ASM
Metals
Reference
Book

Second Edition

Compiled by
The Editorial Staff
Reference Publications
American Society for Metals

AMERICAN SOCIETY FOR METALS
METALS PARK, OHIO 44073

First printing, May 1983
Second printing, September 1983
Third printing, March 1984

Library of Congress Cataloging in Publication Data
Main entry under title:

ASM metals reference book.

1. Metals—Handbooks, manuals, etc.   2. Metal-work—Handbooks,
manuals, etc.   I. American Society for Metals. Reference Publications.
II. Title: A. S. M. metals reference book.   III. Title: Metals reference book.
TA459.A78   1983   620.1'6   82-20660
ISBN 0-87170-156-1

SAN 204-7586

Preface
to the Second Edition

The first edition of the *ASM Metals Reference Book* was designed as a first-stop source for metals data. It was developed in response to persistent demand for a book containing the basic information needed daily by metallurgists and engineers. Data were consolidated on the elements and on the compositions and properties of the major industrial alloys. Expanding on this core of information, additional data areas included crystal structure, frequently used phase diagrams, definitions of terms, and general engineering tables. The book was an immediate success and became one of the most popular titles in the ASM catalog of reference books.

Nevertheless, the first edition of the *ASM Metals Reference Book* was also found wanting in several respects. Although designed for frequent and easy use, the large size, 8½ x 11 inches, was awkward to handle. Therefore, the second edition is a smaller, more compact size that is handier to use.

Another shortcoming was the lack of information on processing. Thus, in the second edition, data on heat treating, forming, joining, and machining supplement the alloy sections. The utility of the book has also been improved through the addition of more data on the elements and conversion factors, plus a new section on testing and inspection containing a sub-stantial amount of information on metallography.

A prime objective in compiling this reference book was to provide source listings to help the reader locate data not included within this volume. Therefore, the final section on organizations and engineering data sources has also been expanded to include information on alloy cross-referencing, automated literature searching, and standards-issuing organizations.

The improvements of the second edition and the quality of the first combine to make this new edition of the *ASM Metals Reference Book* a better and more complete first-stop source for metals engineering data.

Timothy L. Gall
Metals Park, Ohio
September, 1982

Preface

This volume is a helpful handbook of information about metals. The *ASM Metals Reference Book* brings together data from many sources, including major contributions from the Eighth and Ninth Editions of *Metals Handbook*.

The Editors make no claim for completeness; the vast amount of metallurgical data available makes that task difficult. The range of information selected should prove most useful and serves to provide an at-the-fingertips reference volume for engineers, designers, and metallurgists—a one-stop source of engineering data about metals.

The first section of the *ASM Metals Reference Book* is a glossary of terms related to metals and metalworking compiled by the ASM Committee on Definitions of Metallurgical Terms. Metric conversions and hardness-number conversions appear in the second section. The third section contains formulas for length, area, volume and weight, and tables of functions of numbers.

The most extensive collection of information in the *ASM Metals Reference Book* is contained in Sections 4 through 15, which present compositions and properties of the major commercial metals and alloy systems. Mechanical and physical properties of most standard alloys, compositions, standard designations, and processing and fabrication characteristics are described in 175 tables of data.

Two sections contain crystal-structure information and phase diagrams of importance to the industrial metallurgist. A final section provides additional sources of metallurgical information.

The Editorial Staff
Reference Publications

William H. Cubberly
Director
Reference Publications
American Society for Metals

v

Contents

Glossary of Metallurgical and Metalworking Terms

This glossary clarifies more than 3000 specialized technical terms encountered in metallurgical literature. The list includes terms from: materials science; physical metallurgy; heat treating; extractive metallurgy; casting, forging, machining, forming, welding and joining; metal cleaning and finishing; electrometallurgy; powder metallurgy; mechanical testing, inspection and quality control; and metallography, fractography and failure analysis. Terms whose meanings are the same for both technical and non-technical usage, as well as terms having universal scientific meanings, are largely excluded.

Many cross references to preferred terms, alternative terms and closely related terms have been included; these cross references are printed in italics. Also, terms that are obsolete or otherwise inappropriate for use in current writings are so indicated. Many terms can have more than one meaning in metallurgical literature; alternative meanings are identified by parenthetical numbers preceding each alternative. Whenever possible, a general or generic meaning is given before a specific or specialized meaning, but there is no special significance to the order in which alternative meanings are given.

The definitions in this glossary are compatible with (although not necessarily identical to) definitions published by ANSI, ASQC, ASTM, AWS, SAE and other organizations.

A

A_{cm}, A_1, A_3, A_4. Same as Ae_{cm}, Ae_1, Ae_3 and Ae_4.

abrasion. A roughening or scratching of a surface due to *abrasive wear.* On aluminum parts, also known as a rub mark or traffic mark.

abrasive. (1) A hard substance used for *grinding, honing, lapping, superfinishing, polishing,* pressure blasting or *barrel finishing.* It includes natural materials such as garnet, emery, corundum and diamond, and electric-furnace products like aluminum oxide, silicon carbide and boron carbide. (2) Hard particles, such as rocks, sand or fragments of certain hard metals, that wear away a surface when they move across it under pressure.

abrasive belt. A coated abrasive product, in the form of a belt, used in production grinding and polishing.

abrasive blasting. A process for cleaning or finishing by means of an abrasive directed at high velocity against the workpiece.

abrasive disk. (1) A grinding wheel that is mounted on a steel plate, with the exposed flat side being used for grinding. (2) A disk-shaped, coated abrasive product.

abrasive wear. The removal of material from a surface when hard particles slide or roll across the surface under pressure. The particles may be loose or may be part of another surface in contact with the surface being worn. Contrast with *adhesive wear.*

Ac_{cm}, Ac_1, Ac_3, Ac_4. Defined under *transformation temperature.*

accuracy. The closeness of approach of a measurement to the true value of the quantity measured. Since the true value cannot actually be measured, the most probable value from the available data, critically considered for sources of error, is used as "the truth". Contrast with *precision.*

acicular ferrite. A highly substructured nonequiaxed ferrite that forms upon continuous cooling by a mixed diffusion and shear mode of transformation that begins at a temperature slightly higher than the temperature transformation range for upper bainite. It is distinguished from bainite in that it has a limited amount of carbon available; thus, there is only a small amount of carbide present.

acicular ferrite steels. Those steels having a microstructure consisting of either acicular ferrite or a mixture of acicular and equiaxed ferrite.

acid. A chemical substance that yields

The terms were compiled and collected by the ASM Committee on Definitions of Metallurgical Terms for an earlier publication, *Glossary of Metallurgical Terms and Engineering Tables.* Members of the Committee were: Howard E. Boyer, *Chairman,* Consultant; James W. Barr, Aluminum Association; F. W. Boulger, Senior Technical Advisor, Battelle-Columbus Laboratories; Price B. Burgess, Plant Metallurgist, Hayes-Albion Corp.; John R. Cuthill, Assistant Chief, Alloy Physics Section, National Bureau of Standards; Charles C. Dick, Metallurgist, Lindberg Heat Treating Co.; Jon L. Dossett, Division Manager, Lindberg Heat Treating Co.; F. L. Ewald, Assistant Manager, Production Laboratory, Budd Co.; Irving J. Feinberg, Mechanical Properties Section, National Bureau of Standards; Michael Field, President and General Manager, Metcut Research Associates, Inc.; Ronald Gibala, Professor of Metallurgy and Materials Science, Case Western Reserve University; W. C. Harmon, Technical Advisor, Research Center, Republic Steel Corp.; Thomas J. Hughel, General Motors Technical Center, General Motors Corp.; Kenneth J. Humberstone, Vice President, American Tank & Fabricating Co.; Charles G. Interrante, Metallurgist, National Bureau of Standards; F. L. Jamieson, Metallurgical Laboratory, Steel Co. of Canada; Frank La Que, Consultant; Joseph B. Long, Manager, Tin Research Institute; Paul T. Lovejoy, Senior Metallurgist, Research Center, Allegheny Ludlum Steel Co.; W. Stuart Lyman, Manager, Technical and Market Services, Copper Development Association, Inc.; R. C. McMaster, Regents Professor of Welding Engineering, Ohio State University; Joseph T. Michalak, Senior Scientist, Physical Metallurgy, Research Laboratory, United States Steel Corp.; Thomas J. Moore, Lewis Research Center, National Aeronautics and Space Administration; Fielding Ogburn, National Bureau of Standards; Bernard R. Queneau, (retired), formerly General Manager, Product Metallurgy, United States Steel Corp.; Glenn O. Ratliff, President, Shore Metal Treating, Inc.; Kempton H. Roll, Executive Director, Metal Powder Industries Federation; M. T. Rowley, Vice President, Technology, American Foundrymen's Society; John A. Simmons, Microstructure Characterization Section, National Bureau of Standards; Frank Speight, American Welding Society; Harry Turner, Chief, Metallic Materials Branch, Material and Process Development, McDonnell Aircraft Co.; C. F. Walton, Technical Director, Iron Castings Society; William G. Wood, Vice President of Research and Development, Kolene Corp.; David Benjamin, *Secretary,* Senior Editor, Metals Handbook, American Society for Metals

hydrogen ions (H^+) when dissolved in water. Compare with *base* (3).

acid bottom and lining. The inner bottom and lining of a melting furnace consisting of materials like sand, siliceous rock or silica brick that give an acid reaction at the operating temperature.

acid copper. (1) Copper electrodeposited from an acid solution of a copper salt, usually copper sulfate. (2) The solution referred to in (1).

acid embrittlement. A form of *hydrogen embrittlement* that may be induced in some metals by acid treatment.

acid steel. Steel melted in a furnace with an *acid bottom and lining* and under a slag containing an excess of an acid substance such as silica.

activation. The changing of a passive surface of a metal to a chemically active state. Contrast with *passivation*.

activation energy. The energy required for initiating a metallurgical reaction; for example, plastic flow, diffusion, chemical reaction. The activation energy may be calculated from the slope of the line obtained by plotting the natural log of the reaction rate versus the reciprocal of the absolute temperature.

activity. A measure of the chemical potential of a substance, where chemical potential is not equal to concentration, that allows mathematical relations equivalent to those for ideal systems to be used to correlate changes in an experimentally measured quantity to changes in chemical potential.

addition agent. A substance added to a solution for the purpose of altering or controlling a process. Examples: wetting agents in acid pickles; brighteners or antipitting agents in plating solutions; and inhibitors.

adhesion. Force of attraction between the molecules (or atoms) of two different phases. Contrast with *cohesion*.

adhesive bonding. A materials joining process in which an adhesive, placed between faying surfaces, solidifies to bond the surfaces together.

adhesive wear. The removal of material from a surface by the welding together and subsequent shearing of minute areas of two surfaces that slide across each other under pressure. In advanced stages, may lead to *galling* or *seizing*. Contrast with *abrasive wear*.

adjustable bed. Bed of a press designed so that the die space height can be varied conveniently.

Ae$_{cm}$, Ae$_1$, Ae$_3$, Ae$_4$. Defined under *transformation temperature*.

age hardening. Hardening by aging, usually after rapid cooling or cold working. See *aging*.

age softening. Spontaneous decrease of strength and hardness that takes place at room temperature in certain strain hardened alloys, especially those of aluminum.

aging. A change in the properties of certain metals and alloys that occurs at ambient or moderately elevated temperatures after hot working or a heat treatment (quench aging in ferrous alloys, natural or artificial aging in ferrous and nonferrous alloys) or after a cold working operation (strain aging). The change in properties is often, but not always, due to a phase change (precipitation), but never involves a change in chemical composition of the metal or alloy. See also *age hardening, artificial aging, interrupted aging, natural aging, overaging, precipitation hardening, precipitation heat treatment, progressive aging, quench aging, step aging, strain aging*.

air bend die. Angle-forming dies in which the metal is formed without striking the bottom of the die. Metal contact is made at only three points in the cross section: the nose of the male die and the two edges of a V-shape die opening.

air bending. Bending in an *air bend die*.

air classification. The separation of metal powder into particle-size fractions by means of an air stream of controlled velocity; an application of the principle of *elutriation*.

air-hardening steel. A steel containing sufficient carbon and other alloying elements to harden fully during cooling in air or other gaseous mediums from a temperature above its transformation range. The term should be restricted to steels that are capable of being hardened by cooling in air in fairly large sections, about 2 in. or more in diameter. Same as self-hardening steel.

air-lift hammer. A type of gravity drop hammer where the ram is raised for each stroke by an air cylinder. Since length of stroke may be controlled, ram velocity and thus energy delivered to the workpiece may be varied.

alclad. Composite wrought product comprised of an aluminum alloy core having on one or both surfaces a metallurgically bonded aluminum or aluminum alloy coating that is anodic to the core and thus electrically protects the core against corrosion.

alkali metal. A metal in group IA of the periodic system—namely, lithium, sodium, potassium, rubidium, cesium and francium. They form strongly alkaline hydroxides; hence, the name.

alkaline cleaner. A material blended from alkali hydroxides and such alkaline salts as borates, carbonates, phosphates or silicates. The cleaning action may be enhanced by the addition of surface-active agents and special solvents.

alkaline earth metal. A metal in group IIA of the periodic system—namely, beryllium, magnesium, calcium, strontium, barium and radium—so called because the oxides or "earths" of calcium, strontium and barium were found by the early chemists to be alkaline in reaction.

alligatoring. The longitudinal splitting of flat slabs in a plane parallel to the rolled surface. Also called fish-mouthing.

allotriomorphic crystal. A crystal whose lattice structure is normal but whose external surfaces are not bounded by regular crystal faces; rather, the external surfaces are impressed by contact with other crystals or another surface such as a mold wall, or are irregularly shaped because of nonuniform growth. Compare with *idiomorphic crystal*.

allotropy. A near synonym for *polymorphism*. Allotropy is generally restricted to describing polymorphic behavior in elements, terminal phases, and alloys whose behavior closely parallels that of the predominant constituent element.

allowance. The specified difference in limiting sizes (minimum clearance or maximum interference) between mating parts, as computed arithmetically from the specified dimensions and tolerances of each part.

alloy. A substance having metallic properties and being composed of two or more chemical elements of which at least one is a *metal*.

alloying element. An element added to a metal to effect changes in properties and which remains within the metal.

alloy plating. The codeposition of two or more metallic elements.

alloy powder. A powdered metal in

which each particle is composed of the same alloy.

alloy steel. Steel containing specified quantities of alloying elements (other than carbon and the commonly accepted amounts of manganese, copper, silicon, sulfur and phosphorus) within the limits recognized for constructional alloy steels, added to effect changes in mechanical or physical properties.

all-position electrode. In arc welding, a filler-metal electrode for depositing weld metal in the flat, horizontal, overhead and vertical positions.

all-weld-metal test specimen. A test specimen wherein the portion being tested is composed wholly of weld metal.

alpha ferrite. See *ferrite*.

alpha iron. The body-centered cubic form of pure iron, stable below 910 °C (1670 °F).

alsifer. A deoxidizer (20 Al, 40 Si, 40 Fe) used for steel.

alternate-immersion test. A corrosion test in which the specimens are intermittently immersed in and removed from a liquid medium at definite time intervals.

Alumel. A nickel-base alloy containing about 2.5 Mn, 2 Al and 1 Si used chiefly as a component of pyrometric thermocouples.

aluminizing. Forming an aluminum or aluminum alloy coating on a metal by hot dipping, hot spraying or diffusion.

aluminum bomb. A bomb-shaped container used in determining the oxygen content in liquid steel.

amalgam. An alloy of mercury with one or more other metals.

amorphous. Not having a crystal structure; noncrystalline.

amphoteric. Possessing both acidic and basic properties.

anchorite. A zinc-iron phosphate coating for iron and steel.

anelasticity. The property of solids by virtue of which strain is not a single-valued function of stress in the low-stress range where no permanent set occurs.

angle of bite. In rolling metals where all the force is transmitted through the rolls, the maximum attainable angle between the roll radius at the first contact and the line of roll centers. Operating angles less than the angle of bite are called contact angles or rolling angles.

angle of nip. In rolling, the *angle of bite*. In roll, jaw or gyratory crushing,

the entrance angle formed by the tangents at the two points of contact between the working surfaces and the (assumed) spherical particle to be crushed.

angstrom (unit). A unit of linear measurement equal to 10^{-10} m, or 0.1 nm, sometimes used to express small distances such as interatomic distances and some wavelengths.

anion. A negatively charged ion; it flows to the anode in electrolysis.

anisotropy. The characteristic of exhibiting different values of a property in different directions with respect to a fixed reference system in the material.

annealing. A generic term denoting a treatment, consisting of heating to and holding at a suitable temperature followed by cooling at a suitable rate, used primarily to soften metallic materials, but also to simultaneously produce desired changes in other properties or in microstructure. The purpose of such changes may be, but is not confined to: improvement of machinability, facilitation of cold work, improvement of mechanical or electrical properties, and/or increase in stability of dimensions. When the term is used without qualification, full annealing is implied. When applied only for the relief of stress, the process is properly called stress relieving or stress-relief annealing.

In ferrous alloys, annealing usually is done above the upper critical temperature, but the time-temperature cycles vary widely in both maximum temperature attained and in cooling rate employed, depending on composition, material condition, and results desired. When applicable, the following commercial process names should be used: black annealing, blue annealing, box annealing, bright annealing, cycle annealing, flame annealing, full annealing, graphitizing, in-process annealing, isothermal annealing, malleablizing, orientation annealing, process annealing, quench annealing, spheroidizing, subcritical annealing.

In nonferrous alloys, annealing cycles are designed to: (*a*) remove part or all of the effects of cold working (recrystallization may or may not be involved); (*b*) cause substantially complete coalescence of precipitates from solid solution in relatively coarse form; or (*c*) both, depending on composition and material condition. Specific process names in commercial

use are final annealing, full annealing, intermediate annealing, partial annealing, recrystallization annealing, stress-relief annealing, anneal to temper.

annealing carbon. Fine, apparently amorphous carbon particles formed in white cast iron and certain steels during prolonged annealing. Also called temper carbon.

annealing twin. A *twin* formed in a crystal during recrystallization.

anneal to temper. A final partial anneal that softens a cold worked nonferrous alloy to a specified level of hardness or tensile strength.

anode. The electrode where electrons leave an operating system such as a battery, an electrolytic cell, an x-ray tube or a vacuum tube. In the first of these, it is negative; in the other three, positive. In a battery or electrolytic cell, it is the electrode where oxidation occurs. Contrast with *cathode*.

anode compartment. In an electrolytic cell, the enclosure formed by a diaphragm around the anodes.

anode copper. Special-shaped copper slabs, resulting from the refinement of *blister copper* in a reverberatory furnace, used as anodes in electrolytic refinement.

anode corrosion. The dissolution of a metal acting as an anode.

anode effect. The effect produced by polarization of the anode in electrolysis. It is characterized by a sudden increase in voltage and a corresponding decrease in amperage due to the anode becoming virtually separated from the electrolyte by a gas film.

anode efficiency. *Current efficiency* at the anode.

anode film. (1) The portion of solution in immediate contact with the anode, especially if the concentration gradient is steep. (2) The outer layer of the anode itself.

anode mud. Deposit of insoluble residue formed from the dissolution of the anode in commercial electrolysis. Sometimes called anode slime.

anode polarization. See *polarization*.

anodic cleaning. *Electrolytic cleaning* where the work is the anode. It is also called reverse-current cleaning.

anodic coating. A film on a metal surface resulting from an electrolytic treatment at the anode.

anodic pickling. *Electrolytic pickling* where the work is the *anode*.

anodic protection. Imposing an external electrical potential to protect a metal from corrosive attack. (Applicable only to metals that show active-passive behavior.) Contrast with *cathodic protection*.

anodizing. Forming a *conversion coating* on a metal surface by anodic oxidation; most frequently applied to aluminum.

anolyte. The electrolyte adjacent to the anode in an electrolytic cell.

antiferromagnetic material. A material wherein interatomic forces hold the elementary atomic magnets (electron spins) of a solid in alignment, the state being similar to that of a *ferromagnetic material* but with the difference that equal numbers of elementary magnets (spins) face in opposite directions and are antiparallel, causing the solid to be weakly magnetic, that is, paramagnetic, instead of ferromagnetic.

antipitting agent. An *addition agent* for electroplating solutions to prevent the formation of pits or large pores in the electrodeposit.

anvil. (1) In drop forging, the base of the hammer into which the *sow block* and lower die part are set. (2) A block of steel upon which metal is forged.

anvil cap. Same as *sow block*.

apparent density. (1) The weight per unit volume of a metal powder, in contrast to the weight per unit volume of the individual particles. (2) The weight per unit volume of a porous solid, where the unit volume is determined from external dimensions of the mass. Apparent density is always less than the true density of the material itself.

approach distance. The linear distance, in the direction of feed, between the point of initial cutter contact and the point of full cutter contact.

Ar$_{cm}$, Ar$_1$, Ar$_3$, Ar$_4$, Ar', Ar''. Defined under *transformation temperature*.

arbitration bar. A test bar, cast with a heat of material, used to determine chemical composition, hardness, tensile strength, and deflection and strength under transverse loading in order to establish the state of acceptability of the casting.

arbor. (1) In machine grinding, the spindle on which the wheel is mounted. (2) In machine cutting, a shaft or bar for holding and driving the cutter. (3) In founding, a metal shape embedded in green sand or dry sand cores to support the sand or the applied load during casting.

arbor press. A machine used for forcing arbors or mandrels into drilled or bored parts preparatory to turning or grinding. Also used for forcing bushings, shafts or pins into or out of holes.

arbor-type cutters. Cutters having a hole for mounting on an arbor and usually having a keyway for a driving key.

arc blow. The swerving of an electric arc from its normal path because of magnetic forces.

arc brazing. A brazing process in which the heat required is obtained from an electric arc.

arc cutting. A group of cutting processes that melt the metals to be cut with the heat of an arc between an electrode and the base metal. See *carbon-arc cutting, metal-arc cutting, gas tungsten-arc cutting, plasma arc cutting*.

arc furnace. A furnace in which material is heated either directly by an electric arc between an electrode and the work or indirectly by an arc between two electrodes adjacent to the material.

arc gouging. An arc cutting procedure used to form a bevel or groove.

arc melting. Melting metal in an electric arc furnace.

arc of contact. The portion of the circumference of a grinding wheel or cutter touching the work being processed.

arc time. The time the arc is maintained in making an arc weld. Also known as weld time.

arc voltage. The voltage across any electric arc—for example, across a welding arc.

arc welding. A group of welding processes that fuse metals together by heating them with an arc, with or without the application of pressure and with or without the use of filler metal.

artifact. A feature of artificial character (such as a scratch or a piece of dust on a metallographic specimen) that can be erroneously interpreted as a real feature. In inspection, an artifact often produces a *false indication*.

artificial aging. Aging above room temperature. See *aging*. Compare with *natural aging*.

athermal transformation. A reaction that proceeds without benefit of thermal fluctuations; that is, thermal activation is not required. Such reactions are diffusionless and can take place with great speed when the driving force is sufficiently high. For example, many martensitic transformations occur athermally on cooling, even at relatively low temperatures, because of the progressively increasing driving force. In contrast, a reaction that occurs at constant temperature is an *isothermal transformation;* thermal activation is necessary in this case and the reaction proceeds as a function of time.

atmospheric riser. A riser that uses atmospheric pressure to aid feeding. Essentially a *blind riser* into which a small core or rod protrudes, the function of the core or rod being to provide an open passage so that the molten interior of the riser will not be under a partial vacuum when metal is withdrawn to feed the casting, but will always be under atmospheric pressure. Often called Williams riser.

atomic fission. The breakup of the nucleus of an atom in which the combined weight of the fragments is less than that of the original nucleus, the difference being converted to a very large energy release.

atomic hydrogen welding. An arc welding process that fuses metals together by heating them with an electric arc maintained between two metal electrodes enveloped in a stream of hydrogen. Shielding is provided by the hydrogen, which also carries heat by molecular dissociation and subsequent recombination. Pressure may or may not be used and filler metal may or may not be used. (This process is now of limited industrial significance.)

atomic number. The number of protons in an atomic nucleus; determines the individuality of the atom as a chemical element.

atomic percent. The number of atoms of an element in a total of 100 representative atoms of a substance.

atomization. The dispersion of a molten metal into small particles by a rapidly moving stream of gas or liquid.

attenuation. The fractional decrease of the intensity of an energy flux, including the reduction of intensity resulting from geometrical spreading, absorption and scattering.

attritious wear. Wear of abrasive grains in grinding such that the sharp edges gradually become rounded. A grinding wheel that has undergone such wear usually has a glazed appearance.

ausforming. Hot deformation of metastable austenite within controlled ranges of temperature and time that avoids formation of nonmartensitic transformation products.

austempering. A heat treatment for ferrous alloys in which a part is quenched from the austenitizing temperature at a rate fast enough to avoid formation of ferrite or pearlite and then held at a temperature just above M_s until transformation to bainite is complete.

austenite. A solid solution of one or more elements in face-centered cubic iron. Unless otherwise designated (such as nickel austenite), the solute is generally assumed to be carbon.

austenitic grain size. The size attained by the grains of steel when heated to the austenitic region; may be revealed by appropriate etching of cross sections after cooling to room temperature.

austenitic steel. An alloy steel whose structure is normally austenitic at room temperature.

austenitizing. Forming austenite by heating a ferrous alloy into the transformation range (partial austenitizing) or above the transformation range (complete austenitizing). When used without qualification, the term implies complete austenitizing.

autofrettage. Prestressing a hollow metal cylinder by the use of momentary internal pressure exceeding the yield strength.

autogenous weld. A fusion weld made without the addition of filler metal.

automatic brazing. Brazing with equipment that performs the brazing operation without constant observation and adjustment by a brazing operator. The equipment may or may not perform the loading and unloading of the work.

automatic press. A press in which the work is fed mechanically through the press in synchronism with the press action. An automation press is an automatic press that, in addition, is provided with built-in electrical and pneumatic control equipment.

automatic welding. Welding with equipment that performs the welding operation without adjustment of the controls by an operator. The equipment may or may not load and unload the work. Compare with *machine welding.*

automation press. See *automatic press.*

autoradiography. An inspection technique in which radiation sponta-neously emitted by a material is recorded photographically. The radiation is emitted by radioisotopes that are (*a*) produced in a metal by bombarding it with neutrons, (*b*) added to a metal such as by alloying, or (*c*) contained within a cavity in a metal part. The technique serves to locate the position of the radioactive element or compound.

auxiliary anode. In electroplating, a supplementary *anode* placed in a position to raise the current density on a certain area of the cathode to get better plate distribution.

Avogadro's number. The number of atoms (or molecules) in a mole of substance; equals 6.02252×10^{23} per mole.

axial rake. For angular (not helical) flutes, the angle between a plane containing the tooth face and the axial plane through the tooth point. See sketch accompanying *face mill.*

axial relief. The relief or clearance behind the end cutting edge of a milling cutter.

axial runout. For any rotating element, the total variation from a true plane of rotation, taken in a direction parallel to the axis of rotation. Compare with *radial runout.*

axis of weld. A line through the length of a weld perpendicular to the cross section at its geometric center.

B

back draft. A reverse taper on a casting pattern or a forging die that prevents the pattern or forged stock from being removed from the cavity.

back extrusion. See *backward extrusion.*

backfire. The recession of a flame into the tip of a torch followed by immediate reappearance or complete extinction of the flame. See *flashback.*

backhand welding. Welding in which the back of the principal hand (torch or electrode hand) of the welder faces

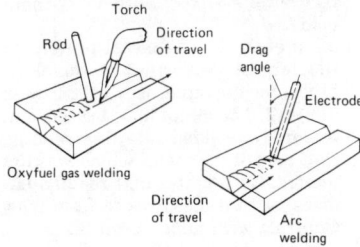

Backhand welding

the direction of travel. It has special significance in oxyfuel gas welding in that the flame is directed backward toward the weld bead, which provides *postheating.* Compare with *forehand welding.*

backing. (1) In grinding, the material (paper, cloth or fiber) that serves as the base for coated abrasives. (2) In welding, a material placed under or behind a joint to enhance the quality of the weld at the root. It may be a metal backing ring or strip; a pass of weld metal; or a nonmetal such as carbon, granular flux or a protective gas.

backlash. Lost motion, play or movement in moving parts such that the driving element (as a gear) can be reversed for some angle or distance before working contact is again made with a driven element.

backoff. A rapid withdrawal of a grinding wheel or cutting tool from contact with workpiece.

back rake. The angle on a single-point turning tool corresponding to axial rake in milling. It is the angle measured between the plane of the tool face and the reference plane and which lies in a plane perpendicular to the axis of the work material and the base of the tool. See sketch accompanying *single-point tool.*

backstep sequence. A longitudinal welding sequence in which the direction of general progress is opposite to that of welding the individual increments.

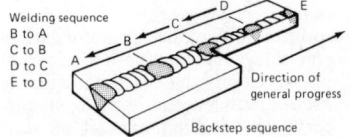

Backstep sequence

backward extrusion. Same as indirect extrusion. See *extrusion.*

back weld. A weld deposited at the back of a single-groove weld.

baghouse. A chamber containing bags for filtering solids out of gases.

bail. Hoop or arched connection between the crane hook and ladle or between crane hook and mold trunnions.

bainite. A metastable aggregate of ferrite and cementite resulting from the transformation of austenite at temperatures below the pearlite range but above M_s. Its appearance is feathery if formed in the upper part of the bainite transformation range; acicular, resembling tempered martensite, if formed in the lower part.

baking. (1) Heating to a low temperature in order to remove gases. (2) Curing or hardening surface coatings such as paints by exposure to heat. (3) Heating to drive off moisture, as in the baking of sand cores after molding.

balance. (1) (dynamic) Condition existing where the principal inertia axis of a body coincides with its rotational axis. (2) (static) Condition existing where the center of gravity of a body lies on its rotational axis.

ball burnishing. (1) Same as *ball sizing*. (2) Removing burrs and polishing small stampings and small machined parts by *tumbling* in the presence of metal balls.

ball mill. A machine consisting of a rotating hollow cylinder partly filled with metal balls (usually hardened steel or white cast iron) or sometimes pebbles; used to pulverize crushed ores or other substances such as pigments.

ball sizing. Sizing and finishing a hole by forcing a ball of suitable size, finish and hardness through the hole or by using a burnishing bar or broach consisting of a series of spherical lands of gradually increasing size coaxially arranged. Also called *ball burnishing,* and sometimes ball broaching.

banded structure. A segregated structure consisting of alternating nearly parallel bands of different composition, typically aligned in the direction of primary hot working.

band mark. An indentation in carbon steel sheet or strip caused by external pressure on the packaging band around cut lengths or coils; it may occur in handling, transit or storage.

bands. (1) Hot rolled steel strip, usually produced for rerolling into thinner sheet or strip. Also known as hot bands or band steel. (2) See *electron bands.*

bar. (1) An obsolete unit of pressure equal to 100 kPa. (2) An elongated rolled metal product that is relatively thick and narrow; most bars have simple, uniform cross sections such as rectangular, square, round, oval or hexagonal. Also known as barstock. Compare with *section* (3).

bare electrode. A filler-metal arc welding electrode, in the form of a wire or rod having no coating other than that incidental to the drawing of

the wire or to its preservation.

bar end. See *end loss.*

bar folder. A machine in which a folding bar or wing is used to bend a metal sheet whose edge is clamped between the upper folding leaf and the lower stationary jaw into a narrow, sharp, close and accurate fold along the edge. It is also capable of making rounded folds such as those used in wiring. A universal folder is more versatile in that it is limited to width only by the dimensions of the sheet.

bark. The decarburized layer just beneath the scale that results from heating steel in an oxidizing atmosphere.

Barkhausen effect. The sequence of abrupt changes in magnetic induction occurring when the magnetizing force acting on a ferromagnetic specimen is varied.

barrel cleaning. Mechanical or electrolytic cleaning of metal in rotating equipment.

barrel finishing. Improving the surface finish of metal objects or parts by processing them in rotating equipment along with abrasive particles that may be suspended in a liquid.

barreling. Convexity of the surfaces of cylindrical or conical bodies, often produced unintentionally during upsetting or as a natural consequence during compression testing.

barrel plating. Plating articles in a rotating container, usually a perforated cylinder that operates at least partially submerged in a solution.

barstock. Same as *bar.*

basal plane. A plane perpendicular to the principal axis (*c* axis) in a tetragonal or hexagonal structure.

base. (1) The surface on which a single-point tool rests when held in a tool post. Also known as heel. See sketch accompanying *single-point tool.* (2) In forging—see *anvil.* (3) A chemical substance that yields hydroxyl ions (OH⁻) when dissolved in water.

base bullion. Crude lead containing recoverable silver, with or without gold.

base metal. (1) The metal present in the largest proportion in an alloy; brass, for example, is a copper-base alloy. (2) The metal to be brazed, cut, soldered or welded. (3) After welding, that part of the metal which was not melted. (4) A metal that readily oxidizes, or that dissolves to form ions. Contrast with *noble metal* (2).

basic bottom and lining. The inner bottom and lining of a melting furnace, consisting of materials such as crushed burned dolomite, magnesite, magnesite bricks or basic slag that give a basic reaction at the operating temperature.

basic steel. Steel melted in a furnace with a *basic bottom and lining* and under a slag containing an excess of a basic substance such as magnesia or lime.

basin. Same as *pouring basin.*

basis metal. The original metal to which one or more coatings are applied.

batch. See *lot.*

Bauschinger effect. For both single-crystal and polycrystalline metals, any change in stress-strain characteristics that can be ascribed to changes in the microscopic stress distribution within the metal, as distinguished from changes caused by strain hardening. In the narrow sense, the process whereby plastic deformation in one direction causes a reduction in yield strength when stress is applied in the opposite direction.

Bayer process. A process for extracting alumina from bauxite ore before the electrolytic reduction. The bauxite is digested in a solution of sodium hydroxide, which converts the alumina to soluble aluminate. After the "red mud" residue has been filtered out, aluminum hydroxide is precipitated, filtered out and calcined to alumina.

beach marks. Progression marks on a fatigue fracture surface that indicate successive positions of the advancing crack front. The classic appearance is of irregular elliptical or semielliptical rings, radiating outward from one or more origins. Beach marks (also known as clamshell marks or tide marks) are typically found on service fractures where the part is loaded randomly, intermittently, or with periodic variations in mean stress or alternating stress.

beaded flange. A flange reinforced by a low ridge, used mostly around a hole.

beading. Raising a ridge or projection on sheet metal.

bead weld. See preferred term *surfacing weld.*

bearing stress. The shear load on a mechanical joint (such as a pinned or riveted joint) divided by the effective bearing area. The effective bearing area of a riveted joint, for example, is the sum of the diameters of all rivets times the thickness of the loaded member.

bed. (1) The stationary portion of a press structure that usually rests on the floor or foundation, forming the support for the remaining parts of the press and the pressing load. The *bolster* and sometimes the lower die are mounted on the top surface of the bed. (2) For machine tools, the portion of the main frame that supports the tools, the work, or both.

Beilby layer. A layer of metal disturbed by mechanical working presumed to be without regular crystalline structure (amorphous); originally applied to grain boundaries.

bel. A unit denoting the ratio of power levels of signals or sound. The number of bels may be given as the common logarithm of the ratio of powers:

$$n = \log (p_1/p_2)$$

where p_1 and p_2 are the initial and final power levels.

belt grinding. Grinding with an *abrasive belt*.

bench press. Any small press that can be mounted on a bench or table.

bend allowance. The length of the arc of the neutral axis between the tangent points of a bend.

bend angle. The angle through which a bending operation is performed.

bender. Term denoting a die impression, tool or mechanical device designed to bend forging stock to conform to the general configuration of die impressions to be subsequently used.

bending brake. A *press brake* used for bending.

bending moment. The algebraic sum of the couples or the moments of the external forces, or both, to the left or right of any section on a member subjected to bending by couples or transverse forces, or both.

bending rolls. Two or three rolls with an adjustment for imparting a desired curvature in sheet or strip metal.

bend radius. (1) The inside radius of a bent section. (2) The radius of a tool around which metal is bent during fabrication.

bend tangent. A tangent point where a bending arc ceases or changes.

bend test. A test for determining relative ductility of metal that is to be formed (usually sheet, strip, plate or wire) for determining soundness and toughness of metal (after welding, for

example). The specimen is usually bent over a specified diameter through a specified angle for a specified number of cycles.

beneficiation. Concentration or other preparation of ore for smelting.

bentonite. A colloidal claylike substance derived from the decomposition of volcanic ash composed chiefly of the minerals of the montmorillonite family. Western bentonite is slightly alkaline; southern bentonite is usually slightly acidic.

bessemer process. A process for making steel by blowing air through molten pig iron contained in a refractory lined vessel so as to remove by oxidation most of the carbon, silicon and manganese. This process is essentially obsolete in the United States.

beta ray. A ray of electrons emitted during the spontaneous disintegration of certain atomic nuclei.

beta structure. A Hume-Rothery designation for structurally analogous body-centered cubic phases (similar to beta brass) or electron compounds that have ratios of three valence electrons to two atoms. Not to be confused with a beta phase on a constitution diagram.

Betts process. A process for the electrolytic refining of lead in which the electrolyte contains lead fluosilicate and fluosilicic acid.

bevel. See preferred term, *corner angle*, and also sketch accompanying *face mill*.

bevel angle. The angle formed between the prepared edge of a member and a plane perpendicular to the surface of the member.

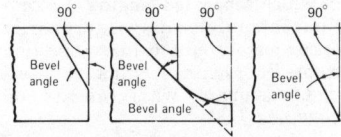

bevel flanging. Same as *flaring*.

biaxiality. In a *biaxial stress* state, the ratio of the smaller to the larger principal stress.

biaxial stress. A state of stress in which only one of the *principal stresses* is zero, the other two usually being in tension.

billet. (1) A solid semifinished round or square product that has been hot worked by forging, rolling or extrusion; usually smaller than a *bloom*. (2) A general term for wrought starting stock used in making forgings or extrusions.

billet mill. A primary rolling mill used to make steel billets.

binary alloy. An alloy containing only two component elements.

binder. (1) In founding, a material, other than water, added to foundry sand to bind the particles together, sometimes with the use of heat. (2) In powder metallurgy, a cementing medium; either a material added to the powder to increase the green strength of the compact and that is expelled during sintering, or a material (usually of relatively low melting point) added to a powder mixture for the specific purpose of cementing together powder particles that alone would not sinter into a strong body.

bipolar electrode. An *electrode* in an electrolytic cell that is not mechanically connected to the power supply, but is so placed in the electrolyte, between the anode and cathode, that the part nearer the anode becomes cathodic and the part nearer the cathode becomes anodic. Also called intermediate electrode.

bipolar field. A longitudinal magnetic field that creates two magnetic poles within a piece of material. Compare with *circular field*.

biscuit. (1) An upset blank for drop forging. (2) A small cake of primary metal (such as uranium made from uranium tetrafluoride and magnesium by bomb reduction). Compare with *derby* and *dingot*.

black annealing. Box annealing or pot annealing ferrous alloy sheet, strip or wire. See *box annealing*.

blackheart malleable. See *malleable cast iron*.

blacking. Carbonaceous materials such as plumbago, graphite or powdered carbon used in coating pouring ladles, molds, runners, pig beds.

black light. Electromagnetic radiation not visible to the human eye. The portion of the spectrum generally used in fluorescent inspection falls in the ultraviolet region between 330 and 400 nm, with the peak at 365 nm.

black oxide. A black finish on a metal produced by immersing it in hot oxidizing salts or salt solutions.

blade-setting angle. See preferred term, *cone angle*.

blank. (1) In forming, a piece of sheet material, produced in cutting dies, that is usually subjected to further press operations. (2) A pressed, presintered or fully sintered powder metallurgy compact, usually in the unfinished condition and requiring cutting, machining or some other operation to produce the final shape. (3) A piece of stock from which a forging is made; often called a slug or multiple.

blank carburizing. Simulating the carburizing operation without introducing carbon. This is usually accomplished by using an inert material in place of the carburizing agent, or by applying a suitable protective coating to the ferrous alloy.

blank holder. The part of a drawing or forming die that holds the workpiece against the draw ring to control metal flow.

blanking. Producing desired shapes from metal to be used for forming or other operations, usually by punching.

blank nitriding. Simulating the nitriding operation without introducing nitrogen. This is usually accomplished by using an inert material in place of the nitriding agent or by applying a suitable protective coating to the ferrous alloy.

blast furnace. A shaft furnace in which solid fuel is burned with an air blast to smelt ore in a continuous operation. Where the temperature must be high, as in the production of pig iron, the air is preheated. Where the temperature can be lower, as in smelting copper, lead and tin ores, a smaller furnace is economical, and preheating of the blast is not required.

blasting. Cleaning or finishing metals by impingement with abrasive particles moving at high speed and usually carried by gas or liquid or thrown centrifugally from a wheel.

blemish. A nonspecific quality control term designating an imperfection that mars the appearance of a part but does not detract from its ability to perform its intended function.

blending. In powder metallurgy, the thorough intermingling of powders of the same nominal composition (not to be confused with *mixing*).

blind riser. A *riser* that does not extend through the top of the mold.

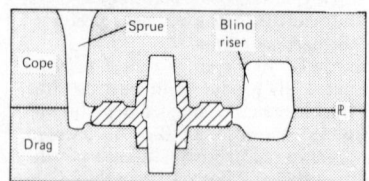

blister. A raised area, often dome shaped, resulting from (*a*) loss of adhesion between a coating or deposit and the basis metal or (*b*) delamination under the pressure of expanding gas trapped in a metal in a near subsurface zone. Very small blisters may be called pinheads or pepper blisters.

blister copper. An impure intermediate product in the refining of copper, produced by blowing copper *matte* in a converter, the name being derived from the large blisters on the cast surface that result from the liberation of SO_2 and other gases.

block brazing. An obsolete brazing process in which the joint was heated using hot blocks.

blocker. The impression in the dies (often one of a series of impressions in a single die set) that imparts to the forging an intermediate shape, preparatory to forging of the final shape. Also called blocking impression.

blocker-type forging. A forging that approximates the general shape of the final part with relatively generous finish allowance and radii. Such forgings are sometimes specified to reduce die costs where only a small number of forgings is desired and the cost of machining each part to its final shape is not excessive.

blocking. In forging, a preliminary operation performed in closed dies, usually hot, to position metal properly so that in the finish operation the dies will be filled correctly.

blocking impression. Same as *blocker*.

block sequence. A welding sequence in which separated lengths of a continuous multiple-pass weld are partly or completely built up in cross section before intervening lengths are deposited. Compare with *cascade sequence*.

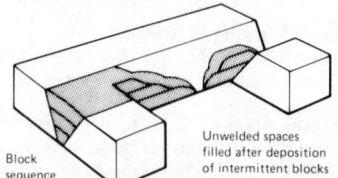

bloom. (1) A semifinished hot rolled product, rectangular in cross section, produced on a blooming mill. See also *billet*. For steel, the width of a bloom is not more than twice the thickness, and the cross-sectional area is usually not less than about 230 cm² (36

in.²). Steel blooms are sometimes made by forging. (2) A visible exudation or efflorescence on the surface of an electroplating bath. (3) A bluish fluorescent cast to a painted surface caused by deposition of a thin film of smoke, dust or oil. (4) A loose, flower-like corrosion product that forms when certain metals are exposed to a moist environment.

bloomer. The mill or other equipment used in reducing steel ingots to blooms.

blooming mill. A primary rolling mill used to make blooms.

blotter. In grinding, a disk of compressible material, usually blotting-paper stock, used between the grinding wheel and its flanges to avoid concentrated stresses.

blowhole. A hole in a casting or a weld caused by gas entrapped during solidification.

blowpipe. A welding or cutting torch.

blue annealing. Heating hot rolled ferrous sheet in an open furnace to a temperature within the transformation range and then cooling in air, in order to soften the metal. The formation of a bluish oxide on the surface is incidental.

blue brittleness. Brittleness exhibited by some steels after being heated to some temperature within the range of about 200 to 370 °C (400 to 700 °F), particularly if the steel is worked at the elevated temperature. Killed steels are virtually free of this kind of brittleness.

blue dip. A solution containing a mercury compound, once widely used to deposit mercury on a metal by immersion, usually prior to silver plating.

bluing. Subjecting the scale-free surface of a ferrous alloy to the action of air, steam or other agents at a suitable temperature, thus forming a thin blue film of oxide and improving the appearance and resistance to corrosion. NOTE: This term is ordinarily applied to sheet, strip or finished parts. It is used also to denote the heating of springs after fabrication in order to improve their properties.

board hammer. A type of forging hammer in which the upper die and ram are attached to "boards" that are raised to the striking position by power-driven rollers and let fall by gravity. See *drop hammer*.

bolster. A plate to which dies may be fastened, the assembly being secured to the top surface of a press bed. In mechanical forging, such a plate is also attached to the ram.

bond. (1) In grinding wheels and other relatively rigid abrasive products, the material that holds the abrasive grains together. (2) In welding, brazing or soldering, the junction of joined parts. Where filler metal is used, it is the junction of the fused metal and the heat-affected base metal. (3) In an adhesive bonded or diffusion bonded joint, the line along which the faying surfaces are joined together.

book mold. A split permanent mold hinged like a book.

bore. A hole or cylindrical cavity produced by a single-point or multipoint tool other than a drill.

boring. A machining method using single-point tools on internal surfaces of revolution.

bort. Industrial diamond.

bosh. (1) The section of a blast furnace extending upward from the tuyeres to the plane of maximum diameter. (2) A lining of quartz that builds up during the smelting of copper ores and that decreases the diameter of the furnace at the tuyeres. (3) A tank, often with sloping sides, used for washing metal parts or for holding cleaned parts.

boss. A relatively short protrusion or projection from the surface of a forging or casting, often cylindrical in shape.

bottom board. A flat base for holding the flask in making sand molds.

bottom drill. A flat-ended twist drill used to convert a cone at the bottom of a drilled hole into a cylinder.

bottoming tap. A tap with a *chamfer* of 1 to 1½ threads in length.

bottom pipe. An oxide-lined fold or cavity at the butt end of a slab, bloom or billet; formed by folding the end of an ingot over on itself during primary rolling. Bottom pipe is not pipe, in that it is not a shrinkage cavity, and in that sense, the term is a misnomer. Bottom pipe is similar to *extrusion pipe.* It is normally discarded when the slab, bloom or billet is cropped following primary reduction.

bowing. Deviation from flatness.

box annealing. Annealing a metal or alloy in a sealed container under conditions that minimize oxidation. In box annealing a ferrous alloy, the charge is usually heated slowly to a temperature below the transformation range, but sometimes above or within it, and is then cooled slowly; this process is also called close an-

nealing or pot annealing. See *black annealing.*

boxing. Continuing a fillet weld around a corner as an extension of the principal weld. Also called an end return.

brake. A device for bending sheet metal to a desired angle.

brale. A diamond penetrator of specified spheroconical shape used with a Rockwell hardness tester. This penetrator is used for the A, C, D and N scales for testing hard metals.

brass. An alloy consisting mainly of copper (over 50%) and zinc, to which smaller amounts of other elements may be added.

braze welding. A method of welding by using a filler metal having a liquidus above 450 °C (840 °F) and below the solidus of the base metals. Unlike *brazing,* in braze welding, the filler metal is not distributed in the joint by capillary attraction.

brazing. A group of welding processes that join solid materials together by heating them to a suitable temperature and by using a filler metal having a liquidus above 450 °C (840 °F) and below the solidus of the base materials. The filler metal is distributed between the closely fitted surfaces of the joint by capillary attraction.

brazing alloy. See preferred term *brazing filler metal.*

brazing filler metal. A nonferrous filler metal used in *brazing* and *braze welding.*

brazing sheet. Brazing filler metal in sheet form or flat-rolled metal clad with brazing filler metal on one or both sides.

breakdown. (1) An initial rolling or drawing operation, or a series of such operations, for the purpose of reducing a casting or extruded shape prior to the finish reduction to desired size. (2) A preliminary press-forging operation.

breaking stress. Same as *fracture stress,* part (1).

breaks. Creases or ridges usually in "untempered" or in aged material where the yield point has been exceeded. Depending on the origin of the break, it may be termed a *cross break,* a *coil break,* an edge break, or a *sticker break.*

bridge die. A two-section extrusion die capable of producing tubing or intricate hollow shapes without the use of a separate mandrel. Metal separates into two streams as it is extruded past a bridge section, which is attached to

the main die section and holds a stub mandrel in the die opening; the metal then is rewelded by extrusion pressure before it enters the die opening. Compare with *porthole die.*

bridging. (1) Premature solidification of metal across a mold section before the metal below or beyond solidifies. (2) Solidification of slag within a cupola at or just above the tuyeres. (3) Welding or mechanical locking of the charge in a downfeed melting or smelting furnace. (4) In powder metallurgy, the formation of arched cavities in a powder mass. (5) In soldering, an unintended solder connection between two or more conductors, either securely or by mere contact. Also called a crossed joint or solder short.

bright annealing. Annealing in a protective medium to prevent discoloration of the bright surface.

bright dip. A solution that produces, through chemical action, a bright surface on an immersed metal.

brightener. An agent or combination of agents added to an electroplating bath to produce a lustrous deposit.

bright finish. A high-quality finish produced on ground and polished rolls. Suitable for electroplating.

bright plate. An electrodeposit that is lustrous in the as-plated condition.

bright range. The range of current densities, other conditions being constant, within which a given electroplating bath produces a bright plate.

Brillouin zones. See *electron bands.*

Brinell hardness test. A test for determining the hardness of a material by forcing a hard steel or carbide ball of specified diameter into it under a specified load. The result is expressed as the Brinell hardness number, which is the value obtained by dividing the applied load in kilograms by the surface area of the resulting impression in square millimetres.

brinelling. Evenly spaced dents in a raceway of a rolling-element bearing that occur when the bearing assembly is subjected to a force or impact load great enough to cause the rolling elements to indent the raceway surface. Also called true brinelling. Compare with *false brinelling.*

brittle crack propagation. A very sudden propagation of a crack with the absorption of no energy except that stored elastically in the body. Microscopic examination may reveal some deformation even though it is not noticeable to the unaided eye.

brittle fracture. Separation of a solid accompanied by little or no macroscopic plastic deformation. Typically, brittle fracture occurs by rapid crack propagation with less expenditure of energy than for ductile fracture.

brittleness. The quality of a material that leads to crack propagation without appreciable plastic deformation.

broach. A bar-shaped cutting tool provided with a series of cutting edges or teeth that increase in size or change in shape from the starting to finishing end. The tool cuts in the axial direction when pushed or pulled and is used to shape either holes or outside surfaces.

bronze. A copper-rich copper-tin alloy with or without small proportions of other elements such as zinc and phosphorus. By extension, certain copper-base alloys containing considerably less tin than other alloying elements, such as manganese bronze (copper-zinc plus manganese, tin and iron) and leaded tin bronze (copper-lead plus tin and sometimes zinc). Also, certain other essentially binary copper-base alloys containing no tin, such as aluminum bronze (copper-aluminum), silicon bronze (copper-silicon) and beryllium bronze (copper-beryllium). Also, trade designations for certain specific copper-base alloys that are actually brasses, such as architectural bronze (57 Cu, 40 Zn, 3 Pb) and commercial bronze (90 Cu, 10 Zn).

bronzing. (1) Applying a chemical finish to copper or copper-alloy surfaces to alter the color. (2) Plating a copper-tin alloy on various materials.

brush anodizing. An *anodizing* process similar to *brush plating*.

brush plating. Plating with a concentrated solution or gel held in or fed to an absorbing medium, pad or brush carrying the anode (usually insoluble). The brush is moved back and forth over the area of the cathode to be plated.

brush polishing (electrolytic). A method of *electropolishing* in which the electrolyte is applied with a pad or brush in contact with the part to be polished.

buckle. (1) A local waviness in metal bar or sheet, usually transverse to the direction of rolling. (2) An indentation in a casting resulting from expansion of molding sand into the mold cavity.

buckling. Producing a bulge, bend, bow, kink or other wavy condition by compressively stressing a beam, column, plate, bar or sheet.

Bucky diaphragm. An x-ray scatter-reducing device originally intended for medical radiography but also applicable to industrial radiography in some circumstances. Thin strips of lead, with their width held parallel to the primary radiation, are used to absorb scattered radiation preferentially; the array of strips is in motion during exposure, to prevent formation of a pattern on the film.

buffer. A substance whose purpose is to maintain a constant hydrogen ion concentration in water solutions, even where acid or alkalis are added. Each buffer has a characteristic limited range of pH over which it is effective.

buffing. Developing a lustrous surface by contacting the *work* with a rotating *buffing wheel*.

buffing wheel. Buff sections assembled to the required face width for use on a rotating shaft between flanges. Sometimes called buff.

buff section. A number of fabric, paper or leather disks with concentric center holes held together by various types of sewing to provide degrees of flexibility or hardness. These sections are assembled to make wheels for polishing.

builder. A material, such as an alkali, a buffer or a water softener, added to soap or synthetic surface-active agent to produce a mixture having enhanced detergency. Examples: (1) alkalis—caustic soda, soda ash and trisodium phosphate; (2) *buffers*—sodium metasilicate and borax; and (3) water softeners—sodium tripolyphosphate, sodium tetraphosphate, sodium hexametaphosphate and ethylene diamine tetraacetic acid.

buildup. Excessive electrodeposition that occurs on high-current-density areas, such as corners or edges.

buildup sequence. The order in which weld beads are deposited, generally designated in cross section as shown in the accompanying illustration.

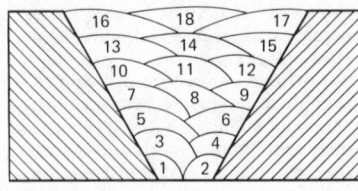

Buildup sequence

built-up edge. Chip material adhering to the tool face adjacent to the cutting edge during cutting.

bulging. Expanding the walls of a cup, shell or tube with an internally expanded segmented punch or a punch composed of air, liquids or semiliquids such as waxes, rubber and other elastomers.

bull block. A machine with a power-driven revolving drum for cold drawing wire through a drawing die as the wire winds around the drum.

bulldozer. A horizontal machine, usually mechanical, having two bull gears with eccentric pins, two connecting links to a ram, and dies to perform bending, forming and punching of narrow plate and bars. Railroad car sills are formed with a bulldozer.

bullion. (1) A semirefined alloy containing sufficient precious metal to make recovery profitable. (2) Refined gold or silver, uncoined.

bull's-eye structure. The microstructure of malleable or ductile cast iron when graphite nodules are surrounded by a ferrite layer in a pearlitic matrix.

bumper. A machine used for packing molding sand in a flask by repeated jarring or jolting.

bumping. (1) Forming a dish in metal by means of many repeated blows. (2) Forming a head. (3) Setting the seams on sheet metal parts. (4) Ramming sand in a flask by repeated jarring and jolting.

burned deposit. A dull, nodular electrodeposit resulting from excessive current density.

burned-on sand. A mixture of sand and cast metal adhering to the surface of a casting. In some instances, may resemble *metal penetration*.

burning. (1) Permanently damaging a metal or alloy by heating to cause either incipient melting or intergranular oxidation. See *overheating*. (2) In grinding, getting the work hot enough to cause discoloration or to change the microstructure by tempering or hardening.

burnishing. Smoothing surfaces through frictional contact between the work and some hard pieces of material such as hardened metal balls.

burn-off. (1) Unintentional removal of an autocatalytic deposit from a nonconducting substrate, during subsequent electroplating operations, owing to the application of excessive current or a poor contact area. (2) Removal of volatile lubricants such as metallic stearates from metal powder compacts by heating immediately prior to sintering. (3) See *melting rate.*

burr. (1) A turned-over edge on work resulting from cutting, punching or grinding. (2) A rotary tool having teeth similar to those on hand files.

burring. Same as *deburring.*

bushing. A bearing or guide.

buster. A pair of shaped dies used to combine preliminary forging operations such as edging and blocking, or to loosen the scale.

butler finish. A semilustrous metal finish composed of fine, uniformly distributed parallel lines, usually produced with a soft abrasive wheel; it is similar in appearance to the traditional hand-rubbed finish on silver.

buttering. A form of surfacing in which one or more layers of weld metal are deposited on the groove face of one member (for example, a high-alloy weld deposit on steel base metal that is to be welded to a dissimilar base metal). The buttering provides a suitable transition weld deposit for subsequent completion of the butt weld.

butt joint. A joint between two abutting members lying approximately in the same plane. A welded butt joint may contain a variety of grooves. See *groove weld.*

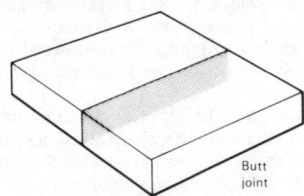

Butt joint

button. (1) A globule of metal remaining in an assaying crucible or cupel after fusion has been completed. (2) That part of a weld that tears out in the destructive testing of spot, seam or projection welded specimens.

butt seam welding. See *seam welding.*

butt welding. Welding a butt joint.

C

cake. (1) A copper or copper alloy casting rectangular in cross section used for rolling into sheet or strip. (2) A coalesced mass of unpressed metal powder.

calcination. Heating ores, concentrates, precipitates or residues to decompose carbonates, hydrates or other compounds.

calomel electrode (calomel half cell). A secondary reference electrode of the composition: $Pt/Hg\text{-}Hg_2Cl_2/KCl$ solution. For $1.0N$ KCl solution, its potential versus a hydrogen electrode at 25 °C and one atmosphere is $+0.281$ V.

calorizing. Imparting resistance to oxidation to an iron or steel surface by heating in aluminum powder at 800 to 1000 °C (1472 to 1832 °F).

camber. (1) Deviation from edge straightness, usually referring to the greatest deviation of side edge from a straight line. (2) Sometimes used to denote crown in rolls where the center diameter has been increased to compensate for deflection caused by the rolling pressure.

cam press. A mechanical press in which one or more of the slides are operated by cams; usually a double-action press in which the blank-holder slide is operated by cams through which the dwell is obtained.

canning. (1) A dished distortion in a flat or nearly flat surface, sometimes referred to as oil canning. (2) Enclosing a highly reactive metal within one relatively inert for the purpose of hot working without undue oxidation of the active metal.

capillary attraction. The combined force of adhesion and cohesion that causes liquids, including molten metals, to flow between very closely spaced solid surfaces, even against gravity.

capped steel. A type of steel similar to rimmed steel, usually cast in a bottle-top ingot mold, in which the application of a mechanical or a chemical cap renders the rimming action incomplete by causing the top metal to solidify. The surface condition of capped steel is much like that of rimmed steel, but certain other characteristics are intermediate between those of *rimmed steel* and those of *semikilled steel.*

capping. The partial or complete separation of a powder metallurgy compact into two or more portions by cracks that originate near the edges of the punch faces and that proceed diagonally into the compact.

carbide. A compound of carbon with one or more metallic elements.

carbide tools. Cutting or forming tools, usually made from tungsten, titanium, tantalum, or niobium carbides, or a combination of them, in a matrix of cobalt, nickel, or other metals. Carbide tools are characterized by high hardnesses and compressive strengths and may be coated to improve wear resistance.

carbon dioxide welding. *Gas metal-arc welding* using carbon dioxide as the shielding gas.

carbon edges. Carbonaceous deposits in a wavy pattern along the edges of a sheet or strip; also known as snaky edges.

carbon electrode. A carbon or graphite rod used in carbon-arc equipment, such as in carbon-arc welding or cutting torches.

carbon equivalent. (1) For cast iron, an empirical relationship of the total carbon, silicon and phosphorus contents expressed by the formula:

$$CE = TC + \tfrac{1}{3}(Si + P)$$

(2) For rating of weldability:

$$CE = C + \frac{Mn}{6} + \frac{Cr + Mo + V}{5} + \frac{Ni + Cu}{15}$$

carbonitriding. A case hardening process in which a suitable ferrous material is heated above the lower transformation temperature in a gaseous atmosphere of such composition as to cause simultaneous absorption of carbon and nitrogen by the surface and, by diffusion, create a concentration gradient. The process is completed by cooling at a rate that produces the desired properties in the workpiece.

carbonization. Conversion of an organic substance into elemental carbon. (Should not be confused with *carburization*).

carbon potential. A measure of the ability of an environment containing active carbon to alter or maintain, under prescribed conditions, the carbon level of the steel. NOTE: In any particular environment, the carbon level attained will depend on such factors as temperature, time and steel composition.

carbon restoration. Replacing the carbon lost in the surface layer from previous processing by carburizing this layer to substantially the original carbon level. Sometimes called recarburizing.

carbon steel. Steel having no specified minimum quantity for any alloying

12

element (other than the commonly accepted amounts of manganese, silicon and copper) and that contains only an incidental amount of any element other than carbon, silicon, manganese, copper, sulfur and phosphorus.

carbonyl powder. A metal powder prepared by the thermal decomposition of a metal carbonyl.

carburizing. Absorption and diffusion of carbon into solid ferrous alloys by heating, to a temperature usually above Ac₃, in contact with a suitable carbonaceous material. A form of *case hardening* that produces a carbon gradient extending inward from the surface, enabling the surface layer to be hardened either by quenching directly from the carburizing temperature or by cooling to room temperature, then reaustenitizing and quenching.

carburizing flame. A gas flame that will introduce carbon into some heated metals, as during a gas welding operation. A carburizing flame is a *reducing flame,* but a reducing flame is not necessarily a carburizing flame.

cascade sequence. A welding sequence in which a continuous multiple-pass weld is built up by depositing weld beads in overlapping layers, usually laid in a *backstep sequence.* Compare with *block sequence.*

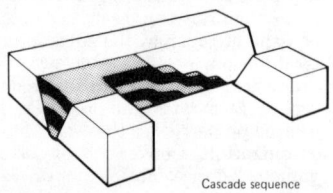

Cascade sequence

case. That portion of a ferrous alloy, extending inward from the surface, whose composition has been altered so that it can be *case hardened.* Typically considered to be the portion of the alloy (*a*) whose composition has been measurably altered from the original composition, (*b*) that appears dark on an etched cross section, or (*c*) that has a hardness, after hardening, equal to or greater than a specified value. Contrast with *core* (2).

case hardening. A generic term covering several processes applicable to steel that change the chemical composition of the surface layer by absorption of carbon, nitrogen, or a mixture of the two and, by diffusion, create a concentration gradient. The processes commonly used are carburizing and quench hardening; cyaniding; nitriding; and carbonitriding. The use of the applicable specific process name is preferred.

cassette. A lighttight holder, used to contain radiographic films during exposure to x-rays or gamma rays, that may or may not contain intensifying or filter screens, or both. A distinction is often made between a cassette, which has positive means for ensuring contact between screens and film and is usually rigid, and an exposure holder, which is rather flexible.

CASS test. Abbreviation for *copper accelerated salt spray test.*

cast. See *die proof.*

cast-alloy tool. A cutting tool made by casting a cobalt-base alloy and used at machining speeds between those for high speed steels and sintered carbides.

casting. (1) An object at or near finished shape obtained by solidification of a substance in a mold. (2) Pouring molten metal into a mold to produce an object of desired shape.

casting copper. Fire-refined tough pitch copper usually cast from melted secondary metal into ingot bars only, and used for making foundry castings but not wrought products.

casting shrinkage. (1) Liquid shrinkage—the reduction in volume of liquid metal as it cools to the liquidus. (2) Solidification shrinkage—the reduction in volume of metal from the beginning to ending of solidification. (3) Solid shrinkage—the reduction in volume of metal from the solidus to room temperature.

casting strains. Strains in a casting caused by *casting stresses* that develop as the casting cools.

casting stresses. Residual stresses set up when the shape of a casting impedes contraction of the solidified casting during cooling.

cast iron. A generic term for a large family of cast ferrous alloys in which the carbon content exceeds the solubility of carbon in austenite at the eutectic temperature. Most cast irons contain at least 2% carbon, plus silicon and sulfur, and may or may not contain other alloying elements. For the various forms *gray cast iron, white cast iron, malleable cast iron* and *ductile cast iron,* the word "cast" is often left out, resulting in "gray iron", "white iron", "malleable iron" and "ductile iron", respectively.

cast steel. Steel in the form of *castings.*

cast structure. The metallographic structure of a *casting* evidenced by shape and orientation of grains and by segregation of impurities.

catalyst. A substance capable of changing the rate of a reaction without itself undergoing any net change.

catastrophic failure. Sudden failure of a component or assembly that frequently results in extensive secondary damage to adjacent components or assemblies.

cathode. The electrode where electrons enter an operating system such as a battery, an electrolytic cell, an x-ray tube or a vacuum tube. In the first of these, it is positive; in the other three, negative. In a battery or electrolytic cell, it is the electrode where reduction occurs. Contrast with *anode.*

cathode compartment. In an electrolytic cell, the enclosure formed by a diaphragm around the cathode.

cathode copper. Copper deposited at the cathode in electrolytic refining.

cathode efficiency. *Current efficiency* at the cathode.

cathode film. The portion of solution in immediate contact with the cathode during electrolysis.

cathodic cleaning. *Electrolytic cleaning* where the work is the cathode.

cathodic pickling. *Electrolytic pickling* where the work is the cathode.

cathodic protection. Partial or complete protection of a metal from corrosion by making it a cathode, using either a galvanic or impressed current. Contrast with *anodic protection.*

catholyte. The electrolyte adjacent to the cathode in an electrolytic cell; in a divided cell, the portion on the cathode side of the diaphragm.

cation. A positively charged ion; it flows to the cathode in electrolysis.

cationic detergent. A detergent in which the *cation* is the active part.

caustic cracking. A form of *stress-corrosion cracking* most frequently encountered in carbon steels or iron-chromium-nickel alloys that are exposed to concentrated hydroxide solutions at temperatures of 200 to 250 °C (400 to 480 °F).

caustic dip. A strongly alkaline solution into which metal is immersed for etching, for neutralizing acid or for removing organic materials such as grease or paints.

cavitation. The formation and instantaneous collapse of innumerable tiny voids or cavities within a liquid sub-jected to rapid and intense pressure changes. Cavitation produced by ul-trasonic radiation is sometimes used to give violent localized agitation. That caused by severe turbulent flow often leads to *cavitation damage*.

cavitation damage. Erosion of a solid surface through the formation and collapse of cavities in an adjacent liq-uid.

cavitation erosion. See preferred term, *cavitation damage*.

cell feed. The material supplied to the cell in the electrolytic production of metals.

cementation. The introduction of one or more elements into the outer por-tion of a metal object by means of diffusion at high temperature.

cement copper. Impure copper recov-ered by *chemical deposition* when iron (most often shredded steel scrap) is brought into prolonged contact with a dilute copper sulfate solu-tion.

cemented carbide. A solid and coher-ent mass made by pressing and sin-tering a mixture of powders of one or more metallic carbides and a much smaller amount of a metal, such as cobalt, to serve as a binder.

cementite. A compound of iron and car-bon, known chemically as iron car-bide and having the approximate chemical formula Fe_3C. It is charac-terized by an orthorhombic crystal structure. When it occurs as a phase in steel, the chemical composition will be altered by the presence of manganese and other carbide-form-ing elements.

center drilling. Drilling a short, coni-cal hole in the end of a workpiece—the hole to be used to center the work-piece for turning on a lathe.

centering plug. A plug fitting both spindle and cutter to ensure concen-tricity of the cutter mounting.

centerless grinding. Grinding the out-side or inside of a workpiece mounted on rollers rather than on centers. The workpiece may be in the form of a cylinder or the frustum of a cone.

centrifugal casting. A casting made by pouring metal into a mold that is rotated or revolved.

ceramic tools. Cutting tools made from fused, sintered or cemented me-tallic oxides.

cereal. An organic *binder,* usually corn flour.

cermet. A powder metallurgy product consisting of ceramic particles

bonded with a metal.

C-frame press. Same as *gap-frame press*.

CG iron. Same as *compacted graphite cast iron*.

chafing fatigue. Fatigue initiated in a surface damaged by rubbing against another body. See *fretting*.

chain-intermittent fillet welding. Depositing a line of intermittent fillet welds on each side of a member at a joint so that the increments on one side are essentially opposite those on the other. Contrast with *staggered-intermittent fillet welding*.

chamfer. (1) A beveled surface to elim-inate an otherwise sharp corner. (2) A relieved angular cutting edge at a tooth corner.

chamfer angle. (1) The angle between a reference surface and the bevel. (2) On a milling cutter, the angle be-tween a beveled surface and the axis of the cutter.

chamfering. Making a sloping surface on the edge of a member. Also called beveling. See *bevel angle*.

chaplet. Metal support that holds a core in place within a mold; molten metal solidifies around a chaplet and fuses it into the finished casting.

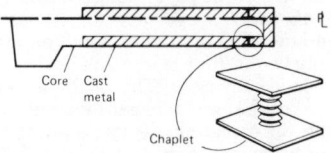

Core Cast
metal

Chaplet

characteristic radiation. High-inten-sity single-wavelength x-rays, char-acteristic of the element emitting the rays, that appear in addition to con-tinuous "white" radiation whenever the element is bombarded with elec-trons whose energy exceeds a specific critical value that is different for each element.

charge. (1) The materials fed into a furnace. (2) Weights of various liquid and solid materials put into a furnace during one feeding cycle.

charging. (1) For a lap, impregnating the surface with fine abrasive. (2) Placing materials into a furnace.

Charpy test. A pendulum-type single-blow impact test in which the speci-men, usually notched, is supported at both ends as a simple beam and bro-ken by a falling pendulum. The en-ergy absorbed, as determined by the subsequent rise of the pendulum, is a measure of impact strength or notch toughness. Contrast with *Izod test*.

chase. To make a series of cuts each, except the first, following in the path of the cut preceding it, as in chasing a thread.

chatter. In machining or grinding, (1) a vibration of the tool, wheel or work-piece producing a wavy surface on the work and (2) the finish produced by such vibration.

checked edges. Sawtooth edges seen after hot rolling and/or cold rolling.

checkers. In a chamber associated with a metallurgical furnace, bricks stacked openly so that heat may be absorbed from the combustion pro-ducts and later transferred to incom-ing air when the direction of flow is reversed.

checks. Numerous, very fine cracks in a coating or at the surface of a metal part. Checks may appear during pro-cessing or during service and are most often associated with thermal treatment or thermal cycling. Also called check marks, checking, *heat checks*.

cheek. The intermediate section of a flask that is used between the *cope* and the *drag* when molding a shape that requires more than one parting plane.

chelating agent. A substance used in metal finishing to control or elimi-nate certain metallic ions present in undesirable quantities.

chemical deposition. The precipita-tion or plating-out of a metal from solutions of its salts through the in-troduction of another metal or re-agent to the solution.

chemically precipitated powder. Metal powder produced as a precipi-tate by chemical displacement.

chemical machining. Removing metal stock by controlled selective chemical dissolution.

chemical metallurgy. See *process met-allurgy*.

chemical polishing. Improving the surface luster of a metal by chemical treatment.

chevron pattern. A fractographic pat-tern of radial marks (shear ledges) that looks like nested letters "V"; sometimes called a herringbone pat-tern. Chevron patterns are typically found on brittle fracture surfaces in parts whose widths are considerably greater than their thicknesses. The points of the chevrons can be traced back to the fracture origin.

chill. (1) A metal or graphite insert embedded in the surface of a sand mold or core or placed in a mold cavity to increase the cooling rate at that point. (2) White iron occurring on a gray or ductile iron casting, such as the chill in the wedge test. Compare with *inverse chill.*

Chinese script. The angular microstructural form suggestive of Chinese writing and characteristic of the constituents α(Al-Fe-Si) and α(Al-Fe-Mn-Si) in cast aluminum alloys. A similar microstructure is found in cast magnesium alloys containing silicon as Mg_2Si.

chip breaker. (1) Notch or groove in the face of a tool parallel to the cutting edge, to break the continuity of the chips. (2) A step formed by an adjustable component clamped to the face of the cutting tool.

chipping. (1) Removing seams and other surface imperfections in metals manually with a chisel or gouge, or by a continuous machine, before further processing. (2) Similarly, removing excessive metal.

chips. Pieces of material removed from a workpiece by cutting tools or an abrasive medium.

chlorination. (1) Roasting ore in contact with chlorine or a chloride salt to produce chlorides. (2) Removing dissolved gases and entrapped oxides by passing chlorine gas through molten metal such as aluminum and magnesium.

chromadizing. Improving paint adhesion on aluminum or aluminum alloys, mainly aircraft skins, by treatment with a solution of chromic acid. Also called chromodizing, chromatizing. Not to be confused with *chromating, chromizing.*

chromate treatment. A treatment of metal in a solution of a hexavalent chromium compound to produce a *conversion coating* consisting of trivalent and hexavalent chromium compounds.

chromating. Performing a *chromate treatment.*

Chromel. (1) A 90Ni-10Cr alloy used in thermocouples. (2) A series of nickel-chromium alloys, some with iron, used for heat-resistant applications.

chrome pickle. (1) Producing a chromate *conversion coating* on magnesium for temporary protection or for a paint base. (2) The solution that produces the conversion coating.

chromizing. A surface treatment at elevated temperature, generally carried out in pack, vapor or salt bath, in which an alloy is formed by the inward diffusion of chromium into the base metal.

chuck. A device for holding work or tools on a machine so that the part can be held or rotated during machining or grinding.

chucking lug. A projection forged or cast onto a part to act as a positive means of driving or locating when the part is being machined.

circle grinding. Either *cylindrical* or *internal grinding,* the preferred terms.

circle shear. A shearing machine with two rotary disk cutters mounted on parallel shafts driven in unison and equipped with an attachment for cutting circles where the desired piece of material is inside the circle. It cannot be employed to cut circles where the desired material is outside the circle.

circular field. The magnetic field that *(a)* surrounds a nonmagnetic conductor of electricity, *(b)* is completely contained within a magnetic conductor of electricity or *(c)* both exists within and surrounds a magnetic conductor. Generally applied to the magnetic field within any magnetic conductor resulting from a current being passed through the part or through a section of the part. Compare with *bipolar field.*

clad metal. A composite metal containing two or three layers that have been bonded together. The bonding may have been accomplished by co-rolling, welding, casting, heavy chemical deposition or heavy electroplating.

clamshell marks. Same as *beach marks.*

classification. (1) The separation of ores into fractions according to size and specific gravity, generally in accordance with Stokes' law of sedimentation. (2) Separation of a metal powder into fractions according to particle size.

clay. An earthy or stony mineral aggregate consisting essentially of hydrous silicates of alumina, and which is plastic when sufficiently pulverized and wetted, rigid when dry, and vitreous when fired at a sufficiently high temperature. Clay minerals most commonly used in the foundry are montmorillonites and kaolinites.

cleanup allowance. See *finish allowance.*

clearance. (1) The gap or space between two mating parts. (2) Space provided between the relief of a cutting tool and the surface that has been cut.

clearance angle. The angle between a plane containing the flank of the tool and a plane passing through the cutting edge in the direction of relative motion between the cutting edge and work. See sketches accompanying *face-mill* and *single-point tool.*

clearance fit. Any of various classes of fit between mating parts where there is a positive allowance (gap) between the parts, even when they are made to the respective extremes of individual tolerances that ensure the tightest fit between the parts. Contrast with *interference fit.*

cleavage. The splitting (fracture) of a crystal on a crystallographic plane of low index.

cleavage fracture. A fracture, usually of a polycrystalline metal, in which most of the grains have failed by cleavage, resulting in bright reflecting facets. It is one type of *crystalline fracture* and is associated with low-energy brittle fracture. Contrast with *shear fracture.*

cleavage plane. A characteristic crystallographic plane or set of planes on which cleavage fracture easily occurs.

climb cutting. Analogous to *climb milling.*

climb milling. Milling in which the cutter moves in the direction of feed at the point of contact.

clip and shave. In forging, a dual operation in which one cutting surface in the clipping die removes the *flash* and then another shaves and sizes the piece.

close annealing. Same as *box annealing.*

closed die forging. See *impression die forging.*

closed dies. Forging or forming impression dies designed to restrict the flow of metal to the cavity within the die set, as opposed to open dies, in which there is little or no restriction to lateral flow.

closed pass. A pass of metal through rolls where the bottom roll has a groove deeper than the bar being rolled and the top roll has a collar fitting into the groove, thus producing the desired shape free from *flash* or fin.

close tolerance forging. A forging held to unusually close dimensional tolerances. Often, little or no machining is required after forging.

cloudburst treatment. A form of *shot peening.*

cluster mill. A rolling mill where each

of the two working rolls of small diameter is supported by two or more backup rolls.

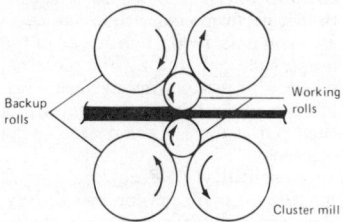

Cluster mill

CO₂ welding. See *carbon dioxide welding.*

coalesced copper. Massive oxygen-free copper made by briquetting ground, brittle cathode copper, then sintering the briquettes in a pressurized reducing atmosphere, followed by hot working.

coalescence. (1) The union of particles of a dispersed phase into larger units, usually effected at temperatures below the fusion point. (2) In welding, brazing or soldering, the union of two or more components into a single body, which usually involves melting of a filler metal or of the base metal.

coarsening. An increase in the grain size, usually, but not necessarily, by *grain growth.*

coated abrasive. An abrasive product, sandpaper being an example, in which a layer of abrasive particles is firmly attached to a paper, cloth or fiber backing by means of glue or synthetic-resin adhesive.

coated electrode. See preferred term, *lightly coated electrode.*

coaxing. Improvement of the fatigue strength of a specimen by the application of a gradually increasing stress amplitude, usually starting below the fatigue limit.

coefficient of elasticity. Same as *modulus of elasticity.*

coercive force. The magnetizing force that must be applied in the direction opposite to that of the previous magnetizing force in order to reduce magnetic flux density to zero; thus, a measure of the magnetic retentivity of magnetic materials.

cogging mill. A *blooming mill.*

coherency. The continuity of lattice of precipitate and parent phase (solvent) maintained by mutual strain and not separated by a phase boundary.

coherent precipitate. A crystalline precipitate that forms from solid solution with an orientation that maintains continuity between the crystal lattice of the precipitate and the lattice of the matrix, usually accompanied by some strain in both lattices. Because the lattices fit at the interface between precipitate and matrix, there is no discernible phase boundary.

cohesion. Force of attraction between the molecules (or atoms) within a single phase. Contrast with *adhesion.*

cohesive strength. (1) The hypothetical stress causing tensile fracture without plastic deformation. (2) The stress corresponding to the forces between atoms. (3) Same as *technical cohesive strength.* (4) Same as *disruptive strength.*

coil breaks. Creases or ridges in sheet or strip that appear as parallel lines across the direction of rolling, and that generally extend the full width of the sheet or strip.

coining. (1) A closed-die squeezing operation, usually performed cold, in which all surfaces of the work are confined or restrained, resulting in a well-defined imprint of the die upon the work. (2) A *restriking* operation used to sharpen or change an existing radius or profile. (3) The final pressing of a sintered powder metallurgy compact to obtain a definite surface configuration (not to be confused with *re-pressing* or *sizing*).

coin silver. An alloy containing 90% silver, with copper being the usual alloying element.

cold chamber machine. A *die-casting* machine where the metal chamber and plunger are not heated.

cold extrusion. See *impact; extrusion.*

cold heading. Working metal at room temperature in such a manner that the cross-sectional area of a portion or all of the stock is increased.

cold inspection. A visual (usually final) inspection of forgings for visible imperfections, dimensions, weight, and surface condition at room temperature. The term may also be used to describe certain nondestructive tests such as magnetic particle, dye penetrant and sonic inspection.

cold lap. Wrinkled markings on the surface of an ingot, caused by incipient freezing of the surface while the liquid is still in motion; results from too low a pouring temperature. See also *cold shut* (1).

cold mill. A mill for cold rolling sheet or strip.

cold pressing. Forming a powder metallurgy *compact* at a temperature low enough to avoid *sintering,* usually room temperature. Contrast with *hot pressing.*

cold rolled sheets. A mill product produced from a hot rolled pickled coil that has been given substantial cold reduction at room temperature. The resulting product usually requires further processing to make it suitable for most common applications. The usual end product is characterized by improved surface, greater uniformity in thickness and improved mechanical properties compared to hot rolled sheet.

cold shortness. Brittleness that exists in some metals at temperatures below the recrystallization temperature.

cold shot. A portion of the surface of an ingot or casting showing premature solidification; caused by splashing of molten metal onto a cold mold wall during pouring.

cold shut. (1) A discontinuity that appears on the surface of cast metal as a result of two streams of liquid meeting and failing to unite. (2) A lap on the surface of a forging or billet that was closed without fusion during deformation. (3) Freezing of the top surface of an ingot before the mold is full.

cold treatment. Exposing to suitable subzero temperatures for the purpose of obtaining desired conditions or properties such as dimensional or microstructural stability. When the treatment involves the transformation of retained austenite, it is usually followed by tempering.

cold trimming. Removing flash or excess metal from the forging in a trimming press when the forging is at room temperature.

cold welding. A solid state welding process in which pressure is used at room temperature to produce coalescence of metals with substantial deformation at the weld. Compare *hot pressure welding, diffusion welding,* and *forge welding.*

cold work. Permanent strain in a metal accompanied by strain hardening.

cold working. Deforming metal plastically under conditions of temperature and strain rate that induce strain hardening. Usually, but not necessarily, conducted at room temperature. Contrast with *hot working.*

collapsibility. The requirement that a

sand mold or core break down under the pressure and temperature of casting in order to avoid hot tears, or to facilitate the separation of sand and casting.

collet. A split sleeve used to hold work or tools during machining or grinding.

color buffing. Producing a final high luster by buffing. Sometimes called "coloring".

coloring. Producing desired colors on metal by a chemical or electrochemical reaction. See also *color buffing.*

columnar structure. A coarse structure of parallel elongated grains formed by unidirectional growth, most often observed in castings, but sometimes in structures resulting from diffusional growth accompanied by a solid-state transformation.

combination die. (1) A die-casting die having two or more different cavities for different castings. (2) For forming, see *compound die.*

combination mill. An arrangement of a continuous mill for roughing, and a *guide mill* or *looping mill* for shaping.

combined carbon. The part of the total carbon in steel or cast iron that is present as other than *free carbon.*

combined cyanide. The cyanide of a metal-cyanide complex ion.

combined stresses. Any state of stress that cannot be represented by a single component of stress; that is, one that is more complicated than simple tension, compression or shear.

comminution. (1) Breaking up or grinding an ore into small fragments. (2) Reducing metal to powder by mechanical means.

commutator-controlled welding. Spot or projection welding in which several electrodes, in simultaneous contact with the work, function progressively under the control of an electrical commutating device.

compact. An object produced by the compression of metal powder, generally while confined in a die, with or without the inclusion of nonmetallic constituents. See also *compound compact* and *composite compact.*

compacted graphite cast iron. Cast iron having a graphite shape intermediate between the flake form typical of gray cast iron and the spherical form of fully spherulitic ductile cast iron. Also known as CG iron or vermicular iron, compacted graphite cast iron is produced in a manner

similar to ductile cast iron, but using a technique that inhibits the formation of fully spherulitic graphite nodules.

complete fusion. Fusion that has occurred over the entire base-metal surfaces exposed for welding.

complexing agent. A substance that is an electron donor and that will combine with a metal ion to form a soluble complex ion.

complex ion. An ion that may be formed by the addition reaction of two or more other ions.

component. (1) One of the elements or compounds used to define a chemical (or alloy) system, including all phases, in terms of the fewest substances possible. (2) One of the individual parts of a vector as referred to a system of coordinates.

composite compact. A powder metallurgy *compact* consisting of two or more adhering layers of different metals or alloys with each layer retaining its original identity.

composite electrode. A welding electrode made from two or more distinct components, at least one of which is filler metal. A composite electrode may exist in any of various physical forms, such as stranded wires, filled tubes or covered wire.

composite joint. A joint in which welding is used in conjunction with mechanical joining.

composite material. A heterogeneous, solid structural material consisting of two or more distinct components that are mechanically or metallurgically bonded together (such as a *cermet,* or boron wire embedded in a matrix of epoxy resin).

composite plate. An electrodeposit consisting of layers of at least two different compositions.

composite structure. A structural member (such as a panel, plate, pipe or other shape) that is built up by bonding together two or more distinct components, each of which may be made of a metal, alloy, nonmetal or *composite material.* Examples of composite structures include: honeycomb panels, clad plate, electrical contacts, sleeve bearings, carbide-tipped drills or lathe tools, and weldments constructed of two or more different alloys.

compound compact. A powder metallurgy *compact* consisting of mixed metals, the particles of which are joined by pressing or sintering, or

both, with each metal particle retaining substantially its original composition.

compound die. Any die so designed that it performs more than one operation on a part with one stroke of the press, such as blanking and piercing, where all functions are performed simultaneously within the confines of the particular blank size being worked.

compressibility. (1) Reciprocal of *bulk modulus.* (2) In powder metallurgy, the reciprocal of the *compression ratio* where a compact is made following a procedure in which the die, the pressure and the pressing speed are specified.

compression ratio. In powder metallurgy, the ratio of the volume of the loose powder to the volume of the compact made from it.

compressive strength. The maximum compressive stress that a material is capable of developing, based on original area of cross section. If a material fails in compression by a shattering fracture, the compressive strength has a very definite value. If a material does not fail in compression by a shattering fracture, the value obtained for compressive strength is an arbitrary value depending upon the degree of distortion that is regarded as indicating complete failure of the material.

concave fillet weld. A fillet weld having a concave face.

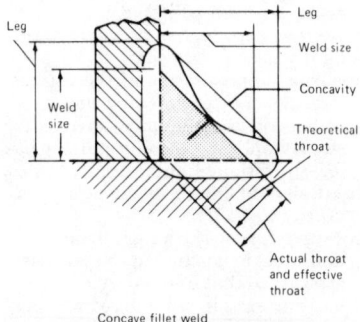

Concave fillet weld

concentration. A process for enrichment of an ore in valuable mineral content by separation and removal of waste material, or *gangue.*

concentration polarization. That part of the total polarization that is caused by changes in the activity of the potential-determining components of the electrolyte.

concurrent heating. Using a second source of heat to supplement the primary heat in cutting or welding.

conditioning heat treatment. A preliminary heat treatment used to prepare a material for a desired reaction to a subsequent heat treatment. For the term to be meaningful, the exact heat treatment must be specified.

congruent melting. An isothermal or isobaric melting in which both the solid and liquid phases have the same composition throughout the transformation.

congruent transformation. An isothermal or isobaric phase change in which both of the phases concerned have the same composition throughout the process.

constantan. A group of copper-nickel alloys containing 45 to 60% copper with minor amounts of iron and manganese and characterized by relatively constant electrical resistivity irrespective of temperature; used in resistors and thermocouples.

constituent. (1) One of the ingredients that make up a chemical system. (2) A phase or combination of phases that occur in a characteristic configuration in an alloy microstructure.

constitution diagram. A graphical representation of the temperature and composition limits of phase fields in an alloy system as they actually exist under the specific conditions of heating or cooling (synonymous with phase diagram). A constitution diagram may be an equilibrium diagram, an approximation to an equilibrium diagram or a representation of metastable conditions or phases. Compare with *equilibrium diagram.*

constraint. Any restriction that occurs to the transverse contraction normally associated with a longitudinal tension, and that hence causes a secondary tension in the transverse direction; usually used in connection with welding. Contrast with *restraint.*

consumable electrode. A general term for any arc-welding electrode made chiefly of filler metal. Use of specific names such as *covered electrode,* bare electrode, flux-cored electrode and *lightly coated* electrode is preferred.

consumable electrode remelting. A process for refining metals in which an electric current passes between an electrode made of the metal to be re-

fined and an ingot of the refined metal, which is contained in a water-cooled mold. As a result of the passage of electric current, droplets of molten metal form on the electrode and fall to the ingot. The refining action occurs from contact with the atmosphere, vacuum or slag through which the drop falls. See *electroslag remelting* and *vacuum arc remelting.*

contact fatigue. Cracking and subsequent pitting of a surface subjected to alternating Hertzian stresses such as those produced under rolling contact or combined rolling and sliding. The phenomenon of contact fatigue is encountered most often in rolling-element bearings or in gears, where the surface stresses are high due to the concentrated loads and are repeated many times during normal operation.

contact plating. A metal plating process where the plating current is provided by galvanic action between the work metal and a second metal, without the use of an external source of current.

contact potential. The potential difference at the junction of two dissimilar substances.

contact scanning. In ultrasonic inspection, a planned systematic movement of the beam relative to the object being inspected, the search unit being in contact with and coupled to this object by a thin film of coupling material.

container. The chamber into which an ingot or billet is inserted prior to extrusion. The container for backward extrusion of cups or cans is sometimes called a die.

continuous casting. A casting technique in which a cast shape is continuously withdrawn through the bottom of the mold as it solidifies, so that its length is not determined by mold dimensions. Used chiefly to produce semifinished mill products such as billets, blooms, ingots, slabs and tubes. See also *strand casting.*

continuous mill. A rolling mill consisting of a number of stands of synchronized rolls (in tandem) in which metal undergoes successive reductions as it passes through the various stands.

continuous phase. In an alloy or portion of an alloy containing more than one phase, the phase that forms the matrix in which the other phase or phases are present as isolated units.

continuous precipitation. Precipita-

tion from a supersaturated solid solution in which the precipitate particles grow by long-range diffusion without recrystallization of the matrix. Continuous precipitates grow from nuclei distributed more or less uniformly throughout the matrix. They usually are randomly oriented, but may form a *Widmanstätten structure.* Also called general precipitation. Compare with *discontinuous precipitation, localized precipitation.*

continuous weld. A weld extending continuously from one end of a joint to the other; where the joint is essentially circular, completely around the joint. Contrast with *intermittent weld.*

contour forming. See *stretch forming, tangent bending, wiper forming.*

contour machining. Machining of irregular surfaces, such as those generated in tracer turning, tracer boring and *tracer milling.*

contour milling. Milling of irregular surfaces. See *tracer milling.*

controlled cooling. Cooling from an elevated temperature in a predetermined manner, to avoid hardening, cracking, or internal damage, or to produce desired microstructure or mechanical properties.

controlled-pressure cycle. A forming cycle during which the hydraulic pressure in the forming cavity is controlled by an adjustable cam that is coordinated with the punch travel.

conventional forging. A forging characterized by design complexity and tolerances that fall within the broad range of general forging practice.

conventional milling. Milling in which the cutter moves in the direction opposite to the feed at the point of contact.

conventional strain. See *strain.*

conventional stress. See *stress.*

conversion coating. A coating consisting of a compound of the surface metal, produced by chemical or electrochemical treatments of the metal. (Examples are chromate coatings on zinc, cadmium, magnesium and aluminum, and oxides and phosphate coatings on steel.)

converter. A furnace in which air is blown through a bath of molten metal or matte, oxidizing the impurities and maintaining the temperature through the heat produced by the oxidation reaction.

convex fillet weld. A fillet weld having a convex face.

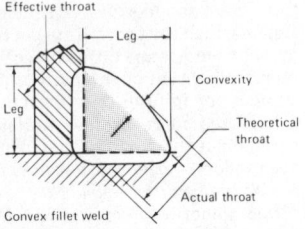

Convex fillet weld

coolant. In metal cutting, the preferred term is *cutting fluid.*

cooling curve. A curve showing the relation between time and temperature during the cooling of a material.

cooling stresses. Residual stresses resulting from nonuniform distribution of temperature during cooling.

cooling table. Same as *hot bed.*

coordination number. (1) Number of atoms or radicals coordinated with the central atom in a complex covalent compound. (2) Number of nearest neighboring atoms to a selected atom in crystal structure.

cope. The upper or topmost section of a flask, mold or pattern.

copper accelerated salt spray test. An accelerated corrosion test for some electrodeposits and for anodic coatings on aluminum. Often referred to as CASS test.

copper brazing. A term improperly used to denote joining with a copper-base filler metal. See preferred terms *brazing* and *braze welding.*

copperhead. A reddish spot in a porcelain enamel coating caused by iron pickup during enameling, iron oxide left on poorly cleaned basis metal, or burrs on iron or steel basis metal that protrude through the coating and are oxidized during firing.

core. (1) A specially formed material inserted in a mold to shape the interior or other part of a casting that cannot be shaped as easily by the pattern. (2) In a ferrous alloy prepared for *case hardening,* that portion of the alloy that is not part of the *case.* Typically considered to be the portion that (*a*) appears light on an etched cross section, (*b*) has an essentially unaltered chemical composition, or (*c*) has a hardness, after hardening, less than a specified value.

core blower. A machine for making foundry cores using compressed air to blow and pack the sand into the core box.

cored bar. A powder metallurgy *compact* of bar shape, the interior of

which has been melted by passage of electricity.

core forging. (1) Displacing metal with a punch to fill a die cavity; (2) the product of such an operation.

core rod. The part of a die used to produce a hole in a powder metallurgy *compact.*

coring. (1) A condition of variable composition between the center and surface of a unit of microstructure (such as a dendrite, grain, carbide particle); results from nonequilibrium solidification, which occurs over a range of temperature. (2) A central cavity at the butt end of rod extrusions, sometimes called *extrusion pipe.*

corner angle. On face milling cutters, the angle between an angular cutting edge of a cutter tooth and the axis of the cutter, measured by rotation into an axial plane. See sketch accompanying *face mill.*

corner joint. A joint between two members located approximately at right angles to each other in the form of an "L".

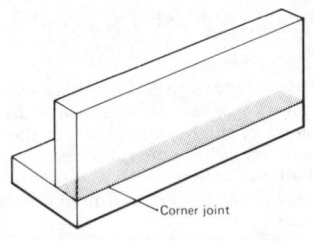

Corner joint

corona. In spot welding, an area sometimes surrounding the nugget at the faying surfaces, where solid state welding occurs. Corona contributes variably to over-all bond strength, depending on the size of the corona and the degree of solid state bonding achieved.

Corrodkote test. An accelerated corrosion test for electrodeposits.

corrosion. The deterioration of a metal by chemical or electrochemical reaction with its environment.

corrosion embrittlement. The severe loss of ductility of a metal resulting from corrosive attack, usually intergranular and often not visually apparent.

corrosion fatigue. Cracking produced by the combined action of repeated or fluctuating stress and a corrosive environment.

corrugating. Forming sheet metal into a series of straight parallel alternate ridges and grooves by using a rolling

mill equipped with matched roller dies or by using a press brake equipped with special-shaped punch and die.

corrugations. Transverse ripples caused by a variation in strip shape during hot or cold reduction.

corundum. Natural abrasive of the aluminum oxide type that has higher purity than emery.

Cottrell process. Removal of solid particulates from gases with electrostatic precipitation.

coulometer. An electrolytic cell arranged to measure the quantity of electricity by the chemical action produced in accordance with Faraday's law.

counterblow hammer. A forging hammer in which both the *ram* and *anvil* are driven simultaneously toward each other by air or steam pistons.

counterboring. Drilling or boring a flat-bottomed hole, often concentric with other holes.

counterlock. A jog in the mating surfaces of dies to prevent lateral die shifting from side thrusts developed in forging irregular-shaped pieces.

countersinking. Forming a flaring depression around the top of a hole for deburring, for receiving the head of a fastener or for receiving a center.

coupling. The degree of mutual interaction between two or more elements resulting from mechanical, acoustical or electrical linkage.

coupon. A piece of metal from which a test specimen is to be prepared—often an extra piece (as on a casting or forging) or a separate piece made for test purposes (such as a test weldment).

covalent bond. A bond between two or more atoms resulting from the completion of shells by the sharing of electrons.

covered electrode. A composite filler-metal welding electrode, consisting of a bare wire or a metal-cored electrode plus a covering sufficient to provide a layer of slag on deposited weld metal. The covering often contains materials that provide shielding during welding, deoxidizers for the weld metal, and arc stabilization; it may also contain alloying elements or other additives for the weld metal.

cover half. The stationary half of a die-casting die.

covering power. The ability of a solution to give a satisfactory plate at very low current densities, a condition that would exist in recesses and pits. This term suggests an ability to cover, but not necessarily to build up, a uniform coating, while *throwing power* suggests ability to get a uniform thickness on an irregularly shaped object.

"C" process. See *Croning process.*

crank press. A mechanical press, the slides of which are actuated by a crankshaft.

crater. (1) In machining, a depression in a cutting tool face eroded by chip contact. (2) In arc welding, a depression at the termination of a bead or in the weld pool beneath the electrode.

crater crack. A crack, often star shaped, that forms in the crater of a weld bead, usually during cooling after welding.

creep. Time-dependent strain occurring under stress. The creep strain occurring at a diminishing rate is called primary creep; that occurring at a minimum and almost constant rate, secondary creep; that occurring at an accelerating rate, tertiary creep.

creep limit. (1) The maximum stress that will cause less than a specified quantity of creep in a given time. (2) The maximum nominal stress under which the creep strain rate decreases continuously with time under constant load and at constant temperature. Sometimes used synonymously with *creep strength.*

creep recovery. Time-dependent strain after release of load in a creep test.

creep-rupture test. Same as *stress-rupture test.*

creep strength. (1) The constant nominal stress that will cause a specified quantity of creep in a given time at constant temperature. (2) The constant nominal stress that will cause a specified rate of secondary creep at constant temperature.

crevice corrosion. A type of concentration-cell corrosion; corrosion caused by the concentration or depletion of dissolved salts, metal ions, oxygen or other gases, and such, in crevices or pockets remote from the principal fluid stream, with a resultant building up of differential cells that ultimately cause deep pitting.

crimping. Forming relatively small corrugations in order to: (1) set down and lock a seam, (2) create an arc in a strip of metal, or (3) reduce an existing arc or diameter.

critical cooling rate. The rate of continuous cooling required to prevent undesirable transformation. For steel, it is the minimum rate at which austenite must be continuously cooled to suppress transformations above the M_s temperature.

critical current density. In an electrolytic process, a current density at which an abrupt change occurs in an operating variable or in the nature of an electrodeposit or electrode film.

critical point. (1) The temperature or pressure at which a change in crystal structure, phase or physical properties occurs. Same as *transformation temperature.* (2) In an equilibrium diagram, that specific value of composition, temperature and pressure, or combinations thereof, at which the phases of a heterogeneous system are in equilibrium.

critical shear stress. The shearing stress required to cause slip in a designated slip direction on a given slip plane. It is called the critical resolved shear stress if the shearing stress is induced by tension or compression forces acting on the crystal.

critical strain. The strain just sufficient to cause *recrystallization;* because the strain is small, usually only a few percent, recrystallization takes place from only a few nuclei, which produces a recrystallized structure consisting of very large grains.

critical temperature. (1) Synonymous with *critical point* if the pressure is constant. (2) The temperature above which the vapor phase cannot be condensed to liquid by an increase in pressure.

critical temperature ranges. Synonymous with *transformation ranges,* which is the preferred term.

Croning process. A *shell molding* process utilizing a phenolic resin binder. Sometimes referred to as "C" process.

crop. (1) An end portion of an ingot that is cut off as scrap. (2) To shear a bar or billet.

crop end. See *end loss.*

cross breaks. Same as *coil breaks.*

cross-country mill. A rolling mill in which the mill stands are so arranged that their tables are parallel with a transfer (or crossover) table connecting them. They are used for rolling structural shapes, rails and any special form of bar stock not rolled in the ordinary bar mill.

crossed joint. See *bridging* (5).

cross forging. Preliminary working of forging stock in flat dies to develop mechanical properties, particularly in the center portions of heavy sections.

cross rolling. The rolling of sheet or plate so that the direction of rolling is about 90° from the direction of a previous rolling.

cross-roll straightener. A machine having paired rolls of special design for straightening round bars or tubes, the pass being made with the work parallel to the axes of the rolls.

cross-wire weld. A weld made at the junction between crossed wires or bars.

crown. (1) A contour on a sheet or roll where the thickness or diameter increases from edge to center. (2) The top section of a press structure where the cylinders and other working parts may be mounted. Also called dome, head or top platen.

crucible. A vessel or pot, made of a refractory substance or of a metal with a high melting point, used for melting metals or other substances.

crush. (1) Buckling or breaking of a section of a casting mold due to incorrect register when the mold is closed. (2) An indentation in the surface of a casting due to displacement of sand when the mold was closed.

crush forming. Shaping a grinding wheel by forcing a rotating metal roll into its face so as to reproduce the desired contour.

crushing test. (1) A radial compressive test applied to tubing, sintered-metal bearings or other similar products for determining radial crushing strength (maximum load in compression). (2) An axial compressive test for determining quality of tubing, such as soundness of weld in welded tubing.

crystal. A solid composed of atoms, ions or molecules arranged in a pattern that is repetitive in three dimensions.

crystalline fracture. A pattern of brightly reflecting crystal facets on the fracture surface of a polycrystalline metal and resulting from cleavage fracture of many individual crystals. Contrast with *fibrous fracture, silky fracture.*

crystallization. (1) The separation, usually from a liquid phase on cooling, of a solid crystalline phase. (2) Sometimes erroneously used to explain fracturing that actually has occurred by fatigue.

crystal orientation. See *orientation.*

cubic plane. A plane perpendicular to any one of the three crystallographic axes of the cubic (isometric) system; the *Miller indices* are {100}.

cup. (1) Sheet-metal part, the product of the first step in deep drawing. (2) Any cylindrical part or shell closed at one end.

cupellation. Oxidation of molten lead containing gold and silver to produce lead oxide, thereby separating the precious metals from the base metal.

cup fracture (cup-and-cone fracture). A mixed-mode fracture, often seen in tensile test specimens of a ductile material, where the central portion undergoes *plane-strain* fracture and the surrounding region undergoes *plane-stress* fracture. It is called a cup fracture (or cup-and-cone fracture) because one of the mating fracture surfaces looks like a miniature cup—that is, it has a central depressed flat-face region surrounded by a shear lip; the other fracture surface looks like a miniature truncated cone.

cupola. A cylindrical vertical furnace for melting metal, especially cast iron, by having the charge come in contact with the hot fuel, usually metallurgical coke.

cupping. (1) The first step in deep drawing. (2) The fracture of severely worked rods or wire where one end has the appearance of a cup and the other that of a cone.

Curie temperature. The temperature of magnetic transformation below which a metal or alloy is ferromagnetic and above which it is paramagnetic.

curling. Rounding the edge of sheet metal into a closed or partly closed loop.

current decay. In spot, seam or projection welding, the controlled reduction of the welding current from its peak amplitude to a lower value to prevent excessively rapid cooling of the weld nugget.

current efficiency. The proportion of current used in a given process to accomplish a desired result; in electroplating, the proportion used in depositing or dissolving metal.

cushion. Same as *die cushion*.

cut. (1) In castings, rough spots or areas of excess metal caused by erosion of the mold or core surface by metal flow. (2) In powder metallurgy, same as *fraction*.

cut-and-carry method. Stamping method where the part remains at-tached to the strip or is forced back into the strip to be fed through the succeeding stations of a progressive die.

cut edge. A mechanically sheared edge obtained by slitting, shearing or blanking.

cutoff wheel. A thin abrasive wheel for severing or slotting any material or part.

cutting down. Removing roughness or irregularities of a metal surface by abrasive action.

cutting edge. The leading edge of a cutting tool (such as a lathe tool, drill or milling cutter) where a line of contact is made with the work during machining. See sketch accompanying *single-point tool*.

cutting fluid. A fluid used in metal cutting to improve finish, tool life or dimensional accuracy. On being flowed over the tool and work, the fluid reduces the friction, the heat generated and the tool wear and prevents galling. It conducts the heat away from the point of generation and also serves to wash the *chips* away.

cutting speed. The linear or peripheral speed of relative motion between the tool and workpiece in the principal direction of cutting.

cutting tip. The part of a cutting torch from which gas issues.

cyanide copper. Copper electrodeposited from an alkali-cyanide solution containing a complex ion made up of univalent copper and the cyanide radical; also, the solution itself.

cyanide slimes. Finely divided metallic precipitates that are formed when precious metals are extracted from their ores using cyanide solutions.

cyaniding. A case hardening process in which a ferrous material is heated above the lower transformation range in a molten salt containing cyanide to cause simultaneous absorption of carbon and nitrogen at the surface and, by diffusion, create a concentration gradient. Quench hardening completes the process.

cycle annealing. An annealing process employing a predetermined and closely controlled time-temperature cycle to produce specific properties or microstructures.

cylindrical grinding. Grinding the outer cylindrical surface of a rotating part.

cylindrical land. *Land* having zero relief.

D

damping capacity. The ability of a material to absorb vibration (cyclical stresses) by internal friction, converting the mechanical energy into heat.

dangler. The flexible electrode used in barrel plating to conduct current to the work.

daylight. The maximum clear distance between the pressing surfaces of a hydraulic press with the surfaces in their usable open position. Where a bolster is supplied, it shall be considered the pressing surface. See also *shut height*.

dc casting. Same as *direct chill casting*.

direct chill casting. A continuous method of making ingots for rolling or extrusion by pouring the metal into a short mold. The base of the mold is a platform that is gradually lowered while the metal solidifies, the frozen shell of metal acting as a retainer for the liquid metal below the wall of the mold. The ingot is usually cooled by the impingement of water directly on the mold or on the walls of the solid metal as it is lowered. The length of the ingot is limited by the depth to which the platform can be lowered; therefore, it is often called semicontinuous casting.

dead center. (1) A stationary center to hold rotating work. (2) Either of the two points in the path of a moving crank or connecting rod that lie at the ends of its stroke.

dead roast. A *roasting* process for complete elimination of sulfur. Also known as *sweet roast*.

dead soft. A *temper* of nonferrous alloys and some ferrous alloys corresponding to the condition of minimum hardness and tensile strength produced by *full annealing*.

deburring. Removing burrs, sharp edges or fins from metal parts by filing, grinding, or rolling the work in a barrel containing abrasives suspended in a suitable liquid medium. Sometimes called burring.

decalescence. A phenomenon, associated with the transformation of alpha iron to gamma iron on the heating (superheating) of iron or steel, revealed by the darkening of the metal surface owing to the sudden **decrease in temperature caused by the fast absorption of the latent heat**

of transformation. Contrast with *recalescence*.

decarburization. Loss of carbon from the surface layer of a carbon-containing alloy due to reaction with one or more chemical substances in a medium that contacts the surface.

decomposition potential. The minimum potential difference necessary to decompose the electrolyte of a cell.

deep drawing. Forming deeply recessed parts by forcing sheet metal to undergo plastic flow between dies, usually without substantial thinning of the sheet.

deep etching. Severe *macroetching*.

defect. A departure of any *quality characteristic* from its intended (usually specified) level that is severe enough to cause the product or service not to fulfill its anticipated function. According to ANSI standards, defects are classified according to severity:

Very serious defects lead directly to severe injury or catastrophic economic loss.

Serious defects lead directly to significant injury or significant economic loss.

Major defects are related to major problems with respect to anticipated use.

Minor defects are related to minor problems with respect to anticipated use.

defective. A quality control term describing a unit of product or service containing at least one *defect*, or having several lesser imperfections that, in combination, cause the unit not to fulfill its anticipated function. NOTE: The term *defective* is not synonymous with *nonconforming* (or rejectable) and should be applied only to those units incapable of performing their anticipated functions.

deformation bands. Parts of a crystal that have rotated differently during deformation to produce bands of varied orientation within individual grains.

degasifier. A substance that can be added to molten metal to remove soluble gases that might otherwise be occluded or entrapped in the metal during solidification.

degassing. Removing gases from liquids or solids.

degreasing. Removing oil or grease from a surface. See *solvent degreasing* and *vapor degreasing*.

degrees of freedom. The number of independent variables (such as temperature, pressure or concentration within the phases present) that may be altered at will without causing a phase change in an alloy system at equilibrium; or the number of such variables that must be fixed arbitrarily to define the system completely.

delayed yield. A phenomenon involving a delay in time between the application of a stress and the occurrence of the corresponding yield point strain.

delta ferrite. See *ferrite*.

dendrite. A crystal that has a treelike branching pattern, being most evident in cast metals slowly cooled through the solidification range. Illustrated below.

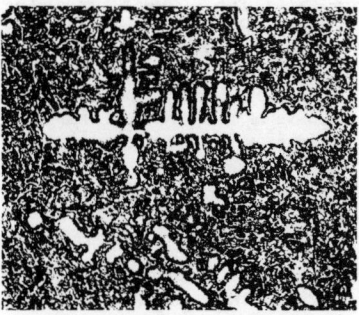

dendritic powder. Particles of metal powder, usually of electrolytic origin, having typical pine-tree structure.

denickelification. Corrosion in which nickel is selectively leached from nickel-containing alloys. Most commonly observed in copper-nickel alloys after extended service in fresh water.

density ratio. The ratio of the determined density of a powder metallurgy compact to the absolute density of metal of the same composition, usually expressed as a percentage.

deoxidized copper. Copper from which cuprous oxide has been removed by adding a deoxidizer, such as phosphorus, to the molten bath.

deoxidizer. A substance that can be added to molten metal to remove either free or combined oxygen.

deoxidizing. (1) The removal of oxygen from molten metals by use of suitable deoxidizers. (2) Sometimes refers to the removal of undesirable elements other than oxygen by the introduction of elements or compounds that readily react with them. (3) In metal

finishing, the removal of oxide films from metal surfaces by chemical or electrochemical reaction.

depolarization. A decrease in the *polarization* of an electrode.

depolarizer. A substance that produces *depolarization*.

deposition efficiency. In welding, the ratio of the weight of deposited weld metal to the net weight of electrodes consumed, exclusive of stubs.

deposition sequence. The order in which increments of weld metal are deposited.

depth of cut. The thickness of material removed from a workpiece in a single machining pass.

depth of fusion. In welding, the distance that fusion extends into the base metal or into a previous pass.

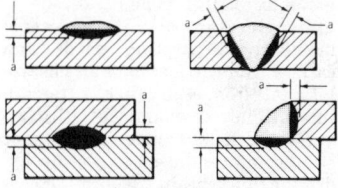

a = depth of fusion

depth of penetration. See *joint penetration* and *root penetration*.

derby. A massive piece (intermediate in size, extending to more than 100 lb, and usually cylindrical) of primary metal made by bomb reduction (such as uranium from uranium tetrafluoride reduced with magnesium). Compare with *biscuit* and *dingot*.

descaling. Removing the thick layer of oxides formed on some metals at elevated temperatures.

deseaming. Analogous to *chipping*, the surface imperfections being removed by gas cutting.

detergent. A chemical substance, generally used in aqueous solution, that removes *soil*.

detritus. Wear debris.

developed blank. A blank that requires little or no trimming when formed.

dewaxing. Removing the expendable wax pattern from an investment mold by heat or solvent.

dezincification. Corrosion in which zinc is selectively leached from zinc-containing alloys. Most commonly found in copper-zinc alloys containing less than 85% copper after extended service in water containing dissolved oxygen.

diamagnetic material. A material

whose specific permeability is less than unity and is therefore repelled weakly by a magnet. Compare with *ferromagnetic material, paramagnetic material.*

diamond boring. Precision boring with a shaped diamond (but not with other tool materials).

diamond pyramid hardness test. See *Vickers hardness test.*

diamond tool. (1) A diamond, shaped or formed to the contour of a single-pointed cutting tool, for use in the precision machining of nonferrous or nonmetallic materials. (2) Sometimes an insert made from multicrystalline diamond compacts.

diamond wheel. A grinding wheel in which crushed and sized industrial diamonds are held in a resinoid, metal or vitrified bond.

diaphragm. (1) A porous or permeable membrane separating anode and cathode compartments of an electrolytic cell from each other or from an intermediate compartment. (2) Universal die member made of rubber or similar material used to contain hydraulic fluid within the forming cavity and transmit pressure to the part being formed.

dichromate treatment. A chromate *conversion coating* produced on magnesium alloys in a boiling solution of sodium dichromate.

didymium. A natural mixture of the rare-earth elements praseodymium and neodymium, often given the quasichemical symbol Di.

die. A tool, usually containing a cavity, that imparts shape to solid, molten or powdered metal primarily because of the shape of the tool itself. Used in many press operations (including blanking, drawing, forging and forming), in die casting and in forming green powder metallurgy compacts. Die-casting and powder-metallurgy dies are sometimes referred to as *molds.*

die block. A block, usually of tool steel, into which the desired impressions are sunk, formed, or machined and from which forgings or die castings are made.

die body. The stationary or fixed part of a powder pressing die.

die casting. (1) A casting made in a die. (2) A casting process where molten metal is forced under high pressure into the cavity of a metal mold.

die clearance. Clearance between a mated punch and die; commonly expressed as clearance per side. Also

called clearance, punch-to-die clearance.

die cushion. A press accessory located beneath or within a *bolster* or *die block* to provide an additional motion or pressure for stamping operations; actuated by air, oil, rubber or springs, or by a combination thereof.

die forging. A forging whose shape is determined by impressions in specially prepared dies.

die forming. The shaping of solid or powdered metal by forcing it into or through the cavity in a die.

die holder. A plate or block, upon which the die block is mounted, having holes or slots for fastening to the bolster or the bed of the press.

die insert. A removable liner or part of a die body or punch.

die layout. The transfer of drawing or sketch dimensions to templates or die surfaces for use in sinking dies.

die life. The productive life of a die impression, usually expressed as the number of units produced before the impression has worn beyond permitted tolerances.

die lines. Lines or markings on formed, drawn or extruded metal parts caused by imperfections in the surface of the die.

die lubricant. A lubricant applied to working surfaces of dies and punches to facilitate drawing, pressing, stamping and/or ejection. In powder metallurgy, the die lubricant is sometimes mixed into the powder before pressing into a compact.

die match. The condition where dies, after having been set up in a press or other equipment, are in proper alignment relative to each other.

die opening. In flash or upset welding, the distance between the electrodes, usually measured with the parts in contact before welding has commenced or immediately upon completion of the cycle but before upsetting.

die proof. A casting of the die impression made to confirm the exactness of the impression. Also called cast.

die radius. The radius on the exposed edge of a drawing die, over which the sheet flows in forming drawn shells.

die scalping. Removing surface layers from bar, rod, wire or tube by drawing through a sharp-edged die to eliminate minor surface defects.

die set. A tool or tool holder consisting of a die base and punch plate for the attachment of a die and punch, respectively.

die shift. A condition requiring correction where, after dies have been set up in the forging equipment, displacement of a point in one die from the corresponding point in the opposite die occurs in a direction parallel to the fundamental parting line of the dies.

die sinking. Forming or machining a depressed pattern in a die.

die welding. Forge welding between dies.

differential coating. A coated product having a specified coating on one surface and a significantly lighter coating on the other surface (such as a hot dip galvanized product or electrolytic tin plate).

differential flotation. Separating a complex ore into two or more valuable minerals and *gangue* by *flotation.* Also called selective flotation.

differential heating. Heating that intentionally produces a temperature gradient within an object such that, after cooling, a desired stress distribution or variation in properties is present within the object.

diffusion. (1) Spreading of a constituent in a gas, liquid or solid, tending to make the composition of all parts uniform. (2) The spontaneous movement of atoms or molecules to new sites within a material.

diffusion aid. A solid filler metal sometimes used in *diffusion welding.*

diffusion bonding. See preferred terms *diffusion welding, diffusion brazing.*

diffusion brazing. A brazing process that joins two or more components by heating them to suitable temperatures and by using a filler metal or an *in situ* liquid phase. The filler metal may be distributed by capillary attraction or may be placed or formed at the faying surfaces. The filler metal is diffused with the base metal to the extent that the joint properties have been changed to approach those of the base metal.

diffusion coating. Any process whereby a basis metal or alloy is either: (1) coated with another metal or alloy and heated to a sufficient temperature in a suitable environment or (2) exposed to a gaseous or liquid medium containing the other metal or alloy, thus causing diffusion of the coating or of the other metal or alloy into the basis metal with resultant change in the composition and properties of its surface.

diffusion coefficient. A factor of proportionality representing the amount of substance diffusing across a unit area through a unit concentration gradient in unit time.

diffusion welding. A high-temperature solid state welding process that permanently joins faying surfaces by the simultaneous application of pressure and heat. The process does not involve macroscopic deformation, melting, or relative motion of parts. A solid filler metal (diffusion aid) may or may not be inserted between the faying surfaces.

digging. A sudden erratic increase in cutting depth or in the load of a cutting tool caused by unstable conditions in the machine setup. Usually the machine is stalled or either the tool or the workpiece is destroyed.

dilatometer. An instrument for measuring the linear expansion or contraction in a metal resulting from changes in such factors as temperature or allotropy.

dimple rupture. A fractographic term describing ductile fracture that occurred through the formation and coalescence of microvoids along the fracture path. The fracture surface of such a ductile fracture appears dimpled when observed at high magnification and usually is most clearly resolved when viewed in a scanning electron microscope.

dimpling. (1) Stretching a relatively small shallow indentation into sheet metal. (2) In aircraft, stretching thin metal into a conical flange for use with a countersunk head rivet.

dingot. An oversized *derby* (possibly a ton or more) of a metal produced in a bomb reaction (such as uranium from uranium tetrafluoride reduced with magnesium). The term "ingot" for these metals is reserved for massive units produced in vacuum melting and casting. See *biscuit* and *derby*.

dinking. Cutting of nonmetallic materials or light-gage soft metals by using a hollow punch with a knifelike edge acting against a wooden fiber or resiliently mounted metal plate.

dip brazing. Brazing by immersing the assembly to be joined in a bath of hot molten chemicals or hot metal. A molten chemical bath may provide brazing flux; molten metal, the filler metal.

diphase cleaning. Removing *soil* by an emulsion that produces two phases in the cleaning tank: a solvent phase and an aqueous phase. Cleaning is effected by both solvent action and emulsification.

dip plating. Same as *immersion plating*.

direct-arc furnace. An electric-arc furnace in which the metallic charge is one of the poles of the arc.

direct-current cleaning. Same as *cathodic cleaning*.

direct extrusion. See *extrusion*.

directional property. Property whose magnitude varies depending on the relation of the test axis to a specific direction within the metal. The variation results from preferred orientation or from fibering of constituents or inclusions.

directional solidification. The solidification of molten metal in such a manner that feed metal is always available for that portion that is just solidifying.

direct quenching. (1) Quenching carburized parts directly from the carburizing operation. (2) Also used for quenching pearlitic malleable parts directly from the malleablizing operation.

discontinuity. Any interruption in the normal physical structure or configuration of a part, such as cracks, laps, seams, inclusions or porosity. A discontinuity may or may not affect the usefulness of a part.

discontinuous precipitation. Precipitation from a supersaturated solid solution in which the precipitate particles grow by short-range diffusion, accompanied by recrystallization of the matrix in the region of precipitation. Discontinuous precipitates grow into the matrix from nuclei near grain boundaries, forming cells of alternate lamellae of precipitate and depleted (and recrystallized) matrix. Often referred to as cellular or nodular precipitation. Compare with *continuous precipitation, localized precipitation*.

discontinuous yielding. The nonuniform plastic flow of a metal exhibiting a yield point in which plastic deformation is inhomogeneously distributed along the gage length. Under some circumstances, it may occur in metals not exhibiting a distinct yield point, either at the onset of or during plastic flow.

dishing. Forming a shallow concave surface, the area being large compared to the depth.

disk grinding. Grinding with the flat side of an abrasive disk or segmented wheel.

dislocation. A linear imperfection in a crystalline array of atoms. Two basic types are recognized: an edge dislocation corresponds to the row of mismatched atoms along the edge formed by an extra, partial plane of atoms within the body of a crystal; a screw dislocation corresponds to the axis of a spiral structure in a crystal, characterized by a distortion that joins normally parallel planes together to form a continuous helical ramp (with a pitch of one interplanar distance) winding about the dislocation. Most prevalent is the so-called mixed dislocation, which is the name given to any combination of an edge dislocation and a screw dislocation.

disordering. Forming a lattice arrangement in which the solute and solvent atoms of a solid solution occupy lattice sites at random. See also *ordering, superlattice*.

dispersing agent. A material that increases the stability of a suspension of particles in a liquid medium by deflocculation of the primary particles.

disruptive strength. The stress at which a metal fractures under hydrostatic tension.

distortion. Any deviation from an original size, shape or contour that occurs because of the application of stress or the release of residual stress.

disturbed metal. The cold worked metal layer formed at a polished surface during the process of mechanical grinding and polishing.

divided cell. A cell containing a diaphragm or other means for physically separating the anolyte from the catholyte.

divorced eutectic. A metallographic appearance in which the two constituents of a eutectic structure appear as massive phases rather than the finely divided mixture characteristic of normal eutectics. Often, one of the constituents of the eutectic is continuous with and indistinguishable from an accompanying proeutectic constituent.

domain. A substructure in a ferromagnetic material within which all the elementary magnets (electron spins) are held aligned in one direction by interatomic forces; if isolated, a domain would be a saturated permanent magnet.

doré silver. Crude silver containing a small amount of gold, obtained after removing lead in a cupelling furnace. Same as doré bullion and doré metal.

double-acting hammer. A forging hammer in which the ram is raised by admitting steam or air into a cylinder below the piston, and the blow intensified by admitting steam or air above the piston on the downward stroke.

double-action die. A die designed to perform more than one operation in a single stroke of the press.

double-action forming. Forming or drawing where more than one action is achieved in a single stroke of the press.

double-action mechanical press. A press having two independent parallel movements by means of two slides, one moving within the other. The inner slide or plunger is usually operated by a crankshaft, whereas the outer or blankholder slide, which dwells during the drawing operation, is usually operated by a toggle mechanism or cams.

double aging. Employment of two different aging treatments to control the type of precipitate formed from a supersaturated matrix in order to obtain the desired properties. The first aging treatment, sometimes referred to as intermediate or stabilizing, is usually carried out at higher temperature than the second.

double-bevel groove weld. A groove weld in which the joint edge of one member is beveled from both sides.

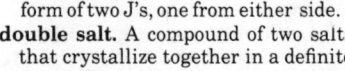
Double-bevel

double-J groove weld. A groove weld in which the joint edge of one member is in the form of two J's, one from either side.

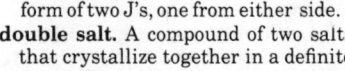
Double-J

double salt. A compound of two salts that crystallize together in a definite proportion.

double tempering. A treatment in which a quench-hardened ferrous metal is subjected to two complete tempering cycles, usually at substantially the same temperature, for the purpose of ensuring completion of the tempering reaction and promoting stability of the resulting microstructure.

double-U groove weld. A groove weld in which each joint edge is in the form of two J's or two half-U's, one from either side of the member.

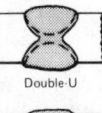
Double-U

double-V groove weld. A groove weld in which each joint

Double-V

edge is beveled from both sides.

double-welded joint. A butt, edge, tee, corner or lap joint in which welding has been done from both sides.

down cutting. See preferred term, *climb cutting.*

downgate. Same as *sprue.*

downhand welding. See *flat-position welding.*

down milling. See preferred term, *climb milling.*

down slope time. In resistance welding, time associated with current decrease using *slope control.*

downsprue. Same as *sprue.*

Dow process. A process for the production of magnesium by electrolysis of molten magnesium chloride.

draft. (1) An angle or taper on the surface of a pattern, core box, punch or die (or of the parts made with them) that makes it easier to remove the parts from a mold or die cavity, or to remove a core from a casting. (2) The change in cross section that occurs during rolling or cold drawing.

drag. The bottom section of a flask, mold or pattern.

drag angle. In welding, the angle between the axis of the electrode or torch and a line normal to the plane of the weld when welding is being done with the torch positioned ahead of the weld puddle. See sketch accompanying *backhand welding.*

drag-in. Water or solution carried into another solution by the work and its associated handling equipment.

dragout. Solution carried out of a bath by the work and its associated handling equipment.

drag technique. A method used in manual arc welding where the electrode is in contact with the assembly being welded without being in short circuit. The electrode is usually used without oscillation.

drawability. A measure of the workability of a metal subject to a drawing process. A term usually expressed to indicate a metal's ability to be deep drawn.

draw bead. (1) A bead or offset used for

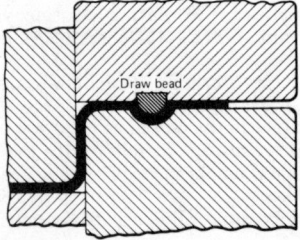
Draw bead

controlling metal flow. (2) Riblike projections on draw rings or hold-down surfaces for controlling metal flow.

drawbench. The stand that holds the die and draw head used in the drawing of wire, rod and tubing.

draw forging. See *radial forging.*

draw forming. A method of curving bars, tubes or rolled or extruded sections, in which the stock is bent around a rotating form block. Stock is bent by clamping it to the form block, then rotating the form block while the stock is pressed between the form block and a pressure die held against the periphery of the form block. Contrast with *wiper forming.*

draw head. Set of rolls or dies mounted on a drawbench for forming a section from strip, tubing or solid stock. See *Turk's-head rolls.*

drawing. (1) Forming recessed parts by forcing the plastic flow of metal in dies. (2) Reducing the cross section of bar stock, wire or tubing by pulling it through a die. (3) A misnomer for *tempering.*

drawing compound. A substance applied to prevent *pickup* and *scoring* during drawing or pressing operations by preventing metal-to-metal contact of the work and die. Also known as *die lubricant.*

drawing out. A stretching operation resulting from forging a series of upsets along the length of the workpiece.

draw marks. See *scoring, galling, pickup, die lines.*

drawn shell. An article formed by drawing sheet metal into a hollow structure having a predetermined geometrical configuration.

draw plate. A circular plate with a hole in the center contoured to fit a forming punch, used to support the blank during the forming cycle.

draw radius. The radius at the edge of a die or punch over which the work is drawn.

draw ring. A ring-shaped die part over the inner edge of which the metal is drawn by the punch.

dresser. A tool used for *truing* and *dressing* a grinding wheel.

dressing. Cutting, breaking down or crushing the surface of a grinding wheel to improve its cutting ability and accuracy.

drift. (1) A flat piece of steel of tapering width used to remove taper shank drills and other tools from their holders. (2) A tapered rod used to force

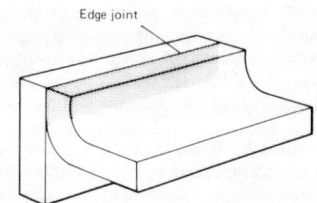
Edge joint

mismated holes in line for riveting or bolting. Sometimes called a drift pin.

drill. A rotary end-cutting tool used for making holes; it has one or more cutting lips and an equal number of helical or straight flutes for the passage of chips and the admission of a cutting fluid.

drive fit. A type of *force fit*.

drop. A casting imperfection due to a portion of the sand dropping from the cope or other overhanging section of the mold.

drop forging. A shallow forging made in impression dies; usually with a drop hammer.

drop hammer. A forging hammer that depends on gravity for its force.

dross. The scum that forms on the surface of molten metals largely because of oxidation but sometimes because of the rising of impurities to the surface.

dry cyaniding. (obsolete) Same as *carbonitriding*.

dry sand mold. A casting mold made of sand and then dried at 100°C (212°F) or above before using. Contrast with *green sand mold*.

ductile cast iron. A *cast iron* that has been treated while molten with an element such as magnesium or cerium to induce the formation of free graphite as nodules or spherulites, which imparts a measurable degree of ductility to the cast metal. Also known as nodular cast iron, spherulitic graphite cast iron and SG iron.

ductile crack propagation. Slow crack propagation that is accompanied by noticeable plastic deformation and requires energy to be supplied from outside the body.

ductile fracture. Fracture characterized by tearing of metal accompanied by appreciable gross plastic deformation and expenditure of considerable energy.

ductility. The ability of a material to deform plastically without fracturing, being measured by elongation or reduction of area in a tensile test, by height of cupping in an Erichsen test or by other means.

dummy block. In extrusion, a thick unattached disk placed between the ram and billet to prevent overheating of the ram.

dummy cathode. (1) A cathode, usually corrugated to give variable current densities, that is plated at low current densities to preferentially remove impurities from a plating solution. (2) A substitute cathode that is used during adjustment of operating conditions.

dummying. Plating with *dummy cathodes*.

duplex coating. See *composite plate*.

duplexing. Any two-furnace melting or refining process. Also called duplex melting or duplex processing.

duplicating. In machining and grinding, reproducing a form from a master with an appropriate type of machine tool, utilizing a suitable tracer or program-controlled mechanism.

duralumin. (obsolete) A term formerly applied to the class of age-hardenable aluminum-copper alloys containing manganese, magnesium or silicon.

Durville process. A casting process that involves rigid attachment of the mold in an inverted position above the crucible. The melt is poured by tilting the entire assembly, causing the metal to flow along a connecting launder and down the side of the mold.

dusting. Applying a powder; as sulfur to molten magnesium, or graphite to a mold surface.

duty cycle. For electric welding equipment, the percentage of time that current flows during a specified period. In arc welding, the specified period is 10 min.

dynamic creep. Creep that occurs under conditions of fluctuating load or fluctuating temperature.

E

earing. The formation of scallops (ears) around the top edge of a drawn part caused by directional differences in the properties of the sheet metal used.

eccentric press. A mechanical press in which the eccentric and strap are used to move the slide, rather than a crankshaft and connection.

ECM. An abbreviation for *electrochemical machining*.

eddy-current testing. An electromagnetic nondestructive testing method in which eddy-current flow is induced in the test object. Changes in the flow caused by variations in the object are reflected into a nearby coil or coils where they are detected and measured by suitable instrumentation.

edge dislocation. See *dislocation*.

edge joint. A joint between the edges of two or more parallel or nearly parallel members.

edger. In forging, the portion of a die that generally distributes the metal in portions required for the shape to be forged, usually a gathering operation. A rolling edger shapes the stock into various solids of revolution; a ball edger forms a ball.

edge strain. Transverse strain lines or Lüders lines ranging from 25 to 300 mm (1 to 12 in.) in from the edges of cold rolled steel sheet or strip.

edging. (1) In forming, reducing the flange radius by retracting the forming punch a small amount after the stroke but prior to releasing the pressure. (2) In forging, removing flash that is directed upward between dies, usually accomplished in a lathe. (3) In rolling, working metal where the axis of the roll is parallel to the thickness dimension. Also called edge rolling.

EDM. An abbreviation for *electrical discharge machining*.

effective rake. The angle between a plane containing a tooth face and the axial plane through the tooth point as measured in the direction of chip flow through the tooth point. Thus, it is the rake resulting from both cutter configuration and direction of chip flow.

ejector. A device mounted in such a way that it removes or assists in removing a formed part from a die.

ejector half. The movable half of a die-casting die containing the ejector pins.

elastic aftereffect. Time-dependent recovery, toward original dimensions, after the load has been reduced or removed from an elastically or plastically strained body. See *anelasticity*.

elastic constants. Factors of proportionality that describe elastic response of a material to applied forces; includes *modulus of elasticity* (either in tension, compression or shear), *Poisson's ratio, compressibility* and bulk modulus.

elastic deformation. A change in dimensions directly proportional to and in phase with an increase or decrease in applied force.

elastic hysteresis. A misnomer for an anelastic strain that lags a change in applied stress, thereby creating energy loss during cyclic loading. More properly termed *mechanical hysteresis.*

elasticity. Ability of a solid to deform in direct proportion to and in phase with increases or decreases in applied force.

elastic limit. The maximum stress to which a material may be subjected without any permanent strain remaining upon complete release of stress.

elastic modulus. Same as *modulus of elasticity.*

elastic ratio. *Yield point* divided by *tensile strength.*

elastic strain. Same as *elastic deformation.*

elastic strain energy. See *strain energy.*

elastic waves. Mechanical vibrations in an elastic medium.

electrical discharge machining. Removal of stock from an electrically conductive material by rapid, repetitive spark discharge through a dielectric fluid flowing between the workpiece and a shaped electrode. Often abbreviated EDM. Variations of the process include electrical discharge grinding and electrical discharge wire cutting.

electrical disintegration. Metal removal by an electrical spark acting in air. It is not subject to precise control, the most common application being the removal of broken tools such as taps and drills; hence the shop name "tap buster".

electrochemical corrosion. Corrosion that is accompanied by a flow of electrons between cathodic and anodic areas on metallic surfaces.

electrochemical equivalent. The weight of an element, compound, radical, or ion involved in a specified electrochemical reaction during the passage of a unit quantity of electricity.

electrochemical machining. Removal of stock from an electrically conductive material by anodic dissolution in an electrolyte flowing rapidly through a gap between the workpiece and a shaped electrode. Often abbreviated ECM. Variations of the process include electrochemical deburring and electrochemical grinding.

electrochemical series. Same as *electromotive series.*

electrode. (1) In arc welding, a current-carrying rod that supports the arc between the rod and work, or between two rods as in twin carbon-arc welding. It may or may not furnish filler metal. See *bare electrode, covered electrode* and *lightly coated electrode.* (2) In resistance welding, a part of a resistance welding machine through which current and, in most instances, pressure are applied directly to the work. The electrode may be in the form of a rotating wheel, rotating roll, bar, cylinder, plate, clamp, chuck or modification thereof. (3) An electrical conductor for leading current into or out of a medium.

electrode cable. Same as *electrode lead.*

electrode deposition. The weight of weld metal deposit obtained from a unit length of electrode.

electrode force. The force between electrodes in spot, seam and projection welding.

electrode lead. The electrical conductor between the source of arc welding current and the electrode holder.

electrodeposition. The deposition of a substance upon an electrode by passing electric current through an electrolyte. Electroplating (plating), electroforming, electrorefining and electrowinning result from electrodeposition.

electrode potential. The potential of a *half cell* as measured against a standard reference half cell.

electrode skid. In spot, seam or projection welding, the sliding of an electrode along the surface of the work.

electroforming. Making parts by electrodeposition on a removable form.

electrogalvanizing. The electroplating of zinc upon iron or steel.

electrogas welding. A process for vertical position welding in which molding shoes confine the molten weld metal. Welding may be done by either *gas metal arc welding* or *flux cored arc welding.*

electroless plating. A process in which metal ions in a dilute aqueous solution are plated out on a substrate by means of autocatalytic chemical reduction.

electrolysis. Chemical change resulting from the passage of an electric current through an electrolyte.

electrolyte. (1) An ionic conductor. (2) A liquid, most often a solution, that will conduct an electric current.

electrolytic brightening. Same as *electropolishing.*

electrolytic cell. An assembly, consisting of a vessel, electrodes and an electrolyte, in which electrolysis can be carried out.

electrolytic cleaning. Removing soil from work by *electrolysis,* the work being one of the electrodes. The electrolyte is usually alkaline.

electrolytic copper. Copper that has been refined by electrolytic deposition, including cathodes that are the direct product of the refining operation, refinery shapes cast from melted cathodes, and, by extension, fabricators' products made therefrom. Usually when this term is used alone, it refers to electrolytic tough pitch copper without elements other than oxygen being present in significant amounts.

electrolytic deposition. Same as *electrodeposition.*

electrolytic grinding. A combination of grinding and machining where a metal-bonded abrasive wheel, usually diamond, is the cathode in physical contact with the anodic workpiece, the contact being made underneath the surface of a suitable electrolyte. The abrasive particles produce grinding and act as nonconducting spacers permitting simultaneous machining through electrolysis.

electrolytic machining. Controlled removal of metal by an applied potential and suitable electrolyte to produce the shapes and dimensions desired.

electrolytic pickling. *Pickling* where electric current is used, the work being one of the electrodes.

electrolytic powder. Metal powder produced by electrolytic deposition or by the pulverization of an electrodeposit, or from metal made by electrodeposition.

electrolytic protection. See the preferred term, *cathodic protection.*

electrometallurgy. Industrial recovery or processing of metals and alloys by electric or electrolytic methods.

electromotive force. Electrical potential; voltage.

electromotive series. A list of elements arranged according to their *standard electrode potentials.* In corrosion studies, the analogous but more practical *galvanic series* of metals is generally used. The relative position of a given metal is not necessarily the same in the two series.

electron bands. Energy states for the free electrons in a metal, as described using the band theory (zone theory) of electron structure. Also called Brillouin zones.

electron beam cutting. A cutting process that uses the heat obtained from a concentrated beam composed primarily of high-velocity electrons, which impinge upon the workpieces to be cut; it may or may not use an externally supplied gas.

electron beam machining. Removing material by melting and vaporizing the workpiece at the point of impingement of a focused high velocity beam of electrons. The machining is done in high vacuum to eliminate scattering of the electrons due to interaction with gas molecules.

electron beam microprobe analyzer. An instrument for selective analysis of a microscopic component or feature in which an electron beam bombards the point of interest in a vacuum at a given energy level. Scanning of a larger area permits determination of the distribution of selected elements. The analysis is made by measuring the wavelengths and intensities of secondary electromagnetic radiation resulting from the bombardment.

electron beam welding. A welding process that produces coalescence of metals with the heat obtained from a concentrated beam composed primarily of high-velocity electrons impinging upon the surfaces to be joined.

electron compound. An intermediate phase on a *constitution diagram,* usually a binary phase, that has the same crystal structure and the same ratio of valence electrons to atoms as intermediate phases in several other systems. An electron compound is often a solid solution of variable composition and good metallic properties. Occasionally, an ordered arrangement of atoms is characteristic of the compound, in which case the range of composition is usually small. Phase stability depends essentially on electron concentration and crystal structure and has been observed at valence-electron-to-atom ratios of $3/2$, $21/13$ and $7/4$.

electrophoresis. The transport of charged colloidal or macromolecular materials in an electric field.

electroplating. Electrodepositing a metal or alloy in an adherent form on an object serving as a cathode.

electropolishing. (1) A technique commonly used to prepare metallographic specimens, in which a high polish is produced by making the specimen the anode in an electrolytic cell, where preferential dissolution at high points smooths the surface. (2) A variation of *chemical machining* wherein electrolytic deplating promotes chemical cutting, especially at surface irregularities.

electrorefining. Using electric or electrolytic methods to convert impure metal to purer metal, or to produce an alloy from impure or partly purified raw materials.

electroslag remelting. A *consumable electrode remelting* process in which heat is generated by the passage of electric current through a conductive slag. The droplets of metal are refined by contact with the slag. Sometimes abbreviated ESR.

electroslag welding. A fusion welding process in which the welding heat is provided by passing an electric current through a layer of molten conductive slag contained in a pocket formed by molding shoes that bridge the gap between the members being welded. The resistance heated slag not only melts filler metal electrodes as they are fed into the slag layer, but also provides shielding for the massive weld puddle characteristic of the process.

electrostrictive effect. The reversible interaction, exhibited by some crystalline materials, between an elastic strain and an electric field. The direction of the strain is independent of the polarity of the field. Compare with *piezoelectric effect.*

electrotinning. Electroplating tin on an object.

electrotyping. The production of printing plates by electroforming.

electrowinning. Recovery of a metal from an ore by means of electrochemical processes.

elongation. In tensile testing, the increase in the gage length, measured after fracture of the specimen within the gage length, usually expressed as a percentage of the original gage length.

elutriation. Separation of metal powder into particle-size fractions by means of a rising stream of gas or liquid.

embossing. Raising a design in relief against a surface.

embossing die. A die used for producing embossed designs.

embrittlement. Reduction in the normal ductility of a metal due to a physical or chemical change. Examples include *blue brittleness, hydrogen embrittlement* and *temper brittleness.*

emery. An impure mineral of the corundum or aluminum oxide type used extensively as an abrasive before the development of electric-furnace products.

emf. An abbreviation for *electromotive force.*

emissivity. Ratio of the amount of energy or of energetic particles radiated from a unit area of a surface to the amount radiated from a unit area of an ideal emitter under the same conditions.

emulsion. A dispersion of one liquid phase in another.

emulsion cleaner. A cleaner consisting of organic solvents dispersed in an aqueous medium with the aid of an emulsifying agent.

enantiotropy. The relation of crystal forms of the same substance in which one form is stable above a certain temperature and the other form stable below that temperature. Ferrite and austenite are enantiotropic in ferrous alloys, for example.

end clearance angle. See *clearance angle,* and also sketches accompanying *face mill* and *single-point tool.*

end cutting edge angle. The angle of concavity between the face cutting edge and the face plane of the cutter. It serves as relief to prevent the face cutting edges from rubbing in the cut. See sketches accompanying *face mill* and *single-point tool.*

end mark. A roll mark caused by the end of a sheet marking the roll during hot or cold rolling.

end milling. A method of machining with a rotating peripheral and end cutting tool. See also *face milling.*

end-quench hardenability test. A laboratory procedure for determining the hardenability of a steel or other ferrous alloy; widely referred to as the Jominy test. Hardenability is determined by heating a standard specimen above the upper critical temperature, placing the hot specimen in a fixture so that a stream of cold water impinges on one end, and, after cooling to room temperature is completed, measuring the hardness near the surface of the specimen at regularly spaced intervals along its length. The data are normally plotted as hardness versus distance from the quenched end.

end relief. Defined by sketch accompanying *single-point tool.*

endurance limit. The maximum stress below which a material can presumably endure an infinite number of stress cycles. If the stress is not completely reversed, the value of the mean stress, the minimum stress or the stress ratio also should be stated. Compare with *fatigue limit.*

endurance ratio. The ratio of the *endurance limit* for completely reversed flexural stress to the tensile strength of a given material.

entry mark (exit mark). A slight corrugation caused by the entry or exit rolls of a roller leveling unit.

epitaxy. Growth of an electrodeposit or vapor deposit in which the orientations of the crystals in the deposit are directly related to crystal orientations in the underlying crystalline substrate.

epsilon structure. A Hume-Rothery designation for structurally analogous close-packed phases or electron compounds like $CuZn_3$ that have ratios of seven valence electrons to four atoms. Not to be confused with the epsilon phase on a constitution diagram.

equiaxed grain structure. A structure in which the grains have approximately the same dimensions in all directions.

equilibrium. A dynamic condition of physical, chemical, mechanical or atomic balance, where the condition appears to be one of rest rather than change.

equilibrium diagram. A graphical representation of the temperature, pressure and composition limits of phase fields in an alloy system as they exist under conditions of complete equilibrium. In metal systems, pressure is usually considered constant.

Erichsen test. A cupping test in which a piece of sheet metal, restrained except at the center, is deformed by a cone-shaped spherical-end plunger until fracture occurs. The height of the cup in millimetres at fracture is a measure of the ductility.

erosion. Destruction of metals or other materials by the abrasive action of moving fluids, usually accelerated by the presence of solid particles or matter in suspension. When *corrosion* occurs simultaneously, the term erosion-corrosion is often used.

erosion-corrosion. See *erosion.*

etchant. A chemical substance or mixture used for *etching.*

etch cleaning. Removing soil by dissolving away some of the underlying metal.

etch cracks. Shallow cracks in hardened steel containing high residual surface stresses, produced on etching in an embrittling acid.

etch figures. Characteristic markings produced on crystal surfaces by chemical attack, usually having facets that are parallel to low-index crystallographic planes.

etching. (1) Subjecting the surface of a metal to preferential chemical or electrolytic attack in order to reveal structural details for metallographic examination. (2) Chemically or electrochemically removing tenacious films from a metal surface to condition the surface for a subsequent treatment, such as painting or electroplating.

eutectic. (1) An isothermal reversible reaction in which a liquid solution is converted into two or more intimately mixed solids on cooling, the number of solids formed being the same as the number of components in the system. (2) An alloy having the composition indicated by the eutectic point on an equilibrium diagram. (3) An alloy structure of intermixed solid constituents formed by a eutectic reaction.

eutectic carbide. Carbide formed during freezing as one of the mutually insoluble phases participating in the eutectic reaction of ferrous alloys.

eutectic melting. Melting of localized microscopic areas whose composition corresponds to that of the eutectic in the system.

eutectoid. (1) An isothermal reversible reaction in which a solid solution is converted into two or more intimately mixed solids on cooling, the number of solids formed being the same as the number of components in the system. (2) An alloy having the composition indicated by the eutectoid point on an equilibrium diagram. (3) An alloy structure of intermixed solid constituents formed by a eutectoid reaction.

exfoliation. A type of corrosion that progresses approximately parallel to the outer surface of the metal, causing layers of the metal to be elevated by the formation of corrosion product.

expanding. A process used to increase the diameter of a cup, shell or tube. See *bulging.*

expansion fit. An interference or force fit made by placing a cold (subzero) inside member into a warmer outside member and allowing an equalization of temperature.

explosion welding. A solid state welding process effected by a controlled detonation, which causes the parts to move together at high velocity.

explosive forming. Shaping metal parts where the forming pressure is generated by an explosive charge.

extensometer. An instrument for measuring changes in length caused by application or removal of a force. Commonly used in tension testing of metal specimens.

extractive metallurgy. The branch of process metallurgy dealing with the *winning* of metals from their ores. Compare with *refining.*

extra hard. A *temper* of nonferrous alloys and some ferrous alloys characterized by tensile strength and hardness about one-third of the way from *full hard* to *extra spring* temper.

extra spring. A *temper* of nonferrous alloys and some ferrous alloys corresponding approximately to a cold worked state above *full hard* beyond which further cold work will not measurably increase the strength and hardness.

extruded hole. A hole formed by a punch that first cleanly cuts a hole and then is pushed farther through to form a flange with an enlargement of the original hole.

extrusion. Conversion of an ingot or billet into lengths of uniform cross section by forcing metal to flow plastically through a die orifice. In direct extrusion (forward extrusion), the die and ram are at opposite ends of the extrusion stock, and the product and ram travel in the same direction. Also, there is relative motion between the extrusion stock and the *container.* In indirect extrusion (backward extrusion), the die is at the ram end of the stock and the product travels in the opposite direction as the ram, either around the ram (as in the impact extrusion of cylinders such as cases for dry cell batteries) or up through the center of a hollow ram.

Impact extrusion is the process (or resultant product) in which a punch strikes a slug (usually unheated) in a confining die. The metal flow may be either between punch and die or through another opening. Impact extrusion of unheated slugs is often called cold extrusion. See also *Hooker process,* which uses a pierced slug.

A stepped extrusion is a single product having one or more abrupt changes in cross section. It is produced by stopping extrusion to change dies. Often, such an extrusion is made in a complex die having a die section that can be freed from the main die and allowed to ride out with the product when extrusion is resumed.

extrusion billet. A metal slug used as *extrusion stock.*

extrusion defect. See preferred term, *extrusion pipe.*

extrusion ingot. A cast metal slug used as *extrusion stock.*

extrusion pipe. A central oxide-lined discontinuity that occasionally occurs in the last 10 to 20% of an extruded bar. It is caused by the oxidized outer surface of the billet flowing around the end of the billet and into the center of the bar during the final stages of extrusion. Also called coring.

extrusion stock. A rod, bar or other section used to make extrusions.

eyeleting. Displacing material about an opening in sheet or plate so that a lip protruding above the surface is formed.

F

face. In a lathe tool, the surface against which the chips bear as they are formed. See sketch accompanying *single-point tool.*

face mill. See definition of nomenclature in accompanying sketch.

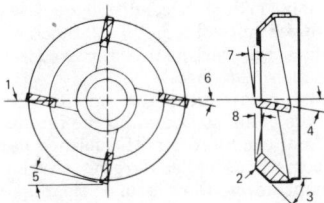

1. Reference plane
2. Tooth point
3. Corner angle (bevel)
4. Axial rake (positive)
5. Peripheral clearance angle
6. Radial rake (negative)
7. End clearance angle
8. End cutting edge angle

Face mill

face milling. Milling a surface that is perpendicular to the cutter axis.

face of weld. The exposed surface of an arc or gas weld on the side from which the welding was done. See sketch accompanying *fillet weld.*

face-type cutters. Cutters that can be mounted directly on and driven from the machine spindle nose.

facing. (1) In machining, generating a surface on a rotating workpiece by the traverse of a tool perpendicular to the axis of rotation. (2) In founding, special sand placed against a pattern to improve the surface quality of the casting. (3) For abrasion resistance, see preferred term *hard facing.*

fagot. In forging work, a bundle of iron bars that will be heated and then hammered and welded to form a single bar.

failure. A general term used to imply that a part in service (*a*) has become completely inoperable, (*b*) is still operable but is incapable of satisfactorily performing its intended function, or (*c*) has deteriorated seriously, to the point that it has become unreliable or unsafe for continued use.

false bottom. An *insert* put in either member of a die set to increase the strength and improve the life of the die.

false brinelling. Evenly spaced depressions in a raceway of a rolling-element bearing caused by fretting that occurs when the bearing is subjected to vibration while it is not rotating. Compare with *brinelling.*

false indication. In nondestructive inspection, an *indication* that may be interpreted erroneously as an *imperfection.* See also *artifact.*

false wiring. Same as *curling.*

fatigue. The phenomenon leading to fracture under repeated or fluctuating stresses having a maximum value less than the tensile strength of the material. Fatigue fractures are progressive, beginning as minute cracks that grow under the action of the fluctuating stress.

fatigue life. The number of cycles of stress that can be sustained prior to failure for a stated test condition.

fatigue limit. The maximum stress that presumably leads to fatigue fracture in a specified number of stress cycles. If the stress is not completely reversed, the value of the mean stress, the minimum stress or the stress ratio also should be stated. Compare with *endurance* limit.

fatigue notch factor (K_f). The ratio of the fatigue strength of an unnotched specimen to the fatigue strength of a notched specimen of the same material and condition; both strengths are determined at the same number of stress cycles.

fatigue notch sensitivity (*q*). An estimate of the effect of a notch or hole on the fatigue properties of a material; measured by $q = (K_f - 1)/(K_t - 1)$. A material is said to be fully notch sensitive if *q* approaches a value of 1.0; it is not notch sensitive if the ratio approaches 0. K_f is the *fatigue notch factor,* and K_t is the *stress-concentration factor,* for a specimen of the material containing a notch or hole of a given size and shape.

fatigue ratio. The *fatigue limit* under completely reversed flexural stress divided by the tensile strength for the same alloy and condition.

fatigue strength. The maximum stress that can be sustained for a specified number of cycles without failure, the stress being completely reversed within each cycle unless otherwise stated.

fatigue-strength reduction factor (K_f). The ratio of the fatigue strength of a member or specimen with no stress concentration to the fatigue strength with stress concentration. K_f has no meaning unless the stress range and the shape, size and material of the member or specimen are stated.

fatigue striations. Parallel lines frequently observed in electron microscope fractographs of fatigue fracture surfaces. The lines are transverse to the direction of local crack propagation; the distance between successive lines represents the advance of the crack front during one cycle of stress variation.

faying surface. The surface of a piece of metal (or a member) in contact with another to which it is or is to be joined.

feed. The rate at which a cutting tool or grinding wheel advances along or into the surface of a workpiece, the direction of advance depending upon the type of operation involved.

feeder (feeder head, feedhead). A *riser.*

feeding. (1) Conveying metal stock or workpieces to a location for use or processing, such as wire to a consumable electrode, strip to a die, or workpieces to an assembler. (2) In casting, providing molten metal to a region undergoing solidification, usually at a rate sufficient to fill the mold cavity ahead of the solidification front and to make up for any shrinkage accompanying solidification.

feed lines. Linear marks on a

machined or ground surface that are spaced at intervals equal to the *feed* per revolution or per stroke.

ferrimagnetic material. A material that macroscopically has properties similar to those of a *ferromagnetic material* but that microscopically also resembles an antiferromagnetic material in that some of the elementary magnetic moments are aligned antiparallel. If the moments are of different magnitudes, the material may still have a large resultant magnetization.

ferrite. (1) A solid solution of one or more elements in body-centered cubic iron. Unless otherwise designated (for instance, as chromium ferrite), the solute is generally assumed to be carbon. On some equilibrium diagrams, there are two ferrite regions separated by an austenite area. The lower area is alpha ferrite; the upper, delta ferrite. If there is no designation, alpha ferrite is assumed. (2) In the field of magnetics, substances having the general formula:

$$M^{++}O \cdot M_2^{+++}O_3$$

the trivalent metal often being iron.

ferrite banding. Parallel bands of free ferrite aligned in the direction of working. Sometimes referred to as ferrite streaks.

ferrite number. An arbitrary, standardized value designating the ferrite content of an austenitic stainless steel weld metal. This value directly replaces percent ferrite or volume percent ferrite and is determined by the magnetic test described in AWS A4.2.

ferrite streaks. Same as *ferrite banding.*

ferritic malleable. See *malleable cast iron.*

ferritizing anneal. A treatment given as-cast gray or ductile (nodular) iron to produce an essentially ferritic matrix. For the term to be meaningful, the final microstructure desired or the time-temperature cycle used must be specified.

ferroalloy. An alloy or iron that contains a sufficient amount of one or more other chemical elements to be useful as an agent for introducing these elements into molten metal, especially into steel or cast iron.

ferrograph. An instrument used to determine the size distribution of wear particles in lubricating oils of mechanical systems.

ferromagnetic material. A material that in general exhibits the phenomena of hysteresis and saturation, and whose permeability is dependent on the magnetizing force. Microscopically, the elementary magnets are aligned parallel in volumes called *domains.* The unmagnetized condition of a ferromagnetic material results from the over-all neutralization of the magnetization of the domains to produce zero external magnetization. Compare with *paramagnetic material, diamagnetic material.*

fiber. (1) The characteristic of wrought metal that indicates *directional properties* and is revealed by the etching of a longitudinal section or is manifested by the fibrous or woody appearance of a fracture. It is caused chiefly by the extension of the constituents of the metal, both metallic and nonmetallic, in the direction of working. (2) The pattern of preferred orientation of metal crystals after a given deformation process, usually wiredrawing. See *preferred orientation.*

fiber stress. Local stress through a small area (a point or line) on a section where the stress is not uniform, as in a beam under a bending load.

fibrous fracture. A fracture where the surface is characterized by a dull gray or silky appearance. Contrast with *crystalline fracture.*

fibrous structure. (1) In forgings, a structure revealed as laminations, not necessarily detrimental, on an etched section or as a ropy appearance on a fracture. It is not to be confused with the silky or ductile fracture of a clean metal. (2) In wrought iron, a structure consisting of slag fibers embedded in ferrite. (3) In rolled steel plate stock, a uniform, fine-grained structure on a fractured surface, free of laminations or shale-type discontinuities. As contrasted with part (1) above, it is virtually synonymous with silky or ductile fracture.

filamentary shrinkage. A fine network of shrinkage cavities, occasionally found in steel castings, that produces a radiographic image resembling lace.

file hardness. Hardness as determined by the use of a file of standardized hardness on the assumption that a material that cannot be cut with the file is as hard as, or harder than, the file. Files covering a range of hardnesses may be employed.

filler. A material used to increase the bulk of a product without adding to its effectiveness in functional performance.

filler metal. Metal added in making a brazed, soldered or welded joint.

fillet. (1) A radius (curvature) imparted to inside meeting surfaces. (2) A concave cornerpiece used on foundry patterns.

fillet weld. A weld, approximately triangular in cross section, joining two surfaces essentially at right angles to each other in a lap, tee or corner joint.

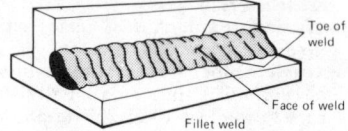

final annealing. An imprecise term used to denote the last anneal given to a nonferrous alloy prior to shipment.

fineness. A measure of the purity of gold or silver expressed in parts per thousand.

fines. (1) The product that passes through the finest screen in sorting crushed or ground material. (2) Sand grains that are substantially smaller than the predominating size in a batch or lot of foundry sand. (3) The portion of a metal powder composed of particles that are smaller than a specified size, currently less than 44 μm. See also *superfines.*

fine silver. Silver with a fineness of 999; equivalent to a minimum content of 99.9% Ag with the remaining content unrestricted.

finish. (1) Surface condition, quality or appearance of a metal. (2) Stock on a forging or casting to be removed when finish machined.

finish allowance. The amount of excess metal surrounding the intended final configuration of a formed part; sometimes called forging envelope, machining allowance, or cleanup allowance.

finished steel. Steel that is ready for the market and has been processed beyond the stages of billets, blooms, sheet bars, slabs and wire rods.

finish grinding. The final grinding action on a workpiece where the objectives are surface finish and dimensional accuracy.

finishing die. The die used to make the final impression on a forging. Sometimes called finisher.

finishing temperature. The temperature at which *hot working* is completed.

finish machining. A machining process analogous to *finish grinding.*

fire-refined copper. Copper that has been refined by the use of a furnace process only, including refinery shapes and, by extension, fabricators' products made therefrom. Usually, when this term is used alone it refers to fire-refined tough pitch copper without elements other than oxygen being present in significant amounts.

fire scale. Intergranular copper oxide remaining below the surface of silver-copper alloys that have been annealed and pickled.

fir-tree crystal. A type of *dendrite.*

fish eyes. Areas on a fractured steel surface having a characteristic white crystalline appearance.

fishmouthing. Same as *alligatoring.*

fishscale. A scaly appearance in a porcelain enamel coating in which the evolution of hydrogen from the basis metal (iron or steel) causes loss of adhesion between the enamel and basis metal. Individual scales are usually small, but have been observed up to 25 mm or more in diameter. The scales are somewhat like blisters that have cracked part way around the perimeter but still remain attached to the coating around the rest of the perimeter; if detached completely, it is one form of *pop-off.*

fishtail. (1) In roll forging, the excess trailing end of a forging. It is often used, before being trimmed off, as a tong hold for a subsequent forging operation. (2) In hot rolling or extrusion, the imperfectly shaped trailing end of a bar or special section that must be cut off and discarded as mill scrap.

fit. The amount of clearance or interference between mating parts is called actual fit. Fit is the preferable term for the range of clearance or interference that may result from the specified limits on dimensions (limits of size). Refer to ANSI standards.

fixed-feed grinding. Grinding where the wheel is fed into the work, or vice versa, by given increments or at a given rate.

fixed-position welding. Welding in which the work is held in a stationary position.

fixture. A positioning device to hold the workpiece only.

flake powder. Metal powder in the form of flat or scalelike particles, relatively thin.

flakes. Short discontinuous internal fissures in ferrous metals attributed to stresses produced by localized transformation and decreased solubility of hydrogen during cooling after hot working. In a fractured surface, flakes appear as bright silvery areas; on an etched surface, they appear as short discontinuous cracks. Also called shatter cracks or snowflakes.

flame annealing. Annealing in which the heat is applied directly by a flame.

flame cleaning. Cleaning metal surfaces of scale, rust, dirt and moisture by use of a gas flame.

flame hardening. A process for hardening the surfaces of hardenable ferrous alloys in which an intense flame is used to heat the surface layers above the upper transformation temperature, whereupon the workpiece is immediately quenched.

flame spraying. *Thermal spraying* in which a coating material is fed into an oxyfuel gas flame, where it is melted. Compressed gas may or may not be used to atomize the coating material and propel it onto the substrate.

flame straightening. Correcting distortion in metal structures by localized heating with a gas flame.

flank. The end surface of a tool that is adjacent to the cutting edge and below it when the tool is in a horizontal position, as for turning. See sketch accompanying *single-point tool.*

flank wear. The loss of relief on the flank of the tool behind the cutting edge due to rubbing contact between the work and the tool during cutting; measured in terms of linear dimension behind the original cutting edge.

flapping. In copper refining, hastening oxidation of molten copper by striking through the slag-covered surface of the melt with a *rabble* just before the bath is poled.

flare test. A test applied to tubing, involving a tapered expansion over a cone. Similar to *pin expansion test.*

flaring. (1) Forming an outward acute-angle flange on a tubular part. (2) Forming a flange by using the head of a hydraulic press.

flash. (1) In forging, excess metal forced out between the upper and lower dies. (2) In casting, a fin of metal that results from leakage between mating mold surfaces. (3) In resistance butt welding, a fin formed perpendicular

to the direction of applied pressure.

flashback. The recession of a flame into or in back of the interior of a torch. See *backfire.*

flash butt welding. See *flash welding.*

flash extension. Portion of flash remaining after trimming. Flash extension is measured from the intersection of the draft and flash at the body of the forging to the trimmed edge of the stock.

flashing. In flash welding, the heating portion of the cycle, consisting of a series of rapidly recurring localized short circuits followed by molten metal expulsions, during which time the surfaces to be welded are moved one toward the other at a predetermined speed.

flash land. Relief at the parting line of a set of closed-die forging dies that is designed either to restrict or to encourage growth of flash, whichever is required to ensure complete filling of the finishing impression.

flash line. The line of location of flash formed around a forging or casting.

flash plate. A very thin final electrodeposited film of metal.

flash welding. A resistance welding process that joins metals by first heating abutting surfaces by passing an electric current across the joint, then forcing the surfaces together by the application of pressure. Flashing and upsetting are accompanied by expulsion of metal from the joint.

flask. A metal or wood frame used for making and holding a sand mold. The upper part is called the cope; the lower, the drag.

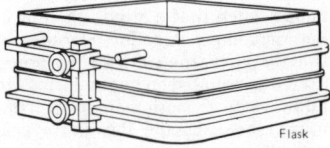
Flask

flat die forging. Forging metal between flat or simple contour dies by repeated strokes and manipulation of the workpiece. Also known as open die forging, hand forging or smith forging.

flat drill. A rotary end-cutting tool constructed from a flat piece of material, provided with suitable cutting lips at the cutting end.

flat edge trimmer. A machine for trimming notched edges on shells. The slide is cam driven to obtain a brief dwell at the bottom of the stroke, at which time the die, sometimes called a shimmy die, oscillates to trim the part.

flat-position welding. Welding from the upper side, the face of the weld being horizontal. Also called downhand welding.

flattening. (1) A preliminary operation performed on forging stock so as to position the metal for a subsequent forging operation. (2) Removing irregularities or distortion in sheets or plates by a method such as *roller leveling* or *stretcher leveling.*

flattening test. A quality test for tubing in which a specimen is flattened to a specified height between parallel plates.

flat wire. A roughly rectangular or square mill product, narrower than *strip*, in which all surfaces are rolled or drawn without any previous slitting, shearing or sawing.

flaw. A nonspecific term often used to imply a cracklike discontinuity. See preferred terms *discontinuity, imperfection, defect.*

flexible cam. An adjustable pressure-control cam of spring steel strips used to obtain varying pressure during a forming cycle.

flex roll. A movable jump roll designed to push up against a sheet as it passes through a roller leveler. The flex roll can be adjusted to deflect the sheet any amount up to the roll diameter.

flex rolling. Passing sheets through a flex-roll unit to minimize yield-point elongation so as to reduce the tendency for *stretcher strains* to appear during forming.

floating die. (1) A die mounted in a die holder or a punch mounted in its holder, such that a slight amount of motion compensates for tolerance in the die parts, the work or the press. (2) A die mounted on heavy springs to allow vertical motion in some trimming, shearing and forming operations.

floating plug. In tube drawing, an unsupported mandrel that locates itself at the die inside the tube causing the reduction of the wall thickness while the die is effecting a reduction of the outside diameter.

floppers. On metals, lines or ridges that are transverse to the direction of rolling and generally confined to the section midway between the edges of a coil as rolled.

flospinning. Forming cylindrical, coni-

cal and curvilinear shaped parts by power spinning over a rotating mandrel.

flotation. The concentration of valuable minerals from ores by agitation of the ground material with water, oil and flotation chemicals. The valuable minerals are generally wetted by the oil, lifted to the surface by clinging air bubbles and then floated off.

flowability. A characteristic of a foundry sand mixture that enables it to move under pressure or vibration so that it makes intimate contact with all surfaces of the pattern or core box.

flow brazing. Brazing by pouring hot molten nonferrous filler metal over a joint until the brazing temperature is attained. The filler metal is distributed in the joint by capillary action.

flow brightening. The melting of an electrodeposit, followed by solidification, especially of tin plate.

flow lines. (1) Texture showing the direction of metal flow during hot or cold working. Flow lines often can be revealed by etching the surface or a section of a metal part (see macrograph on this page). (2) In mechanical metallurgy, paths followed by minute volumes of metal during deformation.

flow stress. The uniaxial true stress at the onset of plastic deformation in a metal.

fluidity. The ability of liquid metal to run into and fill a mold cavity.

fluorescence. The emission of characteristic electromagnetic radiation by a substance as a result of the absorption of electromagnetic or corpuscular radiation having a greater unit energy than that of the fluorescent radiation. It occurs only so long as the stimulus responsible for it is maintained.

fluorescent magnetic-particle in-

spection. Inspection with either dry magnetic particles or those in a liquid suspension, the particles being coated with a fluorescent substance to increase the visibility of the indications.

fluorescent penetrant inspection. Inspection using a fluorescent liquid that will penetrate any surface opening; after wiping the surface clean, the location of any surface flaws may be detected by the fluorescence, under ultraviolet light, of back-seepage of the fluid.

fluoroscopy. An inspection procedure in which the radiographic image of the subject is viewed on a fluorescent screen, normally limited to low-density materials or thin sections of metals because of the low light output of the fluorescent screen at safe levels of radiation.

flute. (1) As applied to drills, reamers and taps, the channels or grooves formed in the body of the tool to provide cutting edges and to permit passage of cutting fluid and chips. (2) As applied to milling cutters and hobs, the chip space between the back of one tooth and the face of the following tooth.

fluting. (1) Forming longitudinal recesses in a cylindrical part, or radial recesses in a conical part. (2) A series of sharp parallel kinks or creases occurring in the arc when sheet metal is roll formed into a cylindrical shape.

flux. (1) In metal refining, a material used to remove undesirable substances, like sand, ash or dirt, as a molten mixture. It is also used as a protective covering for certain molten metal baths. Lime or limestone is generally used to remove sand, as in iron smelting; sand, to remove iron oxide in copper refining. (2) In brazing, cutting, soldering or welding, material used to prevent the formation of, or to dissolve and facilitate removal of, oxides and other undesirable substances.

flux-cored arc welding. An arc welding process that joins metals by heating them with an arc between a continuous tubular filler-metal electrode and the work. Shielding is provided by a flux contained within the consumable tubular electrode. Additional shielding may or may not be obtained from an externally supplied gas or gas mixture. See also *electrogas welding.*

flux density. In magnetism, the number of *flux lines* per unit area passing through a cross section at right angles. It is given by $B = \mu H$, where μ and H are permeability and magnetic field intensity, respectively.

flux lines. Imaginary lines used as a means of explaining the behavior of magnetic and other fields. Their concept is based on the pattern of lines produced when magnetic particles are sprinkled over a permanent magnet. Sometimes called magnetic lines of force.

flux-oxygen cutting. Oxygen cutting with the aid of a flux.

fly ash. A finely divided siliceous material formed during the combustion of coal, coke or other solid fuels.

fly cutting. Cutting with a single-tooth milling cutter.

flying shear. A machine for cutting continuous rolled products to length that does not require a halt in rolling, but rather moves along the runout table at the same speed as the product while performing the cutting, then returns to the starting point in time to cut the next piece.

fog quenching. Quenching in a fine vapor or mist.

foil. Metal in sheet form less than 0.15 mm (0.006 in.) in thickness.

fold. Same as *lap*.

follow board. A board contoured to a pattern to facilitate making a sand mold.

follow die. A *progressive die* consisting of two or more parts in a single holder, used with a separate lower die to perform more than one operation (such as piercing and blanking) on a part at two or more stations.

foot press. A small press with low capacity actuated by foot pressure on a treadle.

force fit. Any of various interference fits between parts assembled under various amounts of force.

forehand welding. Welding in which

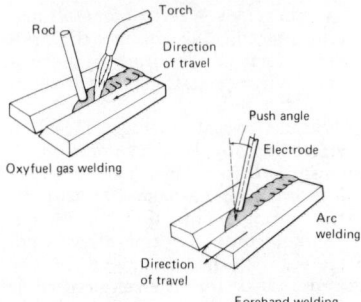

Oxyfuel gas welding

Forehand welding

the palm of the principal hand (torch or electrode hand) of the welder faces the direction of travel. It has special significance in oxyfuel gas welding in that the flame is directed ahead of the weld bead, which provides *preheating*. Contrast with *backhand welding*.

forgeability. Term used to describe the relative ability of material to flow under a compressive load without rupture.

forge delay time. In spot, seam or projection welding, the time between the start of the welding, current or weld interval and the application of forging pressure.

forge welding. Solid state welding in which metals are heated in a forge (in air) then welded together by applying pressure or blows sufficient to cause permanent deformation at the interface.

forging. Plastically deforming metal, usually hot, into desired shapes with compressive force, with or without dies.

forging billet. A wrought metal slug used as *forging stock*.

forging envelope. See *finish allowance*.

forging ingot. A cast metal slug used as *forging stock*.

forging machine. A type of forging equipment, related to the mechanical press, in which the main forming energy is applied horizontally to the workpiece, which is held by dies. Commonly called upsetter or header.

forging plane. In forging, the plane that includes the principal die face and that is perpendicular to the direction of ram travel. When parting surfaces of the dies are flat, the forging plane coincides with the parting line. Contrast *parting plane*.

forging range. Temperature range in which a metal can be forged successfully.

forging rolls. A machine used in *roll forging*. Also called gap rolls.

forging stock. A rod, bar or other section used to make forgings.

formability. The relative ease with which a metal can be shaped through plastic deformation. See *drawability*.

form block. Tooling, usually the male part, used for forming sheet-metal contours, being generally employed in the rubber-pad process.

form cutter. Any cutter, profile sharpened or cam relieved, shaped to pro-

duce a specified form on the work.

form die. A die used to change the shape of a blank with minimum plastic flow.

form grinding. Grinding with a wheel having a contour on its cutting face that is a mating fit to the desired form.

forming. Making a change, with the exception of shearing or blanking, in the shape or contour of a metal part without intentionally altering the thickness.

form-relieved cutter. A cutter so relieved that by grinding only the tooth face the original form is maintained throughout its life.

form rolling. Hot rolling to produce bars having contoured cross sections; not to be confused with roll forming of sheet metal or with roll forging.

form tool. A single-edge, nonrotating cutting tool, circular or flat, that produces its inverse or reverse form counterpart upon a workpiece.

forward extrusion. Same as direct extrusion. See *extrusion*.

foundry. A commercial establishment or building where metal castings are produced.

four-high mill. A type of rolling mill, commonly used for flat-rolled mill products, in which two large-diameter backup rolls are employed to reinforce two smaller working rolls, which are in contact with the product. Either the working rolls or the backup rolls may be driven. Compare with *two-high mill, cluster mill*.

four-point press. A press whose slide is actuated by four connections and four cranks, eccentrics, or cylinders, the chief merit being to equalize the pressure at the corners of the slides.

fraction. In powder metallurgy, the portion of a powder sample that lies between two stated particle sizes. Synonymous with cut.

fractography. Descriptive treatment of fracture, especially in metals, with specific reference to photographs of the fracture surface. Macrofractography involves photographs at low magnification; microfractography, at high magnification.

fracture mechanics. See *linear elastic fracture mechanics*.

fracture stress. (1) The maximum principal true stress at fracture. Usually refers to unnotched tensile specimens. (2) The (hypothetical) true stress that will cause fracture without further deformation at any given strain.

fracture test. Breaking a specimen and examining the fractured surface with the unaided eye or with a low-power microscope to determine such things as composition, grain size, case depth or soundness.

fracture toughness. See *stress-intensity factor*.

fragmentation. The subdivision of a grain into small discrete crystallites outlined by a heavily deformed network of intersecting slip as a result of cold working. These small crystals or fragments differ from one another in orientation and tend to rotate to a stable orientation determined by the slip systems.

freckling. A type of segregation revealed as dark spots on a macroetched specimen of a consumable-electrode vacuum arc remelted alloy.

free carbon. The part of the total carbon in steel or cast iron that is present in elemental form as graphite or temper carbon. Contrast with *combined carbon.*

free ferrite. Ferrite that is formed directly from the decomposition of hypoeutectoid austenite during cooling, without the simultaneous formation of cementite. Also proeutectoid ferrite.

free fit. Various clearance fits for assembly by hand and free rotation of parts. See *running fit.*

free machining. Pertains to the machining characteristics of an alloy to which one or more ingredients have been introduced to give small broken chips, lower power consumption, better surface finish and longer tool life; among such additions are sulfur or lead to steel, lead to brass, lead and bismuth to aluminum, sulfur or selenium to stainless steel.

freezing range. That temperature range between *liquidus* and *solidus* temperatures in which molten and solid constituents coexist.

fretting. A type of wear that occurs between tight-fitting surfaces subjected to cyclic relative motion of extremely small amplitude. Usually, fretting is accompanied by corrosion, especially of the very fine wear debris. Also referred to as fretting corrosion, false brinelling (in rolling-element bearings), friction oxidation, chafing fatigue, molecular attrition and wear oxidation.

fretting fatigue. Fatigue fracture that initiates at a surface area where fretting has occurred.

friction welding. A solid state process in which materials are welded by the heat obtained from rubbing together surfaces that are held against each other under pressure.

full annealing. An imprecise term that denotes an annealing cycle to produce minimum strength and hardness. For the term to be meaningful, the composition and starting condition of the material and the time-temperature cycle used must be stated.

full-automatic plating. Electroplating in which the work is automatically conveyed through the complete cycle.

full center. Mild waviness down the center of a sheet or strip.

fuller. In preliminary forging, the portion of a die that reduces the cross-sectional area between the ends of the stock and permits the metal to move outward.

full hard. A *temper* of nonferrous alloys and some ferrous alloys corresponding approximately to a cold worked state beyond which the material can no longer be formed by bending. In specifications, a full hard temper is commonly defined in terms of minimum hardness or minimum tensile strength (or, alternatively, a range of hardness or strength) corresponding to a specific percentage of cold reduction following a full anneal. For aluminum, a full hard temper is equivalent to a reduction of 75% from *dead soft;* for austenitic stainless steels, a reduction of about 50 to 55%.

furnace brazing. A mass-production *brazing* process in which the filler metal is preplaced on the joint, then the entire assembly is heated to brazing temperature in a furnace. Usually, a protective furnace atmosphere is required, and wetting of the joint surfaces is accomplished without using a brazing flux.

fusion. A change of state from solid to liquid; melting.

fusion face. A surface of the base metal that will be melted during welding.

fusion welding. Any welding process in which filler metal and base metal (substrate), or base metal only, are melted together to complete the weld.

fusion zone. In a weldment, the area of base metal melted as determined on a cross section through the weld.

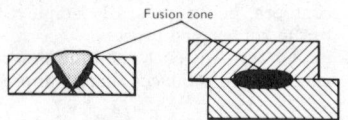

Fusion zone

G

gag. A metal spacer to be inserted so as to render a floating tool or punch inoperative.

gage. (1) The thickness (or diameter) of sheet or wire. The various standards are arbitrary and differ, ferrous from nonferrous products and sheet from wire. (2) An instrument used to measure thickness or length. (3) An aid for visual inspection that enables the inspector to determine more reliably whether the size or contour of a formed part meets dimensional requirements.

gage length. The original length of that portion of the specimen over which strain, change of length and other characteristics are measured.

gagger. An irregular-shaped piece of metal used for reinforcement and support in a sand mold.

galling. A condition whereby excessive friction between high spots results in localized welding with subsequent spalling and a further roughening of the rubbing surfaces of one or both of two mating parts.

galvanic cell. A cell in which chemical change is the source of electrical energy. It usually consists of two dissimilar conductors in contact with each other and with an electrolyte, or of two similar conductors in contact with each other and with dissimilar electrolytes.

galvanic corrosion. Corrosion associated with the current of a galvanic cell consisting of two dissimilar conductors in an electrolyte or two similar conductors in dissimilar electrolytes. Where the two dissimilar metals are in contact, the resulting reaction is referred to as couple action.

galvanic series. A series of metals and alloys arranged according to their relative electrode potentials in a specified environment. Compare with *electromotive series.*

galvanize. To coat a metal surface with zinc using any of various processes.

galvanneal. To produce a zinc-iron alloy coating on iron or steel by keeping the coating molten after hot-dip galvanizing until the zinc alloys completely with the basis metal.

gamma iron. The face-centered cubic form of pure iron, stable from 910 to 1400 °C (1670 to 2550 °F).

gamma ray. Short wavelength electromagnetic radiation, similar to x-rays but of nuclear origin, with a range of wavelengths from about 10^{-14} to 10^{-10} m.

gamma structure. A Hume-Rothery designation for structurally analogous phases or electron compounds that have ratios of 21 valence electrons to 13 atoms; generally, a large complex cubic structure. Not to be confused with gamma phase on a constitution diagram.

gang milling. Milling with several cutters mounted on the same arbor or with workpieces similarly positioned for cutting either simultaneously or consecutively during a single setup.

gang slitter. A machine with a number of pairs of rotary cutters spaced on two parallel shafts, used for slitting sheet metal into strips or for trimming the edges of sheets.

gangue. The worthless portion of an ore that is separated from the desired part before smelting is commenced.

gap. The root opening in a weld joint.

gap-frame press. A general classification of presses in which the uprights or housings are made in the form of a letter "C", thereby making three sides of the die space accessible.

gas cyaniding. A misnomer for *carbonitriding*.

gas holes. Holes in castings or welds that are formed by gas escaping from molten metal as it solidifies. Gas holes may occur individually, in clusters, or distributed throughout the solidified metal.

gas metal-arc welding. A process for welding metals together by heating them with an arc between a continuous filler-metal electrode and the work. Shielding is obtained entirely from an externally supplied gas or gas mixture. Some methods of this process are called MIG or CO_2 welding. See also *electrogas welding, pulsed power welding.*

gas plating. Same as *vapor plating.*

gas pocket. A cavity caused by entrapped gas.

gas porosity. Fine holes or pores within a metal that are caused by entrapped gas or by evolution of dissolved gas during solidification.

gas-shielded arc welding. Arc welding in which the arc and molten metal are shielded from the atmosphere by a stream of gas, such as argon, helium, argon-hydrogen mixtures or carbon dioxide.

gassing. (1) Absorption of gas by a metal. (2) Evolution of gas from a metal during melting operations or on solidification. (3) The evolution of gas from an electrode during electrolysis.

gas tungsten-arc cutting. An arc-cutting process in which metals are severed by melting them with an arc between a single tungsten (nonconsumable) electrode and the work. Shielding is obtained from a gas or gas mixture.

gas tungsten-arc welding. A fusion welding process in which metals are joined by heating them with an electric arc between a nonconsumable tungsten electrode and the work. A gas or gas mixture shields the arc and the weld puddle. Pressure may or may not be applied to the joint, and filler metal may or may not be added. Sometimes referred to as TIG welding.

gas welding. See preferred term, *oxyfuel gas welding.*

gate. The portion of the runner in a mold through which molten metal enters the mold cavity. Sometimes the generic term is applied to the entire network of connecting channels that conduct metal into the mold cavity.

gated pattern. A *pattern* that includes not only the contours of the part to be cast, but also the *gates.*

gathering. A forging operation that increases the cross section of part of the stock; usually a preliminary operation.

gathering stock. Any operation whereby the cross section of a portion of the forging stock is increased above its original size.

geared press. A press whose main crank or eccentric shaft is connected by gears to the driving source.

ghost lines. Lines running parallel to the rolling direction that appear in a panel when it is stretched. These lines may not be evident unless the panel has been sanded or painted. (Not to be confused with leveler lines.)

gibs. Guides that ensure the proper restrained motion of the slide, usually being adjustable to compensate for wear.

glass electrode. A glass membrane electrode used to measure pH or hydrogen-ion activity.

glazing. Dulling the abrasive grains in the cutting face of a wheel during grinding.

glide. (1) Same as *slip.* (2) A noncrys-

tallographic shearing movement, as one grain over another.

globular transfer. In consumable-electrode arc welding, a type of metal transfer in which molten filler metal passes across the arc as large droplets. Compare with *spray transfer, short circuiting transfer.*

gold filled. Covered on one or more surfaces with a layer of gold alloy to form a clad metal. By commercial agreement, a quality mark showing the quantity and fineness of gold alloy may be affixed, indicating the actual proportional weight and karat fineness of the gold alloy cladding. For example, "$^1/_{10}$ 12K Gold Filled" means that the article consists of base metal covered on one or more surfaces with a gold alloy of 12-karat fineness comprising $^1/_{10}$th part by weight of the entire metal in the article. No article having a gold alloy coating of less than 10-karat fineness may have any quality mark affixed. No article having a gold alloy portion of less than $^1/_{20}$th by weight may be marked "Gold Filled", but may be marked "Rolled Gold Plate" provided that the proportional fraction and fineness designation precedes. These standards do not necessarily apply to watch cases.

gooseneck. See *hot chamber machine.*

G-P zone. A *Guinier-Preston zone.*

grain. An individual crystal in a polycrystalline metal or alloy; it may or may not contain twinned regions and subgrains.

grain boundary corrosion. Same as *intergranular corrosion.* See also *interdendritic corrosion.*

grain fineness number. A weighted average grain size of a granular material. The AFS grain fineness number is calculated with prescribed weighting factors from the standard screen analysis.

grain flow. Fiberlike lines appearing on polished and etched sections of forgings, which are caused by orientation of the constituents of the metal in the direction of working during forging. Grain flow produced by proper die design can improve required mechanical properties of forgings.

grain growth. An increase in the average size of the grains in polycrystalline metal, usually as a result of heating at elevated temperature.

grain refiner. A material added to a molten metal to induce a finer than normal grain size in the final structure.

grain size. (1) For metals, a measure of the areas or volumes of grains in a polycrystalline material, usually expressed as an average when the individual sizes are fairly uniform. In metals containing two or more phases, the grain size refers to that of the matrix unless otherwise specified. Grain sizes are reported in terms of number of grains per unit area or volume, average diameter, or as a grain-size number derived from area measurements. (2) For grinding wheels, see preferred term, *grit size*.

granular fracture. A type of irregular surface produced when metal is broken that is characterized by a rough, grainlike appearance as differentiated from a smooth silky, or fibrous, type. It can be subclassified into transgranular and intergranular forms. This type of fracture is frequently called crystalline fracture, but the inference that the metal broke because it "crystallized" is not justified because all metals are crystalline when in the solid state. Contrast with *fibrous fracture, silky fracture*.

granular powder. Particles of metal powder having approximately equidimensional nonspherical shapes.

granulated metal. Small pellets produced by putting liquid metal through a screen or by dropping it onto a revolving disk, and, in both instances, chilling with water.

granulation. The production of coarse metal particles by pouring the molten metal through a screen into water or by agitating the molten metal violently during its solidification.

graphitic carbon. Free carbon in steel or cast iron.

graphitic corrosion. Corrosion of gray iron in which the iron matrix is selectively leached away, leaving a porous mass of graphite behind; it occurs in relatively mild aqueous solutions and on buried pipe and fittings.

graphitic steel. Alloy steel made so that part of the carbon is present as graphite.

graphitization. Formation of graphite in iron or steel. Where graphite is formed during solidification, the phenomenon is called primary graphitization; where formed later by heat treatment, secondary graphitization.

graphitizing. Annealing a ferrous alloy in such a way that some or all of the carbon is precipitated as graphite.

gravity hammer. A class of forging hammer wherein energy for forging is obtained by the mass and velocity of a freely falling ram and the attached upper die. Examples: board hammers and air-lift hammers.

gravity segregation. Variable composition of a casting or ingot caused by the settling out of heavy constituents, or rising of light constituents, before or during solidification.

gray cast iron. A *cast iron* that gives a gray fracture due to the presence of flake graphite. Often called gray iron.

green compact. An unsintered powder metallurgy compact.

green density. Same as *pressed density*.

green rot. A form of high-temperature attack on stainless steels, nickel-chromium alloys and nickel-chromium-iron alloys subjected to simultaneous oxidation and carburization. Basically, attack occurs by first precipitating chromium as chromium carbide, then oxidizing the carbide particles.

green sand. A naturally bonded sand, or a compounded molding sand mixture, that has been "tempered" with water and used while still moist.

green sand core. (1) A *core* made of *green sand* and used as rammed. (2) A sand core that is used in the unbaked condition.

green sand mold. A casting mold composed of moist prepared molding sand. Contrast with *dry sand mold*.

grindability. Relative ease of grinding, analogous to *machinability*.

grindability index. A measure of the grindability of a material under specified grinding conditions, expressed in terms of volume of material removed per unit volume of wheel wear.

grinding. Removing material from a workpiece with a grinding wheel or abrasive belt.

grinding burn. See *burning* (2).

grinding cracks. Shallow cracks formed in the surface of relatively hard materials because of excessive grinding heat or the high sensitivity of the material. See *grinding sensitivity*.

grinding fluid. *Cutting fluid* used in grinding.

grinding oil. An oil-type grinding fluid; it may contain additives, but not water.

grinding relief. A groove or recess located at the boundary of a surface to

permit the corner of the wheel to overhang during grinding.

grinding sensitivity. Susceptibility of a material to surface damage such as grinding cracks; it can be affected by such factors as hardness, microstructure, hydrogen content and residual stress.

grinding stress. *Residual stress*, generated by grinding, in the surface layer of work. It may be tensile, compressive or both.

grinding wheel. A cutting tool of circular shape made of abrasive grains bonded together.

grit blasting. Abrasive blasting with small irregular pieces of steel malleable cast iron or hard nonmetallic materials.

grit size. Nominal size of abrasive particles in a grinding wheel corresponding to the number of openings per linear inch in a screen through which the particles can just pass. Sometimes, but inadvisedly, called grain size.

grizzly. A set of parallel bars (or grating) used for the coarse separation or screening of ores, rock or other material.

groove angle. The total included angle of the groove between parts to be joined. Thus, the sum of two bevel angles, either or both of which may be zero degrees.

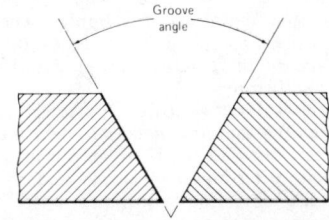

groove face. The portion of a surface or surfaces of a member included in a groove. See sketch accompanying *root of joint*.

groove weld. A weld made in the groove between two members. The standard types are square, single-bevel, single flare-bevel, single flare-V, single-J, single-U, single-V, double-bevel, double flare-bevel, double flare-V, double-J, double-U and double-V.

gross porosity. In weld metal or in a casting, pores, gas holes or globular voids that are larger and in much greater number than obtained in good practice.

ground connection. In arc welding, a

device used for attaching the work lead (ground cable) to the work.

growth. In cast iron, a permanent increase in dimensions resulting from repeated or prolonged heating at temperatures above 480 °C (900 °F) due either to graphitizing of carbides or to oxidation.

guard. (1) A device, often made of sheet metal or wire screening, that prevents accidental contact with moving parts of machinery. (2) In electroplating, same as *robber*.

Guerin forming. A trade-name process. See *rubber-pad forming*.

guided bend test. A test in which the specimen is bent to a definite shape by means of a jig.

guide mill. A small hand mill with several stands in a train and with guides for the work at the entrance to the rolls.

Guinier-Preston (G-P) zone. A small precipitation domain in a supersaturated metallic solid solution. A G-P zone has no well-defined crystalline structure of its own and contains an abnormally high concentration of solute atoms. The formation of G-P zones constitutes the first stage of precipitation and is usually accompanied by a change in properties of the solid solution in which they occur.

gun drill. A drill, usually with one or more flutes and with coolant passages through the drill body, used for deep hole drilling.

gutter. The clearance around the land of a forging die providing space for the flash without trapping it in the dies.

H

habit plane. The plane or system of planes of a crystalline phase along which some phenomenon such as twinning or transformation occurs.

half cell. An electrode immersed in a suitable electrolyte, designed for measurements of electrode potential.

half hard. A *temper* of nonferrous alloys and some ferrous alloys characterized by tensile strength about midway between that of *dead soft* and *full hard* tempers.

Hall process. A commercial process for winning aluminum from alumina by electrolytic reduction of a fused bath of alumina dissolved in cryolite.

hammer forging. Forging in which the work is deformed by repeated blows. Compare with *press forging*.

hammering. Beating metal sheet into a desired shape either over a form or on a high-speed mechanical hammer and a similar anvil to produce the required dishing or thinning.

hammer welding. *Forge welding* by hammering.

hand brake. A small manual folding machine designed to bend sheet metal, being similar in design and purpose to a *press brake*.

hand forging. See *flat die forging*.

handling breaks. Irregular *breaks* caused by improper handling of sheets during processing. These breaks result from the bending or sagging of the sheets while being handled.

Hansgirg process. A process for producing magnesium by the reduction of magnesium oxide with carbon.

hard chromium. Chromium electrodeposited for engineering purposes (such as increasing the wear resistance of sliding metal surfaces) rather than as a decorative coating. It is usually applied directly to basis metal and is customarily thicker than a decorative deposit, but not necessarily harder.

hard drawn. An imprecise term applied to drawn products, such as wire and tubing, that indicates substantial cold reduction without subsequent annealing. Compare with *light drawn*.

hardenability. The relative ability of a ferrous alloy to form martensite when quenched from a temperature above the upper critical temperature. Hardenability is commonly measured as the distance below a quenched surface where the metal exhibits a specific hardness (50 HRC, for example) or a specific percentage of martensite in the microstructure.

hardener. An alloy, rich in one or more alloying elements, added to a melt to permit closer composition control than possible by addition of pure metals or to introduce refractory elements not readily alloyed with the base metal. Sometimes called master alloy or rich alloy.

hardening. Increasing hardness by suitable treatment, usually involving heating and cooling. When applicable, the following more specific terms should be used: *age hardening, case hardening, flame hardening, induction hardening, precipitation hardening* and *quench hardening*.

hard facing. Depositing filler metal on a surface by welding, spraying or braze welding to increase resistance to abrasion, erosion, wear, galling, impact or cavitation damage.

hard head. A hard, brittle, white residue obtained in refining tin by liquation, containing, among other things, tin, iron, arsenic and copper. Also, a refractory lump of ore only partly smelted.

hardness. Resistance of metal to plastic deformation, usually by indentation. However, the term may also refer to stiffness or temper, or to resistance to scratching, abrasion or cutting. Indentation hardness may be measured by various hardness tests, such as *Brinell, Rockwell* and *Vickers*.

hard surfacing. Same as *hard facing*.

hard temper. Same as *full hard* temper.

Haring cell. A four-electrode cell for measurement of electrolyte resistance and electrode polarization during electrolysis.

Hartmann lines. Same as *Lüders lines*.

H-band steel. Alloy steel produced to specified limits of hardenability; the chemical composition range may be slightly different from that of the corresponding grade of ordinary alloy steel.

header. See *upsetter*.

heading. Upsetting wire, rod or bar stock in dies to form parts that usually have some of the cross-sectional area larger than the original.

healed-over scratch. A scratch that occurred in an earlier mill operation and was partially masked in subsequent rolling. It may open up during forming.

hearth. The bottom portion of certain furnaces, such as the blast furnace, air furnace and other reverberatory furnaces, that supports the charge and sometimes collects and holds molten metal.

heat-affected zone. That portion of the base metal that was not melted during brazing, cutting or welding, but whose microstructure and mechanical properties were altered by the heat.

heat check. A pattern of parallel surface cracks that are formed by alternate rapid heating and cooling of the extreme surface metal, sometimes found on forging dies and piercing punches. There may be two sets of parallel cracks, one set perpendicular to the other.

heat-resisting alloy. An alloy developed for very high temperature service where relatively high stresses (tensile, thermal, vibratory or shock) are encountered and where oxidation resistance is frequently required.

heat time. In multiple-impulse or seam welding, the time that the current flows during any one impulse.

heat tinting. Coloration of a metal surface through oxidation by heating to reveal details of the microstructure.

heat treatable alloy. An alloy that can be hardened by heat treatment.

heat treating film. A thin coating or film, usually an oxide, formed on the surface of metals during heat treatment.

heat treatment. Heating and cooling a solid metal or alloy in such a way as to obtain desired conditions or properties. Heating for the sole purpose of hot working is excluded from the meaning of this definition.

heel. Synonymous with *base* (1). See also sketch accompanying *single-point tool*.

hemming. Forming of an edge by bending the metal back on itself.

HERF A common abbreviation for *high-energy-rate forging* or *high-energy-rate forming*.

herringbone pattern. Same as *chevron pattern*.

Heyn stresses. Same as *microscopic stresses*.

high-conductivity copper. Copper that, in the annealed condition, has a minimum electrical conductivity of 100% *IACS* as determined in accordance with ASTM methods of testing.

high-energy-rate forging. Producing forgings at extremely high ram velocities resulting from the sudden release of a compressed gas against a free piston. Forging is usually completed in one blow. Also known as HERF processing, high velocity forging, high speed forging.

high-energy-rate forming. A group of special forming processes in which metal undergoes deformation at high velocity, usually at least ten times the velocity of 0.2 to 6 m/s (0.5 to 20 ft/s) achieved in conventional forming. Commonly abbreviated HERF. Explosive forming, electrohydraulic forming and electromagnetic forming are the most common HERF processes.

high-frequency resistance welding. A resistance-welding process that produces coalescence of metals with the heat generated from the resistance of the workpieces to a high frequency alternating current in the 10 to 500 kHz range and the rapid application of an upsetting force after heating is substantially completed. The path of the current in the workpiece is controlled by the use of the proximity effect (the feed current follows closely the return current conductor).

highlighting. Buffing or polishing selected areas of a complex shape to increase the luster or change the color of those areas.

high residual phosphorus copper. Deoxidized copper with residual phosphorus present in amounts (usually 0.013 to 0.04 %) generally sufficient to decrease appreciably the conductivity of the copper.

hindered contraction. Contraction where the shape will not permit a casting to contract in certain regions in keeping with the coefficient of expansion.

hitch feed. Feed performed by a reciprocating head or slide carrying a gripper shoe that clamps the stock during the feeding movement and releases it on the return stroke.

hob. A rotary cutting tool with its teeth arranged along a helical thread, used for generating gear teeth or other evenly spaced forms on the periphery of a cylindrical workpiece. The hob and the workpiece are rotated in timed relationship to each other while the hob is fed axially or tangentially across or radially into the workpiece. Hobs should not be confused with multiple-thread milling cutters, rack cutters, and similar tools, where the teeth are not arranged along a helical thread.

hogging. Machining a part from bar stock, plate or a simple forging in which much of the original stock is removed.

holddown. A plate, a ring or fingers used to hold work stationary during forming, blanking, piercing or shearing.

holding furnace. A small furnace into which molten metal can be transferred to be held at the proper temperature until it can be used to make castings.

hold time. In resistance welding, the time during which pressure is applied to the work after the current ceases.

hole flanging. Forming an integral collar around the periphery of a previously formed hole. See *extruded hole*.

holidays. Discontinuities in a coating (such as porosity, cracks, gaps and similar flaws) that allow areas of basis metal to be exposed to any corrosive environment that contacts the coated surface.

homogeneous carburizing. Use of a carburizing process to convert a low-carbon ferrous alloy to one of uniform and higher carbon content throughout the section.

homogenizing. Holding at high temperature to eliminate or decrease chemical segregation by diffusion.

honing. A low-speed finishing process used chiefly to produce uniform high dimensional accuracy and fine finish, most often on inside cylindrical surfaces. In honing, very thin layers of stock are removed by simultaneously rotating and reciprocating a bonded abrasive stone or stick that is pressed against the surface being honed with lighter force than is typical of grinding.

hook. Concavity in a tooth face giving a variation in *rake* at different points along the tooth face.

Hooker process. Extrusion of a hollow billet or cup through an annulus formed by the die aperture and the mandrel or pilot to form a tube or long cup.

Hooke's law. Stress is proportional to strain. The law holds only up to the proportional limit.

Hoopes process. An electrolytic refining process for aluminum, using three liquid layers in the reduction cell.

horizontal-position welding. (1) Making a fillet weld on the upper side of the intersection of a vertical surface and a horizontal surface. (2) Making a horizontal groove weld on a vertical surface.

horizontal-rolled-position welding. Topside welding of a butt joint connecting two horizontal pieces of rotating pipe.

horn. In a resistance welding machine, a cylindrical arm or beam that transmits the electrode pressure and usually conducts the welding current.

horn press. A mechanical press equipped with or arranged for a cantilever block or horn that acts as the die or support for the die, used in forming, piercing, setting down, or riveting hollow cylinders and odd-shaped work.

horn spacing. The distance between adjacent surfaces of the horns of a resistance welding machine.

hot bed. An area adjacent to the *runout table* where hot rolled metal is placed to cool. Sometimes called the cooling table.

hot chamber machine. A *die casting* machine in which the metal chamber under pressure is immersed in the molten metal in a furnace. The chamber is sometimes called a gooseneck and the machine, a gooseneck machine.

hot-cold working. (1) A high-temperature thermomechanical treatment consisting of deforming a metal above its transformation temperature and cooling fast enough to preserve some or all of the deformed structure. (2) A general term synonymous with *warm working*.

hot crack. A crack formed in a cast metal because of internal stress developed on cooling following solidification. A hot crack is less open than a *hot tear* and usually exhibits less oxidation and decarburization along the fracture surface.

hot dip coating. A metallic coating obtained by dipping the basis metal into a molten metal.

hot extrusion. Extrusion at elevated temperature that does not cause strain hardening. See also *extrusion*.

hot forming. See *hot working*.

hot isostatic pressing. A process for simultaneously heating and forming a powder metallurgy compact in which metal powder, contained in a sealed flexible mold, is subjected to equal pressure from all directions at a temperature high enough for sintering to take place.

hot isostatic pressure welding. A diffusion-welding method that produces coalescence of materials by heating and applying hot inert gas under pressure.

hot mill. A production line or facility for hot rolling metals.

hot press forging. Plastically deforming metals between dies in presses at temperatures high enough to avoid strain hardening.

hot pressing. Forming a powder metallurgy compact at a temperature high enough to have concurrent *sintering*.

hot pressure welding. A solid state

welding process that produces coalescence materials with heat and application of pressure sufficient to produce macrodeformation of the base material. Vacuum or other shielding media may be used. See also *forge welding* and *diffusion welding*.

hot quenching. An imprecise term used to cover a variety of quenching procedures in which a quenching medium is maintained at a prescribed temperature above 70 °C(160 °F).

hot rod. Same as *wire rod*.

hot shortness. A tendency for some alloys to separate along grain boundaries when stressed or deformed at temperatures near the melting point. Hot shortness is caused by a low-melting constituent, often present only in minute amounts, that is segregated at grain boundaries.

hot tear. A fracture formed in a metal during solidification because of *hindered contraction*. Compare with *hot crack*.

hot top. (1) A reservoir, thermally insulated or heated, to hold molten metal on top of a mold to feed the ingot or casting as it contracts on solidifying to avoid having pipe or voids. See accompanying sketch. (2) A refractory-lined steel or iron casting that is inserted into the tip of the mold and is supported at various heights to feed the ingot as it solidifies.

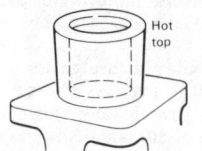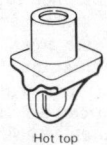

hot trimming. Removing flash or excess metal from a hot part (such as a forging) in a trimming press.

hot working. Deforming metal plastically at such a temperature and strain rate that recrystallization takes place simultaneously with the deformation, thus avoiding any strain hardening.

hubbing. Producing die cavities by pressing a male master plug, known as a hub, into a block of metal.

Hull cell. A special electrodeposition cell giving a range of known current densities for test work.

humidity test. A corrosion test involving exposure of specimens at controlled levels of humidity and temperature. Contrast with *salt-fog test*.

hydraulic press. A press in which fluid

pressure is used to actuate and control the ram.

hydride descaling. *Descaling* by action of a hydride in a fused alkali.

hydrogen brazing. A term sometimes used to denote brazing in a hydrogen-containing atmosphere, usually in a furnace; use of the appropriate process name is preferred.

hydrogen damage. A general term for the embrittlement, cracking, blistering and hydride formation that can occur when hydrogen is present in some metals.

hydrogen embrittlement. A condition of low ductility in metals resulting from the absorption of hydrogen.

hydrogen loss. The loss in weight of metal powder or of a *compact* caused by heating a representative sample for a specified time and temperature in a hydrogen atmosphere. Broadly, a measure of the oxygen content of the sample, when applied to materials containing only such oxides as are reducible with hydrogen and no hydride-forming element.

hydrogen overvoltage (in electroplating). Overvoltage associated with the liberation of hydrogen.

hydrogen-reduced powder. Metal powder produced by the hydrogen reduction of a compound.

hydrometallurgy. Industrial *winning* or *refining* of metals using water or an aqueous solution.

hydrostatic tension. Three equal and mutually perpendicular tensile stresses.

hypereutectic alloy. In an alloy system exhibiting a *eutectic*, any alloy whose composition has an excess of alloying element compared to the eutectic composition, and whose equilibrium microstructure contains some eutectic structure.

hypereutectoid alloy. In an alloy system exhibiting a *eutectoid*, any alloy whose composition has an excess of alloying element compared to the eutectoid composition, and whose equilibrium microstructure contains some eutectoid structure.

hypoeutectic alloy. In an alloy system exhibiting a *eutectic*, any alloy whose composition has an excess of base metal compared to the eutectic composition, and whose equilibrium microstructure contains some eutectic structure.

hypoeutectoid alloy. In an alloy system exhibiting a *eutectoid*, any alloy whose composition has an excess of base metal compared to the eutectoid composition, and whose equilibrium microstructure contains some eutectoid structure.

hysteresis, magnetic. The lag of the magnetization of an iron or steel specimen behind any cyclic variation of the applied magnetizing field.

I

IACS. International annealed copper standard; a standard reference used in reporting electrical conductivity. The conductivity of a material, in %IACS, is equal to 1724.1 divided by the electrical resistivity of the material in nΩ·m.

idiomorphic crystal. An individual crystal that has grown without restraint so that the habit planes are clearly developed. Compare with *allotriomorphic crystal.*

immersion cleaning. Cleaning where the work is immersed in a liquid solution.

immersion coating. A coating produced in a solution by chemical or electrochemical action without the use of external current.

immersion plating. Depositing a metallic coating on a metal immersed in a liquid solution, without the aid of an external electric current. Also called dip plating.

impact energy. The amount of energy required to fracture a material, usually measured by means of an *Izod* or *Charpy* test. The type of specimen and test conditions affect the values and therefore should be specified.

impact extrusion. See *extrusion.*

impact line. A blemish on a drawn sheet-metal part caused by a slight change in metal thickness. The mark is called an impact line when it results from impact of the punch on the blank; it is called a recoil line when it results from transfer of the blank from the die to the punch during forming, or from a reaction to the blank being pulled sharply through the draw ring.

impact strength. Same as *impact energy.*

impact test. A test to determine the behavior of materials when subjected to high rates of loading, usually in bending, tension or torsion. The quantity measured is the energy absorbed in breaking the specimen by a single blow, as in the *Charpy* or *Izod* tests.

imperfection. (1) When referring to the physical condition of a part or metal product, any departure of a quality characteristic from its intended level or state. The existence of an imperfection does not imply *nonconformance,* nor does it have any implication as to the usability of a product or service. An imperfection must be rated on a scale of severity, in accordance with applicable specifications, to establish whether or not the part or metal product is of acceptable quality. (2) Generally, any departure from an ideal design, state or condition. (3) In crystallography, any deviation from an ideal space lattice.

impregnation. (1) The treatment of porous castings with a sealing medium to stop pressure leaks. (2) The process of filling the pores of a sintered compact, usually with a liquid such as a lubricant. (3) The process of mixing particles of a nonmetallic substance in a matrix of metal powder, as in diamond-impregnated tools.

impression die forging. A forging that is formed to the required shape and size by machined impressions in specially prepared dies that exert three-dimensional control on the workpiece.

impurities. Elements or compounds whose presence in a material is undesired.

inclinable press. A press that can be inclined to facilitate handling of the formed parts. See *open-back inclinable press.*

inclusions. Particles of foreign material in a metallic matrix. The particles are usually compounds (such as oxides, sulfides or silicates), but may be of any substance that is foreign to (and essentially insoluble in) the matrix.

indentation. In a spot, seam or projection weld, the depression on the exterior surface of the base metal.

indentation hardness. The resistance of a material to indentation. This is the usual type of hardness test, in which a pointed or rounded indenter is pressed into a surface under a substantially static load.

indication. In inspection, a response to a nondestructive stimulus that implies the presence of an *imperfection.* The indication must be interpreted to determine if (a) it is a true indication or a *false indication* and (b) whether or not a true indication represents an unacceptable deviation.

indicator. A substance that, through some visible change such as color, indicates the condition of a solution or other material as to the presence of free acid, alkali or other substance.

indirect-arc furnace. An electric-arc furnace in which the metallic charge is not one of the poles of the arc.

indirect extrusion. See *extrusion.*

induction brazing. *Brazing* in which the required heat is generated by subjecting the workpiece to electromagnetic induction.

induction furnace. An ac electric furnace in which the primary conductor is coiled and generates, by electromagnetic induction, a secondary current that develops heat within the metal charge.

induction hardening. A surface-hardening process in which only the surface layer of a suitable ferrous workpiece is heated by electromagnetic induction to above the upper critical temperature and immediately quenched.

induction heating. Heating by combined electrical resistance and hysteresis losses induced by subjecting a metal to the varying magnetic field surrounding a coil carrying alternating current.

induction melting. Melting in an *induction furnace.*

induction welding. *Welding* in which the required heat is generated by subjecting the workpiece to electromagnetic induction.

inert anode. An anode that is insoluble in the electrolyte under the conditions prevailing in the electrolysis.

infiltration. The process of filling the pores of a sintered or unsintered powder metallurgy compact with a metal or alloy of lower melting point.

ingate. Same as *gate.*

ingot. A casting of simple shape, suitable for hot working or remelting.

ingot iron. Commercially pure iron.

inhibitor. A substance that retards some specific chemical reaction. Pickling inhibitors retard the dissolution of metal without hindering the removal of scale from steel.

inoculation. The addition of a material to molten metal to form nuclei for crystallization.

insert. (1) A part formed from a second material, usually a metal, which is placed in the mold and appears as an integral structural part of the final casting. (2) A removable portion of a die or mold.

insert die. A relatively small die containing part or all of the impression of a forging, and which is fastened to a master die block.

inserted-blade cutters. Cutters having replaceable blades that are either solid or tipped and are usually adjustable.

intercept method. A quantitative metallographic technique in which the desired quantity (such as grain size or amount of precipitate) is expressed as the number of times per unit length a straight line on a metallographic image crosses particles of the feature being measured.

intercommunicating porosity. In a sintered powder metallurgy compact, a type of porosity in which individual pores are connected in such a way that a fluid may pass from one pore to another throughout the entire compact.

intercrystalline. Between the crystals, or grains, of a metal.

interdendritic corrosion. Corrosive attack that progresses preferentially along interdendritic paths. This type of attack results from local differences in composition, such as coring commonly encountered in alloy castings.

interface. A surface that forms the boundary between phases or systems.

interfacial tension. The contractile force of an interface between two phases.

interference. The difference in lateral dimensions at room temperature between two mating components before assembly by expansion, shrink, or press fitting. Can be expressed in absolute or in relative terms.

interference fit. Any of various classes of fit between mating parts where there is nominally a negative or zero allowance between the parts, and there is either part interference or no gap when the mating parts are made to the respective extremes of individual tolerances that ensure the tightest fit between the parts. Contrast with *clearance fit.*

intergranular corrosion. Corrosion occurring preferentially at grain boundaries, usually with slight or negligible attack on the adjacent grains. See also *interdendritic corrosion.*

intermediate annealing. Annealing wrought metals at one or more stages during manufacture and before final treatment.

intermediate electrode. Same as *bipolar electrode.*

intermediate phase. In an alloy or a chemical system, a distinguishable homogeneous phase whose composition range does not extend to any of the pure components of the system.

intermetallic compound. An intermediate phase in an alloy system, having a narrow range of homogeneity and relatively simple stoichiometric proportions; the nature of the atomic binding can be of various types, ranging from metallic to ionic.

intermittent weld. A weld in which the continuity is broken by recurring unwelded spaces.

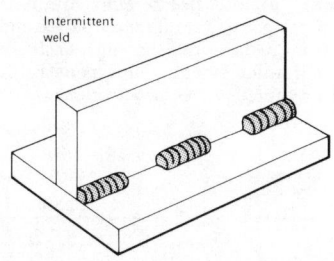

Intermittent weld

internal friction. The conversion of energy into heat by a material subjected to fluctuating stress. In free vibration, the internal friction is measured by the *logarithmic decrement.*

internal oxidation. Preferential in situ oxidation of certain components or phases within the bulk of a solid alloy accomplished by diffusion of oxygen into the body; a form of *subsurface corrosion.*

internal stress. See preferred term, *residual stress.*

interpass temperature. In a multipass weld, the lowest temperature of a *pass* before the succeeding one is commenced.

interrupted aging. Aging at two or more temperatures, by steps, and cooling to room temperature after each step. See *aging,* and compare with *progressive aging* and *step aging.*

interrupted-current plating. Plating in which the flow of current is discontinued for periodic short intervals to decrease anode polarization and elevate the *critical current density.* It is most commonly used in cyanide copper plating.

interrupted quenching. A quenching procedure in which the workpiece is removed from the first quench at a temperature substantially higher than that of the quenchant and is

then subjected to a second quenching system having a different cooling rate than the first.

interstitial solid solution. A solid solution in which the solute atoms occupy positions that do not correspond to lattice points of the solvent. Contrast with *substitutional solid solution.*

intracrystalline. Within or across the crystals or grains of a metal; same as transcrystalline and transgranular.

inverse chill. A condition in an iron casting in which the interior is chilled or white iron while the surfaces are mottled or contain free graphite.

inverse segregation. Segregation in cast metal in which an excess of lower-melting constituents occurs in the earlier freezing portions, apparently the result of liquid metal entering cavities developed in the earlier-solidified metal.

investment casting. (1) Casting metal into a mold produced by surrounding (investing) an expendable pattern with a refractory slurry that sets at room temperature after which the wax, plastic or frozen mercury pattern is removed through the use of heat. Also called precision casting, or lost-wax process. (2) A part made by the investment casting process.

investment compound. A mixture of a graded refractory filler, a binder and a liquid vehicle, used to make molds for *investment casting.*

ion. An atom, or group of atoms, that has gained or lost one or more outer electrons and thus carries an electric charge. Positive ions, or cations, are deficient in outer electrons. Negative ions, or anions, have an excess of outer electrons.

ion exchange. The reversible interchange of ions between a liquid and solid, with no substantial structural changes in the solid.

ionic bond. A bond between two or more atoms that is the result of electrostatic attractive forces between positively and negatively charged ions.

ionic crystal. A crystal in which atomic bonds are *ionic bonds.* This type of atomic linkage, also known as (hetero) polar bonding, is characteristic of many compounds (sodium chloride, for instance).

ionization chamber. An enclosure containing two or more electrodes surrounded by a gas capable of conducting an electric current when it is ionized by x-rays or other ionizing rays. It is commonly used for measuring intensity of such radiation.

iron casting. A part made of *cast iron*.

ironing. Thinning the walls of hollow articles by drawing them between a punch and a die.

iron-powder electrode. A welding electrode with a covering containing up to about 50% iron powder, some of which becomes part of the deposit.

irradiation. The exposure of a material in a field of radiation; the cumulative exposure.

isostatic pressing. A process for forming a powder metallurgy compact by applying pressure equally from all directions to metal powder contained in a sealed flexible mold. See also *hot isostatic pressing.*

isothermal annealing. Austenitizing a ferrous alloy and then cooling to and holding at a temperature at which austenite transforms to a relatively soft ferrite carbide aggregate.

isothermal transformation. A change in phase that takes place at a constant temperature. The time required for transformation to be completed, and in some instances the time delay before transformation begins, depends on the amount of supercooling below (or superheating above) the equilibrium temperature for the same transformation.

isotope. One of several different nuclides of an element having the same number of protons in their nuclei and therefore the same atomic number, but differing in the number of neutrons and therefore in atomic weight.

isotropy. Quality of having identical properties in all directions.

Izod test. A pendulum-type single-blow impact test in which the specimen, usually notched, is fixed at one end and broken by a falling pendulum. The energy absorbed, as measured by the subsequent rise of the pendulum, is a measure of impact strength or notch toughness. Contrast with *Charpy test.*

J

jig. A device to hold a workpiece in place and simultaneously guide the tool in a cutting operation.

jig boring. Boring with a single-point tool where the work is positioned upon a table that can be located so as to bring any desired part of the work under the tool. Thus, holes can be accurately spaced. This type of boring can be done on milling machines or jig borers.

jig grinding. Analogous to *jig boring*, where the holes are ground rather than machined.

joggle. An offset in a flat plane consisting of two parallel bends in opposite directions by the same angle.

joint. The location where two or more members are to be or have been fastened together mechanically or by brazing or welding.

joint efficiency. The strength of a welded joint expressed as a percentage of the strength of the unwelded base metal.

joint penetration. The minimum depth a groove or flange weld extends from its face into the joint, exclusive of reinforcement. Joint penetration may include *root penetration*.

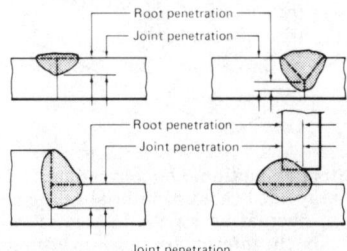

Jominy test. See *end-quench hardenability test.*

K

keel block. A standard test casting, for steel and other high-shrinkage alloys, consisting of a rectangular bar that resembles the keel of a boat, attached to the bottom of a large riser, or shrinkhead. Keel blocks that have only one bar are often called Y-blocks; keel blocks having two bars, double keel blocks. Test specimens are machined from the rectangular bar, and the shrinkhead is discarded.

Kellering. A shop term. See preferred term, *tracer milling.*

kerf. The space that was occupied by the material removed during cutting.

keyhole specimen. A type of specimen containing a hole-and-slot notch, shaped like a keyhole, usually used in impact bend tests. See *Charpy* and *Izod tests.*

killed steel. Steel treated with a strong deoxidizing agent such as silicon or aluminum in order to reduce the oxygen content to such a level that no reaction occurs between carbon and oxygen during solidification.

kiln. A large furnace used for baking, drying or burning fire brick or refractories, or for calcining ores or other substances.

kish. Free graphite that forms in molten hypereutectic cast iron as it cools. In castings, the kish may segregate toward the cope surface, where it lodges at or immediately beneath the casting surface.

knockout. (1) A mechanism for freeing formed parts from a die used for stamping, blanking, drawing, forging or heading operations. (2) A partly pierced hole in a sheet metal part, where the slug remains in the hole and can be forced out by hand if a hole actually is needed. (3) Removing sand cores from a casting. (4) Jarring an investment casting mold to remove the casting and investment from the flask.

Knoop hardness. Microhardness determined from the resistance of metal to indentation by a pyramidal diamond indenter, having edge angles of 172° 30′ and 130°, making a rhombohedral impression with one long and one short diagonal.

knuckle-joint press. A heavy short-stroke press in which the slide is directly actuated by a single toggle joint that is opened and closed by a connection and crank. It is used for embossing, coining, sizing, heading, swaging and extruding.

knurling. Impressing a design into a metallic surface, usually by means of small, hard rollers that carry the corresponding design on their surfaces.

Kroll process. A process for the production of metallic titanium by the reduction of titanium tetrachloride with a more active metal such as magnesium, yielding titanium as granules or powder.

L

ladle. A receptacle used for transferring and pouring molten metal.

laminate. (1) A composite metal, usually in the form of sheet or bar, composed of two or more metal layers so bonded that the composite metal forms a structural member. (2) To form a metallic product of two or more bonded layers.

lamination. (1) A type of discontinuity with separation or weakness generally aligned parallel to the worked surface of a metal. May be the result of pipe, blisters, seams, inclusions or segregation elongated and made directional by working. Laminations may also occur in metal-powder compacts. (2) In electrical products such as motors, a blanked piece of electrical sheet that is stacked up with several other identical pieces to make a stator or rotor.

lancing. (1) A press operation in which a single-line cut is made in strip stock without producing a detached slug. Chiefly used to free metal for forming, or to cut partial contours for blanked parts, particularly in progressive dies. (2) A misnomer for *oxy-fuel gas cutting.*

land. (1) For profile-sharpened milling cutters, the relieved portion immediately behind the cutting edge. (2) For reamers, drills and taps, the solid section between the flutes. (3) On punches, the portion adjacent to the nose that is parallel to the axis and of maximum diameter.

lap. A surface imperfection, appearing as a seam, caused by folding over hot metal, fins or sharp corners and then rolling or forging them into the surface, but not welding them.

lap joint. A joint made with two overlapping members.

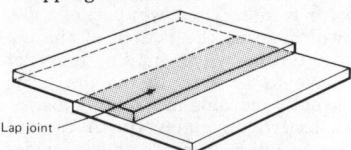

Lap joint

lapping. Finishing surfaces by abrasion with an object, usually made of copper, lead, cast iron or close-grained wood, having very fine abrasive particles rolled into its surface.

laser. A device that emits a concentrated beam of electromagnetic radiation (light). Laser beams are used in metalworking to melt, cut or weld metals; in less concentrated form they are sometimes used to inspect metal parts.

laser-beam cutting. A cutting process that severs materials with the heat obtained by directing a beam from a *laser* against a metal surface. The process can be used with or without an externally supplied shielding gas.

laser-beam machining. Removing material by melting and vaporizing the workpiece at the point of impingement of a highly focused beam of coherent monochromatic light (a laser beam).

laser-beam welding. A welding process that joins metal parts using the heat obtained by directing a beam from a *laser* onto the weld joint.

latent heat. Thermal energy absorbed or released when a substance undergoes a phase change.

lateral extrusion. An operation in which the product is extruded sideways through an orifice in the container wall.

lateral runout. Same as *axial runout.*

lattice constant. See *lattice parameter.*

lattice parameter. The length of any side of a unit cell of a given crystal structure; if the lengths are unequal, all unequal lengths must be given.

launder. (1) A channel for conducting molten metal. (2) A box conduit conveying particles suspended in water.

lay. Direction of predominant surface pattern remaining after cutting, grinding, lapping or other processing.

leaching. Extracting an element or compound from a solid alloy or mixture by preferential dissolution in a suitable liquid.

lead. (1) The axial advance of a helix in one complete turn. (2) The slight bevel at the outer end of a face cutting edge of a face mill.

lead angle. In cutting tools, the helix angle of the flutes.

lead burning. A misnomer for the welding of lead.

lead proof. See *die proof.*

leakage field. The magnetic field that leaves or enters a magnetized part at a magnetic pole.

ledeburite. The eutectic of the iron-carbon system, the constituents being austenite and cementite. The austenite decomposes into ferrite and cementite on cooling below the Ar_1.

left-hand cutting tool. A cutter all of whose flutes twist away in a counterclockwise direction when viewed from either end.

leg of a fillet weld. (1) Actual: The

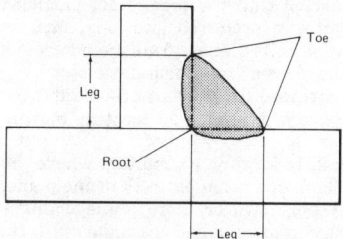

Leg of a fillet weld

distance from the root of the joint to the toe of a fillet weld. See accompanying sketch and sketches of *concave fillet weld* and *convex fillet weld.*

leveler lines. Lines on sheet or strip running transverse to the direction of roller leveling; may be seen on stoning or light sanding after leveling (but before drawing). Usually can be removed by moderate stretching.

leveling. Flattening of rolled sheet, strip or plate by reducing or eliminating distortions. See *stretcher leveling* and *roller leveling.*

leveling action. Action exhibited by a plating solution to give a plate smoother than the basis metal.

levigation. (1) Separating fine powder from coarser material by forming a suspension of the fine material in a liquid. (2) A means of classifying a material as to particle size by the rate of settling from a suspension.

levitation melting. An *induction melting* process in which the metal being melted is suspended by the electromagnetic field and is not in contact with a container.

light drawn. An imprecise term applied to drawn products, such as wire and tubing, that indicates a lesser amount of cold reduction than for *hard drawn* products.

lightly coated electrode. A filler-metal electrode used in arc welding, consisting of a metal wire with a light coating, usually of metal oxides and silicates, applied subsequent to the drawing operation primarily for stabilizing the arc. Contrast with *covered electrode.*

light metal. One of the low-density metals such as aluminum, magnesium, titanium, beryllium or their alloys.

limiting current density. The maximum current density that can be used to get a desired electrode reaction without undue interference such as may come from polarization.

lineage structure. (1) Deviations from perfect alignment of parallel arms of a columnar dendrite as a result of interdendritic shrinkage during solidification from a liquid. This type of deviation may vary in orientation from one area to another from a few minutes to as much as two degrees of arc. (2) A type of substructure consisting of elongated subgrains.

linear elastic fracture mechanics. A method of fracture analysis that can determine the stress (or load) required to induce fracture instability in a structure containing a crack-like flaw of known size and shape. See *stress-intensity factor.*

linear strain. See *strain.*

liner. (1) The slab of coating metal that is placed on the core alloy and is subsequently rolled down to clad sheet as a composite. (2) In extrusion, a removable alloy steel cylindrical chamber, having an outside longitudinal taper firmly positioned in the container or main body of the press, into which the billet is placed for extrusion.

line reaming. Simultaneous reaming of coaxial holes in various sections of a workpiece with a reamer having cutting faces or piloted surfaces with the desired alignment.

lip. For a *milling cutter,* the material included between a relieved land and a tooth face.

lip angle. (1) For a *milling cutter,* the included angle between a tooth face and a relieved land. (2) Defined by sketch accompanying *single-point tool.*

liquation. The partial melting of an alloy, usually as a result of coring or other compositional heterogeneities.

liquation temperature. The lowest temperature at which partial melting can occur in an alloy that exhibits the greatest possible degree of segregation.

liquid honing. Producing a finely polished finish by directing an air-ejected chemical emulsion containing fine abrasives against the surface to be finished.

liquid penetrant inspection. A type of nondestructive inspection that locates discontinuities that are open to the surface of a metal by first allowing a penetrating dye or fluorescent liquid to infiltrate the discontinuity, removing the excess penetrant, and then applying a developing agent that causes the penetrant to seep back out of the discontinuity and register as an indication. Liquid penetrant inspection is suitable for both ferrous and nonferrous materials, but is limited to the detection of open surface discontinuities in nonporous solids.

liquid phase sintering. *Sintering* a powder metallurgy compact under conditions that maintain a liquid metallic phase within the compact during all or part of the sintering schedule. The liquid phase may be derived from a component of the green com-

pact or may be infiltrated into the compact from an outside source.

liquid shrinkage. See *casting shrinkage.*

liquidus. In a constitution or equilibrium diagram, the locus of points representing the temperatures at which the various compositions in the system begin to freeze on cooling or finish melting on heating. See also *solidus.*

liquor finish. A smooth, bright finish characteristic of wet-drawn wire. Formerly produced by using liquor from fermented grain mash as a drawing lubricant.

live center. A lathe or grinder center that holds, yet rotates with, the work. It is used in either the headstock or tailstock of a machine to prevent wear and reduce the driving torque.

loading. (1) In cutting, building up of a cutting tool back of the cutting edge by undesired adherence of material removed from the work. (2) In grinding, filling the pores of a grinding wheel with material from the work, usually resulting in a decrease in production and quality of finish. (3) In powder metallurgy, filling of the die cavity with powder.

loam. A molding material consisting of sand, silt and clay, used over brickwork or other structural backup material for making massive castings, usually of iron or steel.

local action. Corrosion due to the action of "local cells"; that is, *galvanic cells* resulting from inhomogeneities between adjacent areas on a metal surface exposed to an electrolyte.

local cell. A *galvanic cell* resulting from inhomogeneities between areas on a metal surface in an electrolyte. The inhomogeneities may be of physical or chemical nature in either the metal or its environment.

local current density. Current density at a point or on a small area.

localized precipitation. Precipitation from a supersaturated solid solution similar to *continuous precipitation,* except that the precipitate particles form at preferred locations, such as along slip planes, grain boundaries or incoherent twin boundaries.

locational fit. A clearance or interference *fit* intended for locating mating parts.

lock. In forging, a condition where the flash line is not entirely in one plane. Where two or more plane changes occur, it is called a compound lock. Where a lock is placed in the die to

compensate for die shift caused by a steep lock, it is called a counterlock.

longitudinal direction. The principal direction of flow in a worked metal.

longitudinal field. A magnetic field that extends within a magnetized part from one or more poles to one or more other poles and that is completed through a path external to the part.

long transverse. See *transverse.*

looping mill. An arrangement of hot rolling stands such that a hot bar, while it is being discharged from one stand, is fed into a second stand in the opposite direction.

loose metal. Refers to an area in a formed panel that is not stiff enough to hold its shape; may be confused with *oil canning.*

lost-wax process. An *investment casting* process in which a wax pattern is used.

lot. A finite quantity of a given product manufactured under production conditions that are considered uniform. Often used to describe a finite quantity of product submitted for inspection as a single group. For a bulk product (such as a chemical or powdered metal), the term "batch" is often used synonymously with lot.

lower punch. The lower part of a die, which forms the bottom of the die cavity and which may or may not move in relation to the die body; usually movable in a forging die.

low-hydrogen electrode. A covered arc-welding electrode that provides an atmosphere around the arc and molten weld metal that is low in hydrogen.

low-residual-phosphorus copper. Deoxidized copper with residual phosphorus present in amounts (usually 0.004 to 0.012%) generally too small to decrease appreciably the electrical conductivity of the copper.

low shaft furnace. A short shaft-type blast furnace used to produce pig iron and ferroalloys from low-grade ores, using low-grade fuel. The air blast is often enriched with oxygen. It is also used for making a variety of other products such as alumina, cement-making slags and ammonia synthesis gas.

lubricant. Any substance used to reduce friction between two surfaces in contact.

Lüders lines. Elongated surface markings or depressions caused by localized plastic deformation that results from discontinuous (inhomogeneous) yielding. Also known as Lüders bands, Hartmann lines, Piobert lines or *stretcher strains*.

luster finish. A bright as-rolled finish, produced on ground rolls; it is suitable for decorative painting or plating, but usually must undergo additional surface preparation after forming.

lute (1) A mixture of fireclay used to seal cracks between a crucible and its cover or between container and cover when heat is to be applied. (2) To seal with clay or other plastic material.

M

machinability. The relative ease of machining a metal.

machinability index. A relative measure of the machinability of an engineering material under specified standard conditions.

machine forging. Forging performed in upsetters or horizontal forging machines.

machine welding. Welding with equipment that performs under the continual observation and control of a welding operator. The equipment may or may not load and unload the work. Compare with *automatic welding*.

machining. Removing material from a metal part, usually using a cutting tool, and usually using a power-driven machine.

machining allowance. Finish allowance.

machining stress. *Residual stress* caused by machining.

macroetching. *Etching* a metal surface to accentuate gross structural details (such as grain flow, segregation, porosity or cracks) for observation by the unaided eye or at a magnification of ten diameters or less.

macrograph. A graphic reproduction of the surface of a prepared specimen at a magnification not exceeding ten diameters. When photographed, the reproduction is known as a photomacrograph.

macroscopic. Visible at magnifications up to ten diameters.

macroscopic stresses. Residual stresses that vary from tension to compression in a distance (presumably many times the grain size) that is comparable to the gage length in ordinary strain measurements, hence, detectable by x-ray or dissection methods.

macroshrinkage. Isolated, clustered or interconnected voids in a casting that are detectable macroscopically. Such voids are usually associated with abrupt changes in section size and are caused by a lack of adequate feeding to compensate for solidification shrinkage.

macrostress. Same as *macroscopic stress*.

macrostructure. The structure of metals as revealed by macroscopic examination of the etched surface of a polished specimen.

magnesite wheel. A grinding wheel bonded with magnesium oxychloride.

magnetically hard alloy. A ferromagnetic alloy capable of being magnetized permanently because of its ability to retain induced magnetization and magnetic poles after removal of externally applied fields; an alloy with high coercive force. The name is based on the fact that the quality of the early permanent magnets was related to their hardness.

magnetically soft alloy. A ferromagnetic alloy that becomes magnetized readily upon application of a field and that returns to practically a nonmagnetic condition when the field is removed; an alloy with the properties of high magnetic permeability, low coercive force and low magnetic hysteresis loss.

magnetic-analysis inspection. A nondestructive method of inspection to determine the existence of variations in magnetic flux in ferromagnetic materials of constant cross section, such as might be caused by discontinuities and variations in hardness. The variations are usually indicated by a change in pattern on an oscilloscopic screen.

magnetic-particle inspection. A nondestructive method of inspection for determining the existence and extent of surface cracks and similar imperfections in ferromagnetic materials. Finely divided magnetic particles, applied to the magnetized part, are attracted to and outline the pattern of any magnetic-leakage fields created by discontinuities.

magnetic pole. The area on a magnetized part at which the magnetic field leaves or enters the part. It is a point of maximum attraction in a magnet.

magnetic separator. A device used to separate magnetic from less magnetic or nonmagnetic materials. The crushed material is conveyed on a belt past a magnet.

magnetic writing. In magnetic-particle inspection, a *false indication* caused by contact between a magnetized part and another piece of magnetic material.

magnetizing force. A force field, resulting from the flow of electric currents or from magnetized bodies, that produces magnetic induction.

magnetostriction. The characteristic of a material that is manifest by strain when it is subjected to a magnetic field; or the inverse. Some iron-nickel alloys expand; pure nickel contracts.

malleability. The characteristic of metals that permits plastic deformation in compression without rupture.

malleable cast iron. A cast iron made by a prolonged anneal of *white cast iron* in which decarburization or graphitization, or both, take place to eliminate some or all of the cementite. The graphite is in the form of temper carbon. If decarburization is the predominant reaction, the product will have a light fracture, hence, "whiteheart malleable"; otherwise, the fracture will be dark, hence, "blackheart malleable". Ferritic malleable has a predominantly ferritic matrix; pearlitic malleable may contain pearlite, spheroidite or tempered martensite depending on heat treatment and desired hardness.

malleablizing. Annealing *white cast iron* in such a way that some or all of the combined carbon is transformed to graphite or, in some instances, part of the carbon is removed completely.

mandrel. (1) A blunt ended tool or rod used to retain the cavity in hollow metal products during working. (2) A metal bar around which other metal may be cast, bent, formed or shaped. (3) A shaft or bar for holding work to be machined. (4) A form, such as a mold or matrix, used as a cathode in electroforming.

Mannesmann mill. Mill used in *Mannesmann process*.

Mannesmann process. A process used for piercing tube billets in making seamless tubing. The billet is rotated between two heavy rolls mounted at an angle and is forced over a fixed mandrel.

manual welding. Welding wherein the entire welding operation is performed and controlled by hand.

maraging. A precipitation-hardening treatment applied to a special group of iron-base alloys to precipitate one or more intermetallic compounds in a matrix of essentially carbon-free martensite. NOTE: The first developed series of maraging steels contained, in addition to iron, more than 10% nickel and one or more supplemental hardening elements. In this series, aging is done at 480 °C (900 °F).

margin. The cylindrical portion of the *land* of a drill that is not cut away to provide clearance.

marquenching. See *martempering*.

martempering. (1) A hardening procedure in which an austenitized ferrous workpiece is quenched into an appropriate medium whose temperature is maintained substantially at the M_s of the workpiece, held in the medium until its temperature is uniform throughout—but not long enough to permit bainite to form—and then cooled in air. The treatment is frequently followed by tempering. (2) When the process is applied to carburized material, the controlling M_s temperature is that of the case. This variation of the process is frequently called marquenching.

martensite. A generic term for microstructures formed by diffusionless phase transformation in which the parent and product phases have a specific crystallographic relationship. Martensite is characterized by an acicular pattern in the microstructure in both ferrous and nonferrous alloys. In alloys where the solute atoms occupy interstitial positions in the martensitic lattice (such as carbon in iron), the structure is hard and highly strained; but where the solute atoms occupy substitutional positions (such as nickel in iron), the martensite is soft and ductile. The amount of high temperature phase that transforms to martensite on cooling depends to a large extent on the lowest temperature attained, there being a rather distinct beginning temperature (M_s) and a temperature at which the transformation is essentially complete (M_f).

martensite range. The temperature interval between M_s and M_f.

martensitic transformation. A reaction that takes place in some metals on cooling, with the formation of an acicular structure called *martensite*.

mash resistance seam weld. A resistance *seam weld* made in a lap joint, in which the thickness at the lap is reduced plastically to approximately the thickness of one of the lapped parts.

masking tape. A tape used as a *resist* for stopping-off purposes.

master alloy. An alloy, rich in one or more desired addition elements, that can be added to a melt to raise the percentage of a desired constituent.

match. A condition in which a point in one forging-die half is aligned properly with the corresponding point in the opposite die half within specified tolerance.

matched edges. Two edges of a forging-die face that are machined exactly at 90° to each other, and from which all dimensions are taken in laying out the die impression and aligning the dies in the forging equipment.

match lines. Same as *matched edges*.

match plate. A plate of metal or other material on which patterns for metal casting are mounted (or formed as an integral part) so as to facilitate molding. The pattern is divided along its *parting plane* by the plate.

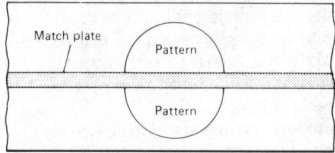

matrix. (1) The principal phase or aggregate in which another constituent is embedded. (2) In electroforming, a form used as a cathode.

matte. An intermediate product of *smelting;* an impure metallic sulfide mixture made by melting a roasted sulfide ore, such as an ore of copper, lead or nickel.

matte dip. An etching solution used to produce a dull finish on metal.

matte finish. (1) A dull texture produced by rolling sheet or strip between rolls that have been roughened by blasting. (2) A dull finish characteristic of some electrodeposits, such as cadmium or tin. Also written mat finish.

McQuaid-Ehn test. A test to reveal grain size after heating into the austenitic temperature range. Eight standard McQuaid-Ehn grain sizes rate the structure, No. 8 being finest, No. 1 coarsest.

mean stress. (1) In fatigue loading, the algebraic mean of the maximum and minimum stress in one cycle. Also called the steady stress component. (2) In any multiaxial stress system, the algebraic mean of three *principal stresses;* more correctly called mean normal stress.

mechanical equation of state. Any equation relating the stress, strain, strain rate and temperature that is based on the concept that the instantaneous value of any one of these quantities is a single-valued function of the others, regardless of the prior history of the deformation.

mechanical hysteresis. Energy absorbed in a complete cycle of loading and unloading within the elastic limit and represented by the closed loop of the stress-strain curves for loading and unloading. Sometimes referred to as elastic, but more properly, mechanical.

mechanical metallurgy. The science and technology dealing with the behavior of metals when subjected to applied forces; often considered restricted to the plastic working or shaping of metals.

mechanical plating. Plating wherein fine metal powders are peened onto the work by *tumbling* or other means.

mechanical press. A press whose slide is operated by a crank, eccentric, cam, toggle links or other mechanical device.

mechanical properties. The properties of a material that reveal its elastic and inelastic behavior when force is applied, thereby indicating its suitability for mechanical applications; for example, modulus of elasticity, tensile strength, elongation, hardness, and fatigue limit. Compare with *physical properties*.

mechanical testing. Determination of *mechanical properties*.

mechanical twin. A *twin* formed in a crystal by simple shear under external loading.

mechanical working. Subjecting metal to pressure, exerted by rolls, hammers or presses, in order to change the metal's shape or physical properties.

melting point. The temperature at which a pure metal, compound or eutectic changes from solid to liquid; the temperature at which the liquid and the solid are in equilibrium.

melting range. The range of temperature over which an alloy other than a compound or eutectic changes from solid to liquid; the range of temperature from *solidus* to *liquidus* at any given composition on a constitution diagram.

melting rate. In electric arc welding, the weight or length of electrode melted in a unit of time. Sometimes called melt-off rate or burn-off rate.

melt-off rate. See *melting rate.*

merchant mill. (obsolete) A mill, consisting of a group of stands of three rolls each arranged in a straight line and driven by one power unit, used to roll rounds, squares or flats of smaller dimensions than would be rolled on the bar mill.

mesh. The screen number of the finest screen of a specified standard screen scale through which almost all the particles of a powder sample will pass. Also called mesh size.

metal. (1) An opaque lustrous elemental chemical substance that is a good conductor of heat and electricity and, when polished, a good reflector of light. Most elemental metals are malleable and ductile and are, in general, denser than the other elemental substances. (2) As to structure, metals may be distinguished from nonmetals by their atomic binding and electron availability. Metallic atoms tend to lose electrons from the outer shells, the positive ions thus formed being held together by the electron gas produced by the separation. The ability of these "free electrons" to carry an electric current, and the fact that this ability decreases as temperature increases, establish the prime distinctions of a metallic solid. (3) From the chemical viewpoint, an elemental substance whose hydroxide is alkaline. (4) An *alloy.*

metal-arc cutting. Any of a group of arc cutting processes that severs metals by melting them with the heat of an arc between a metal electrode and the work.

metal-arc welding. Any of a group of arc welding processes that fuses metals together using the heat of an arc between a metal electrode and the work. Use of the specific process name is preferred.

metal inert-gas welding. *Gas metal-arc welding* using an inert gas such as argon as the shielding gas.

metal leaf. Thin metal sheet, usually thinner than foil, and traditionally produced by beating rather than by rolling.

metallic bond. The principal bond between metal atoms, which arises from the increased spatial extension of valence-electron wave functions when an aggregate of metal atoms is brought close together. See *covalent bond, ionic bond.*

metallic glass. A noncrystalline metal or alloy, commonly produced by drastic supercooling of a molten alloy, by electrodeposition, or by vapor deposition. Also called amorphous alloy.

metallizing. (1) Forming a metallic coating by atomized spraying with molten metal or by *vacuum deposition.* Also called spray metallizing. (2) Applying an electrically conductive metallic layer to the surface of a nonconductor.

metallograph. An optical instrument designed for both visual observation and photomicrography of prepared surfaces of opaque materials at magnifications ranging from about 25 to about 2000 diameters. The instrument consists of a high-intensity illuminating source, a microscope and a camera bellows. On some instruments, provisions are made for examination of specimen surfaces with polarized light, phase contrast, oblique illumination, darkfield illumination and customary brightfield illumination.

metallography. The science dealing with the constitution and structure of metals and alloys as revealed by the unaided eye or by such tools as low-powered magnification, optical microscope, electron microscope and diffraction or x-ray techniques.

metallurgical coke. A coke, usually low in sulfur, having a very high compressive strength at elevated temperatures; used in metallurgical furnaces not only as fuel, but also to support the weight of the charge.

metallurgy. The science and technology of metals and alloys. Process metallurgy is concerned with the extraction of metals from their ores and with the refining of metals; physical metallurgy, with the physical and mechanical properties of metals as affected by composition, processing and environmental conditions; and mechanical metallurgy, with the response of metals to applied forces.

metal penetration. A surface condition in castings in which metal or metal oxides have filled voids between sand grains without displacing them.

metal spraying. Coating metal objects by spraying molten metal against the surface. See *thermal spraying, flame spraying.*

metastable. Refers to a state of pseudoequilibrium that has a higher free energy than the true equilibrium state.

M_f temperature. For any alloy system, the temperature at which martensite formation on cooling is essentially finished. See *transformation temperature* for the definition applicable to ferrous alloys.

microfissure. A crack of microscopic proportions.

micrograph. A graphic reproduction of the surface of a prepared specimen, usually etched, at a magnification greater than ten diameters. If produced by photographic means it is called a photomicrograph (not a microphotograph).

microhardness. The hardness of a material as determined by forcing an indenter such as a Vickers or Knoop indenter into the surface of a material under very light load; usually, the indentations are so small that they must be measured with a microscope. Capable of determining hardnesses of different microconstituents within a structure, or of measuring steep hardness gradients such as those encountered in case hardening.

microprobe. See preferred term, *electron beam microprobe analyzer.*

microradiography. The technique of passing x-rays through a thin section of an alloy in contact with a fine-grained photographic film and then viewing the radiograph at 50 to 100X to observe the distribution of alloying constituents and voids.

microscopic. Visible at magnifications greater than ten diameters.

microscopic stresses. Residual stresses that vary from tension to compression in a distance (presumably approximating the grain size) that is small compared to the gage length in ordinary strain measurements. They are not detectable by dissection methods, but can sometimes be measured from line shift or line broadening in an x-ray diffraction pattern.

microsegregation. *Segregation* within a grain, crystal or small particle. See *coring.*

microshrinkage. A casting imperfection, not detectable microscopically, consisting of interdendritic voids. Microshrinkage results from contraction during solidification where there is not an adequate opportunity to supply filler material to compensate for shrinkage. Alloys with a wide range in solidification temperature are particularly susceptible.

microstress. Same as *microscopic stress.*

microstructure. The structure of metals as revealed by microscopic examination of the etched surface of a polished specimen.

middling. A product intermediate between concentrate and tailing and containing enough of a valuable mineral to make retreatment profitable.

migration. Movement of entities (such as electrons, ions, atoms, molecules, vacancies and grain boundaries) from one place to another under the influence of a driving force (such as an electrical potential or a concentration gradient).

MIG welding. See *metal inert-gas welding.*

mil. One thousandth of an inch (0.001 in.).

mild steel. *Carbon steel* with a maximum of about 0.25% C.

mill. (1) A factory where metals are hot worked, cold worked, or melted and cast into standard shapes suitable for secondary fabrication into commercial products. (2) A production line, usually of four or more *stands,* for hot rolling metal into standard shapes such as bar, rod, plate, sheet or strip. (3) A single machine for hot rolling, cold rolling or extruding metal; examples include blooming mill, *cluster mill, four-high mill,* and *Sendzimer mill.* (4) A shop term for *milling cutter.* (5) A machine or group of machines for grinding or crushing ores and other minerals; see *ball mill, milling* (2).

mill edge. The normal edge produced in hot rolling. This edge is customarily removed when hot rolled sheets are further processed into cold rolled sheets.

Miller indices. A system for identifying planes and directions in any crystal system by means of sets of integers. The indices of a plane are related to the intercepts of that plane with the axes of a unit cell; the indices of a direction, to the multiples of lattice parameter that represent the coordinates of a point on a line parallel to the direction and passing through the arbitrarily chosen origin of a unit cell.

mill finish. A nonstandard (and typically nonuniform) surface finish on mill products that are delivered without subjecting them to a special surface treatment (other than a corrosion preventive treatment) after the final working or heat treating step.

milling. (1) Removing metal with a *milling cutter.* (2) The mechanical treatment of material, as in a *ball mill,* to produce particles or alter their size or shape, or to coat one component of a powder mixture with another.

milling cutter. A rotary cutting tool provided with one or more cutting elements, called teeth, which intermittently engage the workpiece and remove material by relative movement of the workpiece and cutter.

mill product. Any commercial product of a *mill.*

mill scale. The heavy oxide layer formed during hot fabrication or heat treatment of metals.

mineral dressing. Physical and chemical concentration of raw ore into a product from which a metal can be recovered at a profit.

minimized spangle. A hot dip galvanized coating of very small grain size, which makes the spangle less visible when the part is subsequently painted.

minimum bend radius. The minimum radius over which metal products can be bent to a given angle without fracture.

minus sieve. The portion of a sample of a granular substance (such as metal powder) that passes through a standard sieve of specified number. Contrast with *plus sieve.*

mischmetal. A natural mixture of rare-earth elements (atomic numbers 57 through 71) in metallic form. It contains about 50% cerium, the remainder being principally lanthanum and neodymium.

mismatch. Error in register between

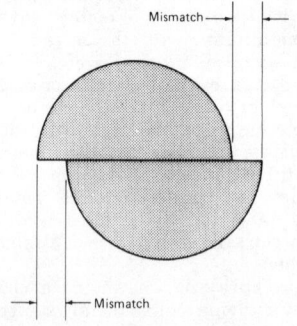

forged surfaces formed by opposing dies.

misrun. A casting not fully formed, resulting from the metal solidifying before the mold is filled.

mixing. In powder metallurgy, the thorough intermingling of powders of two or more different materials (not *blending*).

mixing chamber. The part of a torch or furnace burner in which gases are mixed.

modification. Treatment of molten hypoeutectic (8 to 13% Si) or hypereutectic (13 to 19% Si) aluminum-silicon alloys to improve mechanical properties of the solid alloy by refinement of the size and distribution of the silicon phase. Involves additions of small percentages of sodium or strontium (hypoeutectic alloys) or phosphorus (hypereutectic alloys).

modulus of elasticity. A measure of the rigidity of metal. Ratio of stress, below the proportional limit, to corresponding strain. Specifically, the modulus obtained in tension or compression is Young's modulus, stretch modulus or modulus of extensibility; the modulus obtained in torsion or shear is modulus of rigidity, shear modulus or modulus of torsion; the modulus covering the ratio of the mean normal stress to the change in volume per unit volume is the bulk modulus. The tangent modulus and secant modulus are not restricted within the proportional limit; the former is the slope of the stress-strain curve at a specified point; the latter is the slope of a line from the origin to a specified point on the stress-strain curve. Also called elastic modulus and coefficient of elasticity.

modulus of rigidity. See *modulus of elasticity.*

modulus of rupture. Nominal stress at fracture in a bend test or torsion test. In bending, modulus of rupture is the bending moment at fracture divided by the section modulus. In torsion, modulus of rupture is the torque at fracture divided by the polar section modulus.

modulus of strain hardening. See preferred term, *rate of strain hardening.*

Moh's scale. A scratch hardness test for determining comparative hardness using ten standard minerals from talc (the softest) to diamond (the hardest).

mold. (1) A form made of sand, metal or other material that contains the cavity into which molten metal is poured to produce a casting of definite shape and outline. (2) Same as *die*.

molding machine. A machine for making sand molds by mechanically compacting sand around a pattern.

molding press. A press used to form powder metallurgy *compacts*.

mold jacket. Wood or metal form that is slipped over a sand mold for support during pouring.

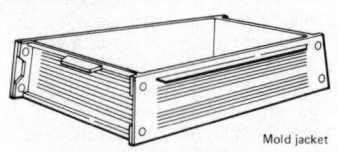
Mold jacket

mold wash. An aqueous or alcoholic emulsion or suspension of various materials used to coat the surface of a mold cavity.

Mond process. A process for extracting and purifying nickel. The main features consist of forming nickel carbonyl by reaction of finely divided reduced metal with carbon monoxide, then decomposing the nickel carbonyl, to deposit purified nickel on small nickel pellets.

monotectic. An isothermal reversible reaction in a binary system, in which a liquid on cooling decomposes into a second liquid of a different composition and a solid. It differs from a *eutectic* in that only one of the two products of the reaction is below its freezing range.

monotron hardness test. A method of determining the *indentation hardness* of metals by measuring the load required to force a spherical penetrator into the metal to a specified depth. Now obsolete.

monotropism. The ability of a solid to exist in two or more forms (crystal structures), but in which one form is the stable modification at all temperatures and pressures. Ferrite and martensite are a monotropic pair below Ac_1 in steels, for example. May also be spelled monotrophism.

mosaic structure. In crystals, a substructure in which neighboring regions have only slightly differing orientations.

M_s temperature. For any alloy system, the temperature at which martensite starts to form on cooling. See *transformation temperature* for the definition applicable to ferrous alloys.

mulling. Mixing sand and clay particles with water by kneading, rolling, rubbing or stirring.

multiaxial stresses. Any stress state in which two or three principal stresses are not zero.

multiple. A piece of stock cut from a longer *mill product* to provide the exact amount of material needed for a single workpiece.

multiple-impulse welding. Spot, projection or upset welding with more than one impulse of current during a single machine cycle. Sometimes called pulsation welding.

multiple-pass weld. A weld made by depositing filler metal with two or more successive passes.

multiple-slide press. A press with individual slides, built into the main slide or connected to individual eccentrics on the main shaft, that can be adjusted so as to give variations in length of stroke and in timing.

multiple spot welding. Spot welding in which several spots are made during one complete cycle of the welding machine.

N

native metal. (1) Any deposit in the earth's crust consisting of uncombined metal. (2) The metal in such a deposit.

natural aging. Spontaneous aging of a supersaturated solid solution at room temperature. See *aging,* and compare with *artificial aging*.

natural strain. See *strain*.

necking. (1) Reducing the cross-sectional area of metal in a localized area by stretching. (2) Reducing the diameter of a portion of the length of a cylindrical shell or tube.

necking down. Localized reduction in area of a specimen during tensile deformation.

necking strain. Same as *uniform strain*.

negative rake. Describes a tooth face in rotation whose cutting edge lags the surface of the tooth face. See sketch accompanying *face mill*.

network structure. A structure in which one constituent occurs primarily at the grain boundaries, thus partially or completely enveloping the grains of the other constituents.

Neumann band. *Mechanical twin* in ferrite.

neutral flame. A gas flame in which there is no excess of either fuel or oxygen in the inner flame. Oxygen from ambient air is used to complete the combustion of CO_2 and H_2 produced in the inner flame.

neutron. Elementary nuclear particle with a mass approximately the same as that of a hydrogen atom and electrically neutral; its mass is 1.008 986 mass units.

neutron embrittlement. Embrittlement resulting from bombardment with neutrons, usually encountered in metals that have been exposed to a neutron flux in the core of a reactor. In steels, neutron embrittlement is evidenced by a rise in the ductile-to-brittle transition temperature.

nibbling. Contour cutting of sheet metal by a rapidly reciprocating punch making numerous small cuts.

nitriding. Introducing nitrogen into the surface layer of a solid ferrous alloy by holding at a suitable temperature (below Ac_1 for ferritic steels) in contact with a nitrogenous material, usually ammonia or molten cyanide of appropriate composition. Quenching is not required to produce a hard case.

nitrocarburizing. Any of several processes in which both nitrogen and carbon are absorbed into the surface layers of a ferrous material at temperatures below the lower critical temperature and, by diffusion, create a concentration gradient. Nitrocarburizing is done mainly to provide an antiscuffing surface layer and to improve fatigue resistance. Compare with *carbonitriding*.

noble metal. (1) A metal whose potential is highly positive relative to the hydrogen electrode. (2) A metal with marked resistance to chemical reaction, particularly to oxidation and to solution by inorganic acids. The term as often used is synonymous with *precious metal*. Contrast with *base metal* (4).

noble potential. The potential for the passive state, if the metal can exist in both the active and passive states in a given medium.

no-draft forging. A forging with extremely close tolerances and little or no draft, requiring a minimum of machining to produce the final part. Mechanical properties can be enhanced by close control of grain flow and retention of surface material in the final component.

nodular cast iron. See preferred term, *ductile cast iron*.

nodular powder. Irregular particles of a metal powder that have knotted, rounded or other similar shapes.

nominal stress. See *stress*.

nonconforming. A quality control

term describing a unit of product or service that does not meet normal acceptance criteria for the specific product or service. A nonconforming unit is not necessarily *defective.*

nondestructive inspection. Inspection by methods that do not destroy the part nor impair its serviceability.

nondestructive testing. Same as *nondestructive inspection,* but implying use of a method in which the part is stimulated and its response measured quantitatively or semiquantitatively.

nonmetallic inclusions. See *inclusions.*

normalizing. Heating a ferrous alloy to a suitable temperature above the transformation range and then cooling in air to a temperature substantially below the transformation range.

normal segregation. Concentration of alloying constituents that have low melting points in those portions of a casting that solidify last. Compare with *inverse segregation.*

normal stress. See *stress.*

nose radius. The radius of the rounded portion of the cutting edge of a tool. See sketch accompanying *single-point tool.*

nosing. Closing in the end of a tubular shape to a desired curved contour.

notch acuity. Relates to the severity of the stress concentration produced by a given notch in a particular structure. If the depth of the notch is very small compared with the width (or diameter) of the narrowest cross section, the acuity may be expressed as the ratio of the notch depth to the notch root radius. Otherwise, the acuity is defined as the ratio of one-half the width (or diameter) of the narrowest cross section to the notch root radius.

notch brittleness. Susceptibility of a material to brittle fracture at points of stress concentration. For example, in a notch tensile test, the material is said to be notch brittle if the *notch strength* is less than the tensile strength of an unnotched specimen. Otherwise, it is said to be notch ductile.

notch depth. The distance from the surface of a test specimen to the bottom of the notch. In a cylindrical test specimen, the percentage of the original cross-sectional area removed by machining an annular groove.

notch ductile. See *notch brittleness.*

notch ductility. The percentage reduction in area after complete separation of the metal in a tensile test of a notched specimen.

notching. Cutting out various shapes from the edge of a strip, blank or part.

notching press. A mechanical press used for notching internal and external circumferences and also for notching along a straight line. These presses are equipped with automatic feeds, because only one notch is made per stroke.

notch rupture strength. The ratio of applied load to original area of the minimum cross section in a stress-rupture test of a notched specimen.

notch sensitivity. A measure of the reduction in strength of a metal caused by the presence of stress concentration. Values can be obtained from static, impact or fatigue tests.

notch sharpness. See *notch acuity.*

notch strength. The maximum load on a notched tensile-test specimen divided by the minimum cross-sectional area (the area at the root of the notch). Also called notch tensile strength.

nucleation. The initiation of a phase transformation at discrete sites, the new phase growing on nuclei. See *nucleus,* (1).

nucleus (1). The first structurally stable particle capable of initiating recrystallization of a phase or the growth of a new phase, and possessing an interface with the parent matrix. The term is also applied to a foreign particle that initiates such action. (2) The heavy central core of an atom, in which most of the mass and the total positive electric charge are concentrated.

nugget. (1) A small mass of metal, such as gold or silver, found free in nature. (2) The weld metal in a spot, seam or projection weld.

O

octahedral plane. In cubic crystals, a plane with equal intercepts on all three axes.

offal. The material trimmed from blanks or formed panels.

offhand grinding. Grinding where the operator manually forces the wheel against the work, or vice versa. It often implies casual manipulation of either grinder or work to achieve the desired result. Dimensions and tolerances frequently are not specified, or

are only loosely specified; the operator relies mainly on visual inspection to determine how much grinding should be done. Contrast with *precision grinding.*

offset. The distance along the strain coordinate between the initial portion of a stress-strain curve and a parallel line that intersects the stress-strain curve at a value of stress that is used as a measure of the *yield strength.* It is used for materials that have no obvious *yield point.* A value of 0.2% is commonly used.

off time. In resistance welding, the time that the electrodes are off the work. This term is generally applied where the welding cycle is repetitive.

oil canning. Same as *canning.*

oilstone. A natural or manufactured abrasive stone, generally impregnated with oil, used for sharpening keen-edged tools.

Olsen ductility test. A cupping test in which a piece of sheet metal, restrained except at the center, is deformed by a standard steel ball until fracture occurs. The height of the cup (in thousandths of an inch) at time of fracture is a measure of the ductility.

open-back inclinable press. A vertical crank press that can be inclined so that the bed will have an inclination generally varying from 0° to 30°. The formed parts slide off through an opening in the back. It is often called an OBI press.

open die forging. Same as *flat die forging.*

open dies. See *closed dies.*

open-gap upset welding. A form of forge welding in which the weld interfaces are heated with a fuel gas flame, then forced into intimate contact by the application of force. Not to be confused with *upset welding,* which is a resistance welding process.

open-hearth furnace. A reverberatory melting furnace with a shallow hearth and a low roof. The flame passes over the charge on the hearth, causing the charge to be heated both by direct flame and by radiation from the roof and sidewalls of the furnace. In ferrous industry, the furnace is regenerative.

open rod press. A hydraulic press in which the slide is guided by vertical, cylindrical rods (usually four) that also serve to hold the crown and bed in position.

operating stress. The stress to which a

structural unit is subjected in service.

optical pyrometer. An instrument for measuring the temperature of heated material by comparing the intensity of light emitted with a known intensity of an incandescent lamp filament.

orange peel. A surface roughening in the form of a pebble-grained pattern where a metal of unusually coarse grain is stressed beyond its elastic limit. Also called pebbles and alligator skin.

ordering. Forming a *superlattice*.

ore. A natural mineral that may be mined and treated for the extraction of any of its components, metallic or otherwise, at a profit.

ore dressing. Same as *mineral dressing*.

orientation. Arrangement in space of the axes of a crystal with respect to a chosen reference or coordinate system. See also *preferred orientation*.

oscillating die press. A small high-speed press in which the die and punch move horizontally with the strip during the working stroke. Through a reciprocating motion, the die and punch return to their original positions to begin the next stroke.

overaging. Aging under conditions of time and temperature greater than those required to obtain maximum change in a certain property, so that the property is altered in the direction of the initial value. See *aging*.

overbending. Bending metal through a greater arc than that required in the finished part, to compensate for springback.

overdraft. A condition where a metal curves upward on leaving the rolls because of the higher speed of the lower roll.

overhauling. Cutting surface layers from castings or slabs to remove scale and surface imperfections. Sometimes called scalping or slab milling.

overhead-drive press. A mechanical press with the driving mechanism mounted in or on the crown or upper parts of the uprights.

overhead-position welding. Welding that is performed from the underside.

overheating. Heating a metal or alloy to such a high temperature that its properties are impaired. When the original properties cannot be restored by further heat treating, by mechanical working or by a combination of working and heat treating, the overheating is known as *burning*.

overlap. (1) Protrusion of weld metal beyond the toe, face or root of a weld. (2) In resistance seam welding, the area in a given weld remelted by the succeeding weld.

oversize powder. Particles of a powdered metal coarser than the maximum permitted by a given specification for particle size.

overstressing. (1) In fatigue testing, cycling at a stress level higher than that used at the end of the test.

overvoltage. The difference between the actual electrode potential when appreciable electrolysis begins and the reversible electrode potential.

oxidation. (1) A reaction in which there is an increase in valence resulting from a loss of electrons. Contrast with *reduction*. (2) A corrosion reaction in which the corroded metal forms an oxide; usually applied to reaction with a gas containing elemental oxygen, such as air.

oxidized surface (on steel). Surface having a thin, tightly adhering, oxidized skin (from straw to blue in color), extending in from the edge of a coil or sheet. Sometimes called annealing border.

oxidizing agent. A compound that causes oxidation, thereby itself becoming reduced.

oxidizing flame. A gas flame produced with excess oxygen in the inner flame.

oxyacetylene cutting. An *oxyfuel gas cutting* process in which the fuel gas is acetylene.

oxyacetylene welding. An *oxyfuel gas welding* process in which the fuel gas is acetylene.

oxyfuel gas cutting. Any of a group of processes used to sever metals by means of chemical reaction between hot base metal and a fine stream of oxygen. The necessary metal temperature is maintained by gas flames resulting from combustion of a specific fuel gas such as acetylene, hydrogen, natural gas or propane. See also *oxygen cutting*.

oxyfuel gas welding. Any of a group of processes used to fuse metals together by heating them with gas flames resulting from combustion of a specific fuel gas such as acetylene, hydrogen, natural gas or propane. The process may be used with or without the application of pressure to the joint, and with or without adding any filler metal.

oxygen cutting. Metal cutting by directing a fine stream of oxygen upon a hot metal. The chemical reaction between oxygen and the base metal furnishes heat for localized melting, hence, cutting.

oxygen deficiency. A form of *crevice corrosion* in which galvanic corrosion proceeds because oxygen is prevented from diffusing into the crevice.

oxygen-free copper. Electrolytic copper free from cuprous oxide, produced without the use of residual metallic or metalloidal deoxidizers.

oxygen gouging. Oxygen cutting in which a chamfer or groove is formed.

oxygen lance. A length of pipe used to convey oxygen, either to the point of cutting in oxygen-lance cutting, or beneath the surface of the melt in a steelmaking furnace.

oxyhydrogen cutting. An *oxyfuel gas cutting* process in which the fuel gas is hydrogen.

oxyhydrogen welding. An *oxyfuel gas welding* process in which the fuel gas is hydrogen.

oxynatural gas cutting. An *oxyfuel gas cutting* process in which the fuel gas is natural gas.

oxynatural gas welding. An *oxyfuel gas welding* process in which the fuel gas is natural gas.

oxypropane cutting. An *oxyfuel gas cutting* process in which the fuel gas is propane.

oxypropane welding. An *oxyfuel gas welding* process in which the fuel gas is propane.

P

packing material. Any material in which powder metallurgy compacts are embedded during the presintering or sintering operations.

pack rolling. Hot rolling a pack of two or more sheets of metal; scale prevents their being welded together.

pancake forging. A rough forged shape, usually flat, that may be obtained quickly with a minimum of tooling. It usually requires considerable machining to attain finish size.

paramagnetic material. A material whose specific permeability is greater than unity and is practically independent of the magnetizing force. Compare with *diamagnetic material, ferromagnetic material*.

Parkes process. A process used to recover precious metals from lead and based on the principle that if 1 to 2%

Zn is stirred into the molten lead, a compound of zinc with gold and silver separates out and can be skimmed off.

partial annealing. An imprecise term used to denote a treatment given cold worked material to reduce the strength to a controlled level or to effect stress relief. To be meaningful, the type of material, the degree of cold work, and the time-temperature schedule must be stated.

particle size. The controlling lineal dimension of an individual particle, such as of a powdered metal, as determined by analysis with screens or other suitable instruments.

particle size distribution. The percentage, by weight or by number, of each fraction into which a powder sample has been classified with respect to sieve number or particle size. Preferred usage: "particle size distribution by weight", or "particle size distribution by frequency".

parting. (1) In the recovery of precious metals, the separation of silver from gold. (2) The zone of separation between cope and drag portions of mold or flask in sand casting. (3) A composition sometimes used in sand molding to facilitate the removal of the pattern. (4) Cutting simultaneously along two parallel lines or along two lines that balance each other in side thrust. (5) A shearing operation used to produce two or more parts from a stamping.

parting line. (1) The intersection of the parting plane of a casting mold or the parting plane between forging dies with the mold or die cavity. (2) A raised line or projection on the surface of a casting or forging that corresponds to said intersection.

parting plane. (1) In forging, the dividing plane between dies. Contrast with *forging plane.* (2) In casting, the dividing plane between mold halves.

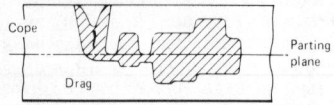

parting sand. Fine sand for dusting on sand mold surfaces that are to be separated.

parts former. A type of upsetter designed to work on short billets instead of bars and tubes, usually for cold forging.

pass. (1) A single transfer of metal through a *stand* of rolls. (2) The open

space between two grooved rolls through which metal is processed. (3) The weld metal deposited in one trip along the axis of a weld.

passivation. The changing of a chemically active surface of a metal to a much less reactive state. Contrast with *activation.*

passivity. A condition in which a piece of metal, because of an impervious covering of oxide or other compound, has a potential much more positive than when the metal is in the active state.

patenting. In wiremaking, a heat treatment applied to medium-carbon or high-carbon steel before the drawing of wire or between drafts. This process consists of heating to a temperature above the transformation range and then cooling to a temperature below Ae_1 in air or in a bath of molten lead or salt.

patent leveling. Same as *stretcher leveling.*

pattern. (1) A form of wood, metal or other material, around which molding material is placed to make a mold for casting metals. (2) A full-scale reproduction of a part used as a guide in cutting.

Pattinson process. A process for separating silver from lead, in which the molten lead is slowly cooled so that crystals poorer in silver solidify out and are removed, leaving the melt richer in silver.

pearlite. A metastable lamellar aggregate of ferrite and cementite resulting from the transformation of austenite at temperatures above the bainite range.

pearlitic malleable. See *malleable cast iron.*

pebbles. Same as *orange peel.*

peeling. The detaching of one layer of a coating from another or from the basis metal, because of poor adherence.

peening. Mechanical working of metal by hammer blows or shot impingement.

penetrant. A liquid with low surface tension used in *liquid penetrant inspection* to flow into surface openings of parts being inspected.

penetrant inspection. See preferred term, *liquid penetrant inspection.*

penetration. (1) In founding, an *imperfection* on a casting surface caused by metal running into voids between sand grains; usually referred to as metal penetration. (2) In welding, the distance from the original surface of

the base metal to that point at which fusion ceased. See *joint penetration.*

penetration hardness. Same as *indentation hardness.*

percussion welding. Resistance welding in which abutting surfaces are heated by an intense spark between them, welding being consummated by applying a hammerlike blow during or immediately after the electrical discharge.

perforating. Piercing holes of desired shapes arranged in a definite pattern in sheets, blanks or formed parts.

periodic reverse. Pertains to periodic changes in direction of flow of the current in electrolysis. It applies to the process and also the machine that controls the time for both directions.

peripheral clearance angle. See *clearance angle,* and also sketch accompanying *face mill.*

peripheral milling. Milling a surface parallel to the axis of the cutter.

peripheral speed. See preferred term, *cutting speed.*

peritectic. An isothermal reversible reaction in which a liquid phase reacts with a solid phase to produce a single (and different) solid phase on cooling.

peritectoid. An isothermal reversible reaction in which a solid phase reacts with a second solid phase to produce a single (and different) solid phase on cooling.

permanent mold. A metal, graphite or ceramic mold (other than an ingot mold) of two or more parts that is used repeatedly for the production of many *castings* of the same form. Liquid metal is poured in by gravity.

permanent set. Plastic deformation that remains upon releasing the stress that produces the deformation.

permeability. (1) In founding, the characteristics of molding materials that permit gases to pass through them. "Permeability number" is determined by a standard test. (2) In powder metallurgy, a property measured as the rate of passage under specified conditions of a liquid or gas through a compact. (3) A general term used to express various relationships between magnetic induction and magnetizing force. These relationships are either "absolute permeability", which is a change in magnetic induction divided by the corresponding change in magnetizing force, or "specific (relative) permeability", the ratio of the absolute permeability to the permeability of free space.

pewter. Any of various alloys in which tin is the chief constituent; especially an alloy of tin and lead formerly used for domestic utensils.

pH. The negative logarithm of the hydrogen ion activity; it denotes the degree of acidity or basicity of a solution. At 25 °C (76 °F), 7.0 is the neutral value. Decreasing values below 7.0 indicate increasing acidity; increasing values above 7.0, increasing basicity.

phase. A physically homogeneous and distinct portion of a material system.

phase diagram. Same as *constitution diagram.*

phosphating. Forming an adherent phosphate coating on a metal by immersion in a suitable aqueous phosphate solution. Also called phosphatizing.

phosphorized copper. General term applied to copper deoxidized with phosphorus. The most commonly used deoxidized copper.

photoelasticity. An optical method for evaluating the magnitude and distribution of stresses, using a transparent model of a part, or a thick film of photoelastic material bonded to a real part.

photomacrograph. See *macrograph.*

photomicrograph. See *micrograph.*

photon. The smallest possible quantity of an electromagnetic radiation that can be characterized by a definite frequency.

physical metallurgy. The science and technology dealing with the properties of metals and alloys, and of the effects of composition, processing and environment on those properties.

physical properties. Properties of a metal or alloy that are relatively insensitive to structure and can be measured without the application of force; for example, density, electrical conductivity, coefficient of thermal expansion, magnetic permeability and lattice parameter. Does not include chemical reactivity. Compare with *mechanical properties.*

physical testing. Determination of *physical properties.*

pickle liquor. A spent acid-pickling bath.

pickle patch. A tightly adhering oxide or scale coating not properly removed during *pickling.*

pickle stain. Discoloration of metal due to chemical cleaning without adequate washing and drying.

pickling. Removing surface oxides from metals by chemical or electrochemical reaction.

pickoff. An automatic device for removing a finished part from the press die after it has been stripped.

pickup. Transfer of metal from tools to part or from part to tools during a forming operation. See *galling.*

Pidgeon process. A process for the production of magnesium by the reduction of magnesium oxide with ferrosilicon.

piezoelectric effect. The reversible interaction, exhibited by some crystalline materials, between an elastic strain and an electric field. The direction of the strain depends on the polarity of the field or vice versa. Compare with *electrostrictive effect.*

pig. A metal casting used in remelting.

pig iron. (1) High-carbon iron made by reduction of iron ore in the blast furnace. (2) Cast iron in the form of *pigs.*

Pilger tube-reducing process. See *tube reducing.*

pinchers. Surface disturbances that result from rolling processes and that ordinarily appear as fernlike ripples running diagonally to the direction of rolling.

pinch pass. A pass of sheet material through rolls to effect a very small reduction in thickness.

pinch trimming. Trimming the edge of a tubular part or shell by pushing or pinching the flange or lip over the cutting edge of a stationary punch or over the cutting edge of a draw punch.

pine-tree crystal. A type of *dendrite.*

pin expansion test. A test for determining the ability of tubes to be expanded or for revealing the presence of cracks or other longitudinal weaknesses, made by forcing a tapered pin into the open end of a tube.

pinhead blister. See *blister.*

pinhole porosity. Porosity consisting of numerous small gas holes distributed throughout the metal; found in weld metal, castings or electrodeposited metal.

pinion. The smaller of two mating gears.

Piobert lines. Same as *Lüders lines.*

pipe. (1) The central cavity formed by contraction in metal, especially ingots, during solidification. See accompanying sketch. (2) An imperfec-

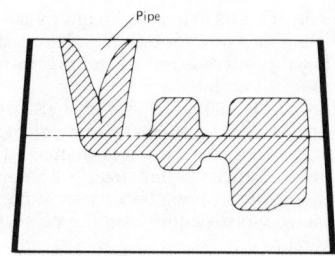
Pipe

tion in wrought or cast products resulting from such a cavity. (3) See *extrusion pipe.* (4) A tubular metal product, cast or wrought.

pipe tap. A *tap* for making internal *pipe threads* within pipe fittings or holes.

pipe threads. Internal or external machine threads, usually tapered, of a design intended for making pressuretight mechanical joints in piping systems.

pitch. See *set.*

pitting. Forming small sharp cavities in a metal surface by nonuniform electrodeposition or by corrosion.

planchet. A metal disk with milled edges, ready for coining.

plane strain. The stress condition in linear elastic fracture mechanics in which there is zero strain in a direction normal to both the axis of applied tensile stress and the direction of crack growth (i.e., parallel to the crack front); most nearly achieved in loading thick plates along a direction parallel to the plate surface. Under plane-strain conditions, the plane of fracture instability is normal to the axis of the principal tensile stress.

plane stress. The stress condition in linear elastic fracture mechanics in which the stress in the thickness direction is zero; most nearly achieved in loading very thin sheet along a direction parallel to the surface of the sheet. Under plane-stress conditions, the plane of fracture instability is inclined 45° to the axis of the principal tensile stress.

planimetric method. A method of measuring grain size in which the grains within a definite area are counted.

planing. Producing flat surfaces by linear reciprocal motion of work and the table to which it is attached, relative to a stationary single-point cutting tool.

planishing. Producing a smooth surface finish on metal by a rapid succession of blows delivered by highly polished dies or by a hammer designed for the purpose, or by rolling in a planishing mill.

plasma-arc cutting. An arc cutting process that severs metals by melting a localized area with heat from a constricted arc and removing the molten metal with a high-velocity jet of hot, ionized gas issuing from the plasma torch.

plasma-arc welding. An arc-welding process that produces coalescence of metals by heating them with a constricted arc between an electrode and the workpiece (transferred arc) or the electrode and the constricting nozzle (nontransferred arc). Shielding is obtained from hot, ionized gas issuing from an orifice surrounding the electrode and may be supplemented by an auxiliary source of shielding gas, which may be an inert gas or a mixture of gases. Pressure may or may not be used, and filler metal may or may not be supplied.

plasma spraying. A *thermal spraying* process in which the coating material is melted with heat from a plasma torch that generates a nontransferred arc (defined in *plasma-arc welding*); molten coating material is propelled against the basis metal by the hot, ionized gas issuing from the torch.

plaster molding. Molding wherein a gypsum-bonded aggregate flour in the form of a water slurry is poured over a pattern, permitted to harden, and, after removal of the pattern, thoroughly dried. The technique is used to make smooth nonferrous castings of accurate size.

plastic deformation. Deformation that does or will remain permanent after removal of the load that caused it.

plastic flow. Same as *plastic deformation*.

plasticity. The ability of a metal to deform nonelastically without rupture.

plate. A flat-rolled metal product of some minimum thickness and width arbitrarily dependent on the type of metal.

platen. (1) Face of a bolster, slide or ram to which a tool assembly is attached. (2) A part of a resistance welding, mechanical testing or other machine with a flat surface to which dies, fixtures, backups or electrode holders are attached and that transmits pressure or force.

plates. Flat particles of metal powder having considerable thickness. Contrast with *flake powder*.

plating. Forming an adherent layer of metal upon an object; often used as a shop term for *electroplating*.

plating rack. A fixture used to hold work and conduct current to it during electroplating.

plating range. The current density range over which a satisfactory electroplate can be deposited.

platinum black. A finely divided form of platinum of a dull black color, usually but not necessarily produced by the reduction of salts in aqueous solution.

plug. (1) A rod or mandrel over which a pierced tube is forced. (2) A rod or mandrel that fills a tube as it is drawn through a die. (3) A punch or mandrel over which a cup is drawn. (4) A protruding portion of a die impression for forming a corresponding recess in the forging. (5) A false bottom in a die. Also called a "peg".

plug tap. A *tap* with *chamfer* extending from three to five threads.

plug weld. A circular weld made by either arc or gas welding through one member of a lap or tee joint. If a hole is used, it may be only partly filled. Neither a fillet-welded hole nor a spot weld is to be construed as a plug weld.

plumbago. A special quality of powdered graphite used to coat molds, and in a mixture with clay, to make crucibles.

plunge grinding. Grinding where the only relative motion of the wheel is radially toward the work.

plus sieve. The portion of a sample of a granular substance (such as metal powder) retained on a standard sieve of specified number. Contrast with *minus sieve*.

plymetal. Sheet consisting of bonded layers of dissimilar metals.

P/M. The acronym for *powder metallurgy*.

point angle. In general, the angle at the point of a cutting tool. Most commonly, the included angle at the point of a twist drill, the general-purpose angle being 118°.

pointing. (1) Reducing the diameter of wire, rod or tubing over a short length at the end by swaging or hammer forging, turning or squeezing to facilitate entry into a drawing die and gripping in the drawhead. (2) The operation in automatic machines of chamfering or rounding the threaded end or the head of a bolt.

Poisson's ratio. The absolute value of the ratio of the transverse strain to the corresponding axial strain, in a body subjected to uniaxial stress; usually applied to elastic conditions.

poke welding. Same as *push welding*.

polar bond. See *ionic bond*.

polarization. A change in the potential of an electrode during electrolysis, such that the potential of an anode becomes more noble and that of a cathode more active than their respective reversible potentials. Often accomplished by the formation of a film on the electrode surface.

polar section modulus. See *section modulus*.

pole. (1) A means of designating the orientation of a crystal plane by stereographically plotting its normal. For example, the north pole defines the equatorial plane. (2) Either of the two regions of a permanent magnet or electromagnet where most of the lines of induction enter or leave.

pole figure. A stereographic projection representing the statistical average distribution of poles of a specific crystalline plane in a polycrystalline metal, with reference to an external system of axes. In an isotropic metal, that is, in one having a completely random distribution of orientations, the pole density is stereographically uniform; preferred orientation is shown by an increased density of poles in certain areas.

poling. A step in the fire refining of copper to reduce the oxygen content to tolerable limits by covering the bath with coal or coke and thrusting green wood poles below the surface. There is a vigorous evolution of reducing gases that combine with the oxygen contained in the metal.

polishing. Smoothing metal surfaces, often to a high luster, by rubbing the surface with a fine abrasive, usually contained in a cloth or other soft lap. Results in microscopic flow of some surface metal together with actual removal of a small amount of surface metal. May be extended to include *electropolishing*. Contrast with *buffing, burnishing*.

polycrystalline. Pertaining to a solid composed of many crystals.

polymorphism. A general term for the ability of a solid to exist in more than one form. In metals, alloys and similar substances, this usually means the ability to exist in two or more crystal structures, or an amorphous state and at least one crystal structure. See also *allotropy, enantiotropy, monotropism.*

pop-off. Loss of small portions of a porcelain enamel coating. The usual cause is outgassing of hydrogen or other gases from the basis metal during firing, but pop-off also may occur because of oxide particles or other debris on the surface of the basis metal. Usually, the pits are minute and cone shaped, but when pop-off is the result of severe *fishscale* the pits may be much larger and irregular.

pores. (1) Small voids in the body of a metal. (2) Minute cavities in a powder metallurgy compact, sometimes intentional. (3) Minute perforations in an electroplated coating.

porosity. Fine holes or pores within a metal.

porthole die. A multiple-section extrusion die capable of producing tubing or intricate hollow shapes without the use of a separate mandrel. Metal is extruded in separate streams through holes in each section and is rewelded by extrusion pressure before it leaves the die. Compare with *bridge die.*

positioned weld. A weld made in a joint that has been oriented to facilitate making the weld.

positive rake. Describes a tooth face in rotation whose cutting edge leads the surface of the tooth face. See sketch accompanying *face mill.*

postheating. Heating weldments immediately after welding, for tempering, for stress relieving, or for providing a controlled rate of cooling to prevent formation of a hard or brittle structure.

pot. (1) A vessel for holding molten metal. (2) The electrolytic reduction cell used to make such metals as aluminum from a fused electrolyte.

pot annealing. Same as *box annealing.*

pot die forming. Forming products from sheet or plate through the use of a hollow die and internal pressure which causes the preformed workpiece to assume the contour of the die.

poultice corrosion. A term used in the automotive industry to describe the corrosion of vehicle body parts due to the collection of road salts and debris on ledges and in pockets that are kept moist by weather and washing.

pouring. Transferring molten metal from a furnace or a ladle to a mold.

pouring basin. A basin on top of a mold to receive the molten metal before it enters the sprue or downgate.

powder. Particles of a solid characterized by small size, nominally within the range of 0.1 to 1000 μm.

powder lubricant. An agent mixed with or incorporated in a powder to facilitate the pressing and ejecting of a powder metallurgy compact.

powder metallurgy. The art of producing metal powders and of utilizing metal powders for the production of massive materials and shaped objects.

powder metallurgy forging. Plastically deforming a powder metallurgy compact or preform into a fully dense finished shape using compressive force; usually done hot, and usually within closed dies.

power reel. A reel that is driven by an electric motor or some other source of power, used to wind or coil strip or wire as it is drawn through a continuous normalizing furnace, through a die, or through rolls as in certain types of cold mills in which the work rolls are not driven.

precharge. In forming, the pressure introduced into the cavity prior to forming the part.

precious metal. One of the relatively scarce and valuable metals: gold, silver and the platinum-group metals.

precipitation hardening. Hardening caused by the precipitation of a constituent from a supersaturated solid solution. See also *age hardening* and *aging.*

precipitation heat treatment. *Artificial aging* in which a constituent precipitates from a supersaturated solid solution.

precision. The closeness of approach of each of a number of similar measurements to the arithmetic mean, the sources of error not necessarily being considered critically. *Accuracy* demands precision, but precision does not ensure accuracy.

precision casting. A metal casting of reproducible accurate dimensions regardless of how it is made.

precision grinding. Machine grinding to specified dimensions and low *tolerances.* Contrast with *offhand grinding.*

precoat. (1) In investment casting, a special refractory slurry applied to a wax or plastic expendable pattern to form a thin coating that serves as a desirable base for application of the main slurry. (2) Making the thin coat-

ing. (3) The thin coating itself.

precoated metal products. Mill products that have a metallic, organic or conversion coating applied to their surfaces before they are fabricated into parts.

preferred orientation. A condition of a polycrystalline aggregate in which the crystal orientations are not random, but rather exhibit a tendency for alignment with a specific direction in the bulk material, commonly related to the direction of working; also called *texture.*

preforming. (1) The initial pressing of a metal powder to form a compact that is to be subjected to a subsequent pressing operation other than coining or sizing. Also, the preliminary shaping of a refractory metal compact after presintering and before the final sintering. (2) Preliminary forming operations, especially for impression die forging.

preheating. Heating before some further thermal or mechanical treatment. For tool steel, heating to an intermediate temperature immediately before final austenitizing. For some nonferrous alloys, heating to a high temperature for a long time, in order to homogenize the structure before working. In welding and related processes, heating to an intermediate temperature for a short time immediately before welding, brazing, soldering, cutting or thermal spraying.

presintering. The heating of a powder metallurgy compact to a temperature lower than the normal temperature for final sintering, usually to increase the ease of handling or forming the compact or to remove a lubricant or binder before sintering.

press. A machine tool having a stationary bed and a slide or ram that has reciprocating motion at right angles to the bed surface, the slide being guided in the frame of the machine.

press brake. An open-frame single-action press used to bend, blank, corrugate, curl, notch, perforate, pierce or punch sheet metal or plate.

pressed density. The density of an unsintered powder metallurgy compact. Sometimes called green density.

press fit. An interference or *force fit* made through the use of a *press.*

press forging. *Forging* metal, usually hot, between dies in a press.

pressing. (1) In metalworking, the product or process of shallow drawing sheet or plate. (2) Forming a powder-metal part with compressive force.

pressing area. The clear distance (left to right) between housings, stops, gibs, gibways or shoulders of strain rods, multiplied by the total distance from front to back on the bed of a *press*. Sometimes called working area.

pressing crack. A rupture in a green powder metallurgy compact that develops during the ejection of the compact from the die; see also *capping* and *lamination*. Sometimes referred to as a slip crack.

pressure casting. (1) Making castings with pressure on the molten or plastic metal, as in injection molding, *die casting, centrifugal casting,* and cold chamber pressure casting. (2) A casting made with pressure applied to the molten or plastic metal.

pressure gas welding. An oxyfuel gas welding process that produces coalescence simultaneously over the entire area of abutting surfaces by heating them with gas flames obtained from the combustion of a fuel gas with oxygen and by the application of pressure, without the use of filler metal.

primary creep. See *creep*.

primary crystal. The first type of crystal that separates from a melt on cooling.

primary current distribution. The current distribution in an electrolytic cell that is free of polarization.

primary metal. Metal extracted from minerals and free of reclaimed metal scrap. Compare with *secondary metal, native metal*.

primary mill. A mill for rolling ingots or the rolled products of ingots to blooms, billets or slabs. This type of mill is often called a *blooming mill* and sometimes a *cogging mill*.

primes. Metal products, principally sheet and plate, of the highest quality and free from blemishes or other visible imperfections.

principal stresses. The normal stresses on three mutually perpendicular planes on which there are no shear stresses.

prismatic plane. In noncubic crystals, any plane that is parallel to the principal axis (*c* axis).

process annealing. An imprecise term denoting various treatments used to improve workability. For the term to be meaningful, the condition of the material and the time-temperature cycle used must be stated.

process metallurgy. The science and technology of winning metals from their ores and purifying metals; sometimes referred to as chemical metallurgy. Its two chief branches are *extractive metallurgy* and *refining*.

process tolerance. The dimensional variations of a part characteristic of a specific process, once the setup is made.

profiling. Any operation that produces an irregular contour on a workpiece, a tracer or template-controlled duplicating equipment usually being employed.

progressive aging. Aging by increasing the temperature in steps or continuously during the aging cycle. See *aging* and compare with *interrupted aging* and *step aging*.

progressive die. A die in which two or more sequential operations are performed at two or more positions, the work being moved from station to station.

progressive forming. Sequential forming at consecutive stations either with a single die or with separate dies.

projection welding. Resistance welding similar to spot welding, but in which the welds are localized at projections, embossments or intersections.

proof. Any reproduction of a die impression in any material, frequently a lead or plaster cast. See *die proof*.

proof load. A predetermined load, generally some multiple of the service load, to which a specimen or structure is submitted before acceptance for use.

proof stress. (1) The stress that will cause a specified small permanent set in a material. (2) A specified stress to be applied to a member or structure to indicate its ability to withstand service loads.

proportional limit. The maximum stress at which strain remains directly proportional to stress.

pseudobinary system. (1) A three-component or ternary alloy system in which an intermediate phase acts as a component. (2) A vertical section through a ternary diagram.

pseudocarburizing. See *blank carburizing*.

pseudonitriding. See *blank nitriding*.

puckering. Wrinkling or buckling in a drawn shell in an area originally inside the draw ring.

pull cracks. In a casting, cracks that

are caused by residual stresses produced during cooling, and that result from the shape of the object.

pulsation welding. Sometimes used as a synonym for *multiple-impulse welding*.

pulsed power welding. Any arc welding process in which the power is cyclically varied to give short duration pulses of either voltage or current that are significantly different from the average value.

pulverization. Synonymous with comminution.

punch. (1) The movable tool that forces material into the die in powder molding and most forming operations. (2) The movable die in a trimming press or a forging machine. (3) The tool that forces the stock through the die in rod and tube extrusion and forms the internal surface in can or cup extrusion.

punching. Producing a hole by die shearing, in which the shape of the hole is controlled by the shape of the punch and its mating die; piercing. Multiple punching of small holes is called perforating.

punch press. (1) In general, any mechanical press. (2) In particular, an endwheel gap-frame press with a fixed bed, used in piercing.

punch radius. The radius on the end of the punch that first contacts the work, sometimes called nose radius.

punch-to-die clearance. See *die clearance*.

push angle. The angle between a welding electrode and a line normal to the face of the weld when the electrode is pointing forward along the weld joint. See sketch accompanying *forehand welding*.

push bench. Equipment used for drawing moderately heavy-gage tubes by cupping sheet and forcing it through a die by pressure exerted against the inside bottom of the cup.

pusher furnace. A type of continuous furnace in which parts to be heated are periodically charged into the furnace in containers, which are pushed along the hearth against a line of previously charged containers thus advancing the containers toward the discharge end of the furnace, where they are removed.

push fit. A loosely defined fit similar to a *snug fit*.

push welding. Spot or projection welding in which the force is applied manually to one electrode, and the work or a backing bar takes the place of the other electrode.

pyramidal plane. In noncubic crystals, any plane that intersects all three axes.

pyrometallurgy. High-temperature *winning* or *refining* of metals.

pyrometer. A device for measuring temperatures above the range of liquid thermometers.

Q

quarter hard. A *temper* of nonferrous alloys and some ferrous alloys characterized by tensile strength about midway between that of *dead soft* and *half hard* tempers.

quality. (1) The totality of features and characteristics of a product or service that bear on its ability to satisfy a given need (fitness-for-use concept of quality). (2) Degree of excellence of a product or service (comparative concept). Often determined subjectively by comparison against an ideal standard or against similar products or services available from other sources. (3) A quantitative evaluation of the features and characteristics of a product or service (quantitative concept).

quality characteristic. Any dimension, mechanical property, physical property, functional characteristic or appearance characteristic that can be used as a basis for measuring the quality of a unit of product or service.

quantitative metallography. Determination of specific characteristics of a microstructure by making quantitative measurements on micrographs or metallographic images. Quantities so measured include volume concentration of phases, grain size, particle size, mean free path between like particles or secondary phases, and surface area to volume ratio of microconstituents, particles or grains.

quasibinary system. In a ternary or higher-order system, a linear composition series between two substances each of which exhibits congruent melting, wherein all equilibriums, at all temperatures or pressures, involve only phases having compositions occurring in the linear series, so that the series may be represented as binary on a phase diagram.

quench-age embrittlement. Embrittlement of low-carbon steel evidenced by a loss of ductility on aging at room temperature following rapid cooling from a temperature below the lower critical temperature.

quench aging. Aging induced by rapid cooling after *solution heat treatment*.

quench annealing. Annealing an austenitic ferrous alloy by *solution heat treatment* followed by rapid quenching.

quench cracking. Fracture of a metal during quenching from elevated temperature. Most frequently observed in hardened carbon steel, alloy steel or tool steel parts of high hardness and low toughness. Cracks often emanate from fillets, holes, corners or other stress raisers and result from high stresses due to the volume changes accompanying transformation to martensite.

quench hardening. (1) Hardening suitable alpha-beta alloys (most often certain copper or titanium alloys) by solution treating and quenching to develop a martensite-like structure. (2) In ferrous alloys, hardening by austenitizing and then cooling at a rate such that a substantial amount of austenite transforms to martensite.

quenching. Rapid cooling. When applicable, the following more specific terms should be used: *direct quenching, fog quenching, hot quenching, interrupted quenching, selective quenching, spray quenching* and *time quenching*.

quench time. In resistance welding, the time from the finish of the weld to the beginning of temper. Also called *chill time*.

quill. (1) A hollow or tubular shaft, designed to slide or revolve, carrying a rotating member within itself. (2) Removable spindle projection for supporting a cutting tool or grinding wheel.

R

rabbit ear. Recess in the corner of a die to allow for wrinkling or folding of the blank.

rabble. A hoelike bladed tool or similar device used for stirring molten metal.

radial draw forming. Forming metals by the simultaneous application of tangential stretch and radial compression forces, the operation being done gradually by tangential contact to the die member. This type of forming is characterized by very close dimensional control.

radial forging. A process utilizing two or more moving anvils or dies for producing shafts with constant or varying diameters along their length or tubes with internal or external variations in diameter; also known as draw forging or rotary swaging.

radial marks. Lines on a fracture surface that radiate from the fracture origin and are visible to the unaided eye or at low magnification. Radial lines result from the intersection and connection of brittle fractures propagating at different levels. Also called shear ledges. See also *chevron pattern*.

radial rake. The angle between the tooth face and a radial line passing through the cutting edge in a plane perpendicular to the cutter axis. See sketch accompanying *face mill*.

radial runout. For any rotating element, the total variation from true radial position, taken in a plane perpendicular to the axis of rotation. Compare with *axial runout*.

radiation damage. A general term for the alteration of properties of a material arising from exposure to ionizing radiation (penetrating radiation) such as x-rays, gamma rays, neutrons, heavy-particle radiation, or fission fragments in nuclear fuel material.

radiation dose. Accumulated exposure to ionizing radiation during a specified period of time.

radiation energy. The energy of a given photon or particle in a beam of radiation, often expressed in electron volts.

radiation gage. An instrument for measuring the intensity and quantity of ionizing radiation.

radiation intensity. In general, the quantity of radiant energy at a specified location passing perpendicularly through unit area in unit time. It may be given as number of particles or photons per square centimetre per second, or in energy units as $J/m^2 \cdot s$ or Rhm.

radiation monitoring. The continuous or periodic measurement of the intensity of radiation received by personnel or present in any particular area.

radiation quality. A term describing roughly the spectrum of radiation produced by a radiation source, with respect to its penetrating power or its suitability for a given application.

radioactive element. An element that has at least one isotope that undergoes spontaneous nuclear disintegration to emit positive alpha particles, negative beta particles, or gamma rays.

radioactive tracer element. A radioactive isotope of an element used to study the movement and behavior of atoms by observing the distribution and intensity of radioactivity.

radioactivity. The spontaneous nuclear disintegration with emission of corpuscular or electromagnetic radiation.

radiograph. A photographic shadow image resulting from uneven absorption of penetrating radiation in a test object.

radiography. A method of nondestructive inspection in which a test object is exposed to a beam of x-rays or gamma rays and the resulting shadow image of the object is recorded on photographic film placed behind the object. Internal discontinuities are detected by observing and interpreting variations in the image caused by differences in thickness, density or absorption within the test object. Variations of radiography include electron radiography, *fluoroscopy,* neutron radiography.

radioisotope. An isotope that emits ionizing radiation during its spontaneous decay.

rake. The angular relationship between the tooth face, or a tangent to the tooth face at a given point, and a given reference plane or line. See sketches accompanying *face mill* and *single-point tool.*

ram. The moving member of a hammer, machine, or press to which a tool is fastened.

ramming. Packing sand, refractory or other material into a compact mass.

ramoff. A casting imperfection resulting from the movement of sand away from the pattern because of improper ramming.

random sequence. A longitudinal welding sequence where the weld-bead increments are deposited at random to minimize distortion.

range. In inspection, the difference between the highest and lowest values of a given *quality characteristic* within a single *sample.*

rare earth metal. One of the group of 15 chemically similar metals with atomic numbers 57 through 71, commonly referred to as the lanthanides.

ratcheting. Progressive cyclic inelastic deformation (growth, for example) that occurs when a component or structure is subjected to a cyclic secondary stress superimposed on a sustained primary stress. The process is called thermal ratcheting when cyclic strain is induced by cyclic changes in temperature, and isothermal ratcheting when cyclic strain is mechanical in origin (even though accompanied by cyclic changes in temperature).

ratchet marks. Lines on a fatigue fracture surface that result from the intersection and connection of fatigue fractures propagating from multiple origins. Ratchet marks are parallel to the over-all direction of crack propagation and are visible to the unaided eye or at low magnification.

rate of strain hardening. Rate of change of true *stress* with respect to true *strain* in the plastic range.

rattail. A surface imperfection on a casting, occurring as one or more irregular lines, caused by expansion of sand in the mold. Compare with *buckle* (2).

RE. Abbreviation for *rare earth* (elements).

reamed extrusion ingot. A cast hollow extrusion ingot that has been machined to remove the original inside surface.

reamer. A rotary cutting tool with one or more cutting elements called teeth, used for enlarging a hole to desired size and contour. It is supported principally by the metal around the hole it cuts.

recalescence. A phenomenon, associated with the transformation of gamma iron to alpha iron on the cooling (supercooling) of iron or steel, revealed by the brightening (reglowing) of the metal surface owing to the sudden increase in temperature caused by the fast liberation of the latent heat of transformation. Contrast with *decalescence.*

recarburize. (1) To increase the carbon content of molten cast iron or steel by adding carbonaceous material, high-carbon pig iron or a high-carbon alloy. (2) To carburize a metal part to return surface carbon lost in processing; also known as carbon restoration.

recess. A groove or depression in a surface.

reclaim rinse. A nonflowing rinse used to recover *dragout.*

recoil line. See *impact line.*

recovery. (1) Reduction or removal of work-hardening effects, without motion of large-angle grain boundaries. (2) The proportion of the desired component obtained by processing an ore, usually expressed as a percentage.

recrystallization. (1) The formation of a new, strain-free grain structure from that existing in cold worked metal, usually accomplished by heating. (2) The change from one crystal structure to another, as occurs on heating or cooling through a critical temperature.

recrystallization annealing. Annealing cold worked metal to produce a new grain structure without phase change.

recrystallization temperature. The approximate minimum temperature at which complete recrystallization of a cold worked metal occurs within a specified time.

recuperator. Equipment for transferring heat from gaseous products of combustion to incoming air or fuel. The incoming material passes through pipes surrounded by a chamber through which the outgoing gases pass.

red mud. A residue, containing a high percentage of iron oxide, obtained in purifying bauxite in the production of alumina in the *Bayer process.*

redrawing. Drawing metal after a previous cupping or drawing operation.

reducing agent. A substance that causes reduction. See *reduction* (3).

reducing flame. A gas flame produced with excess fuel in the inner flame.

reduction. (1) In cupping and deep drawing, a measure of the percentage decrease from blank diameter to cup diameter, or of diameter reduction in redraws. (2) In forging, rolling and drawing, either the ratio of the original to final cross-sectional area or the percentage decrease in cross-sectional area. (3) A reaction in which there is a decrease in valence resulting from a gain in electrons. Contrast with *oxidation.*

reduction cell. A pot or tank in which either a water solution of a salt or a fused salt is reduced electrolytically to form free metals or other substances.

reduction in area. (1) Commonly, the difference, expressed as a percentage of original area, between the original cross-sectional area of a tensile test specimen and the minimum cross-sectional area measured after complete separation. (2) The difference, expressed as a percentage of original area, between original cross-sectional area and that after straining the specimen.

reeding. The operation of forming serrations and corrugations by coining or embossing.

reel. (1) A spool or hub for coiling or feeding wire or strip. (2) To straighten and planish a round bar by passing it between contoured rolls.

reel breaks. Transverse breaks or ridges on successive inner laps of a coil that are the result of crimping the lead end of the coil into a gripping segmented mandrel. Also called reel kinks.

reference plane. (1) The plane that contains the cutter axis and the point of the cutting edge. See sketch accompanying *face mill.* (2) A plane from which measurements are made.

refining. The branch of process metallurgy dealing with the purification of crude or impure metals. Compare with *extractive metallurgy.*

reflector sheet. A clad product, containing a facing layer of high-purity aluminum capable of taking a high polish for reflecting heat or light and a base of commercially pure aluminum or an aluminum-manganese alloy for strength and formability.

reflowing. The melting of an electrodeposit followed by solidification. The surface has the appearance and physical characteristics of being hot dipped (especially tin or tin alloy plates). Also called flow brightening.

refractory. (1) A material of very high melting point with properties that make it suitable for such uses as furnace linings and kiln construction. (2) The quality of resisting heat.

refractory alloy. (1) A heat-resistant alloy. (2) An alloy having an extremely high melting point. See *refractory metal.* (3) An alloy difficult to work at elevated temperatures.

refractory metal. A metal having an extremely high melting point; for example, tungsten, molybdenum, tantalum, niobium (columbium), chromium, vanadium and rhenium. In the broad sense, it refers to metals having melting points above the range of iron, cobalt and nickel.

regenerator. Same as *recuperator* except the gaseous products of combustion heat brick checkerwork in a chamber connected to the exhaust side of the furnace while the incoming air and fuel are being heated by the brick checkerwork in a second chamber, connected to the entrance side. At intervals, the gas flow is reversed so that incoming air and fuel contact hot checkerwork while that in the second chamber is being reheated by exhaust gases.

regulus. The impure button, globule or mass of metal formed beneath the slag in the smelting and reduction of ores. The name was first applied by alchemists to metallic antimony because it readily alloyed with gold.

rejectable. See preferred term, *nonconforming.*

reliability. A quantitative measure of the ability of a product or service to fulfill its intended function for a specified period of time.

relief. The result of the removal of tool material behind or adjacent to the cutting edge to provide clearance and prevent rubbing (heel drag). See sketch accompanying *single-point tool.*

relief angle. The angle formed between a relieved surface and a given plane tangent to a cutting edge or to a point on a cutting edge. See sketch accompanying *single-point tool.*

relieving. Buffing or other abrasive treatment of the high points of an embossed metal surface to produce highlights that contrast to the finish in the recesses.

remanence. The magnetic induction remaining in a magnetic circuit after removal of the applied magnetizing force. Sometimes called remanent induction.

re-pressing. The application of pressure to a previously pressed and sintered powder metallurgy compact, usually for the purpose of improving some physical property.

residual elements. Elements present in an alloy in small quantities, but not added intentionally.

residual field. Same as *residual magnetic field.*

residual magnetic field. The magnetic field that remains in a part after the magnetizing force is removed.

residual method. Method of *magnetic-particle inspection* in which the particles are applied after the magnetizing force has been removed.

residual stress. Stress present in a body that is free of external forces or thermal gradients.

resilience. (1) The amount of energy per unit volume released upon unloading. (2) The capacity of a metal, by virtue of high yield strength and low elastic modulus, to exhibit considerable elastic recovery upon release of load.

resinoid wheel. A grinding wheel bonded with a synthetic resin.

resist. (1) A material applied to a part of a cathode or plating rack to render the surface nonconductive. (2) A material applied to a part of the surface of an article to prevent reaction of metal from that area during chemical or electrochemical processes. (3) A material applied to prevent the flow of brazing filler metal into unwanted areas.

resistance brazing. Brazing by resistance heating, the joint being part of the electrical circuit.

resistance soldering. Soldering in which the joint is heated by electrical resistance. Filler metal is either face fed into the joint or preplaced in the joint.

resistance welding. Welding with resistance heating and pressure, the work being part of the electrical circuit. Examples: resistance *spot welding,* resistance *seam welding, projection welding* and *flash butt welding.*

resistance welding die. The part of a resistance welding machine, usually shaped to the work contour, with which the parts being welded are held and which conducts the welding current.

resolution. The ability to separate closely related items of data or physical features using a given test method; also a quantitative measure of the degree to which they can be discriminated.

restraint. Any external mechanical force that prevents a part from moving to accommodate changes in dimensions due to thermal expansion or contraction. Often applied to weldments made while clamped in a fixture. Compare with *constraint.*

restriking. (1) Striking a trimmed but slightly misaligned or otherwise faulty forging one or more blows to improve alignment, improve surface, maintain close tolerance, increase hardness or effect other improvements. (2) A sizing operation in which coining or stretching is utilized to correct or alter profiles and to counteract distortion.

resultant rake. The angle between the tooth face and an axial plane through the tooth point measured in a plane perpendicular to the cutting edge. The resultant rake of a cutter is a function of three other angles: radial rake, axial rake and corner angle. See sketch accompanying *face mill*.

retentivity. The capacity of a material to retain a portion of the magnetic field set up in it after the magnetizing force has been removed.

retort. A vessel used for the distillation of volatile materials, as in the separation of some metals and in the destructive distillation of coal.

reverberatory furnace. A furnace, with a shallow hearth, usually nonregenerative, having a roof that deflects the flame and radiates heat toward the hearth or the surface of the charge.

reverse-current cleaning. Same as *anodic cleaning*.

reverse drawing. *Redrawing* in a direction opposite to that of the original drawing.

reverse flange. A flange made by shrinking, as opposed to one formed by stretching.

reverse polarity. Direct-current arc welding circuit arrangement in which the electrode is connected to the positive terminal. Contrast with *straight polarity*.

reverse redrawing. A second drawing operation in a direction opposite to that of the original drawing.

rheology. The science of deformation and flow of matter.

rheotropic brittleness. That portion of the brittleness characteristic of non-face-centered cubic metals, when tested in the presence of a stress concentration or at low temperatures or at high strain rates, that may be eliminated by prestraining under milder conditions.

riddle. A sieve used to separate foundry sand or other granular materials into various particle-size grades or to free such a material of undesirable foreign matter.

rigging. The engineering design, layout and fabrication of pattern equipment for producing castings; including a study of the casting solidification program, feeding and gating, risering, skimmers and fitting flasks.

right-hand cutting tool. A cutter all of whose flutes twist away in a clockwise direction when viewed from either end.

rimmed steel. A low-carbon steel containing sufficient iron oxide to give a continuous evolution of carbon monoxide while the ingot is solidifying, resulting in a case or rim of metal virtually free of voids. Sheet and strip products made from the ingot have very good surface quality.

ring and circle shear. A cutting or shearing machine with two rotary-disk cutters driven in unison and equipped with a circle attachment for cutting inside circles or rings from sheet metal, where it is impossible to start the cut at the edge of the sheet. One cutter shaft is inclined to the other to provide cutting clearance so that the outside section remains flat and usable. See *circle shear*.

ringing. The audible or ultrasonic tone produced in a mechanical part by shock, and having the natural frequency or frequencies of the part. The quality, amplitude or decay rate of the tone may sometimes be used to indicate quality or soundness. See also *sonic testing, ultrasonic testing*.

ring riser. A riser block with openings matching those in the press bed.

ring rolling. The process of shaping weldless rings from pierced disks or thick-walled, ring-shaped blanks between rolls that control wall thickness, ring diameter, height and contour.

rinsability. The relative ease of removing a substance from a metal surface with a liquid such as water.

riser. A reservoir of molten metal connected to the casting to provide additional metal to the casting, required as the result of shrinkage before and during solidification.

riser block. (1) Plates or pieces inserted between the top of a press bed or bolster and the die to decrease the height of the die space. (2) Spacers placed between bed and housings to increase *shut height* on a four-piece tie-rod straight-side press.

river pattern. A term used in fractography to describe a characteristic pattern of cleavage steps that run parallel to the local direction of crack propagation on the fracture surface of grains that have separated by cleavage.

riveting. Joining of two or more members of a structure by means of metal rivets, the unheaded end being upset after the rivet is in place.

roasting. Heating an ore to effect some chemical change that will facilitate smelting.

robber. An extra cathode or cathode extension that reduces the current density on what would otherwise be a high-current-density area on work being electroplated.

Rochelle copper. (1) A copper electrodeposit obtained from copper cyanide plating solution to which Rochelle salt (sodium potassium tartrate) has been added for grain refinement, better anode corrosion and cathode efficiency. (2) The solution from which a Rochelle copper electrodeposit is obtained.

rock candy fracture. A fracture that exhibits separated-grain facets, most often used to describe intergranular fractures in large-grained metals.

rocking shear. A type of guillotine shear that utilizes a curved blade to shear sheet metal progressively from side to side by a rocker motion.

Rockrite tube-reducing process. See *tube reducing*.

Rockwell hardness test. An indentation hardness test based on the depth of penetration of a specified penetrator into the specimen under certain arbitrarily fixed conditions.

rod mill. (1) A *hot mill* for rolling rod. (2) A mill for fine grinding, somewhat similar to a *ball mill,* but employing long steel rods instead of balls to effect the grinding.

roll bending. Curving sheets, bars and sections by means of rolls. See *bending rolls*.

roll compacting. The progressive compacting of metal powders by the use of a rolling mill.

rolled gold. Same as *gold filled* except that the proportion of gold alloy to the weight of the entire article may be less than 1/20th. Fineness of the gold alloy may not be less than 10K.

roller leveler breaks. Obvious transverse *breaks* usually about 3 to 6 mm (1/8 to 1/4 in.) apart caused by the sheet fluting during roller leveling. These will not be removed by stretching.

roller leveler lines. Same as *leveler lines*.

roller leveling. *Leveling* by passing flat stock through a machine having a series of small-diameter staggered rolls that are adjusted to produce repeated reverse bending.

roller stamping die. An engraved roller used for impressing designs and markings on sheet metal.

roll flattening. Flattening of sheets that have been rolled in packs by passing them separately through a two-high cold mill, there being virtually no deformation. Not to be confused with *roller leveling.*

roll forging. Forging with rotating dies that are not full round, the desired shape—either straight or tapered—being produced by a groove in the dies.

roll forming. Forming of flat-rolled metal by use of power-driven rolls whose contour determines the shape of the product. Roll forming is used extensively to make metal window frames, drapery rods and similar products from metal strip. The term is sometimes used to describe power spinning.

rolling. Reducing the cross-sectional area of metal stock, or otherwise shaping metal products, through the use of rotating rolls.

rolling mills. Machines used to decrease the cross-sectional area of metal stock and produce certain desired shapes as the metal passes between rotating rolls mounted in a framework comprising a basic unit called a *stand.* Cylindrical rolls produce flat shapes; grooved rolls produce rounds, squares and structural shapes. Among rolling mills may be listed the billet mill, blooming mill, breakdown mill, plate mill, sheet mill, slabbing mill, strip mill and temper mill.

roll resistance spot welding. The making of separated resistance spot welds with one or more rotating circular electrodes. The rotation of the electrodes may or may not be stopped during the making of a weld.

roll straightening. Straightening of metal stock of various shapes by (1) passing it through a series of staggered rolls, the rolls usually being in horizontal and vertical planes; or (2) by reeling in two-roll straightening machines.

roll table. A conveyor table where rolls furnish the contact surface.

roll threading. Making threads by rolling the piece between two grooved die plates, one of which is in motion, or between rotating grooved circular rolls.

roll welding. Solid state welding in which metals are heated, then welded together by applying pressure, with rolls, sufficient to cause deformation at the faying surfaces. See also *forge welding.*

root crack. A crack in either the weld or heat-affected zone at the root of a weld.

root face. The portion of a weld groove face adjacent to the root of the joint.

root of joint. The portion of a weld joint where the members are closest to each other before welding. In cross section, may be a point, a line or an area.

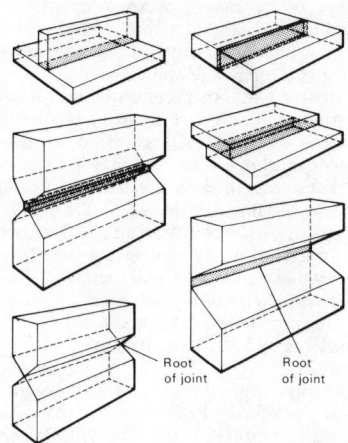

root of weld. The points, as shown in cross section, at which the weld bead intersects the base metal surfaces either nearest to or coincident with the root of joint.

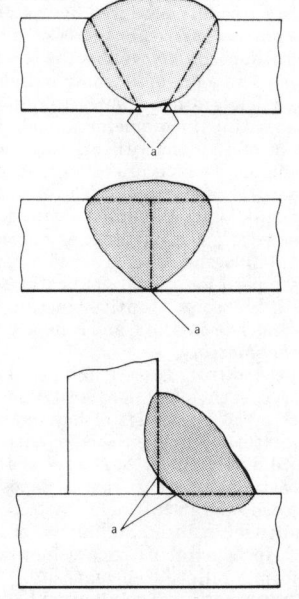

a = root of weld

root opening. In a weldment, the separation between the members at the root of joint prior to welding.

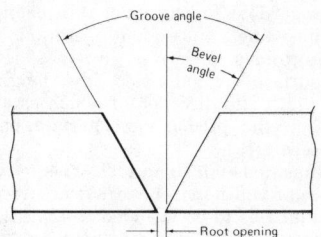

root pass. The first bead of a *multiple-pass weld,* laid in the root of joint.

root penetration. The depth that a weld extends into the root of joint, measured on the centerline of the root cross section. See sketch accompanying *joint penetration.*

rosebuds. Concentric rings of distorted coating, giving the effect of an opened rosebud. Noted only on *minimized spangle.*

rosette. (1) Rounded configuration of microconstituents arranged in whorls or radiating from a center. (2) Strain gages arranged to indicate at a single position strains in three different directions.

rotary forging. A process subjecting the workpiece to pressing between a flat anvil and a swiveling die with a conical working face; the platens move toward each other during forging.

rotary furnace. A circular furnace constructed so that the hearth and workpieces rotate around the axis of the furnace during heating.

rotary shear. A sheet-metal-cutting machine with two rotating-disk cutters mounted on parallel shafts driven in unison.

rotary swager. A swaging machine consisting of a power-driven ring that revolves at high speed causing rollers to engage cam surfaces and force the dies to deliver hammerlike blows upon the work at high frequency. Both straight and tapered sections can be produced.

rouge finish. A highly reflective finish produced with rouge or other very fine abrasive; similar in appearance to the bright polish or mirror finish on sterling silver utensils.

rough grinding. Grinding without regard to finish, usually to be followed by a subsequent operation.

roughing stand. The first stand of rolls through which the reheated billet passes, or the last stand in front of the finishing rolls.

rough machining. Machining without

62

regard to finish, usually to be followed by a subsequent operation.

roughness. Relatively finely spaced surface irregularities, the height, width and direction of which establish the predominant surface pattern.

roughness-width cutoff. The maximum width in inches of surface irregularities to be included in the measurement of roughness height.

rubber blanket. A sheet of rubber or other resilient material used as an auxiliary tool in forming.

rubber forming. Forming where rubber or other resilient material is used as a functional die part. Processes in which rubber is employed only to contain the hydraulic fluid are not classified as rubber forming.

rubber-pad forming. A forming operation for shallow parts where a rubber pad or other resilient material is attached to the press slide and becomes the mating die for a punch, or group of punches, that has been placed on the press bed or plate. Also known as the Guerin process.

rubber wheel. A grinding wheel made with a rubber bond.

rub mark. See *abrasion*.

runner. (1) A channel through which molten metal flows from one receptacle to another. (2) The portion of the gate assembly of a casting that connects the sprue with the gate(s). (3) Parts of patterns and finished castings corresponding to the portion of the gate assembly described in (2).

runner box. A distribution box that divides molten metal into several streams before it enters the mold cavity.

running fit. Any *clearance fit* in the range used for parts that rotate relative to each other. Actual values of clearance resulting from stated shaft and hole tolerances are given in ANSI standards.

runout. (1) The unintentional escape of molten metal from a mold, crucible or furnace. (2) An imperfection in a casting caused by the escape of metal from the mold. (3) See *axial runout* and *radial runout*.

runout table. A *roll table* used to receive a rolled or extruded section.

rust. A corrosion product consisting of hydrated oxides of iron. Applied only to ferrous alloys.

S

sacrificial protection. Reducing the extent of corrosion of a metal in an electrolyte by coupling it to another metal that is electrochemically more active in the environment.

saddling. Forming a seamless ring by forging a pierced disk over a mandrel (or saddle).

sag. An increase or decrease in the section thickness of a casting caused by insufficient strength of the mold sand of the cope or of the core.

salt fog test. An accelerated corrosion test in which specimens are exposed to a fine mist of a solution usually containing sodium chloride but sometimes modified with other chemicals. For testing details see ASTM B117.

salting out. Precipitating a substance in a solution by adding a second substance, usually a salt, without any chemical reaction such as a double decomposition taking place.

salt spray test. More properly, *salt fog test*.

sample. One or more units of product (or a relatively small quantity of a bulk material) that is withdrawn from a *lot* or process stream, and that is tested or inspected to provide information about the properties, dimensions or other quality characteristics of the lot or process stream. Not to be confused with *specimen*.

sand. A granular material, naturally or artificially produced by the disintegration or crushing of rocks or mineral deposits. In casting, the term denotes an aggregate whose individual particle (grain) size is 0.06 to 2 mm ($\frac{1}{400}$ to $\frac{1}{12}$ in.) in diameter, and largely free of finer constituents such as silt and clay, which are often present in natural sand deposits. The most commonly used foundry sand is silica; however, zircon, olivine, chromite, alumina and other crushed ceramics are used for special applications.

sand blasting. Abrasive blasting with sand. See *blasting*, and compare with *shot blasting*.

sand control. Testing and regulation of the chemical, physical and mechanical properties of foundry sand mixtures and their components.

sand hole. A pit in the surface of a sand casting resulting from a deposit of loose sand in the mold cavity.

sandwich rolling. Rolling two or more strips of metal in a pack, sometimes to form a roll-welded composite.

saponification. The alkaline hydrolysis of fats whereby a soap is formed; more generally, the hydrolysis of an ester by an alkali with the formation of an alcohol and a salt of the acid portion.

satin finish. A diffusely reflecting surface finish on metals, lustrous but not mirrorlike. One type is a *butler finish*.

saw gumming. In saw manufacture, the grinding away of punch marks or milling marks in the gullets (spaces between the teeth), and in some cases, simultaneous sharpening of the teeth; in reconditioning worn saws, restoration of the original gullet size and shape.

sawing. Cutting a workpiece with a band, blade or circular disk having teeth.

scab. An imperfection consisting of a thin, flat piece of metal attached to the surface of a sand casting. A sand scab usually is separated from the casting proper by a thin layer of sand and is joined to the casting along one edge. An erosion scab is similar in appearance to a *cut* or wash.

scale pit. (1) A surface depression formed on a forging due to scale remaining in the dies during the forging operation. (2) A pit in the ground in which scale (such as that carried off by cooling water from rolling mills) is allowed to settle out as one step in the treatment of effluent waste water.

scaling. (1) Forming a thick layer of oxidation products on metals at high temperature. (2) Depositing water-insoluble constituents on a metal surface, as in cooling tubes and water boilers.

scalped extrusion ingot. A cast, solid or hollow extrusion ingot that has been machined on the outside surface.

scalping. Removing surface layers from ingots, billets or slabs. See *die scalping*.

scarfing. Cutting surface areas of metal objects, ordinarily by using an oxyfuel gas torch. The operation permits surface imperfections to be cut from ingots, billets or the edges of plate that are to be beveled for butt welding. See *chipping*.

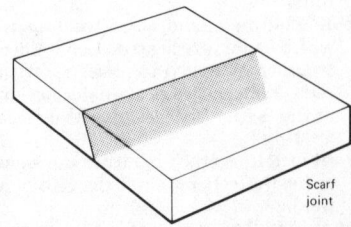

Scarf joint

scarf joint. A butt joint in which the plane of the joint is inclined with respect to the main axis of the members.

Scleroscope test. A hardness test where the loss in kinetic energy of a falling metal "tup", absorbed by indentation upon impact of the tup on the metal being tested, is indicated by the height of rebound.

scorification. The oxidation, in the presence of fluxes, of molten lead containing precious metals to partly remove the lead in order to concentrate the precious metals.

scoring. (1) Marring or scratching of a smooth surface; most often caused by sliding contact with a mating member having a hard projection or embedded particle on its surface. (2) Reducing the thickness of a material along a line to purposely weaken it.

scouring. (1) A wet or dry cleaning process involving mechanical scrubbing. (2) A wet or dry mechanical finishing operation using fine abrasive and low pressure by hand or with a cloth or wire wheel to produce satin or butler-type finishes.

scrap. (1) Products that are discarded because they are defective or otherwise unsuitable for sale. (2) Discarded metallic material, from whatever source, that may be reclaimed through melting and refining.

scratch hardness. The hardness of a metal determined by the width of a scratch made by a cutting point drawn across the surface under a given pressure.

screen. (1) One of a set of sieves, designated by the size of the openings, used to classify granular aggregates such as sand, ore or coke by particle size. (2) A perforated sheet placed in the gating system of a mold to separate dirt from the molten metal.

screw dislocation. See *dislocation*.

screw press. A press whose slide is operated by a screw rather than by a crank or other means.

screw stock. Free-machining bar, rod or wire.

scruff. A mixture of tin oxide and iron-tin alloy formed as dross on a tin-coating bath.

scuffing. A form of *adhesive wear* that produces superficial scratches or a high polish on the rubbing surfaces. It is observed most often on inadequately lubricated parts.

sea coal. Finely ground coal, used as an ingredient in molding sands.

sealing. (1) Closing pores in anodic

coatings to render them less absorbent. (2) Plugging leaks in a casting by introducing thermosetting plastics into porous areas and subsequently setting the plastic with heat.

seal weld. Any weld used primarily to obtain tightness and prevent leakage.

seam. (1) On the surface of metal, an unwelded fold or lap that appears as a crack, usually resulting from a discontinuity obtained in casting or in workpiece.

selective heating. Intentionally heating only certain portions of a workpiece.

selective leaching. Corrosion in which one element is preferentially removed from an alloy, leaving a residue (often porous) of the elements that are more resistant to the particular environment. See also *decarburization, denickelification, dezincification, graphitic corrosion*.

selective quenching. Quenching only certain portions of an object.

self-diffusion. Thermally activated movement of an atom to a new site in a crystal of its own species, as, for example, a copper atom within a crystal of copper.

self-hardening steel. See preferred term, *air-hardening steel*.

semiautomatic plating. *Plating* in which prepared cathodes are mechanically conveyed through the plating baths, with intervening manual transfers.

semiconductor. An electronic conductor whose resistivity at room temperature is in the range of 10^{-7} to $1\ \Omega\cdot m$ and in which the conductivity increases with increasing temperature over some temperature range.

semifinisher. An impression in a forging die that only approximates the finish dimensions of the forging. Semifinishers are often used to extend die life of the finishing impression, ensure proper control of grain flow during forging, and assist in obtaining desired tolerances. Also called semifinishing impression.

semifinishing. Preliminary operations performed prior to finishing.

semikilled steel. Steel that is incompletely deoxidized and contains sufficient dissolved oxygen to react with the carbon to form carbon monoxide to offset solidification shrinkage.

semipermanent mold. A permanent mold in which sand cores are used.

Sendzimir mill. A type of cluster mill

with small-diameter working rolls, larger-diameter backup rolls, backed up by bearings on a shaft mounted eccentrically so that it can be rotated to increase the pressure between the bearings and backup rolls.

sensitivity. The smallest difference in values that can be detected reliably with a given measuring instrument.

sensitization. In austenitic stainless steels, the precipitation of chromium carbides, usually at grain boundaries, upon exposure to temperatures of about 550 to 850 °C (1000 to 1550 °F), leaving the grain boundaries depleted of chromium and therefore susceptible to preferential attack by a corroding (oxidizing) medium.

sequence timer. In resistance welding, a device used for controlling the sequence and duration of any or all of the elements of a complete welding cycle except *weld time* or *heat time*.

sequence weld timer. Same as *sequence timer* except either *weld time* or *heat time,* or both, are also controlled.

sequestering agent. A material that combines with metallic ions to form water-soluble complex compounds.

series welding. Making two or more resistance spot, seam or projection welds simultaneously by a single welding transformer with three or more electrodes forming a series circuit.

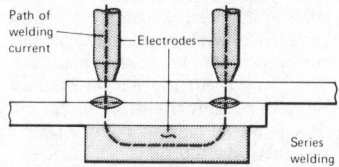

set. The shape of the solidifying surface of a metal, especially copper, with respect to concavity or convexity. May also be called pitch.

set copper. An intermediate copper product containing about 3.5% cuprous oxide, obtained at the end of the oxidizing portion of the fire-refining cycle.

settling. (1) Separation of solids from suspension in a fluid of lower density, solely because of gravitational effects. (2) A process for removing iron from liquid magnesium alloys by holding the melt at a low temperature after manganese has been added to it.

shadowing. (1) Same as *shielding* in electroplating. (2) Directional deposition of carbon or a metallic film on a plastic replica to highlight features to be analyzed by transmission electron microscopy.

shakeout. Removing castings from a sand mold.

shank. (1) The handle for carrying a small ladle or crucible. (2) The portion of a die, tool or forging by which it is held. (3) The main body of a lathe tool. If the tool is an inserted type, the shank is the portion that supports the insert. See sketch accompanying *single-point tool*.

shank-type cutters. Cutters having a straight or tapered shank to fit into a machine-tool spindle or adapter.

shaping. Producing flat surfaces using single-point tools. The work is held in a vise or fixture, or clamped directly to the table. The ram supporting the tool is reciprocated in a linear motion past the work.

shatter cracks. Same as *flakes*.

shaving. (1) As a finishing operation, the accurate removal of a thin layer of a work surface by straight-line motion between a cutter and the surface. (2) Trimming parts such as stampings, forgings and tubes to remove uneven sheared edges or to improve accuracy.

shear. (1) That type of force that causes or tends to cause two contiguous parts of the same body to slide relative to each other in a direction parallel to their plane of contact. (2) A type of cutting tool with which a material in the form of wire, sheet, plate or rod is cut between two opposing blades. (3) The type of cutting action produced by *rake* so that the direction of chip flow is other than at right angles to the cutting edge.

shear angle. The angle that the *shear plane,* in metal cutting, makes with the work surface.

shear fracture. A ductile fracture in which a crystal (or a polycrystalline mass) has separated by sliding or tearing under the action of shear stresses.

shearing strain. See *strain*.

shear ledges. See *radial marks*.

shear lip. A narrow, slanting ridge along the edge of a fracture surface. The term sometimes also denotes a narrow, often crescent-shaped, fibrous region at the edge of a fracture that is otherwise of the cleavage type, even though this fibrous region is in the same plane as the rest of the fracture surface.

shear modulus. See *modulus of elasticity.*

shear plane. A confined zone along which shear takes place in metal cutting. It extends from the cutting edge to the work surface.

shear strain. Same as shearing strain; see *strain*.

shear strength. The stress required to produce fracture in the plane of cross section, the conditions of loading being such that the directions of force and of resistance are parallel and opposite although their paths are offset a specified minimum amount. The maximum load divided by the original cross-sectional area of a section separated by shear.

shear stress. See *stress*.

sheet. A flat-rolled metal product of some maximum thickness and minimum width arbitrarily dependent on the type of metal. It is thinner than plate, and has a width-to-thickness ratio greater than about 50.

sheet separation. In spot, seam or projection welding, the gap that exists between faying surfaces surrounding the weld, after the joint has been welded.

shelf roughness. Roughness on upward facing surfaces where undissolved solids have settled on parts during a plating operation.

shell. (1) A hollow structure or vessel. (2) An article formed by deep drawing. (3) The metal sleeve remaining when a billet is extruded with a dummy block of somewhat smaller diameter. (4) In shell molding, a hard layer of sand and thermosetting plastic or resin formed over a pattern and used as the mold wall. (5) A tubular casting used in making seamless drawn tube. (6) A pierced forging.

shell core. A shell-molded sand core.

shell hardening. A surface-hardening process in which a suitable steel workpiece, when heated through and quench hardened, develops a martensitic layer or shell that closely follows the contour of the piece and surrounds a core of essentially pearlitic transformation product. This result is accomplished by a proper balance among section size, steel hardenability, and severity of quench.

shell molding. Forming a mold from thermosetting resin-bonded sand mixtures brought in contact with preheated (150 to 260 °C, or 300 to 500 °F) metal patterns, resulting in a firm shell with a cavity corresponding to the outline of the pattern. Also called Croning process.

shielded metal-arc welding. Arc welding in which metals are fused together by heating them with an arc between a *covered electrode* and the work. Decomposition of the covering on the consumable electrode provides shielding gas, and the electrode itself provides the filler metal. Pressure is not applied to the joint.

shielding. (1) A material barrier that prevents radiation or a flowing fluid from impinging on an object or a portion of an object. (2) Placing an object in an electrolytic bath so as to alter the current distribution on the cathode. A nonconductor is called a shield; a conductor, a *robber,* thief, or guard.

shift. A casting imperfection caused by mismatch of cope and drag or of cores and mold.

shim. A thin piece of material used between two surfaces to obtain a proper fit, adjustment or alignment.

shimmy die. See *flat edge trimmer.*

shoe. (1) A metal block used in a variety of bending operations to form or support the part being processed. (2) An anvil cap or sow block.

Shore hardness test. Same as *Scleroscope test.*

short circuiting transfer. In consumable-electrode arc welding, a type of metal transfer similar to globular transfer, but in which the drops are so large that the arc is short circuited momentarily during the transfer of each drop to the weld puddle. Compare with *spray transfer, globular transfer.*

shorts. The product that is retained on a specified screen in the screening of a crushed or ground material. See also *plus sieve.*

short transverse. See *transverse.*

shot. Small spherical particles of metal.

shot blasting. *Blasting* with metal *shot;* usually used to remove deposits or mill scale more rapidly or more effectively than can be done by sand blasting.

shot peening. Cold working the surface of a metal by metal-shot impingement.

shotting. The production of shot by pouring molten metal in finely divided streams. Solidified spherical particles are formed during the descent and are cooled in a tank of water.

shrinkage. See *casting shrinkage*.

shrinkage cavity. A void left in cast metals as a result of solidification shrinkage. See *casting shrinkage*.

shrinkage cracks. Hot tears associated with shrinkage cavities.

shrinkage rule. A measuring ruler with graduations expanded to compensate for the change in the dimensions of the solidified casting as it cools in the mold.

shrink fit. An interference fit produced by heating the outside member to a practical temperature to assemble easily. Usually the inside member is kept at or near room temperature. Sometimes the inside member is cooled to increase the ease of assembly.

shrink forming. Forming metal wherein the inner fibers of a cross section undergo a reduction in a localized area by the application of heat, cold upset or mechanically induced pressures.

shut height. For a press, the distance from the top of the bed to the bottom of the slide with the stroke down and adjustment up. In general it is the maximum die height that can be accommodated for normal operation, taking the bolster plate into consideration.

side cutting-edge angle. Defined by sketch accompanying *single-point tool*.

side milling. Milling with cutters having peripheral and side teeth. They are usually profile sharpened but may be form relieved.

side rake. In a single-point turning tool, the angle between the tool face and a reference plane, corresponding to radial rake in milling. It lies in a plane perpendicular to the tool base and parallel to the rotational axis of the work. See sketch accompanying *single-point tool*.

side relief angle. Defined by sketch accompanying *single-point tool*.

sieve analysis. Particle size distribution; usually expressed as the weight percentage retained upon each of a series of standard sieves of decreasing size and the percentage passed by the sieve of finest size. Synonymous with *sieve classification*.

sieve classification. Same as *sieve analysis*.

sieve fraction. The portion of a powder sample that passes through a standard sieve of specified number and is retained by some finer sieve of specified number.

sigma phase. A hard, brittle nonmagnetic intermediate phase with a tetragonal crystal structure, containing 30 atoms per unit cell, space group $P4_2/mnm$, occurring in many binary and ternary alloys of the transition elements. The composition of this phase in the various systems is not the same and the phase usually exhibits a wide range in homogeneity. Alloying with a third transition element usually enlarges the field of homogeneity and extends it deep into the ternary section.

silica flour. A sand additive, containing about 99.5% silica, commonly produced by pulverizing quartz sand in large ball mills to a mesh size of 80 to 325.

siliconizing. Diffusing silicon into solid metal, usually steel, at an elevated temperature.

silky fracture. A metal fracture in which the broken metal surface has a fine texture, usually dull in appearance. Characteristic of tough and strong metals. Contrast with *crystalline fracture, granular fracture*.

silver soldering. Nonpreferred term used to denote brazing with a silver-base filler metal. See preferred terms *furnace brazing, induction brazing*, and *torch brazing*.

single-bevel groove weld. A groove weld in which the joint edge of one member is beveled from one side.

Single-bevel groove weld

single-impulse welding. Spot, projection or upset welding by a single impulse of current. Where alternating current is used, an impulse may be any fraction or number of cycles.

single-J groove weld. A groove weld in which the joint edge of one member is prepared in the form of a J, from one side. See sketch.

Single-J groove weld

single-point tool. See definition of nomenclature in accompanying sketch.

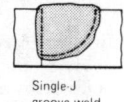

Nomenclature of single-point tool

End cutting-edge angle
Nose
Face
Shank
Cutting edge
Side cutting-edge angle
Side rake angle
Back rake angle
Flank
End relief
Lip
Shank
Side relief
End clearance angle
Base (heel)

single-stand mill. A rolling mill of such design that the product contacts only two rolls at a given moment. Contrast with *tandem mill*.

single-U groove weld. A groove weld in which each joint edge is prepared in the form of a J or half-U from one side. See sketch.

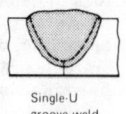
Single-U groove weld

single-V groove weld. A groove weld in which each member is beveled from the same side. See sketch below.

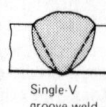
Single-V groove weld

single welded joint. In arc and gas welding, any joint welded from one side only.

sinkhead. Same as *riser*.

sinking. See *tube sinking*.

sinter. To heat a mass of fine particles for a prolonged time below the melting point, usually to cause agglomeration.

sintering. The bonding of adjacent surfaces in a mass of particles by molecular or atomic attraction on heating at high temperatures below the melting temperature of any constituent in the material. Sintering strengthens a powder mass and normally produces densification and, in powdered metals, recrystallization. See also *liquid phase sintering*.

size effect. Effect of the dimensions of a piece of metal upon its mechanical and other properties and upon manufacturing variables such as forging reduction and heat treatment. In general, the mechanical properties are lower for a larger size.

size of weld. (1) The joint penetration in a groove weld. (2) The lengths of the nominal legs of a fillet weld. See illustrations.

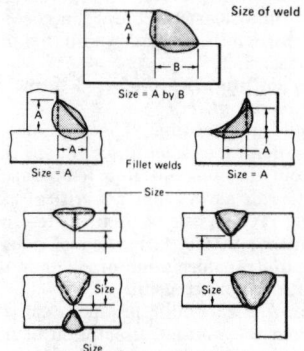

Size of weld
Size = A by B
Fillet welds
Size = A
Size = A
Size
Size
Size
Groove welds

sizing. (1) Secondary forming or squeezing operations, required to square up, set down, flatten or otherwise correct surfaces to produce specified dimensions and tolerances. See *restriking.* (2) Some burnishing, broaching, drawing and shaving operations are also called sizing. (3) A finishing operation for correcting ovality in tubing. (4) Final pressing of a sintered powder metallurgy part.

skelp. The starting stock for making welded pipe or tubing; most often it is strip stock of suitable width, thickness and edge configuration.

skim gate. A gating arrangement designed to prevent the passage of slag and other undesirable materials into a casting.

skimmer. A tool for removing scum, slag and dross from the surface of molten metal.

skin. A thin outside metal layer, not formed by bonding as in cladding or electroplating, that differs in composition, structure or other characteristic from the main mass of metal.

skin lamination. In flat-rolled metals, a surface rupture resulting from the exposure of a subsurface lamination by rolling.

skin pass. See *temper rolling.*

skiving. (1) Removal of a material in thin layers or chips with a high degree of shear or slippage, or both, of the cutting tool. (2) A machining operation in which the cut is made with a form tool with its face so angled that the cutting edge progresses from one end of the work to the other as the tool feeds tangentially past the rotating workpiece.

skull. A layer of solidified metal or dross on the walls of a pouring vessel after the metal has been poured.

slab. A piece of metal, intermediate between ingot and plate, with the width at least twice the thickness.

slabbing mill. A primary mill that produces slabs.

slab milling. See preferred term, *peripheral milling.*

slack quenching. The incomplete hardening of steel due to quenching from the austenitizing temperature at a rate slower than the critical cooling rate for the particular steel, resulting in the formation of one or more transformation products in addition to martensite.

slag. A nonmetallic product resulting from the mutual dissolution of flux and nonmetallic impurities in smelting, refining, and certain welding operations.

slag inclusion. Slag or dross entrapped in a metal.

slant fracture. A type of fracture appearance, typical of plane-stress fractures, in which the plane of metal separation is inclined at an angle (usually about 45°) to the axis of applied stress.

slide. Main reciprocating member of a mechanical press, guided in a press frame and to which the punch or upper die is fastened.

sliding fit. A loosely defined fit similar to a *slip fit.*

slime. (1) A material of extremely fine particle size encountered in ore treatment. (2) A mixture of metals and some insoluble compounds that forms on the anode in electrolysis.

slip. Plastic deformation by the irreversible shear displacement (translation) of one part of a crystal relative to another in a definite crystallographic direction and usually on a specific crystallographic plane. Sometimes called glide.

slip band. A group of parallel slip lines so closely spaced as to appear as a single line when observed under an optical microscope. See *slip line.*

slip direction. The crystallographic direction in which the translation of slip takes place.

slip fit. A loosely defined clearance fit between parts assembled by hand without force, but implying slipping contact.

slip flask. A tapered *flask* that depends on a movable strip of metal to hold the sand in position. After closing the mold, the strip is retracted and the flask can be removed and reused. Molds thus made are usually supported by a *mold jacket* during pouring.

slip-interference theory. Theory involving the resistance to deformation offered by a hard phase dispersed in a ductile matrix.

slip line. The trace of the slip plane on the viewing surface; the trace is (usually) observable only if the surface has been polished before deformation. The usual observation on metal crystals (under the light microscope) is of a cluster of slip lines known as a slip band.

slip plane. The crystallographic plane in which slip occurs in a crystal.

sliver. An imperfection consisting of a very thin elongated piece of metal attached by only one end to the parent metal into whose surface it has been worked.

slope control. Producing electronically a gradual increase or decrease in the welding current between definite limits and within a selected time interval.

slot furnace. A common batch furnace where stock is charged and removed through a slot or opening.

slotting. Cutting a narrow aperture or groove with a reciprocating tool in a vertical shaper or with a cutter, broach or grinding wheel.

slot weld. Similar to *plug weld,* the difference being that the hole is elongated and may extend to the edge of a member without closing.

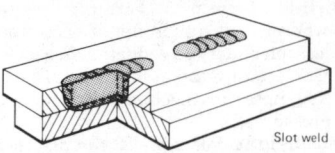
Slot weld

slug. (1) A short piece of metal to be placed in a die for forging or extruding. (2) A small piece of material produced by piercing a hole in sheet material.

slugging. The unsound practice of adding a separate piece of material in a joint before or during welding, resulting in a welded joint in which the weld zone is not entirely built up by adding molten filler metal or by melting and recasting base metal, and therefore does not comply with design, drawing or specification requirements.

slush casting. A hollow casting usually made of an alloy with a low but wide melting temperature range. After the desired thickness of metal has solidified in the mold, the remaining liquid is poured out.

smelting. Thermal processing wherein chemical reactions take place to produce liquid metal from a beneficiated ore.

smith forging. Manual forging with flat or simple-shaped dies that never completely confine the work.

smut. A reaction product sometimes left on the surface of a metal after pickling, electroplating or etching.

snagging. *Offhand grinding* on castings and forgings to remove surplus metal such as gate and riser pads, fins and parting lines.

snake. (1) The product formed by twisting and bending of hot rod prior to its next rolling process. (2) Any crooked surface imperfection in a plate, resembling a snake. (3) A flexible mandrel used in the inside of a shape to prevent flattening or collapse during a bending operation.

snaky edges. See *carbon edges.*

snap flask. A foundry flask hinged on one corner so that it can be opened and removed from the mold for reuse before the metal is poured.

snap temper. A precautionary interim stress-relieving treatment applied to high-hardenability steels immediately after quenching to prevent cracking because of delay in tempering them at the prescribed higher temperature.

***S-N* diagram.** A plot showing the relationship of stress, *S*, and the number of cycles, *N*, before fracture in fatigue testing.

snowflakes. See *flakes.*

snug fit. A loosely defined fit implying the closest clearances that can be assembled manually for firm connection between parts.

soak cleaning. *Immersion cleaning* without electrolysis.

soaking. Prolonged holding at a selected temperature to effect homogenization of structure or composition.

soft soldering. See preferred term, *soldering.*

soft temper. Same as *dead soft* temper.

soil. Undesirable material on a surface and yet not an integral part of the surface. Oil, grease and dirt can be soils; a decarburized skin or excess *hard chromium* are not soils. Loose scale is soil; hard scale may be an integral part of the surface and, hence, not soil.

solderability. The ease with which a surface is wetted by solder.

solder embrittlement. Reduction in mechanical properties of a metal as a result of local penetration of solder along grain boundaries.

soldering. A group of processes that join metals by heating them to a suitable temperature below the solidus of the base metals and applying a filler metal having a liquidus not exceeding 450 °C (840 °F). Molten filler metal is distributed between the closely fitted surfaces of the joint by capillary action.

solder short. See *bridging* (5).

solid cutters. Cutters made of a single piece of material rather than a com-

posite of two or more materials.

solidification. The change in state from liquid to solid on cooling through the melting temperature or melting range.

solidification shrinkage. See *casting shrinkage.*

solid shrinkage. See *casting shrinkage.*

solid solution. A single solid homogeneous crystalline phase containing two or more chemical species.

solid state welding. A group of welding processes that join metals at temperatures essentially below the melting point of the base materials, without the addition of a brazing or soldering filler metal. Pressure may or may not be applied to the joint.

solidus. In a constitution or equilibrium diagram, the locus of points representing the temperatures at which various compositions finish freezing on cooling or begin to melt on heating. See also *liquidus.*

soluble oil. Specially prepared oil whose water emulsion is used as a cutting or grinding fluid.

solute. The component of either a liquid or solid solution that is present to a lesser or minor extent; the component that is dissolved in the *solvent.*

solution heat treatment. Heating an alloy to a suitable temperature, holding at that temperature long enough to cause one or more constituents to enter into solid solution, and then cooling rapidly enough to hold these constituents in solution.

solution potential. *Electrode potential* where the half-cell reaction involves only the metal electrode and its ion.

solvent. The component of either a liquid or solid solution that is present to a greater or major extent; the component that dissolves the *solute.*

solvus. In a constitution or equilibrium diagram, the locus of points representing the temperatures at which the various compositions of the solid phases coexist with other solid phases, that is, the limits of solid solubility.

sonic testing. Any inspection method that uses sound waves (in the audible frequency range, about 20 to 20 000 Hz) to induce a response from a part or test specimen. Sometimes used, but inadvisedly, as a synonym for *ultrasonic testing.*

sorbite. (obsolete) A fine mixture of ferrite and cementite produced either by regulating the rate of cooling of

steel or by tempering steel after hardening. The first type is very fine pearlite difficult to resolve under the microscope; the second type is tempered martensite.

sow block. In forging, a removable block of metal set into the hammer anvil to protect the anvil from shock and wear and occasionally to hold insert dies. Also called an anvil cap or a shoe.

space lattice. A regular, periodic array of points (lattice points) in space that represents the locations of atoms of the same kind in a perfect crystal. The concept may be extended, where appropriate, to crystalline compounds and other substances, in which case the lattice points often represent locations of groups of atoms of identical composition, arrangement and orientation.

spacer strip. A metal strip or bar inserted in the root of a joint prepared for a groove weld, to serve as a backing and to maintain root opening throughout the course of the welding operation.

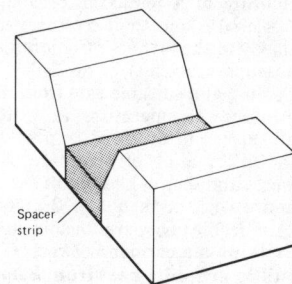

Spacer strip

spade drill. See preferred term, *flat drill.*

spalling. The cracking and flaking of particles out of a surface.

spangle. The characteristic crystalline form in which a hot dipped zinc coating solidifies on steel strip.

spatter. The metal particles expelled during arc or gas welding. They do not form part of the weld.

spatter loss. The metal lost due to *spatter.*

specific energy. In cutting or grinding, the energy expended or work done in removing a unit volume of material.

specific power. Same as *unit power.*

specimen. A test object, often of standard dimensions or configuration, that is used for destructive or nondestructive testing. One or more specimens may be cut from each unit of a *sample*.

speed of travel. In welding, the speed with which a weld is made along its longitudinal axis, usually measured in inches per minute or spots per minute.

speiss. Metallic arsenides and antimonides that result from smelting metal ores such as those of cobalt.

spelter. Crude zinc obtained in smelting zinc ores.

spelter solder. A brazing filler metal of approximately equal parts of copper and zinc.

spheroidite. An aggregate of iron or alloy carbides of essentially spherical shape dispersed throughout a matrix of ferrite.

spheroidizing. Heating and cooling to produce a spheroidal or globular form of carbide in steel. Spheroidizing methods frequently used are

1 Prolonged holding at a temperature just below Ae_1.
2 Heating and cooling alternately between temperatures that are just above and just below Ae_1.
3 Heating to a temperature above Ae_1 or Ae_3 and then cooling very slowly in the furnace or holding at a temperature just below Ae_1.
4 Cooling at a suitable rate from the minimum temperature at which all carbide is dissolved, to prevent the reformation of a carbide network, and then reheating in accordance with method 1 or 2 above. (Applicable to hypereutectoid steel containing a carbide network.)

spherulitic graphite cast iron. Same as *ductile cast iron*.

spider die. Same as *porthole die*.

spiegeleisen (spiegel). A pig iron containing 15 to 30% Mn and 4.5 to 6.5% C.

spindle. (1) Shaft of a machine tool on which a cutter or grinding wheel may be mounted. (2) Metal shaft to which a mounted wheel is cemented.

spinning. Forming a seamless hollow metal part by forcing a rotating blank to conform to a shaped mandrel that rotates concentrically with the blank. In the usual application, a flat-rolled metal blank is forced against the mandrel by a blunt, rounded tool; however, other stock (notably welded or seamless tubing) can be formed, and sometimes the working end of the tool is a roller.

spinodal structure. A fine homogeneous mixture of two phases that form by the growth of composition waves in a solid solution during suitable heat treatment. The phases of a spinodal structure differ in composition from each other and from the parent phase but have the same crystal structure as the parent phase.

spline. Any of a series of longitudinal, straight projections on a shaft that fit into slots on a mating part to transfer rotation to or from the shaft.

split die. Same as *segment die*.

sponge. A form of metal characterized by a porous condition that is the result of the decomposition or reduction of a compound without fusion. The term is applied to forms of iron, titanium, zirconium, uranium, plutonium and the platinum-group metals.

sponge iron. Either porous or powdered iron produced directly without fusion, as by heating high-grade ore with charcoal, or an oxide with a reducing gas.

spot drilling. Making an initial indentation in a work surface, with a drill, to serve as a centering guide in a subsequent machining operation.

spot facing. Machining a flat seat for a bolt head, nut or other similar element at the end of and at right angles to the axis of a previously made hole.

spotting. Fitting one part of a die to another by applying an oil color to the surface of the finished part and bringing this against the surface of the intended mating part, the high spots being marked by the transferred color.

spotting out. Delayed, uneven staining of metal by entrapment of chemicals during the finishing operation.

spot welding. Welding of lapped parts in which fusion is confined to a relatively small circular area. It is generally resistance welding, but may also be gas tungsten-arc, gas metal-arc, or submerged-arc welding.

spray metallizing. See *metallizing*.

spray quenching. Quenching in a spray of liquid.

spray transfer. In consumable-electrode arc welding, a type of metal transfer in which the molten filler metal is propelled across the arc as fine droplets. Compare with *globular transfer, short circuiting transfer*.

springback. (1) The elastic recovery of metal after cold forming. (2) The degree to which metal tends to return to its original shape or contour after undergoing a forming operation. (3) In flash, upset or pressure welding, the deflection in the welding machine caused by the upset pressure.

spring temper. A *temper* of nonferrous alloys and some ferrous alloys characterized by tensile strength and hardness about two-thirds of the way from *full hard* to *extra spring* temper.

sprue. (1) The mold channel that connects the *pouring basin* with the runner or, in the absence of a pouring basin, directly into which molten metal is poured. Sometimes referred to as downsprue or downgate. (2) Sometimes used to mean all gates, risers, runners and similar scrap that are removed from castings after shakeout.

square drilling. Making square holes by means of a specially constructed drill made to rotate and also to oscillate so as to follow accurately the periphery of a square guide bushing or template.

square groove weld. A groove weld in which the abutting surfaces are square.

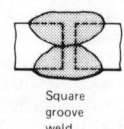

Square groove weld

squaring shear. A machine tool, used for cutting sheet metal or plate, consisting essentially of a fixed cutting knife (usually mounted on the rear of the bed) and another cutting knife mounted on the front of a reciprocally moving crosshead, which is guided vertically in side housings. Corner angles are usually 90°.

squeeze time. In resistance welding, the time between the initial applications of pressure and current.

stabilizing treatment. (1) Before finishing to final dimensions, repeatedly heating a ferrous or nonferrous part to or slightly above its normal operating temperature and then cooling to room temperature to ensure dimensional stability in service. (2) Transforming retained austenite in quenched hardenable steels, usually by *cold treatment*. (3) Heating a solution-treated stabilized grade of austenitic stainless steel to 870 to 900 °C (1600 to 1650 °F) to precipitate all carbon as TiC, NbC, or TaC so that *sensitization* is avoided on subsequent exposure to elevated temperature.

stack cutting. *Oxyfuel gas cutting* of stacked metal plates arranged so that all are severed by a single cut.

stack molding. A molding method that makes use of both faces of a mold section, one face acting as the drag and the other as the cope. Sections, when assembled to other similar sections, form several tiers of mold cavities, all castings being poured together through a common sprue.

stack welding. Resistance *spot welding* of stacked plates, all being joined simultaneously.

staggered-intermittent fillet welding. Making a line of intermittent fillet welds on each side of a joint so that the increments on one side are not opposite those on the other. Contrast with *chain-intermittent fillet welding.*

staggered-tooth cutters. Milling cutters with alternate flutes of oppositely directed helixes.

stainless steel. Any of several steels containing 12 to 30% chromium as the principle alloying element; they usually exhibit *passivity* in aqueous environments.

staking. Fastening two parts together permanently by recessing one part within the other and then causing plastic flow at the joint.

stalagmometer. An apparatus for determining surface tension. The mass of a drop of liquid is measured by weighing a known number of drops or by counting the number of drops obtained from a given volume of the liquid.

stamping. A general term covering almost all press operations. It includes blanking, shearing, hot or cold forming, drawing, bending, coining.

stand. A piece of rolling mill equipment containing one set of working rolls. In the usual sense, any pass of a continuous, looping or cross-country hot rolling mill.

standard electrode potential. The reversible *electrode potential* where all reactants and products are at unit activity.

standard gold. A legally adopted alloy for coinage of gold. In the United States the alloy contains 10% Cu.

stardusting. An extremely fine form of roughness on the surface of a metal deposit.

starting sheet. A thin sheet of metal used as the cathode in electrolytic refining.

state of strain. A complete description of the deformation within a homogeneously deformed volume or at a point. The description requires, in general, the knowledge of six independent components of *strain.*

state of stress. A complete description of the stresses within a homogeneously stressed volume or at a point. The description requires, in general, the knowledge of six independent components of *stress.*

static fatigue. A term sometimes used to identify a form of hydrogen embrittlement in which a metal appears to fracture spontaneously under a steady stress less than the yield stress. There almost always is a delay between the application of stress (or exposure of the stressed metal to hydrogen) and the onset of cracking. More properly referred to as hydrogen-induced delayed cracking.

steadite. A hard structural constituent of cast iron that consists of a binary eutectic of ferrite (containing some phosphorus in solution) and iron phosphide (Fe_3P). The composition of the eutectic is 10.2% P, 89.8% Fe, and the melting temperature is 1050 °C (1920 °F).

Stead's brittleness. A condition of brittleness that causes transcrystalline fracture in the coarse grain structure that results from prolonged annealing of thin sheets of low-carbon steel previously rolled at a temperature below about 705 °C (1300 °F). The fracture usually occurs at about 45° to the direction of rolling.

steadyrest. In cutting or grinding, a stationary support for a long workpiece.

Steckel mill. A cold reducing mill having two working rolls and two backup rolls, none of which is driven. The strip is drawn through the mill by a power reel in one direction as far as the strip will allow and then reversed by a second power reel and so on until the desired thickness is attained.

steel. An iron-base alloy, malleable in some temperature ranges as initially cast, containing manganese, usually carbon, and often other alloying elements. In carbon steel and low-alloy steel, the maximum carbon is about 2.0%; in high-alloy steel, about 2.5%. The dividing line between low-alloy and high-alloy steels is generally regarded as being at about 5% metallic alloying elements.

Steel is to be differentiated from two general classes of "irons": the cast irons, on the high-carbon side, and the relatively pure irons such as ingot iron, carbonyl iron, and electrolytic iron, on the low-carbon side. In some steels containing extremely low carbon, the manganese content is the principle differentiating factor, steel usually containing at least 0.25%; ingot iron, considerably less.

step aging. Aging at two or more temperatures, by steps, without cooling to room temperature after each step. See *aging,* and compare with *interrupted aging* and *progressive aging.*

stepped extrusion. A product with one or more abrupt cross-sectional changes usually obtained by interrupting the extrusion operation and exchanging dies.

stereoradiography. A technique for producing paired radiographs that may be viewed with a stereoscope to exhibit a shadowgraph in three dimensions with various sections in perspective and spatial relation.

sterling silver. A silver alloy containing at least 92.5% Ag, the remainder being unspecified but usually copper.

stick electrode. A shop term for *covered electrode.*

sticker break. Arc-shaped *coil break,* usually located near the center of sheet or strip.

stiffness. The ability of a metal or shape to resist elastic deflection. For identical shapes, the stiffness is proportional to the modulus of elasticity. For a given material, the stiffness increases with increasing moment of inertia, which is computed from cross-sectional dimensions.

stock. A general term for solid starting material that is formed, forged or machined to make parts.

stoking. (obsolete) Presintering, or sintering, in such a way that powder metallurgy compacts are advanced through the furnace at a fixed rate by manual or mechanical means; also called continuous sintering.

stop-off. See *resist.*

stopper rod. A device in a bottom-pour ladle for controlling the flow of metal through the nozzle into a mold. The stopper rod consists of a steel rod, protective refractory sleeves and a graphite stopper head.

stopping off. (1) Applying a *resist.* (2) Depositing a metal (copper, for example) in localized areas to prevent carburization, decarburization or nitriding in those areas. (3) Filling in a portion of a mold cavity to keep out molten metal.

stored-energy welding. Welding with electrical energy accumulated electrostatically, electromagnetically or electrochemically at a relatively low rate and made available at the higher rate required in welding.

straddle milling. Face milling a workpiece on both sides at once using two cutters spaced as required.

straight polarity. Direct-current arc welding circuit arrangement in which the electrode is connected to the negative terminal. Contrast with *reverse polarity.*

strain. A measure of the relative change in the size or shape of a body. Linear strain is the change per unit length of a linear dimension. True strain (or natural strain) is the natural logarithm of the ratio of the length at the moment of observation to the original gage length. Conventional strain is the linear strain over the original gage length. Shearing strain (or shear strain) is the change in angle (expressed in radians) between two lines originally at right angles. When the term "strain" is used alone it usually refers to the linear strain in the direction of applied stress. See also *state of strain.*

strain-age embrittlement. A loss in ductility accompanied by an increase in hardness and strength that occurs with low-carbon steel (especially rimmed or capped steel) is aged following plastic deformation. The degree of embrittlement is a function of aging time and temperature, occurring in a matter of minutes at about 200 °C (400 °F) but requiring a few hours to a year at room temperature.

strain aging. Aging induced by cold working. See *aging.*

strain energy. (1) The work done in deforming a body. (2) The work done in deforming a body within the elastic limit of the material. It is more properly termed elastic strain energy and can be recovered as work rather than heat.

strain hardening. An increase in hardness and strength caused by plastic deformation at temperatures below the recrystallization range.

strain-hardening exponent. A measure of rate of strain hardening. The constant n in the expression:

$$\sigma = \sigma_0 \delta^n$$

where σ is true stress, σ_0 is true stress at unit strain, and δ is true strain.

strain rate. The time rate of straining for the usual tensile test. Strain as measured directly on the specimen gage length is used for determining strain rate. Because strain is dimensionless, the units of strain rate are reciprocal time.

strain-rate sensitivity. Qualitatively, the increase in stress (s) needed to cause a certain increase in plastic strain rate $(\dot\varepsilon)$ at a given level of plastic strain (ε) and a given temperature (T).

Strain-rate sensitivity =
$$m = \frac{\Delta \log s}{\Delta \log \dot\varepsilon}\bigg|_{\varepsilon,T}$$

strain rods. (1) Rods sometimes used on gapframe presses to lessen the frame deflection. (2) Rods used to measure elastic strains, and thus stresses, in frames of presses.

strain state. See *state of strain.*

strand casting. A generic term describing *continuous casting* of one or more elongated shapes such as billets, blooms or slabs; if two or more strands are cast simultaneously, they are often of identical cross section.

stray current. Current flowing in electrodeposition by way of an unplanned and undesired bipolar electrode that may be the tank itself or a poorly connected electrode.

stress. Force per unit area, often thought of as force acting through a small area within a plane. It can be divided into components, normal and parallel to the plane, called normal stress and shear stress, respectively. True stress denotes the stress where force and area are measured at the same time. Conventional stress, as applied to tension and compression tests, is force divided by the original area. Nominal stress is the stress computed by simple elasticity formulas, ignoring stress raisers and disregarding plastic flow; in a notch bend test, for example, it is bending moment divided by minimum section modulus. See also *state of stress.*

stress amplitude. One-half the algebraic difference between the maximum and minimum stress in one cycle of a repetitively varying stress.

stress-concentration factor *(K_t).* A multiplying factor for applied stress that allows for the presence of a structural discontinuity such as a notch or hole; K_t equals the ratio of the greatest stress in the region of the discontinuity to the nominal stress for the entire section.

stress-corrosion cracking. Failure by cracking under combined action of corrosion and stress, either external (applied) stress or internal (residual) stress. Cracking may be either intergranular or transgranular, depending on metal and corrosive medium. See also *season cracking.*

stress-intensity factor. A scaling factor, usually denoted by the symbol K, used in linear-elastic fracture mechanics to describe the intensification of applied stress at the tip of a crack of known size and shape. At the onset of rapid crack propagation in any structure containing a crack, the factor is called the critical stress-intensity factor, or the *fracture toughness.* Various subscripts are used to denote different loading conditions or fracture toughnesses. The most common subscripts, and their meanings, are

K_c. Plane-stress fracture toughness. The value of stress intensity at which crack propagation becomes rapid in sections thinner than those in which plane-strain conditions prevail.

K_I. Stress-intensity factor for a loading condition that displaces the crack faces in a direction normal to the crack plane (also known as the opening mode of deformation).

K_{Ic}. Plane-strain fracture toughness. The minimum value of K_c for any given material and condition, which is attained when rapid crack propagation in the opening mode is governed by plane-strain conditions.

K_{Id}. Dynamic fracture toughness. The fracture toughness determined under dynamic loading conditions; it is used as an approximation of K_{Ic} for very tough materials.

K_{Iscc}. Threshold stress intensity for stress-corrosion cracking. A value of stress intensity characteristic of a specific combination of material, material condition and corrosive environment above which stress-corrosion crack propagation occurs and below which the material is immune from stress-corrosion cracking.

stress raisers. Changes in contour or discontinuities in structure that cause local increases in stress.

stress range. The algebraic difference between the maximum and minimum stress in one cycle of a repetitively varying stress.

stress ratio. In fatigue, the ratio of the minimum stress to the maximum stress in one cycle, considering tensile stresses as positive, compressive stresses as negative.

stress relieving. Heating to a suitable temperature, holding long enough to reduce residual stresses and then cooling slowly enough to minimize the development of new residual stresses.

stress-rupture test. A method of evaluating elevated-temperature durability in which a tension-test specimen is stressed under constant load until it breaks. Data recorded commonly include: initial stress, time to rupture, initial extension, creep extension, reduction of area at fracture. Also known as creep-rupture test.

stress state. See *state of stress*.

stretcher leveling. Leveling a piece of metal (that is, removing warp and distortion) by gripping it at both ends and subjecting it to a stress higher than its yield strength. Sometimes called patent leveling.

stretcher straightening. Straightening rod, tubing or shapes by gripping the stock at both ends and applying tension. The products are elongated a difinite amount to remove warpage.

stretcher strains. Elongated markings that appear on the surface of some materials when deformed just past the yield point. These markings lie approximately parallel to the direction of maximum shear stress and are the result of localized yielding. See also *Lüders lines*.

stretch former. (1) A machine used to perform *stretch forming* operations. (2) A device adaptable to a conventional press for accomplishing stretch forming.

stretch forming. Shaping of a sheet or part, usually of uniform cross section, by first applying suitable tension or stretch and then wrapping it around a die of the desired shape.

stretch wipe forming. Same as *wiper forming*.

striation. A fatigue fracture feature, often observed in electron micrographs, that indicates the position of the crack front after each succeeding cycle of stress. The distance between striations indicates the advance of the crack front across that crystal during one stress cycle, and a line normal to the striations indicates the direction of local crack propagation.

strike. (1) A thin electrodeposited film of metal to be followed by other plated coatings. (2) A plating solution of high covering power and low efficiency designed to electroplate a thin adherent film of metal.

striking. Electrodepositing, under special conditions, a very thin film of metal that will facilitate further plat-

ing with another metal or with the same metal under different conditions.

striking surface. Those areas on the faces of a set of dies that are designed to meet when the upper and lower dies are brought together. Striking surface helps protect impressions from impact shock and aids in maintaining longer die life. Also called beating area.

stringer. In wrought materials, an elongated configuration of microconstituents or foreign material aligned in the direction of working. Commonly, the term is associated with elongated oxide or sulfide inclusions in steel.

stringer bead. A continuous weld bead made without appreciable transverse oscillation. Contrast with *weave bead*.

strip. A flat-rolled metal product of some maximum thickness and width arbitrarily dependent on the type of metal. It is narrower than sheet.

stripper punch. A punch that serves as top or bottom of the die cavity and later moves farther into the die to eject the part or compact. See also *ejector rod, knockout* (1).

stripping. Removing a coating from a metal surface.

structural shape. Piece of metal of any of several designs accepted as standard by the structural branch of the iron and steel industries. See also *section* (3).

stud arc welding. An arc welding process that produces coalescence of metals by heating them with an arc between a metal stud, or similar part, and another part. When the surfaces to be joined are properly heated, they are brought together under pressure. Partial shielding may be obtained by the use of a ceramic ferrule surrounding the stud. Shielding gas or flux may or may not be used.

subboundary structure. A network of low-angle boundaries (usually less than one degree) within the main crystals of a metallographic structure.

subcritical annealing. A process anneal performed on ferrous alloys at a temperature below Ac_1.

subgrain. A portion of a crystal or grain, with an orientation slightly different from the orientation of neighboring portions of the same crystal. Generally, neighboring subgrains are separated by low-angle boundaries such as *tilt boundaries*

and *twist boundaries*.

submerged-arc welding. Arc welding in which the arc, between a bare metal electrode and the work, is shielded by a blanket of granular, fusible material overlying the joint. Pressure is not applied to the joint, and filler metal is obtained from the consumable electrode (and sometimes from a supplementary welding rod).

subsieve analysis. Size distribution of particles all of which will pass through a 44-μm (No. 325) standard sieve, as determined by specified methods.

subsieve fraction. That portion of a powdered sample which will pass through a 44-μm (No. 325) standard sieve.

substitutional solid solution. A solid solution in which the solute atoms are located at some of the lattice points of the solvent, the distribution being random. Contrast with *interstitial solid solution*.

substrate. Layer of metal underlying a coating, regardless of whether the layer is basis metal.

substructure. Same as *subboundary structure*.

subsurface corrosion. Formation of isolated particles of corrosion products beneath a metal surface. This results from the preferential reaction of certain alloy constituents by inward diffusion of oxygen, nitrogen or sulfur.

sulfur dome. An inverted container, holding a high concentration of sulfur dioxide gas, used in die casting to cover a pot of molten magnesium to prevent burning.

sulfur print. A macrographic method of examining for distribution of sulfide inclusions by placing a sheet of wet acidified photographic paper in contact with the polished steel surface to be examined.

superalloy. See *heat-resisting alloy*.

superconductivity. The abrupt and large increase in electrical conductivity exhibited by some metals as the temperature approaches absolute zero.

supercooling. Cooling below the temperature at which an equilibrium phase transformation can take place, without actually obtaining the transformation.

superficial Rockwell hardness test. Form of Rockwell hardness test using relatively light loads that produce minimum penetration by the indenter. Used for determining surface hardness or hardness of thin sections or small parts, or where a large hardness impression might be harmful.

superfines. The portion of a metal powder that is composed of particles smaller than a specified size, usually less than 10 μm.

superfinishing. A form of *honing* in which the abrasive stones are spring supported.

superheating. (1) Heating above the temperature at which an equilibrium phase transformation should occur without actually obtaining the transformation. (2) Heating molten metal above the normal casting temperature so as to obtain more complete refining or greater fluidity.

superlattice. A lattice arrangement in which solute and solvent atoms of a solid solution occupy different preferred sites in the array. Contrast with *disordering*.

superplasticity. The ability of certain metals to undergo unusually large amounts of plastic deformation before local necking occurs.

supersonic. Pertains to phenomena in which the speed is higher than that of sound. Not synonymous with ultrasonic; see *ultrasonic frequency*.

support pins. Rods or pins of accurate length used to support the overhang of irregularly shaped punches.

support plate. A plate that supports the draw ring or draw plate. It also serves as a spacer.

surface checking. Same as *checks*.

surface finish. (1) Condition of a surface as a result of a final treatment. (2) Measured surface profile characteristics, the preferred term being *roughness*.

surface grinding. Producing a plane surface by grinding.

surface hardening. A generic term covering several processes applicable to a suitable ferrous alloy that produces, by quench hardening only, a surface layer that is harder or more wear resistant than the core. There is no significant alteration of the chemical composition of the surface layer. The processes commonly used are induction hardening, flame hardening and shell hardening. Use of the applicable specific process name is preferred.

surface roughness. See *roughness*.

surface tension. Interfacial tension between two phases, one of which is a gas.

surfacing. The deposition of filler metal on a metal surface by welding, spraying or braze welding, to obtain certain desired properties or dimensions. See also *hard facing*.

Surfacing weld. A type of weld composed of one or more stringer or weave beads deposited on an unbroken surface to obtain desired properties or dimensions.

swaging. Tapering bar, rod, wire or tubing by forging, hammering or squeezing; reducing a section by progressively tapering lengthwise until the entire section attains the smaller dimension of the taper.

swarf. Intimate mixture of grinding chips and fine particles of abrasive and bond resulting from a grinding operation.

sweat. Exudation of a low-melting phase during solidification. Also known as sweatback. For tin bronzes, it is called tin sweat.

sweating. A soldering technique in which two or more parts are precoated (tinned), then reheated and joined without adding more solder. Also called sweat soldering.

sweating out. Bringing small globules of one of the low-melting constituents of an alloy to the surface during heat treatment, as lead out of bronze.

sweep. A form or template used for shaping sand molds or cores by hand.

sweeps. Floor and table sweepings containing precious metal particles.

sweet roast. Same as *dead roast*.

swing forging machine. Equipment for continuously hot reducing ingots, blooms, or billets to square flats, rounds, or rectangles by the crank-driven oscillating action of paired dies.

swing-frame grinder. A grinding machine suspended by a chain at the center point so that it may be turned and swung in any direction for the grinding of billets, large castings, or other heavy work. Principal use is removing surface imperfections and roughness.

synchronous timing. In spot, seam or projection welding, a method of regulating the welding transformer primary current so that all the following conditions will prevail: (*a*) The first half-cycle is initiated at the proper time in relation to the voltage to ensure a balanced current wave; (*b*) each succeeding half-cycle is essentially identical to the first; and (*c*) the last half-cycle is of opposite polarity to the first.

syntectic. An isothermal reversible reaction in which a solid phase, on absorption of heat, is converted to two conjugate liquid phases.

synthetic cold rolled sheet. A hot rolled pickled sheet given a sufficient final temper pass to impart a surface approximating that of cold rolled steel.

T

tacking. Making *tack welds*.

tack welds. Small scattered welds made to hold parts of a weldment in proper alignment while the final welds are being made.

taconite. A siliceous iron formation from which certain iron ores of the Lake Superior region are derived; consists chiefly of fine-grain silica mixed with magnetite and hematite.

tailings. The discarded portion of a crushed ore, separated during concentration.

tandem die. Same as *follow die*.

tandem mill. A rolling mill consisting of two or more stands arranged so that the metal being processed travels in a straight line from stand to stand. In continuous rolling, the various stands are synchronized so that the strip may be rolled in all stands simultaneously. Contrast with *single-stand mill*.

tandem welding. Arc welding in which two or more electrodes are in a plane parallel to the line of travel.

tangent bending. Forming one or more identical bends having parallel axes by wiping sheet metal around one or more radius dies in a single operation. The sheet, which may have side flanges, is clamped against the radius die, then made to conform to the radius die by pressure from a rocker-plate die that moves along the periphery of the radius die.

tangent modulus. See *modulus of elasticity*.

tank voltage. The total voltage between the anode and cathode of a plating bath or electrolytic cell during electrolysis. It is equal to the sum of: (*a*) the equilibrium reaction potential, (*b*) the *IR* drop, and (*c*) the electrode potentials.

tap. A cylindrical or conical thread-cutting tool with one or more cutting elements having threads of a desired form on the periphery. By a combination of rotary and axial motions, the leading end cuts an internal thread, the tool deriving its principal support from the thread being produced.

tap density. The apparent density of a metal powder, obtained when the volume receptacle is tapped or vibrated during loading under specified conditions.

tapping. (1) Opening the outlet of a melting furnace to remove molten metal. (2) Removing molten metal from a furnace. (3) Cutting internal threads with a tap.

tarnish. Surface discoloration of a metal caused by formation of a thin film of corrosion product.

Taylor process. A process for making extremely fine wire by inserting a piece of larger diameter wire into a glass tube and stretching the two together at high temperature.

technical cohesive strength. Fracture stress in a notch tensile test. Often used instead of merely "cohesive strength" to avoid confusion among the several definitions of cohesive strength.

tee joint. A joint in which the members are oriented in the form of a T.

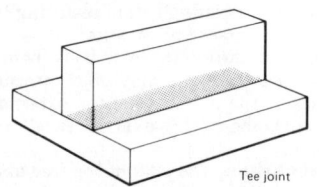

Tee joint

teeming. Pouring molten metal from a ladle into ingot molds. The term applies particularly to the specific operation of pouring either iron or steel into ingot molds.

temper. (1) In heat treatment, reheating hardened steel or hardened cast iron to some temperature below the eutectoid temperature for the purpose of decreasing hardness and increasing toughness. The process also is sometimes applied to normalized steel. (2) In tool steels, temper is sometimes used, but inadvisedly, to denote the carbon content. (3) In nonferrous alloys and in some ferrous alloys (steels that cannot be hardened by heat treatment), the hardness and strength produced by mechanical or thermal treatment, or both, and characterized by a certain structure, mechanical properties, or reduction in area during cold working. (4) To moisten sand for casting molds with water.

temper brittleness. Brittleness that results when certain steels are held within, or are cooled slowly through, a certain range of temperature below the transformation range. The brittleness is manifested as an upward shift in ductile-to-brittle transition temperature, but only rarely produces a low value of reduction of area in a smooth-bar tension test of the embrittled material.

temper carbon. Same as *annealing carbon.*

temper color. A thin, tightly adhering oxide skin (only a few molecules thick) that forms when steel is tempered at a low temperature, or for a short time, in air or a mildly oxidizing atmosphere. The color, which ranges from straw to blue depending on the thickness of the oxide skin, varies with both tempering time and temperature.

temper rolling. Light cold rolling of sheet steel. This operation is performed to improve flatness, minimize the tendency to stretcher strain and flute, and obtain the desired texture and mechanical properties.

temper time. In resistance welding, that part of the postweld interval during which the current is suitable for tempering or heat treatment.

tensile strength. In tensile testing, the ratio of maximum load to original cross-sectional area. Also called *ultimate strength.* Compare with *yield strength.*

terminal phase. A solid solution having a restricted range of compositions, one end of the range being a pure component of an alloy system.

ternary alloy. An alloy that contains three principal elements.

terne. An alloy of lead containing 3 to 15% tin, used as a hot dip coating for steel sheet or plate. Terne coatings, which are smooth and dull in appearance, give the steel better corrosion resistance and enhance its ability to be formed, soldered or painted.

tertiary creep. See *creep.*

texture. In a polycrystalline aggregate, the state of distribution of crystal orientations. In the usual sense, it is synonymous with *preferred orientation.*

thermal analysis. A method for determining transformations in a metal by noting the temperatures at which thermal arrests occur. These arrests are manifested by changes in slope of the plotted or mechanically traced heating and cooling curves. When

such data are secured under nearly equilibrium conditions of heating and cooling, the method is commonly used for determining certain critical temperatures required for the construction of equilibrium diagrams.

thermal electromotive force. The electromotive force generated in a circuit containing two dissimilar metals when one junction is at a temperature different from the other. See also *thermocouple.*

thermal fatigue. Fracture resulting from the presence of temperature gradients that vary with time in such a manner as to produce cyclic stresses in a structure.

thermal shock. The development of a steep temperature gradient and accompanying high stresses within a structure.

thermal spraying. A group of welding or allied processes in which finely divided metallic or nonmetallic materials are deposited in a molten or semimolten condition to form a coating. The coating material may be in the form of powder, ceramic rod, wire or molten materials. See also *flame spraying, plasma spraying.*

thermal stresses. Stresses in metal resulting from nonuniform temperature distribution.

thermit reactions. Strongly exothermic self-propagating reactions such as that where finely divided aluminum reacts with a metal oxide. A mixture of aluminum and iron oxide produces sufficient heat to weld steel, the filler metal being produced in the reaction.

thermit welding. Welding with heat produced by the reaction of aluminum with a metal oxide. Filler metal, if used, is obtained from the reduction of an appropriate oxide.

thermocouple. A device for measuring temperatures, consisting of lengths of two dissimilar metals or alloys that are electrically joined at one end and connected to a voltage-measuring instrument at the other end. When one junction is hotter than the other, a *thermal electromotive force* is produced that is roughly proportional to the difference in temperature between the hot and cold junctions.

thermomechanical working. A general term covering a variety of processes combining controlled thermal and deformation treatments to obtain synergistic effects such as improvement in strength without loss of toughness. Same as thermal-mechanical treatment.

thief. In electroplating, same as robber.

Thomas converter. A Bessemer converter having a basic bottom and lining, usually dolomite, and employing a basic slag.

three-point bending. Bending a piece of metal, or a structural member, in which the object is placed across two supports and force is applied between and in opposition to them. See *V-bend die*.

three-quarters hard. A *temper* of nonferrous alloys and some ferrous alloys characterized by tensile strength and hardness about midway between those of *half hard* and *full hard* tempers.

throat depth. On a resistance-welding machine, the distance from the centerline of the electrodes or platens to the nearest point of interference for flat work.

throat of a fillet weld. (theoretical) The distance from the beginning of the root of the joint perpendicular to the hypotenuse of the largest right triangle that can be inscribed within the fillet-weld cross section. (actual) The shortest distance from the root of a fillet to its face. (effective) The minimum distance from the root of the weld to its face, minus any reinforcement. See sketches accompanying *concave fillet weld, convex fillet weld.*

through weld. A nonpreferred term sometimes used to indicate a weld of substantial length made by melting through one member of a lap or tee joint and into the other member.

throwing power. The ability of a plating solution to produce a uniform metal distribution on an irregularly shaped cathode. Compare with *covering power*.

tiger stripes. Continuous bright lines on sheet or strip in the rolling direction.

tight fit. A loosely defined fit of slight negative allowance requiring a light press or driving force to assemble.

TIG welding. Tungsten inert-gas welding; see preferred term, *gas tungsten-arc welding*.

tilt boundary. A subgrain boundary consisting of an array of edge *dislocations*.

tilt mold. A casting mold, usually a book mold, that rotates from a horizontal to a vertical position during pouring, which reduces agitation and

thus the formation and entrapment of oxides.

tilt mold ingot. An ingot made in a *tilt mold*.

time quenching. Interrupted quenching in which the time in the quenching medium is controlled.

tinning. Coating metal with a very thin layer of molten solder or brazing filler metal.

tin pest. A polymorphic modification of tin that causes it to crumble into a powder known as gray tin. It is generally accepted that the maximum rate of transformation occurs at about $-40\ °C\ (-40\ °F)$, but transformation can occur as high as about 13 °C (55 °F).

tin sweat. See *sweat*.

tin tossing. Oxidizing impurities in molten tin by pouring it from one vessel to another in air, forming a dross that is mechanically separable.

TIR. Abbreviation for *total indicator reading*.

TIV. Abbreviation for *total indicator variation*.

toe crack. A base-metal crack at the *toe of weld*.

toe of weld. The junction between the face of a weld and the base metal. See sketch accompanying *fillet weld*.

toggle press. A mechanical press in which the slide is actuated by one or more toggle links or mechanisms.

tolerance. The specified permissible deviation from a specified nominal dimension, or the permissible variation in size or other quality characteristic of a part.

tolerance limits. The boundaries that define the range of permissible variation in size or other *quality characteristic* of a part.

tong hold. The portion of a forging billet, usually on one end, that is gripped by the operator's tongs. It is removed from the part at the end of the forging operation. Common to drop-hammer and press-type forging.

tool steel. Any of a class of carbon and alloy steels commonly used to make tools. Tool steels are characterized by high hardness and resistance to abrasion, often accompanied by high toughness and resistance to softening at elevated temperature. These attributes are generally attained with high carbon and alloy contents.

tooth. (1) A projection on a multipoint tool (such as on a saw, milling cutter or file) designed to produce cutting.

(2) A projection on the periphery of a wheel or segment thereof (as on a gear, spline or sprocket, for example), designed to engage another mechanism and thereby transmit force or motion, or both. A similar projection on a flat member such as a rack.

tooth point. On a face mill, the chamfered cutting edge of the blade, to which a flat is sometimes added to produce a shaving effect and to improve finish. See sketch accompanying *face mill*.

top-and-bottom process. A process for separating copper and nickel, in which their molten sulfides are separated into two liquid layers by the addition of sodium sulfide. The lower layer holds most of the nickel.

torch. A gas burner used to solder, braze, weld or cut metals. For brazing or welding, it has two gas feed lines: one for fuel, such as acetylene or hydrogen, the other for oxygen. For cutting, there may be an additional feed line for oxygen. See *oxygen cutting*.

torch brazing. Brazing in which the heat is supplied by a fuel gas flame emanating from a *torch*.

torsion. A twisting action resulting in shear stresses and strains.

torsional moment. In a body being twisted, the algebraic sum of the couples or the moments of the external forces about the axis of twist, or both.

total carbon. The sum of the free and combined carbon (including carbon in solution) in a ferrous alloy.

total cyanide. Cyanide content of an electroplating bath (including both simple and complex ions).

total indicator reading. See preferred term, *total indicator variation*.

total indicator variation. The difference between the maximum and minimum indicator readings during a checking cycle.

toughness. Ability of a metal to absorb energy and deform plastically before fracturing. It is usually measured by the energy absorbed in a notch impact test, but the area under the stress-strain curve in tensile testing is also a measure of toughness.

tough pitch copper. Copper containing from 0.02 to 0.05% oxygen, obtained by refining copper in a reverberatory furnace.

traffic mark. See *abrasion*.

tramp alloys. Residual alloying elements that are introduced into steel when unidentified alloy steel is present in the scrap charge to a steelmaking furnace.

transcrystalline. Same as *intracrystalline*.

transference. The movement of ions through the electrolyte associated with the passage of the electric current. Also called transport or migration.

transference number. The proportion of total electroplating current carried by ions of a given kind. Also called transport number.

transformation-induced plasticity. A phenomenon, occurring chiefly in certain highly alloyed steels that have been heat treated to produce metastable austenite or metastable austenite plus martensite, whereby, on subsequent deformation, part of the austenite undergoes strain-induced transformation to martensite. Steels capable of transforming in this manner, commonly referred to as TRIP steels, are highly plastic after heat treatment, but exhibit a very high rate of strain hardening and thus have high tensile and yield strengths after plastic deformation at temperatures between about 20 and 500 °C (70 and 930 °F). Cooling to −195 °C (−320 °F) may or may not be required to complete the transformation to martensite. Tempering usually is done following transformation.

transformation ranges. Those ranges of temperature within which a phase forms during heating and transforms during cooling. The two ranges are distinct, sometimes overlapping but never coinciding. The limiting temperatures of the ranges depend on the composition of the alloy and on the rate of change of temperature, particularly during cooling. See *transformation temperature*.

transformation temperature. The temperature at which a change in phase occurs. The term is sometimes used to denote the limiting temperature of a transformation range. The following symbols are used for iron and steels:

Ac_{cm}. In hypereutectoid steel, the temperature at which the solution of cementite in austenite is completed during heating.

Ac_1. The temperature at which austenite begins to form during heating.

Ac_3. The temperature at which transformation of ferrite to austenite is completed during heating.

Ac_4. The temperature at which austenite transforms to delta ferrite during heating.

Ae_{cm}, Ae_1, Ae_3, Ae_4. The temperatures of phase changes at equilibrium.

Ar_{cm}. In hypereutectoid steel, the temperature at which precipitation of cementite starts during cooling.

Ar_1. The temperature at which transformation of austenite to ferrite or to ferrite plus cementite is completed during cooling.

Ar_3. The temperature at which austenite begins to transform to ferrite during cooling.

Ar_4. The temperature at which delta ferrite transforms to austenite during cooling.

Ar'. The temperature at which transformation of austenite to pearlite starts during cooling.

M_f. The temperature at which transformation of austenite to martensite finishes during cooling.

M_s (or Ar''). The temperature at which transformation of austenite to martensite starts during cooling.

NOTE: All these changes except the formation of martensite occur at lower temperatures during cooling than during heating, and depend on the rate of change of temperature.

transgranular. Same as *intracrystalline*.

transitional fit. A fit that may have either clearance or interference resulting from specified tolerances on hole and shaft.

transition lattice. An unstable crystallographic configuration that forms as an intermediate step in a solid-state reaction such as precipitation from solid solution or eutectoid decomposition.

transition metal. A metal in which the available electron energy levels are occupied in such a way that the *d*-band contains less than its maximum number of ten electrons per atom; for example, iron, cobalt, nickel, and tungsten. The distinctive properties of the transition metals result from the incompletely filled *d*-levels.

transition point. At a stated pressure, the temperature (or at a stated temperature, the pressure) at which two solid phases exist in equilibrium; that is, an allotropic transformation temperature (or pressure).

transition temperature. (1) An arbitrarily defined temperature that lies within the temperature range in which metal fracture characteristics (as usually determined by tests of notched specimens) change rapidly, such as from primarily fibrous (shear) to primarily crystalline (cleavage) fracture. Commonly used definitions are "transition temperature for 50% cleavage fracture", "10 ft·lb transition temperature," and "transition temperature for half maximum energy". (2) Sometimes used to denote an arbitrarily defined temperature within a range in which the ductility changes rapidly with temperature.

transport. See *transference*.

transport number. Same as *transference number*.

transverse. Literally, "across", usually signifying a direction or plane perpendicular to the direction of working. In rolled plate or sheet, the direction across the width is often called long transverse, and the direction through the thickness, short transverse.

transverse rolling machine. Equipment for producing complex preforms or finished forgings from round billets inserted transversely between two or three rolls that rotate in the same direction and drive the billet. The rolls, carrying replaceable die segments with appropriate impressions, make several revolutions for each rotation of the workpiece.

trees. Visible projections of electrodeposited metal formed at sites of high current density.

trepanning. A type of boring where an annular cut is made into a solid material with the coincidental formation of a plug or solid cylinder.

triaxiality. In a *triaxial stress* state, the ratio of the smallest to the largest principal stress, all stresses being tension.

triaxial stress. A state of stress in which none of the three *principal stresses* is zero.

tribology. The science and art concerned with the design, friction, lubrication and wear of contacting surfaces that move relative to each other (as in bearings, cams or gears, for example).

trimmer blades. The portion of trimmers through which a forging is pushed to shear off the flash.

trimmer punch. The upper portion of trimmers, which comes·in contact with a forging and pushes it through the trimmer blades. The lower end of the trimmer punch is generally shaped to fit the surface of the forging against which it pushes.

trimmers. The combination of trimmer punch, trimmer blades and perhaps trimmer shoe used to remove the flash from the forging.

trimming. (1) In drawing, shearing the irregular edge of the drawn part. (2) In forging or die casting, removing any parting-line flash and gates from the part by shearing. (3) In casting, the removal of gates, risers and fins.

trimming shoe. The holder used to support a trimmer. Sometimes called trimming chair.

triple-action press. A mechanical or hydraulic press having three slides with three motions properly synchronized for triple-action drawing, redrawing and forming. Usually, two slides — the blank-holder slide and the plunger — are located above and a lower slide is located within the bed of the press.

triple point. A point on a phase diagram where three phases of a substance coexist in equilibrium.

tripoli. Friable and dustlike silica used as an abrasive.

TRIP steel. A commercial steel product exhibiting *transformation-induced plasticity.*

trommel. A revolving cylindrical screen used in grading coarsely crushed ore.

troostite. (obsolete) A previously unresolvable rapidly etching fine aggregate of carbide and ferrite produced either by tempering martensite at low temperature or by quenching a steel at a rate slower than the critical cooling rate. Preferred terminology for the first product is tempered martensite; for the latter, fine pearlite.

true current density. See preferred term, *local current density.*

true rake. See preferred term, *effective rake.*

true strain. See *strain.*

true stress. See *stress.*

tube reducing. Reducing both the diameter and wall thickness of tubing with a mandrel and a pair of rolls with tapered grooves. The Rockrite process uses a fixed tapered mandrel, and the rolls reciprocate along the tubing with corresponding reversal in rotation. Roll reliefs at the initial and final diameters permit, respectively, advance and rotation of the tubing. The Pilger process uses a uniform rod (broach), which reciprocates with the tubing. The fixed rolls rotate

continuously. During the gap in each revolution, the tubing is advanced and rotated and then, upon roll contact, reduced and partially returned.

tube sinking. Drawing tubing through a die or passing it through rolls without the use of an interior tool (such as a mandrel or plug) to control inside diameter; sinking generally produces a tube of increased wall thickness and length.

tube stock. A semifinished tube suitable for subsequent reduction and finishing.

tumbling. Rotating workpieces, usually castings or forgings, in a barrel partly filled with metal slugs or abrasives, to remove sand, scale or fins. It may be done dry, or with an aqueous solution added to the contents of the barrel. Sometimes called rumbling or rattling.

tungsten inert-gas welding. See preferred term, *gas tungsten-arc welding.*

Turk's-head rolls. Four undriven working rolls, arranged in a square or rectangular pattern, through which strip, wire or tubing is drawn to form square or rectangular sections.

turning. Removing material by forcing a cutting tool (often a *single-point tool*) against the surface of a rotating workpiece. The tool may or may not be moved toward or along the axis of rotation while it cuts away material.

tuyere. An opening in the shell and refractory lining of a furnace, through which air is forced.

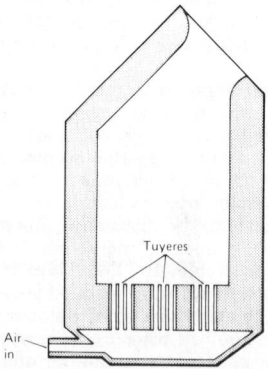

Tuyeres

Air in

twin. Two portions of a crystal having a definite crystallographic relationship; one may be regarded as the parent, the other as the twin. The orientation of the twin is either a mirror

image of the orientation of the parent about a "twinning plane" or an orientation that can be derived by rotating the twin portion about a "twinning axis". See also *annealing twin, mechanical twin.*

twist boundary. A subgrain boundary consisting of an array of screw *dislocations.*

two-high mill. A type of rolling mill in which only two rolls, the working rolls, are contained in a single housing. Compare with *four-high mill, cluster mill.*

type metal. Any of a series of alloys containing 54 to 95% Pb, 2 to 28% Sb and 2 to 20% Sn, used to make printing type.

U

U-bend die. A die, commonly used in press-brake forming, machined horizontally with a square or rectangular cross-sectional opening that provides two edges over which metal is drawn into a channel shape.

Ugine-Sejournet process. A direct extrusion process for metals that uses molten glass to insulate the hot billet and to act as a lubricant.

ultimate strength. The maximum conventional stress (tensile, compressive or shear) that a material can withstand.

ultrasonic beam. A beam of acoustical radiation with a frequency higher than the frequency range for audible sound — that is, above about 20 kHz.

ultrasonic cleaning. Immersion cleaning aided by ultrasonic waves that cause microagitation.

ultrasonic frequency. A frequency, associated with elastic waves, that is greater than the highest audible frequency, generally regarded as being higher than 15 kHz.

ultrasonic machining. A form of abrasive machining in which a tool vibrating at ultrasonic frequency causes a grit-loaded slurry to impinge on the surface of a workpiece, and thereby remove material.

ultrasonic testing. A nondestructive test applied to sound-conductive materials having elastic properties for the purpose of locating inhomogeneities or structural discontinuities within a material by means of an *ultrasonic beam.*

ultrasonic welding. A solid state process in which materials are welded by locally applying high-frequency vibratory energy to a joint held together under pressure.

underbead crack. A subsurface crack in the base metal near a weld.

undercooling. Same as *supercooling*.

undercut. (1) In weldments, a groove melted into the base metal adjacent to the toe of a weld and left unfilled. (2) For castings or forgings, same as *back draft*.

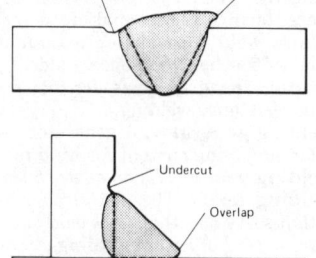

underdraft. A condition wherein a metal curves downward on leaving a set of rolls because of higher speed in the upper roll.

underfill. A portion of a forging that has insufficient metal to give it the true shape of the impression.

understressing. Applying a cyclic stress lower than the *endurance limit*. This may improve fatigue life if the member is later cyclically stressed at levels above the endurance limit.

uniaxial stress. A state of stress in which two of the three principal stresses are zero.

uniform strain. The strain occurring prior to the beginning of localization of strain (necking); the strain to maximum load in the tension test.

unit cell. In crystallography, the fundamental building block of a space lattice. Space lattices are constructed by stacking identical unit cells — that is, parallelepipeds of identical size, shape and orientation, each having a lattice point at every corner — face to face in perfect three-dimensional alignment.

unit die. A *die block* that contains several cavity inserts for making different kinds of castings.

unit power. The net amount of power required during machining to remove a unit volume of metal in unit time.

universal forging mill. A combination of four hydraulic presses arranged in one plane equipped with billet manipulators and automatic controls, used for radial or draw forging.

universal mill. A rolling mill in which rolls with a vertical axis roll the edges of the metal stock between some of the passes through the horizontal rolls.

upset. (1) The localized increase in cross-sectional area of a workpiece or weldment resulting from the application of pressure during mechanical fabrication or welding. (2) That portion of a welding cycle during which the cross-sectional area is increased by the application of pressure.

upset forging. A forging obtained by upset of a suitable length of bar, billet or bloom.

upsetter. A horizontal mechanical press used to make parts from bar stock or tubing by upset forging, piercing, bending or otherwise forming in dies. Also known as a header.

upsetting. Working metal so that the cross-sectional area of a portion or all of the stock is increased. See also *heading*.

upset welding. A resistance welding process in which the weld is produced, simultaneously over the entire area of abutting surfaces or progressively along a joint, by applying mechanical force (pressure) to the joint, then causing electrical current to flow across the joint to heat the abutting surfaces. Pressure is maintained throughout the heating period. See also *open-gap upset welding*.

upslope time. In resistance welding, time associated with current increase using *slope control*.

V

vacancy. A type of lattice imperfection in which an individual atom site is temporarily unoccupied. Diffusion (of other than interstitial solutes) is generally visualized as the shifting of vacancies.

vacuum arc remelting. A *consumable electrode remelting* process in which heat is generated by an electric arc between the electrode and the ingot. The process is performed inside a vacuum chamber. Exposure of the droplets of molten metal to the reduced pressure reduces the amount of dissolved gas in the metal. Sometimes abbreviated VAR.

vacuum deposition. Condensation of thin metal coatings on the cool surface of work in a vacuum.

vacuum fusion. An analytic technique for determining the amount of gases in metals; ordinarily used for hydrogen and oxygen, and sometimes for nitrogen. Applicable to many metals, but not to alkali or alkaline earth metals.

vacuum induction melting. A process for remelting and refining metals in which the metal is melted inside a vacuum chamber by induction heating. The metal may be melted in a crucible, then poured into a mold. The process may also be operated in a configuration similar to that used in *consumable electrode remelting* except that the heat is supplied by an induction heating coil rather than from the passage of electric current through the electrode. Sometimes abbreviated VIM.

vacuum melting. Melting in a vacuum to prevent contamination from air, as well as to remove gases already dissolved in the metal; the solidification may also be carried out in a vacuum or at low pressure.

vapor blasting. Same as *liquid honing*.

vapor degreasing. Degreasing of work in the vapor over a boiling liquid solvent, the vapor being considerably heavier than air. At least one constituent of the soil must be soluble in the solvent. Modifications of this cleaning process include vapor–spray–vapor; warm liquid–vapor; boiling liquid–warm liquid–vapor; and ultrasonic degreasing.

vapor plating. Deposition of a metal or compound on a heated surface by reduction or decomposition of a volatile compound at a temperature below the melting points of the deposit and the base material. The reduction is usually accomplished by a gaseous reducing agent such as hydrogen. The decomposition process may involve thermal dissociation or reaction with the base material. Occasionally used to designate deposition on cold surfaces by vacuum evaporation—see *vacuum deposition*.

V-bend die. A die commonly used in press-brake forming, usually machined with a triangular cross-sectional opening to provide two edges as fulcrums for accomplishing three-point bending.

vector field. Same as *resultant field*.

Vegard's law. The relationship that states that the lattice parameters of substitutional solid solutions vary linearly between the values for the components, with composition expressed in atomic percentage.

veining. A type of subboundary structure that can be delineated because of the presence of a greater than average concentration of precipitate or possibly solute atoms.

vent. A small opening in a mold for the escape of gases.

vermicular iron. Same as *compacted graphite cast iron*.

vertical-position welding. Welding where the axis of the weld is essentially vertical.

vibratory finishing. A process for deburring and surface finishing in which the product and an abrasive mixture are placed in a container and vibrated.

Vickers hardness test. An indentation hardness test employing a 136° diamond pyramid indenter (Vickers) and variable loads enabling the use of one hardness scale for all ranges of hardness from very soft lead to tungsten carbide.

virgin metal. Same as *primary metal*.

voltage efficiency. The ratio, usually expressed as a percentage, of the equilibrium-reaction potential in a given electrochemical process to the bath voltage.

W

Wallner lines. A distinct pattern of intersecting sets of parallel lines, usually producing a set of V-shaped lines, sometimes observed when viewing brittle fracture surfaces at high magnifications in an electron microscope. Wallner lines are attributed to interaction between a shock wave and a brittle crack front propagating at high velocity. Sometimes Wallner lines are misinterpreted as fatigue striations.

wandering sequence. Same as *random sequence*.

warm working. Plastically deforming metal at a temperature above ambient (room) temperature but below the temperature at which the material undergoes recrystallization.

wash. (1) A coating applied to the face of a mold prior to casting. (2) An imperfection at a cast surface similar to a *cut*.

wash metal. Molten metal used to wash out a furnace, ladle or other container.

water break. The appearance of a discontinuous film of water on a surface signifying nonuniform wetting and usually associated with a surface contamination.

waviness. A wave-like variation from a perfect surface, generally much larger and wider than the *roughness* caused by tool or grind marks.

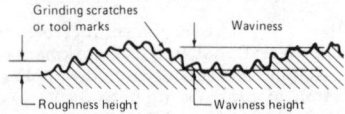

wear pad. In forming, an expendable rubber or rubberlike material of nominal thickness that is placed against the diaphragm to lessen the wear on it. See *diaphragm,* (2).

weave bead. A weld bead made with oscillations transverse to the axis of the weld. Contrast with *stringer bead*.

web. (1) For twist drills and reamers, the central portion of the tool body that joins the lands. (2) In forging, the thin section of metal remaining at the bottom of a cavity or depression or at the location of the top and bottom punches. The former type may be removed by piercing or machining; the latter, by the trim punch. (3) A plate or thin portion between stiffening ribs or flanges, as in an I-beam, H-beam or other similar section.

weight percent. Percentage composition by weight. Contrast with *atomic percent*.

weld. A union made by *welding*.

weldability. A specific or relative measure of the ability of a material to be welded under a given set of conditions. Implicit in this definition is the ability of the completed weldment to fulfill all service designed into the part.

weld bead. A deposit of filler metal from a single welding *pass*.

weld crack. A crack in weld metal.

weld delay time. In spot, seam or projection welding, the time current is delayed with respect to starting the forge delay timer in order to synchronize the forging pressure and the welding heat.

welder. A person who makes welds using manual or semiautomatic equipment. Formerly used as a synonym for *welding machine*.

weld gage. A device for checking shape and size of welds.

welding. (1) Joining two or more pieces of material by applying heat or pressure, or both, with or without filler material, to produce a localized union through fusion or recrystallization across the interface. The thickness of the filler material is much greater than the capillary dimensions encountered in *brazing*. (2) May also be extended to include brazing and soldering.

welding current. The current flowing through a welding circuit during the making of a weld. In resistance welding, the current used during preweld or postweld intervals is excluded.

welding cycle. The complete series of events involved in making a resistance weld. Also applies to semiautomatic mechanized fusion welds.

welding force. Same as *electrode force* in resistance welding.

welding generator. A generator used for supplying current for welding.

welding ground. Same as *work lead*.

welding leads. The electrical cables that serve as either *work lead* or *electrode lead* of an arc welding circuit.

welding machine. Equipment used to perform the welding operation—for example, spot welding machine, arc welding machine, seam welding machine.

welding procedure. The detailed methods and practices, including joint preparation and welding procedures, involved in the production of a *weldment*.

welding rod. Welding or brazing filler metal, usually in rod or wire form, but not a *consumable electrode*. Welding rod does not conduct the electric current to an arc, and may be either fed into the weld puddle or preplaced in the joint.

welding sequence. The order in which the various component parts of a weldment or structure are welded.

welding stress. *Residual stress* caused by localized heating and cooling during welding.

welding technique. The details of a welding operation that, within the limitations of a welding procedure, are performed by the *welder*.

welding tip. (1) A torch tip designed for welding. (2) The electrode tip that contacts the work in resistance spot welding.

weld interval. The total heat and cool times in making one multiple-impulse resistance weld.

weld-interval timer. A device used in resistance welding to control heat and cool times and weld interval when making multiple-impulse welds singly or simultaneously.

weld line. The junction of the weld metal and the base metal, or the junction of the base-metal parts when filler metal is not used.

weldment. An assembly whose component parts are joined by welding.

weld metal. That portion of a weld that has been melted during welding.

weld nugget. The weld metal in spot, seam or projection welding.

weldor. (obsolete) Formerly used to designate a person who makes welds. See preferred term, *welder*.

weld time. In single-impulse and flash welding, the time that the welding current is applied to the work.

weld timer. A device used in resistance welding to control the weld time only.

Wenstrom mill. A rolling mill similar to a universal mill but where the edges and sides of a rolled section are acted on simultaneously.

wet blasting. A process for cleaning or finishing by means of a slurry of abrasive in water directed at high velocity against the workpieces.

wetting. A condition in which the interfacial tension between a liquid and a solid is such that the contact angle is 0° to 90°.

wetting agent. A surface-active agent that produces *wetting* by decreasing the *cohesion* within the liquid.

whiskers. Metallic filamentary growths, often microscopic, sometimes formed during electrodeposition and sometimes spontaneously during storage or service, after finishing.

white cast iron. *Cast iron* that shows a white fracture because the carbon is in combined form.

whiteheart malleable. See *malleable cast iron*.

white metal. (1) A general term covering a group of white-colored metals of relatively low melting points (lead, antimony, bismuth, tin, cadmium and zinc) and of the alloys based on these metals. (2) A copper matte of about 77% Cu obtained from smelting of sulfide copper ores.

white rust. Zinc oxide; the powdery product of corrosion of zinc or zinc-coated surfaces.

Widmanstätten structure. A structure characterized by a geometrical pattern resulting from the formation of a new phase along certain crystallographic planes of the parent solid solution. The orientation of the lattice in the new phase is related crystallographically to the orientation of the lattice in the parent phase. The structure was originally observed in

meteorites, but is readily produced in many other alloys by appropriate heat treatment.

wildness. A condition that exists when molten metal, during cooling, evolves so much gas that it becomes violently agitated, forcibly ejecting metal from the mold or other container.

Williams riser. An *atmospheric riser*.

winning. Recovering a metal from an ore or chemical compound using any suitable hydrometallurgical, pyrometallurgical or electrometallurgical method.

wiped coat. A hot dipped galvanized coating where virtually all free zinc is removed by wiping prior to solidification, leaving only a thin zinc-iron alloy layer.

wiped joint. A joint wherein filler metal is applied in liquid form and distributed by mechanical action.

wiper forming. A method of curving bars, tubes or rolled or extruded sections, in which the stock is bent so that it conforms to a fixed form block. Stock is clamped to the form block, then bent by applying force through a wiper block, shoe or roll that is moved along the periphery of the form block. Sometimes called compression forming. Contrast with *draw forming*.

wiping effect. Activation of a metal surface by mechanically rubbing or wiping to enhance the formation of conversion coatings, such as phosphate coatings.

wire. (1) A thin, flexible, continuous length of metal, usually of circular cross section, and usually produced by drawing through a die. See also *flat wire*. (2) A length of single metallic electrical conductor; it may be of solid, stranded or tinsel construction, and may be either bare or insulated.

wire bar. A cast shape, particularly of tough pitch copper, that has a cross section approximately square with tapered ends, designed for hot rolling to rod for subsequent drawing into wire.

wiredrawing. Reducing the cross section of wire by pulling it through a die. See *Taylor process*.

wire rod. Hot rolled coiled stock that is to be cold drawn into wire.

wiring. Formation of a curl along the edge of a shell, tube or sheet and insertion of a rod or wire within the curl for stiffening the edge. See *curling*.

wood flour. A pulverized wood product used in the foundry to furnish a re-

ducing atmosphere in the mold, help overcome sand expansion, increase flowability, improve casting finish and provide easier shakeout.

woody structure. A macrostructure particulary found in wrought iron and in extruded rods of aluminum alloys that shows elongated surfaces of separation when fractured.

work angle. In arc welding, the angle between the electrode and one member of the joint, taken in a plane normal to the weld axis.

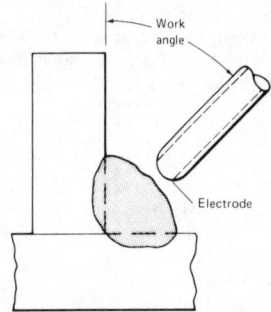

work hardening. Same as *strain hardening*.

work lead. The electrical conductor connecting the source of arc welding current to the work. Also called work connection, welding ground or ground lead.

worm. An exudation (sweat) of molten metal forced through the top crust of solidifying metal by gas evolution. See also *zinc worms*.

wrap forming. See *stretch forming*.

wringing fit. A fit of nominally zero allowance.

wrinkling. A wavy condition obtained in drawing, in the area of the metal that passes over the draw radius. Wrinkling may also occur in other forming operations when unbalanced compressive forces are set up.

wrought iron. A commercial iron consisting of slag (iron silicate) fibers entrained in a ferrite matrix.

X

x-ray. Electromagnetic radiation, of wavelength less than about 50 nm, emitted as the result of deceleration of fast-moving electrons (bremsstrahlung, continuous spectrum) or decay of atomic electrons from excited orbital states *(characteristic radiation)*; specifically, the radiation produced when an electron beam of sufficient energy impinges on a target of suitable material.

Y

Y-block. A single *keel block*.

yield point. The first stress in a material, usually less than the maximum attainable stress, at which an increase in strain occurs without an increase in stress. Only certain metals exhibit a yield point. If there is a decrease in stress after yielding, a distinction may be made between upper and lower yield points.

yield strength. The stress at which a material exhibits a specified devia-

tion from proportionality of stress and strain. An offset of 0.2% is used for many metals. Compare with *tensile strength*.

Young's modulus. See *modulus of elasticity*.

Z

zinc worms. Surface imperfections, characteristic of high-zinc brass castings, that occur when zinc vapor condenses at the mold/metal interface, where it is oxidized and then becomes entrapped in the solidifying metal.

zircon sand. A very refractory mineral, composed chiefly of zirconium

silicate; it has low thermal expansion and high thermal conductivity.

zone melting. Highly localized melting, usually by induction heating, of a small volume of an otherwise solid piece, usually a rod. By moving the induction coil along the rod, the melted zone can be transferred from one end to the other. In a binary mixture where there is a large difference in composition on the liquidus and solidus lines, high purity can be attained by concentrating one of the constituents in the liquid as it moves along the rod.

Physical Data on the Elements and Alloys

Periodic table of the elements

Key to chart

Atomic Number →	**50** +2 ← Oxidation States
Symbol →	**Sn** +4
Atomic Weight →	118.69
	-18-18-4 ← Electron Configuration

Metals — — — Nonmetals

Transition Elements

Group I and II (metals)

Group	Z	Symbol	Oxidation States	Atomic Weight	Electron Configuration	Orbit
Iᵃ	1	H	+1 −1	1.0079	1	K
Iᵃ	3	Li	+1	6.939	2-1	
IIᵃ	4	Be	+2	9.0122	2-2	
Iᵃ	11	Na	+1	22.9898	2-8-1	
IIᵃ	12	Mg	+2	24.312	2-8-2	
Iᵃ	19	K	+1	39.09	-8-8-1	
IIᵃ	20	Ca	+2	40.08	-8-8-2	
Iᵃ	37	Rb	+1	85.467	-18-8-1	
IIᵃ	38	Sr	+2	87.62	-18-8-2	
Iᵃ	55	Cs	+1	132.9054	-18-8-1	
IIᵃ	56	Ba	+2	137.3	-18-8-2	
Iᵃ	87	Fr	+1	(223)	-18-8-1	
IIᵃ	88	Ra	+2	226.0254	-18-8-2	

Transition Elements (IIIb–VIII, Ib, IIb)

Group	Z	Symbol	Oxidation States	Atomic Weight	Electron Configuration
IIIᵇ	21	Sc	+3	44.9559	-8-9-2
IVᵇ	22	Ti	+2 +3 +4	47.9	-8-10-2
Vᵇ	23	V	+2 +3 +4 +5	50.941	-8-11-2
VIᵇ	24	Cr	+2 +3 +6	51.996	-8-13-1
VIIᵇ	25	Mn	+2 +3 +4 +7	54.9380	-8-13-2
VIII	26	Fe	+2 +3	55.847	-8-14-2
VIII	27	Co	+2 +3	58.9332	-8-15-2
VIII	28	Ni	+2 +3	58.71	-8-16-2
Iᵇ	29	Cu	+1 +2	63.54	-8-18-1
IIᵇ	30	Zn	+2	65.38	-8-18-2
IIIᵇ	39	Y	+3	88.9059	-18-9-2
IVᵇ	40	Zr	+4	91.22	-18-10-2
Vᵇ	41	Nb	+3 +5	92.9064	-18-12-1
VIᵇ	42	Mo	+6	95.94	-18-13-1
VIIᵇ	43	Tc	+4 +6 +7	98.9062	-18-13-2
VIII	44	Ru	+3	101.07	-18-15-1
VIII	45	Rh	+3	102.905	-18-16-1
VIII	46	Pd	+2 +4	106.4	-18-18-0
Iᵇ	47	Ag	+1	107.868	-18-18-1
IIᵇ	48	Cd	+2	112.40	-18-18-2
IIIᵇ	57*	La	+3	138.9055	-18-9-2
IVᵇ	72	Hf	+4	178.49	-32-10-2
Vᵇ	73	Ta	+5	180.948	-32-11-2
VIᵇ	74	W	+6	183.85	-32-12-2
VIIᵇ	75	Re	+4 +6 +7	186.207	-32-13-2
VIII	76	Os	+3 +4	190.2	-32-14-2
VIII	77	Ir	+3 +4	192.9	-32-15-2
VIII	78	Pt	+2 +4	195.09	-32-16-2
Iᵇ	79	Au	+1 +3	196.9665	-32-18-1
IIᵇ	80	Hg	+1 +2	200.59	-32-18-2
IIIᵇ	89**	Ac	+3	(227)	-18-9-2
IVᵇ	104	Rf	+4	(261)	-32-10-2
Vᵇ	105	Ha		(262)	-32-11-2
VIᵇ	106			(263)	-32-12-2

Groups IIIa–0 (metals / nonmetals)

Group	Z	Symbol	Oxidation States	Atomic Weight	Electron Configuration	Orbit
0	2	He	0	4.00260	2	K
IIIᵃ	5	B	+3	10.81	2-3	
IVᵃ	6	C	+2 +4 −4	12.011	2-4	
Vᵃ	7	N	+1 +2 +3 +4 +5 −3	14.0067	2-5	
VIᵃ	8	O	−2	15.9994	2-6	
VIIᵃ	9	F	−1	18.998403	2-7	
0	10	Ne	0	10.17₉	2-8	K-L
IIIᵃ	13	Al	+3	26.98154	2-8-3	
IVᵃ	14	Si	+2 +4 −4	28.08	2-8-4	
Vᵃ	15	P	+3 +5 −3	30.97376	2-8-5	
VIᵃ	16	S	+4 +6 −2	32.06	2-8-6	
VIIᵃ	17	Cl	+1 +5 +7 −1	35.453	2-8-7	
0	18	Ar	0	39.948	2-8-8	K-L-M
IIIᵃ	31	Ga	+3	39.72	-8-18-3	
IVᵃ	32	Ge	+2 +4	72.59	-8-18-4	
Vᵃ	33	As	+3 +5 −3	74.9216	-8-18-5	
VIᵃ	34	Se	+4 +6 −2	78.96	-8-18-6	
VIIᵃ	35	Br	+1 +5 −1	79.904	-18-18-7	
0	36	Kr	0	83.80	-8-18-8	L-M-N
IIIᵃ	49	In	+3	114.82	-18-18-3	
IVᵃ	50	Sn	+2 +4	118.69	-18-18-4	
Vᵃ	51	Sb	+3 +5 −3	121.75	-18-18-5	
VIᵃ	52	Te	+4 +6 −2	127.60	-18-18-6	
VIIᵃ	53	I	+1 +5 +7 −1	126.9045	-18-18-7	
0	54	Xe	0	131.30	-18-18-8	M-N-O
IIIᵃ	81	Tl	+1 +3	204.37	-32-18-3	
IVᵃ	82	Pb	+2 +4	207.19	-32-18-4	
Vᵃ	83	Bi	+3 +5	208.980	-32-18-5	
VIᵃ	84	Po	+2 +4	(209)	-32-18-6	
VIIᵃ	85	At	−1	(210)	-32-18-7	
0	86	Rn	0	(222)	-32-18-8	N-O-P

Orbit key: N-O-P / O-P-Q

*Lanthanides

Z	Symbol	Oxidation States	Atomic Weight	Electron Configuration
58	Ce	+3 +4	140.12	-20-8-2
59	Pr	+3	140.9077	-21-8-2
60	Nd	+3	144.24	-22-8-2
61	Pm	+3	147	-23-8-2
62	Sm	+2 +3	150.4	-24-8-2
63	Eu	+2 +3	151.96	-25-8-2
64	Gd	+3	157.25	-25-9-2
65	Tb	+3	158.925	-27-8-2
66	Dy	+3	162.50	-28-8-2
67	Ho	+3	164.9304	-29-8-2
68	Er	+3	167.26	-30-8-2
69	Tm	+3	168.9342	-31-8-2
70	Yb	+2 +3	173.04	-32-8-2
71	Lu	+3	174.967	-32-9-2

**Actinides

Z	Symbol	Oxidation States	Atomic Weight	Electron Configuration
90	Th	+4	232.038	-18-10-2
91	Pa	+5 +4	231.0359	-20-9-2
92	U	+3 +4 +5 +6	238.029	-21-9-2
93	Np	+3 +4 +5 +6	237.0482	-22-9-2
94	Pu	+3 +4 +5 +6	239.052	-24-8-2
95	Am	+3 +4 +5 +6	(243)	-25-8-2
96	Cm	+3	(247)	-25-9-2
97	Bk	+3 +4	(247)	-27-8-2
98	Cf	+3	(251)	-28-8-2
99	Es	+3	(254)	-29-8-2
100	Fm	+3	(257)	-30-8-2
101	Md	+3	(258)	-31-8-2
102	No	+2 +3	(259)	-32-8-2
103	Lr	+3	(260)	-32-9-2

Numbers in parentheses are mass numbers of most stable isotope of that element

Periodic system for ferrous metallurgists

The basic principles of alloying were applied to develop the periodic system shown here, which illustrates the fundamental alloying nature of iron. The solid solubility of each element in iron can be resolved accurately with few exceptions by considering only atomic size. Alloying valence, crystal structure and electronegativity are useful supplementary factors of varying significance. The tendency to form compounds, intermetallic or ionic, is related to the difference in electronegativity which, in general, increases in a sweep from lower left to upper right of the periodic system.

Adapted Primarily for Ferrous Metallurgists

Atomic size factors (in parentheses) are % smaller (–) or larger (+) than gamma (FCC) iron at 75 F. Lattice environment (Coordination No.) is taken into account; CN is 12 except 6 for interstitials H, B, C, N & O. Groups VI, VIb, VII and VIIb form ionic compounds with the metals.

H-1 (−58) ⊗ XX

	0	Ia	IIa	IIIa	IVa	Va	VIa	VIIa	VIII	VIII	VIII	Ib	IIb	IIIb	IVb (IV)	Vb (V)	VIb (VI)	VIIb (VII)
	He-2 FCC (Others)																	
	Ne-10 FCC	Li-3 (+23) ⊗ BCC* HCP†	Be-4 (−11) ● HCP* BCC	B-5 (−29) ▲ XX											C-6 (−34) ▲ XX	N-7 (−36) ▲ XX	O-8 (−33) ▲ XX	F-9
	Ar-18 FCC	Na-11 (+50) ⊗ BCC* HCP	Mg-12 (+27) ⊗ HCP	Al-13 (+14) ⊗ FCC											Si-14 (+7) ⊗ XX	P-15 (+2) ⊗ XX	S-16 (+1) ⊗ XX	Cl-17 XX
	Kr-36 FCC	K-19 (+86) ⊗ BCC	Ca-20 (+56) ⊗ FCC* BCC	Sc-21 (+29) ⊗ HCP* BCC	Ti-22 (+16) ◐ HCP* BCC	V-23 (+6) ● BCC	Cr-24 (+1) ● BCC	Mn-25 (+1) ● XX* FCC‡	Fe-26 (0) ● BCC* FCC	Co-27 (−1) ● HCP* FCC	Ni-28 (−1) ● FCC	Cu-29 (+1) ◐ FCC	Zn-30 (+6) ● HCP	Ga-31 (+12) ⊗ XX	Ge-32 (+9) ⊗ XX	As-33 (+11) ⊗ XX	Se-34 (+11) ⊗ XX	Br-35 XX
	Xe-54 FCC	Rb-37 (+97) ⊗ BCC	Sr-38 (+71) ⊗ FCC* HCP‡	Y-39 (+42) ⊗ HCP* BCC	Zr-40 (+27) ⊗ HCP* BCC	Cb-41 (+15) ⊗ BCC	Mo-42 (+10) ● BCC	Tc-43 (+8) HCP	Ru-44 (+6) HCP	Rh-45 (+6) FCC	Pd-46 (+9) FCC	Ag-47 (+14) ◐ FCC	Cd-48 (+20) ⊗ HCP	In-49 (+25) ⊗ XX	Sn-50 (+23) ⊗ XX	Sb-51 (+27) ⊗ XX	Te-52 (+27) ⊗ XX	I-53 XX
	Rn-86 FCC	Cs-55 (+112) ⊗ BCC	Ba-56 (+76) ⊗ BCC	La-57 (+48) ⊗ HCP* FCC‡	Hf-72 (+26) ⊗ HCP* BCC	Ta-73 (+16) ● BCC	W-74 (+11) ● BCC	Re-75 (+9) HCP	Os-76 (+7) HCP	Ir-77 (+8) FCC	Pt-78 (+10) FCC	Au-79 (+14) ◐ FCC	Hg-80 (+25) ⊗ XX	Tl-81 (+36) ⊗ HCP* BCC	Pb-82 (+39) ⊗ FCC	Bi-83 (+35) ⊗ XX	Po-84 (+40) ⊗ XX	At-85
		Fr-87	Ra-88	Ac-89 (+49) ⊗ FCC														
Alloying Valence		1	2	3	4	5	6	6	6	6	6	5.56	4.56	3.56	2.56 Note 2	1.56 Note 2	(2) Note 3	(1) Note 3

Note 1: The rare-earth (lanthanide, 58-71) and actinide (90-103) series are omitted.
Note 2: Valence is 4 for C; 3 for N and P.
Note 3: (1) and (2) are not alloying valences.

Structure

BCC – BODY CENTERED CUBIC
FCC – FACE CENTERED CUBIC
HCP – HEXAGONAL CLOSE PACKED
XX – NOT BCC, FCC OR HCP
* – USUALLY MORE COMPLEX STRUCTURE AT 75 F
† – ALSO FCC ‡ – ALSO BCC

Substitutional solid solutions

● FAVORABLE SIZE FACTOR: 0 TO ± 13%
◐ BORDERLINE SIZE FACTOR: ± 14 TO ± 16%
⊗ UNFAVORABLE SIZE FACTOR: > ± 16%

Interstitial solid solutions

▲ FAVORABLE SIZE FACTOR: > (−40%)
▲ BORDERLINE SIZE FACTOR: (−30) TO (−40%)
⊿ UNFAVORABLE SIZE FACTOR: < (−30%)

Type of gamma iron (FCC) field if alloyed with iron

◣ GAMMA LOOP, LIKE Cr
◥ LIMITED GAMMA LOOP, LIKE B
▼ OPEN GAMMA REGION, LIKE Ni
⬟ LIMITED GAMMA REGION, LIKE C

General Data

Physical properties of the elements

Element	Atomic No.	Atomic weight	Density(a), g/cm³ (lb/in.³)	Melting point, °C (°F)	Boiling point, °C (°F)	Specific heat(b), cal/g · °C (J/kg · K)	Heat of fusion, cal/g (Btu/lb)
Actinium (Ac)	89	227	... (...)	1050 ± 50 (1920 ± 90)	... (...)	... (...)	... (...)
Aluminum (Al)	13	26.98	2.70 (0.0974)	660 (1220)	2450 (4442)	0.215 (900)	94.5 (170)
Americium (Am)	95	243	11.87 (0.4285)	... (...)	... (...)	... (...)	... (...)
Antimony (Sb)	51	121.76	6.65 (0.240)	630.5 ± 0.1 (1166.9 ± 0.2)	1380 (2516)	0.049 (205)	38.3 (68.9)
Argon (A)	18	39.99	1.784(g) (0.06440)(g)	−189.4 ± 0.2 (−308.9 ± 0.4)	−185.8 (−302.4)	0.125 (523)	6.7 (12)
Arsenic (As)	33	74.91	5.72 (0.206)	817(j) (1503)(j)	613(k) (1135)(k)	0.082 (343)	88.5 (159.3)
Astatine (At)	85	211	... (...)	302(m) (576)(m)	... (...)	... (...)	... (...)
Barium (Ba)	56	137.36	3.6 (0.13)	714 (1317)	1640 (2980)	0.068 (285)	... (...)
Berkelium (Bk)	97	247	... (...)	... (...)	... (...)	... (...)	... (...)
Beryllium (Be)	4	9.01	1.85 (0.0668)	1277 (2332)	2770 (5020)	0.45 (190)	260 (470)
Bismuth (Bi)	83	209.00	9.80 (0.354)	271.3 (520.3)	1560 (2840)	0.0294 (123)	12.5 (22.5)
Boron (B)	5	10.82	2.45 (0.0884)	2030(q) (3690)(q)	... (...)	0.309 (1290)	... (...)
Bromine (Br)	35	79.92	3.12 (0.113)	−7.2 ± 0.2 (19.0 ± 0.4)	58 (136)	0.070 (290)	16.2 (29.2)
Cadmium (Cd)	48	112.41	8.65 (0.312)	320.9 (609.6)	765 (1409)	0.055 (230)	13.2 (23.8)
Calcium (Ca)	20	40.08	1.55 (0.0560)	838 (1540)	1440 (2625)	0.149(u) (624)(u)	52 (93.6)
Californium (Cf)	98	251	... (...)	... (...)	... (...)	... (...)	... (...)
Carbon, graphite (C) ...	6	12.01	2.25 (0.0812)	3727(k) (6740)(k)	4830 (8730)	0.165 (691)	... (...)
Cerium (Ce)	58	140.13	6.77 (0.244)	804 (1479)	3470 (6280)	0.045 (190)	8.5 (15.9)
Cesium (Cs)	55	132.91	1.87 (0.0675)	28.7 (83.6)	690 (1273)	0.04817 (201.7)	3.8 (6.8)
Chlorine (Cl)	17	35.46	3.214(g) (0.1160)(g)	−100.99 (−149.78)	−34.7 (−30.5)	0.116 (486)	21.6 (38.9)
Chromium (Cr)	24	52.01	7.19 (0.260)	1875 (3407)	2665 (4829)	0.11 (460)	96 (173)
Cobalt (Co)	27	58.94	8.85 (0.319)	1495 ± 1 (2723 ± 1.8)	2900 (5250)	0.099 (410)	58.4 (105)
Copper (Cu)	29	63.54	8.96 (0.323)	1083.0 ± 0.1 (1981.4 ± 0.18)	2595 (4703)	0.092 (380)	50.6 (91.1)
Curium (Cm)	96	247	7 (0.3)	... (...)	... (...)	... (...)	... (...)
Dysprosium (Dy)	66	162.51	8.55 (0.309)	1407 (2565)	2330 (4230)	0.041 (170)	25.2 (45.4)
Einsteinium (E)	99	254	... (...)	... (...)	... (...)	... (...)	... (...)
Erbium (Er)	68	167.27	9.15 (0.330)	1497 (2727)	2630 (4770)	0.040 (170)	24.5 (44.1)
Europium (Eu)	63	152.0	5.24 (0.189)	826 (1519)	1490 (2710)	0.039 (160)	16.5 (29.6)
Fermium (Fm)	100	253	... (...)	... (...)	... (...)	... (...)	... (...)
Fluorine (F)	9	19.00	1.696(g) (0.06123)(g)	−219.6 (−363.3)	−188.2 (−306.8)	0.18 (750)	10.1 (18.2)
Francium (Fr)	87	223	... (...)	27(m) (81)(m)	... (...)	... (...)	... (...)
Gadolinium (Gd)	64	157.26	7.86 (0.284)	1312 (2394)	2730 (4950)	0.071 (300)	23.5 (42.4)
Gallium (Ga)	31	69.72	5.91 (0.213)	29.78 (85.60)	2237 (4059)	0.079 (330)	19.16 (34.49)
Germanium (Ge)	32	72.60	5.32 (0.192)	937.4 ± 1.5 (1719.3 ± 2.7)	2830 (5125)	0.073 (310)	... (...)
Gold (Au)	79	197.0	19.3 (0.697)	1063.0 ± 0.0 (1945.4 ± 0.0)	2970 (5380)	0.0312(jj) (131)(jj)	16.1 (29.0)

Physical properties of the elements (continued)

Symbol	Coefficient of linear thermal expansion(c), μin./in. °C (μin./in. °F)	Thermal conductivity(c), cal/cm²/cm/s/°C	Electrical resistivity, μΩ · cm	Modulus of elasticity in tension, 10⁶ psi	Lattice constants(b), Å a	b	c (or axial angle)	Closest approach of atoms
Ac........	... (...)	...	...	...	...	...	...	...
Al........	23.6(d) (13.1)(d)	0.53	2.6548(b)	9	4.0491	...	...	2.862
Am.......	... (...)	...	...	...	...	...	...	...
Sb........	8.5 − 10.8(e) (4.7 − 6)(e)	0.045	39.0(f)	11.3	4.5065	...	57°6.5′	2.904
A	... (...)	0.406×10^{-4}	...	...	5.43(h)	...	...	3.84
As........	4.7 (2.6)	...	33.3(b)	...	4.159	...	53°49′	...
At........	... (...)	...	...	...	...	...	...	...
Ba	... (...)	...	...	...	5.025	...	...	4.348
Bk........	... (...)	...	...	...	...	...	...	...
Be........	11.6(n) (6.4)(n)	0.35	4(b)(p)	40-44	2.2858	...	3.5842	...
Bi........	13.3 (7.4)	0.020	106.8(f)	4.6	4.7457	...	57° 14.2′	3.111
B.........	8.3(r) (4.6)(r)	...	1.8×10^{12}(f)	...	17.89	8.95	10.15	...
Br........	... (...)	...	...	...	4.49(s)	6.68(s)	8.74(s)	2.27
Cd	29.8 (16.55)	0.22	6.83(f)	8(t)	2.9787	...	5.617	...
Ca	22.3(v) (12.4)(v)	0.3	3.91(f)	3.2-3.8(w)	5.582	...	...	...
Cf........	... (...)	...	...	...	...	...	...	...
C.........	0.6-4.3(d) (0.3-2.4)(d)	0.057	1375(f)	0.7	2.4614	...	6.7041	1.42
Ce........	8 (4.44)	0.026(x)	75(y)	6(z)	5.16	...	...	...
Cs........	97(aa) (54)(aa)	...	20(b)	...	6.13(bb)	...	...	...
Cl........	... (...)	0.172×10^{-4}	...	...	8.58(cc)	...	6.13(cc)	1.81
Cr........	6.2 (3.4)	0.16	12.9(f)	36	2.884	...	...	2.498
Co........	13.8 (7.66)	0.165	6.24(b)	30	2.5071	...	4.0686	2.4967
Cu	16.5 (9.2)	0.941 ± 0.005	1.6730(b)	16	3.6153	...	...	2.556
Cm.......	... (...)	...	...	...	...	...	...	...
Dy	9 (5)	0.024(x)	57(y)	10-14(z)	3.59	...	5.65	...
E.........	... (...)	...	...	...	...	...	...	...
Er........	9 (5)	0.023(x)	107(y)	16(z)	3.65	...	5.58	...
Eu........	26 (14.44)	...	90(y)	...	4.58	...	...	...
Fm.......	... (...)	...	...	...	...	...	...	...
F.........	... (...)	...	...	...	...	...	...	...
Fr........	... (...)	...	...	...	...	...	...	...
Gd	4(dd) (2.22)(dd)	0.021(x)	140.5(y)	8-14(z)	3.64	...	5.78	...
Ga	18(ee) (10)(ee)	0.07-0.09(ff)	17.4(gg)	...	4.524(y)	4.523(y)	7.661(y)	2.437
Ge	5.75 (3.19)	0.14	46(hh)	...	5.658	...	...	2.449
Au	14.2 (7.9)	0.71	2.35(b)	11.6	4.078	...	...	2.882

Physical properties of the elements (continued)

Element	Atomic No.	Atomic weight	Density(a), g/cm³ (lb/in.³)	Melting point, °C (°F)	Boiling point, °C (°F)	Specific heat(b), cal/g · °C (J/kg · K)	Heat of fusion, cal/g (Btu/lb)
Hafnium (Hf)	72	178.58	13.1 (0.473)	2222 ± 30 (4032 ± 54)	5400 (9750)	0.0351 (147)	. . . (. . .)
Helium (He)	2	4.00	0.1785(g) (0.006444)(g)	−269.7 (−453.5)	−268.9 (−452.0)	1.25 (5230)	. . . (. . .)
Holmium (Ho)	67	164.94	6.79 (0.245)	1461 (2662)	2330 (4230)	0.039 (160)	24.9 (44.7)
Hydrogen (H)	1	1.008	0.0899(g) (0.00325)(g)	−259.19 (−434.54)	−252.7 (−422.9)	3.45 (14 400)	15.0 (27.0)
Indium (In)	49	114.82	7.31 (0.264)	156.2 (313.1)	2000 (3632)	0.057 (240)	6.8 (12.2)
Iodine (I)	53	126.91	4.94 (0.178)	113.7 (236.7)	183 (361)	0.052 (220)	14.2 (25.6)
Iridium (Ir)	77	192.2	22.65 (0.8177)	2454 ± 3 (4449 ± 5)	5300 (9570)	0.0307 (129)	. . . (. . .)
Iron (Fe)	26	55.85	7.87 (0.284)	1536.5 ± 1 (2797.7 ± 1.8)	3000 ± 150 (5430 ± 270)	0.11 (460)	65.5 (117.9)
Krypton (Kr)	36	83.8	3.743(g) (0.1351)(g)	−157.3 (−251.1)	−152 (−242)	. . . (. . .)	. . . (. . .)
Lanthanum (La)	57	138.92	6.15 (0.222)	920 (1688)	3470 (6280)	0.048 (200)	17.3 (31.1)
Lawrencium (Lw)	103	257	. . . (. . .)	. . . (. . .)	. . . (. . .)	. . . (. . .)	. . . (. . .)
Lead (Pb)	82	207.21	11.34 (0.4094)	327.4258 (621.3664)	1725 (3137)	0.0309(f) (129)f	6.26 (11.27)
Lithium (Li)	3	6.94	0.534 (0.193)	180.54 (356.97)	1330 (2426)	0.79 (3300)	104.2 (187.6)
Lutetium (Lu)	71	174.99	9.85 (0.356)	1652(uu) (3006)(uu)	1930 (3510)	0.037 (150)	26.29 (47.32)
Magnesium (Mg)	12	24.32	1.74 (0.0628)	650 ± 2 (1202 ± 4)	1107 ± 10 (2025 ± 20)	0.245 (1030)	88 ± 2 (158 ± 4)
Manganese (Mn)	25	54.94	7.43 (0.268)	1245 (2273)	2150 (3900)	0.115(xx) (481)(xx)	63.7 (114.7)
Mendelevium (Mv)	101	256	. . . (. . .)	. . . (. . .)	. . . (. . .)	. . . (. . .)	. . . (. . .)
Mercury (Hg)	80	200.61	13.55 (0.4892)	−38.36 (−37.05)	357 (675)	0.033 (140)	2.8 (5.0)
Molybdenum (Mo)	42	95.95	10.2 (0.368)	2610 (4730)	5560 (10 040)	0.066 (280)	69.8(m) (125.6)(m)
Neodymium (Nd)	60	144.27	7.00 (0.253)	1019 (1866)	3180 (5756)	0.045 (190)	11.78 (21.20)
Neon (Ne)	10	20.18	0.8999(g) (0.03249)(g)	−248.6 ± 0.3 (−415.5 ± 0.5)	−246.0 (−410.8)	. . . (. . .)	. . . (. . .)
Neptunium (Np)	93	237	20.5 (0.740)	637 ± 2 (1179 ± 4)	. . . (. . .)	. . . (. . .)	. . . (. . .)
Nickel (Ni)	28	58.71	8.9 (0.32)	1453 (2647)	2730 (4950)	0.105 (440)	73.8 (132.8)
Niobium (Nb)	41	92.91	8.57 (0.309)	2468 ± 10 (4474 ± 18)	4927 (8901)	0.065(f) (270)(f)	69 (124.2)
Nitrogen (N)	7	14.01	1.250(g) (0.04513)(g)	−209.97 (−345.95)	−195.8 (−320.4)	0.247 (1030)	6.2 (11.2)
Nobelium (No)	102	247	. . . (. . .)	. . . (. . .)	. . . (. . .)	. . . (. . .)	. . . (. . .)
Osmium (Os)	76	190.2	22.61 (0.8162)	2700 ± 200(m) (4900 ± 350)(m)	5500 (9950)	0.031 (130)	. . . (. . .)
Oxygen (O)	8	16.00	1.429(g) (0.05159)(g)	−218.83 (−361.89)	−183.0 (−297.4)	0.218 (913)	3.3 (5.9)
Palladium (Pd)	46	106.4	12.02 (0.4339)	1552 (2826)	3980 (7200)	0.0584(f) (245)(f)	34.2 (61.6)
Phosphorus, white (P) ..	15	30.98	1.83 (0.0661)	44.25 (111.65)	280 (536)	0.177 (741)	5.0 (9.0)
Platinum (Pt)	78	195.09	21.45 (0.7743)	1769 (3217)	4530 (8185)	0.0314(f) (131)(f)	26.9 (48.4)
Plutonium (Pu)	94	242	19.4 (0.700)	640 (1184)	3235 (6000)	0.033(qqq) (140)(qqq)	. . . (. . .)
Polonium (Po)	84	210	9.40 (0.339)	254 ± 10 (489 ± 18)	. . . (. . .)	. . . (. . .)	. . . (. . .)
Potassium (K)	19	39.10	0.86 (0.031)	63.7 (146.7)	760 (1400)	0.177 (741)	14.6 (26.3)
Praseodymium (Pr)	59	140.92	6.77 (0.244)	919 (1686)	3020 (5468)	0.045 (188)	11.71 (21.08)
Promethium (Pm)	61	145	. . . (. . .)	1027(m) (1880)(m)	. . . (. . .)	. . . (. . .)	. . . (. . .)

Physical properties of the elements (continued)

Symbol	Coefficient of linear thermal expansion(c), μin./in. °C (μin./in. °F)	Thermal conductivity(c), cal/cm²/cm/s/°C	Electrical resistivity, μΩ · cm	Modulus of elasticity in tension, 10⁶ psi	Lattice constants(b), Å			Closest approach of atoms
					a	*b*	*c* (or axial angle)	
Hf........	519(kk) (288)(kk)	0.223(mm)	35.1(y)	...	3.1883	...	5.0422	...
He	... (...)	3.32×10^{-4}	...	...	3.58(nn)	...	5.84(nn)	3.58
Ho	... (...)	...	87(y)	11(z)	3.58	...	5.62	...
H	... (...)	4.06×10^{-4}	...	...	3.76(pp)	...	6.13(pp)	...
In	33 (18)	0.057	8.37(b)	1.57	4.594	...	4.951	3.25
I	93 (52)	10.4×10^{-4}	1.3×10^{15}(b)	...	4.787	7.266	9.793	2.71
Ir	6.8 (3.8)	0.14	5.3(b)	76	3.8389	...	...	2.714
Fe........	11.76(qq) (6.53)(qq)	0.18(rr)	9.71(b)	28.5 ± 0.5	2.8664(y)	...	...	2.4824
Kr	... (...)	0.21×10^{-4}	...	...	5.69(ss)	...	...	4.03
La........	5 (2.77)	0.033(x)	57(y)	10-11(z)	3.77	...	12.16	...
Lw	... (...)	...	...	...	...	...	...	...
Pb........	29.3(tt) (16.3)(tt)	0.083(f)	20.648(b)	2	4.9489	...	...	3.499
Li	56 (31)	0.17	8.55(f)	...	3.5089	...	...	3.0387
Lu	... (...)	...	79(y)	...	3.50	...	5.50	...
Mg.......	27.1(vv) (15.05)(vv)	0.367	4.45(b)	6.35(ww)	3.2088(y)	...	5.2095(y)	3.196
Mn.......	22(yy) (12.22)(yy)	...	185(zz)	23	8.912	...	...	...
Mv.......	... (...)	...	...	...	...	...	...	...
Hg.......	... (...)	0.0196(f)	98.4(aaa)	...	3.005(bbb)	...	70° 31.7'(bbb)	3.005
Mo.......	4.9(d) (2.7)(d)	0.34	5.2(f)	47	3.1468(y)	...	...	2.725
Nd	6 (3.33)	0.031(ccc)	64(y)	...	3.66	...	11.80	...
Ne	... (...)	0.00011	...	...	4.53(ddd)	...	...	3.21
Np	... (...)	...	...	...	...	...	...	...
Ni........	13.3(u) (7.39)(u)	0.22(y)	6.84(b)	30(eee)	3.5238	...	...	2.491
Nb	7.31 (4.06)	0.125(f)	12.5(f)	...	3.301	...	...	2.859
N	... (...)	0.000060	...	...	4.04(fff)	...	6.60(fff)	...
No	... (...)	...	...	...	...	...	...	...
Os........	4.6(ggg) (2.6)(ggg)	...	9.5(b)	81	2.7341(hhh)	...	4.3197(hhh)	...
O	... (...)	0.000059	...	...	6.84(jjj)	...	...	...
Pd........	11.76 (6.53)	1.68(jj)	10.8(b)	16.3	3.8902	...	...	2.750
P..........	125 (70)	...	1×10^{17}(kkk)	...	7.18(mmm)	...	...	...
Pt........	8.9 (4.9)	0.165(nnn)	10.6(b)	21.3(ppp)	3.9310(y)	...	...	2.775
Pu	55(rrr) (30.55)(rrr)	0.020(y)	141.4(sss)	14(ttt)	6.182(y)	4.826(y)	10.956(y)	...
Po........	... (...)	...	...	...	7.43	4.30	14.13	3.4
K	83 (46)	0.24	6.15(f)	...	5.334	...	...	4.624
Pr........	4 (2.22)	0.028(ccc)	68(y)	7-14(z)	3.67	...	11.84	...
Pm.......	... (...)	...	...	...	...	...	...	...

Physical properties of the elements (continued)

Element	Atomic No.	Atomic weight	Density(a), g/cm³ (lb/in.³)	Melting point, °C (°F)	Boiling point, °C (°F)	Specific heat(b), cal/g · °C (J/kg · K)	Heat of fusion, cal/g (Btu/lb)
Protactinium (Pa)	91	231.1	15.4 (0.556)	1230(m) (2246)(m)	... (...)	... (...)	... (...)
Radium (Ra)	88	226.05	5.0 (0.18)	700 (1292)	... (...)	... (...)	... (...)
Radon (Rn)	86	222	9.960(g) (0.3596)(g)	−71(m) (−96)(m)	−61.8 (−79.2)	... (...)	... (...)
Rhenium (Re)	75	186.22	21.0 (0.76)	3180 ± 20 (5755 ± 35)	5900 (10 650)	0.033 (140)	... (...)
Rhodium (Rh)	45	102.91	12.41 (0.4480)	1966 ± 3 (3571 ± 5)	4500 (8130)	0.059(f) (250)(f)	... (...)
Rubidium (Rb)	37	85.48	1.53 (0.0552)	38.9 (102)	688 (1270)	0.080 (330)	6.5 (11.79)
Ruthenium (Ru)	44	101.07	12.45 (0.4494)	2500 ± 100 (4530 ± 180)	4900 (8850)	0.057(f) (240)(f)	... (...)
Samarium (Sm)	62	150.35	7.49 (0.270)	1072 (1962)	1630 (2966)	0.042(xxx) (180)(xxx)	17.29 (31.12)
Scandium (Sc)	21	44.96	2.9 (0.10)	1539 (2802)	2730 (4946)	0.134 (561)	84.52 (152.14)
Selenium (Se)	34	78.96	4.8 (0.17)	217 (423)	685 ± 1 (1265 ± 2)	0.084(x) (350)(x)	16.4 (29.5)
Silicon (Si)	14	28.09	2.33 (0.0841)	1410 (2570)	2680 (4860)	0.162(f) (678)(f)	432 (778)
Silver (Ag)	47	107.88	10.49 (0.3787)	960.80 (1761.44)	2210 (4010)	0.0559(f) (234)(f)	25 (45)
Sodium (Na)	11	22.99	0.9712 (0.03506)	97.82 (208.08)	892 (1638)	0.295 (1240)	27.5 (49.5)
Strontium (Sr)	38	87.63	2.60 (0.0939)	768 (1414)	1380 (2520)	0.176 (737)	25 (45)
Sulfur, yellow (S)	16	32.07	2.07 (0.0747)	119.0 ± 0.5 (246.2 ± 0.9)	444.6 (832.3)	0.175 (733)	9.3 (16.7)
Tantalum (Ta)	73	180.95	16.6 (0.599)	2996 ± 50 (5425 ± 90)	5425 ± 100 (9800 ± 200)	0.034(y) (140)	38 (68)
Technetium (Tc)	43	98	11.5 (0.415)	2130(m) (3870)(m)	... (...)	... (...)	... (...)
Tellurium (Te)	52	127.61	6.24 (0.225)	449.5 ± 0.3 (841.1 ± 0.5)	989.8 ± 3.8 (1813.6 ± 6.8)	0.047 (200)	32 (58)
Terbium (Tb)	65	158.93	8.25 (0.298)	1356(uu) (2472)(uu)	2530 (4586)	0.044 (180)	24.54 (44.17)
Thallium (Tl)	81	204.39	11.85 (0.4278)	303 (577)	1457 (2655)	0.031 (130)	5.04 (9.07)
Thorium (Th)	90	232.05	11.5 (0.415)	1750 (3182)	3850 ± 350 (7000 ± 600)	0.034 (140)	<19.82 (<35.68)
Thulium (Tm)	69	168.94	9.31 (0.336)	1545 (2813)	1720(www) (3130)(www)	0.038 (160)	26.04 (46.87)
Tin (Sn)	50	118.70	7.30 (0.264)	231.912 ± 0.000 (449.442 ± 0.000)	2270 (4120)	0.054 (230)	14.5 (26.1)
Titanium (Ti)	22	47.90	4.51 (0.163)	1668 ± 10 (3035 ± 18)	3260 (5900)	0.124 (519)	104(m) (188)(m)
Tungsten (W)	74	183.86	19.3 (0.697)	3410 (6170)	5930 (10 706)	0.033 (140)	44 (70)
Uranium (U)	92	238.07	19.07 (0.6884)	1132.3 ± 0.8 (2070.4 ± 1.5)	3818 (6904)	0.02709(jjjj) (113.4)(jjjj)	... (...)
Vanadium (V)	23	50.95	6.11 (0.221)	1900 ± 25 (3450 ± 45)	3400 (6150)	0.119(t) (498)(t)	... (...)
Xenon (Xe)	54	131.30	5.896(g) (0.2128)(g)	−111.9 (−169.4)	−108.0 (−162.4)	... (...)	... (...)
Ytterbium (Yb)	70	173.04	6.96 (0.251)	824 (1515)	1530 (2786)	0.035 (150)	12.71 (22.88)
Yttrium (Y)	39	88.92	4.47 (0.161)	1509(uu) (2748)(uu)	3030 (5490)	0.071 (300)	46 (83)
Zinc (Zn)	30	65.38	7.13 (0.257)	419.5050 (787.1090)	906 (1663)	0.0915 (383)	24.09 (43.36)
Zirconium (Zr)	40	91.22	6.49 (0.234)	1852 (3366)	3580 (6470)	0.067 ± 0.001 (280 ± 4)	60(m) (110)(m)

(a) Density may depend considerably on previous treatment. (b) At 20 °C (68 °F). (c) Near 20 °C (68 °F). (d) From 20 to 100 °C (68 to 212 °F). (e) From 20 to 60 °C (68 to 140 °F). (f) At 0 °C (32 °F). (g) Gas, grams per litre at 20 °C (68 °F) and 760 mm (30 in.). (h) At −233 °C (−387 °F). (j) 28 atm. (k) Sublimes. (m) Estimated. (n) From 25 to 100 °C (77 to 212 °F). (p) Annealed, commercial purity. (q) Approximate. (r) From 20 to 750 °C (68 to 1380 °F). (s) At −150 °C (−238 °F). (t) Sand cast. (u) From 0 to 100 °C (32 to 212 °F). (v) For alpha at 0 to 400 °C (32 to 750 °F). (w) Annealed. (x) At 28 °C (82 °F). (y) At 25 °C (77 °F). (z) Measured from stress-strain relationship on as-cast metal. (aa) From 0 to 26 °C (32 to 70 °F). (bb) At −10 °C (14 °F). (cc) At −185 °C (−300 °F). (dd) Near 40 °C (105 °F); the coefficient of expansion of gadolinium changes rapidly between −100 and +100 °C (−150 and +212 °F). (ee) From 0 to 30 °C (32 to 86 °F). (ff) At melting point. (gg) For a-axis; 8.1 for b-axis and 54.3 for c-axis. (hh) Ohm · cm of intrinsic germanium at 300 K. (jj) At 18 °C (64 °F). (kk) From 20 to 200 °C (68 to 390 °F). (mm) W/cm/°C at 50 °C (120 °F). (nn) At −271.5 °C (−456.7 °F). (pp) At −271 °C (−455.8 °F). (qq) At 25 °C (77 °F) for high-purity k iron. (rr) For ingot iron at 0 °C (32 °F). (ss) At −191 °C (−311.8 °F). (tt) From 17 to 100 °C (63 to 212 °F). (uu) Distilled metal. (vv) Along a-axis; 24.3 along c-axis. (ww) Dynamic; static, 5.77; both for 99.98% magnesium. (xx) For alpha; gamma is 0.120; both at 25.2 °C (77.3 °F). (yy) Alpha; gamma, 14; both from 0 to 100 °C (32 to 212 °F).

Physical properties of the elements (continued)

Symbol	Coefficient of linear thermal expansion(c), μin./in. °C (μin./in. °F)	Thermal conductivity(c), cal/cm²/cm/s/°C	Electrical resistivity, μΩ · cm	Modulus of elasticity in tension, 10⁶ psi	Lattice constants(b), Å			Closest approach of atoms
					a	b	c (or axial angle)	
Pa.........	... (...)	...	...	...	...	...	...	...
Ra	... (...)	...	...	...	...	...	...	...
Rn	... (...)	...	...	...	...	...	...	...
Re.........	6.7(uuu) (3.7)(uuu)	0.17	19.3(b)	66.7(b)	2.760	...	4.458	2.74
Rh	8.3 (4.6)	0.21(nnn)	4.51(b)	42.5(vvv)	3.804	...	...	2.689
Rb	90 (50)	...	12.5(b)	...	5.63(www)	...	...	4.88
Ru	9.1 (5.1)	...	7.6(f)	60(q)	2.7041	...	4.2814	...
Sm	... (...)	...	88(y)	8(z)	8.99	...	23° 13′	...
Sc	... (...)	...	61(yyy)	...	3.31	...	5.27	...
Se.........	37 (21)	7-18.3 × 10⁻⁴	12(f)	8.4	4.346	...	4.954	...
Si	2.8-7.3 (1.6-4.1)	0.20	10(f)	16.35(zzz)	5.428	...	...	2.351
Ag	19.68(u) (10.9)(u)	1.0(f)	1.59(b)	11	4.086 ± 0.0006(jj)	...	...	2.888
Na	71 (39)	0.32	4.2(f)	...	4.289	...	...	3.714
Sr.........	... (...)	...	23(b)	...	6.087	...	...	4.31
S..........	64 (36)	6.31 ×10⁻⁴	2 × 10²³(b)	...	10.50	12.95	24.60	2.12
Ta.........	6.5 (3.6)	0.130	12.45(y)	27(b)	3.303	...	...	2.859
Tc.........	... (...)	...	...	...	...	...	...	...
Te.........	16.75 (9.3)	0.014	436 000(aaaa)	6	4.4570	...	5.9290	2.571
Tb.........	7 (3.88)	...	...	...	3.60	...	5.69	...
Tl.........	28 (16)	0.093	18(f)	...	3.457	...	5.525	3.408
Th	12.5(bbbb) (6.9)(bbbb)	0.090(cccc)	13(f)	...	5.09	...	...	3.60
Tm	... (...)	...	79(y)	...	3.53	...	5.55	...
Sn.........	23(dddd) (13)(dddd)	1.50(e)	11(eeee)	6-6.5(ffff)	5.8314	...	3.1815	...
Ti	8.41 (4.67)	6.6(gggg)	42(b)	16.8	2.95030	...	4.68312	...
W	4.6 (2.55)	0.397(e)	5.65(hhhh)	50	3.158	...	...	2.734
U	6.8-14.1(kkkk) (3.8-7.8)(kkkk)	0.07(mmmm)	30(nnnn)	24	2.8545(y)	5.8681(y)	4.9566(y)	...
V	8.3(pppp) (4.6)(pppp)	0.074(cccc)	24.8-26.0(b)	18-20	3.039	...	...	2.632
Xe	... (...)	1.24 × 10⁻⁴	...	...	6.25(rrrr)	...	...	4.42
Yb	25 (13.9)	...	29(y)	...	5.49	...	...	...
Y	... (...)	0.035(ccc)	57(ssss)	17(z)	3.65	...	5.73	...
Zn.........	39.7(ssss) (22.0)(ssss)	0.27(y)	5.916(b)	(tttt)	2.6649	...	4.9470	2.6648
Zr.........	5.85(uuuu) (3.2)(uuuu)	0.211(vvvv)	40	13.7	3.2312(y)	...	5.1477(y)	3.17

(zz) Alpha at 20 °C (68 °F). (aaa) At 50 °C (122 °F). (bbb) At −50 °C (−58 °F). (ccc) At −2.22 °C (28 °F). (ddd) At −268 °C (−450.4 °F). (eee) At 0 °C (32 °F), unmagnetized. (fff) At −234 °C (−389 °F). (ggg) At 50 °C (122 °F), parallel to a-axis, mean value; parallel to c-axis at 50 °C (122 °F), 5.8 (hhh) At 26 °C (78.8 °F). (jjj) At −225 °C (−373 °F). (kkk) At 11 °C (51.8 °F). (mmm) At −35 °C (−31 °F). (nnn) At 17 °C (63 °F). (ppp) For small cyclic strains. (qqq) For alpha at 25 °C (77 °F). (rrr) From 21 to 104 °C (70 to 219 °F). (sss) At 107 °C (224.6 °F). (ttt) At 25 °C (77 °F), for cast metal. (uuu) From 20 to 500 °C (68 to 930 °F). (vvv) For hard wire. (www) At −173 °C (−279 °F). (xxx) Calculated. (yyy) Average value at 22 °C (72 °F), zone-refined bar. (zzz) Chill cast specimen 90.2 by 24.6 by 24.6 mm (3.55 by 0.97 by 0.97 in.). (aaaa) At 23 °C (73 °F). (bbbb) From 25 to 1000 °C (77 to 1830 °F), for iodide thorium. (cccc) At 100 °C (212 °F). (dddd) From 0 to 100 °C (32 to 212 °F), for polycrystalline metal. (eeee) At 0 °C (32 °F), for white tin. (ffff) Cast tin. (gggg) Btu · ft/h · ft² · °F at −400 °F. (hhhh) At 27 °C (80.6 °F). (jjjj) At 27 °C (80 °F). (kkkk) Rolled rods. (mmmm) At 70 °C (158 °F). (nnnn) Crystallographic average. (pppp) From 23 to 100 °C (73 to 212 °F). (qqqq) At −185 °C (−301 °F). (rrrr) Polycrystalline; c-axis, 135; basal plane, 72. (ssss) From 20 to 250 °C (68 to 480 °F), for polycrystalline metal. (tttt) Pure zinc has no clearly defined molulus of elasticity. (uuuu) Alpha, polycrystalline. (vvvv) W/cm/°C at 27 °C (80.6 °F)

Density of metals and alloys

Metal or alloy	Density g/cm³	lb/in.³
Aluminum and aluminum alloys		
Aluminum (99.996%)	2.6989	0.0975
Wrought alloys		
EC, 1060 alloys	2.70	0.098
1100	2.71	0.098
2011	2.82	0.102
2014	2.80	0.101
2024	2.77	0.100
2218	2.81	0.101
3003	2.73	0.099
4032	2.69	0.097
5005	2.70	0.098
5050	2.69	0.097
5052	2.68	0.097
5056	2.64	0.095
5083	2.66	0.096
5086	2.65	0.096
5154	2.66	0.096
5357	2.70	0.098
5456	2.66	0.096
6061, 6063	2.70	0.098
6101, 6151	2.70	0.098
7075	2.80	0.101
7079	2.74	0.099
7178	2.82	0.102
Casting alloys		
A13	2.66	0.096
43	2.69	0.097
108, A108	2.79	0.101
A132	2.72	0.098
D132	2.76	0.100
F132	2.74	0.099
138	2.95	0.107
142	2.81	0.101
195, B195	2.81	0.101
214	2.65	0.096
220	2.57	0.093
319	2.79	0.101
355	2.71	0.098
356	2.68	0.097
360	2.64	0.095
380	2.71	0.098
750	2.88	0.104
40E	2.81	0.101
Copper and copper alloys		
Wrought coppers		
Pure copper	8.96	0.324
Electrolytic tough pitch copper (ETP)	8.89	0.321
Deoxidized copper, high residual phosphorus (DHP)	8.94	0.323
Free-machining copper		
0.5% Te	8.94	0.323
1.0% Pb	8.94	0.323
Wrought alloys		
Gilding, 95%	8.86	0.320
Commercial bronze, 90%	8.80	0.318
Jewelry bronze, 87.5%	8.78	0.317
Red brass, 85%	8.75	0.316
Low brass, 80%	8.67	0.313
Cartridge brass, 70%	8.53	0.308
Yellow brass	8.47	0.306
Muntz metal	8.39	0.303

Metal or alloy	Density g/cm³	lb/in.³
Leaded commercial bronze	8.83	0.319
Low-leaded brass (tube)	8.50	0.307
Medium-leaded brass	8.47	0.306
High-leaded brass (tube)	8.53	0.308
High-leaded brass	8.50	0.307
Extra-high-leaded brass	8.50	0.307
Free-cutting brass	8.50	0.307
Leaded Muntz metal	8.41	0.304
Forging brass	8.44	0.305
Architectural bronze	8.47	0.306
Inhibited admiralty	8.53	0.308
Naval brass	8.41	0.304
Leaded naval brass	8.44	0.305
Manganese bronze (A)	8.36	0.302
Phosphor bronze, 5% (A)	8.86	0.320
Phosphor bronze, 8% (C)	8.80	0.318
Phosphor bronze, 10% (D)	8.78	0.317
Phosphor bronze, 1.25%	8.89	0.321
Free-cutting phosphor bronze	8.89	0.321
Cupro-nickel, 30%	8.94	0.323
Cupro-nickel, 10%	8.94	0.323
Nickel silver, 65-18	8.73	0.315
Nickel silver, 55-18	8.70	0.314
High-silicon bronze (A)	8.53	0.308
Low-silicon bronze (B)	8.75	0.316
Aluminum bronze, 5% A1	8.17	0.294
Aluminum bronze, (3)	7.78	0.281
Aluminum-silicon bronze	7.69	0.278
Aluminum bronze, (1)	7.58	0.274
Aluminum bronze, (2)	7.58	0.274
Beryllium copper	8.23	0.297
Casting alloys		
Chromium copper (1% Cr)	8.7	0.31
88Cu-10Sn-2Zn	8.7	0.31
88Cu-8Sn-4Zn	8.8	0.32
89Cu-11Sn	8.78	0.317
88Cu-6Sn-1.5Pb-4.5Zn	8.7	0.31
87Cu-8Sn-1Pb-4Zn	8.8	0.32
87Cu-10Sn-1Pb-2Zn	8.8	0.32
80Cu-10Sn-10Pb	8.95	0.323
83Cu-7Sn-7Pb-3Zn	8.93	0.322
85Cu-5Sn-9Pb-1Zn	8.87	0.320
78Cu-7Sn-15Pb	9.25	0.334
70Cu-5Sn-25Pb	9.30	0.336
85Cu-5Sn-5Pb-5Zn	8.80	0.318
83Cu-4Sn-6Pb-7Zn	8.6	0.31
81Cu-3Sn-7Pb-9Zn	8.7	0.31
76Cu-2.5Sn-6.5Pb-15Zn	8.77	0.317
72Cu-1Sn-3Pb-24Zn	8.50	0.307
67Cu-1Sn-3Pb-29Zn	8.45	0.305
61Cu-1Sn-1Pb-37Zn	8.40	0.304
Manganese bronze		
60 ksi	8.2	0.30
65 ksi	8.3	0.30
90 ksi	7.9	0.29
110 ksi	7.7	0.28
Aluminum bronze		
Alloy 9A	7.8	0.28
Alloy 9B	7.55	0.272
Alloy 9C	7.5	0.27
Alloy 9D	7.7	0.28

Metal or alloy	Density g/cm³	lb/in.³
Nickel Silver		
12% Ni	8.95	0.323
16% Ni	8.95	0.323
20% Ni	8.85	0.319
25% Ni	8.8	0.32
Silicon bronze	8.30	0.300
Silicon brass	8.30	0.300
Iron and iron alloys		
Pure iron	7.874	0.2845
Ingot iron	7.866	0.2842
Wrought iron	7.7	0.28
Gray cast iron	7.15(a)	0.258(a)
Malleable iron	7.27(b)	0.262(b)
0.06% C steel	7.871	0.2844
0.23% C steel	7.859	0.2839
0.435% C steel	7.844	0.2834
1.22% C steel	7.830	0.2829
Low-carbon chromium-molybdenum steels		
0.5% Mo steel	7.86	0.283
1Cr-0.5Mo steel	7.86	0.283
1.25Cr-0.5Mo steel	7.86	0.283
2.25Cr-1.0Mo steel	7.86	0.283
5Cr-0.5Mo steel	7.78	0.278
7Cr-0.5Mo steel	7.78	0.278
9Cr-1Mo steel	7.67	0.276
Medium-carbon alloy steels		
1Cr-0.35Mo-0.25V steel	7.86	0.283
H11 die steel (5Cr-1.5Mo-0.4V)	7.79	0.281
Other iron-base alloys		
A-286	7.94	0.286
16-25-6 alloy	8.08	0.292
RA-330	8.03	0.290
Incoloy	8.02	0.290
Incoloy T	7.98	0.288
Incoloy 901	8.23	0.297
T1 tool steel	8.67	0.313
M2 tool steel	8.16	0.295
H41 tool steel	7.88	0.285
20W-4Cr-2V-12Co steel	8.89	0.321
Invar (36% Ni)	8.00	0.289
Hipernik (50% Ni)	8.25	0.298
4% Si	7.6	0.27
10.27% Si	6.97	0.252
Stainless steels and heat-resisting alloys		
Corrosion-resistant steel castings		
CA-15	7.612	0.2750
CA-40	7.612	0.2750
CB-30	7.53	0.272
CC-50	7.53	0.272
CE-30	7.67	0.277
CF-8	7.75	0.280
CF-20	7.75	0.280
CF-8M, CF-12M	7.75	0.280
CF-8C	7.75	0.280
CF-16F	7.75	0.280
CH-20	7.72	0.279
CK-20	7.75	0.280
CN-7M	8.00	0.289

(continued)

Density of metals and alloys (continued)

Metal or alloy	Density g/cm^3	lb/in.^{3}
Heat-resistant alloy castings		
HA	7.72	0.279
HC	7.53	0.272
HD	7.58	0.274
HE	7.67	0.277
HF	7.75	0.280
HH	7.72	0.279
HI	7.72	0.279
HK	7.75	0.280
HL	7.72	0.279
HN	7.83	0.283
HT	7.92	0.286
HU	8.04	0.290
HW	8.14	0.294
HX	8.14	0.294
Wrought stainless and heat-resisting steels		
Type 301	7.9	0.29
Type 302	7.9	0.29
Type 302B	8.0	0.29
Type 303	7.9	0.29
Type 304	7.9	0.29
Type 305	8.0	0.29
Type 308	8.0	0.29
Type 309	7.9	0.29
Type 310	7.9	0.29
Type 314	7.72	0.279
Type 316	8.0	0.29
Type 317	8.0	0.29
Type 321	7.9	0.29
Type 347	8.0	0.29
Type 403	7.7	0.28
Type 405	7.7	0.28
Type 410	7.7	0.28
Type 416	7.7	0.28
Type 420	7.7	0.28
Type 430	7.7	0.28
Type 430F	7.7	0.28
Type 431	7.7	0.28
Types 440A, 440B, 440C	7.7	0.28
Type 446	7.6	0.27
Type 501	7.7	0.28
Type 502	7.8	0.28
19-9DL	7.97	0.29
Precipitation-hardening stainless steels		
PH 15-7 Mo	7.804	0.2819
17-4 PH	7.8	0.28
17-7 PH	7.81	0.282
Nickel-base alloys		
D-979	8.27	0.299
Nimonic 80A	8.25	0.298
Nimonic 90	8.27	0.299
M-252	8.27	0.298
Inconel	8.51	0.307
Inconel "X" 550	8.30	0.300
Inconel 700	8.17	0.295
Inconel "713C"	7.913	0.2859
Waspaloy	8.23	0.296
René 41	8.27	0.298
Hastelloy alloy B	9.24	0.334
Hastelloy alloy C	8.94	0.323
Hastelloy alloy X	8.23	0.297
Udimet 500	8.07	0.291
GMR-235	8.03	0.290
Cobalt-chromium-nickel-base alloys		
N-155 (HS-95)	8.23	0.296
S-590	8.36	0.301

Metal or alloy	Density g/cm^3	lb/in.^{3}
Cobalt-base alloys		
S-816	8.68	0.314
V-36	8.60	0.311
HS-25	9.13	0.330
HS-36	9.04	0.327
HS-31	8.61	0.311
HS-21	8.30	0.300
Molybdenum-base alloy		
Mo-0.5Ti	10.2	0.368
Lead and lead alloys		
Chemical lead (99.90+% Pb)	11.34	0.4097
Corroding lead (99.73+% Pb)	11.36	0.4104
Arsenical lead	11.34	0.4097
Calcium lead	11.34	0.4097
5-95 solder	11.0	0.397
20-80 solder	10.2	0.368
50-50 solder	8.89	0.321
Antimonial lead alloys		
1% antimonial lead	11.27	0.407
Hard lead (96Pb-4Sb)	11.04	0.399
Hard lead (94Pb-6Sb)	10.88	0.393
8% antimonial lead	10.74	0.388
9% antimonial lead	10.66	0.385
Lead-base babbitt alloys		
Lead-base babbitt		
SAE 13	10.24	0.370
SAE 14	9.73	0.352
Alloy 8	10.04	0.363
Arsenical lead		
Babbitt (SAE 15)	10.1	0.365
"G" Babbitt	10.1	0.365
Magnesium and magnesium alloys		
Magnesium (99.8%)	1.738	0.06279
Casting alloys		
AM100A	1.81	0.065
AZ63A	1.84	0.066
AZ81A	1.80	0.065
AZ91A, B, C	1.81	0.065
AZ92A	1.82	0.066
HK31A	1.79	0.065
HZ32A	1.83	0.066
ZH42, ZH62A	1.86	0.067
ZK51A	1.81	0.065
ZE41A	1.82	0.066
EZ33A	1.83	0.066
EK30A	1.79	0.065
EK41A	1.81	0.065
Wrought alloys		
M1A	1.76	0.064
A3A	1.77	0.064
AZ31B	1.77	0.064
PE	1.76	0.064
AZ61A	1.80	0.065
AZ80A	1.80	0.065
ZK60A, B	1.83	0.066
ZE10A	1.76	0.064
HM21A	1.78	0.064
HM31A	1.81	0.065

Metal or alloy	Density g/cm^3	lb/in.^{3}
Nickel and nickel alloys		
Nickel (99.95% Ni + Co)	8.902	0.322
"A" Nickel	8.885	0.321
"D" Nickel	8.78	0.317
Duranickel	8.26	0.298
Cast nickel	8.34	0.301
Monel	8.84	0.319
"K" Monel	8.47	0.306
Monel (cast)	8.63	0.312
"H" Monel (cast)	8.5	0.31
"S" Monel (cast)	8.36	0.302
Inconel	8.51	0.307
Inconel (cast)	8.3	0.30
Ni-o-nel	7.86	0.294
Nickel-molybdenum-chromium-iron alloys		
Hastelloy B	9.24	0.334
Hastelloy C	8.94	0.323
Hastelloy D	7.8	0.282
Hastelloy F	8.17	0.295
Hastelloy N	8.79	0.317
Hastelloy W	9.03	0.326
Hastelloy X	8.23	0.297
Nickel-chromium-molybdenum-copper alloys		
Illium G	8.58	0.310
Illium R	8.58	0.310
Electrical resistance alloys		
80Ni-20Cr	8.4	0.30
60Ni-24Fe-16Cr	8.247	0.298
35Ni-45Fe-20Cr	7.95	0.287
Constantan	8.9	0.32
Tin and tin alloys		
Pure tin	7.3	0.264
Soft solder (30% Pb)	8.32	0.301
Soft solder (37% Pb)	8.42	0.304
Tin babbitt		
Alloy 1	7.34	0.265
Alloy 2	7.39	0.267
Alloy 3	7.46	0.269
Alloy 4	7.53	0.272
Alloy 5	7.75	0.280
White metal	7.28	0.263
Pewter	7.28	0.263
Titanium and titanium alloys		
99.9% Ti	4.507	0.1628
99.2% Ti	4.507	0.1628
99.0% Ti	4.52	0.163
Ti-6Al-4V	4.43	0.160
Ti-5Al-2.5Sn	4.46	0.161
Ti-2Fe-2Cr-2Mo	4.65	0.168
Ti-8Mn	4.71	0.171
Ti-7Al-4Mo	4.48	0.162
Ti-4Al-4Mn	4.52	0.163
Ti-4Al-3Mo-1V	4.507	0.1628
Ti-2.5Al-16V	4.65	0.168
Zinc and zinc alloys		
Pure zinc	7.133	0.2577
AG40A alloy	6.6	0.24
AC41A alloy	6.7	0.24

(continued)

Density of metals and alloys (continued)

Metal or alloy	Density g/cm³	lb/in.³	Metal or alloy	Density g/cm³	lb/in.³	Metal or alloy	Density g/cm³	lb/in.³
Commercial rolled zinc			**Permanent magnet materials**			Ruthenium 12.2		0.441
0.08% Pb 7.14		0.258				Selenium 4.79		0.174
0.06 Pb, 0.06 Cd 7.14		0.258	Cunico 8.30		0.300	Silicon 2.33		0.084
0.3 Pb, 0.3 Cd 7.14		0.258	Cunife 8.61		0.311	Silver 10.49		0.379
Copper-hardened,			Comol 8.16		0.295	Sodium 0.97		0.035
rolled zinc (1% Cu) ..7.18		0.259	Alnico I 6.89		0.249	Tantalum 16.6		0.600
Rolled zinc alloy			Alnico II 7.09		0.256	Thallium 11.85		0.428
(1 Cu, 0.010 Mg) 7.18		0.259	Alnico III 6.89		0.249	Thorium 11.72		0.423
Zn-Cu-Ti alloy			Alnico IV 7.00		0.253	Tungsten 19.3		0.697
(0.8 Cu, 0.15 Ti) 7.18		0.259	Alnico V 7.31		0.264	Uranium 19.07		0.689
			Alnico VI 7.42		0.268	Vanadium 6.1		0.22
			Barium ferrite 4.7		0.17	Zirconium 6.5		0.23
			Vectolite 3.13		0.113			
Precious metals			**Pure metals**			**Rare earth metals**		
Silver 10.49		0.379	Antimony 6.62		0.239	Cerium ....8.23(e)		...
Gold 19.32		0.698	Beryllium 1.848		0.067	Cerium ....6.66(f)		...
70Au-30Pt 19.92		...	Bismuth 9.80		0.354	Cerium ....6.77(g)		...
Platinum 21.45		0.775	Cadmium 8.65		0.313	Dysprosium ....8.55(h)		...
Pt-3.5Rh 20.9		...	Calcium 1.55		0.056	Erbium ....9.15(h)		...
Pt-5Rh 20.65		...	Cesium 1.903		0.069	Europium ....5.245(g)		...
Pt-10Rh 19.97		...	Chromium 7.19		0.260	Gadolinium ....7.86(h)		...
Pt-20Rh 18.74		...	Cobalt 8.85		0.322	Holmium ....6.79(h)		...
Pt-30Rh 17.62		...	Gallium 5.907		0.213	Lanthanum ....6.19(f)		...
Pt-40Rh 16.63		...	Germanium 5.323		0.192	Lanthanum ....6.18(e)		...
Pt-5Ir 21.49		...	Hafnium 13.1		0.473	Lanthanum ....5.97(g)		...
Pt-10Ir 21.53		...	Indium 7.31		0.264	Lutetium ....9.85(h)		...
Pt-15Ir 21.57		...	Iridium 22.5		0.813	Neodymium ....7.00(f)		...
Pt-20Ir 21.61		...	Lithium 0.534		0.019	Neodymium ....6.80(g)		...
Pt-25Ir 21.66		...	Manganese 7.43		0.270	Praseodymium ......6.77(f)		...
Pt-30Ir 21.70		...	Mercury 13.546		0.489	Praseodymium ......6.64(g)		...
Pt-35Ir 21.79		...	Molybdenum 10.22		0.369	Samarium ....7.49(j)		...
Pt-5Ru 20.67		...	Niobium 8.57		0.310	Scandium ....2.99(h)		...
Pt-10Ru 19.94		...	Osmium 22.583		0.816	Terbium ....8.25(h)		...
Palladium 12.02		0.4343	Plutonium 19.84		0.717	Thulium ....9.31(h)		...
60Pd-40Cu 10.6		0.383	Potassium 0.86		0.031	Ytterbium ....6.96(e)		...
95.5Pd-4.5Ru ....12.07(a)		...	Rhenium 21.04		0.756	Yttrium ....4.47(h)		...
95.5Pd-4.5Ru ....11.62(b)		...	Rhodium 12.44		0.447			

(a) 6.95 to 7.35 g/cm² (0.251 to 0.265 lb/in.³). (b) 7.20 to 7.34 g/cm³ (0.260 to 0.265 lb/in.³). (c) Annealed. (d) As cast. (e) Face-centered cubic. (f) Hexagonal. (g) Body-centered cubic. (h) Close-packed hexagonal. (j) Rhombohedral

Linear thermal expansion of metals and alloys

Metal or alloy	Temperature, °C	Coefficient of expansion, µ in./in. · °C	Metal or alloy	Temperature, °C	Coefficient of expansion, µ in./in. · °C	Metal or alloy	Temperature, °C	Coefficient of expansion, µ in./in. · °C
Aluminum and aluminum alloys			6101, 6151 20-100		23.0	750 20-100		23.1
			7075 20-100		23.2	40E 21-93		24.7
Aluminum			7079, 7178 20-100		23.4			
(99.996%) 20-100		23.6				**Copper and copper alloys**		
			Casting alloys					
Wrought alloys						**Wrought coppers**		
			A13 20-100		20.4			
EC, 1060, 1100 20-100		23.6	43 and 108 20-100		22.0	Pure copper 20		16.5
2011, 2014 20-100		23.0	A108 20-100		21.5	Electrolytic tough		
2024 20-100		22.8	A132 20-100		19.0	pitch copper (ETP) .20-100		16.8
2218 20-100		22.3	D132 20-100		20.5	Deoxidized copper,		
3003 20-100		23.2	F132 20-100		20.7	high residual		
4032 20-100		19.4	138 20-100		21.4	phosphorus (DHP) .20-300		17.7
5005, 5050, 5052 20-100		23.8	142 20-300		22.5	Oxygen-free copper. .20-300		17.7
5056 20-100		24.1	195 20-100		23.0	Free-machining		
5083 20-100		23.4	B195 20-100		22.0	copper, 0.5% Te		
5086 60-300		23.9	214 20-100		24.0	or 1% Pb 20-300		17.7
5154 20-100		23.9	220 20-100		25.0			
5357 20-100		23.7	319 20-100		21.5	**Wrought alloys**		
5456 20-100		23.9	355 20-100		22.0			
6061, 6063 20-100		23.4	356 20-100		21.5	Gilding, 95% 20-300		18.1
			360 20-100		21.0	Commercial bronze,		
						90% 20-300		18.4

(continued)

Linear thermal expansion of metals and alloys (continued)

Metal or alloy	Temperature, °C	Coefficient of expansion, μ in./in. · °C
Jewelry bronze, 87.5%	20-300	18.6
Red brass, 85%	20-300	18.7
Low brass, 80%	20-300	19.1
Cartridge brass, 70%	20-300	19.9
Yellow brass	20-300	20.3
Muntz metal	20-300	20.8
Leaded commercial bronze	20-300	18.4
Low-leaded brass	20-300	20.2
Medium-leaded brass	20-300	20.3
High-leaded brass	20-300	20.3
Extra-high-leaded brass	20-300	20.5
Free-cutting brass	20-300	20.5
Leaded Muntz metal	20-300	20.8
Forging brass	20-300	20.7
Architectural bronze	20-300	20.9
Inhibited admiralty	20-300	20.2
Naval brass	20-300	21.2
Leaded naval brass	20-300	21.2
Manganese bronze (A)	20-300	21.2
Phosphor bronze, 5% (A)	20-300	17.8
Phosphor bronze, 8% (C)	20-300	18.2
Phosphor bronze, 10% (D)	20-300	18.4
Phosphor bronze, 1.25%	20-300	17.8
Free-cutting phosphor bronze	20-300	17.3
Cupro-nickel, 30%	20-300	16.2
Cupro-nickel, 10%	20-300	17.1
Nickel silver, 65-18	20-300	16.2
Nickel silver, 55-18	20-300	16.7
Nickel silver, 65-12	20-300	16.2
High-silicon bronze (A)	20-300	18.0
Low-silicon bronze (B)	20-300	17.9
Aluminum bronze (3)	20-300	16.4
Aluminum-silicon bronze	20-300	18.0
Aluminum bronze (1)	20-300	16.8
Beryllium copper	20-300	17.8

Casting alloys

Metal or alloy	Temperature, °C	Coefficient of expansion, μ in./in. · °C
88Cu-8Sn-4Zn	21-177	18.0
89Cu-11Sn	20-300	18.4
88Cu-6Sn-1.5Pb-4.5Zn	21-260	18.5
87Cu-8Sn-1Pb-4Zn	21-177	18.0
87Cu-10Sn-1Pb-2Zn	21-177	18.0
80Cu-10Sn-10Pb	21-204	18.5
78Cu-7Sn-15Pb	21-204	18.5
85Cu-5Sn-5Pb-5Zn	21-204	18.1
72Cu-1Sn-3Pb-24Zn	21-93	20.7
67Cu-1Sn-3Pb-29Zn	21-93	20.2
61Cu-1Sn-1Pb-37Zn	21-260	21.6
Manganese bronze 60 ksi	21-204	20.5
65 ksi	21-93	21.6
110 ksi	21-260	19.8

Metal or alloy	Temperature, °C	Coefficient of expansion, μ in./in. · °C
Aluminum bronze		
Alloy 9A	...	17
Alloy 9B	20-250	17
Alloys 9C, 9D	...	16.2

Iron and iron alloys

Metal or alloy	Temperature, °C	Coefficient of expansion, μ in./in. · °C
Pure iron	20	11.7
Fe-C alloys		
0.06% C	20-100	11.7
0.22% C	20-100	11.7
0.40% C	20-100	11.3
0.56% C	20-100	11.0
1.08% C	20-100	10.8
1.45% C	20-100	10.1
Invar (36% Ni)	20	0-2
13Mn-1.2C	20	18.0
13Cr-0.35C	20-100	10.0
12.3Cr-0.4Ni-0.09C	20-100	9.8
17.7Cr-9.6Ni-0.06C	20-100	16.5
18W-4Cr-1V	0-100	11.2
Gray cast iron	0-100	10.5
Malleable iron (pearlitic)	20-400	12

Lead and lead alloys

Metal or alloy	Temperature, °C	Coefficient of expansion, μ in./in. · °C
Corroding lead (99.73+% Pb)	17-100	29.3
5-95 solder	15-110	28.7
20-80 solder	15-110	26.5
50-50 solder	15-110	23.4
1% antimonial lead	20-100	28.8
Hard lead (96Pb-4Sb)	20-100	27.8
Hard lead (94Pb-6Sb)	20-100	27.2
8% antimonial lead	20-100	26.7
9% antimonial lead	20-100	26.4
Lead-base babbitt SAE 14	20-100	19.6
Alloy 8	20-100	24.0

Magnesium and magnesium alloys

Metal or alloy	Temperature, °C	Coefficient of expansion, μ in./in. · °C
Magnesium (99.8%)	20	25.2

Casting alloys

Metal or alloy	Temperature, °C	Coefficient of expansion, μ in./in. · °C
AM100A	18-100	25.2
AZ63A	20-100	26.1
AZ91A, B, C	20-100	26
AZ92A	18-100	25.2
HZ32A	20-200	26.7
ZH42	20-200	27
ZH62A	20-200	27.1
ZK51A	20	26.1
EZ33A	20-100	26.1
EK30A, EK41A	20-100	26.1

Wrought alloys

Metal or alloy	Temperature, °C	Coefficient of expansion, μ in./in. · °C
M1A, A3A	20-100	26
AZ31B, PE	20-100	26
AZ61A, AZ80A	20-100	26
ZK60A, B	20-100	26
HM31A	20-93	26.1

Nickel and nickel alloys

Metal or alloy	Temperature, °C	Coefficient of expansion, μ in./in. · °C
Nickel (99.95% Ni + Co)	0-100	13.3
Duranickel	0-100	13.0
Monel	0-100	14.0
Monel (cast)	25-100	12.9
Inconel	20-100	11.5
Ni-o-nel	27-93	12.9
Hastelloy B	0-100	10.0
Hastelloy C	0-100	11.3
Hastelloy D	0-100	11.0
Hastelloy F	20-100	14.2
Hastelloy N	21-204	10.4
Hastelloy W	23-100	11.3
Hastelloy X	26-100	13.8
Illium G	0-100	12.19
Illium R	0-100	12.02
80Ni-20Cr	20-1000	17.3
60Ni-24Fe-16Cr	20-1000	17.0
35Ni-45Fe-20Cr	20-500	15.8
Constantan	20-1000	18.8

Tin and tin alloys

Metal or alloy	Temperature, °C	Coefficient of expansion, μ in./in. · °C
Pure tin	0-100	23
Solder (70Sn-30Pb)	15-110	21.6
Solder (63Sn-37Pb)	15-110	24.7

Titanium and titanium alloys

Metal or alloy	Temperature, °C	Coefficient of expansion, μ in./in. · °C
99.9% Ti	20	8.41
99.0% Ti	93	8.55
Ti-5Al-2.5Sn	93	9.36
Ti-8Mn	93	8.64

Zinc and zinc alloys

Metal or alloy	Temperature, °C	Coefficient of expansion, μ in./in. · °C
Pure zinc	20-250	39.7
AG40A alloy	20-100	27.4
AC41A alloy	20-100	27.4
Commercial rolled zinc 0.08 Pb	20-40	32.5
0.3 Pb, 0.3 Cd	20-98	33.9(a)
Rolled zinc alloy (1 Cu, 0.010 Mg)	20-100	34.8(b)
Zn-Cu-Ti alloy (0.8 Cu, 0.15 Ti)	20-100	24.9(c)

Pure metals

Metal or alloy	Temperature, °C	Coefficient of expansion, μ in./in. · °C
Beryllium	25-100	11.6
Cadmium	20	29.8
Calcium	0-400	22.3
Chromium	20	6.2
Cobalt	20	13.8
Gold	20	14.2
Iridium	20	6.8
Lithium	20	56
Manganese	0-100	22
Palladium	20	11.76
Platinum	20	8.9
Rhenium	20-500	6.7
Rhodium	20-100	8.3
Ruthenium	20	9.1
Silicon	0-1400	5
Silver	0-100	19.68
Tungsten	27	4.6
Vanadium	23-100	8.3
Zirconium	...	5.85

(a) With the grain; 23.4 across the grain. (b) With the grain; 21.1 across the grain. (c) With the grain; 19.4 across the grain.

Thermal conductivity of metals and alloys

Metal or alloy	Thermal conductivity near room temperature, cal/cm² · cm · s · °C
Aluminum and aluminum alloys	
Wrought alloys	
EC (O)	0.57
1060 (O)	0.56
1100	0.53
2011 (T3)	0.34
2014 (O)	0.46
2024 (O)	0.45
2218 (T72)	0.37
3003 (O)	0.46
4032 (O)	0.37
5005	0.48
5050 (O)	0.46
5052 (O)	0.33
5056 (O)	0.28
5083	0.28
5086	0.30
5154	0.30
5357	0.40
5456	0.28
6061 (O)	0.41
6063 (O)	0.52
6101 (T6)	0.52
6151 (O)	0.49
7075 (T6)	0.29
7079 (T6)	0.29
7178 (T6)	0.29
Casting alloys	
A13	0.29
43 (F)	0.34
108 (F)	0.29
A108	0.34
A132 (T551)	0.28
D132 (T5)	0.25
F132	0.25
138	0.24
142 (T21, sand)	0.40
195 (T4, T62)	0.33
B195 (T4, T6)	0.31
214	0.33
200 (T4)	0.21
319	0.26
355 (T51, sand)	0.40
356 (T51, sand)	0.40
360	0.35
380	0.23
750	0.44
40E	0.33
Copper and copper alloys	
Wrought coppers	
Pure copper	0.941
Electrolytic tough pitch copper (ETP)	0.934
Deoxidized copper, high residual phosphorus (DHP)	0.81
Free-machining copper (0.5% Te)	0.88
Free-machining copper (1% Pb)	0.92
Wrought alloys	
Gilding, 95%	0.56
Commercial bronze, 90%	0.45
Jewelry bronze, 87.5%	0.41
Red brass, 85%	0.38
Low brass, 80%	0.33

Metal or alloy	Thermal conductivity near room temperature, cal/cm² · cm · s · °C
Cartridge brass, 70%	0.29
Yellow brass	0.28
Muntz metal	0.29
Leaded commercial bronze	0.43
Low-leaded brass (tube)	0.28
Medium-leaded brass	0.28
High-leaded brass (tube)	0.28
High-leaded brass	0.28
Extra-high-leaded brass	0.28
Leaded Muntz metal	0.29
Forging brass	0.28
Architectural bronze	0.29
Inhibited admiralty	0.26
Naval brass	0.28
Leaded naval brass	0.28
Manganese bronze (A)	0.26
Phosphor bronze, 5% (A)	0.17
Phosphor bronze, 8% (C)	0.15
Phosphor bronze, 10% (D)	0.12
Phosphor bronze, 1.25%	0.49
Free-cutting phosphor bronze	0.18
Cupro-nickel, 30%	0.07
Cupro-nickel, 10%	0.095
Nickel silver, 65-18	0.08
Nickel silver, 55-18	0.07
Nickel silver, 65-12	0.10
High-silicon bronze (A)	0.09
Low-silicon bronze (B)	0.14
Aluminum bronze, 5% Al	0.198
Aluminum bronze, (3)	0.18
Aluminum-silicon bronze	0.108
Aluminum bronze, (1)	0.144
Aluminum bronze, (2)	0.091
Beryllium copper	0.20(a)
Casting alloys	
Chromium copper (1% Cr)	0.4(a)
89Cu-11Sn	0.121
88Cu-6Sn-1.5Pb-4.5Zn	(b)
87Cu-8Sn-1Pb-4Zn	(c)
87Cu-10Sn-1Pb-2Zn	(c)
80Cu-10Sn-10Pb	(c)
Manganese bronze, 110 ksi	(d)
Aluminum bronze	
Alloy 9A	(e)
Alloy 9B	(f)
Alloy 9C	(b)
Alloy 9D	(c)
Propeller bronze	(g)
Nickel silver	
12% Ni	(h)
16% Ni	(h)
20% Ni	(j)
25% Ni	(k)
Silicon bronze	(h)
Iron and iron alloys	
Pure iron	0.178
Cast iron (3.16 C, 1.54 Si, 0.57 Mn)	0.112
Carbon steel (0.23 C, 0.64 Mn)	0.124
Carbon steel (1.22 C, 0.35 Mn)	0.108
Alloy steel (0.34 C, 0.55 Mn, 0.78 Cr, 3.53 Ni, 0.39 Mo, 0.05 Cu)	0.079

Metal or alloy	Thermal conductivity near room temperature, cal/cm² · cm · s · °C
Type 410	0.057
Type 304	0.036
T1 tool steel	0.058
Lead and lead alloys	
Corroding lead (99.73+% Pb)	0.083
5-95 solder	0.085
20-80 solder	0.089
50-50 solder	0.111
1% antimonial lead	0.080
Hard lead (96Pb-4Sb)	0.073
Hard lead (94Pb-6Sb)	0.069
8% antimonial lead	0.065
9% antimonial lead	0.064
Lead-base babbitt (SAE 14)	0.057
Lead-base babbitt (alloy 8)	0.058
Magnesium and magnesium alloys	
Magnesium (99.8%)	0.367
Casting alloys	
AM100A	0.17
AZ63A	0.18
AZ81A (T4)	0.12
AZ91A, B, C	0.17
AZ92A	0.17
HK31A (T6, sand cast)	0.22
HZ32A	0.26
ZH42	0.27
ZH62A	0.26
ZK51A	0.26
ZE41A (T5)	0.27
EZ33A	0.24
EK30A	0.26
EK41A (T5)	0.24
Wrought alloys	
M1A	0.33
AZ31B	0.23
AZ61A	0.19
AZ80A	0.18
ZK60A, B (F)	0.28
ZE10A (O)	0.33
HM21A (O)	0.33
HM31A	0.25
Nickel and nickel alloys	
Nickel (99.95% Ni + Co)	0.22
"A" nickel	0.145
"D" nickel	0.115
Monel	0.062
"K" Monel	0.045
Inconel	0.036
Hastelloy B	0.027
Hastelloy C	0.03
Hastelloy D	0.05
Illium G	0.029
Illium R	0.031
60Ni-24Fe-16Cr	0.032
35Ni-45Fe-20Cr	0.031
Constantan	0.051

(continued)

Thermal conductivity of metals and alloys (continued)

Tin and tin alloys

Metal or alloy	Thermal conductivity near room temperature, cal/cm² · cm · s · °C
Pure tin	0.15
Soft solder (63Sn-37Pb)	0.12
Tin foil (92Sn-8Zn)	0.14

Titanium and titanium alloys

Metal or alloy	
Titanium (99.0%)	0.43
Ti-5Al-2.5Sn	0.19
Ti-2Fe-2Cr-2Mo	0.28
Ti-8Mn	0.26

Zinc and zinc alloys

Metal or alloy	
Pure zinc	0.27
AG40A alloy	0.27
AC41A alloy	0.26
Commercial rolled zinc 0.08 Pb	0.257
0.06 Pb, 0.06 Cd	0.257
Rolled zinc alloy (1 Cu, 0.010 Mg)	0.25
Zn-Cu-Ti alloy (0.8 Cu, 0.15 Ti)	0.25

Pure metals

Metal or alloy		Metal or alloy	
Beryllium	0.35	Iridium	0.14
Cadmium	0.22	Lithium	0.17
Chromium	0.16	Molybdenum	0.34
Cobalt	0.165	Niobium	0.13
Germanium	0.14	Palladium	0.168
Gold	0.71	Platinum	0.165
Indium	0.057	Plutonium	0.020
		Rhenium	0.17
		Rhodium	0.21
		Silicon	0.20
		Silver	1.0
		Sodium	0.32
		Tantalum	0.130
		Thallium	0.093
		Thorium	0.090
		Tungsten	0.397
		Uranium	0.071
		Vanadium	0.074
		Yttrium	0.035

(a) Depends on processing. (b) 18% of Cu. (c) 12% of Cu. (d) 9.05% of Cu. (e) 15% of Cu. (f) 16% of Cu. (g) 11% of Cu. (h) 7% of Cu. (j) 6% of Cu. (k) 6.5% of Cu.

Electrical conductivity and resistivity of metals and alloys

Metal or alloy	Conductivity, % IACS	Resistivity, μΩ · cm
Aluminum and aluminum alloys		
Aluminum (99.996%)	64.94	2.65
EC (O, H19)	62	2.8
5052 (O, H38)	35	4.93
5056 (H38)	27	6.4
6101 (T6)	56	3.1
Copper and copper alloys		
Wrought copper		
Pure copper	103.06	1.67
Electrolytic (ETP)	101	1.71
Oxygen-free copper (OF)	101	1.71
Free-machining copper 0.5% Te	95	1.82
1.0% Pb	98	1.76
Wrought alloys		
Cartridge brass, 70%	28	6.2
Yellow brass	27	6.4
Leaded commercial bronze	42	4.1
Phosphor bronze, 1.25%	48	3.6
Nickel silver, 55-18	5.5	31
Low-silicon bronze (B)	12	14.3
Beryllium copper	22-30(a)	5.7-7.8(a)
Casting alloys		
Chromium copper (1% Cr)	80-90(a)	2.10
88Cu-8Sn-4Zn	11	15
87Cu-10Sn-1Pb-2Zn	11	15
Electrical contact materials		
Copper alloys		
0.04 oxide	100	1.72
1.25 Sn + P	48	3.6
5 Sn + P	18	11
8 Sn + P	13	13
15 Zn	37	4.7
20 Zn	32	5.4
35 Zn	27	6.4
2 Be + Ni or Co(b)	17-21	9.6-11.5
Silver and silver alloys		
Fine silver	106	1.59
92.5Ag-7.5Cu	85	2
90Ag-10Cu	85	2
72Ag-28Cu	87	2
72Ag-26Cu-2Ni	60	2.9
85Ag-15Cd	35	4.93
97Ag-3Pt	50	3.5
97Ag-3Pd	60	2.9
90Ag-10Pd	30	5.3
90Ag-10Au	40	4.2
60Ag-40Pd	8	23
70Ag-30Pd	12	14.3
Platinum and platinum alloys		
Platinum	16	10.6
95Pt-5Ir	9	19
90Pt-10Ir	7	25
85Pt-15Ir	6	28.5
80Pt-20Ir	5.6	31
75Pt-25Ir	5.5	33
70Pt-30Ir	5	35
65Pt-35Ir	5	36
95Pt-5Ru	5.5	31.5
90Pt-10Ru	4	43
89Pt-11Ru	4	43
86Pt-14Ru	3.5	46
96Pt-4W	5	36
Palladium and palladium alloys		
Palladium	16	10.8
95.5Pd-4.5Ru	7	24.2
90Pd-10Ru	6.5	27
70Pd-30Ag	4.3	40
60Pd-40Ag	4.0	43
50Pd-50Ag	5.5	31.5
72Pd-26Ag-2Ni	4	43
60Pd-40Cu	5	35(c)
45Pd-30Ag-20Au-5Pt	4.5	39
35Pd-30Ag-14Cu-10Pt-10Au-1Zn	5	35
Gold and gold alloys		
Gold	75	2.35
90Au-10Cu	16	10.8
75Au-25Ag	16	10.8
72.5Au-14Cu-8.5Pt-4Ag-1Zn	10	17
69Au-25Ag-6Pt	11	15
41.7Au-32.5Cu-18.8Ni-7Zn	4.5	39
Electrical heating alloys		
Ni-Cr and Ni-Cr-Fe alloys		
78.5Ni-20Cr-1.5Si (80-20)	1.6	108.05
73.5Ni-20Cr-5Al-1.5Si	1.2	137.97
68Ni-20Cr-8.5Fe-2Si	1.5	116.36
60Ni-16Cr-22.5Fe-1.5Si	1.5	112.20
35Ni-20Cr-43.5Fe-1.5Si	1.7	101.4
Fe-Cr-Al alloys		
72Fe-23Cr-5Al	1.3	138.8
55Fe-37.5Cr-7.5Al	1.2	166.23

(continued)

Electrical conductivity and resistivity of metals and alloys (continued)

Metal or alloy	Conductivity, % IACS	Resistivity, $\mu\Omega \cdot cm$
Pure metals		
Molybdenum 34		5.2
Platinum 16		10.64
Tantalum 13.9		12.45
Tungsten 30		5.65
Nonmetallic heating element materials		
Silicon carbide, SiC ..1-1.7		100-200
Molybdenum disilicide, $MoSi_2$ 4.5		37.24
Graphite		910.1
Instrument and control alloys		
Cu-Ni alloys		
98Cu-2Ni 35		4.99
94Cu-6Ni 17		9.93
89Cu-11Ni 11		14.96
78Cu-22Ni 5.7		29.92
55Cu-45Ni (constantan) 3.5		49.87
Cu-Mn-Ni alloys		
87Cu-13Mn (manganin) 3.5		48.21
83Cu-13Mn-4Ni (manganin) 3.5		48.21
85Cu-10Mn-4Ni (shunt manganin) ...4.5		38.23
70Cu-20Ni-10Mn 3.6		48.88
67Cu-5Ni-27Mn 1.8		99.74
Ni-base alloys		
99.8 Ni 23		7.98
71Ni-29Fe 9		19.95
80Ni-20Cr 1.5		112.2
75Ni-20Cr-3Al + Cu or Fe 1.3		132.98
76Ni-17Cr-4Si-3Mn 1.3		132.98
60Ni-16Cr-24Fe 1.5		112.2
35Ni-20Cr-45Fe 1.7		101.4
Fe-Cr-Al alloy		
72Fe-23Cr-5Al-0.5Co ..1.3		135.48
Pure metals		
Iron (99.99%) 17.75		9.71
Thermostat metals		
75Fe-22Ni-3Cr 3		78.13
72Mn-18Cu-10Ni 1.5		112.2
67Ni-30Cu-1.4Fe-1Mn 3.5		56.52
75Fe-22Ni-3Cr 12		15.79
66.5Fe-22Ni-8.5Cr 3.3		58.18

Metal or alloy	Conductivity, % IACS	Resistivity, $\mu\Omega \cdot cm$
Permanent magnet materials		
Carbon steel (0.65% C) 9.5		18
Carbon steel (1% C) ... 8		20
Chromium steel (3.5% Cr) 6.1		29
Tungsten steel (6% W) 6		30
Cobalt steel (17% Co) ..6.3		28
Cobalt steel (36% Co) ..6.5		27
Intermediate alloys		
Cunico 7.5		24
Cunife 9.5		18
Comol 3.6		45
Alnico alloys		
Alnico I 3.3		75
Alnico II 3.3		65
Alnico III 3.3		60
Alnico IV 3.3		75
Alnico V 3.5		47
Alnico VI 3.5		50
Magnetically soft materials		
Electrical steel sheet		
M-50 .... 9.5		18
M-43 6-9		20-28
M-36 .... 5.5-7.5		24-33
M-27 .... 3.5-5.5		32-47
M-22 .... 3.5-5		41-52
M-19 .... 3.5-5		41-56
M-17 .... 3-3.5		45-58
M-15 .... 3-3.5		45-69
M-14 .... 3-3.5		58-69
M-7 .... 3-3.5		45-52
M-6 .... 3-3.5		45-52
M-5 .... 3-3.5		45-52
Moderately high-permeability materials(d)		
Thermenol 0.5		162
16 Alfenol 0.7		153
Sinimax 2		90
Monimax 2.5		80
Supermalloy 3		65
4-79 Moly Permalloy, Hymu 80 3		58
Mumetal 3		60
1040 alloy 3		56
High Permalloy 49, A-L 4750, Armco 48 ..3.6		48
45 Permalloy 3.6		45

Metal or alloy	Conductivity, % IACS	Resistivity, $\mu\Omega \cdot cm$
High-permeability materials(e)		
Supermendur 4.5		40
2V Permendur 4.5		40
35% Co, 1% Cr 9		20
Ingot iron 17.5		10
0.5% Si steel 6		28
1.75% Si steel 4.6		37
3.0% Si steel 3.6		47
Grain-oriented 3.0% Si steel 3.5		50
Grain-oriented 50% Ni iron 3.6		45
50% Ni iron 3.5		50
Relay steels and alloys after annealing		
Low-carbon iron and steel		
Low-carbon iron 17.5		10
1010 steel 14.5		12
Silicon steels		
1% Si 7.5		23
2.5% Si 4		41
3% Si 3.5		48
3% Si, grain-oriented ...3.5		48
4% Si 3		59
Stainless steels		
Type 410 3		57
Type 416 3		57
Type 430 3		60
Type 443 3		68
Type 446 3		61
Nickel irons		
50% Ni 3.5		48
78% Ni 11		16
77% Ni (Cu, Cr) 3		60
79% Ni (Mo) 3		58
Stainless and heat-resisting alloys		
Type 302 3		72
Type 309 .... 2.5		78
Type 316 .... 2.5		74
Type 317 .... 2.5		74
Type 347 .... 2.5		73
Type 403 3		57
Type 405 3		60
Type 501 .... 4.5		40
HH 2.5		80
HK 2		90
HT 1.7		100

(a) Precipitation hardened; depends on processing. (b) A heat treatable alloy. (c) Annealed and quenched. (d) At low field strength and high electrical resistance. (e) At higher field strength; annealed for optimum magnetic properties.

Vapor pressures of the elements

Element	0.0001		0.001		0.01		0.1		0.50		1.0	
	°C	°F	°C	°F	°C	°F	°C	°F	°C	°F	°C	°F
Aluminum	1110	2030	1263	2305	1461	2662	1713	3115	1940	3524	2056	3733
Antimony	759	1398	872	1602	1013	1855	1196	2185	1359	2478	1440	2624
Arsenic	308	586	363	685	428	802	499	930	578	1072	610	1130
Bismuth	914	1677	1008	1846	1121	2050	1254	2289	1367	2493	1420	2588
Cadmium	307(a)	585(a)	384(b)	723(b)	471	880	594	1101	708	1306	765	1409
Calcium	688	1270	802(c)	1476(c)	958(b)	1756(b)	1175	2147	1380	2516	1487	2709
Carbon	3257	5895	3547	6417	3897	7047	4317	7803	4667	8433	4827	8721
Chromium	1420(a)	2588(a)	1594(b)	2901(b)	1813	3295	2097	3807	2351	4264	2482	4500
Copper	1412	2574	1602	2916	1844	3351	2162	3924	2450	4442	2595	4703
Gallium	1178	2152	1329	2424	1515	2759	1751	3184	1965	3569	2071	3760
Gold	1623	2953	1839	3342	2115	3839	2469	4476	2796	5065	2966	5371
Iron	1564	2847	1760	3200	2004	3639	2316	4201	2595	4703	2735	4955
Lead	815	1499	953	1747	1135	2075	1384	2523	1622	2952	1744	3171
Lithium	592	1098	707	1305	858	1576	1064	1947	1266	2311	1372	2502
Magnesium	516	961	608(a)	1126(a)	725(b)	1337(b)	886	1627	1030	1886	1107	2025
Manganese	1115(d)	2039(d)	1269(b)	2316(b)	1476	2889	1750	3182	2019	3666	2151	3904
Mercury	77.9(b)	172.2(b)	120.8	249.4	176.1	349.0	251.3	484.3	321.5	610.7	357	675
Molybdenum	2727	4941	3057	5535	3477	6291	4027	7281	4537	8199	4804	8679
Nickel	1586	2887	1782	3240	2025	3677	2321	4210	2593	4699	2732	4950
Platinum	2367	4293	2687	4869	3087	5589	3637	6579	4147	7497	4407	7965
Potassium	261	502	332	630	429	804	565	1051	704	1299	774	1425
Rubidium	223	433	288	550	377	711	497	927	617	1143	679	1254
Selenium	282	540	347	657	430	806	540	1004	634	1173	680	1256
Silicon	1572	2862	1707	3105	1867	3393	2057	3735	2217	4023	2287	4149
Silver	1169	2136	1334	2433	1543	2809	1825	3317	2081	3778	2212	4014
Sodium	349	660	429	804	534	993	679	1254	819	1506	892	1638
Strontium	....	....	(a)	(a)	877(b)	1629(b)	1081	1978	1279	2334	1384	2523
Tellurium	(a)	(a)	509(b)	948(b)	632	1170	810	1490	991	1816	1087	1989
Thallium	692	1277	809	1488	962	1764	1166	2131	1359	2478	1457	2655
Tin	....	....	....	....	....	....	1932(b)	3510(b)	2163	3925	2270	4118
Tungsten	3547	6417	3937	7119	4437	8019	5077	9171	5647	10197	5927	10701
Zinc	399(a)	750(a)	477(b)	891(b)	579	1074	717	1323	842	1548	907	1665

(a) In the solid state. (b) In the liquid state. (c) β. (d) γ.
Source: K. K. Kelley, Bureau of Mines Bulletin 383, 1935.

Predominant flame colors of elements

Element	Color
Lithium	Deep red
Strontium	Crimson
Calcium	Yellow-red
Sodium	Bright yellow
Barium	Yellow-green
Molybdenum	Green-yellow
Zinc	Light green
Boron	Green
Tellurium	Deep Green
Thallium	Greenish blue
Antimony	Blue-green
Copper	Green-blue
Arsenic	Light blue
Lead	Light blue
Selenium	Blue
Indium	Deep blue
Potassium	Purple-red
Rubidium	Violet
Cesium	Bluish purple

Energy requirements for production of metal from ore

Metal	Energy				Ratio, actual/theoretical
	Theoretical		Actual		
	10⁶ Btu/ton	TJ/Mg	10⁶ Btu/ton	TJ/Mg	
Aluminum	15.7	18.3	203	236	12:9
Magnesium.......	4.1	4.8	309	359	75:7
Iron.............	3.2	3.7	16.7	19.4	5:2
Steel............	6.0	7	47.5	55	7:9
Copper	1.4	1.6	46.1	54	32:9
Titanium.........	9.9	12	431	501	43:6

Metal melting range and color scale

Color scale: White, Light yellow, Lemon, Orange, Salmon, Bright red, Cherry or dull red, Medium cherry, Dark cherry, Blood red, Faint red

Melting points of metals and alloys: Chromium, Pure iron, Stainless-12% Cr, Mild steel, Cobalt, Nickel, Silicon, Stainless 18-8, Hard steel, Inconel, Monel, Manganese, *Haynes Stellite*, Ambrac, Copper, Gold, Red brass, Everdur, Silver, Yellow brass, Tobin bronze, Manganese bronze, Aluminum, Magnesium, Antimony, Zinc, Lead, Babbit, Tin

Temperature: °C 0–1600; °F 0–3000

Melting point ranges of alloys: Wrought iron, Nickel alloys, Chromium, nickel, Chromium-nickel steels and irons, Cast irons, Chromium-nickel-cast irons, Brasses, Bronzes, Aluminum alloys, Magnesium alloys, Tin alloys, Lead alloys

Standard reduction potentials of metals

Add the correction given at 18 °C (64 °F) if the activity (effective concentration of metallic ions) is $C \times$ standard; if activity is less than standard, $\log_{10} C$ will be negative and the potential will be reduced.

Metal	Ion considered	Standard reduction potential, V	Correction(a)
Gold	Au^{3+}	+1.50	$0.019 \log_{10} C$
Platinum	Pt^{2+}	+1.2	$0.029 \log_{10} C$
Silver	Ag^{+}	+0.799	$0.058 \log_{10} C$
Mercury	$(Hg)_2^{2+}$	+0.789	$0.029 \log_{10} C$
Copper	Cu^{2+}	+0.337	$0.020 \log_{10} C$
Hydrogen (1 atm)	H^{+}	±0.000	$0.058 \log_{10} C$
Lead	Pb^{2+}	−0.126	$0.029 \log_{10} C$
Tin	Sn^{2+}	−0.136	$0.029 \log_{10} C$
Nickel	Ni^{2+}	−0.250	$0.029 \log_{10} C$
Cobalt	Co^{2+}	−0.28	$0.029 \log_{10} C$
Cadmium	Cd^{2+}	−0.403	$0.029 \log_{10} C$
Iron	Fe^{2+}	−0.440	$0.029 \log_{10} C$
Chromium	Cr^{3+}	−0.74	$0.019 \log_{10} C$
Zinc	Zn^{2+}	−0.763	(a)
Manganese	Mn^{2+}	−1.18	(a)
Titanium(b)	Ti^{2+}	−1.63	(a)
Aluminum(b)	Al^{3+}	−1.66	(a)
Magnesium(b)	Mg^{2+}	−2.37	(a)
Sodium(b)	Na^{2+}	−2.71	(a)
Calcium(b)	Ca^{2+}	−2.87	(a)
Potassium(b)	K^{+}	−2.92	(a)
Lithium(b)	Li^{+}	−3.02	(a)

(a) Potential almost independent of original concentration of metal ions. (b) Calculated values, of theoretical interest only. Aluminum, unless amalgamated, gives more positive values owing to the presence of an oxide film. The other metals evolve hydrogen freely, and potential measurements made directly would not represent equilibrium values.

Worldwide distribution of important metals

Element	Source
Manganese	**South Africa** and the **USSR** between them have 83% of the world's high-grade ore reserves (25 to 50% Mn content), with South Africa having 45% and the USSR 38%
Cobalt	Two countries, **Zaire** and **Zambia,** have almost 50% of the world's high-grade ore reserves, with **New Caledonia** and **Australia** having 27%
Chromium	Two countries, **Rhodesia** and **South Africa,** have 97% of the world's high-grade ore reserves and 95% of the low-grade ore reserves. South Africa has 62% and Rhodesia 33%
Aluminum	Four countries have about 75% of the world's high-grade bauxite ores. **Australia** and **Guinea** have approximately 25% each; **Brazil** 15% and **Jamaica** 6%
Tin	**China** has 24% of the world's reserves, **Thailand** 15%, **Malaysia** 12%, and **Bolivia** 10%
Nickel	**New Caledonia** has 44% of the world's reserves, **Canada** 16%, **Australia** and **Indonesia** approximately 9% each, and **Cuba** 6%
Antimony	**China** has 50% of the world's reserves, **Bolivia** 9%, the **USSR** 7%, and **Mexico** 5%
Tungsten	**China** has 54% of the world's reserves, **Canada** 12%, the **USSR** and **North** and **South Korea** approximately 9% each, the **United States** 6%
Iron	The **USSR** has 31% of the world's high-grade ore reserves, **Brazil** 17%, **Canada** 12%, **Australia** 10%, **India** 6%, the **United States** 4%
Uranium	Outside of the Soviet bloc nations, the **United States** has 25% of the world's reserves, **Australia** 22%, **South Africa** 20%, **Canada** 19% (The Soviet bloc nations have not made data on uranium reserves available)
Copper	The **United States** and **Chile** each have 20% of the world's reserves, **Canada** and the **USSR** approximately 9% each, **Peru** and **Zambia** 7% each. However, four countries—Chile, Peru, Zambia, and **Zaire**—account for more than 80% of the world's exports of ore
Molybdenum	The **United States** has 50% of the world's reserves, the **USSR** 15%, **Canada** and **Chile** approximately 14% each
Niobium	**Brazil** has 75% of the world's reserves, **Canada** approximately 8%, the **USSR** 6%
Titanium	Rutile is titanium dioxide ore from which titanium metal is produced. **Brazil** has 65% of the world's reserves, **India** 21%, **Australia** 5%, the **United States** 3%
	Ilmenite is a titanium-iron oxide containing approximately 50% titanium oxide. Much larger resources of ilmenite exist. **India** has 43% of the world's ilmenite, **Canada** 20%, **Norway** 15%, the **United States** 10%
Vanadium	The **USSR** has 75% of the world's reserves, **South Africa** 19%

Source: Bureau of Mines Bulletin 667, **Mineral Facts and Problems,** 1976.

The 45 most abundant elements in the earth's crust

Relative abundance	Element	Abundance, ppm/wt
1	Oxygen	466 000
2	Silicon	277 000
3	Aluminum	81 300
4	Iron	50 000
5	Calcium	36 300
6	Sodium	28 300
7	Potassium	25 900
8	Magnesium	20 900
9	Titanium	4 400
10	Hydrogen	1 400
11	Phosphorus	1 050
12	Manganese	950
13	Fluorine	625
14	Barium	425
15	Strontium	375
16	Sulfur	260
17	Carbon	200
18	Zirconium	165
19	Vanadium	135
20	Chlorine	130
21	Chromium	100
22	Rubidium	90
23	Nickel	75
24	Zinc	70
25	Cerium	60
26	Copper	55
27	Yttrium	33
28	Lanthanum	30
29	Neodymium	28
30	Cobalt	25
31	Scandium	22
32	Lithium	20
33	Niobium	20
34	Nitrogen	20
35	Gallium	15
36	Lead	13
37	Radium	13
38	Boron	10
39	Krypton	9.8
40	Praseodymium	8.2
41	Protoactinium	8.0
42	Thorium	7.2
43	Neon	7.0
44	Samarium	6.0
45	Gadolinium	5.4

Average percentage of metals in igneous rocks

Metal	Percentage
Silicon	27.72
Aluminum	8.13
Iron	5.01
Calcium	3.63
Sodium	2.85
Potassium	2.60
Titanium	0.63
Manganese	0.10
Barium	0.05
Chromium	0.037
Zirconium	0.026
Nickel	0.020
Vanadium	0.017
Rare earths	0.015
Copper	0.010
Tungsten	0.005
Lithium	0.004
Zinc	0.004
Niobium, tantalum	0.003
Hafnium	0.003
Thorium	0.002
Lead	0.002—
Cobalt	0.001
Beryllium	0.001
Strontium	0.001—
Uranium	0.001—

Relative amounts of engineering materials

Metal	Relative amount
Aluminum	4000
Iron	2200
Magnesium	1200
Nickel	10
Copper	1

Source: U.S. Geological Survey

The electrochemical series of elements

In this table, the elements are electropositive to the ones which follow them and will displace them from solutions of their salts

1	Cesium	34	Silver
2	Rubidium	35	Mercury
3	Potassium	36	Palladium
4	Sodium	37	Ruthenium
5	Lithium	38	Rhodium
6	Barium	39	Platinum
7	Strontium	40	Iridium
8	Calcium	41	Osmium
9	Magnesium	42	Gold
10	Beryllium	43	Hydrogen
11	Ytterbium	44	Tin
12	Erbium	45	Silicon
13	Scandium	46	Titanium
14	Aluminum	47	Niobium
15	Zirconium	48	Tantalum
16	Thorium	49	Tellurium
17	Cerium	50	Antimony
18	Didymium	51	Carbon
19	Lanthanum	52	Boron
20	Manganese	53	Tungsten
21	Zinc	54	Molybdenum
22	Iron	55	Vanadium
23	Nickel	56	Chromium
24	Cobalt	57	Arsenic
25	Thallium	58	Phosphorus
26	Cadmium	59	Selenium
27	Lead	60	Iodine
28	Germanium	61	Bromine
29	Indium	62	Chlorine
30	Gallium	63	Fluorine
31	Bismuth	64	Nitrogen
32	Uranium	65	Sulfur
33	Copper	66	Oxygen

Crystal Structures

Crystal structures

Element	Phase	Crystal symmetry (a)	Structure symbol	Structure prototype	Source (b)	Element	Phase	Crystal symmetry (a)	Structure symbol	Structure prototype	Source (b)
Ac (actinium)	...	fcc	A1	Cu	1	Eu (europium)	...	bcc	A2	W	1
Ag (silver)	...	fcc	A1	Cu	1	F (fluorine)	α	mono	...	...	2
Al (aluminum) ...	...	fcc	A1	Cu	1		β	mono	...	...	2
Am (americium)	...	hcp	...	α-La	2	Fe (iron)	α	bcc	A2	W	1
Ar (argon)	...	fcc	A1	Cu	2		γ	fcc	A1	Cu	1
As (arsenic)	α	rhom	A7	α-As	1		δ	bcc	A2	W	1
	β	ortho	...	Black P	1	Fm (fermium)	...	unknown	...	...	...
At (astatine)	...	unknown	...	...	...	Fr (francium)	...	unknown	...	...	...
Au (gold)	...	fcc	A1	Cu	1	Ga (gallium)	...	eco	A11	...	1
B (boron)	α	rhom (LT)	...	...	2	Gd (gadolinium)	α	hcp	A3	Mg	1
	β	rhom (HT)	...	...	2		β	bcc	A2	W	1
	γ	tet	...	...	2	Ge (germanium)	...	fcc	A4	C (diamond)	1
Ba (barium)	...	bcc	A2	W	1						
Be (beryllium)	α	hcp	A3	Mg	1	H (hydrogen)	α	hcp	A3 (?)	Mg	1
	β	bcc	A2	W	1		β	fcc	...	...	1
Bi (bismuth)	...	rhom	A7	α-As	1	He (helium)	α	hcp	A3	Mg	2
Bk (berkelium)	...	double hcp (LT)	...	...	3		β	fcc	A1	Cu	2
	...	fcc (HT)	A1	Cu	3		γ	bcc	A2	W	2
Br (bromine)	...	eco	...	...	2	Hf (hafnium)	α	hcp	A3	Mg	1
C (carbon) .... Graphite	Graphite	hex	A9	C (graphite)	2		β	bcc	A2	W	1
						Hg (mercury)	...	rhom	A10	Hg	1
						Ho (holmium)	α	hcp	A3	Mg	1
	Diamond	fcc	A4	C (diamond)	1		β	bcc	A2	W	1
						I (iodine)	...	eco	...	...	2
						In (indium)	...	bct	A6	In	1
Ca (calcium)	α	fcc	A1	Cu	1	Ir (iridium)	...	fcc	A1	Cu	1
	β	bcc	A2	W	1	K (potassium)	...	bcc	A2	W	1
Cb (columbium)	...	bcc	A2	W	1	Kr (krypton)	...	fcc	A1	Cu	2
Cd (cadmium)	...	hcp	A3	Mg	1	La (lanthanum)	α	hcp	...	α-La	1
Ce (cerium)	α	fcc	A1	Cu	2		β	fcc	A1	Cu	1
	β	hcp	...	α-La	2		γ	bcc	A2	W	1
	γ	fcc	A1	Cu	2	Li (lithium)	α	hcp	A3	Mg	2
	δ	bcc	A2	W	2		β	bcc	A2	W	2
Cf (californium)	...	unknown	...	...	...	Lr (lawrencium)	...	unknown	...	...	...
Cl (chlorine)	...	eco	...	...	2	Lu (lutetium)	...	hcp	A3	Mg	2
Cm (curium)	...	double hcp (LT)	...	...	4	Md (mendelevium) ..	...	unknown	...	...	...
	...	fcc (HT)	A1	Cu	4	Mg (magnesium) ...	...	hcp	A3	Mg	1
Co (cobalt)	α	hcp	A3	Mg	1	Mn (manganese)	α	comp bcc	A12	α-Mn	1
	β	fcc	A1	Cu	1		β	comp cu	A13	β-Mn	1
Cr (chromium)	...	bcc	A2	W	1		γ	fcc	A1	Cu	1
Cs (cesium)	...	bcc	A2	W	1		δ	bcc	A2	W	1
Cu (copper)	...	fcc	A1	Cu	1	Mo (molybdenum) ..	...	bcc	A2	W	1
Dy (dysprosium)	α	hcp	A3	Mg	1	N (nitrogen)	α	cu	...	...	2
	β	bcc	A2	W	1		β	hex	...	...	2
Er (erbium)	...	hcp	A3	Mg	1	Na (sodium)	α	hcp	A3	Mg	1
Es (einsteinium)	...	unknown	...	...	...		β	bcc	A2	W	1

(continued)

(a) bcc = body-centered cubic; bcm = body-centered monoclinic; bct = body-centered tetragonal; comp = complex; cu = cubic; eco = end-centered (base-centered) orthorhombic; fcc = face-centered cubic; fco = face-centered orthorhombic; hcp = hexagonal, close-packed; hex = hexagonal; (HT) = high temperature; (LT) = low temperature; mono = monoclinic; noncu = noncubic; ortho = orthorhombic; rhom = rhombohedral; simp = simple; tet = tetragonal; trig = trigonal. (b) Source 1: W. B. Pearson, "A Handbook of Lattice Spacings and Structures of Metals and Alloys", Vol 1 and 2, Pergamon, 1958, 1967. **Source 2:** R. Hultgren, P. D. Desai, D. T. Hawkins, M. Gleiser, K. K. Kelley and D. D. Wagman, "Selected Values of the Thermodynamic Properties of the Elements", American Society for Metals, 1973. **Source 3:** J. R. Peterson, J. A. Fahey and R. D. Baybarz, *J Inorg Nucl Chem*, Vol 33 (1971), p 3345-3351. **Source 4:** P. K. Smith, W. H. Hale and M. C. Thompson, *J Chem Phys*, Vol 50 (1969), p 5066-5076. **Source 5:** P. G. Pallmer and T. D. Chikalla, *J Less-Common Metals*, Vol 24 (1971), p 233-236. **Source 6:** F. Weigel and A. Trinkl, *Radiochim Acta*, Vol 10 (1967), p 78-82.

Compiled by Donald T. Hawkins, Bell Laboratories, Murray Hill, NJ.

Crystal structures (continued)

Element	Phase	Crystal symmetry (a)	Structure symbol	Structure prototype	Source (b)	Element	Phase	Crystal symmetry (a)	Structure symbol	Structure prototype	Source (b)
Nd (neodymium) ...	α	hcp	···	α-La	1	Sb (antimony)	···	rhom	A7	α-As	1
	β	bcc	A2	W	1	Sc (scandium)	α	hcp	A3	Mg	1
Ne (neon)	···	fcc	A1	Cu	2		β	bcc	A2	W	1
Ni (nickel)	···	fcc	A1	Cu	1	Se (selenium)	α	mono	···	···	1
No (nobelium)	···	unknown	···	···	···		β	mono	···	···	1
Np (neptunium)	α	ortho	···	···	1		γ	hex	A8	γ-Se	1
	β	tet	···	···	1	Si (silicon)	···	fcc	A4	C (diamond)	1
	γ	bcc (?)	A2	W	1						
O (oxygen)	α	eco	···	···	2	Sm (samarium)	α	comp rhom	···	···	1
	β	hex	···	···	2		β	bcc	A2	W	2
	γ	cu	···	···	2	Sn (tin) .... α (gray)		fcc	A4	C (diamond)	1
Os (osmium)	···	hcp	A3	Mg	1						
P (phophorus) ......α (white)		comp cu	···	···	1, 2		β (white)	bct	A5	β-Sn	1
	β (white)	comp noncu	···	···	1, 2	Sr (strontium)	α	fcc	A1	Cu	2
							β	bcc	A2	W	2
	red	unknown	···	···	1, 2	Ta (tantalum)	···	bcc	A2	W	1
	black	eco	···	···	1, 2	Tb (terbium)	α	hcp	A3	Mg	1
Pa (protactinium) ...	···	bct	···	···	1		β	bcc	A2	W	1
Pb (lead)	···	fcc	A1	Cu	1	Tc (technetium)	···	hcp	A3	Mg	1
Pd (palladium)	···	fcc	A1	Cu	1	Te (tellurium)	···	trig	A8	γ-Se	1
Pm (promethium) ...	···	double hcp	···	···	5	Th (thorium)	α	fcc	A1	Cu	1
Po (polonium)	α	simp cu	···	···	1		β	bcc	A2	W	1
	β	rhom	···	···	1	Ti (titanium)	α	hcp	A3	Mg	1
Pr (praseodymium) .	α	hcp	···	α-La	1		β	bcc	A2	W	1
......	β	bcc	A2	W	1	Tl (thallium)	α	hcp	A3	Mg	1
Pt (platinum)	···	fcc	A1	Cu	1		β	bcc	A2	W	1
Pu (plutonium)	α	mono	···	···	2	Tm (thulium)	···	hcp	A3	Mg	1
	β	bcm	···	···	2	U (uranium)	α	eco	A20	α-U	1
	γ	fco	···	···	2		β	comp tet	Like $D8_b$	Like CrFe	1
	δ	fcc	A1	Cu	2						
	δ'	bct	A6	In	2		γ	bcc	A2	W	1
	ε	bcc	A2	W	2	V (vanadium)	···	bcc	A2	W	1
Ra (radium)	···	bcc	A2	W	6	W (tungsten)	···	bcc	A2	W	1
Rb (rubidium)	···	bcc	A2	W	1	Xe (xenon)	···	fcc	A1	Cu	2
Re (rhenium)	···	hcp	A3	Mg	1	Y (yttrium)	α	hcp	A3	Mg	2
Rh (rhodium)	···	fcc	A1	Cu	1		β	bcc (?)	A2	W	2
Rn (radon)	···	unknown	···	···	···	Yb (ytterbium)	α	fcc	A1	Cu	2
Ru (ruthenium)	···	hcp	A3	Mg	1		β	bcc	A2	W	2
S (sulfur)	α	comp fco	···	···	2	Zn (zinc)	···	hcp	A3	Mg	1
	β	comp mono	···	···	2	Zr (zirconium)	α	hcp	A3	Mg	1
	γ	rhom	···	···	2		β	bcc	A2	W	1

(a) bcc = body-centered cubic; bcm = body-centered monoclinic; bct = body-centered tetragonal; comp = complex; cu = cubic; eco = end-centered (base-centered) orthorhombic; fcc = face-centered cubic; fco = face-centered orthorhombic; hcp = hexagonal, close-packed; hex = hexagonal; (HT) = high temperature; (LT) = low temperature; mono = monoclinic; noncu = noncubic; ortho = orthorhombic; rhom = rhombohedral; simp = simple; tet = tetragonal; trig = trigonal. (b) **Source 1:** W. B. Pearson, "A Handbook of Lattice Spacings and Structures of Metals and Alloys", Vol 1 and 2, Pergamon, 1958, 1967. **Source 2:** R. Hultgren, P. D. Desai, D. T. Hawkins, M. Gleiser, K. K. Kelley and D. D. Wagman, "Selected Values of the Thermodynamic Properties of the Elements", American Society for Metals, 1973. **Source 3:** J. R. Peterson, J. A. Fahey and R. D. Baybarz, *J Inorg Nucl Chem*, Vol 33 (1971), p 3345-3351. **Source 4:** P. K. Smith, W. H. Hale and M. C. Thompson, *J Chem Phys*, Vol 50 (1969), p 5066-5076. **Source 5:** P. G. Pallmer and T. D. Chikalla, *J Less-Common Metals*, Vol 24 (1971), p 233-236. **Source 6:** F. Weigel and A. Trinkl, *Radiochim Acta*, Vol 10 (1967), p 78-82.

Compiled by Donald T. Hawkins, Bell Laboratories, Murray Hill, NJ.

Crystal structure of metals

A1 Copper type. Face-centered cubic: $Fm3m$; $cF4$. Four atoms per cell, at $0,0,0$; $\frac{1}{2},0,\frac{1}{2}$; $0,\frac{1}{2},\frac{1}{2}$ and $\frac{1}{2},\frac{1}{2},0$. For Cu, $a = 3.61$ A. **Examples:** Ag, Al, Au, α-Ca, α-Ce, β-Co, Cu, γ-Fe, Ir, Ni, Pb, Pd, Pt, Rh, α-Sr, α-Th.

A2 Tungsten type. Body-centered cubic: $Im3m$; $cI2$. Two atoms per cell, at $0,0,0$ and $\frac{1}{2},\frac{1}{2},\frac{1}{2}$. For W, $a = 3.16$ A. **Examples:** Ba, Cb, Cr, Cs, β Cu-Zn (HT), α-Fe, δ-Fe, K, β-Li, Mo, β-Na, Rb, Ta, V, W.

A3 Magnesium type. Hexagonal close-packed: $P6_3/mmc$; $hP2$. Two atoms per cell, at $0,0,0$ and $\frac{1}{3},\frac{2}{3},\frac{1}{2}$. (The atoms are at the positions of the zinc atoms of the $B4$ structure shown.) For Mg, $a = 3.21$ A and $c = 5.20$ A. **Examples:** α-Be, Cd, α-Co, Mg, α-Ti, Zn, α-Zr.

A4 Carbon (diamond) type. Face-centered cubic: $Fd3m$; $cF8$. Eight atoms per cell, at $0,0,0$; $0,\frac{1}{2},\frac{1}{2}$; $\frac{1}{2},0,\frac{1}{2}$; $\frac{1}{2},\frac{1}{2},0$; $\frac{1}{4},\frac{1}{4},\frac{1}{4}$; $\frac{1}{4},\frac{3}{4},\frac{3}{4}$; $\frac{3}{4},\frac{1}{4},\frac{3}{4}$ and $\frac{3}{4},\frac{3}{4},\frac{1}{4}$. For C (diamond), $a = 3.57$ A. **Examples:** C (diamond), Ge, Si, α-Sn.

A5 Tin type (β-Sn, white). Body-centered tetragonal; $I4_1/amd$; $tI4$. Four atoms per cell, at $0,0,0$; $\frac{1}{2},\frac{1}{2},\frac{1}{2}$; $0,\frac{1}{2},\frac{1}{4}$ and $\frac{1}{2},0,\frac{3}{4}$. For β-Sn, $a = 5.83$ A and $c = 3.18$ A. **Examples:** AlSb II (HP), InSb II (HP), β-Sn (white).

A6 Indium type. Body-centered tetragonal: $I4/mmm$; $tI2$. Two atoms per cell, at $0,0,0$ and $\frac{1}{2},\frac{1}{2},\frac{1}{2}$. It is conventional, however, to use the cell that has four atoms, at $0,0,0$; $0,\frac{1}{2},\frac{1}{2}$; $\frac{1}{2},0,\frac{1}{2}$ and $\frac{1}{2},\frac{1}{2},0$. For In, $a = b = 4.60$ A and $c = 4.95$ A. At room temperature, the unit cell resembles $A1$. **Examples:** δ GaNi$_2$, In, InPd$_3$.

A7 Arsenic type (α-As). Rhombohedral: $R\bar{3}m$; $hR2$. In the cell based on hexagonal axes, there are six atoms, at $0,0,z$; $\frac{1}{3},\frac{2}{3},\frac{2}{3}+z$; $\frac{2}{3},\frac{1}{3},\frac{1}{3}+z$; $0,0,\bar{z}$; $\frac{1}{3},\frac{2}{3},\frac{2}{3}-z$ and $\frac{2}{3},\frac{1}{3},\frac{1}{3}-z$; where $z = 0.226$, $a = 3.76$ A and $c = 10.55$ A for α-As. A cell based on rhombohedral axes contains two atoms, at x,x,x and $\bar{x},\bar{x},\bar{x}$; where $x = 0.276$, $a = 4.13$ A and $\alpha = 54°\,8'$ for α-As. **Examples:** α-As, Bi, Sb.

A8 Selenium type (γ-Se). Hexagonal: $P3_121$ or $P3_221$; $hP3$. Three atoms per cell, at $x,0,\frac{1}{3}$; $0,x,\frac{2}{3}$ and $\bar{x},\bar{x},0$ (or at $x,0,\frac{2}{3}$; $0,x,\frac{1}{3}$ and $\bar{x},\bar{x},0$). For Se, $a = 4.36$ A, $c = 4.96$ A and $x = 0.217$. **Example:** γ-Se.

A9 Carbon (graphite). Hexagonal: $P6_3/mmc$; $hP4$. Four atoms per cell, at $0,0,x$; $0,0,x+\frac{1}{2}$; $\frac{1}{3},\frac{2}{3},y$ and $\frac{2}{3},\frac{1}{3},y+\frac{1}{2}$. For C (graphite), $x = y = 0$, $a = 2.46$ A and $c = 6.70$ A. There is a less common rhombohedral structure in which $a = 2.46$ A and $c = 6.70$ A.

A10 Mercury type. Rhombohedral: $R\bar{3}m$; $hR1$. One atom per cell, at $0,0,0$. For Hg, $a = 3.005$ A and $\alpha = 70°\,32'$. A hexagonal cell, where $a = 3.47$ A and $c = 6.74$ A, has three atoms per cell, at $0,0,0$; $\frac{1}{3},\frac{2}{3},\frac{2}{3}$ and $\frac{2}{3},\frac{1}{3},\frac{1}{3}$. **Example:** Hg.

A11 Gallium type. Orthorhombic: $Cmca$; $oC8$. Eight atoms per cell, at $0,y,z$; $0,\bar{y},\bar{z}$; $\frac{1}{2},y,\frac{1}{2}-z$; $\frac{1}{2},\bar{y},\frac{1}{2}+z$; $\frac{1}{2},\frac{1}{2}+y,z$; $\frac{1}{2},\frac{1}{2}-y,\bar{z}$; $0,\frac{1}{2}+y,\frac{1}{2}-z$ and $0,\frac{1}{2}-y,\frac{1}{2}+z$. For Ga, $a = 2.90$ A, $b = 8.13$ A, $c = 3.17$ A, $y = 0.1549$ and $z = 0.081$.

A12 Alpha-manganese type. Cubic: $I\bar{4}3m$; $cI58$. Fifty-eight atoms per cell. Alpha-manganese appears to be an ordered array of either two or three physically distinguishable types of manganese atoms located on four crystallographically different sets of positions. These sets of positions have an ordered array of atoms in the chi-phase structure (Fe$_{36}$Cr$_{12}$Mo$_{12}$ and Al$_{12}$Mg$_{17}$). Other closely related structures are the mu, P, R and delta phases. **Examples:** γ Al$_{12}$Mg$_{17}$, χ Co$_5$Cr$_3$Si$_2$, CrMn$_2$, Fe$_{36}$Cr$_{12}$Mo$_{12}$, Fe$_5$Si$_2$V$_3$, α-Mn.

A13 Beta-manganese type. Cubic: $P4_132$; $cP20$. Twenty atoms per cell. **Examples:** β Ag$_3$Al, T C-Cr-Fe-W, γ Cu$_5$Si, β-Mn.

A15 W$_3$O or Cr$_3$Si type. Cubic: $Pm3n$; $cP8$. Atom I, two at $0,0,0$ and $\frac{1}{2},\frac{1}{2},\frac{1}{2}$; atom II, six at $\frac{1}{4},0,\frac{1}{2}$; $\frac{1}{2},\frac{1}{4},0$; $0,\frac{1}{2},\frac{1}{4}$; $\frac{3}{4},0,\frac{1}{2}$; $\frac{1}{2},\frac{3}{4},0$ and $0,\frac{1}{2},\frac{3}{4}$. For W$_3$O, $a = 5.04$ A. The prototype structure originally was attributed to β-W. This has since been shown to be the oxide, W$_3$O, having random distribution of atoms. **Examples:** AlV$_3$, AuTi$_3$, CoV$_3$, Cr$_3$O, Cr$_3$Si, Mo$_3$O, V$_3$Si, W$_3$O, W$_3$Si.

A20 Alpha-uranium type. Orthorhombic: $Cmcm$; $oC4$. Four atoms per cell, at $0,y,\frac{1}{4}$; $0,\bar{y},\frac{3}{4}$; $\frac{1}{2},\frac{1}{2}+y,\frac{1}{4}$ and $\frac{1}{2},\frac{1}{2}-y,\frac{3}{4}$. For α-U, $a = 2.85$ A, $b = 5.87$ A, $c = 4.95$ A and $y = 0.1024$.

A$_f$ HgSn$_{10}$ type. Hexagonal: $P6/mmm$. One Hg or Sn atom per cell, at $0,0,0$.

A$_g$ Boron (alpha-boron) type. Tetragonal: $P4_2/nnm$; $tP50$. Fifty atoms per cell.

A$_h$ Alpha-polonium type. Primitive cubic: $Pm3m$; $cP1$. One atom per cell, at $0,0,0$. For α-Po, $a = 3.34$ A. **Examples:** Ag-Te (metastable), Au-Te (metastable), α-Po, Sb II (HP).

A$_i$ Beta-polonium type. Rhombohedral: $R\bar{3}m$; $hR1$. One atom per cell, at $0,0,0$. For β-Po, $a = 3.36$ A and $\alpha = 98°\,13'$. See Fig. 1. A cell based on a hexagonal axis is like the $A10$ (Hg-type) structure.

A$_k$ Alpha-selenium type (a metastable form). Monoclinic: $P2_1/c$; $mP32$. Each cell contains 32 atoms.

··· Samarium type (α-Sm). Rhombohedral: $R\bar{3}m$; $hR3$. In a cell based on a hexagonal axis, there ae nine atoms per cell, at $0,0,0$; $\frac{1}{3},\frac{2}{3},\frac{2}{3}$; $\frac{2}{3},\frac{1}{3},\frac{1}{3}$; $0,0,z$; $0,0,\bar{z}$; $\frac{1}{3},\frac{2}{3},\frac{2}{3}+z$; $\frac{1}{3},\frac{2}{3},\frac{2}{3}-z$; $\frac{2}{3},\frac{1}{3},\frac{1}{3}+z$ and $\frac{2}{3},\frac{1}{3},\frac{1}{3}-z$. For α-Sm, $a = 3.621$ A, $c = 26.25$ A and $z = \frac{2}{9}$. A cell based on rhombohedral axes contains three atoms per cell, at $0,0,0$; x,x,x and $\bar{x},\bar{x},\bar{x}$; with $a = 9.00$ A, $\alpha = 23°\,19'$ and $x = \frac{2}{9}$. **Examples:** α-Sm, Ce-Y, δ Nd-Tm, δ Pr-Y.

··· Lanthanum (α-La) type. Hexagonal: $P6_3/mmc$; $hP4$. Four atoms per cell, at $0,0,0$; $0,0,\frac{1}{2}$ and $\pm(\frac{1}{3},\frac{2}{3},\frac{1}{4})$ or $\pm(\frac{1}{3},\frac{2}{3},\frac{3}{4})$. **Examples:** Am, β-Ce (LT), α-La, α-Nd, α-Pr.

··· Beta-uranium type. Tetragonal: $P4_2/mnm$; $tP30$. Thirty atoms per cell.

··· Alpha-plutonium type. Monoclinic: $P2_1/m$; $mP16$. Sixteen atoms per cell. For α-Pu, $a = 6.183$ A, $b = 4.822$ A, $c = 10.963$ A and $\beta = 101.79°$.

B1 NaCl type. Face-centered cubic: $Fm3m$; $cF8$. Four sodium atoms at $0,0,0$; $0,\frac{1}{2},\frac{1}{2}$; $\frac{1}{2},0,\frac{1}{2}$ and $\frac{1}{2},\frac{1}{2},0$; four chlorine atoms at $\frac{1}{2},\frac{1}{2},\frac{1}{2}$; $\frac{1}{2},0,0$; $0,\frac{1}{2},0$ and $0,0,\frac{1}{2}$. For NaCl, $a = 5.64$ A. **Examples:** BaS, CdO, CdS, CrN, HfC, HfN, NaCl, NiO (HT), PbS, PbSe, TiO, UC, UO, UP, US, VO, ZrO.

Crystal structure of metals (continued)

B2   CsCl or β' Cu-Zn type. Cubic: $Pm3m$; $cP2$. One cesium atom at $0,0,0$; and one chlorine atom at $\frac{1}{2},\frac{1}{2},\frac{1}{2}$. For CsCl, $a = 4.11$ A. **Examples:** AgCd, CoTi, CsCl, FeAl, FeCo, FeTi, FeV, β NiAl, β NiGa, δ NiIn, NiTi, β' Cu-Zn.

B3   ZnS (sphalerite, or zinc blende) type. Face-centered cubic: $F\bar{4}3m$; $cF8$. Four zinc atoms at $0,0,0$; $0,\frac{1}{2},\frac{1}{2}$; $\frac{1}{2},0,\frac{1}{2}$ and $\frac{1}{2},\frac{1}{2},0$; four sulfur atoms at $\frac{1}{4},\frac{1}{4},\frac{1}{4}$; $\frac{1}{4},\frac{3}{4},\frac{3}{4}$; $\frac{3}{4},\frac{1}{4},\frac{3}{4}$ and $\frac{3}{4},\frac{3}{4},\frac{1}{4}$. For ZnS (sphalerite), $a = 5.42$ A. **Examples:** CdS, CdSe, CdTe, CuFeS$_2$ (HT), GaP, GaSb, InAs, InP, InSb, β MnS, β SiC, ZnO, ZnS (sphalerite), ZnSe.

B4   ZnS (wurtzite) type. Hexagonal: $P6_3mc$; $hP4$. Two zinc atoms at $\frac{1}{3}$, $\frac{2}{3},z$, and $\frac{2}{3},\frac{1}{3},\frac{1}{2}+z$ (with $z = 0$); two sulfur atoms at $\frac{1}{3},\frac{2}{3},z$ and $\frac{2}{3},\frac{1}{3},\frac{1}{2}+z$ (with $z = 0.371$). For ZnS (wurtzite), $a = 3.82$ A and $c = 6.26$ A. Equivalent positions for Zn, $0,0,0$ and $\frac{1}{3},\frac{2}{3},\frac{1}{2}$; for S, $0,0,z$ and $\frac{1}{3},\frac{2}{3},\frac{1}{2}+z$ (with $z = 0.371$). **Examples:** AlN, BeO, CdS, CdSe, CuH, InN, InSb, γ MnS, ZnO, ZnS (wurtzite), ZnSe.

B8$_1$ NiAs type. Hexagonal: $P6_3/mmc$; $hP4$. Two nickel atoms at $0,0,0$ and $0,0,\frac{1}{2}$; two arsenic atoms at $\frac{1}{3},\frac{2}{3},\frac{1}{4}$ and $\frac{2}{3},\frac{1}{3},\frac{3}{4}$. **Examples:** CoSb, CoSe, CoTe, CrH, CrSe, a'' FeS, MnSb, NiAs, NiSb, NiTe, TiS, VS, VSb.

B8$_2$ Ni$_2$In type. Hexagonal: $P6_3/mmc$; $hP6$. Two nickel atoms at $0,0,0$ and $0,0,\frac{1}{2}$; two nickel atoms at $\frac{1}{3},\frac{2}{3},\frac{3}{4}$ and $\frac{2}{3},\frac{1}{3},\frac{1}{4}$; two indium atoms at $\frac{1}{3},\frac{2}{3},\frac{1}{4}$ and $\frac{2}{3},\frac{1}{3},\frac{3}{4}$. For Ni$_2$In, $a = 4.18$ A and $c = 5.13$ A. **Examples:** AlZr$_2$, CoNiSn, Cu$_2$In, In$_2$Bi, Ni$_2$In, Ni$_{1.4}$Sn, Ti$_2$Sn.

B10   PbO type. Tetragonal: $P4/nmm$; $tP4$. Two oxygen atoms at $0,0,0$ and $\frac{1}{2},\frac{1}{2},0$; two lead atoms at $0,\frac{1}{2},z$ and $\frac{1}{2},0,\bar{z}$ (with $z = 0.237$). **Examples:** FeS, β FeTe$_{0.9}$, PbO, SnO.

B11   Gamma CuTi type. Tetragonal: $P4/nmm$; $tP4$. Two copper atoms at $0,\frac{1}{2},z$ and $\frac{1}{2},0,\bar{z}$ (with $z = 0.10$); two titanium atoms at $0,\frac{1}{2},z$ and $\frac{1}{2},0,\bar{z}$ (with $z = 0.65$).For γ CuTi, $a = 3.12$ A and $c = 5.92$ A. **Examples:** AgZr, AuTi (LT), γ CuTi.

B19 β' AuCd type. Orthorhombic: $Pmma$; $oP4$. Two gold atoms at $\frac{1}{4}$, $\frac{1}{2},z$ and $\frac{3}{4},\frac{1}{2},\bar{z}$ (with $z = 0.812$); two cadmium atoms at $\frac{1}{4},0,z$ and $\frac{3}{4},0,\bar{z}$ (with $z = 0.313$). For β' AuCd, $a = 4.76$ A, $b = 3.15$ A and $c = 4.86$ A. **Examples:** β'' AgCd, β' AuCd, CdMg, IrMo, IrW.

B20   FeSi type. Cubic: $P2_13$; $cP8$. Four iron atoms at x,x,x; $\frac{1}{2}+x,\frac{1}{2}-x,\bar{x}$; $\bar{x},\frac{1}{2}+x,\frac{1}{2}-x$ and $\frac{1}{2}-x,\bar{x},\frac{1}{2}+x$ (with $x = 0.137$); four silicon atoms at x,x,x; $\frac{1}{2}+x,\frac{1}{2}-x,\bar{x}$; $\bar{x},\frac{1}{2}+x,\frac{1}{2}-x$ and $\frac{1}{2}-x,\bar{x},\frac{1}{2}+x$ (with $x = 0.842$). **Examples:** CoSi, FeSi, MnSi.

B31   MnP type. Orthorhombic: $Pnma$; $oP8$. Four manganese atoms at x, $\frac{1}{4},z$; $\bar{x},\frac{3}{4},\bar{z}$; $\frac{1}{2}-x,\frac{3}{4},\frac{1}{2}+z$ and $\frac{1}{2}+x,\frac{1}{4},\frac{1}{2}-z$ (with $x = 0.20$ and $z = 0.005$); four phosphorus atoms at $x,\frac{1}{4},z$; $\bar{x},\frac{3}{4},\bar{z}$; $\frac{1}{2}-x,\frac{3}{4},\frac{1}{2}+z$ and $\frac{1}{2}+x,\frac{1}{4},\frac{1}{2}-z$ (with $x = 0.57$ and $z = 0.19$). **Examples:** CoP, CrP, FeP, MnP, WP.

B$_h$ WC type. Hexagonal: $P\bar{6}m2$; $hP2$. One tungsten atom at $0,0,0$; one carbon atom at $\frac{1}{3},\frac{2}{3},\frac{1}{2}$ or at $\frac{2}{3},\frac{1}{3},\frac{1}{2}$. For WC, $a = 2.91$ A and $c = 2.84$ A. **Examples:** γ MoC, TiS, WC, WN, Zr$_3$S$_2$.

C1   CaF$_2$ (fluorite) type. Face-centered cubic: $Fm3m$; $cF12$. Four calcium atoms at $0,0,0$; $0,\frac{1}{2},\frac{1}{2}$; $\frac{1}{2},0,\frac{1}{2}$ and $\frac{1}{2},\frac{1}{2},0$; eight fluorine atoms at $\frac{1}{4},\frac{1}{4},\frac{1}{4}$; $\frac{1}{4},\frac{3}{4},\frac{3}{4}$; $\frac{3}{4},\frac{1}{4},\frac{3}{4}$; $\frac{3}{4},\frac{3}{4},\frac{1}{4}$; $\frac{3}{4},\frac{3}{4},\frac{3}{4}$; $\frac{3}{4},\frac{1}{4},\frac{1}{4}$; $\frac{1}{4},\frac{3}{4},\frac{1}{4}$ and $\frac{1}{4},\frac{1}{4},\frac{3}{4}$. For CaF$_2$, $a = 5.46$ A. **Examples:** Be$_2$B, Be$_2$C, CaF$_2$, CoSi$_2$, rare-earth hydrides, K$_2$O, K$_2$S, Mg$_2$Pb, Mg$_2$Si, ζ NiSi$_2$, UN$_2$, UO$_2$.

C2   FeS$_2$ (pyrite) type. Cubic: $Pa3$; $cP12$. Four iron atoms at $0,0,0$; $0,\frac{1}{2},\frac{1}{2}$; $\frac{1}{2},0,\frac{1}{2}$ and $\frac{1}{2},\frac{1}{2},0$; eight sulfur atoms at x, x,x; $\frac{1}{2}+x,\frac{1}{2}-x,\bar{x}$; $\bar{x},\frac{1}{2}+x,\frac{1}{2}-x$; $\frac{1}{2}-x,\bar{x},\frac{1}{2}+x$; $\bar{x},\bar{x},\bar{x}$; $\frac{1}{2}-x,\frac{1}{2}+x,x$; $x,\frac{1}{2}-x,\frac{1}{2}+x$ and $\frac{1}{2}+x,x,\frac{1}{2}-x$. For FeS$_2$ (pyrite), $x = 0.386$ and $a = 5.42$ A. **Examples:** CoPS, CoS$_2$, CoSe$_2$, FeS$_2$ (pyrite), MnS$_2$, MnTe$_2$, NiS$_{2+x}$, NiSe$_2$.

C3   Cu$_2$O type. Cubic: $Pn3m$; $cP6$. Four copper atoms at $\frac{1}{4},\frac{1}{4},\frac{1}{4}$; $\frac{1}{4},\frac{3}{4},\frac{3}{4}$; $\frac{3}{4},\frac{1}{4},\frac{3}{4}$ and $\frac{3}{4},\frac{3}{4},\frac{1}{4}$; two oxygen atoms at $0,0,0$ and $\frac{1}{2},\frac{1}{2},\frac{1}{2}$. For Cu$_2$O, $a = 4.26$ A. **Examples:** Ag$_2$O, Cu$_2$O.

C4   TiO$_2$ (rutile) type. Tetragonal: $P4_2/mnm$; $tP6$. Two titanium atoms at $0,0,0$ and $\frac{1}{2},\frac{1}{2},\frac{1}{2}$; four oxygen atoms at $x,x,0$; $\bar{x},\bar{x},0$; $\frac{1}{2}+x,\frac{1}{2}-x,\frac{1}{2}$ and $\frac{1}{2}-x,\frac{1}{2}+x,\frac{1}{2}$. For TiO$_2$, $x = 0.3056$, $a = 4.59$ A and $c = 2.96$ A. **Examples:** CrO$_2$, β MnO$_2$, PbO$_2$, SnO$_2$, TaO$_2$, TeO$_2$, TiO$_2$ (rutile), VO$_2$ (HT), WO$_2$.

C11$_b$ MoSi$_2$ type. Tetragonal: $I4/mmm$; $tI6$. Two molybdenum atoms at 0, $0,0$ and $\frac{1}{2},\frac{1}{2},\frac{1}{2}$; four silicon atoms at $0,0,z$; $0,0,\bar{z}$; $\frac{1}{2},\frac{1}{2},\frac{1}{2}+z$ and $\frac{1}{2},\frac{1}{2},\frac{1}{2}-z$ (with $z = 0.333$). For MoSi$_2$, $a = 3.20$ A and $c = 7.86$ A. **Examples:** AgZr$_2$, AlCr$_2$, Au$_2$Be, Au$_2$Mn, CuTi$_2$, Hg$_2$Mg, MoSi$_2$, Ni$_2$Ta, Si$_2$W.

C14   MgZn$_2$ type. Hexagonal: $P6_3/mmc$; $hP12$. Four magnesium atoms at $\frac{1}{3},\frac{2}{3},z$; $\frac{2}{3},\frac{1}{3},\bar{z}$; $\frac{2}{3},\frac{1}{3},\frac{1}{2}+z$ and $\frac{1}{3},\frac{2}{3},\frac{1}{2}-z$ (with $z = 0.062$); two zinc atoms at $0,0,0$ and $0,0,\frac{1}{2}$; six zinc atoms at $x,2x,\frac{1}{4}$; $2\bar{x},\bar{x},\frac{1}{4}$; $x,\bar{x},\frac{1}{4}$; $\bar{x},2\bar{x},\frac{3}{4}$; $2x,x,\frac{3}{4}$ and $\bar{x},x,\frac{3}{4}$ (with $x = 0.83$). For MgZn$_2$, $a = 5.18$ A and $c = 8.52$ A. **Examples:** Al$_2$Zr, Be$_2$Mo, CaCd$_2$, CaMg$_2$, CdCu$_2$, Fe$_2$Mo, FeSiW, Fe$_2$Ta, Fe$_2$Ti, Fe$_2$W, MgZn$_2$, TiZn$_2$.

C15   Cu$_2$Mg type. Face-centered cubic: $Fd3m$; $cF24$. Eight magnesium atoms at $0,0,0$; $0,\frac{1}{2},\frac{1}{2}$; $\frac{1}{2},0,\frac{1}{2}$; $\frac{1}{2},\frac{1}{2},0$; $\frac{1}{4},\frac{1}{4},\frac{1}{4}$; $\frac{1}{4},\frac{3}{4},\frac{3}{4}$; $\frac{3}{4},\frac{1}{4},\frac{3}{4}$ and $\frac{3}{4},\frac{3}{4},\frac{1}{4}$; 16 copper atoms at $\frac{5}{8},\frac{5}{8},\frac{5}{8}$; $\frac{5}{8},\frac{1}{8},\frac{1}{8}$; $\frac{1}{8},\frac{5}{8},\frac{1}{8}$; $\frac{1}{8},\frac{1}{8},\frac{5}{8}$; $\frac{5}{8},\frac{7}{8},\frac{7}{8}$; $\frac{5}{8},\frac{3}{8},\frac{3}{8}$; $\frac{1}{8},\frac{7}{8},\frac{3}{8}$; $\frac{1}{8},\frac{3}{8},\frac{7}{8}$; $\frac{7}{8},\frac{5}{8},\frac{7}{8}$; $\frac{7}{8},\frac{1}{8},\frac{3}{8}$; $\frac{3}{8},\frac{5}{8},\frac{3}{8}$; $\frac{3}{8},\frac{1}{8},\frac{7}{8}$; $\frac{7}{8},\frac{7}{8},\frac{5}{8}$; $\frac{7}{8},\frac{3}{8},\frac{1}{8}$; $\frac{3}{8},\frac{7}{8},\frac{1}{8}$ and $\frac{3}{8},\frac{3}{8},\frac{5}{8}$. For Cu$_2$Mg, $a = 7.05$ A. **Examples:** Al$_2$Ca, Al$_2$U, CdCuZn, Co$_2$U, Co$_2$Zr, Cr$_2$Ti, Cu$_2$Mg, FeNiTa, Fe$_2$U, Fe$_2$Zr, MgNiZn, α TiCo$_2$, ZrW$_2$.

C16   CuAl$_2$ type. Body-centered tetragonal: $I4/mcm$; $tI12$. Four copper atoms at $0,0,\frac{1}{4}$; $\frac{1}{2},\frac{1}{2},\frac{3}{4}$; $0,0,\frac{3}{4}$ and $\frac{1}{2},\frac{1}{2},\frac{1}{4}$; eight aluminum atoms at $x,\frac{1}{2}+x,0$; $\bar{x},\frac{1}{2}-x,0$; $\frac{1}{2}+x,\bar{x},0$; $\frac{1}{2}+x,x,0$; $\frac{1}{2}+x,x,\frac{1}{2}$; $\frac{1}{2}-x,x,\frac{1}{2}$; $x,\frac{1}{2}-x,\frac{1}{2}$ and $\bar{x},\frac{1}{2}+x,\frac{1}{2}$ (with $x = 0.158$). For CuAl$_2$, $a = 6.07$ A and $c = 4.87$ A. **Examples:** Co$_2$B, Cr$_2$B, θ CuAl$_2$, Fe$_2$B, FeSn$_2$, Mo$_2$B, Ni$_2$B, W$_2$B.

C18,   FeS$_2$ (marcasite) type. Orthorhombic: $Pnnm$; $oP6$. Two iron atoms at $0,0,0$ and $\frac{1}{2},\frac{1}{2},\frac{1}{2}$; four sulfur atoms at $x,y,0$; $\bar{x},\bar{y},0$; $\frac{1}{2}+x,\frac{1}{2}-y,\frac{1}{2}$ and $\frac{1}{2}-x,\frac{1}{2}+y,\frac{1}{2}$ (with $x = 0.200$ and $y = 0.378$). For FeS$_2$ (marcasite), $a = 4.44$ A, $b = 5.42$ A and $c = 3.39$ A. **Examples:** γ CrSb$_2$, FeP$_2$, FeS$_2$ marcasite), FeSe$_2$, FeTe$_2$, NiSb$_2$.

C22   Fe$_2$P type. Hexagonal: $P\bar{6}2m$; $hP9$. Three iron atoms at $x,0,0$; $0,x,0$ and $\bar{x},\bar{x},0$ (with $x = 0.256$); three iron atoms at $x,0,\frac{1}{2}$; $0,x,\frac{1}{2}$ and $\bar{x},\bar{x},\frac{1}{2}$ (with $x = 0.594$); three phosphorus atoms at $\frac{1}{3},\frac{2}{3},0$ and $\frac{2}{3},\frac{1}{3},0$. For Fe$_2$P, $a = 5.93$ A and $c = 3.45$ A. **Examples:** Fe$_2$P, Mn$_2$P, Ni$_2$P, Pt$_2$Si (HT).

C38   Cu$_2$Sb type. Tetragonal: $P4/nmm$; $tP6$. Four copper atoms at $0,0,0$; $\frac{1}{2},\frac{1}{2},0$; $0,\frac{1}{2},z$ and $\frac{1}{2},0,\bar{z}$ (with $z = 0.27$; two antimony atoms at $0,\frac{1}{2},z$ and $\frac{1}{2},0,\bar{z}$ (with $z = 0.70$). For Cu$_2$Sb, $a = 3.99$ A and $c = 6.09$ A. **Examples:** AlNaSi$_4$, AsCr$_2$, AsMn$_2$, Bi$_2$U, Cu$_2$Sb, Mn$_2$Sb, Pu$_2$U, Sb$_2$U.

Crystal structure of metals (continued)

$D0_3$ BiF_3 or $BiLi_3$ type. Face-centered cubic superlattice: *Fm3m*; *cF*16. Four bismuth atoms at 0,0,0; 0,½,½; ½,0,½ and ½,½,0; 12 fluorine (or lithium) atoms at ½,½,½; ½,0,0; 0,½,0; 0,0,½; ¼,¼,¼; ¼,¾,¾; ¾,¼,¾; ¾,¾,¼; ¾,¾,¾; ¾,¼,¼; ¼,¾,¼ and ¼,¼,¾. For $BiLi_3$, a = 6.71 A. **Examples:** BiF_3, $BiLi_3$, Fe_3Al, γ Cu_3n (HT), α Fe_3Si, Mn_3Si, Ni_3Sn (HT).

$D0_{11}$ Fe_3C (cementite) type. Orthorhombic: *Pnma*; *oP*16. Sixteen atoms per cell. **Examples:** Co_3B, Co_3C, Fe_3C, Mn_3C, Ni_3C, Pd_3P.

$D0_{19}$ Ni_3Sn type. Hexagonal: $P6_3/mmc$; *hP*8. Two tin atoms at ⅓,⅔,¼ and ⅔,⅓,¾; six nickel atoms at x,2x,¼; 2$\bar{x}$,$\bar{x}$,¼; x,$\bar{x}$,¼; $\bar{x}$,2$\bar{x}$,¾; 2x,x,¾ and $\bar{x}$,x,¾ (with x = 0.833). For Ni_3Sn, a = 5.29 A and c = 4.24 A. **Examples:** $AlTi_{2-3}$, Cd_3Mg, $CdMg_3$, Co_3Mo, Co_3W, β'' Fe_3Sn, γ Ni_3In, Ni_3Sn, Ti_4Pb.

$D0_{24}$ Ni_3Ti type. Hexagonal: $P6_3/mmc$; *hP*16. Four titanium atoms at 0,0,0; 0,0,½; ⅓,⅔,¼ and ⅔,⅓,¾; 12 nickel atoms at ½,0,0; 0,½,0; ½,½,0; ½,0,½; 0,½,½; ½,½,½; x,2x,¼; 2$\bar{x}$,$\bar{x}$,¼; x,$\bar{x}$,¼; $\bar{x}$,2$\bar{x}$,¾; 2x,x,¾ and $\bar{x}$,x,¾ (with x = 0.833). For Ni_3Ti, a = 2.55 A and c = 8.31 A. **Examples:** Co_3Ti, Ni_3Ti, Pd_3Zr.

$D5_1$ Alpha Al_2O_3 type. Rhombohedral-hexagonal: $R\bar{3}c$; *hR*10. Ten atoms per unit rhombohedral cell or 30 atoms per unit hexagonal cell. (There are also other structures of alumina.) **Examples:** α Al_2O_3, α Fe_2O_3, Rh_2O_3, Ti_2O_3, V_2O_3 (HT).

$D8_2$ Gamma Cu_5Zn_8 (gamma brass) type. Body-centered cubic: $I\bar{4}3m$; *cI*52. Fifty-two atoms per cell. **Examples:** γ Ag_5Cd_8, γ Ag_5Zn_8, Al_8V_5, δ Cd_8Cu_5, γ Cu_5Zn_8.

$D8_4$ $Cr_{23}C_6$ type. Face-centered cubic: *Fm3m*; *cF*116. One hundred sixteen atoms per cell. **Examples:** $Cr_{23}C_6$, $Fe_{21}Mo_2C_6$, $Fe_{21}W_2C_6$, $Mn_{23}C_6$, $Ni_{19.5}Zr_{3.5}B_6$.

$D8_5$ Fe_7W_6 (mu-phase) type. Rhombohedral: $R\bar{3}m$; *hR*13. Thirteen atoms per cell. **Examples:** μ Co_7Mo_6, Co_7W_6, Fe_7Mo_6, Fe_7W_6, NiTa.

$D8_b$ Sigma FeCr (sigma-phase) type. Tetragonal: $P4_2/mnm$; *tP*30. Thirty atoms per cell. **Examples:** σ CoCr, Co_2Mo_3, $CrMn_3$, σ FeCr, σ FeMo, σ FeV, σ Mn-Mo, TaV.

$E2_1$ $CaTiO_3$ (perovskite) type. Cubic: *Pm3m*; *cP*5. One calcium atom at 0,0,0; one titanium atom at ½,½,½; three oxygen atoms at 0,½,½; ½,0,½ and ½,½,0. **Examples:** $AlCFe_3$, $AlCMn_3$, $AlCTi_3$, $CaTiO_3$, Fe_3C_xIn, Fe_3NNi, Fe_3NPd, Fe_3NSn.

$E9_3$ Fe_3W_3C (eta-carbide) type. Face-centered cubic: *Fd3m*; *cF*112. One hundred twelve atoms per cell. **Examples:** $CoCb_2(C,N,O)_x$, Co_2Mo_4C, Cr_3Cb_3C, η Fe_2Cb_3, Fe_3Mo_3C, Fe_3Mo_3N, Fe_3W_3C, Mn_3Mo_3C, Mo_3Ni_3C, $NiTi_2$, Ni_3W_3C.

$H1_1$ Spinel (Al_2MgO_4 or Fe_3O_4) type. Face-centered cubic: *Fd3m*; *cF*56. Fifty-six atoms per cell. **Examples:** Al_2CrS_4, Al_2MgO_4, Co_2NiS_4, Co_3O_4, Co_3S_4, CuS_4Ti_2, $FeNi_2S_4$, Fe_3O_4, Fe_3S_4 (greigite), Ni_3S_4 (LT).

. . . *R*-phase (in Co-Cr-Mo) type. Rhombohedral-hexagonal: $R\bar{3}$; *hR*53. Fifty-three atoms per cell. See *A*12 type, above. **Examples:** *R*-(Co-Cr-Mo), $Co_3Cr_3Si_2$, *R*-(Co-Mn-Mo), *R*-$Fe_{52}Mn_{16}Mo_{32}$, Fe_2SiV_2, $Mn_{78}Mo_3Si_{19}$, Mn_6Si, Ni_3SiV_6.

. . . *P*-phase (in Mo-Cr-Ni) type. Orthorhombic: *Pbnm*: *oP*56. Fifty-six atoms per cell. See *A*12 type, above. **Examples:** *P*-$Cr_{18}Mo_{42}Ni_{40}$, *P*-(Mo-Fe-Ni), *P*-(Mo-Mn-Co).

$L1_0$ AuCu I type. Tetragonal superlattice: *P4/mmm*; *tP*4. Two gold atoms at 0,0,0 and ½,½,0; two copper atoms at 0,½,½ and ½,0,½. **Examples:** AgTi, AlTi, AuCu I, θ CdPt, FePd, γ'' FePt, θ MnNi, NiPt.

$L1_1$ CuPt type. Rhombohedral superlattice: $R\bar{3}m$; *hR*32. **Example:** CuPt.

$L1_2$ $AuCu_3$ I type. Cubic superlattice: *Pm3m*; *cP*4. One gold atom at 0,0,0; three copper atoms at 0,½,½; ½,0,½ and ½,½,0. **Examples:** α' $AlNi_3$, $AlZr_3$, Au_3Cu, $AuCu_3$ I, $CoPt_3$, Cr_3Pt, Fe_3Ga, $FePd_3$, Ni_3Fe, Ni_3Mn, Sn_3U.

L′2 Martensite (Fe-C) type. Tetragonal: *I4/mmm*. In the unit cell there are iron atoms at 0,0,0 and ½,½,½; the carbon atoms are random, at ½,½,0 and/or 0,0,½, to provide two iron atoms and up to 0.12 carbon atoms per cell. **Examples:** Fe-C martensite, α' Fe-N martensite.

L′3 Fe_2N, or W_2C, type. Hexagonal: $P6_3/mmc$; *hP*3. Two iron atoms at ⅓,⅔,¼ and ⅔,⅓,¾; one nitrogen atom at either 0,0,0 or 0,0,½. **Examples:** Fe_2N, ζ Mn_2N, β Ta_2C, $Ta_{\sim 2}N$, V_2C, W_2C (LT).

Crystal structure of metals (continued)

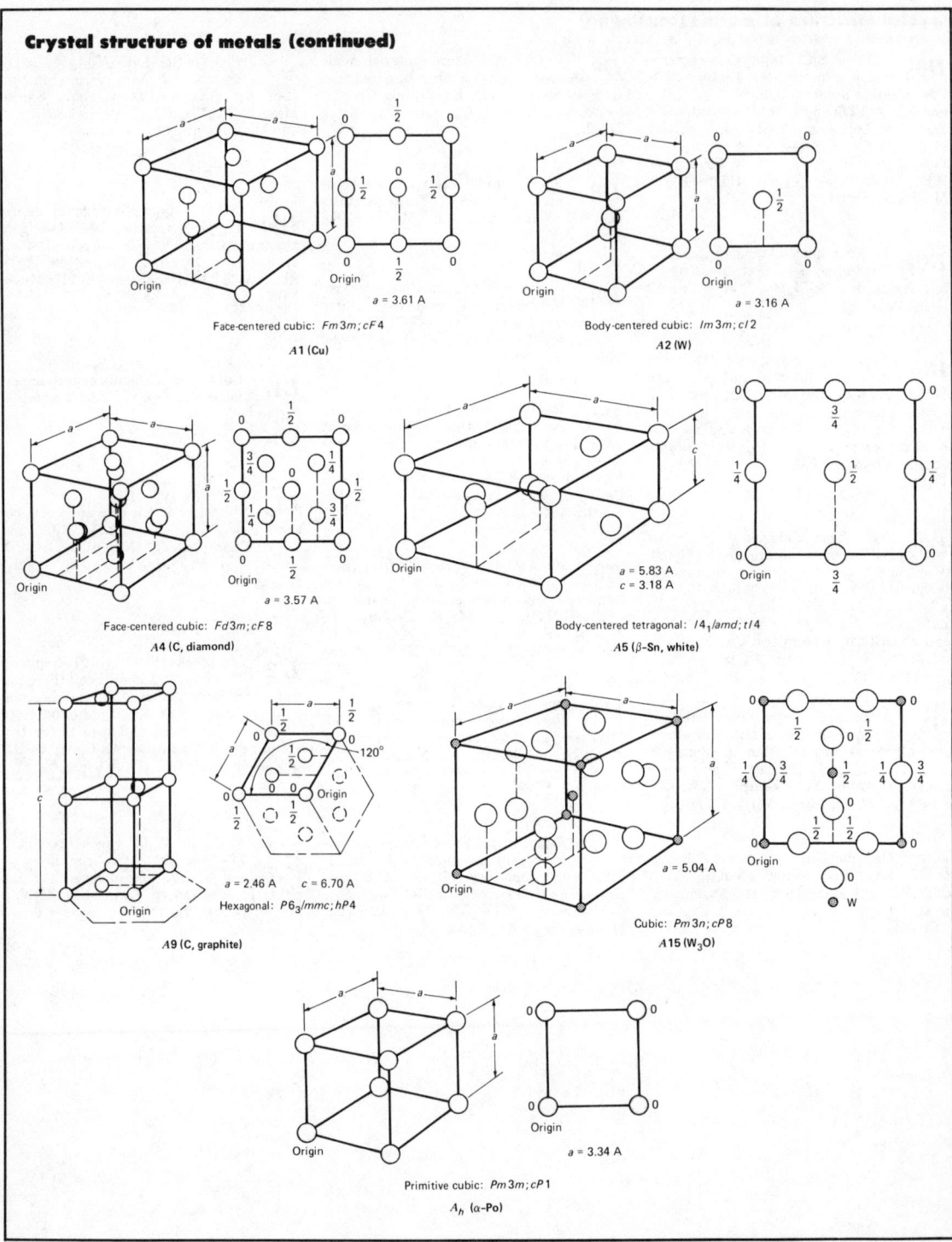

Face-centered cubic: $Fm3m$; $cF4$

A1 (Cu)

$a = 3.61$ A

Body-centered cubic: $Im3m$; $cI2$

A2 (W)

$a = 3.16$ A

Face-centered cubic: $Fd3m$; $cF8$

A4 (C, diamond)

$a = 3.57$ A

Body-centered tetragonal: $I4_1/amd$; $tI4$

A5 (β-Sn, white)

$a = 5.83$ A
$c = 3.18$ A

Hexagonal: $P6_3/mmc$; $hP4$

A9 (C, graphite)

$a = 2.46$ A $c = 6.70$ A

Cubic: $Pm3n$; $cP8$

A15 (W_3O)

$a = 5.04$ A

W

Primitive cubic: $Pm3m$; $cP1$

A_h (α-Po)

$a = 3.34$ A

Crystal structure of metals (continued)

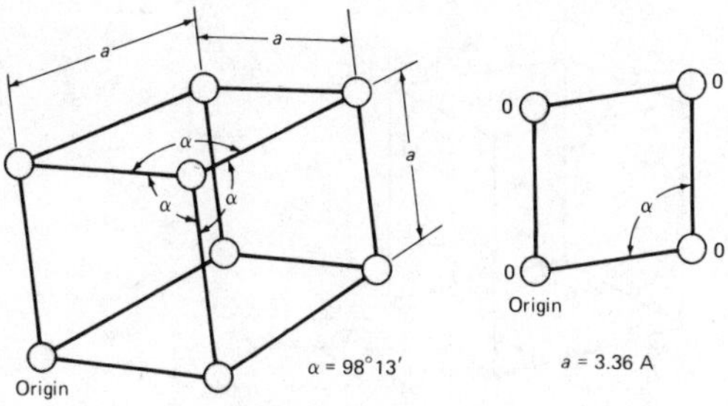

$\alpha = 98°13'$

$a = 3.36$ A

Rhombohedral: $R\overline{3}m; hR1$

A_i (β–Po)

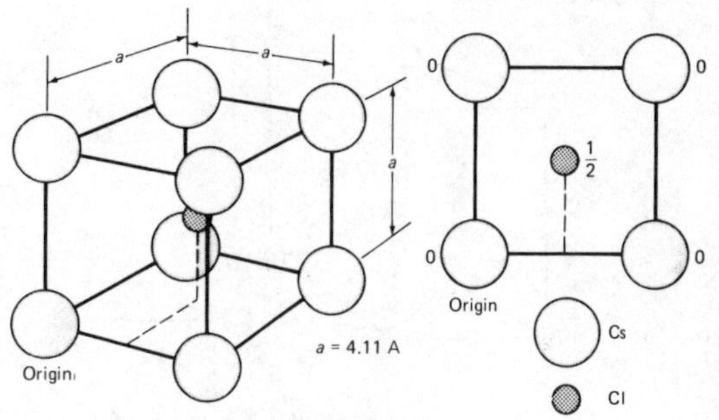

$a = 4.11$ A

Cs

Cl

Cubic: $Pm3m; cP2$

$B2$ (CsCl)

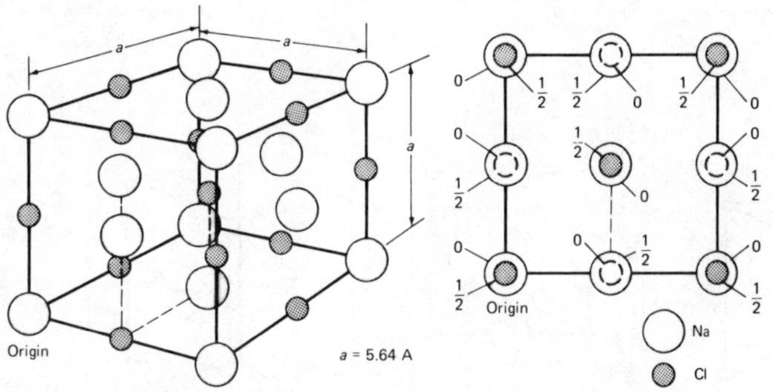

$a = 5.64$ A

Na

Cl

Face-centered cubic: $Fm3m; cF8$

$B1$ (NaCl)

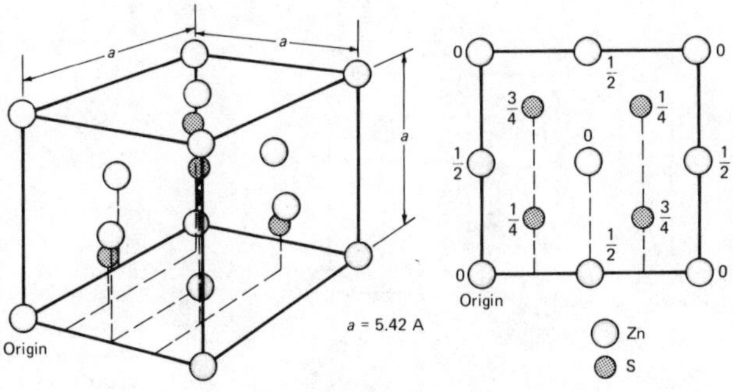

$a = 5.42$ A

Zn

S

Face-centered cubic: $F\overline{4}3m; cF8$

$B3$ (ZnS, sphalerite)

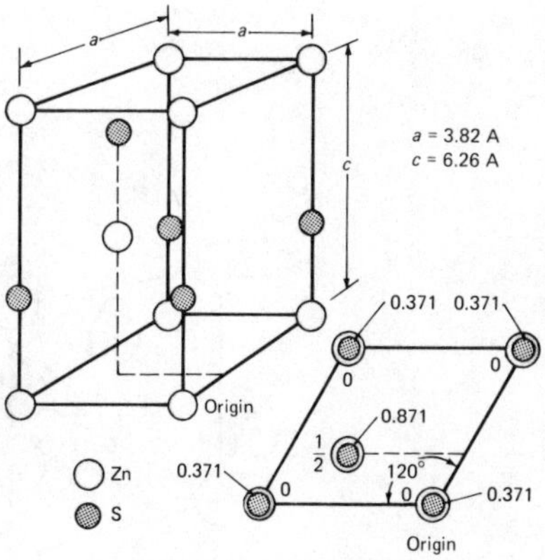

$a = 3.82$ A
$c = 6.26$ A

Zn

S

0.371   0.371

0.871

0.371

0.371

120°

Hexagonal: $P6_3mc; hP4$

$B4$ (ZnS, wurtzite)

Crystal structure of metals (continued)

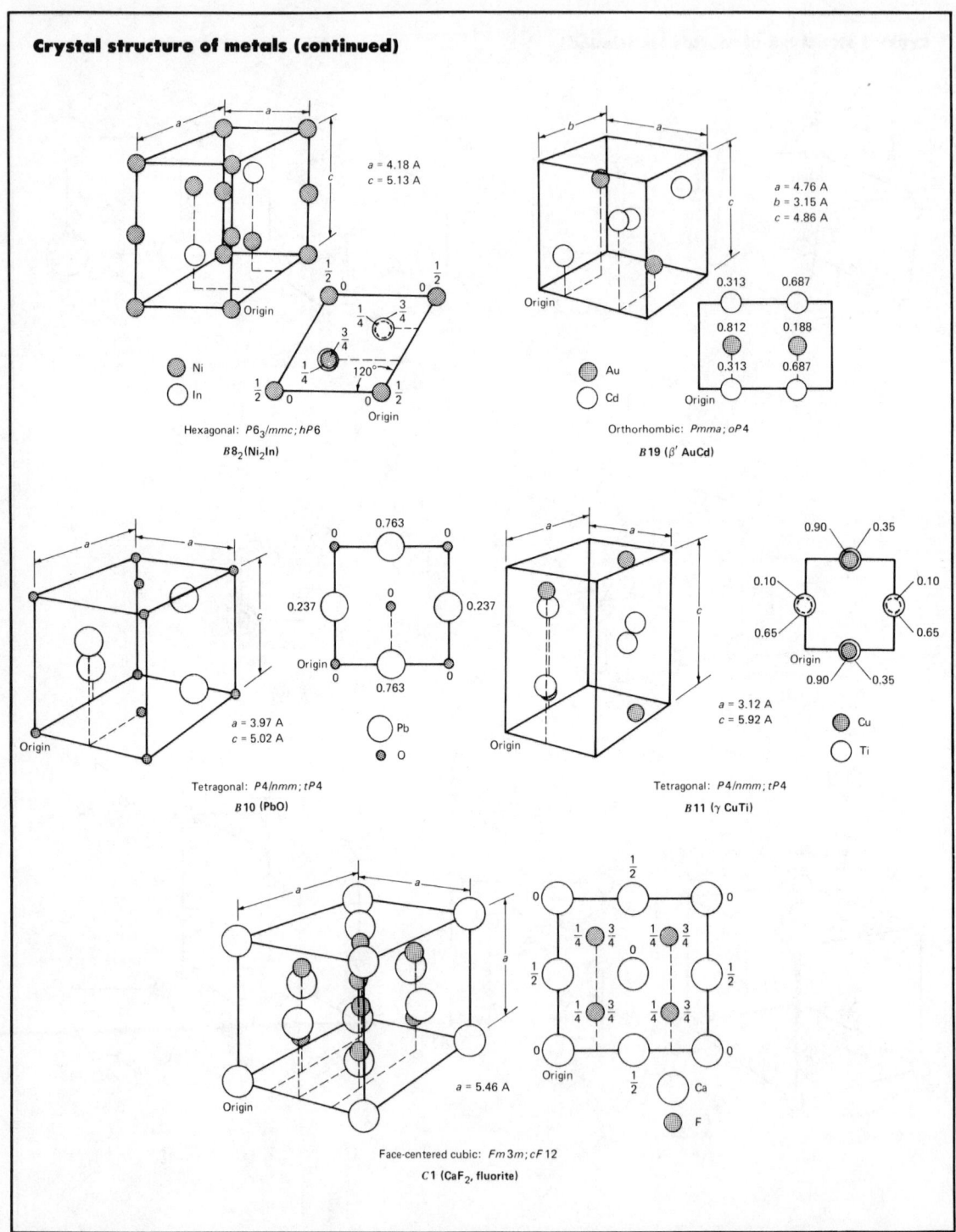

a = 4.18 A
c = 5.13 A

Ni
In

Hexagonal: $P6_3/mmc$; $hP6$
$B8_2$(Ni$_2$In)

a = 4.76 A
b = 3.15 A
c = 4.86 A

Au
Cd

Orthorhombic: $Pmma$; $oP4$
$B19$ (β' AuCd)

a = 3.97 A
c = 5.02 A

Pb
O

Tetragonal: $P4/nmm$; $tP4$
$B10$ (PbO)

a = 3.12 A
c = 5.92 A

Cu
Ti

Tetragonal: $P4/nmm$; $tP4$
$B11$ (γ CuTi)

a = 5.46 A

Ca
F

Face-centered cubic: $Fm3m$; $cF12$
$C1$ (CaF$_2$, fluorite)

Crystal structure of metals (continued)

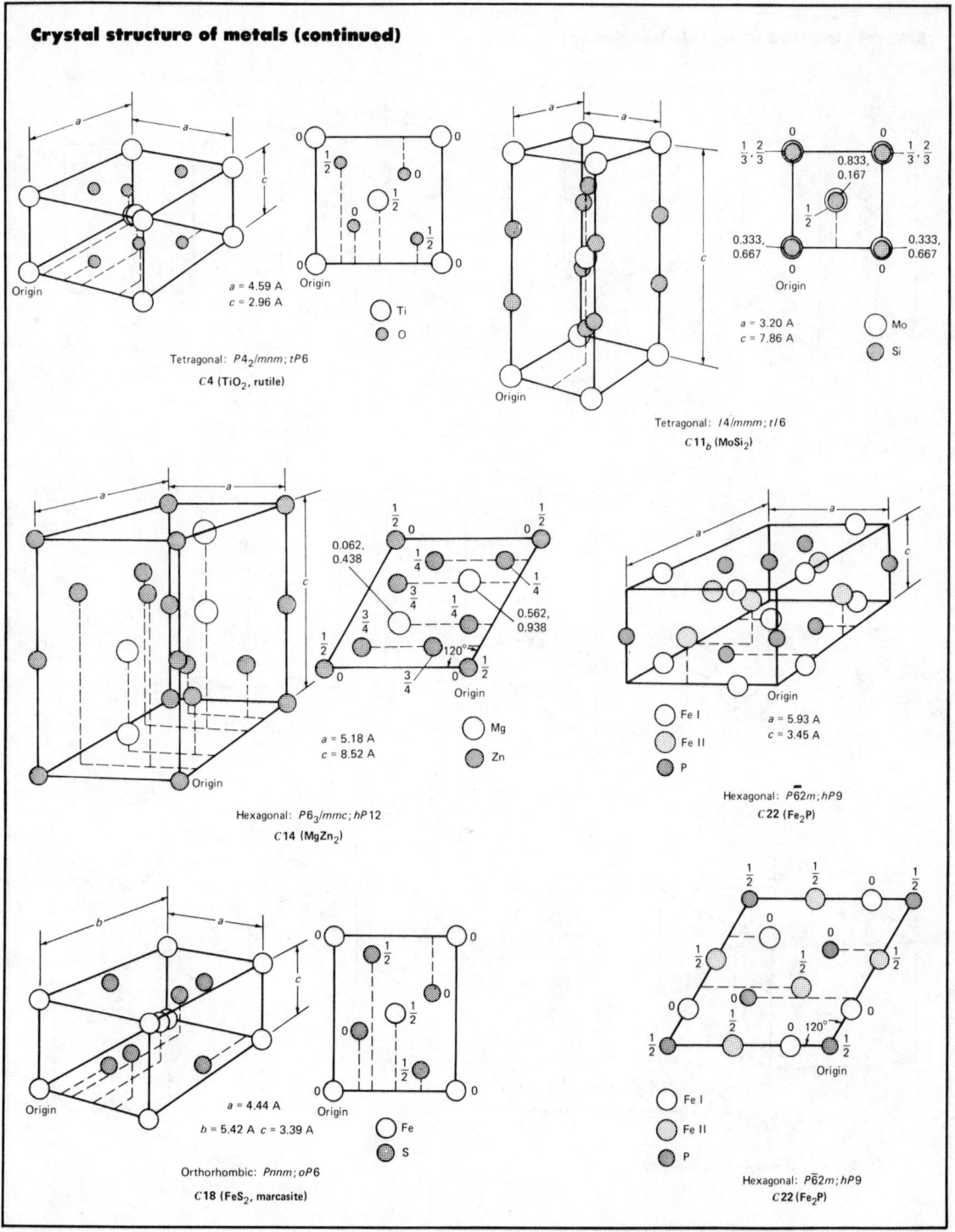

a = 4.59 A
c = 2.96 A

○ Ti
● O

Tetragonal: $P4_2/mnm$; $tP6$

$C4$ (TiO_2, rutile)

a = 3.20 A
c = 7.86 A

○ Mo
● Si

Tetragonal: $I4/mmm$; $tI6$

$C11_b$ ($MoSi_2$)

a = 5.18 A
c = 8.52 A

○ Mg
● Zn

Hexagonal: $P6_3/mmc$; $hP12$

$C14$ ($MgZn_2$)

○ Fe I
● Fe II
● P

a = 5.93 A
c = 3.45 A

Hexagonal: $P\overline{6}2m$; $hP9$

$C22$ (Fe_2P)

a = 4.44 A
b = 5.42 A c = 3.39 A

○ Fe
● S

Orthorhombic: $Pnnm$; $oP6$

$C18$ (FeS_2, marcasite)

○ Fe I
● Fe II
● P

Hexagonal: $P\overline{6}2m$; $hP9$

$C22$ (Fe_2P)

Crystal structure of metals (continued)

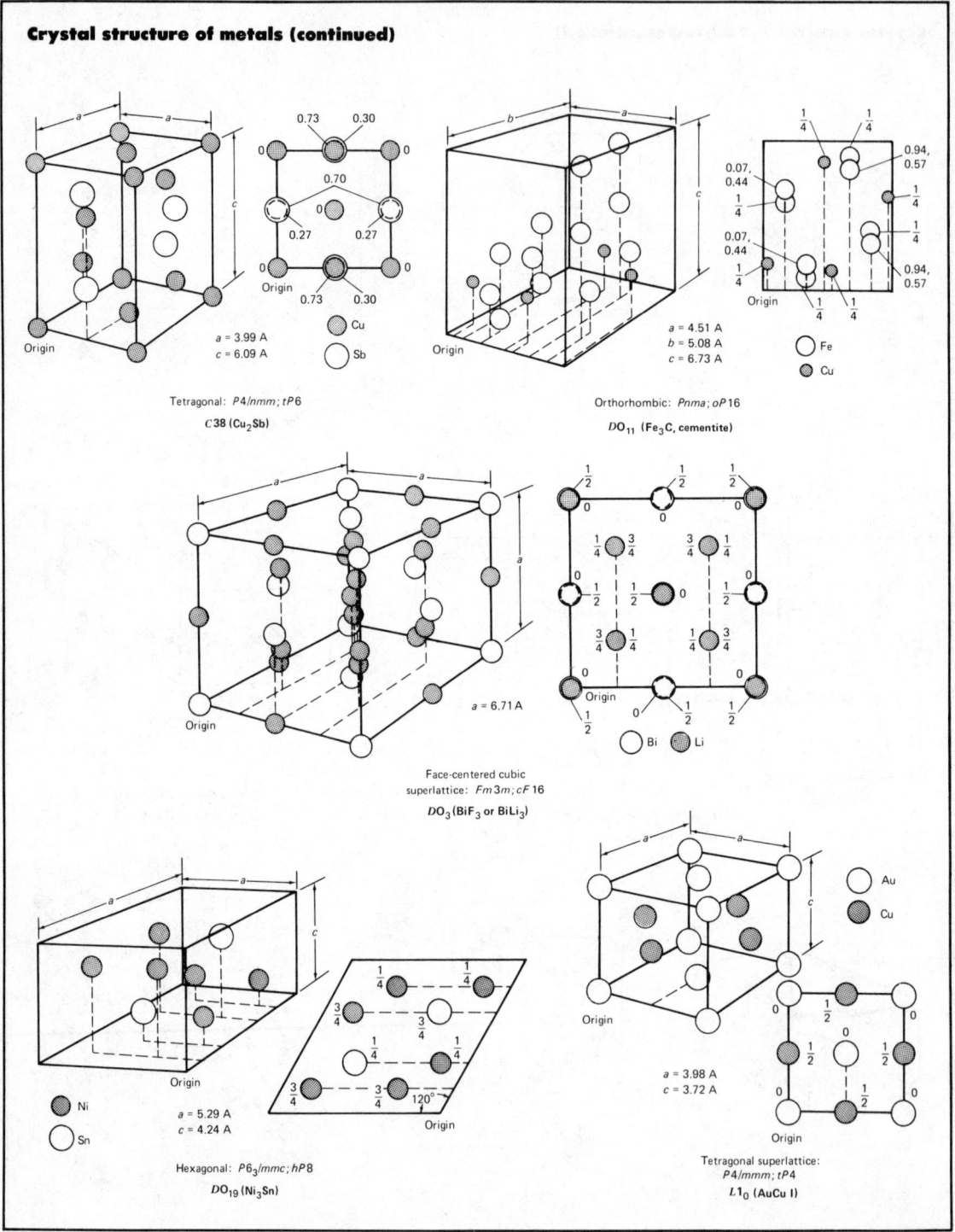

a = 3.99 A
c = 6.09 A

Cu
Sb

Tetragonal: P4/nmm; tP6

C38 (Cu₂Sb)

a = 4.51 A
b = 5.08 A
c = 6.73 A

Fe
Cu

Orthorhombic: Pnma; oP16

DO₁₁ (Fe₃C, cementite)

a = 6.71 A

Bi   Li

Face-centered cubic
superlattice: Fm3m; cF16

DO₃ (BiF₃ or BiLi₃)

Ni
Sn

a = 5.29 A
c = 4.24 A

Hexagonal: P6₃/mmc; hP8

DO₁₉ (Ni₃Sn)

Au
Cu

a = 3.98 A
c = 3.72 A

Tetragonal superlattice:
P4/mmm; tP4

L1₀ (AuCu I)

Crystal structure of metals (continued)

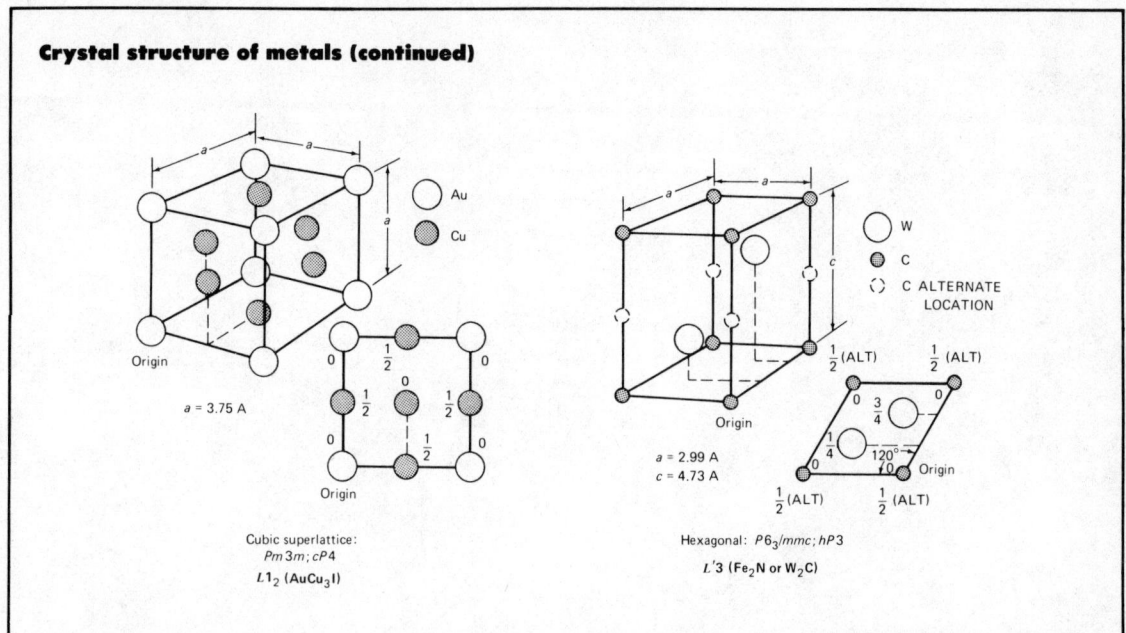

Origin

a = 3.75 A

Origin

Cubic superlattice:
$Pm\,3m$; $cP4$

$L1_2$ (AuCu$_3$I)

Au

Cu

Origin

a = 2.99 A
c = 4.73 A

W

C

C ALTERNATE
LOCATION

$\frac{1}{2}$ (ALT) $\frac{1}{2}$ (ALT)

$\frac{1}{2}$ (ALT) $\frac{1}{2}$ (ALT)

Origin

Hexagonal: $P6_3/mmc$; $hP3$

$L'3$ (Fe$_2$N or W$_2$C)

Testing and Inspection

General Information

Advantages and limitations of nondestructive and destructive testing

Nondestructive testing	Destructive testing

Nondestructive testing

Advantages

1. Can be done directly on production items without regard to part cost or quantity available, and no scrap losses are incurred except for bad parts
2. Can be done on 100% of production or on representative samples
3. Can be used when variability is wide and unpredictable
4. Different tests can be applied to the same item simultaneously or sequentially
5. The same test can be repeated on the same item
6. May be performed on parts in service
7. Cumulative effect of service usage can be measured directly
8. May reveal failure mechanism
9. Little or no specimen preparation is required
10. Equipment is often portable for use in field
11. Labor costs are usually low, especially for repetitive testing of similar parts

Limitations

1. Results often must be interpreted by a skilled, experienced technician
2. In absence of proven correlation, different observers may disagree on meaning and significance of test results
3. Properties are measured indirectly, and often only qualitative or comparative measurements can be made
4. Some nondestructive tests require large capital investments

Destructive testing

Advantages

1. Can often directly and reliably measure response to service conditions
2. Measurements are quantitative, and usually valuable for design or standardization
3. Interpretation of results by a skilled technician usually not required
4. Correlation between tests and service usually direct, leaving little margin for disagreement among observers as to meaning and significance of test results

Limitations

1. Can be applied only to a sample, and separate proof that the sample represents the population is required
2. Tested parts cannot be placed in service
3. Repeated tests of same item are often impossible, and different types of tests may require different samples
4. Extensive testing usually cannot be justified, because of large scrap losses
5. May be prohibited on parts with high material or fabrication costs, or on parts of limited availability
6. Cumulative effect of service usage cannot be measured directly, but only inferred from tests on parts used for different lengths of time
7. Difficult to apply to parts in service, and usually terminates their useful life
8. Extensive machining or other preparation of test specimens is often required
9. Capital investment and manpower costs are often high

Nondestructive methods for evaluating materials

Method	Measures or detects	Applications	Advantages	Limitations
Acoustic emission	Crack initiation and growth rate Internal cracking in welds during cooling Boiling or cavitation Friction or wear Plastic deformation Phase transformations	Pressure vessels Stressed structures Turbine or gear boxes Fracture mechanics research Weldments Sonic signature analysis	Remote and continuous surveillance Permanent record Dynamic (rather than static) detection of cracks Portable Triangulation techniques to locate flaws	Transducers must be placed on part surface Highly ductile materials yield low amplitude emissions Part must be stressed or operating Test system noise needs to be filtered out
Acoustic-impact (tapping)	Debonded areas or delaminations in metal or nonmetal composites or laminates Cracks in turbine wheels or turbine blades Loose rivets or fasteners Crushed core	Brazed or adhesive-bonded structures Bolted or riveted assemblies Turbine blades Turbine wheels Composite structures Honeycomb assemblies	Portable Easy to operate May be automated Permanent record or positive meter readout No couplant required	Part geometry and mass influences test results Impactor and probe must be repositioned to fit geometry of part Reference standards required Pulser impact rate is critical for repeatability
Barkhausen noise analysis	Residual stresses in ferromagnetic steels	Jet engine components such as compressor blades, discs, diffuser cases	Nondestructive stress analysis Permanent record Fully automatic	Expensive Requires reference standard Need trained operator Not yet a production tool

(continued)

Nondestructive methods for evaluating materials (continued)

Method	Measures or detects	Applications	Advantages	Limitations
Eddy current (100 Hz-10 kHz)	Subsurface cracks around fastener holes in aircraft structure	Aluminum and titanium structure	Detect subsurface cracks not detectable by radiography	Part geometry Will not detect short cracks
Eddy current (10 kHz-6 MHz).........	Surface and subsurface cracks and seams Alloy content Heat treatment variations Wall thickness, coating thickness Crack depth Conductivity Permeability	Tubing Wire Ball bearings "Spot checks" on all types of surfaces Proximity gage Metal detector Metal sorting Measure conductivity in % IACS	No special operator skills required High speed, low cost Automation possible for symmetrical parts Permanent record capability for symmetrical parts No couplant or probe contact required	Conductive materials Shallow depth of penetration (thin walls only) Masked or false indications caused by sensitivity to variations, such as part geometry, lift-off Reference standards required Permeability variations
Eddy-sonic..............	Debonded areas in metal-core or metal-faced honeycomb structures Delaminations in metal laminates or composites Crushed core	Metal-core honeycomb Metal-faced honeycomb Conductive laminates such as boron or graphite fiber composites Bonded metal panels	Portable Simple to operate No couplant required May be automated	Specimen or part must contain conductive materials to establish eddy-current field Reference standards required Part geometry
Electric current (direct current conduction method)	Cracks Crack depth Resistivity Wall thickness Corrosion-induced wall-thinning	Metallic materials Electrically conductive materials Train rails Nuclear fuel elements Bars, plates, other shapes	Access to only one surface required Battery or dc source Portable	Edge effect Surface contamination Good surface contact required Difficult to automate Electrode spacing Reference standards required
Electrified particle........	Surface defects in non-conducting material Through-to-metal pinholes on metal-backed material Tension, compression, cyclic cracks Brittle-coating stress cracks	Glass Porcelain enamel Nonhomogeneous materials such as plastic or asphalt coatings Glass-to-metal seals	Portable Useful on materials not practical for penetrant inspection	Poor resolution on thin coatings False indications from moisture streaks or lint Atmospheric conditions High voltage discharge
Exo-electron emission......	Fatigue in metals	Metals	Access to only one surface required Permanent record Quantitative	No surface films or contamination Geometry limitations Skilled technician required
Filtered particle..........	Cracks Porosity Differential absorption	Porous materials such as clay, carbon, powdered metals, concrete Grinding wheels High-tension insulators Sanitary ware	Colored or fluorescent particles Leaves no residue after baking part over 205°C (400°F) Quickly and easily applied Portable	Size and shape of particles must be selected before use Penetrating power of suspension medium is critical Particle concentration must be controlled Skin irritation

(continued)

Nondestructive methods for evaluating materials (continued)

Method	Measures or detects	Applications	Advantages	Limitations
Fluoroscopy (cine-fluorography, kine-fluorography)	Level of fill in containers Foreign objects Internal components Density variations Voids, thickness Spacing or position	Particles in liquid flow Presence of cavitation Operation of valves and switches Burning in small solid-propellant rocket motors	High-brightness images Real-time viewing Image magnification Permanent record Moving subject can be observed	Costly equipment Geometric unsharpness Thick specimens Speed of event to be studied Viewing area
Holography (acoustical-liquid surface levitation)	Lack of bond Delaminations Voids Porosity Resin-rich or resin-starved areas Inclusions Density variations	Metals Plastics Composites Laminates Honeycomb structures Ceramics Biological specimens	No hologram film development required Real-time imaging provided Liquid-surface responds rapidly to ultrasonic energy	Through-transmission techniques only Object and reference beams must superimpose on special liquid surface Immersion test only Laser required
Holography (interferometry)	Strain Plastic deformation Cracks Debonded areas Voids and inclusions Vibration	Bonded and composite structures Automotive or aircraft tires Three-dimensional imaging	Surface of test object can be uneven No special surface preparations or coatings required No physical contact with test specimen	Vibrationfree environment is required Heavy base to dampen vibrations Difficult to identify type of flaw detected
Infrared (radiometers)	Lack of bond Hot spots Heat transfer Isotherms Temperature ranges	Brazed joints Adhesive-bonded joints Metallic platings or coatings; debonded areas or thickness Electrical assemblies Temperature monitoring	Sensitive to 0.85°C (1.5°F) temperature variation Permanent record or thermal picture Quantitative Remote sensing; need not contact part Portable	Emissivity Liquid-nitrogen-cooled detector Critical time-temperature relationship Poor resolution for thick specimens Reference standards required
Leak testing	Leaks: helium, ammonia, smoke, water, air bubbles, radioactive gas, and halogens	Joints: welded, brazed, and adhesive-bonded Sealed assemblies Pressure or vacuum chambers Fuel or gas tanks	High sensitivity to extremely small, tight separations not detectable by other NDT methods Sensitivity related to method selected	Accessibility to both surfaces of part required Smeared metal or contaminants may prevent detection Cost related to sensitivity
Magnetic field	Cracks Wall thickness Hardness Coercive force Magnetic anisotropy Magnetic field Nonmagnetic coating thickness on steel	Ferromagnetic materials Ship degaussing Liquid level control Treasure hunting Wall thickness of nonmetallic materials Material sorting	Measurement of magnetic material properties May be automated Easily detects magnetic objects in nonmagnetic material Portable	Permeability Reference standards required Edge-effect Probe lift-off
Magnetic particle	Surface and slightly subsurface defects; cracks, seams, porosity, inclusions Permeability variations Extremely sensitive for locating small tight cracks	Ferromagnetic materials; bar, forgings, weldments, extrusions, etc.	Advantage over penetrant in that it indicates subsurface defects, particularly inclusions Relatively fast and low cost May be portable	Alignment of magnetic field is critical Demagnetization of parts required after tests Parts must be cleaned before and after inspection Masking by surface coatings

(continued)

Nondestructive methods for evaluating materials (continued)

Method	Measures or detects	Applications	Advantages	Limitations
Magnetic perturbation	Cracks Crack depth Broken strands in steel cables Permeability effects Nonmetallic inclusions Grinding burns and cracks under chromium plating	Ferromagnetic metals Broken steel cables in reinforced concrete	May be automated Easily detects magnetic objects in nonmagnetic materials Detects subsurface defects	Requires reference standard Need trained operator Part geometry Expensive equipment
Microwave (300 MHz-300 GHz)	Cracks, holes, debonded areas, etc. in non-metallic parts Changes in composition, degree of cure, moisture content Thickness measurement Dielectric constant Loss tangent	Reinforced plastics Chemical products Ceramics Resins Rubber Liquids Polyurethane foam Radomes	Between radio waves and infrared in the electromagnetic spectrum Portable Contact with part surface not normally required Can be automated	Will not penetrate metals Reference standards required Horn to part spacing critical Part geometry Wave interference Vibration
Mössbauer effect	Nuclear magnetic resonance in materials, most common being iron-57 Polarization of magnetic domains in steel	Detect and identify iron in specimen or sample Detect iron films on stainless steel Measure retained austenite (2-35%) in steels Determine nitrided surfaces on steel Interaction of domains with dislocation in ferromagnetic materials	Provide unique information about the surroundings of the iron-57 nuclei	Radiation hazard Trained engineers or physicists required Nonportable Precision equipment for vibrating source and spectrum analysis
Neutron activation analysis (reactor, accelerator, or radio-isotope)	Radiation emission resulting from neutron activation Oxygen in steel Nitrogen in food products Silicon in metals and ores	Metallurgical Prospecting Well logging Oceanography On-line process control of liquid or solid materials	Automatic systems Accurate (ppm range) Fast No contact with sample Sample preparation minimal	Radiation hazard Fast decay time Reference standard required Sensitivity varies with irradiation time
Penetrants (dye or fluorescent)	Defects open to surface of parts; cracks, porosity, seams, laps, etc. Through-wall leaks	All parts with non-absorbing surfaces: forgings, weldments, castings, etc. (Note: bleed-out from porous surfaces can mask indications of defects)	Low cost Portable Indications may be further examined visually Results easily interpreted	Surface films, such as coatings, scale, and smeared metal may prevent detection of defects Parts must be cleaned before and after inspection Defect must be open to surface

(continued)

Nondestructive methods for evaluating materials (continued)

Method	Measures or detects	Applications	Advantages	Limitations
Radiography (thermal neutrons from reactor, accelerator, or Californium-252)	Hydrogen contamination of titanium or zirconium alloys Defective or improperly loaded pyrotechnic devices Improper assembly of metal, nonmetal parts Corrosion products	Pyrotechnic devices Metallic, nonmetallic assemblies Biological specimens Nuclear reactor fuel elements and control rods Adhesive bonded structures	High neutron absorption by hydrogen, boron, lithium, cadmium, uranium, plutonium Low neutron absorption by most metals Complement to X-ray or gamma-ray radiography	Very costly equipment Nuclear reactor or accelerator required Trained physicists required Radiation hazard Nonportable Indium or gadolinium screens required
Radiography (gamma rays), Cobalt-60, Iridium-192	Internal defects and variations; porosity, inclusions, cracks, lack of fusion, geometry variations, corrosion thinning Density variations Thickness, gap and position	Usually where X-ray machines are not suitable because source cannot be placed in part with small openings and/or power source not available Panoramic imaging	Low initial cost Permanent records; film Small sources can be placed in parts with small openings Portable Low contrast	One energy level per source Source decay Radiation hazard Trained operators needed Lower image resolution Cost related to source size
Radiography (X-rays, film)	Internal defects and variations; porosity; inclusions; cracks; lack of fusion; geometry variations; corrosion thinning Density variations Thickness, gap and position Misassembly Misalignment	Castings Electrical assemblies Weldments Small, thin, complex wrought products Nonmetallics Solid propellant rocket motors Composites	Permanent records; film Adjustable energy levels (5 kv-25 meV) High sensitivity to density changes No couplant required Geometry variations do not affect direction of X-ray beam	High initial costs Orientation of linear defects in part may not be favorable Radiation hazard Depth of defect not indicated Sensitivity decreases with increase in scattered radiation
Radiometry (X-, gamma-, or beta-ray; transmission or backscatter)	Wall thickness Plating thickness Variations in density or composition Fill level in cans or containers Inclusions or voids	Sheet, plate, foil, strip, tubing Nuclear reactor fuel rods Cans or containers Plated parts Composites	Fully automatic Fast Extremely accurate In-line process control Portable	Radiation hazard Beta-ray useful for ultrathin coatings only Source decay Reference standards required
Sonic (less than 0.1 MHz)	Debonded areas or delaminations in metal or nonmetal composites or laminates Cohesive bond strength under controlled conditions Crushed or fractured core Bond integrity of metal insert fasteners	Metal or nonmetal composite or laminates brazed or adhesive-bonded Plywood Rocket motor nozzles Honeycomb	Portable Easy to operate Locates far-side debonded areas May be automated Access to only one surface required	Surface geometry influences test results Reference standards required Adhesive or core thickness variations influence results

(continued)

Nondestructive methods for evaluating materials (continued)

Method	Measures or detects	Applications	Advantages	Limitations
Thermal (thermochromic paint, liquid crystals)	Lack of bond Hot spots Heat transfer Isotherms Temperature ranges Blockage in coolant passages	Brazed joints Adhesive-bonded joints Metallic platings or coatings Electrical assemblies Temperature monitoring	Very low initial cost Can be readily applied to surfaces which may be difficult to inspect by other methods No special operator skills	Thin-walled surfaces only Critical time-temperature relationship Image retentivity affected by humidity Reference standards required
Thermoelectric probe	Thermoelectric potential Coating thickness Physical properties Thompson effect P-N junctions in semiconductors	Metal sorting Ceramic coating thickness on metals Semiconductors	Portable Simple to operate Access to only one surface required	Hot probe Difficult to automate Reference standards required Surface contaminants Conductive coatings
Tomography	Boundaries Surface reconstruction Crack size, location, and orientation	Metals research Medicine	Pinpoint defect location Image display is computer controlled	Very expensive Need highly trained operator
Ultrasonic (0.1-25 MHz)	Internal defects and variations; cracks, lack of fusion, porosity, inclusions, delaminations, lack of bond, texturing Thickness or velocity Poisson's ratio, elastic modulus	Wrought metals Welds Brazed joints Adhesive-bonded joints Nonmetallics In-service parts	Most sensitive to cracks Test results known immediately Automating and permanent record capability Portable High penetration capability	Couplant required Small, thin, complex parts may be difficult to check Reference standards required Trained operators for manual inspection Special probes
Ultrasonic angle reflectivity	Elastic properties, acoustic attenuation in solids Near-surface metallic property gradients, e.g. carburization in steel Metallic grain structure and size	Metals Nonmetals	Access to only one surface required Permanent record Quantitative No physical contact of sample required Sample preparation minimal	Test parts must be immersed Geometry limitations: test part must have a flat, smooth area Goniometer device required Skilled technician required

Source: Hagemaier, Donald J., Douglas Aircraft Co., McDonnell Douglas Corp., Long Beach, CA

Tension Testing

Percentage reduction of area for tension test specimens

Diam(a), in.	Area, in.²	Reduction of area, % for initial diam of: 0.505 in.	0.506 in.	0.504 in.	Diam(a), in.	Area, in.²	Reduction of area, %, for initial diam of: 0.505 in.	0.506 in.	0.504 in.	Diam(a), in.	Area, in.²	Reduction of area, %, for initial diam of: 0.505 in.	0.506 in.	0.504 in.
0.251 ..	0.0494	75.3	75.4	75.2	0.311 ..	0.0759	62.1	62.2	62.0	0.371 ..	0.1081	46.0	46.2	45.8
0.252 ..	0.0498	75.1	75.2	75.0	0.312 ..	0.0764	61.8	62.0	61.7	0.372 ..	0.1086	45.8	46.0	45.6
0.253 ..	0.0502	74.9	75.0	74.8	0.313 ..	0.0769	61.6	61.7	61.5	0.373 ..	0.1092	45.5	45.7	45.3
0.254 ..	0.0506	74.7	74.8	74.6	0.314 ..	0.0774	61.3	61.5	61.2	0.374 ..	0.1098	45.2	45.4	45.0
0.255 ..	0.0510	74.5	74.6	74.4	0.315 ..	0.0779	61.1	61.2	61.0	0.375 ..	0.1104	44.9	45.1	44.5
0.256 ..	0.0514	74.3	74.4	74.2	0.316 ..	0.0784	60.8	61.0	60.7	0.376 ..	0.1110	44.6	44.8	44.4
0.257 ..	0.0518	74.1	74.2	74.0	0.317 ..	0.0789	60.6	60.7	60.5	0.377 ..	0.1116	44.3	44.5	44.1
0.258 ..	0.0522	73.9	74.0	73.8	0.318 ..	0.0794	60.3	60.5	60.2	0.378 ..	0.1122	44.0	44.2	43.8
0.259 ..	0.0526	73.7	73.8	73.6	0.319 ..	0.0799	60.1	60.2	60.0	0.379 ..	0.1128	43.7	43.9	43.5
0.260 ..	0.0530	73.5	73.6	73.4	0.320 ..	0.0804	59.8	60.0	59.7	0.380 ..	0.1134	43.4	43.6	43.2
0.261 ..	0.0535	73.3	73.4	73.2	0.321 ..	0.0809	59.6	59.8	59.4	0.381 ..	0.1140	43.1	43.3	42.9
0.262 ..	0.0539	73.0	73.2	73.0	0.322 ..	0.0814	59.4	59.5	59.2	0.382 ..	0.1146	42.8	43.0	42.6
0.263 ..	0.0543	72.9	73.0	72.8	0.323 ..	0.0819	59.1	59.3	58.9	0.383 ..	0.1152	42.5	42.7	42.3
0.264 ..	0.0547	72.7	72.8	72.6	0.324 ..	0.0824	58.8	59.0	58.7	0.384 ..	0.1158	42.2	42.4	42.0
0.265 ..	0.0551	72.5	72.6	72.4	0.325 ..	0.0829	58.6	58.8	58.4	0.385 ..	0.1164	41.9	42.1	41.7
0.266 ..	0.0555	72.3	72.4	72.2	0.326 ..	0.0834	58.3	58.5	58.2	0.386 ..	0.1170	41.6	41.8	41.4
0.267 ..	0.0559	71.9	72.2	72.0	0.327 ..	0.0839	58.1	58.3	57.9	0.387 ..	0.1176	41.3	41.5	41.1
0.268 ..	0.0564	71.8	71.9	71.7	0.328 ..	0.0844	57.8	58.0	57.7	0.388 ..	0.1182	41.0	41.2	40.8
0.269 ..	0.0568	71.6	71.7	71.5	0.329 ..	0.0850	57.5	57.7	57.4	0.389 ..	0.1188	40.7	40.9	40.5
0.270 ..	0.0572	71.4	71.5	71.3	0.330 ..	0.0855	57.3	57.5	57.1	0.390 ..	0.1194	40.4	40.6	40.2
0.271 ..	0.0576	71.2	71.3	71.1	0.331 ..	0.0860	57.0	57.2	56.9	0.391 ..	0.1200	40.1	40.3	39.9
0.272 ..	0.0581	71.0	71.1	70.9	0.332 ..	0.0865	56.8	57.0	56.6	0.392 ..	0.1206	39.8	40.0	39.6
0.273 ..	0.0585	70.8	70.9	70.7	0.333 ..	0.0870	56.5	56.7	56.4	0.393 ..	0.1213	39.4	39.7	39.2
0.274 ..	0.0589	70.6	70.7	70.5	0.334 ..	0.0876	56.2	56.4	56.1	0.394 ..	0.1219	39.1	39.4	38.9
0.275 ..	0.0593	70.4	70.5	70.3	0.335 ..	0.0881	56.0	56.2	55.8	0.395 ..	0.1225	38.8	39.1	38.6
0.276 ..	0.0598	70.1	70.2	70.0	0.336 ..	0.0886	55.7	55.9	55.6	0.396 ..	0.1231	38.5	38.8	38.3
0.277 ..	0.0602	69.9	70.0	69.8	0.337 ..	0.0891	55.5	55.7	55.3	0.397 ..	0.1237	38.2	38.5	38.0
0.278 ..	0.0606	69.7	69.9	69.6	0.338 ..	0.0897	55.2	55.4	55.0	0.398 ..	0.1244	37.9	38.1	37.6
0.279 ..	0.0611	69.5	69.6	69.4	0.339 ..	0.0902	54.9	55.1	54.8	0.399 ..	0.1250	37.6	37.8	37.3
0.280 ..	0.0615	69.3	69.4	69.2	0.340 ..	0.0907	54.7	54.9	54.5	0.400 ..	0.1256	37.3	37.5	37.0
0.281 ..	0.0620	69.0	69.2	69.0	0.341 ..	0.0913	54.4	54.6	54.2	0.401 ..	0.1262	37.0	37.2	36.7
0.282 ..	0.0624	68.8	69.0	68.7	0.342 ..	0.0918	54.1	54.3	54.0	0.402 ..	0.1269	36.6	36.9	36.4
0.283 ..	0.0629	68.6	68.7	68.5	0.343 ..	0.0924	53.8	54.0	53.7	0.403 ..	0.1275	36.3	36.6	36.1
0.284 ..	0.0633	68.4	68.5	68.3	0.344 ..	0.0929	53.6	53.8	53.4	0.404 ..	0.1281	36.0	36.3	35.8
0.285 ..	0.0637	68.2	68.3	68.1	0.345 ..	0.0934	53.3	53.5	53.2	0.405 ..	0.1288	35.7	35.9	35.4
0.286 ..	0.0642	67.9	68.1	67.8	0.346 ..	0.0940	53.0	53.2	52.9	0.406 ..	0.1294	35.4	35.6	35.1
0.287 ..	0.0646	67.7	67.9	67.6	0.347 ..	0.0945	52.8	53.0	52.6	0.407 ..	0.1301	35.0	35.3	34.8
0.288 ..	0.0651	67.5	67.6	67.4	0.348 ..	0.0951	52.5	52.7	52.3	0.408 ..	0.1307	34.7	35.0	34.5
0.289 ..	0.0655	67.3	67.4	67.2	0.349 ..	0.0956	52.2	52.4	52.1	0.409 ..	0.1313	34.4	34.7	34.2
0.290 ..	0.0660	67.0	67.2	66.9	0.350 ..	0.0962	51.9	52.1	51.8	0.410 ..	0.1320	34.1	34.3	33.8
0.291 ..	0.0665	66.8	66.9	66.7	0.351 ..	0.0967	51.7	51.9	51.5	0.411 ..	0.1326	33.8	34.0	33.5
0.292 ..	0.0670	66.5	66.7	66.4	0.352 ..	0.0973	51.4	51.6	51.2	0.412 ..	0.1333	33.4	33.7	33.2
0.293 ..	0.0674	66.3	66.5	66.2	0.353 ..	0.0978	51.1	51.3	51.0	0.413 ..	0.1339	33.1	33.4	32.9
0.294 ..	0.0679	66.1	66.2	66.0	0.354 ..	0.0984	50.8	51.0	50.7	0.414 ..	0.1346	32.8	33.0	32.5
0.295 ..	0.0683	65.9	66.0	65.8	0.355 ..	0.0989	50.6	50.8	50.4	0.415 ..	0.1352	32.5	32.8	32.2
0.296 ..	0.0688	65.6	65.8	65.5	0.356 ..	0.0995	50.3	50.5	50.1	0.416 ..	0.1359	32.1	32.4	31.9
0.297 ..	0.0692	65.4	65.6	65.3	0.357 ..	0.1000	50.0	50.2	49.9	0.417 ..	0.1365	31.8	32.1	31.6
0.298 ..	0.0697	65.2	65.3	65.1	0.358 ..	0.1006	49.8	50.0	49.6	0.418 ..	0.1372	31.5	31.7	31.2
0.299 ..	0.0702	64.9	65.1	64.8	0.359 ..	0.1012	49.5	49.7	49.3	0.419 ..	0.1378	31.2	31.4	30.9
0.300 ..	0.0707	64.7	64.8	64.6	0.360 ..	0.1017	49.2	49.4	49.0	0.420 ..	0.1385	30.8	31.1	30.6
0.301 ..	0.0712	64.4	64.6	64.3	0.361 ..	0.1023	48.9	49.1	48.7	0.421 ..	0.1392	30.5	30.7	30.2
0.302 ..	0.0716	64.2	64.4	64.1	0.362 ..	0.1029	48.6	48.8	48.4	0.422 ..	0.1398	30.2	30.4	29.9
0.303 ..	0.0721	64.0	64.1	63.9	0.363 ..	0.1034	48.4	48.6	48.2	0.423 ..	0.1405	29.8	30.1	29.6
0.304 ..	0.0725	63.8	63.9	63.7	0.364 ..	0.1040	48.1	48.3	47.9	0.424 ..	0.1411	29.5	29.8	29.3
0.305 ..	0.0730	63.5	63.7	63.4	0.365 ..	0.1046	47.8	48.0	47.6	0.425 ..	0.1418	29.2	29.5	28.9
0.306 ..	0.0735	63.3	63.4	63.2	0.366 ..	0.1052	47.5	47.7	47.3	0.426 ..	0.1425	28.8	29.1	28.6
0.307 ..	0.0740	63.0	63.2	62.9	0.367 ..	0.1057	47.2	47.4	47.0	0.427 ..	0.1432	28.5	28.8	28.2
0.308 ..	0.0745	62.8	62.9	62.7	0.368 ..	0.1063	46.9	47.1	46.7	0.428 ..	0.1438	28.2	28.5	27.9
0.309 ..	0.0749	62.6	62.7	62.5	0.369 ..	0.1069	46.6	46.8	46.4	0.429 ..	0.1445	27.8	28.1	27.6
0.310 ..	0.0754	62.3	62.5	62.2	0.370 ..	0.1075	46.3	46.5	46.1	0.430 ..	0.1452	27.5	27.8	27.4

(continued)

This table was compiled by Arthur W. F. Green, and was first published in the 1939 *Metals Handbook*. It is designed to save time in computing the percentage reduction of area of standard 0.505-in. diam tension test specimens after testing. The 0.505-in. diam is a nominal dimension; specimens are usually machined to a ±0.001-in. tolerance, hence the utility of a table that gives values for 0.506 and 0.504, as well as the nominal 0.505 in. Percentage reduction of area is obtained from measurement of the necked-down diameter of the specimen (first column of the table) and reading the percentage of reduction of area in the applicable column for initial diameter of test specimen. For example, if the initial diameter of the test specimen were 0.506 in. and the diameter at the fracture after testing were 0.441 in., the reduction of area would be 24.0%, as shown in the table.
(a) At the fracture after testing.

Percentage reduction of area for tension test specimens (continued)

Diam(a), in.	Area, in.²	Reduction of area, % for initial diam of:		
		0.505 in.	0.506 in.	0.504 in.
0.431 ..	0.1458	27.2	27.5	26.9
0.432 ..	0.1465	26.8	27.1	26.6
0.433 ..	0.1472	26.5	26.8	26.2
0.434 ..	0.1479	26.1	26.4	25.9
0.435 ..	0.1486	25.8	26.1	25.5
0.436 ..	0.1493	25.4	25.7	25.2
0.437 ..	0.1499	25.1	25.4	24.9
0.438 ..	0.1506	24.8	25.1	24.5
0.439 ..	0.1513	24.4	24.7	24.2
0.440 ..	0.1520	24.1	24.4	23.9
0.441 ..	0.1527	23.7	24.0	23.5
0.442 ..	0.1534	23.4	23.7	23.2
0.443 ..	0.1541	23.0	23.3	22.8
0.444 ..	0.1548	22.7	23.0	22.5
0.445 ..	0.1555	22.3	22.6	22.1
0.446 ..	0.1562	22.0	22.3	21.7
0.447 ..	0.1569	21.6	21.9	21.4
0.448 ..	0.1576	21.3	21.6	21.0
0.449 ..	0.1583	20.9	21.2	20.7
0.450 ..	0.1590	20.6	20.9	20.3
0.451 ..	0.1597	20.2	20.5	20.0
0.452 ..	0.1604	19.9	20.2	19.6
0.453 ..	0.1611	19.5	19.9	19.2

Diam(a), in.	Area, in.²	Reduction of area, % for initial diam of:		
		0.505 in.	0.506 in.	0.504 in.
0.454 ..	0.1618	19.2	19.5	18.9
0.455 ..	0.1625	18.8	19.2	18.5
0.456 ..	0.1633	18.4	18.8	18.1
0.457 ..	0.1640	18.1	18.4	17.8
0.458 ..	0.1647	17.7	18.1	17.4
0.459 ..	0.1654	17.4	17.7	17.1
0.460 ..	0.1661	17.0	17.4	16.7
0.461 ..	0.1669	16.6	17.0	16.3
0.462 ..	0.1676	16.3	16.6	16.0
0.463 ..	0.1683	15.9	16.3	15.6
0.464 ..	0.1690	15.6	15.9	15.3
0.465 ..	0.1698	15.2	15.5	14.9
0.466 ..	0.1705	14.8	15.2	14.5
0.467 ..	0.1712	14.5	14.8	14.2
0.468 ..	0.1720	14.1	14.4	13.8
0.469 ..	0.1727	13.7	14.1	13.4
0.470 ..	0.1734	13.4	13.7	13.1
0.471 ..	0.1742	13.0	13.3	12.7
0.472 ..	0.1749	12.6	13.0	12.3
0.473 ..	0.1757	12.2	12.6	11.9
0.474 ..	0.1764	11.9	12.2	11.6
0.475 ..	0.1772	11.5	11.8	11.2
0.476 ..	0.1779	11.1	11.5	10.8

Diam(a), in.	Area, in.²	Reduction of area, % for initial diam of:		
		0.505 in.	0.506 in.	0.504 in.
0.477 ..	0.1787	10.7	11.1	10.4
0.478 ..	0.1794	10.4	10.8	10.1
0.479 ..	0.1802	10.0	10.3	9.7
0.480 ..	0.1809	9.6	10.0	9.3
0.481 ..	0.1817	9.2	9.6	8.9
0.482 ..	0.1824	8.9	9.3	8.6
0.483 ..	0.1832	8.5	8.9	8.2
0.484 ..	0.1839	8.1	8.5	7.8
0.485 ..	0.1847	7.7	8.1	7.4
0.486 ..	0.1855	7.3	7.7	7.0
0.487 ..	0.1862	7.0	7.4	6.7
0.488 ..	0.1870	6.6	7.0	6.3
0.489 ..	0.1878	6.2	6.6	5.9
0.490 ..	0.1885	5.8	6.2	5.5
0.492 ..	0.1901	5.0	5.4	4.7
0.494 ..	0.1916	4.3	4.7	4.0
0.496 ..	0.1932	3.5	3.9	3.2
0.498 ..	0.1947	2.7	3.1	2.4
0.500 ..	0.1963	1.9	2.3	1.6
0.502 ..	0.1979	1.1	1.5	0.8
0.504 ..	0.1995	0.4	0.7	0.0
0.505 ..	0.2002	0.0	0.4	..
0.506 ..	0.2010	..	0.0	..

This table was compiled by Arthur W. F. Green, and was first published in the 1939 *Metals Handbook*. It is designed to save time in computing the percentage reduction of area of standard 0.505-in. diam tension test specimens after testing. The 0.505-in. diam is a nominal dimension; specimens are usually machined to a ±0.001-in. tolerance, hence the utility of a table that gives values for 0.506 and 0.504, as well as the nominal 0.505 in. Percentage reduction of area is obtained from measurement of the necked-down diameter of the specimen (first column of the table) and reading the percentage of reduction of area in the applicable column for initial diameter of test specimen. For example, if the initial diameter of the test specimen were 0.506 in. and the diameter at the fracture after testing were 0.441 in., the reduction of area would be 24.0%, as shown in the table.
(a) At the fracture after testing.

Hardness Testing

Equivalent Hardness Numbers and Tensile Strengths for Steel

Vickers hardness numbers (a)

Vickers hardness	Brinell hardness, 3000-kg load, 10-mm ball		Rockwell hardness				Rockwell superficial hardness, superficial Brale indenter			Knoop hardness, 500-g load and greater	Sclero-scope hardness	Tensile strength (approx), ksi	Vickers hardness
	Standard ball	Tungsten carbide ball	A scale, 60-kg load, Brale indenter	B scale, 100-kg load, 1/16-in. diam ball	C scale, 150-kg load, Brale indenter	D scale, 100-kg load, Brale indenter	15N scale, 15-kg load	30N scale, 30-kg load	45N scale, 45-kg load				
940	· · ·	· · ·	85.6	· · ·	68.0	76.9	93.2	84.4	75.4	920	97	· · ·	940
920	· · ·	· · ·	85.3	· · ·	67.5	76.5	93.0	84.0	74.8	908	96	· · ·	920
900	· · ·	· · ·	85.0	· · ·	67.0	76.1	92.9	83.6	74.2	895	95	· · ·	900
880	· · ·	(767)	84.7	· · ·	66.4	75.7	92.7	83.1	73.6	882	93	· · ·	880
860	· · ·	(757)	84.4	· · ·	65.9	75.3	92.5	82.7	73.1	867	92	· · ·	860
840	· · ·	(745)	84.1	· · ·	65.3	74.8	92.3	82.2	72.2	852	91	· · ·	840
820	· · ·	(733)	83.8	· · ·	64.7	74.3	92.1	81.7	71.8	.837	90	· · ·	820
800	· · ·	(722)	83.4	· · ·	64.0	73.8	91.8	81.1	71.0	822	88	· · ·	800
780	· · ·	(710)	83.0	· · ·	63.3	73.3	91.5	80.4	70.2	806	87	· · ·	780
760	· · ·	(698)	82.6	· · ·	62.5	72.6	91.2	79.7	69.4	788	86	· · ·	760
740	· · ·	(684)	82.2	· · ·	61.8	72.1	91.0	79.1	68.6	772	84	· · ·	740
720	· · ·	(670)	81.8	· · ·	61.0	71.5	90.7	78.4	67.7	754	83	· · ·	720
700	· · ·	(656)	81.3	· · ·	60.1	70.8	90.3	77.6	66.7	735	81	· · ·	700
690	· · ·	(647)	81.1	· · ·	59.7	70.5	90.1	77.2	66.2	725	· · ·	· · ·	690
680	· · ·	(638)	80.8	· · ·	59.2	70.1	89.8	76.8	65.7	716	80	355	680
670	· · ·	(630)	80.6	· · ·	58.8	69.8	89.7	76.4	65.3	706	· · ·	348	670
660	· · ·	620	80.3	· · ·	58.3	69.4	89.5	75.9	64.7	697	79	342	660
650	· · ·	611	80.0	· · ·	57.8	69.0	89.2	75.5	64.1	687	78	336	650
640	· · ·	601	79.8	· · ·	57.3	68.7	89.0	75.1	63.5	677	77	328	640
630	· · ·	591	79.5	· · ·	56.8	68.3	88.8	74.6	63.0	667	76	323	630
620	· · ·	582	79.2	· · ·	56.3	67.9	88.5	74.2	62.4	657	75	317	620
610	· · ·	573	78.9	· · ·	55.7	67.5	88.2	73.6	61.7	646	· · ·	310	610
600	· · ·	564	78.6	· · ·	55.2	67.0	88.0	73.2	61.2	636	74	303	600
590	· · ·	554	78.4	· · ·	54.7	66.7	87.8	72.7	60.5	625	73	298	590
580	· · ·	545	78.0	· · ·	54.1	66.2	87.5	72.1	59.9	615	72	293	580
570	· · ·	535	77.8	· · ·	53.6	65.8	87.2	71.7	59.3	604	· · ·	288	570
560	· · ·	525	77.4	· · ·	53.0	65.4	86.9	71.2	58.6	594	71	283	560
550	(505)	517	77.0	· · ·	52.3	64.8	86.6	70.5	57.8	583	70	276	550
540	(496)	507	76.7	· · ·	51.7	64.4	86.3	70.0	57.0	572	69	270	540
530	(488)	497	76.4	· · ·	51.1	63.9	86.0	69.5	56.2	561	68	265	530
520	(480)	488	76.1	· · ·	50.5	63.5	85.7	69.0	55.6	550	67	260	520
510	(473)	479	75.7	· · ·	49.8	62.9	85.4	68.3	54.7	539	· · ·	254	510
500	(465)	471	75.3	· · ·	49.1	62.2	85.0	67.7	53.9	528	66	247	500
490	(456)	460	74.9	· · ·	48.4	61.6	84.7	67.1	53.1	517	65	241	490
480	(448)	452	74.5	· · ·	47.7	61.3	84.3	66.4	52.2	505	64	235	480
470	441	442	74.1	· · ·	46.9	60.7	83.9	65.7	51.3	494	· · ·	228	470
460	433	433	73.6	· · ·	46.1	60.1	83.6	64.9	50.4	482	62	223	460
450	425	425	73.3	· · ·	45.3	59.4	83.2	64.3	49.4	471	· · ·	217	450

(a) For carbon and alloy steels in the annealed, normalized, and quenched-and-tempered conditions: less accurate for cold worked condition and for austenitic steels. The values in bold-faced type correspond to the values in the joint SAE-ASM-ASTM hardness conversions as printed in ASTM E140. The values in parentheses are beyond normal range and are given for information only.

Vickers hardness numbers (a) (continued)

Vickers hardness	Brinell hardness, 3000-kg load, 10-mm ball		Rockwell hardness				Rockwell superficial hardness, superficial Brale indenter			Knoop hardness, 500-g load and greater	Sclero-scope hardness	Tensile strength (approx), ksi	Vickers hardness
	Standard ball	Tungsten carbide ball	A scale, 60-kg load, Brale indenter	B scale, 100-kg load, 1/16-in. diam ball	C scale, 150-kg load, Brale indenter	D scale, 100-kg load, Brale indenter	15N scale, 15-kg load	30N scale, 30-kg load	45N scale, 45-kg load				
440	415	415	72.8	...	44.5	58.8	82.8	63.5	48.4	459	59	212	440
430	405	405	72.3	...	43.6	58.2	82.3	62.7	47.4	447	58	205	430
420	397	397	71.8	...	42.7	57.5	81.8	61.9	46.4	435	57	199	420
410	388	388	71.4	...	41.8	56.8	81.4	61.1	45.3	423	56	193	410
400	379	379	70.8	...	40.8	56.0	80.8	60.2	44.1	412	55	187	400
390	369	369	70.3	...	39.8	55.2	80.3	59.3	42.9	400	...	181	390
380	360	360	69.8	(110.0)	38.8	54.4	79.8	58.4	41.7	389	52	175	380
370	350	350	69.2	...	37.7	53.6	79.2	57.4	40.4	378	51	170	370
360	341	341	68.7	(109.0)	36.6	52.8	78.6	56.4	39.1	367	50	164	360
350	331	331	68.1	...	35.5	51.9	78.0	55.4	37.8	356	48	159	350
340	322	322	67.6	(108.0)	34.4	51.1	77.4	54.4	36.5	346	47	155	340
330	313	313	67.0	...	33.3	50.2	76.8	53.6	35.2	337	46	150	330
320	303	303	66.4	(107.0)	32.2	49.4	76.2	52.3	33.9	328	45	146	320
310	294	294	65.8	...	31.0	48.4	75.6	51.3	32.5	318	...	142	310
300	284	284	65.2	(105.5)	29.8	47.5	74.9	50.2	31.1	309	42	138	300
295	280	280	64.8	...	29.2	47.1	74.6	49.7	30.4	305	...	136	295
290	275	275	64.5	(104.5)	28.5	46.5	74.2	49.0	29.5	300	41	133	290
285	270	270	64.2	...	27.8	46.0	73.8	48.4	28.7	296	...	131	285
280	265	265	63.8	(103.5)	27.1	45.3	73.4	47.8	27.9	291	40	129	280
275	261	261	63.5	...	26.4	44.9	73.0	47.2	27.1	286	39	127	275
270	256	256	63.1	(102.0)	25.6	44.3	72.6	46.4	26.2	282	38	124	270
265	252	252	62.7	...	24.8	43.7	72.1	45.7	25.2	277	...	122	265
260	247	247	62.4	(101.0)	24.0	43.1	71.6	45.0	24.3	272	37	120	260
255	243	243	62.0	...	23.1	42.2	71.1	44.2	23.2	267	...	117	255
250	238	238	61.6	99.5	22.2	41.7	70.6	43.4	22.2	262	36	115	250
245	233	233	61.2	...	21.3	41.1	70.1	42.5	21.1	258	35	113	245
240	228	228	60.7	98.1	20.3	40.3	69.6	41.7	19.9	253	34	111	240
230	219	219	...	96.7	(18.0)	...	...	...	...	243	33	106	230
220	209	209	...	95.0	(15.7)	...	...	...	...	234	32	101	220
210	200	200	...	93.4	(13.4)	...	...	...	...	226	30	97	210
200	190	190	...	91.5	(11.0)	...	...	...	...	216	29	92	200
190	181	181	...	89.5	(8.5)	...	...	...	...	206	28	88	190
180	171	171	...	87.1	(6.0)	...	...	...	...	196	26	84	180
170	162	162	...	85.0	(3.0)	...	...	...	...	185	25	79	170
160	152	152	...	81.7	(0.0)	...	...	...	...	175	23	75	160
150	143	143	...	78.7	...	...	...	...	...	164	22	71	150
140	133	133	...	75.0	...	...	...	...	...	154	21	66	140
130	124	124	...	71.2	...	...	...	...	...	143	20	62	130
120	114	114	...	66.7	...	...	...	...	...	133	18	57	120
110	105	105	...	62.3	...	...	...	...	...	123	...	...	110
100	95	95	...	56.2	...	...	...	...	...	112	...	...	100
95	90	90	...	52.0	...	...	...	...	...	107	...	...	95
90	86	86	...	48.0	...	...	...	...	...	102	...	...	90
85	81	81	...	41.0	...	...	...	...	...	97	...	...	85

(a) For carbon and alloy steels in the annealed, normalized, and quenched-and-tempered conditions: less accurate for cold worked condition and for austenitic steels. The values in bold-faced type correspond to the values in the joint SAE-ASM-ASTM hardness conversions as printed in ASTM E140. The values in parentheses are beyond normal range and are given for information only.

Brinell hardness numbers (a)

| Brinell indentation diam, mm | Brinell hardness (b), 3000-kg load, 10-mm ball | | Vickers hardness | Rockwell hardness | | | | Rockwell superficial hardness, superficial Brale indenter | | | Knoop hardness, 500-g load and greater | Scleroscope hardness | Tensile strength (approx), ksi | Brinell indentation diam, mm |
	Standard ball	Tungsten carbide ball		A scale, 60-kg load, Brale indenter	B scale, 100-kg load, 1/16-in.-diam ball	C scale, 150-kg load, Brale indenter	D scale, 100-kg load, Brale indenter	15N scale, 15-kg load	30N scale, 30-kg load	45N scale, 45-kg load				
2.25	· · ·	(745)	840	84.1	· · ·	65.3	74.8	92.3	82.2	72.2	852	91	· · ·	2.25
2.30	· · ·	(712)	783	83.1	· · ·	63.4	73.4	91.6	80.5	70.4	808	· · ·	· · ·	2.30
2.35	· · ·	(682)	737	82.2	· · ·	61.7	72.0	91.0	79.0	68.5	768	84	· · ·	2.35
2.40	· · ·	(653)	697	81.2	· · ·	60.0	70.7	90.2	77.5	66.5	732	81	· · ·	2.40
2.45	· · ·	627	667	80.5	· · ·	58.7	69.7	89.6	76.3	65.1	703	79	347	2.45
2.50	· · ·	601	640	79.8	· · ·	57.3	68.7	89.0	75.1	63.5	677	77	328	2.50
2.55	· · ·	578	615	79.1	· · ·	56.0	67.7	88.4	73.9	62.1	652	75	313	2.55
2.60	· · ·	555	591	78.4	· · ·	54.7	66.7	87.8	72.7	60.6	626	73	298	2.60
2.65	· · ·	534	569	77.8	· · ·	53.5	65.8	87.2	71.6	59.2	604	71	288	2.65
2.70	· · ·	514	547	76.9	· · ·	52.1	64.7	86.5	70.3	57.6	579	70	273	2.70
2.75	(495)	· · ·	539	76.7	· · ·	51.6	64.3	86.3	69.9	56.9	571	· · ·	269	2.75
	· · ·	495	528	76.3	· · ·	51.0	63.8	85.9	69.4	56.1	558	68	263	
2.80	(477)	· · ·	516	75.9	· · ·	50.3	63.2	85.6	68.7	55.2	545	· · ·	257	2.80
	· · ·	477	508	75.6	· · ·	49.6	62.7	85.3	68.2	54.5	537	66	252	
2.85	(461)	· · ·	495	75.1	· · ·	48.8	61.9	84.9	67.4	53.5	523	· · ·	244	2.85
	· · ·	461	491	74.9	· · ·	48.5	61.7	84.7	67.2	53.2	518	65	242	
2.90	444	· · ·	474	74.3	· · ·	47.2	61.0	84.1	66.0	51.7	499	· · ·	231	2.90
	· · ·	444	472	74.2	· · ·	47.1	60.8	84.0	65.8	51.5	496	63	229	
2.95	429	429	455	73.4	· · ·	45.7	59.7	83.4	64.6	49.9	476	61	220	2.95
3.00	415	415	440	72.8	· · ·	44.5	58.8	82.8	63.5	48.4	459	59	212	3.00
3.05	401	401	425	72.0	· · ·	43.1	57.8	82.0	62.3	46.9	441	58	202	3.05
3.10	388	388	410	71.4	· · ·	41.8	56.8	81.4	61.1	45.3	423	56	193	3.10
3.15	375	375	396	70.6	· · ·	40.4	55.7	80.6	59.9	43.6	407	54	184	3.15
3.20	363	363	383	70.0	· · ·	39.1	54.6	80.0	58.7	42.0	392	52	177	3.20
3.25	352	352	372	69.3	(110.0)	37.9	53.8	79.3	57.6	40.5	379	51	172	3.25
3.30	341	341	360	68.7	(109.0)	36.6	52.8	78.6	56.4	39.1	367	50	164	3.30
3.35	331	331	350	68.1	(108.5)	35.5	51.9	78.0	55.4	37.8	356	48	159	3.35
3.40	321	321	339	67.5	(108.0)	34.3	51.0	77.3	54.3	36.4	345	47	154	3.40
3.45	311	311	328	66.9	(107.5)	33.1	50.0	76.7	53.3	34.4	336	46	149	3.45
3.50	302	302	319	66.3	(107.0)	32.1	49.3	76.1	52.2	33.8	327	45	146	3.50
3.55	293	293	309	65.7	(106.0)	30.9	48.3	75.5	51.2	32.4	318	43	142	3.55

(a) For carbon and alloy steels in the annealed, normalized, and quenched-and-tempered conditions; less accurate for cold worked condition and for austenitic steels. Values in bold-faced type correspond to the values in the joint SAE-ASM-ASTM hardness conversions as printed in ASTM E140. Values in parentheses are beyond normal range and are given for information only. (b) Brinell numbers are based on the diameter of impressed indentation. If the ball distorts (flattens) during test, Brinell numbers will vary in accordance with the degree of such distortion when related to hardnesses determined with a Vickers diamond pyramid, Rockwell Brale, or other indenter that does not sensibly distort. At high hardnesses, therefore, the relationship between Brinell and Vickers or Rockwell scales is affected by the type of ball used. Standard steel balls tend to flatten slightly more than tungsten carbide balls, resulting in a larger indentation and a lower Brinell number than shown by a tungsten carbide ball. Thus, on a specimen of about 539 to 547 HV, a standard ball will leave a 2.75-mm indentation (495 HB), and a tungsten carbide ball a 2.70-mm indentation (514 HB). Conversely, identical indentation diameters for both types of ball will correspond to different Vickers and Rockwell values. Thus, if indentations in two different specimens both are 2.75 mm in diameter (495 HB), the specimen tested with a standard ball has a Vickers hardness of 539, whereas the specimen tested with a tungsten carbide ball has a Vickers hardness of 528.

Brinell hardness numbers (a) (continued)

Brinell indentation diam, mm	Brinell hardness (b), 3000-kg load, 10-mm ball		Vickers hardness	Rockwell hardness				Rockwell superficial hardness, superficial Brale indenter			Knoop hardness, 500-g load and greater	Scleroscope hardness	Tensile strength (approx), ksi	Brinell indentation diam, mm
	Standard ball	Tungsten carbide ball		A scale 60-kg load, Brale indenter	B scale, 100-kg load, 1/16-in.-diam ball	C scale, 150-kg load, Brale indenter	D scale, 100-kg load, Brale indenter	15N scale 15-kg load	30N scale 30-kg load	45N scale 45-kg load				
3.60	285	285	301	65.3	(105.5)	29.9	47.6	75.0	50.3	31.2	310	42	138	3.60
3.65	277	277	292	64.6	(104.5)	28.8	46.7	74.4	49.3	29.9	302	41	134	3.65
3.70	269	269	284	64.1	(104.0)	27.6	45.9	73.7	48.3	28.5	294	40	131	3.70
3.75	262	262	276	63.6	(103.0)	26.6	45.0	73.1	47.3	27.3	286	39	127	3.75
3.80	255	255	269	63.0	(102.0)	25.4	44.2	72.5	46.2	26.0	279	38	123	3.80
3.85	248	248	261	62.5	(101.0)	24.2	43.2	71.7	45.1	24.5	272	37	120	3.85
3.90	241	241	253	61.8	100.0	22.8	42.0	70.9	43.9	22.8	265	36	116	3.90
3.95	235	235	247	61.4	99.0	21.7	41.4	70.3	42.9	21.5	259	35	114	3.95
4.00	229	229	241	60.8	98.2	20.5	40.5	69.7	41.9	20.1	253	34	111	4.00
4.05	223	223	234	⋯	97.3	(19.0)	⋯	⋯	⋯	⋯	247	⋯	107	4.05
4.10	217	217	228	⋯	96.4	(17.7)	⋯	⋯	⋯	⋯	242	33	105	4.10
4.15	212	212	222	⋯	95.5	(16.4)	⋯	⋯	⋯	⋯	237	32	102	4.15
4.20	207	207	218	⋯	94.6	(15.2)	⋯	⋯	⋯	⋯	232	31	100	4.20
4.25	201	201	212	⋯	93.7	(13.8)	⋯	⋯	⋯	⋯	227	⋯	98	4.25
4.30	197	197	207	⋯	92.8	(12.7)	⋯	⋯	⋯	⋯	222	30	95	4.30
4.35	192	192	202	⋯	91.9	(11.5)	⋯	⋯	⋯	⋯	217	29	93	4.35
4.40	187	187	196	⋯	90.9	(10.2)	⋯	⋯	⋯	⋯	212	⋯	90	4.40
4.45	183	183	192	⋯	90.0	(9.0)	⋯	⋯	⋯	⋯	207	28	89	4.45
4.50	179	179	188	⋯	89.0	(8.0)	⋯	⋯	⋯	⋯	202	27	87	4.50
4.55	174	174	182	⋯	88.0	(6.7)	⋯	⋯	⋯	⋯	198	⋯	85	4.55
4.60	170	170	178	⋯	87.0	(5.4)	⋯	⋯	⋯	⋯	194	26	83	4.60
4.65	167	167	175	⋯	86.0	(4.4)	⋯	⋯	⋯	⋯	190	⋯	81	4.65
4.70	163	163	171	⋯	85.0	(3.3)	⋯	⋯	⋯	⋯	186	25	79	4.70
4.75	159	159	167	⋯	83.9	(2.0)	⋯	⋯	⋯	⋯	182	⋯	78	4.75
4.80	156	156	163	⋯	82.9	(0.9)	⋯	⋯	⋯	⋯	178	24	76	4.80
4.85	152	152	159	⋯	81.9	⋯	⋯	⋯	⋯	⋯	174	⋯	75	4.85
4.90	149	149	156	⋯	80.8	⋯	⋯	⋯	⋯	⋯	170	23	73	4.90
4.95	146	146	153	⋯	79.7	⋯	⋯	⋯	⋯	⋯	166	⋯	72	4.95
5.00	143	143	150	⋯	78.6	⋯	⋯	⋯	⋯	⋯	163	22	71	5.00
5.10	137	137	143	⋯	76.4	⋯	⋯	⋯	⋯	⋯	157	21	67	5.10
5.20	131	131	137	⋯	74.2	⋯	⋯	⋯	⋯	⋯	151	⋯	65	5.20
5.30	126	126	132	⋯	72.0	⋯	⋯	⋯	⋯	⋯	145	20	63	5.30
5.40	121	121	127	⋯	69.8	⋯	⋯	⋯	⋯	⋯	140	19	60	5.40
5.50	116	116	122	⋯	67.6	⋯	⋯	⋯	⋯	⋯	135	18	58	5.50
5.60	111	111	117	⋯	65.4	⋯	⋯	⋯	⋯	⋯	131	17	56	5.60

(a) For carbon and alloy steels in the annealed, normalized, and quenched-and-tempered conditions: less accurate for cold worked condition and for austenitic steels. Values in bold-faced type correspond to the values in the joint SAE-ASM-ASTM hardness conversions as printed in ASTM E140. Values in parentheses are beyond normal range and are given for information only. (b) Brinell numbers are based on the diameter of impressed indentation. If the ball distorts (flattens) during test, Brinell numbers will vary in accordance with the degree of such distortion when related to hardnesses determined with a Vickers diamond pyramid, Rockwell Brale, or other indenter that does not sensibly distort. At high hardnesses, therefore, the relationship between Brinell and Vickers or Rockwell scales is affected by the type of ball used. Standard steel balls tend to flatten slightly more than tungsten carbide balls, resulting in a larger indentation and a lower Brinell number than shown by a tungsten carbide ball. Thus, on a specimen of about 539 to 547 HV, a standard ball will leave a 2.75-mm indentation (495 HB), and a tungsten carbide ball a 2.70-mm indentation (514 HB). Conversely, identical indentation diameters for both types of ball will correspond to different Vickers and Rockwell values. Thus, if indentations in two different specimens both are 2.75 mm in diameter (495 HB), the specimen tested with a standard ball has a Vickers hardness of 539, whereas the specimen tested with a tungsten carbide ball has a Vickers hardness of 528.

Rockwell C hardness numbers (a)

Rockwell C-scale hardness	Vickers hardness	Brinell hardness, 3000-kg load, 10-mm ball		Rockwell hardness			Rockwell superficial hardness, superficial Brale indenter			Knoop hardness, 500-g load and greater	Scleroscope hardness	Tensile strength (approx), ksi	Rockwell C-scale hardness
		Standard ball	Tungsten carbide ball	A scale, 60-kg load, Brale indenter	B scale, 100-kg load, 1/16-in.-diam ball	D scale, 100-kg load, Brale indenter	15N scale, 15-kg load	30N scale, 30-kg load	45N scale, 45-kg load				
68	940	· · ·	· · ·	85.6	· · ·	76.9	93.2	84.4	75.4	920	97	· · ·	68
67	900	· · ·	· · ·	85.0	· · ·	76.1	92.9	83.6	74.2	895	95	· · ·	67
66	865	· · ·	· · ·	84.5	· · ·	75.4	92.5	82.8	73.3	870	92	· · ·	66
65	832	· · ·	(739)	83.9	· · ·	74.5	92.2	81.9	72.0	846	91	· · ·	65
64	800	· · ·	(722)	83.4	· · ·	73.8	91.8	81.1	71.0	822	88	· · ·	64
63	772	· · ·	(705)	82.8	· · ·	73.0	91.4	80.1	69.9	799	87	· · ·	63
62	746	· · ·	(688)	82.3	· · ·	72.2	91.1	79.3	68.8	776	85	· · ·	62
61	720	· · ·	(670)	81.8	· · ·	71.5	90.7	78.4	67.7	754	83	· · ·	61
60	697	· · ·	(654)	81.2	· · ·	70.7	90.2	77.5	66.6	732	81	· · ·	60
59	674	· · ·	(634)	80.7	· · ·	69.9	89.8	76.6	65.5	710	80	351	59
58	653	· · ·	615	80.1	· · ·	69.2	89.3	75.7	64.3	690	78	338	58
57	633	· · ·	595	79.6	· · ·	68.5	88.9	74.8	63.2	670	76	325	57
56	613	· · ·	577	79.0	· · ·	67.7	88.3	73.9	62.0	650	75	313	56
55	595	· · ·	560	78.5	· · ·	66.9	87.9	73.0	60.9	630	74	301	55
54	577	· · ·	543	78.0	· · ·	66.1	87.4	72.0	59.8	612	72	292	54
53	560	· · ·	525	77.4	· · ·	65.4	86.9	71.2	58.6	594	71	283	53
52	544	(500)	512	76.8	· · ·	64.6	86.4	70.2	57.4	576	69	273	52
51	528	(487)	496	76.3	· · ·	63.8	85.9	69.4	56.1	558	68	264	51
50	513	(475)	481	75.9	· · ·	63.1	85.5	68.5	55.0	542	67	255	50
49	498	(464)	469	75.2	· · ·	62.1	85.0	67.6	53.8	526	66	246	49
48	484	(451)	455	74.7	· · ·	61.4	84.5	66.7	52.5	510	64	238	48
47	471	442	443	74.1	· · ·	60.8	83.9	65.8	51.4	495	63	229	47
46	458	432	432	73.6	· · ·	60.0	83.5	64.8	50.3	480	62	221	46
45	446	421	421	73.1	· · ·	59.2	83.0	64.0	49.0	466	60	215	45
44	434	409	409	72.5	· · ·	58.5	82.5	63.1	47.8	452	58	208	44
43	423	400	400	72.0	· · ·	57.7	82.0	62.2	46.7	438	57	201	43
42	412	390	390	71.5	· · ·	56.9	81.5	61.3	45.5	426	56	194	42
41	402	381	381	70.9	· · ·	56.2	80.9	60.4	44.3	414	55	188	41
40	392	371	371	70.4	· · ·	55.4	80.4	59.5	43.1	402	54	182	40
39	382	362	362	69.9	· · ·	54.6	79.9	58.6	41.9	391	52	177	39
38	372	353	353	69.4	· · ·	53.8	79.4	57.7	40.8	380	51	171	38
37	363	344	344	68.9	· · ·	53.1	78.8	56.8	39.6	370	50	166	37
36	354	336	336	68.4	(109.0)	52.3	78.3	55.9	38.4	360	49	161	36
35	345	327	327	67.9	(108.5)	51.5	77.7	55.0	37.2	351	48	157	35
34	336	319	319	67.4	(108.0)	50.8	77.2	54.2	36.1	342	47	153	34
33	327	311	311	66.8	(107.5)	50.0	76.6	53.3	34.9	334	46	149	33
32	318	301	301	66.3	(107.0)	49.2	76.1	52.1	33.7	326	44	145	32
31	310	294	294	65.8	(106.0)	48.4	75.6	51.3	32.5	318	43	141	31
30	302	286	286	65.3	(105.5)	47.7	75.0	50.4	31.3	311	42	138	30
29	294	279	279	64.7	(104.5)	47.0	74.5	49.5	30.1	304	41	135	29
28	286	271	271	64.3	(104.0)	46.1	73.9	48.6	28.9	297	40	131	28
27	279	264	264	63.8	(103.0)	45.2	73.3	47.7	27.8	290	39	128	27
26	272	258	258	63.3	(102.5)	44.6	72.8	46.8	26.7	284	38	125	26
25	266	253	253	62.8	(101.5)	43.8	72.2	45.9	25.5	278	38	122	25
24	260	247	247	62.4	(101.0)	43.1	71.6	45.0	24.3	272	37	119	24
23	254	243	243	62.0	100.0	42.1	71.0	44.0	23.1	266	36	117	23
22	248	237	237	61.5	99.0	41.6	70.5	43.2	22.0	261	35	114	22
21	243	231	231	61.0	98.5	40.9	69.9	42.3	20.7	256	35	112	21

(a) For carbon and alloy steels in the annealed, normalized, and quenched-and-tempered conditions; less accurate for cold worked condition and for austenitic steels. The values in bold-faced type correspond to the values in the joint SAE-ASM-ASTM hardness conversions as printed in ASTM E140. The values in parentheses are beyond normal range and are given for information only.

Rockwell B hardness numbers (a)

Rockwell B-scale hardness	Vickers hardness	Brinell hardness, 10-mm-diam ball		Rockwell hardness			Rockwell superficial hardness, 1/16-in.-diam ball			Knoop hardness, 500-g load and greater	Scleroscope hardness	Tensile strength (approx), ksi	Rockwell B-scale hardness
		500-kg load	3000-kg load	A scale, 60-kg load, Brale indenter	C scale, 150-kg load, Brale indenter	F scale, 60-kg load, 1/16-in.-diam ball	15T scale, 15-kg load	30T scale, 30-kg load	45T scale, 45-kg load				
98	228	189	228	60.2	(19.9)	· · ·	92.5	81.8	70.9	241	34	107	98
97	222	184	222	59.5	(18.6)	· · ·	92.1	81.1	69.9	236	33	104	97
96	216	179	216	58.9	(17.2)	· · ·	91.8	80.4	68.9	231	32	102	96
95	210	175	210	58.3	(15.7)	· · ·	91.5	79.8	67.9	226	· · ·	99	95
94	205	171	205	57.6	(14.3)	· · ·	91.2	79.1	66.9	221	31	97	94
93	200	167	200	57.0	(13.0)	· · ·	90.8	78.4	65.9	216	30	94	93
92	195	163	195	56.4	(11.7)	· · ·	90.5	77.8	64.8	211	· · ·	92	92
91	190	160	190	55.8	(10.4)	· · ·	90.2	77.1	63.8	206	29	90	91
90	185	157	185	55.2	(9.2)	· · ·	89.9	76.4	62.8	201	28	88	90
89	180	154	180	54.6	(8.0)	· · ·	89.5	75.8	61.8	196	27	86	89
88	176	151	176	54.0	(6.9)	· · ·	89.2	75.1	60.8	192	· · ·	84	88
87	172	148	172	53.4	(5.8)	· · ·	88.9	74.4	59.8	188	26	82	87
86	169	145	169	52.8	(4.7)	· · ·	88.6	73.8	58.8	184	26	81	86
85	165	142	165	52.3	(3.6)	· · ·	88.2	73.1	57.8	180	25	79	85
84	162	140	162	51.7	(2.5)	· · ·	87.9	72.4	56.8	176	· · ·	78	84
83	159	137	159	51.1	(1.4)	· · ·	87.6	71.8	55.8	173	24	76	83
82	156	135	156	50.6	(0.3)	· · ·	87.3	71.1	54.8	170	24	75	82
81	153	133	153	50.0	· · ·	· · ·	86.9	70.4	53.8	167	· · ·	73	81
80	150	130	150	49.5	· · ·	· · ·	86.6	69.7	52.8	164	23	72	80
79	147	128	147	48.9	· · ·	· · ·	86.3	69.1	51.8	161	· · ·	70	79
78	144	126	144	48.4	· · ·	· · ·	86.0	68.4	50.8	158	22	69	78
77	141	124	141	47.9	· · ·	· · ·	85.6	67.7	49.8	155	22	68	77
76	139	122	139	47.3	· · ·	· · ·	85.3	67.1	48.8	152	· · ·	67	76
75	137	120	137	46.8	· · ·	99.6	85.0	66.4	47.8	150	21	66	75
74	135	118	135	46.3	· · ·	99.1	84.7	65.7	46.8	148	21	65	74
73	132	116	132	45.8	· · ·	98.5	84.3	65.1	45.8	145	· · ·	64	73
72	130	114	130	45.3	· · ·	98.0	84.0	64.4	44.8	143	20	63	72
71	127	112	127	44.8	· · ·	97.4	83.7	63.7	43.8	141	20	62	71
70	125	110	125	44.3	· · ·	96.8	83.4	63.1	42.8	139	· · ·	61	70
69	123	109	123	43.8	· · ·	96.2	83.0	62.4	41.8	137	19	60	69
68	121	107	121	43.3	· · ·	95.6	82.7	61.7	40.8	135	19	59	68
67	119	106	119	42.8	· · ·	95.1	82.4	61.0	39.8	133	19	58	67
66	117	104	117	42.3	· · ·	94.5	82.1	60.4	38.7	131	· · ·	57	66
65	116	102	116	41.8	· · ·	93.9	81.8	59.7	37.7	129	18	56	65
64	114	101	114	41.4	· · ·	93.4	81.4	59.0	36.7	127	18	· · ·	64
63	112	99	112	40.9	· · ·	92.8	81.1	58.4	35.7	125	18	· · ·	63
62	110	98	110	40.4	· · ·	92.2	80.8	57.7	34.7	124	· · ·	· · ·	62
61	108	96	108	40.0	· · ·	91.7	80.5	57.0	33.7	122	17	· · ·	61
60	107	95	107	39.5	· · ·	91.1	80.1	56.4	32.7	120	· · ·	· · ·	60
59	106	94	106	39.0	· · ·	90.5	79.8	55.7	31.7	118	· · ·	· · ·	59
58	104	92	104	38.6	· · ·	90.0	79.5	55.0	30.7	117	· · ·	· · ·	58
57	103	91	103	38.1	· · ·	89.4	79.2	54.4	29.7	115	· · ·	· · ·	57
56	101	90	101	37.7	· · ·	88.8	78.8	53.7	28.7	114	· · ·	· · ·	56
55	100	89	100	37.2	· · ·	88.2	78.5	53.0	27.7	112	· · ·	· · ·	55

(a) For carbon and alloy steels in the annealed, normalized, and quenched-and-tempered conditions; less accurate for cold worked condition and for austenitic steels. The values in bold-faced type correspond to the values in the joint SAE-ASM-ASTM hardness conversions as printed in ASTM E140. The values in parentheses are beyond normal range and are given for information only.

Diamond Pyramid Hardness Numbers

Diamond pyramid hardness numbers are obtained when the 136° diamond pyramid indenter is used on micro-hardness tester at any test load. This indenter is cut in the shape of a square-based pyramid having an apex angle of 136°. Both diagonals of the indentation are measured and the average is used to compute the diamond pyramid number. This number is defined as the load per unit area of surface contact in kilograms per square millimeter and is calculated from the formula:

$$DPN = \frac{2L \sin\frac{a}{2}}{d^2}$$

Where DPN is the diamond pyramid number; L is the load in kilograms applied to the indenter; a is the 136° apex angle; and d is the length of average diagonal in millimeters.

The diamond pyramid number corresponding to a measured average length of diagonal, d, for an applied load of 1 g may be obtained directly from the following table. To obtain the DPN for any other applied load, multiply the number obtained for length, d, for 1 g from the table by the actual applied load in grams used to make the indentation. As an alternative method and in order to avoid the necessity of making interpolations, it is possible to multiply the average length of the diagonals by 10 and then look up the number of the table corresponding to this length. Multiply this number by 100 times the applied load in grams used to make the indentation to find the proper diamond pyramid number.

Example: Direct Procedure. A specimen tested with an applied load of 1000 g is measured and shows an average length for the two diagonals of 50.5 μm. By interpolation a number is obtained from the table of 0.7272. Since an applied load of 1000 g was used while the table was computed using a load of 1 g, multiply this number by 1000 g giving a DPN of 727.2 for the material.

Example: Alternative Procedure. Multiply the length of the average diagonal, 50.5 μm, by 10, giving a length of 505 μm. This length gives a number of 0.00727 on the table and multiplying this number by 100 × 1000 g obtains a DPN of 727.

Diamond pyramid hardness numbers

Computed using a 1-g load

Average diagonal length, μm	0	1	2	3	Applied load, g 4	5	6	7	8	9
00	1854.	463.6	206.0	115.9	74.16	51.51	37.84	28.97	22.89	
10 . . . . . .18.54	15.33	12.88	10.97	9.461	8.242	7.244	6.416	5.723	5.137	
20 . . . . . .4.636	4.205	3.831	3.505	3.219	2.967	2.743	2.544	2.365	2.205	
30 . . . . . . .2.060	1.930	1.811	1.703	1.604	1.514	1.431	1.355	1.284	1.219	
40 . . . . . . .1.159	1.103	1.051	1.003	0.9578	0.9157	0.8764	0.8395	0.8048	0.7723	
50 . . . . . . .0.7416	0.7128	0.6857	0.6600	0.6358	0.6129	0.5912	0.5706	0.5511	0.5326	
60 . . . . . . .0.5150	0.4983	0.4823	0.4671	0.4526	0.4388	0.4256	0.4130	0.4010	0.3894	
70 . . . . . . .0.3784	0.3678	0.3576	0.3479	0.3386	0.3296	0.3210	0.3127	0.3047	0.2971	
80 . . . . . . .0.2897	0.2826	0.2757	0.2691	0.2628	0.2566	0.2507	0.2449	0.2394	0.2341	
90 . . . . . . .0.2289	0.2239	0.2190	0.2144	0.2098	0.2054	0.2012	0.1970	0.1930	0.1892	
100 . . . . . . .0.1854	0.1817	0.1782	0.1748	0.1714	0.1682	0.1650	0.1619	0.1590	0.1560	
110 . . . . . . .0.1533	0.1505	0.1478	0.1452	0.1427	0.1402	0.1378	0.1354	0.1332	0.1310	
120 . . . . . . .0.1288	0.1267	0.1246	0.1226	0.1206	0.1187	0.1168	0.1150	0.1132	0.1115	
130 . . . . . . .0.1097	0.1081	0.1064	0.1048	0.1033	0.1018	0.1003	0.0988	0.0974	0.0960	
140 . . . . . . .0.0649	0.0933	0.0920	0.0907	0.0894	0.0882	0.0870	0.0858	0.0847	0.0835	
150 . . . . . . .0.0824	0.0813	0.0803	0.0792	0.0782	0.0772	0.0762	0.0752	0.0743	0.0734	
160 . . . . . . .0.0724	0.0715	0.0707	0.0698	0.0690	0.0681	0.0673	0.0665	0.0657	0.0647	
170 . . . . . . .0.0642	0.0634	0.0627	0.0620	0.0613	0.0606	0.0599	0.0592	0.0585	0.0579	
180 . . . . . . .0.0572	0.0566	0.0560	0.0554	0.0548	0.0542	0.0536	0.0530	0.0525	0.0519	
190 . . . . . . .0.0514	0.0508	0.0503	0.0498	0.0493	0.0488	0.0483	0.0478	0.0473	0.0468	
200 . . . . . . .0.0464	0.0459	0.0455	0.0450	0.0446	0.0442	0.0437	0.0433	0.0429	0.0425	
210 . . . . . . .0.0421	0.0417	0.0413	0.0409	0.0405	0.0401	0.0397	0.0394	0.0390	0.0387	
220 . . . . . . .0.0383	0.0380	0.0376	0.0373	0.0370	0.0366	0.0363	0.0360	0.0357	0.0354	
230 . . . . . . .0.0351	0.0348	0.0345	0.0342	0.0339	0.0336	0.0333	0.0330	0.0327	0.0325	
240 . . . . . . .0.0322	0.0319	0.0317	0.0314	0.0312	0.0309	0.0306	0.0304	0.0302	0.0299	
250 . . . . . . .0.0297	0.0294	0.0292	0.0289	0.0287	0.0285	0.0283	0.0281	0.0279	0.0276	
260 . . . . . . .0.0274	0.0272	0.0270	0.0268	0.0266	0.0264	0.0262	0.0260	0.0258	0.0256	
270 . . . . . . .0.0254	0.0253	0.0251	0.0249	0.0247	0.0245	0.0243	0.0242	0.0240	0.0238	
280 . . . . . . .0.0236	0.0235	0.0233	0.0232	0.0230	0.0228	0.0227	0.0225	0.0224	0.0222	
290 . . . . . . .0.0221	0.0219	0.0218	0.0216	0.0215	0.0213	0.0212	0.0210	0.0209	0.0207	

(continued)

Diamond pyramid hardness numbers (continued)

Average diagonal length, μm	0	1	2	3	Applied load, g 4	5	6	7	8	9
300	0.0206	0.0205	0.0203	0.0202	0.0201	0.0199	0.0198	0.0197	0.0196	0.0194
310	0.0193	0.0192	0.0191	0.0189	0.0188	0.0187	0.0186	0.0185	0.0183	0.0182
320	0.0181	0.0180	0.0179	0.0178	0.0177	0.0176	0.0175	0.0173	0.0172	0.0171
330	0.0170	0.0169	0.0168	0.0167	0.0166	0.0165	0.0164	0.0163	0.0162	0.0161
340	0.0160	0.0160	0.0159	0.0158	0.0157	0.0156	0.0155	0.0154	0.0153	0.0152
350	0.01514	0.01505	0.01497	0.01488	0.01480	0.01471	0.01463	0.01455	0.01447	0.01439
360	0.01431	0.01423	0.01415	0.01407	0.01400	0.01392	0.01384	0.01377	0.01369	0.01362
370	0.01355	0.01347	0.01340	0.01333	0.01326	0.01319	0.01312	0.01305	0.01298	0.01291
380	0.01284	0.01277	0.01271	0.01264	0.01258	0.01251	0.01245	0.01238	0.01232	0.01226
390	0.01219	0.01213	0.01207	0.01201	0.01195	0.01189	0.01183	0.01177	0.01171	0.01165
400	0.01159	0.01153	0.01148	0.01142	0.01136	0.01131	0.01125	0.01119	0.01114	0.01109
410	0.01103	0.01098	0.01093	0.01087	0.01082	0.01077	0.01072	0.01066	0.01061	0.01056
420	0.01051	0.01046	0.01041	0.01036	0.01031	0.01027	0.01022	0.01017	0.01012	0.01008
430	0.01003	0.00998	0.00994	0.00989	0.00985	0.00980	0.00976	0.00971	0.00967	0.00962
440	0.00958	0.00953	0.00949	0.00945	0.00941	0.00936	0.00932	0.00928	0.00924	0.00920
450	0.00916	0.00912	0.00908	0.00904	0.00900	0.00896	0.00892	0.00888	0.00884	0.00880
460	0.00876	0.00873	0.00869	0.00865	0.00861	0.00858	0.00854	0.00850	0.00847	0.00843
470	0.00840	0.00836	0.00832	0.00829	0.00825	0.00822	0.00818	0.00815	0.00812	0.00808
480	0.00805	0.00802	0.00798	0.00795	0.00792	0.00788	0.00785	0.00782	0.00779	0.00776
490	0.00772	0.00769	0.00766	0.00763	0.00760	0.00757	0.00754	0.00751	0.00748	0.00745
500	0.00742	0.00739	0.00736	0.00733	0.00730	0.00727	0.00724	0.00721	0.00719	0.00716
510	0.00713	0.00710	0.00707	0.00705	0.00702	0.00699	0.00696	0.00694	0.00691	0.00688
520	0.00686	0.00683	0.00681	0.00678	0.00675	0.00673	0.00670	0.00668	0.00665	0.00663
530	0.00660	0.00658	0.00655	0.00653	0.00650	0.00648	0.00645	0.00643	0.00641	0.00638
540	0.00636	0.00634	0.00631	0.00629	0.00627	0.00624	0.00622	0.00620	0.00617	0.00615
550	0.00613	0.00611	0.00609	0.00606	0.00604	0.00602	0.00600	0.00598	0.00596	0.00593
560	0.00591	0.00589	0.00587	0.00585	0.00583	0.00581	0.00579	0.00577	0.00575	0.00573
570	0.00571	0.00569	0.00567	0.00565	0.00563	0.00561	0.00559	0.00557	0.00555	0.00553
580	0.00551	0.00549	0.00547	0.00546	0.00544	0.00542	0.00540	0.00538	0.00536	0.00534
590	0.00533	0.00531	0.00529	0.00527	0.00526	0.00524	0.00522	0.00520	0.00519	0.00517
600	0.00515	0.00513	0.00512	0.00510	0.00508	0.00507	0.00505	0.00503	0.00502	0.00500
610	0.00498	0.00497	0.00495	0.00494	0.00492	0.00490	0.00489	0.00487	0.00486	0.00484
620	0.00482	0.00481	0.00479	0.00478	0.00476	0.00475	0.00473	0.00472	0.00470	0.00469
630	0.00467	0.00466	0.00464	0.00463	0.00461	0.00460	0.00458	0.00457	0.00456	0.00454
640	0.00453	0.00451	0.00450	0.00448	0.00447	0.00446	0.00444	0.00443	0.00442	0.00440
650	0.00439	0.00438	0.00436	0.00435	0.00434	0.00432	0.00431	0.00430	0.00428	0.00427
660	0.00426	0.00424	0.00423	0.00422	0.00421	0.00419	0.00418	0.00417	0.00416	0.00414
670	0.00413	0.00412	0.00411	0.00409	0.00408	0.00407	0.00406	0.00405	0.00403	0.00402
680	0.00401	0.00400	0.00399	0.00398	0.00396	0.00395	0.00394	0.00393	0.00392	0.00391
690	0.00390	0.00388	0.00387	0.00386	0.00385	0.00384	0.00383	0.00382	0.00381	0.00380
700	0.00378	0.00377	0.00376	0.00375	0.00374	0.00373	0.00372	0.00371	0.00370	0.00369
710	0.00368	0.00367	0.00366	0.00365	0.00364	0.00363	0.00362	0.00361	0.00360	0.00359
720	0.00358	0.00357	0.00356	0.00355	0.00354	0.00353	0.00352	0.00351	0.00350	0.00349
730	0.00348	0.00347	0.00346	0.00345	0.00344	0.00343	0.00342	0.00341	0.00340	0.00340
740	0.00339	0.00338	0.00337	0.00336	0.00335	0.00334	0.00333	0.00332	0.00331	0.00331
750	0.00330	0.00329	0.00328	0.00327	0.00326	0.00325	0.00324	0.00324	0.00323	0.00322
760	0.00321	0.00320	0.00319	0.00318	0.00318	0.00317	0.00316	0.00315	0.00314	0.00314
770	0.00313	0.00312	0.00311	0.00310	0.00309	0.00309	0.00308	0.00307	0.00307	0.00306
780	0.00305	0.00304	0.00303	0.00303	0.00302	0.00301	0.00300	0.00299	0.00299	0.00298
790	0.00297	0.00296	0.00296	0.00295	0.00294	0.00293	0.00293	0.00292	0.00291	0.00291
800	0.00290	0.00289	0.00288	0.00288	0.00287	0.00287	0.00286	0.00285	0.00284	0.00283
810	0.00283	0.00282	0.00281	0.00280	0.00280	0.00279	0.00278	0.00278	0.00277	0.00277
820	0.00276	0.00275	0.00274	0.00274	0.00273	0.00273	0.00272	0.00271	0.00270	0.00270
830	0.00269	0.00268	0.00268	0.00267	0.00267	0.00266	0.00265	0.00265	0.00264	0.00263
840	0.00263	0.00262	0.00262	0.00261	0.00260	0.00260	0.00259	0.00258	0.00258	0.00257

(continued)

Diamond pyramid hardness numbers (continued)

Average diagonal length, μm	0	1	2	3	Applied load, g 4	5	6	7	8	9
850	0.00257	0.00256	0.00256	0.00255	0.00254	0.00254	0.00253	0.00253	0.00252	0.00251
860	0.00251	0.00250	0.00250	0.00249	0.00248	0.00248	0.00247	0.00247	0.00246	0.00246
870	0.00245	0.00244	0.00244	0.00243	0.00243	0.00242	0.00242	0.00241	0.00241	0.00240
880	0.00240	0.00239	0.00238	0.00238	0.00237	0.00237	0.00236	0.00236	0.00235	0.00235
890	0.00234	0.00234	0.00233	0.00233	0.00232	0.00232	0.00231	0.00230	0.00230	0.00229
900	0.00229	0.00228	0.00228	0.00227	0.00227	0.00226	0.00226	0.00225	0.00225	0.00224
910	0.00224	0.00223	0.00223	0.00223	0.00222	0.00222	0.00221	0.00221	0.00220	0.00220
920	0.00219	0.00219	0.00218	0.00218	0.00217	0.00217	0.00216	0.00216	0.00215	0.00215
930	0.00214	0.00214	0.00214	0.00213	0.00213	0.00212	0.00212	0.00211	0.00211	0.00210
940	0.00210	0.00209	0.00209	0.00208	0.00208	0.00208	0.00207	0.00207	0.00206	0.00206
950	0.00205	0.00205	0.00205	0.00204	0.00204	0.00203	0.00203	0.00202	0.00202	0.00202
960	0.00201	0.00201	0.00200	0.00200	0.00200	0.00199	0.00199	0.00198	0.00198	0.00198
970	0.00197	0.00197	0.00196	0.00196	0.00196	0.00195	0.00195	0.00194	0.00194	0.00194
980	0.00193	0.00193	0.00192	0.00192	0.00192	0.00191	0.00191	0.00190	0.00190	0.00190
990	0.00189	0.00189	0.00188	0.00188	0.00188	0.00187	0.00187	0.00187	0.00186	0.00186
1000	0.00185									

Source: Torsion Balance Co., Clifton, N. J.

Knoop Hardness Numbers

Knoop hardness numbers are obtained when the Knoop diamond indenter is used with microhardness testers at any test load. These numbers will vary for the same material according to the test load used. Therefore, in all cases the test load must be specified. The Kentron tester will apply loads of from 1 to 10 000 g, but a range of 1 to 1000 g is ordinarily used with the Knoop indenter.

The Knoop indenter is cut in the shape of a diamond-based pyramid giving a diamond-shaped impression in which the long diagonal is very close to seven times the length of the short diagonal. The included longitudinal angle measured from edge to edge is 172° 30' and the transverse angle is 130° 00'. Because of the difference in the lengths of the two diagonals, almost all of the elastic recovery of the indentations made with the Knoop indenter takes place in the transverse direction. Hence, the measurement of the long diagonal together with the computed indenter constant gives a very close approximation of the unrecovered projected area of the indentation in square millimeters. The relationship between the applied load in kilograms and the approximate unrecovered projected area in square millimeters is called the Knoop hardness number for the specimen for that applied load.

The Knoop hardness number may be expressed by the formula:

$$KN = \frac{L}{A_p} - \frac{L}{l^2 C_p}$$

Where KN is the Knoop hardness number; L is the load in kilograms applied to the indenter; A_p is the unrecovered projected area in square millimeters; l is the measured length of the long diagonal of the indentation in millimeters; C_p is the constant relating l to the unrecovered projected area of the indentation. For an indenter with a longitudinal angle of 172° 30' and a transverse angle of 130° 00'. C_p is 7.028×10^{-2}.

The Knoop hardness number corresponding to a measured length l for an applied load of 1 g may be obtained directly from the table. To obtain the Knoop hardness number for any other applied load, multiply the number obtained from the table for 1 g by the actual applied load in grams used to make the indentation.

Example. A specimen tested on a microhardness tester with an applied load of 100 g is measured under the microscope and shows a length of 42 μm for the long diagonal. Reference to the table shows that this length of diagonal would give a Knoop hardness number of 8.066. However, because an applied load of 100 g was used while the table was computed using a load of 1 g, multiply the number obtained from the table by 100 giving a Knoop hardness number of 806.6 for this material for an applied test load of 100 g.

Knoop hardness numbers

Computed using a 1-g load, for a theoretical indenter having a longitudinal angle of 172°30′ and a transverse angle of 130°00′ giving a constant for projected area (C_p) of 7.028 × 10^{-2}; a constant correction is provided to correct the Knoop number obtained to simulate a test made with an indenter with perfect angles.

Length of diagonal, μm	Applied load, g									
	0.0	0.1	0.2	0.3	0.4	0.5	0.6	0.7	0.8	0.9
1.0	14 229	11 759	9 881	8 419	7 260	6 324	5 558	4 923	4 392	3 942
2.0	3 557	3 226	2 940	2 690	2 470	2 277	2 105	1 952	1 815	1 692
3.0	1 581	1 481	1 390	1 307	1 231	1 162	1 098	1 039	985.4	935.5
4.0	889.3	846.5	806.6	769.5	735.0	702.7	672.4	644.1	617.6	592.6
5.0	569.2	547.1	526.2	506.5	488.0	470.4	453.7	437.9	423.0	408.8
6.0	395.2	382.4	370.2	358.5	347.4	336.8	326.7	317.0	307.7	298.9
7.0	290.4	282.3	274.5	267.0	259.8	253.0	246.4	240.0	233.9	228.0
8.0	222.3	216.9	211.6	206.5	201.7	196.9	192.4	188.0	183.7	179.6
9.0	175.7	171.8	168.1	164.5	161.0	157.7	154.4	151.2	148.2	145.2
10.0	142.3	139.5	136.8	134.1	131.6	129.1	126.6	124.3	122.0	119.8
11.0	117.6	115.5	113.4	111.4	109.5	107.6	105.7	103.9	102.2	100.5
12.0	98.81	97.18	95.60	94.05	92.54	91.06	89.62	88.22	86.85	85.50
13.0	84.19	82.91	81.66	80.44	79.24	78.07	76.93	75.81	74.72	73.64
14.0	72.60	71.57	70.57	69.58	68.62	67.68	66.75	65.85	64.96	64.09
15.0	63.24	62.40	61.58	60.78	59.99	59.22	58.47	57.73	57.00	56.28
16.0	55.58	54.89	54.22	53.55	52.90	52.26	51.64	51.02	50.41	49.82
17.0	49.23	48.66	48.10	47.54	47.00	46.64	45.93	45.42	44.91	44.41
18.0	43.92	43.44	42.96	42.49	42.02	41.57	41.13	40.69	40.26	39.83
19.0	39.42	39.00	38.60	38.20	37.81	37.42	37.04	36.66	36.29	35.93
20.0	35.57	35.22	34.87	34.53	34.19	33.86	33.53	33.21	32.89	32.57
21.0	32.27	31.96	31.66	31.36	31.07	30.78	30.50	30.22	29.94	29.67
22.0	29.40	29.13	28.87	28.61	28.36	28.11	27.86	27.61	27.37	27.13
23.0	26.90	26.67	26.44	26.21	25.99	25.77	25.55	25.33	25.12	24.91
24.0	24.70	24.50	24.30	24.10	23.90	23.71	23.51	23.32	23.14	22.95
25.0	22.77	22.59	22.41	22.23	22.05	21.88	21.71	21.54	21.38	21.21
26.0	21.05	20.89	20.73	20.57	20.42	20.26	20.11	19.96	19.81	19.66
27.0	19.52	19.37	19.23	19.09	18.95	18.82	18.68	18.54	18.41	18.28
28.0	18.15	18.02	17.89	17.77	17.64	17.52	17.40	17.27	17.15	17.04
29.0	16.92	16.80	16.69	16.57	16.46	16.35	16.24·	16.13	16.02	15.92
30.0	15.81	15.71	15.60	15.50	15.40	15.30	15.20	15.10	15.00	14.90
31.0	14.81	14.71	14.62	14.52	14.43	14.34	14.25	14.16	14.07	13.98
32.0	13.90	13.81	13.72	13.64	13.55	13.47	13.39	13.31	13.23	13.15
33.0	13.07	12.99	12.91	12.83	12.75	12.68	12.60	12.53	12.45	12.38
34.0	12.31	12.24	12.17	12.09	12.02	11.95	11.89	11.82	11.75	11.68
35.0	11.62	11.55	11.48	11.42	11.35	11.29	11.23	11.16	11.10	11.04
36.0	10.93	10.92	10.86	10.80	10.74	10.68	10.62	10.56	10.51	10.45
37.0	10.39	10.34	10.28	10.23	10.17	10.12	10.06	10.01	9.958	9.906
38.0	9.854	9.802	9.751	9.700	9.650	9.600	9.550	9.501	9.452	9.403
39.0	9.355	9.307	9.260	9.213	9.166	9.120	9.074	9.028	8.983	8.938
40.0	8.893	...	...	...	...	...	...	...	...	...

Length of diagonal, μm	Applied load, g									
	0	1	2	3	4	5	6	7	8	9
10	142.3	117.6	98.81	84.19	72.60	63.24	55.58	49.23	43.92	39.42
20	35.57	32.36	29.40	26.90	24.70	22.77	21.05	19.52	18.15	16.92
30	15.81	14.81	13.90	13.07	12.31	11.62	10.98	10.39	9.854	9.355
40	8.893	8.465	8.066	7.695	7.350	7.027	6.724	6.441	6.176	5.926
50	5.692	5.471	5.262	5.065	4.880	4.704	4.537	4.379	4.230	4.088
60	3.952	3.824	3.702	3.585	3.474	3.368	3.267	3.170	3.077	2.989
70	2.904	2.823	2.745	2.670	2.598	2.530	2.463	2.400	2.339	2.280
80	2.223	2.169	2.116	2.065	2.017	1.969	1.924	1.880	1.837	1.796
90	1.757	1.718	1.681	1.645	1.610	1.577	1.544	1.512	1.482	1.452
100	1.423	1.395	1.368	1.341	1.316	1.291	1.266	1.243	1.220	1.198
110	1.176	1.155	1.134	1.114	1.095	1.076	1.057	1.039	1.022	1.005
120	0.9881	0.9718	0.9560	0.9405	0.9254	0.9107	0.8962	0.8822	0.8685	0.8550
130	0.8419	0.8291	0.8166	0.8044	0.7924	0.7807	0.7693	0.7581	0.7472	0.7364
140	0.7260	0.7157	0.7057	0.6958	0.6862	0.6768	0.6675	0.6585	0.6496	0.6409

(continued)

Knoop hardness numbers (continued)

Length of diagonal, μm	Applied load, g									
	0	1	2	3	4	5	6	7	8	9
150	0.6324	0.6240	0.6158	0.6078	0.5999	0.5922	0.5847	0.5773	0.5700	0.5628
160	8.5558	0.5489	0.5422	0.5355	0.5290	0.5226	0.5164	0.5102	0.5051	0.4982
170	0.4923	0.4866	0.4810	0.4754	0.4700	0.4664	0.4593	0.4542	0.4491	0.4441
180	0.4392	0.4344	0.4296	0.4249	0.4202	0.4157	0.4113	0.4069	0.4026	0.3983
190	0.3942	0.3900	0.3860	0.3820	0.3781	0.3742	0.3704	0.3666	0.3629	0.3593
200	0.3557	0.3522	0.3487	0.3453	0.3419	0.3386	0.3353	0.3321	0.3289	0.3257
210	0.3227	0.3196	0.3166	0.3136	0.3107	0.3078	0.3050	0.3022	0.2994	0.2967
220	0.2940	0.2913	0.2887	0.2861	0.2836	0.2811	0.2786	0.2761	0.2737	0.2713
230	0.2690	0.2667	0.2644	0.2621	0.2599	0.2577	0.2555	0.2533	0.2512	0.2491
240	0.2470	0.2450	0.2430	0.2410	0.2390	0.2371	0.2351	0.2332	0.2314	0.2295
250	0.2277	0.2259	0.2241	0.2223	0.2205	0.2188	0.2171	0.2154	0.2138	0.2121
260	0.2105	0.2089	0.2073	0.2057	0.2042	0.2026	0.2011	0.1996	0.1981	0.1966
270	0.1952	0.1937	0.1923	0.1909	0.1895	0.1882	0.1868	0.1854	0.1841	0.1828
280	0.1815	0.1802	0.1789	0.1777	0.1764	0.1752	0.1740	0.1727	0.1715	0.1704
290	0.1692	0.1680	0.1669	0.1657	0.1646	0.1635	0.1624	0.1613	0.1602	0.1592
300	0.1581	0.1571	0.1560	0.1550	0.1540	0.1530	0.1520	0.1510	0.1500	0.1490
310	0.1481	0.1471	0.1462	0.1452	0.1443	0.1434	0.1425	0.1416	0.1407	0.1398
320	0.1390	0.1381	0.1372	0.1364	0.1355	0.1347	0.1339	0.1331	0.1323	0.1315
330	0.1307	0.1299	0.1291	0.1283	0.1275	0.1268	0.1260	0.1253	0.1245	0.1238
340	0.1231	0.1224	0.1217	0.1209	0.1202	0.1195	0.1189	0.1182	0.1175	0.1168
350	0.1162	0.1155	0.1148	0.1142	0.1135	0.1129	0.1123	0.1116	0.1110	0.1104
360	0.1098	0.1092	0.1086	0.1080	0.1074	0.1068	0.1062	0.1056	0.1051	0.1045
370	0.1039	0.1034	0.1028	0.1023	0.1017	0.1012	0.1006	0.1001	0.09958	0.09906
380	0.09854	0.09802	0.09751	0.09700	0.09650	0.09600	0.09550	0.09501	0.09452	0.09403
390	0.09355	0.09307	0.09260	0.09213	0.09166	0.09120	0.09074	0.09028	0.08983	0.08938
400	0.08893	0.08849	0.08805	0.08761	0.08718	0.08675	0.08632	0.08590	0.08548	0.08506
410	0.08465	0.08423	0.08383	0.08342	0.08302	0.08262	0.08222	0.08183	0.08144	0.08105
420	0.08066	0.08028	0.07990	0.07952	0.07915	0.07878	0.07841	0.07804	0.07768	0.07731
430	0.07695	0.07660	0.07624	0.07589	0.07554	0.07520	0.07485	0.07451	0.07417	0.07383
440	0.07350	0.07316	0.07283	0.07250	0.07218	0.07185	0.07153	0.07121	0.07090	0.07058
450	0.07027	0.06996	0.06965	0.06934	0.06903	0.06873	0.06843	0.06813	0.06783	0.06754
460	0.06724	0.06695	0.06666	0.06638	0.06609	0.06581	0.06552	0.06524	0.06497	0.06469
470	0.06441	0.06414	0.06387	0.06360	0.06333	0.06306	0.06280	0.06254	0.06228	0.06202
480	0.06176	0.06150	0.06125	0.06099	0.06074	0.06049	0.06024	0.06000	0.05975	0.05951
490	0.05926	0.05902	0.05878	0.05854	0.05831	0.05807	0.05784	0.05761	0.05737	0.05714
500	0.05692	0.05669	0.05646	0.05624	0.05602	0.05579	0.05557	0.05536	0.05514	0.05492
510	0.05471	0.05449	0.05428	0.05407	0.05386	0.05365	0.05344	0.05323	0.05303	0.05282
520	0.05262	0.05242	0.05222	0.05202	0.05182	0.05162	0.05143	0.05123	0.05104	0.05085
530	0.05065	0.05046	0.05027	0.05009	0.04990	0.04971	0.04953	0.04934	0.04916	0.04898
540	0.04880	0.04862	0.04844	0.04826	0.04808	0.04790	0.04773	0.04756	0.04738	0.04721
550	0.04704	0.04687	0.04670	0.04653	0.04626	0.04619	0.04603	0.04586	0.04570	0.04554
560	0.04537	0.04521	0.04505	0.04489	0.04473	0.04457	0.04442	0.04426	0.04410	0.04395
570	0.04379	0.04364	0.04349	0.04334	0.04319	0.04304	0.04289	0.04274	0.04259	0.04244
580	0.04230	0.04215	0.04201	0.04186	0.04172	0.04158	0.04144	0.04129	0.04115	0.04101
590	0.04088	0.04074	0.04060	0.04046	0.04033	0.04019	0.04006	0.03992	0.03979	0.03966
600	0.03952	0.03939	0.03926	0.03913	0.03900	0.03887	0.03875	0.03862	0.03849	0.03837
610	0.03824	0.03811	0.03799	0.03787	0.03774	0.03762	0.03750	0.03738	0.03726	0.03714
620	0.03702	0.03690	0.03678	0.03666	0.03654	0.03643	0.03631	0.03619	0.03608	0.03596
630	0.03585	0.03574	0.03562	0.03551	0.03540	0.03529	0.03518	0.03507	0.03496	0.03485
640	0.03474	0.03463	0.03452	0.03442	0.03431	0.03420	0.03410	0.03399	0.03389	0.03378
650	0.03368	0.03357	0.03347	0.03337	0.03327	0.03317	0.03306	0.03296	0.03286	0.03276
660	0.03267	0.03257	0.03247	0.03237	0.03227	0.03218	0.03208	0.03198	0.03189	0.03179
670	0.03170	0.03160	0.03151	0.03142	0.03132	0.03123	0.03114	0.03105	0.03095	0.03086
680	0.03077	0.03068	0.03059	0.03050	0.03041	0.03032	0.03024	0.03015	0.03006	0.02997
690	0.02989	0.02980	0.02971	0.02963	0.02954	0.02946	0.02937	0.02929	0.02921	0.02912
700	0.02904	0.02896	0.02887	0.02879	0.02871	0.02863	0.02855	0.02847	0.02839	0.02831
710	0.02823	0.02815	0.02807	0.02799	0.02791	0.02783	0.02776	0.02768	0.02760	0.02752
720	0.02745	0.02737	0.02730	0.02722	0.02715	0.02707	0.02700	0.02692	0.02685	0.02677
730	0.02670	0.02663	0.02656	0.02648	0.02641	0.02634	0.02627	0.02620	0.02613	0.02605
740	0.02598	0.02591	0.02584	0.02577	0.02571	0.02564	0.02557	0.02550	0.02543	0.02536

(continued)

Knoop hardness numbers (continued)

Length of diagonal, µm	0	1	2	3	Applied load, g 4	5	6	7	8	9
750	0.02530	0.02523	0.02516	0.02509	0.02503	0.02496	0.02490	0.02483	0.02476	0.02470
760	0.02463	0.02457	0.02451	0.02444	0.02438	0.02431	0.02425	0.02419	0.02412	0.02406
770	0.02400	0.02394	0.02387	0.02381	0.02375	0.02369	0.02363	0.02357	0.02351	0.02345
780	0.02339	0.02333	0.02327	0.02321	0.02315	0.02309	0.02303	0.02297	0.02292	0.02286
790	0.02280	0.02274	0.02268	0.02263	0.02257	0.02251	0.02246	0.02240	0.02234	0.02229
800	0.02223	0.02218	0.02212	0.02207	0.02201	0.02196	0.02190	0.02185	0.02179	0.02174
810	0.02169	0.02164	0.02158	0.02153	0.02147	0.02142	0.02137	0.02132	0.02127	0.02121
820	0.02116	0.02111	0.02106	0.02101	0.02096	0.02091	0.02086	0.02080	0.02075	0.02070
830	0.02065	0.02060	0.02056	0.02051	0.02046	0.02041	0.02036	0.02031	0.02026	0.02021
840	0.02017	0.02012	0.02007	0.02002	0.01998	0.01993	0.01988	0.01983	0.01979	0.01974
850	0.01969	0.01965	0.01960	0.01956	0.01951	0.01946	0.01942	0.01937	0.01933	0.01928
860	0.01924	0.01919	0.01915	0.01911	0.01906	0.01902	0.01897	0.01893	0.01889	0.01884
870	0.01880	0.01876	0.01871	0.01867	0.01863	0.01858	0.01854	0.01850	0.01846	0.01842
880	0.01837	0.01833	0.01829	0.01825	0.01821	0.01817	0.01813	0.01809	0.01804	0.01800
890	0.01796	0.01792	0.01788	0.01784	0.01780	0.01776	0.01772	0.01768	0.01764	0.01761
900	0.01757	0.01753	0.01749	0.01745	0.01741	0.01737	0.01733	0.01730	0.01726	0.01722
910	0.01718	0.01714	0.01711	0.01707	0.01703	0.01700	0.01696	0.01692	0.01688	0.01685
920	0.01681	0.01677	0.01674	0.01670	0.01667	0.01663	0.01659	0.01656	0.01652	0.01649
930	0.01645	0.01642	0.01638	0.01635	0.01631	0.01628	0.01624	0.01621	0.01617	0.01614
940	0.01610	0.01607	0.01604	0.01600	0.01597	0.01593	0.01590	0.01587	0.01583	0.01580
950	0.01577	0.01573	0.01570	0.01567	0.01563	0.01560	0.01557	0.01554	0.01550	0.01547
960	0.01544	0.01541	0.01538	0.01534	0.01531	0.01528	0.01525	0.01522	0.01519	0.01515
970	0.01512	0.01509	0.01506	0.01503	0.01500	0.01497	0.01494	0.01491	0.01488	0.01485
980	0.01482	0.01479	0.01476	0.01473	0.01470	0.01467	0.01464	0.01461	0.01458	0.01455
990	0.01452	0.01449	0.01446	0.01443	0.01440	0.01437	0.01434	0.01431	0.01429	0.01426
1000	0.01423	...	...	...	...	...	...	...	...	...

Metallography

General Data

Characteristics of minerals used in coated abrasives

Commercial name	Chemical composition	Mineral name	Origin	Specific gravity	Hardness scale Mohs	Knoop	Grain shape
Flint	SiO_2	Quartz	Natural	2.6	6.8-7.0	820	Light wedges
Emery	Al_2O_3, FeO	Impure corundum	Natural	3.7-4.3	8.5-9.0		Blocky
Garnet	SiO_2, FeO, Al_2O_3 complex	Almandite	Natural	3.4-4.3	7.5-8.5		Light wedges
Crocus	FeO	Iron oxide hematite	Synthetic and natural	4.0-5.3	6.0		Fine milled
Aluminum oxide(a)	Al_2O_3 fused	Corundum (alpha)	Synthetic	3.96	9.4	2050	Heavy wedges
Silicon carbide(a)	SiC	Moissanite (alpha)	Synthetic	3.2	9.6	2480	Sharp wedges, silvery

(a) Used most often in metallographic grinding.

Comparison of manufacturers' brand names

	Manufacturer			
Material	Minnesota Mining & Mfg. Co.	Armour	Behr-Manning	Carborundum
Garnet	3M Garnet	Armour H. T. Garnet	B-M Garnet	Carborundum Garnet
Aluminum oxide metal working	Three-M-ite	Alundum	Metalite	Metal Cloth—ALO Industrial Cloth—ALO
Aluminum oxide metal working fibre and fibre combination discs	3M Type A	Armourclad Fibre combination	Metalite Fibre Comb.	Sander Discs ALO— Closed
	3M Type B	Combination	Openkote Metalite Fibre Combination	Sander Discs ALO— Open
	3M Type C	Armourclad Resin Fibre	Speed-Wet Metalite Fibre	Resin Sander Discs
Aluminum oxide metal working narrow rolls	Three-M-ite Utility Rolls	Economy Rolls	Metalite Handy Rolls	Economy Rolls—ALO
Aluminum oxide woodworking	Production	Garalun	Adalox	Belt Cloth—ALO
Silicon carbide	Tri-M-ite	Crystolon	Durite	Abrasive Cloth—SiC
Electrostatic coating	Elek-Tro-Cut	Electrocoated	Lightning	Electro-coated
Treated glue	Grit-Lok Bond	Armourclad	Durabonded	M (Modified)
Resin over glue	Resinite	...	Resinized	Resin Metal
Resinbond	Resin Bond	...	Resinall	Resin Industrial
Waterproof	Wetordry	Rubwet	Speed-Wet	Waterproof

	Manufacturer			
Material	Clover	Abrasive Products	Midwest	Sandpaper, Inc.
Garnet.........................	Clover Garnet	Jewel Garnet	Garnet	Clipper Garnet
Aluminum oxide metal working	Clover Aluminum Oxide Metal Working	Jewelox	Aluminum Oxide	A/O Metal Cloth
Aluminum oxide metal working fibre and fibre combination discs	Resin-Bonded Fibre Disc	Weldisks (Closed or Open)	Aluminum Oxide Resin Fibre	Phenobond Sandisk
Aluminum oxide metal working narrow rolls	Clover Mechanics Rolls	Jewelox Ready Rolls	Shop Rolls	Bench Rolls
Aluminum oxide woodworking	Clover Aluminum Oxide Woodworking	New Process	Aluminum Oxide	A/O W W Cloth
Silicon carbide	Clover Silicon Carbide	Jewelite	Silicon Carbide	Clipper S/C
Treated Glue	...	...	Nubond	...
Resin over glue	Rez-Size	Resin-Sized	Resin Seal	...
Resinbond	All-Rez	Resin-Bonded	...	Phenobond
Waterproof	All-Rez Waterproof	Waterproof	Watersand	Phenowet

Lubrication selection chart

Type of material sanded	Grease stick	Straight mineral oil	Sulfurized and chlorinated oil	Soluble oil with water(a)	Straight water(a)(b)	10% lard oil
Ferrous metals	X	X	X	X		X
Nonferrous metals(c)	X	X		X		X
Plastics					X	
Glass-stone-marble					X	
Rubber					X	
Nickel and nickel chrome alloys of the heat-resisting type	X		X			

(a) Use waterproof coated abrasives only. (b) Use only nonrecirculating system. (c) Such as brass, bronze, and aluminum.

Grit comparison

Grit	Aluminum oxide silicon carbide	Garnet	Flint	Emery
Very fine	600(a)	...	...	...
	500	...	...	...
	400(a)	400-10/0	...	...
	360	...	...	...
	320(a)	320-9/0	...	...
	280	280-8/0	...	...
	240(a)	240-7/0	...	...
	220	220-6/0	Extra fine	...
Fine	180(a)	180-5/0	...	...
	150	150-4/0	...	...
	120	120-3/0	Fine	Fine
	100	100-2/0	...	...
	80(a)	80-1/0	Medium	...
	60	60-1/2	...	Medium
Coarse	50	50-1	...	...
	40	40-1½	...	...
Very coarse	36	36-2	Extra coarse	...
	30	30-2½	...	Extra coarse
	24	24-3	...	...
	20	20-3½	...	...
	16	16-4	...	...
	12	12-4½	...	...

(a) Recommended for most metallographic grinding.

Electrolytic Polishing

Electrolytes. Table 1 gives the formulas of eight groups of electrolytes, together with conditions for their use in electropolishing of various metals and alloys. Table 2 summarizes the applicability of these electrolytes to electropolishing of specific metals. Preferred (or sometimes required) characteristics of an electrolyte are:

- A somewhat viscous consistency
- Acts as a good solvent for the anode metal (the specimen) during electrolysis conditions
- Does not attack the anode metal when no current is flowing
- Contains one or more ions of large radii, such as $(PO_4)^{-3}$, $(ClO_4)^{-1}$, or $(SO_4)^{-2}$, and sometimes large organic molecules
- Simple to mix, stable, and safe to handle (many effective electrolytes are deficient in these respects)
- Effectively functions at room temperature and not sensitive to temperature changes

Advantages and Limitations of Electropolishing. When properly applied, electropolishing can be a useful tool for the metallographer. The principal advantages of electropolishing are:

- For some metals, electropolishing can produce a high-quality surface finish that is equivalent to the best that can be obtained by mechanical methods

- Once a procedure has been established, good results can be obtained with less operator skill than that required for mechanical polishing
- There can be a marked saving of time if many specimens of the same material are to be polished sequentially
- Electropolishing is especially suited to the softer metals, which may be difficult to polish by mechanical methods
- No scratches are produced in electrolytic polishing—a definite advantage in viewing high-quality electropolished surfaces of optically active materials under polarized light
- Artifacts resulting from mechanical deformation, such as disturbed metal or mechanical twins, which are produced on the surface even by careful grinding and mechanical polishing, do not occur in electropolishing
- Surfaces resulting from electropolishing are completely unworked by the polishing procedure, an important feature in low-load hardness testing or x-ray studies
- In some applications, etching can be accomplished by simply reducing the voltage to approximately one-tenth the potential required for polishing, then continuing electrolysis for a few seconds
- Electropolishing is frequently useful in electron metallography (where high resolution is often important)

because it can produce clean, undistorted metal surfaces

Metallographic preparation by electropolishing is subject to several limitations; these should be recognized to prevent misapplication of the method and disappointment in the results. The principal disadvantages include:

- Because the chemicals and combinations of chemicals used in electropolishing are poisonous and many are highly flammable or potentially explosive, only well-trained personnel who are thoroughly familiar with chemical laboratory procedures should be permitted to handle or mix the chemicals, or to operate the polishing baths
- The conditions and electrolytes required to obtain a satisfactorily polished surface differ for different alloys; hence, when appropriate procedures do not exist, considerable time may be required to develop a procedure for a new alloy, if it can be developed at all
- In multiphase alloys, the rates of polishing of different phases often are not the same. Polishing results depend heavily on whether the second or third phases are strongly cathodic or anodic with respect to the matrix. The matrix is dissolved preferentially if the other phases are to stand in relief. Preferential

Electrolytic Polishing (continued)

attack may also occur at the interface between two phases. These effects are most pronounced when phases other than the matrix are virtually unattacked by the polishing bath and are reversed when the matrix phase is relatively cathodic

- A large number of electrolytes may be needed to polish the variety of metals encountered by a given laboratory

- Plastic or metal mounting materials may react with the electrolyte

- Electropolished surfaces exhibit an undulating rather than a plane surface, and in some cases may not be suited for examination at all magnifications. Under some conditions, furrowing and pitting may be produced

- Edge effects limit applications involving small specimens, surface

phenomena, coatings, interfaces, and cracks

- Attack around nonmetallic particles and adjacent metal, voids, and various inhomogeneities may not be the same as that of the matrix, thus exaggerating the size of the voids and inclusions

- Electropolished surfaces of certain materials may be passive and difficult to etch

Table 1 Electrolytes for electropolishing of various metals and alloys (based on ASTM E3)

Class	Formula	Use	Cell voltage	Time	Notes
Group I—Electrolytes composed of perchloric acid and alcohol with or without organic additions					
I-1	800 ml ethanol (absolute)(a) 140 ml distilled water (optional), 60 ml perchloric acid (60%)	Al and Al alloys with less than 2% Si	30-80	15-60 s	...
		Carbon, alloy and stainless steels	35-65	15-60 s	...
		Pb, Pb-Sn, Pb-Sn-Cd, Pb-Sn-Sb	12-35	15-60 s	...
		Zn, Zn-Sn-Fe, Zn-Al-Cu	20-60	...	...
		Mg and high-Mg alloys	...	...	(b)
I-2	800 ml ethanol (absolute)(a), 200 ml perchloric acid (60%)	Stainless steel; aluminum	35-80	15-60 s	...
I-3	940 ml ethanol (absolute)(a), 6 ml distilled water, 54 ml perchloric acid (70%)	Stainless steel	30-45	15-60 s	...
		Thorium	30-40	15-45 s	...
I-4	700 ml ethanol (absolute)(a), 120 ml distilled water, 100 ml 2-butoxyethanol, 80 ml perchloric acid (60%)	Steel, cast iron, Al, Al alloys, Ni, Sn, Ag, Be, Ti, Zr, U, heat-resisting alloys	30-65	15-60 s	(c)
I-5	700 ml ethanol (absolute)(a), 120 ml distilled water, 100 ml glycerol, 80 ml perchloric acid (60%)	Stainless, alloys and high speed steels; Al, Fe, Fe-Si alloys, Pb, Zr	15-50	15-60 s	(d)
I-6	760 ml ethanol (absolute)(a), 30 ml distilled water, 190 ml ether, 20 ml perchloric acid (60%)	Aluminum, aluminum-silicon alloys, iron-silicon alloys	35-60	15-60 s	(e)
I-7	600 ml methanol (absolute), 370 ml 2-butoxyethanol, 30 ml perchloric acid (60%)	Molybdenum, titanium, zinc, zirconium, uranium-zirconium alloy	60-150	5-30 s	...
I-8	840 ml methanol (absolute), 4 ml distilled water, 125 ml glycerol, 31 ml perchloric acid (70%)	Aluminum, aluminum-silicon alloys, iron-silicon alloys	50-100	5-60 s	...
I-9	590 ml methanol (absolute), 6 ml distilled water, 350 ml 2-butoxyethanol, 54 ml perchloric acid (70%)	Germanium	25-35	30-60 s	...
		Titanium	58-66	45 s	(f)
		Vanadium	30	3 s	(g)
		Zirconium	70-75	15 s	(h)
I-10	950 ml methanol (absolute), 15 ml nitric acid, 50 ml perchloric acid (60%)	Aluminum	30-60	15-60 s	...
Group II—Electrolytes composed of perchloric acid (60%) and glacial acetic acid					
II-1	940 ml acetic acid, 60 ml perchloric acid	Cr, Ti, U, Zr, Fe, cast iron; carbon, alloy and stainless steels	20-60	1-5 min	(j)
II-2	900 ml acetic acid, 100 ml perchloric acid	Zr, Ti, U, steels, superalloys	12-70	½-2 min	...
II-3	800 ml acetic acid, 200 ml perchloric acid	U, Zr, Ti, Al, steels, superalloys	40-100	1-15 min	...
II-4	700 ml acetic acid, 300 ml perchloric acid	Nickel, lead, lead-antimony alloys	40-100	1-5 min	...
II-5	650 ml acetic acid, 350 ml perchloric acid	3% silicon iron	...	5 min	(k)
Group III—Electrolytes composed of phosphoric acid (85%) in water or organic solvent					
III-1	1000 ml phosphoric acid	Cobalt	1.2	3-5 min	...
III-2	175 ml distilled water, 825 ml phosphoric acid	Pure copper	1.0-1.6	10-40 min	(m)
III-3	300 ml water, 700 ml phosphoric acid	Stainless steel, brass, copper and copper alloys except tin-bronze	1.5-1.8	5-15 min	(m)
III-4	600 ml water, 400 ml phosphoric acid	α or α + β brass, Cu-Fe, Cu-Co, Co, Cd	1-2	1-15 min	(n)
III-5	1000 ml water, 580 g pyrophosphoric acid	Copper, copper-zinc	1-2	10 min	(m)
III-6	500 ml diethylene glycol monoethyl ether, 500 ml phosphoric acid	Steel	5-20	5-15 min	(p)

(continued)

Table 1 (continued)

Class	Formula	Use	Cell voltage	Time	Notes
Group III—Electrolytes composed of phosphoric acid (85%) in water or organic solvent (continued)					
III-7	200 ml water, 380 ml ethanol (95%), 400 ml phosphoric acid	Aluminum, magnesium, silver	25-30	4-6 min	(q)
III-8	300 ml ethanol (absolute), 300 ml glycerol (cp), 300 ml phosphoric acid	Uranium	...	...	...
III-9	500 ml ethanol (95%), 250 ml glycerol, 250 ml phosphoric acid	Manganese, manganese-copper alloys	18	...	...
III-10	500 ml distilled water, 250 ml ethanol (95%), 250 ml phosphoric acid	Copper and copper-base alloys	...	1-5 min	...
III-11	Ethanol (absolute) to make 1000 ml of solution; 400 g pyrophosphoric acid	Stainless steel; all austenitic heat-resisting alloys	...	10 min	(r)
III-12	625 ml ethanol (95%), 375 ml phosphoric acid	Magnesium-zinc	1.5-2.5	3-30 min	...
III-13	445 ml ethanol (95%), 275 ml ethylene glycol, 275 ml phosphoric acid	Uranium	18-20	5-15 min	(s)
Group IV—Electrolytes composed of sulfuric acid in water or organic solvent					
IV-1	250 ml water, 750 ml sulfuric acid	Stainless steel	1.5-6	1-2 min	...
IV-2	400 ml water, 600 ml sulfuric acid	Stainless steel, iron, nickel	1.5-6	2-6 min	...
IV-3	750 ml water, 250 ml sulfuric acid	Stainless steel, iron, nickel	1.5-6	2-10 min	...
		Molybdenum	1.5-6	⅓-1 min	(t)
IV-4	900 ml water, 100 ml sulfuric acid	Molybdenum	1.5-6	⅓-2 min	(t)
IV-5	70 ml water, 200 ml glycerol, 720 ml sulfuric acid	Stainless steel	1.5-6	½-5 min	...
IV-6	220 ml water, 200 ml glycerol, 580 ml sulfuric acid	Stainless steel, aluminum	1.5-12	1-20 min	...
IV-7	875 ml methanol (absolute), 125 ml sulfuric acid	Molybdenum	6-18	½-1½ min	(u)
Group V—Electrolytes composed of chromic acid in water					
V-1	830 ml water, 620 g chromic acid	Stainless steel	1.5-9	2-10 min	...
V-2	830 ml water, 170 g chromic acid	Zinc, brass	1.5-12	10-60 s	...
Group VI—Electrolytes composed of mixed acids or salts in water or organic solution					
VI-1	600 ml phosphoric acid (85%), 400 ml sulfuric acid	Stainless steel	...	...	...
VI-2	150 ml water, 300 ml phosphoric acid (85%), 550 ml sulfuric acid	Stainless steel	...	2 min	(v)
VI-3	240 ml water, 420 ml phosphoric acid (85%), 340 ml sulfuric acid	Stainless and alloy steels	...	2-10 min	(w)
VI-4	330 ml water, 550 ml phosphoric acid (85%), 120 ml sulfuric acid	Stainless steel	...	1 min	(x)
VI-5	450 ml water, 390 ml phosphoric acid (85%), 160 ml sulfuric acid	Bronze (to 9% tin)	...	1-5 min	(y)
VI-6	330 ml water, 580 ml phosphoric acid (85%), 90 ml sulfuric acid	Bronze (to 6% tin)	...	1-5 min	(y)
VI-7	140 ml water, 100 ml glycerol, 430 ml phosphoric acid (85%), 330 ml sulfuric acid	Steel	...	1-5 min	(z)
VI-8	200 ml water, 590 ml glycerol, 100 ml phosphoric acid (85%), 110 ml sulfuric acid	Stainless steel	...	5 min	(aa)
VI-9	260 ml water, 175 g chromic acid, 175 ml phosphoric acid (85%), 580 ml sulfuric acid	Stainless steel	...	30 min	(bb)
VI-10	175 ml water, 105 g chromic acid, 460 ml phosphoric acid (85%), 390 ml sulfuric acid	Stainless steel	...	60 min	(cc)
VI-11	240 ml water, 80 g chromic acid, 650 ml phosphoric acid (85%), 130 ml sulfuric acid	Stainless and alloy steels	...	5-60 min	(dd)
VI-12	100 ml hydrofluoric acid, 900 ml sulfuric acid	Tantalum	...	9 min	(ee)
VI-13	210 ml water, 180 ml hydrofluoric acid, 610 ml sulfuric acid	Stainless steel	...	5 min	(ff)

(continued)

Table 1 (continued)

Class	Formula	Use	Cell voltage	Time	Notes
Group VI—Electrolytes composed of mixed acids or salts in water or organic solution (continued)					
VI-14	800 ml water, 100 g chromic acid, 46 ml sulfuric acid, 310 g sodium dichromate, 96 ml acetic acid (glacial)	Zinc	...	...	(gg)
VI-15	260 ml hydrogen peroxide (30%), 240 ml hydrofluoric acid, 500 ml sulfuric acid	Stainless steel	...	5 min	(hh)
VI-16	520 ml water, 80 ml hydrofluoric acid, 400 ml sulfuric acid	Stainless steel	...	½-4 min	(jj)
VI-17	600 ml water, 180 g chromic acid, 60 ml nitric acid, 3 ml hydrochloric acid, 240 ml sulfuric acid	Stainless steel	...	...	...
VI-18	750 ml glycerol, 125 ml acetic acid (glacial), 125 ml nitric acid	Bismuth	12	1-5 min	(kk)
VI-19	900 ml ethylene glycol monoethyl ether, 100 ml hydrochloric acid	Magnesium	50-60	10-30 s	(mm)
VI-20	685 ml methanol (absolute), 225 ml hydrochloric acid, 90 ml sulfuric acid	Molybdenum, sintered and cast	19-35	20-35 s	(nn)
VI-21	855 ml ethanol (absolute), 100 ml n-butyl alcohol, 109 g $AlCl_3 \cdot 6H_2O$, 250 g zinc chloride (anhydrous)	Titanium	30-60	1-6 min	...
VI-22	750 ml acetic acid (glacial), 210 ml distilled water, 180 g chromic acid	Uranium	80	5-30 min	(pp)
VI-23	720 ml ethanol (95%), 90 g $AlCl_3 \cdot 6H_2O$, 225 g zinc chloride (anhydrous), 120 ml distilled water, 80 ml n-butyl alcohol	Pure zinc	25-40	½-3 min	(qq)
VI-24	870 ml glycerol, 43 ml hydrofluoric acid, 87 ml nitric acid	Zirconium(h)	9-12	1-10 min	(rr)
VI-25	980 ml saturated solution of potassium iodide in distilled water, 20 ml hydrochloric acid	Bismuth	7	30 s	(ss)
Group VII—Alkaline electrolytes					
VII-1	Water to make 1000 ml, 80 g potassium cyanide, 40 g potassium carbonate, 50 g gold chloride	Gold, silver	7.5	2-4 min	(tt)
VII-2	Water to make 1000 ml, 100 g sodium cyanide, 100 g potassium ferrocyanide	Silver	2.5	To 1 min	(tt)
VIII-3	Water to make 1000 ml, 400 g potassium cyanide, 280 g silver cyanide, 280 g potassium dichromate	Silver	...	To 9 min	(uu)
VII-4	Water to make 1000 ml, 160 g trisodium phosphate	Tungsten	...	10 min	(vv)
VII-5	Water to make 1000 ml, 100 g sodium hydroxide	Tungsten, lead	...	8-10 min	(ww)
VII-6	Water to make 1000 ml, 200 g potassium hydroxide	Zinc, tin	2-6	15 min	(xx)
Group VIII—electrolyte composed of methanol and nitric acid					
VIII-1	600 ml methanol (absolute), 330 ml nitric acid	Nickel, copper, zinc, Monel, brass, nickel-chrome, stainless steel	40-70	10-60 s	(yy)

Note. Chemical components of electrolytes are listed in the order of mixing. Except where noted otherwise, the electrolytes are intended for use at ambient temperatures, in the approximate range of 18 to 38 °C (65 to 100 °F), and with stainless steel cathodes.

(a) In etchants I-1 through I-6, absolute 5D-3A or SD-30 ethanol can be substituted for absolute ethanol. **(b)** Nickel cathode. **(c)** One of the best electrolytes for universal use. **(d)** Universal electrolyte comparable to I-4. **(e)** Particularly good with aluminum-silicon alloys. **(f)** Polish only. **(g)** 3-s cycles repeated at least seven times to prevent heating. **(h)** Polish and etch simultaneously. **(j)** Good general-purpose electrolyte. **(k)** 0.06 A/cm². **(m)** Copper cathode. **(n)** Copper or stainless steel cathode. **(p)** 49 °C (120 °F). **(q)** Aluminum cathode; 38 to 43 °C (100 to 110 °F). **(r)** 38 °C (100 °F) plus. **(s)** 0.03 A/cm². **(t)** Particularly good for sintered molybdenum; 0 to 27 °C (32 to 80 °F). **(u)** 0 to 27 °C (32 to 80 °F). **(v)** 0.3 A/cm². **(w)** 0.1 to 0.2 A/cm². **(x)** 0.05 A/cm². **(y)** 0.1 A/cm². **(z)** 1 to 5 A/cm²; 38 °C (100 °F) plus. **(aa)** 1 A/cm²; 27 to 49 °C (80 to 120 °F). **(bb)** 0.6 A/cm²; 27 to 49 °C (80 to 120 °F). **(cc)** 0.5 A/cm²; 27 to 49 °C (80 to 120 °F). **(dd)** 0.5 A/cm²; 38 to 54 °C (100 to 130 °F). **(ee)** Graphite cathode; 0.1 A/cm²; 32 to 38 °C (90 to 100 °F). **(ff)** 0.5 A/cm²; 21 to 49 °C (70 to 120 °F). **(gg)** 0.002 A/cm²; 21 to 38 °C (70 to 100 °F). **(hh)** 0.5 A/cm². *Caution:* dangerous. **(jj)** 0.08 to 0.3 A/cm². **(kk)** 0.5 (approx) A/cm². *Caution:* this mixture will decompose vigorously after a short time; do not try to keep. **(mm)** Bath should be stirred. Cool below 2 °C (35 °F) with cracked ice. **(nn)** Mix slowly. Heat is developed. Avoid contamination with water. Use below 2 °C (35 °F). **(pp)** The chromic acid is dissolved in the water, and this solution is then added to the acetic acid. Electrolyte is used below 2 °C (35 °F). **(qq)** Electrolyte is used below 16 °C (60 °F). **(rr)** *Caution:* electrolyte will decompose on standing, and is dangerous if kept too long. **(ss)** Polish 3 s, but allow to remain in electrolyte until brown film is dissolved. **(tt)** Graphite cathode. **(uu)** Graphite cathode; 0.003 to 0.009 A/cm². **(vv)** Graphite cathode; 0.09 A/cm²; 38 to 49 °C (100 to 120 °F). **(ww)** Graphite cathode; 0.03 to 0.06 A/cm². **(xx)** Copper cathode; 0.01 to 0.2 A/cm². **(yy)** An extremely useful electrolyte for certain applications, but dangerous.

Table 2 Applicability of electrolytes in Table 1 to electropolishing of various metals and alloys (based on ASTM E3)

Metal	Electrolyte
Aluminum	I-1, I-2, I-4, I-5, I-6, I-8, I-10, II-3, III-7, IV-6
Aluminum-silicon alloys	I-6, I-8
Antimony	II-4
Beryllium	I-4
Bismuth	VI-18, VI-25
Cadmium	III-4
Cast iron	I-4, II-1
Chromium	II-1, VIII-1
Cobalt	I-5, III-1, III-4
Copper	III-2, III-3, III-4, III-5, III-10, VIII-1
Copper-nickel alloys	III-3, III-10, VIII-1
Copper-tin alloys	III-10, VI-5, VI-6, VIII-1
Copper-zinc alloys	III-3, III-4, III-5, III-10, V-2, VIII-1
Germanium	I-9
Gold	VII-1
Iron, pure	I-5, II-1, IV-2, IV-3
Iron-copper alloys	III-3, III-4
Iron-nickel alloys	I-5, II-1, II-2, II-4, IV-3, VIII-1
Iron-silicon alloys	I-5, I-6, I-8, II-5
Lead	I-1, I-5, IV-4, VI-5
Magnesium	I-1, III-7, III-12, VI-19

Metal	Electrolyte
Manganese	III-9
Molybdenum	I-7, IV-3, IV-4, IV-7, VI-20
Nickel	I-4, II-4, IV-2, VIII-1
Nickel-chromium alloys	II-4, VIII-1
Silver	I-4, III-7, VII-1, VII-2, VII-3
Steel: austenitic, stainless, and superalloys	I-1, I-2, I-3, I-4, I-5, II-1, II-2, II-3, III-3, III-6, III-11, IV-1, IV-2, IV-3, IV-5, IV-6, V-1, VI-1, VI-2, VI-3, VI-4, VI-7, VI-8, VI-9, VI-10, VI-11, VI-13, VI-15, VI-16, VI-17, VIII-1
Steel: carbon and alloy	I-1, I-4, I-5, II-1, II-2, II-3, III-6, VI-3, VI-7, VI-11
Tantalum	VI-12
Thorium	I-3
Tin	I-4, VI-5, VI-6, VII-6
Titanium	I-4, I-7, I-9, II-1, II-2, II-3, VI-21
Tungsten	VII-4, VII-5
Uranium	I-4, I-7, II-1, II-2, II-3, III-8, III-13, VI-22
Vanadium	I-9
Zinc	I-1, I-5, III-12, V-2, VI-14, VI-23, VII-6, VIII-1
Zirconium	I-4, I-5, I-7, I-9, II-1, II-2, II-3, VI-24

Electrolytes and voltages for tampon-type local electropolishing of various metals

Electrolyte composition	Metal	Voltage
9 ml perchloric acid (60%), 91 ml butyl cellosolve	Steel, iron and iron-base alloys	35-40
	Aluminum and aluminum alloys	30-35
	Beryllium and beryllium alloys	43-46
10 ml perchloric acid (60%), 45 ml acetic acid (glacial), 45 ml butyl cellosolve	Steel	30-35
	Chromium-base alloys	32-37
	Nickel and nickel-base alloys	30-40
	Cobalt-base alloys	30-60
54 ml phosphoric acid (85%), 22 ml ethanol (absolute), 3 ml distilled water, 21 ml butyl cellosolve	Copper and copper alloys	4-6
11 ml perchloric acid (60%), 65 ml methanol (absolute), 24 ml butyl cellosolve	Titanium alloys	26-28

Characteristics of pure methanol and ethanol

Name	Active constituent	Nominal composition, vol %(a)
Methanol (methyl alcohol)	CH_3OH	99.5(b)
Methanol (methyl alcohol), 95%	CH_3OH	95(c)
Ethanol (ethyl alcohol), anhydrous	C_2H_5OH	99.5(d)(e)
Ethanol (ethyl alcohol), 95%	C_2H_5OH	95(e)

(a) Nominal percentage of the active constituent; remainder is water, unless otherwise specified. (b) Synthetic methanol; the commercial grade is of high purity and is satisfactory for use in all ordinary metallographic etchants where methanol is specified (wood alcohol has not been manufactured commercially in the United States since 1969). Methanol is available only as an anhydrous (also called absolute) grade containing less than 0.1 or 0.2% water as packaged, and usually not more than about 0.5% water at time of use, depending on storage and handling. (c) Where methanol, 95%, is called for, the ordinary anhydrous grade must be diluted by the user with 5% water by volume. (d) The anhydrous (also called absolute) grade of ethanol is ordinarily used only where no significant amount of water can be tolerated. It contains less than 0.1 or 0.2% water as packaged, and usually not more than about 0.5% water at time of use, depending on storage and handling (e) Available only with special government permit.

Characteristics of aqueous liquid chemicals used in many metallographic etchants

Except for sulfuric acid, all data apply to both laboratory and technical or commercial grades of chemicals

Name	Active constituent	Nominal composition, wt%(a)	Specific gravity	Degrees Baumé(b)
Aqueous acids				
Acetic acid, glacial	$HC_2H_3O_2$	99.5	1.05	7.0
Fluoboric acid	HBF_4	48	1.32	35
Hydrochloric acid(c)	HCl	37	1.18	22
Hydrofluoric acid	HF	48	1.15	19
Lactic acid	$HC_3H_5O_3$	85	1.20	24
Nitric acid	HNO_3	70	1.42	43
Perchloric acid	$NClO_4$	70	1.67	58
		60	1.53	50
Phosphoric acid (ortho)	H_3PO_4	85	1.70	60
Sulfuric acid	H_2SO_4	96(d)	1.84(e)	66(e)
Miscellaneous aqueous chemicals				
Ammonium hydroxide	NH_4OH	28(f)	0.90	26
Hydrogen peroxide	H_2O_2	3(g)	1.01	1.4
		30(h)	1.11	15
		50(j)	1.20	24

(a) Nominal percentage of the active constituent; remainder is water. Reagents made by different manufacturers may differ slightly in nominal concentration and allowable range of concentration. (b) Specific gravity as indicated on the Baumé scale; sometimes used for technical grades and in laboratory measurements. (c) Technical grade is also called muriatic acid. (d) Laboratory grade. Technical grade has concentration of 93%. (e) Specific gravity and degrees Baumé are nearly constant for 93 to 100% sulfuric acid. (f) Percent NH_3. (g) Sometimes called "10 volume". (h) Sometimes called "100 volume". (j) Sometimes called "170 volume".

Nominal compositions of various grades of denatured alcohol (ethanol) used in some metallographic etchants

Component	Parts by volume in specially denatured alcohol(a)					
	Formula SD-1(b)		Formula SD-3A		Formula SD-30	
	Anhydrous	95%(c)	Anhydrous	95%(c)	Anhydrous	95%(c)
Ethanol, anhydrous	100	95	100	95	100	95
Water	...	5	...	5	...	5
Methanol	4	4	5	5	10	10
Methyl isobutyl ketone .	1	1	...	...	...	...

Component	Parts by volume in proprietary solvent(d)		Parts by volume in "reagent" alcohol(d)	
	Anhydrous	95%(c)	Anhydrous	95%(c)
SD-1, anhydrous(b)	100	...	...	...
SD-1, 95%(b)(c)	...	100	...	...
SD-3A, anhydrous	...	...	95	...
SD-3A, 95%(c)	...	...	...	95
Methyl isobutyl ketone	1	1	...	...
Hydrocarbon solvent or gasoline	1	1	...	...
Ethyl acetate	1	1	...	...
Isopropyl alcohol	...	...	5	5

(a) Specially denatured alcohol is available only with special government permit. (b) The formula shown here has replaced the old SD-1 formula in which wood alcohol was specified; wood alcohol has not been manufactured commercially in the United States since 1969. (c) The designation of type of denatured alcohol as 95% means that the denatured product contains 5 parts of water for every 95 parts of anhydrous (absolute) ethanol, plus denaturants as specified. (d) Available without government permit from suppliers of laboratory chemicals, for scientific and general laboratory purposes.

Description of miscellaneous chemicals used in metallographic etchants

aluminum chloride, anhydrous. Solid; $AlCl_3$; reacts violently with water, evolving HCl gas; use of hydrated form, $AlCl_3 \cdot 6H_2O$, is preferred.

ammonium molybdate. Crystals; also called ammonium paramolybdate or heptamolybdate; $(NH_4)_6Mo_7O_{24} \cdot 4H_2O$; can be used inter changeably with "molybdic acid, 85%".

benzalkonium chloride. Crystals; essentially alkyl-dimethyl-benzyl-ammonium chloride. May not be readily available in this form; see *zephiran chloride*.

1-butanol. See *n-butyl alcohol*.

2-butoxyethanol. See *butyl cellosolve*.

n-butyl alcohol. Liquid; normal butyl alcohol; also called butyl alcohol and 1-butanol.

butyl carbitol. Liquid; diethylene glycol monotutyl ether.

butyl cellosolve. Liquid; ethylene glycol monobutyl ether; also called 2-butoxyethanol.

carbitol. Liquid; diethylene glycol monoethyl ether.

cellosolve. Liquid; ethylene glycol monoethyl ether.

chromic acid. Dark-red crystals or flakes; CrO_3; also called chromic anhydride, chromic acid anhydride, and chromium trioxide. See *chromic oxide*, Cr_2O_3.

chromic anhydride. See *chromic acid*.

chromic oxide. Fine green powder; Cr_2O_3; a polishing abrasive. Do not confuse with chromic acid (CrO_3), which is a strong acid and a component of many etchants.

cupric ammonium chloride. Crystals; a double salt, $CuCl_2 \cdot 2NH_4Cl \cdot 2H_2O$. If not available, substitute 0.6 g $CuCl_2 \cdot 2H_2O$ plus 0.4 g NH_4Cl for each gram of the double salt.

diethylene glycol. Syrupy liquid; also called 2,2'-oxydiethanol and dihydroxydiethyl ether; $(HOCH_2CH_2)_2O$. More viscous than ethylene glycol; otherwise similar in behavior.

diethylene glycol monobutyl ether. See *butyl carbitol*.

diethylene glycol monoethyl ether. See *carbitol*.

diethyl ether. See *ether*.

ether. Liquid; also called ethyl ether and diethyl ether; very low flash point, highly explosive; boiling point is 34.4 °C (94 °F).

ethylene glycol. Syrupy liquid; also called 1,2-ethanediol and dihydroxyethane; $(CH_2)_2(OH)_2$. Less viscous than diethylene glycol; otherwise similar in behavior.

ethylene glycol monobutyl ether. Liquid; also called 2-butoxyethanol or butyl cellosolve.

ethylene glycol monoethyl ether. See *cellosolve*.

ethyl ether. See *ether*.

ferric nitrate. Crystals; $Fe(NO_3)_3 \cdot 9H_2O$. There is no anhydrous form of this salt.

fluoboric acid, 48%. Liquid; HBF_4; is not readily available in small quantities, substitute 10.3 ml HF (48%) plus 4.4 g H_3BO_3, for each 10 ml of 48% fluoboric acid specified.

glycerol. Syrupy liquid; also called glycerin or glycereine; $C_3H_5(OH)_3$; contains up to 5% (by weight) water.

molybdic acid, 85%. Crystals or powder containing the equivalent of 85% MoO_3. This misnamed chemical consists mostly of ammonium molybdate (or paramolybdate), which is $(NH_4)_6Mo_7O_{24} \cdot 4H_2O$. The two chemicals can be used interchangeably. See *ammonium molybdate*.

muriatic acid. Liquid; technical grade HCl.

picric acid. Crystals; 2,4,6-trinitrophenol; crystals of laboratory chemical contain 10 to 15% water; explosive; its crystalline metallic salts are even more explosive. Do not use grades that do not have the 10 to 15% water content.

pyrophosphoric acid. Crystals or viscous liquid; $H_4P_2O_7$, anhydrous; hydrolyzes to phosphoric acid (H_3PO_4) slowly in cold water and rapidly in hot water.

zephiran chloride. Aqueous solution; a proprietary material produced in grades containing about 12% and 17% (by weight) benzalkonium chloride (alkyl-dimethyl-benzyl-ammonium chloride) as the active constituent, plus some ammonium acetate; also called sephiran chloride. Available from pharmacies or pharmaceutical distributors. See *benzalkonium chloride*.

Irons and Steels

Etchants and recommendations for macroetching of carbon and alloy steels

Etchant composition(a)	Etching time	Surface required(b)	Purpose, or characteristic revealed
Etchants for use at 71 to 82 °C (160 to 180 °F)			
1 part HCl, 1 part water	15-60 min	A or B	Segregation, porosity, hardness penetration, cracks, inclusions, dendrites, flow lines, soft spots, structure, weld examination
Concentrated HCl	15-60 min	A or B	Segregation, porosity, hardness penetration, cracks, inclusions, dendrites, flow lines, soft spots, structure, weld examination
2 parts H_2SO_4, 1 part HCl, 3 parts water	30-60 min	A	Segregation, porosity, hardness penetration, cracks, inclusions, dendrites, flow lines, soft spots, structure, weld examination
50 parts HCl, 7 parts H_2SO_4, 18 parts water	30-60 min	A	Segregation, porosity, hardness penetration, cracks, inclusions, dendrites, flow lines, soft spots, structure, weld examination
10 to 40 parts HNO_3, 4 to 10 parts HF (48%), 50 to 87 parts water(c)	Until desired etch is obtained	B or C	Segregation, porosity, hardness penetration, cracks, inclusions, dendrites, flow lines, soft spots, structure, weld examination
38 parts HCl, 12 parts H_2SO_4, 50 parts water	30-60 min	B or C	Segregation, porosity, hardness penetration, cracks, inclusions, dendrites, flow lines, soft spots, structure, weld examination
10 parts H_2SO_4, 90 parts water	15-60 min	A	Sulfide and oxide inclusions
Etchants for use at room temperature			
2 to 25% HNO_3 in water or ethanol	5-30 min	B or C	Carburization and decarburization, hardness penetration, cracks, segregation, weld examination
2.5 g $CuCl_2 \cdot 2H_2O$, 20 g $MgCl_2 \cdot 6H_2O$, 10 ml HCl, 500 ml ethanol	Until coppery sheen appears	B or C	Phosphorus-rich areas, banding
50 g $(NH_4)_2S_2O_8$, 500 ml water	Swab until desired etch is obtained	C	Grain size, weld examination
40 g $FeCl_3$, 3 g $CuCl_2$, 40 ml HCl, 500 ml water	15-30 s	B or C	Dendritic structure of cast steel(d)
30 g $FeCl_3$, 1 g $CuCl_2$, 0.5 g $SnCl_2$, 50 ml HCl, 500 ml ethanol, 500 ml water	30 s to 2 min	C	Dendritic structure of cast steel(e)
4 g picric acid in 100 ml methanol	3-5 h	C	Carbon segregation

(a) Parts are listed by volume. All acids listed are of concentrated strength; commercial grades ordinarily can be used instead of laboratory or reagent grades. Water or alcohol should never be poured into an acid; rather, the acid should always be poured and gradually stirred into the other liquid. (b) A indicates a saw-cut or machined surface; B, an average ground surface; C, a polished surface. (c) Ratio of HNO_3 to HF can vary as indicated. (d) Precede use of this etchant with etch in 10% nital for 10 to 20 s. (e) Overetching deposits excessive copper, which may obscure details of structure.

Etchants for microscopic examination of carbon and alloy steels of medium carbon content

Etchant	Purpose, or characteristic revealed
Nital: 1 to 5 ml HNO_3 in 100 ml ethanol (95%) or methanol (95%)	General structure (most-used etchant for routine work)
Picral: saturated solution of picric acid in ethanol (95%) or methanol (95%)	General structure; provides better resolution of certain carbide structures than is obtained with nital
Vilella's reagent: 5 ml HCl, 1 g picric acid, 100 ml ethanol (95%) or methanol (95%)	Reveals outlines of prior austenite grain boundaries in quenched-and-tempered steels
Super picral: picral with a few drops of HCl or zephiran chloride per 25 ml of solution	General structure; for good resolution of carbide structures; often preferred for heat treated structures
Potassium metabisulfite solution: 10 g potassium metabisulfite in 100 ml water	For resolution of hardened structures; use should be preceded by an etch in nital or picral
Howarth's reagent: 10 ml H_2SO_4, 10 ml HNO_3, 80 ml water	Detection of overheating and burning
10 g tartaric acid, 100 ml water	Examination of inclusions
30 g potassium dichromate in 225 ml hot distilled water; add 30 ml acetic acid (glacial)	Reveals lead inclusions, causing them to appear yellow or gold under polarized light
Alkaline chromate solution: 16 g CrO_3 in 145 ml distilled water; add 80 g NaOH (slowly, with constant stirring)	Reveals intergranular oxidation; used for medium-carbon alloy steels that contain nickel

Checklist of principal macroetch observations to be recorded for semifinished steel products

Surface or subsurface	Center or central area	General
(A) Cracks	(a) Pipe	(α) Flakes or cooling cracks
(B) Seams or laps	(b) Porosity	(β) Dendritic pattern
(C) Decarburization	(c) Bursts	(γ) Ingot pattern
(D) Pinholes	(d) Segregations	(δ) Grain size
(E) Segregations		

Etchants for macroscopic examination of cast irons

Etchant	Composition	Etching technique	Application
Stead's reagent	10 g cupric chloride, 40 g magnesium chloride, 20 ml hydrochloric acid, 1 000 ml ethanol(a)	Immersion for up to 3 h	Used to reveal the eutectic cell number in gray cast irons
Rapid cell-etching reagent	10 g cupric chloride, 50 ml water, 100 ml hydrochloric acid	Dip etch for about 60 s	As above, but results are less distinct
Ammonium persulfate	10 g ammonium persulfate, 100 ml water, few drops concentrated H_2SO_4(b)	Immersion and swabbing	Reveals carbide and phosphide distribution
Nital	5 or 10% nitric acid, 95 or 90% ethanol	Dip etch for up to 3 min	Used to reveal macrostructure in white irons
4% picral	4% picric acid, 96% ethanol	Dip etch for up to 3 min	Used to reveal macrostructure in white irons

(a) Dissolve cupric chloride in a minimum quantity of hot water (10-15 ml); add magnesium chloride and dissolve; add ethanol, then hydrochloric acid. (b) Add H_2SO_4 just before use.

Etchants for microscopic examination of cast irons

Etchant	Composition	Etching technique	Applications
Picral	4% picric acid, 96% ethanol	Dip etch for 2-10 s	General-purpose etching of all pearlitic gray, malleable and ductile cast irons; best etchant for pearlite; etches some austenitic cast irons, Ni-Hard and acicular irons
Nital, 5%	5% nitric acid, 95% ethanol	(1) Dip etch for 2-10 s	(1) General-purpose etching of all ferritic gray, malleable and ductile cast irons; etches grain boundaries; etches some austenitic irons and irons containing martensite
		(2) Electrolytic etch(a)	(2) High-chromium irons
Nital, 2%	2% nitric acid, 98% ethanol	Dip etch for 2-10 s	Observation of ferritic grain boundaries at high magnification
Ferric chloride	10 g ferric chloride, 100 ml water	Dip etch for 3-20 s	Austenitic cast irons
Mixed acid in glycerol	10 ml HNO_3, 20 ml HF, 40 ml glycerol	Dip etch for 10-40 s	High-silicon irons (14 to 16% Si)
Vilella's reagent	1 vol HNO_3, 2 vol HCl, 3 vol glycerol	Dip etch for up to 20 s	High-chromium irons
Potassium ferricyanide	10% alkaline aqueous solution of potassium ferricyanide	Dip etch for 5-30 s in etchant at 50 °C (122 °F)	High-chromium irons
Murakami's reagent	10 g KOH, 10 g $K_3Fe(CN)_6$	(1) Dip etch for 2-3 min (2) Dip etch for 10-30 s	(1) 30% chromium irons (2) High-phosphorus irons, to distinguish between iron phosphide and iron carbide
Alkaline sodium picrate	2 g picric acid, 25 g NaOH, 100 ml water; warm to dissolve	(1) Dip etch for 10 s to 2 min at boiling point (2) Electrolytic etch(b)	(1) Blackens cementite (2) Blackens cementite

(a) Specimen is anode; platinum cathode. Current density, 0.13 to 0.31 A/cm^2 (0.5 to 2.0 A/in.^{2}) for up to 2 min. (b) Specimen is anode; stainless steel cathode. Current density, 0.13 to 0.31 A/cm^2 (0.5 to 2.0 A/in.^{2}) for up to 2 min in cold solution.

Etchants for microscopic examination of carbon and alloy steels

Etchant	Purpose, or characteristic revealed
Nital: 1 to 5 ml HNO_3 in 100 ml ethanol (95%) or methanol (95%)	Develops ferrite grain boundaries in low-carbon steels; produces maximum contrast between pearlite and a cementite or ferrite network; develops grain boundaries in 4% silicon steel; develops ferrite boundaries in structures consisting of martensite and ferrite; etches chromium-bearing low-alloy steels resistant to action of picral
Picral; saturated solution of picric acid in ethanol (95%) or methanol (95%)	Reveals maximum detail in pearlite, untempered and tempered martensite, and bainite; reveals undissolved carbide particles in martensite; differentiates ferrite, martensite, and massive carbide by coloration; differentiates bainite and fine pearlite; reveals carbide particles in grain boundaries of low-carbon steel and wrought iron
50 ml 1 to 2% nital, 50 ml 4% picral	Etches some alloy steels, such as 4340
Vilella's reagent: 5 ml HCl, 1 g picric acid, 100 ml ethanol (95%) or methanol (95%)	For contrast etching(a); reveals outlines of prior austenite grains in untempered and tempered martensite, and in austempered steels; reveals pearlite colonies
1 to 1.5 ml HCl (conc), 2 to 3 g picric acid, 100 ml ethanol (95%)	Reveals pearlite colonies(b)
10 g tartaric acid, 100 ml water	For grading inclusions(c)
30 g $K_2Cr_2O_7$ in 225 ml hot distilled water; add 30 ml acetic acid (glacial)	Reveals lead inclusions, causing them to appear yellow or gold when specimen is examined under polarized light(d)
16 g CrO_3 in 145 ml distilled water; add 80 g NaOH(e)	Reveals intergranular oxidation; used for medium-carbon alloy steels that contain nickel(f)
Super picral: picral with a few drops of HCl or zephiran chloride per 25 ml of solution	General structure; for good resolution of carbide structures; often preferred for heat treated structures
10 g potassium metabisulfite, 100 ml water	For resolution of hardened structures; use should be preceded by an etch in nital or picral
Howarth's reagent: 10 ml H_2SO_4, 10 ml HNO_3, 80 ml water	For detection of overheating and burning, and for examination of steel forgings
8 g sodium metabisulfite in 100 ml water	Produces good contrast in as-quenched martensitic structures
1 g KCN in 100 ml water, mixed with 0.25 g diphenylthiocarbazone in 10 ml chloroform	Reveals lead inclusions by coloring them red; coloration is most visible when specimens are viewed under polarized light

(a) Specimen should be tempered 20 to 30 min at 316 °C (600 °F). (b) Immerse specimen 5 to 10 s in solution at room temperature. (c) Immerse specimen for 5 min, rinse in hot water, polish lightly with alumina to remove film, rinse in hot water and dry. (d) Etch 10 to 20 s in solution at room temperature, rinse in hot water and dry. (e) Sodium hydroxide (NaOH) must be added slowly, with constant stirring. (f) Immerse specimen in boiling solution for 10 to 30 min, rinse in hot water, dry in air blast.

Etchants suggested for microscopic examination of wrought stainless steels

Steel	Etchant No.(a)
200 and 300 series:	
General structure	18, 4, 2, 3, 19, 1, 11a, 11b, 20, 25
Sigma	13, 12, 14, 21
Carbides	18, 24
Carbides and sigma	22, 23
400 series:	
General structure	1, 2, 3, 4, 5, 6, 7, 8, 9, 10, 11a, 11b
Sigma	12, 13, 14
Carbides	15
600 series:	
General structure	1, 15, 4, 16, 17, 9
Carbides	15

(a) Numbers correspond to etchants for which compositions and procedures are presented in accompanying table. Where two or more etchants are given, they are listed in order of descending preference.

Compositions and procedures for use of etchants for microscopic examination of wrought stainless steels

Etchant	Composition	Procedure
1	Vilella's reagent: 5 ml HCl, 1 g picric acid, 100 ml ethanol (95%) or methanol (95%)	Immerse or swab specimen for a few seconds to 15 min; reaction may be accelerated by adding a few drops of 3% H_2O_2
2(a)	Glyceregia: 10 ml HNO_3, 20 to 50 ml HCl, 30 ml glycerol	Mix HCl and glycerol thoroughly before adding HNO_3; discard before solution attains a dark orange color; immerse or swab specimen for a few seconds to a few minutes; higher percentage of HCl minimizes pitting
3(a)	10 ml HNO_3, 20 ml HCl, 30 ml water	Immerse specimen for a few seconds to a minute; produces much stronger reaction than etchant 2; discard before solution attains a dark orange color
4(a)	10 ml HNO_3, 10 ml acetic acid, 15 ml HCl, 2 to 5 drops glycerol	Immerse or swab specimen for a few seconds to a few minutes
5	10 ml HNO_3, 20 ml HF, 20 to 40 ml glycerol	Immerse specimen for 2 to 10 s
6	Fry's reagent; 40 ml HCl, 5 g $CuCl_2$, 30 ml water, 25 ml ethanol (95%) or methanol (95%)	Swab specimen for a few seconds to a minute
7	5 g $FeCl_3$, 50 ml HCl, 100 ml water	Immerse or swab for a few seconds to a few minutes; small additions of HNO_3 activate solution and minimize pitting Alternative procedure: Immerse or swab specimen for a few seconds at a time; repeat as necessary
8	5 g $FeCl_3$, 15 ml HCl, 60 ml ethanol (95%) or methanol (95%)	Immerse or swab specimen for a few seconds to a few minutes
9	10 ml HCl, 100 ml ethanol (95%) or methanol (95%)	Immerse specimen for 5 to 30 min, or electrolytic at 6 V for 3 to 5 s
10(a)	Concentrated HNO_3	Electrolytic at 0.2 A/cm^2 for a few seconds
11a	Kalling's reagent 1: 2 g $CuCl_2$, 40 ml HCl, 40 to 80 ml ethanol (95%) or methanol (95%), 40 ml water	Immerse or swab specimen for a few seconds to a few minutes
11b	Kalling's reagent 2: 2 g $CuCl_2$, 40 ml HCl, 40 to 80 ml ethanol (95%) or methanol (95%)	Submerged swabbing for a few seconds to several minutes; attacks ferrite more readily than austenite
12	85 g NaOH, 50 ml water	Electrolytic at 6 V for 5 to 10 s
13	45 g KOH, 60 ml water	Electrolytic at 2.5 V for a few seconds; usually stains sigma and chi yellow to red-brown, ferrite gray to blue-gray, carbides barely touched, austenite not touched
14(a)(b)	Murakami's reagent: 10 g $K_3Fe(CN)_6$, 10 g KOH or NaOH, 100 ml water; use fresh solution	Immerse or swab specimen for 15 to 60 s; stains carbides and sigma(c) Immerse in fresh, hot solution 2 to 20 min; stains carbides dark, ferrite yellow, sigma blue; austenite turns brown on overetching Swab 5 to 60 s; immersion will produce a stain etch Follow with water rinse, alcohol rinse, and dry
15	10 g ammonium persulfate, 100 ml water	Electrolytic at 6 V for a few seconds to a minute
16	25 ml HCl, 3 g ammonium bifluoride, 125 ml water, few grains potassium metabisulfite	Mix fresh; for stock solution, mix first three items; add potassium metabisulfite just before use Immerse specimen for a few seconds to a few minutes
17	10 g $FeCl_3$, 90 ml water	Immerse specimen for a few seconds
18	10 g oxalic acid, 100 ml water	Electrolytic at 6 V for a few to 60 s
19	10 g CrO_3, 100 ml water	Electrolytic at 6 V for 5 to 60 s; attacks carbides
20	2 g CrO_3, 20 ml HCl, 80 ml water	Immerse 5 to 60 s; CrO_3 may be increased up to 20 g for difficult alloys; staining and pitting increase as CrO_3 is increased
21(a)(b)	10 g NaCN, 100 ml water	Electrolytic at 6 V: 5 s for etching sigma, 30 s for ferrite and general structure, and up to 5 min for carbides
22	20 ml HNO_3, 4 ml HCl, 20 ml methanol	Immerse specimen for 10 to 60 s
23	5 ml HNO_3, 45 ml HCl, 50 ml water	Immerse specimen for 10 min or longer
24	Concentrated NH_4OH	Electrolytic at 6 V for 30 to 60 s; attacks carbides only
25	Marble's reagent: 10 g $CuSO_4$, 50 ml HCl, 50 ml water	Immerse or swab specimen for 5 to 60 s

(a) Use exhaust hood; etchant can give off extremely poisonous or noxious fumes. (b) Poisonous by ingestion and by contact. To discard, neutralize or turn basic with ammonia and flush down an acid-disposal drain with a large amount of water. (c) To differentiate, etchant 15, electrolytic at 4 V, will attack sigma but not carbides. If pitting occurs, reduce voltage.

Compositions and applications of etchants for stainless steel casting alloys

Etchant No. and name	Composition
1 Oxalic acid (electrolytic, 6 V)	10 g oxalic acid, 100 ml water
2 Vilella's reagent ...... 5 ml HCl, 1 g picric acid, 100 ml ethanol (95%) or methanol (95%)	
3 Kalling's reagent 2 .... 100 ml HCl, 5 g $CuCl_2$, 100 ml ethanol (95%)	
4 Murakami's reagent (unheated) .... 1-4 g $K_3Fe(CN)_6$, 10 g KOH (or 7 g NaOH), 100 ml water	
5 Murakami's reagent (boiling) .... Same composition as etchant 4, above, but heated to boiling temperature for use	
6 Chromic acid (electrolytic, 6 V) .... 10 g CrO_3, 100 ml water	
7 $10N$ potassium hydroxide (electrolytic, 6 V) ...... 560 g KOH diluted with distilled water to a volume of 1000 ml	
8 HCl, HNO_3, acetic acid .... 15 ml HCl, 10 ml HNO_3, 10 ml acetic acid	
9 Acid ferric chloride .... Saturated solution of $FeCl_3 \cdot 6H_2O$ in concentrated HCl; add a few drops HNO_3	
10 Glyceregia .... 10 ml HNO_3, 20 to 50 ml HCl, 30 ml glycerol	
11 Sodium cyanide (electrolytic, 6 V) .... 10 g NaCN, 90 ml water	

Application of etchants to examination of specific stainless steel casting alloys

Etchant numbers correspond to those assigned in the table on compositions and applications of etchants for stainless steel casting alloys.

Alloy	Normal heat treatment	Etchants for revealing: General microstructure	Ferrite	Carbide	Sigma phase
CA-6NM ...	Hardened and tempered(a)	2	3	4	...
CA-15	Hardened and tempered(a)	2 or 9	3	4	...
CD-4MCu	Annealed(b)	1, 2, or 6	...	...	2 then 7
CE-30	As cast	2	3	4	2 then 7; or 11
CF-3	Annealed(c)	7	3	4	2 then 7
CF-3M	Annealed(c)	8	3	4	2 then 7; or 11
CF-8	Annealed(d)	7 or 10	3	4	2 then 7
CF-8C	Annealed(d)	1 or 6	3	4	2 then 7; or 11
CF-8M	Annealed(d)	9	3	4	2 then 7
CF-20	Annealed(d)	1	3	4	...
CG-8M	Annealed(d)	1	3	4	5; or 7 then 11
CN-7M	Annealed(e)	1 or 6	...	...	...

(a) Heat to 955 °C (1750 °F) min, air cool and temper at 593 °C (1100 °F) min. (b) Heat to 1120 °C (2050 °F) min, furnace cool to 1040 °C (1900 °F), quench in water or oil. (c) Heat to 1040 °C (1900 °F) min, rapid cool. (d) Heat to 1040 °C (1900 °F) min, water quench. (e) Heat to 1120 °C (2050 °F) min, water quench.

Microetching procedures for wrought iron-nickel-chromium heat-resisting alloys

Etchant numbers correspond to those assigned in the table on etchants for microscopic examination of wrought heat-resisting alloys

Etchant no.	Etching method	Etching time, s	Cell voltage	Purpose, or characteristics revealed
Alloy A-286 (AISI 660)				
1	Swab	3-20	...	General structure
2	Swab	5-60	...	General structure; may stain or pit
3	Immerse	10-60	...	General structure
Incoloy 800				
4	Electrolytic	15-30	5-10	General structure; grain boundaries
5	Electrolytic	10-20	5-10	Grain boundaries; carbide particles
6	Electrolytic	10-15	20	Carbide particles
7	Swab	15-30	...	Grain boundaries(a); carbide; no staining
8	Electrolytic	10-30	10	Preferential attack at grain boundaries
Incoloy 825				
4	Electrolytic	15-30	5-10	General structure; grain boundaries
5	Electrolytic	10-20	5-10	Grain boundaries; carbide particles
7	Swab	15-30	...	Grain boundaries; carbide; no staining
9	Swab or immerse	(b)	...	General structure
RA 330				
8	Electrolytic	5-10	5	For etch pitting
10	Electrolytic	2-10	3	General structure; precipitates

(a) Grain boundaries are faint if free of carbide particles. (b) Etching time varies from a few seconds to 12 min.

Microetching procedures for wrought cobalt-base heat-resisting alloys

Etchant numbers correspond to those assigned in the table on etchants for microscopic examination of wrought heat-resisting alloys

Etchant No. (a)	Etching method	Etching time, s	Cell voltage	Purpose or characteristics revealed
Haynes 25 (AISI 670), Haynes 188				
26	Electrolytic	2-5	6	General structure
Stellite 6B				
26	Electrolytic	2-5	6	General structure
27:				
Stage 1	Electrolytic	2-5	6	General structure
Stage 2	Immerse	5-10	...	Carbide particles

Microetching procedures for wrought nickel-base heat-resisting alloys

Etchant numbers correspond to those assigned in the table on etchants for microscopic examination of wrought heat-resisting alloys

Etchant no.	Etching method	Etching time, s	Cell voltage	Purpose, or characteristics revealed
Hastelloy C				
11	Electrolytic	2-10	3	General structure
Hastelloy W				
12	Immerse	...	...	General structure
Hastelloy X (AISI 680)				
13	Electrolytic	2-10	6	General structure; remove stains with HNO$_3$
Inconel 600 and 601				
4	Electrolytic	15-20	5-10	General structure. No pitting
5	Electrolytic	15-20	5-10	General structure; grain boundaries; carbide
6	Electrolytic	15-20	5-10	General structure; excellent for revealing carbide particles
7	Swab or immerse	(a)	...	Grain-boundary contrast fair; carbide
14	Immerse	(b)(c)	...	General structure; carbide particles
Inconel 625				
8	Electrolytic	1-2	50	Grain-boundary films; results vary with thermal history of specimen
15	Electrolytic	10-20	5-10	Grain boundaries; no staining; results vary with thermal history of specimen
16	Electrolytic	15-20	5-10	General structure; grain boundaries
17	Electrolytic	15-20	5-10	General structure
18	Electrolytic	8-20	2-10	Outlines phases; may cause pitting; poor results on cold worked metal
Inconel 706 and Alloy 718				
8	Electrolytic	1-2	50	Grain-boundary films; shows grain boundaries in relief
15	Electrolytic	10-20	5-10	Good for general structure and phase outline; grain boundaries
16	Electrolytic	15-20	5-10	Good for general structure, phase outline, and matrix segregation for most heat treated conditions; grain boundaries
17	Electrolytic	15-20	5-10	General structure; precipitate phases in fully heat treated material
19	Swab or immerse	(a)	...	General structure; microsegregation
20	Immerse	(d)	...	Carbide particles; chromium carbide particles darken faster than nitrides and Laves phase
Inconel X-750 (AISI 688)				
4	Electrolytic	15-20	5-10	General structure; no pitting
5	Electrolytic	15-20	5-10	Grain boundaries; carbide; no pitting
15	Electrolytic	10-20	5-10	Good for revealing grain boundaries and carbide particles
21	Swab	2-10	...	Excellent for showing details of overaged gamma prime
22	Swab	5-60	...	General structure; microsegregation
U-700 (AISI 687)				
23	Swab or immerse	10-20	...	Good for contrast(e)
24	Swab or immerse	(b)	...	General structure; grain boundaries; no staining
28	Electrolytic	5-20	5-10	General structure; grain boundaries
Waspaloy (AISI 685)				
1	Swab	3-20	...	General structure
24	Swab or immerse	(b)	...	General structure; no staining
25	Swab or immerse	5-30	...	General structure

(a) ½ to 5 min. (b) 1 to 5 min. (c) Heat specimen to reduce etching time. (d) 5 to 10 min. (e) Use well-prepared specimen.

Etchants for microscopic examination of wrought heat-resisting alloys

Etchant No. and name	Composition(a)	Remarks on preparation and use
1 HCl, HNO₃, acetic acid	15 ml HCl, 10 ml HNO_3, 10 ml acetic acid	...
2 Chrome regia	2 g CrO_3, 20 ml HCl, 80 ml water	CrO_3 may be increased, but staining may result
3 Ferric chloride—hydrochloric	5 g $FeCl_3$, 15 ml HCl, 100 ml methanol	...
4 Nital	5 ml HNO_3, 95 ml methanol	Use colorless acid and absolute methanol
5 Oxalic acid	10 g oxalic acid, 100 ml water	Can be stored
6 Phosphoric acid	80 ml H_3PO_4, 20 ml water	Change to a 1-to-1 solution for specific results
7 Glyceregia	10 ml HNO_3, 20 ml HCl, 40 ml glycerol	Must be freshly prepared
8 Hydrochloric-methanol	10 ml HCl, 90 ml methanol	Water can be substituted for methanol to show segregation
9 Vilella's reagent	5 ml HCl, 1 g picric acid, 100 ml methanol	A few drops of 3% H_2O_2 will speed etching reaction
10 HCl-H₂O	5 ml HCl, 95 ml water	...
11 Chromic acid	2 to 10 g CrO_3, 100 ml water	...
12 Hydrochloric-chromic	80 ml HCl, 20 ml 50% chromic acid	Use fresh solution
13 Oxalic acid	10 g oxalic acid, 90 ml water	...
14 Nitric-hydrofluoric	20 ml HNO_3, 3 ml HF	Use colorless acids; remove thoroughly by water rinse
15 Chromic-acetic	25 g CrO_3, 7 ml water, 130 ml acetic acid	Can be stored for up to one month
16 Chromic acid	5 g CrO_3, 100 ml water	...
17 47-41-12	47 ml H_2SO_4, 41 ml HNO_3, 12 ml H_3PO_4	Add H_2SO_4 last, and slowly; produces noxious fumes and is highly corrosive
18 Hydrochloric-acetic	10 ml acetic acid, 3 drops HCl, 90 ml water	...
19 Inverted glyceregia	50 ml HCl, 10 ml glycerol, 10 ml HNO_3	...
20 Murakami's reagent	10 g KOH or NaOH, 10 g $K_3Fe(CN)_6$, 100 ml water	Dissolve KOH (or NaOH) and $K_3Fe(CN)_6$ in boiling water; etch specimen in boiling solution; prepare fresh for use
21 Nitric-hydrofluoric	50 ml HNO_3, 50 drops HF	Use colorless acids
22 Hydrochloric-hydrofluoric-nitric	80 ml HCl, 13 ml HF, 7 ml HNO_3	...
23 Marble's reagent	4 g $CuSO_4 \cdot 5H_2O$, 20 ml HCl, 20 ml water	Dissolve $CuSO_4$ in water and add HCl
24 Kalling's reagent	2 g $CuCl_2$, 40 ml HCl, 80 ml methanol	Can be stored
25 92-5-3	92 ml HCl, 5 ml H_2SO_4, 3 ml HNO_3	Must be freshly prepared
26 Hydrochloric—hydrogen peroxide ...	97 ml HCl, 3 ml 3% H_2O_2	Must be freshly prepared
27 Grosbeck's reagent (two-stage)	Stage 1: 2 to 10% CrO_3 in water; stage 2: equal parts 20% $KMnO_4$, 8% NaOH	Mix second stage immediately before use
28 HCl-ethanol-H₂O₂	35 ml HCl, 65 ml ethanol (95%), 7 drops H_2O_2 (30%)	Must be freshly prepared

(a) Use concentrated acids, unless indicated otherwise. Use distilled water to avoid staining.

Compositions of the iron-chromium-nickel heat-resistant casting alloys

ACI designation(a)	C(c)	Composition, %(b) Mn max	Si max	Cr	Ni
HA (d)	0.20 max	0.35-0.65	1.00	8-10	...
HB	0.50 max	1.00	1.00	18-22	2.0 max
HC	0.50 max	1.00	2.00	26-30	4.0 max
HD	0.50 max	1.50	2.00	26-30	4-7
HE	0.20-0.50	2.00	2.00	26-30	8-11
HF	0.20-0.40	2.00	2.00	19-23	9-12
HH (e)	0.20-0.50	2.00	2.00	24-28	11-14
HI	0.20-0.50	2.00	2.00	26-30	14-18
HK	0.20-0.60	2.00	2.00	24-38	18-22
HL	0.20-0.60	2.00	2.00	28-32	18-22
HN	0.20-0.50	2.00	2.00	19-23	23-27
HP	0.38-0.75	2.00	2.50	24-28	34-38
HT	0.35-0.75	2.00	2.50	13-17	33-37
HU	0.35-0.75	2.00	2.50	17-21	37-41
HW	0.35-0.75	2.00	2.50	10-14	58-62
HX	0.35-0.75	2.00	2.50	15-19	64-68

(a) ACI is the Alloy Casting Institute. (b) Iron comprises the remainder of the composition for each alloy, varying according to the quantities of other elements. Compositions also include 0.04% max P and 0.04% max S. Molybdenum is not intentionally added to alloys other than the HA alloys, but is permissible to 0.5% max. (c) Carbon limitations do not apply if a numerical suffix is added to the alloy designation (for example, HF-25). The suffix implies the midpoint of a ±0.05% range. (d) Also contains 0.90 to 1.20% Mo. (e) Also contains 0.2% max N.

Etchants for microscopic examination of iron-chromium-nickel heat-resistant casting alloys

Common name	Composition	Remarks on use
Etchants for delineating general structure		
Aqua regia	20 ml HNO_3, 60 ml HCl	Immerse specimen
Glyceregia	10 ml HNO_3, 20-50 ml HCl, 30 ml glycerol	Immerse specimen; use a hood
Hydrochloric acid (50%)	50 ml HCl, 50 ml water	Outlines ferrite; immerse specimen
Marble's reagent	10 g $CuSO_4$, 50 ml HCl, 50 ml water	Immerse specimen
Vilella's reagent	1 g picric acid, 5 ml HCl, 100 ml ethanol	Immerse specimen
Etchants for staining or film-forming		
Alkaline hydrogen peroxide	25 ml NH_4OH, 50 ml H_2O_2 (3%), 25 ml water	Ordinarily used after a delineating etchant; immerse specimen
Alkaline potassium ferricyanide	10 g $K_3Fe(CN)_6$, 10 g NaOH, 100 ml water	Same as above
Alkaline potassium permanganate	4 g NaOH, 10 g $KMnO_4$, 85 ml water	Same as above
Alkaline sodium picrate	2 g picric acid, 25 g NaOH, 100 ml water	Same as above
Emmanuel's reagent	30 g $K_3Fe(CN)_6$, 30 g KOH, 60 ml water	Attacks sigma phase with little or no effect on carbide particles; immerse specimen
Murakami's reagent	10 g $K_3Fe(CN)_6$, 10 g KOH, 100 ml water	Stains carbide particles without staining sigma phase(a); immerse specimen
Solutions for electrolytic etching		
Ammonium hydroxide	Concentrated NH_4OH	Final electrolytic etch after etching in Vilella's reagent and in 10N KOH (electrolytic)
Cadmium acetate	10 g cadmium acetate, 100 ml water	Attacks $(Cr, Fe)_{23}C_6$ carbide particles
Chromic acid	2-10 g Cr_2O_3, 100 ml water	Outlines carbide particles; extracts sigma phase
Lead acetate (2N)	38 g $Pb(C_2H_3O_2)_2 \cdot 3H_2O$, distilled water to make 100 ml	Stains austenite, then sigma phase, then carbide particles; 1.5 V for 30 s
Oxalic acid	10 g oxalic acid, 100 ml water	Outlines carbide and sigma; 6 V, 1 to 5 s
Potassium hydroxide (1N)	5.6 g KOH, 100 ml water	Blackens sigma phase without outlining other phases; 1.5 V for 1 s
Potassium hydroxide (10N)	56 g KOH, 100 ml water	Intermediate etch between Vilella's and ammonium hydroxide (electrolytic)
Sodium cyanide	10 g NaCN, 100 ml water	Used after glyceregia; outlines carbide particles, stains sigma phase; use at 1 A/in.^{2} for 1 to 5 s, under hood

(a) Sometimes sigma phase is stained. Behavior must be established on a given composition.

Silicon Steels

Pitting etchants for determination of grain orientation in silicon steels by optical microscopy

Etchant	Composition (parts are by volume)	Conditions for use	Purpose
1	1 part HF, 1 part HNO_3, 4 parts water	Immerse for 10 s	Exposes {100} crystallographic faces in (110) [001] (cube-on-edge) oriented 3.25% Si steel
2	2 parts HF, 1 part HNO_3, 3 parts methanol, 4 parts glycerol	Swab for 1 min	Same as for etchant 1
3	A: 6 ml H_2O_2 (30%), 0.1 ml HCl, 100 ml water B: 40 ml $FeCl_3 \cdot 6H_2O$, 40 ml ethanol, 20 ml water	Immerse in A for 10 s, rinse and dry; then immerse in B for 3 s, rinse and dry	Develops etch pits in (110)[001] (cube-on-edge) oriented 3.25% silicon steel
4	100 g ferric sulfate, 100 ml H_2SO_4, 1000 ml water	Immerse for 15 s in solution heated to 80-90 °C (175-195 °F)	Develops etch pits in (100)[001] (cube-on-face) oriented 3.25% silicon steel
5	A: 5 ml HF, 95 ml methanol B: 100 ml H_2O_2 (3%), 100 ml water, 2 drops HCl C: 5 ml HCl, 95 ml methanol	Polish, etch heavily in nital; repolish, etch in nital to reveal grain boundaries; immerse 10 s in A, rinse, dry; immerse 2 s in B, rinse, dry; immerse 30 s in C, rinse, dry	Exposes {100} crystallographic faces in primary recrystallized 3.25% silicon steel and nonoriented silicon steels
6	600 g $FeCl_3$, 10 g ammonium bisulfate, 600 ml HCl, 150 ml HNO_3, 1650 ml water	Immerse for 1 min in solution heated to 49-60 °C (120-140 °F)	Exposes {111} crystallographic faces in secondary recrystallized 50Ni-50Fe

Refractory Metals

Etchants for metallographic specimens of refractory metals

Etchant name or ASTM number (E407)	Composition
Etchants for tungsten and molybdenum and their alloys	
Murakami's reagent (etchant 98c)	10 g $K_3Fe(CN)_6$, 10 g KOH or NaOH, 100 ml water
Murakami's reagent (mod)	15 g $K_3Fe(CN)_6$, 2 g NaOH, 100 ml water
Etchant 131 (electrolytic)	5 ml H_2SO_4, 1 ml HF, 100 ml methanol (95%)
Etchant 132(a)	5 ml HF, 10 ml HNO_3, 30 ml lactic acid
Etchant 209	15 ml HNO_3, 3 ml HF, 80 ml water
Additional etchants for molybdenum and molybdenum alloys	
Etchant 129	10 ml HF, 30 ml HNO_3, 60 ml lactic acid
Etchant 130 (electrolytic)	25 ml HCl, 10 ml H_2SO_4, 75 ml methanol
Etchants for niobium and tantalum and their alloys	
Etchant 66	30 ml HF, 15 ml HNO_3, 30 ml HCl
Etchant 158	10 ml HF, 10 ml HNO_3, 20 ml glycerol
Etchant 159	5 ml HF, 20 ml HNO_3, 50 ml acetic acid
Etchant 161	25 ml HNO_3, 5 ml HF, 50 ml water
Etchant 163	30 ml H_2SO_4, 30 ml HF, 3 to 5 drops H_2O_2 (30%), 30 ml water
Etchant 164	50 ml HNO_3, 30 g ammonium bifluoride, 20 ml water
Additional etchants for niobium and niobium alloys	
Etchant 160	20 ml HF, 15 ml H_2SO_4, 5 ml HNO_3, 50 ml water
Etchant 162B	30 ml lactic acid, 10 ml HNO_3, 10 ml HF
HNO_3-HF-water	20 ml HNO_3, 10 ml HF, 70 ml water
HCl-H_2SO_4-HNO_3-water	15 ml HCl, 15 ml H_2SO_4, 8 ml HNO_3, 62 ml water
Additional etchants for tantalum and tantalum alloys	
Etchant 177	10 g NaOH, 100 ml water
Etchant 178	20 ml HF, 20 ml HNO_3, 60 ml lactic acid
Etchant 179B (electrolytic)	10 ml HF, 90 ml H_2SO_4

(a) Procedure: Swab with heavy pressure for 5 to 10 s, water rinse, alcohol rinse, dry, etch with Murakami's reagent (etchant 98c).

Magnetic Alloys

Electrolytes and conditions for electropolishing of iron-nickel and iron-cobalt magnetic alloys

Alloy	Electrolyte	Conditions for use
Fe-Ni only	135 ml acetic acid (glacial), 25 g CrO_3, 7 ml water	80 V, 0.8 to 1.6 A/cm^2, 5 to 30 s at 7 °C (45 °F) max
Fe-Ni or Fe-Co ...	100 ml acetic acid (glacial), 10 ml perchloric acid	45 V, 0.2 A/cm^2, 3 to 4 min at 24 °C (75 °F)

Etchants for microscopic examination of iron-nickel and iron-cobalt magnetic alloys

Etchant	Composition	Conditions for use(a)
Chemical etching		
1	100 ml HCl, 2 g $CuCl_2$, 7 g $FeCl_3$, 5 ml HNO_3, 200 ml methanol, 100 ml water	Immerse or swab for 10 to 15 s
2(b) ..	15 ml HCl, 5 g $FeCl_3$ (anhydrous), 60 ml ethanol	Immerse for 5 to 10 s
3	3 ml HCl, 1 ml HNO_3, saturated with $CuCl_3$	Swab for 2 to 3 s
4	15 ml HCl, 5 ml HNO_3, 10 ml glycerol	Swab for 10 to 15 s
5	Ammonium persulfate (saturated aqueous solution)	Immerse for 20 to 30 s
6	2 to 10% nital (HNO_3 in ethanol or methanol)	Immerse for 5 to 10 s
7	50 ml HCl, 10 g $CuSO_4$, 50 ml water (Marble's reagent)	Immerse or swab for 5 s
Electrolytic etching		
8	5 to 10 ml HCl, 100 ml water	2-5 s at 250-500 mA/cm^2
9	2 g CrO_3, 100 ml water	2-5 s at 100-200 mA/cm^2
10	3% sulfuric acid	5 to 10 s

(a) All etchants are used at room temperature. (b) Recommended for electron metallography.

Etchants recommended for microscopic examination of iron-nickel and iron-cobalt magnetic alloys

Etchant numbers correspond to those assigned in the table on etchants for microscopic examination of iron-nickel and iron-cobalt magnetic alloys

Etchant	Characteristic revealed
Iron-nickel alloys	
1, 2, 4, 5(a), 6	Grain size, structure
3(b), 7	Grain size
Iron-cobalt alloys	
1, 5, 6, 8, 9, 10	Grain size, structure

(a) For etching 50Fe-50Ni. (b) For etching high-nickel alloys such as Moly Permalloy.

Electrolytes for electrolytic etching of permanent magnet materials other than Alnico alloys and hard ferrites

Magnet material	Electrolyte	Characteristic revealed
Vicalloy	10 ml HNO_3, 90 ml water	General structure
Cunife	10 ml $CuSO_4$, 50 ml HCl, 50 ml water	Grain size and structure of the solid-solution alloy
Rare earth ($Co_3Cu_{1.6}Fe_{0.5}Ce$) ..	Nital (various strengths)	Identification of copper-rich phase
82Co-6Au-12Fe	3 ml HNO_3, 3 ml H_2SO_4, 94 ml water	Grain-boundary precipitates; use after electropolishing

Electrical Contact Materials

Etchants and etching procedures for electrical contact materials

Etchant No.	Composition	Procedure for use
1	20 ml NH_4OH, 10-20 ml H_2O_2 (30%), 10-20 ml water	Swab at room temperature, 3-10 s; use fresh; more water, less H_2O_2 for copper alloys; vice versa for silver alloys
2	2 g $K_2Cr_2O_7$, 1.5 g NaCl, 8 ml H_2SO_4 (conc), 100 ml water	Swab at room temperature, 5-10 s; good for etching hard-to-etch copper alloys
3	50 ml NH_4OH, 10-30 ml H_2O_2 (30%)	Swab at room temperature for 3-10 s; use fresh
4	10 g $FeCl_3$, 90 ml water	Swab or immerse
5	A: 100 ml saturated aqueous solution of $K_2Cr_2O_7$, 2 ml saturated aqueous solution of NaCl, 10 ml H_2SO_4 B: 1 part solution A, 10 parts water C: 98 ml water, 3 g CrO_3, 2 ml H_2SO_4	Use solution A, then solution B, then solution C; swab at room temperature for 15-20 s with each solution; rinse in water between solutions
6	20 g CrO_3, 4.5 g NH_4Cl, 18 ml HNO_3 (conc), 15 ml H_2SO_4 (conc), water to make ½ L (Waterbury reagent)	Dilute 2 to 1 with water at time of use; swab at room temperature for 3-10 s
7	A: 25 ml HNO_3, 1 g $K_2Cr_2O_7$, 100 ml water B: 40 g CrO_3, 3 g Na_2SO_4, 200 ml water	Mix equal parts of A and B; swab at room temperature for 5-10 s
8	20 ml HNO_3 (conc), 20 ml acetic acid (glacial), 20 ml glycerol	Swab at 38-42 °C (100-110 °F) for 3-10 s
9	0.2% CrO_3 and 0.2% H_2SO_4, in water	Swab for 1 min
10	A: 200 ml HNO_3 (50%), 2 g $K_2Cr_2O_7$ B: 20 g CrO_3, 1.5 g Na_2SO_4, 100 ml water	Mix 1 part A with 20 parts B at time of use; swab at room temperature, 3-15 s
11	10 ml $K_3Fe(CN)_6$ (30%), 10 ml NaOH (10%) (Murakami's reagent)	Swab at room temperature for 5-15 s; use at half strength for more control
12	20 ml HNO_3 (conc), 20 ml acetic acid (glacial)	Immerse at room temperature, 10-20 s
13	20 ml KCN (10%), 20 ml $(NH_4)_2S_2O_8$ (10%)	Use in a hood; immerse at room temperature for 10-30 s
14	A: 5% nital B: 5% $FeCl_3$ in methanol	Immerse specimen alternately in A and B
15	10 ml HNO_3, 20 ml HCl, 10 ml glycerol	Swab at room temperature for 3-10 s
16	30 ml HCl, 10 ml water	Electrolytic; up to 5 V dc; 1.5 A/cm^2; room temperature, 1-3 min

Suggested etchants for specific electrical contact materials

Etchant numbers correspond to those assigned in the table on etchants and etching procedures for electrical contact materials

Material	Etchant No.	Material	Etchant No.	Material	Etchant No.
Copper-graphite	4	Silver-graphite	6, 3, 8(a)	Gold-plated nickel-iron	11
Chromium-copper	1	Silver-nickel	9, 10	Gold-silver clad palladium	12
Cadmium-copper	1	Silver-magnesium-nickel	9	Platinum-ruthenium	16
Copper-cobalt-beryllium	1, 2	Silver-molybdenum	11	Platinum-iridium	16
Copper-cobalt-silicon	1, 2	Silver-tungsten	3, 11	Palladium welded to nickel silver	12
Copper-tungsten	1, 11	Silver-tungsten carbide	11	Palladium-ruthenium	13
Silver (99.9% Ag)	3, 4, 5	Tungsten	11	Palladium-copper	13
Silver-copper	6, 7	Tungsten-copper	11	Palladium-silver	13
Silver-copper-cadmium	7	Tungsten-nickel	11	Palladium-platinum-gold-silver-copper-zinc	16
Silver-cadmium	6, 3, 8(a)	Molybdenum	11	Rhodium and gold-plated nickel-iron	14, 15
Silver-cadmium brazed to brass	6, 3, 8(a)	Molybdenum-silver	11		
		Gold-silver-platinum	13		

(a) In the order given.

Sleeve-Bearing Materials

Etchants for microscopic examination of sleeve-bearing materials

Etchant	Some applications
NH_4OH, H_2O_2(a)	Commercial bronze liner Copper-lead alloy liner Copper-lead-tin alloy liner High-leaded tin bronze liner Leaded tin bronze liner Lead-tin-copper overlay on copper-lead alloy liner Nickel bronze infiltrated with lead-base babbitt Nickel-tin bronze infiltrated with lead-base babbitt Silver electroplate on steel Silver-lead alloy electroplate on steel Tin-base babbitt overlay on copper-lead-tin alloy liner Tin bronze infiltrated with lead-base babbitt Tin bronze infiltrated with Teflon Trimetal bearing: lead-tin-copper electroplated overlay, brass electroplated barrier, copper-lead alloy
0.5% HF	Aluminum alloy clad to steel Aluminum-silicon alloy clad to steel High-tin aluminum alloy clad with unalloyed aluminum Lead-tin-copper overlay on aluminum alloy liner Low-tin aluminum alloy clad to steel Trimetal bearing: lead-tin-copper electroplated overlay, copper electroplated barrier, aluminum-silicon-cadmium alloy
5% nital	High-tin aluminum alloy clad to nickel-plated steel Lead-base babbitt liner Tin-base babbitt liner Steel backing of any bearing alloy
Keller's reagent	Lead-tin-copper overlay on aluminum-cadmium alloy
Ferric chloride(b)	Cadmium alloy liner

(a) Equal parts of concentrated NH_4OH and water with 2 to 4 drops of H_2O_2 (30%) per 10 ml of solution. (b) 10 g $FeCl_3$, 90 ml ethanol.

Aluminum

Etchants for use in macroscopic examination of aluminum alloys

Etchant	Composition(a)	Procedure for use
1 (caustic etch)	10 g NaOH to each 90 ml water	Immerse specimen 5-15 min in solution heated to 60-70 °C (140-160 °F)(b), rinse in water, dip in 50% HNO_3 solution to desmut, rinse in water, dry
2 (Tucker's reagent) ...	45 parts HCl(c), 15 parts HNO_3(c), 15 parts HF (48%), 25 parts water	Mix fresh before using; immerse or swab specimen for 10-15 s, rinse in warm water, dry, and examine for desired effect; repeat as necessary until desired effect is obtained
3	1 part HF (48%), 9 parts water	Requires fairly smooth surface; immerse until desired effect is obtained, hot-water rinse, dry
4 (Poulton's reagent) ...	12 parts HCl(c), 6 parts HNO_3(c), 1 part HF (48%), 1 part water	May be premixed and stored(d) for long periods; etch by brief immersion or by swabbing; rinse in cool water, and do not allow either the etchant or the specimen to heat up during etching
5	50 parts HCl(c), 15 parts HNO_3(c), 3 parts HF (48%), 5 parts $FeCl_3$ solution (c)	Mix fresh before use; cool solution to 10-15 °C (50-59 °F) with jacket of cold water; immerse a few seconds, rinse in cold water; repeat immersion and rinsing until desired effect is obtained

(a) Parts are by volume. (b) This etchant may be used without being heated, but the etching action will be slower. (c) Concentrated. (d) Solution should be stored in a vented container, preferably under a fume hood, to prevent buildup of gas pressure. The container should be made of polyethylene or be lined with wax.

Applicability of etchants to macroscopic examination of aluminum alloys

Etchant numbers correspond to those assigned in the table on etchants for use in macroscopic examination of aluminum alloys

Alloy	Etchant
High-purity aluminum	4 or 5
Commercial-purity aluminum:	
1xxx series	4, 2, or 1
All high-copper alloys:	
2xxx series and casting alloys .	1
Al-Mn alloys:	
3xxx series	4, 2, or 1
Al-Si alloys:	
4xxx series and casting alloys(a)	4, 2 or 3
Al-Mg alloys:	
5xxx series and casting alloys .	4, 2, or 1
Al-Mg-Si alloys:	
6xxx series and casting alloys .	4, 2, or 1
Al-Cu-Mg-Zn alloys:	
7xxx series and casting alloys .	1

(a) Also, welds and brazed joints made with the use of these alloys as filler metals.

Applicability of etchants to microscopic examination of aluminum alloys

Etchant numbers correspond to those assigned in the table on etchants for use in microscopic examination of aluminum alloys

Alloy	Etchant	Evidence revealed
Examination for grain size and shape		
1xxx, 3xxx, 5xxx, 6xxx series; most casting alloys	5 or 12	Grain contrast when using crossed polarizers, with or without sensitive tint
2xxx, 7xxx series; Al-Cu or Al-Zn casting alloys	3A or 11	Grain contrast or grain-boundary lines
5xxx series alloys with more than 3% Mg	8 (3-5 min)	Precipitation in grain boundaries
Examination for cold working		
1xxx, 3xxx, 5xxx, 6xxx series alloys	5 or 12	Deformation bands or markings that cause streaked effect when using crossed polarizers
2xxx, 7xxx series alloys	3A or 11	Deformation bands or markings that accompany relatively great amounts of cold working
5xxx series alloys with more than 3% Mg	8 (3-5 min)	Precipitation in bands of slip
Examination for incomplete recrystallization		
1xxx, 3xxx, 5xxx, 6xxx series alloys	5 or 12	Even-toned, well-outlined grains that are recrystallized, otherwise streaked, or banded
2xxx series alloys, hot worked and heat treated	3A or 11	Unrecrystallized grains made up of multiple, very fine subgrains
6xxx series alloys, hot worked and heat treated	9	Unrecrystallized grains made up of multiple, very fine subgrains
7xxx series alloys, hot worked and heat treated	8(3-5 min) or 14	Unrecrystallized grains made up of multiple, very fine subgrains
Examination for preferred orientation		
1xxx, 3xxx, 5xxx, 6xxx series alloys	5 or 12	Predominance of certain gray tones when crossed polarizers are used, lack of randomness
2xxx series alloys in T4 temper	3A or 11	Lack of randomness in grain contrast
Examination for identification of constituents		
1xxx series alloys	1 or 7	(a)
2xxx, 3xxx series; Al-Cu and Al-Mn casting alloys ...	8 (1 min)	(a)
7xxx series; Al-Zn casting alloys	3B	(a)
Examination for overheating (partial melting)		
2xxx series alloys	8 (1 min)	Rosettes and grain-boundary eutectic
6xxx series alloys	2	Grain-boundary eutectic formations
7xxx series alloys	3B	Rosettes and grain-boundary eutectic formations
Examination for general constituent size and distribution		
All wrought alloys and casting alloys	1, 8 (1 min)(b)	Coarse insoluble particles and fine precipitate particles. Longer etching time exaggerates size of fine particles.
Examination for distinction between solution heat treated (T4) and artificially aged (T6) tempers		
2xxx series alloys	3A or 11	Loss of grain contrast, general darkening, in T6 compared with T4
6061	9	Clear outlining of grain boundaries in T6; faint outlining in T4
7075, recrystallized	4	More grain contrast, sharper grain-boundary outlining, in T4
Examination for overaging or poor quench of solution heat treated alloy		
2017 and 2024, in T4 temper	6	Faint dark precipitate at grain boundaries
Examination for cladding thickness		
Alclad 2014, 2024, 7075	3A or 11	Boundary between high grain contrast or outlining of alloy core and lighter-etching cladding
Brazing sheet	1 (swab) or 13	Boundary of high-silicon cladding alloy
Other clad alloys	1 (immerse), 2, 3A, 5 or 11	Any differences in structure that demarcate one layer from another
Examination for solid-solution coring or segregation, and diffusion effects		
3xxx, 5xxx series; Al-Mg casting alloys	10	Interference colors due to differences in thickness of tarnish films laid down on the surface
2xxx series alloys and others with more than 1% Cu .	3A or 11	Brownish-colored films due to redeposition of copper

(a) See table on metallographic identification of phases in aluminum alloys. (b) Or any etchant that does not pit solid-solution matrix.

Metallographic identification of phases in aluminum alloys

Basic and alternative phase designations(a)	Elements that enter in solution	External shape(b)	Appearance before etching(c)	Birefringence(d)	Etchants that aid identification(e)
Si	...	Cubic habit; primary particles form isometric polygons; eutectic may form script, blades, or very fine lamellae	Light bluish-gray	None	Generally best identified without etching; etchant 1 (swab) outlines particles and appears to lighten the color
Mg_2Si	...	Cubic habit; eutectic forms script that easily coalesces on heating	Natural color is darker bluish-gray than silicon, but usually tarnishes to bright blue, black or vari-colored	None (when not roughened or tarnished)	Easily identified without etching; caustic etchant 2 will not attack and may enhance blue color; acid etchants will attack and dissolve readily
$MgZn_2$ or $\eta(Mg\text{-}Zn)$	Isomorphous series with CuMgAl	Usually well-rounded or irregular, except in lamellar eutectic or precipitated from solid solution	White, watery; does not polish in relief	Slight change from light to dark gray	Etchant 3B gives a smooth, dark-gray to black color
$CrAl_7$	Fe as $(Cr,Fe)Al_7$; Mn as $(Cr,Mn)Al_7$	Primary crystals form elongated polygons	Light metallic gray	Weak, but will reveal twinning in large crystals	Resists attack by all common etchants
$CuAl_2$ or $\theta(Al\text{-}Cu)$	...	Usually well-rounded or irregular, except when precipitated from solid solution	Pale pinkish color	Strong, orange to greenish-blue; some orientations show little change	Remains light and clear in etchants 1 (swab), 3A, and 8 (1 min); etchant 6 will darken and is good for detecting barely visible grain-boundary precipitate
$FeAl_3$	Cr as $(Fe,Cr)Al_3$; Mn as $(Fe,Mn)Al_3$; Possibly Cu	Elongated blades or star-shaped clusters when eutectic; resists coalescence	Light metallic gray; slightly darker than Fe_3SiAl_{12}	Weak and not easily detectable	Etchant 7 will dissolve and blacken; in high-copper alloys, etchant 8 (1 min) will color it dark-brown to bluish-black; in Al-Cu-Mg-Zn alloys, etchant 3B will color it medium-brown or gray; rough and outlined
$FeAl_6$	A metastable phase in absence of Mn or Cu (see $MnAl_6$)	Isomorphous with $MnAl_6$, but usually found only under conditions of high solidification rate; forms fine lamellar eutectic	Not easily defined, because of fine particle size	Same as $MnAl_6$	Not attacked by etchant 7 but darkened by etchant 1 (swab)
Mg_2Al_3 or Mg_5Al_8, $\beta(Al\text{-}Mg)$	...	Usually well-rounded or irregular	White; lighter than aluminum but may tarnish to yellow or tan; not in relief	None (when not tarnished)	Caustic etchant such as 2 will not attack or color; acid etchants generally pit and dissolve it with varying rapidity
$MnAl_6$	Fe as $(Fe,Mn)Al_6$; isomorphous with $(Fe,Cu)(Al,Cu)_6$, or $(Fe,Cu)Al_6$	Primary or coarse eutectic forms solid or hollow parallelograms; fine eutectic may form script	Light metallic gray	Strong; light to dark gray; does not twin	Etchant 8 (1 min) will neither attack nor darken this phase; however, it will attack companion phases such as $(Fe,Mn)Al_3$ or $(Fe,Mn)_3SiAl_{12}$
$Cr_2Mg_3Al_{18}$ or $T(Al\text{-}Cr\text{-}Mg)$, $E(Al\text{-}Cr\text{-}Mg)$	...	Usually forms by precipitation or by peritectic reaction from $CrAl_7$	Very light metallic gray; not much in relief	None	Strongly attacked by etchants 6 and 7
$(Fe,Cu)(Al,Cu)_6$ or $(Fe,Cu)Al_6$, $\alpha(Al\text{-}Cu\text{-}Fe)$		——————— (See $MnAl_6$) ———————			
Cu_2FeAl_7 or $\beta(Al\text{-}Cu\text{-}Fe)$, $N(Al\text{-}Cu\text{-}Fe)$	...	Elongated blades when formed eutectically; also forms peritectically from $(Fe,Mn)_3SiAl_{12}$ and other iron-rich phases	Very light metallic gray; only slightly darker than $CuAl_2$	Moderate; light to dark gray	Outlined, but not colored, by etchants 3B and 8 (1 min); hence, can be distinguished from other iron-rich phases with which it is associated

Basic and alternative phase designations(a)	Elements that enter in solution	External shape(b)	Appearance before etching(c)	Birefringence(d)	Etchants that aid identification(e)
$CuMgAl_2$ or $Cu_2Mg_2Al_5$, S(Al-Cu-Mg)	...	Very much resembles $CuAl_2$	Slightly grayer than $CuAl_2$; tarnishes to brown or black very readily during polishing	Very strong; yellowish to purple or greenish-blue	Roughened and darkened to varying degrees by etchants 3B and 8 (1 min), depending on polish; etchant 3A darkens this phase while leaving $CuAl_2$ uncolored; etchant 6 reveals barely visible grain-boundary precipitate
CuMgAl	——————————————— (See $MgZn_2$) ———————————————				
$Cr_4Si_4Al_{13}$ or α(Al-Cr-Si)	——————————————— (See Fe_3SiAl_{12})(f) ———————————————				
$CuMg_4Al_5$ or T(Al-Cu-Mg), c(Al-Cu-Mg)	Isomorphous series with $Mg_3Zn_3Al_2$	Irregular rounded	Very light or slightly yellow	None	Behaves like other Mg-rich phases, attacked rapidly by acidic etchants, not attacked by caustic etchants
Fe_3SiAl_{12} or $Fe_3Si_2Al_{12}$, α(Al-Fe-Si), c(Al-Fe-Si); also $(Fe,Cu)_3SiAl_{12}$ or α(Al-Fe,Cr-Si); $(Fe,Mn)_3SiAl_{12}$ or α(Al-Fe,Mn-Si)	(f); besides the apparent interchangeability of Fe, Cr and Mn, this phase can probably also contain Cu	Usually well-defined script when formed eutectically, especially when silicon is not low; may also form polyhedrons or irregular shapes, or precipitate as Widmanstätten type	Light metallic gray, slightly lighter than either $FeAl_3$ or $Fe_3Si_2Al_9$; often polishes in relief	None	Rarely attacked strongly, but can darken to shades of brown when copper is present, using etchant 8 (1 min); in the absence of copper, etchant 8 (1 min) will roughen and outline it, distinguishing it from $MnAl_6$; chromium makes it more resistant to etching(g)
$Fe_2Si_2Al_9$ or $FeSiAl_5$, β(Al-Fe-Si)	...	Bladelike when formed eutectically; retains flat shape in wrought alloys	Light metallic gray, intermediate between Fe_3SiAl_{12} and silicon	Moderate; light to dark gray	Etchant 1 (immerse) will attack and darken to varying degrees, depending on Fe-Si ratio; etchant 7 will attack and dissolve it out; in both cases, Fe_3SiAl_{12} is outlined but not appreciably darkened
$Mg_3Zn_3Al_2$ or T(Al-Mg-Zn)	——————————————— (See $CuMg_4Al_5$) ———————————————				
Mn_3SiAl_{12} or α(Al-Mn-Si)	——————————————— (See Fe_3SiAl_{12})(f) ———————————————				
$Cu_2Mg_8Si_6Al_5$ or Q(Al-Cu-Mg-Si), λ(Al-Cu-Mg-Si), h(Al-Cu-Mg-Si)	...	This is a true quaternary phase; forms irregular shapes in eutectics	Light metallic gray; darker than $CuAl_2$	Strong; changes from orange to blue	Etchant 8 (1 min) does not attack it, but the color distinction between it and $CuAl_2$ remains the same as when not etched
$FeMg_3Si_6Al_8$ or Q(Al-Fe-Mg-Si), π(Al-Fe-Mg-Si), h(Al-Fe-Mg-Si)	...	This is a true quaternary phase; forms irregular shapes in eutectics; sometimes shows hexagonal symmetry	Very light metallic gray; not much in relief	Strong; changes from yellow to light blue	Not attacked by etchant 1 (immerse), hence distinguished from $Fe_2Si_2Al_9$, with which it is usually associated

Note: There are some phases other than those listed in this table that are less common or that appear in such small amount or as such fine particulate that identification can be made only indirectly. These include $TiAl_3$, AlB_2 and TiB_2, Pb and Bi, $NiAl_3$, Ni_2Al_3, $FeNiAl_9$, Cu_3NiAl_6, and $Cu_2Mn_3Al_{20}$. Other phases that do not normally come into equilibrium with aluminum may occasionally be encountered as a result of incomplete melting or some other abnormality in practice.

(a) There is no widely accepted manner of naming or designating phases as they are encountered in equilibrium phase diagrams or in description of alloy constitution. Even composition formulas are inexact, because many phases have broad homogeneity ranges or their actual composition may not coincide exactly with the ideal atomic arrangement upon which crystal structure is based. Phragmén (Phragmén, G, On the Phases Occurring in Alloys of Aluminum with Copper, Magnesium, Manganese, Iron, and Silicon, *J Inst Metals*, Vol 77, 1950, p 489-552) advocated the use of a lower-case-letter prefix indicating the basic crystal structure (c=cubic, h=hexagonal, and so on). Otherwise, Greek letters and upper-case English letters have been arbitrarily used, although "T" usually denotes a ternary phase and "Q" a quaternary phase. (b) Applies mainly to cast forms or to wrought alloys that have not been extensively worked. However, some iron-rich phases that are resistant to coalescence or spheroidization will retain dimensional ratios that are indicative of crystalline symmetry. (c) Applies to appearance after mechanical polishing. Electrolytic polishing is rarely suitable for making phase identification. (d) An exceptionally good flat polish with no tarnishing is required, because any element of the surface that is not parallel to the plane of the surface (that is, normal to the optical axis) will cause an apparent birefringence that is not due to crystal structure. The sensitivity of this technique will also vary with the quality of the optical system. A rotating stage is necessary. (e) Etchant numbers correspond to those assigned in the table on etchants for use in microscopic examination of aluminum alloys. (f) There are two crystal forms of α(Al-Fe-Si)—namely, Fe_3SiAl_{12} (cubic, also called α_1 and Fe_2SiAl_8) and $Fe_3Si_2Al_{12}$ (hexagonal, also called α_2). It was believed at one time that cubic Fe_3SiAl_{12} was isomorphous with analogous ternary phases $Cr_4Si_4Al_{13}$ and Mn_3SiAl_{12}, but the latter at least has since been found to be hexagonal. Nevertheless, the presence of even very small amounts of Mn, Cr_3 and Cu in α(Al-Fe-Si) seems to favor the cubic form normally encountered in commercial alloys. Metallographic distinction between the cubic and the hexagonal forms is very difficult to detect. When etched in etchant 3B (see the table on etchants for use in microscopic examination of aluminum alloys), complex alloys containing chromium and manganese (such as 5083 and 7075) may show etching contrasts within the scriptlike phase normally taken to be cubic Fe_3SiAl_{12}, but no separate identity has yet been established. (g) This phase and its variants can give a variety of etching responses for a given etch, depending on its composition and that of the matrix.

Etchants for use in microscopic examination of aluminum alloys

Etchant	Composition	Procedure for use
1 (hydrofluoric acid etch)	1 ml HR (48%), 200 ml water	Swab for 15 s or immerse for 30 to 45 s
2	1 g NaOH, 100 ml water	Swab for 5 to 10 s
3A (Keller's reagent)	2 ml HF (48%), 3 ml HCl(a), 5 ml HNO$_3$(a), 190 ml water	Immerse for 8 to 15 s, wash in stream of warm water, blow dry; do not remove etching products from surface
3B (dilute Keller's reagent) ..	20 ml etchant 3A, 80 ml water	Mix fresh before using; immerse specimen for 5 to 10 s
4 (modified Keller's reagent) .	2 ml HF (48%), 3 ml HCl(a), 20 ml HNO$_3$(a), 175 ml water	Immerse for 10 to 60 s, wash in stream of warm water, blow dry; do not remove etching products from surface
5 (Barker's reagent)	4 to 5 ml HBF$_4$ (48%), 200 ml water	Electrolytic: use Al, Pb or stainless steel for cathode, specimen is anode; anodize 40 to 80 s at about 0.2 A/cm^2 (about 20 V dc); check results on microscope with crossed polarizers
6	25 ml HNO$_3$(a), 75 ml water	Immerse in solution at 70 °C (160 °F) for 45 to 60 s
7	20 ml H$_2$SO$_4$(a), 80 ml water	Immerse at 70 °C (160 °F) for 30 s; rinse in cold water
8	10 ml H$_3$PO$_4$ (85%), 90 ml water	Immerse at 50 °C (120 °F) 1 min, or 3 to 5 min
9	5 ml HF (48%), 10 ml H$_2$SO$_4$, 85 ml water	Immerse for 30 s
10	4 g KMnO$_4$, 2 g Na$_2$CO$_3$, 94 ml water, wetting agent	Specimen surface must be well polished, and be precleaned in 20% H$_3$PO$_4$ at 95 °C (205 °F) for uniform wettability(b); after precleaning, rinse in cold water and immediately immerse in etchant for 30 s
11	2 g NaOH, 5 g NaF, 93 ml water	Immerse for 2 to 3 min
12	50 ml Poulton's reagent(b), 25 ml HNO$_3$(a), 40 ml solution of 3 g chromic acid per 10 ml of water	Put a few drops on as-rolled or as-extruded surface for 1 to 4 min, rinse, and swab to desmut; examine on microscope with crossed polarizers to show grains; repeat etching, if necessary; for some 5xxx alloys, increase amount of HNO$_3$ in solution to 50 ml
13	8 ml HNO$_3$(a), 2 ml HCl(a), 45 ml water, 45 ml methanol	Immerse for 10 s
14	5 ml acetic acid (glacial), 1 ml HNO$_3$(a), 94 ml water	Immerse for 20 to 30 min

(a) Concentrated. (b) Etchant 4 in the table on etchants for use in macroscopic examination of aluminum alloys.

Phases that may be present in various aluminum alloy systems

Alloy system	Examples of alloy	Alloy form	Phases
Al-Fe-Si........................	1100, EC	Ingot	FeAl$_3$, FeAl$_6$, Fe$_3$SiAl$_{12}$, Fe$_2$Si$_2$Al$_9$, Si
		Wrought	FeAl$_3$, Fe$_3$SiAl$_{12}$
Al-Fe-Mn-Si	3003	Ingot	(Fe,Mn)Al$_6$, α(Al-Fe,Mn-Si), Si
		Wrought	(Fe,Mn)Al$_6$, α(Al-Fe,Mn-Si)
Al-Fe-Mg-Si (Mg:Si ≈ 1.7:1)	6063	Ingot	FeAl$_3$, FeAl$_6$, Fe$_3$SiAl$_{12}$, Mg$_2$Si
		Wrought	FeAl$_3$, Fe$_3$SiAl$_{12}$, Mg$_2$Si
Al-Fe-Mg-Si (high Si)	356	Cast	Fe$_2$Si$_2$Al$_9$, Mg$_2$Si, Si
Al-Fe-Mg-Si (high Mg)	520	Cast	FeAl$_3$, Fe$_3$SiAl$_{12}$, Mg$_2$Si, Mg$_2$Al$_3$
Al-Cu-Fe-Si	295	Cast	FeAl$_3$, Fe$_3$SiAl$_{12}$, CuAl$_2$, Cu$_2$FeAl$_7$
Al-Fe-Mg-Si-Cr	6061	Ingot	(Fe,Cr)$_3$SiAl$_{12}$, Fe$_2$Si$_2$Al$_9$, FeMg$_3$Si$_6$Al$_8$, Mg$_2$Si, Si
		Wrought	(Fe,Cr)$_3$SiAl$_{12}$, Mg$_2$Si
Al-Cu-Fe-Si-Mg-Mn	2014	Ingot	(Fe,Mn)$_3$SiAl$_{12}$, CuAl$_2$, Cu$_2$Mg$_8$Si$_6$Al$_5$, Si
		Wrought	(Fe,Mn)$_3$SiAl$_{12}$, CuAl$_2$, Cu$_2$Mg$_8$Si$_6$Al$_5$
	2024	Ingot	(Fe,Mn)Al$_6$, (Fe,Mn)Al$_3$, (Fe,Mn)$_3$SiAl$_{12}$, Mg$_2$Si, CuAl$_2$, CuMgAl$_2$, Cu$_2$FeAl$_7$
		Wrought	(Fe,Mn)$_3$SiAl$_{12}$, Mg$_2$Si, CuMgAl$_2$, Cu$_2$FeAl$_7$, Cu$_2$Mn$_3$Al$_{20}$(a)
Al-Cu-Mg-Ni-Fe-Si	2218, 2618	Ingot and wrought	In addition to others, Ni may cause NiAl$_3$, Ni$_2$Al$_3$, Cu$_3$NiAl$_6$ or FeNiAl$_9$ to appear
Al-Fe-Mg-Si-Mn-Cr	5083, 5086, 5456	Ingot	(Fe,Mn,Cr)Al$_6$, (Fe,Mn,Cr)$_3$SiAl$_{12}$, Mg$_2$Al$_3$, (Cr,Mn,Fe)Al$_7$(b)
		Wrought	(Fe,Mn,Cr)$_3$SiAl$_{12}$, Mg$_2$Si, Mg$_2$Al$_3$, Cr$_2$Mg$_3$Al$_{18}$(a)
Al-Cu-Mg-Zn-Fe-Si-Cr............	7075	Ingot	(Fe,Cr)Al$_3$, (Fe,Cr)$_3$SiAl$_{12}$, Mg$_2$Si, Mg(Zn$_2$,AlCu), CrAl$_7$(b)
		Wrought	(Fe,Cr)$_3$SiAl$_{12}$, Cu$_2$FeAl$_7$, Mg$_2$Si, CuMgAl$_2$, Mg(Zn$_2$,AlCu), Cr$_2$Mg$_3$Al$_{18}$(a)

(a) May be identity of fine precipitate which comes out at elevated temperatures; not positively identified. (b) Only when chromium content is near high side of range.

Copper

Etchants for macroscopic examination of coppers and copper alloys
Procedure for use: immerse at room temperature, rinse in warm water, dry

Composition of etchant	Copper or copper alloy	Purpose, or characteristic revealed
50 ml HNO_3, 0.5 g $AgNO_3$, 50 ml water............	All coppers and copper alloys	Produces a brilliant, deep etch
10 ml HNO_3, 90 ml water	Coppers and all brasses	Grains; cracks and other defects
50 ml HNO_3, 50 ml water(a).....................	Coppers, all brasses, aluminum bronze(b)	Grains; cracks and other defects; reveals grain contrast
30 ml HCl, 10 ml $FeCl_3$, 120 ml water or methanol ..	Coppers and all brasses	Grains; cracks and other defects; reveals grain contrast(c)
20 ml acetic acid, 10 ml 5% CrO_3, 5 ml 10% $FeCl_3$, 100 ml water(d)..................................	All brasses	Produces a brilliant, deep etch
2 g $K_2Cr_2O_7$, 4 ml saturated solution of NaCl, 8 ml H_2SO_4, 100 ml water(e).....................	Coppers, high-Cu alloys, phosphor bronze	Grain boundaries, oxide inclusions
40 g CrO_3, 7.5 g NH_4Cl, 50 ml HNO_3, 8 ml H_2SO_4, 100 ml water	Silicon brass, silicon bronze	General macrostructure

(a) Solution should be agitated during etching, to prevent pitting of some alloys. (b) Aluminum bronzes may form smut, which can be removed by brief immersion in concentrated HNO_3. (c) Excellent for grain contrast. (d) Amount of water can be varied as desired. (e) Immerse specimen for 15 to 30 min, then swab with fresh solution.

Electrolytes and conditions for electrolytic polishing of coppers and copper alloys

Electrolyte composition	Voltage	Current density, A/dm^2	Cathode	Time	Copper or copper alloy
825 ml H_3PO_4, 175 ml water................	1.0-1.6	2-10	Copper	10-40 min	Unalloyed copper
250 ml H_3PO_4, 250 ml ethanol, 50 ml propanol, 500 ml distilled water, 3 g urea	3-6	40-80	Stainless steel	50 sec	Coppers and copper alloys
700 ml H_3PO_4, 350 ml water................	1.2-2.0	6-10	Copper	15-30 min	Coppers; alpha, beta, alpha-beta brasses; Al, Si, Sn, and P bronzes; copper with less than 3% Be, Fe, Pb, or Cr
580 g $H_4P_2O_7$, 1000 ml water................	1.2-1.9	8-12	Copper	10-15 min	Coppers, brasses
300 ml HNO_3, 600 ml methanol	20-70 / 30-50	65-310 / 250-310	Stainless / Stainless	10-60 sec / 5-10 sec	Coppers, brasses / Silicon bronze, phosphor bronze
170 g CrO_3, 830 ml water................	1.5-12	95-220	Stainless	10-60 sec	Brasses
400 ml H_3PO_4, 600 ml water................	1.0-2.0	6-15	Copper or stainless	1-15 min	Alpha, alpha-beta brasses; Cu-Fe, Cu-Cr
30 ml HNO_3, 900 ml methanol, 300 g copper nitrate	45-50	105-125	Stainless	15 sec	Bronzes (have tendency to etch)
670 ml H_3PO_4, 100 ml H_2SO_4, 300 ml distilled water.........	2-3	10	Copper	15 min	Copper; Cu-Sn containing up to 6% Sn
470 ml H_3PO_4, 200 ml H_2SO_4, 400 ml distilled water.........	2-2.3	10	Copper	15 min	Cu-Sn up to 9% Sn
350 ml H_3PO_4, 650 ml ethanol	2-5	2-7	Copper	10-15 min	Copper alloys with high lead (up to 30%)
540 ml H_3PO_4, 460 ml water................	2 / 2-2.2	0.65-0.75 / 10-15	Copper / Copper	5-15 min / 15 min	Copper / Nickel silver

Etchants and procedures for microetching of coppers and copper alloys

Etchant composition(a)	Procedure	Copper or copper alloy
20 ml NH_4OH, 0-20 ml water, 8-20 ml H_2O_2 (3%)	Immersion or swabbing for 1 min; H_2O_2 content varies with copper content of alloy to be etched, use fresh H_2O_2 for best results(b)	Coppers and copper alloys; film on etched aluminum bronze can be removed with weak Grard's solution
1 g $Fe(NO_3)_3$, 100 ml water	Immersion	Etching and attack polishing of coppers and copper alloys
25 ml NH_4OH, 25 ml water, 50 ml $(NH_4)_2S_2O_8$ (2.5%)	Immersion	Attack polishing of coppers and some copper alloys
2 g $K_2Cr_2O_7$, 8 ml H_2SO_4, 4 ml NaCl (saturated solution), 100 ml water	Immersion (NaCl replaceable by 1 drop HCl per 25 ml soln; add just before using); follow with $FeCl_3$ or other contrast etch	Coppers; copper alloys of beryllium, manganese and silicon; nickel silver; bronzes; chromium copper
CrO_3 (saturated aqueous solution)	Immersion or swabbing	Coppers, brasses, bronzes, nickel silver
50 ml CrO_3 (10-15%), 1-2 drops HCl	Immersion (add HCl at time of use)	Same as above. Color by electrolytic etching or with $FeCl_3$ etchants
10 g $(NH_4)_2S_2O_8$, 90 ml water	Immersion (use either cold or boiling)	Coppers, brasses, bronzes, nickel silver, aluminum bronze
10% aqueous solution of copper ammonium chloride plus ammonium hydroxide to neutrality or alkalinity	Immersion; Wash specimen thoroughly.	Coppers, brasses, nickel silver; darkening large areas of beta in alpha-beta brass
$FeCl_3$, g HCl, ml Water, ml 5 50 100 20 5 100(c) (d) 25 25 100 1 20 100 8 25 100 5 10 100(e) (f)	Immersion or swabbing; etch lightly or by successive light etches to required results	Coppers, brasses, bronzes, aluminum bronze; darkens beta in brass; gives contrast following dichromate and other etches
5 g $FeCl_3$, 100 ml ethanol, 5-30 ml HCl	Immersion or swabbing for 1 s to several minutes	Coppers and copper alloys
Nitric acid (various concentrations)	Immersion or swabbing; $AgNO_3$ (0.15-0.3%) added to 1:1 solution gives a brilliant, deep etch	Coppers and copper alloys
Ammonium hydroxide (dilute solutions)	Immersion	Attack polishing of brasses and bronzes
50 ml HNO_3, 20 g CrO_3, 30 ml water	Immersion	Aluminum bronze, free-cutting brass; film from polishing can be removed with 10% HF
5 ml HNO_3, 20 g CrO_3, 75 ml water	Immersion	Same as above

(a) The use of concentrated etchants is intended unless otherwise specified. (b) This etchant may be alternated with $FeCl_3$. (c) Grard's No. 1 etchant. (d) Plus 1 g CrO_3. (e) Grard's No. 2 etchant. (f) Plus 1 g $CuCl_2$ and 0.05 g $SnCl_2$.

Electrolytes and operating conditions for electrolytic etching of copper and copper alloys

Electrolyte composition	Operating conditions	Copper or copper alloy
5 to 14% H_3PO_4, 8% (sp gr 1.042) is preferred	1-4 V; etching time, 10 s	Coppers
Remainder, water	1-8 V; etching time, 5-7 s	Cartridge brass, free-cutting brass, admiralty, gilding metal
Struer's D-2: 250 ml H_3PO_4 (85%), 250 ml ethanol (95%), 500 ml water, 2 ml wetting agent	1-3 V; current density, 10-15 A/dm^2; etching time, 30-60 s	Coppers
10 ml $(NH_4)C_2H_3O_2$, 30 ml $Na_2S_2O_3$, 30 ml NH_4OH, 30 ml water	30 A/dm^2 time varies with composition and previous treatment of specimen	Cold worked brasses
30 g $FeSO_4$, 4 g NaOH, 100 ml H_2SO_4, 1900 ml water	0.1 A at 8-10 V; for not over 15 s; do not swab surface after etching	Darkens beta in brasses and gives contrast after H_2O_2-NH_4OH etch; also for nickel silver and bronzes.
1% CrO_3, 99% water	6 V; aluminum cathode; etching time, 3-6 s	Beryllium copper and aluminum bronze
5 ml acetic acid (glacial), 10 ml HNO_3, 30 ml water	½-1 V; current density, 20-50 A/dm^2; etching time, 5-15 s	Copper-nickel alloys; avoiding contrast associated with coring

Lead

Recommended etchants and procedures for macroscopic and microscopic examination of lead and lead alloys

Etchant No.	Composition(a)	Procedure	Use
1	1 part acetic acid (glacial), 1 part nitric acid(b), 4 parts glycerol	Use freshly prepared solution at 80 °C (175 °F); discard after use; for macroetching: etch several minutes, rinse in water; for microetching: etch several seconds; for best results, alternate etching with polishing	Macroetching of lead; development of microstructures and grain boundaries in lead, and in lead-calcium, lead-antimony, and lead-tin (low-tin) alloys
2	100 parts acetic acid (glacial), 10 parts hydrogen peroxide (30%)	Etch for 10-30 min, depending on the depth of the disturbed layer; dry and clean with concentrated nitric acid if required	Microetching of lead-antimony alloys containing up to 2% antimony
3	3 parts acetic acid (glacial), 1 part hydrogen peroxide (30%)	Etch by immersing specimen in solution for 6-15 s; dry with alcohol	Microetching of lead, Pb-Ca alloys, and Pb-Sb alloys containing more than 2% Sb. Also removes disturbed metal
4	Solution A: 15 g ammonium molybdate, 100 ml distilled water / Solution B: 6 parts nitric acid(b), 4 parts distilled water	Mix equal quantities of solutions A and B; etch by alternately swabbing specimen and washing in running water	Macroetching of lead. A very rapid etchant; well suited for removing thick layers of disturbed metal from specimens
5	3 parts acetic acid (glacial), 4 parts nitric acid(b), 16 parts distilled water	Use freshly prepared solution at 40-42 °C (105-110 °F); immerse specimen for 4-30 min until disturbed layer is removed; clean with cotton in running water	Microetching of unalloyed lead, and lead-tin alloys containing up to 3% tin
6	2 parts acetic acid (glacial), 2 parts nitric acid(b), 2 parts hydrogen peroxide (30%), 5 parts distilled water	Etch for 2-10 s by swabbing; rinse specimen in running water and dry with alcohol	Macroetching of unalloyed lead, and of lead-bismuth, lead-tellurium and lead-nickel alloys
7	1 part nitric acid(b), 1 part distilled water	Etch for 5-10 min by immersion; if thick layer of disturbed metal is to be removed, solution can be heated to boiling; rinse in running water, rinse in alcohol and dry	Developing macrostructure of welds and laminations in lead products
8	Solution A: 10% aqueous solution of ammonium persulfate / Solution B: 30% aqueous solution of tartaric acid	Mix 5 ml of solution A with 2 ml of solution B; swab specimen for 5-10 s; rinse in running water	Microetching to distinguish cuboidal SbSn phase from Sb-rich phases in Pb-Sb-Sn alloys such as bearing alloys or type metals. Solution A blackens SbSn phase; solution B etches Sb-rich phases
9	6 parts perchloric acid (70%), 4 parts water	Immerse specimen (cathode) in electrolyte; anode is platinum spiral; etch 45-90 s at 6 V, 4 A, from a rectifier	Electrolytic etching of lead-antimony alloys containing more than 2% antimony
10	1 part hydrochloric acid(b), 9 parts water	Same as for etchant 9	Same as for etchant 9

(a) Parts are by volume. (b) Concentrated

Nickel

Electrolytes and current densities for electropolishing of nickel and nickel-copper alloys

Composition of electrolyte	Applicable alloys	Current density, A/in.^{2}
37 ml H_3PO_4(a), 56 ml glycerol, 7 ml water	Nickel 200	9-10
	Nickel 270	10-12
	Duranickel 301	8-10
	Monel 400	6-7
33 ml HNO_3(a), 66 ml methanol	Monel 400, R-405, K-500	10-15

(a) Concentrated.

Etchants for microscopic examination of nickel and nickel-copper alloys for grain boundaries and general structure

Composition of etchant	Conditions for use
Etchants for Nickel 200 and 270; Duranickel 301; and Monel 400, R-405 and K-500	
1 part 10% aqueous solution of sodium cyanide, 1 part 10% aqueous solution of ammonium persulfate; mix solutions when ready to use	Immerse or swab specimen for 5 to 90 s (*Caution:* use fume hood; solutions release toxic fumes when mixed)
1 part concentrated nitric acid, 1 part acetic acid (glacial); use fresh solution	For revealing grain boundaries; immerse or swab specimen for 5 to 20 s
Alternative etchant for Monel K-500	
Glyceregia: 10 ml concentrated nitric acid, 20 ml concentrated hydrochloric acid, 30-40 ml glycerol	Etch by immersing or swabbing the specimen for 30 s to 5 min

Magnesium

Selected etchants for macroscopic and microscopic examination of magnesium alloys

Etchant No.	Composition	Etching procedure	Characteristics and use
1	*Nital:* 1 to 5 ml HNO_3 (conc), 100 ml ethanol (95%) or methanol (95%)	Swab or immerse specimen for a few seconds to 1 min; wash in water then alcohol and dry	Shows general structure
2	*Glycol:* 1 ml HNO_3 (conc), 24 ml water, 75 ml ethylene glycol	Immerse specimen face up and swab with cotton for 3 to 5 s for as-cast or aged metal, and up to 1 min for heat treated metal; wash in water, then alcohol and dry	Shows general structure; reveals constituents in Mg-rare earth and Mg-Th alloys
3	*Acetic glycol:* 20 ml acetic acid, 1 ml HNO_3 (conc), 60 ml ethylene glycol, 20 ml water	Immerse specimen face up with gentle agitation for 1 to 3 s for as-cast or aged metal, and for 10 s for heat treated metal; wash in water, then alcohol and dry	Shows general structure and grain boundaries in heat treated castings; shows grain boundaries in Mg-rare earth and Mg-Th alloys
4	10 ml HF (48%), 90 ml water	Immerse specimen face up for 1 to 2 s; wash in water, then alcohol and dry	Darkens $Mg_{17}Al_{12}$ phase and leaves $Mg_{32}(Al, Zn)_{49}$ phase unetched and white
5	*Phospho-picral:* 0.7 ml H_3PO_4, 4 to 6 g picric acid, 100 ml ethanol (95%)	Immerse specimen face up for about 10 to 20 s or until polished surface is darkened; wash in alcohol and dry	For estimating the amount of massive phase; stains matrix and leaves phase white; staining improves as magnesium-ion content increases with use
6	*Acetic-picral:* 5 ml acetic acid, 6 g picric acid, 10 ml water, 100 ml ethanol (95%)	Immerse specimen face up with gentle agitation until face turns brown; wash in a stream of alcohol and dry with a blast of air	A universal etchant; defines grain boundaries in most alloys and tempers by etch rate and color of stain; reveals cold work and twinning readily
7	*Acetic-picral:* 20 ml acetic acid, 3 g picric acid, 20 ml water, 50 ml ethanol (95%)	Same as for etchant 6, above, but etch for at least 15 s to develop a heavy film	Orientation of crackled film is parallel to trace of basal plane; film crackles in high-alloy areas; distinguishes between fusion voids surrounded by normal level of alloy and microshrinkage with low alloy content
8	*Acetic-picral:* 10 ml acetic acid, 4.2 g picric acid, 10 ml water, 70 ml ethanol (95%)	Immerse specimen face up with gentle agitation until face turns brown; wash in a stream of alcohol and dry with a blast of air	Reveals grain boundaries more readily than etchant 6, above, especially in dilute alloys
9	0.6 g picric acid, 10 ml ethanol (95%), 90 ml water	Immerse specimen face up for 15 to 30 s; wash in alcohol and dry	Used after HF etchant to darken matrix to give better contrast between matrix and white ternary phase
10	2 ml HF (48%), 2 ml HNO_3 (conc), 96 ml water	Immerse specimen face up with gentle agitation; do not swab	Grain structure and coring in Mg-Zn-Zr alloys

Tin

Etchants for use in microscopic examination of tin and tin alloys

Etchant composition	Uses
5 ml HCl, 2 g $FeCl_3$, 30 ml water, 60 ml absolute alcohol	General use for tin and tin alloys
2 ml HCl, 98 ml methanol (95%) or ethanol (95%)	Grain-boundary etch for pure tin
10 ml HNO_3, 10 ml acetic acid, 80 ml glycerol	Darkens the lead in the eutectic of tin-rich tin-lead alloys
5% silver nitrate in water	Darkens primary and eutectic lead in lead-rich tin-lead alloys
2% nital	Recommended for etching tin-antimony alloys; darkens tin-rich matrix, leaving intermetallic compounds unattacked; often used for etching specimens of babbitted bearings
Picral	For etching tin-coated steel and tin-coated cast iron
1 drop concentrated HNO_3, 2 drops HF, 25 ml glycerol; then picral	For etching tin-coated steel
Dilute ammonium hydroxide with a few drops of 30% hydrogen peroxide	For etching tin-coated copper and copper alloys

Titanium

Etchants for microscopic examination of titanium and titanium alloys

Specimen metal	Composition of etchant	Purpose
Unalloyed titanium	1 to 3 ml HF, 10 ml HNO_3, 30 ml lactic acid	Reveals hydrides
	1 ml HF, 30 ml HNO_3, 30 ml lactic acid	Reveals hydrides
Most titanium alloys	Kroll's reagent: 1-3 ml HF, 2-6 ml HNO_3, water to 1000 ml	General-purpose etch
	10 ml HF, 5 ml HNO_3, 85 ml water	General-purpose etch
	1 ml HF, 2 ml HNO_3, 50 ml H_2O_2, 47 ml water	Removes stain
	10 ml HF, 10 ml HNO_3, 30 ml lactic acid	Chemical polish and etch
	2 ml HF, 98 ml water	Reveals alpha case
	1-2 ml HF, 4-5 ml H_2O_2, water to 1000 ml	Nonstaining etch
Near-alpha titanium alloys	2 ml HF, 98 ml water; then 1 ml HF, 2 ml HNO_3, 97 ml water	General-purpose etch(a)
Alpha-beta titanium alloys	10 ml KOH (40%), 5 ml H_2O_2, 20 ml water	Stains alpha, transformed beta
Ti-Al-Zr and Ti-Si alloys	18.5 g benzalkonium chloride, 33 ml ethanol, 40 ml glycerol, 25 ml HF	General-purpose etch
Ti-3Al-8V-6Cr-4Mo-4Zr	30 ml H_2O_2, 3 drops HF	General-purpose etch
Ti-8Mn; aged Ti-13V-11Cr-3Al	2 ml HF, 4 ml HNO_3, 94 ml water	General-purpose etch
Ti-Si alloys	2 drops HF, 1 drop HNO_3, 3 ml HCl, 25 ml glycerol	General-purpose etch

(a) First etchant stains alpha phase; second etchant removes stain.

Zinc

Etchants for zinc and zinc alloys

Etchant	Composition
1(a)	200 g CrO_3, 15 g Na_2SO_4, 1000 ml water
2(b)	50 g CrO_3, 4 g Na_2SO_4, 1000 ml water
3	200 g CrO_3, 1000 ml water

(a) For rolled zinc-copper alloys, the Na_2SO_4 content can be reduced to 7.5 g. If desired, a smoothly etched surface can be obtained by increasing the Na_2SO_4 to 30 g. (b) This etchant can be made by mixing one part (by volume) of etchant 1 and three parts of water.

Etchants and etching times for zinc and zinc die-casting alloys

Specimen metal	Etchant(a)	Time, s, for examination at: 250 ×	Time, s, for examination at: 1000 ×
Cast or rolled zinc	1	5	1
Alloy AC41A or AG40A	2	1	1

(a) Etchant numbers correspond to those assigned in the table on etchants for zinc and zinc alloys.

Cast Irons

Gray Cast Iron

Typical base compositions and mechanical properties of SAE J431 automotive gray cast irons(a)

UNS	SAE grade	TC	Mn	Si	P	S
F10004	G1800(b)	...3.40 to 3.70	0.50 to 0.80	2.80 to 2.30	0.15	0.15
F10005	G2500(b)	...3.20 to 3.50	0.60 to 0.90	2.40 to 2.00	0.12	0.15
F10006	G3000(c)	...3.10 to 3.40	0.60 to 0.90	2.30 to 1.90	0.10	0.15
F10007	G3500(c)	...3.00 to 3.30	0.60 to 0.90	2.20 to 1.80	0.08	0.15
F10008	G4000(c)	...3.00 to 3.30	0.70 to 1.00	2.10 to 1.80	0.07	0.15

SAE grade	Hardness, HB	Minimum transverse load		Minimum deflection		Minimum tensile strength	
		kg	lb	mm	in.	MPa	ksi
G1800	187 max	780	1720	3.6	0.14	118	18
G2500	170 to 229	910	2000	4.3	0.17	173	25
G3000	187 to 241	1000	2200	5.1	0.20	207	30
G3500	207 to 255	1110	2450	6.1	0.24	241	35
G4000	217 to 269	1180	2600	6.9	0.27	276	40

(a) If either carbon or silicon is on the high side of the range, the other should be on the low side. Properties determined from an as-cast test bar (1.2-in., or 30.5-mm, diam). (b) Ferritic-pearlitic microstructure. (c) Pearlitic microstructure.

Typical base compositions and mechanical properties of SAE J431 automotive gray cast irons for heavy duty service (a)

UNS	SAE grade	TC	Mn	Si	P	S
F10009	G2500a(b)	...3.40 min	0.60 to 0.90	1.60 to 2.10	0.12	0.12
F10010	G3500b(c)	...3.40 min	0.60 to 0.90	1.30 to 1.80	0.08	0.12
F10011	G3500c(c)	...3.50 min	0.60 to 0.90	1.30 to 1.80	0.08	0.12
F10012	G4000d(d)	...3.10 to 3.60	0.60 to 0.90	1.95 to 2.40	0.07	0.12

SAE grade	Hardness, HB	Minimum transverse load		Minimum deflection		Minimum tensile strength	
		kg	lb	mm	in.	MPa	ksi
G2500a	170 to 229	910	2000	4.3	0.17	173	25
G3500b	207 to 255	1090	2400	6.1	0.24	241	35
G3500c	207 to 255	1090	2400	6.1	0.24	241	35
G4000d	241 to 321(e)	1180	2600	6.9	0.27	276	40

(a) If either carbon or silicon is on the high side of the range, the other should be on the low side. Alloying elements not listed in this table may be required. Properties determined from an as-cast test bar (1.2-in., or 30.5-mm, diam). (b) Microstructure: size 2 to 4 type A graphite in a matrix of lamellar pearlite containing not more than 15% free ferrite. (c) Microstructure: size 3 to 5 type A graphite in a matrix of lamellar pearlite containing not more than 5% free ferrite or free carbide. (d) Alloy gray iron containing 0.85 to 1.25 Cr, 0.40 to 0.60 Mo and 0.20 to 0.45 Ni or as agreed. Microstructure: primary carbides and size 4 to 7 type A or E graphite in a matrix of fine pearlite, as determined in a zone at least 1/8 in. (3.2 mm) deep at a specified location on a cam surface. (e) Determined on a specified bearing surface.

Automotive applications of gray cast iron

Grade	Typical uses	Grade	Typical uses
G1800 ..	Miscellaneous soft iron castings (as cast or annealed) in which strength is not a primary consideration		castings, pistons, medium duty brake drums and clutch plates
G2500 ..	Small cylinder blocks, cylinder heads, air-cooled cylinders, pistons, clutch plates, oil pump bodies, transmission cases, gear boxes, clutch housings, and light duty brake drums	G3500 ..	Diesel engine blocks, truck and tractor cylinder blocks and heads, heavy flywheels, tractor transmission cases, heavy gear boxes
G2500a .	Brake drums and clutch plates for moderate service requirements, where high carbon iron is desired to minimize heat checking	G3500b .	Brake drums and clutch plates for heavy duty service where both resistance to heat checking and higher strength are definite requirements
		G3500c .	Brake drums for extra heavy duty service
G3000 ..	Automobile and diesel cylinder blocks, cylinder heads, flywheels, differential carrier	G4000 ..	Diesel engine castings, liners, cylinders, and pistons
		G4000d .	Camshafts

Typical mechanical properties of standard gray iron test bars, as cast

ASTM class	Tensile strength MPa	ksi	Torsional shear strength MPa	ksi	Compressive strength MPa	ksi	Reversed bending fatigue limit MPa	ksi	Transverse load on test bar B kg	lb	Hardness, HB
20	152	22	179	26	572	83	69	10	839	1850	156
25	179	26	220	32	669	97	79	11.5	987	2175	174
30	214	31	276	40	752	109	97	14	1145	2525	210
35	252	36.5	334	48.5	855	124	110	16	1293	2850	212
40	293	42.5	393	57	965	140	128	18.5	1440	3175	235
50	362	52.5	503	73	1130	164	148	21.5	1638	3600	262
60	431	62.5	610	88.5	1293	187.5	169	24.5	1678	3700	302

Typical moduli of elasticity of standard gray iron test bars, as cast

ASTM class	Tensile modulus GPa	10^6 psi	Torsional modulus GPa	10^6 psi
20	66 to 97	9.6 to 14.0	27 to 39	3.9 to 5.6
25	79 to 102	11.5 to 14.8	32 to 41	4.6 to 6.0
30	90 to 113	13.0 to 16.4	36 to 45	5.2 to 6.6
35	100 to 119	14.5 to 17.2	40 to 48	5.8 to 6.9
40	110 to 138	16.0 to 20.0	44 to 54	6.4 to 7.8
50	130 to 157	18.8 to 22.8	50 to 55	7.2 to 8.0
60	141 to 162	20.4 to 23.5	54 to 59	7.8 to 8.5

Ductile Irons

Compositions and general uses for standard grades of ductile irons(a)

Specification No.	Grade or class	UNS	TC(b)	Typical composition, % Si	Mn	P	S	Description	General uses
ASTM A395; ASME SA395	60-40-18	F32800	3.00 min	2.50 max(c)	· · ·	0.08 max	· · ·	Ferritic; annealed	Pressure-containing parts for use at elevated temperatures
ASTM A476; SAE AMS5316 ...	80-60-03	F34100	3.00 min(d)	3.0 max	· · ·	0.08 max	0.05 max	As cast	Paper mill dryer rolls, at temperatures up to 230 °C (450 °F)
ASTM A536; MIL-I-11466B(MR)	60-40-18(e)	F32800						Ferritic; may be annealed	Shock-resistant parts; low-temperature service
	65-45-12(e)	F33100						Mostly ferritic; as cast or annealed	General service
	80-55-06(e)	F33800						Ferritic/pearlitic; as cast	General service
	100-70-03(e)	F34800						Mostly pearlitic; may be normalized	Best combination of strength, wear resistance and response to surface hardening
	120-90-02(e)	F36200						Martensitic; oil quenched and tempered	Highest strength and wear resistance
SAE J434c	D4018(f)	F32800	3.20-4.10	1.80-3.00	0.10-1.00	0.015-0.10	0.005-0.035	Ferritic	Moderately stressed parts requiring good ductility and machinability
	D4512(f)	F33100						Ferritic/pearlitic	Moderately stressed parts requiring moderate machinability
	D5506(f)	F33800						Ferritic/pearlitic	Highly stressed parts requiring good toughness
	D7003(f)	F34800						Pearlitic	Highly stressed parts requiring very good wear resistance and good response to selective hardening

(a) For mechanical properties and typical applications, see the following table. (b) Total carbon. (c) The silicon limit may be increased by 0.08% for each 0.01% reduction in phosphorus content, up to 2.75 Si. (d) Carbon equivalent (CE), 3.8-4.5; CE = TC + 0.3 (Si + P). (e) Composition subordinate to mechanical properties; composition range for any element may be specified by agreement between supplier and purchaser. (f) General composition given under grade D4018 for reference only. Typically, foundries will produce to narrower ranges than those shown and will establish different median compositions for different grades. (g) For castings with sections 13 mm (½ in.) and smaller, may have 2.75 max Si and 0.08 max P, or 3.00 max Si with 0.05 max P; for castings with sections 50 mm (2 in.) and greater, CE must not exceed 4.3. (h) Plus 18-22 Ni and 1.7-2.4 Cr. (j) Plus 20-23 Ni and 0.5 max Cr. (k) Stress relieved at 650°C (1200°F); or solution treated at 950°C (1750°F), if necessary to dissolve carbides

Compositions and general uses for standard grades and ductile irons(a) (continued)

Specification No.	Grade or class	UNS	TC(b)	Typical composition, % Si	Mn	P	S	Description	General uses
	DQ & T(f)	F30000						Martensitic	Highly stressed parts requiring uniformity of microstructure and close control of properties
MIL-I-24137(Ships)	Class A	F33101	3.0 min	2.50 max(g)	· · ·	0.08 max	· · ·	Ferritic; annealed	General shipboard service
	Class B	F43020	2.40-3.00	1.80-3.20	0.80-1.50(h)	0.20 max	· · ·	Austenitic(k)	Shipboard service requiring resistance to corrosion, heat or shock; or requiring nonmagnetic properties
	Class C	F43021	2.70-3.10	2.00-3.00	1.90-2.50(j)	0.15 max	· · ·	Austenitic(k)	

(a) For mechanical properties and typical applications, see the following table. (b) Total carbon. (c) The silicon limit may be increased by 0.08% for each 0.01% reduction in phosphorus content, up to 2.75 Si. (d) Carbon equivalent (CE), 3.8-4.5; CE = TC + 0.3 (Si + P). (e) Composition subordinate to mechanical properties; composition range for any element may be specified by agreement between supplier and purchaser. (f) General composition given under grade D4018 for reference only. Typically, foundries will produce to narrower ranges than those shown and will establish different median compositions for different grades. (g) For castings with sections 13 mm (½ in.) and smaller, may have 2.75 max Si and 0.08 max P, or 3.00 max Si with 0.05 max P; for castings with sections 50 mm (2 in.) and greater, CE must not exceed 4.3. (h) Plus 18-22 Ni and 1.7-2.4 Cr. (j) Plus 20-23 Ni and 0.5 max Cr. (k) Stress relieved at 650°C (1200°F); or solution treated at 950°C (1750°F), if necessary to dissolve carbides.

Mechanical properties and typical applications for standard grades of ductile irons (a)

Specification No.	Grade or class	Hardness, HB(b)	Tensile strength, min(c) MPa	ksi	Yield strength, min(c) MPa	ksi	Elongation in 50 mm, or 2 in., min, %(c)	Typical applications
ASTM A395-76; ASME SA395	60-40-18	143-187	414	60	276	40	18	Valves and fittings for steam and chemical-plant equipment
ASTM A476-70(d); SAE AMS5316	80-60-03	201 min	552	80	414	60	3	Paper-mill dryer rolls
ASTM A536-72, MIL-I-11466B(MR) . .	60-40-18	· · ·	414	60	276	40	18	Pressure-containing parts such as valve and pump bodies
	65-45-12	· · ·	448	65	310	45	12	Machine components subject to shock and fatigue loads
	80-55-06	· · ·	552	80	379	55	6	Crankshafts, gears and rollers
	100-70-03	· · ·	689	100	483	70	3	High-strength gears and machine components
	120-90-02	· · ·	827	120	621	90	2	Pinions, gears, rollers and slides
SAE J434c	D4018	170 max	414	60	276	40	18	Steering knuckles
	D4512	156-217	448	65	310	45	12	Disc-brake calipers
	D5506	187-255	552	80	379	55	6	Crankshafts
	D7003	241-302	689	100	483	70	3	Gears
	DQ & T	(e)	(f)	(f)	(f)	(f)	(f)	Rocker arms
MIL-I-24137(Ships) . .	Class A	190 max	414	60	310	45	15	Electric equipment, engine blocks, pumps, housings, gears, valve bodies, clamps and cylinders
	Class B	190 max	379	55	207	30	7	Pressure parts, machine components and propellers
	Class C	175 max	345	50	172	25	20	Pressure parts, machine components and propellers

(a) For compositions, descriptions and uses, see the preceding table. (b) Measured at a predetermined location on the casting. (c) Determined using a standard specimen taken from a separately cast test block, as set forth in the applicable specification. (d) Reapproved in 1976. (e) Range specified by mutual agreement between producer and purchaser. (f) Value must be compatible with minimum hardness specified for production castings.

Average mechanical properties of ductile irons heat treated to various strength levels (a)

Nearest standard grade	Hardness, HB	Ultimate strength MPa	Ultimate strength ksi	Yield strength MPa	Yield strength ksi	Elongation, %(b)	Modulus GPa	Modulus 10⁶ psi	Poisson's ratio
Tension									
60-40-18	167	461	66.9	329(c)	47.7(c)	15.0	169	24.5	0.29
65-45-12	167	464	67.3	332(c)	48.2(c)	15.0	168	24.4	0.29
80-55-06	192	559	81.1	362(c)	52.5(c)	11.2	168	24.4	0.31
120-90-02	331	974	141.3	864(c)	125.3(c)	1.5	164	23.8	0.28
Compression									
60-40-18	167	· · ·	· · ·	359(c)	52.0(c)	· · ·	164	23.8	0.26
65-45-12	167	· · ·	· · ·	362(c)	52.5(c)	· · ·	163	23.6	0.31
80-55-06	192	· · ·	· · ·	386(c)	56.0(c)	· · ·	165	23.9	0.31
120-90-02	331	· · ·	· · ·	920(c)	133.5(c)	· · ·	164	23.8	0.27
Torsion									
60-40-18	167	472	68.5	195(d)	28.3(d)	· · ·	63 / 65.5(e)	9.1 / 9.5(e)	· · ·
65-45-12	167	475	68.9	297(d)	30.0(d)	· · ·	64 / 65(e)	9.3 / 9.4(e)	· · ·
80-55-06	192	504	73.1	193(d)	28.0(d)	· · ·	62 / 64(e)	9.0 / 9.3(e)	· · ·
120-90-02	331	875	126.9	492(d)	71.3(d)	· · ·	63.4 / 64(e)	9.2 / 9.3(e)	· · ·

(a) Determined for a single heat of ductile iron, heat treated to approximate various standard grades. Properties were obtained using test bars machined from 25-mm (1-in.) keel blocks. (b) In 50 mm, or 2 in. (c) 0.2% offset. (d) 0.0375% offset. (e) Calculated from tensile modulus and Poisson's ratio in tension.

Malleable Irons

Typical composition ranges for ferritic and pearlitic malleable irons, analyzed in the white iron condition

Type of iron	TC	Mn	Si	P	S
Ferritic					
Grade 32510	2.30-2.70	0.25-0.55	1.00-1.75	0.05 max	0.03-0.18
Grade 35018	2.00-2.45	0.25-0.55	1.00-1.35	0.05 max	0.03-0.18
Pearlitic	2.00-2.70	0.25-1.25	1.00-1.75	0.05 max	0.03-0.18

Applications of malleable iron castings(a)

Specification No.	Class or grade	Microstructure	Typical applications
Ferritic			
ASTM A47; ANSI G48.1; FED QQ-I-666C	32510 35018	Temper carbon and ferrite	General engineering service at normal and elevated temperatures for good machinability and excellent shock resistance.
ASTM A338	32510 35018	Temper carbon and ferrite	Flanges, pipe fittings, and valve parts for railroad, marine, and other heavy duty service up to 345 °C (650 °F).
ASTM A197; ANSI G49.1	· · ·	Free of primary graphite	Pipe fittings and valve parts for pressure service.
Pearlitic and Martensitic			
ASTM A220; ANSI G48.2; MIL-I-11444B	40010 45008 45006 50005 60004 70003 80002 90001	Temper carbon in necessary matrix without primary cementite or graphite	General engineering service at normal and elevated temperatures. Dimensional tolerance range for castings is stipulated.
Automotive			
ASTM A602; SAE J158	M3210	Ferritic	For low-stress parts requiring good machinability: steering gear housings, carriers, and mounting brackets.
	M4504	Ferrite and tempered pearlite(b)	Compressor crankshafts and hubs.
	M5003	Ferrite and tempered pearlite(b)	For selective hardening: planet carriers, transmission gears, differential cases.
	M5503	Tempered martensite	For machinability and improved response to induction hardening.
	M7002	Tempered martensite	For high-strength parts: connecting rods and universal joint yokes.
	M8501	Tempered martensite	For high strength plus good wear resistance: certain gears.

(a) For mechanical properties, see the following table. (b) May be all tempered martensite for some applications.

Properties of malleable iron castings(a)

Specification No.	Class or grade	Tensile strength MPa	ksi	Yield strength MPa	ksi	Hardness, HB	Elongation(b), %
Ferritic							
ASTM A47, A338; ANSI G48.1; FED QQ-I-666c	32510	345	50	224	32	156 max	10
	35018	365	53	241	35	156 max	18
ASTM A197	· · ·	276	40	207	30	156 max	5
Pearlitic and Martensitic							
ASTM A220; ANSI G48.2; MIL-I-11444B........	40010	414	60	276	40	149-197	10
	45008	448	65	310	45	156-197	8
	45006	448	65	310	45	156-207	6
	50005	483	70	345	50	179-229	5
	60004	552	80	414	60	197-241	4
	70003	586	85	483	70	217-269	3
	80002	655	95	552	80	241-285	2
	90001	724	105	621	90	269-321	1
Automotive							
ASTM A602; SAE J158	M3210(c)	345	50	224	32	156 max	10
	M4504(d)	448	65	310	45	163-217	4
	M5003(d)	517	75	345	50	187-241	3
	M5503(e)	517	75	379	55	187-241	3
	M7002(e)	621	90	483	70	229-269	2
	M8501(e)	724	105	586	85	269-302	1

(a) For microstructures and typical applications, see the preceding table. (b) Minimum in 50 mm (2 in.). (c) Annealed. (d) Air quenched and tempered. (e) Liquid quenched and tempered.

Alloy Cast Irons: General

Ranges of alloy content for various types of alloy cast irons

Description	TC(b)	Mn	P	S	Composition, wt % (a) Si	Ni	Cr	Mo	Cu	Matrix structure, as-cast(c)
Abrasion-Resistant White Irons										
Low-C white iron(d)	2.2 to 2.8	0.2 to 0.6	0.15	0.15	1.0 to 1.6	1.5	1.0	0.5	(e)	CP
High-C, low-Si white iron	2.8 to 3.6	0.3 to 2.0	0.30	0.15	0.3 to 1.0	2.5	3.0	1.0	(e)	CP
Malleable white iron	2.2 to 2.5	0.3 to 0.5	0.15	0.15	1.0 to 1.6	. . .	. . .	. . .	. . .	CP
Martensitic nickel-chromium iron . .	2.5 to 3.7	1.3	0.30	0.15	0.8	2.7 to 5.0	1.1 to 4.0	1.0	. . .	M, A
Martensitic nickel, high-chromium iron	2.5 to 3.6	1.3	0.10	0.15	1.0 to 2.2	5 to 7	7 to 11	1.0	. . .	M, A
Martensitic chromium-molybdenum iron	2.0 to 3.6	0.5 to 1.5	0.10	0.06	1.0	1.5	11 to 23	0.5 to 3.5	1.2	M, A
High-chromium iron	2.3 to 3.0	0.5 to 1.5	0.10	0.06	1.0	1.5	23 to 28	1.5	1.2	M
Corrosion-Resistant Irons										
High silicon iron(f)	0.4 to 1.1	1.5	0.15	0.15	14 to 17	. . .	5.0	1.0	0.5	F
High chromium iron	1.2 to 4.0	0.3 to 1.5	0.15	0.15	0.5 to 3.0	5.0	12 to 35	4.0	3.0	M, A
Nickel-chromium gray iron(g)	3.0	0.5 to 1.5	0.08	0.12	1.0 to 2.8	13.5 to 36	1.5 to 6.0	1.0	7.0	A
Nickel-chromium ductile iron(h) . . .	3.0	0.7 to 4.5	0.08	0.12	1.0 to 3.0	18 to 36	1.0 to 5.5	1.0	. . .	A
Heat-Resistant Gray Irons										
Medium silicon iron(j)	1.6 to 2.5	0.4 to 0.8	0.30	0.10	4.0 to 7.0	. . .	. . .	. . .	. . .	F
High chromium iron	1.8 to 3.0	0.3 to 1.5	0.15	0.15	0.5 to 2.5	5.0	15 to 35	. . .	. . .	F, CP
Nickle-chromium iron(g)	1.8 to 3.0	0.4 to 1.5	0.15	0.15	1.0 to 2.75	13.5 to 36	1.8 to 6.0	1.0	7.0	A
Nickel-chromium-silicon iron(k)	1.8 to 2.6	0.4 to 1.0	0.10	0.10	5.0 to 6.0	13 to 43	1.8 to 5.5	1.0	10.0	A
High aluminum iron	1.3 to 2.0	0.4 to 1.0	0.15	0.15	1.3 to 6.0	. . .	20 to 25 Al	. . .	. . .	F
Heat-Resistant Ductile Irons										
Medium silicon ductile iron	2.8 to 3.8	0.2 to 0.6	0.08	0.12	2.5 to 6.0	1.5	. . .	. . .	. . .	F
Nickel-chromium ductile iron(h) . . .	3.0	0.7 to 2.4	0.08	0.12	1.75 to 5.5	18 to 36	1.75 to 3.5	1.0	. . .	A

(a) Where a single value is given rather than a range, that value is a maximum limit. (b) Total carbon. (c) CP, coarse pearlite; M, martensite; A, austenite; F, ferrite. (d) May be produced from a malleable-iron base composition. (e) Cu may replace all or part of the Ni. (f) Such as Duriron, Durichlor 51, Superchlor. (g) Such as Ni-Resist austenitic iron (ASTM A436). (h) Such as Ni-Resist austenitic ductile iron (ASTM A439). (j) Such as Silal. (k) Such as Nicrosilal.

Physical properties of selected alloy cast irons

Description(a)	Density Mg/m³	lb/in.³	Coefficient of thermal expansion(b) μm/m · °C	10^{-6} in./in. °F	Electrical resistivity, μΩ · m	Thermal conductivity W/m · K	Btu/ft · h · °F
Abrasion-Resistant White Irons							
Low-C white iron	7.6 to 7.8	0.275 to 0.282	12(c)	6.7(c)	0.53	22(d)	13(d)
Martensitic nickel-chromium iron	7.6 to 7.8	0.275 to 0.282	8 to 9(c)	4.4 to 5(c)	0.80	30(d)	17(d)
Corrosion-Resistant Irons							
High-silicon iron	7.0 to 7.05	0.252 to 0.254	12.4 to 13.1	6.9 to 7.3	0.50	. . .	. . .
High-chromium iron	7.3 to 7.5	0.264 to 0.271	9.4 to 9.9	5.2 to 5.5	. . .	. . .	. . .
High-nickel gray iron	7.4 to 7.6	0.267 to 0.275	8.1 to 19.3	4.5 to 10.7	1.0(d)	38 to 40	22 to 23
High-nickel ductile iron	7.4	0.267	12.6 to 18.7	7.0 to 10.4	1.0(d)	13.4	7.75
Heat-Resistant Gray Irons							
Medium-silicon iron	6.8 to 7.1	0.246 to 0.256	10.8	6.0	. . .	37	21
High-chromium iron	7.3 to 7.5	0.264 to 0.271	9.3 to 9.9	5.2 to 5.5	. . .	20	12
High-nickel iron	7.3 to 7.5	0.264 to 0.271	8.1 to 19.3	4.5 to 10.7	1.4 to 1.7	37 to 40	21 to 23
Nickel-chromium-silicon iron	7.33 to 7.45	0.265 to 0.269	12.6 to 16.2	7.0 to 9.0	1.5 to 1.7	30	17
High-aluminum iron	5.5 to 6.4	0.20 to 0.23	15.3	8.5	2.4	. . .	. . .
Heat-Resistant Ductile Irons							
Medium-silicon ductile iron	7.1	0.257	10.8 to 13.5	6.0 to 7.5	0.58 to 0.87	. . .	. . .
High-nickel ductile (20 Ni)	7.4	0.268	18.7	10.4	1.02	13	7.7
High-nickel ductile (23 Ni)	7.4	0.268	18.4	10.2	1.0(d)	. . .	. . .

(a) For compositions, see the preceding table. (b) At 21 °C (70 °F). (c) 10 to 260 °C (50 to 500 °F). (d) Estimated.

Alloy Cast Irons: Abrasion-Resistant

Chemical composition of standard martensitic white cast irons (a)

Class	Type	Designation	TC(c)	Mn	P	S	Composition, wt %(b) Si	Cr	Ni	Mo	Cu
I	A	Ni-Cr-HC	3.0 to 3.6	1.3	0.30	0.15	0.8	1.4 to 4.0	3.3 to 5.0	1.0	· · ·
I	B	Ni-Cr-LC	2.5 to 3.0	1.3	0.30	0.15	0.8	1.4 to 4.0	3.3 to 5.0	1.0	· · ·
I	C	Ni-Cr-GB	2.9 to 3.7	1.3	0.30	0.15	0.8	1.1 to 1.5	2.7 to 4.0	1.0	· · ·
I	D	Ni-Hi Cr	2.5 to 3.6	1.3	0.30	0.10	1.0 to 2.2	7 to 11	5 to 7	1.0	· · ·
II	A	12% Cr	2.4 to 2.8	0.5 to 1.5	0.10	0.06	1.0	11 to 14	0.5	0.5 to 1.0	1.2
II	B	15% Cr-Mo-LC	2.4 to 2.8	0.5 to 1.5	0.10	0.06	1.0	14 to 18	0.5	1.0 to 3.0	1.2
II	C	15% Cr-Mo-HC	2.8 to 3.6	0.5 to 1.5	0.10	0.06	1.0	14 to 18	0.5	2.3 to 3.5	1.2
II	D	20% Cr-Mo-LC	2.0 to 2.6	0.5 to 1.5	0.10	0.06	1.0	18 to 23	1.5	1.5	1.2
II	E	20% Cr-Mo-HC	2.6 to 3.2	0.5 to 1.5	0.10	0.06	1.0	18 to 23	1.5	1.0 to 2.0	1.2
III	A	25% Cr	2.3 to 3.0	0.5 to 1.5	0.10	0.06	1.0	23 to 28	1.5	1.5	1.2

(a) From ASTM A532-75a. Certain specific compositions of alloys II-B, II-C, II-D and II-E are covered by U. S. Patent No. 3,410,682. (b) Where a single value is given rather than a range, that value is a maximum limit. (c) Total carbon.

Mechanical properties of standard martensitic white cast irons (a)

Class	Type	Designation	Hardness, HB Min value Sand cast	Chill cast	Max value Hardened	Annealed	Typical maximum section thickness mm	in.
I	A	Ni-Cr-HC	550	600	· · ·	· · ·	200	8
I	B	Ni-Cr-LC	550	600	· · ·	· · ·	200	8
I	C	Ni-Cr-GB	550	600	· · ·	· · ·	75(b)	3(b)
I	D	Ni-Hi-Cr	550	500	600	· · ·	300	12
II	A	12% Cr	550	· · ·	600	400	25(b)	1(b)
II	B	15% Cr-Mo-LC	450	· · ·	600	400	100	4
II	C	15% Cr-Mo-HC	550	· · ·	600	400	75	3
II	D	20% Cr-Mo-LC	450	· · ·	600	400	200	8
II	E	20% Cr-Mo-HC	450	· · ·	600	400	300	12
III	A	25% Cr	450	· · ·	600	400	200	8

(a) From ASTM A532-75a; for compositions, see the preceding table. (b) Ball diameter

Hardness conversions for white cast irons (from averaged data)

HB	HV	HRC	Scleroscope	HB	HV	HRC	Scleroscope
High-Chromium Irons							
815	1000	68.5	· · ·	540	600	53	· · ·
800	975	68	· · ·	520	575	51.5	· · ·
790	950	67.5	· · ·	490	550	50	· · ·
775	925	67	· · ·	475	525	48.5	· · ·
760	900	66	· · ·	440	500	47	· · ·
745	875	65.0	· · ·	420	475	45.5	· · ·
730	850	64.5	· · ·	395	450	43.5	· · ·
720	825	63.5	· · ·	370	425	41.7	· · ·
700	800	62.5	· · ·	· · ·	400	40	· · ·
680	775	61.5	· · ·				
660	750	61.0	· · ·	**Chromium-Nickel Irons**			
640	725	59.5	· · ·	750	830-860	· · ·	90-93
625	700	58	· · ·	700	740-770	· · ·	84-87
610	675	57	· · ·	650	690-720	· · ·	79-82
585	650	56	· · ·	600	630-660	· · ·	75-78
				550	570-610	· · ·	70-73
560	625	54.5	· · ·	500	510-540	· · ·	67-70

Transverse strengths and relative toughness of various pearlitic and martensitic white irons (a)

Type of iron	Basic composition	Transverse strength kg	Transverse strength lb	Deflection mm	Deflection in.	Toughness(b) kg · m	Toughness(b) lb · in.
Sand-cast pearlitic	3.2-3.5 C, 1-2 Cr	635 to 815	1400 to 1800	2.0 to 2.3	0.080 to 0.092	1.27 to 1.87	112 to 162
Sand-cast martensitic	2.8-3.6 C, 1.4-4 Cr, 3.3-5 Ni	1810 to 2490	4000 to 5500	2.0 to 3.0	0.08 to 0.12	3.62 to 7.47	320 to 660
	2.5-3.6 C, 7-11 Cr, 4.5-7 Ni	2270 to 2720	5000 to 6000	2.0 to 2.8	0.08 to 0.11	4.54 to 7.62	400 to 660
	2.8-3.4 C, 12-16 Cr, 2-4 Mo	1015 to 1370	2235 to 3015	3.2 to 3.6	0.125 to 0.14	3.25 to 4.93	279 to 422
	3.5-4.1 C, 12-16 Cr, 2.5-3 Mo	800 to 1000	1760 to 2200	2.0 to 2.8	0.08 to 0.110	1.60 to 2.80	140 to 240
Chill-cast martensitic	2.8-3.6 C, 1.4-4 Cr, 3.3-5 Ni	2040 to 3180	4500 to 7000	2.0 to 3.0	0.08 to 0.12	4.08 to 9.54	360 to 840
	2.5-3.6 C, 7-11 Cr, 4.5-7 Ni	2500 to 3180	5500 to 7000	2.5 to 3.8	0.10 to 0.15	6.25 to 12.1	550 to 1050
	3.2-3.4 C, 12-16 Cr, 1.5-3 Mo	1980 to 2300	4360 to 5060	5.1 to 6.5	0.202 to 0.26	10.1 to 15.0	870 to 1320
	3.5-4.1 C, 12-16 Cr, 2.5-3 Mo	1270 to 1570	2800 to 3470	3.6 to 3.8	0.140 to 0.15	4.57 to 5.97	392 to 520

(a) Data from as-cast 30.5-mm (1.2-in.) diam test bars broken over a 457-mm (18-in.) span. (b) Relative toughness evaluated as product of transverse strength times deflection.

Alloy Cast Irons: Corrosion- and Heat-Resistant

Typical mechanical properties of corrosion-resistant cast irons

Type of iron(a)	Hardness, HB	Tensile strength MPa	Tensile strength ksi	Compressive strength MPa	Compressive strength ksi	Impact energy J	Impact energy ft · lb	Transverse breaking load(b) kg	Transverse breaking load(b) lb	Transverse deflection(b) mm	Transverse deflection(b) in.
High-silicon iron	480 to 520	90 to 180	13 to 26	690	100	2.7 to 5.4(c)	2 to 4(c)	545 to 1000	1200 to 2200	0.65	0.026
High-chromium iron	250 to 740	205 to 830	30 to 120	690	100	0.1 to 3(d) 27 to 47(c)	0.1 to 2(d) 20 to 35(c)	910 to 1590	2000 to 3500	1.5 to 3.8	0.06 to 0.15
High-nickel gray iron	120 to 250	170 to 310	25 to 45	690 to 1100	100 to 160	80 to 200(c)	60 to 150(c)	820 to 1590	1800 to 3500	5 to 25	0.20 to 1.00
High-nickel ductile iron	130 to 240	380 to 480	55 to 70	1240 to 1380	180 to 200	14 to 40(d)	10 to 30(d)	· · ·	· · ·	· · ·	· · ·

(a) For composition ranges, see the table in the section Alloy Cast Irons: General. (b) For as-cast 30.5-mm (1.2-in.) diam bar broken over a 457-mm (18-in.) span. (c) Unnotched 30.5-mm diam test bar broken over a 152-mm (6-in.) span in a Charpy testing machine. (d) Standard Charpy.

Typical mechanical properties of heat-resistant alloy cast irons

Type of iron(a)	Hardness, HB	Tensile strength MPa	Tensile strength ksi	Compressive strength MPa	Compressive strength ksi	Impact energy J	Impact energy ft · lb	Transverse breaking load(b) kg	Transverse breaking load(b) lb	Transverse deflection(b) mm	Transverse deflection(b) in.
Medium-silicon gray iron	170 to 250	170 to 310	25 to 45	620 to 1040	90 to 150	20 to 31(c)	15 to 23(c)	455 to 1090	1000 to 2400	4.6 to 8.9	0.18 to 0.35
High-chromium gray iron	250 to 500	210 to 620	30 to 90	690	100	27 to 47(c)	20 to 35(c)	910 to 1590	2000 to 3500	1.5 to 3.8	0.06 to 0.15
High-nickel gray iron	130 to 250	170 to 310	25 to 45	690 to 1100	100 to 160	80 to 200(c)	60 to 150(c)	820 to 1360	1800 to 3000	5 to 25	0.2 to 1.0
Ni-Cr-Si gray iron	110 to 210	140 to 310	20 to 45	480 to 690	70 to 100	110 to 200(c)	80 to 150(c)	820 to 1130	1800 to 2500	7 to 35	0.3 to 1.4
High-aluminum gray iron	180 to 350	235 to 620	34 to 90	· · ·	· · ·	· · ·	· · ·	· · ·	· · ·	· · ·	· · ·
Medium-silicon ductile iron	140 to 300	415 to 690	60 to 100(c)	· · ·	· · ·	7 to 155(d)	5 to 115(d)	· · ·	· · ·	· · ·	· · ·
High-nickel ductile iron (20 Ni)	140 to 200	380 to 415	55 to 60(e)	1240 to 1380	180 to 200	16(f)	12(f)	· · ·	· · ·	· · ·	· · ·
High-nickel ductile iron (23 Ni)	130 to 170	400 to 450	58 to 65(g)	· · ·	· · ·	38(f)	28(f)	· · ·	· · ·	· · ·	· · ·

(a) For composition ranges, see the table in the section Alloy Cast Irons: General. (b) Unnotched 30.5-mm (1.2-in.) diam test bar broken on 152-mm (6-in.) supports in a Charpy testing machine. (c) Yield strength, 310 to 520 MPa (45 to 75 ksi); elongation, 0.2%. (d) Standard Charpy test on 10-mm unnotched specimen. (e) Yield strength, 210 to 240 MPa (30 to 35 ksi); elongation, 8 to 20%. (f) Standard Charpy test on 10-mm notched specimen. (g) Yield strength, 195 to 240 MPa (28 to 35 ksi); elongation, 20 to 40%.

Oxidation of plain and alloy cast irons and one stainless steel

Iron	Composition, %(a) TC	Si	Cr	Ni	Oxide penetration At 760 °C (1400 °F)(b) mm/yr	mil/yr	At 815 °C (1500 °F)(c) mm/yr	mil/yr	Growth at 815 °C (1500 °F)(c) mm/yr	mil/yr
Austenitic(d)	2.69	1.96	2.05	13.96	4.7	184	9.5	374	0.4	15
Austenitic(e)	2.97	2.38	4.87	(14.0)	2.4	96	5.9	232	0.2	8
Austenitic	2.40	1.57	2.98	30.28	2.1	83	6.3	249	0.2	8
Austenitic	(1.8)	(6.0)	(5.0)	(30.0)	0.05	2	1.3	53	0.2	8
Austenitic	(2.8)	(1.7)	(2.0)	(20.0)	4.2	166	7.9	312	0.2	8
Austenitic	(2.7)	(2.5)	(5.0)	(20.0)	1.9	74	3.6	143	0.4	15
Plain ferritic	(3.2)	(2.2)	· · ·	· · ·	>20(f)	>800(f)	>85(f)	>3300(f)	2.0	78
Low-alloy ferritic .	(3.3)	(1.5)	(0.6)	(1.5)	>20(f)	>800(f)	>90(f)	>3500(f)	1.4	54
Low-alloy ferritic .	(3.3)	(2.2)	(1.0)	(1.0)	5.8	228	25.9	1020	1.2	47
Low-alloy ferritic .	(3.1)	(2.2)	(0.9)	(1.5)	7.2	284	29.0	1140	1.6	62
Type 309 stainless	· · ·	· · ·	(25.0)	(12.0)	nil	nil	nil	nil	nil	nil

(a) Figures enclosed in parentheses are estimated values. Phosphorus and sulfur contents in all iron samples were about 0.10%. (b) Exposure of 2000 h in electric furnace at 1400 °F with air atmosphere containing 17 to 19% O. (c) Exposed for 492 h in gas-fired heat treating furnace at 1500 °F. (d) 6.05% copper. (e) 6.0% copper. (f) Specimen completely burned.

Oxidation of ferritic and austenitic cast irons and one stainless steel

Iron	Composition, %(a) TC	Si	Cr	Ni	Growth mm/yr	mil/yr	Oxide penetration mm/yr	mil/yr
After 3723 h at 1375 to 1400 °F in Electric Furnace, Air Atmosphere								
Ferritic	3.05	2.67	0.90	1.55	2.0	78	(b)	(b)
Austenitic	2.97	1.63	1.89	20.02	0.8	31	6.9	270
Austenitic	2.52	2.67	5.16	20.03	nil	nil	0.2	6
Austenitic	2.32	1.86	2.86	30.93	nil	nil	2.0	78
Austenitic	1.86	5.84	5.00	29.63	nil	nil	<0.1	<3
309 stainless . . . · · ·		· · ·	(25.0)	(12.0)	nil	nil	<0.1	<3
After 1677 h at 1500 to 1700 °F in Gas-Fired Furnace, Slightly Reducing Atmosphere								
Ferritic	(3.2)	(2.2)	· · ·	· · ·	3.2	125	(b)	(b)
Austenitic(c) . . .	(3.0)	(2.4)	(5.0)	(14.0)	0.4	15	8.4	330
Austenitic	(2.7)	(2.5)	(5.0)	(20.0)	0.4	15	5.6	220
Austenitic	(2.4)	(1.6)	(3.0)	(30.0)	0.4	15	6.9	270
Austenitic	(1.8)	(6.0)	(5.0)	(30.0)	0.4	15	0.1	5
309 stainless . . . · · ·		· · ·	(25.0)	(12.0)	nil	nil	0.1	5

(a) Figures enclosed in parentheses are estimated values. Phosphorus and sulfur contents in all iron samples were about 0.10%. (b) Sample was completely burned. (c) 6.0% copper (est).

Machining

Machinability of gray iron

Microstructure	ASTM class	Tensile strength MPa	ksi	Hardness, HB	Cutting speed(a) m/min	ft/min
Acicular iron	50	407	59	263	46	150
Fine pearlite, alloy	40	310	45	225	95	310
Ferrite (annealed)	...	108	15.7	100	293	960
Coarse pearlite, no alloy ..	35	241	35	195	99	325

(a) Cutting speed at which removal of 200 in.³ (3280 cm³) produced 0.030-in. (0.75-mm) wear land on single-point carbide tools.

Speeds for machining ductile iron

Ductile iron	High-speed steel tools, ft/min(a) Turning (b)	Drilling (c)	Reaming (d)	Tapping	Thread chasing	Milling (e)	Shaping (f) and planing	Broaching	Cemented carbide tools, ft/min(a) Turning (b)	Reaming (d)	Milling (e)
60-45-10 with full ferritic matrix	50-150	80-130	50-100	20-30	30-70	50-125	40-100	20-35	175-400	75-150	200-400
Semipearlitic matrix	40-90	50-100	40-70	15-20	20-50	35-65	30-75	15-25	100-300	50-90	175-350
80-60-03 with full pearlitic matrix ...	40-90	50-70	40-70	15-20	20-50	35-65	30-75	15-25	100-300	50-90	175-350

(a) Multiply listed values by 0.3 to find recommended speeds in m/min. In all instances, the longest tool life between regrinds will result when tools are operated at minimum to medium speeds for the range given. (b) Feeds of 0.25 to 0.50 mm/rev (0.010 to 0.020 ipr). Maximum speed is for cuts not more than 1.5 mm (¹/₁₆ in.) deep. (c) Feeds should be commensurate with drill diameter—light feeds for small-diameter drills and heavier feeds for larger drills. It is good practice to reduce speed as drill diameter increases; speeds at or near the maximum values given in this table are for drills 12 mm, or ½ in., in diameter or less. (d) Use feeds three to four times those used for drills of similar size. An allowance of 0.3 to 0.4 mm (0.012 to 0.015 in.) is sufficient for reaming. (e) Speeds cited are principally for face milling; however, they may be used for plain milling. (f) Depth of cut and feed vary with sturdiness of the setup. The operating speed for roughing should approach the minimum value cited.

Recommended cutting conditions for turning malleable iron(a)

Grade	Hardness, HB	Type of cut (b)	Feed, in./rev(c)	Depth of cut, in.(c)	Dry, for tool life of: 20 min	30 min	40 min	In soluble oil for tool life of: 20 min	30 min	40 min
32510	109	Rough skin	0.015	0.100	440	390	340	500	410	340
			0.030	0.100	350	300	250	410	360	600
		Coarse underskin	0.015	0.060	600	510	430	840	720	630
			0.030	0.060	440	380	350	600	530	460
		Finish	0.003	0.010	1380	1240	1150	1660	1520	1400
			0.007	0.010	1210	950	700	1380	1220	1080
48004	179	Rough skin	0.015	0.100	230	180	130	270	220	130
			0.030	0.100	160	120	70	200	150	110
		Coarse underskin	0.015	0.060	300	260	230	355	330	280
			0.030	0.060	230	185	150	260	230	210
		Finish	0.003	0.010	590	515	460	700	650	615
			0.007	0.010	510	470	450	670	610	550
60003	230	Rough skin	0.015	0.100	175	140	115	200	165	140
			0.030	0.100	130	115	100	150	120	100
		Coarse underskin	0.015	0.060	245	220	200	285	240	225
			0.030	0.060	170	145	125	210	165	135
		Finish	0.003	0.010	525	495	465	540	510	480
			0.007	0.010	470	430	400	480	440	415
80002	250	Rough skin	0.015	0.100	...	...	...	195	165	135
			0.030	0.100	...	...	...	145	115	90
		Coarse underskin	0.015	0.060	185	160	145	190	170	155
			0.030	0.060	160	125	100	160	125	100
		Finish	0.003	0.010	...	...	...	470	420	385
			0.007	0.010	...	...	...	365	330	305

(a) All tests were made under the direction of the Machining Subcommittee of the Malleable Founders Society (now the Iron Castings Society). Grade C2 cutting tools were used dry or with soluble-oil cutting fluid. Tests were discontinued at a uniform wear land of 0.015 in. Discontinued grades 48004 and 60003 are approximately equivalent to grades 50005 and 60004, respectively. (b) Tool configuration for rough skin and coarse underskin cuts: −5° BR; 15° SCEA; −5° SR; 15° ECEA; 5° relief. Tool configuration for finish cuts: 0° BR; 15° SCEA; 5° SR; 15° ECEA; 5° relief. (c) Multiply tabulated values by 25 to obtain an equivalent value in mm. (d) Multiply tabulated values by 0.3 to obtain an equivalent value in m/min.

Carbon and Alloy Steels

Quality Descriptors

The need for communication among producers and between producers and users has resulted in the development of a group of terms known as "fundamental quality descriptors". These are names applied to various steel products to imply that the particular products possess certain characteristics that make them especially well suited for specific applications or fabrication processes. Some of the fundamental quality descriptors in common use are listed below.

Carbon Steels

Semifinished for forging
Forging quality
Special hardenability
Special internal soundness
Nonmetallic inclusion
requirement
Special surface
Carbon steel structural sections
Structural quality
Carbon steel plates
Regular quality
Structural quality
Cold drawing quality
Cold pressing quality
Cold flanging quality
Forging quality
Pressure vessel quality
Marine quality

Hot rolled carbon steel bars
Merchant quality
Special quality
Special hardenability
Special internal soundness
Nonmetallic inclusion
requirement
Special surface
Scrapless nut quality
Axle shaft quality
Cold extrusion quality
Cold heading and cold
forging quality

Cold finished carbon steel bars
Standard quality
Special hardenability
Special internal soundness
Nonmetallic inclusion
requirement
Special surface
Cold heading and cold
forging quality
Cold extrusion quality

Hot rolled sheets
Commercial quality

Drawing quality
Drawing quality special killed
Structural quality
Cold rolled sheets
Commercial quality
Drawing quality
Drawing quality special killed
Structural quality
Porcelain enameling sheets
Commercial quality
Drawing quality

Long terne sheets
Commercial quality
Drawing quality
Drawing quality special killed
Structural quality

Galvanized sheets
Commercial quality
Drawing quality
Drawing quality special killed
Structural quality
Lock forming quality

Electrolytic zinc-coated sheets
Commercial quality
Drawing quality
Drawing quality special killed
Structural quality

Hot rolled strip
Commercial quality
Drawing quality
Drawing quality special killed
Structural quality

Cold rolled strip
Specific quality descriptors are not
provided in cold rolled strip,
since this product is largely produced for specific end use

Tin mill products
Specific quality descriptors are not
applicable to tin mill products

Carbon steel wire
Industrial quality wire
Cold extrusion wires
Heading, forging and roll thread-
ing wires
Mechanical spring wires
Upholstery spring construction
wires
Welding wire
Carbon steel flat wire
Stitching wire
Stapling wire
Carbon steel pipe
Structural tubing
Line pipe
Oil country tubular goods
Steel specialty tubular products
Pressure tubing

Mechanical tubing
Aircraft tubing
Hot rolled carbon steel wire rods
Industrial quality
Rods for manufacture of wire for
electric welded chain
Rods for heading, forging, and
roll threading wire
Rods for lock washer wire
Rods for scrapless nut wire
Rods for upholstery spring wire
Rods for welding wire

Alloy Steels

Alloy steel plates
Regular quality or structural
quality
Drawing quality
Pressure vessel quality
Structural quality
Aircraft quality
Aircraft physical quality
Hot rolled alloy steel bars
Regular quality
Aircraft structural or steel subject
to magnetic particle inspection
Axle shaft quality
Bearing quality
Cold heading quality
Special cold heading quality
Rifle barrel quality, gun quality,
shell or AP shot quality
Alloy steel wire
Aircraft quality
Bearing quality
Special surface quality
Cold finished alloy steel bars
Regular quality
Aircraft quality or steel subject
to magnetic particle inspection
Axle shaft quality
Bearing shaft quality
Cold heading quality
Special cold heading quality
Rifle barrel quality, gun quality,
shell or AP shot quality
Line pipe
Oil country tubular goods
Steel specialty tubular goods
Pressure tubing
Mechanical tubing
Stainless and heat resisting pipe,
pressure tubing and mechanical
tubing
Aircraft tubing
Pipe

Note: Detailed descriptions of many of the categories listed in this table appear in an appropriate section of the AISI Steel Products Manual.

Composition Ranges and Tolerances

Alloy steel heat composition ranges and limits—bars, blooms, billets and slabs

Element	Limit or max of specified range, %	Range, % Open hearth or basic oxygen steel	Range, % Electric furnace steel
Carbon	To 0.55 incl	0.05	0.05
	Over 0.55 to 0.70 incl	0.08	0.07
	Over 0.70 to 0.80 incl	0.10	0.09
	Over 0.80 to 0.95 incl	0.12	0.11
	Over 0.95 to 1.35 incl	0.13	0.12
Manganese	To 0.60 incl	0.20	0.15
	Over 0.60 to 0.90 incl	0.20	0.20
	Over 0.90 to 1.05 incl	0.25	0.25
	Over 1.05 to 1.90 incl	0.30	0.30
	Over 1.90 to 2.10 incl	0.40	0.35
Sulfur(a)	To 0.050 incl	0.015	0.015
	Over 0.050 to 0.07 incl	0.02	0.02
	Over 0.07 to 0.10 incl	0.04	0.04
	Over 0.10 to 0.14 incl	0.05	0.05
Silicon	To 0.15 incl	0.08	0.08
	Over 0.15 to 0.20 incl	0.10	0.10
	Over 0.20 to 0.40 incl	0.15	0.15
	Over 0.40 to 0.60 incl	0.20	0.20
	Over 0.60 to 1.00 incl	0.30	0.30
	Over 1.00 to 2.20 incl	0.40	0.35
Chromium	To 0.40 incl	0.15	0.15
	Over 0.40 to 0.90 incl	0.20	0.20
	Over 0.90 to 1.05 incl	0.25	0.25
	Over 1.05 to 1.60 incl	0.30	0.30
	Over 1.60 to 1.75 incl	(b)	0.35
	Over 1.75 to 2.10 incl	(b)	0.40
	Over 2.10 to 3.99 incl	(b)	0.50
Nickel	To 0.50 incl	0.20	0.20
	Over 0.50 to 1.50 incl	0.30	0.30
	Over 1.50 to 2.00 incl	0.35	0.35
	Over 2.00 to 3.00 incl	0.40	0.40
	Over 3.00 to 5.30 incl	0.50	0.50
	Over 5.30 to 10.00 incl	1.00	1.00

Element	Limit or max of specified range, %	Range, % Open hearth or basic oxygen steel	Range, % Electric furnace steel
Molybdenum	To 0.10 incl	0.05	0.05
	Over 0.10 to 0.20 incl	0.07	0.07
	Over 0.20 to 0.50 incl	0.10	0.10
	Over 0.50 to 0.80 incl	0.15	0.15
	Over 0.80 to 1.15 incl	0.20	0.20
Tungsten	To 0.50 incl	0.20	0.20
	Over 0.50 to 1.00 incl	0.30	0.30
	Over 1.00 to 2.00 incl	0.50	0.50
	Over 2.00 to 4.00 incl	0.60	0.60
Copper	To 0.60 incl	0.20	0.20
	Over 0.60 to 1.50 incl	0.30	0.30
	Over 1.50 to 2.00 incl	0.35	0.35
Vanadium	To 0.25 incl	0.05	0.05
	Over 0.25 to 0.50 incl	0.10	0.10
Aluminum	Up to 0.10 incl	0.05	0.05
	Over 0.10 to 0.20 incl	0.10	0.10
	Over 0.20 to 0.30 incl	0.15	0.15
	Over 0.30 to 0.80 incl	0.25	0.25
	Over 0.80 to 1.30 incl	0.35	0.35
	Over 1.30 to 1.80 incl	0.45	0.45

Steelmaking Process		Lowest max, % (c)
Phosphorus	Basic open hearth, basic oxygen or basic electric furnace steels	0.035 (d)
	Basic electric furnace "E" steels	0.025
	Acid open hearth or electric furnace steel	0.050
Sulfur	Basic open hearth, basic oxygen or basic electric furnace steels	0.040 (d)
	Basic electric furnace "E" steels	0.025
	Acid open hearth or electric furnace steel	0.050

(a) A range of sulfur content normally indicates a resulfurized steel. (b) Not normally produced by open hearth process. (c) Not applicable to rephosphorized or resulfurized steels. (d) Lower maximum limits on phosphorus and sulfur are required by certain quality descriptors.

Carbon steel heat composition ranges and limits—semifinished products for forging, hot rolled and cold finished bars, wire rod and seamless tubing

Element	Limit or max of specified range, %	Range, %	Element	Limit or max of specified range, %	Range, %
Carbon			Sulfur		
(a)	To 0.25 incl	0.05	(b)	0.050 (c) to 0.09 incl	0.03
	Over 0.25 to 0.40 incl	0.06		Over 0.09 to 0.15	0.05
	Over 0.40 to 0.55 incl	0.07		Over 0.15 to 0.23 incl	0.07
	Over 0.55 to 0.80 incl	0.10		Over 0.23 to 0.35 incl	0.09
	Over 0.80	0.13	Silicon		
Manganese	To 0.40 incl	0.15	(d)	To 0.15 incl	0.08
	Over 0.40 to 0.50 incl	0.20		Over 0.15 to 0.20 incl	0.10
	Over 0.50 to 1.65 incl	0.30		Over 0.20 to 0.30 incl	0.15
Phosphorus				Over 0.30 to 0.60 incl	0.20
(b)	0.040 (c) to 0.08 incl	0.03			
	Over 0.08 to 0.13 incl	0.05			

(a) Add 0.01 to specified carbon ranges for steels with manganese contents exceeding 1.10%. (b) Lower maximum limits on phosphorus and sulfur are required by certain quality descriptors. (c) Lowest permissible maximum for this element. (d) Silicon content not normally specified for acid bessemer steels.

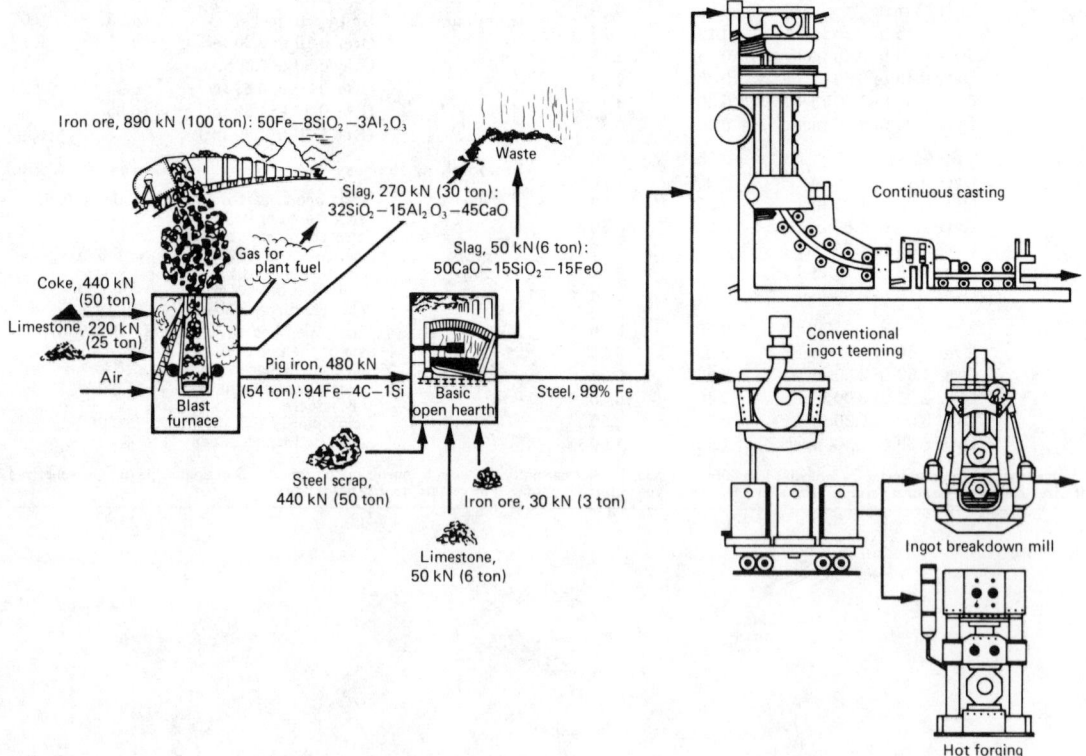

Carbon steel heat composition ranges and limits—structural shapes, plate, strip, sheet and welded tubing

Element	Limit or max of specified range, %	Range, %	Element	Limit or max of specified range, %	Range, %
Carbon			Sulfur		
(a)	0.08 (b)(c) to 0.15 incl	0.05	(d)	0.05 to 0.08 incl	0.03
	Over 0.15 to 0.30 incl	0.06		Over 0.08 to 0.15 incl	0.05
	Over 0.30 to 0.40 incl	0.07		Over 0.15 to 0.23 incl	0.07
	Over 0.40 to 0.60 incl	0.08		Over 0.23 to 0.33 incl	0.10
	Over 0.60 to 0.80 incl	0.11			
	Over 0.80 to 1.35 incl	0.14	Silicon . . .	0.10 to 0.15 incl	0.08
				Over 0.15 to 0.30 incl	0.15
Manganese	0.40 (b) to 0.50 incl	0.20		Over 0.30 to 0.60 incl	0.30
	Over 0.50 to 1.15 incl	0.30			
	Over 1.15 to 1.65 incl	0.35			
Phosphorus					
(d)	0.04 (b) to 0.08 incl	0.03			
	Over 0.08 to 0.15 incl	0.05			

(a) Add 0.01 to specified carbon range for steels with manganese contents exceeding 1.00%. (b) Lowest permissible maximum limit for this element. (c) 0.12% for structural shapes and plate. (d) Lower maximum limits on phosphorus and sulfur are required by certain quality descriptors.

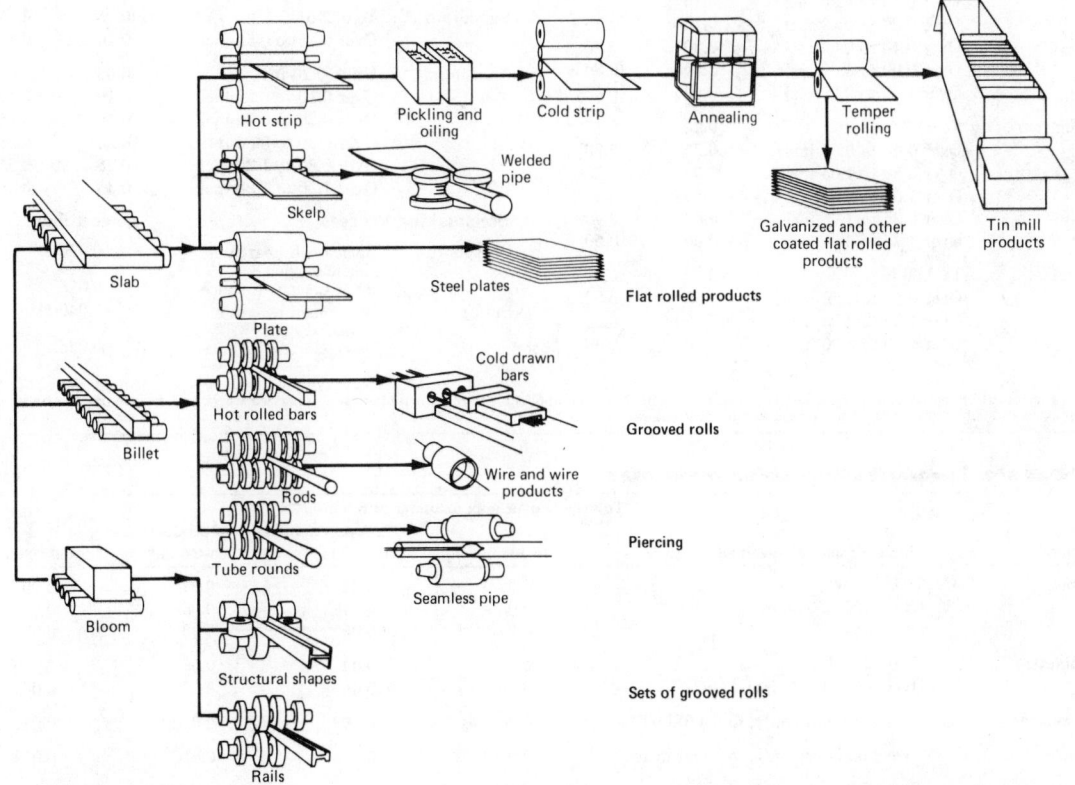

Steel processing flow lines

Flat rolled products are usually rolled from slabs by mills using sets of cylindrical rolls. Grooved rolls squeeze billets into different cross-sections (round, angles, etc.) in a sequence of operations. Piercing is the process used to make seamless pipe and tubing from a semifinished product called tube rounds. Sets of grooved rolls are used to roll blooms into heavy beams for construction or for rails. A small but significant percentage of heated ingot steel is squeezed in forging presses to make large shafts for power plants, nuclear plant components, and other products.
Source: Steel Processing Flow Lines, American Iron and Steel Institute

Alloy steel heat composition ranges and limits—plate

Element	Limit or max of specified range, %	Range, % Open hearth or basic oxygen steels	Electric furnace steels	Element	Limit or max of specified range, %	Range, % Open hearth or basic oxygen steels	Electric furnace steels
Carbon	To 0.25 incl	0.06	0.05	Nickel	Over 3.00 to 5.30 incl	0.50	0.50
	Over 0.25 to 0.40 incl	0.07	0.06		Over 5.30 to 10.00 incl	1.00	1.00
	Over 0.40 to 0.55 incl	0.08	0.07	Molybdenum . .	To 0.10 incl	0.05	0.05
	Over 0.55 to 0.70 incl	0.11	0.10		Over 0.10 to 0.20 incl	0.07	0.07
	Over 0.70	0.14	0.13		Over 0.20 to 0.50 incl	0.10	0.10
Manganese	To 0.45 incl	0.20	0.15		Over 0.50 to 0.80 incl	0.15	0.15
	Over 0.45 to 0.80 incl	0.25	0.20		Over 0.80 to 1.15 incl	0.20	0.20
	Over 0.80 to 1.15 incl	0.30	0.25	Tungsten	To 0.50 incl	0.20	0.20
	Over 1.15 to 1.70 incl	0.35	0.30		Over 0.50 to 1.00 incl	0.30	0.30
	Over 1.70 to 2.10 incl	0.40	0.35		Over 1.00 to 2.00 incl	0.50	0.50
Sulfur(a)	To 0.060 incl	0.02	0.02		Over 2.00 to 4.00 incl	0.60	0.60
	Over 0.060 to 0.100 incl	0.04	0.04	Copper	To 0.60 incl	0.20	0.20
	Over 0.100 to 0.140 incl	0.05	0.05		Over 0.60 to 1.50 incl	0.30	0.30
Silicon	To 0.15 incl	0.08	0.08		Over 1.50 to 2.00 incl	0.35	0.35
	Over 0.15 to 0.20 incl	0.10	0.10	Vanadium	To 0.25 incl	0.05	0.05
	Over 0.20 to 0.40 incl	0.15	0.15		Over 0.25 to 0.50 incl	0.10	0.10
	Over 0.40 to 0.60 incl	0.20	0.20	Aluminum . . .	Up to 0.10 incl	0.05	0.05
	Over 0.60 to 1.00 incl	0.30	0.30		Over 0.10 to 0.20 incl	0.10	0.10
	Over 1.00 to 2.20 incl	0.40	0.35		Over 0.20 to 0.30 incl	0.15	0.15
Chromium	To 0.40 incl	0.20	0.15		Over 0.30 to 0.80 incl	0.25	0.25
	Over 0.40 to 0.80 incl	0.25	0.20		Over 0.80 to 1.30 incl	0.35	0.35
	Over 0.80 to 1.05 incl	0.30	0.25		Over 1.30 to 1.80 incl	0.45	0.45
	Over 1.05 to 1.25 incl	0.35	0.30	**Steelmaking Process**		**Lowest max, %(b)**	
	Over 1.25 to 1.75 incl	0.50	0.40	Phosphorus . . .	Basic open hearth or basic oxygen	0.035(c)	
	Over 1.75 to 3.99 incl	0.60	0.50		Basic electric furnace	0.025	
Nickel	To 0.50 incl	0.20	0.20	Sulfur	Basic open hearth or basic oxygen	0.040(c)	
	Over 0.50 to 1.50 incl	0.30	0.30		Basic electric furnace	0.025	
	Over 1.50 to 2.00 incl	0.35	0.35				
	Over 2.00 to 3.00 incl	0.40	0.40				

(a) A range of sulfur content normally indicates a resulfurized steel. (b) Not applicable to resulfurized or rephosphorized steels. (c) Lower maximum limits on phosphorus and sulfur are required by certain quality descriptors.

Carbon steel product composition tolerances

Element	Limit or max of specified range, %	Tolerance over max or under min limits, % Cross-sectional area of product: to 100 in.²	100-200 in.²	200-400 in.²	400-800 in.²
Carbon	To 0.25 incl	0.02	0.03	0.04	0.05
	Over 0.25 to 0.55 incl	0.03	0.04	0.05	0.06
	Over 0.55	0.04	0.05	0.06	0.07
Manganese	To 0.90 incl	0.03	0.04	0.06	0.07
	Over 0.90 to 1.65 incl	0.06	0.06	0.07	0.08
Phosphorus	Over maximum only, to 0.040 incl	0.008	0.008	0.010	0.015
Sulfur	Over maximum only, to 0.050 incl	0.008	0.010	0.010	0.015
Silicon	To 0.35 incl	0.02	0.02	0.03	0.04
	Over 0.35 to 0.60 incl	0.05	. . .	. . .	. . .
Copper	Under minimum only	0.02	0.03	. . .	. . .
Lead	0.15 to 0.35 incl	0.03	0.03	. . .	. . .

Note: Product composition requirements are not applicable to: (a) rimmed or capped steels; (b) boron content of boron steels; (c) phosphorus and sulfur contents of rephosphorized and resulfurized steels. Product composition tolerances for alloying elements in HSLA steels are given in Table 6.

Alloy steel product composition tolerances—bars, billets, blooms and slabs

Element	Limit or max of specified range, %	Tolerance over max or under min limits, %(a) Cross-sectional area of product:			
		to 100 in.²	100-200 in.²	200-400 in.²	400-800 in.²
Carbon	To 0.30 incl	0.01	0.02	0.03	0.04
	Over 0.30 to 0.75 incl	0.02	0.03	0.04	0.05
	Over 0.75	0.03	0.04	0.05	0.06
Manganese ...	To 0.90 incl	0.03	0.04	0.05	0.06
	Over 0.90 to 2.10 incl	0.04	0.05	0.06	0.07
Phosphorus ...	Over max only	0.005	0.010	0.010	0.010
Sulfur	Over max only	0.005	0.010	0.010	0.010
Silicon	To 0.40 incl	0.02	0.02	0.03	0.04
	Over 0.40 to 2.20 incl	0.05	0.06	0.06	0.07
Chromium	To 0.90 incl	0.03	0.04	0.04	0.05
	Over 0.90 to 2.10 incl	0.05	0.06	0.06	0.07
	Over 2.10 to 3.99 incl	0.10	0.10	0.12	0.14
Nickel	To 1.00 incl	0.03	0.03	0.03	0.03
	Over 1.00 to 2.00 incl	0.05	0.05	0.05	0.05
	Over 2.00 to 5.30 incl	0.07	0.07	0.07	0.07
	Over 5.30 to 10.00 incl	0.10	0.10	0.10	0.10
Molybdenum ..	To 0.20 incl	0.01	0.01	0.02	0.03
	Over 0.20 to 0.40 incl	0.02	0.03	0.03	0.04
	Over 0.40 to 1.15 incl	0.03	0.04	0.05	0.06
Tungsten	To 1.00 incl	0.04	0.05	0.05	0.06
	Over 1.00 to 4.00 incl	0.08	0.09	0.10	0.12
Copper(b)	To 1.00 incl	0.03	...	...	...
	Over 1.00 to 2.00 incl	0.05	...	...	...
Vanadium	To 0.10 incl	0.01	0.01	0.01	0.01
	Over 0.10 to 0.25 incl	0.02	0.02	0.02	0.02
	Over 0.25 to 0.50 incl	0.03	0.03	0.03	0.03
	Min value specified, check under min limit(b)	0.01	0.01	0.01	0.01
Niobium(b) ...	To 0.10 incl	0.01(c)	...	...	...
Titanium(b) ...	To 0.10 incl	0.01(c)	...	...	...
Zirconium(b) ..	To 0.15 incl	0.03	...	...	...
Aluminum(b)	Up to 0.10 incl	0.03	...	...	...
	Over 0.10 to 0.20 incl	0.04	...	...	...
	Over 0.20 to 0.30 incl	0.05	...	...	...
	Over 0.30 to 0.80 incl	0.07	...	...	...
	Over 0.80 to 1.80 incl	0.10	...	...	...
Lead(b)	0.15 to 0.35 incl	0.03	...	...	...
Nitrogen(b) ...	To 0.030 incl	0.005	...	...	...

(a) Product composition requirements are not applicable to boron content of boron steels or sulfur content of resulfurized steels. (b) Tolerances shown apply only to cross-sectional areas of 100 in.² or less. (c) If the minimum of the range is 0.01%, the lower tolerance is 0.005%.

Alloy steel product composition tolerances—plate

Element	Limit or max of specified range, %	Tolerance over max or under min limits, %
Carbon ...	To 0.30 incl	0.02
	Over 0.30 to 0.75 incl	0.03
	Over 0.75	0.04
Manganese	To 0.90 incl	0.04
	Over 0.90 to 2.10 incl	0.05
Phosphorus (a)	Over max only	0.01
Sulfur (a) (b) ...	...	0.01
Silicon ...	To 0.40 incl	0.02
	Over 0.40 to 2.20 incl	0.06
Chromium	To 0.90 incl	0.04
	Over 0.90 to 2.10 incl	0.06
	Over 2.10 to 3.99 incl	0.10
Nickel ...	To 1.00 incl	0.03
	Over 1.00 to 2.00 incl	0.05
	Over 2.00 to 5.30 incl	0.07
	Over 5.30	0.10
Molybde-num	To 0.20 incl	0.01
	Over 0.20 to 0.40 incl	0.03
	Over 0.40 to 1.15 incl	0.04
Tungsten	To 1.00 incl	0.05
	Over 1.00 to 4.00 incl	0.09
Copper ...	To 1.00 incl	0.03
	Over 1.00 to 2.00 incl	0.05
Vanadium	To 0.10 incl	0.01
	Over 0.10 to 0.25 incl	0.02
	Over 0.25 to 0.50 incl	0.03
	Min value specified, check under min limit	0.01
Aluminum	Up to 0.10 incl	0.03
	Over 0.10 to 0.20 incl	0.04
	Over 0.20 to 0.30 incl	0.05
	Over 0.30 to 0.80 incl	0.07
	Over 0.80 to 1.80 incl	0.10

(a) For pressure-vessel quality plate, the specified composition includes product composition tolerances for phosphorus and sulfur. (b) Product composition requirements not applicable to sulfur content of resulfurized steel.

AISI-SAE Designation System

AISI-SAE system of designations

Numerals and digits	Type of steel and nominal alloy content
Carbon Steels	
10XX(a)	Plain carbon (Mn 1.00% max)
11XX	Resulfurized
12XX	Resulfurized and rephosphorized
15XX	Plain carbon (max Mn range— 1.00 to 1.65%)
Manganese Steels	
13XX	Mn 1.75
Nickel Steels	
23XX	Ni 3.50
25XX	Ni 5.00
Nickel-Chromium Steels	
31XX	Ni 1.25; Cr 0.65 and 0.80
32XX	Ni 1.75; Cr 1.07
33XX	Ni 3.50; Cr 1.50 and 1.57
34XX	Ni 3.00; Cr 0.77
Molybdenum Steels	
40XX	Mo 0.20 and 0.25
44XX	Mo 0.40 and 0.52
Chromium-Molybdenum Steels	
41XX	Cr 0.50, 0.80 and 0.95; Mo 0.12, 0.20, 0.25 and 0.30

Numerals and digits	Type of steel and nominal alloy content
Nickel-Chromium-Molybdenum Steels	
43XX	Ni 1.82; Cr 0.50 and 0.80; Mo 0.25
43BVXX	Ni 1.82; Cr 0.50; Mo 0.12 and 0.25; V 0.03 min
47XX	Ni 1.05; Cr 0.45; Mo 0.20 and 0.35
81XX	Ni 0.30; Cr 0.40; Mo 0.12
86XX	Ni 0.55; Cr 0.50; Mo 0.20
87XX	Ni 0.55; Cr 0.50; Mo 0.25
88XX	Ni 0.55; Cr 0.50; Mo 0.35
93XX	Ni 3.25; Cr 1.20; Mo 0.12
94XX	Ni 0.45; Cr 0.40; Mo 0.12
97XX	Ni 0.55; Cr 0.20; Mo 0.20
98XX	Ni 1.00; Cr 0.80; Mo 0.25
Nickel-Molybdenum Steels	
46XX	Ni 0.85 and 1.82; Mo 0.20 and 0.25
48XX	Ni 3.50; Mo 0.25
Chromium Steels	
50XX	Cr 0.27, 0.40, 0.50 and 0.65
51XX	Cr 0.80, 0.87, 0.92, 0.95, 1.00 and 1.05

Numerals and digits	Type of steel and nominal alloy content
Chromium Steels	
50XXX	Cr 0.50 ⎫
51XXX	Cr 1.02 ⎬ C 1.00 min
52XXX	Cr 1.45 ⎭
Chromium-Vanadium Steels	
61XX	Cr 0.60, 0.80 and 0.95; V 0.10 and 0.15 min
Tungsten-Chromium Steel	
72XX	W 1.75; Cr 0.75
Silicon-Manganese Steels	
92XX	Si 1.40 and 2.00; Mn 0.65, 0.82 and 0.85; Cr 0.00 and 0.65
High-Strength Low-Alloy Steels	
9XX	Various SAE grades
Boron Steels	
XXBXX	B denotes boron steel
Leaded Steels	
XXLXX	L denotes leaded steel

(a) XX in the last two digits of these designations indicates that the carbon content (in hundredths of a percent) is to be inserted.

Composition ranges and limits for AISI-SAE standard carbon steels—structural shapes, plate, strip, sheet and welded tubing

AISI-SAE designation	UNS designation	Heat composition ranges and limits, %(a) C	Mn	AISI-SAE designation	UNS designation	Heat composition ranges and limits, %(a) C	Mn	AISI-SAE designation	UNS designation	Heat composition ranges and limits, %(a) C	Mn
1006	G10060	0.08 max	0.25-0.45	1038	G10380	0.34-0.42	0.60-0.90	1090	G10900	0.84-0.98	0.60-0.90
1008	G10080	0.10 max	0.25-0.50	1039	G10390	0.36-0.44	0.70-1.00	1095	G10950	0.90-1.04	0.30-0.50
1009	G10090	0.15 max	0.60 max	1040	G10400	0.36-0.44	0.60-0.90	1524(b)	G15240	0.18-0.25	1.30-1.65
1010	G10100	0.08-0.13	0.30-0.60	1042	G10420	0.39-0.47	0.60-0.90	1527(b)	G15270	0.22-0.29	1.20-1.55
1012	G10120	0.10-0.15	0.30-0.60	1043	G10430	0.39-0.47	0.70-1.00	1536(b)	G15360	0.30-0.38	1.20-1.55
1015	G10150	0.12-0.18	0.30-0.60	1045	G10450	0.42-0.50	0.60-0.90	1541(b)	G15410	0.36-0.45	1.30-1.65
1016	G10160	0.12-0.18	0.60-0.90	1046	G10460	0.42-0.50	0.70-1.00	1548(b)	G15480	0.43-0.52	1.05-1.40
1017	G10170	0.14-0.20	0.30-0.60	1049	G10490	0.45-0.53	0.60-0.90	1552(b)	G15520	0.46-0.55	1.20-1.55
1018	G10180	0.14-0.20	0.60-0.90	1050	G10500	0.47-0.55	0.60-0.90				
1019	G10190	0.14-0.20	0.70-1.00	1055	G10550	0.52-0.60	0.60-0.90				
1020	G10200	0.17-0.23	0.30-0.60	1060	G10600	0.55-0.66	0.60-0.90				
1021	G10210	0.17-0.23	0.60-0.90	1064	G10640	0.59-0.70	0.50-0.80				
1022	G10220	0.17-0.23	0.70-1.00	1065	G10650	0.59-0.70	0.60-0.90				
1023	G10230	0.19-0.25	0.30-0.60	1070	G10700	0.65-0.76	0.60-0.90				
1025	G10250	0.22-0.28	0.30-0.60	1074	G10740	0.69-0.80	0.50-0.80				
1026	G10260	0.22-0.28	0.60-0.90	1078	G10780	0.72-0.86	0.30-0.60				
1030	G10300	0.27-0.34	0.60-0.90	1080	G10800	0.74-0.88	0.60-0.90				
1033	G10330	0.29-0.36	0.70-1.00	1084	G10840	0.80-0.94	0.60-0.90				
1035	G10350	0.31-0.38	0.60-0.90	1085	G10850	0.80-0.94	0.70-1.00				
1037	G10370	0.31-0.38	0.70-1.00	1086	G10860	0.80-0.94	0.30-0.50				

(a) Limits on phosphorus and sulfur contents are given in Table 2; typical limits are 0.040% maximum phosphorus and 0.050% maximum sulfur. When silicon ranges or limits are required, the values in Table 2 apply. Steels listed in this table can be produced with additions of lead or boron. Leaded steels typically contain 0.15 to 0.35% lead and are identified by inserting the letter "L" in the designation—11L17; boron steels can be expected to contain 0.0005 to 0.003% boron and are identified by inserting the letter "B" in the designation—15B41. (b) Formerly designated 10XX grade.

Composition ranges and limits for AISI-SAE standard carbon steels containing less than 1.00% manganese—semifinished products for forging, hot rolled and cold finished bars, wire rod and seamless tubing

AISI-SAE designation	UNS designation	Heat composition ranges and limits, %(a)		AISI-SAE designation	UNS designation	Heat composition ranges and limits, %(a)		AISI-SAE designation	UNS designation	Heat composition ranges and limits, %(a)	
		C	Mn			C	Mn			C	Mn
1005	G10050	0.06 max	0.35 max	1035	G10350	0.32-0.38	0.60-0.90	1074(b) . . .	G10740	0.70-0.80	0.50-0.80
1006	G10060	0.08 max	0.25-0.40	1037	G10370	0.32-0.38	0.70-1.00	1075(b) . . .	G10750	0.70-0.80	0.40-0.70
1008	G10080	0.10 max	0.30-0.50	1038	G10380	0.35-0.42	0.60-0.90	1078	G10780	0.72-0.85	0.30-0.60
1010	G10100	0.08-0.13	0.30-0.60	1039	G10390	0.37-0.44	0.70-1.00	1080	G10800	0.75-0.88	0.60-0.90
1011(b) . . .	G10110	0.08-0.13	0.60-0.90	1040	G10400	0.37-0.44	0.60-0.90	1084	G10840	0.80-0.93	0.60-0.90
1012	G10120	0.10-0.15	0.30-0.60	1042	G10420	0.40-0.47	0.60-0.90	1085(b) . . .	G10850	0.80-0.93	0.70-1.00
1013(b) . . .	G10130	0.11-0.16	0.50-0.80	1043	G10430	0.40-0.47	0.70-1.00	1086	G10860	0.80-0.93	0.30-0.50
1015	G10150	0.13-0.18	0.30-0.60	1044	G10440	0.43-0.50	0.30-0.60	1090	G10900	0.85-0.98	0.60-0.90
1016	G10160	0.13-0.18	0.60-0.90	1045	G10450	0.43-0.50	0.60-0.90	1095	G10950	0.90-1.03	0.30-0.50
1017	G10170	0.15-0.20	0.30-0.60	1046	G10460	0.43-0.50	0.70-1.00				
1018	G10180	0.15-0.20	0.60-0.90	1049	G10490	0.46-0.53	0.60-0.90				
1019	G10190	0.15-0.20	0.70-1.00	1050	G10500	0.48-0.55	0.60-0.90				
1020	G10200	0.18-0.23	0.30-0.60	1053	G10530	0.48-0.55	0.70-1.00				
1021	G10210	0.18-0.23	0.60-0.90	1055	G10550	0.50-0.60	0.60-0.90				
1022	G10220	0.18-0.23	0.70-1.00	1059(c) . . .	G10590	0.55-0.65	0.50-0.80				
1023	G10230	0.20-0.25	0.30-0.60	1060	G10600	0.55-0.65	0.60-0.90				
1025	G10250	0.22-0.28	0.30-0.60	1064	G10640	0.60-0.70	0.50-0.80				
1026	G10260	0.22-0.28	0.60-0.90	1065	G10650	0.60-0.70	0.60-0.90				
1029	G10290	0.25-0.31	0.60-0.90	1069(b) . . .	G10690	0.65-0.75	0.40-0.70				
1030	G10300	0.28-0.34	0.60-0.90	1070	G10700	0.65-0.75	0.60-0.90				

(a) Limits on phosphorus and sulfur contents are given in Table 1; typical limits are 0.040% maximum phosphorus and 0.050% maximum sulfur. When silicon ranges or limits are required, the values in Table 1 apply. Steels listed in this table can be produced with additions of lead or boron. Leaded steels typically contain 0.15 to 0.35% lead and are identified by inserting the letter "L" in the designation—11L17; boron steels can be expected to contain 0.0005 to 0.003% boron and are identified by inserting the letter "B" in the designation—15B41. (b) SAE standard grade only. (c) AISI standard grade only.

Composition ranges and limits for AISI-SAE standard resulfurized carbon steels

AISI-SAE designation	UNS designation	Heat composition ranges and limits, %(a)		
		C	Mn	S
1110	G11100	0.08-0.13	0.30-0.60	0.08-0.13
1117	G11170	0.14-0.20	1.00-1.30	0.08-0.13
1118	G11180	0.14-0.20	1.30-1.60	0.08-0.13
1137	G11370	0.32-0.39	1.35-1.65	0.08-0.13
1139	G11390	0.35-0.43	1.35-1.65	0.13-0.20
1140	G11400	0.37-0.44	0.70-1.00	0.08-0.13
1141	G11410	0.37-0.45	1.35-1.65	0.08-0.13
1144	G11440	0.40-0.48	1.35-1.65	0.24-0.33
1146	G11460	0.42-0.49	0.70-1.00	0.08-0.13
1151	G11510	0.48-0.55	0.70-1.00	0.08-0.13

(a) Limit on phosphorus content is given in Table 1; the typical value is 0.040% maximum phosphorus. Because of the adverse effect of silicon on machinability, steels listed in this table are generally not deoxidized with silicon. Steel listed in this table can be produced as leaded steels, typically containing 0.15 to 0.35% lead and identified by inserting the letter "L" in the designation—11L17.

Composition ranges and limits for AISI-SAE merchant-quality steels

AISI-SAE desig-nation	Heat composition ranges and limits, %			
	C	Mn	P max	S max
M1008 . .	0.10 max	0.25-0.60	0.04	0.05
M1010 . .	0.07-0.14	0.25-0.60	0.04	0.05
M1012 . .	0.09-0.16	0.25-0.60	0.04	0.05
M1015 . .	0.12-0.19	0.25-0.60	0.04	0.05
M1017 . .	0.14-0.21	0.25-0.60	0.04	0.05
M1020 . .	0.17-0.24	0.25-0.60	0.04	0.05
M1023 . .	0.19-0.27	0.25-0.60	0.04	0.05
M1025 . .	0.20-0.30	0.25-0.60	0.04	0.05
M1031 . .	0.26-0.36	0.25-0.60	0.04	0.05
M1044 . .	0.40-0.50	0.25-0.60	0.04	0.05

Composition ranges and limits for AISI-SAE standard carbon H-steels

AISI-SAE designation	UNS designation	Heat composition ranges and limits, %(a)			AISI-SAE designation	UNS designation	Heat composition ranges and limits, %(a)		
		C	Mn	Si			C	Mn	Si
1038H	H10380	0.34-0.43	0.50-1.00	0.15-0.30	15B21H(b) .	H15211	0.17-0.24	0.70-1.20	0.15-0.30
1045H	H10450	0.42-0.51	0.50-1.00	0.15-0.30	15B35H(b) .	H15351	0.31-0.39	0.70-1.20	0.15-0.30
1522H	H15220	0.17-0.25	1.00-1.50	0.15-0.30	15B37H(b) .	H15371	0.30-0.39	1.00-1.50	0.15-0.30
1524H	H15240	0.18-0.26	1.25-1.75	0.15-0.30	15B41H(b,c)	H15411	0.35-0.45	1.25-1.75	0.15-0.30
1526H	H15260	0.21-0.30	1.00-1.50	0.15-0.30	15B48H(b,c)	H15481	0.43-0.53	1.00-1.50	0.15-0.30
1541H	H15410	0.35-0.45	1.25-1.75	0.15-0.30	15B62H(b) .	H15621	0.54-0.67	1.00-1.50	0.40-0.60

(a) Limits on phosphorus and sulfur content are given in Table 1; typical limits are 0.040% maximum phosphorus and 0.050% maximum sulfur. (b) Can be expected to contain 0.0005 to 0.003% boron. (c) AISI grade only.

Composition ranges and limits for AISI-SAE standard resulfurized and rephosphorized carbon steels

AISI-SAE designation	UNS designation	Heat composition ranges and limits, %(a)			
		C max	Mn	P	S
1211	G12110	0.13	0.60-0.90	0.07-0.12	0.10-0.15
1212	G12120	0.13	0.70-1.00	0.07-0.12	0.16-0.23
1213	G12130	0.13	0.70-1.00	0.07-0.12	0.24-0.33
12L14(b)	G12144	0.15	0.85-1.15	0.04-0.09	0.26-0.35
1215	G12150	0.09	0.75-1.05	0.04-0.09	0.26-0.35

(a) Because of the adverse effect of silicon on machinability, steels listed in this table are generally not deoxidized with silicon. (b) Contains 0.15 to 0.35% lead; other steels listed in this table can be produced with the same lead content.

Composition ranges and limits for AISI-SAE standard carbon steels with a maximum manganese content exceeding 1.10%—semifinished products for forging, hot rolled and cold finished bars, wire rod and seamless tubing

AISI-SAE designation	UNS designation	Heat composition ranges and limits, %(a)				Former AISI-SAE designation
		C	Mn	P max	S max	
1513	G15130	0.10-0.16	1.10-1.40	0.040	0.050	...
1518(b)	G15180	0.15-0.21	1.10-1.40	0.040	0.050	...
1522	G15220	0.18-0.24	1.10-1.40	0.040	0.050	...
1524	G15240	0.19-0.25	1.35-1.65	0.040	0.050	1024
1525(b)	G15250	0.23-0.29	0.80-1.10	0.040	0.050	...
1526	G15256	0.22-0.29	1.10-1.40	0.040	0.050	...
1527	G15270	0.22-0.29	1.20-1.50	0.040	0.050	1027
1536(b)	G15360	0.30-0.37	1.20-1.50	0.040	0.050	1036
1541	G15410	0.36-0.44	1.35-1.65	0.040	0.050	1041
1547(b)	G15470	0.43-0.51	1.35-1.65	0.040	0.050	1047
1548	G15480	0.44-0.52	1.10-1.40	0.040	0.050	1048
1551	G15510	0.45-0.56	0.85-1.15	0.040	0.050	1051
1552	G15520	0.47-0.55	1.20-1.50	0.040	0.050	1052
1561	G15610	0.55-0.65	0.75-1.05	0.040	0.050	1061
1566	G15660	0.60-0.71	0.85-1.15	0.040	0.050	1066
1572(b)	G15720	0.65-0.76	1.00-1.30	0.040	0.050	1072

(a) Limits on phosphorus and sulfur contents are given in Table 1; typical limits are 0.040% maximum phosphorus and 0.050% maximum sulfur. Killed steels commonly contain 0.15 to 0.30% silicon; other ranges are negotiable. Steels listed in this table can be produced with additions of lead or boron. Leaded steels typically contain 0.15 to 0.35% lead and are identified by inserting the letter "L" in the designation—11L17; boron steels can be expected to contain 0.0005 to 0.003% boron and are identified by inserting the letter "B" in the designation—15B41. (b) SAE standard grade only.

Composition ranges and limits for AISI-SAE standard alloy steels—bars, billets, blooms and slabs

AISI-SAE designation	UNS designation	Heat composition ranges and limits, %							
		C	Mn	P max(a)	S max(a)	Si	Cr	Ni	Mo
1330	G13300	0.28-0.33	1.60-1.90	0.035	0.040	0.15-0.30	· · ·	· · ·	· · ·
1335	G13350	0.33-0.38	1.60-1.90	0.035	0.040	0.15-0.30	· · ·	· · ·	· · ·
1340	G13400	0.38-0.43	1.60-1.90	0.035	0.040	0.15-0.30	· · ·	· · ·	· · ·
1345	G13450	0.43-0.48	1.60-1.90	0.035	0.040	0.15-0.30	· · ·	· · ·	· · ·
4012	G40120	0.09-0.14	0.75-1.00	0.035	0.040	0.15-0.30	· · ·	· · ·	0.15-0.25
4023	G40230	0.20-0.25	0.70-0.90	0.035	0.040	0.15-0.30	· · ·	· · ·	0.20-0.30
4024	G40240	0.20-0.25	0.70-0.90	0.035	0.035-0.050(b)	0.15-0.30	· · ·	· · ·	0.20-0.30
4027	G40270	0.25-0.30	0.70-0.90	0.035	0.040	0.15-0.30	· · ·	· · ·	0.20-0.30
4028	G40280	0.25-0.30	0.70-0.90	0.035	0.035-0.050(b)	0.15-0.30	· · ·	· · ·	0.20-0.30
4032	G40320	0.30-0.35	0.70-0.90	0.035	0.040	0.15-0.30	· · ·	· · ·	0.20-0.30
4037	G40370	0.35-0.40	0.70-0.90	0.035	0.040	0.15-0.30	· · ·	· · ·	0.20-0.30
4042(c)	G40420	0.40-0.45	0.70-0.90	0.035	0.040	0.15-0.30	· · ·	· · ·	0.20-0.30
4047	G40470	0.45-0.50	0.70-0.90	0.035	0.040	0.15-0.30	· · ·	· · ·	0.20-0.30
4118	G41180	0.18-0.23	0.70-0.90	0.035	0.040	0.15-0.30	0.40-0.60	· · ·	0.08-0.15
4130	G41300	0.28-0.33	0.40-0.60	0.035	0.040	0.15-0.30	0.80-1.10	· · ·	0.15-0.25
4135(c)	G41350	0.33-0.38	0.70-0.90	0.035	0.040	0.15-0.30	0.30-1.10	· · ·	0.15-0.25
4137	G41370	0.35-0.40	0.70-0.90	0.035	0.040	0.15-0.30	0.80-1.10	· · ·	0.15-0.25
4140	G41400	0.38-0.43	0.75-1.00	0.035	0.040	0.15-0.30	0.80-1.10	· · ·	0.15-0.25
4142	G41420	0.40-0.45	0.75-1.00	0.035	0.040	0.15-0.30	0.80-1.10	· · ·	0.15-0.25
4145	G41450	0.43-0.48	0.75-1.00	0.035	0.040	0.15-0.30	0.80-1.10	· · ·	0.15-0.25
4147	G41470	0.45-0.50	0.75-1.00	0.035	0.040	0.15-0.30	0.80-1.10	· · ·	0.15-0.25
4150	G41500	0.48-0.53	0.75-1.00	0.035	0.040	0.15-0.30	0.80-1.10	· · ·	0.15-0.25
4161	G41610	0.56-0.64	0.75-1.00	0.035	0.040	0.15-0.30	0.70-0.90	· · ·	0.25-0.35
4320	G43200	0.17-0.22	0.45-0.65	0.035	0.040	0.15-0.30	0.40-0.60	1.65-2.00	0.20-0.30
4340	G43400	0.38-0.43	0.60-0.80	0.035	0.040	0.15-0.30	0.70-0.90	1.65-2.00	0.20-0.30
E4340(d)	G43406	0.38-0.43	0.65-0.85	0.025	0.025	0.15-0.30	0.70-0.90	1.65-2.00	0.20-0.30
4419(c)	G44190	0.18-0.23	0.45-0.65	0.035	0.040	0.15-0.30	· · ·	· · ·	0.45-0.60
4422(c)	G44220	0.20-0.25	0.70-0.90	0.035	0.040	0.15-0.30	· · ·	· · ·	0.35-0.45
4427(c)	G44270	0.24-0.29	0.70-0.90	0.035	0.040	0.15-0.30	· · ·	· · ·	0.35-0.45
4615	G46150	0.13-0.18	0.45-0.65	0.035	0.040	0.15-0.30	· · ·	1.65-2.00	0.20-0.30
4617(c)	G46170	0.15-0.20	0.45-0.65	0.035	0.040	0.15-0.30	· · ·	1.65-2.00	0.20-0.30
4620	G46200	0.17-0.22	0.45-0.65	0.035	0.040	0.15-0.30	· · ·	1.65-2.00	0.20-0.30
4621(c)	G46210	0.18-0.23	0.70-0.90	0.035	0.040	0.15-0.30	· · ·	1.65-2.00	0.20-0.30
4626	G46260	0.24-0.29	0.45-0.65	0.035	0.04	0.15-0.30	· · ·	0.70-1.00	0.15-0.25
4718(c)	G47180	0.16-0.21	0.70-0.90	· · ·	· · ·	· · ·	0.35-0.55	0.90-1.20	0.30-0.40
4720	G47200	0.17-0.22	0.50-0.70	0.035	0.040	0.15-0.30	0.35-0.55	0.90-1.20	0.15-0.25
4815	G48150	0.13-0.18	0.40-0.60	0.035	0.040	0.15-0.30	· · ·	3.25-3.75	0.20-0.30
4817	G48170	0.15-0.20	0.40-0.60	0.035	0.040	0.15-0.30	· · ·	3.25-3.75	0.20-0.30
4820	G48200	0.18-0.23	0.50-0.70	0.035	0.040	0.15-0.30	· · ·	3.25-3.75	0.20-0.30
5015(e)	G50150	0.12-0.17	0.30-0.50	0.035	0.040	0.15-0.30	0.30-0.50	· · ·	· · ·
50B40(c,e)	G50401	0.38-0.43	0.75-1.00	0.035	0.040	0.15-0.30	0.40-0.60	· · ·	· · ·
50B44(e)	G50441	0.43-0.48	0.75-1.00	0.035	0.040	0.15-0.30	0.40-0.60	· · ·	· · ·
5046(c)	G50460	0.43-0.48	0.75-1.00	0.035	0.040	0.15-0.30	0.20-0.35	· · ·	· · ·
50B46(e)	G50461	0.44-0.49	0.75-1.00	0.035	0.040	0.15-0.30	0.20-0.35	· · ·	· · ·
50B50(e)	G50501	0.48-0.53	0.75-1.00	0.035	0.040	0.15-0.30	0.40-0.60	· · ·	· · ·
5060(c)	G50600	0.56-0.64	0.75-1.00	0.035	0.040	0.15-0.30	0.40-0.60	· · ·	· · ·
50B60(e)	· · ·	0.56-0.64	0.75-1.00	0.035	0.040	0.15-0.30	0.40-0.60	· · ·	· · ·
5115(c)	G51150	0.13-0.18	0.70-0.90	0.035	0.040	0.15-0.30	0.70-0.90	· · ·	· · ·
5117(f)	G51170	0.15-0.20	0.70-0.90	0.035	0.040	0.15-0.30	0.70-0.90	· · ·	· · ·
5120	G51200	0.17-0.22	0.70-0.90	0.035	0.040	0.15-0.30	0.70-0.90	· · ·	· · ·

(continued)

(a) Limits for phosphorus and sulfur are for steel made by open hearth or basic oxygen processes; limits for steels made by other processes are given in Table 3. (b) A range of sulfur content normally indicates a resulfurized steel. (c) SAE standard grade only. (d) Prefix "E" indicates that the steel is made by electric furnace process. (e) Can be expected to contain 0.0005 to 0.003% boron. (f) AISI standard grade only. (g) Contains 0.10 to 0.15% vanadium. (h) Contains 0.15% min vanadium.

Composition ranges and limits for AISI-SAE standard alloy steels—bars, billets, blooms and slabs (continued)

AISI-SAE designation	UNS designation	Heat composition ranges and limits, %							
		C	Mn	P max(a)	S max(a)	Si	Cr	Ni	Mo
5130	G51300	0.28-0.33	0.70-0.90	0.035	0.040	0.15-0.30	0.80-1.10	· · ·	· · ·
5132	G51320	0.30-0.35	0.60-0.80	0.035	0.040	0.15-0.30	0.75-1.00	· · ·	· · ·
5135	G51350	0.33-0.38	0.60-0.80	0.035	0.040	0.15-0.30	0.80-1.05	· · ·	· · ·
5140	G51400	0.38-0.43	0.70-0.90	0.035	0.040	0.15-0.30	0.70-0.90	· · ·	· · ·
5145(c)	G51450	0.43-0.48	0.70-0.90	0.035	0.040	0.15-0.30	0.70-0.90	· · ·	· · ·
5147(c)	G51470	0.46-0.51	0.70-0.95	0.035	0.040	0.15-0.30	0.85-1.15	· · ·	· · ·
5150	G51500	0.48-0.53	0.70-0.90	0.035	0.040	0.15-0.30	0.70-0.90	· · ·	· · ·
5155	G51550	0.51-0.59	0.70-0.90	0.035	0.040	0.15-0.30	0.70-0.90	· · ·	· · ·
5160	G51600	0.56-0.64	0.75-1.00	0.035	0.040	0.15-0.30	0.70-0.90	· · ·	· · ·
51B60(e)	G51601	0.56-0.64	0.75-1.00	0.035	0.040	0.15-0.30	0.70-0.90	· · ·	· · ·
50100	G50986	0.98-1.10	0.25-0.45	0.025	0.025	0.15-0.30	0.40-0.60	· · ·	· · ·
51100	G51986	0.98-1.10	0.25-0.45	0.025	0.025	0.15-0.30	0.90-1.15	· · ·	· · ·
52100	G52986	0.98-1.10	0.25-0.45	0.025	0.025	0.15-0.30	1.30-1.60	· · ·	· · ·
6118(g)	G61180	0.16-0.21	0.50-0.70	0.035	0.040	0.15-0.30	0.50-0.70	· · ·	· · ·
6150(h)	G61500	0.48-0.53	0.70-0.90	0.035	0.040	0.15-0.30	0.80-1.10	· · ·	
8115(c)	G81150	0.13-0.18	0.70-0.90	0.035	0.040	0.15-0.30	0.30-0.50	0.20-0.40	0.08-0.15
81B45(e)	G81451	0.43-0.48	0.75-1.00	0.035	0.040	0.15-0.30	0.35-0.55	0.20-0.40	0.08-0.15
8615	G86150	0.13-0.18	0.70-0.90	0.035	0.040	0.15-0.30	0.40-0.60	0.40-0.70	0.15-0.25
8617	G86170	0.15-0.20	0.70-0.90	0.035	0.040	0.15-0.30	0.40-0.60	0.40-0.70	0.15-0.25
8620	G86200	0.18-0.23	0.70-0.90	0.035	0.040	0.15-0.30	0.40-0.60	0.40-0.70	0.15-0.25
8622	G86220	0.20-0.25	0.70-0.90	0.035	0.040	0.15-0.30	0.40-0.60	0.40-0.70	0.15-0.25
8625	G86250	0.23-0.28	0.70-0.90	0.035	0.040	0.15-0.30	0.40-0.60	0.40-0.70	0.15-0.25
8627	G86270	0.25-0.30	0.70-0.90	0.035	0.040	0.15-0.30	0.40-0.60	0.40-0.70	0.15-0.25
8630	G86300	0.28-0.33	0.70-0.90	0.035	0.040	0.15-0.30	0.40-0.60	0.40-0.70	0.15-0.25
8637	G86370	0.35-0.40	0.75-1.00	0.035	0.040	0.15-0.30	0.40-0.60	0.40-0.70	0.15-0.25
8640	G86400	0.38-0.43	0.75-1.00	0.035	0.040	0.15-0.30	0.40-0.60	0.40-0.70	0.15-0.25
8642	G86420	0.40-0.45	0.75-1.00	0.035	0.040	0.15-0.30	0.40-0.60	0.40-0.70	0.15-0.25
8645	G86450	0.43-0.48	0.75-1.00	0.035	0.040	0.15-0.30	0.40-0.60	0.40-0.70	0.15-0.25
86B45(c,e)	G86451	0.43-0.48	0.75-1.00	0.035	0.040	0.15-0.30	0.40-0.60	0.40-0.70	0.15-0.25
8650(c)	G86500	0.48-0.53	0.75-1.00	0.035	0.040	0.15-0.30	0.40-0.60	0.40-0.70	0.15-0.25
8655	G86550	0.51-0.59	0.75-1.00	0.035	0.040	0.15-0.30	0.40-0.60	0.40-0.70	0.15-0.25
8660(c)	G86600	0.56-0.64	0.75-1.00	0.035	0.040	0.15-0.30	0.40-0.60	0.40-0.70	0.15-0.25
8720	G87200	0.18-0.23	0.70-0.90	0.035	0.040	0.15-0.30	0.40-0.60	0.40-0.70	0.20-0.30
8740	G87400	0.38-0.43	0.75-1.00	0.035	0.040	0.15-0.30	0.40-0.60	0.40-0.70	0.20-0.30
8822	G88220	0.20-0.25	0.75-1.00	0.035	0.040	0.15-0.30	0.40-0.60	0.40-0.70	0.30-0.40
9254(c)	G92540	0.51-0.59	0.60-0.80	0.035	0.040	1.20-1.60	0.60-0.80	· · ·	· · ·
9255(c)	G92550	0.51-0.59	0.70-0.95	0.035	0.040	1.80-2.20	· · ·	· · ·	· · ·
9260	G92600	0.56-0.64	0.75-1.00	0.035	0.040	1.80-2.20	· · ·	· · ·	· · ·
9310(c)	G93106	0.08-0.13	0.45-0.65	0.025	0.025	0.15-0.30	1.00-1.40	3.00-3.50	0.08-0.15
94B15(c,e)	G94151	0.13-0.18	0.75-1.00	0.035	0.040	0.15-0.30	0.30-0.50	0.30-0.60	0.08-0.15
94B17(c)	G94171	0.15-0.20	0.75-1.00	0.035	0.040	0.15-0.30	0.30-0.50	0.30-0.60	0.08-0.15
94B30(e)	G94301	0.28-0.33	0.75-1.00	0.035	0.040	0.15-0.30	0.30-0.50	0.30-0.60	0.08-0.15

(a) Limits for phosphorus and sulfur are for steel made by open hearth or basic oxygen processes; limits for steels made by other processes are given in Table 3. (b) A range of sulfur content normally indicates a resulfurized steel. (c) SAE standard grade only. (d) Prefix "E" indicates that the steel is made by electric furnace process. (e) Can be expected to contain 0.0005 to 0.003% boron. (f) AISI standard grade only. (g) Contains 0.10 to 0.15% vanadium. (h) Contains 0.15% min vanadium.

Composition ranges and limits for AISI-SAE standard alloy steels—plates

AISI-SAE designation	UNS designation	Heat composition ranges and limits, % (a)					
		C	Mn	Si (b)	Cr	Ni	Mo
1330	G13300	0.27-0.34	1.50-1.90	0.15-0.30	. . .	. . .	. . .
1335	G13350	0.32-0.39	1.50-1.90	0.15-0.30	. . .	. . .	. . .
1340	G13400	0.36-0.44	1.50-1.90	0.15-0.30	. . .	. . .	. . .
1345	G13450	0.41-0.49	1.50-1.90	0.15-0.30	. . .	. . .	. . .
4118	G41180	0.17-0.23	0.60-0.90	0.15-0.30	0.40-0.65	. . .	0.08-0.15
4130	G41300	0.27-0.34	0.35-0.60	0.15-0.30	0.80-1.15	. . .	0.15-0.25
4135	G41350	0.32-0.39	0.65-0.95	0.15-0.30	0.08-1.15	. . .	0.15-0.25
4137	G41370	0.33-0.40	0.65-0.95	0.15-0.30	0.80-1.15	. . .	0.15-0.25
4140	G41400	0.36-0.44	0.70-1.00	0.15-0.30	0.08-1.15	. . .	0.15-0.25
4142	G41420	0.38-0.46	0.70-1.00	0.15-0.30	0.80-1.15	. . .	0.15-0.25
4145	G41450	0.41-0.49	0.70-1.00	0.15-0.30	0.80-1.15	. . .	0.15-0.25
4340	G43400	0.36-0.44	0.55-0.80	0.15-0.30	0.60-0.90	1.65-2.00	0.20-0.30
E4340(c)	G43406	0.37-0.44	0.60-0.85	0.15-0.30	0.65-0.90	1.65-2.00	0.20-0.30
4615	G46150	0.12-0.18	0.40-0.65	0.15-0.30	. . .	1.65-2.00	0.20-0.30
4617	G46170	0.15-0.21	0.40-0.65	0.15-0.30	. . .	1.65-2.00	0.20-0.30
4620	G46200	0.16-0.22	0.40-0.65	0.15-0.30	. . .	1.65-2.00	0.20-0.30
5160	G51600	0.54-0.65	0.70-1.00	0.15-0.30	0.60-0.90	. . .	. . .
6150(d)	G61500	0.46-0.54	0.60-0.90	0.15-0.30	0.80-1.15	. . .	. . .
8615	G86150	0.12-0.18	0.60-0.90	0.15-0.30	0.35-0.60	0.40-0.70	0.15-0.25
8617	G86170	0.15-0.21	0.60-0.90	0.15-0.30	0.35-0.60	0.40-0.70	0.15-0.25
8620	G86200	0.17-0.23	0.60-0.90	0.15-0.30	0.35-0.60	0.40-0.70	0.15-0.25
8622	G86220	0.19-0.25	0.60-0.90	0.15-0.30	0.35-0.60	0.40-0.70	0.15-0.25
8625	G86250	0.22-0.29	0.60-0.90	0.15-0.30	0.35-0.60	0.40-0.70	0.15-0.25
8627	G86270	0.24-0.31	0.60-0.90	0.15-0.30	0.35-0.60	0.40-0.70	0.15-0.25
8630	G86300	0.27-0.34	0.60-0.90	0.15-0.30	0.35-0.60	0.40-0.70	0.15-0.25
8637	G86370	0.33-0.40	0.70-1.00	0.15-0.30	0.35-0.60	0.40-0.70	0.15-0.25
8640	G86400	0.36-0.44	0.70-1.00	0.15-0.30	0.35-0.60	0.40-0.70	0.15-0.25
8655	G86550	0.49-0.60	0.70-1.00	0.15-0.30	0.35-0.60	0.40-0.70	0.15-0.25
8742	G87420	0.38-0.46	0.70-1.00	0.15-0.30	0.35-0.60	0.40-0.70	0.20-0.30

(a) Indicated ranges and limits apply to steels made by open hearth or basic oxygen processes; maximum content for phosphorus is 0.035% and for sulfur 0.040%. For steels made by electric furnace process, the ranges and limits are reduced as follows: C—0.01%; Mn—0.05%; Cr—0.05% (under 1.25%), 0.10% (over 1.25%); maximum content for either phosphorus or sulfur is 0.025%. (b) Other silicon ranges may be negotiated. (c) Prefix "E" indicates that the steel is made by electric furnace process. (d) Contains 0.15% minimum vanadium.

Composition ranges and limits for carbon steels formerly listed by SAE

SAE designation	C	Mn	P	S	Year last listed
1009	0.15 max	0.60 max	0.040 max	0.050 max	1965
1033	0.30-0.36	0.70-1.00	0.040 max	0.050 max	1965
1034	0.32-0.38	0.50-0.80	0.040 max	0.050 max	1968
1059	0.55-0.65	0.50-0.80	0.040 max	0.050 max	1968
1062	0.54-0.65	0.85-1.15	0.040 max	0.050 max	1953
1111	0.13 max	0.60-0.90	0.07-0.12	0.10-0.15	1969
1112	0.13 max	0.70-1.00	0.07-0.12	0.16-0.23	1969
1113	0.13 max	0.70-1.00	0.07-0.12	0.24-0.33	1969
1114	0.10-0.16	1.00-1.30	0.040 max	0.08-0.13	1952
1115	0.13-0.18	0.60-0.90	0.040 max	0.08-0.13	1965
1120	0.18-0.23	0.70-1.00	0.040 max	0.08-0.13	1965
1126	0.23-0.29	0.70-1.00	0.040 max	0.08-0.13	1965
1138	0.34-0.40	0.70-1.00	0.040 max	0.08-0.13	1965

Composition ranges and limits for AISI-SAE standard alloy H-steels

AISI-SAE designation	UNS designation	C	Mn	Si	Cr	Ni	Mo
				Heat composition ranges and limits, % (a)			
1330H	H13300	0.27-0.33	1.45-2.05	0.15-0.30	. . .	. . .	. . .
1335H	H13350	0.32-0.38	1.45-2.05	0.15-0.30	. . .	. . .	. . .
1340H	H13400	0.37-0.44	1.45-2.05	0.15-0.30	. . .	. . .	. . .
1345H	H13450	0.42-0.49	1.45-2.05	0.15-0.30	. . .	. . .	. . .
4027H	H40270	0.24-0.30	0.60-1.00	0.15-0.30	. . .	. . .	0.20-0.30
4028H(b)	H40280	0.24-0.30	0.60-1.00	0.15-0.30	. . .	. . .	0.20-0.30
4032H	H40320	0.29-0.35	0.60-1.00	0.15-0.30	. . .	. . .	0.20-0.30
4037H	H40370	0.34-0.41	0.60-1.00	0.15-0.30	. . .	. . .	0.20-0.30
4042H	H40420	0.39-0.46	0.60-1.00	0.15-0.30	. . .	. . .	0.20-0.30
4047H	H40470	0.44-0.51	0.60-1.00	0.15-0.30	. . .	. . .	0.20-0.30
4118H	H41180	0.17-0.23	0.60-1.00	0.15-0.30	0.30-0.70	. . .	0.08-0.15
4130H	H41300	0.27-0.33	0.30-0.70	0.15-0.30	0.75-1.20	. . .	0.15-0.25
4135H	H41350	0.32-0.38	0.60-1.00	0.15-0.30	0.75-1.20	. . .	0.15-0.25
4137H	H41370	0.34-0.41	0.60-1.00	0.15-0.30	0.75-1.20	. . .	0.15-0.25
4140H	H41400	0.37-0.44	0.65-1.10	0.15-0.30	0.75-1.20	. . .	0.15-0.25
4142H	H41420	0.39-0.46	0.65-1.10	0.15-0.30	0.75-1.20	. . .	0.15-0.25
4145H	H41450	0.42-0.49	0.65-1.10	0.15-0.30	0.75-1.20	. . .	0.15-0.25
4147H	H41470	0.44-0.51	0.65-1.10	0.15-0.30	0.75-1.20	. . .	0.15-0.25
4150H	H41500	0.47-0.54	0.65-1.10	0.15-0.30	0.75-1.20	. . .	0.15-0.25
4161H	H41610	0.55-0.65	0.65-1.10	0.15-0.30	0.65-0.95	. . .	0.25-0.35
4320H	H43200	0.17-0.23	0.40-0.70	0.15-0.30	0.35-0.65	1.55-2.00	0.20-0.30
4340H	H43400	0.37-0.44	0.55-0.90	0.15-0.30	0.65-0.95	1.55-2.00	0.20-0.30
E4340H(b)	H43406	0.37-0.44	0.60-0.95	0.15-0.30	0.65-0.95	1.55-2.00	0.20-0.30
4419H(c)	H44190	0.17-0.23	0.35-0.75	0.15-0.30	. . .	. . .	0.45-0.60
4620H	H46200	0.17-0.23	0.35-0.75	0.15-0.30	. . .	1.55-2.00	0.20-0.30
4621H(c)	H46210	0.17-0.23	0.60-1.00	0.15-0.30	. . .	1.55-2.00	0.20-0.30
4626H(d)	H46260	0.23-0.29	0.40-0.70	0.15-0.30	. . .	0.65-1.05	0.15-0.25
4718H(c)	H47180	0.15-0.21	0.60-0.95	0.15-0.30	0.30-0.60	0.85-1.25	0.30-0.40
4720H	H47200	0.17-0.23	0.45-0.75	0.15-0.30	0.30-0.60	0.85-1.25	0.15-0.25
4815H	H48150	0.12-0.18	0.30-0.70	0.15-0.30	. . .	3.20-3.80	0.20-0.30
4817H	H48170	0.14-0.20	0.30-0.70	0.15-0.30	. . .	3.20-3.80	0.20-0.30
4820H	H48200	0.17-0.23	0.40-0.80	0.15-0.30	. . .	3.20-3.80	0.20-0.30
50B40H(e)	H50401	0.37-0.44	0.65-1.10	0.15-0.30	0.30-0.70	. . .	. . .
50B44H(e)	H50441	0.42-0.49	0.65-1.10	0.15-0.30	0.30-0.70	. . .	. . .
5046H	H50460	0.43-0.50	0.65-1.10	0.15-0.30	0.13-0.43	. . .	. . .
50B46H(e)	H50461	0.43-0.50	0.65-1.10	0.15-0.30	0.13-0.43	. . .	. . .
50B50H(e)	H50501	0.47-0.54	0.65-1.10	0.15-0.30	0.30-0.70	. . .	. . .
50B60H(e)	H50601	0.55-0.65	0.65-1.10	0.15-0.30	0.30-0.70	. . .	. . .
5120H	H51200	0.17-0.23	0.60-1.00	0.15-0.30	0.60-1.00	. . .	. . .
5130H	H51300	0.27-0.33	0.60-1.10	0.15-0.30	0.75-1.20	. . .	. . .
5132H	H51320	0.29-0.35	0.50-0.90	0.15-0.30	0.65-1.10	. . .	. . .
5135H	H51350	0.32-0.38	0.50-0.90	0.15-0.30	0.70-1.15	. . .	. . .
5140H	H51400	0.37-0.44	0.60-1.00	0.15-0.30	0.60-1.00	. . .	. . .
5145H(c)	H51450	0.42-0.49	0.60-1.00	0.15-0.30	0.60-1.00	. . .	. . .
5147H(c)	H51470	0.45-0.52	0.60-1.05	0.15-0.30	0.80-1.25	. . .	. . .
5150H	H51500	0.47-0.54	0.60-1.00	0.15-0.30	0.60-1.00	. . .	. . .
5155H	H51550	0.50-0.60	0.60-1.00	0.15-0.30	0.60-1.00	. . .	. . .
5160H	H51600	0.55-0.65	0.65-1.10	0.15-0.30	0.60-1.00	. . .	. . .
51B60H(e)	H51601	0.55-0.65	0.65-1.10	0.15-0.30	0.60-1.00	. . .	. . .
6118H(f)	H61180	0.15-0.21	0.40-0.80	0.15-0.30	0.40-0.80	. . .	. . .
6150H(g)	H61500	0.47-0.54	0.60-1.00	0.15-0.30	0.75-1.20	. . .	. . .
81B45H(e)	H81451	0.42-0.49	0.70-1.05	0.15-0.30	0.30-0.60	0.15-0.45	0.08-0.15

(continued)

(a) Typical limits on phosphorus and sulfur contents are 0.035% maximum phosphorus and 0.040% maximum sulfur. (b) Electric furnace steel. (c) SAE standard grade only. (d) AISI standard grade only. (e) Can be expected to contain 0.0005 to 0.003% boron. (f) Contains 0.10 to 0.15% vanadium. (g) Contains 0.15% minimum vanadium

Composition ranges and limits for AISI-SAE standard alloy H-steels (continued)

AISI-SAE designation	UNS designation	Heat composition ranges and limits, % (a)					
		C	Mn	Si	Cr	Ni	Mo
8617H	H86170	0.14-0.20	0.60-0.95	0.15-0.30	0.35-0.65	0.35-0.75	0.15-0.25
8620H	H86200	0.17-0.23	0.60-0.95	0.15-0.30	0.35-0.65	0.35-0.75	0.15-0.25
8622H	H86220	0.19-0.25	0.60-0.95	0.15-0.30	0.35-0.65	0.35-0.75	0.15-0.25
8625H	H86250	0.22-0.28	0.60-0.95	0.15-0.30	0.35-0.65	0.35-0.75	0.15-0.25
8627H	H86270	0.24-0.30	0.60-0.95	0.15-0.30	0.35-0.65	0.35-0.75	0.15-0.25
8630H	H86300	0.27-0.33	0.60-0.95	0.15-0.30	0.35-0.65	0.35-0.75	0.15-0.25
86B30H(e)	H86301	0.27-0.33	0.60-0.95	0.15-0.30	0.35-0.65	0.35-0.75	0.15-0.25
8637H	H86370	0.34-0.41	0.70-1.05	0.15-0.30	0.35-0.65	0.35-0.75	0.15-0.25
8640H	H86400	0.37-0.44	0.70-1.05	0.15-0.30	0.35-0.65	0.35-0.75	0.15-0.25
8642H	H86420	0.39-0.46	0.70-1.05	0.15-0.30	0.35-0.65	0.35-0.75	0.15-0.25
8645H	H86450	0.42-0.49	0.70-1.05	0.15-0.30	0.35-0.65	0.35-0.75	0.15-0.25
86B45H(e)	H86451	0.42-0.49	0.70-1.05	0.15-0.30	0.35-0.65	0.35-0.75	0.15-0.25
8650H	H86500	0.47-0.54	0.70-1.05	0.15-0.30	0.35-0.65	0.35-0.70	0.15-0.25
8655H	H86550	0.50-0.60	0.70-1.05	0.15-0.30	0.35-0.65	0.35-0.75	0.15-0.25
8660H	H86600	0.55-0.65	0.70-1.05	0.15-0.30	0.35-0.65	0.35-0.75	0.15-0.25
8720H	H87200	0.17-0.23	0.60-0.95	0.15-0.30	0.35-0.65	0.35-0.75	0.20-0.30
8740H	H87400	0.37-0.44	0.70-1.05	0.15-0.30	0.35-0.65	0.35-0.75	0.20-0.30
8822H	H88220	0.19-0.25	0.70-1.05	0.15-0.30	0.35-0.65	0.35-0.75	0.30-0.40
9260H	H92600	0.55-0.65	0.65-1.10	1.70-2.20	. . .	. . .	. . .
9310H(b)	H93100	0.07-0.13	0.40-0.70	0.15-0.30	1.00-1.45	2.95-3.55	0.08-0.15
94B15H(e)	H94151	0.12-0.18	0.70-1.05	0.15-0.30	0.25-0.55	0.25-0.65	0.08-0.15
94B17H(e)	H94171	0.14-0.20	0.70-1.05	0.15-0.30	0.25-0.55	0.25-0.65	0.08-0.15
94B30H(e)	H94301	0.27-0.33	0.70-1.05	0.15-0.30	0.25-0.55	0.25-0.65	0.08-0.15

(a) Typical limits on phosphorus and sulfur contents are 0.035% maximum phosphorus and 0.040% maximum sulfur. (b) Electric furnace steel. (c) SAE standard grade only. (d) AISI standard grade only. (e) Can be expected to contain 0.0005 to 0.003% boron. (f) Contains 0.10 to 0.15% vanadium. (g) Contains 0.15% minimum vanadium

Composition ranges and limits for SAE HSLA steels

SAE designation(b)	Heat composition limits, %(a)			SAE designation(b)	Heat composition limits, %(a)		
	C max	Mn max	P max		C max	Mn max	P max
942X	0.21	1.35	0.04	950D	0.15	1.00	0.15
945A	0.15	1.00	0.04	950X	0.23	1.35	0.04
945C	0.23	1.40	0.04	955X	0.25	1.35	0.04
945X	0.22	1.35	0.04	960X	0.26	1.45	0.04
950A	0.15	1.30	0.04	965X	0.26	1.45	0.04
950B	0.22	1.30	0.04	970X	0.26	1.65	0.04
950C	0.25	1.60	0.04	980X	0.26	1.65	0.04

(a) Maximum contents of sulfur and silicon for all grades: 0.050% S, 0.90% Si. (b) Second and third digits of designation indicate minimum yield strength in ksi. Suffix "X" indicates that the steel contains niobium, vanadium, nitrogen or other alloying elements. A second suffix "K" indicates that the steel is produced fully killed using fine grain practice; otherwise, the steel is produced semikilled.

Composition ranges and limits for SAE experimental steels

SAE designation	Heat composition ranges and limits, %							
	C	Mn	P max	S max	Si	Cr	Ni	Mo
EX 1	0.15-0.21	0.35-0.60	0.040	0.040	0.20-0.35	· · ·	4.80-5.30	0.20-0.30
EX 9	0.19-0.24	0.95-1.25	0.035	0.040	0.050 max	0.25-0.40	0.20-0.40	0.05-0.10
EX 10	0.19-0.24	0.95-1.25	0.035	0.040	0.20-0.35	0.25-0.40	0.20-0.40	0.05-0.10
EX 11(a)	0.38-0.43	0.75-1.00	0.035	0.040	0.050 max	0.25-0.40	0.20-0.40	0.05-0.10
EX 12(a)	0.38-0.43	0.75-1.00	0.035	0.040	0.20-0.35	0.25-0.40	0.20-0.40	0.05-0.10
EX 13	0.66-0.75	0.80-1.05	0.025	0.025	0.050 max	0.25-0.40	0.20-0.40	0.05-0.10
EX 14	0.66-0.75	0.80-1.05	0.025	0.025	0.20-0.35	0.25-0.40	0.20-0.40	0.05-0.10
EX 15	0.18-0.23	0.90-1.20	0.035	0.040	0.20-0.35	0.40-0.60	· · ·	0.13-0.20
EX 16	0.20-0.25	0.90-1.20	0.035	0.040	0.20-0.35	0.40-0.60	· · ·	0.13-0.20
EX 17	0.23-0.28	0.90-1.20	0.035	0.040	0.20-0.35	0.40-0.60	· · ·	0.13-0.20
EX 18	0.25-0.30	0.90-1.20	0.035	0.040	0.20-0.35	0.40-0.60	· · ·	0.13-0.20
EX 19(a)	0.18-0.23	0.90-1.20	0.035	0.040	0.20-0.35	0.40-0.60	· · ·	0.08-0.15
EX 20	0.13-0.18	0.90-1.20	0.035	0.040	0.20-0.35	0.40-0.60	· · ·	0.13-0.20
EX 21	0.15-0.20	0.90-1.20	0.035	0.040	0.20-0.35	0.40-0.60	· · ·	0.13-0.20
EX 24	0.18-0.23	0.75-1.00	0.035	0.040	0.20-0.35	0.45-0.65	· · ·	0.20-0.30
EX 27	0.25-0.30	0.75-1.00	0.035	0.040	0.20-0.35	0.45-0.65	· · ·	0.20-0.30
EX 29	0.18-0.23	0.75-1.00	0.035	0.040	0.20-0.35	0.45-0.65	0.40-0.70	0.30-0.40
EX 30	0.13-0.18	0.70-0.90	0.035	0.040	0.20-0.35	0.45-0.65	0.70-1.00	0.45-0.60
EX 31	0.15-0.20	0.70-0.90	0.035	0.040	0.20-0.35	0.45-0.65	0.70-1.00	0.45-0.60
EX 32	0.18-0.23	0.70-0.90	0.035	0.040	0.20-0.35	0.45-0.65	0.70-1.00	0.45-0.60
EX 33	0.17-0.24	0.85-1.25	0.035	0.040	0.20-0.35	0.20 min	0.20 min	0.05 min
EX 34	0.28-0.33	0.90-1.20	· · ·	· · ·	0.20-0.35	0.40-0.60	· · ·	0.13-0.20
EX 35	0.35-0.40	0.90-1.20	· · ·	· · ·	0.20-0.35	0.45-0.65	· · ·	0.13-0.20
EX 36	0.38-0.43	0.90-1.20	· · ·	· · ·	0.20-0.35	0.45-0.65	· · ·	0.13-0.20
EX 37	0.40-0.45	0.90-1.20	· · ·	· · ·	0.20-0.35	0.45-0.65	· · ·	0.13-0.20
EX 38	0.43-0.48	0.90-1.20	· · ·	· · ·	0.20-0.35	0.45-0.65	· · ·	0.13-0.20
EX 39	0.48-0.53	0.90-1.20	· · ·	· · ·	0.20-0.35	0.45-0.65	· · ·	0.13-0.20
EX 40	0.51-0.59	0.90-1.20	· · ·	· · ·	0.20-0.35	0.45-0.65	· · ·	0.13-0.20
EX 41	0.56-0.64	0.90-1.20	· · ·	· · ·	0.20-0.35	0.45-0.65	· · ·	0.13-0.20
EX 42	0.13-0.18	0.95-1.25	0.035	0.040	0.20-0.35	0.25-0.40	0.20-0.40	0.05-0.10
EX 43(a)	0.13-0.18	0.95-1.25	0.035	0.040	0.20-0.35	0.25-0.40	0.20-0.40	0.05-0.10
EX 44	0.15-0.20	0.95-1.25	0.035	0.040	0.20-0.35	0.25-0.40	0.20-0.40	0.05-0.10
EX 45(a)	0.15-0.20	0.95-1.25	0.035	0.040	0.20-0.35	0.25-0.40	0.20-0.40	0.05-0.10
EX 46	0.20-0.25	0.95-1.25	0.035	0.040	0.20-0.35	0.25-0.40	0.20-0.40	0.05-0.10
EX 47	0.23-0.28	0.95-1.25	0.035	0.040	0.20-0.35	0.25-0.40	0.20-0.40	0.05-0.10
EX 48	0.25-0.30	0.95-1.25	0.035	0.040	0.20-0.35	0.25-0.40	0.20-0.40	0.05-0.10
EX 49	0.28-0.33	0.95-1.25	0.035	0.040	0.20-0.35	0.25-0.40	0.20-0.40	0.05-0.10
EX 50	0.33-0.38	0.95-1.25	0.035	0.040	0.20-0.35	0.25-0.40	0.20-0.40	0.05-0.10
EX 51	0.35-0.40	0.95-1.25	0.035	0.040	0.20-0.35	0.25-0.40	0.20-0.40	0.05-0.10
EX 52	0.38-0.43	0.91-1.25	0.035	0.040	0.20-0.35	0.25-0.40	0.20-0.40	0.05-0.10
EX 53	0.40-0.45	0.95-1.25	0.035	0.040	0.20-0.35	0.25-0.40	0.20-0.40	0.05-0.10
EX 54	0.19-0.25	0.70-1.05	0.035	0.040	0.35 max	0.40-0.70	· · ·	0.05 min
EX 55	0.15-0.20	0.70-1.00	0.035	0.040	0.20-0.35	0.45-0.65	1.65-2.00	0.65-0.80
EX 56	0.08-0.13	0.70-1.00	0.035	0.040	0.20-0.35	0.45-0.65	1.65-2.00	0.65-0.80

(a) Can be expected to contain 0.0005 to 0.003% boron.

Composition ranges and limits for alloy steels formerly listed by SAE

SAE designation	C	Mn	P max	S max	Si	Cr	Ni	Mo	Year last listed
1320	0.18-0.23	1.60-1.90	0.040	0.040	0.20-0.35	· · ·	· · ·	· · ·	1956
2317	0.15-0.20	0.40-0.60	0.040	0.040	0.20-0.35	· · ·	3.25-3.75	· · ·	1956
2330	0.28-0.33	0.60-0.80	0.040	0.040	0.20-0.35	· · ·	3.25-3.75	· · ·	1953
2340	0.38-0.43	0.70-0.90	0.040	0.040	0.20-0.35	· · ·	3.25-3.75	· · ·	1953
2345	0.43-0.48	0.70-0.90	0.040	0.040	0.20-0.35	· · ·	3.25-3.75	· · ·	1952
2512	0.09-0.14	0.45-0.60	0.025	0.025	0.20-0.35	· · ·	4.75-5.25	· · ·	1953
2515	0.12-0.17	0.40-0.60	0.040	0.040	0.20-0.35	· · ·	4.75-5.25	· · ·	1956
2517	0.15-0.20	0.45-0.60	0.025	0.025	0.20-0.35	· · ·	4.75-5.25	· · ·	1959
3115	0.13-0.18	0.40-0.60	0.040	0.040	0.20-0.35	0.55-0.75	1.10-1.40	· · ·	1953
3120	0.17-0.22	0.60-0.80	0.040	0.040	0.20-0.35	0.55-0.75	1.10-1.40	· · ·	1956
3130	0.28-0.33	0.60-0.80	0.040	0.040	0.20-0.35	0.55-0.75	1.10-1.40	· · ·	1956
3135	0.33-0.38	0.60-0.80	0.040	0.040	0.20-0.35	0.55-0.75	1.10-1.40	· · ·	1960
X3140	0.38-0.43	0.70-0.90	0.040	0.040	0.20-0.35	0.70-0.90	1.10-1.40	· · ·	1947
3140	0.38-0.43	0.70-0.90	0.040	0.040	0.20-0.35	0.55-0.75	1.10-1.40	· · ·	1964
3145	0.43-0.48	0.70-0.90	0.040	0.040	0.20-0.35	0.70-0.90	1.10-1.40	· · ·	1952
3150	0.48-0.53	0.70-0.90	0.040	0.040	0.20-0.35	0.70-0.90	1.10-1.40	· · ·	1952
3215	0.10-0.20	0.30-0.60	0.040	0.050	0.15-0.30	0.90-1.25	1.50-2.00	· · ·	1941
3220	0.15-0.25	0.30-0.60	0.040	0.050	0.15-0.30	0.90-1.25	1.50-2.00	· · ·	1941
3230	0.25-0.35	0.30-0.60	0.040	0.050	0.15-0.30	0.90-1.25	1.50-2.00	· · ·	1941
3240	0.35-0.45	0.30-0.60	0.040	0.040	0.15-0.30	0.90-1.25	1.50-2.00	· · ·	1941
3245	0.40-0.50	0.30-0.60	0.040	0.040	0.15-0.30	0.90-1.25	1.50-2.00	· · ·	1941
3250	0.45-0.55	0.30-0.60	0.040	0.040	0.15-0.30	0.90-1.25	1.50-2.00	· · ·	1941
3310	0.08-0.13	0.45-0.60	0.025	0.025	0.20-0.35	1.40-1.75	3.25-3.75	· · ·	1964
3312	0.08-0.13	0.45-0.60	0.025	0.025	0.20-0.35	1.40-1.75	3.25-3.75	· · ·	1948
3316	0.14-0.19	0.45-0.60	0.025	0.025	0.20-0.35	1.40-1.75	3.25-3.75	· · ·	1956
3325	0.20-0.30	0.30-0.60	0.040	0.050	0.15-0.30	1.25-1.75	3.25-3.75	· · ·	1936
3335	0.30-0.40	0.30-0.60	0.040	0.050	0.15-0.30	1.25-1.75	3.25-3.75	· · ·	1936
3340	0.35-0.45	0.30-0.60	0.040	0.050	0.15-0.30	1.25-1.75	3.25-3.75	· · ·	1936
3415	0.10-0.20	0.30-0.60	0.040	0.050	0.15-0.30	0.60-0.95	2.75-3.25	· · ·	1941
3435	0.30-0.40	0.30-0.60	0.040	0.050	0.15-0.30	0.60-0.95	2.75-3.25	· · ·	1936
3450	0.45-0.55	0.30-0.60	0.040	0.050	0.15-0.30	0.60-0.95	2.75-3.25	· · ·	1936
4053	0.50-0.56	0.75-1.00	0.040	0.040	0.20-0.35	· · ·	· · ·	0.20-0.30	1956
4063	0.60-0.67	0.75-1.00	0.040	0.040	0.20-0.35	· · ·	· · ·	0.20-0.30	1964
4068	0.63-0.70	0.75-1.00	0.040	0.040	0.20-0.35	· · ·	· · ·	0.20-0.30	1957
4119	0.17-0.22	0.70-0.90	0.040	0.040	0.20-0.35	0.40-0.60	· · ·	0.20-0.30	1956
4125	0.23-0.28	0.70-0.90	0.040	0.040	0.20-0.35	0.40-0.60	· · ·	0.20-0.30	1950
4317	0.15-0.20	0.45-0.65	0.040	0.040	0.20-0.35	0.40-0.60	1.65-2.00	0.20-0.30	1953
4337	0.35-0.40	0.60-0.80	0.040	0.040	0.20-0.35	0.70-0.90	1.65-2.00	0.20-0.30	1964
4608	0.06-0.11	0.25-0.45	0.040	0.040	0.025 max	· · ·	1.40-1.75	0.15-0.25	1956
46B12(a)	0.10-0.15	0.45-0.65	0.040	0.040	0.20-0.35	· · ·	1.65-2.00	0.20-0.30	1957
X4620	0.18-0.23	0.50-0.70	0.040	0.040	0.20-0.35	· · ·	1.65-2.00	0.20-0.30	1956
4640	0.38-0.43	0.60-0.80	0.040	0.040	0.20-0.35	· · ·	1.65-2.00	0.20-0.30	1952
4812	0.10-0.15	0.40-0.60	0.040	0.040	0.20-0.35	· · ·	3.25-3.75	0.20-0.30	1956
5045	0.43-0.48	0.70-0.90	0.040	0.040	0.20-0.35	0.55-0.75	· · ·	· · ·	1953
5117	0.15-0.20	0.70-0.90	0.040	0.040	0.20-0.35	0.70-0.90	· · ·	· · ·	1956
5152	0.48-0.55	0.70-0.90	0.040	0.040	0.20-0.35	0.90-1.20	· · ·	· · ·	1956
6115(b)	0.10-0.20	0.30-0.60	0.040	0.050	0.15-0.30	0.80-1.10	· · ·	· · ·	1936
6117(c)	0.15-0.20	0.70-0.90	0.040	0.040	0.20-0.35	0.70-0.90	· · ·	· · ·	1956
6120(c)	0.17-0.22	0.70-0.90	0.040	0.040	0.20-0.35	0.70-0.90	· · ·	· · ·	1961
6125(b)	0.20-0.30	0.60-0.90	0.040	0.050	0.15-0.30	0.80-1.10	· · ·	· · ·	1936
6130(b)	0.25-0.35	0.60-0.90	0.040	0.050	0.15-0.30	0.80-1.10	· · ·	· · ·	1936
6135(b)	0.30-0.40	0.60-0.90	0.040	0.050	0.15-0.30	0.80-1.10	· · ·	· · ·	1941

(continued)

(a) Can be expected to contain 0.0005 to 0.003% boron. (b) Contains 0.15% minimum vanadium. (c) Contains 0.10% minimum vanadium. (d) Contains 12.00 to 15.00% tungsten. (e) Contains 15.00 to 18.00% tungsten. (f) Contains 1.50 to 2.00% tungsten. (g) Contains 0.03% minimum vanadium.

Composition ranges and limits for alloy steels formerly listed by SAE (continued)

SAE designation	Heat composition ranges and limits, %							
	C	Mn	P max	S max	Si	Cr	Ni	Mo
6140(b) 0.35-0.45	0.60-0.90	0.040	0.050	0.15-0.30	0.80-1.10	· · ·	· · ·	1936
6145(b) 0.43-0.48	0.70-0.90	0.040	0.050	0.20-0.35	0.80-1.10	· · ·	· · ·	1956
6195(b) 0.90-1.05	0.20-0.45	0.030	0.035	0.15-0.30	0.80-1.10	· · ·	· · ·	1936
71360(d) 0.50-0.70	0.30 max	0.035	0.040	0.15-0.30	3.00-4.00	· · ·	· · ·	1936
71660(e) . . . . .'0.50-0.70	0.30 max	0.035	0.040	0.15-0.30	3.00-4.00		· · ·	1936
7260(f) 0.50-0.70	0.30 max	0.035	0.040	0.15-0.30	0.50-1.00	· · ·	· · ·	1936
8632 . . . . .0.30-0.35	0.70-0.90	0.040	0.040	0.20-0.35	0.40-0.60	0.40-0.70	0.15-0.25	1951
8635 . . . . .0.33-0.38	0.75-1.00	0.040	0.040	0.20-0.35	0.40-0.60	0.40-0.70	0.15-0.25	1956
8641 . . . . .0.38-0.43	0.75-1.00	0.040	0.040-0.060	0.20-0.35	0.40-0.60	0.40-0.70	0.15-0.25	1956
8653 . . . . .0.50-0.56	'0.75-1.00	0.040	0.040	0.20-0.35	0.50-0.80	0.40-0.70	0.15-0.25	1956
8647 . . . . .0.45-0.50	0.75-1.00	0.040	0.040	0.20-0.35	0.40-0.60	0.40-0.70	0.15-0.25	1948
8715 . . . . .0.13-0.18	0.70-0.90	0.040	0.040	0.20-0.35	0.40-0.60	0.40-0.70	0.20-0.30	1956
8717 . . . . .0.15-0.20	0.70-0.90	0.040	0.040	0.20-0.35	0.40-0.60	0.40-0.70	0.20-0.30	1956
8719 . . . . .0.18-0.23	0.60-0.80	0.040	0.040	0.20-0.35	0.40-0.60	0.40-0.70	0.20-0.30	1952
8735 . . . . .0.33-0.38	0.75-1.00	0.040	0.040	0.20-0.35	0.40-0.60	0.40-0.70	0.20-0.30	1952
8742 . . . . .0.40-0.45	0.75-1.00	0.040	0.040	0.20-0.35	0.40-0.60	0.40-0.70	0.20-0.30	1964
8745 . . . . .0.43-0.48	0.75-1.00	0.040	0.040	0.20-0.35	0.40-0.60	0.40-0.70	0.20-0.30	1953
8750 . . . . .0.48-0.53	0.75-1.00	0.040	0.040	0.20-0.35	0.40-0.60	0.40-0.70	0.20-0.30	1956
9250 . . . . .0.45-0.55	0.60-0.90	0.040	0.040	1.80-2.20	· · ·	· · ·	· · ·	1941
9261 . . . . .0.55-0.65	0.75-1.00	0.040	0.040	1.80-2.20	0.10-0.25	· · ·	· · ·	1956
9262 . . . . .0.55-0.65	0.75-1.00	0.040	0.040	1.80-2.20	0.25-0.40			1961
9315 . . . . .0.13-0.18	0.45-0.65	0.025	0.025	0.20-0.35	1.00-1.40	3.00-3.50	0.08-0.15	1959
9317 . . . . .0.15-0.20	0.45-0.65	0.025	0.025	0.20-0.35	1.00-1.40	3.00-3.50	0.08-0.15	1959
9437 . . . . .0.35-0.40	0.90-1.20	0.040	0.040	0.20-0.35	0.30-0.50	0.30-0.60	0.08-0.15	1950
9440 . . . . .0.38-0.43	0.90-1.20	0.040	0.040	0.20-0.35	0.30-0.50	0.30-0.60	0.08-0.15	1950
94B40(a) 0.38-0.43	0.75-1.00	0.040	0.040	0.20-0.35	0.30-0.50	0.30-0.60	0.08-0.15	1964
9442 . . . . .0.40-0.45	0.90-1.20	0.040	0.040	0.20-0.35	0.30-0.50	0.30-0.60	0.08-0.15	1950
9445 . . . . .0.43-0.48	0.90-1.20	0.040	0.040	0.20-0.35	0.30-0.50	0.30-0.60	0.08-0.15	1950
9447 . . . . .0.45-0.50	0.90-1.20	0.040	0.040	0.20-0.35	0.30-0.50	0.30-0.60	0.08-0.15	1950
9747 . . . . .0.45-0.50	0.50-0.80	0.040	0.040	0.20-0.35	0.10-0.25	0.40-0.70	0.15-0.25	1950
9763 . . . . .0.60-0.67	0.50-0.80	0.040	0.040	0.20-0.35	0.10-0.25	0.40-0.70	0.15-0.25	1950
9840 . . . . .0.38-0.43	0.70-0.90	0.040	0.040	0.20-0.35	0.70-0.90	0.85-1.15	0.20-0.30	1964
9845 . . . . .0.43-0.48	0.70-0.90	0.040	0.040	0.20-0.35	0.70-0.90	0.85-1.15	0.20-0.30	1950
9850 . . . . .0.48-0.53	0.70-0.90	0.040	0.040	0.20-0.35	0.70-0.90	0.85-1.15	0.20-0.30	1961
43BV12(a,g) . .0.08-0.13	0.75-1.00	· · ·	· · ·	0.20-0.35	0.40-0.60	1.65-2.00	0.20-0.30	· · ·
43BV14(a,g) . .0.10-0.15	0.45-0.65	· · ·	· · ·	0.20-0.35	0.40-0.60	1.65-2.00	0.08-0.15	· · ·

(a) Can be expected to contain 0.0005 to 0.003% boron. (b) Contains 0.15% minimum vanadium. (c) Contains 0.10% minimum vanadium. (d) Contains 12.00 to 15.00% tungsten. (e) Contains 15.00 to 18.00% tungsten. (f) Contains 1.50 to 2.00% tungsten. (g) Contains 0.03% minimum vanadium.

ASTM Specifications

ASTM specifications that incorporate AISI-SAE designations

A29 . . . . . . Carbon and alloy steel bars, hot rolled and cold finished, generic
A108 . . . . . Standard-quality cold finished carbon steel bars
A295 . . . . . High carbon-chromium ball and roller bearing steel
A304 . . . . . Alloy steel bars having hardenability requirements
A322 . . . . . Hot rolled alloy steel bars
A331 . . . . . Cold finished alloy steel bars
A434 . . . . . Hot rolled or cold finished quenched and tempered alloy steel bars
A505 . . . . . Hot rolled and cold rolled alloy steel sheet and strip, generic
A506 . . . . . Regular-quality hot rolled and cold rolled alloy steel sheet and strip
A507 . . . . . Drawing quality hot rolled and cold rolled alloy steel sheet and strip
A510 . . . . . Carbon steel wire rods and coarse round wire, generic
A534 . . . . . Carburizing steels for antifriction bearings
A535 . . . . . Special-quality ball and roller bearing steel
A544 . . . . . Scrapless nut quality carbon steel wire
A545 . . . . . Cold heading quality carbon steel wire for machine screws
A546 . . . . . Cold heading quality medium-high-carbon steel wire for hexagon-head bolts
A547 . . . . . Cold heading quality alloy steel wire for hexagon head bolts
A548 . . . . . Cold heading quality carbon steel wire for tapping or sheet metal screws
A549 . . . . . Cold heading quality carbon steel wire for wood screws
A575 . . . . . Merchant-quality hot rolled carbon steel bars
A576 . . . . . Special-quality hot rolled carbon steel bars
A634 . . . . . Aircraft-quality hot rolled and cold rolled alloy steel sheet and strip
A646 . . . . . Premium-quality alloy steel blooms and billets for aircraft and aerospace forgings
A659 . . . . . Commercial-quality hot rolled carbon steel sheet and strip
A680 . . . . . Untempered spring quality cold rolled hard carbon steel strip
A682 . . . . . Cold rolled spring quality carbon steel strip, generic
A684 . . . . . Untempered spring quality cold rolled soft carbon steel strip
A689 . . . . . Carbon and alloy steel bars for springs
A711 . . . . . Carbon and alloy steel blooms, billets and slabs for forging
A713 . . . . . High-carbon spring steel wire for heat treated components

Generic ASTM specifications

A6 . . . . Rolled steel structural plate, shapes, sheet piling and bars, generic
A20 . . . . Steel plate for pressure vessels, generic
A29 . . . . Carbon and alloy steel bars, hot rolled and cold finished, generic
A505 . . . . . . Alloy steel sheet and strip, hot rolled and cold rolled, generic

A510 . . . . . . Carbon steel wire rod and coarse round wire, generic
A568 . . . . . . Carbon and HSLA, hot rolled and cold rolled steel sheet and hot rolled strip, generic
A646 . . . . . . Premium-quality alloy steel blooms and billets for aircraft and aerospace forgings
A711 . . . . . . Carbon and alloy steel blooms, billets and slabs for forging

Composition ranges and limits for sheet and strip, plain carbon and HSLA grades (ASTM specifications)

ASTM speci-fication	Description (a)	C max	Mn max	P max	S max	Other	ASTM speci-fication	Descrip-tion	C max	Mn max	P max	S max	Other
A611	CRSQ						A414	Pressure vessel					
	Grades A, B, C	0.20	0.60	0.04	0.04	(b)		Grade A	0.15	0.90	0.035	0.04	(b)
	Grade E	0.20	0.90	0.04	0.04	(b)		Grade B	0.22	0.90	0.035	0.04	(b)
A366	CRCQ	0.15	0.60	0.035	0.04	(b)		Grade C	0.25	0.90	0.035	0.04	(b)
A109	CR strip							Grade D	0.25	1.20	0.035	0.04	(b)
	Tempers 1, 2, 3	0.25	0.60	0.035	0.04	(b)		Grade E	0.27	1.20	0.035	0.04	(b)
	Tempers 4, 5	0.15	0.60	0.035	0.04	(b)		Grade F	0.31	1.20	0.035	0.04	(b)
A619	CRDQ	0.10	0.50	0.025	0.035	(b)		Grade G	0.31	1.35	0.035	0.04	(b)
A620	CR DQSK	0.10	0.50	0.025	0.035	(c)	A606	HSLA	0.22	1.25	· · ·	0.05	(d)
A570	HR SQ						A607	Grade 45	0.22	1.35	0.04	0.05	(e)
	Grades A, B, C	0.25	0.25-0.60	0.04	0.04	(b)		Grade 50	0.23	1.35	0.04	0.05	(e)
	Grades D, E	0.25	0.60-0.90	0.04	0.04	(b)		Grade 55	0.25	1.35	0.04	0.05	(e)
A569	HR CQ	0.15	0.60	0.035	0.04	(b)		Grade 60	0.26	1.50	0.04	0.05	(e)
A621	HR DQ	0.10	0.50	0.025	0.035	· · ·		Grade 65	0.26	1.50	0.04	0.05	(e)
A622	HR DQSK	0.10	0.50	0.025	0.035	(c)		Grade 70	0.26	1.65	0.04	0.05	0.012 max N (e)
							A715	Basic composition	0.15	1.65	0.025	0.035	0.012 max N

Type 1: 0.05 min Ti, 0.10 max Si (f)
Type 2: 0.02 min V, 0.60 max Si(g), 0.005 min N (f)(g)
Type 3: 0.005 min Nb, 0.08 max V(g), 0.60 max Si(g), 0.020 max N(f)(g)
Type 4: 0.05 min Zr, 0.90 max Si, 0.80 max Cr(g), 0.10 max Ti(g), 0.0025 max B(g), 0.005-0.06 Nb(f)(h)

Type 5: 0.03 min Nb(j), 0.20 min Mo(j), 0.30 max Si (f)
Type 6: 0.005-0.10 Nb, 0.90 max Si (f)
Type 7: 0.005 min Nb or V, or both, 0.60 max Si, 0.020 max N(f)

(a) CR, cold rolled; SQ, structural quality; DQ, drawing quality; DQSK, drawing quality special killed. (b) Cu when specified as Cu-bearing steel: 0.20% min. (c) Aluminum as deoxidizer usually exceeds 0.010% in the product. (d) Other elements may be added if necessary to meet mechanical and corrosion requirements. (e) 0.005 min Nb or 0.01 min V for all grades. (f) These elements are added to basic composition. (g) Not added to grades 50 and 60. (h) Might not be added to grade 50. (j) Available as grade 80 only.

Composition ranges and limits for carbon steel structural shapes and plate (ASTM specifications)

ASTM specification	Form, type or grade	UNS designation	C max	Heat composition ranges and limits, % (a) Mn	Si	Cu (b)
A36	Plate	· · ·	0.29	0.80-1.20	· · ·	0.20
	Shapes	K02600	0.26	(c)	(d)	0.20
	Bars	· · ·	0.29	0.60-0.90	· · ·	0.20
A283	Plate	· · ·	· · ·	· · ·	· · ·	0.20
A284	Grade A	K01804	0.24	0.90 max	0.10-0.30	· · ·
	Grade B	K02001	0.24	0.90 max	0.15-0.30	· · ·
	Grade C	K02401	0.36	0.90 max	0.15-0.30	· · ·
	Grade D	K02702	0.35	0.90 max	0.15-0.30	· · ·
A529	Plate, bars and shapes	K02703	0.27	1.20 max	· · ·	0.20
A573	Grade 58	K02301	0.23	0.60-0.90	0.10-0.35	· · ·
	Grade 65	K02404	0.26	0.85-1.20	0.15-0.30	· · ·
	Grade 70	K02701	0.28	0.85-1.20	0.15-0.30	· · ·
A678	Grade A	K01600	0.16	0.90-1.50	0.15-0.50	0.20
	Grade B	K02002	0.20	0.70-1.60	0.15-0.50	0.20
	Grade C	K02204	0.22	1.00-1.60	0.20-0.50	0.20

(a) Limits on phosphorus and sulfur contents are given in Table 2; typical limits are 0.040% maximum phosphorus and 0.050% maximum sulfur. (b) Minimum copper content applicable only if copper-bearing steel is specified. (c) 0.85-1.35% manganese required for shapes heavier than 634 kg/m (426 lb/ft). (d) 0.15-0.30% silicon required for shapes heavier than 634 kg/m (426 lb/ft).

Composition ranges and limits for HSLA and alloy steel plate (ASTM specifications)

ASTM specification	Type or grade	UNS designation	Heat composition ranges and limits, %									
			C	Mn	P max	S max	Si	Cr	Ni	Mo	V	Other
A242	Type 1	K11510	0.15 max	1.00 max	0.45	0.05	...	...	.	...	...	0.20 min Cu
	Type 2	K12010	0.20 max	1.35 max	0.04	0.05	...	...	...	...	...	0.20 min Cu if both 0.5 Si and 0.5 Cr not present
A440	...	K12810	0.28 max	1.10-1.60	0.04	0.05	0.30 max	...	...	...	...	0.20 min Cu
A441	...	K12211	0.22 max	0.85-1.25	0.04	0.05	0.30 max					0.20 min Cu; 0.02 min V
A514	Type A	K11856	0.15-0.21	0.80-1.10	0.035	0.04	0.40-0.80	0.50-0.80	...	0.18-0.28	...	0.05-0.15 Zr; 0.0025 max B
	Type B	K11630	0.12-0.21	0.70-1.00	0.035	0.04	0.20-0.35	0.40-0.65	...	0.15-0.25	0.03-0.08	0.01-0.03 Ti; 0.0005-0.005 B
	Type C	K11511	0.10-0.20	1.10-1.50	0.035	0.04	0.15-0.30	...	...	0.20-0.30	...	0.001-0.005 B
	Type D	K11662	0.13-0.20	0.40-0.70	0.035	0.04	0.20-0.35	0.85-1.20	...	0.15-0.25	...	0.04-0.10 Ti; 0.20-0.40 Cu; 0.0015-0.005 B
	Type E	K21604	0.12-0.20	0.40-0.70	0.035	0.04	0.20-0.35	1.40-2.00	...	0.40-0.60	...	0.04-0.10 Ti; 0.20-0.40 Cu; 0.0015-0.005 B
	Type F	K11576	0.10-0.20	0.60-1.00	0.035	0.04	0.15-0.35	0.40-0.65	0.70-1.00	0.40-0.60	0.03-0.08	0.15-0.50 Cu; 0.0005-0.006 B
	Type G	K11872	0.15-0.21	0.80-1.10	0.035	0.04	0.50-0.90	0.50-0.90	...	0.40-0.60	...	0.05-0.15 Zr; 0.0025 max B
	Type H	K11646	0.12-0.21	0.95-1.30	0.035	0.04	0.20-0.35	0.40-0.65	0.30-0.70	0.20-0.30	0.03-0.08	0.0005-0.005 B
	Type J	K11625	0.12-0.21	0.45-0.70	0.035	0.04	0.20-0.35	...	...	0.50-0.65	...	0.001-0.005 B
	Type K	K11523	0.10-0.20	1.10-1.50	0.035	0.04	0.15-0.30	...	...	0.45-0.55	...	0.001-0.005 B
	Type L	K11682	0.13-0.20	0.40-0.70	0.035	0.04	0.20-0.35	1.15-1.65	...	0.25-0.40	...	0.04-0.10 Ti; 0.20-0.40 Cu; 0.0015-0.005 B
	Type M	K11683	0.12-0.21	0.45-0.70	0.035	0.04	0.20-0.35	...	1.20-1.50	0.45-0.60	...	0.001-0.005 B
	Type N	K11847	0.15-0.21	0.80-1.10	0.035	0.04	0.40-0.90	0.50-0.80	...	0.25 max	...	0.05-0.15 Zr; 0.0005-0.0025 B
	Type P	K21650	0.12-0.21	0.45-0.70	0.035	0.04	0.20-0.35	...	1.20-1.50	...	...	...
A572	Grade 42	...	0.21 max	1.35 max	0.04	0.05	0.30 max	...	...	...	...	0.20 min Cu(a)

(continued)

(a) These grades may contain niobium, vanadium or nitrogen.

Composition ranges and limits for HSLA and alloy steel plate (ASTM specifications) (continued)

ASTM specification	Type or grade	UNS designation	C	Mn	P max	S max	Si	Cr	Ni	Mo	V	Other
A572	Grade 45	· · ·	0.22 max	1.35 max	0.04	0.05	0.30 max	· · ·	· · ·	· · ·	· · ·	0.20 min Cu(a)
	Grade 50	· · ·	0.23 max	1.35 max	0.04	0.05	0.30 max	· · ·		· · ·	· · ·	0.20 min Cu(a)
	Grade 55	· · ·	0.25 max	1.35 max	0.04	0.05	0.30 max	· · ·			· · ·	0.20 min Cu(a)
	Grade 60	· · ·	0.26 max	1.35 max	0.04	0.05	0.30 max	· · ·			· · ·	0.20 min Cu(a)
	Grade 65	· · ·	0.26 max	1.65 max	0.04	0.05	0.30 max	· · ·			· · ·	0.20 min Cu(a)
A588	Grade A	K11430	0.10-0.19	0.90-1.25	0.04	0.05	0.15-0.30	0.40-0.65	· · ·	· · ·	0.02-0.10	0.25-0.40 Cu
	Grade B	K12043	0.20 max	0.75-1.25	0.04	0.05	0.15-0.30	0.40-0.70	0.25-0.50	· · ·	0.01-0.10	0.20-0.40 Cu
	Grade C	K11538	0.15 max	0.80-1.35	0.04	0.05	0.15-0.30	0.30-0.50	0.25-0.50	· · ·	0.01-0.10	0.20-0.50 Cu
	Grade D	K11552	0.10-0.20	0.75-1.25	0.04	0.05	0.50-0.90	0.50-0.90	· · ·	· · ·	· · ·	0.30 max Cu; 0.05-0.15 Zr; 0.04 max Nb
	Grade E	K11567	0.15 max	1.20 max	0.04	0.05	0.15-0.30	· · ·	0.75-1.25	0.10-0.25	0.05 max	0.50-0.80 Cu
	Grade F	K11541	0.10-0.20	0.50-1.00	0.04	0.05	0.30 max	0.30 max	0.40-1.10	0.10-0.20	0.01-0.10	0.30-1.00 Cu
	Grade G	K12040	0.20 max	1.20 max	0.04	0.05	0.25-0.70	0.50-1.00	0.80 max	0.10 max	· · ·	0.30-0.50 Cu; 0.07 max Ti
	Grade H	K12032	0.20 max	1.25 max	0.035	0.040	0.25-0.75	0.10-0.25	0.30-0.60	0.15 max	0.02-0.10	0.20-0.35 Cu; 0.005-0.030 Ti
	Grade J	K12044	0.20 max	0.60-1.00	0.04	0.05	0.30-0.50	· · ·	0.50-0.70	· · ·	· · ·	0.30 min Cu; 0.03-0.05 Ti
A633	Grade A	K01802	0.18 max	1.00-1.35	0.04	0.05	0.15-0.50	· · ·	· · ·	· · ·	· · ·	0.05 max Nb
	Grade B	K01803	0.18 max	1.00-1.35	0.04	0.05	0.15-0.50	· · ·	· · ·	· · ·	0.10 max	· · ·
	Grade C	K12000	0.20 max	1.15-1.50	0.04	0.05	0.15-0.50	· · ·	· · ·	· · ·	· · ·	0.01-0.05 Nb
	Grade D	K02003	0.20 max	0.70-1.60	0.04	0.05	0.15-0.50	0.25 max	0.25 max	0.08 max	· · ·	0.35 max Cu
	Grade E	K12202	0.22 max	1.15-1.50	0.04	0.05	0.15-0.50	· · ·	· · ·	· · ·	0.04-0.11	0.01-0.03 N
A656	Grade 1	K11804	0.18 max	1.60 max	0.040	0.050	0.60 max	· · ·	· · ·	· · ·	0.05-0.15	0.020 min Al; 0.005-0.030 N
	Grade 2	K11503	0.15 max	0.90 max	0.040	0.050	0.10 max	· · ·	· · ·	· · ·	· · ·	0.01 min Al; 0.05-0.50 Ti
A699	· · ·	K10614	0.06 max	1.20-2.20	0.04	0.025	0.35 max	· · ·	· · ·	0.25-0.35	· · ·	0.03-0.09 Nb; 0.20-0.35 Cu optional
A710	Grade A	K20747	0.07 max	0.40-0.70	0.025	0.025	0.35 max	0.60-0.90	0.70-1.00	0.15-0.25	· · ·	1.00-1.30 Cu; 0.02 min Nb
	Grade B	K20622	0.06 max	0.40-0.65	0.025	0.025	0.20-0.35	· · ·	1.20-1.50	· · ·	· · ·	1.00-1.30 Cu; 0.02 min Nb

(a) These grades may contain niobium, vanadium or nitrogen.

Composition ranges and limits for alloy steel for pressure-vessel plate (ASTM specifications)

ASTM specification	Type or grade	UNS designation	Heat composition ranges and limits, %									
			C	Mn	P max	S max	Si	Cr	Ni	Mo	V	Other
A202	Grade A	K11742	0.17 max	1.05-1.40	0.035	0.04	0.60-0.90	0.35-0.60	...	...	...	...
	Grade B	K12542	0.25 max	1.05-1.40	0.035	0.04	0.60-0.90	0.35-0.60	...	...	...	...
A203	Grade A	K21703	0.23 max	0.80 max	0.035	0.04	0.15-0.30	...	2.10-2.50	...	...	...
	Grade B	K22103	0.25 max	0.80 max	0.035	0.04	0.15-0.30	...	2.10-2.50	...	...	...
	Grade D	K31718	0.20 max	0.80 max	0.035	0.04	0.15-0.30	...	3.25-3.75	...	...	...
	Grade E	K32018	0.23 max	0.80 max	0.035	0.04	0.15-0.30	...	3.25-3.75	...	...	...
A204	Grade A	K11820	0.25 max	0.90 max	0.035	0.04	0.15-0.30	...	0.45-0.60	0.45-0.60	...	...
	Grade B	K12020	0.27 max	0.90 max	0.035	0.04	0.15-0.30	...	0.45-0.60	0.45-0.60	...	...
	Grade C	K12320	0.28 max	0.90 max	0.035	0.04	0.15-0.30	...	...	0.45-0.60		
A225	Grade A	K11803	0.18 max	1.45 max	0.035	0.04	0.15-0.30	...	...	...	0.09-0.14	...
	Grade B	K12003	0.20 max	1.45 max	0.035	0.04	0.15-0.30	...	...	...	0.09-0.14	...
	Grade C	K12524	0.25 max	1.60 max	0.035	0.04	0.13-0.32	0.37-0.73	...	...	0.11-0.20	...
A302	Grade A	K12021	0.25 max	0.95-1.30	0.035	0.040	0.15-0.30	...	...	0.45-0.60	...	...
	Grade B	K12022	0.25 max	1.15-1.50	0.035	0.040	0.15-0.30	...	...	0.45-0.60	...	...
	Grade C	K12039	0.25 max	1.15-1.50	0.035	0.040	0.15-0.30	...	0.40-0.70	0.45-0.60	...	...
	Grade D	K12054	0.25 max	1.15-1.50	0.035	0.040	0.15-0.30	...	0.70-1.00	0.45-0.60	...	...
A353	...	K81340	0.13 max	0.90 max	0.035	0.040	0.15-0.30	...	8.50-9.50	...	...	...
A387	Grade 2	K12143	0.21 max	0.55-0.80	0.035	0.040	0.15-0.30	0.50-0.80	...	0.45-0.60	...	
	Grade 12	K11757	0.17 max	0.40-0.65	0.035	0.040	0.15-0.30	0.80-1.15	...	0.45-0.60	...	
	Grade 11	K11789	0.17 max	0.40-0.65	0.035	0.040	0.50-0.80	1.00-1.50	...	0.45-0.65	...	
	Grade 22	K21590	0.15 max	0.30-0.60	0.035	0.035	0.50 max	2.00-2.50	...	0.90-1.10	...	
	Grade 21	K31545	0.15 max	0.30-0.60	0.035	0.035	0.50 max	2.75-3.25	...	0.90-1.10	...	...
	Grade 5	K41545	0.15 max	0.30-0.60	0.040	0.030	0.50 max	4.00-6.00	...	0.45-0.65	...	...
	Grade 7	S50300	0.15 max	0.30-0.60	0.030	0.030	1.00 max	6.00-8.00	...	0.45-0.65	...	
	Grade 9	S50400	0.15 max	0.30-0.60	0.030	0.030	1.00 max	8.00-10.00	...	0.90-1.10	...	
A517	Grade A	K11856	0.15-0.21	0.80-1.10	0.035	0.04	0.40-0.80	0.50-0.80		0.18-0.28	...	0.05-0.15 Zr; 0.0025 max B
	Grade B	K11630	0.12-0.21	0.70-1.00	0.035	0.04	0.20-0.65	0.40-0.65	...	0.15-0.25	0.03-0.08	0.01-0.03 Ti; 0.0005-0.005 B
	Grade C	K11511	0.10-0.20	1.10-1.50	0.035	0.04	0.15-0.30	...	...	0.20-0.30	...	0.001-0.005 B
	Grade D	K11662	0.13-0.20	0.40-0.70	0.035	0.04	0.20-0.35	0.85-1.20	...	0.15-0.25	...	0.04-0.10 Ti; 0.20-0.40 Cu; 0.0015-0.005 B
	Grade E	K21604	0.12-0.20	0.40-0.70	0.035	0.04	0.20-0.35	1.40-2.00	...	0.40-0.60	...	0.04-0.10 Ti; 0.20-0.40 Cu; 0.0015-0.005 B
	Grade F	K11576	0.10-0.20	0.60-1.00	0.035	0.04	0.15-0.35	0.40-0.65	0.70-1.00	0.40-0.60	0.03-0.08	0.15-0.50 Cu; 0.002-0.006 B
	Grade G	K11872	0.15-0.21	0.80-1.10	0.035	0.04	0.50-0.90	0.50-0.90	...	0.40-0.60	...	0.05-0.15 Zr; 0.0025 max B
	Grade H	K11646	0.12-0.21	0.95-1.30	0.035	0.04	0.20-0.35	0.40-0.650	0.30-0.70	0.20-0.30	0.03-0.08	0.0005-0.005 B
	Grade J	K11625	0.12-0.21	0.45-0.70	0.035	0.04	0.20-0.35	...	...	0.50-0.65	...	0.001-0.005 B
	Grade K	K11523	0.10-0.20	1.10-1.50	0.035	0.04	0.15-0.30	...	...	0.45-0.55	...	0.001-0.005 B
	Grade L	K11682	0.13-0.20	0.40-0.70	0.035	0.04	0.20-0.35	1.15-1.65	...	0.25-0.40	...	0.04-0.10 Ti; 0.20-0.40 Cu; 0.0015-0.005 B
	Grade M	K11683	0.12-0.21	0.45-0.70	0.035	0.04	0.20-0.35	...	1.20-1.50	0.45-0.60	...	0.001-0.005 B
	Grade P	K21650	0.12-0.21	0.45-0.70	0.035	0.04	0.20-0.35	0.85-1.20	1.20-1.50	0.45-0.60	...	0.001-0.005 B
A533	Type A	K12521	0.25 max	1.15-1.50	0.035	0.040	0.15-0.30	...	...	0.45-0.60	...	...
	Type B	K12539	0.25 max	1.15-1.50	0.035	0.040	0.15-0.30	...	0.40-0.70	0.45-0.60	...	...
	Type C	K12554	0.25 max	1.15-1.50	0.035	0.040	0.15-0.30	...	0.70-1.00	0.45-0.60	...	...
	Type D	K12529	0.25 max	1.15-1.50	0.035	0.040	0.15-0.30	...	0.20-0.40	0.45-0.60	...	...
A538	Grade A	K92810	0.03 max	0.10 max	0.010	0.010	0.10 max	...	17.0-19.0	4.0-4.5	...	0.10-0.25 Ti; 7.0-8.5 Co; 0.05-0.15 Al; 0.003 B; 0.02 Zr; 0.05 Ca
	Grade B	K92890	0.03 max	0.10 max	0.010	0.010	0.10 max	...	17.0-19.0	4.6-5.1	...	0.30-0.50 Ti; 7.0-8.5 Co; 0.05-0.15 Al

(continued)

Composition ranges and limits for alloy steel for pressure-vessel plate (ASTM specifications) (continued)

ASTM specification	Type or grade	UNS designation	Heat composition ranges and limits, %									
			C	Mn	P max	S max	Si	Cr	Ni	Mo	V	Other
A538	Grade C	K93120	0.03 max	0.10 max	0.010	0.010	0.10 max	· · ·	18.0-19.0	4.6-5.2	· · ·	0.55-0.80 Ti; 8.0-9.5 Co; 0.05-0.15 Al
A542	· · ·	K21590	0.15 max	0.30-0.60	0.035	0.035	0.15-0.30	2.00-2.50	· · ·	0.90-1.10	· · ·	· · ·
A543	Type A	K42338	0.23 max	0.40 max	0.035	0.040	0.20-0.35	1.50-2.00	2.60-4.00	0.45-0.60	0.03 max	· · ·
	Type B	K42339	0.23 max	0.40 max	0.020	0.020	0.20-0.35	1.50-2.00	2.60-4.00	0.45-0.60	0.03 max	· · ·
A553	Type I	K81340	0.13 max	0.90 max	0.035	0.040	0.15-0.30	· · ·	8.50-9.50	· · ·	· · ·	· · ·
	Type II	K71340	0.13 max	0.90 max	0.035	0.040	0.15-0.30	· · ·	7.50-8.50	· · ·	· · ·	· · ·
A562	· · ·	K11224	0.12 max	1.20 max	0.04	0.05	0.15-0.50	· · ·	· · ·	· · ·	· · ·	(4 × %C) Ti; 0.15 max Cu
A590	· · ·	K91890	0.03 max	0.10 max	0.010	0.010	0.10 max	4.50-5.50	11.5-12.5	2.75-3.25	· · ·	0.20-0.35 Ti; 0.40 max Al
A605	· · ·	K91401	0.13 max	0.20-0.40	0.010	0.010	0.10 max	0.65-0.85	8.5-9.5	0.90-1.10	0.06-0.12	4.25-5.00 Co
A645	· · ·	K41583	0.30-0.60	0.30-0.60	0.025	0.025	0.20-0.35	· · ·	4.75-5.25	0.20-0.35	· · ·	0.020 max N; 0.02-0.12 Al
A734	Type A	· · ·	0.17 max	0.45-0.75	0.035	0.015	0.35 max	0.90-1.20	0.90-1.20	0.25-0.40	· · ·	0.06 max Al
	Type B	· · ·	0.17 max	1.60 max	0.035	0.015	0.35 max	0.25 max	· · ·	· · ·	0.11 max	0.35 max Cu; 0.030 max N; 0.050 max Nb; 0.06 max Al
A735	· · ·	· · ·	0.06 max	1.20-2.20	0.04	0.025	0.35 max	· · ·	· · ·	0.23-0.47	· · ·	0.20-0.35 Cu; 0.03-0.09 Nb
A736	· · ·	· · ·	0.07 max	0.40-0.70	0.025	0.025	0.35 max	0.60-0.90	0.70-1.00	0.15-0.25	· · ·	1.00-1.30 Cu; 0.02 min Nb
A737	Grade A	· · ·	0.20 max	1.00-1.35	0.035	0.030	0.15-0.50	· · ·	· · ·	· · ·	0.10 max	· · ·
	Grade B	· · ·	0.20 max	1.15-1.50	0.035	0.030	0.15-0.50	· · ·	· · ·	· · ·	· · ·	0.05 max Nb
	Grade C	· · ·	0.20 max	1.15-1.50	0.035	0.030	0.15-0.50	· · ·	· · ·	· · ·	0.04-0.11	0.03 max N

Composition ranges and limits for carbon steel pressure-vessel plate (ASTM specifications)

Specification	Type or grade	UNS designation	Heat composition ranges and limits, %				
			C max	Mn	P max	S max	Si
A285	Grade A	K01700	0.17	· · ·	0.035	0.045	· · ·
	Grade B	K02200	0.22	0.90 max	0.035	0.045	· · ·
	Grade C	K02801	0.28	0.90 max	0.035	0.045	· · ·
A288	· · ·	K02803	0.30	0.90-1.50	0.035	0.040	0.15-0.30
A442	Grade 55	K02202	0.24	0.60-1.10	0.04	0.05	0.15-0.30
	Grade 60	K02402	0.27	0.60-1.10	0.04	0.05	0.15-0.30
A455	Type I	K03300	0.33	0.85-1.20	0.040	0.050	0.10 max
	Type II	K02802	0.28	0.85-1.20	0.040	0.050	0.15-0.30
A515	Grade 55	K02001	0.28	0.90 max	0.035	0.040	0.15-0.30
	Grade 60	K02401	0.31	0.90 max	0.035	0.040	0.15-0.30
A515	Grade 65	K02800	0.33	0.90 max	0.035	0.040	0.15-0.30
	Grade 70	K03101	0.35	0.90 max	0.035	0.040	0.15-0.30
A516	Grade 55	K01800	0.26	0.60-1.20	0.035	0.04	0.15-0.30
	Grade 60	K02100	0.27	0.60-1.20	0.035	0.04	0.15-0.30
	Grade 65	K02403	0.29	0.85-1.20	0.035	0.04	0.15-0.30
	Grade 70	K02700	0.31	0.85-1.20	0.035	0.04	0.15-0.30
A537	· · ·	K02400	0.24	0.70-1.60	0.035	0.040	0.15-0.30
A612(a)	· · ·	K02900	0.27	1.00-1.50	0.035	0.040	0.15-0.30
A662	Grade A	K01701	0.17	0.90-1.35	0.035	0.040	0.15-0.30
	Grade B	K02203	0.19	0.85-1.50	0.035	0.040	0.15-0.30
A724(a)	Grade A	· · ·	0.18	1.00-1.60	0.035	0.040	0.55 max

(a) Residual alloying elements restricted as follows: 0.35 max Cu; 0.25 max Ni; 0.25 max Cr; 0.08 max Mo; 0.08 max V.

AMS Designations

Product descriptions and nominal compositions for wrought alloy steels

AMS designation	Product form (a)	C	Cr	Ni	Mo	Other	Nearest AISI-SAE grade
6242C	Bars, forgings	0.15-0.20	···	5	···	···	2517
6250F	Bars, forgings, tubing	0.07-0.13	1.5	3.5	···	···	3310
6260G	Bars, forgings, tubing	0.07-0.13	1.2	3.25	0.12	···	9310
6263D	Bars, forgings, tubing	0.11-0.17	1.2	3.25	0.12	···	9315
6264D	Bars, forgings, tubing	0.14-0.20	1.2	3.25	0.12	···	9317
6265C	Bars, forgings, tubing (P,VM)	0.07-0.13	1.2	3.25	0.12	···	9310
6266C	Bars, forgings, tubing	0.08-0.13	0.50	1.85	0.25	0.003 B	43BV12
6267A	Bars, forgings, tubing (P)	0.07-0.13	1.2	3.25	0.12	···	9310
6270G	Bars, forgings, tubing	0.11-0.17	0.50	0.55	0.20	···	8615
6272E	Bars, forgings, tubing	0.15-0.20	0.50	0.55	0.20	···	8617
6274G	Bars, forgings, tubing	0.18-0.23	0.50	0.55	0.20	···	8620
6275B	Bars, forgings, tubing	0.15-0.20	0.40	0.45	0.12	0.003 B	94B17
6276C	Bars, forgings, tubing (P,VM)	0.18-0.23	0.50	0.55	0.2	···	8620
6277A	Bars, forgings, tubing (P)	0.18-0.23	0.50	0.55	0.20	···	8620
6280E	Bars, forgings	0.28-0.33	0.50	0.55	0.20	···	8630
6281C	Tubing	0.28-0.33	0.5	0.55	0.2	···	8630
6282D	Tubing	0.33-0.38	0.50	0.55	0.25	···	8735
6290C	Bars, forgings	0.11-0.17	···	1.8	0.25	···	4615
6292C	Bars, forgings	0.15-0.20	···	1.8	0.25	···	4617
6294C	Bars, forgings	0.17-0.22	···	1.8	0.25	···	4620
6299A	Bars, forgings, tubing	0.17-0.23	0.50	1.8	0.25	···	4320H
6300	Bars, forgings	0.35-0.40	···	···	0.25	···	4037
6302B	Bars, forgings	0.28-0.33	1.25	···	0.50	0.65 Si 0.25 V	···
6303A	Bars, forgings	0.25-0.30	1.25	···	0.5	0.65 Si 0.85 V	···
6304C	Bars, forgings, tubing	0.40-0.50	0.95	···	0.55	0.30 V	···
6312A	Bars, forgings	0.38-0.43	···	1.8	0.25	···	4640
6317B	Bars, forgings (heat treated; 125 ksi tensile strength)	0.38-0.43	···	1.8	0.25	···	4640
6320F	Bars, forgings	0.33-0.38	0.50	0.55	0.25	···	8735
6321A	Bars, forgings, tubing	0.38-0.43	0.43	0.30	0.12	0.003B	81B40
6322F	Bars, forgings	0.38-0.43	0.50	0.55	0.25	···	8740
6323D	Tubing	0.38-0.43	0.50	0.55	0.25	···	8740
6324C	Bars, forgings	0.38-0.43	0.65	0.70	0.25	···	8740 mod
6325D	Bars, forgings (heat treated; 105 ksi tensile strength)	0.38-0.43	0.50	0.55	0.25	···	8740
6327D	Bars, forgings (heat treated; 125 ksi tensile strength)	0.38-0.43	0.50	0.55	0.25	···	8740
6328E	Bars, forgings, tubing	0.48-0.53	0.50	0.55	0.25	···	8750
6330A	Bars, forgings	0.33-0.38	0.6	1.25	···	···	3135
6342D	Bars, forgings, tubing	0.38-0.32	0.80	1.0	0.25	···	9840
6350D	Plate, sheet, strip (annealed)	0.28-0.33	0.95	···	0.20	···	4130
6351A	Plate, sheet, strip (spheroidized)	0.28-0.33	0.95	···	0.20	···	4130
6352B	Plate, sheet, strip (annealed)	0.32-0.39	0.95	···	0.2	···	4135H
6354	Plate, sheet, strip	0.10-0.17	0.6	···	0.2	0.75 Si 0.1 Zr	NAX 9115-AC
6355G	Plate, sheet, strip (annealed)	0.28-0.33	0.50	0.55	0.20	···	8630
6356A	Plate, sheet, strip	0.30-0.35	0.95	···	0.20	···	···
6357D	Plate, sheet, strip	0.33-0.38	0.50	0.55	0.25	···	8735
6358B	Plate, sheet, strip	0.38-0.43	0.50	0.55	0.25	···	8740
6359B	Plate, sheet, strip	0.38-0.43	0.80	1.8	0.25	···	4340
6360F	Tubing, seamless	0.28-0.33	0.95	···	0.20	···	4130
6361	Tubing, seamless (125 ksi tensile strength)	0.27-0.33	0.95	···	0.2	···	4130
6362	Tubing, seamless (150 ksi tensile strength)	0.27-0.33	0.95	···	0.2	···	4130
6365E	Tubing, seamless	0.33-0.38	0.95	···	0.20	···	4135
6370F	Bars, forgings, rings	0.28-0.33	0.95	···	0.2	···	4130
6371D	Tubing	0.28-0.33	0.95	···	0.20	···	4130
6372D	Tubing	0.33-0.38	0.95	···	0.20	···	4135
6373A	Tubing, welded	0.28-0.33	0.95	···	0.20	···	4130

(continued)

(a) P, premium quality; VM, vacuum melted; CM, consumable electrode remelted.

Product descriptions and nominal compositions for wrought alloy steels (continued)

AMS designation	Product form (a)	Nominal composition, %					Nearest AISI-SAE grade
		C	Cr	Ni	Mo	Other	
6378	Bars (die drawn and tempered; 130 ksi yield strength)	0.38-0.45	0.95	⋯	0.2	⋯	4140
6379	Bars (die drawn and tempered; 165 ksi yield strength)	0.40-0.53	0.95	⋯	0.2	⋯	4140
6381B ...	Tubing	0.38-0.43	0.95	⋯	0.20	⋯	4140
6382G ...	Bars, forgings	0.38-0.43	0.95	⋯	0.20	⋯	4140
6385B ...	Plate, sheet, strip	0.27-0.33	1.25	⋯	0.50	0.65 Si 0.25 V	⋯
6386A ...	Plate, sheet (heat treated; 90 and 100 ksi yield strength)	⋯	⋯	⋯	⋯	⋯	⋯
6390A ...	Tubing (special quality)	0.38-0.43	0.95	⋯	0.20	⋯	4140
6395	Plate, sheet, strip	0.38-0.43	0.95	⋯	0.20	⋯	4140
6406A ...	Plate, sheet, strip	0.41-0.46	2.1	⋯	0.58	1.6 Si 0.05 V	⋯
6407B ...	Bars, forgings, tubing	0.27-0.33	1.2	2.05	0.45	⋯	⋯
6411	Bars, forgings, tubing (P, CM)	0.28-0.33	0.85	1.8	0.40	⋯	⋯
6412F ...	Bars, forgings	0.35-0.40	0.80	1.8	0.25	⋯	4337
6413D ...	Tubing	0.35-0.40	0.8	1.8	0.25	⋯	4337
6414A ...	Bars, forgings, tubing (P)	0.38-0.43	0.80	1.8	0.25	⋯	4340
6415G ...	Bars, forgings, tubing	0.38-0.43	0.80	1.8	0.25	⋯	4340
6416	Bars, forgings, tubing	0.41-0.46	0.8	1.8	0.4	1.6 Si 0.07 V	⋯
6417	Bars, forgings, tubing (P, CM)	0.38-0.43	0.82	1.8	0.40	1.6 Si 0.07 V	4340 mod
6418C ...	Bars, forgings, tubing	0.23-0.28	0.30	1.8	0.40	1.3 Mn 1.5 Si	Hy-Tuf
6419	Bars, forgings, tubing (P, CM)	0.41-0.46	0.82	1.8	0.40	1.6 Si 0.07 V	300 M
6421A ...	Bars, forgings, tubing	0.35-0.40	0.80	0.85	0.20	0.003B	Mod 98B37
6422C ...	Bars, forgings, tubing	0.38-0.43	0.80	0.85	0.20	0.003B	Mod 98B40
6423A ...	Bars, forgings, tubing	0.40-0.46	0.92	0.75	0.52	0.003B	⋯
6426A ...	Bars, forgings, tubing (VM)	0.80-0.90	1.0	⋯	0.58	0.75 Si	⋯
6427D ...	Bars, forgings, tubing	0.28-0.33	0.85	1.8	0.40	0.07 V	⋯
6428A ...	Bars, forgings, tubing	0.32-0.38	0.80	1.8	0.35	0.20 V	4335 mod
6429A ...	Bars, forgings, tubing (CM or VM)	0.33-0.38	0.80	1.8	0.35	0.20 V	4335 mod
6430A ...	Bars, forgings, tubing (special grade)	0.33-0.38	0.80	1.8	0.35	0.20 V	4335 mod
6431A ...	Bars, forgings, tubing (P, VM)	0.45-0.50	1.05	0.55	1.0	0.11 V	D6AC
6432	Bars, forgings, tubing	0.43-0.49	1.05	0.55	1.0	0.11 V	D6A
6433A ...	Plate, sheet, strip (special grade)	0.33-0.38	0.80	1.8	0.35	0.20 V	⋯
6434A ...	Plate, sheet, strip	0.31-0.38	0.80	1.8	0.35	0.20 V	4335 mod
6435A ...	Plate, sheet, strip (P, CM) (annealed)	0.33-0.38	0.80	1.8	0.35	0.20 V	4335 mod
6436A ...	Plate, sheet, strip	0.20-0.25	1.25	⋯	0.5	0.65 Si 0.85 V	⋯
6437A ...	Plate, sheet, strip	0.38-0.43	5.0	⋯	1.3	0.5 V	H-11
6438A ...	Plate, sheet, strip (P, CM)	0.45-0.50	1.05	0.55	1.0	0.11 V	D6
6440E ...	Bars, wire, forgings	0.98-1.10	1.45	⋯	⋯	⋯	52100
6441D ...	Tubing (bearing quality)	0.98-1.10	1.45	⋯	⋯	⋯	52100
6442C ...	Bars, wire, forgings	0.98-1.10	0.50	⋯	⋯	⋯	50100
6443B ...	Bars, wire, forgings (P, VM)	0.95-1.1	1.05	⋯	⋯	⋯	51100
6444B ...	Bars, wire, forgings tubing (P, VM)	0.95-1.10	1.45	⋯	⋯	⋯	52100
6445A ...	Bars, wire, forgings, tubing (P, VM)	0.92-1.02	1.05	⋯	⋯	1.1 Mn	51100 mod
6446	Bars, forgings (P)	0.95-1.10	1.05	⋯	⋯	⋯	51100
6447	Bars, forgings, tubing (P)	0.95-1.10	1.45	⋯	⋯	⋯	52100
6448C ...	Bars, forgings	0.48-0.53	0.95	⋯	⋯	0.22 V	6150
6450C ...	Wire (spring-annealed)	0.48-0.53	0.95	⋯	⋯	0.22 V	6150
6455C ...	Plate, sheet, strip (spring)	0.48-0.53	0.95	⋯	⋯	0.22 V	6150
6470F ...	Bars, forgings, tubing for nitriding	0.38-0.43	1.6	⋯	0.35	1.15 Al	⋯
6471	Bars, forgings, tubing for nitriding (P)	0.38-0.43	1.6	⋯	0.35	1.15 Al	⋯

(continued)

(a) P, premium quality; VM, vacuum melted; CM, consumable electrode remelted.

Product descriptions and nominal compositions for wrought alloy steels (continued)

AMS designation	Product form (a)	C	Cr	Ni	Mo	Other	Nearest AISI-SAE grade
6472	Bars, forgings for nitriding (heat treated; 112 ksi tensile strength)	0.38-0.43	1.6	...	0.35	1.13 Al	...
6475C ...	Bars, forgings, tubing for nitriding	0.21-0.26	1.1	3.5	0.25	1.25 Al	...
6485B ...	Bars, forgings	0.38-0.43	5.0	...	1.3	0.50 V	H-11
6487C ...	Bars, forgings (P, VM)	0.38-0.43	5.0	...	1.3	0.50 V	H-11
6488	Bars, forgings (P)	0.38-0.43	5.0	...	1.3	0.50 V	H-11
6490B ...	Bars, forgings, tubing (P, VM)	0.77-0.85	4.0	...	4.25	1.0 V	M-50
6512	Bars, forgings, tubing, rings (CM) (annealed)	...	...	18	4.9	7.8 Co 0.40 Ti 0.10 Al	...
6514	Bars, forgings, tubing, rings (annealed) (CM)	...	...	18.5	4.9	9.0 Co 0.65 Ti 0.10 Al	...
6520	Plate, sheet, strip (solution heat treated) (CM)	...	...	18	4.9	7.8 Co 0.40 Ti 0.10 Al	...
6521	Plate, sheet, strip (solution heat treated) (CM)	...	...	18.5	4.9	9.0 Co 0.65 Ti 0.10 Al	...
6526	Bars, forgings, tubing (annealed)(P, CM)	0.29-0.34	1.0	7.5	1.0	4.5 Co 0.09 V	...
6530E ...	Tubing, seamless	0.28-0.33	0.50	0.55	0.20	...	8630
6535D ...	Tubing, seamless	0.33-0.38	0.50	0.55	0.25	...	8735
6540A ...	Bars, forgings, rings, tubing (annealed)	0.24-0.30	0.48	8.0	0.48	4.0 Co 0.09 V	...
6541A ...	Bars, forgings, tubing, rings (annealed) (P, CM)	0.24-0.30	0.48	8.0	0.48	4.0 Co 0.09 V	...
6542A ...	Bars, forgings, tubing (annealed) (P, CM)	0.42-0.48	0.27	7.75	0.27	4.0 Co 0.09 V	...
6545A ...	Plate, sheet, strip (annealed)	0.24-0.30	0.48	8.0	0.48	4.0 Co 0.09 V	...
6546A ...	Plate, sheet, strip (annealed) (P, CM)	0.24-0.30	0.48	8.0	0.48	4.0 Co 0.09 V	
6550E ...	Tubing, welded	0.28-0.33	0.50	0.55	0.20	...	8630

(a) P, premium quality; VM, vacuum melted; CM, consumable electrode remelted.

Product descriptions and carbon contents for wrought carbon steels

AMS designation	Product form	Carbon content	Nearest AISI-SAE grade	AMS designation	Product form	Carbon content	Nearest AISI-SAE grade
5010E	Bars—screw machine stock	· · ·	1112	5069A	Bars, forgings, tubing	0.15-0.20	1018
5020	Bars, forging, tubing	0.32-0.39	11L37	5070C	Bars, forgings (55 ksi tensile strength)	0.18-0.23	1022
5022G	Bars, forging, tubing	0.14-0.20(a)	1117				
5024D	Bars, forging, tubing	0.32-0.39	1137	5075B	Tubing, seamless (55 ksi tensile strength)	0.22-0.28	1025
5032B	Wire (annealed)	0.18-0.23	1020	5077B	Tubing, welded	0.22-0.28	1025
5040F	Sheet, strip (deep forming grade)	0.15 max	1010	5080D	Bars, forgings, tubing	0.31-0.38	1035
5041	Sheet, strip (cold rolled, extra deep drawing)	0.08 max	1006	5082A	Tubing, seamless (90 ksi tensile strength)	0.31-0.38	1035
5042F	Sheet, strip (forming grade)	0.15 max	1010	5085A	Plate, sheet, strip (annealed)	0.47-0.55	1050
5044D	Sheet, strip (half hard temper)	Low	1010	5110B	Music wire—commercial	· · ·	1080
5045C	Sheet, strip (hard temper)	Low	1020	5112E	Music spring wire—best quality	· · ·	1090
5047A	Sheet, strip (aluminum killed)	Low	1010				
5050F	Tubing, seamless (annealed)	0.15 max	1010	5115C	Wire—spring	0.60-0.75	1070
5053C	Tubing, welded (annealed)	0.13 max	1010	5120F	Strip	0.68-0.80	1074
5060C	Bars, forgings, tubing	0.13-0.18	1015	5121C	Strip, spring	0.89-1.04	1095
5061B	Bars, wire	0.08-0.20	· · ·	5122C	Strip (hard temper)	0.89-1.04	1095
5062B	Bars, forgings, tubing, plate, sheet, strip	0.25 max	· · ·	5132D	Bars	0.90-1.30	1095

(a) Contains 1.2% manganese

Physical Properties

Densities of steels

Nearest AISI-SAE grade	C	Mn	Si	Cr	Ni	Other	Treatment or condition	Density Mg/m³	lb/in.³
1008	0.06	0.38	0.01	...	...	...	Annealed	7.871	0.2844
1024	0.23	0.64	0.11	...	...	...	Annealed	7.859	0.2839
1042	0.44	0.69	0.20	...	...	...	Annealed	7.844	0.2834
(a)	1.22	0.35	0.16	...	...	...	Annealed	7.830	0.2829
5130	0.31	0.74	...	1.00	...	...	Hardened and tempered	7.84	0.283
52100	0.98	0.28	...	1.68	...	...	Annealed	7.81	0.282
(a)	0.51	0.22	...	1.72	3.52	...	Quenched in brine (BQ)	7.79	0.281
							BQ, tempered 190 °C (375 °F)	7.80	0.282
							BQ, tempered 365 °C (690 °F)	7.82	0.283
							BQ, tempered 600 °C (1110 °F)	7.835	0.2831
							Annealed	7.835	0.2831
18Ni250(b)	0.026	0.1	0.11	...	18.5	4.7 Mo 7.0 Co 0.22 Ti 0.003 B	...	8.0	0.289

(a) No AISI-SAE grade of similar composition. (b) Nominal composition.

Specific damping capacity (a)

AISI-SAE grade	Treatment or condition	Specific damping capacity, %	AISI-SAE grade	Treatment or condition	Specific damping capacity, %	AISI-SAE grade	Treatment or condition	Specific damping capacity, %
1018	Normalized	1.5	1095	WQ, temper at 100 °C	0.2	1Cr-3Ni-0.3C	Not known	0.8
1095	Spheroidized	0.8				Gray cast iron	Not known	7 to 20
1095	Water quench from 800 °C	0.5	4140	Not known	0.15			
			1Ni-0.4C	Not known	0.3			

(a) Surface shear stress, 34.5 MPa (5 ksi); data apply to room temperature

Specific heat of steels

Nearest AISI-SAE grade	C	Mn	Cr	Ni	Mo	Other	Treatment or condition	Mean apparent specific heat, J/Kg·K, Temperature ranges, °C												
								50 to 100	150 to 200	200 to 250	250 to 300	300 to 350	350 to 400	450 to 500	550 to 600	650 to 700	700 to 750	750 to 800	850 to 900
1008	0.06	0.38	...	...	...	...	Annealed	481	519	536	553	574	595	662	754	867	1105	875	846
1008	0.08	0.31	...	...	...	...	Annealed	481	523	544	557	569	595	662	741	858	1139	960	...
1010(a) ..	0.10	0.42	...	...	...	0.008 P; 0.028 S	Not known	450	500	520	535	565	590	650	730	825	(a)	(a)	(a)
1025	0.23	0.64	...	...	...	...	Annealed	486	519	532	557	574	599	662	749	846	1432	950	...
1042	0.42	0.64	...	...	...	...	Annealed	486	515	528	548	569	586	649	708	770	1583	624	548
1078	0.80	0.32	...	...	...	...	Annealed	490	532	548	565	586	607	670	712	770	2081	615	...
(b)	1.22	0.35	...	...	...	...	Annealed	486	540	544	557	578	599	636	699	816	2089	649	...
1524	0.23	1.51	...	...	...	0.11 Cu	Annealed	477	511	528	544	565	590	649	741	837	1449	821	536
4130(c) ..	0.3	0.5	0.95	...	0.2	...	Hardened and tempered	477	515	...	544	...	595	657	737	825	...	833	...
4140	0.41	0.67	1.01	...	0.23	...	Hardened and tempered	...	473(d)	...	...	...	519(d)	...	561(d)	...	...	...	...
5132	0.32	0.69	1.09	0.073	...	...	Annealed	494	523	536	553	574	595	657	741	837	1499	934	574
5140	0.39	0.79	1.03	...	...	...	Hardened and tempered	452(d)	473(d)	...	...	...	519(d)	...	561(d)	...	...	...	...
(b)	0.35	0.59	0.88	0.26	0.20	...	Annealed	477	515	528	544	569	595	657	737	825	1616	883	...
(b)	0.33	0.55	0.17	3.47	...	...	Not known	481	523	536	548	569	590	662	749	1637	955	603	640
(b)	0.34	0.55	0.78	3.53	0.39	...	Hardened and tempered	486	523	540	557	582	607	670	770	1051	1662	636	636
(b)	0.49	0.90	...	...	...	1.98 Si; 0.64 Cu	Not known	498	523	540	557	578	603	666	749	829	904	1365	...

(a) See graph of specific heat versus temperature. (b) No equivalent grade. (c) Nominal composition. (d) Value presented is mean value for range of temperatures between room temperature and the higher of the cited temperatures.

Coefficients of linear thermal expansion

Nearest AISI-SAE grade	C	Mn	P	S	Si	Cr	Ni	Mo	Other	Treatment or condition
1008	0.06	0.38	0.017	0.035	0.01	0.02	0.55	0.03	0.08 Cu; 0.001 Al; 0.039 As	Annealed
1008	0.07	0.08	0.01	0.02	0.01	...	...	...	0.02 Cu	Annealed
1010	0.08	0.31	0.029	0.05	0.08	0.04	0.07	0.02	0.002 Al; 0.032 As	Annealed
1010	0.10	0.42	0.008	0.028	...	...	...	...	...	Not known
1015	0.17	0.42	0.012	0.035	...	...	...	...	Rolled	
1020	0.22	0.12	0.01	0.03	0.01	...	...	...	...	Annealed
1022	0.22	0.90	0.01	0.017	0.25	...	...	...	...	Not known
1022	0.23	0.64	0.034	0.034	0.11	Trace	0.07	...	0.13 Cu; 0.010 Al; 0.036 As	Annealed
1035	0.36	0.32	...	...	0.20	...	...	...	...	Annealed
1035	0.33	0.12	0.01	0.03	0.03	...	...	...	0.03 Cu	Annealed
1040	0.40	0.11	0.01	0.03	0.07	...	...	...	0.03 Cu	Annealed
1040	0.42	0.64	0.031	0.029	0.11	Trace	0.06	...	0.12 Cu; 0.006 Al; 0.033 As	Annealed
1045	0.44	0.69	0.037	0.038	0.20	0.03	0.04	...	0.06 Cu; 0.006 Al; 0.024 As	Annealed
1045	0.44	0.57	0.013	0.033	0.16	...	...	...	0.14 V	Annealed
1052	0.49	1.21	0.05	0.05	0.12	...	...	...	...	Annealed
1055	0.56	0.19	Trace	0.023	0.04	...	...	...	0.02 Cu	Annealed
1060	0.59	0.92	0.024	0.033	0.25	...	...	...	...	Annealed
1070	0.66	0.72	0.028	0.016	0.18	...	...	...	...	Rolled
1078	0.80	0.32	...	...	...	...	...	...	...	Annealed
1080	0.81	0.10	Trace	0.025	0.025	...	...	...	0.02 Cu	Annealed
(e)	0.82	1.65	...	...	0.20	0.03	...	...	...	Not known
1085	0.80	0.32	0.008	0.009	0.13	0.11	0.13	...	0.07 Cu; 0.004 Al; 0.021 As	Annealed
1095	1.1	0.3	0.025	0.025	0.2	...	...	...	...	Annealed / Hardened
(e)	1.22	0.35	0.009	0.015	0.16	0.11	0.13	0.01	0.077 Cu; 0.006 Al; 0.025 As	Annealed
1145	0.42	0.64	0.031	0.029	0.11	Trace	0.06	...	0.12 Cu; 0.006 Al; 0.003 As	Annealed
1145	0.44	0.69	0.037	0.038	0.20	0.03	0.04	...	0.06 Cu; 0.006 Al; 0.024 As	Annealed
1524	0.23	1.51	...	...	...	...	...	...	...	Not known
(e)	0.82	1.65	...	...	0.20	0.03	...	...	...	Not known
(e)	0.40	0.67	...	...	...	...	0.80	...	...	Oil hardened, tempered 600 °C (1110 °F)
2330	0.33	0.78	0.014	0.035	0.09	...	3.59	...	...	Annealed
3140	0.40-0.50	0.50-0.80	...	...	...	0.45-0.75	1.00-1.50	...	...	Hardened and tempered
4137	0.39	0.51	0.015	0.029	0.19	0.87	...	0.21	...	Rolled
4140	0.41	0.67	...	...	...	1.01	...	0.23	...	Oil hardened, tempered 600 °C (1110 °F)
4340	0.41	...	...	...	...	1.07	1.43	0.26	...	Oil hardened, tempered 630 °C (1170 °F)
(e)	0.32	...	...	...	...	0.67	2.60	0.51	...	Oil hardened, tempered 650 °C (1200 °F)
4615	...	...	...	...	...	...	1.65	0.30	...	Not known
4617	0.18	...	...	...	...	...	1.76	0.20	...	Carburized and hardened
5140	0.35	0.31	...	...	0.19	0.75	...	...	...	Annealed
(e)	0.37	0.33	...	...	0.21	1.57	...	...	...	Annealed
52100	0.94	0.34	...	...	0.27	0.95	...	...	...	Annealed / Hardened
6150	0.22	0.75	0.019	0.033	0.27	0.96	...	...	0.17 V	Annealed / Hardened, tempered 205 °C (400 °F)
6150	0.35	1.42	0.013	0.057	0.20	1.00	...	...	0.11 V	Annealed
6150	0.53	0.80	0.015	0.020	0.15	1.02	...	...	0.17 V	Annealed / Hardened, tempered 425 °C (800 °F) / Hardened, tempered 650 °C (1200 °F)
18Ni250(e) ...	0.026	0.1	...	...	0.11	...	18.5	4.7	7.0 Co; 0.22 Ti; 0.003 B	Not known

(continued)

(a) To obtain coefficients in μin./in.·°F multiply values in table by 0.556. (b) Stated value represents average coefficient between 0 °C (32 °F) and indicated temperature. (c) 10.3 μm/m·K from − 100 °C to 20 °C, 9.8 μm/m·K from − 150 to 20 °C. (d) Stated value represents average coefficient between 25 °C (75 °F) and indicated temperature. (e) Nominal composition. (f) 11.2 μm/m·K from − 100 to 20 °C; 10.4 μm/m·K from − 150 to 20 °C. (g) Stated value represents average coefficient between 24 and 284 °C (74 and 540 °F).

Coefficients of linear thermal expansion (continued)

Nearest AISI-SAE grade	20 °C to: 68 °F to:	Average coefficients of expansion, μm/m·K(a)									°C °F
		100 212	200 392	300 572	400 752	500 932	600 1112	700 1292	800 1472	1000 1832	
1008		12.6(b)	13.1(b)	13.5(b)	13.8(b)	14.2(b)	14.6(b)	15.0(b)	14.7	13.8	
1008		11.6	12.5	13.0	13.6	14.2	14.6	15.0	. . .	. . .	
1010		12.2(b)	13.0(b)	13.5(b)	13.9(b)	14.3(b)	14.7(b)	15.0(b)	. . .	. . .	
1010		11.9(c)	12.6	13.3	13.8	14.3	14.7	14.9	14.0	. . .	
1015		11.9(b)	12.5(b)	13.0(b)	13.6(b)	14.2(b)	. . .	. . .	. . .	. . .	
1020		11.7	12.1	12.8	13.4	13.9	14.4	14.8	. . .	. . .	
1022		12.5	12.7	. . .	. . .	. . .	. . .	. . .	. . .	. . .	
1022		12.2(b)	12.7(b)	13.1(b)	13.5(b)	13.9(b)	14.4(b)	14.9(b)	12.6	13.4	
1035		. . .	12.6	13.3	13.8	14.3	14.8	15.2	. . .	. . .	
1035		11.1	11.9	12.7	13.4	14.0	14.4	14.8	. . .	. . .	
1040		11.3	12.0	12.5	13.3	13.9	14.4	14.8	. . .	. . .	
		11.2(b)	12.1(b)	13.0(b)	13.6(b)	14.0(b)	14.6(b)	14.8(b)	11.8	13.6	
1045		11.6(b)	12.3(b)	13.1(b)	13.7(b)	14.2(b)	14.7(b)	15.1(b)	. . .	. . .	
		11.2(d)	11.9(d)	12.7(d)	13.5(d)	14.1(d)	14.5(d)	14.8(d)	. . .	. . .	
1052		11.3(d)	11.8(d)	12.7(d)	13.7(d)	14.5(d)	14.7(d)	15.0(d)	. . .	. . .	
1055		11.0	11.8	12.6	13.4	14.0	14.5	14.8	. . .	. . .	
1060		11.1(d)	11.9(d)	12.9(d)	13.5(d)	14.1(d)	14.6(d)	14.9(d)	. . .	. . .	
1070		11.8(b)	12.6(b)	13.3(b)	14.0(b)	. . .	. . .	. . .	. . .	. . .	
1078		11.1	11.7	. . .	13.2	. . .	14.2	. . .	13.8	15.7	
1080		11.0	11.6	12.4	13.2	13.8	14.2	14.7	. . .	. . .	
(e)		8.8(b)	9.8(b)	11.3(b)	12.3(b)	13.1(b)	13.6(b)	14.2(b)	. . .	. . .	
1085		11.1(b)	11.7(b)	12.5(b)	13.2(b)	13.6(b)	14.2(b)	14.7(b)	. . .	. . .	
1095		11.4(b)	. . .	. . .	. . .	. . .	. . .	. . .	. . .	. . .	
		13.0(b)	. . .	. . .	. . .	. . .	. . .	. . .	. . .	. . .	
(e)		10.6(b)	11.2(b)	12.1(b)	12.9(b)	13.5(b)	14.2(b)	14.7(b)	14.3	16.8	
1145		11.2(b)	12.1(b)	13.0(b)	13.6(b)	14.0(b)	14.6(b)	14.8(b)	. . .	. . .	
		11.6(b)	12.3(b)	13.1(b)	13.7(b)	14.2(b)	14.7(b)	15.1(b)	. . .	. . .	
1524		11.9	12.7	. . .	13.9	. . .	14.7	. . .	12.1	13.8	
(e)		8.8(b)	9.8(b)	11.3(b)	12.3(b)	13.1(b)	13.6(b)	14.2(b)	. . .	. . .	
		11.9	12.6	. . .	13.8	. . .	14.5	. . .	. . .	. . .	
2330		10.9(d)	11.2(d)	12.1(d)	12.9(d)	13.4(d)	13.8(d)	. . .	. . .	. . .	
3140		11.8(b)	12.3(b)	12.9(b)	13.4(b)	14.0(b)	. . .	. . .	. . .	. . .	
4137		11.2(b)	11.8(b)	12.4(b)	13.0(b)	13.6(b)	. . .	. . .	. . .	. . .	
4140		12.3	12.7	. . .	13.7	. . .	14.5	. . .	. . .	. . .	
4340		(f)	12.4	. . .	13.6	. . .	14.3	. . .	. . .	. . .	
(e)		. . .	11.6	. . .	13.1	. . .	13.9	. . .	. . .	. . .	
4615		11.5	12.1	12.7	13.2	13.7	14.1	. . .	. . .	. . .	
4617		12.5	13.1	. . .	. . .	. . .	. . .	. . .	. . .	. . .	
5140		. . .	12.6	13.4	13.9	14.3	14.6	15.0	. . .	. . .	
(e)		. . .	12.8	13.4	13.8	14.2	14.4	14.6	. . .	. . .	
52100		11.9(b)	. . .	. . .	. . .	. . .	. . .	. . .	. . .	. . .	
		12.6(b)	. . .	. . .	. . .	. . .	. . .	. . .	. . .	. . .	
6150		12.2	12.7	13.3	13.7	14.1	14.4	. . .	. . .	. . .	
		12.0	12.5	12.9	13.0	13.3	13.7	. . .	. . .	. . .	
6150		12.4(d)	12.6(d)	13.3(d)	13.8(d)	14.2(d)	14.5(d)	14.7(d)	. . .	. . .	
6150		12.4	12.8	13.4	13.9	14.2	14.5	. . .	. . .	. . .	
		11.8	12.4	13.1	13.6	13.9	14.1	. . .	. . .	. . .	
		12.3	12.7	13.4	13.9	14.3	14.7	. . .	. . .	. . .	
18Ni250(e) . . .		. . .	. . .	10.1(g)	. . .	10.1	. . .	. . .	. . .	. . .	

(a) To obtain coefficients in μin./in.·°F multiply values in table by 0.556. (b) Stated value represents average coefficient between 0 °C (32 °F) and indicated temperature. (c) 10.3 μm/m·K from −100 °C to 20 °C, 9.8 μm/m·K from −150 °C to 20 °C. (d) Stated value represents average coefficient between 25 °C (75 °F) and indicated temperature. (e) Nominal composition. (f) 11.2 μm/m·K from −100 to 20 °C; 10.4 μm/m·K from −150 to 20 °C. (g) Stated value represents average coefficient between 24 and 284 °C (74 and 540 °F)

Electrical resistivities of steels

Nearest AISI-SAE grade	Chemical composition, %							Treatment or condition
	C	Mn	Si	Cr	Ni	Mo	Other	
1008	0.06	0.38	. . .	. . .	. . .	. . .	. . .	Annealed
1008	0.08	0.31	. . .	. . .	. . .	. . .	. . .	Annealed
1025	0.23	0.64	. . .	. . .	. . .	. . .	. . .	Annealed
1042	0.42	0.64	. . .	. . .	. . .	. . .	. . .	Annealed
1078	0.80	0.32	. . .	. . .	. . .	. . .	. . .	Annealed
(a)	1.22	0.35	. . .	. . .	. . .	. . .	. . .	Annealed
1524	0.23	1.51	. . .	. . .	. . .	. . .	0.11 Cu	Not known
4130(b)	0.3	0.5	0.3	0.95	. . .	0.2	. . .	Hardened and tempered
4140	0.41	0.67	. . .	1.01	. . .	0.23	. . .	Hardened and tempered
4340	0.41	. . .	. . .	1.07	1.43	0.26	. . .	Hardened and tempered
5132	0.32	0.69	. . .	1.09	0.073	. . .	. . .	Annealed
5140	0.39	0.79	. . .	1.03	. . .	. . .	. . .	Hardened and tempered
(a)	0.35	0.59	. . .	0.88	0.26	0.20	. . .	Annealed
(a)	0.33	0.55	. . .	0.17	3.47	. . .	. . .	Not known
(a)	0.34	0.55	. . .	0.78	3.53	0.39	. . .	Hardened and tempered
(a)	0.49	0.90	1.98	. . .	. . .	. . .	0.64 Cu	Not known
18Ni250(b) .	0.026	0.1	0.11	. . .	18.5	4.7	7.0 Co; 0.22 Ti; 0.003 B	Annealed
								Aged

(continued)

(a) No AISI-SAE standard grade of similar composition. (b) Nominal composition.

Electrical resistivities of steels (continued)

| °C 20 | 100 | 200 | 400 | 600 | 700 | 800 | 900 | 1000 | 1100 | 1200 | 1300 |
°F 68	212	392	752	1112	1292	1472	1652	1832	2012	2192	2372
0.130	0.178	0.252	0.448	0.725	0.898	1.073	1.124	1.160	1.189	1.216	1.241
0.142	0.190	0.263	0.458	0.734	0.905	1.081	1.130	1.165	1.193	1.220	1.244
0.169	0.219	0.292	0.487	0.758	0.925	1.094	1.136	1.167	1.194	1.219	1.239
0.171	0.221	0.296	0.493	0.766	0.932	1.111	1.149	1.179	1.207	1.230	· · ·
0.180	0.232	0.308	0.505	0.772	0.935	1.129	1.164	1.191	1.214	1.231	1.246
0.196	0.252	0.333	0.540	0.802	0.964	1.152	1.196	1.226	1.249	1.271	1.287
0.208	0.259	0.333	0.523	0.786	0.946	1.103	1.143	1.174	1.202	1.227	1.250
0.223	0.271	0.342	0.529	0.786	· · ·	1.103	· · ·	1.171	· · ·	1.222	· · ·
0.222	0.263	0.326	0.475	0.646	· · ·	· · ·	· · ·	· · ·	· · ·	· · ·	· · ·
0.248	0.298	0.367	0.552	0.797	· · ·	· · ·	· · ·	· · ·	· · ·	· · ·	· · ·
0.210	0.259	0.330	0.517	0.778	0.934	1.106	1.145	1.177	1.205	1.230	1.251
0.228	0.281	0.352	0.530	0.785	· · ·	· · ·	· · ·	· · ·	· · ·	· · ·	· · ·
0.223	0.271	0.342	0.529	0.786	0.944	1.103	1.138	1.171	1.200	1.222	1.242
0.271	0.320	0.390	0.567	0.814	0.992	1.122	1.149	1.180	1.204	1.228	1.248
0.289	0.337	0.406	0.582	0.825	0.994	1.114	1.146	1.176	1.199	1.222	1.242
0.429	0.470	0.529	0.685	0.911	1.057	1.173	1.197	1.223	1.249	1.271	1.289
0.6 to 0.7	· · ·	· · ·	· · ·	· · ·	· · ·	· · ·	· · ·	· · ·	· · ·	· · ·	· · ·
0.36 to 0.6	· · ·	· · ·	· · ·	· · ·	· · ·	· · ·	· · ·	· · ·	· · ·	· · ·	· · ·

Resistivity, μΩ·m, at indicated temperature

Thermal conductivities of steels

Nearest AISI-SAE grade	C	Mn	P	S	Si	Cr	Ni	Mo	Other	Treatment or condition
						Chemical composition, %				
1008 ......0.08	0.31	...	...	...	0.045	0.07	0.02	...	...	Not known
1008 ......0.06	0.4	...	...	...	...	...	...	...	...	Annealed
1010 ......0.10	0.42	0.008	0.028	...	...	...	...	...	...	Not known
1025 ......0.23	0.64	...	...	...	Trace	0.074	...	0.13 Cu		Annealed
1042 ......0.42	0.64	...	...	...	Trace	0.063	...	0.12 Cu		Annealed
1078 ......0.80	0.32	...	...	...	0.11	0.13	0.01	0.07 Cu		Annealed
(d) 1.22	0.35	...	...	...	0.11	0.13	0.01	0.08 Cu		Annealed
1524 ......0.23	1.51	0.037	0.038	0.12	0.06	0.04	0.025	0.105 Cu; 0.033 Co; 0.015 Al		Annealed
4037 ......0.37	1.56	...	...	...	...	...	0.26	...		Hardened and tempered
4130(e) 0.3	0.5	...	...	0.3	0.95	...	0.5	...		Hardened and tempered
4140 ......0.41	0.67	...	...	...	1.01	...	0.23	...		Hardened and tempered
5132 ......0.32	0.69	...	...	...	1.09	0.073	0.012	0.07 Cu		Annealed
5140 ......0.39	0.79	...	...	...	1.03	...	...	...		Hardened and tempered
(d) 0.35	0.59	...	...	...	0.88	0.26	0.20	0.12 Cu		Not known
(d) 0.33	0.55	...	...	...	0.17	3.47	0.04	0.09 Cu		Not known
(d) 0.34	0.55	...	...	...	0.78	3.53	0.39	0.05 Cu		Hardened and tempered
(d) 1.22	13.0	...	...	0.22	0.03	0.07	...	0.07 Cu		Not known
18Ni250(e) ..0.026	0.1	...	...	0.11	...	18.5	4.5	7.0 Co; 0.22 Ti; 0.003 B		Not known

Nearest AISI-SAE grade	0 / 32	100 / 212	200 / 392	300 / 572	400 / 752	500 / 932	600 / 1112	700 / 1292	800 / 1472	1000 / 1832	1200 / 2192	°C / °F
					Conductivity, W/m·K (a)							
1008 ....59.5	57.8	53.2	49.4	45.6	41.0	36.8	33.1	28.5	27.6	29.7	
1008 ....65.3(b)	60.3	54.9	...	45.2	...	36.4	...	28.5	27.6	...	
1010 ....65.2	60.2	55.5	50.7	46.0	41.5	36.9	32.9	28.9	...	...	(c)
1025 ....51.9	51.1	49.0	46.1	42.7	39.4	35.6	31.8	26.0	27.2	29.7	
1042 ....51.9	50.7	48.2	45.6	41.9	38.1	33.9	30.1	24.7	26.8	29.7	
1078 ....47.8	48.2	45.2	41.4	38.1	35.2	32.7	30.1	24.3	26.8	30.1	
(d) 45.2	44.8	43.5	41.0	38.5	36.0	33.5	31.0	23.9	26.0	28.5	
1524 ....46.0	45.8	45.0	42.6	40.1	37.4	34.4	30.6	26.6	27.2	...	
4037	48.2	45.6	...	39.4	...	33.9	...	...	...	...	
4130(e)	42.7	...	40.6	...	37.3	...	31.0	...	28.1	30.1	
4140	42.7	42.3	...	37.7	...	33.1	...	...	...	...	
5132 ....48.6	46.5	44.4	42.3	38.5	35.6	31.8	28.9	26.0	28.1	30.1	
5140	44.8	43.5	...	37.7	...	31.4	...	...	...	...	
(d) 42.7	42.7	41.9	40.6	38.9	36.4	33.9	31.0	26.4	28.1	30.1	
(d) 36.4	37.7	38.9	39.4	36.8	35.2	32.7	26.4	25.1	27.6	30.1	
(d) 33.1	33.9	35.2	35.6	35.6	33.5	30.6	28.1	26.8	28.5	30.1	
(d) 13.0	13.8	16.3	18.0	19.3	20.5	21.8	22.6	23.4	25.5	28.1	
18Ni250(e) 19.7(b)	20.9	...	...	...	...	...	...	...	...	...	

(a) To obtain conductivities in BTU/(ft·h·°F), multiply values in table by 0.5778; to obtain conductivities in cal/(cm·s·°C), multiply by 0.002388. (b) Thermal conductivity at 21 °C (70 °F). (c) 70.4 W/m·K at −100 °C (−148 °F). (d) No equivalent grade. (e) Nominal composition.

Mechanical Properties

This guide shows what to expect from a given grade of steel in the indicated condition. Data were obtained from specimens 0.505 in. in diameter which were machined from 1-in. rounds; gage lengths were 2 in. Average properties of hot rolled, normalized, and annealed material are listed, while properties of quenched and tempered grades are for single heats. Sources of the data are Bethlehem Steel Corp. and Republic Steel Corp.

Because of the many variables that affect a steel's properties, however, these listed properties should not be considered either as average or typical. Both strengths and ductilities may range up and down from the values given, depending on the compositions of individual heats of the same grade, section sizes, and internal structures. Properties of carbon steels and many alloy steels are also affected by residual elements (particularly nickel, chromium and molybdenum), even though their amounts are limited to maximums by AISI and SAE specifications.

Fine-grained steels normally have better impact strength than coarse-grained types, a factor which should be considered when reviewing the results of Izod tests. Hardness values are not always related to corresponding tensile strengths. In particular, this effect occurs with carbon steels because they are shallow-hardening. Hardness tests were made on surfaces, and these hardnesses will not reflect the tensile strengths obtained with specimens representing bar centers. (Center hardnesses are usually lower than surface hardnesses.)

Hot rolled properties for alloy steels are not given because these grades are customarily heat treated. Because the samples were small enough to assure full quenching, values indicate strengths and ductilities which may be obtained with hardened, fine-grained steels of a similar section size at room temperatures.

Mechanical properties of selected carbon and alloy steels in the hot rolled, normalized and annealed condition

AISI No. (a)	Treatment	Austenitizing temperature °C	°F	Tensile strength MPa	ksi	Yield strength MPa	ksi	Elongation, %	Reduction in area, %	Hardness, HB	Izod impact strength J	ft·lb
1015 ...	As-rolled	...	...	420.6	61.0	313.7	45.5	39.0	61.0	126	110.5	81.5
	Normalized	925	1700	424.0	61.5	324.1	47.0	37.0	69.6	121	115.5	85.2
	Annealed	870	1600	386.1	56.0	284.4	41.3	37.0	69.7	111	115.0	84.8
1020 ...	As-rolled	...	...	448.2	65.0	330.9	48.0	36.0	59.0	143	86.8	64.0
	Normalized	870	1600	441.3	64.0	346.5	50.3	35.8	67.9	131	117.7	86.8
	Annealed	870	1600	394.7	57.3	294.8	42.8	36.5	66.0	111	123.4	91.0
1022 ...	As-rolled	...	...	503.3	73.0	358.5	52.0	35.0	67.0	149	81.3	60.0
	Normalized	925	1700	482.6	70.0	358.5	52.0	34.0	67.5	143	117.3	86.5
	Annealed	870	1600	429.2	62.3	317.2	46.0	35.0	63.6	137	120.7	89.0
1030 ...	As-rolled	...	...	551.6	80.0	344.7	50.0	32.0	57.0	179	74.6	55.0
	Normalized	925	1700	520.6	75.5	344.7	50.0	32.0	60.8	149	93.6	69.0
	Annealed	845	1550	463.7	67.3	341.3	49.5	31.2	57.9	126	69.4	51.2
1040 ...	As-rolled	...	...	620.5	90.0	413.7	60.0	25.0	50.0	201	48.8	36.0
	Normalized	900	1650	589.5	85.5	374.0	54.3	28.0	54.9	170	65.1	48.0
	Annealed	790	1450	518.8	75.3	353.4	51.3	30.2	57.2	149	44.3	32.7
1050 ...	As-rolled	...	...	723.9	105.0	413.7	60.0	20.0	40.0	229	31.2	23.0
	Normalized	900	1650	748.1	108.5	427.5	62.0	20.0	39.4	217	27.1	20.0
	Annealed	790	1450	636.0	92.3	365.4	53.0	23.7	39.9	187	16.9	12.5
1060 ...	As-rolled	...	...	813.6	118.0	482.6	70.0	17.0	34.0	241	17.6	13.0
	Normalized	900	1650	775.7	112.5	420.6	61.0	18.0	37.2	229	13.2	9.7
	Annealed	790	1450	625.7	90.8	372.3	54.0	22.5	38.2	179	11.3	8.3
1080 ...	As-rolled	...	...	965.3	140.0	586.1	85.0	12.0	17.0	293	6.8	5.0
	Normalized	900	1650	1010.1	146.5	524.0	76.0	11.0	20.6	293	6.8	5.0
	Annealed	790	1450	615.4	89.3	375.8	54.5	24.7	45.0	174	6.1	4.5

(continued)

(a) All grades are fine grained except for those in the 1100 series which are coarse-grained. Heat treated specimens were oil quenched unless otherwise indicated.

Mechanical properties of selected carbon and alloy steels in the hot rolled, normalized and annealed condition (continued)

AISI No. (a)	Treatment	Austenitizing temperature °C	°F	Tensile strength MPa	ksi	Yield strength MPa	ksi	Elongation, %	Reduction in area, %	Hardness, HB	Izod impact strength J	ft·lb
1095 ...	As-rolled	...	...	965.3	140.0	572.3	83.0	9.0	18.0	293	4.1	3.0
	Normalized	900	1650	1013.5	147.0	499.9	72.5	9.5	13.5	293	5.4	4.0
	Annealed	790	1450	656.7	95.3	379.2	55.0	13.0	20.6	192	2.7	2.0
1117 ...	As-rolled	...	...	486.8	70.6	305.4	44.3	33.0	63.0	143	81.3	60.0
	Normalized	900	1650	467.1	67.8	303.4	44.0	33.5	63.8	137	85.1	62.8
	Annealed	855	1575	429.5	62.3	279.2	40.5	32.8	58.0	121	93.6	69.0
1118 ...	As-rolled	...	...	521.2	75.6	316.5	45.9	32.0	70.0	149	108.5	80.0
	Normalized	925	1700	477.8	69.3	319.2	46.3	33.5	65.9	143	103.4	76.3
	Annealed	790	1450	450.2	65.3	284.8	41.3	34.5	66.8	131	106.4	78.5
1137 ...	As-rolled	...	...	627.4	91.0	379.2	55.0	28.0	61.0	192	82.7	61.0
	Normalized	900	1650	668.8	97.0	396.4	57.5	22.5	48.5	197	63.7	47.0
	Annealed	790	1450	584.7	84.8	344.7	50.0	26.8	53.9	174	49.9	36.8
1141 ...	As-rolled	...	...	675.7	98.0	358.5	52.0	22.0	38.0	192	11.1	8.2
	Normalized	900	1650	706.7	102.5	405.4	58.8	22.7	55.5	201	52.6	38.8
	Annealed	815	1500	598.5	86.8	353.0	51.2	25.5	49.3	163	34.3	25.3
1144 ...	As-rolled	...	...	703.3	102.0	420.6	61.0	21.0	41.0	212	52.9	39.0
	Normalized	900	1650	667.4	96.8	399.9	58.0	21.0	40.4	197	43.4	32.0
	Annealed	790	1450	584.7	84.8	346.8	50.3	24.8	41.3	167	65.1	48.0
1340 ...	Normalized	870	1600	836.3	121.3	558.5	81.0	22.0	62.9	248	92.5	68.2
	Annealed	800	1475	703.3	102.0	436.4	63.3	25.5	57.3	207	70.5	52.0
3140 ...	Normalized	870	1600	891.5	129.3	599.8	87.0	19.7	57.3	262	53.6	39.5
	Annealed	815	1500	689.5	100.0	422.6	61.3	24.5	50.8	197	46.4	34.2
4130 ...	Normalized	870	1600	668.8	97.0	436.4	63.3	25.5	59.5	197	86.4	63.7
	Annealed	865	1585	560.5	81.3	360.6	52.3	28.2	55.6	156	61.7	45.5
4140 ...	Normalized	870	1600	1020.4	148.0	655.0	95.0	17.7	46.8	302	22.6	16.7
	Annealed	815	1500	655.0	95.0	417.1	60.5	25.7	56.9	197	54.5	40.2
4150 ...	Normalized	870	1600	1154.9	167.5	734.3	106.5	11.7	30.8	321	11.5	8.5
	Annealed	815	1500	729.5	105.8	379.2	55.0	20.2	40.2	197	24.7	18.2
4320 ...	Normalized	895	1640	792.9	115.0	464.0	67.3	20.8	50.7	235	72.9	53.8
	Annealed	850	1560	579.2	84.0	609.5	61.6	29.0	58.4	163	109.8	81.0
4340 ...	Normalized	870	1600	1279.0	185.5	861.8	125.0	12.2	36.3	363	15.9	11.7
	Annealed	810	1490	744.6	108.0	472.3	68.5	22.0	49.9	217	51.1	37.7
4620 ...	Normalized	900	1650	574.3	83.3	366.1	53.1	29.0	66.7	174	132.9	98.0
	Annealed	855	1575	512.3	74.3	372.3	54.0	31.3	60.3	149	93.6	69.0
4820 ...	Normalized	860	1580	75.0	109.5	484.7	70.3	24.0	59.2	229	109.8	81.0
	Annealed	815	1500	681.2	98.8	464.0	67.3	22.3	58.8	197	92.9	68.5
5140 ...	Normalized	870	1600	792.9	115.0	472.3	68.5	22.7	59.2	229	38.0	28.0
	Annealed	830	1525	572.3	83.0	293.0	42.5	28.6	57.3	167	40.7	30.0
5150 ...	Normalized	870	1600	870.8	126.3	529.5	76.8	20.7	58.7	255	31.5	23.2
	Annealed	825	1520	675.7	98.0	357.1	51.8	22.0	43.7	197	25.1	18.5
5160 ...	Normalized	855	1575	957.0	138.8	530.9	77.0	17.5	44.8	269	10.8	8.0
	Annealed	815	1495	722.6	104.8	275.8	40.0	17.2	30.6	197	10.0	7.4
6150 ...	Normalized	870	1600	939.8	136.3	615.7	89.3	21.8	61.0	269	35.5	26.2
	Annealed	815	1500	667.4	96.8	412.3	59.8	23.0	48.4	197	27.4	20.2
8620 ...	Normalized	915	1675	632.9	91.8	357.1	51.8	26.3	59.7	183	99.7	73.5
	Annealed	870	1600	536.4	77.8	385.4	55.9	31.3	62.1	149	112.2	82.8
8630 ...	Normalized	870	1600	650.2	94.3	429.5	62.3	23.5	53.5	187	94.6	69.8
	Annealed	845	1550	564.0	81.8	372.3	54.0	29.0	58.9	156	95.2	70.2
8650 ...	Normalized	870	1600	1023.9	148.5	688.1	99.8	14.0	40.4	302	13.6	10.0
	Annealed	795	1465	715.7	103.8	386.1	56.0	22.5	46.4	212	29.4	21.7
8740 ...	Normalized	870	1600	929.4	134.8	606.7	88.0	16.0	47.9	269	17.6	13.0
	Annealed	815	1500	695.0	100.8	415.8	60.3	22.2	46.4	201	40.0	29.5
9255 ...	Normalized	900	1650	932.9	135.3	579.2	84.0	19.7	43.4	269	13.6	10.0
	Annealed	845	1550	774.3	112.3	486.1	70.5	21.7	41.1	229	8.8	6.5
9310 ...	Normalized	890	1630	906.7	131.5	570.9	82.8	18.8	58.1	269	119.3	88.0
	Annealed	845	1550	820.5	119.0	439.9	63.8	17.3	42.1	241	78.6	58.0

(a) All grades are fine grained except for those in the 1100 series which are coarse-grained. Heat treated specimens were oil quenched unless otherwise indicated.

Mechanical properties of selected carbon and alloy steels in the quenched and tempered condition

AISI No.(a)	Tempering temperature °C	°F	Tensile strength MPa	ksi	Yield strength MPa	ksi	Elongation, %	Reduction in area, %	Hardness, HB
1030(b)	205	400	848	123	648	94	17	47	495
	315	600	800	116	621	90	19	53	401
	425	800	731	106	579	84	23	60	302
	540	1000	669	97	517	75	28	65	255
	650	1200	586	85	441	64	32	70	207
1040(b)	205	400	896	130	662	96	16	45	514
	315	600	889	129	648	94	18	52	444
	425	800	841	122	634	92	21	57	352
	540	1000	779	113	593	86	23	61	269
	650	1200	669	97	496	72	28	68	201
1040	205	400	779	113	593	86	19	48	262
	315	600	779	113	593	86	20	53	255
	425	800	758	110	552	80	21	54	241
	540	1000	717	104	490	71	26	57	212
	650	1200	634	92	434	63	29	65	192
1050(b)	205	400	1124	163	807	117	9	27	514
	315	600	1089	158	793	115	13	36	444
	425	800	1000	145	758	110	19	48	375
	540	1000	862	125	655	95	23	58	293
	650	1200	717	104	538	78	28	65	235
1050	205	400	...	...	...	...	...	...	...
	315	600	979	142	724	105	14	47	321
	425	800	938	136	655	95	20	50	277
	540	1000	876	127	579	84	23	53	262
	650	1200	738	107	469	68	29	60	223
1060	205	400	1103	160	779	113	13	40	321
	315	600	1103	160	779	113	13	40	321
	425	800	1076	156	765	111	14	41	311
	540	1000	965	140	669	97	17	45	277
	650	1200	800	116	524	76	23	54	229
1080	205	400	1310	190	979	142	12	35	388
	315	600	1303	189	979	142	12	35	388
	425	800	1289	187	951	138	13	36	375
	540	1000	1131	164	807	117	16	40	321
	650	1200	889	129	600	87	21	50	255
1095(b)	205	400	1489	216	1048	152	10	31	601
	315	600	1462	212	1034	150	11	33	534
	425	800	1372	199	958	139	13	35	388
	540	1000	1138	165	758	110	15	40	293
	650	1200	841	122	586	85	20	47	235
1095	205	400	1289	187	827	120	10	30	401
	315	600	1262	183	813	118	10	30	375
	425	800	1213	176	772	112	12	32	363
	540	1000	1089	158	676	98	15	37	321
	650	1200	896	130	552	80	21	47	269
1137	205	400	1082	157	938	136	5	22	352
	315	600	986	143	841	122	10	33	285
	425	800	876	127	731	106	15	48	262
	540	1000	758	110	607	88	24	62	229
	650	1200	655	95	483	70	28	69	197
1137(b)	205	400	1496	217	1165	169	5	17	415
	315	600	1372	199	1124	163	9	25	375
	425	800	1103	160	986	143	14	40	311
	540	1000	827	120	724	105	19	60	262
	650	1200	648	94	531	77	25	69	187

(continued)

(a) All grades are fine-grained except for those in the 1100 series which are coarse-grained. Heat treated specimens were oil quenched unless otherwise indicated. (b) Water quenched.

Mechanical properties of selected carbon and alloy steels in the quenched and tempered condition (continued)

AISI No.(a)	Tempering temperature		Tensile strength		Yield strength		Elongation, %	Reduction in area, %	Hardness, HB
	°C	°F	MPa	ksi	MPa	ksi			
1141	205	400	1634	237	1213	176	6	17	461
	315	600	1462	212	1282	186	9	32	415
	425	800	1165	169	1034	150	12	47	331
	540	1000	896	130	765	111	18	57	262
	650	1200	710	103	593	86	23	62	217
1144	205	400	876	127	627	91	17	36	277
	315	600	869	126	621	90	17	40	262
	425	800	848	123	607	88	18	42	248
	540	1000	807	117	572	83	20	46	235
	650	1200	724	105	503	73	23	55	217
1330(b)	205	400	1600	232	1455	211	9	39	459
	315	600	1427	207	1282	186	9	44	402
	425	800	1158	168	1034	150	15	53	335
	540	1000	876	127	772	112	18	60	263
	650	1200	731	106	572	83	23	63	216
1340	205	400	1806	262	1593	231	11	35	505
	315	600	1586	230	1420	206	12	43	453
	425	800	1262	183	1151	167	14	51	375
	540	1000	965	140	827	120	17	58	295
	650	1200	800	116	621	90	22	66	252
4037	205	400	1027	149	758	110	6	38	310
	315	600	951	138	765	111	14	53	295
	425	800	876	127	731	106	20	60	270
	540	1000	793	115	655	95	23	63	247
	650	1200	696	101	421	61	29	60	220
4042	205	400	1800	261	1662	241	12	37	516
	315	600	1613	234	1455	211	13	42	455
	425	800	1289	187	1172	170	15	51	380
	540	1000	986	143	883	128	20	59	300
	650	1200	793	115	689	100	28	66	238
4130(b)	205	400	1627	236	1462	212	10	41	467
	315	600	1496	217	1379	200	11	43	435
	425	800	1282	186	1193	173	13	49	380
	540	1000	1034	150	910	132	17	57	315
	650	1200	814	118	703	102	22	64	245
4140	205	400	1772	257	1641	238	8	38	510
	315	600	1551	225	1434	208	9	43	445
	425	800	1248	181	1138	165	13	49	370
	540	1000	951	138	834	121	18	58	285
	650	1200	758	110	655	95	22	63	230
4150	205	400	1931	280	1724	250	10	39	530
	315	600	1765	256	1593	231	10	40	495
	425	800	1517	220	1379	200	12	45	440
	540	1000	1207	175	1103	160	15	52	370
	650	1200	958	139	841	122	19	60	290
4340	205	400	1875	272	1675	243	10	38	520
	315	600	1724	250	1586	230	10	40	486
	425	800	1469	213	1365	198	10	44	430
	540	1000	1172	170	1076	156	13	51	360
	650	1200	965	140	855	124	19	60	280
5046	205	400	1744	253	1407	204	9	25	482
	315	600	1413	205	1158	168	10	37	401
	425	800	1138	165	931	135	13	50	336
	540	1000	938	136	765	111	18	61	282
	650	1200	786	114	655	95	24	66	235

(continued)

(a) All grades are fine-grained except for those in the 1100 series which are coarse-grained. Heat treated specimens were oil quenched unless otherwise indicated. (b) Water quenched.

Mechanical properties of selected carbon and alloy steels in the quenched and tempered condition (continued)

AISI No.(a)	Tempering temperature °C	Tempering temperature °F	Tensile strength MPa	Tensile strength ksi	Yield strength MPa	Yield strength ksi	Elongation, %	Reduction in area, %	Hardness, HB
50B46	205	400	...	...	...	...	...	...	560
	315	600	1779	258	1620	235	10	37	505
	425	800	1393	202	1248	181	13	47	405
	540	1000	1082	157	979	142	17	51	322
	650	1200	883	128	793	115	22	60	273
50B60	205	400	...	...	...	...	...	...	600
	315	600	1882	273	1772	257	8	32	525
	425	800	1510	219	1386	201	11	34	435
	540	1000	1124	163	1000	145	15	38	350
	650	1200	896	130	779	113	19	50	290
5130	205	400	1613	234	1517	220	10	40	475
	315	600	1496	217	1407	204	10	46	440
	425	800	1275	185	1207	175	12	51	379
	540	1000	1034	150	938	136	15	56	305
	650	1200	793	115	689	100	20	63	245
5140	205	400	1793	260	1641	238	9	38	490
	315	600	1579	229	1448	210	10	43	450
	425	800	1310	190	1172	170	13	50	365
	540	1000	1000	145	862	125	17	58	280
	650	1200	758	110	662	96	25	66	235
5150	205	400	1944	282	1731	251	5	37	525
	315	600	1737	252	1586	230	6	40	475
	425	800	1448	210	1310	190	9	47	410
	540	1000	1124	163	1034	150	15	54	340
	650	1200	807	117	814	118	20	60	270
5160	205	400	2220	322	1793	260	4	10	627
	315	600	1999	290	1772	257	9	30	555
	425	800	1606	233	1462	212	10	37	461
	540	1000	1165	169	1041	151	12	47	341
	650	1200	896	130	800	116	20	56	269
51B60	205	400	...	...	...	...	...	...	600
	315	600	...	...	...	...	...	...	540
	425	800	1634	237	1489	216	11	36	460
	540	1000	1207	175	1103	160	15	44	355
	650	1200	965	140	869	126	20	47	290
6150	205	400	1931	280	1689	245	8	38	538
	315	600	1724	250	1572	228	8	39	483
	425	800	1434	208	1331	193	10	43	420
	540	1000	1158	168	1069	155	13	50	345
	650	1200	945	137	841	122	17	58	282
81B45	205	400	2034	295	1724	250	10	33	550
	315	600	1765	256	1572	228	8	42	475
	425	800	1407	204	1310	190	11	48	405
	540	1000	1103	160	1027	149	16	53	338
	650	1200	896	130	793	115	20	55	280
8630	205	400	1641	238	1503	218	9	38	465
	315	600	1482	215	1392	202	10	42	430
	425	800	1276	185	1172	170	13	47	375
	540	1000	1034	150	896	130	17	54	310
	650	1200	772	112	689	100	23	63	240
8640	205	400	1862	270	1669	242	10	40	505
	315	600	1655	240	1517	220	10	41	460
	425	800	1379	200	1296	188	12	45	400
	540	1000	1103	160	1034	150	16	54	340
	650	1200	896	130	800	116	20	62	280

(continued)

(a) All grades are fine-grained except for those in the 1100 series which are coarse-grained. Heat treated specimens were oil quenched unless otherwise indicated. (b) Water quenched.

Mechanical properties of selected carbon and alloy steels in the quenched and tempered condition (continued)

AISI No.(a)	Tempering temperature °C	°F	Tensile strength MPa	ksi	Yield strength MPa	ksi	Elongation, %	Reduction in area, %	Hardness, HB
86B45 205		400	1979	287	1641	238	9	31	525
	315	600	1696	246	1551	225	9	40	475
	425	800	1379	200	1317	191	11	41	395
	540	1000	1103	160	1034	150	15	49	335
	650	1200	903	131	876	127	19	58	280
8650 205		400	1937	281	1675	243	10	38	525
	315	600	1724	250	1551	225	10	40	490
	425	800	1448	210	1324	192	12	45	420
	540	1000	1172	170	1055	153	15	51	340
	650	1200	965	140	827	120	20	58	280
8660 205		400	. . .	. . .	. . .	. . .	. . .	. . .	580
	315	600	. . .	. . .	. . .	. . .	. . .	. . .	535
	425	800	1634	237	1551	225	13	37	460
	540	1000	1310	190	1213	176	17	46	370
	650	1200	1068	155	951	138	20	53	315
8740 205		400	1999	290	1655	240	10	41	578
	315	600	1717	249	1551	225	11	46	495
	425	800	1434	208	1358	197	13	50	415
	540	1000	1207	175	1138	165	15	55	363
	650	1200	986	143	903	131	20	60	302
9255 205		400	2103	305	2048	297	1	3	601
	315	600	1937	281	1793	260	4	10	578
	425	800	1606	233	1489	216	8	22	477
	540	1000	1255	182	1103	160	15	32	352
	650	1200	993	144	814	118	20	42	285
9260 205		400	. . .	. . .	. . .	. . .	. . .	. . .	600
	315	600	. . .	. . .	. . .	. . .	. . .	. . .	540
	425	800	1758	255	1503	218	8	24	470
	540	1000	1324	192	1131	164	12	30	390
	650	1200	979	142	814	118	20	43	295
94B30 205		400	1724	250	1551	225	12	46	475
	315	600	1600	232	1420	206	12	49	445
	425	800	1344	195	1207	175	13	57	382
	540	1000	1000	145	931	135	16	65	307
	650	1200	827	120	724	105	21	69	250

(a) All grades are fine-grained except for those in the 1100 series which are coarse-grained. Heat treated specimens were oil quenched unless otherwise indicated. (b) Water quenched.

Monotonic and cyclic stress-strain properties of selected steels

Grade(a)	Orientation(e)	Description(f)	Hardness, HB	Tensile strength MPa	ksi	Reduction in area, %	True strain at fracture	Modulus of elasticity GPa	10⁶ psi	Fatigue strength coefficient MPa	ksi	Fatigue strength exponent	Fatigue ductility coefficient	Fatigue ductility exponent
A538A(b)	L	STA	405	1515	220	67	1.10	185	27	1655	240	−0.065	0.30	−0.62
A538B(b)	L	STA	460	1860	270	56	0.82	185	27	2135	310	−0.071	0.80	−0.71
A538C(b)	L	STA	480	2000	290	55	0.81	180	26	2240	325	−0.07	0.60	−0.75
AM-350(c)	L	HR, A		1315	191	52	0.74	195	28	2800	406	−0.14	0.33	−0.84
AM-350(c)	L	CD	496	1905	276	20	0.23	180	26	2690	390	−0.102	0.10	−0.42
Gainex(c)	LT	HR sheet		530	77	58	0.86	200	29.2	805	117	−0.07	0.86	−0.65
Gainex(c)	L	HR sheet		510	74	64	1.02	200	29.2	805	117	−0.071	0.86	−0.65
H-11	L	Ausformed	660	2585	375	33	0.40	205	30	3170	460	−0.077	0.08	−0.74
RQC-100(c)	LT	HR plate	290	940	136	43	0.56	205	30	1240	180	−0.07	0.66	−0.69
RQC-100(c)	L	HR plate	290	930	135	67	1.02	205	30	1240	180	−0.07	0.66	−0.69
10B62	L	Q&T	430	1640	238	38	0.89	195	28	1780	258	−0.067	0.32	−0.56
1005-1009	LT	HR sheet	90	360	52	73	1.3	205	30	580	84	−0.09	0.15	−0.43
1005-1009	LT	CD sheet	125	470	68	66	1.09	205	30	515	75	−0.059	0.30	−0.51
1005-1009	L	CD sheet	125	415	60	64	1.02	200	29	540	78	−0.073	0.11	−0.41
1005-1009	L	HR sheet	90	345	50	80	1.6	200	29	640	93	−0.109	0.10	−0.39
1015	L	Normalized	80	415	60	68	1.14	205	30	825	120	−0.11	0.95	−0.64
1020	L	HR plate	108	440	64	62	0.96	205	29.5	895	130	−0.12	0.41	−0.51
1040	L	As forged	225	620	90	60	0.93	200	29	1540	223	−0.14	0.61	−0.57
1045	L	Q&T	225	725	105	65	1.04	200	29	1225	178	−0.095	1.00	−0.66
1045	L	Q&T	410	1450	210	51	0.72	200	29	1860	270	−0.073	0.60	−0.70
1045	L	Q&T	390	1345	195	59	0.89	205	30	1585	230	−0.074	0.45	−0.68
1045	L	Q&T	450	1585	230	55	0.81	205	30	1795	260	−0.07	0.35	−0.69
1045	L	Q&T	500	1825	265	51	0.71	205	30	2275	330	−0.08	0.25	−0.68
1045	L	Q&T	595	2240	325	41	0.52	205	30	2725	395	−0.081	0.07	−0.60
1144	L	CDSR	265	930	135	33	0.51	195	28.5	1000	145	−0.08	0.32	−0.58
1144	L	DAT	305	1035	150	25	0.29	200	28.8	1585	230	−0.09	0.27	−0.53
1541F	L	Q&T forging	290	950	138	49	0.68	205	29.9	1275	185	−0.076	0.68	−0.65
1541F	L	Q&T forging	260	890	129	60	0.93	205	29.9	1275	185	−0.071	0.93	−0.65
4130	L	Q&T	258	895	130	67	1.12	220	32	1275	185	−0.083	0.92	−0.63
4130	L	Q&T	365	1425	207	55	0.79	200	29	1695	246	−0.081	0.89	−0.69
4140	L	Q&T, DAT	310	1075	156	60	0.69	200	29.2	1825	265	−0.08	1.2	−0.59
4142	L	DAT	310	1060	154	29	0.35	200	29	1450	210	−0.10	0.22	−0.51
4142	L	DAT	335	1250	181	28	0.34	200	28.9	1250	181	−0.08	0.06	−0.62
4142	L	Q&T	380	1415	205	48	0.66	205	30	1825	265	−0.08	0.45	−0.75
4142	L	Q&T and deformed	400	1550	225	47	0.63	200	29	1895	275	−0.09	0.50	−0.75
4142	L	Q&T	450	1760	255	42	0.54	205	30	2000	290	−0.08	0.40	−0.73
4142	L	Q&T and deformed	475	2035	295	20	0.22	200	29	2070	300	−0.082	0.20	−0.77
4142	L	Q&T and deformed	450	1930	280	37	0.46	200	29	2105	305	−0.09	0.60	−0.76
4142	L	Q&T	475	1930	280	35	0.43	205	30	2170	315	−0.081	0.09	−0.61
4142	L	Q&T	560	2240	325	27	0.31	205	30	2655	385	−0.089	0.07	−0.76
4340	L	HR, A	243	825	120	43	0.57	195	28	1200	174	−0.095	0.45	−0.54
4340	L	Q&T	409	1470	213	38	0.48	200	29	2000	290	−0.091	0.48	−0.60
4340	L	Q&T	350	1240	180	57	0.84	195	28	1655	240	−0.076	0.73	−0.62
5160	L	Q&T	430	1670	242	42	0.87	195	28	1930	280	−0.071	0.40	−0.57
52100	L	SH, Q&T	518	2015	292	11	0.12	205	30	2585	375	−0.09	0.18	−0.56
9262	L	A	260	925	134	14	0.16	205	30	1040	151	−0.071	0.16	−0.47
9262	L	Q&T	280	1000	145	33	0.41	195	28	1220	177	−0.073	0.41	−0.60
9262	L	Q&T	410	1565	227	32	0.38	200	29	1855	269	−0.057	0.38	−0.65
950C(d)	LT	HR plate	159	565	82	64	1.03	205	29.6	1170	170	−0.12	0.95	−0.61
950C(d)	L	HR bar	150	565	82	69	1.19	205	30	970	141	−0.11	0.85	−0.59
950X(d)	L	Plate channel	150	440	64	65	1.06	205	30	625	91	−0.075	0.35	−0.54
950X(d)	L	HR plate	156	530	77	72	1.24	205	29.5	1005	146	−0.10	0.85	−0.61
980X(d)	L	Plate channel	225	695	101	68	1.15	195	28.2	1055	153	−0.08	0.21	−0.53

(a) AISI/SAE grade, unless otherwise indicated. (b) ASTM designation. (c) Proprietary designation. (d) SAE HSLA grade. (e) Orientation of axis of specimen, relative to rolling direction; L is longitudinal (parallel to rolling direction); LT is long transverse (perpendicular to rolling direction). (f) STA, solution treated and aged; HR, hot rolled; CD, cold drawn; Q&T, quenched and tempered; CDSR, cold drawn strain relieved; DAT, drawn at temperature; A, annealed.

Low-temperature impact properties of quenched and tempered alloy steels(a)

AISI No.	C	Mn	Composition, % Ni	Cr	Mo	Quenching temperature(b) °C	°F	Tempering temperature °C	°F
2320(c)	0.18	0.91	3.47	. . .	. . .	900	1650	150	300
								540	1000
2330(c)	0.33	0.86	3.50	. . .	. . .	855	1575	425	800
								540	1000
2340(c)	0.43	0.83	3.38	. . .	. . .	845	1550	425	800
								540	1000
2360(c)(d)	0.57	0.91	3.50	. . .	. . .	800	1475	425	800
								540	1000
2380(c)(d)	0.75	1.00	3.50	. . .	. . .	790	1450	425	800
								540	1000
3120(c)	0.23	0.75	1.26	0.58	. . .	900	1650	150	300
								540	1000
								650	1200
3140(c)	0.38	0.77	1.26	0.64	. . .	845	1550	425	800
								650	1200
3180(c)(d)	0.72	0.92	1.34	0.63	. . .	790	1450	425	800
								650	1200
4320	0.21	0.74	1.53	1.09	0.19	900	1650	205	400
								540	1000
								650	1200
4330	0.30	0.84	1.69	1.10	0.20	855	1575	205	400
								425	800
								540	1000
								650	1200
4340	0.38	0.77	1.65	0.93	0.21	845	1550	205	400
								425	800
								540	1000
								650	1200
4360(d)	0.57	0.87	1.62	1.08	0.22	800	1475	425	800
								650	1200
4380(d)	0.76	0.91	1.67	1.11	0.21	790	1450	425	800
								540	1000
								650	1200
4620(c)	0.20	0.67	1.85	0.30	0.18	900	1650	150	300
								540	1000
								650	1200
4640(c)	0.43	0.69	1.78	0.29	0.20	845	1550	425	800
								540	1000
								650	1200
4680(c)(d)	0.74	0.77	1.81	0.30	0.21	790	1450	425	800
								540	1000
								650	1200
8620	0.20	0.89	0.60	0.68	0.20	900	1650	150	300
								425	800
								650	1200
8630	0.34	0.77	0.66	0.62	0.22	855	1575	425	800
								540	1000
								650	1200
8640	0.45	0.78	0.65	0.61	0.20	845	1550	425	800
								540	1000
								650	1200
8660(d)	0.56	0.81	0.70	0.56	0.25	800	1475	425	800
								540	1000
								650	1200
8680(d)	0.76	0.81	0.67	0.60	0.22	790	1450	425	800
								540	1000
								650	1200

Source: International Nickel Co. Inc. (a) Induction furnace laboratory heats normalized before hardening and tempering. (b) Specimens 1.14-cm- (0.45-in.-) square bars were quenched in oil. (c) Not currently standard. (d) Higher than standard carbon content. (e) Charpy V-notch values scaled from curves.

Low-temperature impact properties of quenched and tempered alloy steels(a) (continued)

Hardness, HRC	−185 °C (−300 °F) J	ft·lb	−130 °C (−200 °F) J	ft·lb	−75 °C (−100 °F) J	ft·lb	−20 °C (0 °F) J	ft·lb	40 °C (100 °F) J	ft·lb	Transition temperature (50% brittle) °C	°F
43.5	23	17	33	24	41	30	41	30	41	30	· · ·	· · ·
22	26	19	33	24	88	65	102	75	102	75	−95	−140
35	20	15	27	20	41	30	56	41	62	46	−80	−115
26	14	10	33	24	66	49	80	59	81	60	−105	−160
39	14	10	24	18	33	24	41	30	47	35	−55	−65
30	20	15	26	19	50	37	60	44	61	45	−100	−150
40	7	5	12	9	14	10	26	19	33	24	· · ·	· · ·
32	11	8	15	11	24	18	41	30	47	35	−55	−65
45	7	5	8	6	14	10	16	12	19	14	· · ·	· · ·
35	11	8	14	10	16	12	19	14	27	20	5	40
43	14	10	20	15	31	23	38	28	41	30	· · ·	· · ·
26	7	5	18	13	45	33	103	76	106	78	· · ·	· · ·
19	18	13	47	35	129	95	130	96	130	96	· · ·	· · ·
43	14	10	20	15	28	21	35	26	41	30	−120	−185
25	19	14	34	25	75	55	85	63	89	66	−130	−200
47	5	4	7	5	14	10	19	14	19	14	· · ·	· · ·
31	· · ·	· · ·	24	18	26	19	54	40	61	45	· · ·	· · ·
43	12	9	23	17	34	25	37	27	38	28	· · ·	· · ·
33	11	8	22	16	38	28	54	40	61	45	−55	−65
21	26	19	47	35	108	80	122	90	122	90	−115	−175
50	7	5	19	14	22	16	27	20	27	20	· · ·	· · ·
43	8	6	14	10	19	14	26	19	28	21	20	70
35	23	17	23	17	31	23	41	30	46	34	−80	−110
27	23	17	47	35	61	45	72	53	76	56	−120	−185
52	15	11	20	15	27	20	28	21	28	21	· · ·	· · ·
44	12	9	18	13	22	16	28	21	34	25	· · ·	· · ·
38	20	15	24	18	38	28	47	35	49	36	−90	−130
30	20	15	38	28	75	55	75	55	75	55	−120	−185
48	7	5	7	5	14	10	15	11	19	14	· · ·	· · ·
30	16	12	20	15	34	25	57	42	58	43	−80	−110
49	5	4	7	5	11	8	12	9	14	10	· · ·	· · ·
42	11	8	11	8	14	10	16	12	20	15	15	60
31	7	5	15	11	26	19	45	33	52	38	−45	−50
42	19	14	27	20	38	28	47	35	47	35	· · ·	· · ·
29	22	16	46	34	75	55	106	78	106	78	· · ·	· · ·
19	23	17	65	48	140	103	156	115	159	117	· · ·	· · ·
42	22	16	23	17	27	20	34	25	37	27	· · ·	· · ·
37	23	17	30	22	47	35	53	39	53	39	−125	−190
29	23	17	41	30	75	55	91	67	91	67	−120	−180
46	7	5	11	8	18	13	20	15	22	16	· · ·	· · ·
41	15	11	16	12	20	15	26	19	30	22	· · ·	· · ·
31	15	11	18	13	23	17	53	39	58	43	· · ·	· · ·
43	15	11	22	16	31	23	47	35	47	35	· · ·	· · ·
36	11	8	18	13	27	20	47	35	61	45	−30	−20
21	14	10	115	85	145	107	156	115	159	117	−125	−195
41	9	7	16	12	23	17	34	25	42	31	−20	0
34	15	11	27	20	58	43	72	53	73	54	−105	−155
27	24	18	38	28	100	74	108	80	111	82	−110	−165
46	7	5	14	10	19	14	27	20	31	23	· · ·	· · ·
38	15	11	20	15	33	24	54	40	54	40	−80	−110
30	24	18	30	22	66	49	85	63	89	66	−95	−140
47	5	4	8	6	14	10	18	13	22	16	· · ·	· · ·
41	14	10	16	12	20	15	27	20	41	30	−25	−10
30	22	16	24	18	34	25	73	54	81	60	−70	−90
50	4	3	5	4	7	5	12	9	14	10	· · ·	· · ·
42	5	4	7	5	14	10	19	14	23	17	· · ·	· · ·
32	4	3	8	6	15	11	34	25	54	40	−35	−30

Source: International Nickel Co. Inc. (a) Induction furnace laboratory heats, normalized before hardening and tempering. (b) Specimens 1.14-cm- (0.45-in.-) square bars were quenched in oil. (c) Nor currently standard. (d) Higher than standard carbon content. (e) Charpy V-notch values scaled from curves.

Typical mechanical properties of normalized alloy steel sheet

Grade	Thickness mm	in.	Tensile strength MPa	ksi	Yield strength(a) MPa	ksi	Elonga-tion(b), %	Hardness, HRC
4130	4.9	0.193..........	835	121	585	85	14	25
4335(c)	4.6	0.180..........	1725	250	1240	180	8	48
4340(c)	2.0	0.080..........	1860	270	1345	195	7	50

(a) At 0.2% offset. (b) In 50 mm or 2 in. (c) Modified: 0.40% Mo, 0.20% V.

Heat Treating

Effect of mass on hardness of normalized carbon and alloy steels

All data are based on single heats. Sources: data for 3310, 3140 and 4063 are from *Modern Steels and Their Properties,* 6th Ed., Bethlehem Steel Corp., 1966; all other data are from *Modern Steels and Their Properties* (Handbook 3310), Bethlehem Steel Corp., Sept 1978.

Grade	Normalizing temperature °C	°F	Hardness, HB, for bar with diameter, mm (in.), of: 13(½)	25(1)	50(2)	100(4)
Carbon steels, carburizing grades						
1015..........	925	1700	126	121	116	116
1020..........	925	1700	131	131	126	121
1022..........	925	1700	143	143	137	131
1117..........	900	1650	143	137	137	126
1118..........	925	1700	156	143	137	131
Carbon steels, direct-hardening grades						
1030..........	925	1700	156	149	137	137
1040..........	900	1650	183	170	167	167
1050..........	900	1650	223	217	212	201
1060..........	900	1650	229	229	223	223
1080..........	900	1650	293	293	285	269
1095..........	900	1650	302	293	269	255
1137..........	900	1650	201	197	197	192
1141..........	900	1650	207	201	201	201
1144..........	900	1650	201	197	192	192
Alloy steels, carburizing grades						
3310..........	890	1630	269	262	262	248
4118..........	910	1670	170	156	143	137
4320..........	895	1640	248	235	212	201
4419..........	955	1750	149	143	143	143
4620..........	900	1650	192	174	167	163
4820..........	860	1580	235	229	223	212
8620..........	915	1675	197	183	179	163
9310..........	890	1630	285	269	262	255
Alloy steels, direct-hardening grades						
1340..........	870	1600	269	248	235	235
3140..........	870	1600	302	262	248	241
4027..........	905	1660	179	179	163	156
4063..........	870	1600	285	285	285	277
4130..........	870	1600	217	197	167	163
4140..........	870	1600	302	302	285	241
4150..........	870	1600	375	321	311	293
4340..........	870	1600	388	363	341	321
5140..........	870	1600	235	229	223	217
5150..........	870	1600	262	255	248	241
5160..........	860	1575	285	269	262	255
6150..........	870	1600	285	269	262	255
8630..........	870	1600	201	187	187	187
8650..........	870	1600	363	302	293	285
8740..........	870	1600	269	269	262	255
9255..........	900	1650	277	269	269	269

Typical normalizing temperatures for standard carbon and alloy steels

Based on production experience, normalizing temperature may vary from as much as 27 °C (50 °F) below to as much as 55 °C (100 °F) above indicated temperature. The steel should be cooled in still air from indicated temperature

Grade	Temperature °C	°F	Grade	Temperature °C	°F
Plain carbon steels			**Standard alloy steels (continued)**		
1015	915	1675	4817	925	1700
1020	915	1675	4820	925	1700
1022	915	1675	5046	870	1600
1025	900	1650	5120	925	1700
1030	900	1650	5130	900	1650
1035	885	1625	5132	900	1650
1040	860	1575	5135	870	1600
1045	860	1575	5140	870	1600
1050	860	1575	5145	870	1600
1060	830	1525	5147	870	1600
1080	830	1525	5150	870	1600
1090	830	1525	5155	870	1600
1095	845	1550	5160	870	1600
1117	900	1650	6118	925	1700
1137	885	1625	6120	925	1700
1141	860	1575	6150	900	1650
1144	860	1575	8617	925	1700
			8620	925	1700
			8622	925	1700
Standard alloy steels			8625	900	1650
1330	900	1650	8627	900	1650
1335	870	1600	8630	900	1650
1340	870	1600	8637	870	1600
3135	870	1600	8640	870	1600
3140	870	1600	8642	870	1600
3310	925	1700	8645	870	1600
4027	900	1650	8650	870	1600
4028	900	1650	8655	870	1600
4032	900	1650	8660	870	1600
4037	870	1600	8720	925	1700
4042	870	1600	8740	925	1700
4047	870	1600	8742	870	1600
4063	870	1600	8822	925	1700
4118	925	1700	9255	900	1650
4130	900	1650	9260	900	1650
4135	870	1600	9262	900	1650
4137	870	1600	9310	925	1700
4140	870	1600	9840	870	1600
4142	870	1600	9850	870	1600
4145	870	1600	50B40	870	1600
4147	870	1600	50B44	870	1600
4150	870	1600	50B46	870	1600
4320	925	1700	50B50	870	1600
4337	870	1600	60B60	870	1600
4340	870	1600	81B45	870	1600
4520	925	1700	86B45	870	1600
4620	925	1700	94B15	925	1700
4621	925	1700	94B17	925	1700
4718	925	1700	94B30	900	1650
4720	925	1700	94B40	900	1650
4815	925	1700			

Fe-Fe₃C phase diagram, showing the temperature range of interest for annealing plain carbon steels

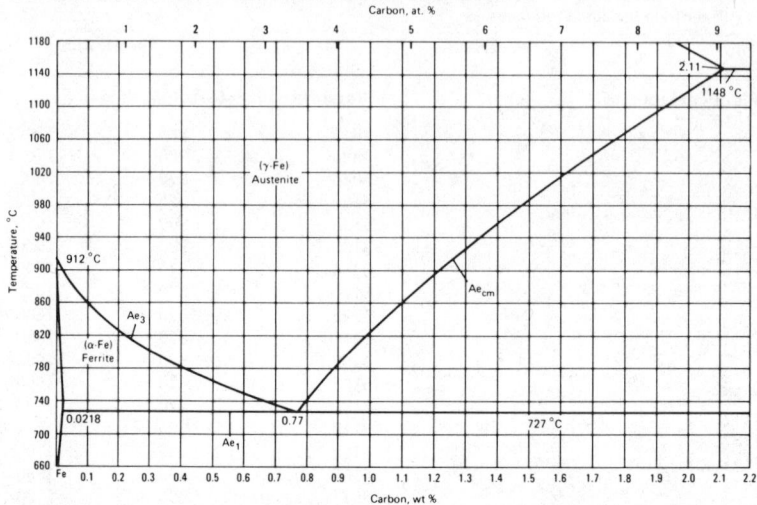

The equilibrium transformation temperatures Ae₁, Ae₃ and Ae$_{cm}$ are labeled on the diagram.

Approximate critical temperatures for selected carbon and low-alloy steels

Steel	Critical temperatures on heating at 28 °C/h (50 °F/h)				Critical temperatures on cooling at 28 °C/h (50 °F/h)			
	Ac₁		Ac₃		Ar₃		Ar₁	
	°C	°F	°C	°F	°C	°F	°C	°F
1010	724	1335	877	1610	849	1560	682	1260
1020	724	1335	846	1555	816	1500	682	1260
1030	727	1340	813	1495	788	1450	677	1250
1040	727	1340	793	1460	757	1395	671	1240
1050	727	1340	768	1415	741	1365	682	1260
1060	727	1340	746	1375	727	1340	685	1265
1070	727	1340	732	1350	710	1310	691	1275
1080	729	1345	735	1355	699	1290	693	1280
1340	716	1320	777	1430	721	1330	621	1150
3140	735	1355	766	1410	721	1330	660	1220
4027	727	1340	807	1485	760	1400	671	1240
4042	727	1340	793	1460	732	1350	654	1210
4130	757	1395	810	1490	754	1390	693	1280
4140	732	1350	804	1480	743	1370	679	1255
4150	743	1370	766	1410	729	1345	671	1240
4340	724	1335	774	1425	710	1310	654	1210
4615	727	1340	810	1490	760	1400	649	1200
5046	716	1320	771	1420	732	1350	682	1260
5120	766	1410	838	1540	799	1470	699	1290
5140	738	1360	788	1450	727	1340	693	1280
5160	710	1310	766	1410	716	1320	677	1250
52100	727	1340	768	1415	716	1320	688	1270
6150	749	1380	788	1450	743	1370	693	1280
8115	721	1330	838	1540	788	1450	671	1240
8620	732	1350	829	1525	768	1415	660	1220
8640	732	1350	779	1435	727	1340	666	1230
9260	743	1370	816	1500	749	1380	713	1315

Recommended temperatures and cooling cycles for full annealing of small carbon steel forgings

Data are for forgings up to 75 mm (3 in.) in section thickness. Time at temperature usually is a minimum of 1 h for sections up to 25 mm (1 in.) thick; ½ h is added for each additional 25 mm (1 in.) of thickness

Steel	Annealing temperature °C	°F	Cooling cycle(a) °C From	To	°F From	To	Hardness range, HB
1018	855–900	1575–1650	855	705	1575	1300	111–149
1020	855–900	1575–1650	855	700	1575	1290	111–149
1022	855–900	1575–1650	855	700	1575	1290	111–149
1025	855–900	1575–1650	855	700	1575	1290	111–187
1030	845–885	1550–1625	845	650	1550	1200	126–197
1035	845–885	1550–1625	845	650	1550	1200	137–207
1040	790–870	1450–1600	790	650	1450	1200	137–207
1045	790–870	1450–1600	790	650	1450	1200	156–217
1050	790–870	1450–1600	790	650	1450	1200	156–217
1060	790–845	1450–1550	790	650	1450	1200	156–217
1070	790–845	1450–1550	790	650	1450	1200	167–229
1080	790–845	1450–1550	790	650	1450	1200	167–229
1090	790–830	1450–1525	790	650	1450	1200	167–229
1095	790–830	1450–1525	790	655	1450	1215	167–229

(a) Furnace cooling at 28 °C/h (50 °F/h).

Recommended annealing temperatures for alloy steels (furnace cooling)

Steel	Annealing temperature °C	°F	Hardness (max), HB
1330	845–900	1550–1650	179
1335	845–900	1550–1650	187
1340	845–900	1550–1650	192
1345	845–900	1550–1650	...
3140	815–870	1500–1600	187
4037	815–855	1500–1575	183
4042	815–855	1500–1575	192
4047	790–845	1450–1550	201
4063	790–845	1450–1550	223
4130	790–845	1450–1550	174
4135	790–845	1450–1550	...
4137	790–845	1450–1550	192
4140	790–845	1450–1550	197
4145	790–845	1450–1550	207
4147	790–845	1450–1550	...
4150	790–845	1450–1550	212
4161	790–845	1450–1550	...
4337	790–845	1450–1550	...
4340	790–845	1450–1550	223
50B40	815–870	1500–1600	187
50B44	815–870	1500–1600	197
5046	815–870	1500–1600	192
50B46	815–870	1500–1600	192
50B50	815–870	1500–1600	201
50B60	815–870	1500–1600	217
5130	790–845	1450–1550	170
5132	790–845	1450–1550	170
5135	815–870	1500–1600	174
5140	815–870	1500–1600	187
5145	815–870	1500–1600	197
5147	815–870	1500–1600	197
5150	815–870	1500–1600	201
5155	815–870	1500–1600	217
5160	815–870	1500–1600	223
51B60	815–870	1500–1600	223
50100	730–790	1350–1450	197
51100	730–790	1350–1450	197
52100	730–790	1350–1450	207
6150	845–900	1550–1650	201
81B45	845–900	1550–1650	192
8627	815–870	1500–1600	174
8630	790–845	1450–1550	179
8637	815–870	1500–1600	192
8640	815–870	1500–1600	197
8642	815–870	1500–1600	201
8645	815–870	1500–1600	207
86B45	815–870	1500–1600	207
8650	815–870	1500–1600	212
8655	815–870	1500–1600	223
8660	815–870	1500–1600	229
8740	815–870	1500–1600	202
8742	815–870	1500–1600	...
9260	815–870	1500–1600	229
94B30	790–845	1450–1550	174
94B40	790–845	1450–1550	192
9840	790–845	1450–1550	207

Recommended temperatures and time cycles for annealing of alloy steels

Steel	Austenitizing temperature °C	°F	Conventional cooling(a) Temperature °C From	°C To	°F From	°F To	Cooling rate °C/h	°F/h	Time, h	Isothermal method(b) Cool to °C	°F	Hold, h	Hardness (approx), HB
To obtain a predominantly pearlitic structure(c)													
1340	830	1525	735	610	1350	1130	11	20	11	620	1150	4.5	183
2340	800	1475	655	555	1210	1030	8.3	15	12	595	1100	6	201
2345	800	1475	655	550	1210	1020	9.1	15	12.7	595	1100	6	201
3120(d)	885	1625	. . .	. . .	. . .	. . .	. . .	. . .	. . .	650	1200	4	179
3140	830	1525	735	650	1350	1200	11	20	7.5	660	1225	6	187
3150	830	1525	705	645	1300	1190	11	20	5.5	660	1225	6	201
3310(e)	870	1600	. . .	. . .	. . .	. . .	. . .	. . .	. . .	595	1100	14	187
4042	830	1525	745	640	1370	1180	11	20	9.5	660	1225	4.5	197
4047	830	1525	735	630	1350	1170	11	20	9	660	1225	5	207
4062	830	1525	695	630	1280	1170	8.3	15	7.3	660	1225	6	223
4130	855	1575	765	665	1410	1230	20	35	5	675	1250	4	174
4140	845	1550	755	665	1390	1230	14	25	6.4	675	1250	5	197
4150	830	1525	745	670	1370	1240	8.4	15	8.6	675	1250	6	212
4320(d)	885	1625	. . .	. . .	. . .	. . .	. . .	. . .	. . .	660	1225	6	197
4340	830	1525	705	565	1300	1050	8.3	15	16.5	650	1200	8	223
4620(d)	885	1625	. . .	. . .	. . .	. . .	. . .	. . .	. . .	650	1200	6	187
4640	830	1525	715	600	1320	1110	7.6	15	15	620	1150	8	197
4820(d)	. . .	. . .	. . .	. . .	. . .	. . .	. . .	. . .	. . .	605	1125	4	192
5045	830	1525	755	665	1390	1230	11	20	8	660	1225	4.5	192
5120(d)	885	1625	. . .	. . .	. . .	. . .	. . .	. . .	. . .	690	1275	4	179
5132	845	1550	755	670	1390	1240	11	20	7.5	675	1250	6	183
5140	830	1525	740	670	1360	1240	11	20	6	675	1250	6	187
5150	830	1525	705	650	1300	1200	11	20	5	675	1250	6	201
52100(f)	. . .	. . .	. . .	. . .	. . .	. . .	. . .	. . .	. . .	. . .	. . .	. . .	. . .
6150	830	1525	760	675	1400	1250	8.4	15	10	675	1250	6	201
8620(d)	885	1625	. . .	. . .	. . .	. . .	. . .	. . .	. . .	660	1225	4	187
8630	845	1550	735	640	1350	1180	11	20	8.5	660	1225	6	192
8640	830	1525	725	640	1340	1180	11	20	8	660	1225	6	197
8650	830	1525	710	650	1310	1200	8.4	15	7.2	650	1200	8	212
8660	830	1525	700	655	1290	1210	8.4	15	8	650	1200	8	229
8720(d)	885	1625	. . .	. . .	. . .	. . .	. . .	. . .	. . .	660	1225	4	187
8740	830	1525	725	645	1340	1190	11	20	7.5	660	1225	7	201
8750	830	1525	720	630	1330	1170	8.4	15	10.7	660	1225	7	217
9260	860	1575	760	705	1400	1300	8.4	15	6.7	660	1225	6	229
9310(e)	870	1600	. . .	. . .	. . .	. . .	. . .	. . .	. . .	595	1100	14	187
9840	830	1525	695	640	1280	1180	8.4	15	6.6	650	1200	6	207
9850	830	1525	700	645	1290	1190	8.4	15	6.7	650	1200	8	223
To obtain a predominantly ferritic and spheroidized carbide structure													
1320(d)	805	1480	. . .	. . .	. . .	. . .	. . .	. . .	. . .	650	1200	8	170
1340	750	1380	735	610	1350	1130	5.5	10	22	640	1180	8	174
2340	715	1320	655	555	1210	1030	5.5	10	18	605	1125	10	192
2345	715	1320	655	550	1210	1020	5.5	10	19	605	1125	10	192
3120(d)	790	1450	. . .	. . .	. . .	. . .	. . .	. . .	. . .	650	1200	8	163
3140	745	1370	735	650	1350	1200	5.5	10	15	660	1225	10	174
3150	750	1380	705	645	1300	1190	5.5	10	11	660	1225	10	187
9840	745	1370	695	640	1280	1180	5.5	10	11	650	1200	10	192
9850	745	1370	700	645	1290	1190	5.5	10	11	650	1200	12	207

(a) The steel is cooled in the furnace at the indicated rate through the temperature range shown. (b) The steel is cooled rapidly to the temperature indicated and is held at that temperature for the time specified. (c) In isothermal annealing to obtain pearlitic structure, steels may be austenitized at temperatures up to 70 °C (125 °F) higher than temperatures listed. (d) Seldom annealed. Structures of better machinability are developed by normalizing or by transforming isothermally after rolling or forging. (e) Annealing is impractical by the conventional process of continuous slow cooling. The lower transformation temperature is markedly depressed, and excessively long cooling cycles are required to obtain transformation to pearlite. (f) Predominantly pearlitic structures are seldom desired in this steel.

Austenitizing temperatures for direct-hardening carbon and alloy steels (SAE)

Steel	°C	°F
Carbon steels		
1025	855-900	1575-1650
1030	845-870	1550-1600
1035	830-855	1525-1575
1037	830-855	1525-1575
1038(a)	830-855	1525-1575
1039(a)	830-855	1525-1575
1040(a)	830-855	1525-1575
1042	800-845	1475-1550
1043(a)	800-845	1475-1550
1045(a)	800-845	1475-1550
1046(a)	800-845	1475-1550
1050(a)	800-845	1475-1550
1055	800-845	1475-1550
1060	800-845	1475-1550
1065	800-845	1475-1550
1070	800-845	1475-1550
1074	800-845	1475-1550
1078	790-815	1450-1500
1080	790-815	1450-1500
1084	790-815	1450-1500
1085	790-815	1450-1500
1086	790-815	1450-1500
1090	790-815	1450-1500
1095	790-815(a)	1450-1500(b)
Free-cutting carbon steels		
1137	830-855	1525-1575
1138	815-845	1500-1550
1140	815-845	1500-1550
1141	800-845	1475-1550
1144	800-845	1475-1550
1145	800-845	1475-1550
1146	800-845	1475-1550
1151	800-845	1475-1550
1536	815-845	1500-1550
1541	815-845	1500-1550
1548	815-845	1500-1550
1552	815-845	1500-1550
1566	855-885	1575-1625
Alloy steels		
1330	830-855	1525-1575
1335	815-845	1500-1550
1340	815-845	1500-1550
1345	815-845	1500-1550
3140	815-845	1500-1550
4037	830-855	1525-1575
4042	830-855	1525-1575
4047	815-855	1500-1575
4063	800-845	1475-1550
4130	815-870	1500-1600
4135	845-870	1550-1600
4137	845-870	1550-1600
4140	845-870	1550-1600
4142	845-870	1550-1600
4145	815-845	1500-1550
4147	815-845	1500-1550
4150	815-845	1500-1550
4161	815-845	1500-1550
4337	815-845	1500-1550
4340	815-845	1500-1550
50B40	815-845	1500-1550
50B44	815-845	1500-1550
5046	815-845	1500-1550
50B46	815-845	1500-1550
50B50	800-845	1475-1550
50B60	800-845	1475-1550
5130	830-855	1525-1575
5132	830-855	1525-1575
5135	815-845	1500-1550
5140	815-845	1500-1550
5145	815-845	1500-1550
5147	800-845	1475-1550
5150	800-845	1475-1550
5155	800-845	1475-1550
5160	800-845	1475-1550
51B60	800-845	1475-1550
50100	775-800(c)	1425-1475(c)
51100	775-800(c)	1425-1475(c)
52100	775-800(c)	1425-1475(c)
6150	845-885	1550-1625
81B45	815-855	1500-1575
8630	830-870	1525-1600
8637	830-855	1525-1575
8640	830-855	1525-1575
8642	815-855	1500-1575
8645	815-855	1500-1575
86B45	815-855	1500-1575
8650	815-855	1500-1575
8655	800-845	1475-1550
8660	800-845	1475-1550
8740	830-855	1525-1575
8742	830-855	1525-1575
9254	815-900	1500-1650
9255	815-900	1500-1650
9260	815-900	1500-1650
94B30	845-885	1550-1625
94B40	845-885	1550-1625
9840	830-855	1525-1575

(a) Commonly used on parts where induction hardening is employed. All steels from SAE 1030 up may have induction hardening applications. (b) This temperature range may be employed for 1095 steel that is to be quenched in water, brine or oil. For oil quenching, 1095 steel may alternatively be austenitized in the range 815 to 870 °C (1500 to 1600 °F). (c) This range is recommended for steel that is to be water quenched. For oil quenching, steel should be austenitized in the range 815 to 870 °C (1500 to 1600 °F).

Reheating (austenitizing) temperatures for hardening of carburized(a) carbon and alloy steels (SAE)

Steel	°C	°F
Carbon steels		
1010	760-790	1400-1450
1012	760-790	1400-1450
1015	760-790	1400-1450
1016	760-790	1400-1450
1017	760-790	1400-1450
1018	760-790	1400-1450
1019	760-790	1400-1450
1020	760-790	1400-1450
1022	760-790	1400-1450
1513	760-790	1600-1650
1518	760-790	1600-1650
1522	760-790	1600-1650
1524	760-790	1600-1650
1525	760-790	1600-1650
1526	760-790	1600-1650
1527	760-790	1600-1650
Free-cutting carbon steels		
1109	760-790	1400-1450
1115	760-790	1400-1450
1117	760-790	1400-1450
1118	760-790	1400-1450
Alloy steels		
3310	790-830	1450-1525
4320	830-845	1525-1550
4615	815-845	1500-1550
4617	815-845	1500-1550
4620	815-845	1500-1550
4621	815-845	1500-1550
4626	815-845	1500-1550
4718	815-845	1500-1550
4720	815-845	1500-1550
4815	800-830	1475-1525
4817	800-830	1475-1525
4820	800-830	1475-1525
8115	845-870	1550-1600
8615	845-870	1550-1600
8617	845-870	1550-1600
8620	845-870	1550-1600
8622	845-870	1550-1600
8625	845-870	1550-1600
8627	845-870	1550-1600
8720	845-870	1550-1600
8822	845-870	1550-1600
9310	790-830	1450-1525

(a) Carburizing is commonly carried out at 900 to 925 °C (1650 to 1700 °F); slow cooled and reheated to given austenizing temperature.

Typical hardnesses of various carbon and alloy steels after tempering

Data were obtained on 25-mm (1-in.) bars adequately quenched to develop full hardness

Grade	Carbon content, %		°C: 205 °F: 400	260 500	315 600	370 700	425 800	480 900	540 1000	595 1100	650 1200	Heat treatment
Carbon steels, water hardening												
1030	0.30	.....	50	45	43	39	31	28	25	22	95(a)	Normalized at 900 °C (1650 °F); water quenched
1040	0.40	.....	51	48	46	42	37	30	27	22	94(a)	from 830-845 °C (1525-1550 °F); average dew
1050	0.50	.....	52	50	46	44	40	37	31	29	22	point, 16 °C (60 °F)
1060	0.60	.....	56	55	50	42	38	37	35	33	26	Normalized at 885 °C (1625 °F); water quenched
1080	0.80	.....	57	55	50	43	41	40	39	38	32	from 800-815 °C (1475-1500 °F); average dew
1095	0.95	.....	58	57	52	47	43	42	41	40	33	point, 7 °C (45 °F)
1137	0.40	.....	44	42	40	37	33	30	27	21	91(a)	Normalized at 900 °C (1650 °F); water quenched
1141	0.40	.....	49	46	43	41	38	34	28	23	94(a)	from 830-855 °C (1525-1575 °F); average dew
1144	0.40	.....	55	50	47	45	39	32	29	25	97(a)	point, 13 °C (55 °F)
Alloy steels, water hardening												
1330	0.30	.....	47	44	42	38	35	32	26	22	16	Normalized at 900 °C (1650 °F); water quenched
2330	0.30	.....	47	44	42	38	35	32	26	22	16	from 800-815 °C (1475-1500 °F); average dew
3130	0.30	.....	47	44	42	38	35	32	26	22	16	point, 16 °C (60 °F)
4130	0.30	.....	47	45	43	42	38	34	32	26	22	Normalized at 885 °C (1625 °F); water quenched
5130	0.30	.....	47	45	43	42	38	34	32	26	22	from 800-855 °C (1475-1575 °F); average dew
8630	0.30	.....	47	45	43	42	38	34	32	26	22	point, 16 °C (60 °F)
Alloy steels, oil hardening												
1340	0.40	.....	57	53	50	46	44	41	38	35	31	Normalized at 870 °C (1600 °F); oil quenched
3140	0.40	.....	55	52	49	47	41	37	33	30	26	from 830-845 °C (1525-1550 °F); average dew
4140	0.40	.....	57	53	50	47	45	41	36	33	29	point, 16 °C (60 °F)
4340	0.40	.....	55	52	50	48	45	42	39	34	31	Normalized at 870 °C (1600 °F); oil quenched
4640	0.40	.....	52	51	50	47	42	40	37	31	27	from 830-845 °C (1525-1575 °F); average dew
8740	0.40	.....	57	53	50	47	44	41	38	35	22	point, 13 °C (55 °F)
4150	0.50	.....	56	55	53	51	47	46	43	39	35	Normalized at 870 °C (1600 °F); oil quenched
5150	0.50	.....	57	55	52	49	45	39	34	31	28	from 830-870 °C (1525-1600 °F); average dew
6150	0.50	.....	58	57	53	50	46	42	40	36	31	point, 13 °C (55 °F)
8650	0.50	.....	55	54	52	49	45	41	37	32	28	Normalized at 870 °C (1600 °F); oil quenched
8750	0.50	.....	56	55	52	51	46	44	39	34	32	from 815-845 °C (1500-1550 °F); average dew
9850	0.50	.....	54	53	51	48	45	41	36	33	30	point, 13 °C (55 °F)

(a) Hardness, HRB

Machining

Machinability ratings of plain carbon steels, percent of cutting speed for B1112/1212

Grade(a)	Machinability rating, %	HB	Grade(a)	Machinability rating, %	HB
1005	45	· · ·	1050	45	197
1006	50	95		55(c)	189
1008	55	95	1053	55	· · ·
1010	55	105	1055	55(c)	197
1011	53	· · ·	1059	52	· · ·
1012	55	105	1060	60(d)	183
1013	53	· · ·	1064	60(d)	183
1015	60	111	1065	60(d)	187
1016	70	121	1066	48	· · ·
1017	65	116	1069	48	· · ·
1018	70	126	1070	55(d)	192
1019	70	131	1074	55(d)	192
1020	65	121	1075	48	· · ·
1021	70	131	1078	55(d)	192
1022	70	137	1080	45(d)	192
1023	65	121	1084	45(d)	192
1025	65	126	1085	45(d)	192
1026	75	143	1086	45(d)	192
1029	68	· · ·	1090	45(d)	197
1030	70	149	1095	45(d)	197
1035	65	163	1524	60	163
1037	65	167	1527	65	163
1038	65	163	1536	55	187
1039	60	179	1541	45	207
1040	60	170		60(c)	184
1042	60	179	1547	40	207
	70(b)	179		45(c)	187
1043	60	179	1548	45	217
	70(b)	179		50(c)	192
1045	55	179	1552	50(c)	193
	65(c)	170			
1046	55	187			
	65(c)	179			
1049	45	197			
	55(c)	187			

(a) Unless otherwise indicated, all values are for cold drawn steels. (b) Normalized and cold drawn. (c) Annealed and cold drawn. (d) Spheroidized and cold drawn.

Machinability ratings and recommended feeds and speeds for cold drawn carbon steel bars (SI units) (a)

Steel grade	Machinability rating(b), %	Form turning			Single-point turning			Drilling		
		Width of cut, mm	Speed, mm/s	Feed, mm/rev	Depth of cut, mm	Speed, mm/s	Feed, mm/rev	Size of hole, mm	Speed, mm/s	Feed, mm/rev
12L14	158	12.7	1320	0.084	3.18	1320	0.24	6.35	735	0.15
		25.4	1220	0.071	6.35	1220	0.22	12.7	735	0.17
		38.1	1220	0.069	9.53	1195	0.18	19.05	815	0.20
		50.8	1195	0.051	12.7	1170	0.15	25.4	815	0.22
		63.5	1170	0.041	· · ·	· · ·	· · ·	31.75	840	0.25
1213	136	12.7	1145	0.076	3.18	1145	0.22	6.35	635	0.14
		25.4	1065	0.064	6.35	1065	0.20	12.7	635	0.15
		38.1	1065	0.064	9.53	1040	0.17	19.05	710	0.18
		50.8	1040	0.046	12.7	1015	0.14	25.4	710	0.20
		63.5	1015	0.038	· · ·	· · ·	· · ·	31.75	735	0.23
1119, 1212	100	12.7	840	0.064	3.18	840	0.18	6.35	535	0.12
		25.4	815	0.051	6.35	815	0.17	12.7	535	0.13
		38.1	815	0.046	9.53	785	0.14	19.05	585	0.15
		50.8	785	0.038	12.7	760	0.11	25.4	585	0.18
		63.5	760	0.030	· · ·	· · ·	· · ·	31.75	610	0.20
1211	94	12.7	785	0.058	3.18	785	0.17	6.35	505	0.11
		25.4	760	0.048	6.35	760	0.15	12.7	505	0.12
		38.1	760	0.043	9.53	740	0.13	19.05	550	0.14
		50.8	740	0.036	12.7	715	0.11	25.4	550	0.17
		63.5	715	0.028	· · ·	· · ·	· · ·	31.75	575	0.19
1117, 1118	91	12.7	760	0.056	3.18	760	0.16	6.35	485	0.10
		25.4	735	0.046	6.35	735	0.15	12.7	485	0.11
		38.1	735	0.041	9.53	715	0.13	19.05	535	0.14
		50.8	715	0.036	12.7	690	0.10	25.4	535	0.16
		63.5	690	0.028	· · ·	· · ·	· · ·	31.75	605	0.19
1144, annealed	85	12.7	710	0.053	3.18	710	0.15	6.35	450	0.10
		25.4	690	0.043	6.35	690	0.14	12.7	450	0.11
		38.1	690	0.038	9.53	670	0.12	19.05	500	0.14
		50.8	670	0.033	12.7	645	0.10	25.4	500	0.16
		63.5	645	0.025	· · ·	· · ·	· · ·	31.75	520	0.18
1141, annealed	81	12.7	685	0.050	3.18	685	0.14	6.35	435	0.10
		25.4	660	0.043	6.35	660	0.13	12.7	435	0.11
		38.1	660	0.038	9.53	645	0.11	19.05	480	0.14
		50.8	645	0.031	12.7	620	0.094	25.4	480	0.16
		63.5	620	0.025	· · ·	· · ·	· · ·	31.75	500	0.18
1016, 1018, 1022	78	12.7	660	0.048	3.18	660	0.14	6.35	415	0.097
		25.4	635	0.041	6.35	635	0.13	12.7	415	0.11
		38.1	635	0.036	9.53	615	0.11	19.05	455	0.13
		50.8	615	0.031	12.7	595	0.089	25.4	455	0.15
		63.5	595	0.023	· · ·	· · ·	· · ·	31.75	480	0.17
1144	76	12.7	635	0.048	3.18	635	0.13	6.35	400	0.094
		25.4	615	0.038	6.35	615	0.12	12.7	400	0.11
		38.1	615	0.036	9.53	595	0.10	19.05	440	0.13
		50.8	595	0.028	12.7	575	0.086	25.4	440	0.15
		63.5	575	0.023	· · ·	· · ·	· · ·	31.75	460	0.17
1020, 1137, 1045, annealed . .	72	12.7	610	0.046	3.18	610	0.13	6.35	385	0.089
		25.4	585	0.036	6.35	585	0.12	12.7	385	0.10
		38.1	585	0.033	9.53	570	0.10	19.05	420	0.12
		50.8	570	0.028	12.7	550	0.081	25.4	420	0.14
		63.5	550	0.023	· · ·	· · ·	· · ·	31.75	435	0.16

(continued)

(a) All cutting speeds and feeds based on cutting with high-speed steel tools. (b) Based on a machinability rating of 100% for 1212 steel.

Machinability ratings and recommended feeds and speeds for cold drawn carbon steel bars (SI units) (a) (continued)

Steel grade	Machinability rating(b), %	Form turning			Single-point turning			Drilling		
		Width of cut, in.	Speed, ft/min	Feed, in./rev	Depth of cut, in.	Speed, ft/min	Feed, in./rev	Size of hole, in.	Speed, ft/min	Feed, in./rev
1035, 1141, 1050, annealed ..	70	12.7	585	0.043	3.18	585	0.12	6.35	370	0.086
		25.4	570	0.036	6.35	570	0.11	12.7	370	0.097
		38.1	570	0.033	9.53	550	0.097	19.05	405	0.11
		50.8	550	0.028	12.7	535	0.079	25.4	405	0.13
		63.5	535	0.020	· · ·	· · ·	· · ·	31.75	425	0.16
1040	64	12.7	535	0.038	3.18	535	0.11	6.35	340	0.081
		25.4	515	0.030	6.35	515	0.10	12.7	340	0.089
		38.1	515	0.028	9.53	500	0.086	19.05	370	0.11
		50.8	500	0.023	12.7	485	0.071	25.4	370	0.12
		63.5	485	0.018	· · ·	· · ·	· · ·	31.75	385	0.14
1045	57	12.7	485	0.036	3.18	485	0.10	6.35	305	0.071
		25.4	460	0.030	6.35	460	0.094	12.7	305	0.079
		38.1	460	0.025	9.53	445	0.079	19.05	330	0.094
		50.8	445	0.023	12.7	430	0.066	25.4	330	0.11
		63.5	430	0.018	· · ·	· · ·	· · ·	31.75	345	0.13
1050	54	12.7	455	0.036	3.18	455	0.097	6.35	290	0.071
		25.4	440	0.028	6.35	440	0.089	12.7	290	0.079
		38.1	440	0.025	9.53	425	0.076	19.05	315	0.094
		50.8	425	0.020	12.7	410	0.061	25.4	315	0.11
		63.5	410	0.018	· · ·	· · ·	· · ·	31.75	330	0.13

(a) All cutting speeds and feeds based on cutting with high-speed steel tools. (b) Based on a machinability rating of 100% for 1212 steel.

Machinability ratings and recommended feeds and speeds for cold drawn carbon steel bars (customary U.S. units) (a)

Steel grade	Machinability rating(b), %	Form turning			Single-point turning			Drilling		
		Width of cut, in.	Speed, ft/min	Feed, in./rev	Depth of cut, in.	Speed, ft/min	Feed, in./rev	Size of hole, in.	Speed, ft/min	Feed, in./rev
12L14	158	0.500	260	0.0033	0.125	260	0.0093	0.250	145	0.0060
		1.000	240	0.0028	0.250	240	0.0088	0.500	145	0.0066
		1.500	240	0.0027	0.375	235	0.0071	0.750	160	0.0077
		2.000	235	0.0020	0.500	230	0.0060	1.000	160	0.0088
		2.500	230	0.0016	· · ·	· · ·	· · ·	1.250	165	0.0099
1213	136	0.500	225	0.0030	0.125	225	0.0085	0.250	125	0.0054
		1.000	210	0.0025	0.250	210	0.0080	0.500	125	0.0060
		1.500	210	0.0025	0.375	205	0.0065	0.750	140	0.0070
		2.000	205	0.0018	0.500	200	0.0055	1.000	140	0.0080
		2.500	200	0.0015	· · ·	· · ·	· · ·	1.250	145	0.0090
1119, 1212	100	0.500	165	0.0025	0.125	165	0.0070	0.250	105	0.0045
		1.000	160	0.0020	0.250	160	0.0065	0.500	105	0.0050
		1.500	160	0.0018	0.375	155	0.0055	0.750	115	0.0060
		2.000	155	0.0015	0.500	150	0.0045	1.000	115	0.0070
		2.500	150	0.0012	· · ·	· · ·	· · ·	1.250	120	0.0080
1211	94	0.500	155	0.0023	0.125	155	0.0066	0.250	99	0.0042
		1.000	150	0.0019	0.250	150	0.0061	0.500	99	0.0047
		1.500	150	0.0017	0.375	146	0.0052	0.750	108	0.0056
		2.000	146	0.0014	0.500	141	0.0042	1.000	108	0.0066
		2.500	141	0.0011	· · ·	· · ·	· · ·	1.250	113	0.0076
1117, 1118	91	0.500	150	0.0022	0.125	150	0.0064	0.250	95	0.0041
		1.000	145	0.0018	0.250	145	0.0059	0.500	95	0.0045
		1.500	145	0.0016	0.375	141	0.0050	0.750	105	0.0055
		2.000	141	0.0014	0.500	136	0.0041	1.000	105	0.0064
		2.500	136	0.0011	· · ·	· · ·	· · ·	1.250	119	0.0073
1144, annealed	85	0.500	140	0.0021	0.125	140	0.0059	0.250	89	0.0040
		1.000	136	0.0017	0.250	136	0.0055	0.500	89	0.0045
		1.500	136	0.0015	0.375	132	0.0047	0.750	98	0.0055
		2.000	132	0.0013	0.500	127	0.0040	1.000	98	0.0064
		2.500	127	0.0010	· · ·	· · ·	· · ·	1.250	102	0.0070
1141, annealed	81	0.500	135	0.0020	0.125	135	0.0057	0.250	86	0.0040
		1.000	130	0.0017	0.250	130	0.0053	0.500	86	0.0045
		1.500	130	0.0015	0.375	127	0.0045	0.750	94	0.0054
		2.000	127	0.0012	0.500	122	0.0037	1.000	94	0.0063
		2.500	122	0.0010	· · ·	· · ·	· · ·	1.250	98	0.0072
1016, 1018, 1022	78	0.500	130	0.0019	0.125	130	0.0055	0.250	82	0.0038
		1.000	125	0.0016	0.250	125	0.0051	0.500	82	0.0043
		1.500	125	0.0014	0.375	121	0.0043	0.750	90	0.0052
		2.000	121	0.0012	0.500	117	0.0035	1.000	90	0.0060
		2.500	117	0.0009	· · ·	· · ·	· · ·	1.250	94	0.0068
1144	76	0.500	125	0.0019	0.125	125	0.0052	0.250	79	0.0037
		1.000	121	0.0015	0.250	121	0.0049	0.500	79	0.0042
		1.500	121	0.0014	0.375	117	0.0041	0.750	87	0.0050
		2.000	117	0.0011	0.500	113	0.0034	1.000	87	0.0058
		2.500	113	0.0009	· · ·	· · ·	· · ·	1.250	91	0.0066

(continued)

(a) All cutting speeds and feeds based on cutting with high-speed steel tools.(b) Based on a machinability rating of 100% of 1212 steel.

Machinability ratings and recommended feeds and speeds for cold drawn steel bars (customary U.S. units) (a) (continued)

Steel grade	Machinability rating(b), %	Form turning			Single-point turning			Drilling		
		Width of cut, mm	Speed, mm/s	Feed, mm/rev	Depth of cut, mm	Speed, mm/s	Feed, mm/rev	Size of hole, mm	Speed, mm/s	Feed, mm/rev
1020, 1137, 1045, annealed ..	72	0.500	120	0.0018	0.125	120	0.0050	0.250	76	0.0035
		1.000	115	0.0014	0 250	115	0.0047	0.500	76	0.0040
		1.500	115	0.0013	0.375	112	0.0040	0.750	83	0.0047
		2.000	112	0.0011	0.500	108	0.0032	1.000	83	0.0055
		2.500	108	0.0009	...	...	...	1.250	86	0.0064
1035, 1141, 1050, annealed ..	70	0.500	115	0.0017	0.125	115	0.0049	0.250	73	0.0034
		1.000	112	0.0014	0.250	112	0.0045	0.500	73	0.0038
		1.500	112	0.0013	0.375	108	0.0038	0.750	80	0.0045
		2.000	108	0.0011	0.500	105	0.0031	1.000	80	0.0053
		2.500	105	0.0008	...	...	...	1.250	84	0.0062
1040	64	0.500	105	0.0015	0.125	105	0.0044	0.250	67	0.0032
		1.000	101	0.0012	0.250	101	0.0041	0.500	67	0.0035
		1.500	101	0.0011	0.375	98	0.0034	0.750	73	0.0042
		2.000	98	0.0009	0.500	95	0.0028	1.000	73	0.0049
		2.500	95	0.0007	...	...	...	1.250	76	0.0056
1045	57	0.500	95	0.0014	0.125	95	0.0040	0.250	60	0.0028
		1.000	91	0.0012	0.250	91	0.0037	0.500	60	0.0031
		1.500	91	0.0010	0.375	88	0.0031	0.750	65	0.0037
		2.000	88	0.0009	0.500	85	0.0026	1.000	65	0.0044
		2.500	85	0.0007	...	...	...	1.250	68	0.0050
1050	54	0.500	90	0.0014	0.125	90	0.0038	0.250	57	0.0028
		1.000	87	0.0011	0.250	87	0.0035	0.500	57	0.0031
		1.500	87	0.0010	0.375	84	0.0030	0.750	62	0.0037
		2.000	84	0.0008	0.500	81	0.0024	1.000	62	0.0044
		2.500	81	0.0007	...	...	...	1.250	65	0.0050

(a) All cutting speeds and feeds based on cutting with high-speed steel tools. (b) Based on a machinability rating of 100% for 1212 steel.

Machinability ratings for alloy steels

Grade	Machinability rating (a)	Range of typical hardnesses, HB	Grade	Machinability rating (a)	Range of typical hardnesses, HB	Grade	Machinability rating (a)	Range of typical hardnesses, HB
1330	55 (b)	179-235	4626	70 (c)	170-212	8615	70 (c)	179-235
1335	55 (b)	179-235				8617	70 (c)	179-235
1340	50 (b)	183-241	4718	60 (c)	187-229	8620	65 (c)	179-235
1345	45 (c)	183-241	4720	65 (c)	187-229			
						86L20	85 (f)	...
4012	70 (c)	149-196	4815	50 (e)	187-229	8622	65 (c)	179-235
			4817	50 (e)	187-229	8625	60 (b)	179-223
4023	70 (c)	156-207	4820	50 (e)	187-229	8627	60 (b)	170-223
4024	75 (c)	156-207				8630	70 (b)	179-229
4027	70 (c)	167-212	5015	65 (c)	156-196			
4028	75 (c)	167-212	50B40	65 (b)	174-223	8637	65 (b)	179-229
4032	70 (b)	174-217	50B44	65 (b)	174-223	8640	65 (b)	184-229
			5046	60 (b)	174-223	8642	65 (b)	184-229
4037	70 (b)	174-217	50B46	60 (b)	174-223	8645	65 (b)	184-235
4042	65 (b)	179-229				86B45	65 (b)	184-235
4047	65 (b)	179-229	50B50	55 (b)	183-235			
			5060	55 (d)	170-212	8650	60 (b)	187-248
4118	60 (c)	170-207	50B60	55 (d)	170-212	8655	55 (b)	187-248
4130	70 (b)	187-229				8660	55 (d)	179-217
4135	70 (b)	187-229	5115	65 (c)	163-201			
4137	70 (b)	187-229	5120	70 (c)	163-201	8720	65 (c)	179-235
4140	65 (b)	187-229				8740	65 (b)	184-235
41L40	85 (b)	...	5130	70 (b)	174-212			
4142	65 (b)	187-229	5132	70 (b)	174-212	8822	55 (d)	179-223
			5135	70 (b)	179-217			
4145	60 (b)	187-229	5140	65 (b)	179-217	9254	45 (d)	187-241
4147	60 (b)	187-235	5145	65 (b)	179-229	9255	40 (d)	179-229
4150	55 (b)	187-241				9260	40 (d)	184-235
4161	50 (d)	187-241	5147	65 (b)	179-229			
			5150	60 (b)	183-235	9310	50 (e)	184-229
4320	60 (e)	187-229	5155	55 (b)	183-235			
			5160	55 (d)	179-217	94B15	70 (c)	163-202
4340	50 (d)	187-241	51B60	55 (d)	179-217	94B17	70 (c)	163-202
E4340	50 (d)	187-241				94B30	70 (b)	170-223
			50100	40 (d)	183-241			
4419	65 (b)	170-212	51100	40 (d)	183-241			
4422	65 (c)	170-212	52100	40 (d)	183-241			
4427	65 (b)	170-212						
			6118	60 (c)	179-217			
4615	65 (c)	174-223	6150	55 (d)	183-241			
4617	65 (c)	174-223						
4620	65 (c)	183-229	8115	65 (c)	163-202			
4621	60 (c)	183-229	81B45	65 (b)	179-223			

(a) All ratings apply to cold-finished bars. (b) Microstructure composed primarily of lamellar pearlite and ferrite. (c) Microstructure composed primarily of blocky or acicular pearlite and bainite, as found in hot rolled steels. (d) Microstructure compound primarily of spheroidite. (e) Microstructure resulting from subcritical anneal. (f) Microstructure not specified.

Rankings for machinability of several steels (a)

Operation	Type of tool	Steel and hardness range, HB										
		12L13 150-200	41L40 200-250	4140R 200-250	1030 175-225	8620 175-225	1060 175-225	4140 175-225	4340 175-225	H11 225-300	4340 275-325	4340 375-425
Broaching	HSS	11	10	6	6	6	6	6	6	6	2	1
Tapping	HSS	11	9½	7	9½	7	7	4½	4½	3	2	1
Deep drilling ..	HSS	11	10	9	8	5	9	5	5	3	2	1
	Carbide	11	10	6	6	6	6	6	6	6	2	1
Boring	HSS	11	10	7	9	7	7	4½	4½	2½	2½	1
	Carbide	11	10	8	9	6	7	½	½	3	2	1
Form cutting ..	HSS	11	9½	9½	8	6	7	4½	4½	2½	2½	1
	Carbide	11	10	9	8	6	7	4½	4½	2½	2½	1
Drilling	HSS	11	10	9	8	5	7	5	5	3	2	1
Planing	HSS	11	10	7½	9	7½	3	5	5	5	2	1
	Carbide	10½	9	10½	7	7	7	4½	4½	3	2	1
Turning	HSS	11	10	4½	8	9	4½	4½	4½	3	2	1
	Carbide	11	10	8	6	9	4	6	6	3	2	1
End milling ...	HSS	11	10	9	8	4	6½	4	4	6½	2	1
	Carbide	11	10	8	8	6	8	4½	4½	3	2	1

(a) Higher numbers indicate steels easier to machine. Rankings based on suggested cutting speeds in previous table.

Machinability ratings for cold drawn alloy steel bars (a)

Steel grade	Condition before drawing	Machinability rating(b), %	Steel grade	Condition before drawing	Machinability rating(b), %	Steel grade	Condition before drawing	Machinability rating(b), %
1330	Annealed	55	4620	Hot rolled	65	6118	Hot rolled	60
1335	Annealed	55	4626	Hot rolled	70	6150	Annealed	55
1340	Annealed	50	4720	Hot rolled	65	81B45	Annealed	65
1345	Annealed	45	4815	Annealed	50	8615	Hot rolled	70
4023	Hot rolled	70	4817	Annealed	50	8617	Hot rolled	70
4024	Hot rolled	75	4820	Annealed	50	8620	Hot rolled	65
4027	Annealed	70	5015	Hot rolled	65	8622	Hot rolled	65
4028	Annealed	75	50B44	Annealed	65	8625	Annealed	60
4037	Annealed	70	50B46	Annealed	60	8627	Annealed	60
4047	Annealed	65	50B50	Annealed	55	8630	Annealed	70
4118	Hot rolled	60	50B60	Spheroidized	55	8637	Annealed	65
4130	Annealed	70	5120	Hot rolled	70	8640	Annealed	65
4137	Annealed	70	5130	Annealed	70	8642	Annealed	65
4140	Annealed	65	5132	Annealed	70	8645	Annealed	65
4142	Annealed	65	5135	Annealed	70	8655	Annealed	55
4145	Annealed	60	5140	Annealed	65	8720	Hot rolled	65
4147	Annealed	60	5150	Annealed	60	8740	Annealed	65
4150	Annealed	55	5155	Annealed	55	8822	Hot rolled	55
4161	Spheroidized	50	5160	Spheroidized	55	9260	Spheroidized	40
4320	Annealed	60	51B60	Spheroidized	55	94B17	Hot rolled	70
4340	Annealed	50	51100	Spheroidized	40	94B30	Annealed	70
4615	Hot rolled	65	52100	Spheroidized	40			

(a) Source: AISI Committee of Hot Rolled and Cold Finished Bar Producers. (b) Based on cutting with high-speed tool steels and a machinability rating of 100% for 1212 steel.

Conditions recommended for various machining operations on wrought steels of different hardness levels (SI units)

Grade(a)	Material condition(b)	Hardness, HB	Type of tool	Broach	Tap	Deep drill	Bore	Form cut	Drill	Plane	Turn	End mill
12L13	HR, N, A, CD	150-200	HSS	200(d)	380	785(e)	840(f)	835(d)	660(h)	480(i)	990(i)	940(l)
			Carbide			1395(d)	3125(h)	2670(f)		1525(k)	3555(l)	2490(h)
41L40	HR, N, A, CD	200-260	HSS	125(d)	330	560(e)	510(f)	455(d)	455(f)	330(i)	610(j)	710(g)
			Carbide			1270(e)	1930(h)	1575(g)		1477(k)	2415(l)	2030(l)
1030	HR, N, A, CD	175-225	HSS	100(d)	330	455(h)	480(f)	430(d)	380(h)	305(i)	485(j)	610(g)
			Carbide			1015(d)	1780(h)	1395(g)		1270(k)	2030(l)	1905(l)
4140R	HR, N, A, CD	200-250	HSS	100(d)	280	480(h)	430(f)	380(d)	405(f)	280(i)	455(j)	635(g)
			Carbide			1015(h)	1750(h)	1370(g)		1525(k)	2185(l)	1905(l)
8620	HR, A, CD	175-225	HSS	100(d)	280	405(h)	430(f)	380(d)	330(h)	280(i)	560(j)	455(g)
			Carbide			1015(d)	1625(h)	1270(g)		1270(k)	2235(l)	1855(l)
1060	HR, N, A, CD	175-225	HSS	100(d)	280	430(h)	430(f)	405(d)	355(h)	305(m)	455(j)	585(g)
			Carbide			1015(d)	1650(h)	1345(g)		1270(k)	2005(h)	1905(l)
4140	HR, A, CD	175-225	HSS	100(d)	230	405(h)	380(f)	330(d)	330(h)	255(i)	455(j)	455(g)
			Carbide			1015(d)	1500(h)	1145(g)		1220(k)	2030(l)	1780(l)
4340	HR, A, CD	175-225	HSS	100(d)	230	405(h)	380(f)	330(d)	330(h)	255(i)	455(j)	455(g)
			Carbide			1015(d)	1500(h)	1145(g)		1220(k)	2030(l)	1780(l)
H11	A	225-300	HSS	100(d)	150	330(h)	280(g)	255(n)	255(f)	230(i)	330(h)	585(g)
			Carbide			1015(d)	1320(h)	890(d)		1015(k)	1780(l)	1525(l)
4340	N, Q & T	275-325	HSS	75(d)	125	280(f)	280(g)	255(n)	230(g)	180(m)	305(j)	330(g)
			Carbide			760(d)	1220(f)	890(d)		915(i)	1675(j)	1220(g)
4340	Q & T	375-425	HSS	40(d)	60	150(d)	150(d)	150(n)	125(d)	···	180(h)	230(d)
			Carbide	···	···	510(n)	760(g)	560(d)	···	···	1145(j)	710(d)

(a) L, lead-bearing grade; R, resulfurized steel. (b) HR, hot rolled; N, normalized; A, annealed; CD, cold drawn; Q & T, quenched and tempered. (c) Dimensions of cut as follows: tap, deep drill and drill: 12.7-mm diam hole; bore and end mill: 1.3-mm deep cut; form cut: 12.7-mm wide cut: plane: 2.5-mm deep cut; turn: 3.8-mm deep cut. (d) Chip load in broaching or feed in milling: 0.076 to 0.10 mm/tooth; feed in other operations: 0.076 to .10 mm/rev. (e) Feed: 0.30 mm/rev. (f) Feed: 0.18 to 0.20 mm/rev. (g) Feed: 0.13 mm/stroke. (h) Feed: 0.23 to 0.25 mm/rev. (i) Feed: 1.3 mm/stroke. (j) Feed: 0.38 mm/rev. (k) Feed: 2.0 mm/stroke. (l) Feed: 0.51 mm/rev. (m) Feed: 76 mm/stroke. (n) Feed: 0.025 to 0.051 mm/rev.

Conditions recommended for various machining operations on wrought steels of different hardness levels (customary U.S. units)

Grade(a)	Material condition(b)	Hardness, HB	Type of tool	Broach	Tap	Deep drill	Bore	Form cut	Drill	Plane	Turn	End mill
12L13	HR, N, A, CD	150-200	HSS	40(d)	75	155(e)	165(f)	165(d)	130(h)	95(i)	195(j)	185(f)
			Carbide			275(d)	615(h)	525(f)		300(k)	700(l)	490(h)
41L40	HR, N, A, CD	200-260	HSS	25(d)	65	110(e)	100(f)	90(d)	90(f)	65(i)	120(j)	140(g)
			Carbide			250(e)	380(h)	310(g)		290(k)	475(l)	400(f)
1030	HR, N, A, CD	175-225	HSS	20(d)	65	90(h)	95(f)	85(d)	75(h)	60(i)	95(j)	120(g)
			Carbide			200(d)	350(h)	275(g)		250(k)	400(l)	375(f)
4140R	HR, N, A, CD	200-250	HSS	20(d)	55	95(h)	85(f)	75(d)	80(f)	55(i)	90(j)	125(g)
			Carbide			200(h)	345(h)	270(g)		300(k)	430(l)	375(f)
8620	HR, A, CD	175-225	HSS	20(d)	55	80(h)	85(f)	75(d)	65(h)	55(i)	110(j)	90(g)
			Carbide			200(d)	320(h)	250(g)		250(k)	440(l)	365(f)
1060	HR, N, A, CD	175-225	HSS	20(d)	55	85(h)	85(f)	80(d)	70(h)	60(m)	90(j)	115(g)
			Carbide			200(d)	325(h)	265(g)		250(k)	395(h)	375(f)
4140	HR, A, CD	175-225	HSS	20(d)	45	80(h)	75(f)	65(d)	65(h)	50(i)	90(j)	90(g)
			Carbide			200(d)	295(h)	225(g)		240(k)	400(l)	350(f)
4340	HR, A, CD	175-225	HSS	20(d)	45	80(h)	75(f)	65(d)	65(h)	50(i)	90(j)	90(g)
			Carbide			200(d)	295(h)	225(g)		240(k)	400(l)	350(f)
H11	A	225-300	HSS	20(d)	30	65(h)	55(g)	50(n)	50(f)	45(i)	65(h)	115(g)
			Carbide			200(d)	260(h)	175(d)		200(k)	350(l)	300(f)
4340	N, Q & T	275-325	HSS	15(d)	25	55(f)	55(g)	50(n)	45(g)	35(m)	60(j)	65(g)
			Carbide			150(d)	240(f)	175(d)		180(i)	330(j)	240(g)
4340	Q & T	375-425	HSS	8(c)	12	30(d)	30(d)	30(n)	25(d)	...	35(h)	45(d)
			Carbide	...	...	100(h)	150(g)	110(d)	...	...	225(j)	140(d)

(a) L, lead-bearing grade; R, resulfurized steel. (b) HR, hot rolled; N, normalized; A, annealed; CD, cold drawn; Q & T, quenched and tempered. (c) Dimensions of cut as follows: tap, deep drill and drill: 0.5-in. diam hole; bore and end mill: 0.05-in. deep cut; form cut: 0.5-in. wide cut; plane: 0.10-in. deep cut turn: 0.15-in. deep cut. (d) Chip load in broaching or feed in milling: 0.003 to 0.004 in./tooth; feed in other operations: 0.003 to 0.004 in./rev. (e) Feed: 0.012 in./rev. (f) Feed: 0.007 to 0.008 in./rev. (g) Feed: 0.005 to 0.006 in./rev. (h) Feed: 0.009 to 0.010 in./rev. (i) Feed: 0.05 in./stroke. (j) Feed: 0.015 in./rev. (k) Feed: 0.08 in./stroke. (l) Feed: 0.020 in./rev. (m) Feed: 0.03 in./stroke. (n) Feed: 0.001 to 0.002 in./rev.

Forming

Typical mechanical properties of hot rolled steel sheet

Type or quality	Special feature	Yield strength MPa	ksi	Tensile strength MPa	ksi	Elongation in 50 mm or 2 in., %	Hardness, HRB	Strain-hardening exponent, n	Plastic strain ratio, r_m
Commercial	Standard properties	262	38	359	52	30	55	0.15	0.9
Drawing (rimmed)	Improved properties	241	35	345	50	35	50	0.18	1.0
Drawing (special killed)	Non-aging	241	35	345	50	40	50	0.20	1.0
Medium strength	Inclusion shape control	345	50	414	60	25	70	0.15	0.9
High strength	Inclusion shape control	552	80	620	90	15	90	...	...

Typical mechanical properties of cold rolled steel sheet

Type or quality	Special feature	Yield strength MPa	ksi	Tensile strength MPa	ksi	Elongation in 50 mm or 2 in., %	Hardness	Strain-hardening exponent, n	Plastic strain ratio, r_m
Commercial	Improved finish	234	34	317	46	35	45 HRB	0.20	1.0
Drawing (rimmed)	Stretchability	207	30	310	45	45	40 HRB	0.24	1.2
Drawing (special killed)	Deep drawing	172	25	296	43	40	40 HRB	0.22	1.6
Interstitial free	Extra deep drawing	152	22	317	46	45	45 HRB	0.24	2.0
Medium strength	Formable	414	60	483	70	25	85 HRB	0.20	1.2
High strength	Moderately formable	689	100	724	105	10	25 HRC	...	...

Minimum bend radii for selected plain carbon and low-alloy steel sheet materials
. INSIDE RADIUS

Product	Quality temper or strength level	Thickness	Parallel to rolling direction	Across rolling direction
Cold rolled				
1008/1010	CQ	. . .	0.25 mm (0.01 in.)	. . .
1008/1010	DQ	. . .	0.25 mm (0.01 in.)	. . .
1008/1010	No. 3(b)	. . .	1t	0.5t
1008/1010	No. 2(c)	. . .	NR	1t
1008/1010	No. 1(d)	. . .	NR	NR
Hot rolled				
1008/1010	CQ	<2.25 mm (<0.09 in.)	0.75t	0.5t
		>2.25 mm (>0.09 in.)	1.5t	1t
1008/1010	DQ	<2.25 mm (<0.09 in.)	0.5t	0.25t
		>2.25 mm (>0.09 in.)	0.75t	0.5t
Annealed				
1020/1025	. . .	. . .	1t to 2t	. . .
4130, 8630	. . .	. . .	1.5t to 2t	. . .
1070, 1095	. . .	. . .	2t to 3t	. . .
ASTM A607 (HSLA)	345 MPa (50 ksi)	. . .	1.5t	1t
	415 MPa (60 ksi)	. . .	3t	2t
	480 MPa (70 ksi)	. . .	4t	3t

(a) CR, cold rolled; HR, hot rolled; CQ, commercial quality; DQ, drawing quality; t, sheet thickness; NR, not recommended; HSLA, high-strength low-alloy. (b) Quarter hard. (c) Half hard. (d) Full hard.

Suggested minimum inside bend radii for SAE J410c steels of various strengths and thicknesses

Grade	<4.6 mm (<0.180 in.)	4.6 to 6.4 mm (0.180 to 0.250 in.)	6.4 to 12.7 mm (0.250 to 0.500 in.)
942X	. . .	t	2t
945A, 945C	t	2t	2.5t
945X	t	t	2t
950A, 950B, 950C, 950D . . .	t	2t	3t
950X	1.5t	2.5t	2.5t
955X	2t	3t	3t
960X	2.5t	3.5t	3.5t
965X	3t	4t	4t
970X .:	3.5t	4.5t	4.5t
980X(a)	3.5t	4.5t	4.5t

(a) Available only in thicknesses to 9.5 mm (0.375 in.)

Specified bend test radii for inclusion shape controlled ASTM A715 steel sheet

Grade	Transverse bends(a)	Longitudinal bends(a)
50	0.5t	0
60	0.5t	0
70	0.75t	0.5t
80	0.75t	0.5t

(a) For sheet thicknesses up to 5.84 mm (0.2299 in.)

Tool Steels

Composition Limits

Compositions of tool steels no longer in common use

Type	C	W	Mo	Cr	V	Others
High speed steels						
M8 . . . . 0.80	5.00	5.00	4.00	1.50	1.25 Nb
M15 . . . . 1.50	6.50	3.50	4.00	5.00	5.00 Co
M35 . . . . 0.80	6.00	5.00	4.00	2.00	5.00 Co
M45 . . . . 1.25	8.00	5.00	4.25	1.60	5.50 Co
T3 . . . . 1.05	18.00	. . .	4.00	3.00	. . .
T7 . . . . 0.75	14.00	. . .	4.00	2.00	. . .
T9 . . . . 1.20	18.00	. . .	4.00	4.00	. . .
Hot work steels						
H15 . . . . 0.40	. . .	5.00	5.00	. . .	. . .
H16 . . . . 0.55	7.00	. . .	7.00	. . .	. . .
H20 . . . . 0.35	9.00	. . .	2.00	. . .	. . .
H41 . . . . 0.65	1.50	8.00	4.00	1.00	. . .
H43 . . . . 0.55	. . .	8.00	4.00	2.00	. . .
Cold work steels						
D1 . . . . 1.00	. . .	1.00	12.00	. . .	. . .
D6 (a)						
A5 . . . . 1.00	. . .	1.00	1.00	. . .	3.00 Mn
Shock-resisting steels						
S3 . . . . 0.50	1.00	. . .	0.74	. . .	. . .
S4 . . . . 0.55	. . .	. . .	. . .	. . .	2.00 Si, 0.80 Mn
Mold steel						
P1 . . . . 0.10	. . .	. . .	. . .	. . .	. . .
Special-purpose tool steels						
L1 . . . . 1.00	. . .	. . .	1.25	. . .	. . .
L3 . . . . 1.00	. . .	. . .	1.50	0.20	. . .
L4 . . . . 1.00	. . .	. . .	1.50	0.25	0.60 Mn
L5 . . . . 1.00	. . .	0.25	1.00	. . .	1.00 Mn
L7 . . . . 1.00	. . .	0.40	1.40	. . .	0.35 Mn
F1 . . . . 1.00	1.25	. . .	. . .	. . .	. . .
F2 . . . . 1.25	3.50	. . .	. . .	. . .	. . .
F3 . . . . 1.25	3.50	. . .	0.75	. . .	. . .
Water-hardening tool steels						
W3 . . . . 1.00	. . .	. . .	. . .	0.50	. . .
W4 . . . 0.60/1.40(b)	. . .	. . .	0.25	. . .	. . .
W6 . . . . 1.00	. . .	. . .	0.25	0.25	. . .
W7 . . . . 1.00	. . .	. . .	0.50	0.20	. . .

(a) Now included with D3 in the following table. (b) Various carbon contents were available.

Composition limits of principal types of tool steels

AISI	SAE	UNS	C	Mn	Si	Cr	Ni	Mo	W	V	Co
Molybdenum high speed steels											
M1	M1	T11301	0.78-0.88	0.15-0.40	0.20-0.50	3.50-4.00	0.30 max	8.20-9.20	1.40-2.10	1.00-1.35	...
M2	M2	T11302	0.78-0.88; 0.95-1.05	0.15-0.40	0.20-0.45	3.75-4.50	0.30 max	4.50-5.50	5.50-6.75	1.75-2.20	...
M3, class 1	M3	T11313	1.00-1.10	0.15-0.40	0.20-0.45	3.75-4.50	0.30 max	4.75-6.50	5.00-6.75	2.25-2.75	...
M3, class 2	M3	T11323	1.15-1.25	0.15-0.40	0.20-0.45	3.75-4.50	0.30 max	4.75-6.50	5.00-6.75	2.75-3.75	...
M4	M4	T11304	1.25-1.40	0.15-0.40	0.20-0.45	3.75-4.75	0.30 max	4.25-5.50	5.25-6.50	3.75-4.50	...
M6		T11306	0.75-0.85	0.15-0.40	0.20-0.45	3.75-4.50	0.30 max	4.50-5.50	3.75-4.75	1.30-1.70	11.00-13.00
M7		T11307	0.97-1.05	0.15-0.40	0.20-0.55	3.50-4.00	0.30 max	8.20-9.20	1.40-2.10	1.75-2.25	...
M10		T11310	0.84-0.94; 0.95-1.05	0.10-0.40	0.20-0.45	3.75-4.50	0.30 max	7.75-8.50	...	1.80-2.20	...
M30		T11330	0.75-0.85	0.15-0.40	0.20-0.45	3.50-4.25	0.30 max	7.75-9.00	1.30-2.30	1.00-1.40	4.50-5.50
M33		T11333	0.85-0.92	0.15-0.40	0.15-0.50	3.50-4.00	0.30 max	9.00-10.00	1.30-2.10	1.00-1.35	7.75-8.75
M34		T11334	0.85-0.92	0.15-0.40	0.20-0.45	3.50-4.00	0.30 max	7.75-9.20	1.40-2.10	1.90-2.30	7.75-8.75
M36		T11336	0.80-0.90	0.15-0.40	0.20-0.45	3.75-4.50	0.30 max	4.50-5.50	5.50-6.50	1.75-2.25	7.75-8.75
M41		T11341	1.05-1.15	0.20-0.60	0.15-0.50	3.75-4.50	0.30 max	3.25-4.25	6.25-7.00	1.75-2.25	4.75-5.75
M42		T11342	1.05-1.15	0.15-0.40	0.15-0.65	3.50-4.25	0.30 max	9.00-10.00	1.15-1.85	0.95-1.35	7.75-8.75
M43		T11343	1.15-1.25	0.20-0.40	0.15-0.65	3.50-4.25	0.30 max	7.50-8.50	2.25-3.00	1.50-1.75	7.75-8.75
M44		T11344	1.10-1.20	0.20-0.40	0.30-0.55	4.00-4.75	0.30 max	6.00-7.00	5.00-5.75	1.85-2.20	11.00-12.25
M46		T11346	1.22-1.30	0.20-0.40	0.40-0.65	3.70-4.20	0.30 max	8.00-8.50	1.90-2.20	3.00-3.30	7.80-8.80
M47		T11347	1.05-1.15	0.15-0.40	0.20-0.45	3.50-4.00	0.30 max	9.25-10.00	1.30-1.80	1.15-1.35	4.75-5.25
Tungsten high speed steels											
T1	T1	T12001	0.65-0.80	0.10-0.40	0.20-0.40	3.75-4.00	0.30 max	...	17.25-18.75	0.90-1.30	...
T2	T2	T12002	0.80-0.90	0.20-0.40	0.20-0.40	3.75-4.50	0.30 max	1.00 max	17.50-19.00	1.80-2.40	...
T4	T4	T12004	0.70-0.80	0.10-0.40	0.20-0.40	3.75-4.50	0.30 max	0.40-1.00	17.50-19.00	0.80-1.20	4.25-5.75
T5	T5	T12005	0.75-0.85	0.20-0.40	0.20-0.40	3.75-5.00	0.30 max	0.50-1.25	17.50-19.00	1.80-2.40	7.00-9.50
T6		T12006	0.75-0.85	0.20-0.40	0.20-0.40	4.00-4.75	0.30 max	0.40-1.00	18.50-21.00	1.50-2.10	11.00-13.00
T8	T8	T12008	0.75-0.85	0.20-0.40	0.20-0.40	3.75-4.50	0.30 max	0.40-1.00	13.25-14.75	1.80-2.40	4.25-5.75
T15		T12015	1.50-1.60	0.15-0.40	0.15-0.40	3.75-5.00	0.30 max	1.00 max	11.75-13.00	4.50-5.25	4.75-5.25
Chromium hot work steels											
	H10	T20810	0.35-0.45	0.25-0.70	0.80-1.20	3.00-3.75	0.30 max	2.00-3.00	...	0.25-0.75	...
H11	H11	T20811	0.33-0.43	0.20-0.50	0.80-1.20	4.75-5.50	0.30 max	1.10-1.60	...	0.30-0.60	...
H12	H12	T20812	0.30-0.40	0.20-0.50	0.80-1.20	4.75-5.50	0.30 max	1.25-1.75	1.00-1.70	0.50 max	...
H13	H13	T20813	0.32-0.45	0.20-0.50	0.80-1.20	4.75-5.50	0.30 max	1.10-1.75	...	0.80-1.20	...
H14		T20814	0.35-0.45	0.20-0.50	0.80-1.20	4.75-5.50	0.30 max	...	4.00-5.25	...	...
H19		T20819	0.32-0.45	0.20-0.50	0.20-0.50	4.00-4.75	0.30 max	0.30-0.55	3.75-4.50	1.75-2.20	4.00-4.50
Tungsten hot work steels											
H21	H21	T20821	0.26-0.36	0.15-0.40	0.15-0.50	3.00-3.75	0.30 max	...	8.50-10.00	0.30-0.60	...
	H22	T20822	0.30-0.40	0.15-0.40	0.15-0.40	1.75-3.75	0.30 max	...	10.00-11.75	0.25-0.50	...
		T20823	0.25-0.35	0.15-0.40	0.15-0.60	11.00-12.75	0.30 max	...	11.00-12.75	0.75-1.25	...
		T20824	0.42-0.53	0.15-0.40	0.15-0.40	2.50-3.50	0.30 max	...	14.00-16.00	0.40-0.60	...
		T20825	0.22-0.32	0.15-0.40	0.15-0.40	3.75-4.50	0.30 max	...	14.00-16.00	0.40-0.60	...
		T20826	0.45-0.55(b)	0.15-0.40	0.15-0.40	3.75-4.50	0.30 max	...	17.25-19.00	0.75-1.25	...
Molybdenum hot work steels											
	H42	T20842	0.55-0.70(b)	0.15-0.40	...	3.75-4.50	0.30 max	4.50-5.50	5.50-6.75	1.75-2.20	...

(a) All steels except group W contain 0.25 max Cu, 0.03 max P and 0.03 max S; group W steels contain 0.20 max Cu, 0.025 max P and 0.025 max S. Where specified, sulfur may be increased to 0.06 to 0.15% to improve machinability of group H, M and T steels. (b) Available in several carbon ranges. (c) Contains free graphite in the microstructure. (d) Optional. (e) Specified carbon ranges are designated by suffix numbers.

Composition limits of principal types of tool steels (continued)

AISI	Designations SAE	UNS	C	Mn	Si	Cr	Ni	Mo	W	V	Co
						Composition(a), %					
Air-hardening medium-alloy cold work steels											
A2	A2	T30102	0.95-1.05	1.00 max	0.50 max	4.75-5.50	0.30 max	0.90-1.40	...	0.15-0.50	...
A3	...	T30103	1.20-1.30	0.40-0.60	0.50 max	4.75-5.50	0.30 max	0.90-1.40	...	0.80-1.40	...
A4	...	T30104	0.95-1.05	1.80-2.20	0.50 max	0.90-2.20	0.30 max	0.90-1.40	...	...	...
A6	...	T30106	0.65-0.75	1.80-2.50	0.50 max	0.90-1.20	0.30 max	0.90-1.40	...	...	...
A7	...	T30107	2.00-2.85	0.80 max	0.50 max	5.00-5.75	0.30 max	0.90-1.40	0.50-1.50	3.90-5.15	...
A8	...	T30108	0.50-0.60	0.50 max	0.75-1.10	4.75-5.50	0.30 max	1.15-1.65	1.00-1.50	...	...
A9	...	T30109	0.45-0.55	0.50 max	0.95-1.15	4.75-5.50	1.25-1.75	1.30-1.80	...	0.80-1.40	...
A10	...	T30110	1.25-1.50(c)	1.60-2.10	1.00-1.50	...	1.55-2.05	1.25-1.75	...	...	...
High-carbon, high-chromium cold work steels											
D2	D2	T30402	1.40-1.60	0.60 max	0.60 max	11.00-13.00	0.30 max	0.70-1.20	...	1.10 max	1.00 max
D3	D3	T30403	2.00-2.35	0.60 max	0.60 max	11.00-13.50	0.30 max	...	1.00 max	1.00 max	...
D4	...	T30404	2.05-2.40	0.60 max	0.60 max	11.00-13.00	0.30 max	0.70-1.20	...	1.00 max	...
D5	D5	T30405	1.40-1.60	0.60 max	0.60 max	11.00-13.00	0.30 max	0.70-1.20	...	1.00 max	2.50-3.50
D7	D7	T30407	2.15-2.50	0.60 max	0.60 max	11.50-13.50	0.30 max	0.70-1.20	...	3.80-4.40	...
Oil-hardening cold work steels											
O1	O1	T31501	0.85-1.00	1.00-1.40	0.50 max	0.40-0.60	0.30 max	...	0.40-0.60	0.30 max	...
O2	O2	T31502	0.85-0.95	1.40-1.80	0.50 max	0.35 max	0.30 max	0.30 max	...	0.30 max	...
O6	O6	T31506	1.25-1.55(c)	0.30-1.10	0.55-1.50	0.30 max	0.30 max	0.20-0.30	...	...	...
O7	...	T31507	1.10-1.30	1.00 max	0.60 max	0.35-0.85	0.30 max	0.30 max	1.00-2.00	0.40 max	...
Shock-resisting steels											
S1	S1	T41901	0.40-0.55	0.10-0.40	0.15-1.20	1.00-1.80	0.30 max	0.50 max	1.50-3.00	0.15-0.30	...
S2	S2	T41902	0.40-0.55	0.30-0.50	0.90-1.20	...	0.30 max	0.30-0.60	...	0.50 max	...
S5	S5	T41905	0.50-0.65	0.60-1.00	1.75-2.25	0.35 max	...	0.20-1.35	...	0.35 max	...
S6	...	T41906	0.40-0.50	1.20-1.50	2.00-2.50	1.20-1.50	...	0.30-0.50	...	0.20-0.40	...
S7	...	T41907	0.45-0.55	0.20-0.80	0.20-1.00	3.00-3.50	...	1.30-1.80	...	0.20-0.30(d)	...
Low-alloy special-purpose tool steels											
L2	...	T61202	0.45-1.00(b)	0.10-0.90	0.50 max	0.70-1.20	0.30 max	0.25 max	...	0.10-0.30	...
L6	L6	T61206	0.65-0.75	0.25-0.80	0.50 max	0.60-1.20	1.25-2.00	0.50 max	...	0.20-0.30(d)	...
Low-carbon mold steels											
P2	...	T51602	0.10 max	0.10-0.40	0.10-0.40	0.75-1.25	0.10-0.50	0.15-0.40	...	...	...
P3	...	T51603	0.10 max	0.20-0.60	0.40 max	0.40-0.75	1.00-1.50	...	...	...	...
P4	...	T51604	0.12 max	0.20-0.60	0.10-0.40	4.00-5.25	...	0.40-1.00	...	...	...
P5	...	T51605	0.10 max	0.20-0.60	0.40 max	2.00-2.50	0.35 max	...	...	...	...
P6	...	T51606	0.05-0.15	0.35-0.70	0.10-0.40	1.25-1.75	3.25-3.75	...	...	...	...
P20	...	T51620	0.28-0.40	0.60-1.00	0.20-0.80	1.40-2.00	...	0.30-0.55	...	...	...
P21	...	T51621	0.18-0.22	0.20-0.40	0.20-0.40	0.20-0.30	3.90-4.25	...	...	...	1.05-1.25Al
Water-hardening tool steels											
W1	W108,W109, W110,W112	T72301	0.70-1.50(e)	0.10-0.40	0.10-0.40	0.15 max	0.20 max	0.10 max	0.15 max	0.10 max	...
W2	W209,W210	T72302	0.85-1.50(e)	0.10-0.40	0.10-0.40	0.15 max	0.20 max	0.10 max	0.15 max	0.15-0.35	...
W5	...	T72305	1.05-1.15	0.10-0.40	0.10-0.40	0.40-0.60	0.20 max	0.10 max	0.15 max	0.10 max	...

(a) All steels except group W contain 0.25 max Cu, 0.03 max P and 0.03 max S; group W steels contain 0.20 max Cu, 0.025 max P and 0.025 max S. Where specified, sulfur may be increased to 0.06 to 0.15% to improve machinability of group H, M and T steels. (b) Available in several carbon ranges. (c) Contains free graphite in the microstructure. (d) Optional. (e) Specified carbon ranges are designated by suffix numbers.

Properties and Characteristics

Nominal room-temperature mechanical properties of group L and group S tool steels

Type	Condition	Tensile strength MPa	ksi	0.2% yield strength MPa	ksi	Elongation(a), %	Reduction in area, %	Hardness, HRC	Impact energy J	ft·lb
L2	Annealed.........................	710	103	510	74	25	50	96 HRB	...	...
	Oil quenched from 855 °C (1575 °F) and single tempered at:									
	205 °C (400 °F)...................	2000	290	1790	260	5	15	54	28(b)	21(b)
	315 °C (600 °F)...................	1790	260	1655	240	10	30	52	19(b)	14(b)
	425 °C (800 °F)...................	1550	225	1380	200	12	35	47	26(b)	19(b)
	540 °C (1000 °F)..................	1275	185	1170	170	15	45	41	39(b)	29(b)
	650 °C (1200 °F)..................	930	135	760	110	25	55	30	125(b)	92(b)
L6	Annealed.........................	655	95	380	55	25	55	93 HRB	...	...
	Oil quenched from 845 °C (1550 °F) and single tempered at:									
	315 °C (600 °F)...................	2000	290	1790	260	4	9	54	12(b)	9(b)
	425 °C (800 °F)...................	1585	230	1380	200	8	20	46	18(b)	13(b)
	540 °C (1000 °F)..................	1345	195	1100	160	12	30	42	23(b)	17(b)
	650 °C (1200 °F)..................	965	140	830	120	20	48	32	81(b)	60(b)
S1	Annealed.........................	690	100	415	60	24	52	96 HRB	...	...
	Oil quenched from 930 °C (1700 °F) and single tempered at:									
	205 °C (400 °F)...................	2070	300	1895	275	...	...	57.5	249(c)	184(c)
	315 °C (600 °F)...................	2030	294	1860	270	4	12	54	233(c)	172(c)
	425 °C (800 °F)...................	1790	260	1690	245	5	17	50.5	203(c)	150(c)
	540 °C (1000 °F)..................	1680	244	1525	221	9	23	47.5	230(c)	170(c)
	650 °C (1200 °F)..................	1345	195	1240	180	12	37	42	...	...
S5	Annealed.........................	725	105	440	64	25	50	96 HRB	...	...
	Oil quenched from 870 °C (1600 °F) and single tempered at:									
	205 °C (400 °F)...................	2345	340	1930	280	5	20	59	206(c)	152(c)
	315 °C (600 °F)...................	2240	325	1860	270	7	24	58	232(c)	171(c)
	425 °C (800 °F)...................	1895	275	1690	245	9	28	52	243(c)	179(c)
	540 °C (1000 °F)..................	1520	220	1380	200	10	30	48	188(c)	139(c)
	650 °C (1200 °F)..................	1035	150	1170	170	15	40	37	...	...
S7	Annealed.........................	640	93	380	55	25	55	95 HRB	...	...
	Fan cooled from 940 °C (1725 °F) and single tempered at:									
	205 °C (400 °F)...................	2170	315	1450	210	7	20	58	244(c)	180(c)
	315 °C (600 °F)...................	1965	285	1585	230	9	25	55	309(c)	228(c)
	425 °C (800 °F)...................	1895	275	1410	205	10	29	53	243(c)	179(c)
	540 °C (1000 °F)..................	1820	264	1380	200	10	33	51	324(c)	239(c)
	650 °C (1200 °F)..................	1240	180	1035	150	14	45	39	358(c)	264(c)

(a) In 50 mm or 2 in. (b) Charpy V-notch. (c) Charpy unnotched.

242

Density and thermal expansion of selected tool steels

| | Density | | | | Thermal expansion | | | | | | | |
| | | | μm/m·K from 20 °C to | | | | | μin./in. °F from 68 °F to | | | | |
Type	Mg/m³	lb/in.³	100 °C	200 °C	425 °C	540 °C	650 °C	200 °F	400 °F	800 °F	1000 °F	1200 °F
W1	7.84	0.282	10.4	11.0	13.1	13.8(a)	14.2(b)	5.76	6.13	7.28	7.64(a)	7.90(b)
W2	7.85	0.283	...	...	14.4	14.8	14.9	...	...	8.0	8.2	8.3
S1	7.88	0.255	12.4	12.6	13.5	13.9	14.2	6.9	7.0	7.5	7.7	7.9
S2	7.79	0.281	10.9	11.9	13.5	14.0	14.2	6.0	6.6	7.5	7.8	7.9
S5	7.76	0.280	...	...	12.6	13.3	13.7	...	...	7.0	7.4	7.6
S6	7.75	0.279	...	...	12.6	13.3	...	...	...	7.0	7.4	...
S7	7.76	0.280	...	12.6	13.3	13.7(a)	13.3	...	7.0	7.4	7.6(a)	7.4
O1	7.85	0.283	...	10.6(c)	12.8	14.0(d)	14.4(d)	...	5.9(c)	7.1	7.8(d)	8.0(d)
O2	7.66	0.277	11.2	12.6	13.9	14.6	15.1	6.2	7.0	7.7	8.1	8.4
O7	7.8	0.282	...	...	...	...	...	...	...	...	...	...
A2	7.86	0.284	10.7	10.6(c)	12.9	14.0	14.2	5.96	5.91(c)	7.2	7.8	7.9
A6	7.84	0.283	11.5	12.4	13.5	13.9	14.2	6.4	6.9	7.5	7.7	7.9
A7	7.66	0.277	...	...	12.4	12.9	13.5	...	...	6.9	7.2	7.5
A8	7.87	0.284	...	...	12.0	12.4	12.6	...	...	6.7	6.9	7.0
A9	7.78	0.281	...	...	12.0	12.4	12.6	...	...	6.7	6.9	7.0
D2	7.70	0.278	10.4	10.3	11.9	12.2	12.2	5.8	5.7	6.6	6.8	6.8
D3	7.70	0.278	12.0	11.7	12.9	13.1	13.5	6.7	6.5	7.2	7.3	7.5
D4	7.70	0.278	...	...	12.4	...	...	...	...	6.9	...	...
D5	...	...	...	...	...	12.0	...	...	...	...	6.7	...
H10	7.81	0.281	...	...	12.2	13.3	13.7	...	...	6.8	7.4	7.6
H11	7.75	0.280	11.9	12.4	12.8	12.9	13.3	6.6	6.9	7.1	7.2	7.4
H13	7.76	0.280	10.4	11.5	12.2	12.4	13.1	5.8	6.4	6.8	6.9	7.3
H14	7.89	0.285	11.0	...	...	...	...	6.1	...	...	...	...
H19	7.98	0.288	11.0	11.0	12.0	12.4	12.9	6.1	6.1	6.7	6.9	7.2
H21	8.28	0.299	12.4	12.6	12.9	13.5	13.9	6.9	7.0	7.2	7.5	7.7
H22	8.36	0.302	11.0	...	11.5	12.0	12.4	6.1	...	6.4	6.7	6.9
H26	8.67	0.313	...	...	...	12.4	...	...	...	...	6.9	...
H42	8.15	0.295	...	...	...	11.9	...	...	...	...	6.6	...
T1	8.67	0.313	...	9.7	11.2	11.7	11.9	...	5.4	6.2	6.5	6.6
T2	8.67	0.313	...	...	...	...	...	...	...	...	...	...
T4	8.68	0.313	...	...	...	11.9	...	...	...	...	6.6	...
T5	8.75	0.316	11.2	...	...	11.5	...	6.2	...	...	6.4	...
T6	8.89	0.321	...	...	...	...	...	...	...	...	...	...
T8	8.43	0.305	...	...	...	...	...	...	...	...	...	...
T15	8.19	0.296	...	9.9	11.0	11.5	...	...	5.5(c)	6.1	6.4	...
M1	7.89	0.285	...	10.6(c)	11.3	12.0	12.4	...	5.9(c)	6.3	6.7	6.9
M2	8.16	0.295	10.1	9.4(c)	11.2	11.9	12.2	5.6	5.2(c)	6.2	6.6	6.8
M3, class 1	8.15	0.295	...	...	11.5	12.0	12.2	...	...	6.4	6.7	6.8
M3, class 2	8.16	0.295	...	...	11.5	12.0	12.8	...	...	6.4	6.7	7.1
M4	7.97	0.288	...	9.5(c)	11.2	12.0	12.2	...	5.3(c)	6.2	6.7	6.8
M7	7.95	0.287	...	9.5(c)	11.5	12.2	12.4	...	5.3(c)	6.4	6.8	6.9
M10	7.88	0.255	...	...	11.0	11.9	12.4	...	...	6.1	6.6	6.9
M30	8.01	0.289	...	...	11.2	11.7	12.2	...	...	6.2	6.5	6.8
M33	8.03	0.290	...	...	11.0	11.7	12.0	...	...	6.1	6.5	6.7
M36	8.18	0.296	...	...	...	...	...	...	...	...	...	...
M41	8.17	0.295	...	9.7	10.4	11.2	...	...	5.4	5.8	6.2	...
M42	7.98	0.288	...	...	...	...	...	...	...	...	...	...
M46	7.83	0.283	...	...	...	...	...	...	...	...	...	...
M47	7.96	0.288	10.6	11.0	11.9	...	12.6	5.9	6.1	6.6	...	7.0
L2	7.86	0.284	...	...	14.4	14.6	14.8	...	...	8.0	8.1	8.2
L6	7.86	0.284	11.3	12.6	12.6	13.5	13.7	6.3	7.0	7.0	7.5	7.6
P2	7.86	0.284	...	...	13.7	...	...	...	...	7.6	...	...
P5	7.80	0.282	...	...	...	...	...	...	...	...	...	...
P6	7.85	0.284	...	...	...	...	...	...	...	...	...	...
P20	7.85	0.284	...	...	12.8	13.7	14.2	...	...	7.1	7.6	7.9

(a) From 20 °C to 500 °C (68 °F to 930 °F). (b) From 20 °C to 600 °C (68 °F to 1110 °F). (c) From 20 °C to 260 °C (68 °F to 500 °F). (d) From 38 °C (100 °F).

Processing and service characteristics of tool steels

AISI designation	Resistance to decarburization	Hardening and tempering					Machinability	Toughness	Fabrication and service	
		Hardening response	Amount of distortion(a)	Resistance to cracking	Approximate hardness(b), HRC				Resistance to softening	Resistance to wear

Molybdenum high speed steels

AISI designation	Resistance to decarburization	Hardening response	Amount of distortion(a)	Resistance to cracking	Approximate hardness(b), HRC	Machinability	Toughness	Resistance to softening	Resistance to wear
M1	Low	Deep	A or S, low; O, medium	Medium	60-65	Medium	Low	Very high	Very high
M2	Medium	Deep	A or S, low; O, medium	Medium	60-65	Medium	Low	Very high	Very high
M3 (class 1 and class 2)	Medium	Deep	A or S, low; O, medium	Medium	61-66	Medium	Low	Very high	Very high
M4	Medium	Deep	A or S, low; O, medium	Medium	61-66	Low to medium	Low	Very high	Highest
M6	Low	Deep	A or S, low; O, medium	Medium	61-66	Medium	Low	Highest	Very high
M7	Low	Deep	A or S, low; O, medium	Medium	61-66	Medium	Low	Very high	Very high
M10	Low	Deep	A or S, low; O, medium	Medium	60-65	Medium	Low	Very high	Very high
M30	Low	Deep	A or S, low; O, medium	Medium	60-65	Medium	Low	Highest	Very high
M33	Low	Deep	A or S, low; O, medium	Medium	60-65	Medium	Low	Highest	Very high
M34	Low	Deep	A or S, low; O, medium	Medium	60-65	Medium	Low	Highest	Very high
M36	Low	Deep	A or S, low; O, medium	Medium	60-65	Medium	Low	Highest	Very high
M41	Low	Deep	A or S, low; O, medium	Medium	65-70	Medium	Low	Highest	Very high
M42	Low	Deep	A or S, low; O, medium	Medium	65-70	Medium	Low	Highest	Very high
M43	Low	Deep	A or S, low; O, medium	Medium	65-70	Medium	Low	Highest	Very high
M44	Low	Deep	A or S, low; O, medium	Medium	62-70	Medium	Low	Highest	Very high
M46	Low	Deep	A or S, low; O, medium	Medium	67-69	Medium	Low	Highest	Very high
M47	Low	Deep	A or S, low; O, medium	Medium	65-70	Medium	Low	Highest	Very high

Tungsten high speed steels

AISI designation	Resistance to decarburization	Hardening response	Amount of distortion(a)	Resistance to cracking	Approximate hardness(b), HRC	Machinability	Toughness	Resistance to softening	Resistance to wear
T1	High	Deep	A or S, low; O, medium	High	60-65	Medium	Low	Very high	Very high
T2	High	Deep	A or S, low; O, medium	High	61-66	Medium	Low	Very high	Very high
T4	Medium	Deep	A or S, low; O, medium	Medium	62-66	Medium	Low	Highest	Very high
T5	Low	Deep	A or S, low; O, medium	Medium	60-65	Medium	Low	Highest	Very high
T6	Low	Deep	A or S, low; O, medium	Medium	60-65	Low to medium	Low	Highest	Very high
T8	Medium	Deep	A or S, low; O, medium	Medium	60-65	Medium	Toughness	Highest	Very high
T15	Medium	Deep	A or S, low; O, medium	Medium	63-68	Low to medium	Low	Highest	Highest

Adapted from *Tool Steels*, American Iron and Steel Institute. (a) A, air cool; B, brine quench; O, oil quench; S, salt bath quench; W, water quench. (b) After tempering in temperature range normally recommended for this steel. (c) Carburized case hardness. (d) After aging at 510 to 550 °C (950 to 1025 °F). (e) Toughness decreases with increasing carbon content and depth of hardening.

244

Processing and service characteristics of tool steels (continued)

AISI designation	Resistance to decarbur- ization	Harden- ing response	Amount of distortion(a)	Resistance to cracking	Approxi- mate hard- ness(b), HRC	Machin- ability	Tough- ness	Resistance to softening	Resistance to wear
Chromium hot work steels									
H10	Medium	Deep	Very low	Highest	39-56	Medium to high	High	High	Medium
H11	Medium	Deep	Very low	Highest	38-54	Medium to high	Very high	High	Medium
H12	Medium	Deep	Very low	Highest	38-55	Medium to high	Very high	High	Medium
H13	Medium	Deep	Very low	Highest	38-53	Medium to high	Very high	High	Medium
H14	Medium	Deep	Low	Highest	40-47	Medium	High	High	Medium
H19	Medium	Deep	A, low; O, medium	High	40-57	Medium	High	High	Medium to high
Tungsten hot work steels									
H21	Medium	Deep	A, low; O, medium	High	36-54	Medium	High	High	Medium to high
H22	Medium	Deep	A, low; O, medium	High	39-52	Medium	High	High	Medium to high
H23	Medium	Deep	Medium	High	34-47	Medium	Medium	Very high	Medium to high
H24	Medium	Deep	A, low; O, medium	High	45-55	Medium	Medium	Very high	High
H25	Medium	Deep	A, low; O, medium	High	35-44	Medium	High	Very high	Medium
H26	Medium	Deep	A or S, low; O, medium	High	43-58	Medium	Medium	Very high	High
Molybdenum hot work steels									
H42	Medium	Deep	A or S, low; O, medium	Medium	50-60	Medium	Medium	Very high	High
Air-hardening medium-alloy cold work steels									
A2	Medium	Deep	Lowest	Highest	57-62	Medium	Medium	High	High
A3	Medium	Deep	Lowest	Highest	57-65	Medium	Medium	High	Very high
A4	Medium to high	Deep	Lowest	Highest	54-62	Low to medium	Medium	Medium	Medium to high
A6	Medium to high	Deep	Lowest	Highest	54-60	Low to medium	Medium	Medium	Medium to high
A7	Medium	Deep	Lowest	Highest	57-67	Low	Low	High	Highest
A8	Medium	Deep	Lowest	Highest	50-60	Medium	High	High	Medium to high
A9	Medium	Deep	Lowest	Highest	35-56	Medium	High	High	Medium to high
A10	Medium to high	Deep	Lowest	Highest	55-62	Medium to high	Medium	Medium	High
High-carbon, high-chromium cold work steels									
D2	Medium	Deep	Lowest	Highest	54-61	Low	Low	High	High to very high
D3	Medium	Deep	Very low	High	54-61	Low	Low	High	Very high
D4	Medium	Deep	Lowest	Highest	54-61	Low	Low	High	Very high
D5	Medium	Deep	Lowest	Highest	54-61	Low	Low	High	High to very high
D7	Medium	Deep	Lowest	Highest	58-65	Low	Low	High	Highest
Oil-hardening cold work steels									
O1	High	Medium	Very low	Very high	57-62	High	Medium	Low	Medium
O2	High	Medium	Very low	Very high	57-62	High	Medium	Low	Medium
O6	High	Medium	Very low	Very high	58-63	Highest	Medium	Low	Medium
O7	High	Medium	W, high; O, very low	W, low; O, very high	58-64	High	Medium	Low	Medium

Adapted from *Tool Steels*, American Iron and Steel Institute. (a) A, air cool; B, brine quench; O, oil quench; S, salt bath quench; W, water quench. (b) After tempering in temperature range normally recommended for this steel. (c) Carburized case hardness. (d) After aging at 510 to 550 °C (950 to 1025 °F). (e) Toughness decreases with increasing carbon content and depth of hardening.

Processing and service characteristics of tool steels (continued)

AISI designation	Resistance to decarbur-ization	Hardening and tempering			Approxi-mate hard-ness(b), HRC	Machin-ability	Fabrication and service		
		Harden-ing response	Amount of distortion(a)	Resistance to cracking			Tough-ness	Resistance to softening	Resistance to wear
Shock-resisting steels									
S1	Medium	Medium	Medium	High	40-58	Medium	Very high	Medium	Low to medium
S2	Low	Medium	High	Low	50-60	Medium to high	Highest	Low	Low to medium
S5	Low	Medium	Medium	High	50-60	Medium to high	Highest	Low	Low to medium
S6	Low	Medium	Medium	High	54-56	Medium	Very high	Low	Low to medium
S7	Medium	Deep	A, lowest; O, low	A, high-est; O, high	45-57	Medium	Very high	High	Low to medium
Low-alloy special-purpose steels									
L2	High	Medium	W, low; O, medium	W, high; O, medi-um	45-63	High	Very high(c)	Low	Low to medium
L6	High	Medium	Low	High	45-62	Medium	Very high	Low	Medium
Low-carbon mold steels									
P2	High	Medium	Low	High	58-64(c)	Medium to high	High	Low	Medium
P3	High	Medium	Low	High	58-64(c)	Medium	High	Low	Medium
P4	High	High	Very low	High	58-64(c)	Low to medium	High	Medium	High
P5	High	. . .	W, high; O, low	High	58-64(c)	Medium	High	Low	Medium
P6	High	. . .	A, very low; O, low	High	58-61(c)	Medium	High	Low	Medium
P20	High	Medium	Low	High	28-37	Medium to high	High	Low	Low to medium
P21	High	Deep	Lowest	Highest	30-40(d)	Medium	Medium	Medium	Medium
Water-hardening steels									
W1	Highest	Shallow	High	Medium	50-64	Highest	High(e)	Low	Low to medium
W2	Highest	Shallow	High	Medium	50-64	Highest	High(e)	Low	Low to medium
W5	Highest	Shallow	High	Medium	50-64	Highest	High(e)	Low	Low to medium

Adapted from *Tool Steels*, American Iron and Steel Institute. (a) A, air cool; B, brine quench; O, oil quench; S, salt bath quench; W, water quench. (b) After tempering in temperature range normally recommended for this steel. (c) Carburized case hardness. (d) After aging at 510 to 550 °C (950 to 1025 °F). (e) Toughness decreases with increasing carbon content and depth of hardening.

Properties of refractory metal carbides(a)

Carbide	Hardness(b), HV	Crystal system	Melting point		Theoretical density, Mg/m³	Modulus of elasticity	
			°C	°F		GPa	10⁶psi
TiC	3200	Cubic	3065 ± 15	5550 ± 30	4.92	448	65
VC	2950	Cubic	2730 ± 75	4950 ± 150	5.48	434	63
HfC	2700	Cubic	3925 ± 50	7100 ± 100	12.67	. . .	. .
ZrC	2600	Cubic	3440 ± 20	6225 ± 40	6.56	474	68.8
NbC	2400	Cubic	3500 ± 75	6330 ± 135	7.82	~290	~42
Cr_3C_2	2280	Ortho-rhombic	~1900	~3440	6.68	386	56
WC	2080	Hexagonal	~2800	~5030	15.8	669	97
Mo_2C	1950	Hexagonal	2490-2520	4510-4570	9.12	227	33
TaC	1790	Cubic	3915 ± 50	7080 ± 100	14.50	276	40

(a) Data from Ref 2 unless otherwise indicated. (b) Data from Ref 3.

SAE J1072 system for classification of superhard tool materials

Basic classification

Material classification	Designation

Material compound

Nitride	1
Carbide	2
Oxide	3
Other	9(a)

Binder material

None	0
Nickel	1
Iron	2
Cobalt	3
Other	9(a)

Base metal

None	0
Niobium	1
Tungsten	2
Titanium	3
Tantalum	4
Chromium	5
Aluminum	6
Other	9(a)

Suffixes

Material property	Identifier(b)
Binder metal quantity (wt% to nearest 0.1%)	A
Base metal quantity (wt% to nearest 0.1%)	B
Hardness(c) (HRA to nearest 0.1)	C
Specific gravity(c) (to nearest 0.1)	D
Grain size(c) (maximum amount of each type)	E
Apparent porosity(c) (the first digit indicates the amount of type A, the second the amount of type B and the third the amount of type C porosity)	F
Transverse rupture strength(d) (minimum, in ksi)	G
Other properties (written description required)	Z

(a) Material in this category shall be described by suffix Z. (b) Complete description consists of the letter identifier followed by one to three digits that express a quantitative value for the specific property. (c) Determined according to procedures outlined in SAE J439. (d) Determined according to procedures outlined in ASTM B406.

C-grade classification system for cemented carbides

C grade	Application category

Machining of cast iron, nonferrous and nonmetallic materials

C-1	Roughing
C-2	General-purpose machining
C-3	Finishing
C-4	Precision finishing

Machining of carbon and alloy steels

C-5	Roughing
C-6	General-purpose machining
C-7	Finishing
C-8	Precision finishing

Wear-surface applications

C-9	No shock
C-10	Light shock
C-11	Heavy shock

Impact applications

C-12	Light impact
C-13	Medium impact
C-14	Heavy impact

Miscellaneous applications

C-15	Hot weld-flash removal, light cuts
C-15A	Hot weld-flash removal, heavy cuts
C-16	Rock bits
C-17	Cold header dies
C-18	Wear at elevated temperatures and/or resistance to chemicals
C-19	Radioactive shielding, counterbalances and kinetic-energy devices

Typical application of cobalt-bonded cemented carbides

Grade	Grain size	Application
Straight grades		
97WC-3Co	Medium	Machining of cast iron, nonferrous metals and nonmetallic materials; excellent abrasion resistance and low shock resistance; the most wear resistant of the straight WC-Co grades; maintains a sharp cutting edge and makes long finishing cuts to close tolerances possible; also used for fine wire dies and small nozzles
94WC-6Co	Fine	Machining nonferrous and high-temperature alloys
94WC-6Co	Medium	General-purpose machining of work materials other than steel; also used for small and medium size compacting dies, coating dies, burnishing rings and nozzles
94WC-6Co	Coarse	Machining of cast iron, nonferrous metals and nonmetallic materials; also used for small wire-drawing dies, compacting dies, small drawing dies and caps and rings. The hardest grade used in mining applications where impact is encountered, as in rotary percussive bits
90WC-10Co	Fine	Machining steel and milling high-temperature metals (including titanium and its alloys) at low feeds and speeds: face mills, end mills, form tools, cutoff tools and screw-machine tools
90WC-10Co	Coarse	Primarily used for mining roller bits and percussive drilling bits
84WC-16Co	Fine	Primarily used for mining and metal-forming components
84WC-16Co	Coarse	Metal-forming and mining components: medium and large dies where great toughness is required, blanking dies for punch presses, and large mandrels
75WC-25Co	Medium	Metal-forming components for heavy impact applications, such as heading dies, cold extrusion dies, and punches and dies for blanking heavy stock
Complex grades		
71-74.5WC-10-12.5TiC-11-12.0TaC-4.5Co	Medium	Finishing, semifinishing and light roughing operations on plain carbon and alloy steels and alloy cast irons
72-73WC-7-8TiC-11.5-12TaC-8-8.5Co	Medium	Tough, wear-resistant grade for heavy-duty roughing cuts. Successfully withstands high temperatures encountered in heavy-duty machining, interrupted turning, scale cuts and milling of plain carbon and alloy steels and alloy cast irons
64TiC-28WC-2TaC-2Cr$_3$C$_2$-4Co	Medium	High-speed finishing of steels and cast irons
57WC-27TaC-16Co	Coarse	Cutting hot flash formed in the manufacture of welded tubing; also used to make dies for hot extrusion of aluminum wirebar and tubing

Thermal conductivity of selected tool steels

Temperature °C	°F	Thermal conductivity W/m·K	Btu/ft·h·°F
Type W1			
100	200	48.3	27.9
260	500	41.5	24.0
400	750	38.1	22.0
540	1000	34.6	20.0
675	1250	29.4	17.0
815	1500	24.2	14.0
Type H11			
100	200	42.2	24.4
260	500	36.3	21.0
400	750	33.4	19.3
540	1000	31.5	18.2
675	1250	30.1	17.4
815	1500	28.6	16.5
Type H13			
215	420	28.6	16.5
350	660	28.4	16.4
475	890	28.4	16.4
605	1120	28.7	16.6
Type H21			
100	200	27.0	15.6
260	500	29.8	17.2
400	750	29.8	17.2
540	1000	29.4	17.0
675	1250	29.1	16.8
Type T1			
100	200	19.9	11.5
260	500	21.6	12.5
400	750	23.2	13.4
540	1000	24.7	14.3
Type T15			
100	200	20.9	12.1
200	500	24.1	13.9
400	750	25.4	14.7
540	1000	26.3	15.2
Type M2			
100	200	21.3	12.3
200	500	23.5	13.6
400	750	25.6	14.8
540	1000	27.0	15.6
675	1250	28.9	16.7

Properties of representative cobalt-bonded cemented carbides

Nominal composition	Grain size	Hardness, HRA	Density Mg/m³	lb/in.³	Transverse strength MPa	ksi	Compressive strength MPa	ksi	Proportional limit, compression MPa	ksi	Modulus of elasticity GPa	10⁶ psi
97WC-3Co	Medium	92.5–93.2	15.3	0.55	1590	230	5860	850	2410	350	641	93
94WC-6Co	Fine	92.5–93.1	15.0	0.54	1790	260	5930	860	2550	370	614	89
	Medium	91.7–92.2	15.0	0.54	2000	290	5450	790	1930	280	648	94
	Coarse	90.5–91.5	15.0	0.54	2210	320	5170	750	1450	210	641	93
90WC-10Co	Fine	90.7–91.3	14.6	0.53	3100	450	5170	750	1590	230	620	90
	Coarse	87.4–88.2	14.5	0.52	2760	400	4000	580	1170	170	552	80
84WC-16Co	Fine	89	13.9	0.50	3380	490	4070	590	970	140	524	76
	Coarse	86.0–87.5	13.9	0.50	2900	420	3860	560	700	100	524	76
75WC-25Co	Medium	83–85	13.0	0.47	2550	370	3100	450	410	60	483	70
71WC-12.5TiC-12TaC-4.5Co	Medium	92.1–92.8	12.0	0.43	1380	200	5790	840	1170	170	565	82
72WC-8TiC-11.5TaC-8.5Co	Medium	90.7–91.5	12.6	0.45	1720	250	5170	750	1720	250	558	81
64TiC-28WC-2TaC-2Cr₂C₃-4.0Co	Medium	94.5–95.2	6.6	0.24	690	100	4340	630	...	...	...	...
57WC-27TaC-16Co	Coarse	84.0–86.0	13.7	0.49	2690	390	3720	540	1170	170	441	64

(a) Based on a value of 100 for the most abrasion-resistant grade.

Properties of representative cobalt-bonded cemented carbides (continued)

Nominal composition	Tensile strength MPa	ksi	Impact strength J	in.·lb	Relative abrasion resistance(a)	Thermal expansion μm/m·°C at 200 °C	at 1000 °C	μin./in.·°F at 400 °F	at 1800 °F	Thermal conductivity, W/m·K	Electrical conductivity, % IACS
97WC-3Co	...	...	1.13	10	100	4.0	...	2.2	...	121	5.3
94WC-6Co	...	...	1.02	9	100	4.3	5.9	2.4	3.3	...	...
	1450	210	1.36	12	58	4.3	5.4	2.4	3.0	100	7.8
	1520	220	1.36	12	25	4.3	5.6	2.4	3.1	121	10.0
90WC-10Co	...	...	1.69	15	22	...	...	...	...	...	...
	1340	195	2.03	18	7	5.2	...	2.9	...	112	11.4
84WC-16Co	...	...	3.05	27	5	...	...	...	...	...	...
	1860	270	2.83	25	5	5.8	7.0	3.2	3.9	88	9.2
75WC-25Co	1380	200	3.05	27	3	6.3	...	3.5	...	71	9.8
71WC-12.5TiC-12TaC-4.5Co	...	...	0.79	7	11	5.2	6.5	2.9	3.6	35	4.3
72WC-8TiC-11.5TaC-8.5Co	...	...	0.90	8	13	5.8	6.8	3.2	3.8	50	5.2
64TiC-28WC-2TaC-2Cr₂C₃-4.0Co	...	...	...	...	8	...	...	...	...	...	...
57WC-27TaC-16Co	...	...	2.03	18	3	5.9	7.7	3.3	4.3	...	...

(a) Based on a value of 100 for the most abrasion-resistant grade.

Test methods for determining properties of cemented carbides

Property	ASTM/ANSI	CCPA	ISO	SAE
Abrasive wear resistance	B611	P112	(a)	· · ·
Apparent grain size	B390	M203	· · ·	J439
Apparent porosity.	B276	M201	4505	J439
Axial load fatigue.	· · ·	· · ·	(a)	· · ·
Coefficient of sliding friction	· · ·	P111	· · ·	· · ·
Coercive force	· · ·	· · ·	3326	· · ·
Compressive strength	E9(b)	P104	4506	· · ·
Density.	B311	P101	3369	J439
Diametral compression testing	B485	P115	· · ·	· · ·
Electrical resistivity	B421	P107	· · ·	· · ·
Fracture toughness	(a)	· · ·	· · ·	· · ·
Hardness, HRA.	B294	P103	3738	J439
Hardness, HV	E92	· · ·	3878	· · ·
Linear thermal expansion	B95	P108	· · ·	· · ·
Magnetic permeability	A342	P109	· · ·	· · ·
Metallographic preparation of samples.	(a)	· · ·	· · ·	· · ·
Microstructure	B657	M202	4499	· · ·
Poisson's ratio.	E132	P105	· · ·	· · ·
Powder sampling and testing.	· · ·	· · ·	4884	· · ·
Sampling and testing	· · ·	· · ·	4889	· · ·
Tensile testing	B437(c)	P113	· · ·	· · ·
Thermal shock resistance	· · ·	P110	· · ·	· · ·
Transverse rupture strength	B406	P102	3327	· · ·
Young's modulus	E111	P106	3312	· · ·

(a) In preparation. (b) A procedure derived from ISO 4506 is in preparation. (c) Being withdrawn.

Compositions and selected properties of three principal grades of steel-bonded titanium carbide

Characteristic	Grade C	Grade CM	Grade SK
Composition:			
Titanium carbide.	45 vol%	45 vol%	40 vol%
Steel matrix	0.6C-3Cr-3Mo	0.85C-10Cr-3Mo	0.40C-5Cr-4Mo-0.50Ni
Hardness:			
Annealed HRC.	40	45	37
Hardened(a) HRC	70	69	65
Max service temperature: °C	200	540	540
°F	400	1000	1000
Density: Mg/m^3	6.60	6.45	6.80
lb/in.^{3}	0.238	0.232	0.245
Transverse strength: MPa	2070	2140	2070
ksi	300	310	300
Modulus of elasticity: GPa.	305	305	270
10^6 psi	44	44	39
Thermal expansion: μm/m·°C	7.83(b)	8.3(c)	9.47(c)
μin./in.·°F	4.35	4.6	5.26
Electrical conductivity, % IACS . . .	3.2	2.8	3.0

(a) Grade C: austenitized at 950 °C (1750 °F), oil quenched and tempered 1 h at 190 °C (375 °F). Grade CM: austenitized at 1100 °C (2000 °F), oil quenched (gas quenched if heat treated in vacuum), and double tempered at 525 °C (975 °F). Grade SK: austenitized at 1000 °C (1850 °F), oil quenched (gas quenched if heat treated in vacuum), and double tempered at 525 °C. (b) At 21 to 200 °C (70 to 400 °F). (c) At 21 to 540 °C (70 to 1000 °F).

Typical elevated-temperature hardness and strength of groups A-2 and A-3 ceramic tool materials

See Table 13 for other typical properties

Temperature		Hardness, HV		Transverse rupture strength			
°C	°F	Group A-2	Group A-3	Group A-2		Group A-3	
				MPa	ksi	MPa	ksi
RT	RT	2100	2400	690	100	735	107
480	900	2000	2000	690	100	715	104
650	1200	1950	1850	...	...	...	...
815	1500	1850	1700	...	...	...	...
980	1800	1700	1500	700	101	700	102
1200	2200	1400	1400	610	90	690	100

Properties and uses of nickel-bonded titanium carbides

Property	High TiC, plus Mo$_2$C; low nickel	High TiC, plus Mo$_2$C; low nickel	TiC, plus Mo$_2$C; intermediate nickel	Lower TiC, plus Mo$_2$C; high nickel
Grain size	Fine	...	Fine	Fine
Hardness: HRA	93.3	93.0	91.7	90.5
HV	1970	1890	1600	1440
Density: Mg/m^3	5.50	5.63	5.71	5.82
lb/in.^{3}	0.198	0.203	0.206	0.210
Transverse strength: MPa	1170	1380	1720	1890
ksi	170	200	250	275
Compressive strength: MPa	3585	3450	3270	2960
ksi	520	500	475	430
Tensile strength: MPa	970	1100	1170	1240
ksi	140	160	170	180
Modulus of elasticity: GPa	462	448	414	379
10^6 psi	67	65	60	55
Impact strength: J	0.79	0.90	1.02	1.24
in.·lb	7	8	9	11
Thermal expansion(a): μm/m·°C	7.5	7.8	8.4	9.1
μin./in.·°F.....	4.2	4.3	4.7	5.1
Thermal conductivity(b): W/m·K	16.7	16.7	16.7	16.7
Btu/ft·h·°F ..	9.6	9.6	9.6	9.6
Typical classification for machining use: C-grade	C-8	C-7	C-6	C-6
ISO...............	P01	P10	P20	P30

(a) At 21 to 650 °C (70 to 1200 °F). (b) At 100 to 300 °C (200 to 575 °F).

Selected properties of cubic boron nitride

Crystal structure........... Zinc blende
(F $\bar{4}$3m)
Density .... 3.48 Mg/m^3
(0.125 lb/in.^{3})
Hardness:
At 20 °C (70 °F).............. 4000 HV
At 1000 °C (1800 °F)........ ~4000 HV
Temperature for
dislocation mobility .... 1300-1400 °C
(2400-2550 °F)
Melting point
(triple point) .... 3500 K
Thermal conductivity
(theoretical) .... 13 W/m·K
(7.5 Btu/ft·h·°F)
Thermal stability:
Limit of oxidation
resistance in air .... ~1300 °C
(~2400 °F)
Metastable reversion
temperature .... ~1500 °C
(~2700 °F)
Thermal expansion,
21 to 500 °C
(70 to 900 °F)............. 4.8 μm/m·°C
(2.7 μin./in.·°F)

Mutual indentation hot hardness of cast Tantung G

Temperature °C	°F	Hardness, HB(a)	HRA	Equivalent hardness(b) HRC	HRB
RT	RT	 654	81.3	60.1	...
425	800	 479	75.7	49.8	...
650	1200	 479	75.7	49.8	...
870	1600	 267	63.8	27.1	104
980	1800	 114	...	...	66.7

(a) 3000 kg load, applied for 30 s. (b) Converted values.

Typical properties of ceramic tool materials

Property	Group A-1 General	Group A-1 Example(a)	Group A-2	Group A-3
Hardness: HRA	93–94	93–94	93–94	93–94
Density:				
Mg/m^3.....................	3.96–3.98	4.1	4.0	4.24
lb/in.^{3}	0.142–0.143	0.148	0.144	0.153
Transverse strength:				
MPa	480–690	620	640	760
ksi	70–100	90	92.5	110
Compressive strength:				
MPa	3790–4480	2140(b)	4140	3930–4070
ksi	550–650	310(b)	600	570–590
Modulus of elasticity:				
GPa	390	400	390	...
10^6 psi	57	58	57	...
Impact strength:				
J	...	0.23	...	...
in.·lb	...	2	...	...
μm/m·°C	...	6.1(d)	7.2	7.7
Thermal expansion(c): μin./in.·°F ..	...	3.4(d)	4.0	4.3
Thermal conductivity:				
At room temperature:				
W/m·K	29	...	...	17–21
Btu/ft·h·°F	17	...	...	10–12
At 100 °C (212 °F):				
W/m·K	22	...	29	...
Btu/ft·h·°F	13	...	17	...
At 450 °C (850 °F):				
W/m·K	11	...	...	...
Btu/ft·h·°F	6.5	...	...	...
At 600 °C (1100 °F):				
W/m·K	...	...	...	14.7
Btu/ft·h·°F	...	...	...	8.4

(a) 89Al_2O_3-11TiO, cold pressed and sintered. See Table 14 for hot hardness data. (b) Proportional limit.
(c) At 21 to 200 °C (70 to 400 °F). (d) 8.3 μm/m·°C (4.6 μin./in.·°F) at 21 to 980 °C (70 to 1800 °F).

Typical properties of cast Tantung G

Property	Chill cast	Refractory mold cast
Melting temperature:		
°C	1150–1200	
°F	2100–2200	
Casting temperature:		
°C	1370	
°F	2500	
Density:		
Mg/m^3	8.3	8.3
lb/in.^{3}	0.30	0.30
Thermal expansion:		
μm/m·°C	4.2	4.2
μin./in.·°F	2.3	2.3
Thermal conductivity:		
W/m·K	26.8	26.8
Btu/ft·h·°F	15.5	15.5
Hardness: HRC ...	60–63	53–58
Transverse strength:		
MPa	2240	1030–1200
ksi	325	150–175
Modulus of elasticity:		
GPa	265	...
10^6 psi	41	...
Tensile strength:		
MPa	585–620	450
ksi	85–90	65
Compressive strength:		
MPa	2760	2930
ksi	400	425
Impact strength:		
J	6.1	6.1
ft·lb	4.5	4.5

Heat Treating

Normalizing and annealing temperatures of tool steels

Type	Normalizing treatment/ temperature(a) °C	°F	Annealing(b) Temperature °C	°F	Rate of cooling, max °C/h	°F/h	Hardness, HB
Molybdenum high speed steels							
M1, M10 ....Do not normalize			815 to 870	1500 to 1600	22	40	207 to 235
M2 ....Do not normalize			870 to 900	1600 to 1650	22	40	212 to 241
M3, M4 ....Do not normalize			870 to 900	1600 to 1650	22	40	223 to 255
M6 ....Do not normalize			870	1600	22	40	248 to 277
M7 ....Do not normalize			815 to 870	1500 to 1600	22	40	217 to 255
M30, M33, M34, M36, M41, M42, M46, M47 ....Do not normalize			870 to 900	1600 to 1650	22	40	235 to 269
M43 ....Do not normalize			870 to 900	1600 to 1650	22	40	248 to 269
M44 ....Do not normalize			870 to 900	1600 to 1650	22	40	248 to 293
Tungsten high speed steels							
T1 ....Do not normalize			870 to 900	1600 to 1650	22	40	217 to 255
T2 ....Do not normalize			870 to 900	1600 to 1650	22	40	223 to 255
T4 ....Do not normalize			870 to 900	1600 to 1650	22	40	229 to 269
T5 ....Do not normalize			870 to 900	1600 to 1650	22	40	235 to 277
T6 ....Do not normalize			870 to 900	1600 to 1650	22	40	248 to 293
T8 ....Do not normalize			870 to 900	1600 to 1650	22	40	229 to 255
T15 ....Do not normalize			870 to 900	1600 to 1650	22	40	241 to 277
Chromium hot work steels							
H10, H11, H12, H13 ....Do not normalize			845 to 900	1550 to 1650	22	40	192 to 229
H14 ....Do not normalize			870 to 900	1600 to 1650	22	40	207 to 235
H19 ....Do not normalize			870 to 900	1600 to 1650	22	40	207 to 241
Tungsten hot work steels							
H21, H22, H25 ....Do not normalize			870 to 900	1600 to 1650	22	40	207 to 235
H23 ....Do not normalize			870 to 900	1600 to 1650	22	40	212 to 255
H24, H26 ....Do not normalize			870 to 900	1600 to 1650	22	40	217 to 241
Molybdenum hot work steels							
H41, H43 ....Do not normalize			815 to 870	1500 to 1600	22	40	207 to 235
H42 ....Do not normalize			845 to 900	1550 to 1650	22	40	207 to 235
High-carbon high-chromium cold work steels							
D2, D3, D4 ....Do not normalize			870 to 900	1600 to 1650	22	40	217 to 255
D5 ....Do not normalize			870 to 900	1600 to 1650	22	40	223 to 255
D7 ....Do not normalize			870 to 900	1600 to 1650	22	40	235 to 262
Medium-alloy air-hardening cold work steels							
A2 ....Do not normalize			845 to 870	1550 to 1600	22	40	201 to 229
A3 ....Do not normalize			845 to 870	1550 to 1600	22	40	207 to 229
A4 ....Do not normalize			740 to 760	1360 to 1400	14	25	200 to 241
A6 ....Do not normalize			730 to 745	1350 to 1375	14	25	217 to 248
A7 ....Do not normalize			870 to 900	1600 to 1650	14	25	235 to 262
A8 ....Do not normalize			845 to 870	1550 to 1600	22	40	192 to 223
A9 ....Do not normalize			845 to 870	1550 to 1600	14	25	212 to 248
A10	790	1450	765 to 795	1410 to 1460	8	15	235 to 269
Oil-hardening cold work steels							
O1	870	1600	760 to 790	1400 to 1450	22	40	183 to 212
O2	845	1550	745 to 775	1375 to 1425	22	40	183 to 212
O6	870	1600	765 to 790	1410 to 1450	11	20	183 to 217
O7	900	1650	790 to 815	1450 to 1500	22	40	192 to 217

(continued)

(a) Time held at temperature varies from 15 min for small sections to 1 h for large sizes. Cooling is done in still air. Normalizing should not be confused with low-temperature annealing. (b) The upper limit of ranges should be used for large sections and the lower limit for smaller sections. Time held at temperature varies from 1 h for light sections to 4 h for heavy sections and large furnace charges of high-alloy steel. (c) For 0.25 Si type, 183 to 207 HB; for 1.00 Si type, 207 to 229 HB. (d) Temperature varies with carbon content: 0.60 to 0.75 C, 815 °C (1500 °F); 0.75 to 0.90 C, 790 °C (1450 °F); 0.90 to 1.10 C, 870 °C (1600 °F); 1.10 to 1.40 C, 870 to 925 °C (1600 to 1700 °F). (e) Temperature varies with carbon content: 0.60 to 0.90 C, 740 to 790 °C (1360 to 1450 °F); 0.90 to 1.40 C, 760 to 790 °C (1400 to 1450 °F).

Normalizing and annealing temperatures of tool steels (continued)

Type	Normalizing treatment/ temperature(a) °C	°F	Annealing(b) Temperature °C	°F	Rate of cooling, max °C/h	°F/h	Hardness, HB
Shock-resisting steels							
S1	Do not normalize		790 to 815	1450 to 1500	22	40	183 to 229(c)
S2	Do not normalize		760 to 790	1400 to 1450	22	40	192 to 217
S5	Do not normalize		775 to 800	1425 to 1475	14	25	192 to 229
S7	Do not normalize		815 to 845	1500 to 1550	14	25	187 to 223
Mold steels							
P2	Not required		730 to 815	1350 to 1500	22	40	103 to 123
P3	Not required		730 to 815	1350 to 1500	22	40	109 to 137
P4	Do not normalize		870 to 900	1600 to 1650	14	25	116 to 128
P5	Not required		845 to 870	1550 to 1600	22	40	105 to 116
P6	Not required		845	1550	8	15	183 to 217
P20	900	1650	760 to 790	1400 to 1450	22	40	149 to 179
P21	900	1650	Do not anneal				
Low-alloy special-purpose steels							
L2	871 to 900	1600 to 1650	760 to 790	1400 to 1450	22	40	163 to 197
L3	900	1650	790 to 815	1450 to 1500	22	40	174 to 201
L6	870	1600	760 to 790	1400 to 1450	22	40	183 to 212
Carbon-tungsten special-purpose steels							
F1	900	1650	760 to 800	1400 to 1475	22	40	183 to 207
F2	900	1650	790 to 815	1450 to 1500	22	40	207 to 235
Water-hardening steels							
W1, W2	790 to 925(d)	1450 to 1700(d)	740 to 790(e)	1360 to 1450(e)	22	40	156 to 201
W5	870 to 925	1600 to 1700	760 to 790	1400 to 1450	22	40	163 to 201

(a) Time held at temperature varies from 15 min for small sections to 1 h for large sizes. Cooling is done in still air. Normalizing should not be confused with low-temperature annealing. (b) The upper limit of ranges should be used for large sections and the lower limit for smaller sections. Time held at temperature varies from 1 h for light sections to 4 h for heavy sections and large furnace charges of high-alloy steel. (c) For 0.25 Si type, 183 to 207 HB; for 1.00 Si type, 207 to 229 HB. (d) Temperature varies with carbon content: 0.60 to 0.75 C, 815 °C (1500 °F); 0.75 to 0.90 C, 790 °C (1450 °F); 0.90 to 1.10 C, 870 °C (1600 °F); 1.10 to 1.40 C, 870 to 925 °C (1600 to 1700 °F). (e) Temperature varies with carbon content: 0.60 to 0.90 C, 740 to 790 °C (1360 to 1450 °F); 0.90 to 1.40 C, 760 to 790 °C (1400 to 1450 °F).

Hardening and tempering of tool steels

Type	Rate of heating	Preheat temperature °C	Preheat temperature °F	Hardening temperature °C	Hardening temperature °F	Time at temperature, min	Quenching medium(a)	Tempering temperature °C	Tempering temperature °F
Molybdenum high speed steels									
M1, M7, M10	Rapidly from preheat	730 to 845	1350 to 1550	1175 to 1220	2150 to 2225(b)	2 to 5	O, A or S	540 to 595(c)	1000 to 1100(c)
M2	Rapidly from preheat	730 to 845	1350 to 1550	1190 to 1230	2175 to 2250(b)	2 to 5	O, A or S	540 to 595(c)	1000 to 1100(c)
M3, M4, M30, M33, M34 ..	Rapidly from preheat	730 to 845	1350 to 1550	1205 to 1230(b)	2200 to 2250(b)	2 to 5	O, A or S	540 to 595(c)	1000 to 1100(c)
M6	Rapidly from preheat	790	1450	1175 to 1205(b)	2150 to 2200(b)	2 to 5	O, A or S	540 to 595(c)	1000 to 1100(c)
M36	Rapidly from preheat	730 to 845	1350 to 1550	1220 to 1245(b)	2225 to 2275(b)	2 to 5	O, A or S	540 to 595(c)	1000 to 1100(c)
M41	Rapidly from preheat	730 to 845	1350 to 1550	1190 to 1215(b)	2175 to 2220(b)	2 to 5	O, A or S	540 to 595(d)	1000 to 1100(d)
M42	Rapidly from preheat	730 to 845	1350 to 1550	1190 to 1210(b)	2175 to 2210(b)	2 to 5	O, A or S	510 to 595(d)	950 to 1100(d)
M43	Rapidly from preheat	730 to 845	1350 to 1550	1190 to 1215(b)	2175 to 2220(b)	2 to 5	O, A or S	510 to 595(d)	950 to 1100(d)
M44	Rapidly from preheat	730 to 845	1350 to 1550	1200 to 1225(b)	2190 to 2240(b)	2 to 5	O, A or S	540 to 625(d)	1000 to 1160(d)
M46	Rapidly from preheat	730 to 845	1350 to 1550	1190 to 1220(b)	2175 to 2225(b)	2 to 5	O, A or S	525 to 565(d)	975 to 1050(d)
M47	Rapidly from preheat	730 to 845	1350 to 1550	1180 to 1205(b)	2150 to 2200(b)	2 to 5	O, A or S	525 to 595(d)	975 to 1100(d)
Tungsten high speed steels									
T1, T2, T4, T8	Rapidly from preheat	815 to 870	1500 to 1600	1260 to 1300(b)	2300 to 2375(b)	2 to 5	O, A or S	540 to 595(c)	1000 to 1100(c)
T5, T6.......	Rapidly from preheat	815 to 870	1500 to 1600	1275 to 1300(b)	2325 to 2375(b)	2 to 5	O, A or S	540 to 595(c)	1000 to 1100(c)
T15	Rapidly from preheat	815 to 870	1500 to 1600	1205 to 1260(b)	2200 to 2300(b)	2 to 5	O, A or S	540 to 650(d)	1000 to 1200(d)
Chromium hot work steels									
H10........	Moderately from preheat	815	1500	1010 to 1040	1850 to 1900	15 to 40(e)	A	540 to 650	1000 to 1200
H11, H12....	Moderately from preheat	815	1500	995 to 1025	1825 to 1875	15 to 40(e)	A	540 to 650	1000 to 1200
H13........	Moderately from preheat	815	1500	995 to 1040	1825 to 1900	15 to 40(e)	A	540 to 650	1000 to 1200
H14........	Moderately from preheat	815	1500	1010 to 1065	1850 to 1950	15 to 40(e)	A	540 to 650	1000 to 1200
H19........	Moderately from preheat	815	1500	1095 to 1205	2000 to 2200	2 to 5	A or O	540 to 705	1000 to 1300
Molybdenum hot work steels									
H41, H43....	Rapidly from preheat	730 to 845	1350 to 1550	1095 to 1190	2000 to 2175	2 to 5	O, A or S	565 to 650	1050 to 1200
H42........	Rapidly from preheat	730 to 845	1350 to 1550	1120 to 1220	2050 to 2225	2 to 5	O, A or S	565 to 650	1050 to 1200

(continued)

(a) O, oil quench; A, air cool; S, salt bath quench; W, water quench; B, brine quench. (b) When the high-temperature heating is carried out in a salt bath, the range of temperatures should be about 15 °C (25 °F) lower than given in this line. (c) Double tempering recommended for not less than 1 h at temperature each time. (d) Triple tempering recommended for not less than 1 h at temperature each time. (e) Times apply to open-furnace heat treatment. For pack hardening, a common rule is to heat 30 min/in. of cross section of the pack. (f) Preferable for large tools to minimize decarburization. (g) Carburizing temperature. (h) After carburizing. (j) Carburized case hardness. (k) P21 is a precipitation-hardening steel having a thermal treatment which involves solution treating and aging rather than hardening and tempering. (m) Recommended for large tools and tools with intricate sections

Hardening and tempering of tool steels (continued)

Type	Rate of heating	Preheat temperature °C	Preheat temperature °F	Hardening temperature °C	Hardening temperature °F	Time at temperature, min	Quenching medium(a)	Tempering temperature °C	Tempering temperature °F
Tungsten hot work steels									
H21, H22....	Rapidly from preheat	815	1500	1095 to 1205	2000 to 2200	2 to 5	A or O	595 to 675	1100 to 1250
H23.........	Rapidly from preheat	845	1550	1205 to 1260	2200 to 2300	2 to 5	O	650 to 815	1200 to 1500
H24.........	Rapidly from preheat	815	1500	1095 to 1230	2000 to 2250	2 to 5	O	565 to 650	1050 to 1200
H25.........	Rapidly from preheat	815	1500	1150 to 1260	2100 to 2300	2 to 5	A or O	565 to 675	1050 to 1250
H26.........	Rapidly from preheat	870	1600	1175 to 1260	2150 to 2300	2 to 5	O, A or S	565 to 675	1050 to 1250
Medium-alloy air-hardening cold work steels									
A2.........	Slowly	790	1450	925 to 980	1700 to 1800	20 to 45	A	175 to 540	350 to 1000
A3.........	Slowly	790	1450	955 to 980	1750 to 1800	25 to 60	A	175 to 540	350 to 1000
A4.........	Slowly	675	1250	815 to 870	1500 to 1600	20 to 45	A	175 to 425	350 to 800
A6.........	Slowly	650	1200	830 to 870	1525 to 1600	20 to 45	A	150 to 425	300 to 800
A7.........	Very slowly	815	1500	955 to 980	1750 to 1800	30 to 60	A	150 to 540	300 to 1000
A8.........	Slowly	790	1450	980 to 1010	1800 to 1850	20 to 45	A	175 to 595	350 to 1100
A9.........	Slowly	790	1450	980 to 1025	1800 ro 1875	20 to 45	A	510 to 620	950 to 1150
A10........	Slowly	650	1200	790 to 815	1450 to 1500	30 to 60	A	175 to 425	350 to 800
Oil-hardening cold work steels									
O1.........	Slowly	650	1200	790 to 815	1450 to 1500	10 to 30	O	175 to 260	350 to 500
O2.........	Slowly	650	1200	760 to 800	1400 to 1475	5 to 20	O	175 to 260	350 to 500
O6.........	Slowly	...	...	790 to 815	1450 to 1500	10 to 30	O	175 to 315	350 to 600
O7.........	Slowly	650	1200	790 to 830; 845 to 885	W:1450 to 1525; O:1550 to 1625	10 to 30	O or W	175 to 290	350 to 550
Shock-resisting steels									
S1.........	Slowly	...	...	900 to 955	1650 to 1750	15 to 45	O	205 to 650	400 to 1200
S2.........	Slowly	650(f)	1200(f)	845 to 900	1550 to 1650	5 to 20	B or W	175 to 425	350 to 800
S5.........	Slowly	760	1400	870 to 925	1600 to 1700	5 to 20	0	175 to 425	350 to 800
S7.........	Slowly	650 to 705	1200 to 1300	925 to 955	1700 to 1750	15 to 45	A or O	205 to 620	400 to 1150

(continued)

(a) O, oil quench; A, air cool; S, salt bath quench; W, water quench; B, brine quench. (b) When the high-temperature heating is carried out in a salt bath, the range of temperatures should be about 15 °C (25 °F) lower than given in this line. (c) Double tempering recommended for not less than 1 h at temperature each time. (d) Triple tempering recommended for not less than 1 h at temperature each time. (e) Times apply to open-furnace heat treatment. For pack hardening, a common rule is to heat 30 min/in. of cross section of the pack. (f) Preferable for large tools to minimize decarburization. (g) Carburizing temperature. (h) After carburizing. (j) Carburized case hardness. (k) P21 is a precipitation-hardening steel having a thermal treatment which involves solution treating and aging rather than hardening and tempering. (m) Recommended for large tools and tools with intricate sections

Hardening and tempering of tool steels (continued)

Type	Rate of heating	Hardening Preheat temperature °C	°F	Hardening temperature °C	°F	Time at temperature, min	Quenching medium(a)	Tempering temperature °C	°F
Mold steels									
P2	. . .	900 to 925(g)	1650 to 1700(g)	830 to 845(h)	1525 to 1550(h)	15	O	175 to 260	350 to 500
P3	. . .	900 to 925(g)	1650 to 1700(g)	800 to 830(h)	1475 to 1525(h)	15	O	175 to 260	350 to 500
P4	. . .	970 to 995(g)	1775 to 1825(g)	970 to 995(h)	1775 to 1825(h)	15	A	175 to 480	350 to 900
P5	. . .	900 to 925(g)	1650 to 1700(g)	845 to 870(h)	1550 to 1600(h)	15	O or W	175 to 260	350 to 500
P6	. . .	900 to 925(g)	1650 to 1700(g)	790 to 815(h)	1450 to 1500(h)	15	A or O	175 to 230	350 to 450
P20	. . .	870 to 900(h)	1600 to 1650(h)	815 to 870	1500 to 1600	15	O	480 to 595(j)	900 to 1100(j)
P21(k)	Slowly	Do not preheat		705 to 730	1300 to 1350	60 to 180	A or O	510 to 550	950 to 1025
Low-alloy special-purpose steels									
L2	Slowly	. . .	. . .	W: 790 to 845 O: 845 to 925	W: 1450 to 1550 O: 1550 to 1700	10 to 30	O or W	175 to 540	350 to 1000
L3	Slowly	. . .	. . .	W: 775 to 815 O: 815 to 870	W: 1425 to 1500 O: 1500 to 1600	10 to 30	O or W	175 to 315	350 to 600
L6	Slowly	. . .	. . .	790 to 845	1450 to 1550	10 to 30	O	175 to 540	350 to 1000
Carbon-tungsten special-purpose steels									
F1, F2	Slowly	650	1200	790 to 870	1450 to 1600	15	W or B	175 to 260	350 to 500
Water-hardening steels									
W1, W2, W3	Slowly	565 to 650(m)	1050 to 1200(m)	760 to 815	1400 to 1550	10 to 30	B or W	175 to 345	350 to 650
High-carbon, high-chromium cold work steels									
D1, D5	Very slowly	815	1500	980 to 1025	1800 to 1875	15 to 45	A	205 to 540	400 to 1000
D3	Very slowly	815	1500	925 to 980	1700 to 1800	15 to 45	O	205 to 540	400 to 1000
D4	Very slowly	815	1500	970 to 1010	1775 to 1850	15 to 45	A	205 to 540	400 to 1000
D7	Very slowly	815	1500	1010 to 1065	1850 to 1950	30 to 60	A	150 to 540	300 to 1000

(a) O, oil quench; A, air cool; S, salt bath quench; W, water quench; B, brine quench. (b) When the high-temperature heating is carried out in a salt bath, the range of temperatures should be about 15 °C (25 °F) lower than given in this line. (c) Double tempering recommended for not less than 1 h at temperature each time. (d) Triple tempering recommended for not less than 1 h at temperature each time. (e) Times apply to open-furnace heat treatment. For pack hardening, a common rule is to heat 30 min/in. of cross section of the pack. (f) Preferable for large tools to minimize decarburization. (g) Carburizing temperature. (h) After carburizing. (j) Carburized case hardness. (k) P21 is a precipitation-hardening steel having a thermal treatment which involves solution treating and aging rather than hardening and tempering. (m) Recommended for large tools and tools with intricate sections

Microconstituents in four tool steels after hardening

Steel	Hardening treatment	As-quenched hardness, HRC		Martensite, vol %	Retained austenite, vol %	Undissolved carbides, vol %
W1	790 °C (1450 °F), 30 min; WQ	67.0	..	88.5	9	2.5
L3	840 °C (1550 °F), 30 min; OQ	66.5	..	90	7	3.0
M2	1225 °C (2235 °F), 6 min; OQ	64	..	71.5	20	8.5
D2	1040 °C (1900 °F), 30 min; AC	62	..	45	40	15

Typical dimensional changes in hardening and tempering

Tool steel	Hardening treatment °C	°F	Quenching medium	Total change in linear dimensions, % after quenching	°C 150 °F 300	205 400	260 500	315 600	370 700	425 800	480 900	510 950	540 1000	565 1050	595 1100
O1	816	1500	Oil	0.22	0.17	0.16	0.18	...	...	...	...	...	...	...	...
O1	788	1450	Oil	0.18	0.09	0.12	0.13	...	...	...	...	...	...	...	...
O6	788	1450	Oil	0.12	0.07	0.10	0.14	0.10	0.00	−0.05	−0.06	...	−0.07	...	...
A2	954	1750	Air	0.09	0.06	0.06	0.08	0.07	...	0.05	0.04	...	0.06	...	...
A10	788	1450	Air	0.04	0.00	0.00	0.08	0.08	0.01	0.01	0.02	...	0.01	...	0.02
D2	1010	1850	Air	0.06	0.03	0.03	0.02	0.00	...	−0.01	−0.02	...	0.06	...	...
D3	954	1750	Oil	0.07	0.04	0.02	0.01	−0.02	...	...	...	...	...	...	...
D4	1038	1900	Air	0.07	0.03	0.01	−0.01	−0.03	...	−0.4	−0.03	...	0.05	...	...
D5	1010	1850	Air	0.07	0.03	0.02	0.01	0.00	...	0.3	0.03	...	0.05	...	...
H11	1010	1850	Air	0.11	0.06	0.07	0.08	0.08	...	0.3	0.01	...	0.12	...	...
H13	1010	1850	Air	−0.01	...	...	...	...	...	...	0.00	...	0.06	...	...
M2	1210	2210	Oil	−0.02	...	...	...	...	...	...	...	−0.06	0.10	0.14	0.16
M41	1210	2210	Oil	−0.16	...	...	...	...	...	...	...	−0.17	0.08	0.21	0.23

Machining

Approximate machinability ratings for annealed tool steels

Type	Machinability rating
O6	125
W1, W2, W5	100(a)
A10	90
P2, P3, P4, P5, P6	75 to 90
P20, P21	65 to 80
L2, L6	65 to 75
S1, S2, S5, S6, S7	60 to 70
H10, H11, H13, H14, H19	60 to 70(b)
O1, O2, O7	45 to 60
A2, A3, A4, A6, A8, A9	45 to 60
H21, H22, H24, H25, H26, H42	45 to 55(b)
T1	40 to 50
M2	40 to 50
T4	35 to 40
M3 (class 1)	35 to 40
D2, D3, D4, D5, D7, A7	30 to 40
T15	25 to 30
M15	25 to 30

(a) Equivalent to approximately 30% of the machinability of B1112. (b) For hardness range 150 to 200 HB.

Typical grinding ratios for high speed steels

Type	Hardness, HRC	Grinding ratio(a) 32A46-H8VBE	32A60-H8VBE	32A80-H8VBE
T15	65.7	0.49	0.62	0.51
M44	67.7	0.97	0.99	0.88
M41	68.7	1.2	1.6	1.4
M43	67.5	1.4	2.2	1.7
M42	68.8	4.8	6.5	3.8
M2	64.9	6.1	7.2	6.7
M1	64.9	7.8	8.0	11.9

(a) For the following conditions: work, 152 mm (6 in.) long by 38 mm (1.5 in.) wide; wheel size, 200 mm (8 in.) in diameter by 13 mm (0.5 in.) wide; wheel speed (idling), 30 m/s (6000 ft/min); table speed, 0.3 m/s (60 ft/min); unit crossfeed, 1.27 mm (0.050 in.) after each table traverse; unit downfeed, 0.025 mm (0.001 in.) after each complete crossfeed; total downfeed, 0.25 mm (0.010 in.) preceded by four unit downfeeds to break wheel in after dressing with a diamond tool; grinding fluid, 1.25% water emulsion of general-purpose soluble oil.

Standard machining allowances for hot rolled square and flat bars

Specified width mm	in.	Top and bottom surfaces mm	in.	Edges mm	in.
Specified thickness, up to 12.7 mm (½ in.)					
0 to 12.7	0 to ½	0.64	0.025	0.64	0.025
>12.7 to 25.4	>½ to 1	0.64	0.025	0.89	0.035
>25.4 to 50.8	>1 to 2	0.76	0.030	1.02	0.040
>50.8 to 76.2	>2 to 3	0.89	0.035	1.27	0.050
>76.2 to 101.6	>3 to 4	1.02	0.040	1.65	0.065
>101.6 to 127.0	>4 to 5	1.14	0.045	2.03	0.080
>127.0 to 152.4	>5 to 6	1.27	0.050	2.41	0.095
>152.4 to 177.8	>6 to 7	1.40	0.055	2.67	0.105
>177.8 to 203.2	>7 to 8	1.52	0.060	3.05	0.120
>203.2 to 228.6	>8 to 9	1.52	0.060	3.30	0.130
>228.6 to 304.8	>9 to 12	1.52	0.060	3.56	0.140
Specified thickness, >12.7 to 25.4 mm (>½ to 1 in.)					
>12.7 to 25.4	>½ to 1	1.14	0.045	1.14	0.045
>25.4 to 50.8	>1 to 2	1.14	0.045	1.27	0.050
>50.8 to 76.2	>2 to 3	1.27	0.050	1.52	0.060
>76.2 to 101.6	>3 to 4	1.40	0.055	1.90	0.075
>101.6 to 127.0	>4 to 5	1.52	0.060	2.41	0.095
>127.0 to 152.4	>5 to 6	1.65	0.065	2.92	0.115
>152.4 to 177.8	>6 to 7	1.78	0.070	3.30	0.130
>177.8 to 203.2	>7 to 8	1.90	0.075	3.81	0.150
>203.2 to 228.6	>8 to 9	1.90	0.075	3.94	0.155
>228.6 to 304.8	>9 to 12	1.90	0.075	3.94	0.155
Specified thickness, >25.4 to 50.8 mm (>1 to 2 in.)					
>25.4 to 50.8	>1 to 2	1.65	0.065	1.65	0.065
>50.8 to 76.2	>2 to 3	1.65	0.065	1.78	0.070
>76.2 to 101.6	>3 to 4	1.78	0.070	2.16	0.085
>101.6 to 127.0	>4 to 5	1.78	0.070	2.67	0.105
>127.0 to 152.4	>5 to 6	1.90	0.075	3.18	0.125
>152.4 to 177.8	>6 to 7	2.03	0.080	3.68	0.145
>177.8 to 203.2	>7 to 8	2.03	0.080	4.19	0.165
>203.2 to 228.6	>8 to 9	2.41	0.095	4.32	0.170
>228.6 to 304.8	>9 to 12	2.54	0.100	4.32	0.170
Specified thickness, >50.8 to 76.2 mm (>2 to 3 in.)					
>50.8 to 76.2	>2 to 3	2.16	0.085	2.16	0.085
>76.2 to 101.6	>3 to 4	2.16	0.085	2.54	0.100
>101.6 to 127.0	>4 to 5	2.16	0.085	3.05	0.120
>127.0 to 152.4	>5 to 6	2.16	0.085	3.43	0.135
>152.4 to 177.8	>6 to 7	2.29	0.090	3.94	0.155
>177.8 to 203.2	>7 to 8	2.54	0.100	4.32	0.170
>203.2 to 228.6	>8 to 9	2.54	0.100	4.83	0.190
>228.6 to 304.8	>9 to 12	2.54	0.100	4.83	0.190
Specified thickness, >76.2 to 101.6 mm (>3 to 4 in.)					
>76.2 to 101.6	>3 to 4	2.92	0.115	2.92	0.115
>101.6 to 127.0	>4 to 5	2.92	0.115	3.18	0.125
>127.0 to 152.4	>5 to 6	2.92	0.115	3.56	0.140
>152.4 to 177.8	>6 to 7	2.92	0.115	4.32	0.170
>177.8 to 203.2	>7 to 8	3.18	0.125	4.83	0.190
>203.2 to 228.6	>8 to 9	3.18	0.125	4.83	0.190
>228.6 to 304.8	>9 to 12	3.18	0.125	4.83	0.190

(a) Minimum allowance per side for machining prior to heat treatment. Maximum decarburization limit, 80% of machining allowance.

ISO R513 classification of carbides according to use for machining

Designation(a)	Groups of application	
	Material to be machined	Use and working conditions
P 01	Steel, steel castings	Finish turning and boring; high cutting speeds, small chip section, accuracy of dimensions and fine finish, vibration-free operation
P 10	Steel, steel castings	Turning, copying, threading and milling; high cutting speeds, small or medium chip sections
P 20	Steel, steel castings Malleable cast iron with long chips	Turning, copying, milling, medium cutting speeds and chip sections; planing with small chip sections
P 30	Steel, steel castings Malleable cast iron with long chips	Turning, milling, planing, medium or low cutting speeds, medium or large chip sections, and machining in unfavorable conditions(b)
P 40	Steel Steel castings with sand inclusion and cavities	Turning, planing, slotting, low cutting speeds, large chip sections with the possibility of large cutting angles for machining in unfavorable conditions(b) and work on automatic machines
P 50	Steel Steel castings of medium or low tensile strength, with sand inclusion and cavities	For operations demanding very tough carbide: turning, planing, slotting, low cutting speeds, large chip sections, with the possibility of large cutting angles for machining in unfavorable conditions(b) and work on automatic machines
M 10	Steel, steel castings, manganese steel Grey cast iron, alloy cast iron	Turning, medium or high cutting speeds. Small or medium chip sections
M 20	Steel, steel castings, austenitic or manganese steel, grey cast iron	Turning, milling. Medium cutting speeds and chip sections
M 30	Steel, steel castings, austenitic steel, grey cast iron, high temperature resistant alloys	Turning, milling, planing. Medium cutting speeds, medium or large chip sections
M 40	Mild free cutting steel, low tensile steel Nonferrous metals and light alloys	Turning, parting off, particularly on automatic machines
K 01	Very hard grey cast iron, chilled castings of over 85 scleroscope hardness, high silicon aluminum alloys, hardened steel, highly abrasive plastics, hard cardboard, ceramics	Turning, finish turning, boring, milling, scraping
K 10	Grey cast iron over 220 HB malleable cast iron with short chips, hardened steel, silicon aluminum alloys, copper alloys, plastics, glass, hard rubber, hard cardboard, porcelain, stone	Turning, milling, drilling, boring, broaching, scraping
K 20	Grey cast iron up to 220 HB nonferrous metals: copper, brass, aluminum	Turning, milling, planing, boring, broaching, demanding very tough carbide
K 30	Low hardness grey cast iron, low tensile steel, compressed wood	Turning, milling, planing, slotting, for machining in unfavorable conditions(b) and with the possibility of large cutting angles
K 40	Soft wood or hard wood Nonferrous metals	Turning, milling, planing, slotting, for machining in unfavorable conditions(b) and with the possibility of large cutting angles

(a) In each letter category, low designation numbers are for high speeds and light feeds, higher numbers for slower speeds and/or heavier feeds. Also, increasing designation numbers imply increasing toughness and decreasing wear resistance of the cemented carbide materials. (b) Unfavorable conditions include: shapes that are awkward to machine; material having a casting or forging skin; material having variable hardness; and machining that involves variable depth of cut, interrupted cut or moderate to severe vibrations.

Wrought Stainless Steels

Family Relationships

Family relationships for standard austenitic stainless steels

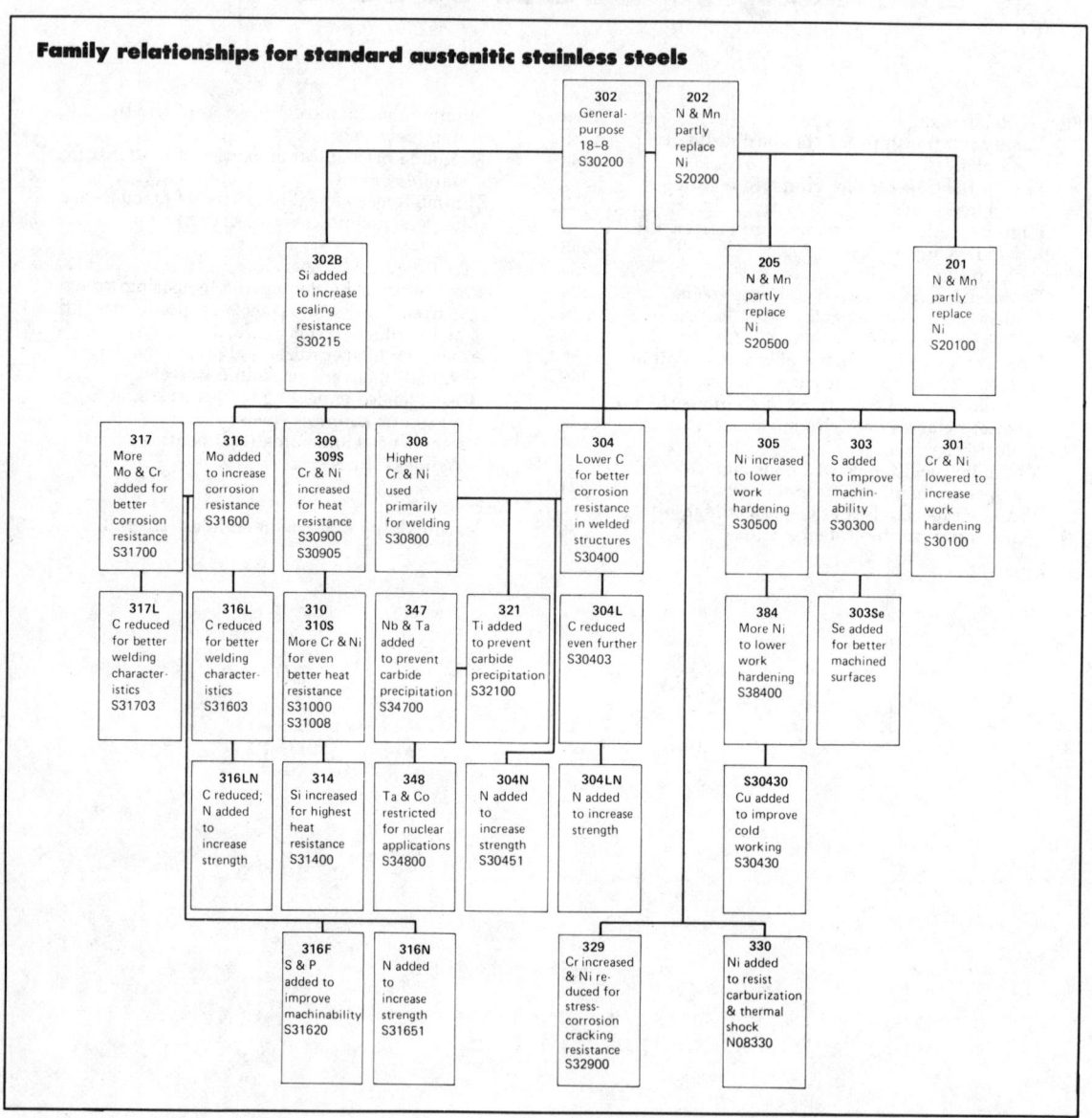

Family relationships for standard ferritic stainless steels

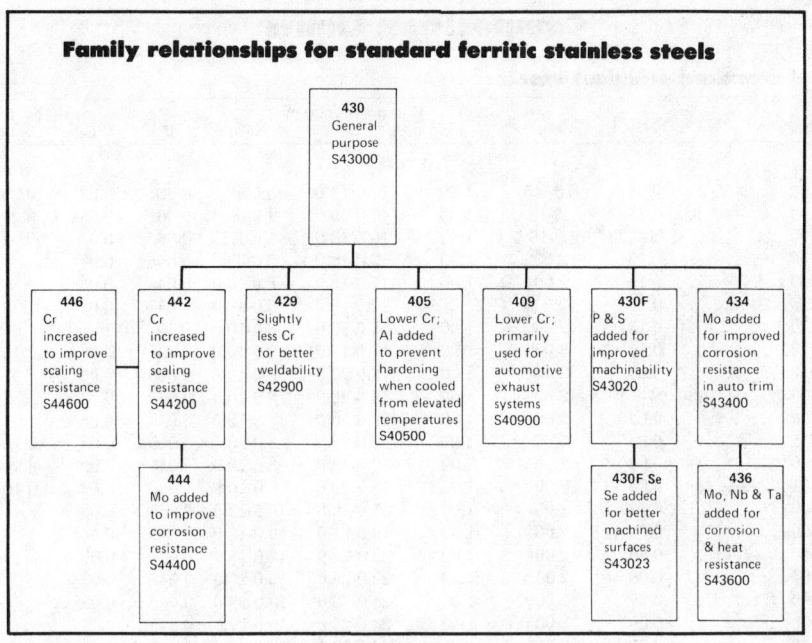

Family relationships for standard martensitic stainless steels

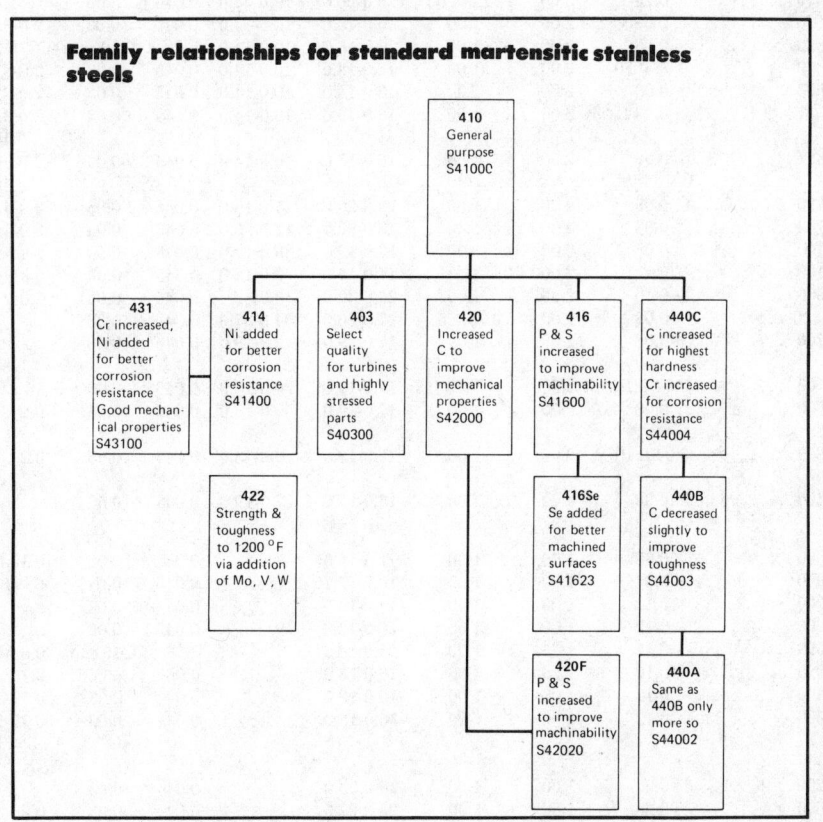

Composition Limits

Compositions of standard stainless steels

Type	UNS number	C	Mn	Si	Cr	Ni(b)	P	S	Others
Austenitic types									
201	S20100	0.15	5.5-7.5	1.00	16.0-18.0	3.5-5.5	0.06	0.03	0.25 N
202	S20200	0.15	7.5-10.0	1.00	17.0-19.0	4.0-6.0	0.06	0.03	0.25 N
205	S20500	0.12-0.25	14.0-15.5	1.00	16.5-18.0	1.0-1.75	0.06	0.03	0.32-0.40 N
301	S30100	0.15	2.00	1.00	16.0-18.0	6.0-8.0	0.045	0.03	....
302	S30200	0.15	2.00	1.00	17.0-19.0	8.0-10.0	0.045	0.03	....
302B	S30215	0.15	2.00	2.0-3.0	17.0-19.0	8.0-10.0	0.045	0.03	....
303	S30300	0.15	2.00	1.00	17.0-19.0	8.0-10.0	0.20	0.15 min	0.6 Mo(c)
303Se	S30323	0.15	2.00	1.00	17.0-19.0	8.0-10.0	0.20	0.06	0.15 min Se
304	S30400	0.08	2.00	1.00	18.0-20.0	8.0-10.5	0.045	0.03	....
304H	S30409	0.04-0.10	2.00	1.00	18.0-20.0	8.0-10.5	0.045	0.03	....
304L	S30403	0.03	2.00	1.00	18.0-20.0	8.0-12.0	0.045	0.03	....
304LN	....	0.03	2.00	1.00	18.0-20.0	8.0-10.5	0.045	0.03	0.10-0.15 N
S30430	S30430	0.08	2.00	1.00	17.0-19.0	8.0-10.0	0.045	0.03	3.0-4.0 Cu
304N	S30451	0.08	2.00	1.00	18.0-20.0	8.0-10.5	0.045	0.03	0.10-0.16 N
305	S30500	0.12	2.00	1.00	17.0-19.0	10.5-13.0	0.045	0.03	....
308	S30800	0.08	2.00	1.00	19.0-21.0	10.0-12.0	0.045	0.03	....
309	S30900	0.20	2.00	1.00	22.0-24.0	12.0-15.0	0.045	0.03	....
309S	S30908	0.08	2.00	1.00	22.0-24.0	12.0-15.0	0.045	0.03	....
310	S31000	0.25	2.00	1.50	24.0-26.0	19.0-22.0	0.045	0.03	....
310S	S31008	0.08	2.00	1.50	24.0-26.0	19.0-22.0	0.045	0.03	....
314	S31400	0.25	2.00	1.5-3.0	23.0-26.0	19.0-22.0	0.045	0.03	....
316	S31600	0.08	2.00	1.00	16.0-18.0	10.0-14.0	0.045	0.03	2.0-3.0 Mo
316F	S31620	0.08	2.00	1.00	16.0-18.0	10.0-14.0	0.20	0.10 min	1.75-2.5 Mo
316H	S31609	0.04-0.10	2.00	1.00	16.0-18.0	10.0-14.0	0.045	0.03	2.0-3.0 Mo
316L	S31603	0.03	2.00	1.00	16.0-18.0	10.0-14.0	0.045	0.03	2.0-3.0 Mo
316LN	....	0.03	2.00	1.00	16.0-18.0	10.0-14.0	0.045	0.03	2.0-3.0 Mo; 0.10-0.30 N
316N	S31651	0.08	2.00	1.00	16.0-18.0	10.0-14.0	0.045	0.03	2.0-3.0 Mo; 0.10-0.16 N
317	S31700	0.08	2.00	1.00	18.0-20.0	11.0-15.0	0.045	0.03	3.0-4.0 Mo
317L	S31703	0.03	2.00	1.00	18.0-20.0	11.0-15.0	0.045	0.03	3.0-4.0 Mo
321	S32100	0.08	2.00	1.00	17.0-19.0	9.0-12.0	0.045	0.03	5×%C min Ti
321H	S32109	0.04-0.10	2.00	1.00	17.0-19.0	9.0-12.0	0.045	0.03	5×%C min Ti
329	S32900	0.10	2.00	1.00	25.0-30.0	3.0-6.0	0.045	0.03	1.0-2.0 Mo
330	N08330	0.08	2.00	0.75-1.5	17.0-20.0	34.0-37.0	0.04	0.03	....
347	S34700	0.08	2.00	1.00	17.0-19.0	9.0-13.0	0.045	0.03	10×%C min Nb + Ta(c)
347H	S34709	0.04-0.10	2.00	1.00	17.0-19.0	9.0-13.0	0.045	0.03	10×%C min Nb + Ta
348	S34800	0.08	2.00	1.00	17.0-19.0	9.0-13.0	0.045	0.03	0.2 Cu; 10×%C min Nb + Ta(c)
348H	S34809	0.04-0.10	2.00	1.00	17.0-19.0	9.0-13.0	0.045	0.03	0.2 Cu; 10×%C min Nb + Ta(c)
384	S38400	0.08	2.00	1.00	15.0-17.0	17.0-19.0	0.045	0.03	....
Ferritic types									
405	S40500	0.08	1.00	1.00	11.5-14.5	....	0.04	0.03	0.10-0.30 Al
409	S40900	0.08	1.00	1.00	10.5-11.75	....	0.045	0.045	6×% C min Ti(e)
429	S42900	0.12	1.00	1.00	14.0-16.0	....	0.04	0.03	....
430	S43000	0.12	1.00	1.00	16.0-18.0	....	0.04	0.03	....
430F	S43020	0.12	1.25	1.00	16.0-18.0	....	0.06	0.15 min	0.6 Mo(c)
430FSe	S43023	0.12	1.25	1.00	16.0-18.0	....	0.06	0.06	0.15 min Se
434	S43400	0.12	1.00	1.00	16.0-18.0	....	0.04	0.03	0.75-1.25 Mo
436	S43600	0.12	1.00	1.00	16.0-18.0	....	0.04	0.03	0.75-1.25 Mo; 5×%C min Nb + Ta(f)
442	S44200	0.20	1.00	1.00	18.0-23.0	....	0.04	0.03	....
446	S44600	0.20	1.50	1.00	23.0-27.0	....	0.04	0.03	0.25 N

(continued)

(a) Single values are maximum values unless otherwise indicated. (b) For some tubemaking processes, the nickel content of certain austenitic types must be slightly higher than shown. (c) Optional. (d) 0.10% max Ta. (e) 0.75% maximum. (f) 0.70% maximum.

Compositions of standard stainless steels (continued)

Type	UNS number	C	Mn	Si	Cr	Ni(b)	P	S	Others
Martensitic types									
403	S40300	0.15	1.00	0.50	11.5-13.0	....	0.04	0.03	....
410	S41000	0.15	1.00	1.00	11.5-13.0	....	0.04	0.03	....
414	S41400	0.15	1.00	1.00	11.5-13.5	1.25-2.50	0.04	0.03	....
416	S41600	0.15	1.25	1.00	12.0-14.0	....	0.04	0.03	0.6 Mo(c)
416Se	S41623	0.15	1.25	1.00	12.0-14.0	....	0.06	0.06	0.15 min Se
420	S42000	0.15 min	1.00	1.00	12.0-14.0	....	0.04	0.03	....
420F	S42020	0.15 min	1.25	1.00	12.0-14.0	....	0.06	0.15 min	0.6 Mo(c)
422	S42200	0.20-0.25	1.00	0.75	11.0-13.0	0.5-1.0	0.025	0.025	0.75-1.25 Mo; 0.75-1.25 W; 0.15-0.3 V
431	S43100	0.20	1.00	1.00	15.0-17.0	1.25-2.50	0.04	0.03	....
440A	S44002	0.60-0.75	1.00	1.00	16.0-18.0	....	0.04	0.03	0.75 Mo
440B	S44003	0.75-0.95	1.00	1.00	16.0-18.0	....	0.04	0.03	0.75 Mo
440C	S44004	0.95-1.20	1.00	1.00	16.0-18.0	....	0.04	0.03	0.75 Mo
501	S50100	0.10 min	1.00	1.00	4.0-6.0	....	0.04	0.03	0.40-0.65 Mo
501A	S50300	0.15	0.30-0.60	0.50-1.00	6.0-8.0	....	0.03	0.03	0.45-0.65 Mo
501B	S50400	0.15	0.30-0.60	0.50-1.00	8.0-10.0	....	0.03	0.03	0.9-1.1 Mo
502	S50200	0.10	1.00	1.00	4.0-6.0	....	0.04	0.03	0.40-0.65 Mo
503	S50300	0.15	1.00	1.00	6.0-8.0	....	0.04	0.04	0.45-0.65 Mo
504	S50400	0.15	1.00	1.00	8.0-10.0	....	0.04	0.04	0.9-1.1 Mo
Precipitation-hardening types									
PH 13-8 Mo	S13800	0.05	0.10	0.10	12.25-13.25	7.5-8.5	0.01	0.008	2.0-2.5 Mo; 0.90-1.35 Al; 0.01 N
15-5 PH	S15500	0.07	1.00	1.00	14.0-15.5	3.5-5.5	0.04	0.03	2.5-4.5 Cu; 0.15-0.45 Nb + Ta
17-4 PH	S17400	0.07	1.00	1.00	15.5-17.5	3.0-5.0	0.04	0.03	3.0-5.0 Cu; 0.15-0.45 Nb + Ta
17-7 PH	S17700	0.09	1.00	1.00	16.0-18.0	6.5-7.75	0.04	0.03	0.75-1.5 Al

(a) Single values are maximum values unless otherwise indicated. (b) For some tubemaking processes, the nickel content of certain austenitic types must be slightly higher than shown. (c) Optional. (d) 0.10% max Ta. (e) 0.75% maximum. (f) 0.70% maximum

Compositions of nonstandard stainless steels

Designation(a)	UNS number	C	Mn	Si	Composition(b), % Cr	Ni	P	S	Others
Austenitic stainless steels									
Type 216 (XM-17)	S21600	0.08	7.5-9.0	1.00	17.5-22.0	5.0-7.0	0.045	0.03	2.0-3.0 Mo; 0.25-0.50 N
Type 304HN	S30452	0.04-0.10	2.00	1.00	18.0-20.0	8.0-10.5	0.045	0.03	0.10-0.16 N
Type 308	S30800	0.08	2.00	1.00	19.0-21.0	10.0-12.0	0.045	0.03	....
Type 308L	...	0.03	2.00	1.00	19.0-21.0	10.0-12.0	0.045	0.03	....
Type 309S	S30908	0.08	2.00	1.00	22.0-24.0	12.0-15.0	0.045	0.03	....
Type 309S Cb	S30940	0.08	2.00	1.00	22.0-24.0	12.0-15.0	0.045	0.03	8 × %C min Nb
Type 309 Cb + Ta	...	0.08	2.00	1.00	22.0-24.0	12.0-15.0	0.045	0.03	8 × %C min Nb + Ta
Type 312	...	0.15	2.00	1.00	30.0 nom	9.0 nom	0.045	0.03	....
Type 317LM	...	0.03	2.00	1.00	18.0-20.0	12.0-16.0	0.045	0.03	4.0-5.0 Mo
Type 330HC	...	0.40	1.50	1.25	19.0 nom	35.0 nom	....	....	....
Type 332	...	0.04	1.00	0.50	21.5 nom	32.0 nom	0.045	0.03	....
Type 385	...	0.08	2.00	1.00	11.5-13.5	14.0-16.0	0.045	0.03	....
904L	N08904	0.02	2.00	1.00	19.0-23.0	23.0-28.0	0.045	0.035	4.0-5.0 Mo; 1.0-2.0 Cu
18-18-2 (XM-15)	S38100	0.08	2.00	1.5-2.5	17.0-19.0	17.5-18.5	0.03	0.03	0.08-0.18 N
18-18 Plus	S28200	0.15	17.0-19.0	1.00	17.5-19.5	....	0.045	0.03	0.5-1.5 Mo; 0.5-1.5 Cu; 0.4-0.6 N
20Cb-3	N08020	0.07	2.00	1.00	19.0-21.0	32.0-38.0	0.045	0.035	2.0-3.0 Mo; 3.0-4.0 Cu; 8 × %C min Nb(c)
AL-6X	N08366	0.03	2.00	0.75	20.0-22.0	23.5-25.5	0.030	0.003	6.0-7.0 Mo
303 Plus X (XM-5)	...	0.15	2.5-4.5	1.00	17.0-19.0	7.0-10.0	0.20	0.25 min	0.6 Mo
HNM (d)	...	0.30	3.5	0.5	18.5	9.5	0.25	....	....
Crutemp 25 (d)	...	0.05	1.5	0.4	25.0	25.0	....	....	....
JS-700	N08700	0.04	2.00	1.00	19.0-23.0	24.0-26.0	0.04	0.03	4.3-5.0 Mo; 0.5 Cu; 8 × %C min Nb(e); 0.005 Pb; 0.035 Sn
JS-777	...	0.04	2.00	1.00	19.0-23.0	24.0-26.0	0.045	0.035	4.0-5.0 Mo; 1.9-2.5 Cu
Nitronic 32 (d)	S24100	0.10	12.0	0.5	18.0	1.6	....	....	0.35 N
Nitronic 33 (d)	S24000	0.06	13.0	0.5	18.0	3.0	....	....	0.30 N
Nitronic 40 (21-6-9) (XM-10)	S21900	0.08	8.0-10.0	1.00	18.0-20.0	5.0-7.0	0.06	0.03	0.15-0.40 N
Nitronic 50 (22-13-5) (XM-19)	S20910	0.06	4.0-6.0	1.00	20.5-23.5	11.5-13.5	0.04	0.03	1.5-3.0 Mo; 0.2-0.4 N; 0.1-0.3 Nb; 0.1-0.3 V
Nitronic 60	S21800	0.10	7.0-9.0	3.5-4.5	16.0-18.0	8.0-9.0	0.04	0.03	....
Tenelon (XM-31)	S21400	0.12	14.5-16.0	0.3-1.0	17.0-18.5	0.75	0.045	0.03	0.35 N
Cryogenic Tenelon (XM-14)	S21460	0.12	14.0-16.0	1.00	17.0-19.0	5.0-6.0	0.06	0.03	0.35-0.50 N

(continued)

(a) XM designations in this column are ASTM designations for the listed alloy. Type numbers in parentheses are obsolete AISI designations. (b) Single values are maximum values unless otherwise indicated. (c) 1.00% maximum. (d) Nominal composition; composition limits not available. (e) 0.50% maximum. (f) 0.75% maximum. (g) 0.80 maximum. (h) 7 N minimum.

Compositions of nonstandard stainless steels (continued)

Designation(a)	UNS number	Composition(b), % C	Mn	Si	Cr	Ni	P	S	Others
Ferritic stainless steels									
Type 404	...	0.05	1.00	0.50	11.0-12.5	1.25-2.00	0.03	0.03	...
Type 430Ti	S43036	0.10	1.00	1.00	16.0-19.5	0.75	0.04	0.03	5×%C min Ti(f)
Type 444 (18-2)	S44400	0.025	1.00	1.00	17.5-19.5	1.00	0.04	0.03	1.75-2.5 Mo; 0.035 max N; 0.2 + 4(%C + %N) min (Ti + Nb)
18SR (d)	...	0.04	0.3	1.00	18.0	...	...	...	2.0 Al; 0.4 Ti
18-2 FM	S18200	0.08	2.50	1.00	17.5-19.5	...	0.04	0.15 min	...
E Brite 26-1 (XM-27)	S44625	0.01	0.40	0.40	25.0-27.5	0.50	0.02	0.02	0.75-1.5 Mo; 0.015 N; 0.2 Cu; 0.5 Ni + Cu
26-1 Ti (XM-33)	S44626	0.06	0.75	0.75	25.0-27.0	0.50	0.04	0.02	0.75-1.5 Mo; 0.04 N; 0.2 Cu; 0.2-1.0 Ti(g)
29-4	S44700	0.010	0.30	0.20	28.0-30.0	0.15	0.025	0.02	3.5-4.2 Mo
29-4-2	S44800	0.010	0.30	0.20	28.0-30.0	2.0-2.5	0.025	0.02	3.5-4.2 Mo
Monit	S44635	0.25	1.00	0.75	24.5-26.0	3.5-4.5	0.04	0.03	3.5-4.5 Mo; 0.3-0.6 (Ti + Nb)
Sea-cure/Sc-1	S44660	0.025	1.00	0.75	25.0-27.0	1.5-3.5	0.04	0.03	2.5-3.5 Mo; 0.2 + 4(%C + %N) min (Ti + Nb)
Martensitic stainless steels									
Type 410Cb (XM-30)	S41040	0.18	1.00	1.00	11.5-13.5	...	0.04	0.03	0.05-0.30 Nb
Type 410S	S41008	0.08	1.00	1.00	11.5-13.5	0.60	0.04	0.03	...
Type 414L	...	0.06	0.50	0.15	12.5-13.0	2.5-3.0	0.04	0.03	0.5 Mo; 0.03 Al
416 Plus X (XM-6)	S41610	0.15	1.5-2.5	1.00	12.0-14.0	...	0.06	0.15 min	0.6 Mo
Precipitation-hardening stainless steels									
AM-350 (Type 633)	S35000	0.07-0.11	0.5-1.25	0.50	16.0-17.0	4.0-5.0	0.04	0.03	2.5-3.25 Mo; 0.07-0.13 N
AM-355 (Type 634)	S35500	0.10-0.15	0.5-1.25	0.50	15.0-16.0	4.0-5.0	0.04	0.03	2.5-3.25 Mo
AM-363 (d)	...	0.04	0.15	0.05	11.0	4.0	...	...	0.25 Ti
Custom 450 (XM-25)	S45000	0.05	1.00	1.00	14.0-16.0	5.0-7.0	0.03	0.03	1.25-1.75 Cu; 0.5-1.0 Mo; 8×%C min Nb
Custom 455 (XM-16)	S45500	0.05	0.50	0.50	11.0-12.5	7.5-9.5	0.04	0.03	0.5 Mo; 1.5-2.5 Cu; 0.8-1.4 Ti; 0.1-0.5 Nb
PH 15-7 Mo (Type 632)	S15700	0.09	1.00	1.00	14.0-16.0	6.5-7.75	0.04	0.03	2.0-3.0 Mo; 0.75-1.5 Al
Stainless W (Type 635)	S17600	0.08	1.00	1.00	16.0-17.5	6.0-7.5	0.04	0.03	0.4 Al; 0.4-1.2 Ti
17-10 P (d)	...	0.07	0.75	0.5	17.0	10.5	0.28	...	...

(a) XM designations in this column are ASTM designations for the listed alloy. Type numbers in parentheses are obsolete AISI designations. (b) Single values are maximum values unless otherwise indicated. (c) 1.00% maximum. (d) Nominal composition; composition limits not available. (e) 0.50% maximum. (f) 0.75% maximum. (g) 0.80 maximum. (h) 7 N minimum

Physical Properties

Typical physical properties of wrought stainless steels, annealed condition

Type	UNS designation	Density Mg/m³	Density lb/in.³	Elastic modulus GPa	Elastic modulus 10⁶ psi	Mean coefficient of thermal expansion μm/m·°C 0 °C to: 32 °F to: 100 212	315 600	538 1000	μin./in.·°F 100 212	315 600	538 1000
201	S20100 .... 7.8		0.28	197	28.6	15.7	17.5	18.4	8.7	9.7	10.2
202	S20200 .... 7.8		0.28	...	...	17.5	18.4	19.2	9.7	10.2	10.7
205	S20500 .... 7.8		0.28	197	28.6	...	17.9	19.1	...	9.9	10.6
301	S30100 .... 8.0		0.29	193	28.0	17.0	17.2	18.2	9.4	9.6	10.1
302	S30200 .... 8.0		0.29	193	28.0	17.2	17.8	18.4	9.6	9.9	10.2
302B	S30215 .... 8.0		0.29	193	28.0	16.2	18.0	19.4	9.0	10.0	10.8
303	S30300 .... 8.0		0.29	193	28.0	17.2	17.8	18.4	9.6	9.9	10.2
304	S30400 .... 8.0		0.29	193	28.0	17.2	17.8	18.4	9.6	9.9	10.2
304L	S30403 .... 8.0		0.29	...	...	...	...	...	...	...	...
S30430	S30430 .... 8.0		0.29	193	28.0	17.2	17.8	...	9.6	9.9	...
304N	S30451 .... 8.0		0.29	196	28.5	...	...	...	...	...	...
305	S30500 .... 8.0		0.29	193	28.0	17.2	17.8	18.4	9.6	9.9	10.2
308	S30800 .... 8.0		0.29	193	28.0	17.2	17.8	18.4	9.6	9.9	10.2
309	S30900 .... 8.0		0.29	200	29.0	15.0	16.6	17.2	8.3	9.2	9.6
310	S31000 .... 8.0		0.29	200	29.0	15.9	16.2	17.0	8.8	9.0	9.4
314	S31400 .... 7.8		0.28	200	29.0	...	15.1	...	...	8.4	...
316	S31600 .... 8.0		0.29	193	28.0	15.9	16.2	17.5	8.8	9.0	9.7
316L	S31603 .... 8.0		0.29	...	...	...	...	...	...	...	...
316N	S31651 .... 8.0		0.29	196	28.5	...	...	...	...	...	...
317	S31700 .... 8.0		0.29	193	28.0	15.9	16.2	17.5	8.8	9.0	9.7
317L	S31703 .... 8.0		0.29	200	29.0	16.5	...	18.1	9.2	...	10.1
321	S32100 .... 8.0		0.29	193	28.0	16.6	17.2	18.6	9.2	9.6	10.3
329	S32900 .... 7.8		0.28	...	...	...	...	...	...	...	...
330	N08330 .... 8.0		0.29	196	28.5	14.4	16.0	16.7	8.0	8.9	9.3
347	S34700 .... 8.0		0.29	193	28.0	16.6	17.2	18.6	9.2	9.6	10.3
384	S38400 .... 8.0		0.29	193	28.0	17.2	17.8	18.4	9.6	9.9	10.2
405	S40500 .... 7.8		0.28	200	29.0	10.8	11.6	12.1	6.0	6.4	6.7
409	S40900 .... 7.8		0.28	...	...	11.7	...	...	6.5	...	...
410	S41000 .... 7.8		0.28	200	29.0	9.9	11.4	11.6	5.5	6.3	6.4
414	S41400 .... 7.8		0.28	200	29.0	10.4	11.0	12.1	5.8	6.1	6.7
416	S41600 .... 7.8		0.28	200	29.0	9.9	11.0	11.6	5.5	6.1	6.4
420	S42000 .... 7.8		0.28	200	29.0	10.3	10.8	11.7	5.7	6.0	6.5
422	S42200 .... 7.8		0.28	...	...	11.2	11.4	11.9	6.2	6.3	6.6
429	S42900 .... 7.8		0.28	200	29.0	10.3	...	...	5.7	...	...
430	S43000 .... 7.8		0.28	200	29.0	10.4	11.0	11.4	5.8	6.1	6.3
430F	S43020 .... 7.8		0.28	200	29.0	10.4	11.0	11.4	5.8	6.1	6.3
431	S43100 .... 7.8		0.28	200	29.0	10.2	12.1	...	5.7	6.7	...
434	S43400 .... 7.8		0.28	200	29.0	10.4	11.0	11.4	5.8	6.1	6.3
436	S43600 .... 7.8		0.28	200	29.0	9.3	...	...	5.2	...	...
440A	S44002 .... 7.8		0.28	200	29.0	10.2	...	...	5.7	...	...
440C	S44004 .... 7.8		0.28	200	29.0	10.2	...	...	5.7	...	...
444	S44400 .... 7.8		0.28	200	29.0	10.0	10.6	11.4	5.6	5.9	6.3
446	S44600 .... 7.5		0.27	200	29.0	10.4	10.8	11.2	5.8	6.0	6.2
PH 13-8 Mo	S13800 .... 7.8		0.28	203	29.4	10.6	11.2	11.9	5.9	6.2	6.6
15-5 PH	S15500 .... 7.8		0.28	196	28.5	10.8	11.4	...	6.0	6.3	...
17-4 PH	S17400 .... 7.8		0.28	196	28.5	10.8	11.6	...	6.0	6.4	...
17-7 PH	S17700 .... 7.8		0.28	204	29.5	11.0	11.6	...	6.1	6.4	...

(continued)

(a) At 0 to 100 °C (32 to 212 °F). (b) Approximate values

Typical physical properties of wrought stainless steels, annealed condition (continued)

Type	Thermal conductivity W/m·K °C: 100 °F: 212	500 932	Btu/h·ft·°F 100 212	500 932	Specific heat(a) J/kg·K	Btu/lb·°F	Electrical resistivity, nΩ·m	Magnetic permeability(b)	Melting range °C	°F
201	16.2	21.5	9.4	12.4	500	0.12	69	1.02	1400-1450	2550-2650
202	16.2	21.6	9.4	12.5	500	0.12	69	1.02	1400-1450	2550-2650
205	...	...	...	...	500	0.12	...	...	...	...
301	16.2	21.5	9.4	12.4	500	0.12	72	1.02	1400-1420	2550-2590
302	16.2	21.5	9.4	12.4	500	0.12	72	1.02	1400-1420	2550-2590
302B	15.9	21.6	9.2	12.5	500	0.12	72	1.02	1375-1400	2500-2550
303	16.2	21.5	9.4	12.4	500	0.12	72	1.02	1400-1420	2550-2590
304	16.2	21.5	9.4	12.4	500	0.12	72	1.02	1400-1450	2550-2650
304L	...	...	...	...	...	...	...	1.02	1400-1450	2550-2650
S30430	11.2	21.5	6.5	12.4	500	0.12	72	1.02	1400-1450	2550-2650
304N	...	...	...	...	500	0.12	72	1.02	1400-1450	2550-2650
305	16.2	21.5	9.4	12.4	500	0.12	72	1.02	1400-1450	2550-2650
308	15.2	21.6	8.8	12.5	500	0.12	72	...	1400-1420	2550-2590
309	15.6	18.7	9.0	10.8	500	0.12	78	1.02	1400-1450	2550-2650
310	14.2	18.7	8.2	10.8	500	0.12	78	1.02	1400-1450	2550-2650
314	17.5	20.9	10.1	12.1	500	0.12	77	1.02	...	...
316	16.2	21.5	9.4	12.4	500	0.12	74	1.02	1375-1400	2500-2550
316L	...	...	...	...	...	...	...	1.02	1375-1400	2500-2550
316N	...	...	...	...	500	0.12	74	1.02	1375-1400	2500-2550
317	16.2	21.5	9.4	12.4	500	0.12	74	1.02	1375-1400	2500-2550
317L	14.4	...	8.3	...	500	0.12	79	...	1375-1400	2500-2550
321	16.1	22.2	9.3	12.8	500	0.12	72	1.02	1400-1425	2550-2600
329	...	...	...	...	460	0.11	75	...	...	...
330	...	...	...	...	460	0.11	102	1.02	1400-1425	2550-2600
347	16.1	22.2	9.3	12.8	500	0.12	73	1.02	1400-1425	2550-2600
384	16.2	21.5	9.4	12.4	500	0.12	79	1.02	1400-1450	2550-2650
405	27.0	...	15.6	...	460	0.11	60	...	1480-1530	2700-2790
409	...	...	...	...	...	...	...	...	1480-1530	2700-2790
410	24.9	28.7	14.4	16.6	460	0.11	57	700-1000	1480-1530	2700-2790
414	24.9	28.7	14.4	16.6	460	0.11	70	...	1425-1480	2600-2700
416	24.9	28.7	14.4	16.6	460	0.11	57	700-1000	1480-1530	2700-2790
420	24.9	...	14.4	...	460	0.11	55	...	1450-1510	2650-2750
422	23.9	27.3	13.8	15.8	460	0.11	...	...	1470-1480	2675-2700
429	25.6	...	14.8	...	460	0.11	59	...	1450-1510	2650-2750
430	26.1	26.3	15.1	15.2	460	0.11	60	600-1100	1425-1510	2600-2750
430F	26.1	26.3	15.1	15.2	460	0.11	60	...	1425-1510	2600-2750
431	20.2	...	11.7	...	460	0.11	72	...	...	...
434	...	26.3	...	15.2	460	0.11	60	600-1100	1425-1510	2600-2750
436	23.9	26.0	13.8	15.0	460	0.11	60	600-1100	1425-1510	2600-2750
440A	24.2	...	14.0	...	460	0.11	60	...	1370-1480	2500-2700
440C	24.2	...	14.0	...	460	0.11	60	...	1370-1480	2500-2700
444	26.8	...	15.5	...	420	0.10	62	...	...	...
446	20.9	24.4	12.1	14.1	500	0.12	67	400-700	1425-1510	2600-2750
PH 13-8 Mo	14.0	22.0	8.1	12.7	460	0.11	102	...	1400-1440	2560-2625
15-5 PH	17.8	23.0	10.3	13.1	420	0.10	77	95	1400-1440	2560-2625
17-4 PH	18.3	23.0	10.6	13.1	460	0.11	80	95	1400-1440	2560-2625
17-7 PH	16.4	21.8	9.5	12.6	460	0.11	83	...	1400-1440	2560-2625

(a) At 0 to 100 °C (32 to 212 °F). (b) Approximate values

Resistance of standard types of stainless steel to various classes of environments

Austenitic stainless steels

Type	Mild atmospheric and fresh water	Atmospheric		Salt water	Chemical		
		Industrial	Marine		Mild	Oxidizing	Reducing
201	x	x	x		x	x	
202	x	x	x		x	x	
205	x	x	x		x	x	
301	x	x	x		x	x	
302	x	x	x		x	x	
302B	x	x	x		x	x	
303	x	x			x	x	
303Se	x	x			x	x	
304	x	x	x		x	x	
304H	x	x	x		x	x	
304L	x	x	x		x	x	
304N	x	x	x		x	x	
S30430	x	x	x		x	x	
305	x	x	x		x	x	
308	x	x	x		x	x	
309	x	x	x		x	x	
309S	x	x	x		x	x	
310	x	x	x		x	x	
310S	x	x	x		x	x	
314	x	x	x		x	x	
316	x	x	x	x	x	x	x
316F	x	x	x	x	x	x	x
316H	x	x	x	x	x	x	x
316L	x	x	x	x	x	x	x
316N	x	x	x	x	x	x	x
317	x	x	x	x	x	x	x
317L	x	x	x	x	x	x	x
321	x	x	x		x	x	
321H	x	x	x		x	x	
329	x	x	x	x	x	x	x
330	x	x	x	x	x	x	x
347	x	x	x		x	x	
347H	x	x	x		x	x	
348	x	x	x		x	x	
348H	x	x	x		x	x	
384	x	x	x		x	x	

Ferritic stainless steels

Type	Mild atmospheric and fresh water	Atmospheric		Salt water	Chemical		
		Industrial	Marine		Mild	Oxidizing	Reducing
405	x				x	x	
409	x				x	x	
429	x	x			x	x	
430	x	x			x	x	
430F	x	x			x	x	
430FSe	x	x			x	x	
434	x	x	x		x	x	
436	x	x	x		x	x	
442	x	x			x	x	
446	x	x	x		x	x	

Martensitic stainless steels

Type	Mild atmospheric and fresh water	Atmospheric		Salt water	Chemical		
		Industrial	Marine		Mild	Oxidizing	Reducing
403	x				x	x	
410	x				x	x	
414	x				x	x	
416	x				x	x	
416Se	x				x	x	
420	x				x	x	
420F	x				x	x	
422	x				x	x	
431	x	x	x		x	x	
440A	x				x	x	
440B	x				x	x	
440C	x				x	x	
501					x	x	
502					x	x	
503					x	x	
504					x	x	

Precipitation-hardening stainless steels

Type	Mild atmospheric and fresh water	Atmospheric		Salt water	Chemical		
		Industrial	Marine		Mild	Oxidizing	Reducing
PH 13-8 Mo	x	x			x	x	
15-5 PH	x	x	x		x	x	
17-4 PH	x	x	x		x	x	
17-7 PH	x	x	x		x	x	

An "x" notation above indicates that the specific type is considered resistant to the corrosive environment.

Minimum room-temperature mechanical properties of austenitic stainless steels

Product form(a)	Condition	Tensile strength MPa	ksi	0.2% yield strength MPa	ksi	Elongation, %	Reduction in area, %	Hardness, HRB	ASTM specification
Type 301 (UNS S30100)									
B, W, P, Sh, St	Annealed	515	75	205	30	40	...	88 max	A167
Sh, St	¼ hard	860	125	515	75	25	...	...	A177
Sh, St	½ hard	1030	150	760	110	18	...	...	A177
Sh, St	¾ hard	1210	175	930	135	12	...	...	A177
Sh, St	Full hard	1280	185	965	140	9	...	...	A177
Type 302 (UNS S30200)									
B	Hot finished and annealed	515	75	205	30	40	50	...	A276
B	Cold finished and annealed (b)	620	90	310	45	30	40	...	A276
B	Cold finished and annealed (c)	515	75	205	30	30	40	...	A276
W	Annealed	515	75	205	30	...	...	...	A580
W	Cold finished	620	90	310	45	...	...	...	A580
P, Sh, St	Annealed	515	75	205	30	40	...	88 max	A167
P, Sh, St	High tensile, grade B	585	85	310	45	40	...	...	A666
P, Sh, St	High tensile, grade C	860	125	515	75	...	...	...	A666
P, Sh, St	High tensile, grade D	1030	150	760	110	...	...	...	A666
B, W	High tensile(d)	2240	325	...	...	...	...	...	A313
Type 302B (UNS S30215)									
B	Hot finished and annealed	515	75	205	30	40	50	...	A276
B	Cold finished and annealed (b)	620	90	310	45	30	40	...	A276
B	Cold finished and annealed (c)	515	75	205	30	30	40	...	A276
W	Annealed	515	75	205	30	...	...	...	A580
W	Cold finished	620	90	310	45	...	...	...	A580
P, Sh, St	Annealed	515	75	205	30	...	...	...	A167
Type 302Cu (UNS S30430)									
B	Annealed	450 to 585	65 to 85	...	...	...	...	...	A493
B	Lightly drafted	485 to 620	70 to 90	...	...	...	...	...	A493
W (e)	Annealed	485 to 620	70 to 90	...	...	...	...	...	A493
W (e)	Lightly drafted	485 to 620	70 to 90	...	...	...	...	...	A493
W (f)	Annealed	485 to 690	70 to 100	...	...	...	...	...	A493
W (f)	Lightly drafted	520 to 725	75 to 105	...	...	...	...	...	A493

(continued)

(a) B, bar; W, wire; P, plate; Sh, sheet; St, strip. (b) Up to 13 mm (0.5 in.) thick. (c) Over 13 mm (0.5 in.) thick. (d) Depending on size and amount of cold reduction. (e) 4 mm (0.156 in.) in diameter and over. (f) Under 4 mm (0.156 in.) in diameter. (g) Values given are typical. (h) Not a basis for acceptance or rejection

Minimum room-temperature mechanical properties of austenitic stainless steels (continued)

Product form(a)	Condition	Tensile strength MPa	ksi	0.2% yield strength MPa	ksi	Elongation, %	Reduction in area, %	Hardness, HRB	ASTM specification
Types 303 (UNS S30300) and 303Se (UNS S30323)									
B	Annealed	585(g)	85(g)	240(g)	35(g)	50(g)	55(g)	...	A581
W	Annealed	585 to 860	85 to 125	...	...	...	...	...	A581
W	Cold worked	790 to 1000	115 to 145	...	...	...	...	...	A581
Type 304 (UNS S30400)									
B	Hot finished and annealed	515	75	205	30	40	50	...	A276
B	Cold finished and annealed(b)	620	90	310	45	30	40	...	A276
B	Cold finished and annealed(c)	515	75	205	30	30	40	...	A276
W	Annealed	515	75	205	30	...	...	...	A580
W	Cold finished	620	90	310	45	...	...	...	A580
P, Sh, St	Annealed	515	75	205	30	40	...	88 max	A167
Sh, St	High tensile, grade B	550	80	310	45	...	...	...	A666
Sh, St	High tensile, grade C	860	125	515	75	...	...	...	A666
Sh, St	High tensile, grade D	1030	150	690	110	...	...	...	A666
B, W	High tensile(d)	2240	325	...	...	...	...	...	A313
Type 304L (UNS S30403)									
B	Hot finished and annealed	480	70	170	25	40	50	...	A276
B	Cold finished and annealed(b)	620	90	310	45	30	40	...	A276
B	Cold finished and annealed(c)	480	70	170	25	30	40	...	A276
W	Annealed	480	70	170	25	...	...	...	A580
W	Cold finished	620	90	310	45	...	...	...	A580
P, Sh, St	Annealed	480	70	170	25	40	...	88 max	A167
Types 304N (UNS S30451) and 316N (UNS S31651)									
B	Annealed	550	80	240	35	30	...	...	A276
Type 304LN									
B	Annealed	515	75	205	30	...	...	...	...

(continued)

(a) B, bar; W, wire; P, plate; Sh, sheet; St, strip. (b) Up to 13 mm (0.5 in.) thick. (c) Over 13 mm (0.5 in.) thick. (d) Depending on size and amount of cold reduction. (e) 4 mm (0.156 in.) in diameter and over. (f) Under 4 mm (0.156 in.) in diameter. (g) Values given are typical. (h) Not a basis for acceptance or rejection

Minimum room-temperature mechanical properties of austenitic stainless steels (continued)

Product form(a)	Condition	Tensile strength MPa	ksi	0.2% yield strength MPa	ksi	Elongation, %	Reduction in area, %	Hardness, HRB	ASTM specification
Type 305 (UNS S30500)									
B	Hot finished and annealed	515	75	205	30	40	50	...	A276
B	Cold finished and annealed(b)	620	90	310	45	30	40	...	A276
B	Cold finished and annealed(c)	515	75	205	30	30	40	...	A276
W	Annealed	515	75	205	30	...	...	...	A580
W	Cold finished	620	90	310	45	...	...	...	A580
P, Sh, St	Annealed	480	70	170	25	40	...	88 max	A167
B, W	High tensile(d)	1690	245	...	...	...	...	...	...
Types 308 (UNS S30800), 321 (UNS S32100), 347 (UNS S34700) and 348 (UNS S34800)									
B	Hot finished and annealed	515	75	205	30	40	50	...	A276
B	Cold finished and annealed(b)	620	90	310	45	30	40	...	A276
B	Cold finished and annealed(c)	515	75	205	30	30	40	...	A276
W	Annealed	515	75	205	30	...	...	...	A580
W	Cold finished	620	90	310	45	...	...	...	A580
P, Sh, St	Annealed	515	75	205	30	40	...	88 max	A167
Type 308L									
B	Annealed	550(g)	80(g)	207(g)	30(g)	60(g)	70(g)	...	...
Types 309 (UNS S30900), 309S (UNS S30908), 310 (UNS S31000) and 310S (UNS S31008)									
B	Hot finished and annealed	515	75	205	30	40	50	...	A276
B	Cold finished and annealed(b)	620	90	310	45	30	40	...	A276
B	Cold finished and annealed(c)	515	75	205	30	30	40	...	A276
W	Annealed	515	75	205	30	...	...	...	A580
W	Cold finished	620	90	310	45	...	...	...	A580
P, Sh, St	Annealed	515	75	205	30	40	...	95 max	A167
Type 312									
Weld metal	...	655	95	...	...	20	...	...	MIL-E-19933
Type 314 (UNS S31400)									
B	Hot finished and annealed	515	75	205	30	40	50	...	A276
B	Cold finished and annealed(b)	620	90	310	45	30	40	...	A276
B	Cold finished and annealed(c)	515	75	205	30	30	40	...	A276
W	Annealed	515	75	205	30	...	...	...	A580
W	Cold finished	620	90	310	45	...	...	...	A580

(continued)

(a) B, bar; W, wire; P, plate; Sh, sheet; St, strip. (b) Up to 13 mm (0.5 in.) thick. (c) Over 13 mm (0.5 in.) thick. (d) Depending on size and amount of cold reduction. (e) 4 mm (0.156 in.) in diameter and over. (f) Under 4 mm (0.156 in.) in diameter. (g) Values given are typical. (h) Not a basis for acceptance or rejection

Minimum room-temperature mechanical properties of austenitic stainless steels (continued)

Product form(a)	Condition	Tensile strength MPa	ksi	0.2% yield strength MPa	ksi	Elongation, %	Reduction in area, %	Hardness, HRB	ASTM specification
Type 316 (UNS S31600)									
B	Hot finished and annealed	515	75	205	30	40	50	...	A276
B	Cold finished and annealed(b)	620	90	310	45	30	40	...	A276
B	Cold finished and annealed(c)	515	75	205	30	30	40	...	A276
W	Annealed	515	75	205	30	...	...	...	A580
W	Cold finished	620	90	310	45	...	...	...	A580
P, Sh, St	Annealed	515	75	205	30	40	...	95 max	A580
B, W	High tensile(d)	1690	245	...	...	...	...	...	A167
Type 316F (UNS S31620)									
B	Annealed	585(g)	85(g)	240(g)	35(g)	40(g)	55(g)	...	...
Type 316L (UNS S31603)									
B	Hot finished and annealed	480	70	170	25	40	50	...	A276
B	Cold finished and annealed(b)	620	90	310	45	30	40	...	A276
B	Cold finished and annealed(c)	480	70	170	25	30	40	...	A276
W	Annealed	480	70	170	25	...	...	...	A580
W	Cold finished	620	90	310	45	...	...	...	A580
Type 316LN									
B	Annealed	515(g)	75(g)	205(g)	30(g)	60(g)	70(g)	...	...
Type 317 (UNS S31700)									
B	Hot finished and annealed	515	75	205	30	40	50	...	A276
B	Cold finished and annealed(b)	620	90	310	45	30	40	...	A276
B	Cold finished and annealed(c)	515	75	205	30	30	40	...	A276
W	Annealed	515	75	205	30	...	...	...	A580
W	Cold finished	620	90	310	45	...	...	...	A580
P, Sh, St	Annealed	515	75	205	30	35	...	95 max	A167
Type 317L (UNS S31703)									
B	Annealed	585(g)	85(g)	240(g)	35(g)	55(g)	65(g)	85 max(g)	A167
P, Sh, St	Annealed	515	75	205	30	35	...	95 max	A167
Type 317LM									
B, P, Sh, St	Annealed	515	75	205	30	35	50	95 max	...

(continued)

(a) B, bar; W, wire; P, plate; Sh, sheet; St, strip. (b) Up to 13 mm (0.5 in.) thick. (c) Over 13 mm (0.5 in.) thick. (d) Depending on size and amount of cold reduction. (e) 4 mm (0.156 in.) in diameter and over. (f) Under 4 mm (0.156 in.) in diameter. (g) Values given are typical. (h) Not a basis for acceptance or rejection

Minimum room-temperature mechanical properties of austenitic stainless steels (continued)

Product form(a)	Condition	Tensile strength MPa	ksi	0.2% yield strength MPa	ksi	Elongation, %	Reduction in area, %	Hardness, HRB	ASTM specification
Type 329 (UNS S32900)									
B	Annealed	724(g)	105(g)	550(g)	80(g)	25(g)	50(g)	...	...
Type 330 (UNS N08330)									
B	Annealed	480	70	210	30	30	...	...	B511
P, Sh, St	Annealed	480	70	210	30	30	...	75 to 85(h)	B536
Type 330HC									
B, W, St	Annealed	585(g)	85(g)	290(g)	42(g)	45(g)	65(g)	...	...
Type 332									
B, W, Sh, St	Annealed	550(g)	80(g)	240(g)	35(g)	45(g)	70(g)	...	...
Types 384 (UNS S38400) and 385 (UNS S38500)									
B	Annealed	415 to 550	60 to 80	...	...	...	...	...	A493
B	Lightly drafted	450 to 585	65 to 85	...	...	...	...	...	A493
W (e)	Annealed	450 to 585	65 to 85	...	...	...	...	...	A493
W (e)	Lightly drafted	485 to 620	70 to 90	...	...	...	...	...	A493
W (f)	Annealed	450 to 655	65 to 95	...	...	...	...	...	A493
W (f)	Lightly drafted	485 to 690	70 to 100	...	...	...	...	...	A493
904L (UNS N08904)									
B, P, Sh, St	Annealed	490	71	220	31	35	...	95 max	B625
AL-6X (UNS N08366)									
Sh, St	Annealed	515	75	205	30	30	...	...	B676
18-18-2 (UNS S38100)									
P, Sh, St	Annealed	515	75	205	30	40	...	96 max	A167
Crutemp 25									
P, Sh, St	Annealed	615(g)	89(g)	275(g)	40(g)	40(g)	...	...	...
JS-700 (UNS N08700)									
P, Sh, St	Annealed	550	80	205	30	30	40	...	B599
JS-777									
B, P, Sh, St	Annealed	550	80	240	35	30	40	95 max	...
20Cb-3 (UNS N08020)									
B	Annealed	585	85	240	35	30	50	...	B473
Shapes	Cold finished and annealed	585	85	240	35	15	50	...	B473
W	Annealed	620 to 825	90 to 120	240	35	...	...	...	B473
P, Sh, St	Annealed	585	85	275	40	30	...	95 max	B463

(a) B, bar; W, wire; P, plate; Sh, sheet; St, strip. (b) Up to 13 mm (0.5 in.) thick. (c) Over 13 mm (0.5 in.) thick. (d) Depending on size and amount of cold reduction. (e) 4 mm (0.156 in.) in diameter and over. (f) Under 4 mm (0.156 in.) in diameter. (g) Values given are typical. (h) Not a basis for acceptance or rejection

Minimum mechanical properties of high-nitrogen austenitic stainless steels

Product form(a)	Condition	Tensile strength MPa	ksi	0.2% yield strength MPa	ksi	Elongation, %	Reduction in area, %	Hardness, HRB	ASTM specification
Type 201 (UNS S20100)									
B	Annealed...............	515	75	275	40	40	45	···	A276
W, P, Sh, St	Annealed...............	655	95	310	45	40	···	···	A276, A412
Sh, St	¼ hard.................	860	125	515	75	20	···	···	A412
Sh, St	½ hard.................	1030	150	760	110	10	···	···	A412
Sh, St	¾ hard.................	1210	175	930	135	7	···	···	A412
Sh, St	Full hard...............	1280	185	965	140	5	···	···	A412
Type 202 (UNS S20200)									
B	Annealed...............	515	75	275	40	40	···	···	A276
W, P, Sh, St	Annealed...............	655	95	310	45	40	···	···	A412
Sh, St	¼ hard.................	860	125	515	75	12	···	···	A412
Sh, St	½ hard.................	1030	150	760	110	10	···	···	A666
Type 205 (UNS S20500)									
P	Annealed(b).............	830(b)	120(b)	475(b)	69(b)	58(b)	62(b)	98 max(b)	
Type 216 (UNS S21600)									
Sh, St	Annealed...............	690	100	415	60	40	···	100 max	A240
P	Annealed...............	620	90	345	50	40	···	100 max	A240
Type 304N (UNS S30451)									
B	Annealed...............	550	80	240	35	30	···	···	A276
P, Sh, St	Annealed...............	550	80	240	35	30	···	88 max	A240
Type 304HN (UNS S30452)									
B	Annealed...............	620	90	345	50	30	50		
Sh, St	Annealed...............	620	90	345	50	30	···	100 max	A240
P	Annealed...............	585	85	275	40	30	···	100 max	A240
Type 316N (UNS S31651)									
B	Annealed...............	550	80	240	35	30	···	···	A276
P, Sh, St	Annealed...............	550	80	240	35	30	···	95 max	A240
Nitronic 32 (UNS S24100)									
B	Annealed...............	690	100	380	55	30	50	···	A276
W	Annealed...............	690	100	380	55	30	50	···	A580
Nitronic 33 (UNS S24000)									
B	Annealed...............	690	100	380	55	30	50	···	A276
W	Annealed...............	690	100	380	55	30	50	···	A580
Sh, St	Annealed...............	690	100	415	60	40	···	···	A412
P	Annealed...............	690	100	380	55	40	···	···	A412
Nitronic 40 (UNS S21900)									
B	Annealed...............	550	80	345	50	45	···	···	A276
W	Annealed...............	620	90	345	50	45	60	···	A580
Sh, St	Annealed...............	690	100	415	60	40	···	···	A412
Sh, St	10% cold rolled..........	895	130	795	115	15	···	···	A412
P	Annealed...............	620	90	345	50	45	···	···	A412
Nitronic 50 (UNS S20910)									
B	Annealed...............	690	100	380	55	35	55	···	A276
W	Annealed...............	690	100	380	55	35	55	···	A580
Sh, St	Annealed...............	825	120	515	75	30	···	···	A412
P	Annealed...............	690	100	380	55	35	···	···	A412

(continued)

(a) B, bar; W, wire; P, plate; Sh, sheet; St, strip. (b) Typical values

Minimum mechanical properties of high-nitrogen austenitic stainless steels (continued)

Product form(a)	Condition	Tensile strength MPa	ksi	0.2% yield strength MPa	ksi	Elonga- tion, %	Reduction in area, %	Hardness, HRB	ASTM specification
Nitronic 60 (UNS S21800)									
B	Annealed	655	95	345	50	35	55	· · ·	A276
W	Annealed	655	95	345	50	35	55	· · ·	A580
18-18 Plus (UNS S28200)									
B	Annealed	825(b)	120(b)	450(b)	65(b)	60(b)	70(b)	95 min(b)	
B	Annealed	760	110	415	60	35	55	· · ·	A276
W	Annealed	760 to 930	110 to 135	· · ·	· · ·	· · ·	· · ·	· · ·	A493
Tenelon (UNS S21400)									
Sh	Annealed	860	125	485	70	40	· · ·	· · ·	A240
St	Annealed	725	105	380	55	40	· · ·	· · ·	A240

(a) B, bar; W, wire; P, plate; Sh, sheet; St, strip. (b) Typical values

Minimum mechanical properties of ferritic stainless steels

Product form(a)	Condition	Tensile strength MPa	ksi	0.2% yield strength MPa	ksi	Elongation, %	Reduction in area, %	Hardness, HRB	ASTM specification
Type 405 (UNS S40500)									
W	Annealed	480	70	275	40	20	45	· · ·	A580
W	Annealed, cold finished	480	70	275	40	16	45	· · ·	A580
P, Sh, St	Annealed	415	60	170	25	20	· · ·	88 max	A176
Type 409 (UNS S40900)									
B	Annealed	450(b)	65(b)	240(b)	35(b)	25(b)	· · ·	75 max(b)	
P, Sh, St	Annealed	415	60	205	30	22(c)	· · ·	80 max	A176
Type 429 (UNS S42900)									
B	Annealed	490(b)	71(b)	310(b)	45(b)	30(b)	65(b)	· · ·	
P, Sh, St	Annealed	450	65	205	30	22(c)	· · ·	88 max	A176
Type 430 (UNS S43000)									
B	Annealed, hot finished	480	70	275	40	20	45	· · ·	A276
B	Annealed, cold finished	480	70	275	40	16	45	· · ·	A276
W	Annealed	480	70	275	40	20	45	· · ·	A580
W	Annealed, cold finished	480	70	275	40	16	45	· · ·	A580
P, Sh, St	Annealed	450	65	205	30	22(c)	· · ·	88 max	A176
Type 430F (UNS S43020)									
W	Annealed	585 to 860	85 to 125	· · ·	· · ·	· · ·	· · ·	· · ·	A581
Type 430Ti (UNS S43036)									
B	Annealed	515(b)	75(b)	310(b)	45(b)	30(b)	65(b)	· · ·	· · ·
Type 434 (UNS S43400)									
W	Annealed	545(b)	79(b)	415(b)	60(b)	33(b)	78(b)	90 max(b)	· · ·
Sh	Annealed	530(b)	77(b)	365(b)	53(b)	23(b)	· · ·	83 max(b)	· · ·
Type 436 (UNS S43600)									
Sh, St	Annealed	530(b)	77(b)	365(b)	53(b)	23(b)	· · ·	83 max(b)	· · ·
Type 442 (UNS S44200)									
B	Annealed	550(b)	80(b)	310(b)	45(b)	20(b)	40(b)	90 max(b)	· · ·
P, Sh, St	Annealed	515	75	275	40	20	· · ·	95 max	A176
		(continued)							

(a) B, bar; W, wire; P, plate; Sh, sheet; St, strip. (b) Typical values. (c) 20% reduction for 1.3 mm (0.050 in.) in thickness and under

Minimum mechanical properties of ferritic stainless steels (continued)

Product form(a)	Condition	Tensile strength MPa	ksi	0.2% yield strength MPa	ksi	Elongation, %	Reduction in area, %	Hardness, HRB	ASTM specification
Type 444 (UNS S44400)									
P, Sh, St	Annealed 415		60	275	40	20	· · ·	95 max	A176
Type 446 (UNS S44600)									
B	Annealed, hot finished 480		70	275	40	20	45	· · ·	A276
B	Annealed, cold finished 480		70	275	40	16	45	· · ·	A276
W	Annealed 480		70	275	40	20	45	· · ·	A580
W	Annealed, cold finished 480		70	275	40	16	45	· · ·	A580
P, Sh, St	Annealed 515		75	275	40	20	· · ·	95 max	A176
18 SR									
Sh, St	Annealed . . . . 620(b)		90(b)	450(b)	65(b)	25(b)	· · ·	90 min(b)	
E-Brite 26-1 (UNS S44625)									
B	Annealed, hot finished 450		65	275	40	20	45	· · ·	A276
B	Annealed, cold finished 450		65	275	40	16	45	· · ·	A276
P, Sh, St	Annealed 450		65	275	40	22(c)	· · ·	90 max	A176
26-1 Ti (UNS S44626)									
P, Sh, St	Annealed 470		68	310	45	20	· · ·	95 max	A176
MONIT (UNS S44635)									
B, P, Sh, St	Annealed 650		94	550	80	20	· · ·	100 max	A176
Sea-cure/SC-1 (UNS S44600)									
B, P, Sh, St	Annealed 550		80	380	55	20	· · ·	100 max	A176
29-4 (UNS S44700)									
B, P, Sh, St	Annealed 550		80	415	60	20	· · ·	98 max	A176
29-4-2 (UNS S44800)									
B, P, Sh, St	Annealed 550		80	415	60	20	· · ·	98 max	A176

(a) B, bar; W, wire; P, plate; Sh, sheet; St, strip. (b) Typical values. (c) 20% reduction for 1.3 mm (0.050 in.) in thickness and under

Minimum mechanical properties of martensitic stainless steels

Product form(a)	Condition	Tensile strength		0.2% yield strength		Elongation, %	Reduction in area, %	Rockwell hardness	ASTM specification
		MPa	ksi	MPa	ksi				
Type 403 (UNS S40300)									
B	Annealed, hot finished	485	70	275	40	20	45	· · ·	A276
B	Annealed, cold finished	485	70	275	40	16	45	· · ·	A276
B	Intermediate temper, hot finished	690	100	550	80	15	45	· · ·	A276
B	Intermediate temper, cold finished	690	100	550	80	12	40	· · ·	A276
B	Hard temper, hot finished	825	120	620	90	12	40	· · ·	A276
B	Hard temper, cold finished	825	120	620	90	12	40	· · ·	A276
W	Annealed	485	70	275	40	20	45	· · ·	A580
W	Annealed, cold finished	485	70	275	40	16	45	· · ·	A580
W	Intermediate temper, cold finished	690	100	550	80	12	40	· · ·	A580
W	Hard temper, cold finished	825	120	620	90	12	40	· · ·	A580
P, Sh, St	Annealed	485	70	205	30	25(b)	· · ·	88 HRB max	A176
Type 404 (UNS S40400)									
B	Tempered 260 °C (500 °F)	1120(c)	162(c)	910(c)	132(c)	15(c)	50(c)	35 HRC(c)	· · ·
B	Tempered 595 °C (1100 °F)	745(c)	108(c)	655(c)	95(c)	23(c)	70(c)	20 HRC(c)	· · ·
Type 410 (UNS S41000)									
B	Annealed, hot finished	485	70	275	40	20	45	· · ·	A276
B	Annealed, cold finished	485	70	275	40	16	45	· · ·	A276
B	Intermediate temper, hot finished	690	100	550	80	15	45	· · ·	A276
B	Intermediate temper, cold finished	690	100	550	80	12	40	· · ·	A276
B	Hard temper, hot finished	825	120	620	90	12	40	· · ·	A276
B	Hard temper, cold finished	825	120	620	90	12	40	· · ·	A276
W	Annealed	485	70	275	40	20	45	· · ·	A580
W	Annealed, cold finished	485	70	275	40	16	45	· · ·	A580
W	Intermediate temper, cold finished	690	100	550	80	12	40	· · ·	A580
W	Hard temper, cold finished	825	120	620	90	12	40	· · ·	A580
P, Sh, St	Annealed	450	65	205	30	22(b)	· · ·	95 HRB max	A176
Type 410S (UNS S41008)									
P, Sh, St	Annealed	415	60	205	30	22	· · ·	95 HRB max	A176

(continued)

(a) B, bar; W, wire; P, plate; Sh, sheet; St, strip. (b) 20% elongation for 1.3 mm (0.050 in.) and under in thickness. (c) Typical values. (d) Heat treated for high-temperature service

Minimum mechanical properties of martensitic stainless steels (continued)

Product form(a)	Condition	Tensile strength MPa	ksi	0.2% yield strength MPa	ksi	Elongation, %	Reduction in area, %	Rockwell hardness	ASTM specification
Type 410Cb (UNS S41040)									
B	Annealed, hot finished	485	70	275	40	13	45	· · ·	A276
B	Annealed, cold finished	485	70	275	40	12	35	· · ·	A276
B	Intermediate temper, hot finished	860	125	690	100	13	45	· · ·	A276
B	Intermediate temper, cold finished	860	125	690	100	12	35	· · ·	A276
Type 414 (UNS S41400)									
B	Intermediate temper, hot finished	795	115	620	90	15	45	· · ·	A276
B	Intermediate temper, cold finished	795	115	620	90	15	45	· · ·	A276
W	Annealed, cold finished	1030 max	150 max	· · ·	· · ·	· · ·	· · ·	· · ·	A580
Type 414L									
B	Annealed	795(c)	115(c)	550(c)	80(c)	20(c)	60(c)	· · ·	· · ·
Types 416 (UNS S41600) and 416Se (UNS S41623)									
W	Annealed	585 to 860	85 to 125	· · ·	· · ·	· · ·	· · ·	· · ·	A581
W	Intermediate temper	795 to 1000	115 to 145	· · ·	· · ·	· · ·	· · ·	· · ·	A581
W	Hard temper	965 to 1210	140 to 175	· · ·	· · ·	· · ·	· · ·	· · ·	A581
Type 416 Plus X									
B	Annealed	515	75	275	40	30	60	· · ·	· · ·
Type 418 (UNS S41800)									
B	Tempered 260 °C (500 °F)	1450(c)	210(c)	1210(c)	175(c)	18(c)	52(c)	· · ·	· · ·
B	Tempered 650 °C (1200 °F)	930(c)	135(c)	725(c)	105(c)	20(c)	60(c)	· · ·	· · ·
Type 420 (UNS S42000)									
B	Tempered 205 °C (400 °F)	1720	250	1480(c)	215(c)	8(c)	25(c)	52 HRC(c)	· · ·
W	Annealed, cold finished	860 max	125 max	· · ·	· · ·	· · ·	· · ·	· · ·	A580
Type 422 (UNS S42200)									
B	Intermediate and hard tempers(d)	965	140	760	110	13	30	· · ·	A565
Type 431 (UNS S43100)									
B	Tempered 260 °C (500 °F)	1370(c)	198(c)	1030(c)	149(c)	16(c)	55(c)	· · ·	· · ·
B	Tempered 595 °C (1100 °F)	965(c)	140(c)	795(c)	115(c)	19(c)	57(c)	· · ·	· · ·
W	Annealed, cold finished	965 max	140 max	· · ·	· · ·	· · ·	· · ·	· · ·	A580
Type 440A (UNS S44002)									
B	Annealed	725(c)	105(c)	415(c)	60(c)	20(c)	· · ·	95 HRB(c)	· · ·
B	Tempered 315 °C (600 °F)	1790(c)	260(c)	1650(c)	240(c)	5(c)	20(c)	51 HRC(c)	· · ·
W	Annealed, cold finished	965 max	140 max	· · ·	· · ·	· · ·	· · ·	· · ·	A580

(continued)

(a) B, bar; W, wire; P, plate; Sh, sheet; St, strip. (b) 20% elongation for 1.3 mm (0.050 in.) and under in thickness. (c) Typical values. (d) Heat treated for high-temperature service

Minimum mechanical properties of martensitic stainless steels (continued)

Product form(a)	Condition	Tensile strength MPa	ksi	0.2% yield strength MPa	ksi	Elongation, %	Reduction in area, %	Rockwell hardness	ASTM specification
Type 440B (UNS S44003)									
B	Annealed	740(c)	107(c)	425(c)	62(c)	18(c)	· · ·	96 HRB(c)	· · ·
B	Tempered 315 °C (600 °F)	1930(c)	280(c)	1860(c)	270(c)	3(c)	15(c)	55 HRC(c)	· · ·
W	Annealed, cold finished	965 max	140 max	· · ·	· · ·	· · ·	· · ·	· · ·	A580
Type 440C (UNS S44004)									
B	Annealed	760(c)	110(c)	450(c)	65(c)	14(c)	· · ·	97 HRB(c)	· · ·
B	Tempered 315 °C (600 °F)	1970(c)	285(c)	1900(c)	275(c)	2(c)	10(c)	57 HRC(c)	· · ·
W	Annealed, cold finished	965 max	140 max	· · ·	· · ·	· · ·	· · ·	· · ·	A580
Type 501 (UNS S50100)									
B, P	Annealed	485(c)	70(c)	205(c)	30(c)	28(c)	65(c)	· · ·	· · ·
B, P	Tempered 540 °C (1000 °F)	1210(c)	175(c)	965(c)	140(c)	15(c)	50(c)	· · ·	· · ·
Type 502 (UNS S50200)									
B, P	Annealed	485(c)	70(c)	205(c)	30(c)	30(c)	70(c)	· · ·	· · ·

(a) B, bar; W, wire; P, plate; Sh, sheet; St, strip. (b) 20% elongation for 1.3 mm (0.050 in.) and under in thickness. (c) Typical values. (d) Heat treated for high-temperature service

Minimum mechanical properties of precipitation-hardening stainless steels

Product form(a)	Condition	Tensile strength MPa	ksi	Yield strength MPa	ksi	Elongation, %	Reduction in area(b), %	Hardness(c), HRC Min	Max
PH 13-8 Mo (UNS S13800)									
B, P, Sh, St	H950	1520	220	1410	205	6-10(c)	45	45	· · ·
B, P, Sh, St	H1000	1380	200	1310	190	6-10(c)	45	43	· · ·
15-5 PH (UNS S15500) and 17-4 PH (UNS S17400)									
B, P, Sh, St	H900	1310	190	1170	170	10(d)	35(e)	40	48
B, P, Sh, St	H925	1170	170	1070	155	10(d)	38(e)	38	47(d)
B, P, Sh, St	H1025	1070	155	1000	145	12(d)	45(e)	35(d)	42(d)
B, P, Sh, St	H1075	1000	145	860	125	13(d)	45(e)	32(d)	38(d)
B, P, Sh, St	H1100	965	140	795	115	14(d)	45(e)	31(d)	38(d)
B, P, Sh, St	H1150	930	135	725	105	16(d)	50(e)	28(d)	36(d)
B, P, Sh, St	H1150M	795	115	515	75	18(d)	55(e)	24(d)	34(d)
17-7 PH (UNS S17700)									
B	RH950	1275	185	1030	150	6	10	41	· · ·
B	TH1050	1170	170	965	140	6	25	38	· · ·
P, Sh, St	RH950	1450	210	1310	190	1-6(d)	· · ·	41(d)	44(d)
P, Sh, St	TH1050	1240	180	1030	150	3-7(d)	· · ·	38	· · ·
P, Sh, St	Cold rolled (condition C)	1380	200	1210	175	1	· · ·	41	· · ·
P, Sh, St	CH900	1650	240	1590	230	1	· · ·	46	· · ·
Custom 450 (UNS S45000)									
B	Annealed	860	125	655	95	10	40	· · ·	33
P, Sh, St	Annealed	895	130	620	90	4	· · ·	25	· · ·
B, P, Sh, St	H900	1240	180	1170	170	10	40	40	· · ·
B, P, Sh, St	H1000	1100	160	1030	150	12	45	36	· · ·
B, P, Sh, St	H1150	860	125	515	75	15	50	26	· · ·

(continued)

(a) B, bar; P, plate; Sh, sheet; St, strip; W, wire; F, forgings. (b) Values are for bar products. (c) Where minimum value is also given, maximum value applies only to flat-rolled products. Both max and min values may vary with thickness for flat-rolled products. (d) Value varies with thickness for flat-rolled products. (e) Value generally lower for flat-rolled products and varies with thickness. (f) Values are typical. (g) Rockwell B hardness

Minimum mechanical properties of precipitation-hardening stainless steels (continued)

Product form(a)	Condition	Tensile strength MPa	ksi	Yield strength MPa	ksi	Elongation, %	Reduction in area(b), %	Hardness(c), HRC Min	Max
Custom 455 (UNS S45500)									
B	H900............ 1620		235	1520	220	8	30	47	...
B	H950............ 1520		220	1410	205	10	40	44	...
P, Sh, St	H950............ 1530		222	1410	205	Up to 4	...	44	...
AM-350 (UNS S35000)									
P, Sh, St	H850............ 1275		185	1030	150	2-8	...	42	...
P, Sh, St	H1000........... 1140		165	1000	145	2-8	...	36	...
AM-355 (UNS S35500)									
B	Equalize plus overtemper 540 °C (1000 °F).. 1170		170	1070	155	12	25	39	...
P, Sh, St	H850............ 1310		190	1140	165	10	...	37	...
P, Sh, St	H1000........... 1170		170	1030	150	12	...	28(f)	...
AM-363									
St	Annealed .... 850(f)		123(f)	730(f)	106(f)	12(f)	...	...	
Stainless W (UNS S17600)									
B, P, Sh, St	H950............ 1310		190	1170	170	8	25	39	...
B, P, Sh, St	H1000........... 1240		180	1100	160	8	30	37	...
B, P, Sh, St	H1050........... 1170		170	1070	150	10	40	35	...
PH 15-7 Mo (UNS S15700)									
B	RH950 1380		200	1210	175	7	25	...	...
B	TH1050.......... 1240		180	1100	160	8	25	...	...
P, Sh, St	RH950 1550		225	1380	200	1-5(d)	...	43(d)	46(d)
P, Sh, St	TH1050.......... 1310		190	1170	170	2-5(d)	...	38(d)	40(d)
P, Sh, St	Cold rolled 1380		200	1210	175	1	...	41	...
P, Sh, St	Cold rolled and aged......... 1650		240	1590	230	1	...	46	...
17-10 P									
B	Annealed .... 615(f)		89(f)	255(f)	37(f)	70(f)	76(f)	82(f)(g)	...
B	Aged .... 945(f)		137(f)	605(f)	88(f)	25(f)	39(f)	30(f)	...
HNM									
B, W, F	Aged at 705 °C (1300 °F).. 825		120	550	80	18	30	...	...

(a) B, bar; P, plate; Sh, sheet; St, strip; W, wire; F, forgings. (b) Values are for bar products. (c) Where minimum value is also given, maximum value applies only to flat-rolled products. Both max and min values may vary with thickness for flat-rolled products. (d) Value varies with thickness for flat-rolled products. (e) Value generally lower for flat-rolled products and varies with thickness. (f) Values are typical. (g) Rockwell B hardness

Heat Treating

Procedures for hardening and tempering wrought martensitic stainless steels to specific strength and hardness levels

Type	Austenitizing(a) Temperature(b) °C	°F	Quenching medium(c)	Tempering temperature(d) °C min	°C max	°F min	°F max	Tensile strength MPa	ksi	Hardness, HRC
403, 410	925-1010	1700-1850	Air or oil	565	605	1050	1125	760-965	110-140	25-31
				205	370	400	700	1105-1515	160-220	38-47
414	925-1050	1700-1925	Air or oil	595	650	1100	1200	760-965	110-140	25-31
				230	370	450	700	1105-1515	160-220	38-49
416, 416(Se)	925-1010	1700-1850	Oil	565	605	1050	1125	760-965	110-140	25-31
				230	370	450	700	1105-1515	160-220	35-45
420	985-1065	1800-1950	Air or oil(e)	205	370	400	700	1550-1930	225-280	48-56
431	985-1065	1800-1950	Air or oil(e)	565	605	1050	1125	860-1035	125-150	26-34
				230	370	450	700	1210-1515	175-220	40-47
440A	1010-1065	1850-1950	Air or oil(e)	150	370	300	700	. . .	. . .	49-57
440B	1010-1065	1850-1950	Air or oil(e)	150	370	300	700	. . .	. . .	53-59
440C, 440F	1010-1065	1850-1950	Air or oil(e)	. . .	160	. . .	325	. . .	. . .	60 min
				. . .	190	. . .	375	. . .	. . .	58 min
				. . .	230	. . .	450	. . .	. . .	57 min
				. . .	355	. . .	675	. . .	. . .	52-56

(a) Preheating to a temperature within the process annealing range (see Table 5) is recommended for thin-gage parts, heavy sections, previously hardened parts, parts with extreme variations in section or with sharp re-entrant angles, and parts that have been straightened or heavily ground or machined, to avoid cracking and minimize distortion, particularly for types 420, 431, and 440A, B, C and F. (b) Usual time at temperature ranges from 30 to 90 min. The low side of the austenitizing range is recommended for all types subsequently tempered to 25 to 31 HRC; generally, however, corrosion resistance is enhanced by quenching from the upper limit of the austenitizing range. (c) Where air or oil is indicated, oil quenching should be used for parts more than 6.4 mm ($^1/_4$ in.) thick; martempering baths at 150 to 400 °C (300 to 750 °F) may be substituted for an oil quench. (d) Generally, the low end of the tempering range of 150 to 370 °C (300 to 700 °F) is recommended for maximum hardness, the middle for maximum toughness, and the high end for maximum yield strength. Tempering in the range of 370 to 565 °C (700 to 1050 °F) is not recommended, because it results in low and erratic impact properties and poor resistance to corrosion and stress corrosion. (e) For minimum retained austenite and maximum dimensional stability, a subzero treatment −75 °C ±10 °C (−100 °F ±20 °F) is recommended; this should incorporate continuous cooling from the austenitizing temperature to the cold transformation temperature.

Annealing temperatures and procedures for wrought martensitic stainless steels

Type	Process (subcritical) annealing Temperature(a), °C	Hardness	Full annealing Temperature(b)(c), °C	Hardness	Isothermal annealing(c) Procedure (d)	Hardness
403, 410	650-760	86-92 HRB	830-885	75-85 HRB	Heat to 830 to 885 °C; hold 6 h at 705 °C	85 HRB
414	650-730	99 HRB-24 HRC	Not recommended		Not recommended	
416, 416(Se)	650-760	86-92 HRB	830-885	75-85 HRB	Heat to 830 to 885 °C; hold 2 h at 720 °C	85 HRB
420	675-760	94-97 HRB	830-885	86-95 HRB	Heat to 830 to 885 °C; hold 2 h at 705 °C	95 HRB
431	620-705	99 HRB-30 HRC	Not recommended		Not recommended	
440A	675-760	90 HRB-22 HRC	845-900	94-98 HRB	Heat to 845 to 900 °C; hold 4 h at 690 °C	98 HRB
440B	675-760	98 HRB-23 HRC	845-900	95 HRB-20 HRC	Same as 440A	20 HRC
440C, 440F	675-760	98 HRB-23 HRC	845-900	98 HRB-25 HRC	Same as 440A	25 HRC

(a) Air cool from temperature; maximum softness is obtained by heating to temperature at high end of range. (b) Soak thoroughly at temperature within range indicated; furnace cool to 790 °C; continue cooling at 15 to 25 °C/h to 595 °C; air cool to room temperature. (c) Recommended for applications in which full advantage may be taken of the rapid cooling to the transformation temperature and from it to room temperature. (d) Preheating to a temperature within the process annealing range is recommended for thin-gage parts, heavy sections, previously hardened parts, parts with extreme variations in section or with sharp re-entrant angles, and parts that have been straightened or heavily ground or machined to avoid cracking and minimize distortion, particularly for types 420 and 431, and 440A, B, C and F.

Recommended annealing temperatures for austenitic stainless steels

UNS No.	Designation	Temperature(a) °C	°F
Conventional grades			
S30100, S30200, S30215 ..	301, 302, 302B	1010 to 1120	1850 to 2050
S30300, S30323	303, 303Se	1010 to 1120	1850 to 2050
S30400, S30500, S30800 ..	304, 305, 308	1010 to 1120	1850 to 2050
S30900, S30908	309, 309S	1040 to 1120	1900 to 2050
S31000, S31008	310, 310S	1040 to 1065	1900 to 1950
S31600	316	1040 to 1120	1900 to 2050
S31700	317	1065 to 1120	1950 to 2050
Stabilized grades			
S32100	321	955 to 1065	1750 to 1950
S34700, S34800	347, 348	980 to 1065	1800 to 1950
N08020	Carpenter 20Cb-3	925 to 955	1700 to 1750
Low-carbon grades			
S30403	304L, 304LN	1010 to 1120	1850 to 2050
S31603, S31703	316L, 316LN, 317L	1040 to 1110	1900 to 2025
High-nitrogen grades			
S20100, S20200	201, 202	1010 to 1120	1850 to 2050
S30451	304N	1010 to 1120	1850 to 2050
S31651	316N	1010 to 1120	1850 to 2050
S24100	Nitronic 32, Carpenter 18Cr-2Ni-12Mn	1010 to 1065	1850 to 1950
S24000	Nitronic 33	1040 to 1095	1900 to 2000
S21904	Nitronic 40, Carpenter 21Cr-6Ni-9Mn	980 to 1175	1800 to 2150
S20910	Nitronic 50, Carpenter 22Cr-13Ni-5Mn	1065 to 1120	1950 to 2050
S21800	Nitronic 60	1040 to 1095	1900 to 2000
S28200	Carpenter 18-18 PLUS	1040 to 1095	1900 to 2000
Highly alloyed grades			
···	317LM, 317LX, 317L PLUS, 317LMO, 7L4	1120 to 1150	2050 to 2100
···	JS700, JS777	1065 to 1150	1950 to 2100
N08904	904L, AL-4X, 2RK65	1075 to 1125	1965 to 2055
N08028	Sanicro 28	···	···
N08366	AL-6X	1205 to 1230	2200 to 2250
S31254	254 SMO	1150 to 1205	2100 to 2200

(a) Temperatures given are for annealing a composite structure. Time at temperature and method of cooling depend on thickness. Light sections may be held at temperature for 3 to 5 min per 2.5 mm (0.10 in.) of thickness, followed by rapid air cooling. Thicker sections are water quenched. For many of these grades, a postweld heat treatment is not necessary. For proprietary alloys, alloy producers may be consulted for details. Although cooling from the annealing temperature must be rapid, it must also be consistent with limitations of distortion.

Recommended annealing treatments for ferritic stainless steels

UNS No.	Designation	Treatment temperature	
		°C	°F
Conventional ferritic grades			
S40500 405		650 to 815	1200 to 1500
S40900 409		870 to 900	1600 to 1650
S43000 430		705 to 790	1300 to 1450
S43020 . . . . 430F		705 to 790	1300 to 1450
S43400 434		705 to 790	1300 to 1450
S44600 446		760 to 830	1400 to 1525
Low-interstitial ferritic grades			
S43035 439		870 to 925	1600 to 1700
S44400 444		955 to 1010	1750 to 1850
S44626 . . . . E-BRITE		760 to 955	1400 to 1750
S44660 . . . . SEA-CURE, SC-1		1010 to 1065	1850 to 1950
. . . . AL 29-4C		1010 to 1065	1850 to 1950
S44800 . . . . Al 29-4-2		1010 to 1065	1850 to 1950
S44635 . . . . MONIT		1010 to 1065	1850 to 1950

Postweld heat treating of the low interstitial ferritic stainless steels is generally unnecessary and frequently undesirable. Any annealing of these grades should be followed by water quenching or very rapid cooling.

Machining

Machinability of wrought stainless steels

	Machining speeds, ft/min(a), for type							
Operation or type of cutting machine	403 (b), 405, 410 (b): 180 to 240 HB	416 (b): 180 to 240 HB	420, 420F (c): 180 to 230 HB	430: 170 to 230 HB	430F: 170 to 230 HB	414 (b), 431: 230 to 280 HB	440A, 440B, 440C, 440F (c): 200 to 265 HB	446: 170 to 230 HB
Automatic screw machine(d)	90 to 100	120 to 150	80 to 100	90 to 100	120 to 150	80 to 100	60 to 80	80 to 100
Heavy-duty single or multiple spindle(d)	80 to 100	110 to 130	60 to 80	80 to 100	110 to 130	70 to 90	50 to 70	60 to 80
Turret lathe(d)	80 to 100	100 to 130	60 to 80	80 to 100	110 to 130	70 to 90	50 to 70	60 to 80
Automatic screw machine(e)	110 to 140	120 to 150	90 to 120	110 to 140	120 to 150	100 to 140	60 to 100	100 to 140
Milling(f)	40 to 60	50 to 80	30 to 50	40 to 60	50 to 80	40 to 60	30 to 50	40 to 60
Reaming(f)								
Smooth finish	15 to 40	15 to 40	15 to 40	15 to 40	15 to 40	15 to 40	15 to 40	15 to 40
Work sizing	40 to 120	40 to 120	40 to 120	40 to 120	40 to 120	40 to 120	40 to 120	40 to 120
Threading(g)	10 to 25	10 to 25	10 to 25	10 to 25	10 to 25	10 to 25	10 to 25	10 to 25
Tapping(g)	10 to 25	10 to 25	10 to 25	10 to 25	10 to 25	10 to 25	10 to 25	10 to 25
Drill press(g)	40 to 80	60 to 90	30 to 50	40 to 80	60 to 90	40 to 60	30 to 50	40 to 60
Single-point turning:								
Carbide tooling:								
Roughing	150 to 200	150 to 200	100 to 150	150 to 200	150 to 200	140 to 180	100 to 150	140 to 180
Finishing	200 to 400	200 to 400	150 to 250	200 to 400	200 to 400	150 to 350	150 to 200	150 to 350
High-cobalt or cast alloy tooling:								
Roughing	100 to 130	100 to 150	80 to 100	100 to 130	100 to 150	90 to 120	60 to 80	100 to 130
Finishing	100 to 150	150 to 200	100 to 150	100 to 150	150 to 200	90 to 140	80 to 100	100 to 150
High speed steel tooling:								
Roughing	80 to 100	80 to 100	60 to 80	80 to 100	80 to 100	60 to 80	40 to 60	60 to 90
Finishing	80 to 130	100 to 150	80 to 120	80 to 130	100 to 150	80 to 100	60 to 80	90 to 120

	Machining speeds, ft/min(a), for type						
Operation or type of cutting machine	301, 302, 304, 304L: 150 to 250 HB	303: 150 to 240 HB	309, 309S, 310, 310S, 316, 316L: 150 to 240 HB	321, 347: 150 to 240 HB	347F: 150 to 240 HB	17-4 PH: 300 to 360 HB	17-7 PH: 150 to 240 HB
Automatic screw machine(d)	70 to 90	100 to 130	60 to 80	70 to 90	90 to 110	60 to 80	60 to 80
Heavy-duty single or multiple spindle(d)	60 to 80	90 to 120	60 to 80	60 to 80	80 to 100	50 to 70	50 to 70
Turret lathe(d)	60 to 80	90 to 120	60 to 80	60 to 80	80 to 100	50 to 70	50 to 70
Automatic screw machine(e)	80 to 120	110 to 130	80 to 120	80 to 120	100 to 120	80 to 120	80 to 120
Milling(f)	40 to 60	40 to 60	30 to 50	40 to 60	40 to 60	40 to 60	40 to 60
Reaming(f)							
Smooth finish	15 to 40	15 to 40	15 to 40	15 to 40	15 to 40	15 to 40	15 to 40
Work sizing	40 to 80	40 to 120	40 to 80	40 to 80	40 to 80	40 to 80	40 to 80
Threading(g)	10 to 25	10 to 25	10 to 25	10 to 25	10 to 25	10 to 25	10 to 25
Tapping(g)	10 to 25	10 to 25	10 to 25	10 to 25	10 to 25	10 to 25	10 to 25
Drill press(g)	30 to 50	50 to 80	30 to 50	30 to 50	30 to 50	40 to 60	40 to 60
Single-point turning:							
Carbide tooling:							
Roughing	130 to 180	150 to 250	130 to 180	130 to 180	150 to 250	130 to 180	130 to 180
Finishing	150 to 300	200 to 400	150 to 300	150 to 300	200 to 400	150 to 300	150 to 300
High-cobalt or cast alloy tooling:							
Roughing	100 to 130	100 to 150	100 to 130	100 to 130	100 to 140	100 to 130	100 to 130
Finishing	100 to 150	150 to 200	100 to 150	100 to 150	140 to 190	100 to 150	100 to 150
High speed steel tooling:							
Roughing	60 to 90	70 to 90	60 to 90	60 to 90	60 to 90	60 to 90	60 to 90
Finishing	100 to 120	100 to 140	100 to 120	100 to 120	100 to 130	100 to 120	100 to 120

(a) To obtain equivalent values in m/min, multiply listed values by 0.3. (b) Harder stock in the 260 to 320 HB range may be machined by reducing these speeds approximately 20%. (c) When using an automatic screw machine, cutting speeds may be increased about 10% over those shown. (d) Based on tungsten or molybdenum high speed steel tooling. Rates may be increased 15 to 30% with high-cobalt or cast alloys. (e) Based on the use of tools made of cemented carbide or cast cobalt-chromium-tungsten alloy. (f) Based on tungsten or molybdenum high speed steel tooling. Greatly increased speeds can be used with carbide tooling. (g) Based on tungsten or molybdenum high speed steel tooling.

Aluminum and Aluminum Alloys

Wrought Aluminum

Temper designations for aluminum alloys

F   **As fabricated.** Applies to products of shaping processes in which no special control over thermal conditions or strain-hardening is employed.

H   **Strain-hardened (wrought products only).** Applies to products which have their strength increased by strain-hardening, with or without supplementary thermal treatments to produce some reduction in strength.

H1   **Strain-hardened only.** Applies to products which are strain-hardened to obtain the desired strength without supplementary thermal treatment.

H111   Applies to products which are strain-hardened less than the amount required for a controlled H11 temper.

H112   Applies to products which acquire some temper from shaping processes not having special control over the amount of strain-hardening or thermal treatment, but for which there are mechanical property limits.

H2   **Strain-hardened and partially annealed.** Applies to products which are strain-hardened more than the desired final amount, and then reduced in strength to the desired level by partial annealing. For alloys that age-soften at room temperature, the H2 tempers have the same minimum tensile strength as the corresponding H3 tempers. For other alloys, the H2 tempers have the same minimum tensile strength as the corresponding H1 tempers and slightly higher elongation. The number following this designation indicates the degree of strain-hardening remaining after the product has been partially annealed.

H3   **Strain-hardened and stabilized.** Applies to products which are strain-hardened and whose mechanical properties are stabilized by a low temperature thermal treatment which results in slightly lowered tensile strength and improved ductility. This designation is applicable only to those alloys which, unless stabilized, gradually age-soften at room temperature. The number following this designation indicates the degree of strain-hardening before the stabilization treatment.

H311   Applies to products which are strain-hardened less than the amount required for a controlled H31 temper.

H321   Applies to products which are strain-hardened less than the amount required for a controlled H32 temper.

H323
H343   Applies to products which are specially fabricated to have acceptable resistance to stress corrosion cracking.

O   **Annealed.** Applies to wrought products which are annealed to obtain the lowest strength temper, and to cast products which are annealed to improve ductility and dimensional stability.

T   **Thermally treated to produce stable tempers other than F, O or H.** Applies to products which are thermally treated, with or without supplementary strain-hardening, to produce stable tempers.

T1   **Cooled from an elevated temperature shaping process and naturally aged to a substantially stable condition.** Applies to products which are not cold worked after cooling from an elevated temperature shaping process, or in which the effect of cold work in flattening or straightening may not be recognized in mechanical property limits.

T2   **Cooled from an elevated temperature shaping process, cold worked, and naturally aged to a substantially stable condition.** Applies to products which are cold worked to improve strength after cooling from an elevated temperature shaping process, or in which the effect of cold work in flattening or straightening is recognized in mechanical property limits.

T3   **Solution heat-treated, cold worked, and naturally aged to a substantially stable condition.** Applies to products which are cold worked to improve strength after solution heat-treatment, or in which the effect of cold work in flattening or straightening is recognized in mechanical property limits.

T4   **Solution heat-treated and naturally aged to a substantially stable condition.** Applies to products which are not cold worked after solution heat-treatment, or in which the effect of cold work in flattening or straightening may not be recognized in mechanical property limits. (T42 indicates material is solution heat-treated from the O or F temper to demonstrate response to heat-treatment, and naturally aged to a substantially stable condition.)

T5   **Cooled from an elevated temperature shaping process and then artificially aged.** Applies to products which are not cold worked after cooling from an elevated temperature shaping process, or in which the effect of cold work in flattening or straightening may not be recognized in mechanical property limits.

T51   **Stress relieved by stretching.** Applies to the following products when stretched the indicated amounts after solution heat-treatment or cooled from an elevated temperature shaping process.
Plate .... 1½ to 3% permanent set
Rod, bar, shapes, extruded tube.... 1 to 3% permanent set

(continued)

Temper designations for aluminum alloys (continued)

Drawn tube .... ½ to 3% permanent set

Applies directly to plate and rolled or cold-finished rod and bar, which receive no further straightening after stretching. Applies to extruded rod, bar, shapes, tubing, and to drawn tubing when designated as follows:

T510 Products that receive no further straightening after stretching.

T511 Products that may receive minor straightening after stretching to comply with standard tolerances.

T52 Stress-relieved by compressing. Applies to products which are stress-relieved by compressing after solution heat-treatment, or cooled from an elevated temperature shaping process to produce a permanent set of 1 to 5%.

T54 Stress-relieved by combined stretching and compressing.

Applies to die forgings which are stress relieved by restriking cold in the finish die.

T6 Solution heat-treated and then artificially aged. Applies to products which are not cold worked after solution heat-treatment, or in which the effect of cold work in flattening or straightening may not be recognized in mechanical property limits. (T62 indicates material is solution heat-treated from the O or F temper to demonstrate response to heat-treatment, and artificially aged.)

T7 Solution heat-treated and stabilized. Applies to products which are stabilized after solution heat-treatment to carry them beyond the point of maximum strength to provide control of some special characteristic.

T8 Solution heat-treated, cold worked, and artificially aged. Applies to products which are cold worked to improve strength, or in which the effect of cold work in flattening or straightening is recognized in mechanical property limits.

T9 Solution heat-treated, artificially aged, and cold worked. Applies to products which are cold worked to improve strength.

T10 Cooled from an elevated temperature shaping process, cold worked, and artificially aged. Applies to products which are cold worked to improve strength, or in which the effect of cold work in flattening or straightening is recognized in mechanical property limits.

W Solution heat-treated. An unstable temper applicable only to alloys which spontaneously age at room temperature after solution heat-treatment.

Source: Aluminum Association

Approximate correlation between Brinell and Rockwell hardness of wrought aluminum alloys

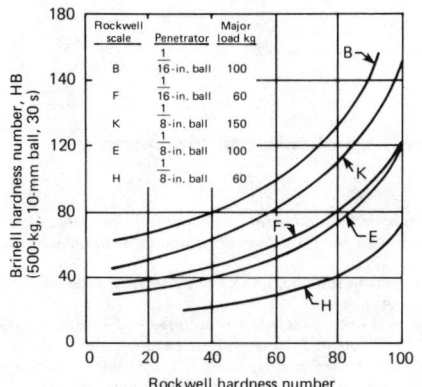

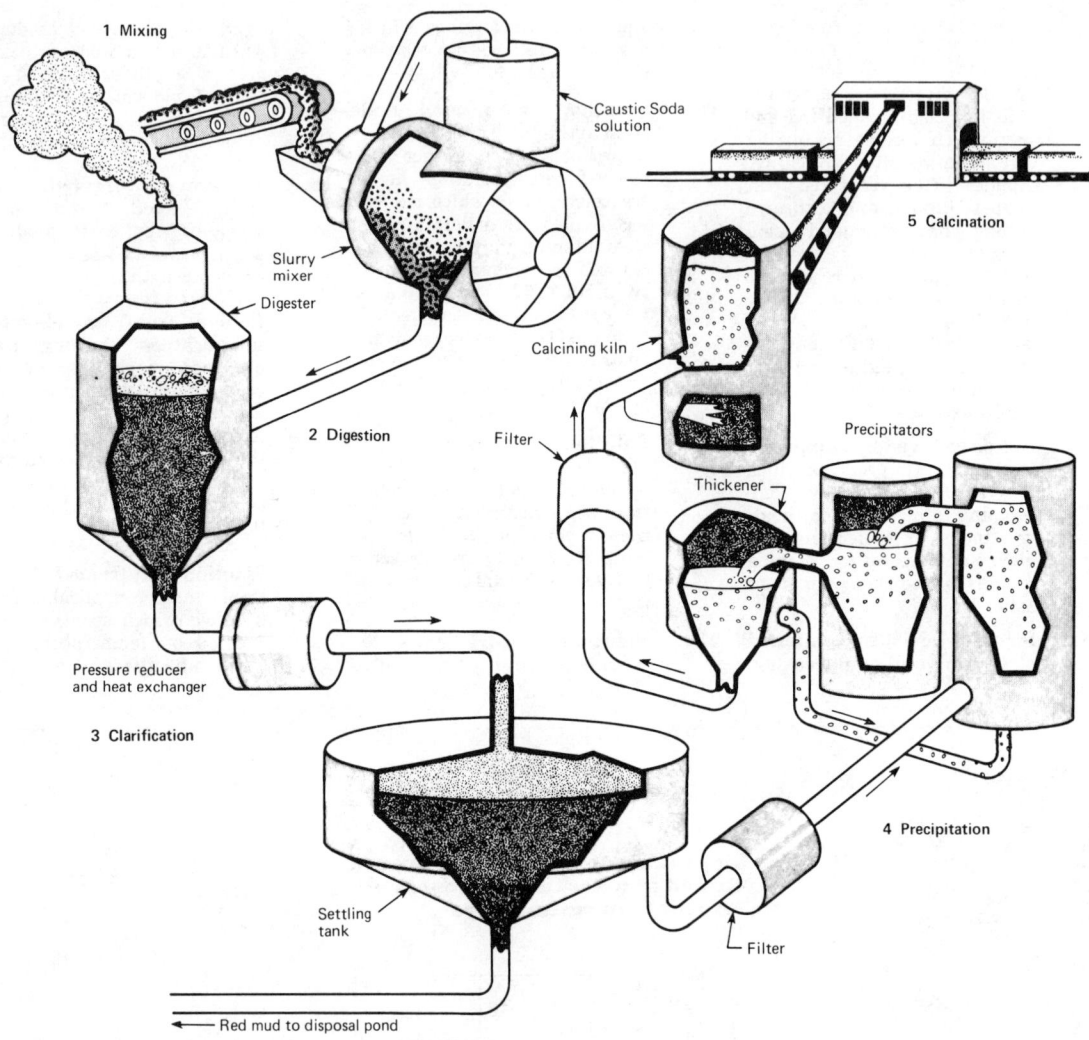

1 Mixing

Caustic Soda solution

5 Calcination

Slurry mixer

Digester

Calcining kiln

2 Digestion

Filter

Precipitators

Thickener

Pressure reducer and heat exchanger

3 Clarification

Settling tank

4 Precipitation

Filter

Red mud to disposal pond

Refining bauxite to alumina; smelting alumina to aluminum

1 **Mixing:** crushed and mixed with caustic soda, bauxite (aluminum ore) is pumped into huge digesters

2 **Digestion:** under high pressure and heat, the caustic soda dissolves the alumina, or aluminum oxide, in the bauxite to form sodium aluminate

3 **Clarification:** while the sodium aluminate remains in solution, iron oxides and other solid impurities drop to the bottom of the settling tank, where, as red mud, they are pumped to a disposal pond

4 **Precipitation:** after the liquid sodium aluminate is further cooled, it is agitated and seeded with aluminum hydroxide crystals. These form larger crystals, which gradually settle out of the solution. Seed crystals and sodium aluminate remaining in solution are recirculated

5 **Calcination:** The aluminum hydroxide crystals are roasted at more than 980 °C (1800 °F) to remove the water. A fine white powder, alumina, remains—half aluminum and half oxygen—ready for transport to a smelter

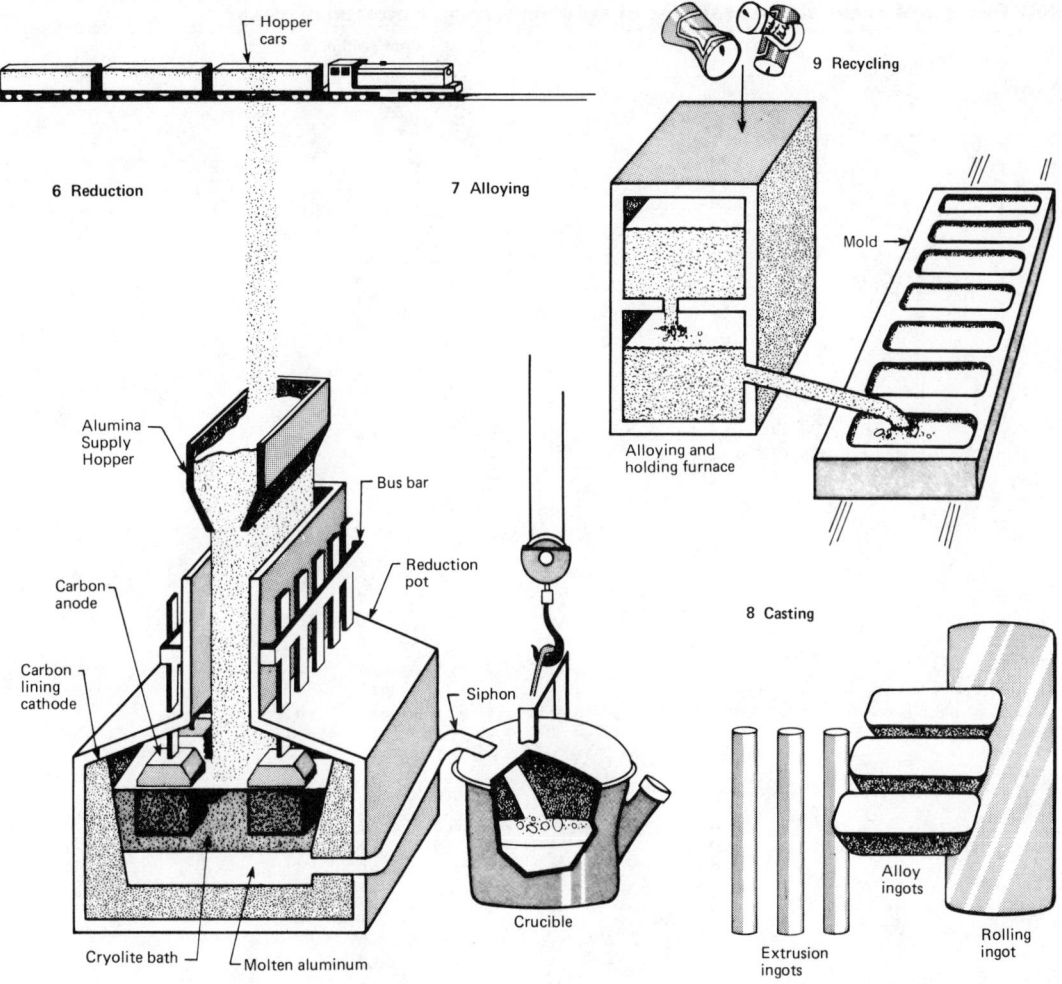

6 Reduction

Hopper cars

7 Alloying

9 Recycling

Mold

Alumina Supply Hopper

Bus bar

Reduction pot

Carbon anode

Carbon lining cathode

Siphon

Alloying and holding furnace

8 Casting

Crucible

Cryolite bath

Molten aluminum

Extrusion ingots

Alloy ingots

Rolling ingot

6 Reduction: the alumina is dissolved in a molten cryolite bath that acts as an electrolyte, in which a powerful electric current wrests aluminum from the oxygen. Molten metal settles to the bottom of the pot
7 Alloying: carried by crucible to a furnace, aluminum is alloyed with small amounts of other metals. Copper adds strength; magnesium imparts additional marine-corrosion resistance
8 Casting: molten aluminum is cast into various shapes, from 180-kN (20-ton) rolling ingots for sheet metal to 20-N (4-lb) alloy ingots for further casting
9 Recycling: nearly indestructible, aluminum can be remelted over and over. Depending on energy used to collect and transport cans and scrap, recycling saves up to 95% of the energy used to make aluminum from bauxite
Source: National Geographic Society

Product forms and nominal compositions of common wrought aluminum alloys

AA number	Product(a)	Composition, % Al	Si	Cu	Mn	Mg	Cr	Zn	Others
1050	DT	99.50 min	...	...	...	...	...	...	...
1060	S, P, ET, DT	99.60 min	...	...	...	...	...	...	...
1100	S, P, F, E, ES, ET, C, DT, FG	99.00 min	...	0.12	...	...	...	...	...
1145	S, P, F	99.45 min	...	...	...	...	...	...	...
1199	F	99.99 min	...	...	...	...	...	...	...
1350	S, P, E, ES, ET, C	99.50 min	...	...	...	...	...	...	...
2011	E, ES, ET, C, DT	93.7	...	5.5	...	...	...	...	0.4Bi; 0.4Pb
2014	S, P, E, ES, ET, C, DT, FG	93.5	0.8	4.4	0.8	0.5	...	...	...
2024	S, P, E, ES, ET, C, DT	93.5	...	4.4	0.6	1.5	...	...	...
2036	S	96.7	...	2.6	0.25	0.45	...	...	...
2048	S, P	94.8	...	3.3	0.4	1.5	...	...	...
2124	P	93.5	...	4.4	0.6	1.5	...	...	...
2218	FG	92.5	...	4.0	...	1.5	...	...	2.0Ni
2219	S, P, E, ES, ET, C, FG	93.0	...	6.3	0.3	...	...	...	0.06Ti; 0.10V; 0.18Zr
2319	C	93.0	...	6.3	0.3	...	...	...	0.18Zn, 0.15Ti, 0.10V
2618	FG	93.7	0.18	2.3	...	1.6	...	...	1.1Fe; 1.0Ni; 0.07Ti
3003	S, P, F, E, ES, ET, C, DT, FG	98.6	...	0.12	1.2	...	...	...	...
3004	S, P, ET, DT	97.8	...	...	1.2	1.0	...	...	...
3105	S	99.0	...	...	0.55	0.50	...	...	...
4032	FG	85.0	12.2	0.9	...	1.0	...	...	0.9Ni
4043	C	94.8	5.2	...	...	...	...	...	...
5005	S, P, C	99.2	...	...	...	0.8	...	...	...
5050	S, P, C, DT	98.6	...	...	...	1.4	...	...	...
5052	S, P, F, C, DT	97.2	...	...	...	2.5	0.25	...	...
5056	F, C	95.0	...	...	0.12	5.0	0.12	...	...
5083	S, P, E, ES, ET, FG	94.7	...	...	0.7	4.4	0.15	...	...
5086	S, P, E, ES, ET, DT	95.4	...	...	0.4	4.0	0.15	...	...
5154	S, P, E, ES, ET, C, DT	96.2	...	...	...	3.5	0.25	...	...
5182	S	95.2	...	...	0.35	4.5	...	...	...
5252	S	97.5	...	...	...	2.5	...	...	...
5254	S, P	96.2	...	...	...	3.5	0.25	...	...
5356	C	94.6	...	...	0.12	5.0	0.12	...	0.13Ti
5454	S, P, E, ES, ET	96.3	...	...	0.8	2.7	0.12	...	...
5456	S, P, E, ES, ET, DT, FG	93.9	...	...	0.8	5.1	0.12	...	...
5457	S	98.7	...	...	0.3	1.0	...	...	...
5652	S, P	97.2	...	...	...	2.5	0.25	...	...
5657	S	99.2	...	...	...	0.8	...	...	...
6005	E, ES, ET	98.7	0.8	...	...	0.5	...	...	...
6009	S	97.7	0.8	0.35	0.5	0.6	...	...	...
6010	S	97.3	1.0	0.35	0.5	0.8	...	...	...
6061	S, P, E, ES, ET, C, DT, FG	97.9	0.6	0.28	...	1.0	0.2	...	...
6063	E, ES, ET, DT	98.9	0.4	...	...	0.7	...	...	...
6066	E, ES, ET, DT, FG	95.7	1.4	1.0	0.8	1.1	...	...	...
6070	E, ES, ET	96.8	1.4	0.28	0.7	0.8	...	...	...
6101	E, ES, ET	98.9	0.5	...	...	0.6	...	...	...
6151	FG	98.2	0.9	...	...	0.6	0.25	...	...
6201	C	98.5	0.7	...	...	0.8	...	...	...
6205	E, ES, ET	98.4	0.8	...	0.1	0.5	0.1	...	0.1Zr
6262	E, ES, ET, C, DT	96.8	0.6	0.28	...	1.0	0.09	...	0.6Bi; 0.6Pb
6351	E, ES	97.8	1.0	...	0.6	0.6	...	...	...
6463	E, ES	98.9	0.4	...	...	0.7	...	...	...
7005	E, ES	93.3	...	...	0.45	1.4	0.13	4.5	0.04Ti; 0.14Zr
7049	P, E, ES, FG	88.2	...	1.5	...	2.5	0.15	7.6	...
7050	P, E, ES, FG	89.0	...	2.3	...	2.3	...	6.2	0.12Zr
7072	S, F	99.0	...	...	...	...	...	1.0	...
7075	S, P, E, ES, ET, C, DT, FG	90.0	...	1.6	...	2.5	0.23	5.6	...
7175	S, P, FG	90.0	...	1.6	...	2.5	0.23	5.6	...
7178	S, P, E, ES, C	88.1	...	2.0	...	2.7	0.26	6.8	...
7475	S, P, FG	90.3	1.5	...	...	2.3	0.22	5.7	...

(a) S = sheet; P = plate; F = foil; E = extruded rod, bar and wire; ES = extruded shapes; ET = extruded tubes; C = cold finished rod, bar and wire; DT = drawn tube; FG = forgings.

Typical physical properties of wrought aluminum alloys

Alloy	Temper	Electrical conductivity(a) Volume	Weight	Electrical resistivity(b) nΩm	ohms(d)	Thermal conductivity(c) W/m·K	Btu/ ft·h·°F
1050	O	61	190	28	17	231	133
1060	O	62	204	28	17	234	135
	H18	61	201	28	17	234	135
1100	O	59	194	29	18	222	128
	H18	57	187	30	18	218	126
1145	O	61	202	28	17	230	133
	H18	60	198	29	18	227	131
1199	O	65	215	27	16	243	140
1350	O	62	204	28	17	234	135
	H1x(e)	61	201	28	17	230	133
2011	T3, T4	39	123	44	27	152	88
	T8	45	142	38	23	173	100
2014	O	50	159	34	21	192	111
	T3, T4	34	108	51	31	134	77
	T6	40	127	43	26	155	89
2024	O	50	160	34	21	190	110
	T3, T4	30	96	57	35	120	69
	T6	37	119	46	28	145	84
	T8	39	125	44	27	152	88
2036	O	52	169	33	20	198	114
	T4	41	135	42	25	159	92
2048	T851	42	137	40	24	159	92
2124	O	50	161	35	21	191	110
	T851	39	126	44	27	152	88
2218	T61	38	121	45	27	148	86
	T72	40	128	43	26	155	90
2219	O	44	138	39	24	170	98
	T31, T37, T351	28	88	62	37	116	67
	T62, T81, T87, T851	30	95	57	35	130	75
2319	O	44	139	39	24	170	98
2618	T61	37	120	47	28	146	84
3003	O	47	154	37	22	180	104
	H12	42	138	41	25	162	94
	H14	41	134	42	25	159	92
	H18	40	130	43	26	155	90
3004	O (all)	42	137	41	25	162	94
3105	O (all)	45	148	38	23	173	100
4032	O	40	132	43	26	155	90
	T6	36	120	48	29	141	82
4043	O	42	140	41	25	163	94
5005	O, H38	52	172	33	20	205	118
5050	O, H38	50	165	34	21	191	110
5052	O, H38	35	116	49	30	137	79
5056	O	29	98	59	36	120	69
	H38	27	91	64	38	112	65
5083	All	29	98	60	36	120	69
5086	All	31	104	56	33	127	73
5154	All	32	108	54	32	127	73
5182	All	31	105	56	33	123	71
5252	All	35	117	49	30	138	80
5254	All	32	107	54	32	127	73
5356	O	29	98	59	36	116	67
5454	All	34	113	51	31	134	77
5456	All	29	98	60	36	116	67
5457	All	46	153	38	23	177	102
5652	All	35	116	49	30	137	79
5657	All	54	179	32	19	⋯	⋯

(continued)

294

Typical physical properties of wrought aluminum alloys (continued)

Alloy	Temper	Electrical conductivity(a) Volume	Weight	Electrical resistivity(b) nΩm	ohms(d)	Thermal conductivity(c) W/m·K	Btu/ ft·h·°F
6005	T5	49	162	35	21	167	97
6009	O	54	184	32	19	205	118
	T4	44	150	39	24	172	99
	T6	47	160	37	22	180	104
6010	O	53	175	33	20	202	117
	T4	39	129	44	27	151	87
	T6	44	146	39	24	180	104
6061	O	47	155	37	22	180	104
	T4	40	132	43	26	154	89
	T6	43	142	40	24	167	97
6063	O	58	191	30	18	218	126
	T1	50	165	35	21	193	112
	T5	55	181	32	19	209	121
	T6	53	175	33	20	201	116
6066	O	40	132	43	26	147	85
	T6	37	122	47	28	147	85
6070	T6	44	145	39	24	172	99
6101	T6	57	188	30	18	218	138
	T8	54	178	32	19	218	138
6151	O	54	178	32	19	205	118
	T4	42	138	41	25	163	94
	T6	45	148	38	23	175	101
6201	T81	54	179	32	19	205	118
6205	T1	45	149	37	22	172	99
	T5	49	162	35	21	188	109
6262	T9	44	145	39	24	172	99
6351	T6	46	152	38	23	176	102
6463	T1	50	165	34	21	192	111
	T5	55	181	31	19	209	121
	T6	53	175	33	20	201	116
7005	O	43	138	40	24	166	96
	T53	38	122	45	27	148	86
	T6	35	113	49	30	137	79
	T63	38	122	45	27	148	86
7049	T73	38	120	43	27	154	89
7050	O	47	148	37	22	180	104
	T73	40	127	43	26	157	91
	T76	40	125	44	26	154	89
7072	O	60	197	29	17	227	131
7075	T6	33	105	52	31	130	75
	T73	40	128	43	26	155	90
	T76	38	123	45	27	150	87
7175	O	46	147	38	23	177	102
	T66	36	115	48	29	142	82
	T73	40	128	43	26	155	90
7475	O	46	147	38	23	177	102
	T6	36	115	48	29	142	82
	T7351	40	128	43	26	155	90
	T76	42	134	41	25	163	94

(a) % IACS at 20 °C (68 °F). (b) At 20 °C (68 °F). (c) At 25 °C (77 °F). (d) Per circular mil/ft. (e) All H1x-type tempers.

Comparative corrosion and fabrication characteristics and typical applications of wrought aluminum alloys

Alloy temper	Resistance to corrosion General(a)	Stress-corrosion cracking(b)	Workability (cold)(e)	Machinability(e)	Weldability(f) Gas	Arc	Resistance spot and seam	Brazeability(f)	Solderability(g)	Some typical applications of alloys
1050 0	A	A	A	E	A	A	B	A	A	Chemical equipment, railroad tank cars
H12	A	A	A	E	A	A	A	A	A	
H14	A	A	A	D	A	A	A	A	A	
H16	A	A	B	D	A	A	A	A	A	
H18	A	A	B	D	A	A	A	A	A	
1060 0	A	A	A	E	A	A	B	A	A	Chemical equipment, railroad tank cars
H12	A	A	A	E	A	A	A	A	A	
H14	A	A	A	D	A	A	A	A	A	
H16	A	A	B	D	A	A	A	A	A	
H18	A	A	B	D	A	A	A	A	A	
1100 0	A	A	A	E	A	A	B	A	A	Sheet-metal work, spun hollowware, fin stock
H12	A	A	A	E	A	A	A	A	A	
H14	A	A	A	D	A	A	A	A	A	
H16	A	A	B	D	A	A	A	A	A	
H18	A	A	C	D	A	A	A	A	A	
1145 0	A	A	A	E	A	A	B	A	A	Foil, fin stock
H12	A	A	A	E	A	A	A	A	A	
H14	A	A	A	D	A	A	A	A	A	
H16	A	A	B	D	A	A	A	A	A	
H18	A	A	B	D	A	A	A	A	A	
1199 0	A	A	A	E	A	A	B	A	A	Electrolytic capacitor foil, chemical equipment, railroad tank cars
H12	A	A	A	E	A	A	A	A	A	
H14	A	A	A	D	A	A	A	A	A	
H16	A	A	B	D	A	A	A	A	A	
H18	A	A	B	D	A	A	A	A	A	
1350 0	A	A	A	E	A	A	B	A	A	Electrical conductors
H12, H111	A	A	A	E	A	A	A	A	A	
H14, H24	A	A	A	D	A	A	A	A	A	
H16, H26	A	A	B	D	A	A	A	A	A	
H18	A	A	B	D	A	A	A	A	A	
2011 T3	D(c)	D	C	A	D	D	D	D	C	Screw-machine products
T4, T451	D(c)	D	B	A	D	D	D	D	C	
T8	D	B	D	A	D	D	D	D	C	
2014 0	...	...	...	D	D	D	B	D	C	Truck frames, aircraft structures
T3, T4, T451	D(c)	C	C	D	B	B	B	D	C	
T6, T651, T6510, T6511	D	C	D	B	D	B	B	D	C	
2024 0	...	...	...	D	D	D	D	D	C	Truck wheels, screw-machine products, aircraft structures
T4, T3, T351, T3510, T3511	D(c)	C	C	B	C	B	B	D	C	
T361	D(c)	C	D	B	D	C	B	D	C	
T6	D	B	C	B	D	C	B	D	C	
T861, T81, T851, T8510, T8511	D	B	D	B	D	C	B	D	C	
T72	...	...	...	B	...	...	...	...	...	
2036 T4	C	...	B	C	...	B	B	D	...	Auto-body panel sheet
2124 T851	D	B	D	B	D	C	B	D	C	Military supersonic aircraft
2218 T61	D	C	...	...	...	...	C	...	C	Jet engine impellers and rings
T72	D	C	...	B	D	C	B	D	C	
2219 0	...	...	...	...	D	A	B	D		Structural uses at high temperatures (to 315 °C or 600 °F) high-strength weldments
T31, T351, T3510, T3511	D(c)	C	C	B	A	A	A	D	NA	
T37	D(c)	C	D	B	A	A	A	D		
T81, T851, T8510, T8511	D	B	D	B	A	A	A	D		
T87	D	B	D	B	A	A	A	D		
2618 T61	D	C	...	B	D	C	B	D	NA	Aircraft engines

(continued)

Comparative corrosion and fabrication characteristics and typical applications of wrought aluminum alloys (continued)

	General(a)	Stress-corrosion cracking(b)	Workability (cold)(e)	Machinability(e)	Gas	Arc	Resistance spot and seam	Brazeability(f)	Solderability(g)	Some typical applications of alloys
	Resistance to corrosion				Weldability(f)					
3003 0	A	A	A	E	A	A	B	A	A	Cooking utensils, chemical equipment, pressure vessels, sheet-metal work, builder's hardware, storage tanks
H12	A	A	A	E	A	A	A	A	A	
H14	A	A	B	D	A	A	A	A	A	
H16	A	A	C	D	A	A	A	A	A	
H18	A	A	C	D	A	A	A	A	A	
H25	A	A	B	D	A	A	A	A	A	
3004 0	A	A	A	D	B	A	B	B	B	Sheet-metal work, storage tanks
H32	A	A	B	D	B	A	A	B	B	
H34	A	A	B	C	B	A	A	B	B	
H36	A	A	C	C	B	A	A	B	B	
H38	A	A	C	C	B	A	A	B	B	
3105 0	A	A	A	E	B	A	B	B	B	Residential siding, mobile homes, rain-carrying goods, sheet-metal work
H12	A	A	B	E	B	A	A	B	B	
H14	A	A	B	D	B	A	A	B	B	
H16	A	A	C	D	B	A	A	B	B	
H18	A	A	C	D	B	A	A	B	B	
H25	A	A	B	D	B	A	A	B	B	
4032 T6	C	B	...	B	D	B	C	D	NA	Pistons
4043	B	A	NA	C	NA	NA	NA	NA	NA	Welding electrode
5005 0	A	A	A	E	A	A	B	B	B	Appliances, utensils, architectural, electrical conductors
H12	A	A	A	E	A	A	A	B	B	
H14	A	A	B	D	A	A	A	B	B	
H16	A	A	C	D	A	A	A	B	B	
H18	A	A	C	D	A	A	A	B	B	
H32	A	A	B	E	A	A	A	B	B	
H34	A	A	C	D	A	A	A	B	B	
H36	A	A	C	D	A	A	A	B	B	
H38	A	A		D	A	A	A	B	B	
5050 0	A	A	A	E	A	A	B	B	C	Builders' hardware, refrigerator trim, coiled tubes
H32	A	A	A	D	A	A	A	B	C	
H34	A	A	B	D	A	A	A	B	C	
H36	A	A	C	C	A	A	A	B	C	
H38	A	A	C	C	A	A	A	B	C	
5052 0	A	A	A	D	A	A	B	C	D	Sheet-metal work, hydraulic tube, appliances
H32	A	A	B	D	A	A	A	C	D	
H34	A	A	B	C	A	A	A	C	D	
H36	A	A	C	C	A	A	A	C	D	
H38	A	A	C	C	A	A	A	C	D	
5056 0	A(d)	B(d)	A	D	C	A	B	D	D	Cable sheathing, rivets for magnesium, screen wire, zippers
H111	A(d)	B(d)	A	D	C	A	A	D	D	
H12, H32	A(d)	B(d)	B	D	C	A	A	D	D	
H14, H34	A(d)	B(d)	B	C	C	A	A	D	D	
H18, H38	A(d)	C(d)	C	C	C	A	A	D	D	
H192	B(d)	D(d)	D	B	C	A	A	D	D	
H392	B(d)	D(d)	D	B	C	A	A	D	D	
5083 0	A(d)	B(d)	B	D	C	A	B	D	D	Unfired, welded pressure vessels, marine, auto aircraft cryogenics, TV towers, drilling rigs, transportation equipment, missile components
H321, H116	A(d)	B(d)	C	D	C	A	A	D	D	
H323	A(d)	B(d)	C	D	C	A	A	D	D	
H343	A(d)	B(d)	C	C	C	A	A	D	D	
H111	A(d)	B(d)	C	D	C	A	A	D	D	

(continued)

Comparative corrosion and fabrication characteristics and typical applications of wrought aluminum alloys (continued)

	Resistance to corrosion		Workability (cold)(e)	Machinability(e)	Weldability(f)			Brazeability(f)	Solderability(g)	Some typical applications of alloys
	General(a)	Stress-corrosion cracking(b)			Gas	Arc	Resistance spot and seam			
5086 0	A(d)	A(d)	A	D	C	A	B	D	D	
H32, H116	A(d)	A(d)	B	D	C	A	A	D	D	
H34	A(d)	B(d)	B	C	C	A	A	D	D	
H36	A(d)	B(d)	C	C	C	A	A	D	D	
H38	A(d)	B(d)	C	C	C	A	A	D	D	
H111	A(d)	A(d)	B	D	C	A	A	D	D	
5154 0	A(d)	A(d)	A	D	C	A	B	D	D	Welded structures, storage tanks, pressure vessels, salt-water service
H32	A(d)	A(d)	B	D	C	A	A	D	D	
H34	A(d)	A(d)	B	C	C	A	A	D	D	
H36	A(d)	A(d)	C	C	C	A	A	D	D	
H38	A(d)	A(d)	C	C	C	A	A	D	D	
5182 0	A	A(d)	A	D	C	A	B	D	D	Automobile body sheet, can ends
H19	A	A(d)	D	B	C	A	A	D	D	
5252 H24	A	A	B	D	A	A	A	C	D	Automotive and appliance trim
H25	A	A	B	C	A	A	A	C	D	
H28	A	A	C	C	A	A	A	C	D	
5254 0	A(d)	A(d)	A	D	C	A	B	D	D	Hydrogen peroxide and chemical storage vessels
H32	A(d)	A(d)	B	D	C	A	A	D	D	
H34	A(d)	A(d)	B	C	C	A	A	D	D	
H36	A(d)	A(d)	C	C	C	A	A	D	D	
H38	A(d)	A(d)	C	C	C	A	A	D	D	
5356	A	A	NA	B	NA	NA	NA	NA	NA	Welding electrode
5454 0	A	A	A	D	C	A	B	D		Welded structures, pressure vessels, marine service
H32	A	A	B	D	C	A	A	D	NA	
H34	A	A	B	C	C	A	A	D		
H111	A	A	B	D	C	A	A	D		
5456 0	A(d)	B(d)	B	D	C	A	B	D		High-strength welded structures, storage tanks, pressure vessels, marine applications
H111	A(d)	B(d)	C	D	C	A	A	D	NA	
H321, H115	A(d)	B(d)	C	D	C	A	A	D		
H323	A(d)	B(d)	C	D	C	A	A	D		
H343	A(d)	B(d)	C	C	C	A	A	D		
5457 0	A	A	A	E	A	A	B	B	B	
5652 0	A	A	A	D	A	A	B	C	D	Hydrogen peroxide and chemical storage vessels
H32	A	A	B	D	A	A	A	C	D	
H34	A	A	B	C	A	A	A	C	D	
H36	A	A	C	C	A	A	A	C	D	
H38	A	A	C	C	A	A	A	C	D	
5657 H241	A	A	A	D	A	A	A	B		Anodized auto and appliance trim
H25	A	A	B	D	A	A	A	B	NA	
H26	A	A	B	D	A	A	A	B		
H28	A	A	C	D	A	A	A	B		
6005 T5	B	A	C	C	A	A	A	A	NA	Heavy-duty structures requiring good corrosion resistance applications, truck and marine, railroad cars, furniture, pipelines
6009 T4	A	A	A	C	A	A	A	A	B	Automobile body sheet
6010 T4	A	A	B	C	A	A	A	A	B	Automobile body sheet
6061 0	B	A	A	D	A	A	B	A	B	Heavy-duty structures requiring good corrosion resistance, truck and marine, railroad cars, furniture, pipelines
T4, T451, T4510, T4511	B	B	B	C	A	A	A	A	B	
T6, T651, T652, T6510, T6511	B	A	C	C	A	A	A	A	B	

(continued)

Comparative corrosion and fabrication characteristics and typical applications of wrought aluminum alloys (continued)

Alloy and temper	Resistance to corrosion — General(a)	Stress-corrosion cracking(b)	Workability (cold)(e)	Machinability(e)	Weldability(f) — Gas	Arc	Resistance spot and seam	Brazeability(f)	Solderability(g)	Some typical applications of alloys
6063 T1	A	A	B	D	A	A	A	A	B	Pipe railing, furniture, architectural extrusions
T4	A	A	B	D	A	A	A	A	B	
T5, T52	A	A	B	C	A	A	A	A	B	
T6	A	A	C	C	A	A	A	A	B	
T83, T831, T832	A	A	C	C	A	A	A	A	B	
6066 0	C	A	B	D	D	B	B	D		Forgings and extrusions for welded structures
T4, T4510, T4511	C	B	C	C	D	B	B	D	NA	
T6, T6510, T6511	C	B	C	B	D	B	B	D		
6070 T4, T4511	B	B	B	C	A	A	A	B	NA	Heavy-duty welded structures, pipelines
T6	B	B	C	C	A	A	A	B		
6101 T6, T63	A	A	C	C	A	A	A	A	NA	High-strength bus conductors
T61, T64	A	A	B	D	A	A	A	A		
6151 T6, T652	...	...	...	...	...	...	...	...	B	Moderate-strength, intricate forgings for machine and auto parts
6201 T81	A	A	...	C	A	A	A	A	NA	High-strength electric conductor wire
6262 T6, T651, T6510, T6511	B	A	C	B	A	A	A	A	NA	Screw-machine products
T9	B	A	D	B	A	A	A	A		
6351 T5, T6	B	A	C	C	A	A	A	A	B	Heavy-duty structures requiring good corrosion resistance, truck and tractor extrusions
6463 T1	A	A	B	D	A	A	A	A		Extruded architectural and trim sections
T5	A	A	B	C	A	A	A	A	NA	
T6	A	A	C	C	A	A	A	A		
7005 T53, T63	B	B	C	A	B	B	B	B	B	Heavy-duty structures requiring good corrosion resistance, trucks, trailers, dump bodies
7049 T73, T7351, T7352	C	B	D	B	D	C	B	D	D	Aircraft and other structures
T76, T7651	C	B	D	B	D	C	B	D	D	
7050 T736, T73651, T73652	C	B	D	B	D	C	B	D	D	Aircraft and other structures
T76, T761	C	B	D	B	D	C	B	D	D	
7072	A	A	A	D	A	A	A	A	A	Fin stock, cladding alloy
7075 0	...	...	D	D	C	B	D	D		Aircraft and other structures
T6, T651, T652, T6510, T6511	C(c)	C	D	B	D	C	B	D	D	
T73, T7351	C	B	D	B	D	C	B	D	D	
7175 T736, T73652	C	B	D	B	D	C	B	D	D	Aircraft and other structures, forgings
7178 0	...	...	...	D	C	B	D	D		Aircraft and other structures
T6, T651, T6510, T6511	C(c)	C	D	B	D	C	B	D	D	
7475 T6, T651	C	C	D	B	D	C	B	D	D	Aircraft and other structures
T73, T7351, T7352	C	B	D	B	D	C	B	D	D	
T76, T7651	C	B	D	B	D	C	B	D	D	

(a) Ratings A through E are relative ratings in decreasing order of merit, based on exposures to sodium chloride solution by intermittent spraying or immersion. Alloys with A and B ratings can be used in industrial and seacoast atmospheres without protection. Alloys with C, D and E ratings generally should be protected at least on faying surfaces. (b) Stress-corrosion cracking ratings are based on service experience and on laboratory tests of specimens exposed to the 3.5% sodium chloride alternate immersion test. A = No known instance of failure in service or in laboratory tests. B = No known instance of failure in service; limited failures in laboratory tests of short transverse specimens. C = Service failures with sustained tension stress acting in short transverse direction relative to grain structure; limited failures in laboratory tests of long transverse specimens. D = Limited service failures with sustained longitudinal or long transverse stress. (c) In relatively thick sections the rating would be E. (d) This rating may be different for material held at elevated temperature for long periods. (e) Ratings A through D for workability (cold), and A through E for machinability, are relative ratings in decreasing order of merit. (f) Ratings A through D for weldability and brazeability are relative ratings defined as follows: A = Generally weldable by all commercial procedures and methods. B = Weldable with special techniques or for specific applications which justify preliminary trials or testing to develop welding procedure and weld performance. C = Limited weldability because of crack sensitivity or loss in resistance to corrosion and mechanical properties. D = No commonly used welding methods have been developed. (g) Ratings A through D and NA for solderability are relative ratings defined as follows: A = Excellent. B = Good. C = Fair. D = Poor. NA = Not applicable.

Typical mechanical properties of wrought aluminum alloys

Alloy	Temper	Tensile strength MPa	ksi	Yield strength MPa	ksi	Elongation(a), % (b)	(c)	Hardness(d)	Shear strength MPa	ksi	Fatigue strength(e) MPa	ksi
1050	O	76	11	28	4	...	...	...	62	9	...	...
	H14	110	16	105	15	...	...	...	69	10	...	...
	H16	130	19	125	18	...	...	...	76	11	...	...
	H18	160	23	145	21	...	...	...	83	12	...	...
1060	O	69	10	28	4	43	...	19	48	7	21	3
	H12	83	12	76	11	16	...	23	55	8	28	4
	H14	97	14	90	13	12	...	26	62	9	34	5
	H16	110	16	105	15	8	...	30	69	10	45	6.5
	H18	130	19	125	18	6	...	35	76	11	45	6.5
1100	O	90	13	34	5	35	45	23	62	9	34	5
	H12	110	16	105	15	12	25	28	69	10	41	6
	H14	125	18	115	17	9	20	32	76	11	48	7
	H16	145	21	140	20	6	17	38	83	12	62	9
	H18	165	24	150	22	5	15	44	90	13	62	9
1350	O	83	12	28	4	23(f)	...	...	55	8	...	...
	H12	97	14	83	12	...	...	...	62	9	...	...
	H14	110	16	97	14	...	...	...	69	10	...	...
	H16	125	18	110	16	...	...	...	76	11	...	...
	H19	185	27	165	24	1.5(f)	...	...	105	15	48	7
2011	T3	380	55	295	43	...	15	95	220	32	125	18
	T8	405	59	310	45	...	15	100	240	35	125	18
2014	O	185	27	97	14	...	18	45	125	18	90	13
	T4	425	62	290(h)	42(h)	...	20	105	260	38	140	20
	T6(g)	485	70	415	60	...	13	135	290	42	125	18
Alclad 2014	O	170	25	69	10	21	...	...	125	18	...	...
	T3	435	63	275	40	20	...	...	255	37	...	...
	T4	420	61	255	37	22	...	...	255	37	...	...
	T6	470	68	415	60	10	...	...	285	41	...	...
2024	O	185	27	76	11	20	22	47	125	18	90	13
	T3	485	70	345	50	18	...	120	285	11	140	20
	T4, T351	470	68	325	47	20	19	120	285	41	140	20
	T361	495	72	395	57	13	...	130	290	42	125	18
Alclad 2024	O	180	26	76	11	20	...	...	125	18	...	...
	T	450	65	310	45	18	...	...	275	40	...	...
	T4, T351	440	64	290	42	19	...	...	275	40	...	...
	T361	460	67	365	53	11	...	...	285	41	...	...
	T81, T851	450	65	415	60	6	...	...	275	40	...	...
	T861	485	70	455	66	6	...	...	290	42	...	...
2036	T4	340	49	195	28	24	...	...	...	...	125(j)	18(j)
2048		455	66	415	60	8.3	...	...	...	...	220(k)	32(k)
2124	T851	490	71	440	64	9.4	...	...	...	...	...	...
2218	T61	405	59	305	44	...	13	115	...	...	...	...
	T71	345	50	275	40	...	11	105	...	...	...	...
	T72	330	48	255	37	...	11	95	205	30	...	...
2219	O	170	25	76	11	18	...	...	...	...	...	...
	T42	360	52	185	27	20	...	...	...	...	...	...
	T31, T351	360	52	250	36	17	...	...	...	...	...	...
	T37	395	57	315	46	11	...	...	...	...	...	...
	T62	415	60	290	42	10	...	...	...	...	105	15
	T81, T851	455	66	350	51	10	...	...	...	...	105	15
	T87	475	69	395	57	10	...	...	...	...	105	15
2618	All	440	64	370	54	10(c)	...	...	260	38	125	18
3003 and Alclad 3003(m)	O	110	16	42	6	30	40	28	76	11	48	7
	H12	130	19	125	18	10	20	35	83	12	55	8
	H14	150	22	145	21	8	16	40	97	14	62	9
	H16	180	26	170	25	5	14	47	105	15	69	10
	H18	200	29	185	27	4	10	55	110	16	69	10

(continued)

Typical mechanical properties of wrought aluminum alloys (continued)

Alloy	Temper	Tensile strength MPa	ksi	Yield strength MPa	ksi	Elonga-tion(a), % (b)	(c)	Hard-ness(d)	Shear strength MPa	ksi	Fatigue strength(e) MPa	ksi
3004 and Alclad 3004(m)	O	180	26	69	10	20	25	45	110	16	97	14
	H32	215	31	170	25	10	17	52	115	17	105	15
	H34	240	35	200	29	9	12	63	125	18	105	15
	H36	260	38	230	33	5	9	70	140	20	110	16
	H38	285	41	250	36	5	6	77	145	21	110	16
3105	O	115	17	55	8	24	…	…	83	12	…	…
	H12	150	22	130	19	7	…	…	97	14	…	…
	H14	170	25	150	22	5	…	…	105	15	…	…
	H16	195	28	170	25	4	…	…	110	16	…	…
	H18	215	31	195	28	3	…	…	115	17	…	…
	H25	180	26	160	23	8	…	…	105	15	…	…
4032	T6	380	55	315	46	…	9	120	260	38	110	16
4043	O	145	21	69	10	22	…	…	…	…	…	…
	H18	285	41	270	39	0.5	…	…	…	…	…	…
5005	O	125	18	41	6	25	…	28	76	11	…	…
	H12	140	20	130	19	10	…	…	97	14	…	…
	H14	160	23	150	22	6	…	…	97	14	…	…
	H16	180	26	170	25	5	…	…	105	15	…	…
	H18	200	29	195	28	4	…	…	110	16		
	H32	140	20	115	17	11	…	36	97	14	…	…
	H34	160	23	140	20	8	…	41	97	14	…	…
	H36	180	26	165	24	6	…	46	105	15	…	…
	H38	200	29	185	27	5	…	51	110	16	…	…
5050	O	145	21	55	8	24	…	36	105	15	83	12
	H32	170	25	145	21	9	…	46	115	17	90	13
	H34	195	28	165	24	8	…	53	125	18	90	13
	H36	205	30	180	26	7	…	58	130	19	97	14
	H38	220	32	200	29	6	…	63	140	20	97	14
5052	O	195	28	90	13	25	27	47	125	18	110	16
	H32	230	33	195	28	12	16	60	140	20	115	17
	H34	260	38	215	31	10	12	68	145	21	125	18
	H36	275	40	240	35	8	9	73	160	23	130	19
	H38	290	42	255	37	7	7	77	165	24	140	20
5056	O	290	42	150	22	…	35	65	180	26	140	20
	H18	435	63	405	59	…	10	105	235	34	150	22
	H38	415	60	345	50	…	15	100	220	32	150	22
5083	O	290	42	145	21	…	22	…	170	25	…	…
	H112	305	44	195	28	…	16	…	…	…	…	…
	H113	315	46	230	33	…	16	…	…	…	…	…
	H321	315	46	230	33	…	16	…	…	…	160	23
	H323, H32	325	47	250	36	…	10	…	…	…	…	…
	H343, H34	345	50	285	41	…	9	…	…	…	…	…
5086	O	260	38	115	17	22	…	…	160	23	…	…
	H32, H116, H117	290	42	205	30	12	…	…	…	…	…	…
	H34	325	47	255	37	10	…	…	185	27	…	…
	H112	270	39	130	19	14	…	…	…	…	…	…
5154	O	240	35	115	17	27	…	58	150	22	115	17
	H32	270	39	205	30	15	…	67	150	22	125	18
	H34	290	42	230	33	13	…	73	165	24	130	19
	H36	310	45	250	36	12	…	78	180	26	140	20
	H38	330	48	270	39	10	…	80	195	28	145	21
	H112	240	35	115	17	25	…	63	…	…	115	17
5182	O	275	40	140	19	25	…	58	150	22	140	20
	H32	315	46	235	34	12	…	…	…	…	…	…
	H34	340	49	285	41	10	…	…	…	…	…	…
	H19(n)	420	61	395	57	4	…	…	…	…	…	…
5252	H25	235	34	170	25	11	…	68	145	21	…	…
	H28, H38	285	41	240	35	5	…	75	160	23	…	…
5254	O	240	35	115	17	27	…	58	150	22	115	17

(continued)

Typical mechanical properties of wrought aluminum alloys (continued)

Alloy	Temper	Tensile strength MPa	ksi	Yield strength MPa	ksi	Elonga-tion(a), % (b)	(c)	Hard-ness(d)	Shear strength MPa	ksi	Fatigue strength(e) MPa	ksi
5254	H32............	270	39	205	30	15	...	67	150	22	125	18
	H34............	290	42	230	33	13	...	73	165	24	130	19
	H36............	310	45	250	36	12	...	78	180	26	140	20
	H38............	330	48	270	39	10	...	80	195	28	145	21
	H112............	240	35	115	17	25	...	63	...	...	115	17
5454	O...............	250	36	115	17	22	...	62	160	23	...	...
	H32............	275	40	205	30	10	...	73	165	24	...	...
	H34............	305	44	240	35	10	...	81	180	26	...	...
	H36............	340	49	275	40	8	...	...	...	...	...	...
	H38............	370	54	310	45	8	...	...	...	...	...	...
	H111............	260	38	180	26	14	...	70	160	23	...	...
	H112............	250	36	125	18	18	...	62	160	23	...	...
	H311............	260	38	180	26	18	...	70	160	23	...	...
5456	O...............	310	45	160	23	...	24	...	...	...	...	...
	H111............	325	47	230	33	...	18	...	...	...	...	...
	H112............	310	45	165	24	...	22	...	...	...	...	...
	H321, H116......	350	51	255	27	...	16	90	205	30	...	...
5457	O...............	130	19	48	7	22	...	32	83	12	...	...
	H25............	180	26	160	23	12	...	48	110	16	...	...
	H28, H38........	205	30	185	27	6	...	55	125	18	...	...
5652	O...............	195	28	90	13	25	30	47	125	18	110	16
	H32............	230	33	195	28	12	18	60	140	20	115	17
	H34............	260	38	215	31	10	14	68	145	21	125	18
	H36............	275	40	240	35	8	10	73	160	23	130	19
	H38............	290	42	255	34	7	8	77	165	24	140	20
5657	H25............	160	23	140	20	12	...	40	97	14	...	...
	H28, H38........	195	28	165	24	7	...	50	105	15	...	...
6005	T1.............	170	25	105	15	16	...	...	...	...	97	14
	T5.............	260	38	240	35	8	10	95	205	30	97	14
6009	T4.............	235	34	130	19	24	...	70(p)	150	22	115	17
	T6.............	345	50	325	47	12	...	...	...	...	...	...
6010	T4.............	255	37	170	25	24	...	76(p)	...	...	115	17
6061	O...............	125	18	55	8	25	30	30	83	12	62	9
	T4, T451........	240	35	145	21	22	25	65	165	24	97	14
	T6, T651........	310	45	275	40	12	17	95	205	30	97	14
Alclad 6061	O...............	115	17	48	7	25	...	...	76	11	...	...
	T4, T451........	230	33	130	19	22	...	...	150	22	...	...
	T6, T651........	290	42	255	37	12	...	...	185	27	...	...
6063	O...............	90	13	48	7	...	...	25	69	10	55	8
	T1.............	150	22	90	13	20	...	42	97	14	62	9
	T4.............	170	25	90	13	22	...	...	...	...	...	...
	T5.............	185	27	145	21	12	...	60	115	17	69	10
	T6.............	240	35	215	31	12	...	73	150	22	69	10
	T83.............	255	37	240	35	9	...	82	150	22	...	...
	T831.............	205	30	185	27	10	...	70	125	18	...	...
	T832.............	290	42	270	39	12	...	95	185	27	...	...
6066	O...............	150	22	83	12	...	18	43	97	14	...	...
	T4, T451........	360	52	205	30	...	18	90	200	29	...	...
	T6, T651........	395	57	360	52	...	12	120	235	34	110	16
6070	O...............	145	21	69	10	20	...	35	97	14	62	9
	T4.............	315	46	170	25	20	...	90	205	30	90	13
	T6.............	380	55	350	51	10	...	120	235	34	97	14
6101	H111............	97	14	76	11	...	...	...	...	...	...	...
6151	T6.............	220	32	195	28	15(q)	...	71	140	20	...	...
6201	T6.............	330	48	300	43	17	...	90	...	...	...	...
	T81.............	330	48	310	45	6(f)	...	...	...	...	...	...
6205	T1.............	260	38	140	20	19	...	65	...	...	...	...
	T5.............	310	45	290	42	11	...	95	205	30	105	15
6262	T9.............	400	58	380	55	...	10	120	240	35	90	13

(continued)

Typical mechanical properties of wrought aluminum alloys (continued)

Alloy	Temper	Tensile strength MPa	ksi	Yield strength MPa	ksi	Elongation(a), % (b)	(c)	Hardness(d)	Shear strength MPa	ksi	Fatigue strength(e) MPa	ksi
6351	T4	250	36	150	22	20	...	...	...	...	...	...
	T6	310	45	285	41	14	...	95	200	29	90	13
6463	T1	150	22	90	13	20	...	42	97	14	69	10
	T5	185	27	145	21	12	...	60	115	17	69	10
	T6	240	35	215	31	12	...	74	150	22	69	10
7005	O	193	28	83	12	20	...	...	117	17	...	...
	T53	393	57	345	50	15	...	...	221	32	140	20
	T6, T63, T6351	372	54	315	46	12	...	...	214	31	125	18
7049	T73	...	...	...	...	...	...	135	...	...	295(k)	43(k)
7050	T736	515	75	455	66	11	15	...	...	...	240(k)	35(k)
7072	O	...	...	...	...	...	...	20	55	8	...	...
	H12	...	...	...	...	...	...	28	62	9	...	...
	H14	...	...	...	...	...	...	32	69	10	...	...
7075	O	230	38	105	15	17	16	60	150	22	...	...
	T6, T651	570	83	505	73	11	11	150	330	48	160	23
	T73	505	73	435	63	13	...	...	...	...	...	...
Alclad 7075	O	220	32	95	14	17	...	...	150	22	...	...
	T6, T651	525	76	460	67	11	...	...	315	46	...	...
7175	T66	595	86	525	76	11	...	150	325	47	160	23
	T736	525	76	455	66	14	...	145	290	42	160	23
7475	T61	525	76	460	67	12	...	...	...	...	...	...
	T651	...	...	...	...	...	...	...	295	43	...	...
	T7351	...	...	...	...	...	...	...	270	9	220	32
	T7651	...	...	...	...	...	...	...	270	39	...	...

(a) In 50 mm or 2 in. (b) Specimen 1.6 mm (1/16 in.) thick. (c) Specimen 12.5 mm (0.50 in.) in diameter. (d) 500-kg load; 10-mm ball. (e) In 5×10^8 cycles; R. R. Moore — type test. (f) In 254 mm or 10 in. (g) Extruded products more than 19 mm (¾ in.) thick are 15 to 20% higher in strength. (h) Die forgings are about 20% lower in yield strength. (j) In 10^7 cycles using flexural-type testing of sheet specimens. (k) In 10^7 cycles; axially loaded specimens tested at $R = 0.1$. (m) No shear-strength or fatigue-strength values for Alclad. (n) Properties for this temper are those of container end stock 0.25 to 0.38 mm (0.010 to 0.015 in.) thick. (p) HR15T. (q) Specimen 6.35 mm (0.25 in.) thick.

Cast Aluminum

Designations and nominal compositions of common aluminum alloys used for casting

AA number	Alloys Former AA designation	Former ASTM number	Product(a)	Cu	Mg	Mn	Si	Others
201.0	...	...	S	4.6	0.35	0.35	...	0.7 Ag, 0.25 Ti
206.0	...	...	S or P	4.6	0.25	0.35	0.10(b)	0.22 Ti, 0.15 Fe(b)
A206.0....	...	...	S or P	4.6	0.25	0.35	0.05(b)	0.22 Ti, 0.10 Fe(b)
208.0	108	CS43A	S	4.0	...	...	3.0	...
242.0	142	CN42A	S or P	4.0	1.5	...	...	2.0 Ni
295.0	195	C4A	S	4.5	...	...	0.8	...
296.0 ..... B295.0, B195		...	P	4.5	...	...	2.5	...
308.0	A108	SC64A	S or P	4.5	...	...	5.5	...
319.0 ..... 319, Allcast		SC64D	S or P	3.5	...	...	6.0	...
336.0 ..... A332.0, A132		SN122A	P	1.0	1.0	...	12.0	2.5 Ni
354.0	354	SC92A	P	1.8	0.50	...	9.0	...
355.0	355	SC51A	S or P	1.2	0.50	0.50(b)	5.0	0.6 Fe(b), 0.35 Zn(b)
C355.0....	C355	SC51B	S or P	1.2	0.50	0.10(b)	5.0	0.20 Fe(b), 0.10 Zn(b)
356.0	356	SG70A	S or P	0.25(b)	0.32	0.35(b)	7.0	0.6 Fe(b), 0.35 Zn(b)
A356.0....	A356	SG70B	S or P	0.20(b)	0.35	0.10(b)	7.0	0.20 Fe(b), 0.10 Zn(b)
357.0	357	...	S or P	...	0.50	...	7.0	...
A357.0....	A357	...	S or P	...	0.6	...	7.0	0.15 Ti, 0.005 Be
359.0	359	SG91A	S or P	...	0.6	...	9.0	...
360.0	360	SG100B	D	...	0.50	...	9.5	2.0 Fe(b)
A360.0....	A360	SG100A	D	...	0.50	...	9.5	1.3 Fe(b)
380.0	380	SC84B	D	3.5	...	...	8.5	2.0 Fe(b)
A380.0....	A380	SC84A	D	3.5	...	...	8.5	1.3 Fe(b)
383.0	...	SC102A	D	2.5	...	...	10.5	...
384.0	384	SC114A	D	3.8	...	...	11.2	3.0 Zn(b)
A384.0....	384	SC114A	D	3.8	...	...	11.2	1.0 Zn(b)
390.0	390	...	D	4.5	0.6	...	17.0	1.3 Zn(b)
A390.0....	A390	...	S or P	4.5	0.6	...	17.0	0.5 Zn(b)
413.0	13	S12B	D	...	...	...	12.0	2.0 Fe(b)
A413.0....	A13	S12A	D	...	...	...	12.0	1.3 Fe(b)
4430......	43	S5B	S	0.6(b)	...	...	5.2	...
A443.0....	43	...	S	0.30(b)	...	...	5.2	...
B443.0....	43	S5A	S or P	0.15(b)	...	...	5.2	...
C443.0....	A43	S5C	D	0.6(b)	...	...	5.2	2.0 Fe(b)
514.0	214	G4A	S	...	4.0	...	...	...
518.0	218	G8A	D	...	8.0	...	...	...
520.0	220	G10A	S	...	10.0	...	...	...
535.0	Almag 35	GM70B	S	...	6.8	0.18	...	0.18 Ti
A535.0....	A218	...	S	...	7.0	0.18	...	...
B535.0....	B218	...	S	...	7.0	...	...	0.18 Ti
712.0 ..... D712.0, D612, 40E		ZG61A	S or P	...	0.6	...	...	5.8 Zn, 0.5 Cr, 0.20 Ti
713.0	613, Tenzaloy	ZC81A,B	S or P	0.7	0.35	...	...	7.5 Zn, 0.7 Cu
771.0	Precedent 71A	ZG71B	S	...	0.9	...	...	7.0 Zn, 0.13 Cr, 0.15 Ti
850.0	750	...	S or P	1.0	...	...	...	6.2 Sn, 1.0 Ni

(a) S = sand casting, P = permanent mold casting, D = die casting. (b) Maximum.

Characteristics of common aluminum alloys used in sand and permanent mold castings(a)(b)

Alloy	Type of mold(c)	Fluidity	Resistance to hot cracking	Pressure tightness	Heat treatment	Strength at elevated temperatures	General corrosion resistance	Machining	Polishing	Anodizing Appearance	Weldability
208.0	S	2	2	2	Optional	3	4	3	3	3	2
213.0	P	2	3	3	No	3	4	2	2	3	3
222.0	S or P	3	3	3	Yes	1	4	1	2	3	3
242.0	S or P	3	4	4	Yes	1	4	2	2	3	4
295.0	S	3	4	4	Yes	3	4	2	2	2	3
296.0	P	3	4	3	Yes	2	4	3	2	3	3
308.0	P	2	2	2	No	3	3	3	3	4	2
319.0	S or P	2	2	2	Optional	3	3	3	4	4	2
328.0	S	1	1	2	Optional	2	3	3	3	4	1
332.0	P	1	2	2	Yes	1	3	4	4	4	2
333.0	P	1	2	2	Yes	2	3	3	3	4	3
336.0	P	1	2	2	Yes	1	3	4	4	4	3
354.0	P	1	1	1	Yes	2	3	4	4	4	3
355.0	S or P	1	1	1	Yes	2	3	3	3	4	1
C355.0	S or P	1	1	1	Yes	2	3	3	3	4	1
356.0	S or P	1	1	1	Yes	3	2	3	4	4	1
A356.0	S or P	1	1	1	Yes	3	2	3	4	4	1
357.0	S or P	1	1	1	Yes	3	2	3	4	4	1
A357.0	S or P	1	1	1	Yes	2	2	3	4	4	1
359.0	S or P	1	2	2	Yes	2	2	4	4	4	1
B443.0	S or P	1	1	1	No	4	2	5	4	4	1
512.0	S	3	3	4	No	3	1	2	2	2	3
513.0	P	4	4	4	No	3	1	1	1	1	3
514.0	S	4	4	5	No	3	1	1	1	1	3
520.0	S	4	4	5	Yes	5	1	1	1	1	4
535.0	S	5	4	5	Optional	3	1	1	1	1	4
705.0	S or P	4	4	4	No	4	2	1	2	2	4
707.0	S or P	4	4	4	No	4	2	1	2	2	4
710.0	S	4	5	4	No	4	4	1	2	2	4
711.0	S	3	5	4	No	4	3	1	2	2	4
713.0	S or P	3	4	4	No	4	3	1	1	1	4
771.0	S	3	4	4	Yes	4	3	1	1	1	4
850.0	S or P	4	5	5	Yes	5	4	1	3	...	5
851.0	S or P	4	5	5	Yes	5	4	1	3	...	5
852.0	S or P	4	5	5	Yes	5	4	1	3	...	5

(a) From Standards for Aluminum Sand and Permanent Mold Castings, The Aluminum Association, 1977. (b) Characteristics are comparatively rated from 1 to 5; 1 is the highest or best possible rating. (c) S = sand; P = permanent.

Characteristics of aluminum die casting alloys(a)

Alloy	Approximate melting temperature, °C	Hot cracking	Resistance to: Die soldering	Corrosion	Die filling capacity	Machining	Polishing	Electroplating	Anodized surface Appearance	Protection	Elevated temperature strength	Pressure tightness
360.0	557-596	1	2	2	3	3	3	2	3	3	1	2
A360.0	557-596	1	2	2	3	3	3	2	3	3	1	2
380.0	538-593	2	1	4	2	3	3	1	3	4	3	2
A380.0	583-593	2	1	4	2	3	3	1	3	4	3	2
383.0	516-582	1	2	3	1	2	3	1	3	4	2	2
384.0	516-582	2	2	5	1	3	3	2	4	5	2	2
413.0	574-582	1	1	2	1	4	5	3	5	3	3	1
A413.0	574-582	1	1	2	1	4	5	3	5	3	3	1
C443.0	574-632	3	4	2	4	5	4	2	2	2	5	3
518.0	535-621	5	5	1	4	1	1	5	1	1	4	5

(a) From ASTM B85. Relative rating of die casting alloys from 1 to 5; 1 is the highest or best possible rating. A rating of 5 in one or more categories does not rule an alloy out of commercial use if other attributes are favorable; however, ratings of 5 may present manufacturing difficulties.

Factors affecting selection of casting process for aluminum alloys

Factor	Sand casting	Casting process Permanent mold casting	Die casting
Cost of equipment	Lowest cost if only a few items required	Less than die casting	Highest
Casting rate	Lowest rate	11 kg/h (25 lb/h) common; higher rates possible	4.5 kg/h (10 lb/h) common; 45 kg/h (100 lb/h) possible
Size of casting	Largest of any casting method	Limited by size of machine	Limited by size of machine
External and internal shape	Best suited for complex shapes where coring required	Simple sand cores can be used, but more difficult to insert than in sand castings	Cores must be able to be pulled because they are metal; undercuts can be formed only by collapsing cores or loose pieces
Minimum wall thickness	3.0-5.0 mm (0.125-0.200 in.) required; 4.0 mm (0.150 in.) normal	3.0-5.0 mm (0.125-0.200 in.) required; 3.5 mm (0.140 in.) normal	1.0-2.5 mm (0.100-0.040 in.); depends on casting size
Type of cores	Complex baked sand cores can be used	Reuseable cores can be made of steel, or nonreuseable baked cores can be used	Steel cores; must be simple and straight so they can be pulled
Tolerance obtainable	Poorest; best linear tolerance is 300 mm/m (300 mils/in.)	Best linear tolerance is 10 mm/m (10 mils/in.)	Best linear tolerance is 4 mm/m (4 mils/in.)
Surface finish	6.5-12.5 μm (250-500 μin.)	4.0-10 μm (150-400 μin.)	1.5 μm (50 μin.); best finish of the three casting processes
Gas porosity	Lowest porosity possible with good technique	Best pressure tightness; low porosity possible with good technique	Porosity may be present
Cooling rate	0.1-0.5 °C/s (0.2-0.9 °F/s)	0.3-1.0 °C/s (0.5-1.8 °F/s)	50-500 °C/s (90-900 °F/s)
Grain size	Coarse	Fine	Very fine on surface
Strength	Lowest	Excellent	Highest, usually used in the "as cast" condition
Fatigue properties	Good	Good	Excellent
Wear resistance	Good	Good	Excellent
Over-all quality	Depends on foundry technique	Highest quality	Tolerance and repeatability very good
Remarks	Very versatile as to size, shape, internal configurations	...	Excellent for fast production rates

Typical tensile properties for separately cast test bars of common aluminum casting alloys

Alloy	Product(a)	Temper	Tensile strength MPa	ksi	Yield strength(b) MPa	ksi	Elongation(c), %
201.0	S	T4	365	53	215	31	20
	S	T6	485	70	435	63	7
	S	T7	460	67	415	60	4.5
206.0, A206.0	S	T7	435	63	345	50	11.7
208.0	S	F	145	21	97	14	2.5
242.0	S	T21	185	27	125	18	1.0
	S	T571	220	32	205	30	0.5
	S	T77	205	30	160	23	2.0
	P	T571	275	40	235	34	1.0
	P	T61	325	47	290	42	0.5
295.0	S	T4	220	32	110	16	8.5
	S	T6	250	36	165	24	5.0
	S	T62	285	41	220	32	2.0
296.0	P	T4	255	37	130	19	9.0
	P	T6	275	40	180	26	5.0
	P	T7	270	39	140	20	4.5
308.0	P	F	195	28	110	16	2.0
319.0	S	F	185	27	125	18	2.0
	S	T6	250	36	165	24	2.0
	P	F	235	34	130	19	2.5
	P	T6	280	40	185	27	3.0
336.0	P	T551	250	36	195	28	0.5
	P	T65	325	47	295	43	0.5
354.0	P	T61	380	55	285	41	6.0
355.0	S	T51	195	28	160	23	1.5
	S	T6	240	35	175	25	3.0
	S	T61	270	39	240	35	1.0
	S	T7	265	38	250	36	0.5
	S	T71	175	35	200	29	1.5
	P	T51	210	30	165	24	2.0
	P	T6	290	42	190	27	4.0
	P	T62	310	45	280	40	1.5
	P	T7	280	40	210	30	2.0
	P	T71	250	36	215	31	3.0
356.0	S	T51	175	25	140	20	2.0
	S	T6	230	33	165	24	3.5
	S	T7	235	34	210	30	2.0
	S	T71	195	28	145	21	3.5
	P	T6	265	38	185	27	5.0
	P	T7	220	32	165	24	6.0
357.0, A357.0	S	T62	360	52	290	42	8.0
359.0	P	T61	330	48	255	37	6.0
		T62	345	50	290	42	5.5
360.0	D	F	325	47	170	25	3.0
A360.0	D	F	320	46	165	24	5.0
380.0	D	F	330	48	165	24	3.0
383.0	D	F	310	45	150	22	3.5
384.0, A384.0	D	F	330	48	165	24	2.5
390.0	D	F	280	41	240	35	1.0
	D	T5	300	43	260	38	1.0
A390.0	S	F, T5	180	26	180	26	<1.0
	S	T6	280	40	280	40	<1.0
	S	T7	250	36	250	36	<1.0
	P	F, T5	200	29	200	29	1.0
	P	T6	310	45	310	45	<1.0
	P	T7	260	38	260	38	<1.0
413.0	D	F	300	43	140	21	2.5
A413.0	D	F	290	42	130	19	3.5

(continued)

Typical tensile properties for separately cast test bars of common aluminum casting alloys (continued)

Alloy	Product(a)	Temper	Tensile Strength		Yield strength(b)		Elonga-tion(c), %
			MPa	ksi	MPa	ksi	
443.0 . . . . S		F	130	19	55	8	8.0
B443.0 . . . . P		F	159	23	62	9	10.0
C443.0 . . . . D		F	228	33	110	16	9.0
514.0 . . . . S		F	170	25	85	12	9.0
518.0 . . . . D		F	310	45	190	28	5.0-8.0
520.0 . . . . S		T4	330	48	180	26	16
535.0 . . . . S		F	275	40	140	20	13
712.0 . . . . S		F	240	35	170	25	5.0
713.0 . . . . S		T5	210	30	150	22	3.0
	P	T5	220	32	150	22	4.0
771.0 . . . . S		T6	345	50	275	40	9.0
850.0 . . . . P		T5	160	23	75	11	10.0

(a) S = sand casting, P = permanent mold casting, D = die casting. (b) 0.2% offset. (c) With 12.7-mm (½-in.) diam specimen.

Extraction of aluminum from bauxite ore

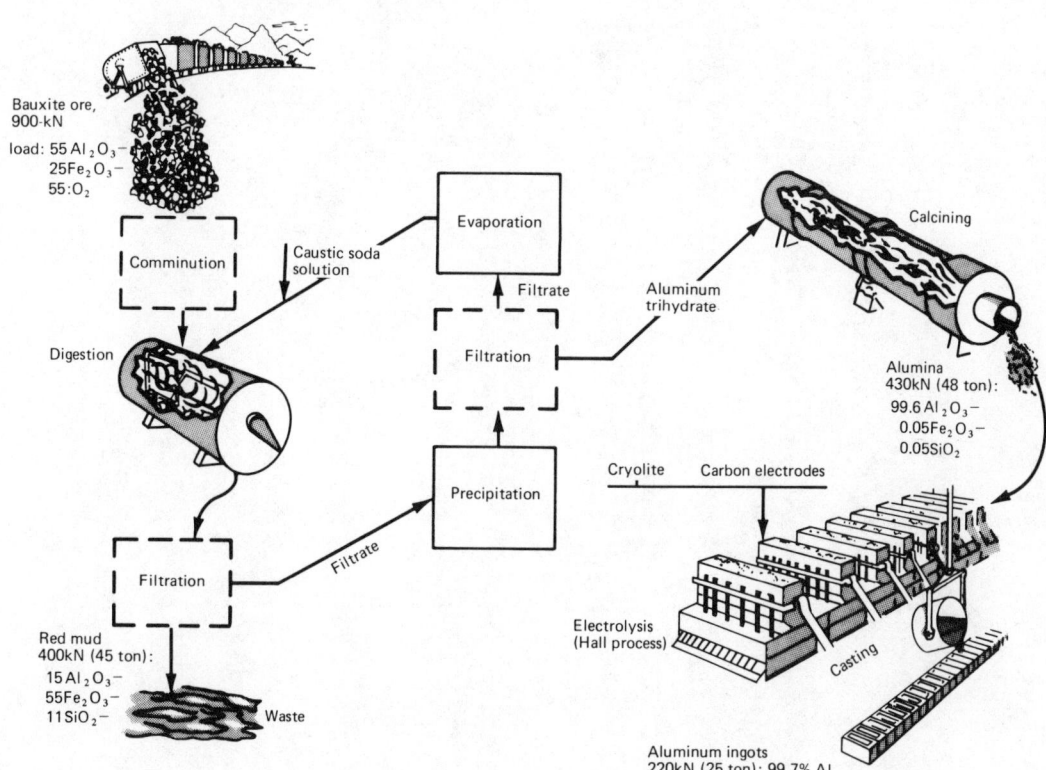

Typical mechanical properties of several aluminum sheet alloys (a)

Alloy and temper	Thickness		Ultimate		Tensile-test values(b) Yield					Cup test values		Bend diameter	
	mm	in.	MPa	ksi	MPa	ksi	$\bar{E}$, %	$\bar{n}$	$\bar{r}$	Olsen, O_d	Swift, LDR	Longi-tudinal	Trans-verse
2036-T4	0.99	0.039	349	50.6	203	29.5	23	0.226	0.75	0.306	2.10	1t	2t
3003-O	0.89	0.035	107	15.5	52	7.6	33	0.235	0.66	0.300	2.04	0	0
5052-O	0.86	0.034	211	30.6	97	14.0	22	0.282	0.62	0.316	2.08	0	0
5056-O	0.86	0.034	283	41.0	147	21.3	27	0.279	0.79	0.325	2.16	0	0
5086-O	1.63	0.064	284	41.2	136	19.7	24	0.359	0.68	0.365	2.06	2t	2t
5086-H32	1.63	0.064	303	43.9	203	29.5	16	0.291	0.71	0.340	2.01	2t	2t
5182-O	1.02	0.040	280	40.6	132	19.2	26	0.340	0.78	0.340	2.09	0	½t
5252-H25	0.79	0.031	247	35.8	192	27.8	14	0.140	1.04	0.216	1.98	· · ·	· · ·
6009-T4	1.32	0.052	261	37.8	150	21.7	24	0.264	0.68	0.330	2.04	0	½t
6010-T4	0.81	0.032	296	42.9	170	24.6	27	0.260	0.61	0.320	2.04	½t	½t
6061-T4	1.02	0.040	301	43.6	193	28.0	23	0.194	0.71	0.290	2.08	2t	2t
6151-T4	1.02	0.040	279	40.5	180	26.1	21	0.195	0.62	0.305	2.07	2t	2t
7021-O	2.36	0.093	195	28.3	132	19.1	20	0.202	0.990	0.345	2.15	2t	2t
7021-O	4.75	0.187	215	31.2	140	20.3	20	0.176	0.372	· · ·	· · ·	2t	2t
7029-O	2.56	0.101	273	39.6	123	17.9	22	0.298	0.611	0.293	2.15	2t	2t
7029-O	5.08	0.200	228	33.0	110	15.9	25	0.264	0.544	0.377	· · ·	2t	2t
7029-W	2.54	0.100	222	32.2	79	11.4	26	0.428	· · ·	0.387	2.30	2t	2t
7146-O	4.72	0.186	174	25.2	133	19.3	21	0.122	0.680	0.370	· · ·	2t	2t
7146-O	2.54	0.100	169	24.5	137	19.8	19	0.123	0.642	0.347	2.26	2t	2t

(a) See text for explanation of column headings. (b) Values shown are planar averages of tensile strength ($\bar{T}$), yield strength ($\bar{Y}$), elongation ($\bar{E}$), strain-hardening exponent ($\bar{n}$) and plastic strain ratio ($\bar{r}$).

Heat Treating

Typical solution and precipitation heat treatments for aluminum alloy mill products

These times and temperatures are typical for various forms, sizes and methods of manufacture and may not exactly describe optimum treatments for specific items.

Alloy	Product form	Solution heat treatment(a) Metal temperature(b) °C	°F	Temper designation	Precipitation heat treatment Metal temperature(b) °C	°F	Time(c), h	Temper designation
2011	Rolled or cold finished rod and bar 525		975	T3(d)	160	320	14	T8(d)
				T4	...	...	...	...
				T451(e)	...	...	...	...
2014(f)	Flat sheet 500		935	T3(d)	160	320	18	T6
				T42	160	320	18	T62
	Coiled sheet 500		935	T4	160	320	18	T6
				T42	160	320	18	T62
	Plate................. 500		935	T42	160	320	18	T62
				T451(e)	160	320	18	T651(e)
	Rolled or cold finished wire, rod and bar 500		935	T4	160(g)	320(g)	18	T6
				T42	160(g)	320(g)	18	T62
				T451(e)	160(g)	320(g)	18	T651(e)
	Extruded rod, bar, shapes and tube.............. 500		935	T4	160(g)	320(g)	18	T6
				T42	160(g)	320(g)	18	T62
				T4510(e)	160(g)	320(g)	18	T6510(e)
				T4511(e)	160(g)	320(g)	18	T6511(e)
	Drawn tube........... 500		935	T4	160(g)	320(g)	18	T6
				T42	160(g)	320(g)	18	T62
	Die forgings .... 500(h)		935(h)	T4	170	340	10	T6
	Hand forgings and rolled rings................ 500(h)		935(h)	T4	170	340	10	T6
				T452(j)	170	340	10	T652(j)
2017	Rolled or cold finished wire, rod and bar 500		935	T4	...	...	...	...
				T42	...	...	...	...
				T451(e)	...	...	...	...
2018	Die forgings .... 510(k)		950(k)	T4	170	340	10	T61
2024(f)	Flat sheet 495		920	T3(d)	190	375	12	T81(d)
				T361(d)	190	375	8	T861(d)
				T42	190	375	9	T62
					190	375	16	T72

(continued)

(a) Material should be quenched from the solution-treating temperature as rapidly as possible and with minimum delay after removal from the furnace. When material is quenched by total immersion in water, unless otherwise indicated, the water should be at room temperature, and should be suitably cooled so that it remains below 38 °C (100 °F) during the quenching cycle. Use of high-velocity, high-volume jets of cold water also is effective for some materials. (b) The nominal temperatures listed should be attained as rapidly as possible and maintained within ± 6 °C (± 10 °F) of nominal during the time at temperature. (c) Approximate time at temperature. The specific time will depend on the time required for the load to reach temperature. The times shown are based on rapid heating, with soak time measured from the time the load reaches a temperature within 6 °C (10 °F) of the applicable temperature. (d) Cold working subsequent to solution heat treatment and prior to any precipitation heat treatment is necessary to attain the specified properties for this temper. (e) Stress relieved by stretching to produce a specified amount of permanent set subsequent to solution heat treatment and prior to any precipitation heat treatment. (f) These heat treatments also apply to alclad sheet and plate in these alloys. (g) An alternative treatment of 8 h at 177 °C (350 °F) also may be used. (h) Solution heat treatment is followed by quenching in water 60 to 82 °C (140 to 180 °F). (j) Stress relieved by 1 to 5% cold reduction subsequent to solution heat treatment and prior to precipitation heat treatment. (k) Solution heat treatment is followed by quenching in water at 100 °C (212 °F). (m) Solution heat treatment is followed by quenching in room-temperature air blast. (n) By suitable control of extrusion temperature, product may be quenched directly from extrusion press to provide specified properties for this temper. Some products may be adequately quenched in room-temperature air blast. (p) See U.S. Patent 4 082 578. (q) Applicable to tread plate only. (r) An alternative treatment of 8 h at 171 °C (340 °F) also may be used. (s) Cold working subsequent to precipitation heat treatment is necessary to attain the specified properties for this temper. (t) An alternative treatment of 3 h at 182 °C (360 °F) also may be used. (u) An alternative treatment of 6 h at 182 °C (360 °F) also may be used. (v) No solution heat treatment; 72 h at room temperature following press quench, followed by two-stage precipitation heat treatment comprised of 8 h at 107 °C (225 °F) plus 16 h at 149 °C (300 °F). (w) Aging practice varies with product, size, nature of equipment, loading procedures and furnace-control capabilities. The optimum practice for a specific item can be ascertained only by actual trial treatment of the item under specific conditions. Typical procedures involve a two-stage treatment comprised of 3 to 30 h at 121 °C (250 °F) followed by 15 to 18 h at 163 °C (325 °F) for extrusions. An alternative two-stage treatment of 8 h at 99 °C (210 °F) followed by 24 to 28 h at 163 °C (325 °F) also may be used. (x) Aging of aluminum alloys 7050, 7075, 7175 and 7475 from any temper to the T73 or T76 temper series requires closer-than-normal controls on aging variables such as time, temperature, heatup rate, etc., for any given item. In addition, when material in a T6-type temper is reaged to a T73- or T76-type temper, the specific condition of the T6 material (such as property levels and other effects of processing variables) is extremely important and will affect the capability of the reaged material to conform to the requirements specified for the applicable T73- or T76-type temper. (y) Two-stage treatment comprised of 6 to 8 h at 107 °C (225 °F) followed by: 24 to 30 h at 163 °C (325 °F) for sheet and plate; 8 to 10 h at 177 °C (350 °F) for rolled or cold finished rod and bar; 6 to 8 h at 177 °C (350 °F) for extrusions and tube; 8 to 10 h at 177 °C (350 °F) for forgings in the T73 temper and 6 to 8 h at 177 °C (350 °F) for forgings in the T7352 temper. (z) An alternative two-stage treatment comprised of 4 h at 96 °C (205 °F) followed by 8 h at 157 °C (315 °F) also may be used. (aa) For sheet, plate, tube and extrusions, an alternative two-stage treatment comprised of 6 to 8 h at 107 °C (225 °F) followed by 14 to 18 h at 168 °C (335 °F) may be used, provided that a heatup rate of approximately 14 °C/h (25 °F/h) is employed. For rolled or cold finished rod and bar, the alternative treatment is 10 h at 177 °C (350 °F). (bb) An alternative three-stage treatment comprised of 5 h at 99 °C (210 °F), 4 h at 121 °C (250 °F) and then 4 h at 149 °C (300 °F) also may be used. (cc) 7175-T736 and -T73652 heat treatments are directed to specific results, may vary from supplier to supplier and are either proprietary or patented. (dd) Must be preceded by soak at 466 to 477 °C (870 to 890 °F). See U.S. Patent 3 791 880.

Typical solution and precipitation heat treatments for aluminum alloy mill products (continued)

Alloy	Product form	Solution heat treatment(a) Metal temperature(b) °C	°F	Temper designation	Precipitation heat treatment Metal temperature(b) °C	°F	Time(c), h	Temper designation
2024(f)	Coiled sheet 495		920	T4	...	...	...	...
				T42	190	375	9	T62
					190	375	16	T72
	Plate.................. 495		920	T351(e)	190	375	12	T851(e)
				T361(d)	190	375	8	T861(d)
				T42	190	375	9	T62
	Rolled or cold finished wire, rod and bar 495		920	T4	190	375	12	T6
				T351(e)	190	375	12	T851(e)
				T36(d)	190	375	8	T86(d)
				T42	190	375	16	T62
	Extruded rod, bar, shapes and tube............. 495		920	T3	190	375	12	T81
				T3510(e)	190	375	12	T8510(e)
				T3511(e)	190	375	12	T8511(e)
				T42	190	375	16	T62
	Drawn tube........... 495		920	T3(d)	...	...	...	...
				T42	...	...	...	...
2025	Die forgings 515		960	T4	170	340	10	T6
2036	Sheet 500		930	T4	...	...	...	...
2117	Rolled or cold finished wire and rod 500		935	T4				
				T42	...	...	...	...
2218	Die forgings .... 510(k)		950(k)	T4	170	340	10	T61
		510(m)	950(m)	T41	240	460	6	T72
2219(f)	Flat sheet 535		995	T31(d)	175	350	18	T81(d)
				T37(d)	165	325	24	T87(d)
				T42	190	375	36	T62
	Plate.................. 535		995	T31(d)	175	350	18	T81(d)
				T37(d)	175	350	18	T87(d)
				T351(e)	175	350	18	T851(e)
				T42	190	375	36	T62
	Rolled or cold finished wire, rod and bar 535		995	T351(e)	190	375	18	T851(e)

(continued)

(a) Material should be quenched from the solution-treating temperature as rapidly as possible and with minimum delay after removal from the furnace. When material is quenched by total immersion in water, unless otherwise indicated, the water should be at room temperature, and should be suitably cooled so that it remains below 38 °C (100 °F) during the quenching cycle. Use of high-velocity, high-volume jets of cold water also is effective for some materials. (b) The nominal temperatures listed should be attained as rapidly as possible and maintained within ± 6 °C (± 10 °F) of nominal during the time at temperature. (c) Approximate time at temperature. The specific time will depend on the time required for the load to reach temperature. The times shown are based on rapid heating, with soak time measured from the time the load reaches a temperature within 6 °C (10 °F) of the applicable temperature. (d) Cold working subsequent to solution heat treatment and prior to any precipitation heat treatment is necessary to attain the specified properties for this temper. (e) Stress relieved by stretching to produce a specified amount of permanent set subsequent to solution heat treatment and prior to any precipitation heat treatment. (f) These heat treatments also apply to alclad sheet and plate in these alloys. (g) An alternative treatment of 8 h at 177 °C (350 °F) also may be used. (h) Solution heat treatment is followed by quenching in water 60 to 82 °C (140 to 180 °F). (j) Stress relieved by 1 to 5% cold reduction subsequent to solution heat treatment and prior to precipitation heat treatment. (k) Solution heat treatment is followed by quenching in water at 100 °C (212 °F). (m) Solution heat treatment is followed by quenching in room-temperature air blast. (n) By suitable control of extrusion temperature, product may be quenched directly from extrusion press to provide specified properties for this temper. Some products may be adequately quenched in room-temperature air blast. (p) See U.S. Patent 4 082 578. (q) Applicable to tread plate only. (r) An alternative treatment of 8 h at 171 °C (340 °F) also may be used. (s) Cold working subsequent to precipitation heat treatment is necessary to attain the specified properties for this temper. (t) An alternative treatment of 3 h at 182 °C (360 °F) also may be used. (u) An alternative treatment of 6 h at 182 °C (360 °F) also may be used. (v) No solution heat treatment; 72 h at room temperature following press quench, followed by two-stage precipitation heat treatment comprised of 8 h at 107 °C (225 °F) plus 16 h at 149 °C (300 °F). (w) Aging practice varies with product, size, nature of equipment, loading procedures and furnace-control capabilities. The optimum practice for a specific item can be ascertained only by actual trial treatment of the item under specific conditions. Typical procedures involve a two-stage treatment comprised of 3 to 30 h at 121 °C (250 °F) followed by 15 to 18 h at 163 °C (325 °F) for extrusions. An alternative two-stage treatment of 8 h at 99 °C (210 °F) followed by 24 to 28 h at 163 °C (325 °F) also may be used. (x) Aging of aluminum alloys 7050, 7075, 7175 and 7475 from any temper to the T73 or T76 temper series requires closer-than-normal controls on aging variables such as time, temperature, heatup rate, etc., for any given item. In addition, when material in a T6-type temper is reaged to a T73- or T76-type temper, the specific condition of the T6 material (such as property levels and other effects of processing variables) is extremely important and will affect the capability of the reaged material to conform to the requirements specified for the applicable T73- or T76-type temper. (y) Two-stage treatment comprised of 6 to 8 h at 107 °C (225 °F) followed by: 24 to 30 h at 163 °C (325 °F) for sheet and plate; 8 to 10 h at 177 °C (350 °F) for rolled or cold finished rod and bar; 6 to 8 h at 177 °C (350 °F) for extrusions and tube; 8 to 10 h at 177 °C (350 °F) for forgings in the T73 temper and 6 to 8 h at 177 °C (350 °F) for forgings in the T7352 temper. (z) An alternative two-stage treatment comprised of 4 h at 96 °C (205 °F) followed by 8 h at 157 °C (315 °F) also may be used. (aa) For sheet, plate, tube and extrusions, an alternative two-stage treatment comprised of 6 to 8 h at 107 °C (225 °F) followed by 14 to 18 h at 168 °C (335 °F) may be used, provided that a heatup rate of approximately 14 °C/h (25 °F/h) is employed. For rolled or cold finished rod and bar, the alternative treatment is 10 h at 177 °C (350 °F). (bb) An alternative three-stage treatment comprised of 5 h at 99 °C (210 °F), 4 h at 121 °C (250 °F) and then 4 h at 149 °C (300 °F) also may be used. (cc) 7175-T736 and -T73652 heat treatments are directed to specific results, may vary from supplier to supplier and are either proprietary or patented. (dd) Must be preceded by soak at 466 to 477 °C (870 to 890 °F). See U.S. Patent 3 791 880.

Typical solution and precipitation heat treatments for aluminum alloy mill products (continued)

Alloy	Product form	Solution heat treatment(a) Metal temperature(b) °C	°F	Temper designation	Precipitation heat treatment Metal temperature(b) °C	°F	Time(c), h	Temper designation
2219(f)	Extruded rod, bar, shapes and tube............. 535		995	T31(d)	190	375	18	T81(d)
				T3510(e)	190	375	18	T8510(e)
				T3511(e)	190	375	18	T8511(e)
				T42	190	375	36	T62
	Die forgings and rolled rings................ 535		995	T4	190	375	26	T6
	Hand forgings......... 535		995	T4	190	375	26	T6
				T352(j)	175	350	18	T852(j)
2618	Forgings and rolled rings................ 530		985	T4	200	390	20	T61
4032	Die forgings .... 510(h)		950(h)	T4	170	340	10	T6
6005	Extruded rod, bar, shapes and tube............. 530(n)		985(n)	T1	175	350	8	T5
6009(p)	Coiled sheet 555		1030	T4	175	350	8	T6
6010(p)	Coiled sheet 565		1050	T4	175	350	8	T6
6053	Die forgings 520		970	T4	170	340	10	T6
6061(f)	Sheet 530		985	T4	160	320	18	T6
				T42	160	320	18	T62
	Plate................. 530		985	T4(q)	160	320	18	T6(q)
				T42	160	320	18	T62
				T451(e)	160	320	18	T651(e)
	Rolled or cold finished wire, rod and bar 530		985	T4	160(r)	320(r)	18	T6
					160(r)	320(r)	18	T89(d)
					160(r)	320(r)	18	T93(s)
					160(r)	320(r)	18	T913(s)
					160(r)	320(r)	18	T94(s)
				T42	160(r)	320(r)	18	T62
				T451(e)	160(r)	320(r)	18	T651(e)
	Extruded rod, bar, shapes and tube............. 530(n)		985(n)	T4	175	350	8	T6
				T4510(e)	175	350	8	T6510(e)
				T4511(e)	175	350	8	T6511(e)
		530	985	T42	175	350	8	T62

(continued)

(a) Material should be quenched from the solution-treating temperature as rapidly as possible and with minimum delay after removal from the furnace. When material is quenched by total immersion in water, unless otherwise indicated, the water should be at room temperature, and should be suitably cooled so that it remains below 38 °C (100 °F) during the quenching cycle. Use of high-velocity, high-volume jets of cold water also is effective for some materials. (b) The nominal temperatures listed should be attained as rapidly as possible and maintained within ± 6 °C (± 10 °F) of nominal during the time at temperature. (c) Approximate time at temperature. The specific time will depend on the time required for the load to reach temperature. The times shown are based on rapid heating, with soak time measured from the time the load reaches a temperature within 6 °C (10 °F) of the applicable temperature. (d) Cold working subsequent to solution heat treatment and prior to any precipitation heat treatment is necessary to attain the specified properties for this temper. (e) Stress relieved by stretching to produce a specified amount of permanent set subsequent to solution heat treatment and prior to any precipitation heat treatment. (f) These heat treatments also apply to alclad sheet and plate in these alloys. (g) An alternative treatment of 8 h at 177 °C (350 °F) also may be used. (h) Solution heat treatment is followed by quenching in water 60 to 82 °C (140 to 180 °F). (j) Stress relieved by 1 to 5% cold reduction subsequent to solution heat treatment and prior to precipitation heat treatment. (k) Solution heat treatment is followed by quenching in water at 100 °C (212 °F). (m) Solution heat treatment is followed by quenching in room-temperature air blast. (n) By suitable control of extrusion temperature, product may be quenched directly from extrusion press to provide specified properties for this temper. Some products may be adequately quenched in room-temperature air blast. (p) See U.S. Patent 4 082 578. (q) Applicable to tread plate only. (r) An alternative treatment of 8 h at 171 °C (340 °F) also may be used. (s) Cold working subsequent to precipitation heat treatment is necessary to attain the specified properties for this temper. (t) An alternative treatment of 3 h at 182 °C (360 °F) also may be used. (u) An alternative treatment of 6 h at 182 °C (360 °F) also may be used. (v) No solution heat treatment; 72 h at room temperature following press quench, followed by two-stage precipitation heat treatment comprised of 8 h at 107 °C (225 °F) plus 16 h at 149 °C (300 °F). (w) Aging practice varies with product, size, nature of equipment, loading procedures and furnace-control capabilities. The optimum practice for a specific item can be ascertained only by actual trial treatment of the item under specific conditions. Typical procedures involve a two-stage treatment comprised of 3 to 30 h at 121 °C (250 °F) followed by 15 to 18 h at 163 °C (325 °F) for extrusions. An alternative two-stage treatment of 8 h at 99 °C (210 °F) followed by 24 to 28 h at 163 °C (325 °F) also may be used. (x) Aging of aluminum alloys 7050, 7075, 7175 and 7475 from any temper to the T73 or T76 temper series requires closer-than-normal controls on aging variables such as time, temperature, heatup rate, etc., for any given item. In addition, when material in a T6-type temper is reaged to a T73- or T76-type temper, the specific condition of the T6 material (such as property levels and other effects of processing variables) is extremely important and will affect the capability of the reaged material to conform to the requirements specified for the applicable T73- or T76-type temper. (y) Two-stage treatment comprised of 6 to 8 h at 107 °C (225 °F) followed by: 24 to 30 h at 163 °C (325 °F) for sheet and plate; 8 to 10 h at 177 °C (350 °F) for rolled or cold finished rod and bar; 6 to 8 h at 177 °C (350 °F) for extrusions and tube; 8 to 10 h at 177 °C (350 °F) for forgings in the T73 temper and 6 to 8 h at 177 °C (350 °F) for forgings in the T7352 temper. (z) An alternative two-stage treatment comprised of 4 h at 96 °C (205 °F) followed by 8 h at 157 °C (315 °F) also may be used. (aa) For sheet, plate, tube and extrusions, an alternative two-stage treatment comprised of 6 to 8 h at 107 °C (225 °F) followed by 14 to 18 h at 168 °C (335 °F) may be used, provided that a heatup rate of approximately 14 °C/h (25 °F/h) is employed. For rolled or cold finished rod and bar, the alternative treatment is 10 h at 177 °C (350 °F). (bb) An alternative three-stage treatment comprised of 5 h at 99 °C (210 °F), 4 h at 121 °C (250 °F) and then 4 h at 149 °C (300 °F) also may be used. (cc) 7175-T736 and -T73652 heat treatments are directed to specific results, may vary from supplier to supplier and are either proprietary or patented. (dd) Must be preceded by soak at 466 to 477 °C (870 to 890 °F). See U.S. Patent 3 791 880.

Typical solution and precipitation heat treatments for aluminum alloy mill products (continued)

Alloy	Product form	Solution heat treatment(a) Metal temperature(b) °C	°F	Temper designation	Precipitation heat treatment Metal temperature(b) °C	°F	Time(c), h	Temper designation
6061(f)	Drawn tube	530	985	T4	160(r)	320(r)	18	T6
				T42	160(r)	320(r)	18	T62
	Die and hand forgings	530	985	T4	175	350	8	T6
	Rolled rings	530	985	T4	175	350	8	T6
				T452(j)	175	350	8	T652(j)
6063	Extruded rod, bar, shapes and tube	(n)	(n)	T1	205(t)	400(t)	1	T5
		520(n)	970(n)	T4	175(u)	350(u)	8	T6
		520	970	T42	175(u)	350(u)	8	T62
	Drawn tube	520	970	T4	175	350	8	T6
					175	350	8	T83(d)(n)
					175	350	8	T831(d)(n)
					175	350	8	T832(d)(n)
				T42	175	350	8	T62
6066	Extruded rod, bar, shapes and tube	530	990	T4	175	350	8	T6
				T42	175	350	8	T62
				T4510(e)	175	350	8	T6510(e)
				T4511(e)	175	350	8	T6511(e)
	Drawn tube	530	990	T4	175	350	8	T6
				T42	175	350	8	T62
	Die forgings	530	990	T4	175	350	8	T6
6070	Extruded rod, bar, shapes and tube	545(n)	1015(n)	T4	160	320	18	T6
				T42	160	320	18	T62
6151	Die forgings	515	960	T4	170	340	10	T6
	Rolled rings	515	960	T4	170	340	10	T6
				T452(j)	170	340	10	T652(j)
6262	Rolled or cold finished wire, rod and bar	540	1000	T4	170	340	8	T6
					170	340	12	T9(s)
				T451	170	340	8	T651(e)
				T42	170	340	8	T62

(continued)

(a) Material should be quenched from the solution-treating temperature as rapidly as possible and with minimum delay after removal from the furnace. When material is quenched by total immersion in water, unless otherwise indicated, the water should be at room temperature, and should be suitably cooled so that it remains below 38 °C (100 °F) during the quenching cycle. Use of high-velocity, high-volume jets of cold water also is effective for some materials. (b) The nominal temperatures listed should be attained as rapidly as possible and maintained within ± 6 °C (± 10 °F) of nominal during the time at temperature. (c) Approximate time at temperature. The specific time will depend on the time required for the load to reach temperature. The times shown are based on rapid heating, with soak time measured from the time the load reaches a temperature within 6 °C (10 °F) of the applicable temperature. (d) Cold working subsequent to solution heat treatment and prior to any precipitation heat treatment is necessary to attain the specified properties for this temper. (e) Stress relieved by stretching to produce a specified amount of permanent set subsequent to solution heat treatment and prior to any precipitation heat treatment. (f) These heat treatments also apply to alclad sheet and plate in these alloys. (g) An alternative treatment of 8 h at 177 °C (350 °F) also may be used. (h) Solution heat treatment is followed by quenching in water 60 to 82 °C (140 to 180 °F). (j) Stress relieved by 1 to 5% cold reduction subsequent to solution heat treatment and prior to precipitation heat treatment. (k) Solution heat treatment is followed by quenching in water at 100 °C (212 °F). (m) Solution heat treatment is followed by quenching in room-temperature air blast. (n) By suitable control of extrusion temperature, product may be quenched directly from extrusion press to provide specified properties for this temper. Some products may be adequately quenched in room-temperature air blast. (p) See U.S. Patent 4 082 578. (q) Applicable to tread plate only. (r) An alternative treatment of 8 h at 171 °C (340 °F) also may be used. (s) Cold working subsequent to precipitation heat treatment is necessary to attain the specified properties for this temper. (t) An alternative treatment of 3 h at 182 °C (360 °F) also may be used. (u) An alternative treatment of 6 h at 182 °C (360 °F) also may be used. (v) No solution heat treatment; 72 h at room temperature following press quench, followed by two-stage precipitation heat treatment comprised of 8 h at 107 °C (225 °F) plus 16 h at 149 °C (300 °F). (w) Aging practice varies with product, size, nature of equipment, loading procedures and furnace-control capabilities. The optimum practice for a specific item can be ascertained only by actual trial treatment of the item under specific conditions. Typical procedures involve a two-stage treatment comprised of 3 to 30 h at 121 °C (250 °F) followed by 15 to 18 h at 163 °C (325 °F) for extrusions. An alternative two-stage treatment of 8 h at 99 °C (210 °F) followed by 24 to 28 h at 163 °C (325 °F) also may be used. (x) Aging of aluminum alloys 7050, 7075, 7175 and 7475 from any temper to the T73 or T76 temper series requires closer-than-normal controls on aging variables such as time, temperature, heatup rate, etc., for any given item. In addition, when material in a T6-type temper is reaged to a T73- or T76-type temper, the specific condition of the T6 material (such as property levels and other effects of processing variables) is extremely important and will affect the capability of the reaged material to conform to the requirements specified for the applicable T73- or T76-type temper. (y) Two-stage treatment comprised of 6 to 8 h at 107 °C (225 °F) followed by: 24 to 30 h at 163 °C (325 °F) for sheet and plate; 8 to 10 h at 177 °C (350 °F) for rolled or cold finished rod and bar; 6 to 8 h at 177 °C (350 °F) for extrusions and tube; 8 to 10 h at 177 °C (350 °F) for forgings in the T73 temper and 6 to 8 h at 177 °C (350 °F) for forgings in the T7352 temper. (z) An alternative two-stage treatment comprised of 4 h at 96 °C (205 °F) followed by 8 h at 157 °C (315 °F) also may be used. (aa) For sheet, plate, tube and extrusions, an alternative two-stage treatment comprised of 6 to 8 h at 107 °C (225 °F) followed by 14 to 18 h at 168 °C (335 °F) may be used, provided that a heatup rate of approximately 14 °C/h (25 °F/h) is employed. For rolled or cold finished rod and bar, the alternative treatment is 10 h at 177 °C (350 °F). (bb) An alternative three-stage treatment comprised of 5 h at 99 °C (210 °F), 4 h at 121 °C (250 °F) and then 4 h at 149 °C (300 °F) also may be used. (cc) 7175-T736 and -T73652 heat treatments are directed to specific results, may vary from supplier to supplier and are either proprietary or patented. (dd) Must be preceded by soak at 466 to 477 °C (870 to 890 °F). See U.S. Patent 3 791 880.

313

Typical solution and precipitation heat treatments for aluminum alloy mill products (continued)

Alloy	Product form	Solution heat treatment(a) Metal temperature(b) °C	°F	Temper designation	Precipitation heat treatment Metal temperature(b) °C	°F	Time(c), h	Temper designation
6262	Extruded rod, bar, shapes and tube	540(n)	1000(n)	T4	175	350	12	T6
				T4510(e)	175	350	12	T6510(e)
				T4511(e)	175	350	12	T6511(e)
		540	1000	T42	175	350	12	T62
	Drawn tube	540	1000	T4	170	340	8	T6
					170	340	8	T9(s)
				T42	170	340	8	T62
6463	Extruded rod, bar, shapes and tube	(n)	(n)	T1	205(t)	400(t)	1	T5
		520(n)	970(n)	T4	175(u)	350(u)	8	T6
		520	970	T42	175(u)	350(u)	8	T62
6951	Sheet	530	985	T4	160	320	18	T6
				T42	160	320	18	T62
7001	Extruded rod, bar, shapes and tube	465	870	W	120	250	24	T6
					120	250	24	T62
				W510(e)	120	250	24	T6510(e)
				W511(e)	120	250	24	T6511(e)
7005	Extruded rod, bar and shapes	...	...	...	...	...	...	T53(v)
7050	Plate	475	890	W51(e)	(w)	(w)	(w)	T7651(x)
					(y)	(y)	(y)	T73651(x)
	Extrusions	475	890	W510(e)	(w)	(w)	(w)	T76510(x)
				W511(e)	(w)	(w)	(w)	T76511(x)
	Die and hand forgings	475	890	W	(y)	(y)	(y)	T736(x)
				W52(e)	(y)	(y)	(y)	T73652(x)
7075(f)	Sheet	480	900	W	120(z)	250(z)	24	T6
					120(z)	250(z)	24	T62
					(w)	(w)	(w)	T76(x)
					(y)(aa)	(y)(aa)	(y)(aa)	T73(x)
	Plate	480	900	W	120(z)	250(z)	24	T62
				W51(e)	(y)(aa)	(y)(aa)	(y)(aa)	T7351(e)(x)
					120(z)	250(z)	24	T651(e)
					(w)	(w)	(w)	T7651(x)

(continued)

(a) Material should be quenched from the solution-treating temperature as rapidly as possible and with minimum delay after removal from the furnace. When material is quenched by total immersion in water, unless otherwise indicated, the water should be at room temperature, and should be suitably cooled so that it remains below 38 °C (100 °F) during the quenching cycle. Use of high-velocity, high-volume jets of cold water also is effective for some materials. (b) The nominal temperatures listed should be attained as rapidly as possible and maintained within ± 6 °C (± 10 °F) of nominal during the time at temperature. (c) Approximate time at temperature. The specific time will depend on the time required for the load to reach temperature. The times shown are based on rapid heating, with soak time measured from the time the load reaches a temperature within 6 °C (10 °F) of the applicable temperature. (d) Cold working subsequent to solution heat treatment and prior to any precipitation heat treatment is necessary to attain the specified properties for this temper. (e) Stress relieved by stretching to produce a specified amount of permanent set subsequent to solution heat treatment and prior to any precipitation heat treatment. (f) These heat treatments also apply to alclad sheet and plate in these alloys. (g) An alternative treatment of 8 h at 177 °C (350 °F) also may be used. (h) Solution heat treatment is followed by quenching in water 60 to 82 °C (140 to 180 °F). (j) Stress relieved by 1 to 5% cold reduction subsequent to solution heat treatment and prior to precipitation heat treatment. (k) Solution heat treatment is followed by quenching in water at 100 °C (212 °F). (m) Solution heat treatment is followed by quenching in room-temperature air blast. (n) By suitable control of extrusion temperature, product may be quenched directly from extrusion press to provide specified properties for this temper. Some products may be adequately quenched in room-temperature air blast. (p) See U.S. Patent 4 082 578. (q) Applicable to tread plate only. (r) An alternative treatment of 8 h at 171 °C (340 °F) also may be used. (s) Cold working subsequent to precipitation heat treatment is necessary to attain the specified properties for this temper. (t) An alternative treatment of 3 h at 182 °C (360 °F) also may be used. (u) An alternative treatment of 6 h at 182 °C (360 °F) also may be used. (v) No solution heat treatment; 72 h at room temperature following press quench, followed by two-stage precipitation heat treatment comprised of 8 h at 107 °C (225 °F) plus 16 h at 149 °C (300 °F). (w) Aging practice varies with product, size, nature of equipment, loading procedures and furnace-control capabilities. The optimum practice for a specific item can be ascertained only by actual trial treatment of the item under specific conditions. Typical procedures involve a two-stage treatment comprised of 3 to 30 h at 121 °C (250 °F) followed by 15 to 18 h at 163 °C (325 °F) for extrusions. An alternative two-stage treatment of 8 h at 99 °C (210 °F) followed by 24 to 28 h at 163 °C (325 °F) also may be used. (x) Aging of aluminum alloys 7050, 7075, 7175 and 7475 from any temper to the T73 or T76 temper series requires closer-than-normal controls on aging variables such as time, temperature, heatup rate, etc., for any given item. In addition, when material in a T6-type temper is reaged to a T73- or T76-type temper, the specific condition of the T6 material (such as property levels and other effects of processing variables) is extremely important and will affect the capability of the reaged material to conform to the requirements specified for the applicable T73- or T76-type temper. (y) Two-stage treatment comprised of 6 to 8 h at 107 °C (225 °F) followed by: 24 to 30 h at 163 °C (325 °F) for sheet and plate; 8 to 10 h at 177 °C (350 °F) for rolled or cold finished rod and bar; 6 to 8 h at 177 °C (350 °F) for extrusions and tube; 8 to 10 h at 177 °C (350 °F) for forgings in the T73 temper and 6 to 8 h at 177 °C (350 °F) for forgings in the T7352 temper. (z) An alternative two-stage treatment comprised of 4 h at 96 °C (205 °F) followed by 8 h at 157 °C (315 °F) also may be used. (aa) For sheet, plate, tube and extrusions, an alternative two-stage treatment comprised of 6 to 8 h at 107 °C (225 °F) followed by 14 to 18 h at 168 °C (335 °F) may be used, provided that a heatup rate of approximately 14 °C/h (25 °F/h) is employed. For rolled or cold finished rod and bar, the alternative treatment is 10 h at 177 °C (350 °F). (bb) An alternative three-stage treatment comprised of 5 h at 99 °C (210 °F), 4 h at 121 °C (250 °F) and then 4 h at 149 °C (300 °F) also may be used. (cc) 7175-T736 and -T73652 heat treatments are directed to specific results, may vary from supplier to supplier and are either proprietary or patented. (dd) Must be preceded by soak at 466 to 477 °C (870 to 890 °F). See U.S. Patent 3 791 880.

Typical solution and precipitation heat treatments for aluminum alloy mill products (continued)

Alloy	Product form	Solution heat treatment(a) Metal temperature(b) °C	°F	Temper designation	Precipitation heat treatment Metal temperature(b) °C	°F	Time(c), h	Temper designation
7075(f)	Rolled or cold finished wire, rod and bar	490	915	W	120	250	24	T6
					120	250	24	T62
					(y)(aa)	(y)(aa)	(y)(aa)	T73(x)
				W51(e)	120	250	24	T651(e)
					(y)(aa)	(y)(aa)	(y)(aa)	T7351(e)(x)
	Extruded rod, bar, shapes and tube.............	465	870	W	120(bb)	250(bb)	24	T6
					120(bb)	250(bb)	24	T62
					(y)(aa)	(y)(aa)	(y)(aa)	T73(x)
					(w)	(w)	(w)	T76(x)
				W510(e)	120(bb)	250(bb)	24	T6510(e)
					(y)(aa)	(y)(aa)	(y)(aa)	T73510(e)(x)
					(w)	(w)	(w)	T76510(x)
				W511(e)	120(bb)	250(bb)	24	T6511(e)
					(y)(aa)	(y)(aa)	(y)(aa)	T73511(e)(x)
					(w)	(w)	(w)	T76511(x)
	Drawn tube...........	465	870	W	120	250	24	T6
					120	250	24	T62
					(y)(aa)	(y)(aa)	(y)(aa)	T73(x)
	Die forgings	470(h)	880(h)	W	120	250	24	T6
					(y)	(y)	(y)	T73(x)
				W52(j)	(y)	(y)	(y)	T7352(j)(x)
	Hand forgings.........	470(h)	880(h)	W	120	250	24	T6
					(y)	(y)	(y)	T73(x)
				W52(j)	120	250	24	T652(j)
					(y)	(y)	(y)	T7352(j)(x)
	Rolled rings...........	470	880	W	120	250	24	T6
7175	Die forgings	(cc)	(cc)	W	(cc)	(cc)	(cc)	T66(cc)
		(cc)	(cc)	W	(cc)	(cc)	(cc)	T736(x)(cc)
		(cc)	(cc)	W52(j)	(cc)	(cc)	(cc)	T73652(j)(x)(cc)
	Hand forgings.........	(cc)	(cc)	W	(cc)	(cc)	(cc)	T736(x)(cc)
		(cc)	(cc)	W52(j)	(cc)	(cc)	(cc)	T7365(j)(x)(cc)

(continued)

(a) Material should be quenched from the solution-treating temperature as rapidly as possible and with minimum delay after removal from the furnace. When material is quenched by total immersion in water, unless otherwise indicated, the water should be at room temperature, and should be suitably cooled so that it remains below 38 °C (100 °F) during the quenching cycle. Use of high-velocity, high-volume jets of cold water also is effective for some materials. (b) The nominal temperatures listed should be attained as rapidly as possible and maintained within ± 6 °C (± 10 °F) of nominal during the time at temperature. (c) Approximate time at temperature. The specific time will depend on the time required for the load to reach temperature. The times shown are based on rapid heating, with soak time measured from the time the load reaches a temperature within 6 °C (10 °F) of the applicable temperature. (d) Cold working subsequent to solution heat treatment and prior to any precipitation heat treatment is necessary to attain the specified properties for this temper. (e) Stress relieved by stretching to produce a specified amount of permanent set subsequent to solution heat treatment and prior to any precipitation heat treatment. (f) These heat treatments also apply to alclad sheet and plate in these alloys. (g) An alternative treatment of 8 h at 177 °C (350 °F) also may be used. (h) Solution heat treatment is followed by quenching in water 60 to 82 °C (140 to 180 °F). (j) Stress relieved by 1 to 5% cold reduction subsequent to solution heat treatment and prior to precipitation heat treatment. (k) Solution heat treatment is followed by quenching in water at 100 °C (212 °F). (m) Solution heat treatment is followed by quenching in room-temperature air blast. (n) By suitable control of extrusion temperature, product may be quenched directly from extrusion press to provide specified properties for this temper. Some products may be adequately quenched in room-temperature air blast. (p) See U.S. Patent 4 082 578. (q) Applicable to tread plate only. (r) An alternative treatment of 8 h at 171 °C (340 °F) also may be used. (s) Cold working subsequent to precipitation heat treatment is necessary to attain the specified properties for this temper. (t) An alternative treatment of 3 h at 182 °C (360 °F) also may be used. (u) An alternative treatment of 6 h at 182 °C (360 °F) also may be used. (v) No solution heat treatment; 72 h at room temperature following press quench, followed by two-stage precipitation heat treatment comprised of 8 h at 107 °C (225 °F) plus 16 h at 149 °C (300 °F). (w) Aging practice varies with product, size, nature of equipment, loading procedures and furnace-control capabilities. The optimum practice for a specific item can be ascertained only by actual trial treatment of the item under specific conditions. Typical procedures involve a two-stage treatment comprised of 3 to 30 h at 121 °C (250 °F) followed by 15 to 18 h at 163 °C (325 °F) for extrusions. An alternative two-stage treatment of 8 h at 99 °C (210 °F) followed by 24 to 28 h at 163 °C (325 °F) also may be used. (x) Aging of aluminum alloys 7050, 7075, 7175 and 7475 from any temper to the T73 or T76 temper series requires closer-than-normal controls on aging variables such as time, temperature, heatup rate, etc., for any given item. In addition, when material in a T6-type temper is reaged to a T73- or T76-type temper, the specific condition of the T6 material (such as property levels and other effects of processing variables) is extremely important and will affect the capability of the reaged material to conform to the requirements specified for the applicable T73- or T76-type temper. (y) Two-stage treatment comprised of 6 to 8 h at 107 °C (225 °F) followed by: 24 to 30 h at 163 °C (325 °F) for sheet and plate; 8 to 10 h at 177 °C (350 °F) for rolled or cold finished rod and bar; 6 to 8 h at 177 °C (350 °F) for extrusions and tube; 8 to 10 h at 177 °C (350 °F) for forgings in the T73 temper and 6 to 8 h at 177 °C (350 °F) for forgings in the T7352 temper. (z) An alternative two-stage treatment comprised of 4 h at 96 °C (205 °F) followed by 8 h at 157 °C (315 °F) also may be used. (aa) For sheet, plate, tube and extrusions, an alternative two-stage treatment comprised of 6 to 8 h at 107 °C (225 °F) followed by 14 to 18 h at 168 °C (335 °F) may be used, provided that a heatup rate of approximately 14 °C/h (25 °F/h) is employed. For rolled or cold finished rod and bar, the alternative treatment is 10 h at 177 °C (350 °F). (bb) An alternative three-stage treatment comprised of 5 h at 99 °C (210 °F), 4 h at 121 °C (250 °F) and then 4 h at 149 °C (300 °F) also may be used. (cc) 7175-T736 and -T73652 heat treatments are directed to specific results, may vary from supplier to supplier and are either proprietary or patented. (dd) Must be preceded by soak at 466 to 477 °C (870 to 890 °F). See U.S. Patent 3 791 880.

Typical solution and precipitation heat treatments for aluminum alloy mill products (continued)

Alloy	Product form	Solution heat treatment(a) Metal temperature(b) °C	°F	Temper designation	Precipitation heat treatment Metal temperature(b) °C	°F	Time(c), h	Temper designation
7475	Sheet	515(dd)	960(dd)	W	120	250	3	
				plus 155	315	3	T61(dd)
					(w)	(w)	(w)	T761(x)(dd)
	Plate.................	510(dd)	950(dd)	W51(e)	120	250	24	T651(dd)
					(w)	(w)	(w)	T7651(x)(dd)
					(y)	(y)	(y)	T7351(x)(dd)
Alclad 7475	Sheet	495	920	W	120	250	3	
				plus 155	315	3	T61(dd)
					(w)	(w)	(w)	T761(x)(dd)

(a) Material should be quenched from the solution-treating temperature as rapidly as possible and with minimum delay after removal from the furnace. When material is quenched by total immersion in water, unless otherwise indicated, the water should be at room temperature, and should be suitably cooled so that it remains below 38 °C (100 °F) during the quenching cycle. Use of high-velocity, high-volume jets of cold water also is effective for some materials. (b) The nominal temperatures listed should be attained as rapidly as possible and maintained within ± 6 °C (± 10 °F) of nominal during the time at temperature. (c) Approximate time at temperature. The specific time will depend on the time required for the load to reach temperature. The times shown are based on rapid heating, with soak time measured from the time the load reaches a temperature within 6 °C (10 °F) of the applicable temperature. (d) Cold working subsequent to solution heat treatment and prior to any precipitation heat treatment is necessary to attain the specified properties for this temper. (e) Stress relieved by stretching to produce a specified amount of permanent set subsequent to solution heat treatment and prior to any precipitation heat treatment. (f) These heat treatments also apply to alclad sheet and plate in these alloys. (g) An alternative treatment of 8 h at 177 °C (350 °F) also may be used. (h) Solution heat treatment is followed by quenching in water 60 to 82 °C (140 to 180 °F). (j) Stress relieved by 1 to 5% cold reduction subsequent to solution heat treatment and prior to precipitation heat treatment. (k) Solution heat treatment is followed by quenching in water at 100 °C (212 °F). (m) Solution heat treatment is followed by quenching in room-temperature air blast. (n) By suitable control of extrusion temperature, product may be quenched directly from extrusion press to provide specified properties for this temper. Some products may be adequately quenched in room-temperature air blast. (p) See U.S. Patent 4 082 578. (q) Applicable to tread plate only. (r) An alternative treatment of 8 h at 171 °C (340 °F) also may be used. (s) Cold working subsequent to precipitation heat treatment is necessary to attain the specified properties for this temper. (t) An alternative treatment of 3 h at 182 °C (360 °F) also may be used. (u) An alternative treatment of 6 h at 182 °C (360 °F) also may be used. (v) No solution heat treatment; 72 h at room temperature following press quench, followed by two-stage precipitation heat treatment comprised of 8 h at 107 °C (225 °F) plus 16 h at 149 °C (300 °F). (w) Aging practice varies with product, size, nature of equipment, loading procedures and furnace-control capabilities. The optimum practice for a specific item can be ascertained only by actual trial treatment of the item under specific conditions. Typical procedures involve a two-stage treatment comprised of 3 to 30 h at 121 °C (250 °F) followed by 15 to 18 h at 163 °C (325 °F) for extrusions. An alternative two-stage treatment of 8 h at 99 °C (210 °F) followed by 24 to 28 h at 163 °C (325 °F) also may be used. (x) Aging of aluminum alloys 7050, 7075, 7175 and 7475 from any temper to the T73 or T76 temper series requires closer-than-normal controls on aging variables such as time, temperature, heatup rate, etc., for any given item. In addition, when material in a T6-type temper is reaged to a T73- or T76-type temper, the specific condition of the T6 material (such as property levels and other effects of processing variables) is extremely important and will affect the capability of the reaged material to conform to the requirements specified for the applicable T73- or T76-type temper. (y) Two-stage treatment comprised of 6 to 8 h at 107 °C (225 °F) followed by: 24 to 30 h at 163 °C (325 °F) for sheet and plate; 8 to 10 h at 177 °C (350 °F) for rolled or cold finished rod and bar; 6 to 8 h at 177 °C (350 °F) for extrusions and tube; 8 to 10 h at 177 °C (350 °F) for forgings in the T73 temper and 6 to 8 h at 177 °C (350 °F) for forgings in the T7352 temper. (z) An alternative two-stage treatment comprised of 4 h at 96 °C (205 °F) followed by 8 h at 157 °C (315 °F) also may be used. (aa) For sheet, plate, tube and extrusions, an alternative two-stage treatment comprised of 6 to 8 h at 107 °C (225 °F) followed by 14 to 18 h at 168 °C (335 °F) may be used, provided that a heatup rate of approximately 14 °C/h (25 °F/h) is employed. For rolled or cold finished rod and bar, the alternative treatment is 10 h at 177 °C (350 °F). (bb) An alternative three-stage treatment comprised of 5 h at 99 °C (210 °F), 4 h at 121 °C (250 °F) and then 4 h at 149 °C (300 °F) also may be used. (cc) 7175-T736 and -T73652 heat treatments are directed to specific results, may vary from supplier to supplier and are either proprietary or patented. (dd) Must be preceded by soak at 466 to 477 °C (870 to 890 °F). See U.S. Patent 3 791 880.

Typical heat treatments for aluminum alloy sand and permanent mold castings

Alloy	Temper	Type of casting(a)	Solution heat treatment(b) Temperature(c) °C	°F	Time, h	Aging treatment Temperature(c) °C	°F	Time, h
201.0	T6	S	510-515; 525-530	950-960; 980-990	2 14-20	155	310	20
	T7	S	510-515; 525-530	950-960; 980-990	2 14-20	190	370	5
204.0	T4	S or P	520	970	10	...	...	...
208.0	T55	S	...	...	...	155	310	16
222.0	O(d)	S	...	...	...	315	600	3
	T61	S	510	950	12	155	310	11
	T551	P	...	...	...	170	340	16-22
	T65	...	510	950	4-12	170	340	7-9
242.0	O(e)	S	...	...	...	345	650	3
	T571	S	...	...	...	205	400	8
		P	...	...	...	165-170	330-340	22-26
	T77	S	515	960	5(f)	330-355	625-675	2 (min)
	T61	S or P	515	960	4-12(f)	205-230	400-450	3-5
295.0	T4	S	515	960	12	...	...	...
	T6	S	515	960	12	155	310	3-6
	T62	S	515	960	12	155	310	12-24
	T7	S	515	960	12	260	500	4-6
296.0	T4	P	510	950	8	...	...	...
	T6	P	510	950	8	155	310	1-8
	T7	P	510	950	8	260	500	4-6
319.0	T5	S	...	...	...	205	400	8
	T6	S	505	940	12	155	310	2-5
		P	505	940	4-12	155	310	2-5
328.0	T6	S	515	960	12	155	310	2-5
332.0	T5	P	...	...	...	205	400	7-9
333.0	T5	P	...	...	...	205	400	7-9
	T6	P	505	940	6-12	155	310	2-5
	T7	P	505	940	6-12	260	500	4-6
336.0	T551	P	...	...	...	205	400	7-9
	T65	P	515	960	8	205	400	7-9
354.0	...	(g)	525-535	980-995	10-12	(h)	(h)	(h)
355.0	T51	S or P	...	...	...	225	440	7-9
	T6	S	525	980	12	155	310	3-5
		P	525	980	4-12	155	310	2-5
	T62	P	525	980	4-12	170	340	14-18
	T7	S	525	980	12	225	440	3-5
		P	525	980	4-12	225	440	3-9
	T71	S	525	980	12	245	475	4-6
		P	525	980	4-12	245	475	3-6
C355.0	T6	S	525	980	12	155	310	3-5
	T61	P	525	980	6-12	Room temperature 155	310	8 (min) 10-12
356.0	T51	S or P	...	...	...	225	440	7-9
	T6	S	540	1000	12	155	310	3-5
		P	540	1000	4-12	155	310	2-5
	T7	S	540	1000	12	205	400	3-5
		P	540	1000	4-12	225	440	7-9
	T71	S	540	1000	10-12	245	475	3
		P	540	1000	4-12	245	475	3-6

(continued)

(a) S, sand; P, permanent mold. (b) Unless otherwise indicated, solution treating is followed by quenching in water at 65 to 100 °C (150 to 212 °F). (c) Except where ranges are given, listed temperatures are ±6 °C or ±10 °F. (d) Stress relieve for dimensional stability as follows: hold 5 h at 413 ± 14 °C (775 ± 25 °F); furnace cool to 345 °C (650 °F) over a period of 2 h or more; furnace cool to 230 °C (450 °F) over a period of not more than ½ h; furnace cool to 120 °C (250 °F) over a period of approximately 2 h; cool to room temperature in still air outside the furnace. (e) No quench required; cool in still air outside furnace. (f) Air-blast quench from solution-treating temperature. (g) Casting process varies (sand, permanent mold or composite) depending on desired mechanical properties. (h) Solution heat treat as indicated, then artificially age by heating uniformly at the temperature and for the time necessary to develop the desired mechanical properties. (j) Quench in water at 65 to 100 °C (150 to 212 °F) for 10 to 20 s only. (k) Cool to room temperature in still air outside furnace.

Typical heat treatments for aluminum alloy sand and permanent mold castings (continued)

Alloy	Temper	Type of casting(a)	Solution heat treatment(b) Temperature(c) °C	°F	Time, h	Aging treatment Temperature(c) °C	°F	Time, h
A356.0	T6	S.......	540	1000	12	155	310	3-5
	T61	P.......	540	1000	6-12	Room temperature		8 (min)
						155	310	6-12
357.0	T6	P.......	540	1000	8	175	350	6
	T61	S.......	540	1000	10-12	155	310	10-12
A357.0	···	(g).......	540	1000	8-12	(h)	(h)	(h)
359.0	···	(g).......	540	1000	10-14	(h)	(h)	(h)
A444.0	T4	P.......	540	1000	8-12	···	···	···
520.0	T4	S.......	430	810	18(j)	···	···	···
535.0	T5(d)	S.......	400	750	5	···	···	···
705.0	T5	S.......	···	···	···	Room temperature or		21 days
						100	210	8
		P.......	···	···	···	Room temperature or		21 days
						100	210	10
707.0	T5	S.......	···	···	···	155	310	3-5
		P.......	···	···	···	Room temperature or		21 days
						100	210	8
	T7	S.......	530	990	8-16	175	350	4-10
		P.......	530	990	4-8	175	350	4-10
710.0	T5	S.......	···	···	···	Room temperature		21 days
711.0	T1	P.......	···	···	···	Room temperature		21 days
712.0	T5	S.......	···	···	···	Room temperature or		21 days
						155	315	6-8
713.0	T5	S or P.....	···	···	···	Room temperature or		21 days
						120	250	16
771.0	T53(d)	S.......	415(k)	775(k)	5(k)	180(k)	360(k)	4(k)
	T5	S.......	···	···	···	180(k)	355(k)	3-5(k)
	T51	S.......	···	···	···	205	405	6
	T52	S.......	···	···	···	(d)	(d)	(d)
	T6	S.......	590(k)	1090(k)	6(k)	130	265	3
	T71	S.......	590(e)	1090(e)	6(e)	140	285	15
850.0	T5	S or P.....	···	···	···	220	430	7-9
851.0	T5	S or P.....	···	···	···	220	430	7-9
	T6	P.......	480	900	6	220	430	4
852.0	T5	S or P.....	···	···	···	220	430	7-9

(a) S, sand; P, permanent mold. (b) Unless otherwise indicated, solution treating is followed by quenching in water at 65 to 100 °C (150 to 212 °F). (c) Except where ranges are given, listed temperatures are ±6 °C or ±10 °F. (d) Stress relieve for dimensional stability as follows: hold 5 h at 413 ± 14 °C (775 ± 25 °F); furnace cool to 345 °C (650 °F) over a period of 2 h or more; furnace cool to 230 °C (450 °F) over a period of not more than 1/2 h; furnace cool to 120 °C (250 °F) over a period of approximately 2 h; cool to room temperature in still air outside the furnace. (e) No quench required; cool in still air outside furnace. (f) Air-blast quench from solution-treating temperature. (g) Casting process varies (sand, permanent mold or composite) depending on desired mechanical properties. (h) Solution heat treat as indicated, then artificially age by heating uniformly at the temperature and for the time necessary to develop the desired mechanical properties. (j) Quench in water at 65 to 100 °C (150 to 212 °F) for 10 to 20 s only. (k) Cool to room temperature in still air outside furnace.

Effects of annealing treatments on ductility of 7075-O sheet

Annealing treatment	Elongation in tension(a), % in 50 mm or 2 in., for thickness of:			Bend angle(b), degrees, for thickness of:		Elongation in bending(c), % in 50 mm or 2 in., for thickness of:	
	0.5 mm (0.020 in.)	1.6 mm (0.064 in.)	2.6 mm (0.102 in.)	1.6 mm (0.064 in.)	2.6 mm (0.102 in.)	1.6 mm (0.064 in.)	2.6 mm (0.102 in.)
Treatment 1(d) 12		12	12	82	73	48	50
Treatment 2(e) 14		14	14	91	76	58	57
Treatment 3(f) 16		16	...	92.5	84	56	60

(a) Uniform elongation of gridded tension specimens. (b) Bend angle at first fracture. (c) Elongation in bend test for 1.3-mm (0.05-in.) gage spanning fracture. (d) Soak 2 h at 415 ± 14 °C (775 ± 25 °F); furnace cool to 260 °C (500 °F) at 30 °C/h (50 °F/h); air cool. (e) Soak 2 h at 425 °C (800 °F), air cool; soak 2 h at 230 °C (450 °F), air cool. (f) Soak 1 h at 425 °C (800 °F); furnace cool to 230 °C (450 °F) at 30 °C/h (50 °F/h); soak 6 h at 230 °C (450 °F), air cool

Typical full annealing treatments for some common wrought aluminum alloys

These treatments, which anneal the material to the "O" temper, are typical for various sizes and methods of manufacture and may not exactly describe optimum treatments for specific items.

Alloy	Metal temperature °C	°F	Approximate time at temperature, h	Alloy	Metal temperature °C	°F	Approximate time at temperature, h
1060 345		650	(a)	5457 345		650	(a)
1100 345		650	(a)	5652 345		650	(a)
1350 345		650	(a)	6005 .... 415(b)		775(b)	2-3
2014 .... 415(b)		775(b)	2-3	6009 .... 415(b)		775(b)	2-3
2017 .... 415(b)		775(b)	2-3	6010 .... 415(b)		775(b)	2-3
2024 .... 415(b)		775(b)	2-3	6053 .... 415(b)		775(b)	2-3
2036 .... 385(b)		725(b)	2-3	6061 .... 415(b)		775(b)	2-3
2117 .... 415(b)		775(b)	2-3	6063 .... 415(b)		775(b)	2-3
2124 .... 415(b)		775(b)	2-3	6066 .... 415(b)		775(b)	2-3
2219 .... 415(b)		775(b)	2-3	7001 .... 415(c)		775(c)	2-3
3003 415		775	(a)	7005 .... 345(d)		650(d)	2-3
3004 345		650	(a)	7049 .... 415(c)		775(c)	2-3
3105 345		650	(a)	7050 .... 415(c)		775(c)	2-3
5005 345		650	(a)	7075 .... 415(c)		775(c)	2-3
5050 345		650	(a)	7079 .... 415(c)		775(c)	2-3
5052 345		650	(a)	7178 .... 415(c)		775(c)	2-3
5056 345		650	(a)	7475 .... 415(c)		775(c)	2-3
5083 345		650	(a)	**Brazing sheet**			
5086 345		650	(a)	No. 11			
5154 345		650	(a)	and 12 345		650	(a)
5182 345		650	(a)	No. 21			
5254 345		650	(a)	and 22 345		650	(a)
5454 345		650	(a)	No. 23			
5456 345		650	(a)	and 24 345		650	(a)

(a) Time in the furnace need not be longer than necessary to bring all parts of the load to annealing temperature. Cooling rate is unimportant. (b) These treatments are intended to remove the effects of solution treatment and include cooling at a rate of about 30 °C/h (50 °F/h) from the annealing temperature to 260 °C (500 °F). Rate of subsequent cooling is unimportant. Treatment at 345 °C (650 °F), followed by uncontrolled cooling, may be used to remove the effects of cold work, or to partly remove the effects of heat treatment. (c) These treatments are intended to remove the effects of solution treatment and include cooling at an uncontrolled rate to 205 °C (400 °F) or less, followed by reheating to 230 °C (450 °F) for 4 h. Treatment at 345 °C (650 °F), followed by uncontrolled cooling, may be used to remove the effects of cold work, or to partly remove the effects of heat treatment. (d) Cooling rate to 205 °C (400 °F) or below is less than or equal to 30 °C/h (50 °F/h).

Typical acceptable hardness values for wrought aluminum alloys

Acceptable hardness does not guarantee acceptable properties; acceptance should be based on acceptable hardness plus written evidence of compliance with specified heat treating procedures. Hardness values higher than the listed maximums are acceptable provided that the material is positively identified as the correct alloy.

Alloy and temper	Product form(a)	Hardness HRB	HRE	HRH	HR15T
2014-T3, -T4, -T42	All	65-70	87-95	...	...
2014-T6, -T62, -T65	Sheet(b)	80-90	103-110	...	...
	All others	81-90	104-110	...	...
2014-T61	All	...	100-109	...	...
2024-T3	Not clad(c)	69-83	97-106	111-118	82.5-87.5
	Clad, thru 1.60 mm (0.063 in.)	52-71	91-100	109-116	80-84.5
	Clad, over 1.60 mm (0.063 in.)	52-71	93-102	109-116	...
2024-T36	All	76-90	100-110	...	85-90
2024-T4, -T42(d)	Not clad	69-83	97-106	111-118	82.5-87.5
	Clad, thru 1.60 mm (0.063 in.)	52-71	91-100	109-116	80-84.5
	Clad, over 1.60 mm (0.063 in.)	52-71	93-102	109-116	...
2024-T6, -T62	All	74.5-83.5	99-106	...	84-88
2024-T81	Not clad	74.5-83.5	99-106	...	84-88
	Clad	...	99-106	...	...
2024-T86	All	83-90	105-110	...	87.5-90
6053-T6	All	...	79-87	...	74.5-78.5
6061-T4(d)	Sheet	...	60-75	88-100	64-75
	Extrusions; bar	...	70-81	82-103	67-78
6061-T6	Not clad, 0.41 mm (0.016 in.)	...	...	...	75-84
	Not clad, 0.51 mm (0.020 in.) and over	47-72	85-97	...	78-84
	Clad	...	84-96	...	...
6063-T5	All	...	55-70	89-97	62.5-70
6063-T6	All	...	70-85	...	...
6151-T6	All	...	91-102	...	...
7075-T6, -T65	Not clad(e)	85-94	106-114	...	87.5-92
	Clad: Thru 0.91 mm (0.036 in.)	...	102-110	...	86-90
	Over 0.91 thru 1.27 mm (over 0.036 thru 0.050 in.)	78-90	104-110	...	...
	Over 1.27 thru 1.57 mm (over 0.050 thru 0.062 in.)	76-90	104-110	...	...
	Over 1.57 thru 1.78 mm (over 0.062 thru 0.070 in.)	76-90	102-110	...	...
	Over 1.78 mm (0.070 in.)	73-90	102-110	...	...
7079-T6, -T65	All(e)	81-93	104-114	...	87.5-92
7178-T6	Not clad(f)	85 min	105 min	...	88 min
	Clad: Thru 0.91 mm (0.036 in.)	...	102 min	...	86 min
	Over 0.91 thru 1.57 mm (over 0.036 thru 0.062 in.)	85 min	...	...	...
	Over 1.57 mm (0.062 in.)	88 min	...	...	...

(a) Minimum hardness values shown for clad products are valid for thicknesses up to and including 2.31 mm (0.091 in.); for heavier-gage material, cladding should be locally removed for hardness testing or test should be performed on edge of sheet. (b) 126 to 158 HB (10-mm ball, 500-kg load). (c) 100 to 130 HB (10-mm ball, 500-kg load). (d) Alloys 2024-T4, 2024-T42 and 6061-T4 should not be rejected for low hardness until they have remained at room temperature for at least three days following solution treatment. (e) 136 to 164 HB (10-mm ball, 500-kg load). (f) 136 HB min (10-mm ball, 500-kg load)

Machining
Machinability ratings of aluminum alloys

Alloy(a)	Temper(a)	Rating(b)
Casting Alloys		
208.0	F	C
242.0	T21, T571, T61, T77	B
295.0	T4, T6, T7, T62	B
308.0	F	C
319.0	F	C
	T5, T6, T7	C
354.0	T61, T62	C
355.0	T51, T6, T61, T62, T7, T71	C
C355.0	T61	B
356.0	T51, T6, T7, T71	C
A356.0, A357.0	T61	C
357.0	T6	C
359.0	T61, T62	C
360.0	F	C
A360.0	F	C
380.0, A380.0	F, T5	B
390.0, A390.0	T5	E
413.0, 443.0	F	E
514.0	F	A
518.0, B535.0	F	B
520.0	T4	A
712.0	F	A
713.0	F	A
850.0	T5	A
Wrought Alloys		
1100	O, H112, H12	E
	H14 to H18	D
1350	O, H111, H112, H12	E
	H14 to H19	D
2011	T3, T4, T6, T8	A
2014(c), 2124	O	C
	T3, T4, T6	B
2017	O	C
	T4	B
2024(c)	T3, T4, T6, T8	B
2036	T4	C
2219	T3, T6, T8	B
2618	T61	B
3003(c)	O, H112, H12	E
	H14 to H18	D
3004(c)	O, H112, H32	D
	H34 to H38	C
5005	O, H112, H12, H32	E
	H14 to H18	D
	H34 to H38	D
5050	O, H112, H32	D
	H34 to H38	C
5052	O, H112, H32	D
	H34 to H38	C
5056	O	D
	H18, H38	C
5083	O, H111, H116, H321, H323	D
	H131, H343	C
5086, 5154	O, H111, H116, H32	D
	H34 to H38	C
5182	O	D
	H19	C
5454, 5456	O, H112, H311	D
	H343	C

(continued)

Machinability ratings of aluminum alloys (continued)

Alloy(a)	Temper(a)	Rating(b)
5457	O	D
	H25, H28, H38	C
5657	O	E
	H25, H28, H38	D
6061(c)	O	D
	T4, T6	C
6009, 6010	T4	D
6063	O, T2, T4	D
	T5, T6, T8	C
6262	T4, T9	B
6463	O, T1	D
	T4, T5, T6	C
7049, 7050	T73, T76	B
	T736, T76	B
7075(c)	T6, T73, T76	B
7175	T736	B
7178(c)	T6	B
7475	T6, T73, T76	B

(a) Alloys and tempers are those commonly used. Alloy modifications designated by other second digits and temper variations designated by added numerals will have the same ratings. (b) A, B, C, D and E are relative ratings in increasing order of chip length and decreasing order of quality of finish and are defined as: A—Free cutting, very small broken chips and excellent finish; B—Curled or easily broken chips and good-to-excellent finish; C—Continuous chips and good finish; D—Continuous chips and satisfactory finish; E—Optimum tool design and machine settings required to obtain satisfactory control of chip and finish. (c) Includes clad alloys and tempers.

Joining

Weldability of aluminum alloys by the gas metal-arc and gas tungsten-arc processes

Readily Weldable

Wrought alloys

Unalloyed aluminum, 1060, 1100, 1350

2219

3003, 3004, 3105

5005, 5050, 5052, 5056, 5083, 5086, 5154, 5252, 5254, 5454, 5456, 5457, 5652, 5657

6061, 6063, 6070, 6101, 6201, 6262, 6463

7005

Casting alloys

328.0, 355.0, C355.0, 356.0, A356.0, 357.0, A357.0, 359.0

443.0, A443.0, B443.0

Weldable in Most Applications(a)

Wrought alloys: 2014, 4032, 6066

Casting alloys

208.0, 308.0, 319.0, 332.0

413.0, 712.0

Limited Weldability(b)

Wrought alloys: 2024, 2218, 2618

Casting alloys

213.0, 222.0, 295.0, 296.0

333.0, 336.0, 354.0

512.0, 513.0, 514.0

Die casting alloys

Welding Not Recommended

Wrought alloys: 2011, 7075, 7178

Casting alloys

242.0, 520.0, 535.0

705.0, 707.0, 710.0, 711.0, 713.0, 771.0

(a) May require special techniques for some applications. (b) Require special techniques.

Minimum expected properties at room temperature for butt welded aluminum alloys(a)(b)

Alloy and temper	Filler wire	Product forms	Thickness range, mm	Tensile strength Ultimate MPa	ksi	Yield(c) MPa	ksi	Compressive yield strength(c) MPa	ksi	Shear strength Ultimate MPa	ksi	Yield MPa	ksi	Bearing strength Ultimate MPa	ksi	Yield MPa	ksi
1100-H12, H14, H16, H18	1100	All	All	76	11	31	4.5	31	4.5	55	8	17	2.5	160	23	55	8
3003-H12, H14, H16, H18	1100	All	All	97	14	48	7	48	7	69	10	28	4	205	30	83	12
Alclad 3003-H12, H14, H16, H18	1100	All	All	90	13	41	6	41	6	69	10	24	3.5	205	30	76	11
3004-H32, H34, H36, H38	4043	All	All	150	22	76	11	76	11	97	14	45	6.5	315	46	140	20
Alclad 3004-H32, H34, H36, H38, H14, H16	4043	All	All	145	21	76	11	76	11	90	13	45	6.5	305	44	130	19
3003-H25	1100	Sheet	All	115	17	62	9	62	9	83	12	34	5	250	36	105	15
5005-H12, H14, H32, H34	4043	All	All	97	14	48	7	48	7	62	9	28	4	195	28	69	10
5050-H32, H34	4043	All	All	125	18	55	8	55	8	83	12	30	4.5	250	36	83	12
5052-H32, H34	5654	All	All	170	25	90	13	90	13	110	16	52	7.5	345	50	130	19
5083-H111	5183	Extrusions	All	270	39	145	21	140	20	160	23	83	12	540	78	220	32
H321	5183	Sheet and plate	4.7-38.1	275	40	165	24	165	24	165	24	97	14	550	80	250	36
H321	5183	Plate	38.1-76.2	270	39	160	23	165	23	160	24	90	13	540	78	235	34
H323, H343	5183	Sheet	All	275	40	165	24	165	24	165	24	97	14	550	80	250	36
5086-H111	5356	Extrusions	All	240	35	125	18	115	17	145	21	69	10	485	70	195	28
H112	5356	Plate	6.4-12.7	240	35	115	17	115	17	145	21	66	9.5	485	70	195	28
H112	5356	Plate	12.7-25.4	240	35	110	16	110	16	145	21	62	9	485	70	195	28
H112	5356	Plate	25.4-50.8	240	35	97	14	97	14	145	21	55	8	485	70	195	28
H32, H34	5356	Sheet and plate	All	240	35	130	19	130	19	145	21	76	11	485	70	195	28
5154-H38	5654	Sheet	All	205	30	105	15	105	15	130	19	59	8.5	415	60	160	23
5454-H111	5554	Extrusions	All	215	31	110	16	105	15	130	19	66	9.5	425	62	165	24
H112	5554	Extrusions	All	215	31	83	12	83	12	130	19	49	7	425	62	165	24
H32, H34	5554	Sheet and plate	All	215	31	110	16	110	16	130	19	66	9.5	425	62	165	24
5456-H111	5556	Extrusions	All	285	41	165	24	150	22	165	24	97	14	565	82	260	38
H112	5556	Extrusions	All	285	41	130	19	130	19	165	24	76	11	565	82	260	38
H321	5556	Sheet and plate	4.7-38.1	290	42	180	26	165	24	170	25	105	15	580	84	260	38
H321	5556	Plate	38.1-76.2	285	41	165	24	160	23	170	25	97	14	565	82	260	38
H323, H343	5556	Sheet	All	290	42	180	26	180	26	170	25	105	15	580	84	260	38
6061-T6, T651, T6510, T6511(d)	4043	All	Over 9.5	165	24	140	20	140	20	105	15	83	12	345	50	205	30
6061-T6, T651, T6510, T6511(e)	4043	All	Over 9.5	165	24	105	15	105	15	105	15	62	9	345	50	205	30
6063-T5, T6	4043	All	Over 9.5	115	17	76	11	76	11	76	11	45	6.5	235	34	150	22
6351-T5(d)	(d)	Extrusions	Over 9.5	165	24	140	20	140	20	105	15	83	12	345	50	205	30
T5(e)	(e)	Extrusions	Over 9.5	165	24	105	15	105	15	105	15	62	9	345	50	205	30

(a) Gas tungsten-arc or gas metal-arc welding with no postweld heat treatment. (b) Ultimate tensile values are ASME weld-qualification-test values. (c) 0.2% offset in 250-mm (10-in.) gage length across a butt weld. (d) Values are for welding with 5183, 5356 or 5556 filler wire, regardless of thickness. Values also apply to thicknesses less than 9.5 mm (0.375 in.) when welding is done with 4043, 5554, or 5654 filler wire. (e) Values are for welding with 4043, 5554 or 5654 filler wire.

Copper and Copper Alloys

Wrought Copper Alloys

Standard color controlled wrought copper alloys

UNS number	Common name	Color description
C11000	Electrolytic tough pitch copper	Soft pink
C21000	Gilding, 95%	Red brown
C22000	Commercial bronze, 90%	Bronze gold
C23000	Red brass, 85%	Tan gold
C26000	Cartridge brass, 70%	Green gold
C28000	Muntz metal, 60%	Light brown gold
C61200	Aluminum bronze	Brown gold
C65500	High-silicon bronze, A	Lavender-brown
C70600	Copper-nickel, 10%	Soft lavender
C74500	Nickel silver, 65-10	Gray white
C75200	Nickel silver, 65-18	Silver

Classification of copper and copper alloys

Family	Principal alloying element	Solid solubility, at. %(a)	UNS numbers(b)
Coppers, high copper alloys	(c)	...	C10000
Brasses	Zn	37	C20000, C30000, C40000, C66400 to C69800
Phosphor bronzes	Sn	9	C50000
Aluminum bronzes	Al	19	C60600 to C64200
Silicon bronzes	Si	8	C64700 to C66100
Copper nickels, nickel silvers	Ni	100	C70000

(a) At 20 °C (68 °F). (b) Wrought alloys. (c) Various elements having less than 8 at. % solid solubility at 20 °C (68 °F).

ASTM B601 temper designation codes for copper and copper alloys

Temper designation	Temper name or material condition	Temper designation	Temper name or material condition
Cold Worked Tempers		**Annealed Tempers(a) (continued)**	
H00	1/8 hard	O50	Light annealed
H01	1/4 hard	O60	Soft annealed
H02	1/2 hard	O61	Annealed
H03	3/4 hard	O65	Drawing annealed
H04	Hard	O68	Deep drawing annealed
H06	Extra hard	O70	Dead soft annealed
H08	Spring	O80	Annealed to temper—1//8 hard
H10	Extra spring	O81	Annealed to temper—1/4 hard
H12	Special spring	O82	Annealed to temper—1/2 hard
H13	Ultra spring	**Annealed Tempers(c)**	
H14	Super spring	OS005	Average grain size 0.005 mm
H50	Extruded and drawn	OS010	Average grain size 0.010 mm
H52	Pierced and drawn	OS015	Average grain size 0.015 mm
H55	Light drawn; light cold rolled	OS025	Average grain size 0.025 mm
H58	Drawn general purpose	OS035	Average grain size 0.035 mm
H60	Cold heading; forming	OS050	Average grain size 0.050 mm
H63	Rivet	OS070	Average grain size 0.070 mm
H64	Screw	OS100	Average grain size 0.100 mm
H66	Bolt	OS120	Average grain size 0.120 mm
H70	Bending	OS150	Average grain size 0.150 mm
H80	Hard drawn	OS200	Average grain size 0.200 mm
H85	Medium-hard-drawn electrical wire	**Solution-treated Temper**	
H86	Hard-drawn electrical wire	TB00	Solution heat treated
Cold Worked and Stress-relieved Tempers		**Solution-treated and Cold Worked Tempers**	
HR01	H01 and stress relieved	TD00	TB00 cold worked to 1/8 hard
HR02	H02 and stress relieved	TD01	TB00 cold worked to 1/4 hard
HR04	H04 and stress relieved	TD02	TB00 cold worked to 1/2 hard
HR06	H06 and stress relieved	TD03	TB00 cold worked to 3/4 hard
HR08	H08 and stress relieved	TD04	TB00 cold worked to full hard
HR10	H10 and stress relieved	**Precipitation-hardened Temper**	
HR50	Drawn and stress relieved	TF00	TB00 and precipitation hardened
Cold Worked and Order-strengthened Tempers		**Cold Worked and Precipitation-hardened Tempers**	
HT04	H04 and order heat treated	TH01	TD01 and precipitation hardened
HT06	H06 and order heat treated	TH02	TD02 and precipitation hardened
HT08	H08 and order heat treated	TH03	TD03 and precipitation hardened
As-manufactured Tempers		TH04	TD04 and precipitation hardened
M01	As sand cast	**Precipitation-hardened and Cold Worked Tempers**	
M02	As centrifugal cast	TL00	TF00 cold worked to 1/8 hard
M03	As plaster cast	TL01	TF00 cold worked to 1/4 hard
M04	As pressure die cast	TL02	TF00 cold worked to 1/2 hard
M05	As permanent mold cast	TL04	TF00 cold worked to full hard
M06	As investment cast	TL08	TF00 cold worked to spring
M07	As continuous cast	TL10	TF00 cold worked to extra spring
M10	As hot forged and air cooled		
M11	As hot forged and quenched		
M20	As hot rolled		
M30	As hot extruded		
M40	As hot pierced		
M45	As hot pierced and rerolled		
Annealed Tempers(a)			
O10	Cast and annealed(b)		
O20	Hot forged and annealed		
O25	Hot rolled and annealed		
O30	Hot extruded and annealed		
O40	Hot pierced and annealed		

(continued)

ASTM B601 temper designation codes for copper and copper alloys (continued)

Temper designation	Temper name or material condition	Temper designation	Temper name or material condition
Precipitation-hardened and Cold Worked Tempers (cont.)		**Tempers of Welded Tubing(d)**	
TR01	TL01 and stress relieved	WH00	Welded and drawn to 1/8 hard
TR02	TL02 and stress relieved	WH01	Welded and drawn to 1/4 hard
TR04	TL04 and stress relieved	WM00	As welded from H00 strip
Mill-hardened Tempers		WM01	As welded from H01 strip
		WM02	As welded from H02 strip
TM00	AM	WM03	As welded from H03 strip
TM01	1/4 HM	WM04	As welded from H04 strip
TM02	1/2 HM	WM06	As welded from H06 strip
TM04	HM	WM08	As welded from H08 strip
TM06	XHM	WM10	As welded from H10 strip
TM08	XHMS	WM15	WM50 and stress relieved
Quench-hardened Tempers		WM20	WM00 and stress relieved
		WM21	WM01 and stress relieved
TQ00	Quench hardened	WM22	WM02 and stress relieved
TQ50	Quench hardened and temper annealed	WM50	As welded from O60 strip
TQ75	Interrupted quench hardened	WO50	Welded and light annealed
		WR00	WM00; drawn and stress relieved
		WR01	WM01; drawn and stress relieved

(a) To produce specified mechanical properties. (b) Homogenization anneal. (c) To produce prescribed average grain size. (d) Tempers of fully finished tubing that has been drawn or annealed to produce specified mechanical properties or that has been annealed to produce a prescribed average grain size are commonly identified by the appropriate H, O, or OS temper designation.

Extraction of copper from a low-grade ore

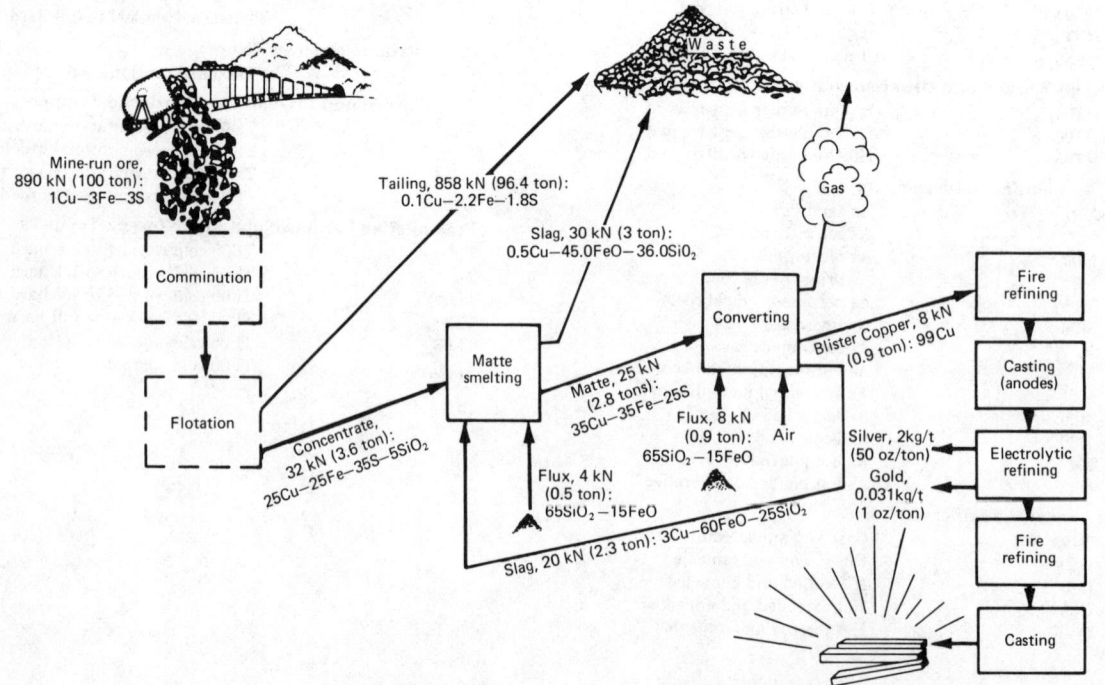

Wire bar copper, 8 kN (0.9 ton): 99.95 Cu

Properties and applications of wrought coppers and copper alloys

Name and number	Nominal composition, %	Commercial forms(a)	Tensile strength MPa	ksi	Yield strength MPa	ksi	Elongation in 2 in., %	Corrosion resistance (c)	Machinability rating (d)	Fabricating characteristics and typical applications
C10100 Oxygen-free electronic	99.99 Cu	F, R, W, T, P, S	221-455	32-66	69-365	10-53	55-4	G-E	20	Excellent hot and cold workability; good forgeability. Fabricated by coining, coppersmithing, drawing and upsetting, hot forging and pressing, spinning, swaging, stamping. Uses: busbars, bus conductors, waveguides, hollow conductors, lead-in wires and anodes for vacuum tubes, vacuum seals, transistor components, glass to metal seals, coaxial cables and tubes, klystrons, microwave tubes, rectifiers.
C10200 Oxygen-free copper	99.95 Cu	F, R, W, T, P, S	221-455	32-66	69-365	10-53	55-4	G-E	20	Fabricating characteristics same as C10100. Uses: busbars, waveguides.
C10300 Oxygen-free, extra-low phosphorus	99.95 Cu, 0.003 P	F, R, T, P, S	221-379	32-55	69-345	10-50	50-6	G-E	20	Fabricating characteristics same as C10100. Uses: busbars, electrical conductors, tubular bus and applications requiring good conductivity and welding or brazing properties.
C10400, C10500, C10700 Oxygen-free, silver-bearing	99.95 Cu(e)	F, R, W, S	221-455	32-66	69-365	10-53	55-4	G-E	20	Fabricating characteristics same as C10100. Uses: auto gaskets, radiators, busbars, conductivity wire, contacts, radio parts, winding, switches, terminals, commutator segments; chemical process equipment, printing rolls, clad metals, printed circuit foil.
C10800 Oxygen-free, low phosphorus	99.95 Cu, 0.009 P	F, R, T, P	221-379	32-55	69-345	10-50	50-4	G-E	20	Fabricating characteristics same as C10100. Uses: refrigerators, air conditioners, gas and heater lines, oil burner tubes, plumbing pipe and tube, brewery tubes, condenser and heat exchanger tubes, dairy and distiller tubes, pulp and paper lines, tanks, air, gasoline, hydraulic and oil lines.
C11000 Electrolytic tough pitch copper	99.90 Cu, 0.04 O	F, R, W, T, P, S	221-455	32-66	69-365	10-53	55-4	G-E	20	Fabricating characteristics same as C10100. Uses: downspouts, gutters, roofing, gaskets, auto radiators, busbars, nails, printing rolls, rivets, radio parts.
C11100 Electrolytic tough pitch, anneal resistant	99.90 Cu, 0.04 O, 0.01 Cd	W	455	66	· · ·	· · ·	1.5 in 60 in.	G-E	20	Fabricating characteristics same as C10100. Uses: electrical power transmission where resistance to softening under overloads is desired.

(continued)

(a) F, flat products; R, rod; W, wire; T, tube; P, pipe; S, shapes. (b) Softest to hardest commercial forms. The strength of the standard copper alloys depends on the temper (annealed grain size or degree of cold work) and the section thickness of the mill product. Ranges cover standard tempers for each alloy. (c) E, excellent; G, good; F, fair. (d) Based on 100% for C360000. (e) C10400, 8 oz/ton Ag; C10500, 10 oz/ton; C10700, 25 oz/ton. (f) C11300, 8 oz/ton Ag; C11400, 10 oz/ton; C11500, 16 oz/ton; C11600, 25 oz/ton. (g) C12000, 0.008 P; C12100, 0.008 P and 4 oz/ton Ag. (h) C12700, 8 oz/ton Ag; C12800, 10 oz/ton; C12900, 16 oz/ton; C13000, 25 oz/ton. (i) 8.30 oz/ton Ag. (j) C18200, 0.9 Cr; C18400, 0.8 Cr; C18500, 0.7 Cr. (k) Rod, 61.0 Cu min. Compiled by Copper Development Assn. Inc., New York.

Properties and applications of wrought coppers and copper alloys (continued)

Name and number	Nominal composition, %	Commercial forms(a)	Tensile strength MPa	ksi	Yield strength MPa	ksi	Elongation in 2 in., %	Corrosion resistance (c)	Machinability rating (d)	Fabricating characteristics and typical applications
C11300, C11400, C11500, C11600 Silver-bearing tough pitch copper	99.90 Cu, 0.04 O, Ag(f)	F, R, W, T, S	221-455	32-66	69-365	10-53	55-4	G-E	20	Fabricating characteristics same as C10100. Uses: gaskets, radiators, busbars, windings, switches, chemical process equipment, clad metals, printed circuit foil.
C12000, C12100	99.9 Cu(g)	F, T, P	221-393	32-57	69-365	10-53	55-4	G-E	20	Fabricating characteristics same as C10100. Uses: busbars, electrical conductors, tubular bus, and applications requiring welding or brazing.
C12200 Phosphorus deoxidized copper, high residual phosphorus	99.90 Cu, 0.02 P	F, R, T, P	221-379	32-55	69-345	10-50	45-8	G-E	20	Fabricating characteristics same as C10100. Uses: gas and heater lines; oil burner tubing; plumbing pipe and tubing; condenser, evaporator, heat exchanger, dairy, and distiller tubing; steam and water lines; air, gasoline, and hydraulic lines.
C12500, C12700, C12800, C12900, C13000 Fire-refined tough pitch with silver	99.88 Cu(h)	F, R, W, S	221-462	32-67	69-365	10-53	55-4	G-E	20	Fabricating characteristics same as C10100. Uses: same as C11000.
C14200 Phosphorus deoxidized, arsenical	99.68 Cu, 0.3 As, 0.02 P	F, R, T	221-379	32-55	69-345	10-50	45-8	G-E	20	Fabricating characteristics same as C10100. Uses: plates for locomotive fireboxes, staybolts, heat exchanger and condenser tubes.
C19200	98.97 Cu, 1.0 Fe, 0.03 P	F, T	255-531	37-77	76-510	11-74	40-2	G-E	20	Excellent hot and cold workability. Uses: automotive hydraulic brake lines, flexible hose, electrical terminals, fuse clips, gaskets, gift hollow ware, applications requiring resistance to softening and stress corrosion, air conditioning and heat exchanger tubing.
C14300	99.9 Cu, 0.1 Cd	F	221-400	32-58	76-386	11-56	42-1	G-E	20	Fabricating characteristics same as C10100. Uses: anneal resistant electrical applications requiring thermal softening and embrittlement resistance, lead frames, contacts, terminals, solder-coated and solder-fabricated parts, furnace-brazed assemblies and welded components, cable wrap.
C14310	99.8 Cu, 0.2 Cd	F	221-400	32-58	76-386	11-56	42-1	G-E	20	Same as C14300.

(continued)

(a) F, flat products; R, rod; W, wire; T, tube; P, pipe; S, shapes. (b) Softest to hardest commercial forms. The strength of the standard copper alloys depends on the temper (annealed grain size or degree of cold work) and the section thickness of the mill product. Ranges cover standard tempers for each alloy. (c) E, excellent; G, good; F, fair. (d) Based on 100% for C36000. (e) C10400, 8 oz/ton Ag; C10500, 10 oz/ton; C10700, 25 oz/ton. (f) C11300, 8 oz/ton Ag; C11400, 10 oz/ton; C11500, 16 oz/ton; C11600, 25 oz/ton. (g) C12000, 0.008 P; C12100, 0.008 P and 4 oz/ton Ag. (h) C12700, 8 oz/ton Ag; C12800, 10 oz/ton; C12900, 16 oz/ton; C13000, 25 oz/ton. (i) 8.30 oz/ton Ag. (j) C18200, 0.9 Cr; C18400, 0.8 Cr; C18500, 0.7 Cr. (k) Rod, 61.0 Cu min. Compiled by Copper Development Assn. Inc., New York.

Properties and applications of wrought coppers and copper alloys (continued)

Name and number	Nominal composition, %	Commercial forms(a)	Mechanical properties(b) Tensile strength MPa	ksi	Yield strength MPa	ksi	Elongation in 2 in., %	Corrosion resistance (c)	Machinability rating (d)	Fabricating characteristics and typical applications
C14500 Phosphorus deoxidized, tellurium bearing	99.5 Cu, 0.50 Te, 0.008 P	F, R, W, T	221-386	32-56	69-352	10-51	50-3	G-E	85	Fabricating characteristics same as C10100. Uses: forgings and screw machine products, and parts requiring high conductivity, extensive machining, corrosion resistance, copper color, or a combination of these; electrical connectors, motor and switch parts, plumbing fittings, soldering coppers, welding torch tips, transistor bases and furnace-brazed articles.
C14700 Sulfur-bearing	99.6 Cu, 0.40 S	R, W	221-393	32-57	69-379	10-55	52-8	G-E	85	Fabricating characteristics same as C10100. Uses: screw machine products and parts requiring high conductivity, extensive machining, corrosion resistance, copper color, or a combination of these; electrical connectors, motor and switch components, plumbing fittings, cold headed and machined parts, cold forgings, furnace brazed articles, screws, soldering coppers, rivets and welding torch tips.
C15000 Zirconium copper	99.8 Cu, 0.15 Zr	R, W	200-524	29-76	41-496	6-72	54-1.5	G-E	20	Fabricating characteristics same as C10100. Uses: switches, high-temperature circuit breakers; commutators, stud bases for power transmitters, rectifiers, soldering welding tips.
C15500	99.75 Cu, 0.06 P, 0.11 Mg, Ag(i)	F	276-552	40-80	124-496	18-72	40-3	G-E	20	Fabricating characteristics same as C10100. Uses: high-conductivity light-duty springs, electrical contacts, fittings, clamps, connectors, diaphragms, electronic components, resistance welding electrodes.
C15710	99.8 Cu, 0.2 Al$_2$O$_3$	R, W	324-724	47-105	268-689	39-100	20-10	· · ·	· · ·	Excellent cold workability. Fabricated by extrusion, drawing, rolling, impacting, heading, swaging, bending, machining, blanking, roll threading. Uses: electrical connectors, light-duty current-carrying springs, inorganic insulated wire, thermocouple wire, lead wire, resistance welding electrodes for aluminum, heat sinks.

(continued)

(a) F, flat products; R, rod; W, wire; T, tube; P, pipe; S, shapes. (b) Softest to hardest commercial forms. The strength of the standard copper alloys depends on the temper (annealed grain size or degree of cold work) and the section thickness of the mill product. Ranges cover standard tempers for each alloy. (c) E, excellent; G, good; F, fair. (d) Based on 100% for C360000. (e) C10400, 8 oz/ton Ag; C10500, 10 oz/ton; C10700, 25 oz/ton. (f) C11300, 8 oz/ton Ag; C11400, 10 oz/ton; C11500, 16 oz/ton; C11600, 25 oz/ton. (g) C12000, 0.008 P; C12100, 0.008 P and 4 oz/ton Ag. (h) C12700, 8 oz/ton Ag; C12800, 10 oz/ton; C12900, 16 oz/ton; C13000, 25 oz/ton. (i) 8.30 oz/ton Ag. (j) C18200, 0.9 Cr; C18400, 0.8 Cr; C18500, 0.7 Cr. (k) Rod, 61.0 Cu min. Compiled by Copper Development Assn. Inc., New York.

Properties and applications of wrought coppers and copper alloys (continued)

Name and number	Nominal composition, %	Commercial forms(a)	Mechanical properties(b) Tensile strength		Yield strength		Elongation in 2 in., %	Corrosion resistance (c)	Machinability rating (d)	Fabricating characteristics and typical applications
			MPa	ksi	MPa	ksi				
C15720	99.6 Cu, 0.4 Al$_2$O$_3$	F, R	462-614	67-89	365-586	53-85	20-3.5	· · ·	· · ·	Excellent cold workability. Fabricated by extrusion, drawing, rolling, impacting, heading, swaging, machining, blanking. Uses: relay and switch springs, lead frames, contact supports, heat sinks, circuit breaker parts, rotor bars, resistance welding electrodes and wheels, connectors, high-strength high-temperature parts.
C15735	99.3 Cu, 0.7 Al$_2$O$_3$	R	483-586	70-85	414-565	60-82	16-10	· · ·	· · ·	Excellent cold workability. Fabricated by extrusion, drawing, heading, impacting, machining. Uses: resistance welding electrodes, circuit breakers, feed-through conductors, heat sinks, motor parts, high-strength high-temperature parts.
C15760	98.9 Cu, 1.1 Al$_2$O$_3$	F, R	483-648	70-94	386-552	56-80	20-8	· · ·	· · ·	Excellent cold workability. Fabricated by extrusion and drawing. Uses: resistance welding electrodes, circuit breakers, electrical connectors, wire feed contact tips, plasma spray nozzles, high-strength high-temperature parts.
C16200 Cadmium copper	99.0 Cu, 1.0 Cd	F, R, W	241-689	35-100	48-476	7-69	57-1	G-E	20	Excellent cold workability; good hot formability. Uses: trolley wire, heating pad, electric-blanket elements, spring contacts, railbands, high-strength transmission lines, connectors, cable wrap, switch gear components and waveguide cavities.
C16500	98.6 Cu, 0.8 Cd, 0.6 Sn	F, R, W	276-655	40-95	97-490	14-71	53-1.5	G-E	20	Fabricating characteristics same as C16200. Uses: electrical springs and contacts, trolley wire, clips, flat cable, resistance welding electrodes.
C17000 Beryllium copper	99.5 Cu, 1.7 Be, 0.20 Co	F, R	483-1310	70-190	221-1172	32-170	45-3	G-E	20	Fabricating characteristics same as C16200. Commonly fabricated by blanking, forming and bending, turning, drilling, tapping. Uses: bellows, bourdon tubing, diaphragms, fuse clips, fasteners, lock-washers, springs, switch parts, roll pins, valves, welding equipment.
C17200 Beryllium copper	99.5 Cu, 1.9 Be, 0.20 Co	F, R, W, T, P, S	469-1462	68-212	172-1344	25-195	48-1	G-E	20	Similar to C17000, particularly for its nonsparking characteristics.

(continued)

(a) F, flat products; R, rod; W, wire; T, tube; P, pipe; S, shapes. (b) Softest to hardest commercial forms. The strength of the standard copper alloys depends on the temper (annealed grain size or degree of cold work) and the section thickness of the mill product. Ranges cover standard tempers for each alloy. (c) E, excellent; G, good; F, fair. (d) Based on 100% for C360000. (e) C10400, 8 oz/ton Ag; C10500, 10 oz/ton; C10700, 25 oz/ton. (f) C11300, 8 oz/ton Ag; C11400, 10 oz/ton; C11500, 16 oz/ton; C11600, 25 oz/ton. (g) C12000, 0.008 P; C12100, 0.008 P and 4 oz/ton Ag. (h) C12700, 8 oz/ton Ag; C12800, 10 oz/ton; C12900, 16 oz/ton; C13000, 25 oz/ton. (i) 8.30 oz/ton Ag. (j) C18200, 0.9 Cr; C18400, 0.8 Cr; C18500, 0.7 Cr. (k) Rod, 61.0 Cu min. Compiled by Copper Development Assn. Inc., New York.

Properties and applications of wrought coppers and copper alloys (continued)

Name and number	Nominal composition, %	Commercial forms(a)	Mechanical properties(b) Tensile strength MPa	ksi	Yield strength MPa	ksi	Elongation in 2 in., %	Corrosion resistance (c)	Machinability rating (d)	Fabricating characteristics and typical applications
C17300 Beryllium copper	99.5 Cu, 1.9 Be, 0.40 Pb	R	469-1479	68-200	172-1255	25-182	48-3	G-E	50	Combines superior machinability with good fabricating characteristics of C17200.
C17500 Copper-cobalt-beryllium alloy	99.5 Cu, 2.5 Co, 0.6 Be	F, R	310-793	45-115	172-758	25-110	28-5	G-E	...	Fabricating characteristics same as C16200. Uses: fuse clips, fasteners, springs, switch and relay parts, electrical conductors, welding equipment.
C18200, C18400, C18500 Chromium copper	99.5 Cu(j)	F, W, R, S, T	234-593	34-86	97-531	14-77	40-5	G-E	20	Excellent cold workability, good hot workability. Uses: resistance welding electrodes, seam welding wheels, switch gear, electrode holder jaws, cable connectors, current carrying arms and shafts, circuit breaker parts, molds, spot welding tips, flash welding electrodes, electrical and thermal conductors requiring strength, switch contacts.
C18700 Leaded copper	99.0 Cu, 1.0 Pb	R	221-379	32-55	69-345	10-50	45-8	G-E	85	Good cold workability; poor hot formability. Uses: connectors, motor and switch parts, screw machine parts requiring high conductivity.
C18900	98.75 Cu, 0.75 Sn, 0.3 Si, 0.20 Mn	R, W	262-655	38-95	62-359	9-52	48-14	G-E	20	Fabricating characteristics same as C10100. Uses: welding rod and wire for inert gas tungsten arc and metal arc welding and oxyacetylene welding of copper.
C19000 Copper-nickel-phosphorus alloy	98.7 Cu, 1.1 Ni, 0.25 P	F, R, W	262-793	38-115	138-552	20-80	50-2	G-E	30	Fabricating characteristics same as C10100. Uses: springs, clips, electrical connectors, power tube and electron tube components, high-strength electrical conductors, bolts, nails, screws, cotter pins, and parts requiring some combination of high-strength, high-electrical or thermal conductivity, high resistance to fatigue and creep, and good workability.
C19100 Copper-nickel-phosphorus-tellurium alloy	98.15 Cu, 1.1 Ni, 0.50 Te, 0.25 P	R, F	248-717	36-104	69-634	10-92	27-6	G-E	75	Good hot and cold workability. Uses: forgings and screw machine parts requiring high strength, hardenability, extensive machining, corrosion resistance, copper color, good conductivity, or a combination of these; bolts, bushings, electrical connectors, gears, marine hardware, nuts, pinions, tie rods, turnbuckle barrels, welding torch tips.

(continued)

(a) F, flat products; R, rod; W, wire; T, tube; P, pipe; S, shapes. (b) Softest to hardest commercial forms. The strength of the standard copper alloys depends on the temper (annealed grain size or degree of cold work) and the section thickness of the mill product. Ranges cover standard tempers for each alloy. (c) E, excellent; G, good; F, fair. (d) Based on 100% for C360000. (e) C10400, 8 oz/ton Ag; C10500, 10 oz/ton; C10700, 25 oz/ton. (f) C11300, 8 oz/ton Ag; C11400, 10 oz/ton; C11500, 16 oz/ton; C11600, 25 oz/ton. (g) C12000, 0.008 P; C12100, 0.008 P and 4 oz/ton Ag. (h) C12700, 8 oz/ton Ag; C12800, 10 oz/ton; C12900, 16 oz/ton; C13000, 25 oz/ton. (i) 8.30 oz/ton Ag. (j) C18200, 0.9 Cr; C18400, 0.8 Cr; C18500, 0.7 Cr. (k) Rod, 61.0 Cu min. Compiled by Copper Development Assn. Inc., New York.

Properties and applications of wrought coppers and copper alloys (continued)

Name and number	Nominal composi- tion, %	Com- mercial forms(a)	Mechanical properties(b) Tensile strength		Yield strength		Elon- gation in 2 in., %	Corro- sion resis- tance (c)	Machin- ability rating (d)	Fabricating characteristics and typical applications
			MPa	ksi	MPa	ksi				
C19400	97.5 Cu, 2.4 Fe, 0.13 Zn, 0.03 P	F	310-524	45-76	165-503	24-73	32-2	G-E	20	Excellent hot and cold work- ability. Uses: circuit breaker components, contact springs, electrical clamps, electrical springs, electrical terminals, flexible hose, fuse clips, gas- kets, gift hollow ware, plug contacts, rivets, and welded condenser tubes.
C19500	97.0 Cu, 1.5 Fe, 0.6 Sn, 0.10 P, 0.80 Co	F	552-669	80-97	448-655	65-95	15-2	G-E	20	Excellent hot and cold work- ability. Uses: electrical springs, sockets, terminals, connectors, clips and other current carrying parts having strength.
C21000 Gilding, 95%	95.0 Cu, 5.0 Zn	F, W	234-441	34-64	69-400	10-58	45-4	G-E	20	Excellent cold workability, good hot workability for blanking, coining, drawing, piercing and punching, shearing, spinning, squeez- ing and swaging, stamping. Uses: coins, medals, bullet jackets, fuse caps, primers, plaques, jewelry base for gold plate.
C22000 Commercial bronze, 90%	90.0 Cu, 10.0 Zn	F, R, W, T	255-496	37-72	69-427	10-62	50-3	G-E	20	Fabricating characteristics same as C21000, plus head- ing and upsetting, roll threading and knurling, hot forging and pressing. Uses: etching bronze, grillwork, screen cloth, weatherstrip- ping, lipstick cases, compacts, marine hardware, screws, rivets.
C22600 Jewelry bronze, 87.5%	87.5 Cu, 12.5 Zn	F, W	269-669	39-97	76-427	11-62	46-3	G-E	30	Fabricating characteristics same as C21000, plus head- ing and upsetting, roll threading and knurling. Uses: angles, channels, chain, fasteners, costume jewelry, lipstick cases, com- pacts, base for gold plate.
C23000 Red brass, 85%	85.0 Cu, 15.0 Zn	F, W, T, P	269-724	39-105	69-434	10-63	55-3	G-E	30	Excellent cold workability; good hot formability. Uses: weatherstripping, conduit, sockets, fasteners, fire extinguishers, condenser and heat exchanger tubing, plumbing pipe, radiator cores.
C24000 Low brass, 80%	80.0 Cu, 20.0 Zn	F, W	290-862	42-125	83-448	12-65	55-3	F-E	30	Excellent cold workability. Fabricating characteristics same as C23000. Uses: battery caps, bellows, musical instruments, clock dials, pump lines, flexible hose.

(continued)

(a) F, flat products; R, rod; W, wire; T, tube; P, pipe; S, shapes. (b) Softest to hardest commercial forms. The strength of the standard copper alloys depends on the temper (annealed grain size or degree of cold work) and the section thickness of the mill product. Ranges cover standard tempers for each alloy. (c) E, excellent; G, good; F, fair. (d) Based on 100% for C360000. (e) C10400, 8 oz/ton Ag; C10500, 10 oz/ton; C10700, 25 oz/ton. (f) C11300, 8 oz/ton Ag; C11400, 10 oz/ton; C11500, 16 oz/ton; C11600, 25 oz/ton. (g) C12000, 0.008 P; C12100, 0.008 P and 4 oz/ton Ag. (h) C12700, 8 oz/ton Ag; C12800, 10 oz/ton; C12900, 16 oz/ton; C13000, 25 oz/ton. (i) 8.30 oz/ton Ag. (j) C18200, 0.9 Cr; C18400, 0.8 Cr; C18500, 0.7 Cr. (k) Rod, 61.0 Cu min. Compiled by Copper Development Assn. Inc., New York.

Properties and applications of wrought coppers and copper alloys (continued)

Name and number	Nominal composition, %	Commercial forms(a)	Mechanical properties(b) Tensile strength MPa	ksi	Yield strength MPa	ksi	Elongation in 2 in., %	Corrosion resistance (c)	Machinability rating (d)	Fabricating characteristics and typical applications
C26000 Cartridge brass, 70%	70.0 Cu, 30.0 Zn	F, R, W, T	303-896	44-130	76-448	11-65	66-3	F-E	30	Excellent cold workability. Fabricating characteristics same as C23000, except for coining, roll threading, and knurling. Uses: radiator cores and tanks, flashlight shells, lamp fixtures, fasteners, locks, hinges, ammunition components, plumbing accessories, pins, rivets.
C26800, C27000 Yellow brass	65.0 Cu, 35.0 Zn	F, R, W	317-883	46-128	97-427	14-62	65-3	F-E	30	Excellent cold workability. Fabricating characteristics same as C23000. Uses: same as C26000 except not used for ammunition.
C28000 Muntz metal	60.0 Cu, 40.0 Zn	F, R, T	372-510	54-74	145-379	21-55	52-10	F-E	40	Excellent hot formability and forgeability for blanking, forming and bending, hot forging and pressing, hot heading and upsetting, shearing. Uses: architectural, large nuts and bolts, brazing rod, condenser plates, heat exchanger and condenser tubing, hot forgings.
C31400 Leaded commercial bronze	89.0 Cu, 1.75 Pb, 9.25 Zn	F, R	255-414	37-60	83-379	12-55	45-10	G-E	80	Excellent machinability. Uses: screws, machine parts, pickling crates.
C31600 Leaded commercial bronze, nickel-bearing	89.0 Cu, 1.9 Pb, 1.0 Ni, 8.1 Zn	F, R	255-462	37-67	83-407	12-59	45-12	G-E	80	Good cold workability; poor hot formability. Uses: electrical connectors, fasteners, hardware, nuts, screws, screw machine parts.
C33000 Low-leaded brass tube	66.0 Cu, 0.5 Pb, 33.5 Zn	T	324-517	47-75	103-414	15-60	60-7	F-E	60	Combines good machinability and excellent cold workability. Fabricated by forming and bending, machining, piercing and punching. Uses: pump and power cylinders and liners, ammunition primers, plumbing accessories.
C33200 High-leaded brass tube	66.0 Cu, 1.6 Pb, 32.4 Zn	T	359-517	52-75	138-414	20-60	50-7	F-E	80	Excellent machinability. Fabricated by piercing, punching, and machining. Uses: general-purpose screw machine parts.
C33500 Low-leaded brass	65.0 Cu, 0.5 Pb, 34.5 Zn	F	317-510	46-74	97-414	14-60	65-8	F-E	60	Similar to C33200. Commonly fabricated by blanking, drawing, machining, piercing and punching, stamping. Uses: butts, hinges, watch backs.

(continued)

(a) F, flat products; R, rod; W, wire; T, tube; P, pipe; S, shapes. (b) Softest to hardest commercial forms. The strength of the standard copper alloys depends on the temper (annealed grain size or degree of cold work) and the section thickness of the mill product. Ranges cover standard tempers for each alloy. (c) E, excellent; G, good; F, fair. (d) Based on 100% for C360000. (e) C10400, 8 oz/ton Ag; C10500, 10 oz/ton; C10700, 25 oz/ton. (f) C11300, 8 oz/ton Ag; C11400, 10 oz/ton; C11500, 16 oz/ton; C11600, 25 oz/ton. (g) C12000, 0.008 P; C12100, 0.008 P and 4 oz/ton Ag. (h) C12700, 8 oz/ton Ag; C12800, 10 oz/ton; C12900, 16 oz/ton; C13000, 25 oz/ton. (i) 8.30 oz/ton Ag. (j) C18200, 0.9 Cr; C18400, 0.8 Cr; C18500, 0.7 Cr. (k) Rod, 61.0 Cu min. Compiled by Copper Development Assn. Inc., New York.

Properties and applications of wrought coppers and copper alloys (continued)

Name and number	Nominal composi- tion, %	Com- mercial forms(a)	Mechanical properties(b) Tensile strength MPa	ksi	Yield strength MPa	ksi	Elon- gation in 2 in., %	Corro- sion resis- tance (c)	Machin- ability rating (d)	Fabricating characteristics and typical applications
C34000 Medium- leaded brass	65.0 Cu, 1.0 Pb, 34.0 Zn	F, R, W, S	324-607	47-88	103-414	15-60	60-7	F-E	70	Similar to C33200. Fabri- cated by blanking, heading and upsetting, machining, piercing and punching, roll threading and knurling, stamping. Uses: butts, gears, nuts, rivets, screws, dials, engravings, instrument plates.
C34200 High- leaded brass	64.5 Cu, 2.0 Pb, 33.5 Zn	F, R	338-586	49-85	117-427	17-62	52-5	F-E	90	Combines excellent machin- ability with moderate cold workability. Uses: clock plates and nuts, clock and watch backs, gears, wheels and channel plate.
C34900	62.2 Cu, 0.35 Pb, 37.45 Zn	R, W	365-469	53-68	110-379	16-55	72-18	F-E	50	Good cold workability, fair hot workability for bending and forming, heading and upsetting, machining, roll threading and knurling. Uses: building hardware, rivets and nuts, plumbing goods, and parts requiring moderate cold working com- bined with some machining.
C35000 Medium- leaded brass	62.5 Cu, 1.1 Pb, 36.4 Zn	F, R	310-655	45-95	90-483	13-70	66-1	F-E	70	Fair cold workability; poor hot formability. Uses: bear- ing cages, book dies, clock plates, engraving plates, gears, hinges, hose cou- plings, keys, lock parts, lock tumblers, meter parts, nuts, sink strainers, strike plates, templates, type characters, washers, wear plates.
C35300 High-leaded brass	62.0 Cu, 1.8 Pb, 36.2 Zn	F, R	338-586	49-85	117-427	17-62	52-5	F-E	90	Similar to C34200.
C35600 Extra-high- leaded brass	63.0 Cu, 2.5 Pb, 34.5 Zn	F	338-510	49-74	117-414	17-60	50-7	F-E	100	Excellent machinability. Fabricated by blanking, machining, piercing and punching, stamping. Uses: same as C34200 and C35300.
C36000 Free- cutting brass	61.5 Cu, 3.0 Pb, 35.5 Zn	F, R, S	338-469	49-68	124-310	18-45	53-18	F-E	100	Excellent machinability. Fabricated by machining, roll threading and knurling. Uses: gears, pinions, auto- matic high-speed screw machine parts.
C36500 to C36800 Leaded Muntz metal	60.0 Cu(k), 0.6 Pb, 39.4 Zn	F	372	54 (As hot rolled)	138	20	45	F-E	60	Combines good machin- ability with excellent hot formability. Uses: condenser tube plates.

(continued)

(a) F, flat products; R, rod; W, wire; T, tube; P, pipe; S, shapes. (b) Softest to hardest commercial forms. The strength of the standard copper alloys depends on the temper (annealed grain size or degree of cold work) and the section thickness of the mill product. Ranges cover standard tempers for each alloy. (c) E, excellent; G, good; F, fair. (d) Based on 100% for C360000. (e) C10400, 8 oz/ton Ag; C10500, 10 oz/ton; C10700, 25 oz/ton. (f) C11300, 8 oz/ton Ag; C11400, 10 oz/ton; C11500, 16 oz/ton; C11600, 25 oz/ton. (g) C12000, 0.008 P; C12100, 0.008 P and 4 oz/ton Ag. (h) C12700, 8 oz/ton Ag; C12800, 10 oz/ton; C12900, 16 oz/ton; C13000, 25 oz/ton. (i) 8.30 oz/ton Ag. (j) C18200, 0.9 Cr; C18400, 0.8 Cr; C18500, 0.7 Cr. (k) Rod, 61.0 Cu min. Compiled by Copper Development Assn. Inc., New York.

Properties and applications of wrought coppers and copper alloys (continued)

Name and number	Nominal composition, %	Commercial forms(a)	Tensile strength MPa	Tensile strength ksi	Yield strength MPa	Yield strength ksi	Elongation in 2 in., %	Corrosion resistance (c)	Machinability rating (d)	Fabricating characteristics and typical applications
C37000 Free-cutting Muntz metal	60.0 Cu, 1.0 Pb, 39.0 Zn	T	372-552	54-80	138-414	20-60	40-6	F-E	70	Fabricating characteristics similar to C36500 to C36800. Uses: automatic screw machine parts.
C37700 Forging brass	59.0 Cu, 2.0 Pb, 39.0 Zn	R, S	359	52	138 (As extruded)	20	45	F-E	80	Excellent hot workability. Fabricated by heading and upsetting, hot forging and pressing, hot heading and upsetting, machining. Uses: forgings and pressings of all kinds.
C38500 Architectural bronze	57.0 Cu, 3.0 Pb, 40.0 Zn	R, S	414	60	138 (As extruded)	20	30	F-E	90	Excellent machinability and hot workability. Fabricated by hot forging and pressing, forming, bending and machining. Uses: architectural extrusions, store fronts, thresholds, trim, butts, hinges, lock bodies and forgings.
C40500	95 Cu, 1 Sn, 4 Zn	F	269-538	39-78	83-483	12-70	49-3	G-E	20	Excellent cold workability. Fabricated by blanking, forming and drawing. Uses: meter clips, terminals, fuse clips, contact and relay springs, washers.
C40800	95 Cu, 2 Sn, 3 Zn	F	290-545	42-79	90-517	13-75	43-3	G-E	20	Excellent cold workability. Fabricated by blanking, stamping and shearing. Uses: electrical connectors.
C41100	91 Cu, 0.5 Sn, 8.5 Zn	F, W	269-731	39-106	76-496	11-72	13-2	G-E	20	Excellent cold workability, good hot formability. Fabricated by blanking, forming and drawing. Uses: bushings, bearing sleeves, thrust washers, terminals, connectors, flexible metal hose, electrical conductors.
C41300	90.0 Cu, 1.0 Sn, 9.0 Zn	F, R, W	283-724	41-105	83-565	12-82	45-2	G-E	20	Excellent cold workability; good hot formability. Uses: plater bar for jewelry products, flat springs for electrical switchgear.
C41500	91 Cu, 1.8 Sn, 7.2 Zn	F	317-558	46-81	117-517	17-75	44-2	G-E	30	Excellent cold workability. Fabricated by blanking, drawing, bending, forming, shearing and stamping. Uses: spring applications for electrical switches.
C42200	87.5 Cu, 1.1 Sn, 11.4 Zn	F	296-607	43-88	103-517	15-75	46-2	G-E	30	Excellent cold workability; good hot formability. Fabricated by blanking, piercing, forming and drawing. Uses: sash chains, fuse clips, terminals, spring washers, contact springs, electrical connectors.
C42500	88.5 Cu, 2.0 Sn, 9.5 Zn	F	310-634	45-92	124-524	18-76	49-2	G-E	30	Excellent cold workability. Fabricated by blanking, piercing, forming and drawing. Uses: electrical switches, springs, terminals, connectors, fuse clips, pen clips, weather stripping.

(continued)

(a) F, flat products; R, rod; W, wire; T, tube; P, pipe; S, shapes. (b) Softest to hardest commercial forms. The strength of the standard copper alloys depends on the temper (annealed grain size or degree of cold work) and the section thickness of the mill product. Ranges cover standard tempers for each alloy. (c) E, excellent; G, good; F, fair. (d) Based on 100% for C360000. (e) C10400, 8 oz/ton Ag; C10500, 10 oz/ton; C10700, 25 oz/ton. (f) C11300, 8 oz/ton Ag; C11400, 10 oz/ton; C11500, 16 oz/ton; C11600, 25 oz/ton. (g) C12000, 0.008 P; C12100, 0.008 P and 4 oz/ton Ag. (h) C12700, 8 oz/ton Ag; C12800, 10 oz/ton; C12900, 16 oz/ton; C13000, 25 oz/ton. (i) 8.30 oz/ton Ag. (j) C18200, 0.9 Cr; C18400, 0.8 Cr; C18500, 0.7 Cr. (k) Rod, 61.0 Cu min. Compiled by Copper Development Assn. Inc., New York.

Properties and applications of wrought coppers and copper alloys (continued)

Name and number	Nominal composi- tion, %	Com- mercial forms(a)	Tensile strength MPa	ksi	Yield strength MPa	ksi	Elon- gation in 2 in., %	Corro- sion resis- tance (c)	Machin- ability rating (d)	Fabricating characteristics and typical applications
C43000	87.0 Cu, 2.2 Sn, 10.8 Zn	F	317-648	46-94	124-503	18-73	55-3	G-E	30	Excellent cold workability; good hot formability. Fabri- cated by blanking, coining, drawing, forming, bending, heading, and upsetting. Uses: same as C42500.
C43400	85.0 Cu, 0.7 Sn, 14.3 Zn	F	310-607	45-88	103-517	15-75	49-3	G-E	30	Excellent cold workability. Fabricated by blanking, drawing, bonding, forming, stamping and shearing. Uses: electrical switch parts, blades, relay springs, contacts.
C43500	81.0 Cu, 0.9 Sn, 18.1 Zn	F, T	317-552	46-80	110-469	16-68	46-7	G-E	30	Excellent cold workability for fabrication by forming and bending. Uses: bourdon tubing and musical instruments.
C44300, C44400, C44500 Inhibited admiralty	71.0 Cu, 28.0 Zn, 1.0 Sn	F, W, T	331-379	48-55	124-152	18-22	65-60	G-E	30	Excellent cold workability for forming and bending. Uses: condenser, evaporator and heat exchanger tubing, condenser tubing plates, distiller tubing, ferrules.
C46400 to C46700 Naval brass	60.0 Cu, 39.25 Zn, 0.75 Sn	F, R, T, S	379-607	55-88	172-455	25-66	50-17	F-E	30	Excellent hot workability and hot forgeability. Fabri- cated by blanking, drawing, bending, heading and upset- ting, hot forging, pressing. Uses: aircraft turnbuckle barrels, balls, bolts, marine hardware, nuts, propeller shafts, rivets, valve stems, condenser plates, welding rod.
C48200 Naval brass, medium- leaded	60.5 Cu, 0.7 Pb, 0.8 Sn, 38.0 Zn	F, R, S	386-517	56-75	172-365	25-53	43-15	F-E	50	Good hot workability for hot forging, pressing, and machining operations. Uses: marine hardware, screw ma- chine products, valve stems.
C48500 Leaded naval brass	60.0 Cu, 1.75 Pb, 37.5 Zn, 0.75 Sn	F, R, S	379-531	55-77	172-365	25-53	40-15	F-E	70	Combines excellent hot forgeability and machin- ability. Fabricated by hot forging and pressing, ma- chining. Uses: marine hard- ware, screw machine parts, valve stems.
C50500 Phosphor bronze, 1.25% E	98.75 Cu, 1.25 Sn, trace P	F, W	276-545	40-79	97-345	14-50	48-4	G-E	20	Excellent cold workability; good hot formability. Fabri- cated by blanking, bending, heading and upsetting, shearing and swaging. Uses: electrical contacts, flexible hose, pole-line hardware.
C51000 Phosphor bronze, 5% A	95.0 Cu, 5.0 Sn, trace P	F, R, W, T	324-965	47-140	131-552	19-80	64-2	G-E	20	Excellent cold workability. Fabricated by blanking, drawing, bending, heading and upsetting, roll thread- ing and knurling, shearing, stamping. Uses: bellows, bourdon tubing, clutch discs, cotter pins, diaphragms, fasteners, lock washers, wire brushes, chemical hardware, textile machinery, welding rod.

(continued)

(a) F, flat products; R, rod; W, wire; T, tube; P, pipe; S, shapes. (b) Softest to hardest commercial forms. The strength of the standard copper alloys depends on the temper (annealed grain size or degree of cold work) and the section thickness of the mill product. Ranges cover standard tempers for each alloy. (c) E, excellent; G, good; F, fair. (d) Based on 100% for C360000. (e) C10400, 8 oz/ton Ag; C10500, 10 oz/ton; C10700, 25 oz/ton. (f) C11300, 8 oz/ton Ag; C11400, 10 oz/ton; C11500, 16 oz/ton; C11600, 25 oz/ton. (g) C12000, 0.008 P; C12100, 0.008 P and 4 oz/ton Ag. (h) C12700, 8 oz/ton Ag; C12800, 10 oz/ton; C12900, 16 oz/ton; C13000, 25 oz/ton. (i) 8.30 oz/ton Ag. (j) C18200, 0.9 Cr; C18400, 0.8 Cr; C18500, 0.7 Cr. (k) Rod, 61.0 Cu min. Compiled by Copper Development Assn. Inc., New York.

Properties and applications of wrought coppers and copper alloys (continued)

Name and number	Nominal composition, %	Commercial forms(a)	Mechanical properties(b)				Elongation in 2 in., %	Corrosion resistance (c)	Machinability rating (d)	Fabricating characteristics and typical applications
			Tensile strength		Yield strength					
			MPa	ksi	MPa	ksi				
C51100	95.6 Cu, 4.2 Sn, 0.2 P	F	317-710	46-103	345-552	50-80	48-2	G-E	20	Excellent cold workability. Uses: bridge bearing plates, locator bars, fuse clips, sleeve bushings, springs, switch parts, truss wire, wire brushes, chemical hardware, perforated sheets, textile machinery, welding rod.
C52100 Phosphor bronze, 8% C	92.0 Cu, 8.0 Sn, trace P	F, R, W	379-965	55-140	165-552	24-80	70-2	G-E	20	Good cold workability for blanking, drawing, forming and bending, shearing, stamping. Uses: generally for more severe service conditions than C51000.
C52400 Phosphor bronze, 10% D	90.0 Cu, 10.0 Sn, trace P	F, R, W	455-1014	66-147	193 (Annealed)	28	70-3	G-E	20	Good cold workability for blanking, forming and bending, shearing. Uses: heavy bars and plates for severe compression, bridge and expansion plates and fittings, articles requiring good spring qualities, resiliency, fatigue resistance, good wear and corrosion resistance.
C54400 Free-cutting phosphor bronze	88.0 Cu, 4.0 Pb, 4.0 Zn, 4.0 Sn	F, R	303-517	44-75	131-434	19-63	50-15	G-E	80	Excellent machinability; good cold workability. Fabricated by blanking, drawing, bending, machining, shearing, stamping. Uses: bearings, bushings, gears, pinions, shafts, thrust washers, valve parts.
C60800 Aluminum bronze, 5%	95.0 Cu, 5.0 Al	T	414	60	186	27	55	G-E	20	Good cold workability; fair hot formability. Uses: condenser, evaporator and heat exchanger tubes, distiller tubes, ferrules.
C61000	92.0 Cu, 8.0 Al	R, W	483-552	70-80	207-379	30-55	65-25	G-E	20	Good hot and cold workability. Uses: bolts, pump parts, shafts, tie rods, overlay on steel for wearing surfaces.
C61300	92.65 Cu, 0.35 Sn, 7.0 Al	F, R, T, P, S	483-586	70-85	207-400	30-58	42-35	G-E	30	Good hot and cold formability. Uses: nuts, bolts, stringers and threaded members, corrosion-resistant vessels and tanks, structural components, machine parts, condenser tube and piping systems, marine protective sheathing and fastening, munitions mixing troughs and blending chambers.
C61400 Aluminum bronze, D	91.0 Cu, 7.0 Al, 2.0 Fe	F, R, W, T, P, S	524-614	76-89	228-414	33-60	45-32	G-E	20	Similar to C61300.

(continued)

(a) F, flat products; R, rod; W, wire; T, tube; P, pipe; S, shapes. (b) Softest to hardest commercial forms. The strength of the standard copper alloys depends on the temper (annealed grain size or degree of cold work) and the section thickness of the mill product. Ranges cover standard tempers for each alloy. (c) E, excellent; G, good; F, fair. (d) Based on 100% for C36000. (e) C10400, 8 oz/ton Ag; C10500, 10 oz/ton; C10700, 25 oz/ton. (f) C11300, 8 oz/ton Ag; C11400, 10 oz/ton; C11500, 16 oz/ton; C11600, 25 oz/ton. (g) C12000, 0.008 P; C12100, 0.008 P and 4 oz/ton Ag. (h) C12700, 8 oz/ton Ag; C12800, 10 oz/ton; C12900, 16 oz/ton; C13000, 25 oz/ton. (i) 8.30 oz/ton Ag. (j) C18200, 0.9 Cr; C18400, 0.8 Cr; C18500, 0.7 Cr. (k) Rod, 61.0 Cu min. Compiled by Copper Development Assn. Inc., New York.

Properties and applications of wrought coppers and copper alloys (continued)

Name and number	Nominal composition, %	Commercial forms(a)	Tensile strength MPa	ksi	Yield strength MPa	ksi	Elongation in 2 in., %	Corrosion resistance (c)	Machinability rating (d)	Fabricating characteristics and typical applications
C61500	90.0 Cu, 8.0 Al, 2.0 Ni	F	483-1000	70-145	152-965	22-140	55-1	G-E	30	Good hot and cold workability. Fabricating characteristics similar to C52100. Uses: hardware, decorative metal trim, interior furnishings and other articles requiring high tarnish resistance.
C61800	89.0 Cu, 1.0 Fe, 10.0 Al	R	552-586	80-85	269-293	39-42.5	28-23	G-E	40	Fabricated by hot forging and hot pressing. Uses: bushings, bearings, corrosion-resistant applications, welding rods.
C61900	86.5 Cu, 4.0 Fe, 9.5 Al	F	634-1048	92-152	338-1000	49-145	30-1	G-E	...	Excellent hot formability for fabricating by blanking, forming, bending, shearing, and stamping. Uses: springs, contacts, and switch components.
C62300	87.0 Cu, 3.0 Fe, 10.0 Al	F, R	517-676	75-98	241-359	35-52	35-22	G-E	50	Good hot and cold formability. Fabricated by bending, hot forging, hot pressing, forming, and welding. Uses: bearings, bushings, valve guides, gears, valve seats, nuts, bolts, pump rods, worm gears, and cams.
C62400	86.0 Cu, 3.0 Fe, 11.0 Al	F, R	621-724	90-105	276-359	40-52	18-14	G-E	50	Excellent hot formability for fabrication by hot forging and hot bending. Uses: bushings, gears, cams, wear strips, nuts, drift pins, tie rods.
C62500	82.7 Cu, 4.3 Fe, 13.0 Al	F, R	689	100	379 (As extruded)	55	1	G-E	20	Excellent hot formability for fabrication by hot forging and machining. Uses: guide bushings, wear strips, cams, dies, forming rolls.
C63000	82.0 Cu, 3.0 Fe, 10.0 Al, 5.0 Ni	F, R	621-814	90-118	345-517	50-75	20-15	G-E	30	Good hot formability. Fabricated by hot forming and forging. Uses: nuts, bolts, valve seats, plunger tips, marine shafts, valve guides, aircraft parts, pump shafts, structural members.
C63200	82.0 Cu, 4.0 Fe, 9.0 Al, 5.0 Ni	F, R	621-724	90-105	310-365	45-53	25-20	G-E	30	Good hot formability. Fabricated by hot forming and welding. Uses: nuts, bolts, structural pump parts, shafting requiring corrosion resistance.
C63600	95.5 Cu, 3.5 Al, 1.0 Si	R, W	414-579	60-84	...	...	64-29	G-E	40	Excellent cold workability; fair hot formability. Fabricated by cold heading. Uses: components for pole line hardware, cold-headed nuts for wire and cable connectors, bolts and screw products.

(continued)

(a) F, flat products; R, rod; W, wire; T, tube; P, pipe; S, shapes. (b) Softest to hardest commercial forms. The strength of the standard copper alloys depends on the temper (annealed grain size or degree of cold work) and the section thickness of the mill product. Ranges cover standard tempers for each alloy. (c) E, excellent; G, good; F, fair. (d) Based on 100% for C360000. (e) C10400, 8 oz/ton Ag; C10500, 10 oz/ton; C10700, 25 oz/ton. (f) C11300, 8 oz/ton Ag; C11400, 10 oz/ton; C11500, 16 oz/ton; C11600, 25 oz/ton. (g) C12000, 0.008 P; C12100, 0.008 P and 4 oz/ton Ag. (h) C12700, 8 oz/ton Ag; C12800, 10 oz/ton; C12900, 16 oz/ton; C13000, 25 oz/ton. (i) 8.30 oz/ton Ag. (j) C18200, 0.9 Cr; C18400, 0.8 Cr; C18500, 0.7 Cr. (k) Rod, 61.0 Cu min. Compiled by Copper Development Assn. Inc., New York.

Properties and applications of wrought coppers and copper alloys (continued)

Name and number	Nominal composition, %	Commercial forms(a)	Mechanical properties(b) Tensile strength MPa	ksi	Yield strength MPa	ksi	Elongation in 2 in., %	Corrosion resistance (c)	Machinability rating (d)	Fabricating characteristics and typical applications
C63800	99.5 Cu, 2.8 Al, 1.8 Si, 0.40 Co	F	565-896	82-130	372-786	54-114	36-4	G-E	...	Excellent cold workability and hot formability. Uses: springs, switch parts, contacts, relay springs, glass sealing and porcelain enameling.
C64200	91.2 Cu, 7.0 Al	F, R	517-703	75-102	241-469	35-68	32-22	G-E	60	Excellent hot formability. Fabricated by hot forming, forging, machining. Uses: valve stems, gears, marine hardware, pole-line hardware, bolts, nuts, valve bodies and components.
C65100 Low-silicon bronze, B	98.5 Cu, 1.5 Si	R, W, T	276-655	40-95	103-476	15-69	55-11	G-E	30	Excellent hot and cold workability. Fabricated by forming and bending, heading and upsetting, hot forging and pressing, roll threading and knurling, squeezing and swaging. Uses: hydraulic pressure lines, anchor screws, bolts, cable clamps, cap screws, machine screws, marine hardware, nuts, pole-line hardware, rivets, U-bolts, electrical conduits, heat exchanger tubing, welding rod.
C65500 High-silicon bronze, A	97.0 Cu, 3.0 Si	F, R, W, T	386-1000	56-145	145-483	21-70	63-3	G-E	30	Excellent hot and cold workability. Fabricated by blanking, drawing, forming and bending, heading and upsetting, hot forging and pressing, roll threading and knurling, shearing, squeezing and swaging. Uses: similar to C65100 including propeller shafts.
C66700 Manganese brass	70.0 Cu, 28.8 Zn, 1.2 Mn	F, W	315-689	45.8-100	83-638	12-92.5	60-2	G-E	30	Excellent cold formability. Fabricated by blanking, bending, forming, stamping, welding. Uses: brass products resistance welded by spot, seam, and butt welding.
C67400	58.5 Cu, 36.5 Zn, 1.2 Al, 2.8 Mn, 1.0 Sn	F, R	483-634	70-92	234-379	34-55	28-20	F-E	25	Excellent hot formability. Fabricated by hot forging and pressing, machining. Uses: bushings, gears, connecting rods, shafts, wear plates.
C67500 Manganese bronze, A	58.5 Cu, 1.4 Fe, 39.0 Zn, 1.0 Sn, 0.1 Mn	R, S	448-579	65-84	207-414	30-60	33-19	F-E	30	Excellent hot workability. Fabricated by hot forging and pressing, hot heading and upsetting. Uses: clutch discs, pump rods, shafting, balls, valve stems and bodies.
C68700 Aluminum brass, arsenical	77.5 Cu, 20.5 Zn, 2.0 Al, 0.1 As	T	414	60	186	27	55	G-E	30	Excellent cold workability for forming and bending. Uses: condenser, evaporator and heat exchanger tubing, condenser tubing plates, distiller tubing, ferrules.

(continued)

(a) F, flat products; R, rod; W, wire; T, tube; P, pipe; S, shapes. (b) Softest to hardest commercial forms. The strength of the standard copper alloys depends on the temper (annealed grain size or degree of cold work) and the section thickness of the mill product. Ranges cover standard tempers for each alloy. (c) E, excellent; G, good; F, fair. (d) Based on 100% for C360000. (e) C10400, 8 oz/ton Ag; C10500, 10 oz/ton; C10700, 25 oz/ton Ag. (f) C11300, 8 oz/ton Ag; C11400, 10 oz/ton; C11500, 16 oz/ton; C11600, 25 oz/ton. (g) C12000, 0.008 P; C12100, 0.008 P and 4 oz/ton Ag. (h) C12700, 8 oz/ton Ag; C12800, 10 oz/ton; C12900, 16 oz/ton; C13000, 25 oz/ton. (i) 8.30 oz/ton Ag. (j) C18200, 0.9 Cr; C18400, 0.8 Cr; C18500, 0.7 Cr. (k) Rod, 61.0 Cu min. Compiled by Copper Development Assn. Inc., New York.

Properties and applications of wrought coppers and copper alloys (continued)

Name and number	Nominal composition, %	Commercial forms(a)	Mechanical properties(b) Tensile strength MPa	ksi	Yield strength MPa	ksi	Elongation in 2 in., %	Corrosion resistance (c)	Machinability rating (d)	Fabricating characteristics and typical applications
C68800	73.5 Cu, 22.7 Zn, 3.4 Al, 0.40 Co	F	565-889	82-129	379-786	55-114	36-2	G-E	· · ·	Excellent hot and cold formability. Fabricated by blanking, drawing, forming and bending, shearing and stamping. Uses: springs, switches, contacts, relays, drawn parts.
C69000	73.3 Cu, 3.4 Al, 0.6 Ni, 22.7 Zn	F	496-896	72-130	345-807	50-117	40-2	G-E	· · ·	Fabricating characteristics same as C68800. Uses: wiring devices, relays, switches, springs, high-strength shells.
C69400 Silicon red brass	81.5 Cu, 14.5 Zn, 4.0 Si	R	552-689	80-100	276-393	40-57	25-20	G-E	30	Excellent hot formability for fabrication by forging, screw machine operations. Uses: valve stems where corrosion resistance and high strength are critical.
C70400	92.4 Cu, 1.5 Fe, 5.5 Ni, 0.6 Mn	F, T	262-531	38-77	276-524	40-76	46-2	G-E	20	Excellent cold workability; good hot formability. Fabricated by forming, bending and welding. Uses: condensers, evaporators, heat exchangers, ferrules, salt water piping, lithium bromide absorption tubing, shipboard condenser intake systems.
C70600 Copper nickel, 10%	88.7 Cu, 1.3 Fe, 10.0 Ni	F, T	303-414	44-60	110-393	16-57	42-10	E	20	Good hot and cold workability. Fabricated by forming and bending, welding. Uses: condensers, condenser plates, distiller tubing, evaporator and heat exchanger tubing, ferrules, salt water piping.
C71000 Copper nickel, 20%	79.0 Cu, 21.0 Ni	F, W, T	338-655	49-95	90-586	13-85	40-3	E	20	Good hot and cold formability. Fabricated by blanking, forming and bending, welding. Uses: communication relays, condensers, condenser plates, electrical springs, evaporator and heat exchanger tubes, ferrules, resistors.
C71500 Copper nickel, 30%	70.0 Cu, 30.0 Ni	F, R, T	372-517	54-75	138-483	20-70	45-15	E	20	Similar to C70600.
C71700	67.8 Cu, 0.7 Fe, 31.0 Ni, 0.5 Be	F, R, W	483-1379	70-200	207-1241	30-180	40-4	G-E	20	Good hot and cold formability. Uses: high-strength constructional parts for sea water corrosion resistance, hydrophone cases, mooring cable wire, springs, retainer rings, bolts, screws, pins for ocean telephone cable applications.

(continued)

(a) F, flat products; R, rod; W, wire; T, tube; P, pipe; S, shapes. (b) Softest to hardest commercial forms. The strength of the standard copper alloys depends on the temper (annealed grain size or degree of cold work) and the section thickness of the mill product. Ranges cover standard tempers for each alloy. (c) E, excellent; G, good; F, fair. (d) Based on 100% for C360000. (e) C10400, 8 oz/ton Ag; C10500, 10 oz/ton; C10700, 25 oz/ton. (f) C11300, 8 oz/ton Ag; C11400, 10 oz/ton; C11500, 16 oz/ton; C11600, 25 oz/ton. (g) C12000, 0.008 P; C12100, 0.008 P and 4 oz/ton Ag. (h) C12700, 8 oz/ton Ag; C12800, 10 oz/ton; C12900, 16 oz/ton; C13000, 25 oz/ton. (i) 8.30 oz/ton Ag. (j) C18200, 0.9 Cr; C18400, 0.8 Cr; C18500, 0.7 Cr. (k) Rod, 61.0 Cu min. Compiled by Copper Development Assn. Inc., New York.

Properties and applications of wrought coppers and copper alloys (continued)

Name and number	Nominal composition, %	Commercial forms(a)	Mechanical properties(b) Tensile strength MPa	ksi	Yield strength MPa	ksi	Elongation in 2 in., %	Corrosion resistance (c)	Machinability rating (d)	Fabricating characteristics and typical applications
C72500	88.2 Cu, 9.5 Ni, 2.3 Sn	F, R, W, T	379-827	55-120	152-745	22-108	35-1	E	20	Excellent cold and hot formability. Fabricated by blanking, brazing, coining, drawing, etching, forming and bending, heading and upsetting, roll threading and knurling, shearing, spinning, squeezing, stamping and swaging. Uses: relay and switch springs, connectors, brazing alloy, lead frames, control and sensing bellows.
C73500	72.0 Cu, 10.0 Zn, 18.0 Ni	F, R, W, T	345-758	50-110	103-579	15-84	37-1	E	20	Fabricating characteristics same as C74500. Uses: holloware, medallions, jewelry, base for silver plate, cosmetic cases, musical instruments, name plates, contacts.
C74500 Nickel silver, 65-10	65.0 Cu, 25.0 Zn, 10.0 Ni	F, W	338-896	49-130	124-524	18-76	50-1	E	20	Excellent cold workability. Fabricated by blanking, drawing, etching, forming and bending, heading and upsetting, roll threading and knurling, shearing, spinning, squeezing and swaging. Uses: rivets, screws, slide fasteners, optical parts, etching stock, hollow ware, nameplates, platers' parts.
C75200 Nickel silver, 65-18	65.0 Cu, 17.0 Zn, 18.0 Ni	F, R, W	386-710	56-103	172-621	25-90	45-3	E	20	Fabricating characteristics similar to C74500. Uses: rivets, screws, table flatware, truss wire, zippers, bows, camera parts, core bars, temples, base for silver plate, costume jewelry, etching stock, hollow ware, nameplates, radio dials.
C75400 Nickel silver, 65-15	65.0 Cu, 20.0 Zn, 15.0 Ni	F	365-634	53-92	124-545	18-79	43-2	E	20	Fabricating characteristics similar to C74500. Uses: camera parts, optical equipment, etching stock, jewelry.
C75700 Nickel silver, 65-12	65.0 Cu, 23.0 Zn, 12.0 Ni	F, W	359-641	52-93	124-545	18-79	48-2	E	20	Fabricating characteristics similar to C74500. Uses: slide fasteners, camera parts, optical parts, etching stock, nameplates.

(continued)

(a) F, flat products; R, rod; W, wire; T, tube; P, pipe; S, shapes. (b) Softest to hardest commercial forms. The strength of the standard copper alloys depends on the temper (annealed grain size or degree of cold work) and the section thickness of the mill product. Ranges cover standard tempers for each alloy. (c) E, excellent; G, good; F, fair. (d) Based on 100% for C360000. (e) C10400, 8 oz/ton Ag; C10500, 10 oz/ton; C10700, 25 oz/ton. (f) C11300, 8 oz/ton Ag; C11400, 10 oz/ton; C11500, 16 oz/ton; C11600, 25 oz/ton. (g) C12000, 0.008 P; C12100, 0.008 P and 4 oz/ton Ag. (h) C12700, 8 oz/ton Ag; C12800, 10 oz/ton; C12900, 16 oz/ton; C13000, 25 oz/ton. (i) 8.30 oz/ton Ag. (j) C18200, 0.9 Cr; C18400, 0.8 Cr; C18500, 0.7 Cr. (k) Rod, 61.0 Cu min. Compiled by Copper Development Assn. Inc., New York.

Properties and applications of wrought coppers and copper alloys (continued)

Name and number	Nominal composition, %	Commercial forms(a)	Mechanical properties(b) Tensile strength MPa	ksi	Yield strength MPa	ksi	Elongation in 2 in., %	Corrosion resistance (c)	Machinability rating (d)	Fabricating characteristics and typical applications
C76200	59.0 Cu, 29.0 Zn, 12.0 Ni	F, T	393-841	57-122	145-758	21-110	50-1	G-E	· · ·	Fabricating characteristics same as C77000. Uses: electrical terminals, contact springs, release brackets, ornamental bits and spurs, optical parts, surgical instruments, electrical contacts.
C77000 Nickel silver, 55-18	55.0 Cu, 27.0 Zn, 18.0 Ni	F, R, W	414-1000	60-145	186-621	27-90	40-2	E	30	Good cold workability. Fabricated by blanking, forming and bending, and shearing. Uses: optical goods, springs and resistance wire.
C72200	82.0 Cu, 16.0 Ni, 0.5 Cr, 0.8 Fe, 0.5 Mn	F, T	317-483	46-70	124-455	18-66	46-6	G-E	· · ·	Good hot and cold workability. Fabricated by forming, bending and welding. Uses: condenser and heat exchanger tubing, salt water piping.
C78200 Leaded nickel silver, 65-8-2	65.0 Cu, 2.0 Pb, 25.0 Zn, 8.0 Ni	F	365-627	53-91	159-524	23-76	40-3	E	60	Good cold formability. Fabricated by blanking, milling and drilling. Uses: key blanks, watch plates, watch parts.

(a) F, flat products; R, rod; W, wire; T, tube; P, pipe; S, shapes. (b) Softest to hardest commercial forms. The strength of the standard copper alloys depends on the temper (annealed grain size or degree of cold work) and the section thickness of the mill product. Ranges cover standard tempers for each alloy. (c) E, excellent; G, good; F, fair. (d) Based on 100% for C360000. (e) C10400, 8 oz/ton Ag; C10500, 10 oz/ton; C10700, 25 oz/ton. (f) C11300, 8 oz/ton Ag; C11400, 10 oz/ton; C11500, 16 oz/ton; C11600, 25 oz/ton. (g) C12000, 0.008 P; C12100, 0.008 P and 4 oz/ton Ag. (h) C12700, 8 oz/ton Ag; C12800, 10 oz/ton; C12900, 16 oz/ton; C13000, 25 oz/ton. (i) 8.30 oz/ton Ag. (j) C18200, 0.9 Cr; C18400, 0.8 Cr; C18500, 0.7 Cr. (k) Rod, 61.0 Cu min. Compiled by Copper Development Assn. Inc., New York.

Scheme of temper designations for wrought copper and brass based on cold reduction

Nominal temper designation	Increase in B&S gage numbers	Rolled sheet Reduction in thickness and area, %	True strain(a)	Drawn wire Reduction in diameter, %	Reduction in area, %	True strain(a)
1/4 hard	1	10.9	0.116	10.9	20.7	0.232
1/2 hard	2	20.7	0.232	20.7	37.1	0.463
3/4 hard	3	29.4	0.347	29.4	50.1	0.694
Hard	4	37.1	0.463	37.1	60.5	0.926
Extra hard	6	50.1	0.696	50.1	75.1	1.39
Spring	8	60.5	0.928	60.5	84.4	1.86
Extra spring	10	68.6	1.16	68.6	90.2	2.32
Special spring	12	75.1	1.39	75.1	93.8	2.78
Super spring	14	80.3	1.62	80.3	96.1	3.25

(a) True strain equals ln A_0/A, where A_0 is initial cross-sectional area and A is final area.

Copper tube alloys and typical applications

UNS number	Alloy type	ASTM specifications	Typical uses
C10200	Oxygen-free copper	B68, B75, B88, B111, B188, B280, B359, B372, B395, B447	Bus tube, conductors, wave guides
C12200	Phosphorus deoxidized copper	B68, B75, B88, B111, B280, B306, B359, B360, B395, B447, B543	Water tubes; condenser, evaporator and heat exchanger tubes; air conditioning and refrigeration, gas, heater and oil burner lines; plumbing pipe and steam tubes; brewery and distillery tubes; gasoline, hydraulic and oil lines; rotating bands
C19200	Copper	B111, B359, B395, B469	Automotive hydraulic brake lines; flexible hose
C23000	Red brass, 85%	B111, B135, B359, B395, B543	Condenser and heat exchanger tubes, flexible hose; plumbing pipe; pump lines
C26000	Cartridge brass, 70%	B135	Plumbing brass goods
C33000	Low-leaded brass (tube)	B135	Pump and power cylinders and liners; plumbing brass goods
C36000	Free-cutting brass		Screw machine parts; plumbing goods
C43500	Tin brass		Bourdon tubes; musical instruments
C44300	Inhibited admiralty metal	B111, B359, B395	Condenser, evaporator and heat exchanger tubes; distiller tubes
C44400			
C44500			
C46400	Naval brass		Marine hardware, nuts
C46500			
C46600			
C46700			
C60800	Aluminum bronze, 5%	B111, B359, B395	Condenser, evaporator and heat exchanger tubes; distiller tubes
C65100	Silicon bronze B	B315	Heater exchanger tubes; electrical conduits
C65500	Silicon bronze A	B315	Chemical equipment, heat exchanger tubes; piston rings
C68700	Arsenical aluminum brass	B111, B359, B395	Condenser, evaporator and heat exchanger tubes; distiller tubes
C70600	Copper nickel, 10%	B111, B359, B395, B466, B467, B543, B552	Condenser, evaporator and heat exchanger tubes; salt water piping; distiller tubes
C71500	Copper nickel, 30%	B111, B359, B395, B446, B467, B543, B552	Condenser, evaporator and heat exchanger tubes; distiller tubes; salt water piping

Typical mechanical properties for copper alloy tube(a)

Temper	Tensile strength MPa	ksi	Yield strength(b) MPa	ksi	Elongation(c), %
C10200					
OS050......... 220		32	69	10	45
OS025......... 235		34	76	11	45
H55 275		40	220	32	25
H80 380		55	345	50	8
C12200					
OS050......... 220		32	69	10	45
OS025......... 235		34	76	11	45
H55 275		40	220	32	25
H80 380		55	345	50	8
C19200					
H55(d)......... 290		42	205(e)	30(e)	35
C23000					
OS050......... 275		40	83	12	55
OS015......... 305		44	125	18	45
H55 345		50	275	40	30
H80 485		70	400	58	8
C26000					
OS050......... 325		47	105	15	65
OS025......... 360		52	140	20	55
H80 540		78	440	64	8
C33000					
OS050......... 325		47	105	15	60
OS025......... 360		52	140	20	50
H80 515		75	415	60	7
C43500					
OS035......... 315		46	110	16	46
H80 515		75	415	60	10
C44300, C44400, C44500					
OS025......... 365		53	150	22	65
C46400, C46500, C46600, C46700(f)					
H80 605		88	455	66	18
C60800					
OS025......... 415		60	185	27	55
C65100					
OS015......... 310		45	140	20	55
H80 450		65	275	40	20
C65500					
OS050......... 395		57	...	...	70
H80 640		93	...	...	22
C68700					
OS025......... 415		60	185	27	55
C70600					
OS025......... 305		44	110	16	42
H55 415		60	395	57	10
C71500					
OS025......... 415		60	170	25	45

(a) Tube size: 25 mm (1 in.) OD by 1.65 mm (0.065 in.) wall. (b) 0.5% extension under load. (c) In 50 mm or 2 in. (d) Tube size: 4.8 mm (0.1875 in.) OD by 0.76 mm (0.030 in.) wall. (e) 0.2% offset. (f) Tube size: 9.5 mm (0.375 in.) OD by 2.5 mm (0.097 in.) wall.

Characteristics of solid round copper wire: ASTM B1, B3, B258

Conductor size, AWG	Conductor diameter, mils	Conductor area, circular mils	Net weight, lb/1000 ft	Soft (annealed) wire		Hard drawn wire		
				Minimum elongation(a), %	Nominal resistance, Ω/1000 ft	Nominal breaking strength, lb	Nominal tensile strength, ksi	Nominal resistance, Ω/1000 ft
4/0 . . . . 460.0		211 600	640.5	35	0.0491	8143	49.0	0.05044
3/0 . . . . 409.6		167 800	507.8	35	0.06180	6720	51.0	0.06361
2/0 . . . . 364.8		133 100	402.8	35	0.07791	5519	52.8	0.08019
1.0 . . . . 324.9		105 600	319.5	35	0.09821	4518	54.5	0.1011
1 . . . . 289.3		83 690	253.3	30	0.1239	3888	56.1	0.1289
2 . . . . 257.6		66 360	200.9	30	0.1563	3002	57.6	0.1625
3 . . . . 229.4		52 620	159.3	30	0.1971	2439	59.0	0.2050
4 . . . . 204.3		41 740	126.3	30	0.2485	1970	60.1	0.2584
5 . . . . 181.9		33 090	100.2	30	0.3134	1590	61.2	0.3259
6 . . . . 162.0		26 240	79.44	30	0.3952	1280	62.1	0.4110
7 . . . . 144.3		20 820	63.03	30	0.4981	1030	63.1	0.5180
8 . . . . 128.5		16 510	49.98	30	0.6281	826.1	63.7	0.6532
9 . . . . 114.4		13 090	39.62	30	0.7923	660.9	64.3	0.8239
10 . . . . 101.9		10 380	31.43	25	0.9991	529.3	64.9	1.039
11 . . . . 90.7		8 230	24.9	25	1.26	423	65.4	1.31
12 . . . . 80.8		6 530	19.8	25	1.59	337	65.7	1.65
13 . . . . 72.0		5 180	15.7	25	2.00	268	65.9	2.08
14 . . . . 64.1		4 110	12.4	25	2.52	214	66.2	2.62
15 . . . . 57.1		3 260	9.87	25	3.18	170	66.4	3.31
16 . . . . 50.8		2 580	7.81	25	4.02	135	66.6	4.18
17 . . . . 45.3		2 050	6.21	25	5.06	108	66.8	5.26
18 . . . . 40.3		1 620	4.92	25	6.40	85.5	67.0	6.66
19 . . . . 35.9		1 290	3.90	25	8.04	68.0	67.2	8.36
20 . . . . 32.0		1 020	3.10	25	10.2	54.2	67.4	10.6
21 . . . . 28.5		812	2.46	25	12.8	43.2	67.7	13.3
22 . . . . 25.3		640	1.94	25	16.2	34.1	67.9	16.9
23 . . . . 22.6		511	1.55	25	20.3	27.3	68.1	21.1
24 . . . . 20.1		404	1.22	20	25.7	21.7	68.3	26.7
25 . . . . 17.9		320	0.970	20	32.4	17.3	68.6	33.7
26 . . . . 15.9		253	0.765	20	41.0	13.7	68.8	42.6
27 . . . . 14.2		202	0.610	20	51.4	10.9	69.0	53.4
28 . . . . 12.6		159	0.481	20	65.2	8.64	69.3	67.8
29 . . . . 11.3		128	0.387	20	81.0	6.96	69.4	84.3
30 . . . . 10.0		100	0.303	15	104.0	5.47	69.7	108.0
31 . . . . 8.9		79.2	0.240	15	131.0	4.35	69.9	136.0
32 . . . . 8.0		64.0	0.194	15	162.0	3.53	70.2	169.0
33 . . . . 7.1		50.4	0.153	15	206.0	2.79	70.4	214.0
34 . . . . 6.3		39.7	0.120	15	261.0	2.20	70.6	272.0
35 . . . . 5.6		31.4	0.0949	15	330.0	1.75	70.9	343.0
36 . . . . 5.0		25.0	0.0757	15	415.0	1.40	71.1	431.0
37 . . . . 4.5		20.2	0.0613	15	513.0	1.13	71.3	534.0
38 . . . . 4.0		16.0	0.0484	15	648.0	0.898	71.5	674.0
39 . . . . 3.5		12.2	0.0371	15	850.0	0.691	71.8	884.0
40 . . . . 3.1		9.61	0.0291	15	1079.0	0.543	72.0	1122.0
41 . . . . 2.8		7.84	0.0237	15	1323.0	0.443	72.0	1376.0
42 . . . . 2.5		6.25	0.0189	15	1659.0	0.353	72.0	1726.0
43 . . . . 2.2		4.48	0.0147	15	2143.0	0.274	72.0	2228.0
44 . . . . 2.0		4.00	0.0121	15	2593.0	0.226	72.0	2696.0

(a) In 10 in.

Copper Casting Alloys

Nominal compositions of principal copper casting alloys

UNS number	Common name	Previous ASTM designation	Cu	Sn	Pb	Zn	Fe	Al	Others
ASTM B22									
C86300	Manganese bronze	B22-E	63	...	...	25	3	6	3 Mn
C90500	Tin bronze	B22-D	88	10	...	2	...	...	...
C91100	Tin bronze	B22-B	84	16	...	...	...	...	...
C91300	Tin bronze	B22-A	81	19	...	...	...	...	...
C93700	High-lead tin bronze	B22-C	80	10	10	...	...	...	...
ASTM B61									
C92200	Valve bronze	...	88	6	1.5	4	...	...	1 max Ni
ASTM B62									
C83600	Leaded red brass	...	85	5	5	5	...	...	...
ASTM B66									
C93800	High-lead tin bronze	...	78	7	15	...	...	...	...
C94300	High-lead tin bronze	...	70	5	25	...	...	...	...
C94400	Leaded phosphor bronze	...	81	8	11	...	...	...	0.35 P
C94500	High-lead tin bronze	...	73	7	19	1	...	...	...
ASTM B67									
C94100	High-lead tin bronze	...	70	5.5	18.5	3 max	...	...	...
ASTM B148									...
C95200	Aluminum bronze	B148-9A	88	...	...	...	3	9	...
C95300	Aluminum bronze	B148-9B	89	...	...	...	1	10	...
C95400	Aluminum bronze	B148-9C	85	...	...	...	4	11	...
C95500	Nickel-aluminum bronze	B148-9D	81	...	...	...	4	11	4 Ni
C95800	Nickel-aluminum bronze	...	81.3	...	...	...	4	9	4.5 Ni, 1.2 Mn
ASTM B176 (Die Casting Alloys)									
C85800	Yellow brass	Z30A	58	1	1	40	...	...	...
C87800	Silicon brass	ZS144A	82	...	...	14	...	...	4 Si
C87900	Silicon brass	ZS331A	65	...	...	33	...	...	1 Si
ASTM B584									
C83600	Leaded red brass	B145-4A	85	5	5	5	...	...	...
C83800	Leaded red brass	B145-4B	83	4	6	7	...	...	...
C84400	Leaded semi-red brass	B145-5A	81	3	7	9	...	...	...
C84800	Leaded semi-red brass	B145-5B	76	3	6	15	...	...	...
C85200	Leaded yellow brass	B146-6A	72	1	3	24	...	...	...
C85400	Leaded yellow brass	B146-6B	67	1	3	29	...	...	...
C85700	Leaded naval brass	B146-6C	63	1	1	34	7	0.03	...
C86200	High-strength manganese bronze	B147-8B	64	...	...	26	3	4	3 Mn
C86300	High-strength manganese bronze	B147-8C	63	...	...	25	3	6	3 Mn
C86400	Leaded manganese bronze	B147-7A or B132-A	59	...	1	38	1	0.5	0.5 Mn
C86500	Manganese bronze	B147-8A	58	...	...	39	1	1	1 Mn
C86700	Leaded manganese bronze	B132-B	58	1	1	34	2	2	2 Mn
C87200	Silicon bronze	B198-12A	Several nominal compositions available						
C87400	Silicon brass	B198-13A	82	...	0.5	14	...	...	3.5 Si
C87500	Silicon brass	B198-13B	82	...	...	14	...	...	4 Si

(continued)

Nominal compositions of principal copper casting alloys (continued)

UNS number	Common name	Previous ASTM designation	Cu	Sn	Pb	Zn	Fe	Al	Others
C87600	Silicon brass	B198-13C	89	...	...	6	...	...	5 Si
C90300	Modified G bronze	B143-1B	88	8	...	4	...	...	...
C90500	G bronze	B143-1A	88	10	...	2	...	...	...
C92200	Steam bronze (a)	B143-2A	88	6	1.5	4.5	...	...	...
C92300	Leaded tin bronze	B143-2B	87	8	1	4	...	...	...
C93200	High-lead tin bronze	B144-3B	83	7	7	3	...	...	...
C93500	High-lead tin bronze	B144-3C	85	5	9	1	...	...	...
C93700	High-lead tin bronze	B144-3A	80	10	10	...	...	...	...
C93800	High-lead tin bronze	B144-3D	78	7	15	...	...	...	...
C94300	High-lead tin bronze	B144-3E	70	5	25	...	...	...	...
C94700	Nickel-tin bronze	B292-A	88	5	...	2	...	...	5 Ni
C94800	Leaded nickel-tin bronze	B292-B	87	5	1	2	...	...	5 Ni
C94900	Leaded nickel-tin bronze	...	80	5	5	5	...	...	5 Ni
C97300	Leaded nickel silver	B149-10A	56	2	10	20	...	...	12 Ni
C97600	Leaded nickel silver	B149-11A	64	4	4	8	...	...	20 Ni
C97800	Leaded nickel silver	B149-11B	66	5	2	2	...	...	25 Ni

(a) Also known as valve bronze or Navy M bronze.

Foundry properties for principal copper alloys for sand casting

UNS number	Common name	Shrinkage allowance, %	Approx liquidus temperature °C	Approx liquidus temperature °F	Castability rating (a)	Fluidity rating (a)
C83600	Leaded red brass	5.7	1010	1850	2	6
C84400	Leaded semi-red brass	2.0	980	1795	2	6
C84800	Leaded semi-red brass	1.4	955	1750	2	6
C85400	Leaded yellow brass	1.5 to 1.8	940	1725	4	4
C85800	Yellow brass	2.0	925	1700	4	4
C86300	Manganese bronze	2.3	920	1690	6	2
C86500	Manganese bronze	1.9	880	1615	6	2
C87200	Silicon bronze	1.8 to 2.0			8	3
C87500	Silicon brass	1.9	915	1680	7	1
C90300	Tin bronze	1.5 to 1.8	980	1795	3	6
C92200	Leaded tin bronze	1.5	990	1810	3	6
C93700	High-lead tin bronze	2.0	930	1705	1	6
C94300	High-lead tin bronze	1.5	925	1700	1	6
C95300	Aluminum bronze	1.6	1045	1910	8	5
C95800	Aluminum bronze	1.6	1060	1940	8	5
C97600	Nickel silver	2.0	1145	2090	5	7
C97800	Nickel silver	1.6	1180	2160	5	7

(a) Relative rating for casting in sand molds. The alloys are ranked from 1 to 8 in both over-all castability and fluidity; 1 is the highest or best possible rating.

Composition and typical properties of heat treated copper casting alloys of high strength and conductivity

UNS number	Nominal composition	Tensile strength MPa	Tensile strength ksi	Yield strength MPa	Yield strength ksi	Elongation, %	Hardness	Electrical conductivity, %IACS
C81400	99Cu-0.8Cr-0.06Be	365	53	250	36	11	69 HRB	70
C81500	99Cu-1Cr	350	51	275	40	17	105 HB	85
C81800	97Cu-1.5Co-1Ag-0.4Be	705	102	515	75	8	96 HRB	48
C82000	97Cu-2.5Co-0.5Be	660	96	515	75	6	96 HRB	48
C82200	98Cu-1.5Ni-0.5Be	655	95	515	75	7	96 HRB	48
C82500	97Cu-2Be-0.5Co-0.3Si	1105	160	1035	150	1	43 HRC	20
C82800	96.6Cu-2.6Be-0.5Co-0.3Si	1140	165	1070	155	1	46 HRC	18

Typical properties of copper casting alloys

UNS number	Tensile strength MPa	ksi	Yield strength(a) MPa	ksi	Compressive yield strength(b) MPa	ksi	Elongation, %	Hardness HB(c)	Electrical conductivity, % IACS
ASTM B22									
C86300	820	119	570(d)	82(d)	490	71	18	177	9.05
C90500	275-345	40-50	140-160	20-23	...	...	24-43	75-85	10.5-11.5
C93700	270	39	125	18	125	18	30	67	10.0
ASTM B61									
C92200	280	41	110	16	105	15	45	64	14.5
ASTM B62									
C83600	240	35	105	15	100	14	32	62	15.0
ASTM B66									
C94300	160-205	23-30	75-105	11-15	80-95	12-14	7-16	42-55	...
ASTM B147									
C86200	625-670	91-97	315-345	46-50	345	50	19-25	170-195(e)	7-8
C86300	820	119	570(d)	82(d)	490	71	18	177	9.0
C86400	415-540	60-78	170-275	25-40	140-180	20-26	15-30	80-95	20-24
C86500	490	71	180	26	165	24	40	98	20.5
ASTM B148									
C95200	480-600	70-87	170-205	25-30	185-215	27-31	22-38	110-140(e)	12-14
C95300(f)	480-585	70-85	205-240	30-35	110-140	16-20	20-35	110-160(e)	12-15
C95300(g)	550-655	80-95	275-380	40-55	240-275	35-45	12-16	160-225(e)	13.8
C95400(f)	515-655	75-95	205-285	30-41	...	...	12-20	150-185(e)	13-15
C95400(g)	620-690	90-100	310-360	45-52	...	...	6-15	190-235(e)	...
C95500(f)	620-725	90-105	275-345	40-50	...	...	7-20	175-210(e)	8-9.5
C95500(g)	760-855	110-124	415-550	60-80	...	...	5-12	215-260(e)	...
ASTM B176									
C85800(h)	380	55	205(d)	30(d)	...	...	15	...	22
C87800(h)	620	90	205(d)	30(d)	...	...	25	...	...
C87900(h)	400	58	205(d)	30(d)	...	...	15	...	...
ASTM B584									
C83600	243	35	105	15	100	14	32	62	15
C83800	205-260	30-38	85-115	12-17	76-83	11-12	15-27	50-60	...
C84400	200-270	29-39	90-115	13-17	...	...	18-30	50-60	18
C84800	260	38	105	15	85	12	37	59	16.5
C85200	240-275	35-40	85-95	12-14	55-70	8-10	25-40	40-55	15-22
C85400	205-260	30-38	75-105	11-15	62	9	20-35	40-60	18-25
C85700	275-310	40-45	95-140	14-20	...	...	15-25	50-75	20-26
C86200	625-670	91-97	315-345	46-50	345	50	19-25	170-195(e)	7-8
C86300	820	119	573	82(d)	490	71	18	177	9.0
C86400	415-540	60-78	170-275	25-40	...	20-26	15-30	80-95	20-24
C86500	490	71	180	26	140-180	24	40	98	20.5
C86700	550	80	220	32	165	...	15	...	...
C87200	380-450	55-65	150-205	22-30	105-150	15-22	25-55	85-120	4.5-6.4
C87400	345-485	50-70	145-225	21-33	...	...	20-50	70-130	...
C87500	470	68	207	30	185	27	17	115	6.0
C87600	414 min	60 min	207 min	30 min	...	...	16 min	...	...
C90300	275-345	40-50	125-150	18-22	...	...	25-50	60-75	12-13
C90500	275-345	40-50	140-160	20-23	...	...	24-43	75-85	10.5-11.5
C92200	280	41	110	16	105	15	45	64	14.5
C92300	225-295	33-43	110-165	16-24	62-76	9-11	18-30	60-75	10-12
C93200	205-260	30-38	115-145	17-21	...	...	12-20	55-65	...
C93500	195-240	28-35	83-105	12-15	90	13	20-35	55-65	15
C93700	270	39	125	18	125	18	30	67	10.0
C93800	170-225	25-33	95-140	14-20	90-110	13-16	10-18	50-60	...
C94300	160-205	23-30	76-105	11-15	83-97	12-14	7-16	42-55	...
C94700(f)	310	45	140	20	...	...	25	...	...
C94700(j)	515	75	345	50	...	...	5	...	...
C94800(f)	275	40	140	20	...	...	20	...	...
C94900	262 min	38 min	97 min	15 min	...	...	15 min	...	...
C97300	205-275	30-40	105-140	15-20	...	...	10-25	50-60	5.7
C97600	325	47	180	26	168	24	22	85	4.8
C97800	345-450	50-65	180-275	26-40	...	...	15-25	120-150	4-5

(a) At 0.5% extension under load. (b) At permanent set of 0.1%. (c) 500 kg load; 10 mm diam ball. (d) At 0.2% offset. (e) 3000 kg load. (f) MO1 temper. (g) TQ00 temper. (h) MO4 temper. (j) TF00 temper.

Corrosion ratings of cast copper metals in various media

Corrosive medium	Copper	Tin bronze	Leaded tin bronze	High-leaded tin bronze	Leaded red brass	Leaded semi-red brass	Leaded yellow brass	Leaded high-strength yellow brass	High-strength yellow brass	Aluminum bronze	Leaded nickel brass	Leaded nickel bronze	Silicon bronze	Silicon brass
Acetate solvents	B	A	A	A	A	A	B	A	A	A	A	A	A	B
Acetic acid														
20%	A	C	B	C	B	C	C	C	C	A	C	A	A	B
50%	A	C	B	C	B	C	C	C	C	A	C	B	A	B
Glacial	A	A	A	C	A	C	C	C	C	A	B	B	A	A
Acetone	A	A	A	A	A	A	A	A	A	A	A	A	A	A
Acetylene (a)	C	C	C	C	C	C	C	C	C	C	C	C	C	C
Alcohols (b)	A	A	A	A	A	A	A	A	A	A	A	A	A	A
Aluminum chloride	C	C	C	C	C	C	C	C	C	C	B	C	C	C
Aluminum sulfate	B	B	B	B	B	C	C	C	C	A	C	C	A	A
Ammonia, moist gas	C	C	C	C	C	C	C	C	C	C	C	C	C	C
Ammonia, moisture-free	A	A	A	A	A	A	A	A	A	A	A	A	A	A
Ammonium chloride	C	C	C	C	C	C	C	C	C	C	C	C	C	C
Ammonium hydroxide	C	C	C	C	C	C	C	C	C	C	C	C	C	C
Ammonium nitrate	C	C	C	C	C	C	C	C	C	C	C	C	C	C
Ammonium sulfate	B	B	B	B	B	C	C	C	C	A	C	C	A	A
Aniline and aniline dyes	C	C	C	C	C	C	C	C	C	B	C	C	C	C
Asphalt	A	A	A	A	A	A	A	A	A	A	A	A	A	A
Barium chloride	A	A	A	A	A	C	C	C	C	A	A	A	A	C
Barium sulfide	C	C	C	C	C	C	C	B	C	C	C	C	C	C
Beer (b)	A	A	B	B	B	C	C	C	C	A	A	A	A	B
Beet sugar syrup	A	A	B	B	B	A	A	A	B	A	A	A	B	B
Benzine	A	A	A	A	A	A	A	A	A	A	A	A	A	A
Benzol	A	A	A	A	A	A	A	A	A	A	A	A	A	A
Boric acid	A	A	A	A	A	A	B	A	A	A	A	A	A	A
Butane	A	A	A	A	A	A	A	A	A	A	A	A	A	A
Calcium bisulfite	A	A	B	B	B	C	C	C	C	A	B	A	A	B
Calcium chloride (acid)	B	B	B	B	B	B	C	C	C	A	C	C	A	C
Calcium chloride (alkaline)	C	C	C	C	C	C	C	C	C	A	C	A	C	B
Calcium hydroxide	C	C	C	C	C	C	C	C	C	B	C	C	C	C
Calcium hypochlorite	C	C	B	B	B	C	C	C	C	B	C	C	C	C
Cane sugar syrups	A	A	B	A	B	A	A	A	A	A	A	A	A	B
Carbonated beverages (b)	A	C	C	C	C	C	C	C	C	A	C	C	A	C
Carbon dioxide, dry	A	A	A	A	A	A	A	A	A	A	A	A	A	A
Carbon dioxide, moist (b)	B	B	B	C	B	C	C	C	C	A	C	A	A	B
Carbon tetrachloride, dry	A	A	A	A	A	A	A	A	A	A	A	A	A	A
Carbon tetrachloride, moist	B	B	B	B	B	B	B	B	B	B	B	A	A	A
Chlorine, dry	A	A	A	A	A	A	A	A	A	A	A	A	A	A
Chlorine, moist	C	C	B	B	B	C	C	C	C	C	C	C	C	C
Chromic acid	C	C	C	C	C	C	C	C	C	C	C	C	C	C
Citric acid	A	A	A	A	A	A	A	A	A	A	A	A	A	A
Copper sulfate	B	A	A	A	A	C	C	C	C	B	B	B	B	A
Cottonseed oil (b)	A	A	A	A	A	A	A	A	A	A	A	A	A	A
Creosote	B	B	B	B	B	C	C	C	C	A	B	B	B	B
Ethers	A	A	A	A	A	A	A	A	A	A	A	A	A	A
Ethylene glycol	A	A	A	A	A	A	A	A	A	A	A	A	A	A
Ferric chloride, sulfate	C	C	C	C	C	C	C	C	C	C	C	C	C	C
Ferrous chloride, sulfate	C	C	C	C	C	C	C	C	C	C	C	C	C	C

(continued)

Corrosion ratings of cast copper metals in various media (continued)

Corrosive medium	Copper	Tin bronze	Leaded tin bronze	High-leaded tin bronze	Leaded red brass	Leaded semi-red brass	Leaded yellow brass	Leaded high-strength yellow brass	High-strength yellow brass	Aluminum bronze	Leaded nickel brass	Leaded nickel bronze	Silicon bronze	Silicon brass
Formaldehyde	A	A	A	A	A	A	A	A	A	A	A	A	A	A
Formic acid	A	A	A	A	B	B	B	B	A	B	A	B	B	C
Freon	A	A	A	A	A	A	A	A	A	A	A	A	A	B
Fuel oil	A	A	A	A	A	A	A	A	A	A	A	A	A	A
Furfural	A	A	A	A	A	A	A	A	A	A	A	A	A	A
Gasoline	A	A	A	A	A	A	A	A	A	A	A	A	A	A
Gelatin (b)	A	A	A	A	A	A	A	A	A	A	A	A	A	A
Glucose	A	A	A	A	A	A	A	A	A	A	A	A	A	A
Glue	A	A	A	A	A	A	A	A	A	A	A	A	A	A
Glycerin	A	A	A	A	A	A	A	A	A	A	A	A	A	A
Hydrochloric or muriatic acid	C	C	C	C	C	C	C	C	C	B	C	C	C	C
Hydrofluoric acid	B	B	B	B	B	B	B	B	B	A	B	B	B	B
Hydrofluosilicic acid	B	B	B	B	B	C	C	C	C	B	C	C	B	C
Hydrogen	A	A	A	A	A	A	A	A	A	A	A	A	A	A
Hydrogen peroxide	C	C	C	C	C	C	C	C	C	C	C	C	C	C
Hydrogen sulfide, dry	C	C	C	C	C	C	C	C	C	B	C	C	B	C
Hydrogen sulfide, moist	C	C	C	C	C	C	C	C	C	B	C	C	C	C
Lacquers	A	A	A	A	A	A	A	A	A	A	A	A	A	A
Lacquer thinners	A	A	A	A	A	A	A	A	A	A	A	A	A	A
Lactic acid	A	A	A	A	A	C	C	C	C	A	C	C	A	C
Linseed oil	A	A	A	A	A	A	A	A	A	A	A	A	A	A
Liquors														
Black liquor	B	B	B	B	B	C	C	C	C	B	C	C	B	B
Green liquor	C	C	C	C	C	C	C	C	C	B	C	C	C	B
White liquor	C	C	C	C	C	C	C	C	C	A	C	C	C	B
Magnesium chloride	A	A	A	A	A	C	C	C	C	A	C	C	A	B
Magnesium hydroxide	B	B	B	B	B	B	B	B	B	A	B	B	B	B
Magnesium sulfate	A	A	A	A	B	C	C	C	C	A	C	B	A	B
Mercury, mercury salts	C	C	C	C	C	C	C	C	C	C	C	C	C	C
Milk (b)	A	A	A	A	A	A	A	A	A	A	A	A	A	A
Molasses (b)	A	A	A	A	A	A	A	A	A	A	A	A	A	A
Natural gas	A	A	A	A	A	A	A	A	A	A	A	A	A	A
Nickel chloride	A	A	A	A	A	C	C	C	C	B	C	C	A	C
Nickel sulfate	A	A	A	A	A	C	C	C	C	A	C	C	A	C
Nitric acid	C	C	C	C	C	C	C	C	C	C	C	C	C	C
Oleic acid	A	A	B	B	B	C	C	C	C	A	C	A	A	B
Oxalic acid	A	A	B	B	B	C	C	C	C	A	C	A	A	B
Phosphoric acid	A	A	A	A	A	C	C	C	C	A	C	A	A	A
Picric acid	C	C	C	C	C	C	C	C	C	C	C	C	C	C
Potassium chloride	A	A	A	A	A	C	C	C	C	A	C	C	A	C
Potassium cyanide	C	C	C	C	C	C	C	C	C	C	C	C	C	C

(continued)

Corrosion ratings of cast copper metals in various media (continued)

Corrosive medium	Copper	Tin bronze	Leaded tin bronze	High-leaded tin bronze	Leaded red brass	Leaded semi-red brass	Leaded yellow brass	Leaded high-strength yellow brass	High-strength yellow brass	Aluminum bronze	Leaded nickel brass	Leaded nickel bronze	Silicon bronze	Silicon brass
Potassium hydroxide	C	C	C	C	C	C	C	C	C	A	C	C	C	C
Potassium sulfate	A	A	A	A	A	C	C	C	C	A	C	C	A	C
Propane gas	A	A	A	A	A	A	A	A	A	A	A	A	A	A
Sea water	A	A	A	A	A	C	C	C	C	A	C	C	B	B
Soap solutions	A	A	A	A	B	C	C	C	C	A	C	C	A	C
Sodium bicarbonate	A	A	A	A	A	A	A	A	A	A	A	A	A	B
Sodium bisulfate	C	C	C	C	C	C	C	C	C	A	C	C	C	C
Sodium carbonate	C	A	A	A	C	C	C	C	C	A	C	C	C	A
Sodium chloride	A	A	A	A	A	B	C	C	C	A	C	C	A	C
Sodium cyanide	C	C	C	C	C	C	C	C	C	B	C	C	C	C
Sodium hydroxide	C	C	C	C	C	C	C	C	C	A	C	C	C	C
Sodium hypochlorite	C	C	C	C	C	C	C	C	C	C	C	C	C	C
Sodium nitrate	B	B	B	B	B	B	B	B	B	A	B	B	A	A
Sodium peroxide	B	B	B	B	B	B	B	B	B	B	B	B	B	B
Sodium phosphate	A	A	A	A	A	A	A	A	A	A	A	A	A	A
Sodium sulfate, silicate	A	A	B	B	B	B	C	C	C	A	C	C	A	B
Sodium sulfide, thiosulfate	C	C	C	C	C	C	C	C	C	B	C	C	C	C
Stearic acid	A	A	A	A	A	A	A	A	A	A	A	A	A	A
Sulfur, solid	C	C	C	C	C	C	C	C	C	A	C	C	C	C
Sulfur chloride	C	C	C	C	C	C	C	C	C	C	C	C	C	C
Sulfur dioxide, dry	A	A	A	A	A	A	A	A	A	A	A	A	A	A
Sulfur dioxide, moist	A	A	A	B	B	C	C	C	C	A	C	C	A	B
Sulfur trioxide, dry	A	A	A	A	A	A	A	A	A	A	A	A	A	A
Sulfuric acid														
78% or less	B	B	B	B	B	C	C	C	C	A	C	C	B	B
78% to 90%	C	C	C	C	C	C	C	C	C	B	C	C	C	C
90% to 95%	C	C	C	C	C	C	C	C	C	B	C	C	C	C
Fuming	C	C	C	C	C	C	C	C	C	A	C	C	C	C
Tannic acid	A	A	A	A	A	A	A	A	A	A	A	A	A	A
Tartaric acid	B	A	A	A	A	A	A	A	A	A	A	A	A	A
Toluene	B	B	A	A	A	B	B	B	B	B	B	B	B	A
Trichlorethylene, dry	A	A	A	A	A	A	A	A	A	A	A	A	A	A
Trichlorethylene, moist	A	A	A	A	A	A	A	A	A	A	A	A	A	A
Turpentine	A	A	A	A	A	A	A	A	A	A	A	A	A	A
Varnish	A	A	A	A	A	A	A	A	A	A	A	A	A	A
Vinegar	A	A	B	B	B	C	C	C	C	B	C	C	A	B
Water, acid mine	C	C	C	C	C	C	C	C	C	C	C	C	C	C
Water, condensate	A	A	A	A	A	A	A	A	A	A	A	A	A	A
Water, potable	A	A	A	A	A	B	B	B	A	A	A	A	A	A
Whiskey (b)	A	A	C	C	C	C	C	C	C	A	C	C	A	C
Zinc chloride	C	C	C	C	C	C	C	C	C	B	C	C	B	C
Zinc sulfate	A	A	A	A	A	C	C	C	C	B	C	A	A	C

Note: ratings: A, recommended; B, acceptable; C, not recommended.

(a) Acetylene forms an explosive compound with copper when moist or when certain impurities are present and the gas is under pressure. Alloys containing less than 65% Cu are satisfactory under this use. When gas is not under pressure other copper alloys are satisfactory. (b) Copper and copper alloys resist corrosion by most food products. Traces of copper may be dissolved and affect taste or color. In such cases, copper metals are often tin coated.

Properties and applications of cast coppers and copper alloys

UNS designation(a)	Nominal composition, %(a)	Typical mechanical properties, as cast (heat treated)(b)								Machinability rating(c)	Casting types(e)	Typical applications
		Tensile strength		Yield strength		Elongation in 2 in., %	Hardness					
		MPa	ksi	MPa	ksi		Rockwell	Brinell 500 kg	3 000 kg			
C80100	99.95 Cu + Ag min, 0.05 others max	172	25	62	9	40	...	44	...	10	C, T, I, M, P, S	Electrical and thermal conductors; corrosion and oxidation resistant applications.
C80300	99.95 Cu + Ag min, 0.034 Ag min, 0.05 others max	172	25	62	9	40	...	44	...	10	C, T, I, M, P, S	Electrical and thermal conductors; corrosion and oxidation resistant applications.
C80500	99.75 Cu + Ag min, 0.034 Ag min, 0.02 B max, 0.23 others max	172	25	62	9	40	...	44	...	10	C, T, I, M, P, S	Electrical and thermal conductors; corrosion and oxidation resistant applications.
C80700	99.75 Cu + Ag min, 0.02 B max, 0.23 others max	172	25	62	9	40	...	44	...	10	C, T, I, M, P, S	Electrical and thermal conductors; corrosion and oxidation resistant applications.
C80900	99.70 Cu + Ag min, 0.034 Ag min, 0.30 others max	172	25	62	9	40	...	44	...	10	C, T, I, M, P, S	Electrical and thermal conductors; corrosion and oxidation resistant applications.
C81100	99.70 Cu + Ag min, 0.30 others max	172	25	62	9	40	...	44	...	10	C, T, I, M, P, S	Electrical and thermal conductors; corrosion and oxidation resistant applications.
High-copper alloys												
C81300	98.5 Cu min, 0.06 Be, 0.80 Co, 0.40 others max	(365)	(53)	(248)	(36)	(11)	...	(39)	...	20	C, T, I, M, P, S	Higher hardness electrical and thermal conductors.
C81400	98.5 Cu min, 0.06 Be, 0.80 Cr, 0.40 others max	(365)	(53)	(248)	(36)	(11)	(B 69)	...	...	20	C, T, I, M, P, S	Higher hardness electrical and thermal conductors.
C81500	98.0 Cu min, 1.0 Cr, 0.50 others max	(352)	(51)	(276)	(40)	(17)	...	(105)	...	20	C, T, I, M, P, S	Electrical and/or thermal conductors used as structural members where strength and hardness greater than that of C80100-81100 are required.
C81700	94.25 Cu min, 1.0 Ag, 0.4 Be, 0.9 Co, 0.9 Ni	(634)	(92)	(469)	(68)	(8)	...	...	(217)	30	C, T, I, M, P, S	Electrical and/or thermal conductors used as structural members where strength and hardness greater than that of C80100-81100 are required. Also used in place of C81500 where electrical and/or thermal conductivities can be sacrificed for hardness and strength.
High-copper alloys												
C81800	95.6 Cu min, 1.0 Ag, 0.4 Be, 1.6 Co	345 (703)	50 (102)	172 (517)	25 (75)	20 (8)	B 55 (B 96)	...	...	20	C, T, I, M, P, S	Resistance welding electrodes, dies.
C82000	96.8 Cu, 0.6 Be, 2.6 Co	345 (689)	50 (100)	138 (517)	20 (75)	20 (8)	B 55 (B 95)	...	(195)	20	C, T, I, M, P, S	Current carrying parts, contact and switch blades, bushings and bearings, soldering iron and resistance welding tips.

(continued)

(a) Nominal composition, unless otherwise noted. For seldom-used alloys, only compositions are available. (b) Values for C82700, 84200, 96200, 96300 are minimum, not typical. As-cast values are for sand casting except C93900, continuous cast; and C85800, 87800, 87900, die cast. Heat treated values, in parentheses, indicate that the alloy responds to heat treatment. If heat treated values are not shown, the copper or copper alloy does not respond. (c) Free cutting brass = 100. (d) As-heat treated value for C94700, 20; for C94800, 40. (e) C, centrifugal; T, continuous; D, die; I, investment; M, permanent mold; P, plaster; S, sand. (Note: C82000, 82400, 82500, 82600, 82800 are also pressure cast.) **Source:** Copper Development Assn. Inc., New York. (Revised March 1980)

Properties and applications of cast coppers and copper alloys (continued)

UNS designation(a)	Nominal composition, %(a)	Tensile strength MPa	ksi	Yield strength MPa	ksi	Elongation in 2 in, %	Hardness Rockwell	Brinell 500 kg	Brinell 3 000 kg	Machinability rating(c)	Casting types(e)	Typical applications
High-copper alloys (continued)												
C82100	97.7 Cu, 0.5 Be, 0.9 Co, 0.9 Ni	(634)	(92)	(469)	(68)	(8)	...	...	(217)	30	C, T, I, M, P, S	Electrical and/or thermal conductors used as structural members where strength and hardness greater than that of C80100-81100 are required. Also used in place of C81500 where electrical and/or thermal conductivities can be sacrificed for hardness and strength.
C82200	96.5 Cu min, 0.6 Be, 1.5 Ni	393 (655)	57 (95)	207 (517)	30 (75)	20 (8)	B 60 (B 96)	...	...	20	C, T, I, M, P, S	Clutch rings, brake drums, seam welder electrodes, projection welding dies, spot welding tips, beam welder shapes, bushings, water-cooled holders.
C82400	96.4 Cu min, 1.70 Be, 0.25 Co	496 (1034)	72 (150)	255 (965)	37 (140)	20 (1)	B 78 (C 38)	...	...	20	C, I, M, P, S	Safety tools, molds for plastic parts, cams, bushings, bearings, valves, pump parts, gears.
C82500	97.2 Cu, 2.0 Be, 0.5 Co, 0.25 Si	552 (1103)	80 (160)	310	45	20 (1)	B 82 (C 40)	...	...	20	C, I, M, P, S	Safety tools, molds for plastic parts, cams, bushings, bearings, valves, pump parts.
C82600	95.2 Cu min, 2.3 Be, 0.5 Co, 0.25 Si	565 (1138)	82 (165)	324 (1069)	47 (155)	20 (1)	B 83 (C 43)	...	...	20	C, I, M, P, S	Bearings and molds for plastic parts.
C82700	96.3 Cu, 2.45 Be, 1.25 Ni	(1069)	(155)	(896)	(130)	(0)	(C 39)	...	...	20	C, I, M, P, S	Bearings and molds for plastic parts.
C82800	96.6 Cu, 2.6 Be, 0.5 Co, 0.25 Si	669 (1138)	97 (165)	379 (1000)	55 (145)	20 (1)	B 85 (C 45)	...	...	10	C, I, M, P, S	Molds for plastic parts, cams, bushings, bearings, valves, pump parts, sleeves.
Red brasses and leaded red brasses												
C83300	93 Cu, 1.5 Sn, 1.5 Pb, 4 Zn	221	32	69	10	35	...	35	...	35	S	Terminal ends for electrical cables.
C83400	90 Cu, 10 Zn	241	35	69	10	30	F 50	...	...	60	C, S	Moderate strength, moderate conductivity castings; rotating bands.
C83600	85 Cu, 5 Sn, 5 Pb, 5 Zn	255	37	117	17	30	...	60	...	84	C, T, I, S	Valves, flanges, pipe fittings, plumbing goods, pump castings, water pump impellers and housings, ornamental fixtures, small gears.
C83800	83 Cu, 4 Sn, 6 Pb, 7 Zn	241	35	110	16	25	...	60	...	90	C, T, S	Low-pressure valves and fittings, plumbing supplies and fittings, general hardware, air-gas-water fittings, pump components, railroad catenary fittings.
Semi-red brasses and leaded semi-red brasses												
C84200	80 Cu, 5 Sn, 2.5 Pb, 12.5 Zn	193	28	103	15	27	...	60	...	80	C, T, S	Pipe fittings, elbows, T's, couplings, bushings, locknuts, plugs, unions.

(continued)

(a) Nominal composition, unless otherwise noted. For seldom-used alloys, only compositions are available. (b) Values for C82700, 84200, 96200, 96300 are minimum, not typical. As-cast values are for sand casting except C93900; continuous cast; and C85800, 87800, 87900, die cast. Heat treated values, in parentheses, indicate that the alloy responds to heat treatment. If heat treated values are not shown, the copper or copper alloy does not respond. (c) Free cutting brass = 100. (d) As-heat treated value for C94700, 20; for C94800, 40. (e) C, centrifugal; T, continuous; D, die; I, investment; M, permanent mold; P, plaster; S, sand. (Note: C82000, 82400, 82500, 82600, 82800 are also pressure cast.) **Source:** Copper Development Assn. Inc., New York. (Revised March 1980)

Properties and applications of cast coppers and copper alloys (continued)

UNS designation(a)	Nominal composition, %(a)	Tensile strength MPa	Tensile strength ksi	Yield strength MPa	Yield strength ksi	Elongation in 2 in., %	Hardness Rockwell	Hardness Brinell 500 kg	Hardness Brinell 3 000 kg	Machinability rating(c)	Casting types(e)	Typical applications
Semi-red brasses and leaded semi-red brasses (continued)												
C84400	81 Cu, 3 Sn, 7 Pb, 9 Zn	234	34	103	15	26	...	55	...	90	C, T, S	General hardware, ornamental castings, plumbing supplies and fixtures, low-pressure valves and fittings.
C84500	78 Cu, 3 Sn, 7 Pb, 12 Zn	241	35	97	14	28	...	55	...	90	C, T, S	Plumbing fixtures, cocks, faucets, stops, waste, air and gas fittings, low-pressure valve fittings.
C84800	76 Cu, 3 Sn, 6 Pb, 15 Zn	248	36	97	14	30	...	55	...	90	C, S	Plumbing fixtures, cocks, faucets, stops, waste, air, and gas, general hardware, and low-pressure valve fittings.
Yellow brasses and leaded yellow brasses												
C85200	72 Cu, 1 Sn, 3 Pb, 24 Zn	262	38	90	13	35	...	45	...	80	C, T	Plumbing fittings and fixtures, ferrules, valves, hardware, ornamental brass, chandeliers, and irons.
C85400	67 Cu, 1 Sn, 3 Pb, 29 Zn	234	34	83	12	35	...	50	...	80	C, T, M, P, S	General purpose yellow casting alloy not subject to high internal pressure. Furniture hardware, ornamental castings, radiator fittings, ship trimmings, cocks, battery clamps, valves and fittings.
C85500	61 Cu, 0.8 Al, bal Zn	414	60	159	23	40	B 55	85	...	80	C, S	Ornamental castings.
C85700	63 Cu, 1 Sn, 1 Pb, 34.7 Zn, 0.3 Al	345	50	124	18	40	...	75	...	80	C, M, P, S	Bushings, hardware fittings, ornamental castings.
C85800	58 Cu, 1 Sn, 1 Pb, 40 Zn	379	55	207	30	15	B 55	...	...	80	D	General purpose die casting alloy having moderate strength.
Manganese and leaded manganese bronze alloys												
C86100	67 Cu, 21 Zn, 3 Fe, 5 Al, 4 Mn	655	95	345	50	20	...	...	180	30	C, I, P, S	Marine castings, gears, gun mounts, bushings and bearings, marine racing propellers.
C86200	64 Cu, 26 Zn, 3 Fe, 4 Al, 3 Mn	655	95	331	48	20	...	...	180	30	C, T, D, I, P, S	Marine castings, gears, gun mounts, bushings and bearings.
C86300	63 Cu, 25 Zn, 3 Fe, 6 Al, 3 Mn	793	115	572	83	15	...	...	225	8	C, I, P, S	Extra-heavy duty, high-strength alloy. Large valve stems, gears, cams, slow-speed heavy-load bearings, screwdown nuts, hydraulic cylinder parts.
C86400	59 Cu, 1 Pb, 40 Zn	448	65	172	25	20	...	90	105	65	C, D, M, P, S	Free-machining manganese bronze. Valve stems, marine fittings, lever arms, brackets, light-duty gears.
C86500	58 Cu, 0.5 Sn, 39.5 Zn, 1 Fe, 1 Al	490	71	193	28	30	...	100	130	26	C, I, P, S	Machinery parts requiring strength and toughness, lever arms, valve stems, gears.
C86700	58 Cu, 1 Pb, 41 Zn	586	85	290	42	20	B 80	...	155	55	C, S	High strength, free-machining manganese bronze. Valve stems.
C86800	55 Cu, 37 Zn, 3 Ni, 2 Fe, 3 Mn	565	82	262	38	22	...	80	80	30	S	Marine fittings, marine propellers.

(continued)

(a) Nominal composition, unless otherwise noted. For seldom-used alloys, only compositions are available. (b) Values for C82700, 84200, 96200, 96300 are minimum, not typical. As-cast values are for sand casting except C93900, continuous cast; and C85800, 87800, 87900, die cast. Heat treated values, in parentheses, indicate that the alloy responds to heat treatment. If heat treated values are not shown, the copper or copper alloy does not respond. (c) Free cutting brass = 100. (d) As-heat treated value for C94700, 20; for C94800, 40. (e) C, centrifugal; T, continuous; D, die; I, investment; M, permanent mold; P, plaster; S, sand. (Note: C82000, 82400, 82500, 82600, 82800 are also pressure cast.) **Source:** Copper Development Assn. Inc., New York. (Revised March 1980)

Properties and applications of cast coppers and copper alloys (continued)

UNS designation(a)	Nominal composition, %(a)	Typical mechanical properties, as cast (heat treated)(b)								Machinability rating(c)	Casting types(e)	Typical applications
		Tensile strength		Yield strength		Elongation in 2 in., %	Hardness					
		MPa	ksi	MPa	ksi		Rockwell	Brinell 500 kg	3 000 kg			
Silicon bronzes and silicon brasses												
C87200	89 Cu min, 4 Si	379	55	172	25	30	...	85	...	40	C, I, M, P, S	Bearings, bells, impellers, pump and valve components, marine fittings, corrosion-resistant castings.
C87400	83 Cu, 14 Zn, 3 Si	379	55	165	24	30	...	70	100	50	C, D, I, M, P, S	Bearings, gears, impellers, rocker arms, valve stems, clamps.
C87500	82 Cu, 14 Zn, 4 Si	462	67	207	30	21	...	115	134	50	C, D, I, M, P, S	Bearings, gears, impellers, rocker arms, valve stems, small boat propellers.
C87600	90 Cu, 5.5 Zn, 4.5 Si	455	66	221	32	20	B 76	110	135	40	S	Valve stems.
C87800	82 Cu, 14 Zn, 4 Si	586	85	345	50	25	B 85	...	...	40	D	High-strength, thin-wall die castings; brush holders, lever arms, brackets, clamps, hexagonal nuts.
C87900	65 Cu, 34 Zn, 1 Si	483	70	241	35	25	B 70	...	...	80	D	General purpose die casting alloy having moderate strength.
Tin bronzes												
C90200	93 Cu, 7 Sn	262	38	110	16	30	...	70	...	20	C, S	Bearings and bushings.
C90300	88 Cu, 8 Sn, 4 Zn	310	45	145	21	30	...	70	...	30	C, T, I, P, S	Bearings, bushings, pump impellers, piston rings, valve components, seal rings, steam fittings, gears.
C90500	88 Cu, 10 Sn, 2 Zn	310	45	152	22	25	...	75	...	30	C, T, I, S	Bearings, bushings, pump impellers, piston rings, valve components, steam fittings, gears.
C90700	89 Cu, 11 Sn	303 (379)	44 (55)	152 (207)	22 (30)	20 (16)	...	80 (102)	...	20	C, T, I, M, S	Gears, bearings, bushings.
C90800	87 Cu, 12 Sn	276	40	138	20	15	...	90	...	20	C, S	Bearings and bushings.
C90900	87 Cu, 13 Sn	221	32	172	25	2	...	105	...	20	C, T, I, S	Piston rings and bearings.
C91000	85 Cu, 14 Sn, 1 Zn											
C91100	84 Cu, 16 Sn	241	35	172	25	2	...	...	135	10	S	Piston rings, bearings, bushings, bridge plates.
C91300	81 Cu, 19 Sn	241	35	207	30	0.5	...	...	170	10	S	Piston rings, bearings, bushings, bridge plates, bells.
C91600	88 Cu, 10.5 Sn, 1.5 Ni	303 (414)	44 (60)	152 (221)	22 (32)	16 (16)	...	85 (106)	...	20	C, T, M, S	Gears.
C91700	86.5 Cu, 12 Sn, 1.5 Ni	303 (414)	44 (60)	152 (221)	22 (32)	16 (16)	...	85 (106)	...	20	C, T, I, M, S	Gears.
Leaded tin bronzes												
C92200	88 Cu, 6 Sn, 1.5 Pb, 4.5 Zn	276	40	138	20	30	...	65	...	42	C, T, I, M, P, S	Valves, fittings, and pressure-containing parts for use up to 550 °F.
C92300	87 Cu, 8 Sn, 4 Zn	276	40	138	20	25	...	70	...	42	C, T, S	Valves, pipe fittings, and high-pressure steam castings. Superior machinability to C90300.

(continued)

(a) Nominal composition, unless otherwise noted. For seldom-used alloys, only compositions are available. (b) Values for C82700, 84200, 96200, 96300 are minimum, not typical. As-cast values are for sand casting except C93900, continuous cast; and C85800, 87800, 87900, die cast. Heat treated values, in parentheses, indicate that the alloy responds to heat treatment. If heat treated values are not shown, the copper or copper alloy does not respond. (c) Free cutting brass = 100. (d) As-heat treated value for C94700, 20; for C94800, 40. (e) C, centrifugal; T, continuous; D, die; I, investment; M, permanent mold; P, plaster; S, sand. (Note: C82000, 82400, 82500, 82600, 82800 are also pressure cast.) **Source:** Copper Development Assn. Inc., New York. (Revised March 1980)

Properties and applications of cast coppers and copper alloys (continued)

UNS designation(a)	Nominal composition, %(a)	Tensile strength MPa	Tensile strength ksi	Yield strength MPa	Yield strength ksi	Elongation in 2 in., %	Hardness Rockwell	Hardness Brinell 500 kg	Hardness Brinell 3 000 kg	Machinability rating(c)	Casting types(e)	Typical applications
Leaded tin bronzes (continued)												
C92400	88 Cu, 10 Sn, 2 Pb, 2 Zn	303	44	138	20	20	...	80	...	30	C, T, M, S	Gears, automotive synchronizer rings.
C92500	87 Cu, 11 Sn, 1 Pb, 1 Ni	303	44	138	20	30	F 78	70	...	40	C, T, S	Bearings, bushings, pump impellers, piston rings, valve components, steam fittings, and gears. Superior machinability to C90500.
C92600	87 Cu, 10 Sn, 1 Pb, 2 Zn											
C92700	88 Cu, 10 Sn, 2 Pb	290	42	145	21	20	...	77	...	45	C, T, S	Bearings, bushings, pump impellers, piston rings, valve components, steam fittings, and gears. Superior machinability to C90500.
C92800	79 Cu, 16 Sn, 5 Pb	276	40	207	30	1	B 80	...	...	70	C, S	Piston rings.
C92900	84 Cu, 10 Sn, 2.5 Pb, 3.5 Ni	324 (324)	47 (47)	179 (179)	26 (26)	20 (20)	...	80 (80)	...	40	C, T, M, S	Gears, wear plates, guides, cams, parts requiring machinability superior to that of C91600 or 91700.
High-leaded tin bronzes												
C93200	83 Cu, 7 Sn, 7 Pb, 3 Zn	241	35	124	18	20	...	65	...	70	C, T, M, S	General-utility bearings and bushings.
C93400	84 Cu, 8 Sn, 8 Pb	221	32	110	16	20	...	60	...	70	C, T, S	Bearings and bushings.
C93500	85 Cu, 5 Sn, 9 Pb	221	32	110	16	20	...	60	...	70	C, T, S	Small bearings and bushings, bronze backing for babbit-lined automotive bearings.
C93700	80 Cu, 10 Sn, 10 Pb	241	35	124	18	20	...	60	...	80	C, T, M, S	Bearings for high speed and heavy pressures, pumps, impellers, corrosion-resistant applications, pressure tight castings.
C93800	78 Cu, 7 Sn, 15 Pb	207	30	110	16	18	...	55	...	80	C, T, M, S	Bearings for general service and moderate pressures, pump impellers and bodies for use in acid mine water.
C93900	79 Cu, 6 Sn, 15 Pb	221	32	152	22	7	...	63	...	80	T	Continuous castings only. Bearings for general service, pump bodies and impellers for mine waters.
C94000	70.5 Cu, 13.0 Sn, 15.0 Pb, 0.50 Zn, 0.75 Ni, 0.25 Fe, 0.05 P, 0.35 Sb(h)											
C94100	70.0 Cu, 5.5 Sn, 18.5 Pb, 3.0 Zn, 1.0 others max											
C94300	70 Cu, 5 Sn, 25 Pb	186	27	90	13	15	...	48	...	80	C, S	High-speed bearings for light loads.

(continued)

(a) Nominal composition, unless otherwise noted. For seldom-used alloys, only compositions are available. (b) Values for C82700, 84200, 96200, 96300 are minimum, not typical. As-cast values are for sand casting except C93900, continuous cast; and C85800, 87800, 87900, die cast. Heat treated values, in parentheses, indicate that the alloy responds to heat treatment. If heat treated values are not shown, the copper or copper alloy does not respond. (c) Free cutting brass = 100. (d) As-heat treated value for C94700, 20; for C94800, 40. (e) C, centrifugal; T, continuous; D, die; I, investment; M, permanent mold; P, plaster; S, sand. (Note: C82000, 82400, 82500, 82600, 82800 are also pressure cast.) **Source:** Copper Development Assn. Inc., New York. (Revised March 1980)

Properties and applications of cast coppers and copper alloys (continued)

UNS designation(a)	Nominal composition, %(a)	Tensile strength		Yield strength		Elongation in 2 in, %	Hardness			Machinability rating(c)	Casting type(e)	Typical applications
		MPa	ksi	MPa	ksi		Rockwell	Brinell 500 kg	3 000 kg			
High-leaded tin bronzes (continued)												
C94400	81 Cu, 8 Sn, 11 Pb	221	32	110	16	18	...	55	...	80	C, T, S	General-utility alloy for bushings and bearings.
C94500	73 Cu, 7 Sn, 20 Pb	172	25	83	12	12	...	50	...	80	C, S	Locomotive wearing parts, high-speed low-load bearings.
Nickel-tin bronzes												
C94700	88 Cu, 5 Sn, 2 Zn, 5 Ni	345 (586)	50 (85)	159 (414)	23 (60)	35 (10)	...	85	(180)	30 (d)	C, T, I, M, S	Valve stems and bodies, bearings, wear guides, shift forks, feeding mechanisms, circuit breaker parts, gears, piston cylinders, nozzles.
C94800	87 Cu, 5 Sn, 5 Ni	310 (414)	45 (60)	159 (207)	23 (30)	35 (8)	...	80 (120)	...	50 (d)	M, S	Structural castings, gear components, motion translation devices, machinery parts, bearings.
C94900	80 Cu, 5 Sn, 5 Pb, 5 Zn, 5 Ni											
Aluminum bronzes												
C95200	88 Cu, 3 Fe, 9 Al	552	80	186	27	35	...	...	125	50	C, T, M, P, S	Acid-resisting pumps, bearings, gears, valve seats, guides, plungers, pump rods, bushings.
C95300	89 Cu, 1 Fe, 10 Al	517 (586)	75 (85)	186 (290)	27 (42)	25 (15)	...	...	140 (174)	55	C, T, M, P, S	Pickling baskets, nuts, gears, steel mill slippers, marine equipment, welding jaws.
C95400	85 Cu, 4 Fe, 11 Al	586 (724)	85 (105)	241 (372)	35 (54)	18 (8)	...	...	170 (195)	60	C, T, M, P, S	Bearings, gears, worms, bushings, valve seats and guides, pickling hooks.
C95410	85 Cu, 4 Fe, 11 Al, 2 Ni											
C95500	81 Cu, 4 Ni, 4 Fe, 11 Al	689 (827)	100 (120)	303 (469)	44 (68)	12 (10)	...	...	192 (230)	50	C, T, M, P, S	Valve guides and seats in aircraft engines, corrosion-resistant parts, bushings, gears, worms, pickling hooks and baskets, agitators.
C95600	91 Cu, 7 Al, 2 Si	517	75	234	34	18	...	...	140	60	C, T, M, P, S	Cable connectors, terminals, valve stems, marine hardware, gears, worms, pole-line hardware.
C95700	75 Cu, 2 Ni, 3 Fe, 8 Al, 12 Mn	655	95	310	45	26	...	...	180	50	C, T, M, P, S	Propellers, impellers, stator clamp segments, safety tools, welding rods, valves, pump casings.
C95800	81 Cu, 5 Ni, 4 Fe, 9 Al, 1 Mn	655	95	262	38	25	...	...	159	50	C, T, M, P, S	Propeller hubs, blades, and other parts in contact with salt water.
Copper-nickels												
C96200	88.6 Cu, 10 Ni, 1.4 Fe	310	45	172	25	20	...	...	...	10	C, S	Components of items being used for sea water corrosion resistance.

(continued)

(a) Nominal composition, unless otherwise noted. For seldom-used alloys, only compositions are available. (b) Values for C82700, 84200, 96200, 96300 are minimum, not typical. As-cast values are for sand casting except C93900, continuous cast; and C85500, 87800, 87900, die cast. Heat treated values, in parentheses, indicate that the alloy responds to heat treatment. If heat treated values are not shown, the copper or copper alloy does not respond. (c) Free cutting brass = 100. (d) As-heat treated value for C94700, 20; for C94800, 40. (e) C, centrifugal; T, continuous; D, die; I, investment; M, permanent mold; P, plaster; S, sand. (Note: C82000, 82400, 82500, 82600, 82800 are also pressure cast.) **Source:** Copper Development Assn. Inc., New York. (Revised March 1980)

Properties and applications of cast coppers and copper alloys (continued)

Typical mechanical properties, as cast (heat treated)(b)

UNS designation(a)	Nominal composition, %(a)	Tensile strength MPa	ksi	Yield strength MPa	ksi	Elongation in 2 in., %	Hardness Rockwell	500 kg	Brinell 3 000 kg	Machinability rating(c)	Casting types(e)	Typical applications
Copper-nickels (continued)												
C96300	79.3 Cu, 20 Ni, 0.7 Fe	517	75	379	55	10	...	150	...	15	C, S	Centrifugally cast tailshaft sleeves.
C96400	69.1 Cu, 30 Ni, 0.9 Fe	469	68	255	37	28	...	...	140	20	C, T, S	Valves, pump bodies, flanges, elbows used for sea-water corrosion resistance.
C96600	68.5 Cu, 30 Ni, 1 Fe, 0.5 Be	(758)	(110)	(482)	(70)	(7)	...	...	(230)	20	C, T, I, M, S	High-strength constructional parts for sea-water corrosion resistance.
C96700	67.6 Cu, 30 Ni, 0.9 Fe, 1.15 Be, 0.15 Zr, 0.15 Ti	(1207)	(175)	(552)	(80)	(10)	C26	...	...	40	I, M, S	Corrosion-resistant molds for plastics, high-strength constructional parts for sea-water use.
Nickel silvers												
C97300	56 Cu, 2 Sn, 10 Pb, 12 Ni, 20 Zn	241	35	117	17	20	...	55	...	70	I, M, S	Hardware fittings, valves and valve trim, statuary, ornamental castings.
C97400	59 Cu, 3 Sn, 5 Pb, 17 Ni, 16 Zn	262	38	117	17	20	...	70	...	60	C, I, S	Valves, hardware, fittings, ornamental castings.
C97600	64 Cu, 4 Sn, 4 Pb, 20 Ni, 8 Zn	310	45	165	24	20	...	80	...	70	C, I, S	Marine castings, sanitary fittings, ornamental hardware, valves, pumps.
C97800	66 Cu, 5 Sn, 2 Pb, 25 Ni, 2 Zn	379	55	207	30	15	...	...	130	60	I, M, S	Ornamental and sanitary castings, valves and valve seats, musical instrument components.
Leaded coppers												
C98200	76.0 Cu, 24.0 Pb											
C98400	70.5 Cu, 28.5 Pb, 1.5 Ag											
C98600	65.0 Cu, 35.0 Pb, 1.5 Ag											
C98800	59.5 Cu, 40.0 Pb, 5.5 Ag											
Special alloys												
C99300	71.8 Cu, 15 Ni, 0.7 Fe, 11 Al, 1.5 Co	655	95	379	55	2	...	200	20	20	T, S	Glass making molds, plate glass rolls, marine hardware.
C99400	90.4 Cu, 2.2 Ni, 2.0 Fe, 1.2 Al, 1.2 Si, 3.0 Zn	455 (545)	66 (79)	234 (372)	34 (54)	25	...	...	125 (170)	50	C, T, S	Valve stems, marine and other uses requiring resistance to dezincification and dealuminification, propeller wheels, electrical parts, mining equipment gears.
C99500	87.9 Cu, 4.5 Ni, 4.0 Fe, 1.2 Al, 1.2 Si, 1.2 Zn	483	70	276	40	12	...	145	50	50	C, T, S	Same as C99400, but where higher yield strength is required.
C99600	58 Cu, 2 Al, 40 Mn	558 (558)	81 (81)	248 (303)	36 (44)	34 (27)	B 72	...	130	...	C, T, M, S	Damping alloy to reduce noise and vibration.
C99700	56.5 Cu, 1 Al, 1.5 Pb, 12 Mn, 5 Ni, 24 Zn	379	55	172	25	25	...	...	110	80	C, D, I, M, P, S	
C99750	58 Cu, 1 Al, 1 Pb, 20 Mn, 20 Zn	448 (517)	65 (75)	221 (276)	32 (40)	30 (20)	B 77 (B 82)	110 (119)	...	...	D, I, M, P, S	

(a) Nominal composition, unless otherwise noted. For seldom-used alloys, only compositions are available. Values for C82700, 84200, 96200, 96300 are minimum, not typical. As-cast values are for sand casting except C93900, continuous cast; and C85800, 87800, 87900, die cast. Heat treated values, in parentheses, indicate that the alloy responds to heat treatment. If heat treated values are not shown, the copper or copper alloy does not respond. (b) ... (c) Free cutting brass = 100. (d) As-heat treated value for C94700, 20; for C94800, 40. (e) C, centrifugal; T, continuous; D, die; I, investment; M, permanent mold; P, plaster; S, sand. (Note: C82000, 82400, 82500, 82600, 82800 are also pressure cast.) **Source:** Copper Development Assn. Inc., New York. (Revised March 1980)

Heat Treating

Typical stress-relieving temperatures for 19 wrought copper alloys

Alloy	Common name	Stress-relieving temperature(a) °C	°F
C21000	Gilding metal	190	375
C22000	Commercial bronze	205	400
C23000	Red brass	230	450
C24000	Low brass	260	500
C26000	Cartridge brass	260	500
C27000	Yellow brass	260	500
C28000	Muntz metal	205	400
C36000	Free-cutting brass	245	475
C44300, C44400, C44500	Inhibited admiralty	290	550
C51000, C52100	Phosphor bronze	205	400
C61300, C61400	Aluminum bronze	345	650
C65500	High-silicon bronze	345	650
C70600, C71500	Copper nickel	260	500
C75200	Nickel silver	260	500

(a) Time at temperature, 1 h

Typical heat treatments and resulting properties for several low-temperature-hardening alloys

Alloy	Solution-treating temperature(a) °C	°F	Aging treatment Temperature °C	°F	Time, h	Hardness	Electrical conductivity, % IACS
Precipitation hardening							
C15000	980	1795	500-550	930-1025	3	30 HRB	87-95
C17000, C17200, C17300	760-800	1400-1475	300-350	575-660	1-3	35-44 HRC	22
C17500, C17600	900-950	1650-1740	455-490	850-915	1-4	95-98 HRB	48
C18000(b), C81540	900-930	1650-1705	425-540	800-1000	2-3	92-96 HRB	42-48
C18200, C18400, C18500, C81500	980-1000	1795-1830	425-500	800-930	2-4	68 HRB	80
C94700	775-800	1425-1475	305-325	580-620	5	180 HB	15
C99400	885	1625	482	900	1	170 HB	17
Spinodal hardening							
C71900	900-950	1650-1740	425-760	800-1400	1-2	86 HRC	4-4
C72800	815-845	1500-1550	350-360	660-680	4	32 HRC	...

(a) Solution treating is followed by water quenching. (b) Alloy C18000 (81540) must be double aged—typically, 3 h at 540 °C (1000 °F) followed by 3 h at 425 °C (800 °F) (U.S. Patent No. 4 191 601)—to develop the higher levels of electrical conductivity and hardness.

Recommended precipitation-hardening schedules and resulting properties for solution-treated copper-beryllium castings

Alloy	Solution treatment Temperature °C	°F	Time, min	Aging treatment Temperature °C	°F	Time, min	Tensile strength MPa	ksi	Yield strength(a) MPa	ksi	Elongation(b), %	Hardness	Electrical conductivity, % IACS
C81300	980–1010	1800–1850	60	480	900	120	365	53	250	36	11	89 HB(c)	60
C81700	900–925	1650–1700	60	455	850	180	635	92	470	68	8	217 HB(d)	48
C81800	900–925	1650–1700	60	480	900	180	705	102	515	75	8	92 HRB	45
C82000	900–925	1650–1700	180	480	900	180	690	100	515	75	8	195 HB(d)	45
C82100	900–925	1650–1700	60	455	850	180	635	92	470	68	8	217 HB(d)	48
C82200	900–925	1650–1700	60	445–455	835–850	120	655	95	515	75	8	96 HRB	45
C82400	785–850	1450–1560	60	345	650	180	1035	150	965	140	1	34 HRC	25
C82500	785–800	1450–1475	60	345	650	180	1105	160	795	115	1	40 HRC	20
C82600	785–800	1450–1475	60	345	650	180	1105	160	1035	150	1	40 HRC	19
C82700	785–800	1450–1475	180	345	650	180	1070	155	895	130	0	39 HRC	20
C82800	785–800	1450–1475	60	345	650	180	1140	165	1000	145	1	42 HRC	18

(a) At 0.2% extension under load. (b) In 50 mm or 2 in. (c) 500-kg load. (d) 3000-kg load

Annealing temperatures for cold worked coppers and copper alloys

Alloy	Common name	Annealing temperature	
		°C	°F
Wrought coppers			
C10200	Oxygen-free copper	425-650	800-1200
C11000	Electrolytic tough pitch copper	250-650	500-1200
C11300, C11400, C11500, C11600	Silver-bearing tough pitch copper	400-475	750-900
C12000	Phosphorus-deoxidized copper, low residual phosphorus	325-650	600-1200
C12200	Phosphorus-deoxidized copper, high residual phosphorus	375-650	700-1200
C14500	Phosphorus-deoxidized, tellurium-bearing copper	425-650	800-1200
Wrought copper alloys			
C17000, C17200, C17500	Beryllium copper	775-925(a)	1425-1700(a)
C21000	Gilding metal	425-800	800-1450
C22000	Commercial bronze	425-800	800-1450
C22600	Jewelry bronze	425-750	800-1400
C23000	Red brass	425-725	800-1350
C24000	Low brass	425-700	800-1300
C26000	Cartridge brass	425-750	800-1400
C26800, C27000, C27400	Yellow brass	425-700	800-1300
C28000	Muntz metal	425-600	800-1100
C31400	Leaded commercial bronze	425-650	800-1200
C33000, C33500	Low-leaded brass	425-650	800-1200
C33200, C34200, C35300	High-leaded brass	425-650	800-1200
C34000, C35000	Medium-leaded brass	425-650	800-1200
C35600	Extra-high-leaded brass	425-650	800-1200
C36000	Free-cutting brass	425-600	800-1100
C36500, C36600, C36700, C36800	Leaded Muntz metal	425-600	800-1100
C37000	Free-cutting Muntz metal	425-650	800-1200
C37700	Forging brass	425-600	800-1100
C38500	Architectural bronze	425-600	800-1100
C44300, C44400, C44500	Inhibited admiralty	425-600	800-1100
C46200	Naval brass	425-600	800-1100
C48200, C48500	Leaded naval brass	425-600	800-1100
C50500	Phosphor bronze	475-650	900-1200
C51000, C52100, C54200	Phosphor bronze	475-675	900-1250
C53200, C53400, C54400	Free-cutting phosphor bronze	475-675	900-1250
C60600, C60800	Aluminum bronze	550-650	1000-1200
C61000	Aluminum bronze	600-675	1100-1250
C61300, C61400	Aluminum bronze	750-875	1400-1600
C61800, C61900 C62400	Aluminum bronze	600-650(b)	1100-1200(b)
C63000	Aluminum bronze	650-700(c)	1200-1300(c)
C63200	Aluminum bronze	675-725(c)	1250-1350(c)
C64200	Aluminum bronze	Above 650	Above 1200
C65100	Low-silicon bronze	475-675	900-1250
C65500	High-silicon bronze	475-700	900-1300
C67000, C67500	Manganese bronze	425-600	800-1100
C68700	Aluminum brass	425-600	800-1100
C70600	Copper nickel, 10%	600-825	1100-1500
C71500	Copper nickel, 30%	650-825	1200-1500
C75200, C75700, C77000	Nickel silver	600-825	1100-1500

(a) Solution-treating temperature; see Table 5 for temperatures for specific alloys. (b) Cool rapidly (cooling method important in determining result of annealing). (c) Air cool (cooling method important in determining result of annealing).

Effects of special precipitation-hardening treatments on mechanical properties and electrical conductivity of Cu-1.9Be strip

Initial condition	Aging treatment Time, min	Temperature °C	°F	Tensile strength MPa	ksi	Yield strength(a) MPa	ksi	Elongation(b), %	Electrical conductivity, % IACS	Fatigue strength (c) MPa	ksi	Modulus of elasticity GPa	10⁶ psi
Alloy C17200													
Annealed ...... None	None	...	...	465	67.5	250	36	49	18.0	205	30	115	16.5
	5	370	700	855	124	695	101	18	19.5	...	...	120	17.5
	15	370	700	1195	173	1055	153	10	22.0	...	...	125	18.0
	30	370	700	1260	182.5	1060	153.5	6	23.0	...	...	125	18.0
	60	370	700	1240	180	1055	153	5	25.5	255	37	130	18.5
	120	370	700	1195	173.5	1040	151	6	26.0	...	...	130	18.5
	240	370	700	1150	167	980	142	6	26.5	...	...	130	19.0
¼ hard ...... None	None	...	...	570	82.5	485	70.5	21	17.0	220	32	115	17.0
	5	370	700	1115	162	945	137	9	18.5	...	...	125	18.0
	15	370	700	1250	181	1115	162	6	20.5	...	...	130	18.5
	30	370	700	1290	187	1125	163.5	4	23.5	290	42	130	18.5
	60	370	700	1230	178.5	1060	154	3	25.5	...	...	130	18.5
	120	370	700	1185	172	1000	145	4	26.5	...	...	130	19.0
	240	370	700	1155	167.5	970	141	6	27.0	...	...	130	19.0
½ hard ...... None	None	...	...	605	87.5	555	80.5	17	16.0	230	33	115	17.0
	3	370	700	1010	146.5	885	128	11	18.0	230	33	125	18.0
	5	370	700	1280	186	1110	161	3	21.0	295	43	125	18.0
	15	370	700	1310	190	1175	170.5	2	23.0	305	44	130	18.5
	30	370	700	1325	192.5	1180	171	2	24.5	305	44	130	18.5
	60	370	700	1280	185.5	1105	160	2	25.0	295	43	130	18.5
	120	370	700	1200	174	1040	150.5	3	26.0	275	40	130	18.5
	240	370	700	1185	172	1035	150	3	27.0	275	40	130	19.0
	420	370	700	1010	146.5	860	125	10	27.0	200	29	130	19.0
Hard ...... None	None	...	...	730	106	690	100	5	15.0	270	39	120	17.5
	5	370	700	1300	188.5	1125	163	3	18.0	...	...	125	18.0
	15	370	700	1360	197	1195	173	2	21.0	...	...	130	18.5
	30	370	700	1310	190	1170	170	1	24.5	315	46	130	19.0
	60	370	700	1295	188	1105	160	2	26.5	...	...	130	19.0
	120	370	700	1240	180	1090	158	2	27.5	...	...	130	19.0
	240	370	700	1215	176	1055	153	2	27.5	...	...	130	19.0
Alloy C17500													
Annealed ...... None	None	...	...	350	51	170	25	30	25	...	...	110	16.3
	120	425	800	805	117	625	91	14	44	...	...	135	19.3
	120	455	850	835	121	675	98	14	48	...	...	140	20.0
	120	480	900	805	116.5	625	91	14	48	215	31	140	20.0
	120	510	950	795	115	600	87	16	48.5	...	...	140	20.0
Hard ...... None	None	...	...	440	63.5	425	61.5	2	27.8	...	...	125	18.3
	120	425	800	985	142.5	860	125	11	44.0	...	...	140	20.0
	120	455	850	915	133	800	116	13	45.0	...	...	140	20.0
	120	480	900	850	123	760	110.5	13	47.5	250	36	140	20.0
	120	510	950	800	116	705	102	12	49.0	...	...	140	20.0

(a) At 0.2% offset. (b) In 50 mm or 2 in. (c) In 10^7 cycles

Properties and precipitation treatments usually specified for copper-beryllium alloys

Initial condition	Standard aging treatment Time, h	Temperature °C	°F	Tensile strength MPa	ksi	Yield strength(a) MPa	ksi	Elongation(b), %	Hardness(c)	Electrical conductivity, % IACS
C17200										
Flat products:										
Annealed . . . . . . None	. . .	. . .	415-540	60-78	195-380	28-55	35-60	45-78 HRB	17-19
¼ hard . . . . None	. . .	. . .	515-605	75-88	415-550	60-80	10-40	68-90 HRB	16-18
½ hard . . . . None	. . .	. . .	585-690	85-100	515-655	75-95	10-25	88-96 HRB	15-17
Hard . . . . None	. . .	. . .	690-825	100-120	620-770	90-112	2-8	96-102 HRB	15-17
Annealed(d) 3	315	600	1140-1345	165-195	965-1205	140-175	4-10	35-40 HRC	22-25
Annealed . . . . . . ½	370	700	1105-1310	160-190	895-1205	130-175	3-10	34-40 HRC	22-25
¼ hard(d) 2	315	600	1205-1415	175-205	1035-1275	150-185	3-6	37-42 HRC	22-25
¼ hard . . . . ⅓	370	700	1170-1380	170-200	965-1275	140-185	2-6	36-42 HRC	22-25
½ hard(d) 2	315	600	1275-1485	185-215	1105-1345	160-195	2-5	39-44 HRC	22-25
½ hard . . . . ¼	370	700	1240-1450	180-210	1070-1345	155-195	2-5	38-44 HRC	22-25
Hard(d) 2	315	600	1310-1575	190-220	1140-1415	165-205	1-4	40-45 HRC	22-25
Hard . . . . ¼	370	700	1275-1480	185-215	1105-1415	160-205	1-4	39-45 HRC	22-25
Rod, bar and plate:										
Annealed . . . . . . None	. . .	. . .	415-585	60-85	185-205	20-30	35-60	45-85 HRB	17-19
Hard . . . . None	. . .	. . .	585-895	85-130	515-725	75-105	10-20	88-103 HRB	15-17
Annealed(d) 3	315	600	1140-1345	165-200	1000-1205	145-175	3-10	36-41 HRC	22-25
Hard(d) 2	315	600	1205-1550	175-225	1035-1380	150-200	2-5	39-45 HRC	22-25
Wire(e):										
Annealed . . . . . . None	. . .	. . .	450-590	65-85	185-240	20-35	35-55	. . .	17-19
¼ hard . . . . None	. . .	. . .	620-795	90-115	485-655	70-95	10-35	. . .	15-17
½ hard . . . . None	. . .	. . .	760-930	110-135	620-760	90-110	4-10	. . .	15-17
¾ hard . . . . None	. . .	. . .	895-1070	130-155	760-930	110-135	2-8	. . .	15-17
Annealed(d) 3	315	600	1140-1310	165-190	1000-1205	145-175	3-8	. . .	22-25
Annealed . . . . . . ½	370	700	1105-1310	160-190	930-1205	135-175	3-8	. . .	22-25
¼ hard(d) 2	315	600	1205-1415	175-205	1105-1310	160-190	2-5	. . .	22-25
¼ hard . . . . ¼	370	700	1170-1415	170-205	1035-1310	150-190	2-5	. . .	22-25
½ hard(d) 1½	315	600	1310-1480	190-215	1205-1380	175-200	1-3	. . .	22-25
½ hard . . . . ¼	370	700	1275-1480	185-215	1170-1380	170-200	1-3	. . .	22-25
¾ hard(d) 1	315	600	1345-1585	195-230	1245-1415	180-205	1-3	. . .	22-25
¾ hard . . . . ¼	370	700	1310-1585	190-230	1205-1415	175-205	1-3	. . .	22-25

(continued)

(a) At 0.2% offset. (b) In 50 mm or 2 in. (c) Rockwell B and C hardness values are accurate only if metal is at least 1 mm (0.040 in.) thick. (d) Heat treatment that provides optimum strength. (e) For wire diameters greater than 1.3 mm (0.050 in.).

Properties and precipitation treatments usually specified for copper-beryllium alloys (continued)

Initial condition	Standard aging treatment Time, h	Temperature °C	°F	Tensile strength MPa	ksi	Yield strength(a) MPa	ksi	Elonga-tion(b), %	Hardness(c)	Electrical conduc-tivity, % IACS
C17000										
Flat products:										
Annealed	None	...	...	415-540	60-78	170-365	25-55	35-60	45-78 HRB	17-19
¼ hard	None	...	...	515-605	75-88	310-515	45-75	10-40	68-90 HRB	16-18
½ hard	None	...	...	585-690	85-100	450-620	65-90	10-25	88-96 HRB	15-17
Hard	None	...	...	690-825	100-120	550-760	80-110	2-8	96-102 HRB	15-17
Annealed	3	315	600	1035-1240	150-180	895-1105	130-165	4-10	33-39 HRC	22-25
Annealed(d)....	3	345	650	1105-1275	160-185	860-1140	125-165	4-10	34-40 HRC	22-25
¼ hard	2	315	600	1105-1310	160-190	860-1140	135-170	3-6	34-40 HRC	22-25
¼ hard(d)......	3	330	625	1170-1345	170-195	895-1170	130-170	3-6	36-41 HRC	22-25
½ hard	2	315	600	1170-1380	170-200	895-1170	145-175	2-5	36-41 HRC	22-25
½ hard(d)......	2	330	625	1240-1380	180-200	965-1240	140-180	2-5	38-42 HRC	22-25
Hard	2	315	600	1240-1450	180-210	965-1240	155-180	2-5	38-42 HRC	22-25
Hard(d)........	2	330	625	1275-1415	185-205	1070-1345	155-195	2-5	39-43 HRC	22-25
Rod and bar:										
Annealed	None	...	...	415-585	60-85	185-205	20-30	35-60	45-85 HRB	17-19
Hard	None	...	...	585-895	85-130	515-725	75-105	10-20	88-103 HRB	15-17
Annealed	3	315	600	1035-1240	150-180	860-1070	125-155	4-10	32-39 HRC	22-25
Annealed(d)....	3	345	650	1105-1275	160-185	930-1140	135-165	4-10	34-40 HRC	22-25
Hard	2	315	600	1140-1380	165-200	930-1140	135-165	2-5	36-41 HRC	22-25
Hard(d)........	2	345	650	1205-1415	175-205	965-1170	140-170	2-5	38-42 HRC	22-25
C17500, C17510										
Rod, bar, plate and flat products:										
Annealed	None	...	...	240-380	35-55	185-205	20-30	20-35	20-43 HRB	25-30
Hard	None	...	...	515-585	75-85	380-550	55-80	3-10	78-88 HRB	20-30
Annealed	3	480	900	690-760	100-120	550-690	80-100	10-20	92-100 HRB	45-60
Annealed(d)....	3	455	850	725-825	105-120	550-725	80-105	8-12	93-100 HRB	45-52
Hard	2	480	900	760-860	110-130	690-825	100-120	8-15	95-103 HRB	45-60
Hard(d)........	2	455	850	795-930	115-135	725-860	105-125	5-8	97-104 HRB	45-52

(a) At 0.2% offset. (b) In 50 mm or 2 in. (c) Rockwell B and C hardness values are accurate only if metal is at least 1 mm (0.040 in.) thick. (d) Heat treatment that provides optimum strength. (e) For wire diameters greater than 1.3 mm (0.050 in.)

Typical effects of heat treatment and cold work on properties of copper-1% chromium alloys

Condition	Ultimate tensile strength MPa	ksi	Yield strength(a) MPa	ksi	Elongation(b), %	Hardness	Electrical conductivity, % IACS
Alloy C18200							
Solution treated................	240	35	105	15	42	50 HRF	35-42
Solution treated and aged......................	350	51	275	40	15	90 HB(c)	75-82
Solution treated and drawn 40%...............	415	60	310	45	15	65 HRB	40
Solution treated, hard drawn and aged..........	435	63	385	56	18	68-75 HRB	80
Solution treated, aged and drawn 30%...............	480	70	425	62	18	75-80 HRB	80
Alloy C81500							
Cast, solution treated and aged......................	350	51	275	40	17	105 HB(c)	75-80

(a) At 0.5% extension under load. (b) In 50 mm or 2 in. (c) 500-kg load

Effect of heat treatment and cold work on properties of copper-zirconium alloy C15000

Solution treating temperature(a) °C	°F	Amount of cold work, %	Aging Temperature °C	°F	Time, h	Tensile strength MPa	ksi	Yield strength MPa	ksi	Elonga-tion(b), %	Hard-ness, HRB	Electrical conduc-tivity, % IACS
900	1650.....	20	475	885	1	310	45	260	38	25	48	85 min
900	1650.....	80	425	795	1	425	62	380	55	12	64	85 min
980	1795.....	None	...	...	...	200	29	41(c)	6(c)	54	...	64
980	1795.....	20	...	...	...	270	39	250(c)	36(c)	26	37	64
980	1795.....	80	...	...	...	440	64	420(c)	61(c)	19	73	64
980	1795.....	None	500	930	3	205	30	90	13	51	...	87
980	1795.....	None	550	1025	3	205	30	90	13	49	...	95
980	1795.....	20	400	750	3	330	48	260	38	31	50	80
980	1795.....	20	450	840	3	330	48	275	40	28	57	92
980	1795.....	85	400	750	3	495	72	440	64	24	79	85
980	1795.....	85	450	840	3	470	68	425	62	23	74	91

(a) Hold 30 min, water quench. (b) In 50 mm or 2 in. (c) 0.5% extension under load

Typical heat treating schedules and resulting properties for precipitation hardened miscellaneous alloys

Alloy	Solution treatment Temperature °C	°F	Time, min	Tempering treatment Temperature °C	°F	Time, min	Tensile strength MPa	ksi	Yield strength(a) MPa	ksi	Elonga-tion(b), %	Hard-ness, HB(c)
C94700.....	775-800	1425-1475	120	305-325	580-620	300	585	85	415	60	10	180
C94800.......	...	...	...	305-325	580-620	360-1000	415	60	205	30	8	120
C96600.......	995	1825	60	510	950	180	760	110	485	70	7	230
C99400.......	885	1625	60	480	900	60	545	79	370	54	...	170
C99500.......	885	1625	60	480	900	60	595	86	425	62	8	196

(a) At 0.2% extension under load for C96600; at 0.5% extension under load for all other alloys. (b) In 50 mm or 2 in. (c) 3000-kg load

Typical strengths and recommended aging times for various spinodal alloys(a)

Alloy	CDA No.	Solution treated and cold worked temper	Aging cycle	Tensile strength MPa	ksi	Yield strength MPa	ksi	Elonga-tion, %
Cu-4Ni-4Sn	C72600	TD 02(½H)	90 min at 350 °C (660 °F)	635-690	92-100	495-570	72-83 (0.05)	12
Cu-4Ni-4Sn	C72600	TD 06(XH)	90 min at 350 °C (660 °F)	690-725	100-105	565-620	82-90 (0.05)	9
Cu-4Ni-4Sn	C72600	TD 08(S)	90 min at 350 °C (660 °F)	705-795	102-115	565-655	82-95 (0.05)	7
Cu-9Ni-6Sn	C72700	TD 04(H)	90 min at 350 °C (660 °F)	860-1035	125-150	760-895	110-130 (0.05)	8
Cu-9Ni-6Sn	C72700	TD 14(SS)	90 min at 350 °C (660 °F)	1055-1145	153-166	930-985	135-143 (0.05)	. . .
Cu-10Ni-8Sn-0.2Nb . .	C72800	TB 00 cast and solution treated	4-6 h at 350 °C (660 °F)	830-965	120-140	550-690	80-100 (0.01)	3
Cu-10Ni-8Sn-0.2Nb . .	C72800	TB 00 hot work and solution treated	3-5 h at 350 °C (660 °F)	965-1070	140-155	690-825	100-120 (0.01)	6-14
Cu-10Ni-8Sn-0.2Nb . .	C72800	TD 01(¼H)	3 h at 350 °C (660 °F)	1140-1240	165-180	895-930	130-135 (0.01)	7
Cu-10Ni-8Sn-0.2Nb . .	C72800	TD 04(H)	3 h at 350 °C (660 °F)	1205-1380	175-200	930-1000	135-145 (0.01)	7
Cu-10Ni-8Sn-0.2Nb . .	C72800	TD 06(XH)	3 h at 350 °C (660 °F)	1205-1380	175-200	965-1035	140-150 (0.01)	5
Cu-10Ni-8Sn-0.2Nb . .	C72800	TD 08(S)	3 h at 350 °C (660 °F)	1240-1380	180-200	1000-1070	145-155 (0.01)	4
Cu-10Ni-8Sn-0.2Nb . .	C72800	TD 14(SS)	90 min at 350 °C (660 °F)	1240-1380	180-200	1070-1140	155-165 (0.01)	2.5
Cu-15Ni-8Sn	C72900	TD 14(SS)	90 min at 350 °C (660 °F)	1140-1380	165-200	1035-1170	150-170 (0.05)	3
Cu-30Ni-3Cr	C71900	Hot extruded	90 min at 760 °C (405 °F)	550	80	345	50 (0.20)	25

(a) At 350 °C (660 °F)

Typical heat treatments and resulting properties for complex (alpha-beta) aluminum bronzes

Alloy	Typical condition(a)	Tensile strength MPa	ksi	Yield strength(b) MPa	ksi	Elonga-tion(c), %	Hard-ness, HB
C62400	As forged or extruded	620-690	90-100	240-260	35-38	14-16	163-183
	Solution treated at 870 °C (1600 °F) and quenched, tempered 2 h at 620 °C (1150 °F)	675-725	98-105	345-385	50-56	8-14	187-202
C63000	As forged or extruded	730	106	365	53	13	187
	Solution treated at 855 °C (1575 °F) and quenched, tempered 2 h at 650 °C (1200 °F)	760	110	425	62	13	212
C95300	As cast	495-530	72-77	185-205	27-30	27-30	137-140
	Solution treated at 855 °C (1575 °F) and quenched, tempered 2 h at 620 °C (1150 °F)	585	85	290	42	14-16	159-179
C95400	As cast	585-690	85-100	240-260	35-38	14-18	156-179
	Solution treated at 870 °C (1600 °F) and quenched, tempered 2 h at 620 °C (1150 °F)	655-725	95-105	330-370	48-54	8-14	187-202
C95500	As cast	640-710	93-103	290-310	42-45	10-14	183-192
	Solution treated at 855 °C (1575 °F) and quenched, tempered 2 h at 650 °C (1200 °F)	775-800	112-116	440-470	64-68	10-14	217-234

(a) As-cast condition is typical for moderate sections shaken out at temperatures above 540 °C (1000 °F) and fan cooled; or mold cooled, annealed at 620 °C (1150 °F) and fan (rapid) cooled. (b) At 0.5% extension under load. (c) In 50 mm or 2 in.

Machining

Machinability ratings of several copper casting alloys

UNS number	Common name	Machinability rating (a), %
Group 1—Free-cutting Alloys		
C83600	Leaded red brass	90
C83800	Leaded red brass	90
C84400	Leaded semi-red brass	90
C84800	Leaded semi-red brass	90
C94300	High-leaded tin bronze	90
C85200	Leaded yellow brass	80
C85400	Leaded yellow brass	80
C93700	High-leaded tin bronze	80
C93800	High-leaded tin bronze	80
C93200	High-leaded tin bronze	70
C93500	High-leaded tin bronze	70
C97300	Leaded nickel brass	70
Group 2—Moderately Machinable Alloys		
C86400	Leaded high-strength manganese bronze	60
C92200	Leaded tin bronze	60
C92300	Leaded tin bronze	60
C90300	Tin bronze	50
C90500	Tin bronze	50
C95600	Silicon-aluminum bronze	50
C95300	Aluminum bronze	35
C86500	High-strength manganese bronze	30
Group 3—Hard-to-Machine Alloys		
C86300	High-strength manganese bronze	20
C95200	9% aluminum bronze	20
C95400	11% aluminum bronze	20
C95500	Nickel-aluminum bronze	20

(a) Machinability rating expressed as a percentage of the machinability of C36000, free-cutting brass. The rating is based on relative speed for equivalent tool life. For instance, a material having a rating of 50 should be machined at about half the speed that would be used to make a similar cut in C36000.

Joining

Relative solderability of various types of copper metals

Type of copper metal	Solderability and remarks
Coppers(a)	Excellent. Need only rosin or other noncorrosive flux
Copper-tin alloys	Good. Easily soldered with activated rosin and intermediate fluxes
Copper-zinc alloys	Good. Easily soldered with activated rosin and intermediate fluxes
Copper-nickel alloys	Good. Easily soldered with intermediate and corrosive type fluxes
Chromium copper and beryllium copper	Good. Require intermediate and corrosive type fluxes
Copper-silicon alloys	Fair. Silicon produces refractory oxides that require use of corrosive fluxes
Copper-aluminum alloys	Difficult. May be soldered with help of very corrosive fluxes
High-strength manganese bronze	Not recommended. Should be plated to ensure consistent solderability

(a) Includes tough-pitch, oxygen-free, phosphorized, arsenical, silver-bearing, leaded, tellurium, and selenium coppers.

Comparative behavior of copper metals in resistance welding

Alloy type	Common name	Conductivity, % IACS	Rating
Cu-Si	Silicon bronzes	7 to 12	Excellent
Cu-Ni	Copper nickels	4 to 10	Excellent
Cu-Ni-Zn	Nickel silvers	5 to 9	Good
Cu-Al	Aluminum bronzes	7 to 18	Good
Cu-Sn	Phosphor bronzes	10 to 22	Fair
Cu-Zn-Mn	Manganese brass	15 to 16	Fair
Cu-Zn-Si	Silicon red brass	15 to 16	Fair
Cu-Zn (high Zn)	Yellow brasses	22 to 28	Fair
Cu-Zn-Mn-Fe	Manganese bronzes	22 to 28	Fair
Cu-Zn (low Zn)	Red brasses	32 to 43	Poor
Cu	High-copper alloys	50 to 65	Poor

Joining-process selection guide for copper alloys based on service requirements

Service requirement	Gas metal-arc	Gas tungsten-arc	Plasma-arc	Shielded metal-arc	Electron beam	Laser beam	Oxyfuel gas	Resistance spot	Resistance seam	Projection	Flash	Diffusion	Explosive	Ultrasonic	Friction	Brazing	Soldering	Adhesive bonding	Mechanical fastening
Primary structural																			
Elevated temperature	B	A	E	B	A	E	C	A	A	D	A	A	C	C	D	B	D	C	B
Ambient temperature	A	A	E	B	A	E	C	A	A	D	A	A	C	C	D	A	C	A	A
Cryogenic	B	A	E	C	A	E	D	A	A	D	A	A	C	C	D	B	D	D	B
Vacuum	B	A	E	D	A	E	D	D	B	D	C	A	C	C	D	B	D	C	C
Atmospheric pressure	A	A	E	B	A	E	C	B	A	B	A	C	C	D	A	D	C	C	C
High pressure	B	A	E	D	A	E	D	D	C	D	C	A	C	C	D	C	D	D	C
Secondary structural	A	A	E	A	A	E	B	A	A	C	A	A	C	C	C	A	D	A	A
Noncritical	A	A	C	A	A	E	B	A	A	A	A	A	C	B	B	A	B	A	A
Dissimilar metal joining	C	C	C	B	C	E	C	D	D	D	C	A	B	A	B	B	B	A	A

Note: A, most satisfactory; B, satisfactory; C, restricted use; D, prohibited use; E, experimental.

Joining-process selection guide for copper alloys based on joint configuration

Joint configuration	Gas metal-arc	Gas tungsten-arc	Plasma-arc	Shielded metal-arc	Electron beam	Laser beam	Oxyfuel gas	Resistance spot	Resistance seam	Projection	Flash	Diffusion	Explosive	Ultrasonic	Friction	Brazing	Soldering	Adhesive bonding	Mechanical fastening
Butt joint	A	A	A	A	A	A,E	B	D	D	D	A	C	D	D	B	D	D	D	D
Tee-joint	A	A	B	A	C	C,E	B	D	D	D	C	B	D	D	D	B	B	C	D
Edge joint	A	A	B	B	A	A,E	B	D	D	D	D	C	D	D	D	C	D	D	D
Corner joint	A	A	B	B	A	A,E	B	D	D	D	D	C	D	D	D	B	B	D	D
Flange joint	A	A	B	B	A	A,E	B	D	D	C	D	C	D	D	B	C	C	C	A
Scarf joint	D	D	D	D	C	D	D	D	D	D	C	B,E	D	B	B	B	B	B	B
Strap butt joint (splice joint)	C	C	C	C	C	B,E	D	B	B	C	D	C	C	B	D	A	B	A	A
Lap joint																			
Shear load	B	B	B	B	A	A,E	D	A	A	C	D	A	B	B	D	A	B	A	A
Tensile load	B	B	B	B	A	A,E	D	B	B	C	D	A	B	B	D	D	D	C	A

Note: A, most satisfactory; B, satisfactory; C, restricted use; D, prohibited use; E, experimental.

Joining-process selection guide for copper alloys based on thickness of parts being joined

| Metal thickness | | Fusion welding | | | | | | | Resistance welding | | | | Solid-state welding | | | | | | | |
| | | Arc welding | | | | Other welding processes | | | | | | | | | | | | | | | |
mm	in.	Gas metal-arc	Gas tungsten-arc	Plasma-arc	Shielded metal-arc	Electron beam	Laser beam	Oxyfuel gas	Resistance spot	Resistance seam	Projection	Flash	Diffusion	Explosive	Ultrasonic	Friction	Brazing	Soldering	Adhesive bonding	Mechanical fastening
0.025–0.25	0.001–0.010	D	B	A	D	A	B,E	D	C	C	D	D	A	D	A	D	B	C	B	C
0.25–0.50	0.010–0.020	D	A	B	D	A	C,E	D	A	A	B	D	A	D	A	D	B	B	A	B
0.50–1.25	0.020–0.050	C	A	B	D	A	D	B	A	A	A	D	A	D	C	B	B	A	A	A
1.25–2.50	0.050–0.100	A	A	A	B	A	D	B	A	A	A	A	D	B	B	C	A	A	A	A
2.50–3.75	0.100–0.150	A	A	A	A	A	D	B	A	A	C	A	A	C	D	B	B	C	A	A
3.75–6.25	0.150–0.250	A	A	A	A	A	D	C	C	C	C	A	A	C	D	B	C	D	B	A
6.25–12.50	0.250–0.500	A	A	C	A	A	D	D	D	D	D	A	A	C	D	B	C	D	C	A
12.50–25.0	0.500–1.00	A	A	D	A	A	D	D	D	D	D	A	A	C	D	B	D	D	D	A
25.0–62.5	1.00–2.50	B	B	D	A	A	D	D	D	D	D	C	A	C	D	C	D	D	D	A
Over 62.5	Over 2.50	C	B	D	A	A	D	D	D	D	D	C	A	C	D	C	D	D	D	A
Thick to thin		A	A	C	B	A	B,E	D	C	C	C	C	A	A	A	···	B	A	A	A

Note: A, most satisfactory; B, satisfactory; C, restricted use; D, prohibited use; E, experimental.

Joining-process selection guide for copper alloys based on alloy composition

| Copper alloy family | Fusion welding | | | | | | | | | Resistance welding | | | | | | | |
| | Arc welding | | | | | Other welding processes | | | | | | | | | | | |
	Gas metal-arc	Gas tungsten-arc	Submerged arc	Shielded metal-arc	Stud	Electron beam	Electroslag	Laser beam	Oxyfuel gas	Spot	Seam	Projection	Flash butt	Brazing	Soldering	Adhesive bonding	Mechanical fastening
Coppers																	
Oxygen free	B	B	D	D	D	B,E	D	E	C	D	D	D	B	A	A	C	A
Deoxidized	A	A	D	D	D	B,E	D	E	B	D	D	D	B	A	A	C	A
Tough pitch	C	C	D	D	D	B,E	D	E	D	D	D	D	B	B	A	C	A
High Copper Alloys																	
Cadmium	B	B	D	D	D	···	D	···	B	D	D	D	B	A	A	C	A
Beryllium	B	B	D	B	D	···	D	···	D	B	C	B	C	B	B	C	A
Chromium	B	B	D	D	D	···	D	···	D	D	D	D	C	B	B	C	A
Leaded	D	D	D	D	D	···	D	···	D	D	D	D	C	B	A	C	A
Oxide dispersion strengthened	D	D	D	D	D	···	D	···	D	C	C	C	C	B	A	C	A
Brasses																	
Red	B	B	D	D	D	···	D	···	B	C	D	C	B	A	A	C	A
Yellow	C	C	D	D	D	···	D	···	B	B	D	B	B	A	A	C	A
Leaded	D	D	D	D	D	···	D	···	D	D	D	D	C	B	A	C	A
Tin	B	B	D	C	D	···	D	···	B	B	C	B	B	A	A	C	A
Bronzes																	
Phosphor	B	B	D	C	D	···	D	···	C	B	C	B	A	A	A	C	A
Leaded phosphor	D	D	D	D	D	···	D	···	D	D	D	D	C	B	A	C	A
Aluminum	A	A	D	B	D	···	D	···	D	B	B	B	C	C	C	C	A
Silicon	A	A	D	C	D	···	D	···	B	A	A	A	A	A	A	C	A
Copper Nickels																	
10% Ni	A	A	D	B	D	···	D	···	C	B	B	B	A	A	A	C	A
30% Ni	C	C	D	C	D	···	D	···	B	A	A	A	A	A	A	C	A
Nickel silvers	C	C	D	D	D	···	D	···	B	B	C	B	B	A	A	C	A

Note: A, most satisfactory; B, satisfactory; C, restricted use; D, prohibited use; E, experimental.

Comparison of soldering with other bonding methods for electrical applications

Factor	Soldering	Metallurgical bonding Brazing	Welding	Crimp-ing	Mechanical bonding Screw-ing	Wrap-ping	Adhesive bonding Conductive cement
Temperature limit of joint, °C (°F)	73-460 (100-800)	460-900 (800-1600)	Conductor melting temperature	No limit except wire	No limit except wire	No limit except wire	100-180 (160-300)
Heating effect on assembly.	Small	Large	Small (quick)	None	None	None	Cures at ambient to 120 °C (250 °F)
Ease of rework and rebonding	Simple	Simple	Not practical	Not practical	Simple	Simple	Not practical
Process economy Equipment cost	Low	Medium	High	Low	Low	Low	Low
Ease of automation.	Easiest	More difficult	More difficult	More difficult	More difficult	More difficult	More difficult
Extra hardware.	No	No	No	Yes	Yes	No	No
Joint stable under vibration.	Yes	Yes	Yes	Yes	No	Yes	Yes
Joint stable in oxidizing environment	Yes	Yes	Yes	No	No	Yes	Yes

Representative list of solders used for joining copper metals

ASTM classification	Nominal composition	Solidus °C	°F	Liquidus °C	°F	Melting range °C	°F	Density Mg/m³	lb/in.³	Typical applications
Tin-Lead(a)										
5A	95Pb-5Sn	308	586	312	594	4	8	11.3	0.408	Coating and joining
10B	90Pb-10Sn	268(b)	514(b)	301	573	33	59	10.8	0.389	Coating and joining
15B	85Pb-15Sn	225(b)	437(b)	290	553	65	116	10.5	0.379	Coating and joining
20A	80Pb-20Sn	183	361	280	535	97	174	10.2	0.368	Coating and joining
25A	75Pb-25Sn	183	361	267	511	84	150	9.99	0.361	Machine and torch soldering
30A	70Pb-30Sn	183	361	255	491	72	130	9.69	0.350	Machine and torch soldering
35A	65Pb-35Sn	183	361	247	477	64	116	9.69	0.350	General purpose; wiping
40A	60Pb-40Sn	183	361	235	455	52	94	9.27	0.335	Wiping; auto radiators
45A	55Pb-45Sn	183	361	228	441	45	80	8.97	0.324	Auto radiators; roofing seams
50A	50Pb-50Sn	183	361	217	421	34	60	8.83	0.319	General purpose; most widely used on copper
60A	40Pb-60Sn	183	361	190	374	7	13	8.64	0.312	"Fine solder"; general purpose, especially where low soldering temperature is essential
63A	37Pb-63Sn	183	361	183	361	0	0	8.40	0.303	"Eutectic solder"; lowest-melting lead-tin solder
70A	30Pb-70Sn	183	361	192	378	9	17	8.32	0.301	
Tin-Lead-Antimony and Tin-Antimony										
20C	79Pb-20Sn-1.0Sb	184	363	270	517	86	154	10.2	0.367	Machine soldering and coating(c)
25C	73.7Pb-25Sn-1.3Sb	185	364	262	504	77	140	9.94	0.359	Torch and machine soldering(c)
30C	68.4Pb-30Sn-1.6Sb	185	364	250	482	65	118	9.63	0.348	Torch and machine soldering(c)
35C	63.2Pb-35Sn-1.8Sb	186	365	243	470	57	105	9.44	0.341	Wiping(c)
40C	58Pb-40Sn-2.0Sb	186	365	231	448	45	83	9.22	0.333	General purpose(c)
95TA	95Sn-5Sb	232	450	240	464	8	14	7.80	0.260	Copper joints in electrical, plumbing and heating systems
Lead-Silver, Lead-Tin-Silver and Tin-Silver										
2.5S	97.5Pb-2.5Ag	304	579	579	579	275	0	11.3	0.409	Torch soldering of copper and brass
5.5S	94.5Pb-5.5Ag	304	579	343	689	39	110	. . .	. . .	Torch soldering of copper and brass
1.5S	97.5Pb-1.0Sn-1.5Ag	313	588	313	588	0	0	11.3	0.409	Torch soldering of copper and brass
. . . .	97Pb-2.5Sn-0.5Ag	303	577	310	590	7	13	. . .	. . .	Torch soldering of copper and brass
. . . .	94.5Pb-5Sn-0.5Ag	294	561	301	574	7	13	. . .	. . .	Torch soldering of copper and brass
. . . .	36Pb-62Sn-2Ag	180	354	190	372	10	18	. . .	. . .	. . .
96TS	96Sn-4Ag	221	430	221	430	0	0	10.4	0.375	Delicate instruments; electrical conductors for use at high temperature

(a) Most class A solders also available as antimonial class B solders (0.20 to 0.50% Sb). (b) This alloy has virtually no strength at temperatures above 183 °C (361 °F). (c) Not recommended for soldering to galvanized iron.

Lead, Tin and Zinc

Lead

Composition limits of pig leads (ASTM B29)

Element	Composition(a), % Corroding lead(b)	Common lead(c)	Chemical lead(d)	Acid-copper lead(e)
Silver, max	0.0015	0.005	0.020	0.002
Silver, min	...	...	0.002	...
Copper, max	0.0015	0.0015	0.080	0.080
Copper, min	...	...	0.040	0.040
Silver + copper, max	0.0025	...	...	...
Arsenic + antimony + tin, max..	0.002	0.002	0.002	0.002
Zinc, max	0.001	0.001	0.001	0.001
Iron, max	0.002	0.002	0.002	0.002
Bismuth, max..................	0.050	0.050(f)	0.005	0.025
Lead(g), min..................	99.94	99.94	99.90	99.90

(a) By agreement between the purchaser and the supplier, analyses may be required and limits established for elements (or compounds) not specified here. (b) "Corroding lead" is a designation used in the trade to describe lead that has been refined to a high degree of purity. (c) "Common lead" is fully refined desilverized lead. (d) "Chemical lead" designates the undesilverized lead produced from southeastern Missouri ores. (e) "Copper-bearing lead" is made by adding copper to fully refined lead. (f) By agreement between the purchaser and the supplier, bismuth levels of up to 0.150% may be allowed. (g) By difference.

Tin and Tin Alloys

Designations, chemical compositions and applications of commercially pure tins

	Grade designation		Composition, % (a)											
ASTM B339	Designation	Class	Sn max	Sb max	As max	Bi max	Cd max	Cu max	Fe max	Pb max	Ni + Co max	S max	Zn max	General applications
AAA	Electrolytic	Extra-high purity	99.98	0.008	0.0005	0.001	0.001	0.002	0.005	0.010	0.005	0.002	0.001	Analytical standards, research
AA	Electrolytic	High purity	99.95	0.02	0.01	0.01	0.001	0.02	0.01	0.02	0.01	0.01	0.005	Research, pharmaceuticals, fine chemicals
A(b)	A, Straits	High purity; commercial	99.80	0.04	0.05	0.015	0.001	0.04	0.015	0.05	0.01	0.01	0.005	Tinplate, foil, collapsible tubes, block tin products, pewter
B(c)	B	General purpose	99.80	⋯	0.05	⋯	⋯	⋯	⋯	⋯	⋯	⋯	⋯	Less exacting, general purpose
C	C	Intermediate grade	99.65	⋯	⋯	⋯	⋯	⋯	⋯	⋯	⋯	⋯	⋯	General purpose alloys
D	D	Lower intermediate grade	99.50	⋯	⋯	⋯	⋯	⋯	⋯	⋯	⋯	⋯	⋯	General purpose alloys
E	E	Common	99.00	(d)	⋯	⋯	⋯	(d)	⋯	(d)	⋯	⋯	⋯	Cast bronze, bearing metal, general purpose alloys, lead base alloys

(a) The maximum impurity limits listed below, which are from ASTM Standard Classification B339, are not specification limits, but simply guides to the maximum impurity contents commonly found in the various brands of tin that fall into these grades. (b) ASTM Grade A includes about 80 to 90% of the refined tin produced. (c) Grade B is intended for those uses where the specific impurity limitations of Grade A are not critical. (d) Limits of these impurities may be specified for some uses.

Tensile properties of commercially pure tin

Temperature °C	°F	Yield strength MPa	ksi	Elongation in 25 mm or 1 in., %	Reduction in area, %
Strained at 0.2 mm/m · min (0.0002 in./in. · min)					
−200	−328	36.2	5.25	6	6
−160	−256	90.3	13.10	15	10
−120	−184	87.6	12.71	60	97
−80	−112	38.9	5.64	89	100
−40	−40	20.1	2.92	86	100
0	32	12.5	1.81	64	100
23	73	11.0	1.60	57	100
Strained at 0.4 mm/m · min (0.0004 in./in. · min)					
15	59	14.5	2.10	75	⋯
50	122	12.4	1.80	85	⋯
100	212	11.0	1.60	55	⋯
150	302	7.6	1.10	55	⋯
200	392	4.5	0.65	45	⋯

Note: It is uncertain if the inconsistencies among these data are due to differences in purity or the difference in straining rate.

Zinc and Zinc Alloys

Grades and compositions of slab zinc (ASTM B6)

Grade	Pb, max	Composition(a), % Fe, max	Cd, max	Zn, min (by difference)
Special High Grade(b)........0.003		0.003	0.003	99.990
High Grade 0.03		0.02	0.02	99.90
Prime Western(c)...........1.4		0.05	0.20	98.0

(a) When specified for use in manufacture of rolled zinc or brass, aluminum is held to 0.005% max.
(b) Tin in Special High Grade Zinc is held to 0.001% max. (c) Aluminum in Prime Western Zinc is held to 0.05% max.

Designations of zinc die casting alloys

Alloy	UNS number	SAE number	Government specification	ASTM specification
AG40A....	Z33520	903	QQ-Z-363	B240 B86
AC41A....	Z35530	925	QQ-Z-363	B240 B86
Alloy 7....	...	...	...	...
ILZRO 16..	...	...	...	...

Classification of wrought zinc alloys

Pb	Fe	Cd	Composition, % Cu	Mg	Al	Other	Characteristics	Typical uses
0.05 to 0.10	0.012 max	0.005 max	0.001 max	...	...	...	High ductility with low hardness and stiffness. Very little work hardening possible	Drawn battery cans, eyelets, fuse links, and a variety of articles drawn, formed and spun. Address plates
0.05 to 0.10	0.012 max	0.06	0.005 max	...	...	...	High ductility with low hardness. Can be work hardened slightly	Drawn battery cans, eyelets and grommets. Extruded battery cans. Address plates, laundry tags
0.15 to 0.35	0.017 max	0.15 to 0.30	0.005 max	...	...	...	High hardness and stiffness. Uniform etching quality. Can be work hardened	Soldered battery cans, photoengraver's plate, lithographer sheet, boiler and ship plates, weather-strips
0.05 to 0.10	0.012 max	0.005 max	0.85 to 1.25	...	...	...	High hardness and stiffness. Good ductility. Good creep resistance. Work hardens easily	Weatherstrips and drawn and formed articles requiring stiffness
0.05 to 0.10	0.015 max	0.005 max	0.85 to 1.25	0.007 to 0.02	...	...	High stiffness and creep resistance. Can be severely work hardened	Flat or formed articles requiring high stiffness and strength
0.005 to 0.10	0.012 max	0.05 max	0.50 to 1.50	...	...	0.12 to 1.50 Ti(a)	Outstanding creep resistance. Can be severely work hardened. Lowest thermal expansivity with the grain. Very high resistance to grain growth during annealing	Corrugated roofing, leaders and gutters, and other uses requiring maximum creep resistance
0.15 to 0.35	0.014 to 0.025	0.15 to 0.30	0.005 max	0.005 to 0.025	...	...	High hardness. Can be baked without severe softening. Good etching characteristics	Photoengraver's sheet
...	...	...	...	0 to 0.025	0.25 to 0.60	...	High hardness. Can be baked without severe softening. Good etching characteristics	Photoengraver's sheet
0.007 max	0.10 max	0.007 max	0 to 3.5	0.02 to 0.10	3.5 to 4.5	0.005 max Sn	High strength and hardness	Shearing and forming dies. Extruded rod, tube and moldings
0.005 to 0.10	0.012 max	0.05 max	0.50 to 1.5	...	...	0.12 to 0.50 Ti	Good creep resistance	Corrugated roofing, leaders and gutters, and formed articles requiring maximum creep resistance

(a) U.S. Patent 2472402.

Compositions of zinc die casting alloys

Alloy	Form	Cu	Al	Mg	Pb, max	Cd, max	Sn, max	Fe, max	Others	Zn
AG40A	Ingot	0.10 max	3.9-4.3	0.025-0.05	0.004	0.003	0.002	0.075	(a)	rem
AC41A	Die castings	0.25 max(b)	3.5-4.3	0.020-0.05(c)	0.005	0.004	0.003	0.100	(a)	rem
Alloy 7	Ingot	0.75-1.25	3.9-4.3	0.03-0.06	0.004	0.003	0.002	0.075	(a)	rem
ILZRO 16 ...	Die castings	0.75-1.25	3.5-4.3	0.03-0.08(c)	0.005	0.004	0.003	0.100	(a)	rem
	Die castings	0.25 max	3.5-4.3	0.010-0.02	0.0020	0.0020	0.0010	0.050	...	rem
	Die castings	1.0 to 1.5	0.01-0.04	...	...	...	...	...	(d)	rem

(a) May contain nickel, chromium, silicon and manganese in amount of 0.02, 0.02, 0.035 and 0.5%, respectively. No harmful effects have ever been noted due to the presence of these elements in these concentrations; therefore, analyses are not required for these elements. (b) For the majority of commercial applications, a copper content in the range of 0.25 to 0.75% will not adversely affect the serviceability of die castings and should not serve as a basis for rejection. (c) Magnesium content may be as low as 0.015% provided that lead, cadmium and tin contents do not exceed 0.003, 0.003 and 0.001%, respectively. (d) 0.15 to 0.25 Ti, 0.10 to 0.20 Cr, 0.30 to 0.40 Ti + Cr.

Average properties of zinc die casting alloys

Properties	ASTM AG40A; SAE 903	ASTM AC41A; SAE 925	Alloy 7	ILZRO 16
Mechanical Properties				
Charpy impact strength, ¼-by-¼-in. bar:				
As cast, J(ft·lb)	58 (43)	65 (48)	...	...
After aging indoors 10 yr, J(ft·lb)	56 (41)	54 (40)	275 (40)	...
Tensile strength:				
As cast, MPa (ksi)	285(41)	330 (47.6)	285 (41)	230 to 235 (33 to 34)
After aging indoors 10 yr, MPa (ksi)	240 (35)	270 (39.3)	...	...
Elongation, % in 50 mm or 2 in.:				
As cast...................	10	7	14	5
After aging indoors 10 yr.....	16	13	...	...
Expansion, after aging indoors 10 yr at room temperature, μm/m......................	80	70	...	...
Other Properties and Constants of As-Cast Alloys				
Brinell hardness (HB)	82	91	76	75 to 77
Compressive strength, MPa (ksi)	415 (60)	600 (87)	...	...
Electrical conductivity, % IACS	27.5	26.5	...	...
Liquidus temperature, °C (°F)......................	387 (728)	386 (727)	...	417 (785)
Solidus temperature, °C (°F)....	381 (717)	380 (716)	...	415 (780)
Modulus of rupture, MPa (ksi)	655 (95)	725 (105)	...	...
Shear strength, MPa (ksi)	215 (31)	260 (38)	...	...
Specific heat, J/kg (Btu/lb)	420 (0.10)	420 (0.10)	...	...
Thermal conductivity, W/m·K (Btu/ft·h·°F)(a)	113 (65.3)	109 (62.9)	...	...
Thermal expansion, μm/m·K (μin./in.·°F)	27.4 (15.2)	27.4 (15.2)	...	...
Transverse deflection, mm (in.)	6.9 (0.27)	4.1 (0.16)	...	...
Density, Mg/m³ (lb/in.³)........	6.6 (0.238)	6.7 (0.242)	...	...

(a) At 18 °C (64 °F).

Magnesium and Magnesium Alloys

Compositions and Properties

Nominal compositions and typical room-temperature mechanical properties of magnesium alloys

Alloy	Al	Mn(a)	Composition Th	Zn	Zr	Others	Tensile strength MPa	ksi	Yield strength Tensile MPa	ksi	Compressive MPa	ksi	Bearing MPa	ksi	Elongation in 50 mm or 2 in., %	Shear strength MPa	ksi	Hardness, HRB(b)
Sand and Permanent Mold Castings																		
AM100A-T61	10.0	0.1	···	···	···	···	275	40	150	22	150	22	···	···	1	···	···	69
AZ63A-T6	6.0	0.15	···	3.0	···	···	275	40	130	19	130	19	360	52	5	145	21	73
AZ81A-T4	7.6	0.13	···	0.7	···	···	275	40	83	12	83	12	305	44	15	125	18	55
AZ91C-T6	8.7	0.13	···	0.7	···	···	275	40	195	21	145	21	360	52	6	145	21	66
AZ92A-T6	9.0	0.10	···	2.0	···	···	275	40	150	22	150	22	450	65	3	150	22	84
EZ33A-T5	···	···	···	2.7	0.6	3.3 RE	160	23	110	16	110	16	275	40	2	145	21	50
HK31A-T6	···	···	3.3	···	0.7	···	220	32	105	15	105	15	275	40	8	145	21	55
HZ32A-T5	···	···	3.3	2.1	0.7	···	185	27	90	13	90	13	255	37	4	140	20	57
K1A-F	···	···	···	···	0.7	···	180	26	55	8	···	···	125	18	1	55	8	···
QE22A-T6	···	···	···	···	0.7	2.5 Ag, 2.1 Di	260	38	195	28	195	28	···	···	3	···	···	80
QH21A-T6	···	···	60	···	0.7	2.5 Ag, 1.0 Di	275	40	205	30	···	···	···	···	4	···	···	···
ZE41A-T5	···	···	···	4.2	0.7	1.2 RE	205	30	140	20	140	20	350	51	3.5	160	23	62
ZE63A-T6	···	···	···	5.8	0.7	2.6 RE	300	44	190	28	195	28	···	···	10	···	···	60-85
ZH62A-T5	···	···	1.8	5.7	0.7	···	240	35	170	25	170	25	340	49	4	165	24	70
ZK51A-T5	···	···	···	4.6	0.7	···	205	30	165	24	165	24	325	47	3.5	160	23	65
ZK61A-T5	···	···	···	6.0	0.7	···	310	45	185	27	185	27	···	···	···	170	25	68
ZK61A-T6	···	···	···	6.0	0.7	···	310	45	195	28	195	28	···	···	10	180	26	70
Die Castings																		
AM60A-F	6.0	0.13	···	···	···	···	205	30	115	17	115	17	···	···	6	···	···	···
AS41A-F(d)	4.3	0.35	···	···	···	1.0 Si	220	32	150	22	150	22	···	···	4	···	···	···
AZ91A and B-F(e)	9.0	0.13	···	0.7	···	···	230	33	150	22	165	24	···	···	3	140	20	63
Extruded Bars and Shapes																		
AZ10A-F	1.2	0.2	···	0.4	···	···	240	35	145	21	69	10	···	···	10	···	···	···
AZ21X1-F	1.8	0.02	···	1.2	···	···	···								···	···	···	···
AZ31 B and C-F(d)	3.0	···	···	1.0	···	···	260	38	200	29	97	14	230	33	15	130	19	49
AZ61A-F	6.5	···	···	1.0	···	···	310	45	230	33	130	19	285	41	16	140	20	60
AZ80A-T5	8.5	···	···	0.5	···	···	380	55	275	40	240	35	···	···	7	165	24	82
HM31A-F	···	1.2	3.0	···	···	···	290	42	230	33	185	27	345	50	10	150	22	···
M1A-F	···	1.2	···	···	···	···	255	37	180	26	83	12	195	28	12	125	18	44
ZK21A-F	···	···	···	2.3	0.45(a)	···	260	38	195	28	135	20	···	···	4	···	···	···
ZK40A-T5	···	···	···	4.0	0.45(a)	···	276	40	255	37	140	20	···	···	4	···	···	···
ZK60A-T5	···	···	···	5.5	0.45(a)	···	365	53	305	44	250	36	405	59	11	180	26	88
Sheet and Plate																		
AZ31B-H24	3.0	···	···	1.0	···	···	290	42	220	32	180	26	325	47	15	160	23	73
HK31A-H24	···	···	3.0	···	0.6	···	255	33	200	29	160	23	285	41	9	140	20	68
HM21A-T8	···	0.6	2.0	···	···	···	235	34	170	25	130	19	270	39	11	125	18	···
PE(f)	3.3	···	···	0.7	···	···	···								···	···	···	···

(a) Minimum. (b) 500-kg load, 10-mm ball. (c) A and B are identical except that 0.30% max residual Cu is allowable in AZ91B. (d) For battery applications. (e) Properties of B and C are identical, but AZ31C has 0.15 min Mn, 0.1 max Cu and 0.03 max Ni. (f) Photoengraving grade.

Some low-temperature tensile properties of various magnesium alloys

Alloy	Thickness mm	in.	Tensile strength MPa	ksi	Yield strength MPa	ksi	Elongation, %
Transverse Tests of Plate Alloys at 24 °C (75 °F)							
HK31A-H24	6.35	0.250	240	35.2	180	25.9	21.0
HK31A-O	6.35	0.250	200	29.0	125	18.0	30.5
HM21A-T8	6.35	0.250	240	35.0	170	24.8	13.7
Longitudinal Tests of Sheet and Plate Alloys at 24 °C (75 °F)							
HK31A-H24	1.63	0.064	250	36.3	200	29.0	7.5
HK31A-H24	6.35	0.250	240	34.5	190	27.3	14.2
Welded(a)	6.35	0.250	200	28.8	150	21.7	2.4
HK31A-O	1.63	0.064	205	29.7	125	17.9	27.5
HK31A-O	6.35	0.250	200	28.9	120	17.7	29.7
Welded(a)	6.35	0.250	160	23.4	120	17.3	3.2
HM21A-T5	(b)	(b)	210	30.4	105	15.5	8.0
HM21A-T8	1.63	0.064	220	32.2	160	23.1	7.2
HM21A-T8	6.35	0.250	235	32.4	175	25.1	5.6
Welded(a)	6.35	0.250	195	28.6	130	18.6	2.7
Longitudinal Tests of Sheet and Plate Alloys at −54 °C (−65 °F)							
HK31A-H24	1.63	0.064	300	43.3	220	32.0	5.0
....	6.35	0.250	280	40.8	230	33.4	9.0
HK31A-O	1.63	0.064	275	39.9	150	21.4	20.7
....	6.35	0.250	265	38.3	150	21.5	18.0
HM21A-T5	(b)	(b)	270	39.5	110	15.8	9.3
HM21A-T8	1.63	0.064	275	39.6	175	25.6	6.2
....	6.35	0.250	265	38.4	205	29.7	4.7
Longitudinal Tests of Sheet and Plate Alloys at −72 °C (−98 °F)							
HK31A-H24	1.63	0.064	295	42.7	210	30.6	4.2
HK31A-H24	6.35	0.250	290	42.4	235	33.8	11.5
Welded(a)	6.35	0.250	195	28.1	165	23.6	0.5
HK31A-O	1.63	0.064	285	41.3	145	21.1	17.5
HK31A-O	6.35	0.250	275	40.2	150	21.9	20.2
Welded(a)	6.35	0.250	205	29.4	145	21.0	2.2
HM21A-T5	(b)	(b)	275	40.0	110	16.3	8.3
HM21A-T8	1.63	0.064	280	40.8	150	22.1	17.5
....	6.35	0.250	275	40.1	215	31.3	5.0
....	6.35	0.250	200	29.2	120	17.5	1.5
Longitudinal Tests of Sheet and Plate Alloys at −196 °C (−320 °F)							
HK31A-H24	1.63	0.064	370	54.0	225	33.0	6.2
HK31A-H24	6.35	0.250	365	52.9	240	34.7	8.0
Welded(a)	6.35	0.250	230	33.7	180	25.9	1.5
HK31A-O	1.63	0.064	330	47.9	170	24.3	12.7
HK31A-O	6.35	0.250	325	47.2	170	24.7	12.5
Welded(a)	6.35	0.250	205	29.7	150	21.6	2.2
HM21A-T5		(b)	320	46.6	125	18.1	8.0
HM21A-T8	1.63	0.064	330	47.6	170	24.9	4.0
HM21A-T8	6.35	0.250	325	47.3	210	30.6	4.2
Welded(a)	6.35	0.250	330	33.1	145	20.9	1.5

(a) Welding rod was EZ33A; weld bead intact. (b) Specimen machined from a forging.
NOTE: Values for wrought alloys are averages of two to four tests at room temperature (2-in. gage length). Values of duplicate tests at low temperatures are also averages (1-in. gage length). Values for cast alloys are averages of two to four tests on separately cast bars.

Some low-temperature tensile properties of various magnesium alloys (continued)

Alloy	Tensile strength MPa	ksi	Yield strength MPa	ksi	Elongation, %	Charpy impact (a) J	ft·lb	(b) J	ft·lb
Cast Alloys at 24 °C (75 °F)									
AZ91C-T6 ...	290	41.8	130	19.2	6.3	7.96	5.87	1.36	1.00
AZ92A-T6 ...	290	41.8	160	23.4	4.0	7.62	5.62	0.68	0.50
EZ33A-T5 ...	190	27.5	115	16.9	7.6	7.46	5.50	0.84	0.62
HK31A-T6 ...	225	32.7	110	16.3	9.5	16.61	12.25	3.80	2.81
ZH62A-T5 ...	275	39.9	190	27.9	5.7	15.02	11.08	1.02	0.75
Cast Alloys at −78 °C (−109 °F)									
AZ91C-T6 ...	305	44.3	150	21.6	5.1	6.26	4.62	1.36	1.00
AZ92A-T6 ...	295	42.7	170	24.6	2.3	6.44	4.75	0.76	0.56
EZ33A-T5 ...	190	27.6	125	18.0	3.1	4.83	3.56	0.68	0.50
HK31A-T6 ...	300	43.3	120	17.5	8.6	16.43	12.12	3.21	2.37
ZH62A-T5 ...	330	47.6	200	29.2	2.7	18.99	14.00	1.02	0.75
Cast Alloys at −196 °C (−321 °F)									
AZ91C-T6 ...	310	44.9	180	26.0	1.7	4.06	3.00	1.02	0.75
AZ92A-T6 ...	320	46.5	195	28.5	0.8	4.57	3.37	0.68	0.50
EZ33A-T5 ...	200	29.0	140	20.3	2.2	5.00	3.69	0.68	0.50
HK31A-T6 ...	330	48.1	135	19.6	6.1	13.72	10.12	3.05	2.25
ZH62A-T5 ...	320	46.6	235	34.1	1.0	8.56	6.31	1.02	0.75

(a) Unnotched specimens. (b) Notched specimens.
NOTE: Values for wrought alloys are averages of two to four tests at room temperature (2-in. gage length). Values of duplicate tests at low temperatures are also averages (1-in. gage length). Values for cast alloys are averages of two to four tests on separately cast bars.

Effect of elevated temperature on values of creep stress and elastic modulus for magnesium alloys

Alloy	Creep stress(a) at 205 °C (400 °F) MPa	ksi	315 °C (600 °F) MPa	ksi	Elastic modulus at 205 °C (400 °F) GPa	10^6 psi	315 °C (600 °F) GPa	10^6 psi
Castings								
AZ92A-T6	3.4	0.5	...	...	31	4.5	21	3.0
EZ33A-T5	38	5.5	6.9	1.0	40	5.8	38	5.5
HK31A-T6	64	9.3	14	2.0	40	5.8	39	5.6
HZ32A-T5	52	7.5	22	3.2	40	5.8	39	5.6
ZH62A-T5	17	2.5	...	...	40	5.8	38	5.5
Extrusions								
ZK60A-T5	7	1.0(b)	...	...	...	...	...	...
HM31A-F	83	12.0	41	6.0	40	5.8	38	5.5
Sheet								
AZ31B-H24 ...	7	1.0(b)	...	...	30	4.3	17	2.5
HK31A-T6	69	10.0	17	2.5	40	5.8	25	3.6
HM21A-T8	76	11.0	34	5.0	40	5.8	34	5.0

(a) Stress to produce 0.2% total extension in 1000 h for cast alloys and 100 h for wrought alloys.
(b) Tested at 150 °C (300 °F).

Effect of elevated temperature on tensile strength of magnesium alloys

Alloy	20 °C MPa	(70 °F) ksi	Exposed 10 min at 150 °C MPa	(300 °F) ksi	315 °C MPa	(600 °F) ksi	205 °C MPa	(400 °F) ksi	315 °C MPa	(600 °F) ksi	205 °C MPa	(400 °F) ksi	Exposed 1000 h at 315 °C MPa	(600 °F) ksi
					Tested at exposure temperature								Tested at room temperature	
									Exposed 1000 h at				Exposed 1000 h at	
Castings														
AZ63A-T6	275	40	165	24	55	8	110	16	...	...	255	37	...	...
AZ92A-T6	275	40	195	28	55	8	115	17	...	...	270	39	...	...
EZ33A-T5	160	23	145	21	83	12	130	19	76	11	170	25	180	26
HK31A-T6	215	31	195	28	125	18	180	26	62	9	240	35	180	26
HZ32A-T5	200	29	145	21	83	12	115	17	76	11	220	32	235	34
ZH62A-T5	290	42	195	28	69	10	...	...	...	...	...	...	...	...
QH21A-T6	275	40	235	34	97	14	...	...	...	...	...	...	...	...
Extrusions														
AZ80A-T5	380	55	235	34	69	10	...	...	...	...	...	...	...	...
ZK60A-T5	365	53	180	26	41	6	...	...	...	...	315	46	315	46
HM31A-F	275	40	195(a)	28(a)	115	17	...	...	...	...	...	...	...	...
Sheet														
AZ31B-H24	285	41	145	21	48	7	90	13	62(a)	9(a)	255	37	260	38
HK31A-T6	255	37	180	26	115	17	...	...	55	8	255	37	215	31
HM21A-T8	235	34	140	20	97	14	...	...	...	...	...	...	...	...

(a) Tested at 260 °C (500 °F).

Heat Treating

Annealing temperatures for wrought magnesium alloys

Alloy	Original temper	Annealing temperature(a) °C	°F
AZ31B	F, H10, H11, H23, H24, H26	345	650
AZ31C	F	345	650
AZ61A	F	345	650
AZ80A	F, T5, T6	385	725
HK31A	H24	400	750
HM21A	T5, T8, T81	455	850
HM31A	T5	455	850
ZK60A	F, T5, T6	290	550

(a) Time at temperature, 1 h or more

Recommended stress-relieving treatments for wrought magnesium alloys

Alloy	Annealed Temperature °C	°F	Time, min	Hard rolled Temperature °C	°F	Time, min	Extrusions and forgings Temperature °C	°F	Time, min
				Sheet					
AZ31B	345	650	120	150	300	60	...	...	...
AZ31B-F	...	...	...	...	...	...	260	500	15
AZ61A	345	650	120	205	400	60	...	...	...
AZ61A-F	...	...	...	...	...	...	260	500	15
AZ80A-F	...	...	...	...	...	...	260	500	15
AZ80A-T5	...	...	...	...	...	...	205	400	60
HK31A	345	650	60	290	550	30	...	...	...
HM21A-T5	...	...	...	...	...	...	370	700	30
HM21A-T8	...	...	...	370	700	30	...	...	...
HM21A-T81	...	...	...	400	750	30	...	...	...
HM31A-T5	...	...	...	...	...	...	425	800	60
ZK60A-F	230	450	180	...	...	...	260	500	15
ZK60A-T5	...	...	...	...	...	...	150	300	60

Note: Stress relieving after welding, to prevent stress-corrosion cracking, is necessary only for alloys that contain more than 1.5% aluminum.

Heat treatments commonly applied to magnesium alloys

Alloy	Heat treatment(a)	Alloy	Heat treatment(a)
Casting alloys			
AM100A	T4, T5, T6, T61(b)	ZE41A	T5
		ZE63A	T6(c)
AZ63A	T4, T5, T6	ZH62A	T5
AZ81A	T4	ZK51A	T5
AZ91C	T4, T6	ZK61A	T4, T6
AZ92A	T4, T6	**Wrought alloys**	
EZ33A	T5	AZ80A	T5
HK31A	T6	HM21A	T5, T8, T81(d)
HZ32A	T5		
QE22A	T6	HM31A	T5
QH21A	T6	ZK60A	T5

(a) Indicated by temper designations (see Table 1). (b) Same as T6 except aged for longer time to increase yield strength. (c) Thermal treatment must include hydriding. (d) Mill modification of T8 to improve mechanical properties

Recommended solution-treating and aging schedules for magnesium alloy castings

For castings up to 51 mm (2 in.) in section thickness; heavier sections may require longer times at temperature.

Alloy	Final temper	Aging(a) Temperature °C, ±6(b)	°F, ±10(b)	Time, h	Solution treating(c) Temperature °C, ±6(b)	°F, ±10(b)	Time, h	Maximum temperature °C	°F	Aging after solution treating Temperature °C, ±6(b)	°F, ±10(b)	Time, h
Magnesium-aluminum-zinc alloys(d)												
AM100A	T5	232	450	5	...	...	...	...	...	...	...	...
	T4	...	...	...	424(e)	795(e)	16–24(e)	432	810	...	...	...
	T6	...	...	...	424(e)	795(e)	16–24(e)	432	810	232	450	5
	T61	...	...	...	424(e)	795(e)	16–24(e)	432	810	218	425	25
AZ63A	T5	260(f)	500(f)	4(f)	...	...	...	...	...	...	...	...
	T4	...	...	...	385	725	10–14	391	735	...	...	...
	T6	...	...	...	385	725	10–14	391	735	218(f)	425(f)	5(f)
AZ81A	T4	...	...	...	413(e)	775(e)	16–24(e)	418	785	...	...	...
AZ91C	T5	168(g)	335(g)	16(g)	...	...	...	...	...	...	...	...
	T4	...	...	...	413(e)	775(e)	16–24(e)	418	785	...	...	...
	T6	...	...	...	413(e)	775(e)	16–24(e)	418	785	168(h)	335(h)	16(h)
AZ92A	T5	260	500	4	...	...	...	...	...	...	...	...
	T4	...	...	...	407(j)	765(j)	16–24(j)	413	775	...	...	...
	T6	...	...	...	407(j)	765(j)	16–24(j)	413	775	218	425	5
Magnesium-zirconium alloys												
EZ33A	T5	216(k)	420(k)	5(k)	...	...	...	...	...	...	...	...
HK31A(m)	T6	...	...	...	566	1050	2	571	1060	204	400	16
HZ32A	T5	316	600	16	...	...	...	...	...	...	...	...
QE22A(n)	T6	...	...	...	527	980	4–8	538	1000	204	400	8
QH21A(n)	T6	...	...	...	527	980	4–8	538	1000	204	400	8
ZE41A	T5	329(p)	625(p)	2(p)	...	...	...	...	...	...	...	...
ZE63A(q)	T6	...	...	...	480	895	10–72	491	915	141	285	48
ZH62A	T5	329	625	2	...	...	...	...	...	...	...	...
	plus:	177	350	16	...	...	...	...	...	...	...	...
ZK51A	T5	177(r)	350(r)	12(r)	...	...	...	...	...	...	...	...
ZK61A	T5	149	300	48	...	...	...	...	...	...	...	...
	T6	...	...	...	499(s)	930(s)	2(s)	502	935	129	265	48

(a) Aging of castings to the T5 temper is done from the as-cast condition. (b) Except where quoted differently. (c) After solution treatment and before subsequent aging, castings are cooled to room temperature by fast fan cooling, except where otherwise indicated. Use carbon dioxide or sulfur dioxide atmosphere above 400 °C (750 °F). (d) For solution treating, Mg-Al-Zn alloys are loaded into the furnace at 260 °C (500 °F) and brought to temperature over a 2-h period at a uniform rate of temperature increase. (e) Alternative treatment, to prevent germination (excessive grain growth): 6 h at 413 ± 6 °C (775 ± 10 °F), 2 h at 352 ± 6 °C (665 ± 10 °F), 10 h at 413 ± 6 °C (775 ± 10 °F). (f) Alternative treatment: 5 h at 232 ± 6 °C (450 ± 10 °F). (g) Alternative treatment: 4 h at 216 ± 6 °C (420 ± 10 °F). (h) Alternative treatment: 5–6 h at 216 ± 6 °C (420 ± 10 °F). (j) Alternative treatment, to prevent germination (excessive grain growth): 6 h at 407 ± 6 °C (765 ± 10 °F), 2 h at 352 ± 6 °C (665 ± 10 °F), 10 h at 407 ± 6 °C (765 ± 10 °F). (k) Alternative treatment, which can be used where maximum resistance to creep at elevated temperature is not of prime importance: 2 h at 343 ± 6 °C (650 ± 10 °F). (m) Alloy HK31A castings must be loaded into the furnace already at temperature and brought back to temperature as quickly as possible. (n) Quench from solution-treating temperature either in water at 65 °C (150 °F) or in other suitable quenching medium. (p) This treatment is adequate for development of satisfactory properties; it may be followed by 16 h at 177 ± 6 °C (350 ± 10 °F), to provide very slight improvements in mechanical properties. (q) Alloy ZE63A must be solution treated in a special hydrogen atmosphere, because its mechanical properties are developed through hydriding of some of its alloying elements. Hydriding time depends on section thickness; as a guide, 6.4-mm (1/4-in.) sections require approximately 10 h, and 19-mm (3/4-in.) sections require about 72 h. Following solution treatment, ZE63A should be quenched in oil, water spray or air blast. (r) Alternative treatment: 8 h at 218 ± 6 °C (425 ± 10 °F). (s) Alternative treatment: 10 h at 482 ± 6 °C (900 ± 10 °F).

Forming

Recommended minimum radii for 90° bends in magnesium sheet(a)

Alloy and temper	Forming temperature(b)							
	20 °C (70 °F)	95 °C (200 °F)	150 °C (300 °F)	205 °C (400 °F)	260 °C (500 °F)	315 °C (600 °F)	370 °C (700 °F)	425 °C (800 °F)
AZ31B-O 5.5*t*	5.5*t*	4*t*	3*t*	2*t*	...	...	...
AZ31B-H24 .... 8*t*	8*t*	6*t*	3*t*	2*t*	...	...	...
HK31A-O 6*t*	6*t*	6*t*	5*t*	4*t*	3*t*	2*t*	1*t*
HK31A-H24 ... 13*t*	13*t*	13*t*	9*t*	8*t*	5*t*	3*t*	...
HM21A-T8 ..... 9*t*	9*t*	9*t*	9*t*	9*t*	8*t*	6*t*	4*t*

(a) Numerical values of bend radii are given as multiples of sheet thickness.

Joining

Weldability of magnesium alloys

Alloy	Thickness		Welding rod	Joint efficiency, %	Joint ductility(a)
	mm	in.			
AZ31B-O	1.63	0.064	AZ61A, AZ92A	97	12.0
AZ31B-H24	1.63	0.064	AZ61A, AZ92A	88	10.0
ZE10A-O	1.63	0.064	AZ61A, AZ92A	94	7.0
ZE10A-H24	1.63	0.064	AZ61A, AZ92A	87	3.0
M1A-F	3.17	0.125	M1A	55	2.0
AZ31B-F	3.17	0.125	AZ61A, AZ92A	92	12.0
AZ61A-F	3.17	0.125	AZ61A, AZ92A	89	8.0
AZ80A-F	3.17	0.125	AZ61A, AZ92A	86	4.0
AZ63A-F	12.70	0.5	AZ63A	83	2.5
AZ63A-T4	12.70	0.5	AZ63A	70	5.0
AZ63A-T6	12.70	0.5	AZ63A	75	2.0
AZ92A-F	12.70	0.5	AZ92A	100	2.5
AZ92A-T4	12.70	0.5	AZ92A	70	4.0
AZ92A-T6	12.70	0.5	AZ92A	75	2.0
AZ91C-F	12.70	0.5	AZ92A	100	2.5
AZ91C-T4	12.70	0.5	AZ92A	78	4.0
AZ91C-T6	12.70	0.5	AZ92A	75	2.0
AZ81A-F	12.70	0.5	AZ92A	100	2.5
AZ81A-T4	12.70	0.5	AZ92A	85	8.0
EK41A-T5	12.70	0.5	EK41A	100	1.0
EK41A-T6	12.70	0.5	EK41A	93	6.2
EZ33A-T5	12.70	0.5	EZ33A	100	1.1
HK31A-T6	12.70	0.5	HK31A	100	9.5
HK31A-H24	...	...	EZ33A	83	1.0
HZ32A-T5	12.70	0.5	HZ32A	93	3.8
HM21A-T8	1.63	0.064	EZ33A	88	1.5
	...	...	HM31A	74	1.5
HM31A-F	15.88	0.625	EZ33A	71	1.8
			HM31A	58	2.5

(a) Percentage elongation across the weld over a 50-mm or 2-in. gage length from tension tests.

Titanium and Titanium Alloys

Compositions and Properties

Summary of commercial and semicommercial grades and alloys of titanium

Designation	Tensile strength (min) MPa	ksi	0.2% yield strength (min) MPa	ksi	N (max)	C (max)	H (max)	Fe (max)	O (max)	Al	Sn	Zr	Mo	Others
Unalloyed grades														
ASTM Grade 1	240	35	170	25	0.03	0.10	0.015	0.20	0.18	...	...	...	...	...
ASTM Grade 2	340	50	280	40	0.03	0.10	0.015	0.30	0.25	...	...	...	...	...
ASTM Grade 3	450	65	380	55	0.05	0.10	0.015	0.30	0.35	...	...	...	...	...
ASTM Grade 4	550	80	480	70	0.05	0.10	0.015	0.50	0.40	...	...	...	...	...
ASTM Grade 7	340	50	280	40	0.03	0.10	0.015	0.30	0.25	...	...	...	...	0.2Pd
Alpha and near-alpha alloys														
Ti Code 12	480	70	380	55	0.03	0.10	0.015	0.30	0.25	...	...	...	0.3	0.8Ni
Ti-5Al-2.5Sn	790	115	760	110	0.05	0.08	0.02	0.50	0.20	5	2.5	...	...	...
Ti-5Al-2.5Sn-ELI	690	100	620	90	0.07	0.08	0.0125	0.25	0.12	5	2.5	...	...	...
Ti-8Al-1Mo-1V	900	130	830	120	0.05	0.08	0.015	0.30	0.12	8	...	...	1	1V
Ti-6Al-2Sn-4Zr-2Mo	900	130	830	120	0.05	0.05	0.0125	0.25	0.15	6	2	4	2	...
Ti-6Al-2Nb-1Ta-0.8Mo	790	115	690	100	0.02	0.03	0.0125	0.12	0.10	6	...	...	1	2Nb, 1Ta
Ti-2.25Al-11Sn-5Zr-1Mo	1000	145	900	130	0.04	0.04	0.008	0.12	0.17	2.25	11.0	5.0	1.0	0.2Si
Ti-5Al-5Sn-2Zr-2Mo(a)........	900	130	830	120	0.03	0.05	0.0125	0.15	0.13	5	5	2	2	0.25Si
Alpha-beta alloys														
Ti-6Al-4V(b)	900	130	830	120	0.05	0.10	0.0125	0.30	0.20	6.0	...	...	...	4.0V
Ti-6Al-4V-ELI(b).............	830	120	760	110	0.05	0.08	0.0125	0.25	0.13	6.0	...	...	...	4.0V
Ti-6Al-6V-2Sn(b).............	1030	150	970	140	0.04	0.05	0.015	1.0	0.20	6.0	2.0	...	...	0.75Cu, 6.0V
Ti-8Mn(b)	860	125	760	110	0.05	0.08	0.015	0.50	0.20	...	...	...	...	8.0Mn
Ti-7Al-4Mo(b)	1030	150	970	140	0.05	0.10	0.013	0.30	0.20	7.0	...	...	4.0	...
Ti-6Al-2Sn-4Zr-6Mo(c)........	1170	170	1100	160	0.04	0.04	0.0125	0.15	0.15	6.0	2.0	4.0	6.0	...
Ti-5Al-2Sn-2Zr-4Mo-4Cr(a)(c)..	1125	163	1055	153	0.04	0.05	0.0125	0.30	0.13	5.0	2.0	2.0	4.0	4.0Cr
Ti-6Al-2Sn-2Zr-2Mo-2Cr(a)(b) .	1030	150	970	140	0.03	0.05	0.0125	0.25	0.14	5.7	2.0	2.0	2.0	2.0Cr, 0.25Si
Ti-10V-2Fe-3Al(a)(c)	1170	170	1100	160	0.05	0.05	0.015	2.5	0.16	3.0	...	...	...	10.0V
Ti-3Al-2.5V(d)...............	620	90	520	75	0.015	0.05	0.015	0.30	0.12	3.0	...	...	...	2.5V
Beta alloys														
Ti-13V-11Cr-3Al(c)...........	1170	170	1100	160	0.05	0.05	0.025	0.35	0.17	3.0	...	...	...	11.0Cr, 13.0V
Ti-8Mo-8V-2Fe-3Al(a)(c)	1170	170	1100	160	0.05	0.05	0.015	2.5	0.17	3.0	...	...	8.0	8.0V
Ti-3Al-8V-6Cr-4Mo-4Zr(a)(b) ..	900	130	830	120	0.03	0.05	0.020	0.25	0.12	3.0	...	4.0	4.0	6.0Cr, 8.0V
Ti-11.5Mo-6Zr-4.5Sn(b)	690	100	620	90	0.05	0.10	0.020	0.35	0.18	...	4.5	6.0	11.5	...

(a) Semicommercial alloy; mechanical properties and composition limits subject to negotiation with suppliers. (b) Mechanical properties given for annealed condition; may be solution treated and aged to increase strength. (c) Mechanical properties given for solution treated and aged condition; alloy not normally applied in annealed condition. Properties may be sensitive to section size and processing. (d) Primarily a tubing alloy; may be cold drawn to increase strength.

AMS specifications for titanium and titanium alloys

AMS No.	Mill form	Condition	Alloy	Similar MIL specification
4900	Plate, sheet, strip	Annealed	Unalloyed; 55-ksi YS	MIL-T-9046
4901	Plate, sheet, strip	Annealed	Unalloyed; 70-ksi YS	MIL-T-9046
4902	Plate, sheet, strip	Annealed	Unalloyed; 40-ksi YS	MIL-T-9046
4906	Sheet, strip; continuously rolled	Annealed	Ti-6Al-4V	. . .
4907	Plate, sheet, strip	Annealed	Ti-6Al-4V-ELI	MIL-T-9046
4908	Sheet, strip	Annealed	Ti-8Mn; 110-ksi YS	MIL-T-9046
4909	Plate, sheet, strip	Annealed	Ti-5Al-2.5Sn-ELI	MIL-T-9046
4910	Plate, sheet, strip	Annealed	Ti-5Al-2.5Sn	MIL-T-9046
4911	Plate, sheet, strip	Annealed	Ti-6Al-4V	MIL-T-9046
4915	Plate, sheet, strip	Single annealed	Ti-8Al-1Mo-1V	MIL-T-9046
4916	Plate, sheet, strip	Duplex annealed	Ti-8Al-1Mo-1V	MIL-T-9046
4917	Plate, sheet, strip	Solution treated	Ti-13V-11Cr-3Al	MIL-T-9046
4918	Plate, sheet, strip	Annealed	Ti-6Al-6V-2Sn	MIL-T-9046
4921	Bar, forgings, rings	Annealed	Unalloyed; 70-ksi YS	MIL-T-9047
4924	Bar, forgings, rings	Annealed	Ti-5Al-2.5Sn-ELI; 90-ksi YS	MIL-T-9047
4926	Bar, rings	Annealed	Ti-5Al-2.5Sn; 110-ksi YS	MIL-T-9047
4928	Bar, forgings	Annealed	Ti-6Al-4V; 120-ksi YS	MIL-T-9047
4930	Bar, forgings, rings	Annealed	Ti-6Al-4V-ELI	MIL-T-9047
4935	Extrusions	Annealed	Ti-6Al-4V	. . .
4936	Extrusions	. . .	Ti-6Al-6V-2Sn	. . .
4941	Tubing, welded	Annealed	Unalloyed; 40-ksi YS	. . .
4942	Tubing, seamless	Annealed	Unalloyed; 40-ksi YS	. . .
4943	Tubing, seamless	Annealed	Ti-3Al-2.5V	. . .
4944	Tubing, seamless hydraulic	Cold worked and stress relieved	Ti-3Al-2.5V	. . .
4951	Wire, welding	. . .		. . .
4953	Wire, welding	Annealed	Ti-5Al-2.5Sn	. . .
4954	Wire, welding	. . .	Ti-6Al-4V	. . .
4955	Wire, welding	. . .	Ti-8Al-1Mo-1V	. . .
4956	Wire, welding	. . .	Ti-6Al-4V-ELI	. . .
4965	Bar, forgings, rings	Precipitation heat treated	Ti-6Al-4V	. . .
4966	Forgings	Annealed	Ti-5Al-2.5Sn; 110-ksi YS	MIL-F-83142
4967	Bar, forgings	Annealed	Ti-6Al-4V	MIL-T-9047
4970	Bar, forgings	Precipitation heat treated	Ti-7Al-4Mo	MIL-T-9047
4971	Bar, forgings, rings	Annealed	Ti-6Al-6V-2Sn	MIL-T-9047, MIL-F-83142
4972	Bar, rings	Solution treated and stabilized	Ti-8Al-1Mo-1V	. . .
4973	Forgings	Solution treated and stabilized	Ti-8Al-1Mo-1V	. . .
4974	Bar, forgings	Precipitation heat treated	Ti-11Sn-5Zr-2.3Al-1Mo-0.21Si	. . .
4975	Bar, rings	Precipitation heat treated	Ti-6Al-2Sn-4Zr-2Mo	MIL-T-9047
4976	Forgings	Precipitation heat treated	Ti-6Al-2Sn-4Zr-2Mo	. . .
4977	Bar, wire	Solution treated	Ti-11.5Mo-6Zr-4.5Sn	MIL-T-9047
4978	Bar, forgings, rings	Annealed	Ti-6Al-6V-2Sn; 140-ksi YS	MIL-T-9047, MIL-F-83142
4979	Bar, forgings, rings	Precipitation heat treated	Ti-6Al-6V-2Sn	MIL-T-9047, MIL-F-83142
4980	Bar, wire	Solution treated at 745 °C (1375 °F)	Ti-11.5Mo-6Zr-4.5Sn	. . .
4981	Bar, forgings	Precipitation heat treated	Ti-6Al-2Sn-4Zr-6Mo	MIL-T-9047

ASTM specifications for titanium and titanium alloys

Specification(b)	Grade	Alloy	Min 0.2% yield strength(a) MPa	ksi	Similar AMS specification(b)
Plate, sheet and strip					
B265	1	Unalloyed	170	25	. . .
	2	Unalloyed	280	40	4902
	3	Unalloyed	380	55	4900
	4	Unalloyed	480	70	4901
	5	Ti-6Al-4V	830	120	4911
	6	Ti-5Al-2.5Sn	790	115	4910
	7	Ti-0.2Pd	280	40	
	10	Ti-4.5Sn-11.5Mo-6Zr	620	90	4977
	11	Ti-0.2Pd	170	25	
Seamless and welded pipe					
B337	1	Unalloyed	170	25	. . .
	2	Unalloyed	280	40	4941(c), 4942(d)
	3	Unalloyed	380	55	. . .
	7	Ti-0.2Pd	280	40	. . .
	9	Ti-3Al-2.5V	480	70	4943
	9	Ti-3Al-2.5V	720(e)	105(e)	4943
	10	Ti-11.5Mo-6Zr-4.5Sn	620(f)	90(f)	4977, 4980
	11	Ti-0.2Pd	170	25	. . .
Seamless and welded tube for condensers and heat exchangers					
B338	1	Unalloyed	170	25	. . .
	2	Unalloyed	280	40	. . .
	3	Unalloyed	380	55	. . .
	7	Ti-0.2Pd	280	40	. . .
	9	Ti-3Al-2.5V	720	105	4943, 4944
	10	Ti-11.5Mo-6Zr-4.5Sn	620	90	4977, 4980
	11	Ti-0.2Pd	170	25	. . .
Bar and billet					
B348	1	Unalloyed	170	25	. . .
	2	Unalloyed	280	40	. . .
	3	Unalloyed	380	55	. . .
	4	Unalloyed	480	70	4921(g)
	5	Ti-6Al-4V	830	120	4928(g)
	6	Ti-5Al-2.5Sn	790	115	4926(g)
	7	Ti-0.2Pd	280	40	. . .
	10	Ti-4.5Sn-11.5Mo-6Zr	620	90	4977, 4980
	11	Ti-0.2Pd	170	25	. . .
Castings					
B367	C-1	Unalloyed	170	25	. . .
	C-2	Unalloyed	280	40	. . .
	C-3	Unalloyed	380	55	. . .
	C-4	Unalloyed	480	70	. . .
	C-5	Ti-6Al-4V	830	120	. . .
	C-6	Ti-5Al-2.5Sn	720	105	. . .
	C-7A	Ti-0.2Pd	170	25	. . .
	C-7B	Ti-0.2Pd	280	40	. . .
	C-8A	Ti-0.2Pd	380	55	. . .
	C-8B	Ti-0.2Pd	480	70	. . .
Forgings					
B381	F1	Unalloyed	170	25	. . .
	F2	Unalloyed	280	40	. . .
	F3	Unalloyed	380	55	. . .
	F4	Unalloyed	480	70	4921
	F5	Ti-6Al-4V	830	120	4928
	F6	Ti-5Al-2.5Sn	790	115	4966
	F7	Ti-0.2Pd	280	40	. . .
	F11	Ti-0.2Pd	170	25	. . .

(a) Annealed. (b) Interstitial and impurity levels, and mechanical property requirements, may show minor differences compared with ASTM specifications. (c) Welded tubing. (d) Seamless tubing. (e) Cold worked and stress relieved. (f) Solution treated. (g) AMS specifications cover bar and forgings but not billet.

Specifications and applications of wrought titanium alloys

Nominal composition, %	Condition	ASTM No.	Military	Bars and forgings	Sheet and plate	Tubing	Wire	Applications and characteristics
Commercially pure								
99.5 Ti	Annealed	B 265 (Gr. 1) B 348 (Gr. 1) B 381 (Gr. 1)	··· ··· ···	··· ··· ···	··· ··· ···	··· ··· ···	··· ··· ···	Airframes; chemical, desalination, and marine parts; plate-type heat exchangers; cold spun or pressed parts; platinized anodes; high formability.
99.2 Ti	Annealed	B 265 (Gr. 2) B 348 (Gr. 2) B 381 (Gr. 2)	MIL-T-9046 ··· ···	··· ··· ···	4902 ··· ···	4941 4942 ···	4951 ··· ···	Airframes; aircraft engines; marine and chemical parts; heat exchangers; condenser and evaporator tubing; high formability.
99.1 Ti	Annealed	B 265 (Gr. 3) B 348 (Gr. 3) B 381 (Gr. 3)	MIL-T-9046 ··· ···	··· ··· ···	4900 ··· ···	··· ··· ···	··· ··· ···	Chemical, marine, airframe, and aircraft engine parts which require formability strength, weldability, and corrosion resistance.
99.0 Ti	Annealed	B-265 (Gr. 4) B-348 (Gr. 4) B-381 (Gr. 4)	MIL-T-9046 MIL-T-9047 ···	4921 ··· ···	4901 ··· ···	··· ··· ···	··· ··· ···	Chemical, marine, airframe, and aircraft engine parts; surgical implants; high speed fans; gas compressors; good formability and corrosion resistance, high strength.
99.2 Ti(a)	Annealed	B-265 (Gr. 7) B-348 (Gr. 7) B-381 (Gr. 7)	··· ··· ···	··· ··· ···	··· ··· ···	··· ··· ···	··· ··· ···	Good corrosion resistance for chemical industry applications where media is mildly reducing or varies between oxidizing and reducing.
98.9(b)	Annealed	···	···	···	···	···	···	Same as 0.2 Pd alloy (above).
Alpha alloys								
5 Al, 2.5 Sn	Annealed	B-265 (Gr. 6) B-348 (Gr. 6) B-381 (Gr. 6)	MIL-T-9046 MIL-T-9047 ···	4926 4966 ···	4910 ··· ···	··· ··· ···	4953 ··· ···	Weldable alloy for forgings and sheet metal parts such as aircraft engine compressor blades and ducting; steam turbine blades; good oxidation resistance and strength at 315 to 595 °C (600 to 1100 °F); good stability at elevated temperatures.
5 Al, 2.5 Sn (low O₂)	Annealed		MIL-T-9046 MIL-T-9047	4924 ···	4909 ···	··· ···	··· ···	Special grade for high-pressure cryogenic vessels operating down to −255 °C (−423 °F).
Near alpha								
8 Al, 1 Mo, 1 V	Duplex annealed	··· ···	MIL-T-9046 MIL-T-9047	4972 4973	4915 4916	··· ···	4955 ···	Airframe and jet engine parts requiring high strength to 455 °C (850 °F); good creep and toughness properties; good weldability.
11 Sn, 1 Mo, 2.25 Al, 5.0 Zr, 1 Mo, 0.2 Si		···	MIL-T-9047	4974	···	···	···	Airframes; blades, discs, wheels, spacers, and fasteners for turbine engines.
6 Al, 2 Sn, 4 Zr, 2 Mo		···	MIL-T-9046	4975	···	···	···	Parts and cases for jet-engine compressors; airframe skin components.
5 Al, 5 Sn, 2 Zr, 2 Mo, 0.25 Si	975 °C (1785 °F) (½ h), AC + 595 °C (1100 °F) (2 h), AC	···	···	···	···	···	···	Jet engine parts; high creep strength to 540 °C (1000 °F).
6 Al, 2 Cb, 1 Ta, 1 Mo	As rolled (1 in. plate)	···	MIL-T-9046	···	···	···	···	High toughness; moderate strength; good resistance to seawater and hot-salt stress corrosion; good weldability.
6 Al, 2 Sn, 1.5 Zr, 1 Mo, 0.35 Bi, 0.1 Si	Beta forge + duplex anneal	···		···	···	···	···	Jet engine discs and blades requiring extra creep resistance and stability.

(continued)

(a) Also contains 0.2 Pd. (b) Also contains 0.8 Ni and 0.3 Mo.
Source: Titanium Metals Corp. of America and RMI Co.

Specifications and applications of wrought titanium alloys (continued)

Nominal composition, %	Condition	ASTM No.	Military	Aerospace material specifications				Applications and characteristics
				Bars and forgings	Sheet and plate	Tubing	Wire	
Alpha-beta alloys								
8 Mn	Annealed	⋯	MIL-T-9046	⋯	4908	⋯	⋯	Aircraft sheet components, structural sections, and skins; good formability, moderate strength.
3 Al, 2.5 V	Annealed	⋯	⋯	⋯	⋯	4943 4944	⋯	Aircraft hydraulic tubing, foil; combines strength, weldability, and formability.
6 Al, 4 V	Annealed	B-265 (Gr. 5) B-348 (Gr. 5) B-381 (Gr. 5)	MIL-T-9046 MIL-T-9047 ⋯	4928 4965	4911 4906	⋯ ⋯ ⋯	4954 ⋯ ⋯	Rocket motor cases; blades and discs for aircraft turbines and compressors; structural forgings and fasteners; pressure vessels; gas and chemical pumps; cryogenic parts; ordnance equipment; marine components; steam-turbine blades.
	Solution + age							
6 Al, 4 V (low O_2)	Annealed	⋯ ⋯	MIL-T-9046 MIL-T-9047	4930 ⋯	4907 ⋯	⋯ ⋯	4956 ⋯	High pressure cryogenic vessels operating down to −195 °C (−320 °F).
6 Al, 6 V, 2 Sn	Annealed Solution + age	⋯ ⋯ ⋯	MIL-T-9046 MIL-T-9047 ⋯	4971 4978 4979	4918 ⋯ ⋯	⋯ ⋯ ⋯	⋯ ⋯ ⋯	Rocket motor cases; ordnance components; structural aircraft parts and landing gears; responds well to heat treatments; good hardenability.
7 Al, 4 Mo	Solution + age	⋯	MIL-T-9047	4970	⋯	⋯	⋯	Airframes and jet engine parts for operation at up to 425 °C (800 °F); missile forgings; ordnance equipment.
6 Al, 2 Sn, 4 Zr, 6 Mo	Solution + age	⋯	⋯	4981	⋯	⋯	⋯	Components for advanced jet engines.
6 Al, 2 Sn, 2 Zr, 2 Mo, 2 Cr, 0.25 Si	Solution + age	⋯	⋯	⋯	⋯	⋯	⋯	Strength, fracture toughness in heavy sections; landing gear wheels.
10 V, 2 Fe, 3 Al	Solution + age	⋯	⋯	⋯	⋯	⋯	⋯	Heavy airframe structural components requiring toughness at high strengths.
Beta alloys								
	Solution + age							High strength fasteners.
13 V, 11 Cr, 3 Al	Solution + age	⋯ ⋯	MIL-T-9046 MIL-T-9047	⋯ ⋯	4917 ⋯	⋯ ⋯	⋯ ⋯	High strength fasteners, aerospace components, honeycomb panels; good formability, heat treatable.
8 Mo, 8 V, 2 Fe, 3 Al	Solution + age	⋯	⋯	⋯	⋯	⋯	⋯	High-strength, tough airframe sheet, plate, fasteners, and forged components.
3 Al, 8 V, 6 Cr, 4 Mo, 4 Zr	Solution + age	⋯	⋯	⋯	⋯	⋯	⋯	High strength fasteners, torsion bars, aerospace components.
	Annealed							Parts requiring formability and corrosion resistance.
11.5 Mo, 6 Zr, 4.5 Sn	Solution + age	⋯	MIL-T-9047	4977	⋯	⋯	4980	High-strength fasteners, high-strength aircraft sheet parts.

(a) Also contains 0.2 Pd. (b) Also contains 0.8 Ni and 0.3 Mo.
Source: Titanium Metals Corp. of America and RMI Co.

Mechanical properties of wrought titanium alloys

Nominal composition, %	Condition	Room temperature Tensile strength MPa	ksi	Yield strength MPa	ksi	Elongation, %	Reduction in area, %	Test temperature °C	°F	Tensile strength MPa	ksi	Yield strength MPa	ksi	Elongation, %	Reduction in area, %	Charpy impact strength J	ft·lb	Hardness
Commercially pure																		
99.5 Ti	Annealed	331	48	241	35	30	55	315	600	152	22	97	14	32	80	...	...	120 HB
99.2 Ti	Annealed	434	63	345	50	28	50	315	600	193	28	117	17	35	75	43	32	200 HB
99.1 Ti	Annealed	517	75	448	65	25	45	315	600	234	34	138	20	34	75	38	28	225 HB
99.0 Ti	Annealed	662	96	586	85	20	40	315	600	310	45	172	25	25	70	20	15	265 HB
99.2 Ti(a)	Annealed	434	63	345	50	28	50	315	600	186	27	110	16	37	75	43	32	200 HB
98.9(b)	Annealed	517	75	448	65	25	42	205	400	345	50	248	36	37	...	...	...	...
								315	600	324	47	207	30	32	...	...	...	...
Alpha alloys																		
5 Al, 2.5 Sn	Annealed	862	125	807	117	16	40	315	600	565	82	448	65	18	45	26	19	36 HRC
5 Al, 2.5 Sn (low O$_2$)	Annealed	807	117	745	108	16	...	−195	−320	1241	180	1158	168	16	...	27	20	35 HRC
								−255	−423	1579	229	1420	206	15	...	...	...	...
Near alpha																		
8 Al, 1 Mo, 1 V	Duplex annealed	1000	145	951	138	15	28	315	600	793	115	621	90	20	38	32	24	35 HRC
								425	800	738	107	565	82	20	44	...	...	...
								540	1000	621	90	517	75	25	55	...	...	...
11 Sn, 1 Mo, 2.25 Al, 5.0 Zr, 1 Mo, 0.2 Si	Duplex annealed	1103	160	993	144	15	35	315	600	896	130	758	110	20	44	...	...	36 HRC
								425	800	827	120	676	98	22	48	...	...	...
								540	1000	758	110	586	85	24	50	...	...	...
6 Al, 2 Sn, 4 Zr, 2 Mo	Duplex annealed	979	142	896	130	15	35	315	600	772	112	586	85	16	42	...	...	32 HRC
								425	800	703	102	586	85	21	55	...	...	...
								540	1000	648	94	489	71	26	60	...	...	...
5 Al, 5 Sn, 2 Zr, 2 Mo, 0.25 Si	975 °C (1785 °F) (½ h), AC + 595 °C (1100 °F) (2 h), AC	1048	152	965	140	13	...	315	600	793	115	565	82	15	...	...	...	...
								425	800	779	113	531	77	17	...	...	...	...
								540	1000	689	100	503	73	19	...	...	...	...
6 Al, 2 Cb, 1 Ta, 1 Mo	As rolled 2.5 cm (1 in.) plate	855	124	758	110	13	34	315	600	586	85	462	67	20	...	31	23	30 HRC
								425	800	517	75	414	60	20	...	...	...	...
								540	1000	483	70	379	55	20	...	...	...	...
6 Al, 2 Sn, 1.5 Zr, 1 Mo, 0.35 Bi, 0.1 Si	Beta forge + duplex anneal	1014	147	945	137	11	...	480	900	724	105	586	85	15	...	...	...	...

(continued)

(a) Also contains 0.2 Pd. (b) Also contains 0.8 Ni and 0.3 Mo.
Source: Titanium Metals Corp. of America and RMI Co.

Mechanical properties of wrought titanium alloys (continued)

		Room temperature						Average mechanical properties — Extreme temperatures									
Nominal composition, %	Condition	Tensile strength MPa	ksi	Yield strength MPa	ksi	Elongation, %	Reduction in area, %	temperature °C	°F	Tensile strength MPa	ksi	Yield strength MPa	ksi	Elongation, %	Reduction in area, %	Charpy impact strength J	ft·lb	Hardness
Alpha-beta alloys																		
8 Mn	Annealed	945	137	862	125	15	32	315	600	717	104	565	82	18	...	...	...	...
3 Al, 2.5 V	Annealed	689	100	586	85	20	...	315	600	483	70	345	50	25	...	...	...	...
6 Al, 4 V	Annealed	993	144	924	134	14	30	315	600	724	105	655	95	14	35	19	14	36 HRC
								425	800	669	97	572	83	18	40	...	...	...
								540	1000	531	77	427	62	35	50	...	...	...
	Solution + age	1172	170	1103	160	10	25	315	600	862	125	703	102	10	28	...	...	41 HRC
								425	800	800	116	621	90	12	35	...	...	...
								540	1000	655	95	483	70	22	45	...	...	...
6 Al, 4 V (low O$_2$)	Annealed	896	130	827	120	15	35	160	320	1517	220	1413	205	14	...	24	18	35 HRC
6 Al, 6 V, 2 Sn	Annealed	1069	155	1000	145	14	30	315	600	931	135	807	117	18	42	18	13	38 HRC
	Solution + age	1276	185	1172	170	10	20	315	600	979	142	896	130	12	28	...	...	42 HRC
7 Al, 4 Mo	Solution + age	1103	160	1034	150	16	22	315	600	876	127	745	108	18	50	18	13	38 HRC
								425	800	848	123	717	104	20	55	...	...	42 HRC
6 Al, 2 Sn, 4 Zr, 6 Mo	Solution + age	1269	184	1172	170	10	23	315	600	1020	148	841	122	18	55	...	...	...
								425	800	951	138	758	110	19	67	...	...	...
								540	1000	848	123	655	95	19	70	...	...	...
6 Al, 2 Sn, 2 Zr, 2 Mo, 2 Cr, 0.25 Si	Solution + age	1276	185	1138	165	11	33	315	600	979	142	807	117	14	27	...	...	...
10 V, 2 Fe, 3 Al	Solution + age	1276	185	1200	174	10	19	205	400	1117	162	1048	152	13	33	...	...	...
								315	600	1103	160	979	142	13	42	...	...	...
Beta alloys																		
13 V, 11 Cr, 3 Al	Solution + age	1220	177	1172	170	8	...	315	600	883	128	793	115	19	...	11	8	40 HRC
	Solution + age	1276	185	1207	175	8	...	425	800	1103	160	827	120	12	...	...	...	...
8 Mo, 8 V, 2 Fe, 3 Al	Solution + age	1310	190	1241	180	8	...	315	600	1131	164	979	142	15	...	...	...	40 HRC
3 Al, 8 V, 6 Cr, 4 Mo, 4 Zr	Solution + age	1448	210	1379	200	7	...	315	600	1034	150	896	130	20	...	10	7.5	42 HRC
								425	800	938	136	758	110	17	...	...	...	...
11.5 Mo, 6 Zr, 4.5 Sn	Annealed	883	128	834	121	15	...	315	600	724	105	655	95	22	...	...	...	...
	Solution + age	1386	201	1317	191	11	...	315	600	903	131	848	123	16	...	...	...	...

(a) Also contains 0.2 Pd. (b) Also contains 0.8 Ni and 0.3 Mo.
Source: Titanium Metals Corp. of America and RMI Co.

Physical properties of wrought titanium alloys

Nominal composition, %	Coefficient of linear thermal expansion, μm/m · K (μin./in./°F) 20-100 °C (70-212 °F)	20-205 °C (70-400 °F)	20-315 °C (70-600 °F)	20-425 °C (70-800 °F)	20-540 °C (70-1000 °F)	20-650 °C (70-1200 °F)	20-815 °C (70-1500 °F)	Modulus of elasticity(a) GPa	10⁶ psi	Modulus of rigidity(a) GPa	10⁶ psi	Poisson's ratio (a)	Density(a) Mg/m³	lb/in.³
Commercially pure														
99.5 Ti	8.6 (4.8)	· · ·	9.2 (5.1)	· · ·	9.7 (5.4)	10.1 (5.6)	10.1 (5.6)	102.7	14.9	38.6	5.6	0.34	4.51	0.163
99.2 Ti	8.6 (4.8)	· · ·	9.2 (5.1)	· · ·	9.7 (5.4)	10.1 (5.6)	10.1 (5.6)	102.7	14.9	38.6	5.6	0.34	4.51	0.163
99.1 Ti	8.6 (4.8)	· · ·	9.2 (5.1)	· · ·	9.7 (5.4)	10.1 (5.6)	10.1 (5.6)	103.4	15.0	38.6	5.6	0.34	4.51	0.163
99.0 Ti	8.6 (4.8)	· · ·	9.2 (5.1)	· · ·	9.7 (5.4)	10.1 (5.6)	10.1 (5.6)	104.1	15.1	38.6	5.6	0.34	4.51	0.163
99.2 Ti(b)	8.6 (4.8)	· · ·	9.2 (5.1)	· · ·	9.7 (5.4)	10.1 (5.6)	10.1 (5.6)	102.7	14.9	38.6	5.6	0.34	4.51	0.163
98.9(c)	· · ·	· · ·	· · ·	· · ·	· · ·	· · ·	· · ·	· · ·	· · ·	102.7	14.9	· · ·	4.54	0.164
Alpha alloys														
5 Al, 2.5 Sn	9.4 (5.2)	· · ·	9.5 (5.3)	· · ·	9.5 (5.3)	9.7 (5.4)	10.1 (5.6)	110.3	16.0	· · ·	· · ·	· · ·	4.48	0.162
5 Al, 2.5 Sn (low O₂)	9.4 (5.2)	· · ·	9.5 (5.3)	· · ·	9.7 (5.4)	9.9 (5.5)	10.1 (5.6)	110.3	16.0	· · ·	· · ·	· · ·	4.48	0.162
Near alpha														
8 Al, 1 Mo, 1 V	8.5 (4.7)	· · ·	9.0 (5.0)	· · ·	10.1 (5.6)	10.3 (5.7)	· · ·	124.1	18.0	46.9	6.8	0.32	4.37	0.158
11 Sn, 1 Mo, 2.25 Al, 5.0 Zr, 1 Mo, 0.2 Si	8.5 (4.7)	· · ·	9.2 (5.1)	· · ·	9.4 (5.2)	· · ·	· · ·	113.8	16.5	· · ·	· · ·	· · ·	4.82	0.174
6 Al, 2 Sn, 4 Zr, 2 Mo	7.7 (4.3)	· · ·	8.1 (4.5)	· · ·	8.1 (4.5)	· · ·	· · ·	113.8	16.5	· · ·	· · ·	· · ·	4.54	0.164
5 Al, 5 Sn, 2 Zr, 2 Mo, 0.25 Si	· · ·	· · ·	· · ·	· · ·	· · ·	· · ·	10.3 (5.7)	113.8	16.5	· · ·	· · ·	0.326	4.51	0.163
6 Al, 2 Cb, 1 Ta, 1 Mo	· · ·	· · ·	· · ·	· · ·	· · ·	9.0 (5.0)	· · ·	113.8	17.5	· · ·	· · ·	· · ·	4.48	0.162
6 Al, 2 Sn, 1.5 Zr, 1 Mo, 0.35 Bi, 0.1 Si	· · ·	· · ·	· · ·	· · ·	· · ·	· · ·	· · ·	· · ·	· · ·	· · ·	· · ·	· · ·	· · ·	· · ·
Alpha-beta alloys														
8 Mn	8.6 (4.8)	9.2 (5.1)	9.7 (5.4)	10.3 (5.7)	10.8 (6.0)	11.7 (6.5)	12.6 (7.0)	113.1	16.4	48.3	7.0	· · ·	4.73	0.171
3 Al, 2.5 V	9.5 (5.3)	· · ·	9.9 (5.5)	· · ·	9.9 (5.5)	· · ·	· · ·	106.9	15.5	· · ·	· · ·	· · ·	4.48	0.162
6 Al, 4 V	8.6 (4.8)	9.0 (5.0)	9.2 (5.1)	9.4 (5.2)	9.5 (5.3)	9.7 (5.4)	· · ·	113.8	16.5	42.1	6.1	0.342	4.43	0.160
6 Al, 4 V (low O₂)	8.6 (4.8)	9.0 (5.0)	9.2 (5.1)	9.4 (5.2)	9.5 (5.3)	9.7 (5.4)	· · ·	113.8	16.5	42.1	6.1	0.342	4.43	0.160
6 Al, 6 V, 2 Sn	9.0 (5.0)	· · ·	9.4 (5.2)	· · ·	9.5 (5.3)	· · ·	· · ·	110.3	16.0	· · ·	· · ·	· · ·	4.54	0.164
7 Al, 4 Mo	9.0 (5.0)	9.2 (5.1)	9.4 (5.2)	9.7 (5.4)	10.1 (5.6)	10.4 (5.8)	11.2 (6.2)	113.8	16.5	44.8	6.5	· · ·	4.48	0.162
6 Al, 2 Sn, 4 Zr, 6 Mo	9.0 (5.0)	9.2 (5.1)	9.4 (5.2)	9.5 (5.3)	(5.3)	· · ·	· · ·	113.8	16.5	· · ·	· · ·	· · ·	4.65	0.168
6 Al, 2 Sn, 2 Zr, 2 Mo, 2 Cr, 0.25 Si	· · ·	· · ·	9.2 (5.1)	· · ·	· · ·	· · ·	· · ·	122.0	17.7	46.2	6.7	0.327	4.57	0.165
10 V, 2 Fe, 3 Al	· · ·	· · ·	· · ·	· · ·	· · ·	· · ·	· · ·	111.7	16.2	· · ·	· · ·	· · ·	4.65	0.168
Beta alloys														4.84	0.175
13 V, 11 Cr, 3 Al	9.4 (5.2)	· · ·	10.1 (5.6)	· · ·	10.6 (5.9)	· · ·	· · ·	101.4	14.7	42.7	6.2	0.304	4.84	0.175
8 Mo, 8 V, 2 Fe, 3 Al	· · ·	· · ·	· · ·	· · ·	· · ·	· · ·	· · ·	106.9	15.5	· · ·	· · ·	· · ·	4.84	0.175
3 Al, 8 V, 6 Cr, 4 Mo, 4 Zr	· · ·	· · ·	· · ·	9.68 (5.38) (to 900 °F)	· · ·	· · ·	· · ·	105.5	15.3	· · ·	· · ·	· · ·	4.82	0.174
11.5 Mo, 6 Zr, 4.5 Sn	· · ·	· · ·	· · ·	· · ·	· · ·	· · ·	· · ·	103.4	15.0	· · ·	· · ·	· · ·	· · ·	· · ·

(a) Room temperature. (b) Also contains 0.2 Pd. (c) Also contains 0.8 Ni and 0.3 Mo.
Source: Titanium Metals Corp. of America and RMI Co.

Heat Treating

Recommended stress-relief treatments for titanium and titanium alloys

Parts can be cooled from stress relief by either air cooling or slow cooling

Alloy	Temperature °C	°F	Time, h
Commercially pure Ti (all grades)	480 to 595	900 to 1100	1/4 to 4
Alpha or near-alpha titanium alloys			
Ti-5Al-2.5Sn	540 to 650	1000 to 1200	1/4 to 4
Ti-8Al-1Mo-1V	595 to 705	1100 to 1300	1/4 to 4
Ti-6Al-2Sn-4Zr-2Mo	595 to 705	1100 to 1300	1/4 to 4
Ti-6Al-2Cb-1Ta-0.8Mo	595 to 650	1100 to 1200	1/4 to 2
Ti-0.3Mo-0.8Ni (Ti Code 12)	480 to 595	900 to 1100	1/4 to 4
Alpha-beta titanium alloys			
Ti-6Al-4V	480 to 650	900 to 1200	1 to 4
Ti-6Al-6V-2Sn (Cu + Fe)	480 to 650	900 to 1200	1 to 4
Ti-3Al-2.5V	540 to 650	1000 to 1200	1/2 to 2
Ti-6Al-2Sn-4Zr-6Mo	595 to 705	1100 to 1300	1/4 to 4
Ti-5Al-2Sn-4Mo-2Zr-4Cr (Ti-17)	480 to 650	900 to 1200	1 to 4
Ti-7Al-4Mo	480 to 705	900 to 1300	1 to 8
Ti-6Al-2Sn-2Zr-2Mo-2Cr-0.25Si	480 to 650	900 to 1200	1 to 4
Ti-8Mn	480 to 595	900 to 1100	1/4 to 2
Beta or near-beta titanium alloys			
Ti-13V-11Cr-3Al	705 to 730	1300 to 1350	1/12 to 1/4
Ti-11.5Mo-6Zr-4.5Sn (Beta III)	720 to 730	1325 to 1350	1/12 to 1/4
Ti-3Al-8V-6Cr-4Zr-4Mo (Beta C)	705 to 760	1300 to 1400	1/6 to 1/2
Ti-10V-2Fe-3Al	675 to 705	1250 to 1300	1/2 to 2
Ti-15V-3Al-3Cr-3Sn	790 to 815	1450 to 1500	1/12 to 1/4

Recommended annealing treatments for titanium and titanium alloys

Alloy	Temperature °C	°F	Time, h	Cooling method
Commercially pure Ti (all grades)	650 to 760	1200 to 1400	1/10 to 2	Air
Alpha or near-alpha titanium alloys				
Ti-5Al-2.5Sn	720 to 845	1325 to 1550	1/6 to 4	Air
Ti-8Al-1Mo-1V	790(a)	1450(a)	1 to 8	Air or furnace
Ti-6Al-2Sn-4Zr-2Mo	900(b)	1650(b)	1/2 to 1	Air
Ti-6Al-2Cb-1Ta-0.8Mo	790 to 900	1450 to 1650	1 to 4	Air
Alpha-beta titanium alloys				
Ti-6Al-4V	705 to 790	1300 to 1450	1 to 4	Air or furnace
Ti-6Al-6V-2Sn (Cu + Fe)	705 to 815	1300 to 1500	3/4 to 4	Air or furnace
Ti-3Al-2.5V	650 to 760	1200 to 1400	1/2 to 2	Air
Ti-6Al-2Sn-4Zr-6Mo	(c)	(c)	...	...
Ti-5Al-2Sn-4Mo-2Zr-4Cr (Ti-17)	(c)	(c)	...	...
Ti-7Al-4Mo	705 to 790	1300 to 1450	1 to 8	Air
Ti-6Al-2Sn-2Zr-2Mo-2Cr-0.25Si	705 to 815	1300 to 1500	1 to 2	Air
Ti-8Mn	650 to 760	1200 to 1400	1/2 to 1	(d)
Beta or near-beta titanium alloys				
Ti-13V-11Cr-3Al	705 to 790	1300 to 1450	1/6 to 1	Air or water
Ti-11.5Mo-6Zr-4.5Sn (Beta III)	690 to 760	1275 to 1400	1/6 to 1	Air or water
Ti-3Al-8V-6Cr-4Zr-4Mo (Beta C)	790 to 815	1450 to 1500	1/4 to 1	Air or water
Ti-10V-2Fe-3Al	(c)	(c)	...	...
Ti-15V-3Al-3Cr-3Sn	790 to 815	1450 to 1500	1/12 to 1/4	Air

(a) For sheet and plate, follow by 1/4 h at 790 °C (1450 °F), then air cool. (b) For sheet, follow by 1/4 h at 790 °C (1450 °F), then air cool (plus 2 h at 595 °C or 1100 °F, then air cool, in certain applications). For plate, follow by 8 h at 595 °C (1100 °F), then air cool. (c) Not normally supplied or used in annealed condition (see Table 3). (d) Furnace or slow cool to 540 °C (1000 °F), then air cool.

Recommended solution treating and aging (stabilizing) treatments for titanium alloys

Alloy	Solution temperature °C	°F	Solution time, h	Cooling rate	Aging temperature °C	°F	Aging time, h
Alpha or near-alpha alloys							
Ti-8Al-1Mo-1V....................	980 to 1010(a)	1800 to 1850(a)	1	Oil or water	565 to 595	1050 to 1100	...
Ti-6Al-2Sn-4Zr-2Mo	955 to 980	1750 to 1800	1	Air	595	1100	8
Alpha-beta alloys							
Ti-6Al-4V	955 to 970(b)(c)	1750 to 1775(b)(c)	1	Water	480 to 595	900 to 1100	4 to 8
	955 to 970	1750 to 1775	1	Water	705 to 760	1300 to 1400	2 to 4
Ti-6Al-6V-2Sn (Cu + Fe)...........	885 to 910	1625 to 1675	1	Water	480 to 595	900 to 1100	4 to 8
Ti-6Al-2Sn-4Zr-6Mo	845 to 890	1550 to 1650	1	Air	580 to 605	1075 to 1125	4 to 8
Ti-5Al-2Sn-2Zr-4Mo-4Cr	845 to 870	1550 to 1600	1	Air	580 to 605	1075 to 1125	4 to 8
Ti-6Al-2Sn-2Zr-2Mo-2Cr-0.25Si	870 to 925	1600 to 1700	1	Water	480 to 595	900 to 1100	4 to 8
Beta or near-beta alloys							
Ti-13V-11Cr-3Al	775 to 800	1425 to 1475	¼ to 1	Air or water	425 to 480	800 to 900	4 to 100
Ti-11.5Mo-6Zr-4.5Sn (Beta III)	690 to 790	1275 to 1450	⅛ to 1	Air or water	480 to 595	900 to 1100	8 to 32
Ti-3Al-8V-6Cr-4Mo-4Zr (Beta C)	815 to 925	1500 to 1700	1	Water	455 to 540	850 to 1000	8 to 24
Ti-10V-2Fe-3Al	760 to 780	1400 to 1435	1	Water	495 to 525	925 to 975	8
Ti-15V-3Al-3Cr-3Sn	790 to 815	1450 to 1500	¼	Air	510 to 595	950 to 1100	8 to 24

(a) For certain products, use solution temperature of 890 °C (1650 °F) for 1 h, then air cool or faster. (b) For thin plate or sheet, solution temperature can be used down to 890 °C (1650 °F) for 6 to 30 min, then water quench. (c) This treatment is used to develop maximum tensile properties in this alloy.

Minimum metal removal after thermal exposure of titanium alloys

Heat treating temperature °C	°F	Time at temperature, h	Minimum stock removal per surface(a) mm	in.
480 to 593	900 to 1100	Up to 12	0.005	0.0002
594 to 648	1101 to 1200	Up to 4	0.008	0.0003
		4 to 12	0.015	0.0006
649 to 704	1201 to 1300	Up to 1	0.013	0.0005
		1 to 8	0.020	0.0008
		8 to 12	0.025	0.0010
705 to 760	1301 to 1400	Up to 1	0.025	0.0010
		1 to 4	0.036	0.0014
		4 to 8	0.038	0.0015
		8 to 12	0.043	0.0017
761 to 787	1401 to 1450	Up to 1	0.030	0.0012
		1 to 2	0.038	0.0015
		2 to 4	0.046	0.0018
		4 to 8	0.051	0.0020
		8 to 12	0.056	0.0022
788 to 815	1451 to 1500	Up to ½	0.036	0.0014
		½ to 1	0.041	0.0016
		1 to 2	0.051	0.0020
816 to 871	1501 to 1600	Up to ½	0.058	0.0023
		½ to 1	0.066	0.0026
		1 to 2	0.076	0.0030
872 to 898	1601 to 1650	Up to ½	0.058	0.0023
		½ to 1	0.081	0.0032
		1 to 2	0.089	0.0035
899 to 926	1651 to 1700	Up to ½	0.086	0.0034
		½ to 1	0.091	0.0036
		1 to 2	0.107	0.0042
927 to 954	1701 to 1750	Up to ½	0.097	0.0038
		½ to 1	0.107	0.0042
		1 to 2	0.122	0.0048

(a) Values shown are typical; actual values may vary with alloy type.

Variation of tensile properties of Ti-6Al-4V bar stock with solution-treating temperature

Solution-treating temperature		Room-temperature tensile properties(a)				Elongation in 4D,
		Tensile strength		Yield strength(b)		
°C	°F	MPa	ksi	MPa	ksi	%
845	1550	1025	149	980	142	18
870	1600	1060	154	985	143	17
900	1650	1095	159	995	144	16
925	1700	1110	161	1000	145	16
940	1725	1140	165	1055	153	16

(a) Properties determined on 13-mm (1/2-in.) bar after solution treating, quenching and aging. Aging treatment: 8 h at 480 °C (900 °F), air cool. (b) At 0.2% offset

Relation of tensile strength of solution treated and aged titanium alloys to size

Alloy	Tensile strength of square bar in section size of:											
	13 mm (1/2 in.)		25 mm (1 in.)		50 mm (2 in.)		75 mm (3 in.)		100 mm (4 in.)		150 mm (6 in.)	
	MPa	ksi	MPa	ksi	MPa	ksi	MPa	ksi	MPa	ksi	MPa	ksi
Ti-6Al-4V	1105	160	1070	155	1000	145	930	135	...	...	...	...
Ti-6Al-6V-2Sn (Cu + Fe)........	1205	175	1205	175	1070	155	1035	150	...	...	...	...
Ti-6Al-2Sn-4Zr-6Mo	1170	170	1170	170	1170	170	1140	165	1105	160	...	...
Ti-5Al-2Sn-2Zr-4Mo-4Cr (Ti-17) ...	1170	170	1170	170	1170	170	1105	160	1105	160	1105	160
Ti-10V-2Fe-3Al	1240	180	1240	180	1240	180	1240	180	1170	170	1170	170
Ti-13V-11Cr-3Al	1310	190	1310	190	1310	190	1310	190	1310	190	1310	190
Ti-11.5Mo-6Zr-4.5Sn (Beta III)	1310	190	1310	190	1310	190	1310	190	1310	190	...	...
Ti-3Al-8V-6Cr-4Zr-4Mo (Beta C) ..	1310	190	1310	190	1240	180	1240	180	1170	170	1170	170

Beta transformation temperatures of titanium alloys

Alloy	Beta transus	
	°C, ±15	°F, ±25
Commercially pure Ti, 0.25 max O_2.....................	910	1675
Commercially pure Ti, 0.40 max O_2.....................	945	1735
Alpha and near-alpha alloys		
Ti-5Al-2.5Sn..	1050	1925
Ti-8Al-1Mo-1V.....................................	1040	1900
Ti-6Al-2Sn-4Zr-2Mo................................	995	1820
Ti-6Al-2Cb-1Ta-0.8Mo	1015	1860
Ti-0.3Mo-0.8Ni (Ti code 12)	880	1615
Alpha-beta alloys		
Ti-6Al-4V	1000(a)	1830(b)
Ti-6Al-6V-2Sn (Cu + Fe)............................	945	1735
Ti-3Al-2.5V..	935	1715
Ti-6Al-2Sn-4Zr-6Mo................................	940	1720
Ti-5Al-2Sn-2Zr-4Mo-4Cr (Ti-17)	900	1650
Ti-7Al-4Mo..	1000	1840
Ti-6Al-2Sn-2Zr-2Mo-2Cr-0.25Si.....................	970	1780
Ti-8Mn...	800(c)	1475(d)
Beta or near-beta alloys		
Ti-13V-11Cr-3Al	720	1330
Ti-11.5Mo-6Zr-4.5Sn (Beta III)	760	1400
Ti-3Al-8V-6Cr-4Zr-4Mo (Beta C)	795	1460
Ti-10V-2Fe-3Al.....................................	805	1480
Ti-15V-3Al-3Cr-3Sn.................................	760	1400

(a) ±20. (b) ±30. (c) ±35. (d) ±50.

Joining

Typical tensile, bend and hardness data for as-welded titanium and several titanium alloys

Material condition	Tensile strength MPa	ksi	Yield strength MPa	ksi	Elonga- tion, %	Minimum bend radius	Hardness Knoop	Rockwell
Ti Grade 1								
Unwelded sheet	315	46	215	31	50.4	0.7t	140	63.5 HRB
Single-bead weld	345	50	255	37	37.5	1.0t	140	55.8 HRB
Multiple-bead weld	365	53	270	39	37.7	...	...	...
Transverse weld	325	47(a)	...	...	...	...	...	...
Ti Grade 2								
Unwelded sheet	460	67	325	47	26.2	2.9t	165	80.6 HRB
Single-bead weld	505	73	380	55	18.3	2.9t	175	83.1 HRB
Multiple-bead weld	510	74	385	56	13.3	...	...	...
Transverse weld	475	69(a)	...	...	...	...	...	...
Ti Grade 3								
Unwelded sheet	545	79	395	57	25.9	1.9t	175	94.4 HRB
Single-bead sheet	605	88	475	69	15.5	4.7t	220	92.4 HRB
Multiple-bead weld	615	89	480	70	14.7	...	...	...
Transverse weld	560	81(a)	...	...	...	...	...	...
Ti Grade 4								
Unwelded sheet	660	96	530	77	22.3	3.2t	215	23.4 HRC
Single-bead weld	695	101	580	84	16.4	5.6t	240	21.2 HRC
Multiple-bead weld	710	103	585	85	16.0	...	...	...
Transverse weld	660	96(a)	...	...	...	...	...	...
Ti-5Al-2.5Sn-ELI								
Unwelded sheet	850	123	805	117	15.7	3.8t	265	33.2 HRC
Single-bead weld	920	133	770	112	9.8	5.9t	310	28.0 HRC
Multiple-bead weld	935	136	820	119	7.5	...	...	...
Transverse weld	850	123(a)	...	...	...	...	...	...
Ti-6Al-2Cb-1Ta-1Mo								
Unwelded sheet	895	130	855	124	9.7	2.8t	275	29.6 HRC
Single-bead weld	930	135	800	116	5.9	7.7t	300	27.7 HRC
Multiple-bead weld	945	137	815	118	5.7	...	...	...
Transverse weld	890	129(a)	...	...	...	...	...	...
Ti-3Al-2.5V								
Unwelded sheet	705	102	670	97	15.2	4.0t	230	23.6 HRC
Single-bead weld	705	102	600	87	12.7	5.4t	250	19.6 HRC
Multiple-bead weld	745	108	625	91	11.2	...	...	...
Transverse weld	710	103(a)	...	...	...	...	...	...
Ti-6Al-4V								
Unwelded sheet	1000	145	945	137	11.0	2.6t	320	32.2 HRC
Single-bead weld	1060	154	920	133	3.5	10.5t	350	35.9 HRC
Multiple-bead weld	1090	158	945	137	3.2	...	...	...
Transverse weld	1015	147(a)	...	...	...	...	...	...
Ti-8Al-1Mo-1V								
Unwelded sheet	1060	154	1020	148	15.0	2.9t	325	36.0 HRC
Single-bead weld	1085	157	930	135	5.5	7.0t	345	35.2 HRC
Multiple-bead weld	1115	162	960	139	3.2	...	...	...
Transverse weld	1060	154(a)	...	...	...	...	...	...
Ti-6Al-6V-2Sn								
Unwelded sheet	1060	154	1005	146	9.8	2.8t	350	34.0 HRC
Single-bead weld	1295	188	1255	182	0.3	25.6t	420	46.8 HRC
Multiple-bead weld	1280	186	...	...	0.1	...	...	...
Transverse weld	1103	160(a)	...	...	...	...	...	...
Ti-13V-11Cr-3Al								
Unwelded sheet	965	140	910	132	13.9	2.7t	300	30.6 HRC
Single-bead weld	950	138	925	134	11.6	2.7t	320	30.1 HRC
Multiple-bead weld	925	134	875	127	9.1	...	...	...
Transverse weld	950	138(a)	...	...	...	...	...	...

(a) Fracture occurred in base metal.

Heat-and Corrosion-Resistant Alloys

Nominal Compositions

Nominal compositions of wrought iron-base heat-resistant alloys

Designation	UNS number	C	Cr	Ni	Mo	N	Nb	Ti	Others
Ferritic stainless steels									
405	S40500	0.15 max	13.0	...	...	...	...	...	0.2 Al
406	...	0.15 max	13.0	...	...	...	...	...	4.0 Al
409	S40900	0.08 max	11.0	0.5	...	...	...	6 × C min	...
430	S43000	0.12 max	16.0	...	...	...	...	...	...
434	S43400	0.12 max	17.0	...	1.0	...	...	...	...
439	S43027	0.07 max	18.25	...	...	...	...	0.2 + 4 (C + N)	...
18 SR	...	0.05	18.0	0.5	...	...	...	0.40	2.0 Al
18Cr-2Mo	...	...	18.0	...	2.0	...	...	...	...
446	S44600	0.20 max	25.0	...	...	0.25	...	...	...
E-Brite 26-1	S44627	0.01 max	26.0	...	1.0	0.015 max	0.1	...	...
26-1Ti	...	0.04	26.0	...	1.0	...	...	10 × C min	...
29Cr-4Mo	...	0.01 max	29.0	...	4.0	0.02 max	...	...	...
Quenched and tempered martensitic stainless steels									
403	S40300	0.15 max	12.0	...	...	...	...	...	...
410	S41000	0.15 max	12.5	...	...	...	...	...	...
416	S41600	0.15 max	13.0	...	0.6(a)	...	...	...	0.15 min S
422	S42200	0.20	12.5	0.75	1.0	...	...	...	1.0 W, 0.22 V
H-46	...	0.12	10.75	0.50	0.85	0.07	0.30	...	0.20 V
Moly Ascoloy	...	0.14	12.0	2.4	1.80	0.05	...	...	0.35 V
Greek Ascoloy	...	0.15	13.0	2.0	...	...	...	...	3.0 W
Jethete M-152	...	0.12	12.0	2.5	1.7	...	...	...	0.30 V
Almar 363	...	0.05	11.5	4.5	...	...	...	10 × C min	...
431	S43100	0.20 max	16.0	2.0	...	...	...	...	...
Precipitation-hardening martensitic stainless steels									
Custom 450	...	0.05 max	15.5	6.0	0.75	...	8 × C min	...	1.5 Cu
Custom 455	...	0.03	11.75	8.5	...	...	0.30	1.2	2.25 Cu
15-5 PH	S15500	0.07	15.0	4.5	...	...	0.30	...	3.5 Cu
17-4 PH	S17400	0.04	16.5	4.25	...	...	0.25	...	3.6 Cu
PH 13-8 Mo	S13800	0.05	12.5	8.0	2.25	...	...	...	1.1 Al
Precipitation-hardening semiaustenitic stainless steels									
AM-350	S35000	0.10	16.5	4.25	2.75	0.10	...	...	...
AM-355	S35500	0.13	15.5	4.25	2.75	0.10	...	...	...
17-7 PH	S17700	0.07	17.0	7.0	...	...	...	...	1.15 Al
PH 15-7 Mo	S15700	0.07	15.0	7.0	2.25	...	...	...	1.15 Al

(continued)

Nominal compositions of wrought iron-base heat-resistant alloys (continued)

Designation	UNS number	C	Cr	Ni	Mo	N	Nb	Ti	Others
						Composition, %			
Austenitic stainless steels									
304	S30400	0.08 max	19.0	10.0	...	...	...	...	...
304L	S30403	0.03 max	19.0	10.0	...	...	...	...	...
304N	S30451	0.08 max	19.0	9.25	...	0.13	...	...	...
309	S30900	0.20 max	23.0	13.0	...	...	...	...	...
310	S31000	0.25 max	25.0	20.0	...	...	...	...	...
316	S31600	0.08 max	17.0	12.0	2.5	...	...	...	...
316L	S31603	0.03 max	17.0	12.0	2.5	...	...	...	...
316N	S31651	0.08 max	17.0	12.0	2.5	0.13	...	...	...
317	S31700	0.08 max	19.0	13.0	3.5	...	...	...	...
321	S32100	0.08 max	18.0	10.0	...	...	...	5 × C min	...
347	S34700	0.08 max	18.0	11.0	...	...	10 × C min	...	...
19-9 DL	K63198	0.30	19.0	9.0	1.25	...	0.4	0.3	1.25 W
19-9 DX	K63199	0.30	19.2	9.0	1.5	...	...	0.55	1.2 W
17-14-CuMo	...	0.12	16.0	14.0	2.5	...	0.4	0.3	3.0 Cu
202	S20200	0.09	18.0	5.0	...	0.10	...	...	8.0 Mn
216	S21600	0.05	20.0	6.0	2.5	0.35	...	...	8.5 Mn
21-6-9	S21900	0.04 max	20.25	6.5	...	0.30	...	...	9.0 Mn
Nitronic 32	...	0.10	18.0	1.6	...	0.34	...	...	12.0 Mn
Nitronic 33	...	0.08 max	18.0	3.0	...	0.30	...	...	13.0 Mn
Nitronic 50	...	0.06 max	21.0	12.0	2.0	0.30	0.20	...	5.0 Mn
Nitronic 60	...	0.10 max	17.0	8.5	2.0	...	...	...	8.0 Mn, 0.20 V, 4.0 Si
Carpenter 18-18 Plus	...	0.10	18.0	<0.50	1.0	0.50	...	...	16.0 Mn, 0.40 Si, 1.0 Cu

(a) Optional.

Nominal compositions of wrought superalloys

Alloy	UNS number	Composition, %										
		Cr	Ni	Co	Mo	W	Nb	Ti	Al	Fe	C	Other
Iron-base solid-solution alloys												
16-25-6	...	16.0	25.0	...	6.00	...	...	...	...	50.7	0.06	1.35 Mn; 0.70 Si; 0.15 N
17-14CuMo	...	16.0	14.0	...	2.50	...	0.4	0.3	...	62.4	0.12	0.75 Mn; 0.50 Si; 3.0 Cu
19-9DL	K63198	19.0	9.0	...	1.25	1.25	0.4	0.3	...	66.8	0.30	1.10 Mn; 0.60 Si
Carpenter 20Cb-3	N08020	20.0	34.0	...	2.50	...	1.0 max	...	...	42.4	0.07 max	3.5 Cu
Incoloy 800	N08800	21.0	32.5	...	...	...	...	0.38	0.38	45.7	0.05	...
Incoloy 801	N08801	20.5	32.0	...	...	...	...	1.13	...	46.3	0.05	...
Incoloy 802	...	21.0	32.5	...	...	...	...	0.75	0.58	44.8	0.35	...
N-155	R30155	21.0	20.0	20.0	3.00	2.5	1.0	...	...	32.2	0.15	0.15 N; 0.02 La; 0.02 Zr
RA330	N08330	19.0	36.0	...	...	...	...	...	...	45.1	0.05	...
Cobalt-base solid-solution alloys												
Haynes 25 (L-605)	R30605	20.0	10.0	50.0	...	15.0	...	...	...	3.0	0.10	1.5 Mn
Haynes 188	R30188	22.0	22.0	37.0	...	14.5	...	...	...	3.0 max	0.10	0.90 La
S-816	R30816	20.0	20.0	42.0	4.0	4.0	4.0	...	...	4.0	0.38	...
Stellite 6B	...	30.0	1.0	61.5	...	4.5	...	...	...	1.0	1.0	...
UMCo-50	...	28.0	...	49.0	...	...	...	...	...	21.0	0.12 max	...
Nickel-base solid-solution alloys												
Hastelloy B	N10001	1.0 max	63.0	2.5 max	28.0	...	...	...	...	5.0	0.05 max	0.03 V
Hastelloy B-2	N10665	1.0 max	69.0	1.0 max	28.0	...	...	...	...	2.0 max	0.02 max	...
Hastelloy C	N10002	16.5	56.0	...	17.0	4.5	...	...	...	6.0	0.15 max	...
Hastelloy C-4	N06455	16.0	63.0	2.0 max	15.5	...	...	0.7 max	...	3.0 max	0.015 max	...
Hastelloy C-276	N10276	15.5	59.0	...	16.0	3.7	...	...	...	5.0	0.02 max	...
Hastelloy N	N10003	7.0	72.0	...	16.0	...	...	...	...	5.0 max	0.06	...
Hastelloy S	...	15.5	67.0	...	15.5	...	...	0.5 max	0.2	1.0	0.02 max	0.02 La
Hastelloy W	N10004	5.0	61.0	2.5 max	24.5	...	...	...	...	5.5	0.12 max	0.6 V
Hastelloy X	N06002	22.0	49.0	1.5 max	9.0	0.6	...	...	2.0	15.8	0.15	...
Inconel 600	N06600	15.5	76.0	...	...	...	...	...	...	8.0	0.08	0.25 max Cu
Inconel 601	N06601	23.0	60.5	...	...	...	...	...	1.35	14.1	0.05	0.5 max Cu
Inconel 604	...	16.0	74.0	...	...	...	2.25	...	...	7.5	0.02	0.03 max Cu
Inconel 617	...	22.0	55.0	12.5	9.0	...	...	...	1.0	...	0.07	...
Inconel 625	N06625	21.5	61.0	...	9.0	...	3.6	0.2	0.2	2.5	0.05	...
NA-224	...	27.0	48.0	...	...	6.0	...	...	...	18.5	0.50	...
Nimonic 75	...	19.5	75.0	...	...	...	...	0.4	0.15	2.5	0.12	0.25 max Cu
RA-333	N06333	25.0	45.0	3.0	3.0	3.0	...	...	...	18.0	0.05	...
Iron-base precipitation-hardening alloys												
A-286	K66286	15.0	26.0	...	1.25	...	...	2.0	0.2	55.2	0.04	0.005 B; 0.3 V
Discaloy	K66220	14.0	26.0	...	3.0	...	...	1.7	0.25	55.0	0.06	...
Haynes 556	...	22.0	21.0	20.0	3.0	2.5	0.1	...	0.3	29.0	0.10	0.50 Ta; 0.02 La; 0.002 Zr
Incoloy 903	...	0.1 max	38.0	15.0	0.1	0.1	3.0	1.4	0.7	41.0	0.04	...
Pyromet CTX-1	...	0.1 max	37.7	16.0	0.1	...	3.0	1.7	1.0	39.0	0.03	0.002 Zr

(continued)

(a) No longer active, but shown here for reference. (b) Also known as Rolls Royce C-263.

Nominal compositions of wrought superalloys (continued)

Alloy	UNS number	Cr	Ni	Co	Mo	W	Nb	Ti	Al	Fe	C	Other
V-57	...	14.8	27.0	...	1.25	...	...	3.0	0.25	48.6	0.08 max	0.01 B; 0.5 max V
W-545	K66545	13.5	26.0	...	1.5	...	...	2.85	0.2	55.8	0.08	0.05 B
Cobalt-base precipitation-hardening alloys												
AR-213	...	19.0	0.5 max	65.0	...	4.5	...	...	3.5	0.5 max	0.17	6.5 Ta; 0.15 Zr; 0.1 Y
MP-35N	R30035	20.0	35.0	35.0	10.0	...	...	...	...	...	...	...
MP-159		19.0	25.0	36.0	7.0	...	0.6	3.0	0.2	9.0	...	...
Nickel-base precipitation-hardening alloys												
Astroloy	...	15.0	56.5	15.0	5.25	...	...	3.5	4.4	<0.3	0.06	0.03 B; 0.06 Zr
D-979	N09979; K66979(a)	15.0	45.0	...	4.0	4.0	...	3.0	1.0	27.0	0.05	0.01 B
IN 100	N13100	10.0	60.0	15.0	3.0	...	...	4.7	5.5	<0.6	0.15	1.0 V; 0.06 Zr; 0.015 B
IN 102	N06102	15.0	67.0	...	2.9	3.0	2.9	0.5	0.5	7.0	0.06	0.005 B; 0.02 Mg; 0.03 Zr
Incoloy 901	N09901	12.5	42.5	...	6.0	...	...	2.7	...	36.2	0.10 max	...
Inconel 706	N09706	16.0	41.5	...	...	...	...	1.75	0.2	37.5	0.03	2.9 (Nb + Ta); 0.15 max Cu
Inconel 718	N07718	19.0	52.5	...	3.0	...	5.1	0.9	0.5	18.5	0.08 max	0.15 max Cu
Inconel 751	...	15.5	72.5	...	...	...	1.0	2.3	1.2	7.0	0.05	0.25 max Cu
Inconel X750	N07750	15.5	73.0	...	...	...	1.0	2.5	0.7	7.0	0.04	0.25 max Cu
M252	N07252	19.0	56.5	10.0	10.0	...	...	2.6	1.0	<0.75	0.15	0.005 B
Nimonic 80A	N07080	19.5	73.0	1.0	...	...	...	2.25	1.4	1.5	0.05	0.10 max Cu
Nimonic 90	N07090	19.5	55.5	18.0	...	...	...	2.4	1.4	1.5	0.06	...
Nimonic 95		19.5	53.5	18.0	...	...	...	2.9	2.0	5.0 max	0.15 max	+B; +Zr
Nimonic 100		11.0	56.0	20.0	5.0	...	...	1.5	5.0	2.0 max	0.30 max	+B; +Zr
Nimonic 105		15.0	54.0	20.0	5.0	...	...	1.2	4.7	...	0.08	0.005 B
Nimonic 115		15.0	55.0	15.0	4.0	...	...	4.0	5.0	1.0	0.20	0.04 Zr
Nimonic 263(b)		20.0	51.0	20.0	5.9	...	...	2.1	0.45	0.7 max	0.06	...
Pyromet 860		13.0	44.0	4.0	6.0	...	...	3.0	1.0	28.9	0.05	0.01 B
Refractory 26		18.0	38.0	20.0	3.2	...	...	2.6	0.2	16.0	0.03	0.015 B
René 41	N07041	19.0	55.0	11.0	10.0	...	...	3.1	1.5	<0.3	0.09	0.01 B
René 95		14.0	61.0	8.0	3.5	3.5	3.5	2.5	3.5	<0.3	0.16	0.01 B; 0.05 Zr
René 100		9.5	61.0	15.0	3.0	...	...	4.2	5.5	1.0 max	0.16	0.015 B; 0.06 Zr; 1.0 V
Udimet 500	N07500	19.0	48.0	19.0	4.0	...	...	3.0	3.0	4.0 max	0.08	0.005 B
Udimet 520		19.0	57.0	12.0	6.0	1.0	...	3.0	2.0	...	0.08	0.005 B
Udimet 630		17.0	50.0	...	3.0	3.0	6.5	1.0	0.7	18.0	0.04	0.004 B
Udimet 700		15.0	53.0	18.5	5.0	...	...	3.4	4.3	<1.0	0.07	0.03 B
Udimet 710		18.0	55.0	14.8	3.0	1.5	...	5.0	2.5	...	0.07	0.01 B
Unitemp AF2-1DA		12.0	59.0	10.0	3.0	6.0	...	3.0	4.6	<0.5	0.35	1.5 Ta; 0.015 B; 0.1 Zr
Waspaloy	N07001	19.5	57.0	13.5	4.3	...	...	3.0	1.4	2.0 max	0.07	0.006 B; 0.09 Zr

(a) No longer active, but shown here for reference. (b) Also known as Rolls Royce C-263.

Compositions of ACI heat-resistant casting alloys

ACI designation	UNS number	ASTM specifications(a)	C	Cr	Ni	Si (max)
HA	...	A217	0.20 max	8 to 10	...	1.00
HC	J92605	A297, A608	0.50 max	26 to 30	4 max	2.00
HD	J93005	A297, A608	0.50 max	26 to 30	4 to 7	2.00
HE	J93403	A297, A608	0.20 to 0.50	26 to 30	8 to 11	2.00
HF	J92603	A297, A608	0.20 to 0.40	19 to 23	9 to 12	2.00
HH	J93503	A297, A608	0.20 to 0.50	24 to 28	11 to 14	2.00
HI	J94003	A297, A567, A608	0.20 to 0.50	26 to 30	14 to 18	2.00
HK	J94224	A297, A351, A567, A608	0.20 to 0.60	24 to 28	18 to 22	2.00
HL	J94604	A297, A608	0.20 to 0.60	28 to 32	18 to 22	2.00
HN	J94213	A297, A608	0.20 to 0.50	19 to 32	23 to 27	2.00
HP	...	A297	0.35 to 0.75	24 to 28	33 to 37	2.00
HP-50WZ (c)	...	...	0.45 to 0.55	24 to 28	33 to 37	2.50
HT	J94605	A297, A351, A567, A608	0.35 to 0.75	13 to 17	33 to 37	2.50
HU	...	A297, A608	0.35 to 0.75	17 to 21	37 to 41	2.50
HW	...	A297, A608	0.35 to 0.75	10 to 14	58 to 62	2.50
HX	...	A297, A608	0.35 to 0.75	15 to 19	64 to 68	2.50

(a) ASTM designations are same as ACI designations. (b) Rem Fe in all compositions. Manganese content: 0.35 to 0.65% for HA, 1% for HC, 1.5% for HD and 2% for the other alloys. Phosphorus and sulfur contents: 0.04% max for all but HP-50WZ. Molybdenum is intentionally added only to HA, which has 0.90 to 1.20% Mo; maximum for other alloys is set at 0.5% Mo. HH also contains 0.2% max N. (c) Also contains 4 to 6% W, 0.1 to 1.0% Zr, and 0.035% max S and P.

Compositions of nickel-base heat-resistant casting alloys

Alloy designation	C	Ni	Cr	Co	Mo	Fe	Al	B	Ti	W	Zr	Others
B-1900	0.1	64	8	10	6	...	6	0.015	1	...	0.10	4Ta(a)
Hastelloy X	0.1	50	21	1	9	18	...	...	...	1	...	...
IN-100	0.18	60.5	10	15	3	...	5.5	0.01	5	...	0.06	1V
IN-738X	0.17	61.5	16	8.5	1.75	...	3.4	0.01	3.4	2.6	0.1	1.75Ta, 0.9Nb
IN-792	0.2	60	13	9	2.0	...	3.2	0.02	4.2	4	0.1	4Ta
Inconel 713C	0.12	74	12.5	...	4.2	...	6	0.012	0.8	...	0.1	2Nb
Inconel 713LC	0.05	75	12	...	4.5	...	6	0.01	0.6	...	0.1	2Nb
Inconel 718	0.04	53	19	...	3	18	0.5	...	0.9	...	...	0.1Cu, 5Nb
Inconel X-750	0.04	73	15	...	...	7	0.7	...	2.5	...	...	0.25Cu, 0.9Nb
M-252	0.15	56	20	10	10	...	1	0.005	2.6	...	...	...
MAR-M 200	0.15	59	9	10	...	1	5	0.015	2	12.5	0.05	1Nb(b)
MAR-M 246	0.15	60	9	10	2.5	...	5.5	0.015	1.5	10	0.05	1.5Ta
MAR-M 247	0.15	59	8.25	10	0.7	0.5	5.5	0.015	1	10	0.05	1.5Hf, 3Ta
NX 188 (DS)	0.04	74	...	...	18	...	8	...	...	...	...	...
René 77	0.07	58	15	15	4.2	...	4.3	0.015	3.3	...	0.04	...
René 80	0.17	60	14	9.5	4	...	3	0.015	5	4	0.03	...
René 100	0.18	61	9.5	15	3	...	5.5	0.015	4.2	...	0.06	1V
TRW-NASA VIA	0.13	61	6	7.5	2	...	5.5	0.02	1	6	0.13	0.4Hf, 0.5Nb, 0.5Re, 9Ta
Udimet 500	0.1	53	18	17	4	2	3	...	3	...	...	...
Udimet 700	0.1	53.5	15	18.5	5.25	...	4.25	0.03	3.5	...	...	...
Udimet 710	0.13	55	18	15	3	...	2.5	...	5	1.5	0.08	...
Waspaloy	0.07	57.5	19.5	13.5	4.2	1	1.2	0.005	3	...	0.09	...
WAZ-20 (DS)	0.20	72	...	...	...	...	6.5	...	...	20	1.5	...

(a) B-1900 + Hf also contains 1.5% Hf. (b) MAR-M 200 + Hf also contains 1.5% Hf.

Compositions of cobalt-base heat-resistant casting alloys

Alloy designation	C	Co	Cr	Ni	Al	B	Fe	Ta	W	Zr	Others
						Nominal composition, %					
AiResist 13	0.45	62	21	...	3.4	...	...	2	11	...	0.1Y
AiResist 213	0.20	64	20	0.5	3.5	...	0.5	6.5	4.5	0.1	0.1Y
AiResist 215	0.35	63	19	0.5	4.3	...	0.5	7.5	4.5	0.1	0.1Y
Haynes 21	0.25	64	27	3	...	...	1	...	...	...	5 Mo
Haynes 25; L-605	0.1	54	20	10	...	...	1	...	15	...	...
Haynes 151(a)	0.48	65	20	...	...	0.03	...	...	12.8	...	3 max Fe + Ni
J-1650	0.20	36	19	27	...	0.02	...	2	12	...	3.8 Ti
MAR-M 302	0.85	58	21.5	...	...	0.005	0.5	9	10	0.2	...
MAR-M 322	1.0	60.5	21.5	...	...	...	0.5	4.5	9	2	0.75 Ti
MAR-M 509	0.6	54.5	23.5	10	...	...	...	3.5	7	0.5	0.2 Ti
MAR-M 918	0.05	52	20	20	...	...	...	7.5	...	0.1	...
NASA Co-W-Re	0.40	67.5	3	...	...	...	...	...	25	1	2 Re, 1 Ti
S-816	0.4	42	20	20	...	...	4	4	4	...	4 Mo, 4 Nb, 1.2 Mn, 0.4 Si
V-36	0.27	42	25	20	...	...	3	2	2	...	4 Mo, 2 Nb, 1 Mn, 0.4 Si
WI-52	0.45	63.5	21	...	...	...	2	...	11	...	2 Nb+Ta
X-40	0.50	57.5	22	10	...	...	1.5	...	7.5	...	0.5 Mn, 0.5 Si

(a) Obsolete alloy, included for reference purposes.

Standard designations and composition ranges for corrosion-resistant steel castings

ACI type(a)	Wrought alloy type(b)	C (max)	Mn (max)	Si (max)	Composition, % Cr	Ni	Others(c)
CA-6NM	···	0.06	1.00	1.00	11.5 to 14.0	3.5 to 4.5	0.40 to 1.0 Mo
CA-15	410	0.15	1.00	1.50	11.5 to 14.0	1.0 max	0.5 max Mo(d)
CA-40	420	0.40	1.00	1.50	11.5 to 14.0	1.0 max	0.5 max Mo(d)
CB-7Cu-1	···	0.07	0.70	1.00	15.5 to 17.7	3.6 to 4.6	2.5 to 3.2 Cu; 0.20 to 0.35 Nb; 0.05 max N
CB-7Cu-2	···	0.07	0.70	1.00	14.0 to 15.5	4.5 to 5.5	2.5 to 3.2 Cu; 0.20 to 0.35 Nb; 0.05 max N
CB-30	431	0.30	1.00	1.50	18.0 to 22.0	2.0 max	···
CC-50	446	0.50	1.00	1.50	26.0 to 30.0	4.0 max	···
CD-4MCu	···	0.04	1.00	1.00	25.0 to 26.5	4.75 to 6.0	1.75 to 2.25 Mo; 2.75 to 3.25 Cu
CE-30	312	0.30	1.50	2.00	26.0 to 30.0	8.0 to 11.0	···
CF-3 (e)	304L	0.03	1.50	2.00	17.0 to 21.0	8.0 to 12.0	···
CF-3M (e)	316L	0.03	1.50	2.00	17.0 to 21.0	8.0 to 12.0	2.0 to 3.0 Mo
CF-8 (e)	304	0.08	1.50	2.00	18.0 to 21.0	8.0 to 11.0	···
CF-8C	347	0.08	1.50	2.00	18.0 to 21.0	9.0 to 12.0	Nb(f)
CF-8M	316	0.08	1.50	2.00	18.0 to 21.0	9.0 to 12.0	2.0 to 3.0 Mo
CF-12M	316	0.12	1.50	2.00	18.0 to 21.0	9.0 to 12.0	2.0 to 3.0 Mo
CF-16F	303	0.16	1.50	2.00	18.0 to 21.0	9.0 to 12.0	1.50 max Mo; 0.20 to 0.35 Se
CF-20	302	0.20	1.50	2.00	18.0 to 21.0	8.0 to 11.0	···
CG-8M	317	0.08	1.50	1.50	18.0 to 21.0	9.0 to 13.0	3.0 to 4.0 Mo
CH-20	309	0.20	1.50	2.00	22.0 to 26.0	12.0 to 15.0	···
CK-20	310	0.20	2.00	2.00	23.0 to 27.0	19.0 to 22.0	···
CN-7M	···	0.07	1.50	1.50	19.0 to 22.0	27.5 to 30.5	2.0 to 3.0 Mo; 3.0 to 4.0 Cu
CN-7MS	···	0.07	1.50	3.50(g)	18.0 to 20.0	22.0 to 25.0	2.5 to 3.0 Mo; 1.5 to 2.0 Cu

(a) Most of these standard grades are covered by ASTM A743 and A744. (b) Type numbers of wrought alloys are listed only for nominal identification of corresponding wrought and cast grades. Composition ranges of cast alloys are not the same as for corresponding wrought alloys; cast alloy designations should be used for castings only. (c) Phosphorus content is 0.04% max except in CF-16F, which has 0.17% max P; sulfur content is 0.04% max in all grades. (d) Molybdenum not intentionally added. (e) CF-3A, CF-3MA and CF-8A have the same composition ranges as CF-3, CF-3M and CF-8, respectively, but have balanced compositions so that ferrite contents are at levels that permit higher mechanical-property specifications than those for related grades. They are covered by ASTM A351. (f) Nb, 8 × %C min (1.0% max); or Nb + Ta, 9 × %C (1.1% max). (g) For CN-7MS, silicon ranges from 2.50 to 3.50.

Physical and Mechanical Properties

Physical properties of selected superalloys

Alloy	Density, Mg/m³	Melting temperatures Liquidus °C	Liquidus °F	Solidus °C	Solidus °F	Specific heat (a) J/kg	Specific heat (a) Btu/lb	Electrical conductivity, % IACS	Electric resistivity, nΩ·m	Magnetic permeability	Curie temperature °C	Curie temperature °F
Iron-base alloys												
Carpenter 20-Cb3	8.055	1425	2600	1370	2500	...	...	...	1040	...	...	...
Haynes 556	8.23	...	...	...	...	472	0.113	...	970	...	...	...
Incoloy 800	7.94	1385	2525	1355	2475	502	0.117	1.7	989	1.0092	...	...
Incoloy 801	7.94	1385	2525	1355	2475	452	0.105	1.7	1012	...	...	...
Incoloy 825	8.14	1400	2500	1370	2500	...	...	1.5	1127	1.005	<−196	<−520
Cobalt-base alloys												
Haynes 25 (L-605)	9.13	1410	2570	1329	2425	374	0.090	...	890	<1.00	...	...
Haynes 188	9.13	1398	2550	1302-1330	2375-2425	423(b)	0.101(b)	...	...	1.01	...	...
Stellite 6B	8.38	1354	2470	1265	2310	421	0.101	...	910	<1.2	...	...
UMCo 50	8.05	1395	2540	1380	2515	...	...	...	825	...	...	...
Nickel-base alloys												
Hastelloy B-2	9.21	...	...	...	...	389(b)	0.093(b)	...	1380(b)	...	...	...
Hastelloy C-4	8.64	...	...	...	...	426(b)	0.102(b)	...	1250	...	...	...
Hastelloy C-276	8.90	1371	2500	1323	2415	427	0.102	...	1330	...	...	...
Hastelloy N	8.93	...	...	...	...	419(b)	0.100(b)	...	1200(b)	...	...	...
Hastelloy S	8.76	1380	2516	1335	2435	427(b)	0.102(b)	...	...	...	...	...
Hastelloy W	9.03	1315	2400	...	...	486	0.116	...	1180	<1.002(c)	...	...
Hastelloy X	8.23	1290	2350	1250	2280	...	...	...	...	1.010	−124	−192
Inconel 600	8.42	1415	2575	1354	2470	444	0.103	1.7	1030	...	...	...
Inconel 617	...	...	...	1333	2430	410	0.095	...	...	1.006	−196	−320
Inconel 625	8.44	1350	2460	1290	2350	456	0.106	1.3	1290	...	...	...
Inconel 671	7.86	1350	2460	1305	2385	...	...	2.0	869	...	...	...
Inconel 690	8.03	1375	2510	1345	2450	431	0.103	1.5	148	...	...	...
Inconel X750	8.25	1425	2600	1393	2540	...	...	...	1215	1.0020	−143	−225
Nimonic 75	...	...	...	1380	2515	...	...	...	...	...	...	...
Nimonic 80A	...	...	...	1360	2480	...	...	...	...	...	...	...
Nimonic 90	...	1310	...	1310	2390	...	...	...	...	...	...	...
Nimonic 100	...	...	...	1290	2256	...	...	...	...	...	...	...
Nimonic 105	...	1290	...	1232	2250	452	0.108	...	...	...	...	...
René 41	8.25	1371	2500	1232	2250	...	...	...	1308	1.002	...	...
Udimet 500	8.14	1345	2450	1260	2300	...	...	...	1203	...	...	...
Udimet 700	7.92	1345	2450	1216	2220	...	...	...	...	...	...	...
Waspaloy	8.20	1355	2475	1339	2425	523(c)	0.125(c)	...	1240	...	...	...

(a) At room temperature. (b) At 100 °C (212 °F). (c) At 93 °C (200 °F).

Typical mechanical properties of cobalt-base and nickel-base superalloys

Temperature		Tensile strength		Yield strength		Elongation,
°C	°F	MPa	ksi	MPa	ksi	%
Cobalt-base alloys						
Haynes 25 (L-605) sheet						
21	70	1010	146	460	67	64
540	1000	800	116	250	36	59
650	1200	710	103	240	35	35
760	1400	455	66	260	38	12
870	1600	325	47	240	35	30
Haynes 188, sheet						
21	70	960	139	485	70	56
540	1000	740	107	305	44	70
650	1200	710	103	305	44	61
760	1400	635	92	290	42	43
870	1600	420	61	260	38	73
S-816, bar						
21	70	965	140	385	56	30
540	1000	840	122	310	45	27
650	1200	765	111	305	44	25
760	1400	650	94	285	41	21
870	1600	360	52	240	35	16
Nickel-base alloys						
Astroloy, bar						
21	70	1410	205	1050	152	16
540	1000	1240	180	965	140	16
650	1200	1310	190	965	140	18
760	1400	1160	168	910	132	21
870	1600	770	112	690	100	25
D-979, bar						
21	70	1410	204	1010	146	15
540	1000	1300	188	925	134	15
650	1200	1100	160	890	129	21
760	1400	7	104	655	95	17
870	1600	345	50	305	44	18
Hastelloy X, sheet						
21	70	785	114	360	52	43
540	1000	650	94	290	42	45
650	1200	570	83	275	40	37
760	1400	435	63	260	38	37
870	1600	255	37	180	26	50
IN-102, bar						
21	70	960	139	505	73	47
540	1000	825	120	400	58	48
650	1200	710	103	400	58	64
760	1400	440	64	385	56	110
870	1600	215	31	200	29	110
Inconel 600, bar						
21	70	620	90	250	36	47
540	1000	580	84	195	28	47
650	1200	450	65	180	26	39
760	1400	185	27	115	17	46
870	1600	105	15	62	9	80

(continued)

Typical mechanical properties of cobalt-base and nickel-base superalloys (continued)

Temperature		Tensile strength		Yield strength		Elongation,
°C	°F	MPa	ksi	MPa	ksi	%

Inconel 601, sheet

21	70	740	107	340	49	45
540	1000	725	105	150	22	38
650	1200	525	76	180	26	45
760	1400	290	42	200	29	73
870	1600	160	23	140	20	92

Inconel 625, bar

21	70	855	124	490	71	50
540	1000	745	108	405	59	50
650	1200	710	103	420	61	35
760	1400	505	73	420	61	42
870	1600	285	41	475	40	125

Inconel 706, bar

21	70	1300	188	980	142	19
540	1000	1120	163	895	130	19
650	1200	1010	147	825	120	21
760	1400	690	100	675	98	32

Inconel 718, bar

21	70	1430	208	1190	172	21
540	1000	1280	185	1060	154	18
650	1200	1230	178	1020	148	19
760	1400	950	138	740	107	25
870	1600	340	49	330	48	88

Inconel 718, sheet

21	70	1280	185	1050	153	22
540	1000	1140	166	945	137	26
650	1200	1030	150	870	126	15
760	1400	675	98	625	91	8

Inconel X 750, bar

21	70	1120	162	635	92	24
540	1000	965	140	580	84	22
650	1200	825	120	565	82	9
760	1400	485	70	455	66	9
870	1600	235	34	165	24	47

M-252, bar

21	70	1240	180	840	122	16
540	1000	1230	178	765	111	15
650	1200	1160	168	745	108	11
760	1400	945	137	715	104	10
870	1600	510	74	485	70	18

Nimonic 75, bar

21	70	750	109	...	...	41
540	1000	635	92	...	...	41
650	1200	538	78	...	...	42
760	1400	290	42	...	...	70
870	1600	145	21	...	...	68

Nimonic 80A, bar

21	70	1240	179	620	90	24
540	1000	1100	160	530	77	24
650	1200	1000	145	550	80	18
760	1400	760	110	505	73	20
870	1600	400	58	260	38	34

(continued)

Typical mechanical properties of cobalt-base and nickel-base superalloys (continued)

Temperature °C	°F	Tensile strength MPa	ksi	Yield strength MPa	ksi	Elongation, %
Nimonic 90, bar						
21	70 1240		180	805	117	23
540	1000 1100		160	725	105	23
650	1200 1030		150	685	99	20
760	1400 825		120	540	78	10
870	1600 430		62	260	38	16
Nimonic 105, bar						
21	70 1140		166	815	118	12
540	1000 1100		160	775	112	18
650	1200 1080		156	800	116	24
760	1400 965		140	655	95	22
870	1600 605		88	365	53	25
Nimonic 115, bar						
21	70 1240		180	860	125	25
540	1000 1090		158	795	115	26
650	1200 1120		163	815	118	25
760	1400 1080		157	800	116	22
870	1600 825		120	550	80	18
Pyromet 860, bar						
21	70 1300		188	835	121	22
540	1000 1250		182	840	122	15
650	1200 1110		161	850	123	17
760	1400 910		132	835	121	18
René 41, bar						
21	70 1420		206	1060	154	14
540	1000 1400		203	1010	147	14
650	1200 1340		194	1000	145	14
760	1400 1100		160	940	136	11
870	1600 620		90	550	80	19
René 95, bar						
21	70 1620		235	1310	190	15
540	1000 1540		224	1250	182	12
650	1200 1460		212	1220	177	14
760	1400 1170		170	1100	160	15
Udimet 500, bar						
21	70 1310		190	840	122	32
540	1000 1240		180	795	115	28
650	1200 1210		176	760	110	28
760	1400 1040		151	730	106	39
870	1600 640		93	495	72	20
Udimet 520, bar						
21	70 1310		190	860	125	21
540	1000 1240		180	825	120	20
650	1200 1170		170	795	115	17
760	1400 725		105	725	105	15
870	1600 515		75	515	75	20
Udimet 700, bar						
21	70 1410		204	965	140	17
540	1000 1280		185	895	130	16
650	1200 1240		180	855	124	16
760	1400 1030		150	825	120	20
870	1600 690		100	635	92	27

(continued)

Typical mechanical properties of cobalt-base and nickel-base superalloys (continued)

Temperature		Tensile strength		Yield strength		Elongation,
°C	°F	MPa	ksi	MPa	ksi	%
Udimet 710, bar						
21	70	1190	172	910	132	7
540	1000	1150	167	850	123	10
650	1200	1290	187	860	125	15
760	1400	1020	148	815	118	25
870	1600	705	102	635	92	29
Unitemp AF2-1DA, bar						
21	70	1290	187	1050	152	10
540	1000	1340	194	1080	157	13
650	1200	1360	197	1080	157	13
760	1400	1150	167	1010	146	8
870	1600	830	120	715	104	8
Waspaloy, bar						
21	70	1280	185	795	115	25
540	1000	1170	170	725	105	23
650	1200	1120	162	690	100	34
760	1400	795	115	675	98	28
870	1600	525	76	515	75	35

Typical rupture strengths of selected superalloys

Temperature		For stress rupture at:			
		100 h		1000 h	
°C	°F	MPa	ksi	MPa	ksi
Incoloy 800					
650	1200 220		32	145	21
760	1400 115		17	69	10
870	1600 45		6.5	33	4.8
Incoloy 801					
650	1200 250		36	...	...
730	1350 145		21	...	...
815	1500 62		9	...	...
Incoloy 802					
650	1200 240		35	170	24
760	1400 145		21	105	15
870	1600 97		14	62	9
Inconel 600					
815	1500 55		8	39	5.6
870	1600 37		5.3	24	3.5
Inconel 601(a)					
540	1000		...	400	58
870	1600 48		7	30	4.3
980	1800 23		3.4	14	2.1
Inconel 617(b)					
815	1500 140		20	97	14
925	1700 62		9	...	5.5
980	1800 41		6	...	3.5
Inconel 625(a)					
650	1200 440		64	370	54
815	1500 130		19	93	13.5
870	1600 72		10.5	48	7
Inconel 718(c)					
540	1000		...	951	138
595	1100 860		125	760	110
650	1200 690		100	585	85
Inconel 751(d)					
815	1500 200		29	125	185
870	1600 120		175	69	10
Inconel X750(e)					
540	1000		...	827	120
870	1600 83		12	45	6.5
925	1700 58		8.4	21	3.1
N-155, bar(f)					
650	1200 360		52	295	43
730	1350 195		28	150	22
870	1600 97		14	66	9.5
N-155(g)					
650	1200 380		55	290	42

(continued)

(a) Solution treated 1150 °C (2100 °F). (b) Solution treated 1175 °C (2150 °F). (c) Heat treated to 980 °C (1800 °F) plus 720 °C (1325 °F) hold for 8 h, F.C. to 620 °C (1150 °F) hold for 8 h. (d) 730 °C (1350 °F) hold for 2 h. (e) Heat treat to 1150 °C (2100 °F) plus 840 °C (1550 °F) hold for 24 h, plus 705 °C (1300 °F) hold for 20 h. (f) Solution treated and aged. (g) Stress-relieved forging. (h) Heat treat to 1050 °C (1920 °F) hold for 1 h. (j) Heat treat to 1080 °C (1975 °F) hold for 8 h, plus 700 °C (1290 °F) hold for 16 h. (k) Heat treat to 1150 °C (2100 °F) hold for 4 h, plus 1050 °C (1920 °F) hold for 16 h, plus 850 °C (1560 °F) hold for 16 h. (m) Heat treat to 1190 °C (2175 °F) hold for 1.5 h, plus 1100 °C (2010 °F) hold for 6 h. (n) Heat treat to 1150 °C (2100 °F) hold for 2 h, W.Q., plus 800 °C (1475 °F) hold for 8 h.

Typical rupture strengths of selected superalloys (continued)

| Temperature | | For stress rupture at: | | | |
| | | 100 h | | 1000 h | |
°C	°F	MPa	ksi	MPa	ksi
N-155, sheet(f)					
980	1800	39	5.6	20	2.9
Nimonic 75(h)					
815	1500	38	5.5	24	3.5
870	1600	23	3.4	15	2.2
925	1700	14	2.1	10	1.5
980	1800	...	...	7.6	1.1
Nimonic 80A(j)					
540	1000	...	...	825	120
815	1500	185	27	115	17
870	1600	105	15	...	...
Nimonic 90(j)					
815	1500	240	35	155	22.5
870	1600	150	22	69	10
925	1700	69	10	...	...
Nimonic 105(k)					
815	1500	325	47	225	32
870	1600	210	30.2	135	19
Nimonic 115(m)					
815	1500	425	62	315	46
870	1600	315	46	205	30
925	1700	205	30	130	18.5
Nimonic 263(n)					
815	1500	170	24.5	105	15
870	1600	93	13.5	46	6.7
925	1700	45	6.5	...	...

(a) Solution treated 1150 °C (2100 °F). (b) Solution treated 1175 °C (2150 °F). (c) Heat treated to 980 °C (1800 °F) plus 720 °C (1325 °F) hold for 8 h, F.C. to 620 °C (1150 °F) hold for 8 h. (d) 730 °C (1350 °F) hold for 2 h. (e) Heat treat to 1150 °C (2100 °F) plus 840 °C (1550 °F) hold for 24 h, plus 705 °C (1300 °F) hold for 20 h. (f) Solution treated and aged. (g) Stress-relieved forging. (h) Heat treat to 1050 °C (1920 °F) hold for 1 h. (j) Heat treat to 1080 °C (1975 °F) hold for 8 h, plus 700 °C (1290 °F) hold for 16 h. (k) Heat treat to 1150 °C (2100 °F) hold for 4 h, plus 1050 °C (1920 °F) hold for 16 h, plus 850 °C (1560 °F) hold for 16 h. (m) Heat treat to 1190 °C (2175 °F) hold for 1.5 h, plus 1100 °C (2010 °F) hold for 6 h. (n) Heat treat to 1150 °C (2100 °F) hold for 2 h, W.Q., plus 800 °C (1475 °F) hold for 8 h.

Typical room-temperature properties of ACI heat-resistant casting alloys

Alloy	Condition	Tensile strength MPa	ksi	Yield strength MPa	ksi	Elonga- tion, %	Hardness, HB
HC	As cast........... 760		110	515	75	19	223
	Aged(a) 790		115	550	80	18	...
HD	As cast........... 585		85	330	48	16	90
HE	As cast........... 655		95	310	45	20	200
	Aged(a) 620		90	380	55	10	270
HF	As cast........... 635		92	310	45	38	165
	Aged(a) 690		100	345	50	25	190
HH, type 1	As cast........... 585		85	345	50	25	185
	Aged(a) 595		86	380	55	11	200
HH, type 2	As cast........... 550		80	275	40	15	180
	Aged(a) 635		92	310	45	8	200
HI	As cast........... 550		80	310	45	12	180
	Aged(a) 620		90	450	65	6	200
HK	As cast........... 515		75	345	50	17	170
	Aged(b) 585		85	345	50	10	190
HL	As cast........... 565		82	360	52	19	192
HN	As cast........... 470		68	260	38	13	160
HP	As cast........... 490		71	275	40	11	170
HT	As cast........... 485		70	275	40	10	180
	Aged(b) 515		75	310	45	5	200
HU	As cast........... 485		70	275	40	9	170
	Aged(c).......... 505		73	295	43	5	190
HW	As cast........... 470		68	250	36	4	185
	Aged(d) 580		84	360	52	4	205
HX	As cast........... 450		65	250	36	9	176
	Aged(c).......... 505		73	305	44	9	185

(a) Aging treatment: 24 h at 760 °C (1400 °F), furnace cool. (b) Aging treatment: 24 h at 760 °C (1400 °F), air cool. (c) Aging treatment: 48 h at 980 °C (1800 °F), air cool. (d) Aging treatment: 48 h at 980 °C (1800 °F), furnace cool.

Approximate rates of corrosion for ACI heat-resistant casting alloys in air and in flue gas

Alloy	Oxidation rate in air, mils/yr(a) 870 °C (1600 °F)	980 °C (1800 °F)	1090 °C (2000 °F)	Corrosion rate, mils/yr(a)(b), in flue gas with sulfur content of: 0.12 g/m^3 Oxidizing	Reducing	2.3 g/m^3 Oxidizing	Reducing
HB.......	25 –	250 –	500 –	100 +	500 –	250 –	500
HC.......	10	50	50	25 –	25 +	25	25 –
HD.......	10 –	50 –	50 –	25 –	25 –	25 –	25 –
HE.......	5 –	25 –	35 –	25 –	25 –	25 –	25 –
HF.......	5 –	50 +	100	50 +	100 +	50 +	250 –
HH.......	5 –	25 –	50	25 –	25	25	25 –
HI	5 –	10 +	35 –	25 –	25 –	25 –	25 –
HK.......	10 –	10 –	35 –	25 –	25 –	25 –	25 –
HL.......	10 +	25 –	35	25 –	25 –	25 –	25 –
HN.......	5	10 +	50 –	25 –	25 –	25	25
HP.......	25 –	25	50	25 –	25 –	25 –	25 –
HT.......	5 –	10 +	50	25	25 –	25	100
HU.......	5 –	10 –	35 –	25 –	25 –	25 –	25
HW	5 –	10 –	35	25	25 –	50 –	250
HX.......	5 –	10 –	35 –	25 –	25 –	25 –	25 –

(a) Data based on 100-h tests. To convert to μm/yr, multiply by 25. (b) At 980 °C (1800 °F).

Results of in-plant corrosion testing of CF-8, CF-8M and CN-7M alloys

Type and composition of corroding solution	Temperature of solution, °C	°F	Alloy	Metal loss on surface µm/yr	mils/yr	Surface condition by visual examination	Remarks
Neutralizer after formation of ammonium sulfate: ammonium sulfate plus small excess of sulfuric acid, ammonia vapor and steam	100	212	CF-8 CF-8M CN-7M	665 28 18	26.2 1.1 0.7	Very heavy etch(a) Light tarnish(b) Bright	CF-8M was installed for low-corrosion-tolerance equipment in this service and performed satisfactorily.
Settling tank after neutralizer: ammonium sulfate plus excess of sulfuric acid	50	122	CF-8 CF-8M CN-7M	385 10 2.5	15.2 0.4 0.1	Very heavy etch(a) Slight tarnish Bright(b)	CF-8 in service showed excessive corrosion rate plus heavy concentration-cell attack.
Ammonium sulfate processing solution: ammonium sulfate at pH of 8.0	50	122	CF-8 CF-8M CN-7M	685 175 5	27.0 6.8 2.0	Heavy etch Moderate etch Light etch	CF-8M had too high a corrosion rate in service for good valve life, although suitable for equipment of greater corrosion tolerance. CN-7M was installed in this service.
99 to 100% fuming nitric acid	20	68	CF-8 CN-7M CF-8M	245 79 345	9.6 3.1 13.5	Moderate etch Light etch Moderate etch	CF-8 was satisfactory except for low-tolerance equipment such as valves. CN-7M valves performed satisfactorily in service.
Saturated solution of sodium chloride plus 15% sodium sulfate; pH, 4.5	60	140	CF-8M CF-8	2.5 240	0.1 9.5	Bright Concentration-cell corrosion at various small areas of specimen	CF-8M was installed for valves in service.

(a) Concentration-cell attack under insulating washer. (b) Slight concentration-cell attack under insulating washer.

Heat Treating

Typical solution treating and aging cycles for heat-resisting casting alloys

| | Solution treating | | | | Aging | | | |
| | Temperature(a) | | Time, h | Cooling procedure | Temperature(b) | | Time, h | Cooling procedure |
Alloy	°C	°F			°C	°F		
A-286	1095	2000	2	Rapid cool	720	1325	16	Air cool
B-1900	As cast		...	...	...	...	...	...
FSX-414	1150	2100	4	Rapid cool	980	1800	4	Air cool
Hastelloy B	1175	2150	2	Rapid cool	(c)	(c)	...	...
Hastelloy C	1220	2225	1	Rapid cool	(c)	(c)	...	...
HS-31 (X-40)	As cast		...	...	...	...	...	...
IN-100	As cast		...	...	...	...	...	...
IN-713C	As cast		...	...	...	...	...	...
IN-738	1120	2050	2	Air cool	845	1550	24	Air cool
IN-792	1120	2050	2	Air cool	845	1550	24	Air cool
IN-939	1160	2120	4	Air cool	850	1560	16	Air cool
Inconel 718	1095	2000	1	Air cool	620	1150	10	Air cool
MAR-M 200	...	...	...	...	870	1600	50	Air cool
MAR-M 200 DS	1230	2250	4	Air cool	870	1600	32	Air cool
MAR-M 246	...	...	...	...	845	1550	50	Air cool
MAR-M 247	...	...	...	...	870	1600	16	Air cool
MAR-M 302	As cast		...	...	...	...	...	...
MAR-M 509	As cast		...	...	...	...	...	...
René 41	1095	2000	½	Rapid cool	900	1650	4	Air cool
René 80	1220	2225	2	Air cool	1095	2000	4	Air cool
					1055	1925	4	Air cool
					845	1550	16	Air cool
Udimet 700	1150	2100	2	Air cool	760	1400	16	Air cool

(a) Furnace temperature tolerance of ± 15 °C (± 25 °F) is satisfactory. (b) Furnace temperature tolerance of ± 10 °C (± 15 °F) is recommended. (c) Aging occurs in service at elevated temperature. Use a vacuum or protective atmosphere for heat treating at temperatures above 1040 °C (1900 °F) and subsequent cooling.

Prepared atmospheres suitable for annealing of nickels and nickel alloys

Atmospheres 2 through 7 can be used for bright annealing of nickel, modified nickels, and nickel-copper alloys; atmosphere 4 or atmosphere 7 must be used for bright annealing of nickel alloys that contain chromium or molybdenum, or both.

| Atmosphere | Air-to-gas ratio(a) | Composition, vol % | | | | | | Dew point (approx) | |
		H_2	CO	CO_2	CH_4	O_2	N_2	°C	°F
1 Completely burned fuel, lean atmosphere	10 to 1	0.5	0.5	10.0	0.0	0.0	89.0	Saturated(b)	
2 Partially burned fuel, medium-rich atmosphere	6 to 1	15.0	10.0	5.0	1.0	0.0	69.0	Saturated(b)	
3 Reacted fuel, rich atmosphere	3 to 1	38.0	19.0	1.0	2.0	0.0	40.0	+ 20	+ 70
4 Dissociated ammonia (complete dissociation)	No air	75.0	0.0	0.0	0.0	0.0	25.0	− 55 to − 75	− 70 to − 100
5 Dissociated ammonia, partially burned	1.25 to 1(c)	15.0	0.0	0.0	0.0	0.0	85.0	Saturated(b)	
6 Dissociated ammonia, completely burned	1.8 to 1(c)	1.0	0.0	0.0	0.0	0.0	99.0	Saturated(b)	
7 Electrolytic hydrogen, dried(d)	No air	100.0	0.0	0.0	0.0	0.0	0.0	− 55 to − 75	− 70 to − 100

(a) Based on use of natural gas containing nearly 100% methane and rated at 37 MJ/m³ (1000 Btu/ft³). For high-hydrogen manufactured gas (20 MJ/m³, or 550 Btu/ft³), ratios are about 50% of values listed. For manufactured gas with lower hydrogen and high carbon monoxide contents (17 MJ/m³, or 450 Btu/ft³), ratios are about 40% of values listed. For propane, ratios are about twice those listed. For butane, multiply listed values by three. (b) When atmosphere is cooled by tap-water heat exchangers, dew point will be about 6 to 8 °C (10 to 15 °F) above the temperature of the tap water. Dew point may be reduced to about 5 °C (40 °F) by refrigeration equipment, and to − 55 °C (− 70 °F) or lower by activated-absorption equipment. (c) Ratio of air to dissociated ammonia. (d) Dried to a dew point of − 55 to − 75 °C (− 70 to − 100 °F) by alumina plus molecular sieve.

Typical solution treating and aging cycles for wrought heat-resisting alloys

Alloy	Solution treating Temperature °C	°F	Time, h	Cooling procedure	Aging Temperature °C	°F	Time, h	Cooling procedure
Iron-base alloys								
A-286	980	1800	1	Oil quench	720	1325	16	Air cool
Discaloy	1010	1850	2	Oil quench	730	1350	20	Air cool
					650	1200	20	Air cool
N-155	1175	2150	1	Water quench	815	1500	4	Air cool
Nickel-base alloys								
Astroloy	1175	2150	4	Air cool	845	1550	24	Air cool
	1080	1975	4	Air cool	760	1400	16	Air cool
Hastelloy B	1175	2150	1/2	(a)	(b)	(b)	. . .	. . .
Hastelloy B-2	1065	1950	1/2	Rapid quench	. . .	. . .	. . .	. . .
Hastelloy C-4	1065	1950	1/2	Rapid quench	. . .	. . .	. . .	. . .
Hastelloy C-276	1120	2050	1/2	Rapid quench	. . .	. . .	. . .	. . .
Hastelloy N	1175	2150	1/2	Rapid quench	. . .	. . .	. . .	. . .
Hastelloy S	1065	1950	1/2	Rapid quench	. . .	. . .	. . .	. . .
Hastelloy C	1220	2225	1	(a)	(b)	(b)	. . .	. . .
Hastelloy W	1175	2150	1	(a)	(b)	(b)	. . .	. . .
Hastelloy X	1175	2150	1	(a)	. . .	. . .	. . .	. . .
Inconel 901	1095	2000	2	Water quench	790	1450	2	Air cool
					720	1325	24	Air cool
Inconel 600	1120	2050	2	Air cool	. . .	. . .	. . .	. . .
Inconel 601	1150	2100	1	Air cool	. . .	. . .	. . .	. . .
Inconel 617	1175	2150	2	(a)	. . .	. . .	. . .	. . .
Inconel 625	1150	2100	2	(a)	. . .	. . .	. . .	. . .
Inconel 706	925-1010	1700-1850	. . .	. . .	845	1550	3	Air cool
					720	1325	8	Furnace cool
					620	1150	8	Air cool
	925-1010	1700-1850	. . .	. . .	730	1350	8	Furnace cool
					620	1150	8	Air cool
Inconel 718	980	1800	1	Air cool	720	1325	8	Furnace cool
					620	1150	8	Air cool
Inconel X-750 (AMS 5667)	855	1625	24	Air cool	705	1300	20	Air cool
Inconel X-750 (AMS 5668)	1150	2100	2	Air cool	845	1550	24	Air cool
					705	1300	20	Air cool
Nimonic 80A	1080	1975	8	Air cool	705	1300	16	Air cool
Nimonic 90	1080	1975	8	Air cool	705	1300	16	Air cool
René 41	1065	1950	1/2	Air cool	760	1400	16	Air cool
Udimet 500	1080	1975	4	Air cool	845	1550	24	Air cool
					760	1400	16	Air cool
Udimet 700	1175	2150	4	Air cool	845	1550	24	Air cool
	1080	1975	4	Air cool	760	1400	16	Air cool
Waspaloy	1080	1975	4	Air cool	845	1550	24	Air cool
					760	1400	16	Air cool
Cobalt-base alloys								
Haynes 25; L-605	1230	2250	1	Rapid air cool	(b)	(b)	. . .	. . .
Haynes 188	1175	2150	1/2	Rapid air cool	. . .	. . .	. . .	. . .
Haynes 556	1175	2150	1/2	Rapid air cool	. . .	. . .	. . .	. . .
S-816	1175	2150	1	(a)	760	1400	12	Air cool
Stellite 6B	1230	2250	1	Air cool	. . .	. . .	. . .	. . .

Note: Alternate treatments may be used to improve specific properties. (a) To provide an adequate quench after solution treating, it is necessary to cool below about 540 °C (1000 °F) rapidly enough to prevent precipitation in the intermediate temperature range. For sheet metal parts of most alloys, rapid air cooling will suffice. Oil or water quenching is frequently required for heavier sections that are not subject to cracking. (b) Aging occurs in service at elevated temperatures.

Typical stress relieving and annealing cycles for wrought heat-resisting alloys

Alloy	Stress relieving Temperature °C	°F	Holding time per inch of section, h	Annealing(a) Temperature °C	°F	Holding time per inch of section, h
Iron-base and iron-nickel-chromium alloys						
RA-330	900	1650	1(b)	1110(c)	2025(c)	¼(d)
19-9 DL	675(e)	1250(e)	4	980	1800	1
A-286	(f)	(f)	...	980	1800	1
Discaloy	(f)	(f)	...	1035	1900	1
Nickel-base alloys						
Astroloy	(f)	(f)	...	1135	2075	4
Hastelloy B	(f)	(f)	...	1175	2150	1
Hastelloy C	(f)	(f)	...	1215	2225	1
Hastelloy W	(f)	(f)	...	1175	2150	1
Hastelloy X	(f)	(f)	...	1175	2150	1
Incoloy 800	870	1600	1½	980	1800	¼
Incoloy 800H	...	...	...	1175	2150	...
Incoloy 825	...	...	...	980	1800	...
Incoloy 901	(f)	(f)	...	1095	2000	2
Inconel 600	900	1650	1	1010	1850	¼(d)
Inconel 601	...	...	...	980	1800	...
Inconel 625	870	1600	1	980	1800	1
Inconel 690	...	...	...	1040	1900	½
Inconel 718	(f)	(f)	...	955	1750	1
Inconel X-750	880(g)	1625(g)	...	1035	1900	½
Nimonic 80A	(f)	(f)	...	1080	1975	2
Nimonic 90	(f)	(f)	...	1080	1975	2
René 41	(f)	(f)	...	1080	1975	2
Udimet 500	(f)	(f)	...	1080	1975	4
Udimet 700	(f)	(f)	...	1135	2075	4
Waspaloy	(f)	(f)	...	1010	1850	4
Cobalt-chromium-nickel-base alloys						
L-605 (HS-25)	(h)	(h)	...	1230	2250	1
N-155 (HS-95)	(h)	(h)	...	1175	2150	...
S-816	(h)	(h)	...	1205	2200	1
Refractory metals(j)						
Ta-10W	1205(k)	2200(k)	1	1425(k)	2600(k)	1
FS-80	1095(k)	2000(k)	1	1315(k)	2400(k)	1
FS-82	1095(k)	2000(k)	1	1315(k)	2400(k)	1
Mo-0.5 Ti	1095(m)	2000(m)	½	1315(m)(n)	2400(m)(n)	1
TZM	1205(m)	2200(m)	1	1425(m)(n)(p)	2600(m)(n)(p)	1

(a) Minimum hardness is achieved by cooling rapidly from the annealing temperature, to prevent precipitation of hardening phases. Water quenching is preferred, and is usually necessary for heavy sections; air cooling is preferred for heavy sections of Waspaloy, Udimet 500, Udimet 700 and Inconel X-750, because water quenching causes cracking. However, for complex shapes subject to excessive distortion, oil quenching is often adequate and more practical. Rapid air cooling usually is adequate for parts formed from strip or sheet. Rapid cooling from the annealing or solution treating temperature does not suppress the aging reaction of some alloys, such as Astroloy; these alloys become harder and stronger. (b) Time given is minimum; some plants use as long as 3 h per inch. (c) Nominal temperature; 1035 to 1175 °C (1900 to 2150 °F) is commonly used. (d) Short time is required for prevention of grain coarsening. (e) Nominal temperature; 650 to 705 °C (1200 to 1300 °F) is permissible. (f) Full annealing is recommended, because intermediate temperatures cause aging. (g) Used only for stress equalizing of warm worked grades. (h) Full annealing is recommended if further fabrication is performed; otherwise, material can be stress relieved at approximately 55 °C (100 °F) below annealing temperature. (j) Annealing temperatures depend on prior plastic deformation, degree of cold work, alloy content and interstitial purity. Annealing temperatures given are those most frequently used for cold worked sheet or plate; in many instances, more precise determination of the recrystallization temperature is necessary for a specific application. (k) Heat and cool in vacuum or inert-gas atmosphere. (m) Heat and cool in hydrogen or vacuum. (n) Seldom used as finished product in annealed condition, because recrystallization raises the ductile-brittle transition temperature, resulting in brittleness at low temperatures. (p) For vacuum-arc-cast material with a minimum of 50% cold work.

Machining

Machining data for ACI heat-resistant casting alloys

ACI designation	Typical hardness, HB	Rough turning(a) Speed, sfm(b)	Rough turning(a) Feed, ipr(c)	Finishing Speed, sfm(b)	Finishing Feed, ipr(c)	Drilling speed(d), sfm(b)
HA	220	40 to 50	0.010 to 0.030	80 to 100	0.005 to 0.010	35 to 70
HC	220	40 to 50	0.025 to 0.035	80 to 100	0.010 to 0.015	40 to 60
HD	190	40 to 50	0.025 to 0.035	80 to 100	0.010 to 0.015	40 to 60
HE	270	30 to 40	0.020 to 0.025	60 to 80	0.005 to 0.010	30 to 60
HF.	190	25 to 35	0.020 to 0.025	50 to 70	0.005 to 0.010	20 to 40
HH	200	25 to 35	0.015 to 0.020	50 to 70	0.005 to 0.010	20 to 40
HI	200	25 to 35	0.015 to 0.020	50 to 70	0.005 to 0.010	20 to 40
HK	190	25 to 35	0.020 to 0.025	50 to 70	0.005 to 0.010	20 to 40
HL.	190	30 to 40	0.020 to 0.025	60 to 80	0.005 to 0.010	30 to 60
HN	160	35 to 45	0.020 to 0.025	70 to 90	0.005 to 0.010	40 to 60
HP.	. . .	35 to 45	0.020 to 0.025	70 to 90	0.005 to 0.010	40 to 60
HT.	200	40 to 45	0.025 to 0.035	80 to 90	0.005 to 0.010	40 to 60
HU	190	40 to 45	0.025 to 0.035	80 to 90	0.010 to 0.015	40 to 60
HW	200	40 to 45	0.025 to 0.035	80 to 90	0.010 to 0.015	40 to 60
HX	185	40 to 45	0.025 to 0.035	80 to 90	0.010 to 0.015	40 to 60

(a) Single-point high speed steel tools usually are ground to 4 to 10° side and back rake, 4 to 7° side relief, 7 to 10° end relief, 8 to 15° end cutting-edge angle, 10 to 15° side cutting-edge angle, and 1/32- to 1/8-in. nose radius. (b) To convert to m/s, multiply by 0.005. (c) To convert to mm/rev, multiply by 25. (d) Recommended drilling feeds are as follows: for drill diameters up to 1/8 in., 0.001 to 0.002 ipr; 1/8 to 1/4 in., 0.002 to 0.004 ipr; 1/4 to 1/2 in., 0.004 to 0.007 ipr; 1/2 to 1 in., 0.007 to 0.015 ipr; over 1 in., 0.015 to 0.025 ipr. Tapping speeds recommended for HA, HC, HD, HE and HL are 10 to 25 sfm; for HF, HH, HI, and HK, 10 to 20 sfm; and for HN, HT, HU, HW and HX, 5 to 15 sfm.

Machinability ratings of resulfurized and rephosphorized carbon steels, percent of cutting speed for B1112/1212

Grade(a)	Machinability rating, %	HB
1117	90	137
1118	85	143
1137	70	197
1140	70	170
1141	70	212
1144	80	217
1146	70	187
1151	65	207
1212	100	. . .
1213	136	. . .
1215	136	. . .
12L14	160	163
12L14(b)	190	137
12L14(c)	235	137
12L14(d)	295	137

(a) All values are for cold drawn steels. (b) Proprietary free-machining variant of 12L14. (c) Proprietary free-machining variant of 12L14 containing bismuth. (d) Proprietary free-machining variants of 12L14 containing bismuth, selenium or tellurium.

Speeds and feeds for machining corrosion-resistant steel castings

Operation	Approximate feed, in./rev(a)	CF-20 CF-8	CF-16F	CE-30 CF-8M CH-20 CK-20	CF-8C	CN-7M	CA-6NM CA-15	CA-40	CB-30	CC-50
Broaching	0.001 to 0.005	8 to 15	10 to 20	8 to 15	8 to 15	8 to 15	10 to 20	8 to 15	10 to 20	10 to 20
Tapping	0.003 to 0.007	10 to 20	15 to 30	10 to 25	10 to 25	12 to 20	10 to 25	10 to 20	10 to 25	10 to 25
Threading	0.003 to 0.008	10 to 20	10 to 25	10 to 25	10 to 25	10 to 20	10 to 20	10 to 20	10 to 25	10 to 25
Reaming	0.003 to 0.008	20 to 60	30 to 100	40 to 80	40 to 80	20 to 60	20 to 60	20 to 60	40 to 120	40 to 120
Drilling	0.003 to 0.007	15 to 40	35 to 85	30 to 50	30 to 50	30 to 60	35 to 75	30 to 60	40 to 60	40 to 60
Turret lathe	0.003 to 0.008	60 to 90	90 to 130	60 to 80	60 to 90	60 to 80	80 to 110	60 to 100	70 to 100	60 to 100
Milling	0.003 to 0.008	35 to 65	75 to 110	30 to 50	40 to 50	35 to 70	70 to 105	35 to 70	40 to 60	40 to 60
Turning, boring	0.003 to 0.008	40 to 85	85 to 120	60 to 120	60 to 120	60 to 80	80 to 115	40 to 80	60 to 120	60 to 120
Screw machine	0.003 to 0.008	60 to 90	90 to 130	60 to 80	60 to 90	60 to 80	80 to 110	60 to 100	70 to 100	60 to 100

Hack sawing . . . . Use a coarse-tooth blade (not over 10 teeth per in.) at about 50 strokes per minute with positive pressure.

(a) For feeds in mm/rev, multiply listed values by 25. (b) For speeds in m/min, multiply listed values by 0.3.

Powder Metals

Properties and applications of nonferrous powder metal parts

Material and specification designation	Type	Con-dition	Den-sity, g/cm³	Tensile strength MPa	ksi	Compres-sive yield strength(a) MPa	ksi	Ap-parent hard-ness	Young's modulus 10⁶ GPa	psi	Impact strength (b) J	ft·lb	Elon-ga-tion, %	Application
Copper alloys														
P/M bronze(c): 86.3-90.5 Cu, 9.5-10.5 Sn, 1.75 C max(d), 1 Fe max	N	Sintered	5.8	55	8	48	7	···	···	···	···	···	1.0	Bearings or mechanical components resistant to atmospheric corro-sion. Sleeve bearings, flange bearings, thrust washers, load-carrying bearing plates
MPIF CT-0010 (N, R, S, T) ASTM B438 (N, R), B255 (S) SAE 840 (N), 841 (R), 842 (S) MIL-B-5687-C (R, S)	R	Sintered	6.6	97	14	76	11	···	···	···	···	···	1.0	
	S	Sintered	7.0	124	18	120.7	17.5	···	···	···	···	···	2.5	
P/M brass: 88.0-91.0 Cu, 8.3-12.0 Zn, 0-0.3 Fe MPIF CZ-0010 (T, U)	T	Sintered	7.4	138	20	62	9	57 HB	···	···	···	···	13.0	Mechanical components requiring corrosion resistance and a pleasing appearance
	U	Sintered	7.8	186	27	69	10	70 HB	···	···	···	···	10.0	
P/M brass (leaded): 88.0-91.0 Cu, 1.0-2.0 Pb, bal Zn MPIF CZP-0010 (T, U)	T	Sintered	7.4	124	18	48	7	46 HB	···	···	···	···	14.0	Same as MPIF CZ-0010; free-machining quality
	U	Sintered	7.8	175.8	25.5	55	8	60 HB	···	···	···	···	20.0	
P/M brass (leaded): 77.0-80.0 Cu, 1.0-2.0 Pb, 0.3 Fe max, 0.1 Sn max, bal Zn MPIF CZP-0218 (T, U, W) ASTM B282 (T, U, W) SAE 890 (T), 891 (U) MIL-B-12128 C (T, U)	T	Sintered	7.4	165	24	83	12	55 HB	82.7	12.0	13.6	10.0	13.0	Mechanical components resistant to atmo-spheric corrosion. Ordnance components, builders' hardware, lock parts, housing, nuts, gears
	U	Sintered	7.8	193	28	97	14	68 HB	89.6	13.0	20.3	15.0	19.0	
	W	Sintered	8.2	221	32	110	16	75 HB	96.5	14.0	28.5	21.0	23.0	
P/M brass: 68.5-71.5 Cu, 27.8-31.5 Zn, 0-0.3 Fe MPIF CZ-0030 (T, U)	T	Sintered	7.4	214	31	90	13	76 HB	···	···	···	···	20.0	Mechanical components requiring corrosion resistance and a pleasing appearance
	U	Sintered	7.8	255	37	103	15	85 HB	···	···	···	···	26.0	
P/M brass (leaded): 68.5-71.5 Cu, 1.0-2.0 Pb, bal Zn MPIF CZP-0030 (T, U)	T	Sintered	7.4	193	28	76	11	65 HB	···	···	···	···	22.0	Same as MPIF CZ-0030; free-machining quality
	U	Sintered	7.8	234	34	90	13	76 HB	···	···	···	···	27.0	
P/M nickel silver: 62.5-65.5 Cu, 16.5-19.5 Ni, bal Zn MPIF CZN-1818 (U, W) ASTM B458 (U, W)	U	Sintered	7.8	207	30	110	16	75 HB	96.5	14.0	13.6	10.0	10.0	Mechanical compo-nents, corrosion resist-ing. Gears, levers, chuck jaws, electrical components, parts for marine exposure
	W	Sintered	8.2	255	37	124	18	85 HB	96.5	14.0	17.6	13.0	12.0	
P/M nickel silver (leaded): 62.5-65.5 Cu, 16.5-19.5 Ni, 1.0-1.8 Pb, bal Zn MPIF CZNP-1618 (U, W) ASTM B458 (U, W)	U	Sintered	7.8	207	30	110	16	75 HB	89.6	13.0	12.2	9.0	10.0	Same as MPIF CZN-1818; free-machining quality
	W	Sintered	8.2	241	35	117	17	85 HB	96.5	14.0	16.2	12.0	12.0	

(continued)

(a) 1% offset. (b) Unnotched Charpy. (c) At 300X, microstructure should be substantially alpha brass without visible free tin. (d) Commonly graphite; up to 1.7% of another type of solid lubricant can be added as authorized by the producer. (e) 1202 composition: 99.4 Al, 0.3 Al_2O_3, 0.15 Fe, 0.07 Si, bal other metallics. (f) Solutioned at 520 °C (970 °F), 505 °C (940 °F) for Grade 201AB and 201AC) for 30 min. quenched in cold water, and aged at 160 °C (320 °F) for 18 h to the T6 condition. (g) Solutioned at 540 °C (1000 °F) for 30 min, quenched in cold water, and aged at 150 °C (300 °F) for 20 h to the T6 condi-tion. (h) Grade number includes suffix: FF = premix with 1.5% lubricant; NL = premix without lubricant. (i) Solutioned at 520 °C (970 °F) for 30 min, quenched in water, and aged at 150 °C (300 °F) for 18 h. (j) Solutioned at 930 F for 60 min, quenched in water, and aged at 150 °C (300 °F) for 18 h. (k) Solutioned at 475 °C (890 °F) for 60 min, quenched in water, and aged at 125 °C (260 °F) for 18 h. (m) Breaking strength in pounds (lb).
Source: Metal Powder Industries Federation (copper alloys); Aluminum Co. of America and Alcan Ingot & Powders, Div. Alcan Aluminum Corp. (aluminum alloys).

Properties and applications of nonferrous powder metal parts (continued)

Material and specification designation	Treatment	Density, g/cm³	Tensile strength MPa	Tensile strength ksi	Yield strength MPa	Yield strength ksi	Hardness	Elongation, %	Application
Aluminum alloys									
601AB (Alcoa): 0.25 Cu, 0.6 Si, 1.0 Mg, 1.5 lubricant, bal 1202(e)	Sintered (T1)	2.55	124	18	58.6	8.5	65-70 HB	8.0	Similar to wrought 6061; strength, ductility, corrosion resistance
	Heat treated(f)	2.55	248	36	241	35	80-85 Re	2.0	
201AB (Alcoa): 4.4 Cu, 0.8 Si, 0.5 Mg, 1.5 lubricant, bal 1202(e)	Sintered (T1)	2.64	208.9	30.3	180.6	26.2	70-75 Re	3.0	Similar to wrought 2014 but without manganese. Good strength properties
	Heat treated(f)	2.64	331.6	48.1	327.5	47.5	70-75 Re	2.0	
202AB (Alcoa): 4.0 Cu, 1.5 lubricant, bal 1202(e)	Sintered (T1)	2.49	160.0	23.2	75.2	10.9	55-60 HB	10.0	Good ductility. Suitable for cold formed parts
	Heat treated(g)	2.49	228	33	146.9	21.3	45-50 Re	7.3	
602AB (Alcoa): 0.4 Si, 0.6 Mg, 1.5 lubricant, bal 1202(e)	Sintered (T1)	2.55	131	19	62	9	55-60 HB	9.0	Good electrical conductivity (from 42.0 to 48.5% IACS, depending on treatment), ductility, and finishability
	Heat treated(f)	2.55	186	27	172	25	65-70 Re	3.0	
601AC (Alcoa): 0.25 Cu, 0.6 Si, 1.0 Mg, bal 1202(e)	Sintered (T1)	2.55	133.8	19.4	146.9	7.7	· · ·	10.7	Same as 601AB, without lubricant; for isostatic compacting
	Heat treated(f)	2.55	269	39	261.3	37.9	· · ·	2.0	
201AC (Alcoa): 4.4 Cu, 0.8 Si, 0.5 Mg, bal 1202(e)	Sintered (T1)	2.64	204.8	29.7	146.9	21.3	· · ·	4.0	Same as 201AB, without lubricant; for isostatic compacting
	Heat treated(f)	2.64	382.0	55.4	368.9	53.5	· · ·	2.0	
22 (Alcan)(h): 2.0 Cu, 1.0 Mg, 0.3 Si, bal Al	Sintered	2.53	165	24	110	16	83 HB	6.0	Has good mechanical properties in sintered or heat treated forms
	Heat treated(i)	2.53	262	38	200	29	74 Re	3.0	
24 (Alcan)(h) (2014): 4.4 Cu, 0.5 Mg, 0.9 Si, 0.4 Mn, bal Al	Sintered	2.54	165	24	97	14	80 HB	5.0	Properties resemble those of its wrought counterpart, 2014. Good mechanical properties
	Heat treated(j)	2.54	241	35	193	28	72 Re	3.0	
67 (Alcan)(h): 0.5 Cu, bal Al	Sintered (T1)	2.52	103	15	55	8	60 HB	12.0	High electrical conductivity (48% IACS) and ductility. Similar to wrought 1100
68 (Alcan)(h): 0.6 Mg, 0.4 Si, bal Al	Sintered (T1)	2.52	117	17	62	9	64 HB	9.0	Good surface finish; high ductility and conductivity (42% IACS). Similar to wrought 6101
69 (Alcan)(h) (6061): 0.25 Cu, 1.0 Mg, 0.6 Si, 0.10 Cr	Sintered	2.50	127.6	18.5	69	10	66 HB	10.0	Properties are similar to those of 6061. Good strength, corrosion resistance, ductility, and conductivity (40% IACS)
	Heat treated(i)	2.50	207	30	193	28	71 Re	2.0	
76 (Alcan)(h) (7075): 1.6 Cu, 2.5 Mg, 0.20 Cr, 5.6 Zn	Sintered	2.51	207	30	152	22	90 HB	3.0	Properties are similar to those of 7075. High strength and hardness
	Heat treated(k)	2.51	310	45	276	40	80 Re	2.0	
91 (Alcan)(h): 26.3 tribaloy	Sintered (T1)	3.05	93.8	13.6	· · ·	· · ·	· · ·	2.0	Excellent wear resistance
	Heat treated	3.05	106.2	15.4	· · ·	· · ·	· · ·	1.0	
4040 (Alcan): 1.0 Cu, 1.0 Si, bal Al −150 + 325 mesh	Loose sintered (T1)	1.40	689.5	100(m)	· · ·	· · ·	· · ·	· · ·	High porosity parts for controlling contamination, pressure, sound, catalytic reactions, etc.
4090 (Alcan): 1.0 Cu, 1.0 Si, bal Al −60 + 150 mesh	Loose sintered (T1)	1.35	689.5	100(m)	· · ·	· · ·	· · ·	· · ·	
4160 (Alcan): 1.0 Cu, 1.0 Si, bal Al −30 + 60 mesh	Loose sintered (T1)	1.30	689.5	100(m)	· · ·	· · ·	· · ·	· · ·	

(a) 1% offset. (b) Unnotched Charpy. (c) At 300X, microstructure should be substantially alpha brass without visible free tin. (d) Commonly graphite; up to 1.7% of another type of solid lubricant can be added as authorized by the producer. (e) 1202 composition: 99.4 Al, 0.3 Al_2O_3, 0.15 Fe, 0.07 Si, bal other metallics. (f) Solutioned at 520 °C (970 °F), 505 °C (940 °F) for Grade 201AB and 201AC for 30 min, quenched in cold water, and aged at 160 °C (320 °F) for 18 h to the T6 condition. (g) Solutioned at 540 °C (1000 °F) for 30 min, quenched in cold water, and aged at 150 °C (300 °F) for 20 h to the T6 condition. (h) Grade number includes suffix: FF = premix with 1.5% lubricant; NL = premix without lubricant. (i) Solutioned at 520 °C (970 °F) for 30 min, quenched in water, and aged at 150 °C (300 °F) for 18 h. (j) Solutioned at 930 °F for 60 min, quenched in water, and aged at 150 °C (300 °F) for 18 h. (k) Solutioned at 475 °C (890 °F) for 60 min, quenched in water, and aged at 125 °C (260 °F) for 18 h. (m) Breaking strength in pounds (lb).
Source: Metal Powder Industries Federation (copper alloys); Aluminum Co. of America and Alcan Ingot & Powders, Div. Alcan Aluminum Corp. (aluminum alloys).

Properties and applications of ferrous powder metal parts

Material and specification designation	Type	Condition	Density, g/cm³	Tensile strength MPa	ksi	Yield strength MPa	ksi	Apparent hardness	Young's modulus GPa	10⁶ psi	Elongation, %	Impact strength (a) J	ft·lb	Fatigue strength (b) MPa	ksi	Application
P/M iron F-0000-N through T: 0.3 C max	N	As sintered	5.8	110	16	76	11	10 HB	72.4	10.5	2.0	4.1	3.0	41.92	6.08	Structural (lightly loaded gears); magnetic (motor pole pieces); self-lubricating bearings; structural wear-resisting (small levers and cams) as carbonitrided
	P	As sintered	6.2	131	19	97	14	70 HB	89.6	13.0	2.5	6.1	4.5	49.78	7.22	
	R	As sintered	6.6	165	24	110	16	80 HB	110.3	16.0	5	12.9	9.5	62.88	9.12	
	S	As sintered	7.0	207	30	148.2	21.5	15 HRB	131.0	19.0	9	20	15	78.6	11.4	
	T	As sintered	7.4	276	40	179	26	30 HRB	158.6	23.0	15	34	25	104.8	15.2	
P/M steel F-0005-N through T: 0.3 to 0.6 C	N	As sintered	5.8	124	18	100.0	14.5	5 HRB	72.4	10.5	1.0	3.4	2.5	47.16	6.84	Structural (moderately loaded gears, levers, cams); structural (moderately loaded gears, levers, and cams requiring wear resistance) as heat treated
	P	As sintered	6.2	172	25	138	20	20 HRB	89.6	13.0	1.5	4.7	3.5	65.5	9.5	
	R	As sintered	6.6	221	32	159	23	45 HRB	110.3	16.0	2.5	6.8	5.0	83.84	12.16	
	R	Heat treated	6.6	414	60	393	57	100 HRB	110.3	16.0	0.5	...	...	157.2	22.8	
	S	As sintered	7.0	296	43	193	28	60 HRB	131.0	19.0	3.5	12.2	9.0	112.66	16.34	
	S	Heat treated	7.0	552	80	517	75	25 HRC	131.0	19.0	0.5	...	...	209.6	30.4	
P/M steel F-0008-N through T: 0.6 to 1.0 C	N	As sintered	5.8	200	29	172	25	35 HRB	72.4	10.5	0.5	2.7	2.0	75.98	11.02	Structural (moderately loaded gears, levers, cams); structural (moderately loaded gears, levers, and cams requiring wear resistance) as heat treated
	N	Heat treated	5.8	290	42	...	...	90 HRB	72.4	10.5	<0.5	...	...	110.04	15.96	
	P	As sintered	6.2	241	35	207	30	50 HRB	89.6	13.0	1.0	4.1	3.0	91.7	13.3	
	P	Heat treated	6.2	400	58	...	...	100 HRB	89.6	13.0	<0.5	...	...	151.96	22.04	
	R	As sintered	6.6	290	42	248	36	65 HRB	110.3	16.0	1.5	4.7	3.5	94.32	13.68	
	R	Heat treated	6.6	510	74	...	...	25 HRC	110.3	16.0	<0.5	...	...	193.88	28.12	
	S	As sintered	7.0	393	57	276	40	75 HRB	131.0	19.0	2.5	9.5	7.0	149.34	21.66	
	S	Heat treated	7.0	648	94	627	91	30 HRC	131.0	19.0	<0.5	...	...	246.28	35.72	
P/M copper iron FC-0200-P through S: 1.5 to 3.9 Cu, 0.3 C max	P	As sintered	6.2	159	23	117	17	80 HB	89.6	13.0	2.5	7.5	5.5	60.26	8.74	Bearings or mechanical components
	R	As sintered	6.6	207	30	145	21	15 HRB	110.3	16.0	4.0	9.5	7.0	78.6	11.4	Mechanical components
	S	As sintered	7.0	255	37	159	23	30 HRB	131.0	19.0	7.0	23.0	17.0	96.94	14.06	Mechanical components
P/M copper steel FC-0205-P through S: 1.5 to 3.9 Cu, 0.3 to 0.6 C	P	As sintered	6.2	276	40	234	34	35 HRB	89.6	13.0	1.0	4.7	3.5	104.8	15.2	Bearings or mechanical components
	R	As sintered	6.6	345	50	262	38	60 HRB	110.3	16.0	1.5	7.5	5.5	131.0	19.0	Mechanical components
	R	Heat treated	6.6	586	85	558	81	30 HRC	110.3	16.0	<0.5	...	...	212.22	30.78	Mechanical components
	S	As sintered	7.0	427	62	310	45	75 HRB	131.0	19.0	3.0	12.9	9.5	162.44	23.56	Mechanical components
	S	Heat treated	7.0	689	100	655	95	35 HRC	131.0	19.0	<0.5	...	...	262.0	38.0	Mechanical components
P/M copper steel FC-0208-N through S: 1.5 to 3.9 Cu, 0.6 to 1.0 C	N	As sintered	5.8	228	33	207	30	45 HRB	72.4	10.5	<0.5	3.4	2.5	86.46	12.54	Bearings or mechanical components
	N	Heat treated	5.8	296	43	...	...	95 HRB	72.4	10.5	<0.5	...	...	112.66	16.34	Bearings or mechanical components
	P	As sintered	6.2	310	45	283	41	50 HRB	89.6	13.0	<0.5	4.1	3.0	117.9	17.1	Bearings or mechanical components
	P	Heat treated	6.2	379	55	...	...	25 HRC	89.6	13.0	<0.5	...	...	144.1	20.9	Bearings or mechanical components
P/M copper steel FC-0505-N through S: 4.0 to 6.0 Cu, 0.3 to 0.6 C	N	As sintered	5.8	241	35	207	30	50 HRB	72.4	10.5	0.5	4.1	3.0	91.7	13.3	Mechanical components
	N	Heat treated(c)	5.8	...	...	...	...	90 HRB	72.4	10.5	1.0	...	...	131.0	19.0	Mechanical components
	P	As sintered	6.2	345	50	290	42	60 HRB	89.6	13.0	1.0	6.1	4.5	131.0	19.0	Mechanical components
	P	Heat treated(c)	6.2	...	...	...	...	95 HRB	89.6	13.0	...	...	...	...	...	Mechanical components
	R	As sintered	6.6	455	66	379	55	75 HRB	110.3	16.0	1.5	6.8	5.0	172.9	25.08	Mechanical components
	R	Heat treated(c)	6.6	...	...	...	...	25 HRC	110.3	16.0	...	...	...	...	...	Mechanical components

(continued)

(a) Unnotched Charpy. (b) F-0000-N to FC-0508-R, 0.38% of tensile strength; all others, determined at 10^7 cycles in rotating beam test. (c) Generally heat treated for wear resistance rather than strength. (d) Compositions correspond to AISI types 303, 316, and 410, except nickel in type 303 is higher. **Source:** Metal Powder Industries Federation

Properties and applications of ferrous powder metal parts (continued)

Material and specification designation	Type	Condition	Density, g/cm³	Tensile strength MPa	ksi	Yield strength MPa	ksi	Apparent hardness	Young's modulus GPa	10^6 psi	Elongation, %	Impact strength (a) J	ft·lb	Fatigue strength (b) MPa	ksi	Application
P/M copper steel FC-0508-N through R: 4.0 to 6.0 Cu, 0.6 to 1.0 C	N	As sintered	5.8	241	35	207	30	50 HRB	72.4	10.5	0.5	4.1	3.0	91.7	13.3	Mechanical components
	N	Heat treated(c)	5.8	...	...	...	...	90 HRB	...	...	...	...	...	...	...	Mechanical components
	P	As sintered	6.2	345	50	290	42	60 HRB	89.6	13.0	1.0	6.1	4.5	131.0	19	Mechanical components
	P	Heat treated(c)	6.2	...	...	...	...	95 HRB	...	...	...	...	...	...	...	Mechanical components
	R	As sintered	6.6	455	66	379	55	75 HRB	110.3	16.0	1.5	6.8	5.0	172.92	25.08	Mechanical components
	R	Heat treated(c)	6.6	...	...	...	...	25 HRC	...	...	...	...	...	...	...	Mechanical components
P/M copper steel FC-0808-N: 6 to 11 Cu, 0.6 to 1.0 C	N	As sintered	5.8	248	36	...	...	55 HRB	...	...	0-0.5	...	...	...	...	Mechanical components
P/M copper iron FC-1000-N: 9.5 to 10.5 Cu, 0.3 C max	N	As sintered	5.8	207	30	...	...	70 HRF	...	...	0.5	...	...	...	...	Bearings or mechanical components
P/M iron-nickel FN-0200-R through T: 1 to 3 Ni, 2.5 Cu max, 0.3 C max	R	As sintered	6.6	193	28	124	18	38 HRB	117.2	17.0	4.0	19.0	14.0	75	11	Mechanical components (can be case hardened)
	S	As sintered	7.0	262	38	172	25	42 HRB	144.8	21.0	7.0	43.4	32.0	103	15	Mechanical components (can be case hardened)
	T	As sintered	7.4	310	45	207	30	51 HRB	158.6	23.0	10.5	67.8	50.0	124	18	Mechanical components (can be case hardened)
P/M nickel steel FN-0205-R through T: 1 to 3 Ni, 0.3 to 0.6 C, 2.5 Cu max	R	As sintered	6.6	255	37	159	23	50 HRB	117.2	17.0	3.0	13.6	10.0	103	15	Structural (couplings) as sintered; structural, wear resisting (oil pump gears and heavily loaded support brackets) as heat treated; structural, wear and impact resisting (oil pump gears to 3000 psi and heavily loaded transmission gears)
	R	Heat treated	6.6	565	82	448	65	32 HRC	117.2	17.0	0.5	8.1	6.0	228	33	
	S	Sintered and sized	7.0	345	50	214	31	70 HRB	144.8	21.0	3.5	24.4	18.0	138	20	
	S	Heat treated	7.0	758	110	607	88	42 HRC	144.8	21.0	1.0	21.7	16.0	303	44	
	T	Sintered and sized	7.4	421	61	255	37	85 HRB	158.6	23.0	4.5	43.4	32.0	165	24	
	T	Heat treated	7.4	924	134	724	105	46 HRC	158.6	23.0	2.0	38.0	28.0	372	54	
P/M nickel steel FN-0208-R through T: 1 to 3 Ni, 0.6 to 0.9 C, 2.5 Cu max	R	As sintered	6.6	331	48	207	30	62 HRB	117.2	17.0	2.0	10.8	8.0	131	19	Mechanical components
	R	Heat treated	6.6	689	100	648	94	34 HRC	117.2	17.0	0.5	8.1	6.0	276	40	Mechanical components
	S	As sintered	7.0	448	65	283	41	79 HRB	144.8	21.0	3.0	19.0	14.0	179	26	Mechanical components
	S	Heat treated	7.0	931	135	883	128	45 HRC	144.8	21.0	0.5	16.2	12.0	372	54	Mechanical components
	T	As sintered	7.4	545	79	345	50	87 HRB	158.6	23.0	3.5	29.8	22.0	221	32	Mechanical components
	T	Heat treated	7.4	1103	160	1069	155	47 HRC	158.6	23.0	0.5	24.4	18.0	414	60	Mechanical components
P/M iron-nickel FN-0400-R through T: 3 to 5.5 Ni, 0.3 to 0.6 C, 2.0 Cu max	R	As sintered	6.6	248	36	152	22	40 HRB	117.2	17.0	5.0	21.7	16.0	97	14	Mechanical components
	S	As sintered	7.0	338	49	207	30	60 HRB	144.8	21.0	6.0	47.4	35.0	138	20	Mechanical components (can be case hardened)
	T	As sintered	7.4	400	58	248	36	67 HRB	158.6	23.0	6.5	67.8	50.0	159	23	Mechanical components (can be case hardened)
P/M nickel steel FN-0405-R through T: 3 to 5 Ni, 0.3 to 0.6 C, 2.0 Cu max	R	As sintered	6.6	310	45	179	26	63 HRB	117.2	17.0	3.0	13.6	10.0	124	18	Mechanical components
	R	Heat treated	6.6	772	112	648	94	27 HRC	117.2	17.0	0.5	8.1	6.0	310	45	Mechanical components
	S	As sintered	7.0	427	62	241	35	72 HRB	144.8	21.0	4.5	20.3	15.0	165	24	Mechanical components
	S	Heat treated	7.0	1062	154	883	128	39 HRC	144.8	21.0	1.0	13.6	10.0	414	60	Mechanical components
	T	As sintered	7.4	510	74	296	43	80 HRB	158.6	23.0	6.0	40.7	30.0	207	30	Mechanical components
	T	Heat treated	7.4	1241	180	1062	154	44 HRC	158.6	23.0	1.5	19.0	14.0	448	65	Mechanical components
P/M nickel steel FN-0408-R through T: 3.0 to 5.0 Ni, 0.6 to 0.9 C, 2.0 Cu max	R	As sintered	6.6	393	57	290	42	72 HRB	117.2	17.0	1.5	8.1	6.0	159	23	Structural, wear resisting, high stress (planetary differential and transmission gears up to 6 hp) as heat treated; structural, wear resisting, high stress and requiring welded assembly (welded assembly of pinion and sprocket) as carbonitrided
	S	As sintered	7.0	531	77	393	57	88 HRB	144.8	21.0	3.0	13.6	10.0	214	31	
	T	As sintered	7.4	641	93	469	68	95 HRB	158.6	23.0	4.5	21.7	16.0	255	37	

(continued)

(a) Unnotched Charpy. (b) F-0000-N to FC-0508-R, 0.38% of tensile strength; all others, determined at 10^7 cycles in rotating beam test. (c) Generally heat treated for wear resistance rather than strength. (d) Compositions correspond to AISI types 303, 316, and 410, except nickel in type 303 is higher.
Source: Metal Powder Industries Federation

Properties and applications of ferrous powder metal parts (continued)

Material and specification designation	Type	Condition	Density, g/cm³	Tensile strength MPa	ksi	Yield strength MPa	ksi	Apparent hardness	Young's modulus GPa	10⁶ psi	Elongation, %	Impact strength (a) J	ft·lb	Fatigue strength (b) MPa	ksi	Application
P/M iron-nickel FN-0700-R through T: 6 to 8 Ni, 0.3 C max, 2.0 Cu max	R	As sintered	6.6	359	52	207	30	60 HRB	117.2	17.0	2.5	16.2	12.0	145	21	Mechanical components (can be case hardened)
	S	As sintered	7.0	490	71	276	40	72 HRB	144.8	21.0	4.0	28.4	21.0	193	28	Mechanical components (can be case hardened)
	T	As sintered	7.4	586	85	331	48	83 HRB	158.6	23.0	6.0	35.3	26.0	234	34	Mechanical components (can be case hardened)
P/M nickel steel FN-0705-R through T: 6 to 8 Ni, 0.3 to 0.6 C, 2.0 Cu max	R	As sintered	6.6	372	54	241	35	69 HRB	117.2	17.0	2.0	12.2	9.0	152	22	Mechanical components
	R	Heat treated	6.6	703	102	552	80	24 HRC	117.2	17.0	0.5	10.8	8.0	283	41	Mechanical components
	S	As sintered	7.0	524	76	331	48	83 HRB	144.8	21.0	3.5	23.0	17.0	207	30	Mechanical components
	S	Heat treated	7.0	965	140	758	110	38 HRC	144.8	21.0	1.0	20.3	15.0	386	56	Mechanical components
	T	As sintered	7.4	621	90	393	57	90 HRB	158.6	23.0	5.0	32.5	24.0	248	36	Mechanical components
	T	Heat treated	7.4	1158	168	896	130	40 HRC	158.6	23.0	1.5	27.1	20.0	448	65	Mechanical components
P/M nickel steel FN-0708-R through T: 6 to 8 Ni, 0.6 to 0.9 C, 2.0 Cu max	R	As sintered	6.6	393	57	283	41	75 HRB	117.2	17.0	1.5	8.1	6.0	159	23	Mechanical components
	S	As sintered	7.0	552	80	379	55	88 HRB	144.8	21.0	2.5	16.2	12.0	221	32	Mechanical components
	T	As sintered	7.4	655	95	455	66	96 HRB	158.6	23.0	3.0	21.7	16.0	262	38	Mechanical components
P/M infiltrated steel FX-1005-T: 8 to 14.9 Cu, 0.3 to 0.6 C	T	As sintered	7.4	572	83	441	64	75 HRB	137.9	20.0	4.0	19.0	14.0	...	...	Mechanical components (special shapes)
	T	Heat treated	7.4	827	120	738	107	35 HRC	137.9	20.0	1.0	9.5	7.0	...	...	Mechanical components (special shapes)
FX-1008-T: 8 to 14.9 Cu, 0.6 to 1.0 C	T	As sintered	7.4	621	90	517	75	80 HRB	137.9	20.0	2.5	16.3	12.0	...	...	Mechanical components (special shapes)
	T	Heat treated	7.4	896	130	724	105	40 HRC	137.9	20.0	<0.5	9.5	7.0	...	...	Mechanical components (special shapes)
FX-2000-T: 15 to 25 Cu, 10.3 C max	T	As sintered	7.4	448	65	...	...	60 HRB	137.9	20.0	1.0	20.3	15.0	...	...	Mechanical components
FX-2005-T: 15 to 25 Cu, 0.3 to 0.6 C	T	As sintered	7.4	517	75	345	50	75 HRB	124.1	18.0	1.5	12.9	9.5	...	...	Mechanical components
	T	Heat treated	7.4	793	115	655	95	30 HRC	124.1	18.0	<0.5	8.1	6.0	...	...	Mechanical components
FX-2008-T: 15 to 25 Cu, 0.6 to 1.0 C	T	As sintered	7.4	586	85	517	75	80 HRB	124.1	18.0	1.0	13.6	10.0	...	...	Mechanical components
	T	Heat treated	7.4	862	125	738	107	42 HRC	124.1	18.0	<0.5	6.8	5.0	...	...	Mechanical components
P/M austenitic stainless steel(d) SS-303-P	P	As sintered	6.2	241	35	221	32	...	...	...	1.0	...	...	...	...	Type 303, mechanical components requiring secondary machining; type 316, structural, corrosion resisting, nonmagnetic (small gears, levers, cams, and other parts for exposure to salt water and specific industrial acids); type 410, structural, corrosion resisting, (small gears, levers, cams, and other parts where applications require heat treating for wear resistance.)
SS-303-R	R	As sintered	6.6	359	52	324	47	...	...	...	2.0	...	...	...	...	
SS-316-P	P	As sintered	6.2	262	38	221	32	...	...	...	2.0	...	...	...	...	
SS-316-R	R	As sintered	6.6	372	54	276	40	...	...	...	4.0	...	...	...	...	
SS-410-N	N	As sintered	5.8	290	42	283	41	...	...	...	<1.0	...	...	...	...	
SS-410-P	P	As sintered	6.2	379	55	372	54	...	...	...	<1.0	...	...	...	...	

(a) Unnotched Charpy. (b) F-0000-N to FC-0508-R, 0.38% of tensile strength; all others, determined at 10⁷ cycles in rotating beam test. (c) Generally heat treated for wear resistance rather than strength. (d) Compositions correspond to AISI types 303, 316, and 410, except nickel in type 303 is higher. **Source:** Metal Powder Industries Federation

Alloy Phase Diagrams

Edited by Hugh Baker
Manager
ASM/NBS Data Program for Alloy Phase Diagrams

Ferrous Binary Alloy Systems

General References

1 L. A. Willey, Appendix 1, Phase Diagrams, in *Aluminum*, Vol I, K. R. Van Horn, Ed., American Society for Metals, 1967, p 359–381
2 W. B. Pearson, *Handbook of Lattice Spacings and Structures of Metals*, Vol 1 and 2, Pergamon, 1958 and 1967

3 M. Hansen, *Constitution of Binary Alloys*, 2nd ed., prepared with the cooperation of K. Anderko, McGraw-Hill, 1958
4 R. P. Elliott, *Constitution of Binary Alloys, First Supplement*, McGraw-Hill, 1965
5 F. A. Shunk, *Constitution of Binary Alloys, Second Supplement*, McGraw-Hill, 1969

6 R. Hultgren, P. D. Desai, D. T. Hawkins, M. Gleiser and K. K. Kelley, *Selected Values of the Thermodynamic Properties of Binary Alloys*, American Society for Metals, 1973
7 E. Rudy, Compendium of Phase Diagram Data, Part V, Air Force Materials Laboratory Report AFML-TR-65-2, 1969

Al-Fe (Aluminum-Iron)

By L. A. Willey

The diagram and crystal-structure data are from Hultgren *et al* (Gen Ref 6). Additional information is found in Willey (Gen Ref 1):

Phase	Formula	Symmetry	Symbol	Prototype
β_1	$AlFe_3$	ord fcc	DO_3	BiF_3
β_2	AlFe	ord bcc	$B2$	CsCl
ε		comp bcc	...	...
ζ	ζ Al_2Fe	comp rhom	...	...
η	η Al_5Fe_2	eco	...	...
θ	θ Al_3Fe	very comp ecm	...	...

B-Fe (Boron-Iron)

By A. R. Marder

The diagram is from Hansen (Gen Ref 3), Elliott (Gen Ref 4), and Shunk (Gen Ref 5), with the addition of data from K. I. Portnoi, M. Kh. Levinskaya and V. M. Romashov, *Poroshkovaya Met*, Akad Nauk Ukr SSR, No. 8, 1969, p 66–70. Crystal-structure data are from Pearson (Gen Ref 2):

Phase	Formula	Symmetry	Symbol	Prototype
ε	Fe_2B	bct	$C16$	Al_2Cu
ζ, ζ'	FeB	ortho	$B27$	FeB
η	FeB_{19}	unknown	...	...

C-Fe (Carbon-Iron)

By John Chipman

The diagrams and tables are from a review by Chipman (1). Many iron carbides have been reported, but only those that have been studied under stable or metastable equilibrium conditions are considered here.

The major lines of the diagrams are based on direct observations and are drawn to conform to the known thermodynamic properties of iron, from Orr and Chipman (2), and to experimentally determined activity coefficients. Lines showing composition-temperature relations at several pressures were calculated from *(a)* the known vapor pressure of liquid iron, from Hultgren *et al* (3); and *(b)* the activity of iron in liquid-iron-plus-carbon solution, from Chipman (1) and Syu *et al* (4). The liquidus curve for cementite (Fe_3C) has never been observed but was calculated by Chipman (1, 5) from the free energy of cementite and from that of the equilibrium liquid as extrapolated into the metastable range below the graphite liquidus. Many iron carbides have been reported in the literature. Most of them are metastable, and are not shown on the diagrams. Crystal-structure data that appear to have been confirmed are listed below; see Pearson (Gen Ref 2) and Hultgren *et al* (Gen Ref 6):

Phase	Formula	Symmetry	Symbol	Prototype
... .	Fe_4C	cu	...	...
Cementite..........	Fe_3C	ortho	DO_{11}	Fe_3C
ε carbide..........	ε $Fe_{2.3}C$	hex	...	Ni_3N
χ (Hägg carbide)......	$Fe_{2.2}C$ or Fe_5C_2	ecm	...	Mn_5C_2
... .	Fe_7C_3	hex	$D10_2$	Fe_3Th_7
... .	$Fe_{20}C_9$	ortho	...	...
... .	Fe_2C (?)	hex or ortho	...	...

The Austenite Field. The properties and boundary lines of the face-centered cubic austenite (γ-Fe) solid-solution area are well known. The activity of carbon as a function of temperature and composition has been determined by many observers, chiefly through studies of the equilibria: C (in Fe) + $2H_2$ = CH_2 and C (in Fe) + CO_2 = $2CO$. The earlier investigations of the latter reaction, by Dünwald and Wagner (6) and by Smith (7), are in agreement with the more recent work of Scheil *et al* (8)—except at the highest carbon concentrations—and of Ban-ya *et al* (9). Studies of the former reaction have been subject to errors, especially at low carbon levels, because of the reaction of methane with residual gas impurities. These errors may have caused the differences observed by Smith (7) between activity coefficients determined by the two equilibria. Studies based on the latter reaction were subject to similar but generally smaller errors, which increased with increasing temperature and carbon content.

The solubility of graphite in austenite (γ-Fe) was calculated from the data on activity of carbon in austenite, using the equations presented by Chipman (1). These solubility values are in rather good agreement with the direct measurements of Wells (10) and of Gurry (11) and with a downward extrapolation of the solidus line of Benz and Elliott (12) to the eutectic temperature. Average values of these three determinations are listed in Table 1.

Table 1 Solubility of graphite and cementite in γ-Fe (austenite)

Temperature		Graphite		Cementite	
°C	°F	C:Fe(a)	C, wt %	C:Fe(a)	C, wt %
727(b)	1341	...	...	0.0356	0.77
738(c)	1360	0.0320	0.68	...	...
800	1472	0.0408	0.87	0.0442	0.94
900	1652	0.0561	1.19	0.0580	1.22
1000	1832	0.0725	1.53	0.0730	1.55
1100	2012	0.0896	1.89	0.0910	1.92
1148(d)	2098	...	...	0.1000	2.11
1154(e)	2109	0.0990	2.08	...	...

(a) Ratio of number of carbon atoms to number of iron atoms. (b) Cementite eutectoid. (c) Graphite eutectoid. (d) Cementite eutectic. (e) Graphite eutectic

Table 2 Solid-liquid equilibria for γ-Fe

Temperature		γ-Fe Solidus		Liquidus	
°C	°F	C:Fe(a)	C, wt %	C:Fe(a)	C, wt %
1148(b)	2098	0.1000	2.11	0.2092	4.30
1154(c)	2109	0.0900	2.08	0.2072	4.26
1200	2192	0.0877	1.85	0.1906	3.93
1250	2282	0.0718	1.59	0.1689	3.50
1300	2372	0.0613	1.30	0.1450	3.02
1350	2462	0.0475	1.01	0.1179	2.47
1400	2552	0.0333	0.71	0.0891	1.88
1450	2642	0.0196	0.42	0.0570	1.21
1495(d)	2723	0.0079	0.17	0.0248	0.53
1527(e)	2781	0.0000	0.00	0.0000	0.00
1538(f)	2800	...	...	...	...

(a) Ratio of number of carbon atoms to number of iron atoms. (b) Cementite eutectic. (c) Graphite eutectic. (d) Peritectic. (e)Metastable melting point of γ-Fe. (f) Melting point of δ-Fe

The (γ-Fe)-Liquid Equilibrium. The austenite solidus line of Benz and Elliott and a portion of the austenite liquidus line of Buckley and Hume-Rothery (13) are shown in the diagram. The solidus line was given a slight inflection with downward curvature near its lower end to conform to the data of Ban-ya *et al*. Both lines are superior in accuracy to those of earlier investigators and are supported by data of Adcock (14). When the liquidus and solidus lines are extended, as shown by Chipman (1), they intersect at the metastable melting point of γ-Fe (1527 °C; 2781 °F) as calculated by Orr and Chipman. Values from the curves in their entirety are listed in Table 2.

The Liquid Phase. The best data on the activity of carbon in liquid Fe-C alloys were determined by Richardson and Dennis (15) at 1560, 1660 and 1760 °C (2840, 3020 and 3200 °F), using the equilibrium: C (in Fe) + CO_2 = $2CO$. For lower temperatures, Chipman (1) used the activity of carbon in austenite at points along the solidus to determine the activity coefficient in the liquid for compositions along the liquidus.

The solubility of graphite in liquid iron has been measured by many investigators. Up to 1800 °C (3272 °F), excellent agreement is found among Ruer and Biren (16), Chipman *et al* (17) and Kitchener *et al* (18). Data for temperatures up to 2500 °C (4532 °F) were reported by Ruer and Biren, for up to 2875 °C (5207 °F) by Cahill *et al* (19), and for 2050 to 2375 °C (3722 to 4307 °F) by Vertman *et al*

Table 3 Solubility of graphite in liquid iron

Temperature			Temperature		
°C	°F	C, wt %	°C	°F	C, wt %
1154	2109	4.26 ± 0.02	2100	3812	7.05 ± 0.2
1200	2192	4.37	2200	3992	7.56 ± 0.3
1300	2372	4.63	2300	4172	8.10
1400	2552	4.88	2400	4352	8.68
1500	2732	5.14	2500	4532	9.28 ± 0.4
1600	2912	5.40 ± 0.03	2600	4712	9.87
1700	3092	5.66	2700	4892	10.50
1800	3272	5.94 ± 0.05	2800	5072	11.12
1900	3452	6.26 ± 0.10	2900	5252	11.75 ± 0.5
2000	3632	6.63 ± 0.10			

Table 4 Solubility of graphite, Fe_3C and $Fe_{2.2}C$ in α-Fe (ferrite)

Temperature, °C	Solubility, ppm			Temperature, °C	Solubility, ppm		
	Graphite(a)	Fe_3C(b)	$Fe_{2.2}C$(c)		Graphite(a)	Fe_3C(b)	$Fe_{2.2}C$(c)
738	206	⋯	⋯	500	4.3	13	⋯
727	⋯	218	⋯	450	1.35	5.7	⋯
700	127	160	⋯	400	0.37	2.3	⋯
650	63	102	⋯	350	0.081	0.75	1.3
600	28	57	⋯	300	0.013	0.21	0.30
550	11.7	28	⋯	250	⋯	0.045	0.050
				200	⋯	0.007	0.0055

(a) Data derived from equation in Chipman (1). (b) Calculated from observed value at 727 °C (1341 °F) and free energy of formation. (c) From free energy of formation

(20). Interpolated data and estimates of probable accuracy are given in Table 3. The selected eutectic at 1154 °C (2109 °F) and 4.24 to 4.28 wt % C has been confirmed by Ruth and Turpin (21).

The δ Phase and Peritectic. The δ-Fe region is shown on the diagram with the peritectic at 1495 °C (2723 °F), as recommended by Buckley and Hume-Rothery. The liquidus of δ-Fe phase is shown as a straight line from the melting point to 0.53 wt % C at the peritectic. The compositions of the peritectic are: δ-Fe, 0.09 wt % C; γ-Fe, 0.17 wt % C; liquid, 0.53 wt % C. Compositions of δ-Fe and liquid at the peritectic are averages of several published values that have been selected to agree with the heat effects at 1394 °C (2541 °F), and at the melting point, from Orr and Chipman, and with the activity of iron in the δ-Fe and liquid solutions from Chipman (5) and Syu et al.

The α-γ Equilibrium. The (γ-Fe)-(α-Fe + γ-Fe) phase boundary is based almost entirely on the data of Mehl and Wells (22), corrected by 1 °C (1.8 °F) at the pure iron end with negligible correction at the eutectoid end. The intersections of Chipman's solubility lines (1) place the Fe-graphite eutectoid at 738 °C (1360 °F) and 0.68 wt % C, and the Fe-cementite eutectoid at 727 °C (1341 °F) and 0.77 wt % C. The latter temperature agrees with Smith and Darken (23) and is 4 °C (7 °F) higher than that of Mehl and Wells.

The (α-Fe)-(α-Fe + γ-Fe) phase boundary, which is based on data at 800 and 750 °C (1472 and 1382 °F) by Smith (7), is extrapolated to 0.0206 and 0.0218 wt % C at the graphite and cementite eutectoid temperatures.

Solubility of Graphite in α-Fe. The activity of carbon in body-centered-cubic α-Fe was determined by Smith (7, 24) and Swartz (25) at 590 to 800 °C (1094 to 1472 °F). The activity in body-centered-cubic δ-Fe at the peritectic temperature (1495 °C; 2723 °F) is known from the activity in the γ-Fe and liquid phases. These activity data, and the activity of carbon at the peritectic temperature, were used to calculate the values for the solubility of graphite in α-Fe that are given in Table 4.

The Iron Carbides. Numerous carbides of iron are reported, ranging in composition from FeC to Fe_4C. Only two of these have been studied under metastable equilibrium conditions, and it is only for these two that thermodynamic data are available.

Cementite, sometimes called θ carbide, is usually assigned the formula Fe_3C, but its exact conformance to the stoichiometric composition has not been proved. A slight iron deficiency above 800 °C (1472 °F) was suggested by Benz et al (26). Several observers have reported variations in composition and in the lattice parameter.

The existence of Fe_2C was suggested by Glud et al (27). An x-ray diffraction pattern by Hofmann and Groll (28) for iron carburized at temperatures below 400 °C (752 °F) showed an unknown carbide and Fe_3C; in iron carburized above 400 °C (752 °F), only Fe_3C was found. Experiments using a hydrogen-reduction method for determination of carbidic and free carbon contents showed that when iron was carburized with CO at 225 °C (437 °F), the product contained 9.7 wt % C, the carbon content of Fe_2C. At higher temperatures, a mixture of Fe_2C and Fe_3C was formed; at 400 °C (752 °F) and above, only Fe_3C was formed. The x-ray diffraction pattern for the new carbide was determined by Hägg (29). This carbide is frequently called Hägg or χ carbide and has a slightly variable composition approximating $Fe_{2.2}C$.

The solubilities of Fe_3C and $Fe_{2.2}C$ in α-Fe (ferrite) are given in Table 4.

Cementite (θ Carbide). The (γ-Fe)-(γ-Fe + cementite) phase boundary is from Ban-ya et al, and is based on the direct measurements of solubility by Smith (30) and on the CO-CO_2 equilibrium measurements by Scheil et al. The Fe_3C liquidus line and the melting point (1227 °C; 2241 °F) were calculated by Chipman (1). Values from the phase boundary are listed in Table 1.

The solubility of cementite in α-Fe (ferrite) was calculated from the free energy of formation of cementite as determined by the gas-solid equilibrium studies of Scheil et al. Within the temperature range shown in the diagram, the result is in fair agreement with studies by Swartz of the solubility of stress-free cementite using internal-friction methods.

Hägg (χ) Carbide. The carbide prepared by Browning et al (31) by treatment of hydrogen-reduced iron with butane at 275 °C (527 °F) was identified by its diffraction pattern as the same as that prepared by Hägg at 225 °C (437 °F). Both observers found that this carbide was converted to Fe_3C by heating to 500 °C (932 °F). The structure was listed as orthorhombic with the added notation that a monoclinic or triclinic structure was not excluded. Recent studies show a monoclinic structure isotypic with Mn_5C_2. The χ carbide is represented by the formula $Fe_{2.2}C$ (or Fe_5C_2) and has a Curie temperature of 247 °C (477 °F).

The transformation indicated on the diagram at 230 °C (446 °F) is uncertain but is based on observations of gas equilibria with the respective carbides by Browning *et al.* Below a temperature estimated at 230 to 350 °C (446 to 662 °F), χ carbide is more stable than cementite.

The ε carbide occurs as a transition phase in the tempering and aging of steel. It has not been isolated, and its thermodynamic properties are unknown. Hofer (32) described it as close-packed hexagonal with a Curie temperature of 370 °C (698 °F) and a variable composition—commonly about $Fe_{2.4}C$.

REFERENCES

1 J. Chipman, *Met Trans,* Vol 3, 1972, p 55-64
2 R. L. Orr and J. Chipman, *Trans Met Soc AIME,* Vol 239, 1967, p 630-633
3 R. Hultgren, P. D. Desai, D. T. Hawkins, M. Gleiser and K. K. Kelley, "Selected Values of The Thermodynamic Properties of Binary Alloys", American Society for Metals, 1973
4 T. Syu, A. V. Polyakov and A. M. Samarin, *Izv Vyssh Ucheb Zaved Chern Met,* No. 11, 1959, p 3-12
5 J. Chipman, *Met Trans,* Vol 1, 1970, p 2163-2168
6 H. Dünwald and C. Wagner, *Z Anorg Allgem Chem,* Vol 199, 1931, p 321-346
7 R. P. Smith, *J Am Chem Soc,* Vol 68, 1946, p 1163-1175
8 E. Scheil, T. Schmidt and J. Wünning, *Arch Eisenhuetteow,* Vol 32, 1961, p 251-260
9 S. Ban-ya, J. F. Elliott and J. Chipman, *Met Trans,* Vol 1, 1970, p 1313-1320
10 C. Wells, *Trans Am Soc Met,* Vol 26, 1938, p 289-344
11 R. W. Gurry, *Trans AIME,* Vol 150, 1942, p 147-153
12 M. G. Benz and J. F. Elliott, *Trans Met Soc AIME,* Vol 221, 1961, p 323-331 and p 888
13 R. A. Buckley and W. Hume-Rothery, *J Iron Steel Inst,* Vol 196, 1960, p 403-406; Vol 200, 1962, p 142-143
14 F. Adcock, *J Iron Steel Inst,* Vol 135, 1937, p 281-287
15 F. D. Richardson and W. E. Dennis, *Trans Faraday Soc,* Vol 49, 1953, p 171-180
16 R. Ruer, J. Biren, *Z Anorg Allgem Chem,* Vol 113, 1920, p 98-112
17 J. Chipman, R. M. Alfred, L. W. Gott, R. B. Small, D. M. Wilson, C. N. Thomson, D. L. Guernsey and J. C. Fulton, *Trans Am Soc Met,* Vol 44, p 1215-1230
18 J. A. Kitchener, J. O'M. Bockris and D. A. Spratt, *Trans Faraday Soc,* Vol 48, 1952, p 608-617
19 J. A. Cahill, A. D. Kirshenbaum and A. V. Grosse, *Trans Am Soc Met,* Vol 57, 1964, p 417-426
20 A. A. Vertman, V. K. Grigorovich, N.A. Nedumov and A. M. Samarin, *Dokl Akad Nauk SSSR,* Vol 159, 1964, p 121-124
21 J. C. Ruth and M. Turpin, *C R Acad Sci Paris,* Vol 265, 1967, p 786-788
22 R. F. Mehl and C. Wells, *Trans AIME,* Vol 125, 1937, p 429-469
23 R. P. Smith and L. S. Darken, *Trans AIME,* Vol 215, 1959, p 727-728
24 R. P. Smith, *Trans Met Soc AIME,* Vol 224, 1962, p 105-111
25 J. C. Swartz, *Trans Met Soc AIME,* Vol 239, 1967, p 68-75; Vol 245, 1969, p 1083-1092
26 R. Benz, J. F. Elliott and J. Chipman, *Met Trans,* Vol 4, 1973
27 W. Glud, K. V. Otto and H. Ritter, *Ber Ges Kohlentech,* Vol 3, 1929, p 40
28 U. Hofmann, E. Groll, *Z Anorg Allgem Chem,* Vol 191, 1930, p 914
29 G. Hägg, *Z Kristallogr,* Vol 89, 1934, p 92
30 R. P. Smith, *Trans Met Soc AIME,* Vol 215, 1959, p 954-957
31 L. C. Browning, T. W. DeWitt and P. H. Emmett, *J Am Chem Soc,* Vol 72, 1950, p 4211-4217
32 L. J. E. Hofer, *U S Bur Mines Bull,* No 631, 1966

Co-Fe (Cobalt-Iron)

By Theodore E. Torok

The diagram is from Hansen (Gen Ref 3). Elliott (Gen Ref 4) proposes a diagram with additional intermetallic compounds at 580 °C (1076 °F), 26% Co, and 500 °C (932 °F), 76% Co, based on electrical-resistivity measurements. Metallographic and dilatometric studies could not confirm their presence; hence, the older diagram given by Hansen was preferred. Crystal-structure data are from Pearson (Gen Ref 2):

Phase	Formula	Symmetry	Symbol	Prototype
α1	CoFe	ord bcc	B2	CsCl

Cr-Fe (Chromium-Iron)

By M. Vikram Rao

The diagram is from Hansen (Gen Ref 3) and Elliott (Gen Ref 4). Calculations using the heat of fusion of Fe predict that the liquidus-solidus gap in the region 0 to 20% Cr is smaller than indicated (1). Crystal-structure data are from Pearson (Gen Ref 2):

Phase	Formula	Symmetry	Symbol	Prototype
σ	CrFe	tet	$D8_b$	CrFe

REFERENCE

1 M. V. Rao, private communication to D. T. Hawkins

Cu-Fe (Copper-Iron)

By Robert E. Johnson

The lower temperature range of the iron-rich end of the diagram is from Speich, Gula and Fisher (1); the higher temperature range is from Hellawell and Hume-Rothery (2). The remainder of the diagram is from Hansen (Gen Ref 3).

REFERENCES

1 G. R. Speich, J. A. Gula and R. M. Fisher, Diffusivity and Solubility Limit of Copper in Alpha and Gamma Iron, p 525-542 in "Electron Microprobe", Wiley, 1966
2 A. Hellawell and W. Hume-Rothery, *Phil Trans Roy Soc London,* Ser A, Vol 249, 1957, p 417-459

Fe-Mn (Iron-Manganese)

By Malcolm J. Roberts

The diagram is from Hansen (Gen Ref 3) and Elliott (Gen Ref 4).

Fe-Mo (Iron-Molybdenum)

By Donald T. Hawkins

From 0 to 65% Mo, the diagram is from Sinha, Buckley and Hume-Rothery (1); at higher molybdenum compositions, the diagram is from Hansen (Gen Ref 3). Rawlings and Newey (2) present a similar diagram except they did not observe the β phase. The γ-loop determination of Hillert, Wada and Wada (3) agrees with that of Sinha, Buckley and Hume-Rothery. Abrahamson and Lopata (4) report a much lower solubility of molybdenum in α-Fe than other workers. Crystal-structure data are from Hultgren *et al* (Gen Ref 6):

Phase	Formula	Symmetry	Symbol	Prototype
β........	Fe_2Mo	hex	C14	$MgZn_2$
δ........	...	comp rhom	...	Mn_6Si
ε........	Fe_7Mo_6	rhom	$D8_5$	Fe_7W_6
σ........	FeMo	tet	$D8_b$	CrFe

REFERENCES

1   A. K. Sinha, R. A. Buckley and W. Hume-Rothery, *J Iron Steel Inst,* Vol 205, 1967, p 191-195
2   R. D. Rawlings and C. W. A. Newey, *J Iron Steel Inst,* Vol 206, 1968, p 723
3   M. Hillert, T. Wada and H. Wada, *J Iron Steel Inst,* Vol 205, 1967, p 539-546
4   E. P. Abrahamson III and S. L. Lopata, U S Dept of Commerce Rept AD-639663, 1966

Fe-Ni (Iron-Nickel)

By J. I. Goldstein

The liquidus and (δ-Fe) region are from Elliott (Gen Ref 4). Equilibria in the solid state are very difficult to determine because of the long times needed to reach equilibrium. Using diffusion couples and long annealing times, Goldstein and Ogilvie (1) have determined the solubility limits of (α-Fe) and (γ-Fe). Below 500 °C (932 °F), these solubility limits have been calculated thermodynamically by Goldstein and Ogilvie, and Hillert, Wada and Wada (2). The phase relations in the vicinity of the ordered state are from Viting (3). Crystal-structure data are from Pearson (Gen Ref 2):

Phase	Formula	Symmetry	Symbol	Prototype
γ'........	$FeNi_3$	ord fcc	$L1_2$	$AuCu_3I$

REFERENCES

1   J. I. Goldstein and R. E. Ogilvie, *Trans Met Soc AIME,* Vol 233, 1965, p 2083-2087
2   M. Hillert, T. Wada and H. Wada, *J Iron Steel Inst,* Vol 205, 1967, p 539-546
3   L. M. Viting, *Zh Neorgan Khim,* Vol 2, 1957, p 367-374

Fe-Nb (Iron-Niobium)

By L. R. Woodyatt

The portion of the diagram below 45% Nb is from Hansen (Gen Ref 3), Elliott (Gen Ref 4), and Shunk (Gen Ref 5). Little is known of the remainder of the diagram, except the existence of the μ phase as reported by Raman (1) and confirmed by Kripyakevich, Gladyshevskii and Skolozdra (2). Crystal-structure data are from Pearson (Gen Ref 2):

Phase	Formula	Symmetry	Symbol	Prototype
ε........	Fe_2Nb	hex	C14	$MgZn_2$
μ........	Fe_7Nb_6	rhom	$D8_5$	Fe_7W_6

REFERENCES

1   A. Raman, *Proc Indian Acad Sci,* Section A, Vol 65, 1965, p 256-264
2   P. I. Kripyakevich, E. I. Gladyshevskii and R. V. Skolozdra, *Sov Phys-Cryst,* Vol 12, 1968, p 525-527

Fe-P (Iron-Phosphorus)

By Donald T. Hawkins

The diagram is from Hansen (Gen Ref 3), Elliott (Gen Ref 4), and Shunk (Gen Ref 5). Crystal-structure data are from Pearson (Gen Ref 2):

Phase	Formula	Symmetry	Symbol	Prototype
δ........	Fe_3P	bct	$D0_e$	Ni_3P
ε........	Fe_2P	hex	C22	Fe_2P
ζ........	FeP	ortho	B31	MnP
η........	FeP_2	ortho	C18	FeS_2 (marcasite)

Fe-Pb (Iron-Lead)

By Donald T. Hawkins

The diagram is from Lord and Parlee (1) for the iron-rich region and from Miller and Elliott (2) and Stevenson and Wulff (3) for the lead-rich region. Elliott (Gen Ref 4) states that it is difficult to judge which work relating to the lead-rich region is more accurate.

REFERENCES

1   A. E. Lord and N. A. Parlee, *Trans Met Soc AIME,* Vol 218, 1960, p 644-646
2   K. O. Miller and J. F. Elliott, *Trans Met Soc AIME,* Vol 218, 1960, p 900-910
3   D. A. Stevenson and J. Wulff, *Trans Met Soc AIME,* Vol 221, 1961, p 271-275

Fe-S (Iron-Sulfur)

By John Chipman

The diagram is similar to that of Hansen (Gen Ref 3) with certain modifications. Sulfur vapor pressures are from Rosenqvist (1), Urbain *et al* (2) and Nagamori and Kameda (3). Liquidus lines and melting point (1190 °C; 2174 °F) of the ε phase are from Jensen (4). Other boundaries of the ε phase are from Arnold (5, 6), Yund and Hall (7, 8), and Andresen and Torbo (9). Kullerud and Yoker (10) report the melting point of FeS_2 (pyrite) as 743 °C (1369 °F) at 10 bars. The inset showing solubility of sulfur in solid iron is from Rosenqvist and Dunicz (11), Barloga *et al* (12), Turkdogan (13), and Herrnstein *et al* (14).

Crystal-structure data are from Pearson (Gen Ref 2):

Phase	Formula	Symmetry	Symbol	Prototype
ζ	FeS	hex	. . .	. . .
ε	$Fe_{1-x}S$	mono	. . .	. . .
ε'	$Fe_{1-x}S$	unknown (metastable)	. . .	. . .
η	FeS_2 (pyrite)	cu	C2	FeS_2 (pyrite)
·	FeS_2 (marcasite)	ortho	B18	FeS_2 (marcasite)

REFERENCES

1 T. Rosenqvist, *J Iron Steel Inst,* Vol 176, 1954, p 37-57
2 G. Urbain, W. Burgmann and M. G. Frohberg, *Compt Rend,* Ser C, Vol 263, 1966, p 595
3 M. Nagamori and M. Kameda, *Trans Japan Inst Metals,* Vol 9, 1968, p 187-194
4 E. Jensen, *Am J Sci,* Vol 240, 1942, p 695-709
5 R. G. Arnold, *Econ Geol,* Vol 57, 1962, p 72-90
6 R. G. Arnold, *Econ Geol,* Vol 64, 1969, p 405-419
7 R. A. Yund and H. T. Hall, *Mat Res Bull,* Vol 3, 1968, p 779-783
8 R. A. Yund and H. T. Hall, *Econ Geol,* Vol 64, 1969, p 420-434
9 A. F. Andresen and P. Torbo, *Acta Chem Scand,* Vol 21, 1967, p 2841
10 G. Kullerud and H. S. Yoker, *Econ Geol,* Vol 54, 1959, p 533-572
11 T. Rosenqvist and B. L. Dunicz, *Trans AIME,* Vol 194, 1952, p 604-608
12 A. M. Barloga, K. R. Bock and N. Parlee, *Trans Met Soc AIME,* Vol 221, 1961, p 173-179
13 E. T. Turkdogan, *Trans Met Soc AIME,* Vol 242, 1968, p 1665-1672
14 W. H. Herrnstein III, F. H. Beck and M. G. Fontana, *Trans Met Soc AIME,* Vol 242, 1968, p 1049-1956

Fe-Si (Iron-Silicon)

By John Chipman

The diagram and crystal-structure data are from Hultgren *et al* (Gen Ref 6):

Phase	Formula	Symmetry	Symbol	Prototype
α_1	α' Fe_3Si	ord bcc	$D0_3$	BiF_3
α_2 . . . . · · ·		ord bcc	B2	CsCl
β	Fe_2Si	bcc	A2	W
η	Fe_5Si_3	ord hex	$D8_8$	Mn_5Si_3
ε	FeSi	ord cu	B20	FeSi
ζ_1	$FeSi_{2.33}$	eco (a)	. . .	. . .
ζ_2	$FeSi_2$	eco	. . .	. . .

(a) With 13 to 23% vacancies on Fe sites

Fe-V (Iron-Vanadium)

By E. C. Oren

The solid phase relations were determined by Hellawell and Hume-Rothery (1); the remainder of the diagram is from Hansen (Gen Ref 3). There is disagreement as to the maximum extent of the γ loop; Lisnik and Skvorchuk (2) have suggested it is 1.0 to 1.2% V. Crystal-structure data are from Pearson (Gen Ref 2):

Phase	Formula	Symmetry	Symbol	Prototype
σ . . . . · · ·		tet	$D8_b$	CrFe

REFERENCES

1 A. Hellawell and W. Hume-Rothery, *Phil Trans Roy Soc (London),* Vol A249, 1957, p 417-459
2 A. G. Lisnik and V. P. Skvorchuk, *Dopovidi Akad Nauk Ukr SSR,* No. 10, 1960, p 1408-1419

Fe-W (Iron-Tungsten)

By E. T. Stephenson

The α solvus between 600 and 900 °C (1112 and 1652 °F) is from Abrahamson and Lopata (1). The remainder of the diagram from 0 to 72 wt % W is from Sinha and Hume-Rothery (2), except the γ loop, which is from Hillert *et al* (3), and is in good agreement with thermodynamic calculations by Sinha and Hume-Rothery. Crystal-structure data are from Pearson (Gen Ref 2):

Phase	Formula	Symmetry	Symbol	Prototype
δ	Fe_2W	hex	C14	$MgZn_2$
ε	Fe_7W_6	rhom	$D8_5$	Fe_7W_6

REFERENCES

1 E. P. Abrahamson II and S. L. Lopata, *Trans Met Soc AIME,* Vol 236, 1966, p 76-87
2 A. K. Sinha and W. Hume-Rothery, *J Iron Steel Inst,* Vol 205, 1967, p 1145-1149
3 M. Hillert, T. Wada and H. Wada, *J Iron Steel Inst,* Vol 205, 1967, p 539-546

Ferrous Binary Alloy Phase Diagrams

Al-Fe (Aluminum-Iron)

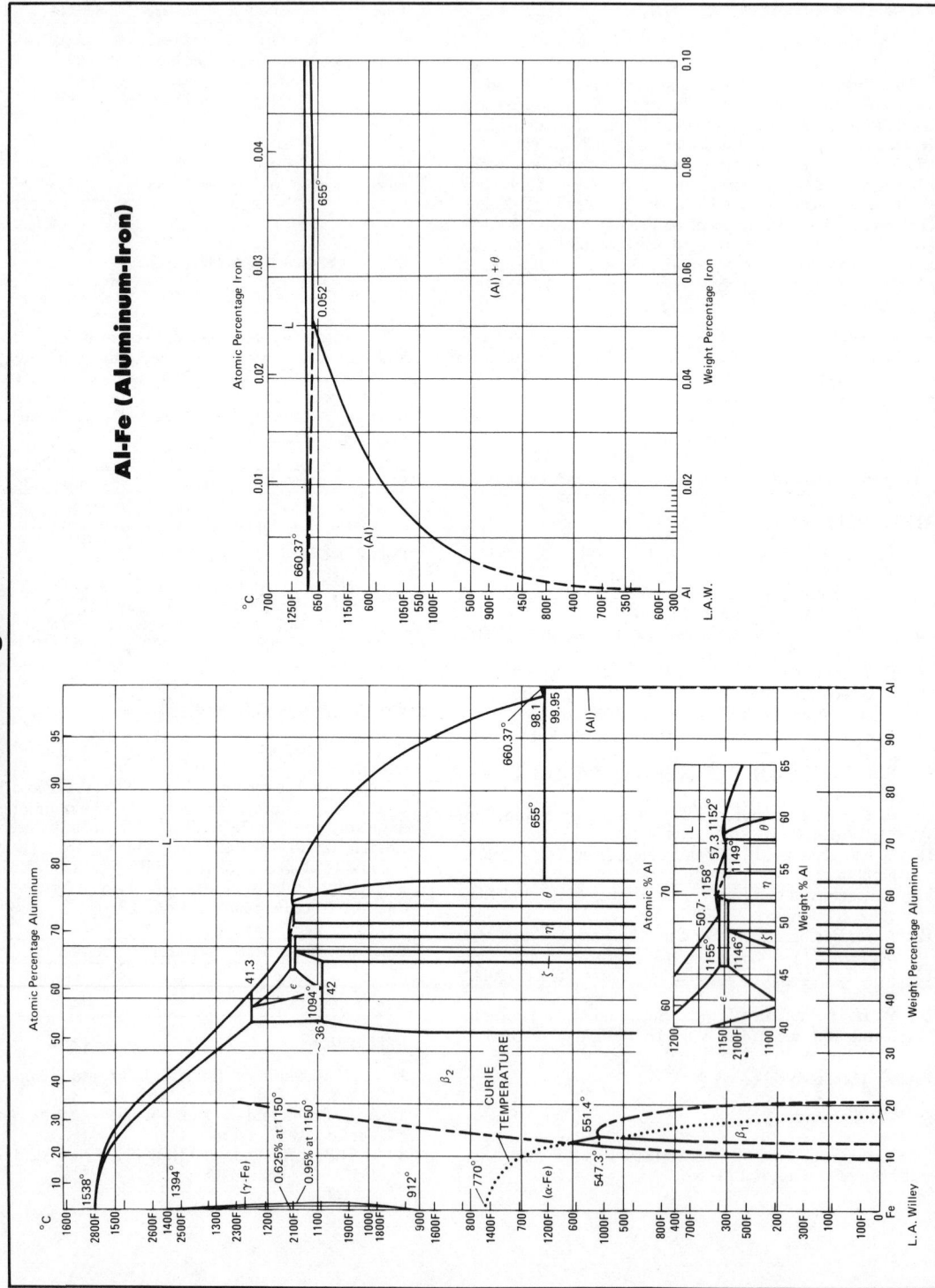

L.A.W.

L. A. Willey

B-Fe (Boron-Iron)

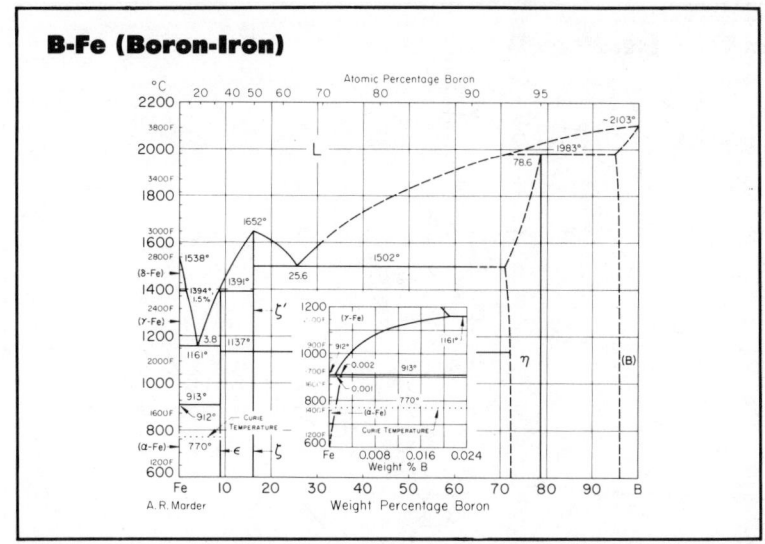

A. R. Marder

C-Fe (Carbon-Iron)

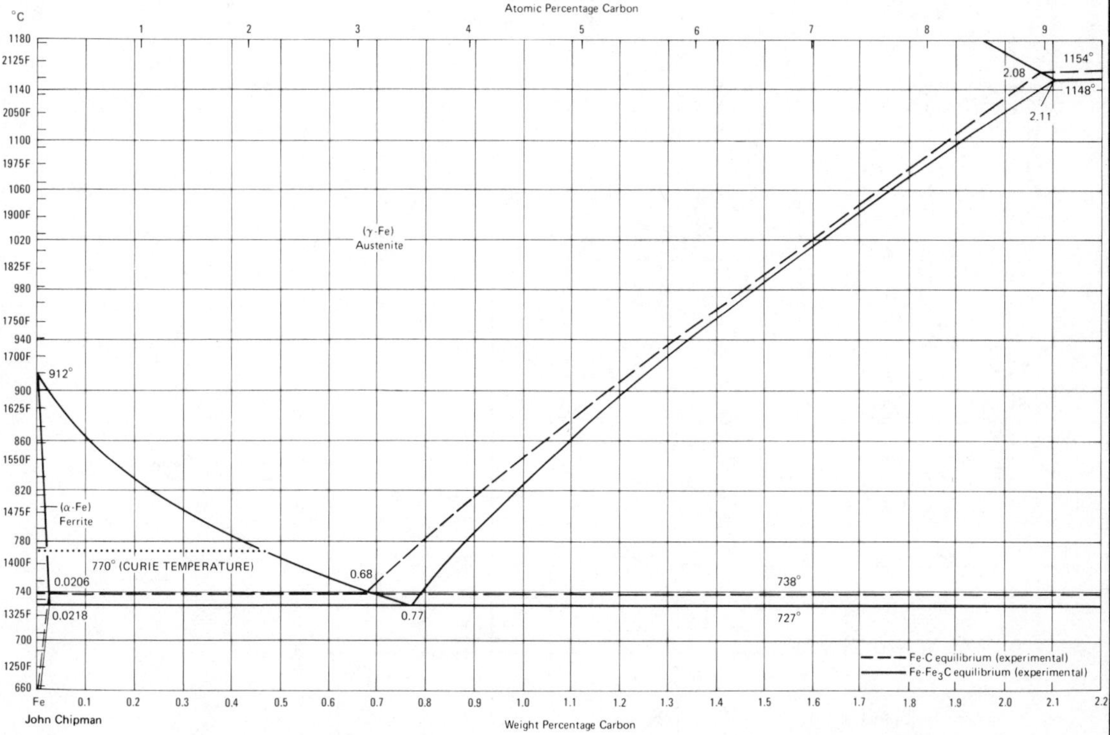

John Chipman

- - - - Fe-C equilibrium (experimental)
———— Fe-Fe₃C equilibrium (experimental)

C-Fe (Carbon-Iron) (continued)

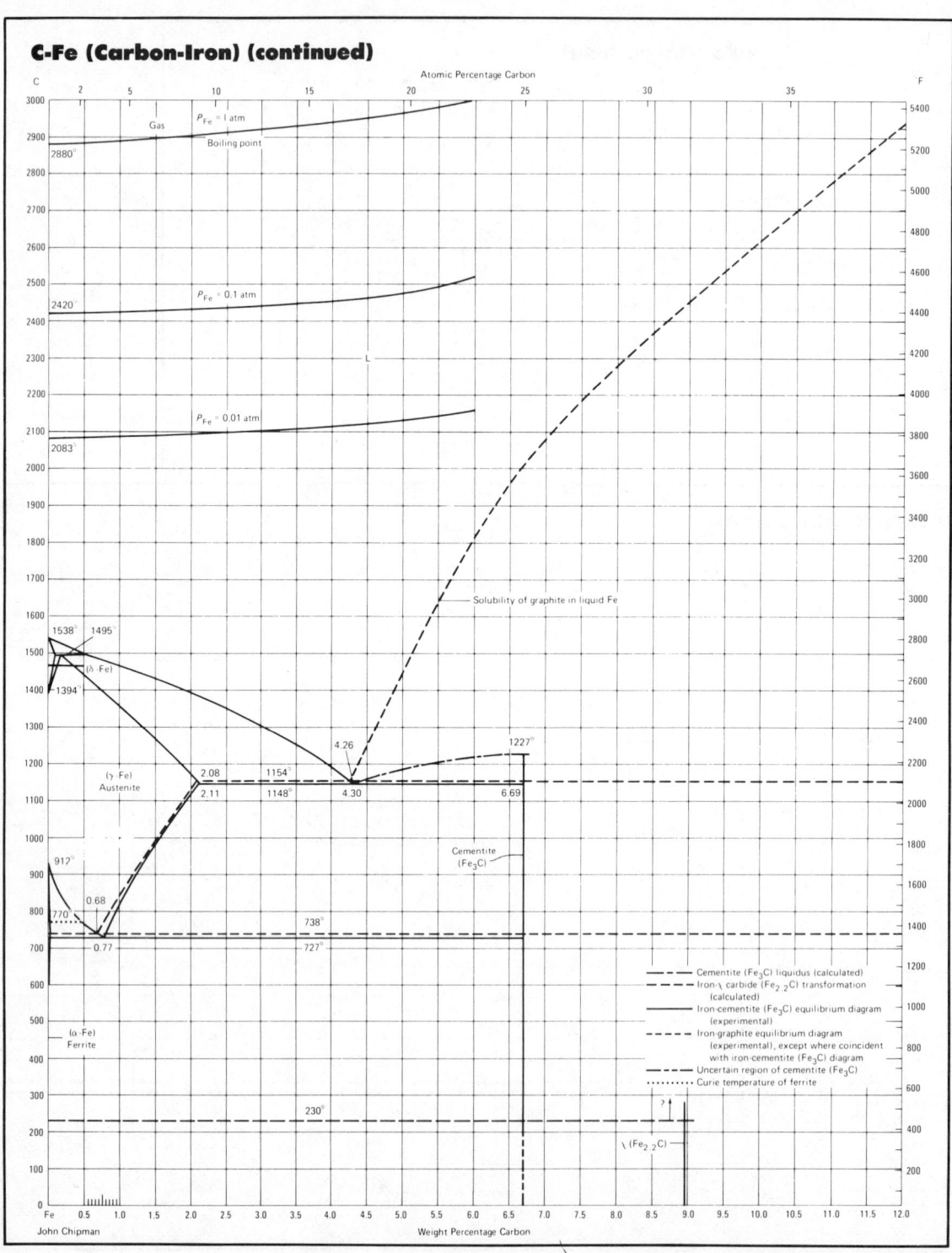

John Chipman

C-Fe (Carbon-Iron) (continued)

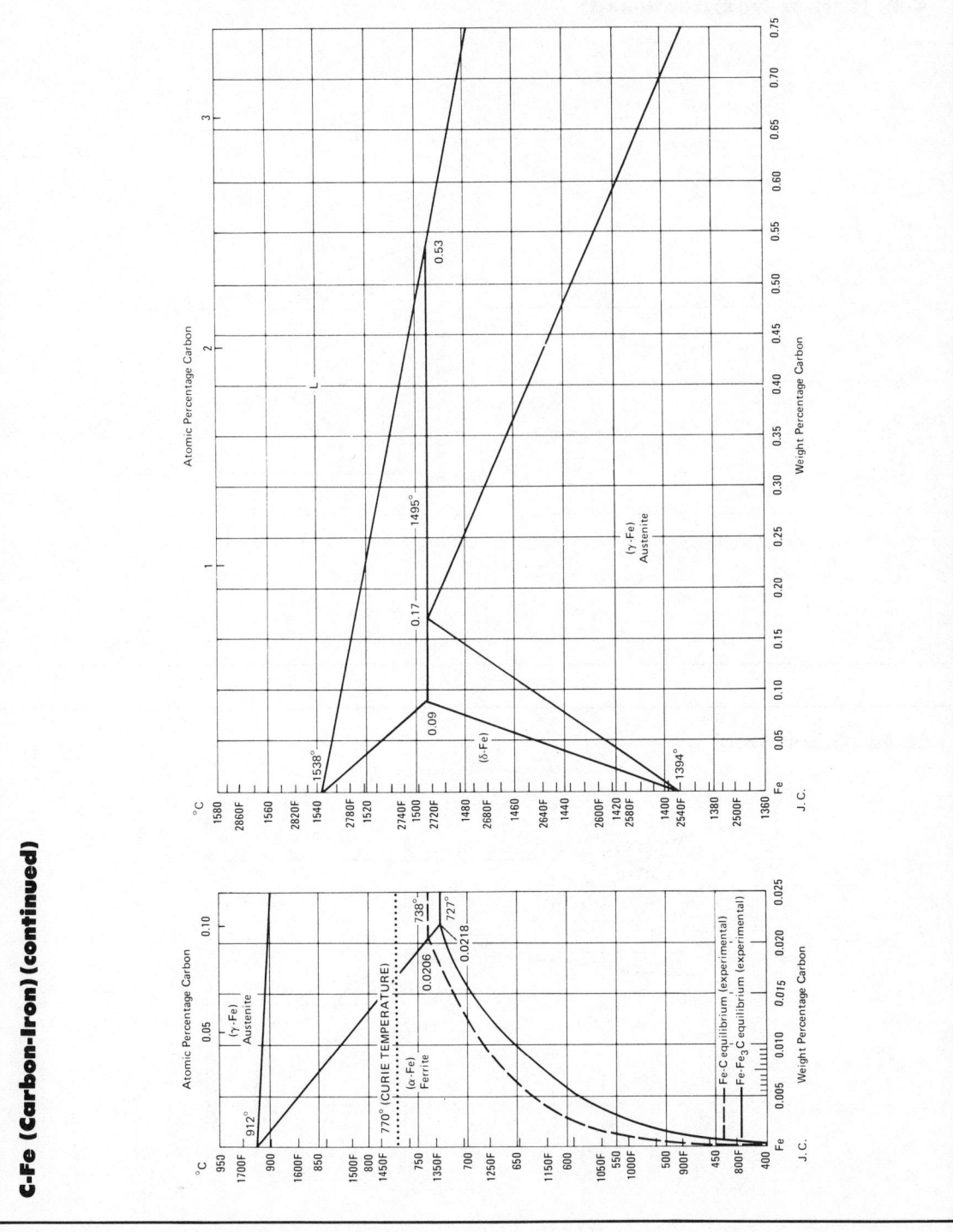

C-Fe (Carbon-Iron)(continued)

Two-phase fields involving graphite or iron carbide

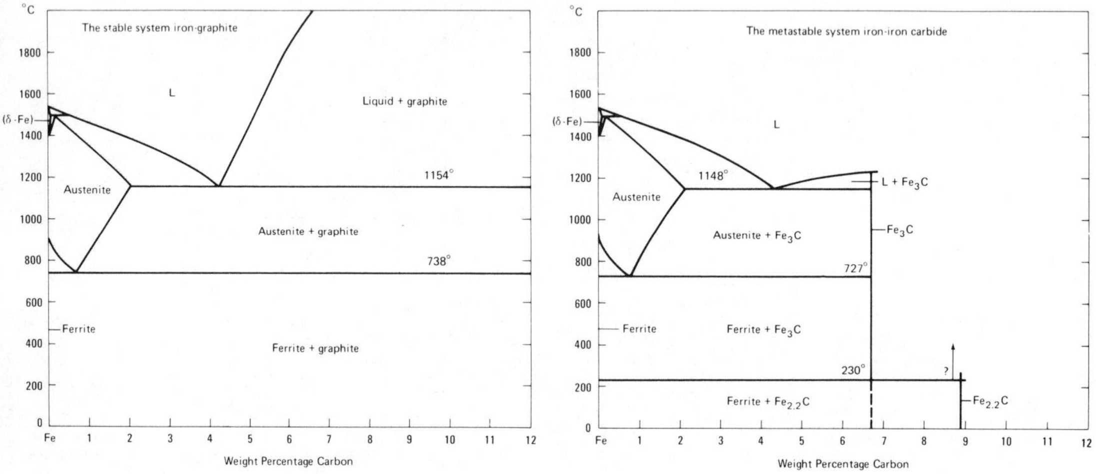

The stable system iron-graphite

°C

1800

1600

(δ-Fe) — 1400

L

1200 — 1154°

Austenite

1000

800 — 738°

Austenite + graphite

600

Ferrite

400

Ferrite + graphite

200

Liquid + graphite

0
Fe 1 2 3 4 5 6 7 8 9 10 11 12

Weight Percentage Carbon

The metastable system iron-iron carbide

°C

1800

1600

(δ-Fe) — 1400

L

1200 — 1148°   L + Fe₃C

Austenite   Fe₃C

1000

800 — Austenite + Fe₃C 727°

600

Ferrite   Ferrite + Fe₃C

400

200 — 230° ?

Ferrite + Fe₂.₂C   Fe₂.₂C

0
Fe 1 2 3 4 5 6 7 8 9 10 11 12

Weight Percentage Carbon

Co-Fe (Cobalt-Iron)

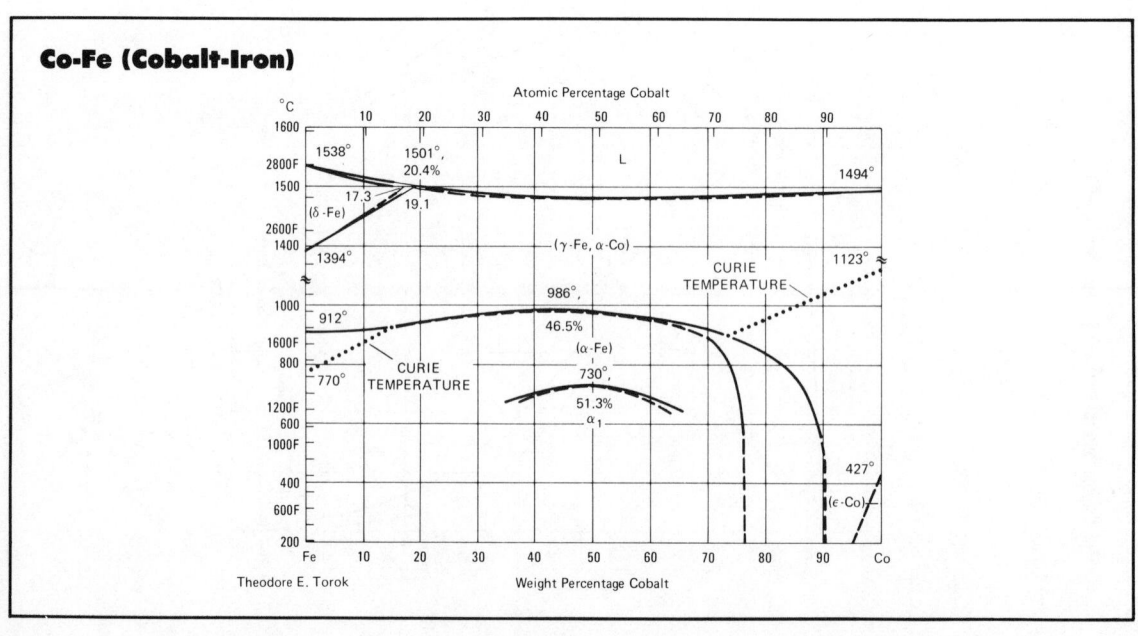

Theodore E. Torok

Cr-Fe (Chromium-Iron)

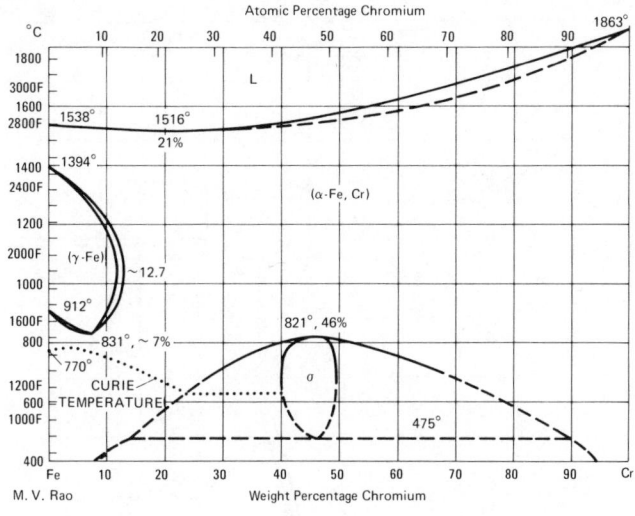

M. V. Rao

Cu-Fe (Copper-Iron)

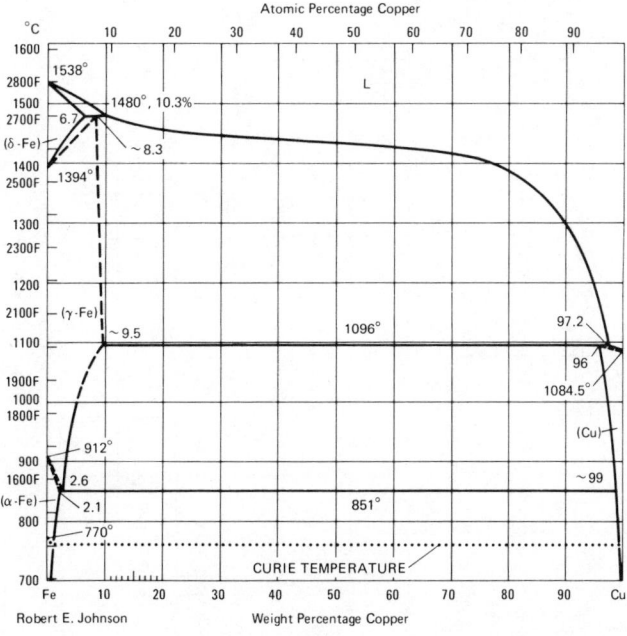

Robert E. Johnson

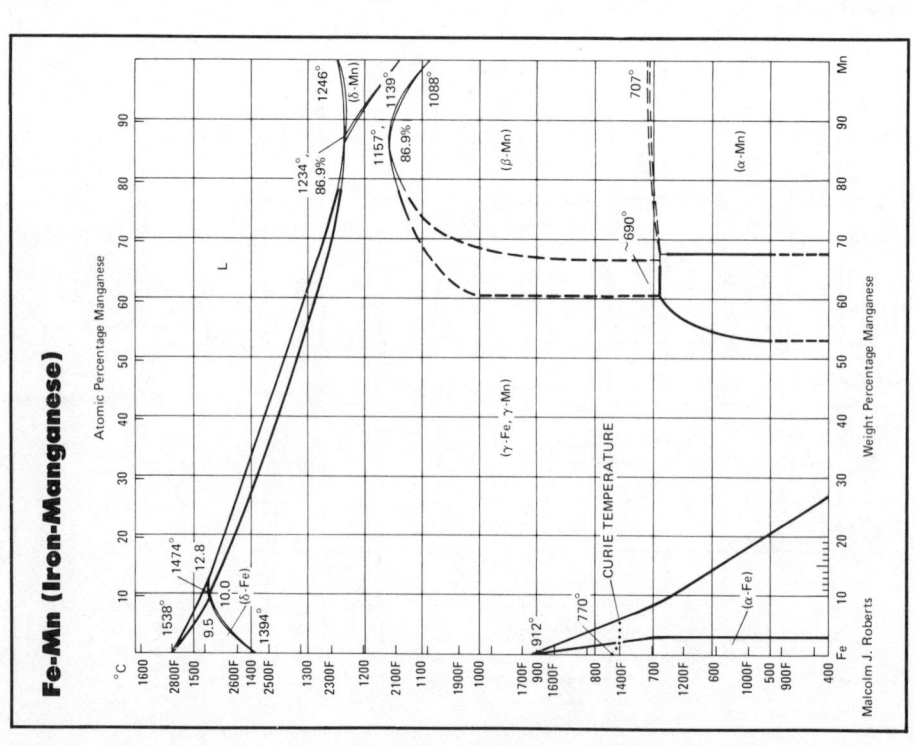

Fe-Mo (Iron-Molybdenum)

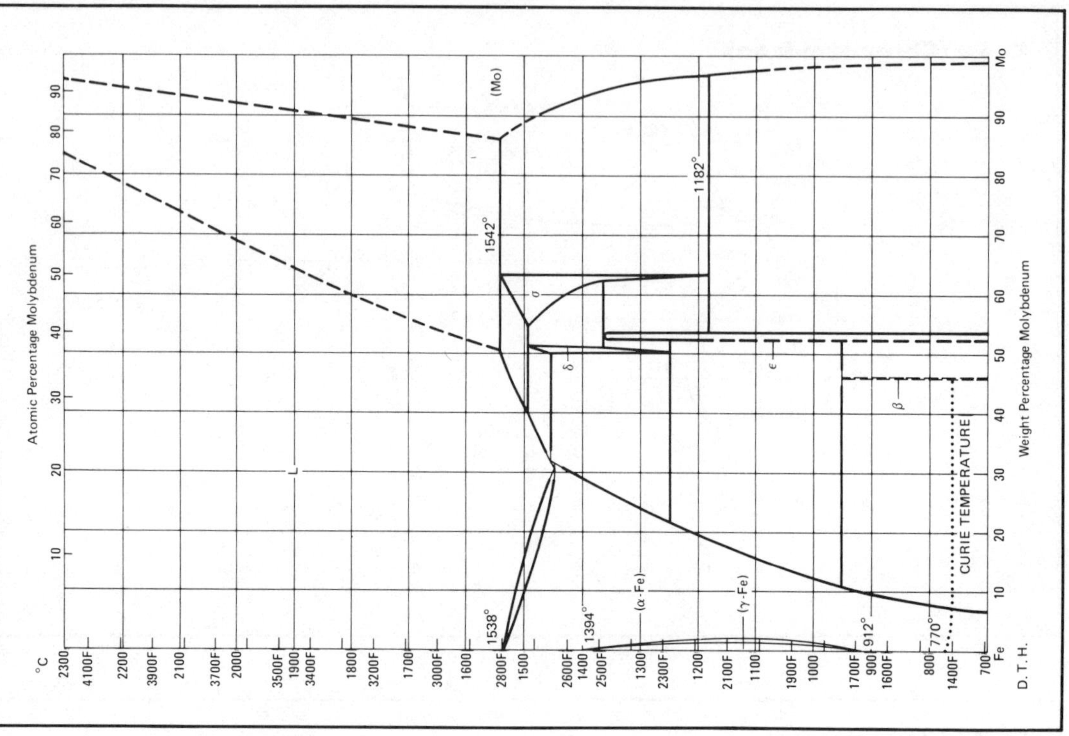

Atomic Percentage Molybdenum

°C
2300
2200
2100
2000
1900
1800
1700
1600
1538°
1500
1400 1394°
1300
1200
1100
1000
900 912°
800 770°
700

4100F
3900F
3700F
3500F
3400F
3200F
3000F
2800F
2600F
2500F
2300F
2100F
1900F
1700F
1600F
1400F

1542°
1182°
(Mo)
σ
δ
ε
β
(α-Fe)
(γ-Fe)
CURIE TEMPERATURE

Weight Percentage Molybdenum

D. T. H.

Fe-Mn (Iron-Manganese)

Atomic Percentage Manganese

°C
1600
1538°
1500
1474°
1400 1394°
1300
1200
1100
1000
912°
900
800 770°
700
600
500
400

2800F
2600F
2500F
2300F
2100F
1900F
1700F
1600F
1400F
1200F
1000F
900F

L
12.8
10.0
9.5
(δ-Fe)
1234° 86.9%
1246°
(δ-Mn)
1157°
1139°
86.9%
1088°
(β-Mn)
707°
690°
(α-Mn)
(γ-Fe, γ-Mn)
CURIE TEMPERATURE
(α-Fe)

Weight Percentage Manganese

Malcolm J. Roberts

Fe-Ni (Iron-Nickel)

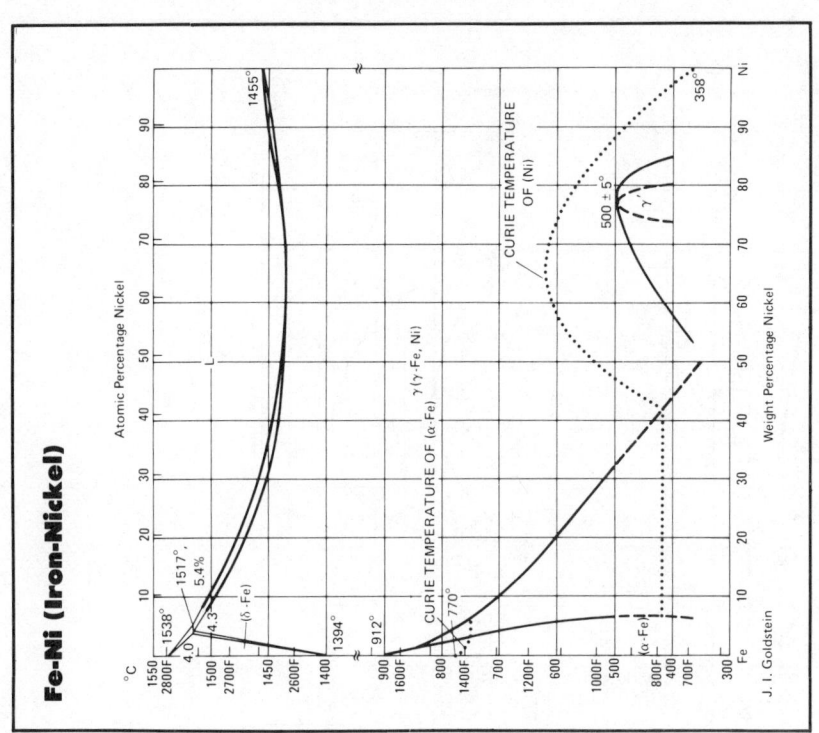

Fe-Nb (Iron-Niobium)

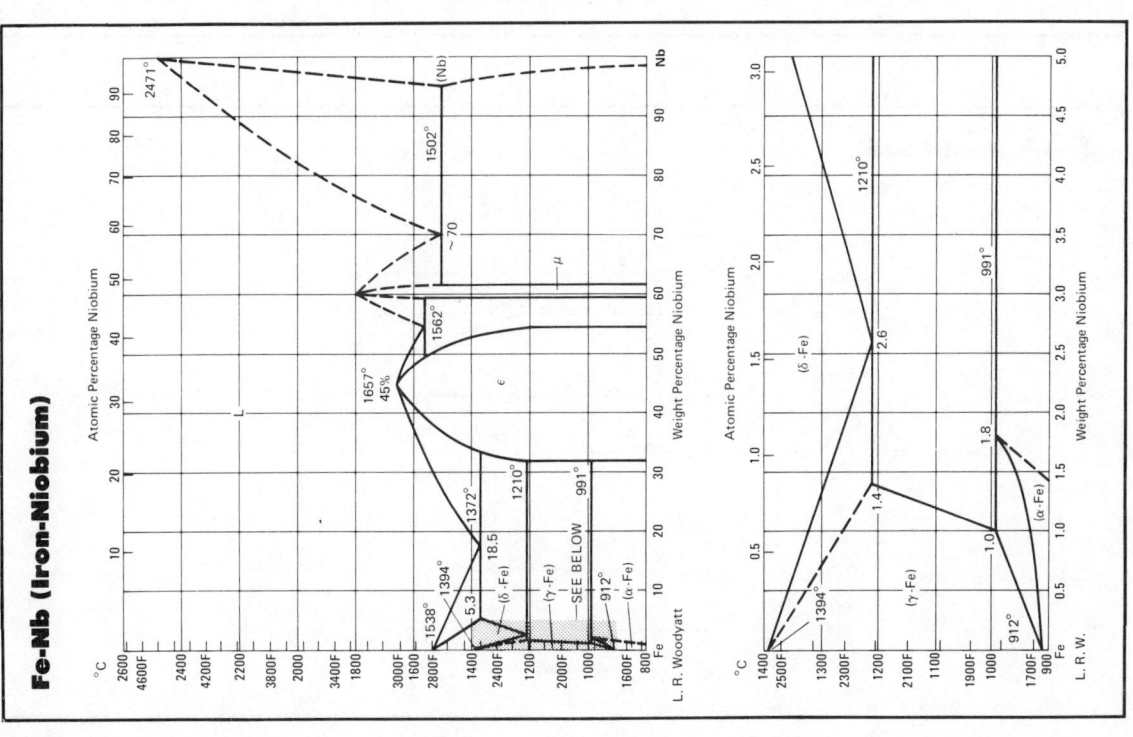

Fe-P (Iron-Phosphorus)

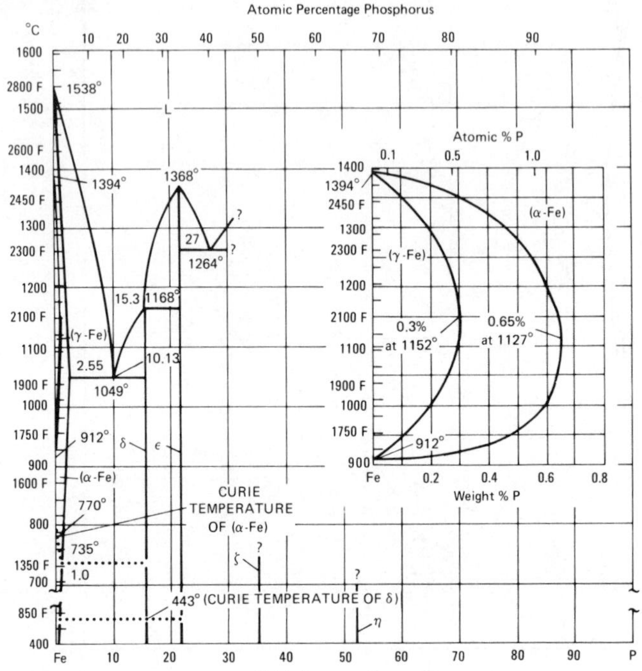

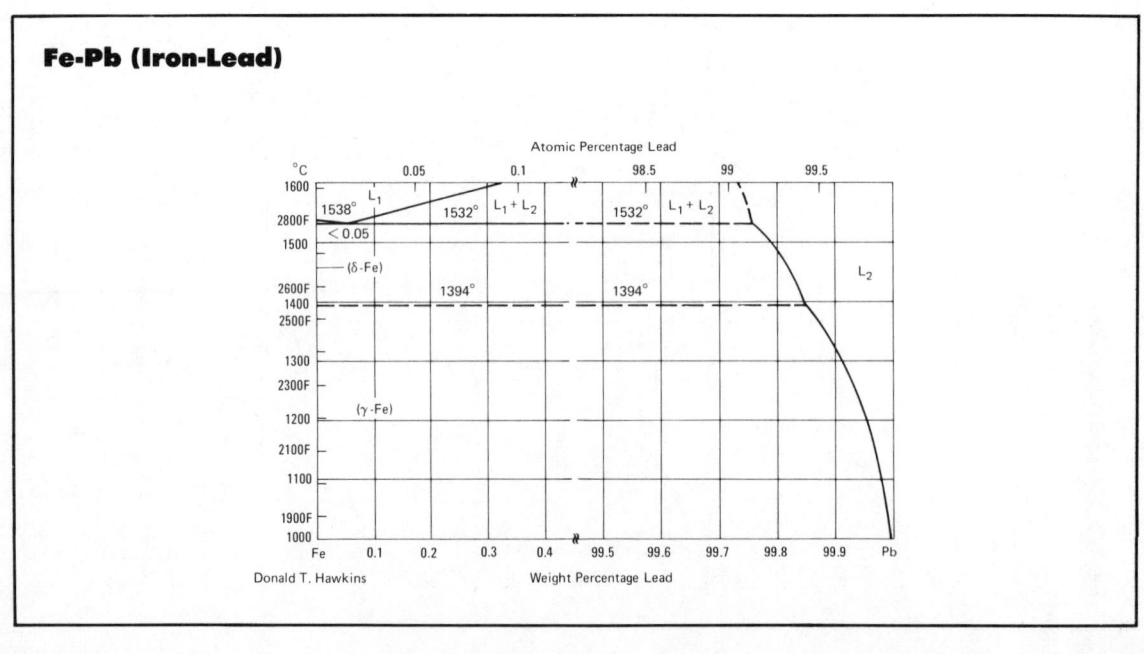

Donald T. Hawkins

Fe-Pb (Iron-Lead)

Donald T. Hawkins

Fe-S (Iron-Sulfur)

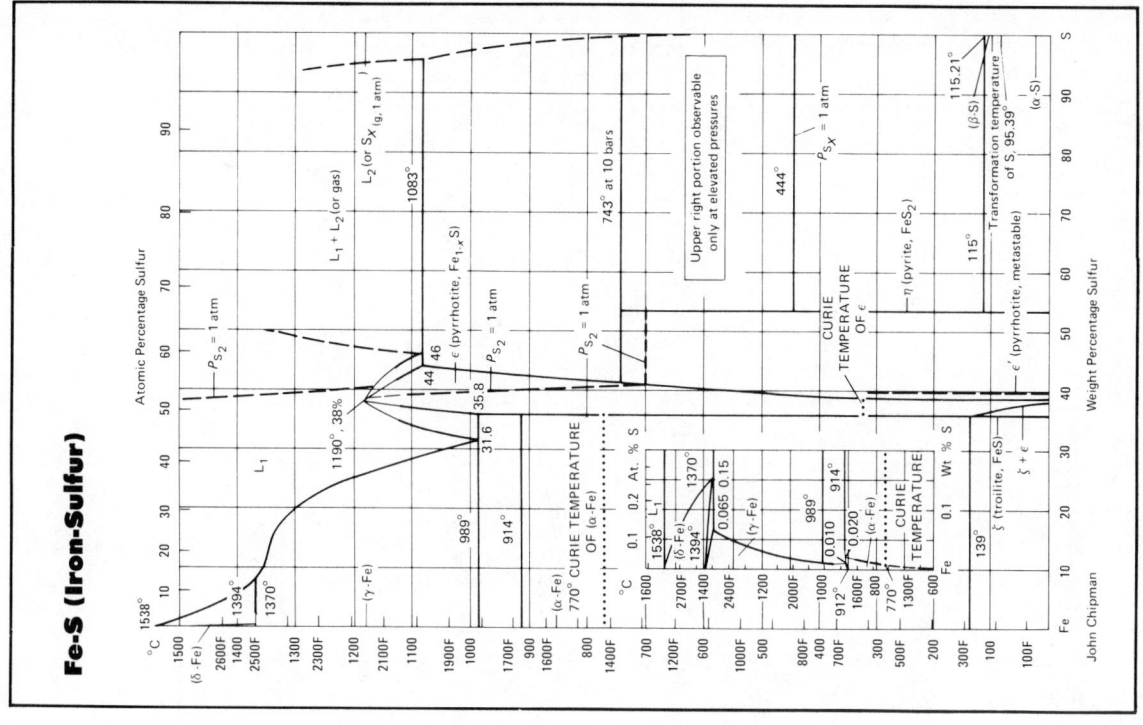

John Chipman

Fe-Si (Iron-Silicon)

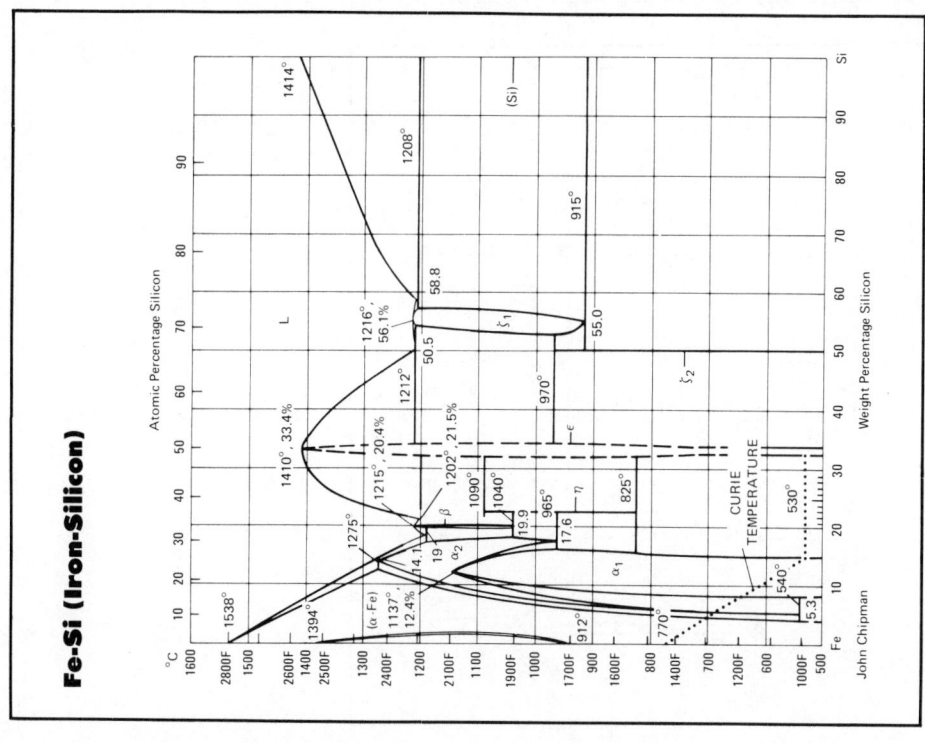

John Chipman

Fe-V (Iron-Vanadium)

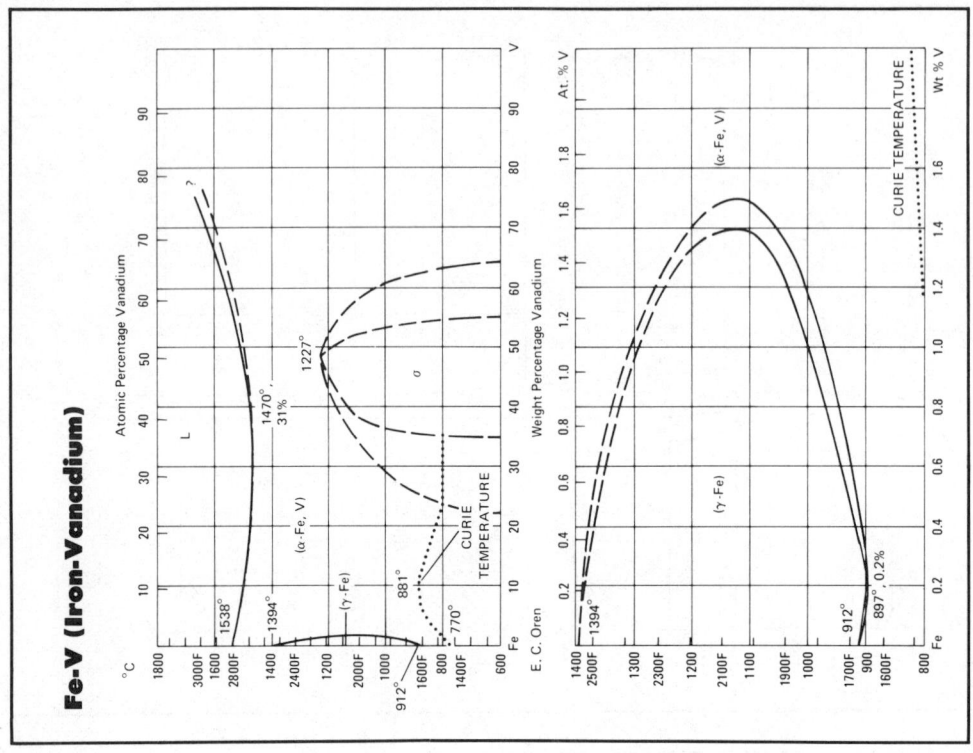

E. C. Oren

Fe-W (Iron-Tungsten)

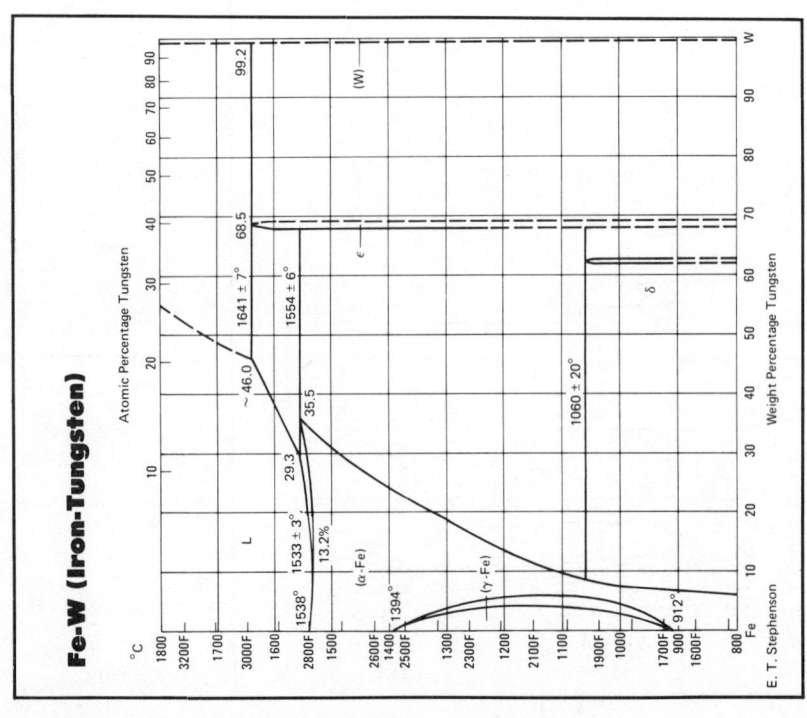

E. T. Stephenson

Ferrous Ternary Alloy Systems

General References

1 L. A. Willey, Appendix 1, Phase Diagrams, in *Aluminum*, Vol I, K. R. Van Horn, Ed., American Society for Metals, 1967, p 359–381

2 W. B. Pearson, *Handbook of Lattice Spacings and Structures of Metals*, Vol 1 and 2, Pergamon, 1958 and 1967

3 M. Hansen, *Constitution of Binary Alloys*, 2nd ed., prepared with the cooperation of K. Anderko, McGraw-Hill, 1958

4 R. P. Elliott, *Constitution of Binary Alloys, First Supplement*, McGraw-Hill, 1965

5 F. A. Shunk, *Constitution of Binary Alloys, Second Supplement*, McGraw-Hill, 1969

6 R. Hultgren, P. D. Desai, D. T. Hawkins, M. Gleiser and K. K. Kelley, *Selected Values of the Thermodynamic Properties of Binary Alloys*, American Society for Metals, 1973

7 E. Rudy, Compendium of Phase Diagram Data, Part V, Air Force Materials Laboratory Report AFML-TR-65-2, 1969

Al-Cr-Fe (Aluminum-Chromium-Iron)

By L. A. Willey

The liquidus diagram is based on data by Mondolfo (1) and on unpublished data by Willey (Alcoa Research Laboratories). On the solvus diagram, the isotherms at 600 and 425 °C (1112 and 797 °F) are from observations by Pratt and Raynor (2).

No ternary phases were found in equilibrium with the aluminum-rich solid solution. However, the $CrAl_7$ phase (sometimes called θ Al-Cr) will dissolve up to approximately 5.5 wt % Fe, and the Cr_2Al_{11} phase will dissolve up to approximately 13.5 wt % Fe. The $FeAl_3$ phase will dissolve a maximum of 4 wt % Cr.

Additional information on the Al-Cr-Fe system can be found in Ref 3.

REFERENCES

1 L. F. Mondolfo, p 70-71 in "Metallography of Aluminum Alloys", Wiley, 1943

2 J. N. Pratt, G. V. Raynor, *J Inst Metals*, Vol 80, 1951-52, p 449-458

3 H. W. L. Phillips, p 78 in "Equilibrium Diagrams of Aluminum Alloy Systems", Aluminium Development Assn. (London), 1961

B-C-Fe (Boron-Carbon-Iron)

By L. Brewer and S-G. Chang

Nicholson (1) found that at 1000 °C (1832 °F) up to 80 wt % of the carbon in cementite, Fe_3C, could be replaced by boron to give $Fe_3C_{0.2}B_{0.8}$. As boron replaces carbon, lattice parameters of the phase change, with contraction along the a-axis and c-axis and expansion along the b-axis of the orthorhombic unit cell. Also, the saturation magnetic moment and the Curie temperature increase with increase in boron content.

Carroll *et al* (2) found a ternary iron borocarbide, $Fe_{23}(C,B)_6$, that is isomorphous with cubic $Cr_{23}C_6$ ($D8_4$-type structure). The $Fe_{23}(C,B)_6$ phase has been more fully investigated by Stadelmaier and Gregg (3), and Lucco Borlera and Pradelli (4). Stadelmaier and Gregg report that $Fe_{23}(C,B)_6$ melts congruently and that lattice constant a varies from 10.594 to 10.628 A from carbon-rich to boron-rich compositions, after quenching from 800 °C (1472 °F).

The isothermal sections at 700, 800, 900 and 1000 °C (1292, 1472, 1652 and 1832 °F) herewith are from Lucco Borlera and Pradelli (4). At 700 °C (1292 °F), the $Fe_{23}(C,B)_6$ phase has compositions from $Fe_{23}(C_{0.73}B_{0.27})_6$, which is in equilibrium with $Fe_3(C,B)$ (borocementite) and (α-Fe), to $Fe_{23}(C_{0.44}B_{0.56})_6$, which is in equilibrium with Fe_2B and $Fe_3(C,B)$. At 800 °C (1472 °F), $Fe_{23}(C,B)_6$ is present within larger limits of carbon and boron—between $Fe_{23}(C_{0.38}B_{0.62})_6$ and $Fe_{23}(C_{0.77}B_{0.23})_6$—and is in equilibrium with both (γ-Fe) and $Fe_3(C,B)$. The $Fe_{23}(C,B)_6$ phase is stable up to 965 ± 5 °C (1769 ± 9 °F).

REFERENCES

1 M. E. Nicholson, *Trans Met Soc AIME*, Vol 209, 1957, p 1-6

2 K. G. Carroll, L. S. Darken, R. M. Fisher and L. Zwell, *Nature (London)*, Vol 174, 1954, p 978

3 H. H. Stadelmaier and R. A. Gregg, *Metall*, Vol 17, 1963, p 412-414

4 M. Lucco Borlera and G. Pradelli, *Met Ital*, Vol 59, 1967, p 907-916

C-Cr-Fe (Carbon-Chromium-Iron)

By W. D. Forgeng and W. D. Forgeng, Jr.

The general features of the C-Cr-Fe system presented here are well established (1 to 10). The region between Fe_3C and Cr_7C_3 is apparently a quasibinary system, and the isothermal sections have been constructed on this basis.

The carbide phases have the following formulas and composition limits:

M_3C, cementite, containing up to 15 wt % Cr

M_7C_3, trigonal chromium carbide, containing up to 55 wt % Fe

$M_{23}C_6$, cubic chromium carbide, containing up to 30 wt % Fe

M_3C_2, orthorhombic chromium carbide, containing up to 20 wt % Fe.

The other phases have been designated as follows: Graphite; L, liquid; α, ferrite; (Cr), chromium ferrite; γ, austenite; and σ, sigma (FeCr).

The ternary liquidus projection, from Griffing et al (8), indicates the liquidus-field boundaries of the primary phases formed during solidification. The arrows along the boundaries indicate the direction of composition change of the remaining liquid with decreasing temperature. Each intersection of three interval boundaries is the composition of liquid involved in isothermal equilibrium with the three primary phases, the liquidus fields of which touch there. The general topography of the liquidus surface consists of a eutectic valley sloping down from the Cr-C edge at 3.39 wt % C to the Fe-C edge at 4.26 wt % C. The liquidus surface rises in two directions from this valley, one slope terminating at the Cr-Fe edge, and the opposing slope rising toward the carbon (graphite) corner. Note that M_3C can solidify as a stable (equilibrium) phase in this system when the liquid contains more than about 3 wt % Cr, whereas in plain Fe-C alloys, Fe_3C (cementite) occurs only as a metastable (nonequilibrium) phase. In the three isothermal sections, the metastable Fe_3C is the terminal of the M_3C phase along the Fe-C axis, shown in this way for simplicity.

Six of the seven four-phase equilibria in this system, and their approximate temperatures, are indicated in the liquidus-projection diagram. The other nonvariant equilibrium (not shown in any of the four vertical sections) is of considerable practical importance; it occurs at about 800 °C (1472 °F) and involves the α, γ, M_3C and M_7C_3 phases. The coordinates of the α and γ corners of this quadrangle are at 0.05 wt % C, 2.6 wt % Cr, and at 0.7 wt % C, 2 wt % Cr, respectively. Therefore, M_7C_3 is stable in medium-carbon steels containing more than 2 wt % Cr, and the behavior of such steels during heat treatment is different from that of plain carbon steels.

Alloys with less than 2 wt % Cr are similar in behavior to plain iron-carbon alloys, differing only in that three-phase regions can exist over a range of temperatures and compositions rather than at a fixed point. As the chromium content is increased, the single-phase γ region becomes smaller; it disappears entirely at about 20 wt % Cr. In addition, the temperature range of the γ region becomes narrower at higher chromium contents. Thus, from a practical standpoint, steels containing 10 to 18 wt % Cr must be heated at temperatures between 900 and 1100 °C (1652 and 2012 °F) if full hardening is to be obtained during subsequent quenching, and for maximum hardening by transformation of the γ phase, the carbon content must lie within the limits of the γ region for a given chromium content. Also, the carbon content of the eutectoid is decreased by increasing the chromium content, and at 12 wt % Cr, the eutectoid contains only about 0.35 wt % C. The α solid solution in the equilibrium α + M_7C_3 + $M_{23}C_6$ becomes richer in iron at lower temperatures; consequently, $M_{23}C_6$ may appear in alloys with less than 10 wt % Cr.

At 20 to 30 wt % Cr, hardening by the transformation of the γ phase becomes less pronounced. Because medium-carbon alloys with 30 wt % Cr contain only α, M_7C_3 and $M_{23}C_6$ phases, such alloys cannot be hardened by heat treatment.

The solubility of carbon in the α phase of iron-rich alloys at temperatures just below the solidus increases from about 0.05 wt % C at 12 wt % Cr to about 0.1 wt % C at 20 to 30 wt % Cr. Cost and Blazek (11) report that the solubility of carbon in α phase at lower temperatures and lower chromium contents (below 10 wt % Cr) is markedly reduced by the presence of chromium, being only about 10 ppm at 700 °C (1292 °F) in an alloy containing 2 wt % Cr, compared with about 200 ppm in pure iron at this temperature.

In alloys containing 15 to 40 wt % Cr (or more—depending on carbon content), the α solid solution is gradually converted to the hard, brittle σ phase by holding within the temperature range of 550 to 820 °C (1022 to 1508 °F) for long periods. The formation of a chromium-rich ferrite (Cr) precipitate below 520 °C (968 °F) in low-carbon alloys is reported by Williams and Paxton (12), and by Williams (13), and explains a phenomenon of embrittlement in stainless steels in this temperature range that was formerly attributed to the formation of σ. The formation of the (Cr) phase in higher-carbon iron-chromium alloys has not been studied, and a similar precipitation behavior in these alloys cannot be inferred.

REFERENCES

1 T. Hatsuta, *Technol Rept Tohoku Imp Univ,* Vol 10, 1931-1932, p 680
2 A. Westgren, *Nature,* Vol 132, 1933, p 480
3 W. Tofaute, A. Sponheuer and H. Bennek, *Arch Eisenhuettenw,* Vol 8, 1935, p 499
4 W. Tofaute, C. Küttner and A. Büttinghaus, *Arch Eisenhuettenw,* Vol 9, 1936, p 606
5 A. B. Kinzel and W. Crafts, "The Alloys of Iron and Chromium", Vol I, McGraw-Hill, 1937
6 K. Bungardt, E. Kunze and E. Horn, *Arch Eisenhuettenw,* Vol 29, 1958, p 193-203
7 L. Messulam and A. S. Appleton, *Trans AIME,* Vol 236, 1966, p 222-224
8 N. R. Griffing, W. D. Forgeng and G. W. Healy, *Trans AIME,* Vol 224, 1962, p 148-159
9 T. Wada, H. Wada, J. F. Elliott and J. Chipman, *Met Trans,* Vol 3, 1972, p 2865-2872
10 R. S. Jackson, *J Iron Steel Inst (London),* Vol 208, 1970, p 163-167
11 J. R. Cost and K. E. Blazek, *Met Trans* in manuscript as of June 1973, preparatory to publication
12 R. O. Williams and H. W. Paxton, *J Iron Steel Inst (London),* Vol 185, 1957, p 358-374
13 R. O. Williams, *Trans AIME,* Vol 212, 1958, p 497-502

C-Cu-Fe (Carbon-Copper-Iron)

By L. Brewer and S-G. Chang

The phases that are present in the C-Cu-Fe system are: L_1, iron-rich liquid; L_2, copper-rich liquid; α, γ and δ solid solutions of iron; ε solid solution of copper; and iron carbide, Fe_3C. Stogoff and Messkin (1) reported evidence of graphite at high concentrations of carbon and copper.

Maddocks and Claussen (2) determined the intersection of the two-liquids region with the liquidus surface. Their work shows that the liquid immiscibility line does not intersect the plane of the iron-copper binary diagram. This is confirmed by the work of Iwasé et al (3). The liquid-solubility diagrams at 1450 and 1540 °C (2642 and 2804 °F) are derived from the work of Iwasé *et al* (3), Schenck and Perbix (4), and Chipman (5). The experimental tie lines are shown in the region of the two-liquids phase. The composition of L_2 in equilibrium with L_1 and carbon at 1450 and 1540 °C (2642 and 2804 °F) was extrapolated to 4 wt % Fe; because of the uncertainties of the thermodynamic data, this composition could be as high as 6 wt % Fe. According to Iwasé *et al,* about 0.02 wt % C is sufficient to introduce liquid immiscibility in alloys containing 50 wt % Fe and 50 wt % Cu. Mutual liquid solubility increases with temperature.

The vertical section through the 50Fe-50Cu line, obtained by Maddocks and Claussen, shows that the monotectic and peritectic four-phase planes exist separately, the former at 1150 °C (2102 °F) and the latter at 1094 °C (2001 °F). The iron-carbon eutectic temperature is raised to 1150 °C (2102 °F) by the addition of copper; the monotectic four-phase point is at about 4.2 wt % C, 2.5 wt % Cu.

The δ-γ transformation temperature of the iron-carbon system (1394 °C, 2541 °F) is raised by the addition of copper, according to Andrew *et al* (6).

The ε solid solution of copper probably dissolves some carbon, although quantitative data are lacking. Bever and Floe (7) reported that copper dissolves 0.0001 wt % C at 1100 °C (2012 °F) and 0.003 wt % C at 1700 °C (3092 °F).

The boundaries of the γ phase have not been accurately established. According to Ishiwara *et al* (8), the maximum extent of the γ solid solution is at 1.7 wt % C, about 6 wt % Cu. The γ region is limited on its lower side by the three surfaces representing primary separation of the α phase, ε phase and cementite. The ternary eutectoid point, according to Ishiwara, is about 700 °C (1292 °F), and its composition is about 0.9 wt % C, 1.9 wt % Cu.

REFERENCES

1 A. F. Stogoff and W. S. Messkin, *Arch Eisenhuettenw,* Vol 2, 1928, p 321
2 W. R. Maddocks and G. E. Claussen, *Iron Steel Inst (London) Spec Rept No. 14,* 1936, p 97
3 K. Iwasé, M. Okamoto and T. Ameniya, *Sci Rept Tohoku Imp Univ,* Vol 26, 1938, p 618
4 H. Schenck and G. Perbix, *Arch Eisenhuettenw,* Vol 32, 1961, p 123-127
5 J. Chipman, *Met Trans,* Vol 3, 1972, p 55-64
6 J. H. Andrew, G. T. C. Bottomley, W. R. Maddocks and R. T. Percival, *Iron Steel Inst (London) Spec Rept No. 23,* 1938, p 5
7 M. B. Bever and C. F. Floe, *Trans Inst Met Div AIME,* Vol 166, 1946, p 128-141
8 T. Ishiwara, T. Yonekura and T. Ishigaki, *Sci Rept Tohoku Imp Univ,* Vol 15, 1926, p 81

C-Fe-Mn (Carbon-Iron-Manganese)

By L. Brewer, J. Chipman and S-G. Chang

The isothermal sections for the C-Fe-Mn system at 1100, 1000, 900, 800 and 600 °C (2012, 1832, 1652, 1472 and 1112 °F) are from Benz, Elliott and Chipman (1), who derived their diagrams from experimental results and from reports of earlier investigators (2-5).

At 1100 °C (2012 °F), a small liquid field is present. The region of the liquid moves to higher manganese concentrations as temperature is decreased. Liquid persists down to 1045 °C (1913 °F), where it freezes, probably by the eutectic reaction L → ε + M_3C + γ. At 1013 °C (1855 °F), an invariant point of four-phase equilibrium occurs, with coexistence of ε, M_3C, $M_{23}C_6$ and γ. Above the invariant point, there are two three-phase fields: ε + M_3C + γ, and ε + $M_{23}C_6$ + M_3C. Below the invariant point, the three-phase fields are $M_{23}C_6$ + M_3C + γ, and ε + $M_{23}C_6$ + γ. Several other three-phase areas are shown in the diagrams. The ε phase has variable composition and is sometimes designated M_4C_{1+x}.

The carbide M_3C is stable over a wide range of manganese and iron contents, extending, at 1000 °C (1832 °F), from Mn_3C to nearly pure Fe_3C. Fe_3C is metastable with respect to graphite and γ (austenite); this metastability is overcome by less than 1 wt % Mn in the metal phase (see Smith, Ref 6). The face-centered-cubic γ phase extends from the Fe-C side to the Mn-C side of the system at temperatures above 891 °C (1636 °F). Other phases that may be in equilibrium with γ include: L at or above 1045 °C (1913 °F), ε from 991 to 1260 °C (1816 to 2300 °F), and $M_{23}C_6$ at and below 1000 °C (1832 °F). $M_{23}C_6$ also occurs in equilibrium with (α-Mn).

The portion of the system with <2.0 wt % C and <20 wt % Mn has been studied by many investigators (6-10). Purdy *et al* (10), in contrast to the early work of Wells (9), indicate that the surface between the α + γ region and the γ region is concave, as shown in the vertical section for 2.5 wt % Mn.

The vertical sections at constant manganese (2.5, 4.5 and 13 wt % Mn) are from Walters and Wells (2) and Wells (9).

REFERENCES

1 R. Benz, J. F. Elliott and J. Chipman, *Met Trans,* Vol 4, 1973, p 1449-1452
2 F. M. Walters, Jr. and C. Wells, *Trans Am Soc Metals,* Vol 24, 1936, p 359-374
3 R. Vögel and W. Döring, *Arch Eisenhuettenw,* Vol 9, 1935, p 247
4 M. Isobe, *Sci Rept Tohoku Imp Univ,* Vol A3, 1951, p 540
5 K. Kuo and L. E. Persson, *J Iron Steel Inst (London),* Vol 178, 1954, p 39-44
6 R. P. Smith, *J Am Chem Soc,* Vol 70, 1948, p 2724-2729
7 E. C. Bain, E. S. Davenport and W. S. N. Waring, *Trans AIME,* Vol 100, 1932, p 228-249
8 W. Tofaute and K. Linden, *Arch Eisenhuettenw,* Vol 10, 1936-1937, p 515
9 C. Wells, p 1251-1252 in "Metals Handbook", T. Lyman, Editor, American Society for Metals, 1948
10 G. P. Purdy, D. H. Weichert and J. S. Kirkaldy, *Trans Met Soc AIME,* Vol 230, 1964, p 1025-1034

C-Fe-Mo (Carbon-Iron-Molybdenum)

By T. Wada

Since the publication of the phase diagrams for the C-Fe-Mo system in the 1948 edition of Metals Handbook, which were based on a review by Marsh (1) of Takei's data (2), revised diagrams have been reported by Kuo (3), Sato *et al* (4), Campbell *et al* (5), Nishizawa (6), Jellinghaus (7), and Harvig and Uhrenius (8). Campbell *et al* were con-

Table 1 Compositions and properties of phases in the C-Fe-Mo system

Characteristic composition	Properties
(α-Fe); (δ-Fe)	Solid solution of C and Mo in bcc Fe
(γ-Fe)	Solid solution of C and Mo in fcc Fe
Fe₃C	Mo may dissolve in Fe₃C (cementite): up to ~6 wt % at 1000 C (1832 F) (Ref 6 and 8) and ~3 wt % at 700 C (1292 F) (Ref 9 and 10).
Fe₂MoC	Orthorhombic structure (Ref 11); 40 to 42 wt % Mo (Ref 8)
M₂₃C₆	Face-centered-cubic structure (Ref 3). The composition is close to Fe₂₁Mo₂C₆, ranging from 2 to 8 wt % Mo (Ref 3 and 9).
M₆C	Face-centered-cubic structure (Ref 12). Most probable composition is 58 to 63 wt % Mo (Ref 8).
M₂C	Hexagonal structure. Two modifications of arrangement of C atoms have been reported (Ref 13, 14).
(Mo)	Solid solution of C and Fe in Mo
Fe₃Mo₂	Corresponds crystallographically to Fe₇Mo₆ (Ref 15)
Fe₂Mo	A Laves phase has been reported (Ref 15 and 16), but others were unable to find it (Ref 17).
MoC₀.₆₄(a)	Stable temperature range: 1658 to 2550 C (3016 to 4622 F)
MoC₀.₆₂(a)	Stable temperature range: 1963 to 2605 C (3565 to 4721 F)
Fe₅Mo₃(a)	Stable temperature range: 1247 to 1493 C (2277 to 2719 F)
FeMo(a)	Stable temperature range: 1182 to 1542 C (2160 to 2808 F)

(a) These phases do not appear in the diagrams because the temperatures and compositions at which they are stable are not included.

cerned mainly with alloys with up to 6 wt % Mo; Nishizawa, and Harvig and Uhrenius, determined the isothermal section of the diagram at 1000 °C (1832 °F).

The solid phases (in addition to graphite) that form in the C-Fe-Mo system are listed in Table 1, together with properties of these phases.

The liquidus surface is based on data by Jellinghaus (7). The isothermal section at 1000 °C (1832 °F) is mainly based on the report of Harvig and Uhrenius (8). The most probable diagram of the isotherm at 700 °C (1292 °F) is shown.

The principal difference between the 700 and 1000 °C (1292 and 1832 °F) isotherms is the presence of M₂₃C₆ in the 700 °C (1292 °F) isotherm. Kuo (3) and Jellinghaus (7) report that this carbide may be metastable, but Sato et al (4), Bowman and Parke (9), Bowman (10) and Aldén et al (18) report that it exists in C-Fe-Mo alloys of a certain composition range even after tempering for hundreds of hours at temperatures around 700 °C (1292 °F). Only Campbell et al (5) found M₂₃C₆ at 1000 °C (1832 °F); consequently, the highest temperature at which M₂₃C₆ might exist is assumed to be about 900 °C (1652 °F).

According to Kuo (3), Harvig and Uhrenius (8), and Ridal and Quarrell (19), the two carbides Fe₂MoC and M₂C may also be metastable at 700 °C (1292 °F), but Dyson and

Andrews (11) and Aldén et al (18) have reported their presence in steels after tempering for about 1000 h. Kuo (3) has reported that MoC is a stable carbide in steel, but there has been no further support for this observation.

The isotherm at 1000 °C (1832 °F) for the iron-rich corner of the system is based on data of Harvig and Uhrenius (8). Both Campbell et al (5) and Bungardt et al (20) report lower solubilities of carbides in (γ-Fe), but these are considered less reliable. The results of Nishizawa (6) are in agreement with those of Harvig and Uhrenius, which in turn are supported by the results of Wada et al (21). For simplicity, the (γ-Fe) + graphite region that is shown in this diagram has been omitted from the more complete isotherm at 1000 °C (1832 °F).

The pseudobinary diagrams showing sections at 2, 4, 10 and 20% Mo are revisions of diagrams by Marsh (1), based on information from more current isothermal diagrams and from thermodynamic requirements. Some phase boundaries, for instance those in the L + γ + carbides region of the sections at 10 and 20% Mo, are uncertain.

REFERENCES

1 J. S. Marsh, Appendix 1 in "Alloys of Iron and Molybdenum", Engineering Foundation, McGraw-Hill, 1932
2 T. Takei, *Kinzoku-no-Kenkyu,* Vol 9, 1932, p 97-124, 142-173
3 K. Kuo, *J Iron Steel Inst (London),* Vol 173, 1953, p 363-375
4 T. Sato, T. Nishizawa and K. Tamaki, *Nippon Kinzoku Gakkaishi,* Vol 24, 1960, p 395-399
5 R. F. Campbell, S. H. Reynolds, L. W. Ballard and K. G. Carroll, *Trans Met Soc AIME,* Vol 218, 1960, p 723-732
6 T. Nishizawa, *Scand J Met,* Vol 1, 1972, p 41-48
7 W. Jellinghaus, *Arch Eisenhuettenw,* Vol 39, 1968, p 705-718
8 H. Harvig and B. Uhrenius, Rept TRITA-MAC-0008, Materials Center, Royal Inst Technology, Stockholm, 1971
9 F. E. Bowman and R. M. Parke, *Trans Am Soc Metals,* Vol 33, 1944, p 481-493
10 F. E. Bowman, *Trans Am Soc Metals,* Vol 36, 1946, p 61-80
11 D. J. Dyson and K. W. Andrews, *J Iron Steel Inst (London),* Vol 202, 1964, p 325-329
12 A. C. Fraker and H. H. Stadelmaier, *Trans Met Soc AIME,* Vol 245, 1969, p 847-850
13 M. Hansen, "Constitution of Binary Alloys", 2nd Ed. (prepared with the cooperation of K. Anderko), McGraw-Hill, 1958
14 F. A. Shunk, "Constitution of Binary Alloys, Second Supplement", McGraw-Hill, 1969
15 A. K. Sinha, R. A. Buckley and W. Hume-Rothery, *J Iron Steel Inst (London),* Vol 205, 1967, p 191-195
16 J. P. Michel, *Mem Sci Rev Met,* Vol 68, 1971, p 785-792
17 R. D. Rawlings and C. W. A. Newey, *J Iron Steel Inst (London),* Vol 206, 1968, p 723
18 M. Aldén, S. Asplund and B. Aronsson, *J Iron Steel Inst (London),* Vol 207, 1969, p 235-236
19 K. A. Ridal and A. G. Quarrell, *J Iron Steel Inst (London),* Vol 200, 1962, p 359-365

20 K. Bungardt, E. Schürmann, H. Preisendanz, P. Schüler and H. J. Osing, *DEW Tech Ber,* Vol 9, 1969, p 439-462

21 T. Wada, H. Wada, J. F. Elliott and J. Chipman, *Met Trans,* Vol 3, 1972, p 2865-2872

Carbon, wt %	A	B	Carbon, wt %	A	B
1.2.....	817 ± 110	2.352 ± 0.09	0.50....	682 ± 80	2.143 ± 0.60
1.0.....	769 ± 60	2.272 ± 0.05	0.00....	656 ± 25	2.090 ± 0.02
0.66....	673 ± 80	2.155 ± 0.06			

C-Fe-N (Carbon-Iron-Nitrogen)

By L. Brewer and S-G. Chang

A tentative isothermal section at 450 °C (842 °F) for the C-Fe-N system was constructed by Jack (1). Scheil, Mayr and Müller (2) investigated the system up to 1.4 wt % C and 2.4 wt % N at 600 to 700 °C (1112 to 1292 °F). They found that, with increasing temperature, (γ-Fe) dissolves more carbon and also extends nearer to the iron corner.

The isothermal sections at 500, 565, 575, 600 and 700 °C (932, 1049, 1067, 1112 and 1292 °F) shown here are from Naumann and Langenscheid (3). Cementite dissolves little nitrogen. The (α-Fe) + Fe₃C two-phase region does not extend beyond 0.1 wt % N. The χ carbide has been found to dissolve up to 0.5 wt % N at 500 °C (932 °F). Above 650 °C (1202 °F), χ carbide no longer exists. The γ' nitride dissolves no more than 0.2 wt % C, and is not stable above 680 °C (1256 °F). At 500 °C (932 °F), ε nitride dissolves no more than 4.1 wt % C. The ε nitride can dissolve as much as 3.8, 3.4 and 3.0 wt % C at 550, 600 and 700 °C (1022, 1112 and 1292 °F), respectively. As temperature is increased, ε nitride extends nearer to the iron corner of the phase diagram, reaching 5.0, 3.6 and 3.0 wt % N at 500, 600 and 700 °C (932, 1112 and 1292 °F), respectively. The (γ-Fe) appears above 580 °C (1076 °F). The stabilization temperatures for the three-phase regions up to 11 wt % C plus N have been determined by Naumann and Langenscheid (3):

Three-phase region	Stabilization temperature
(α-Fe)-(γ-Fe)-Fe₃C	>560 °C (1040 °F)
(α-Fe)-(γ-Fe)-γ'....	Between 565 and 585 °C (1049 and 1085 °F)
(γ-Fe)-γ'-Fe₃C.....	Between 565 and 575 °C (1049 and 1067 °F)
(γ-Fe)-γ'-ε........	Between 575 and 650 °C (1067 and 1202 °F)
(γ-Fe)-ε-Fe₃C	>575 °C (1067 °F)
γ'-ε-Fe₃C.........	>575 °C (1067 °F)
ε-χ-Fe₃C	>650 °C (1202 °F)
(α-Fe)-γ'-Fe₃C.....	>565 °C (1049 °F)

The four-phase regions (γ-Fe) -γ'-ε-Fe₃C and (α-Fe) - (γ-Fe) -γ'-Fe₃C occur at 575 and 565 °C (1067 and 1049 °F), respectively.

The vertical section at 91 wt % Fe, from Naumann and Langenscheid (3) and Jack (1), shows that solubility of carbon in ε nitride increases as temperature decreases.

Solubility of nitrogen in Fe-C alloys was determined by Milinskaya and Tomilin (4) at 910 to 1060 °C (1670 to 1940 °F) and by Svyazhin *et al* (5) at 1750, 1960, and 2150 °C (3182, 3560 and 3902 °F). The solubility is proportional to the square root of the pressure of nitrogen. Milinskaya and Tomilin presented the results in the form of an equation: $\log [(\text{wt \% N}) / (P_{N_2})^{1/2}] = A/T - B$. The constants A and B were:

The solubility of nitrogen in carbon-saturated liquid iron alloys was studied by Opravit and Pehlke (6). It increases with temperature and increasing manganese, and decreases with increasing silicon.

Levy, Libsch and Wood (7) found that γ' is the harder of the two carbonitride phases, ε and γ'. The hardness of the ε and γ' phases in the as-nitrided condition decreases as the ratio C/N increases. Both γ' and ε carbonitride are very brittle.

REFERENCES

1 K. H. Jack, *Proc Roy Soc,* A195, 1948, p 34-61
2 E. Scheil, W. Mayr and J. Müller, *Arch Eisenhuettenw,* Vol 33, 1962, p 385-392
3 F. K. Naumann and G. Langenscheid, *Arch Eisenhuettenw,* Vol 36, 1965, p 677-682
4 I. N. Milinskaya and I. A. Tomilin, *Zh Fiz Khim,* Vol 43, 1969, p 2355-2356
5 A. G. Svyazhin, G. M. Chursin, A. F. Vishkarev and V. I. Yavoiskii, *Izv Vysshikh Uchebn Zavedenii, Chernaya Met,* No. 5, 1969, p 43-46
6 O. Opravil and R. D. Pehlke, *Cast Metals Res J,* Vol 5, 1969, p 197-199
7 S. A. Levy, J. F. Libsch and J. D. Wood, *Trans Met Soc AIME,* Vol 245, 1969, p 753-758

C-Fe-Ni (Carbon-Iron-Nickel)

By Rodney P. Elliott

The problems associated with the determination of the C-Fe-Ni system pointed out by Marsh (1), after reviewing the literature, are: (a) thermodynamically unstable Fe₃C is rendered even less stable by the graphitizing effect of Ni, which gives rise to experimental and interpretative difficulties; and (b) the transformation at low temperatures is so sluggish that establishment of equilibrium in a reasonable time is either difficult or impossible.

The liquidus surface was determined by Kase (2) using thermal analysis techniques on 126 alloys; starting materials contained some silicon and manganese. Kase showed that there was no ternary eutectic, but rather a eutectic trough originating in the C-Ni system and sweeping to the C-Fe system. The proeutectic phase for hypereutectic compositions was assumed by Kase to be a continuously miscible carbide (Fe,Ni)₃C. Schichtel and Piwowarsky (3) verified that the ternary system contains a eutectic that collapses with decreasing temperature into the C-Fe system, and they positioned the eutectic trough at higher-carbon compositions than Kase had. However, Schichtel and Piwowarsky showed that only for ternary compositions containing less than 4% Ni is the metastable austenite-Fe₃C equilibrium achieved; for compositions containing >4 to 32% Ni (the maximum investigated), the stable austenite-graphite system was achieved.

Marsh, using Kase's data, hypothesized the ternary system assuming various forms of equilibrium for the quasi-binary section Fe_3C-Ni_3C. Constructions of Marsh are useful to the extent that they hypothesize the course of the eutectoid into the ternary prism. However, some areas are not valid, because, as Schichtel and Piwowarsky established, the metastable equilibrium is seldom achieved.

Buckley and Hume-Rothery (4) determined the liquidus and solidus surfaces for the Fe-rich corner of the system to 1.2% C, 40% Ni. The peritectic reaction $L + \delta \rightarrow \gamma$ originates in the Fe-Ni system at 1517 C (2763 F) and terminates in the peritectic reaction in the C-Fe system at 1495 °C (2723 °F) without involving any other phases. Isothermal saturation of graphite in molten Fe-Ni alloys was determined by Turkdogan et al (5) at 1350 and 1550 °C (2462 and 2822 °F), by Ward and Wright (6) at 1350 °C (2462 °F), and by Miller and Elliott (7) at 1300 to 1460 °C (2372 to 2660 °F).

The liquidus-surface diagram is based on the data of Kase (2) as reworked by Marsh (1), on data by Buckley and Hume-Rothery (4), and on summarized data by Miller and Elliott (7). Experimental data have been modified where necessary to be consistent with the bounding binary phase diagrams as re-evaluated for this volume. The combined data of these multiple investigations yield a diagram that is qualitatively consistent, but that can be considered definitive only for the Fe-rich portion with liquidus temperatures above 1450 °C (2642 °F). A major criticism is that the hypereutectic liquidus surface constructed from the 1350 and 1550 °C (2462 and 2822 °F) isotherms, together with the course of the eutectic trough, is concave for the Fe-rich side of the system; such a shape is inconsistent with the bounding C-Fe binary system.

Soehnchen and Piwowarsky (8) measured the solubility of graphite in Fe-rich Fe-Ni austenite. Chipman and Brush (9), using activity data of Smith (10), calculated the solubility at 1000 °C (1832 °F) with poor agreement with measurements by Soehnchen and Piwowarsky. Fray and Chipman (11) redetermined the solubility at 1000 °C (1832 °F), which was in good agreement with the theoretical calculations of Chipman and Brush. Greenbank (12) formulated expressions by which the solubility of graphite in Fe-Ni austenite can be calculated at all temperatures.

Wada et al (13) determined the activity of graphite in Fe-Ni austenite at 800, 1000 and 1200 °C (1472, 1832 and 2192 °F) and the solubility at 800 and 1000 °C (1472 and 1832 °F). The solubility data for graphite in Fe-Ni austenite, as given in the diagram, is based on the data of Wada et al. The saturation at 1200 °C (2192 °F) was determined by extrapolating activity data to unit activity; the 900 and 1100 °C (1652 and 2012 °F) isotherms were interpolated. Because the austenite-graphite eutectic in the C-Fe binary system is at 1154 °C (2109 °F), a portion of the 1200 °C (2192 °F) isotherm near the C-Fe binary system must be theoretical, because it will be cut off by the eutectic reaction.

The more complex equilibria for austenite—alpha iron—Fe_3C (or graphite) have not been established; schematic representations of probable equilibrium involving these phases have been made by Marsh and by Lange and Mathieu (14).

REFERENCES

1 J. S. Marsh, p 56-82 in "The Alloys of Iron and Nickel", Vol 1, Engineering Foundation, McGraw-Hill, 1938

2 T. Kase, *Sci Rept Sendai Univ,* Vol 14, 1925, p 173-217

3 K. Schichtel and E. Piwowarsky, *Arch Eisenhuettenw,* Vol 3, 1929, p 139-147

4 R. A. Buckley and W. Hume-Rothery, *J Iron Steel Inst (London),* Vol 202, 1964, p 895-898

5 E. T. Turkdogan, R. A. Hancock, S. I. Herlitz and J. Dentan, *J Iron Steel Inst (London),* Vol 183, 1956, p 69-72

6 R. G. Ward and J. A. Wright, *J Iron Steel Inst (London),* Vol 194, 1960, p 304-306

7 K. O. Miller and J. F. Elliott, *Trans Met Soc AIME,* Vol 218, 1960, p 900-910

8 E. Soehnchen and E. Piwowarsky, *Arch Eisenhuettenw,* Vol 5, 1931, p 111-121

9 J. Chipman and E. F. Brush, *Trans Met Soc AIME,* Vol 242, 1968, p 35-41

10 R. P. Smith, *Trans Met Soc AIME,* Vol 218, 1960, p 62-64

11 D. J. Fray and J. Chipman, *Trans Met Soc AIME,* Vol 245, 1969, p 1143-1144

12 J. C. Greenbank, *J Iron Steel Inst (London),* Vol 209, 1971, p 819-825

13 T. Wada, H. Wada, J. F. Elliott and J. Chipman, *Met Trans,* Vol 2, 1971, p 2199-2208

14 H. Lange and K. Mathieu, *Mitt K-W Inst Eisenforsch,* Vol 20, 1938, p 125-134

C-Fe-Si (Carbon-Iron-Silicon)

By L. Brewer, J. Chipman and S-G. Chang

The addition of silicon to the C-Fe binary system destabilizes the cementite phase (already metastable) and promotes graphitization. Furthermore, silicon acts as a ferrite stabilizer, counteracting carbon by constricting the γ field and moving it to higher silicon contents so as to join the (α-Fe) and (δ-Fe) fields. This is shown by the sections of the metastable C-Fe-Si diagram for 2.3, 3.5, 5.2 and 7.9 wt % silicon, which are from Hilliard and Owen (1), slightly modified to match the isothermal section at 1000 °C (1832 °F).

The phases designated on the diagrams are graphite, α (ferrite), γ (austenite), C_1 (cementite, Fe_3C), C_2 (iron silicocarbide), C_3 (silicon carbide, SiC), and L (liquid).

The isothermal section of the stable iron-rich C-Fe-Si system at 1000 °C (1832 °F) is from Chipman (2), Smith (3), Chipman and Brush (4), and Wada et al (5). A similar section at the same temperature for the metastable C-Fe-Si system is from Chipman (2), Wada et al (5), and Malinochka and Dolinskaya (6). According to Malinochka and Dolinskaya, the composition of iron silicocarbide (C_2) is 3.4 wt % C, 8.9 wt % Si and it has the formula $Fe_{5.55}CSi_{1.12}$. Shevchuk and Gurinovich (7) reported a range of Fe_4CSi to Fe_6CSi for the formula. Earlier values proposed for the point for iron silicocarbide in the C-Fe-Si metastable system vary from a binary composition to several ternary compositions: 6.7 wt % C, 0 wt % Si by Scheil (8); 7.0 wt % C, 4.0 wt % Si by Kriz and Poboril (9); and 6.5 wt % C, 9.0 wt % Si by Humphreys and Owen (10).

The isothermal section of the iron-rich C-Fe-Si system at 1300 °C (2372 °F) is also shown. The solubility line for graphite is from Chipman *et al* (11,12) and Rein and Chipman (13). The solid phases are based on Hilliard and Owen (1), except for the Fe-C boundary and the Fe-Si boundary, which are from the respective binary diagrams in this volume.

Addition of silicon to the Fe-C eutectic raises its temperature slightly and lowers drastically the carbon content of L and γ, according to Hilliard and Owen (1) and Greiner *et al* (14).

REFERENCES

1 J. E. Hilliard and W. S. Owen, *J Iron Steel Inst,* Vol 172, 1952, p 268-282
2 J. Chipman, *Met Trans,* Vol 3, 1972, p 55-64
3 R. P. Smith, *J Am Chem Soc,* Vol 70, 1948, p 2724-2729
4 J. Chipman and E. F. Brush, *Trans Met Soc AIME,* Vol 242, p 35-41, 1968
5 T. Wada, H. Wada, J. F. Elliott and J. Chipman, *Met Trans,* Vol 3, 1972, p 1657-1662
6 Ya. N. Malinochka and V. Z. Dolinskaya, *Liteinoe Proizv,* Vol 7, 1970, p 26-27
7 L. A. Shevchuk and V. I. Gurinovich, *Vestsi Akad Navuk Belarusk SSR, Ser Fiz Tekhn Navuk,* Vol 1, 1971, p 28-32
8 E. Scheil, *Stahl Eisen,* Vol 50, 1930, p 1725
9 A. Kriz and F. Poboril, *J Iron Steel Inst,* No. II, 1930, p 191-210
10 J. G. Humphreys and W. S. Owen, *J Iron Steel Inst (London),* Vol 198, 1961, p 38-45
11 J. Chipman, R. M. Alfred, L. W. Gott, R. B. Small, D. M. Wilson, C. N. Thomson, D. L. Guernsey and J. C. Fulton, *Trans Am Soc Metals,* Vol 44, 1952, p 1215-1230
12 J. Chipman, J. C. Fulton, N. Gokcen and G. R. Caskey, Jr., *Acta Met,* Vol 2, 1954, p 439-450
13 R. H. Rein and J. Chipman, *J Phys Chem,* Vol 67, 1963, p 839
14 E. S. Greiner, J. S. Marsh and B. Stoughton, "Alloys of Iron and Silicon", Engineering Foundation, McGraw-Hill, 1933

C-Fe-V (Carbon-Iron-Vanadium)

By Charles E. Dremann

Diagrams of the C-Fe-V system are important to an understanding of tool steels and alloy steels containing vanadium, and yet only small portions of the system are well known.

All of the phases in this ternary system are present in the individual binary systems except the η phase, a complex carbide discovered by Flender and Wever (1) and identified as Fe_3V_3C by Borusevich *et al* (2) and Grdina *et al* (3). The η phase has the face-centered-cubic $E9_3$ type of structure, isotypic with Fe_3W_3C. Lattice parameter a = 10.877 or 10.897 A, according to Borusevich *et al* and Grdina *et al*, respectively. Kuo (4) found Fe_3W_3C to be a stable double carbide; therefore, the η phase is probably stable.

The vertical sections of the C-Fe-V system shown on page 1256 of the 1948 edition of *Metals Handbook* were from Wever *et al* (5). The phases were indicated by C_1 and C_2 and were considered to be Fe_3C and V_4C_3, respectively.

Later investigations have shown that the formula for C_2 is VC. (Although the formula VC correctly describes the structure of the compound, all of the lattice sites are not occupied; this results in a defect structure corresponding to the formula range of $VC_{0.73}$ to $VC_{0.88}$ and has prompted using other formulas to describe the composition, such as VC_{1-x}, V_4C_3 and V_6C_5.)

V_2C and η were not considered in the diagrams shown in the 1948 edition of *Metals Handbook*. Both diagrams and the reactions should be redetermined to include these phases.

Because Fe_3C is metastable, it can be replaced by graphite, the stable phase. According to Goldschmidt (6), the reaction γ → α + VC + Fe_3C occurs at 690 °C (1274 °F).

The 500 °C (932 °F) isothermal section presented here, from Grdina *et al* (7), indicates a high solubility of Fe in VC and gives the tentative location of the η phase. The metastable Fe_3C phase exists up to about 30 wt % V before being replaced by graphite. Grdina *et al* (3,7) also found an X phase (face-centered-cubic, with a = 4.33 A), which they believe is an allotrope of V_2C.

Many investigators (1,8,9,10,11) have determined the iron-rich corner of the isothermal diagram at 1000 °C (1832 °F). The diagram shown here was taken mainly from data by Ebeling and Wever (10). The dashed line shows the solubility of VC in austenite (γ) as found by Wada *et al* (11); they found that the solubility is less than that proposed by Ebeling and Wever, but in agreement with Savostyanova and Shvartsmann (12). The results of Wada *et al* also suggest that an extension of the dashed line would intersect the γ − (α + γ) boundary to form a ternary region, α + γ + C.

The temperature dependence of the solubility of VC in ferrite is reported by Alisova and Budberg (13) and in austenite by Bungardt *et al* (14).

The influence of vanadium content on the solubility of carbon in molten iron was determined at 1530 °C (2786 °F) by Schenck *et al* (15) and at 1350 °C (2462 °F) and 1550 °C (2822 °F) by Schurmann and Kramer (16).

REFERENCES

1 H. Flender and H. Wever, *Arch Eisenhuettenw,* Vol 34, 1963, p 727-732
2 L. K. Borusevich, E. I. Gladyshevskii, T. F. Fedorov and N. M. Popova, *Zh Strukt Khim,* Vol 6(2), 1965, p 313-314
3 Yu. V. Grdina, I. D. Lykhin and T. F. Fedorov, *Izv Vysshikh Uchebn Zavedenii, Chernaya Met,* Vol 4, 1966, p 124-127
4 K. Kuo, *J Iron Steel Inst (London),* Vol 173, 1953, p 368-375
5 F. Wever, A. Rose and H. Eggers, *Mitt Kaiser-Wilhelm Inst Eisenforsch,* Vol 18, 1936, p 239-246
6 H. J. Goldschmidt, *J Iron Steel Inst (London),* Vol 160, 1948, p 345-362
7 Yu. V. Grdina, I. D. Lykhin and T. F. Fedorov, *Izv Vysshikh Uchebn Zavedenii, Chernaya Met,* Vol 6, 1966, p 156-160
8 R. R. Zupp and D. A. Stevenson, *Trans Met Soc AIME,* Vol 236, 1966, p 1316-1323
9 B. M. Mogutnov, I. A. Tomilin and L. A. Shvartsmann, *Zh Fiz Khim,* Vol 45(8), 1971, p 2047

10 R. Ebeling and H. Wever, *Arch Eisenhuettenw,* Vol 40(7), 1969, p 551-555

11 T. Wada, H. Wada, J. F. Elliott and J. Chipman, *Met Trans,* Vol 3, 1972, p 2865-2872

12 N. A. Savostyanova and L. A. Shvartsmann, *Phys Metals Metallog (USSR) (English Transl),* Vol 9(4), 1969, p 35

13 "Phase Diagrams in Metallic Systems" (S. P. Alisova and P. B. Budberg, Editors), Academy of Sciences, USSR, 1968, p 156

14 K. Bungardt, K. Kind and W. Oelsen, *Arch Eisenhuettenw,* Vol 27, 1956, p 61-66

15 H. Schenck, M. Gloz and E. Steinmetz, *Arch Eisenhuettenw,* Vol 41, 1970, p 1-3

16 E. Schurmann and D. Kramer, *Giessereiforschung,* Vol 21 (1), 1969

C-Fe-W (Carbon-Iron-Tungsten)

By R. A. Cary and R. J. Henry

The early investigations of the C-Fe-W system by Honda and Murakami (1), Hultgren (2), Takeda (3), Westgren and Phragmén (4), and Westgren (5) resulted in disagreement as to the nature and composition of the carbides present. Recent work by Jellinghaus (6), Pollock and Stadelmaier (7), and Harvig and Uhrenius (8) established the identity of some constituents. Fundamental differences in phase equilibria still persist. For example, Pollock and Stadelmaier found ferrite in equilibrium with WC and M_6C, whereas Harvig and Uhrenius found austenite in equilibrium with WC and M_6C.

The η carbide (M_6C) has been shown to have limited solubility at 1000 °C (1832 °F) by Jellinghaus (6) and by Harvig and Uhrenius (8), ranging from about $Fe_{3.4}W_{2.6}C$ to $Fe_{2.9}W_{3.1}C$. Harvig and Uhrenius have shown up to 5 at. % Fe dissolved in WC. Furthermore, no $M_{23}C_6$ carbide was found at 1000 °C (1832 °F), substantiating the work of Krainer (9), who demonstrated the decomposition of $M_{23}C_6$ to WC + Fe_3C at temperatures above 700 °C (1292 °F). The effect of tungsten in increasing the stability of Fe_3C, first shown by Kuo and Hultgren (10), has been substantiated by Harvig and Uhrenius, who set the solubility limit at 0.46 at. % W.

Pollock and Stadelmaier (7) confirmed the identity of a second η carbide ($M_{12}C$), with limited solubility and corresponding to the composition of Fe_6W_6C. It had been previously reported by Leciejewicz (11).

Although Orton (12) has shown a eutectoidal decomposition of W_2C to W + WC below 1200 °C (2192 °F), Pollock and Stadelmaier have found W_2C in equilibrium with FeW_3C at 1000 °C (1832 °F). Telegus *et al* (13) also found W_2C to be stable below 1200 °C (2192 °F), to as low as 800 °C (1472 °F). However, the hexagonal structure transforms to an orthorhombic structure below 1100 °C (2012 °F). The transformation is similar to that of Mo_2S at 1432 °C (2610 °F). Yvon *et al* (14) confirmed that the low-temperature form of W_2C was of the Fe_2N type of orthorhombic structure. The isothermal section here is a best approximation.

The liquidus projection is from Jellinghaus (6), and the schematic binary showing the effect of tungsten on the gamma field is from Gobin and Bouchy (15).

REFERENCES

1 Honda and Murakami, *Sci Rept Tohoku Imp Univ,* Vol 6, No. 5, 1918, p 235

2 A. Hultgren, *Jernkontorets Ann,* Vol 105, 1921, p 499

3 S. Takeda, *Technol Rept Tohoku Imp Univ,* Vol 10, 1931, p 42-92

4 A. Westgren and G. Phragmén, *Jernkontorets Ann,* Vol 111, 1927, p 535

5 A. Westgren, *Jernkontorets Ann,* Vol 117, 1933, p 1

6 W. Jellinghaus, *Arch Eisenhuettenw,* Vol 39, 1968, p 705-718

7 C. B. Pollock and H. H. Stadelmaier, *Met Trans,* Vol 1, 1970, p 767-770

8 H. Harvig and B. Uhrenius, TRITA-MAC-0008, Royal Institute of Technology (Stockholm), 1971

9 H. Krainer, *Arch Eisenhuettenw,* Vol 21, 1950, p 33-38

10 K. Kuo and A. Hultgren, *Kgl Vetensk Acad Handl,* Vol 4(3), 1953, p 22

11 J. Leciejewicz, *J Less-Common Metals,* Vol 7, 1964, p 318

12 G. W. Orton, *Trans Met Soc AIME,* Vol 230, 1964, p 600-602

13 V. S. Telegus, E. I. Gladyshevsky and P. I. Kripyakevich, *Kristallografiya,* Vol 12, 1967, p 936-939

14 K. Yvon, H. Nowotny and F. Benesovsky, *Monatsh Chem,* Vol 99, 1968, p 726-729

15 P. Gobin and C. Bouchy, *Mem Sci Rev Met,* Vol 63, No. 1, 1966, p 75

Co-Cr-Fe (Cobalt-Chromium-Iron)

By J. M. Drapier and D. Coutsouradis

A prominent feature of the Co-Cr-Fe system is the formation of a σ solid-solution phase between the σ phases of the Co-Cr and Cr-Fe binary systems, as reported by Köster (1). The diagram showing phases in an isothermal section at 1200 °C (2192 °F) by Rideout *et al* (2) is dominated by two solid-solution fields: that occupied by the face-centered-cubic (γ-Fe,α-Co) and that occupied by the body-centered-cubic (α-Fe-Cr). Alloys within area A on the diagram transform completely to body-centered-cubic (α-Fe) (ferrite) on quenching to room temperature, whereas alloys within area B do not decompose on quenching, apart from the precipitation of σ phase.

A comparison of the extent of the ternary σ phase at 1200 °C (2192 °F) and that at 800 °C (1472 °F), at which temperature the Fe-Cr σ phase is stable, is shown in the sigma-phase diagram, from Rideout *et al* (2).

The isothermal sections at 700 °C (1292 °F) and 600 °C (1112 °F) show that the phases in the (α-Fe) + (α-Co) + σ area have the following compositions:

	Composition, wt % at:					
	700 °C (1292 °F)			600 °C (1112 °F)		
Phase	Fe	Co	Cr	Fe	Co	Cr
(α-Fe)	72	8	20	78	10	12
(α-Co)	60	20	20	29	54	17
σ	24	33	43	30	27	43

Köster and Hofmann (3) report that many alloys in the area contain (α-Fe), (ε-Co) and σ phases, instead of the expected mixture of (α-Fe), (α-Co) and σ. This must mean that there is an invariant reaction at approximately the temperature of the annealing experiments, such that (α-Co) + σ $\rightleftharpoons$ (α-Fe) + (ε-Co); there would thus be three-phase areas containing (α-Fe) + (ε-Co) + (α-Co) and (α-Fe) + (ε-Co) + σ in the isothermal sections at temperatures lower than 600 °C (1112 °F), in contrast to the (α-Fe) + (α-Co) + σ and (α-Co) + (ε-Co) + σ areas characteristic of higher temperatures.

The peritectic reactions at 1501 °C (2734 °F) and 1472 °C (2681 °F) in the Co-Fe and Co-Cr binary systems, respectively, are reported by Köster and Hofmann to extend into the ternary system to give rise to an invariant reaction at 1410 °C (2570 °F), at which L + (α-Fe) → (α-Co) + δ, the high-temperature modification of the ternary σ phase. The binary eutectic temperature of 1402 °C (2556 °F) in the Co-Cr binary system is raised by iron to the invariant reaction temperature of 1410 °C (2570 °F). The (α-Co) + δ → σ reaction in the Co-Cr binary system is depressed from 1312 °C (2394 °F) to 1290 °C (2354 °F) in the ternary system, and at this lower temperature, the reaction (α-Co) + δ → (α-Fe) + σ takes place.

On cooling to room temperature, the constitution of the system can be represented as in the isothermal section at 20 °C (68 °F). This diagram does not correspond with equilibrium, but shows the metastable phase fields likely to be encountered after furnace cooling.

The effect of additions of up to 6 wt % Fe on the transformation and precipitation reactions in Co-rich Co-Cr alloys has been studied by Elsea and McBride (4). It has been reported that nitrogen tends to promote the formation of the σ phase in the Co-Cr-Fe alloys. The Co-Cr-Fe-C system was studied by Coutsouradis and Habraken (5).

REFERENCES

1 W. Köster, *Arch Eisenhuettenw,* Vol 6, 1932, p 113-116

2 S. Rideout, W. D. Manly, E. L. Kamen, B. S. Lement and P. A. Beck, *Trans AIME,* Vol 191, 1951, p 872-876

3 W. Köster and G. Hofmann, *Arch Eisenhuettenw,* Vol 30, 1959, p 249-251

4 A. R. Elsea and C. C. McBride, *Trans AIME,* Vol 188, 1950, p 154-161

5 D. Coutsouradis and L. Habraken, *Cobalt,* Vol 13, Dec 1961, p 4-23

Cr-Fe-N (Chromium-Iron-Nitrogen)

By L. Brewer and S-G. Chang

The Cr-Fe-N system has been extensively studied by Imai, Masumoto and Maeda (1). The isothermal sections at 1300, 1100, 1000, 800 and 700 °C (2372, 2012, 1832, 1472 and 1292 °F), as well as the vertical sections at 0.1 and 0.3 wt % N, and at 7, 18 and 26 wt % Cr, are from their results. At high temperatures, with increasing nitrogen, the γ-phase region expands to higher chromium. The solubility of nitrogen in (α-Fe) increases with the increase of chromium content. As the temperature decreases, the solubility of nitrogen in (α-Fe) and γ is reduced, narrowing the

(α-Fe), (α-Fe) + γ and γ regions; also, the (α-Fe) + Cr_2N region expands, and the (α-Fe) + γ + Cr_2N region shifts to lower chromium. The solubility of nitrogen in (α-Fe) is remarkably reduced between 1100 and 1000 °C (2012 and 1832 °F), becoming less than 0.05 wt % N at 1000 °C (1832 °F). At 800 °C (1472 °F), (α-Fe) in the three-phase region (α-Fe) + γ + Cr_2N contains about 7 wt % Cr, and the (α-Fe) + γ region connects with the (α-Fe) + γ region in the Fe-N binary system. On cooling, the γ phase in the γ and (α-Fe) + γ regions undergoes a phase transition into (α-Fe) and nitrides (Fe_4N, CrN and Cr_2N). Below 700 °C (1292 °F), the γ phase exists only in the low-chromium, high-nitrogen region.

Okamoto and Naito (2) constructed an isothermal section at 1250 °C (2282 °F) in the composition range of 12 to 60 wt % Cr and up to about 2 wt % N. They studied the liquidus surface and presented six vertical sections (0.2 and 0.3 wt % N; 25, 30, 35 and 40 wt % Cr) between 1300 and 1550 °C (2372 and 2822 °F). Okamoto and Naito report that a ternary peritecto-eutectic reaction L + (α-Fe) → γ + Cr_2N occurs at 1328 °C (2422 °F) and that the compositions of the four phases participating in this invariant reaction are as follows: (α-Fe) : 37 wt % Cr, 0.05 wt % N, remainder Fe; γ: 15 wt % Cr, 0.4 wt % N, remainder Fe; L: 16 wt % Cr, 0.9 wt % N, remainder Fe; Cr_2N: 88.4 wt % Cr, 11.6 wt % N. The results of Imai et al (1) show that the ternary peritecto-eutectic reaction occurs at about 1330 °C (2426 °F) and that the compositions of the (α-Fe) and γ phases in the invariant reaction are: (α-Fe): 39 wt % Cr, 0.25 wt % N, remainder Fe; γ: 35 wt % Cr, 1.3 wt % N, remainder Fe, which differ considerably from the values obtained by Okamoto and Naito. The (α-Fe) + γ + Cr_2N region shown by Okamoto and Naito in their isothermal section at 1250 °C (2282 °F) is very broad and thus narrows the (α-Fe) + γ region, which is contradictory to what has been determined by Imai et al. These differences seem to be the result of Okamoto and Naito's including the nonequilibrium phase, Cr_2N, precipitated from (α-Fe) during cooling, in their equilibrium diagram. The phase boundaries of (α-Fe), (α-Fe) + γ and γ reported by Turkdogan and Ignatowicz (3) at 1200 to 1370 °C (2192 to 2498 °F) and the phase boundary of γ determined by Tavadze and Gogvadze (4) at 1250 °C (2282 °F) agree reasonably well with those shown by Imai et al. Imai et al also confirmed that there are two other invariant reaction planes: a ternary peritecto-eutectoid reaction γ + Cr_2N → (α-Fe) + CrN at about 790 °C (1454 °F) and a ternary peritecto-eutectoid reaction γ + CrN → (α-Fe) + Fe_4N at about 760 °C (1400 °F); however, the CrN phase is a nonequilibrium phase that precipitates from (α-Fe) during cooling.

According to Wada et al (5), the solubility of nitrogen in liquid iron-chromium alloys increases with increasing chromium content, up to 4.1 wt % Cr at 1900 °C (3452 °F), and decreases with increasing temperature; however, according to Tavadze et al (6), the solubility of nitrogen in liquid iron-chromium alloys changes very little up to 10 to 12 wt % Cr; it changes significantly with more than 10 to 12 wt % Cr, and it changes vigorously at 22 to 25 wt % Cr.

REFERENCES

1 Y. Imai, T. Masumoto and K. Maeda, *Sci Rep Res Inst Tohoku Imp Univ Ser A,* Vol 19, 1967, p 35-49

2 M. Okamoto and T. Naito, *J Iron Steel Inst (Japan)*, Vol 49, 1963, p 1915-1921

3 E. T. Turkdogan and S. Ignatowicz, *J Iron Steel Inst*, Vol 199, 1961, p 287-296

4 F. N. Tavadze and T. I. Gogvadze, *Vop Metalloved Korroz Metal*, No. 2, 1971, p 29-31

5 H. Wada, K. Gunji and T. Wada, *Nippon Kinzoku Gakkaishi*, Vol 32, 1968, p 933-938

6 F. N. Tavadze, M. A. Nabichvrishvili, V. A. Pirtskhalaishvili and T. I. Gogvadze, *Vop Metalloved Korroz Metal* (F. N. Tavadze, Editor), Metsniereba, Tbilisi, USSR, 1968, p 54-58

Cr-Fe-Ni (Chromium-Iron-Nickel)

By G. R. Speich

For the Cr-Fe-Ni system, most of the studies have been restricted to the iron-rich and nickel-rich regions, because most stainless steels and the high-temperature nickel-base alloys containing chromium are associated with these regions. The principal features of the Cr-Fe-Ni system include the phase equilibria resulting from the high-temperature face-centered-cubic structure of γ-Fe (austenite) and of nickel, which are completely miscible in each other; the low-temperature body-centered-cubic structure of α-Fe (ferrite) and of chromium, which are completely miscible above 821 °C (1510 °F); and the formation of sigma phase at higher chromium contents, at temperatures below 821 °C (1510 °F).

The maximum separation of liquidus and solidus in the chromium-rich ternary alloys reaches 100 °C (180 °F) in a 48 wt % Cr, 30 wt % Fe, 22 wt % Ni alloy; in iron-rich alloys of the stainless steel type, the separation does not exceed 40 °C (72 °F), according to Jenkins *et al* (1). A minimum in the liquidus surface occurs below 1300 °C (2372 °F) at 49 wt % Cr, 8 wt % Fe, 43 wt % Ni. Masing and Rogers (2) reported that the solidus and liquidus surfaces are typical of those found in a transition from a peritectic binary (in the Fe-Ni system) to a eutectic binary (in the Cr-Ni system) to complete solid solution (in the Cr-Fe system).

The ferrite-stabilizing influence of chromium is predominant at high and low temperatures (see the Cr-Fe binary phase diagram), whereas the austenite-stabilizing influence of nickel is predominant at intermediate temperatures (see the Fe-Ni binary phase diagram). The outstanding feature, however, is the pronounced reluctance of metastable austenite to transform when once established at high temperatures. Aborn and Bain (3) and Schafmeister and Ergang (4) show that the temperature range in which the stable γ region is broadest lies between 900 and 1300 °C (1652 and 2372 °F), but the range for a particular alloy is often considerably narrower and depends on composition.

Studies by Schafmeister and Ergang (4), Pugh and Nisbet (5), Baerlecken and Hirsch (6), and Marcinkowski *et al* (7) show that the addition of nickel appears to raise the upper temperature limit of stability of sigma phase (σ CrFe), the brittle iron-chromium compound, from 821 °C (1510 °F) to about 900 °C (1652 °F); in so doing, as much as 12 wt % Ni may be dissolved. Although σ is shown in equilibrium with γ within certain composition limits, σ usually forms from α. The σ phase in Cr-Fe-Ni alloys forms even more sluggishly than in Cr-Fe alloys, so that the presence of σ is uncertain below 650 °C (1202 °F). However, the Cr-Fe binary phase diagram in this volume shows that σ decomposes by a eutectoid reaction into an iron-rich α and a chromium-rich α below 475 °C (887 °F); therefore, σ disappears and these phases would appear in isothermal sections below 475 °C (887 °F). This decomposition reaction is extremely sluggish, as is the formation reaction, and requires long-time annealing before it can be detected. Also, below 500 °C (932 °F), ordered Ni_3Fe and ordered Ni_3Cr phases occur in Fe-Ni and Cr-Ni binary alloys; these phases would appear in isothermal sections below 500 °C (932 °F).

The transformation of metastable austenitic iron-rich Cr-Fe-Ni alloys has been studied by Otte (8), Breedis (9), and Ludwigson and Brickner (10) because of its importance in the strength and formability characteristics of stainless steels. The diagram of the structure of alloys quenched from 1100 °C (2012 °F) shows the various transformation products. Nomenclature adopted for the structures (see Ref 11 and 12) is as follows:

α_m	Ferrite formed by massive $\gamma \rightarrow \alpha$ transformation
M_L	Martensite in which units are small laths
M_P	Martensite in which units are large plates
M_ε	Hexagonal close-packed martensite
γ_u	Unstable austenite (may transform if cold worked)
γ_s	Stable austenite

Of particular interest is the hexagonal close-packed phase, ε, that appears in some of the alloys that transform martensitically during quenching or by cold working.

As shown by Houdremont (13), many commercial alloys, particularly of the stainless steel type, contain 0.02 to 0.10 wt % C, and so chromium carbides may be present in these alloys in addition to the phases shown in the isothermal sections.

REFERENCES

1 C. H. M. Jenkins, E. H. Bucknall, C. R. Austin and G. A. Mellor, *J Iron Steel Inst (London)*, Vol 136, 1937, p 187-222

2 G. Masing and B. A. Rogers, p 79-81 in "Ternary Systems", Reinhold, 1944

3 R. H. Aborn and E. C. Bain, *Trans Am Soc Steel Treating*, Vol 18, 1930, p 837-873

4 P. Schafmeister and R. Ergang, *Arch Eisenhuettenw*, Vol 12, 1938-1939, p 459-464

5 J. W. Pugh and J. D. Nisbet, *J Metals*, Vol 2, 1950, p 268-276

6 E. Baerlecken and W. Hirsch, *Stahl Eisen*, Vol 75, 1955, p 570-579

7 M. J. Marcinkowski, R. M. Fisher and A. Szirmae, *Trans Met Soc AIME*, Vol 230, 1964, p 676-689

8 H. M. Otte, *Acta Met*, Vol 5, 1957, p 614-627

9 J. F. Breedis, *Trans Met Soc AIME*, Vol 230, 1964, p 1583-1596

10 D. C. Ludwigson and K. G. Brickner, *Sheet Metal Ind*, Vol 42, 1965, p 245-254

11 G. Krauss and A. R. Marder, *Met Trans*, Vol 2, 1971, p 2343-2357

12 R. P. Reed and F. J. Breedis, p 60-132 in "Behavior of Materials at Cryogenic Temperatures", *ASTM Spec Tech Publ 387*, 1966

13 E. Houdremont, p 644-653 in Vol 1 of "Handbuch der Sonderstahlkunde", Springer Verlag (Berlin), 1956

Fe-Mo-Ni (Iron-Molybdenum-Nickel)

By T. Wada

The isothermal section at 1200 °C (2192 °F) is based on the diagram proposed by D. K. Das, S. P. Rideout and P. A. Beck, *Trans AIME,* Vol 194, 1952, p 1071-1075. The phases appearing in the isothermal section are:

(Fe,Mo) . . . . A solid solution of iron and molybdenum containing nickel, and having a body-centered-cubic structure

(Fe,Ni) . . . . A solid solution of iron and nickel containing molybdenum, and having a face-centered-cubic structure. Alloys in region A, near the Fe corner, partly or fully transform to (Fe,Mo) phase on quenching. Alloys in region B can be quenched to room temperature without transforming.

MoNi . . . . An intermetallic compound, which is designated as the β phase in the Mo-Ni binary diagram in this volume. Iron can be dissolved in this phase to some extent.

P . . . . A hard, brittle intermetallic phase. The crystal structure is unknown, but the phase is isomorphous with the phase of the same designation in the Cr-Mo-Ni system.

$(Fe, Ni)_3Mo_2$ . . . An intermetallic phase with a composition close to Fe_3Mo_2 in the Fe-Mo binary system. It extends toward the Mo-Ni side of the diagram. In some systems, including the Fe-Mo binary system, this phase is designated ε phase.

The region that has a higher molybdenum content than that of the $(Fe,Ni)_3Mo_2$-P-MoNi line has not been determined.

Ferrous Ternary Alloy Phase Diagrams

Al-Cr-Fe (Aluminum-Chromium-Iron)

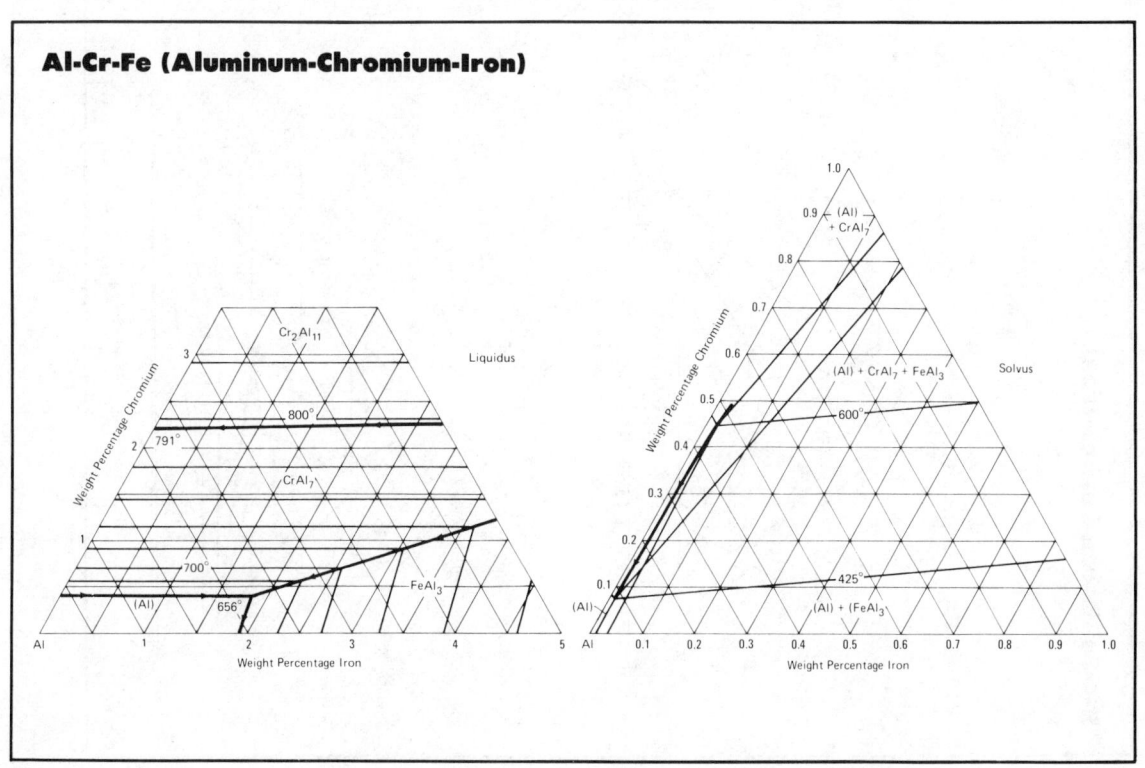

B-C-Fe (Boron-Carbon-Iron)

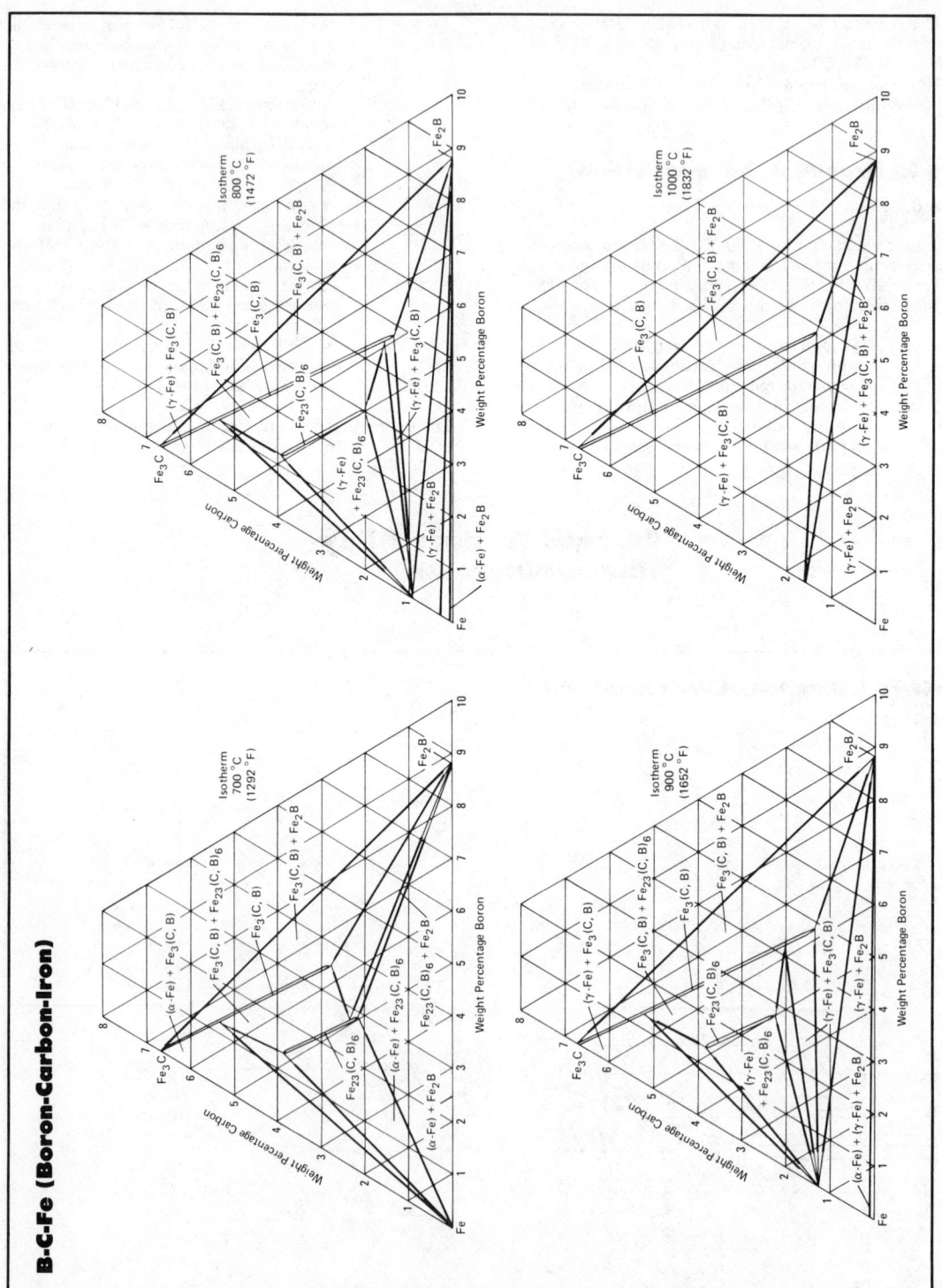

C-Cr-Fe (Carbon-Chromium-Iron)

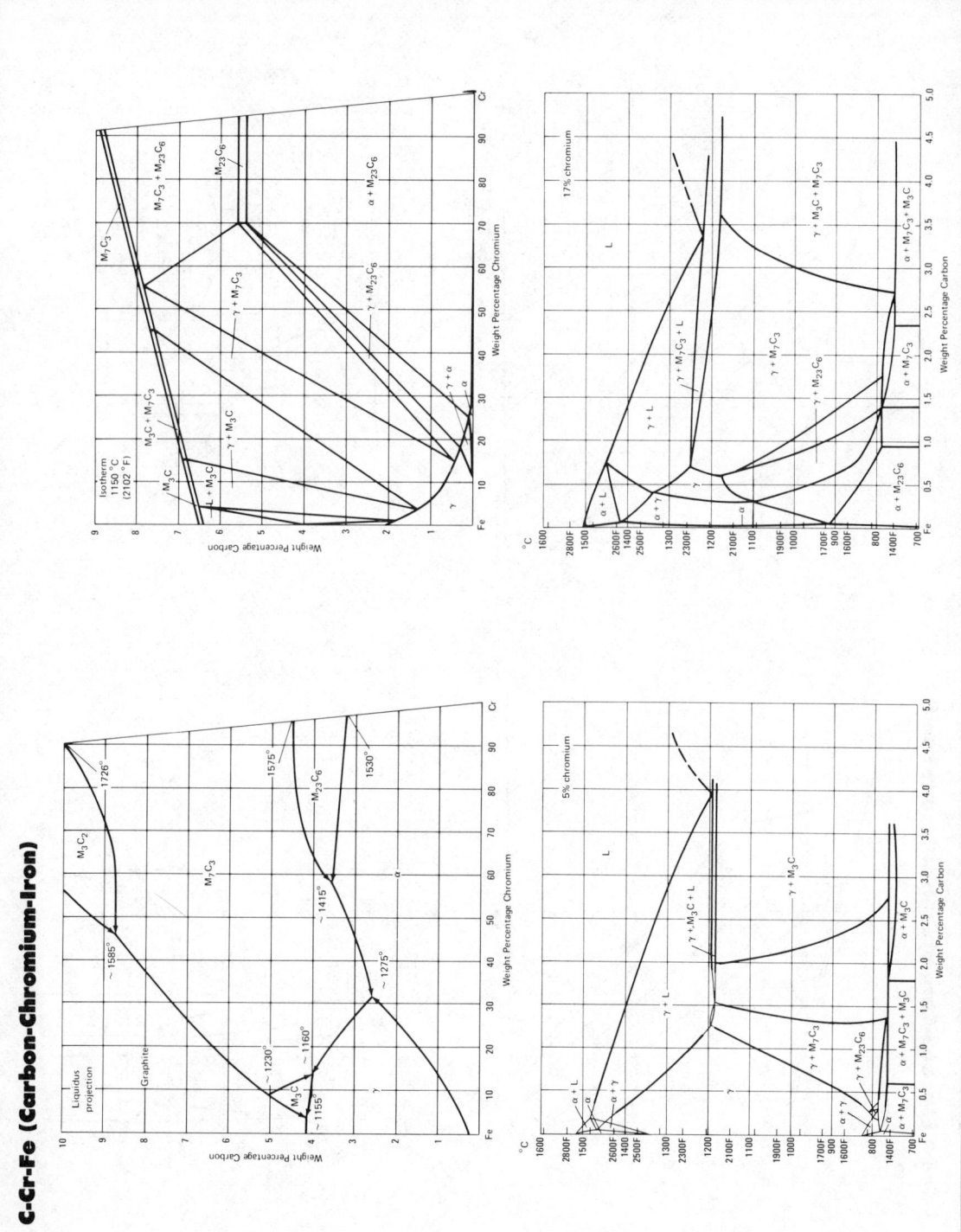

C-Cr-Fe (Carbon-Chromium-Iron) (continued)

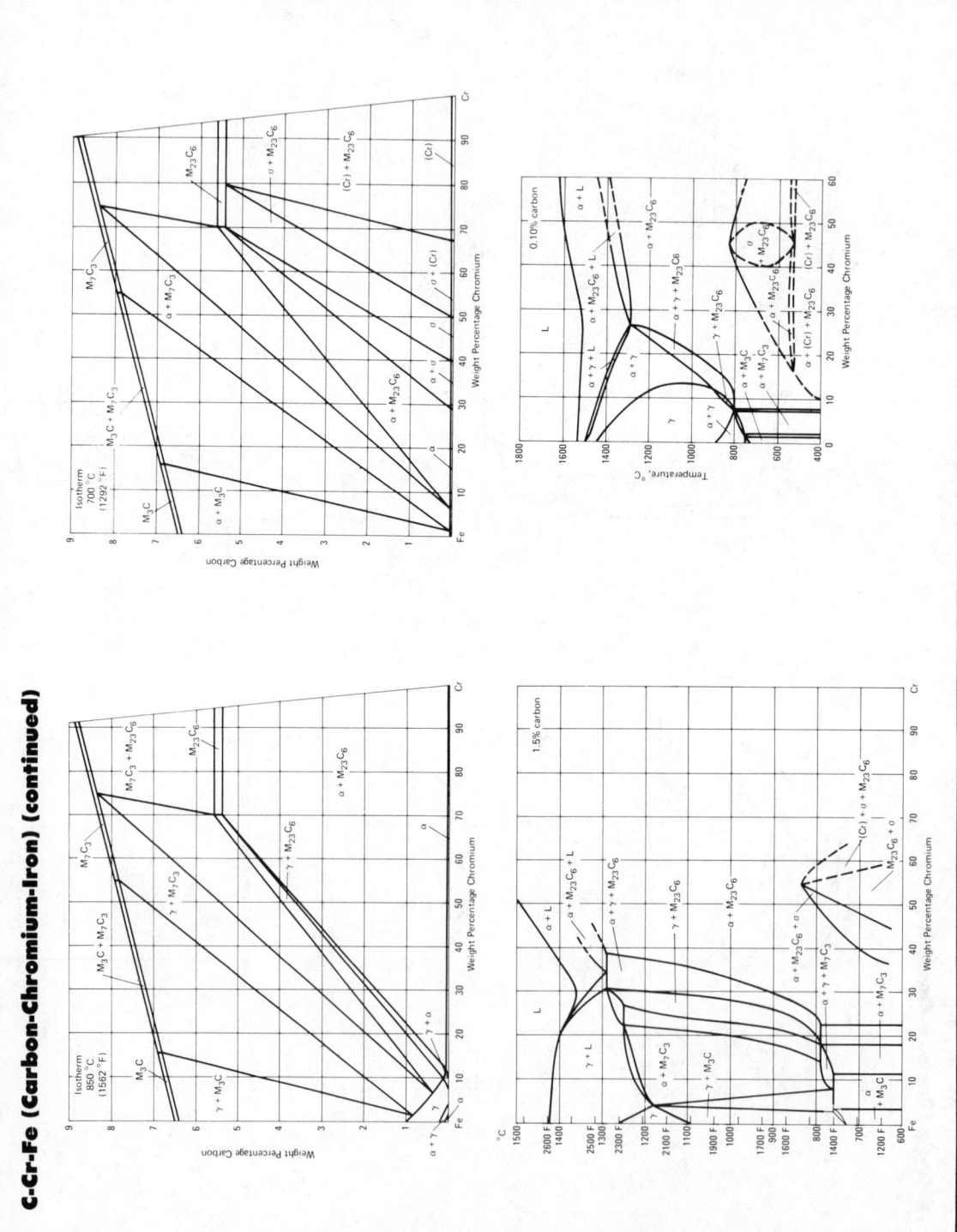

C-Cu-Fe (Carbon-Copper-Iron)

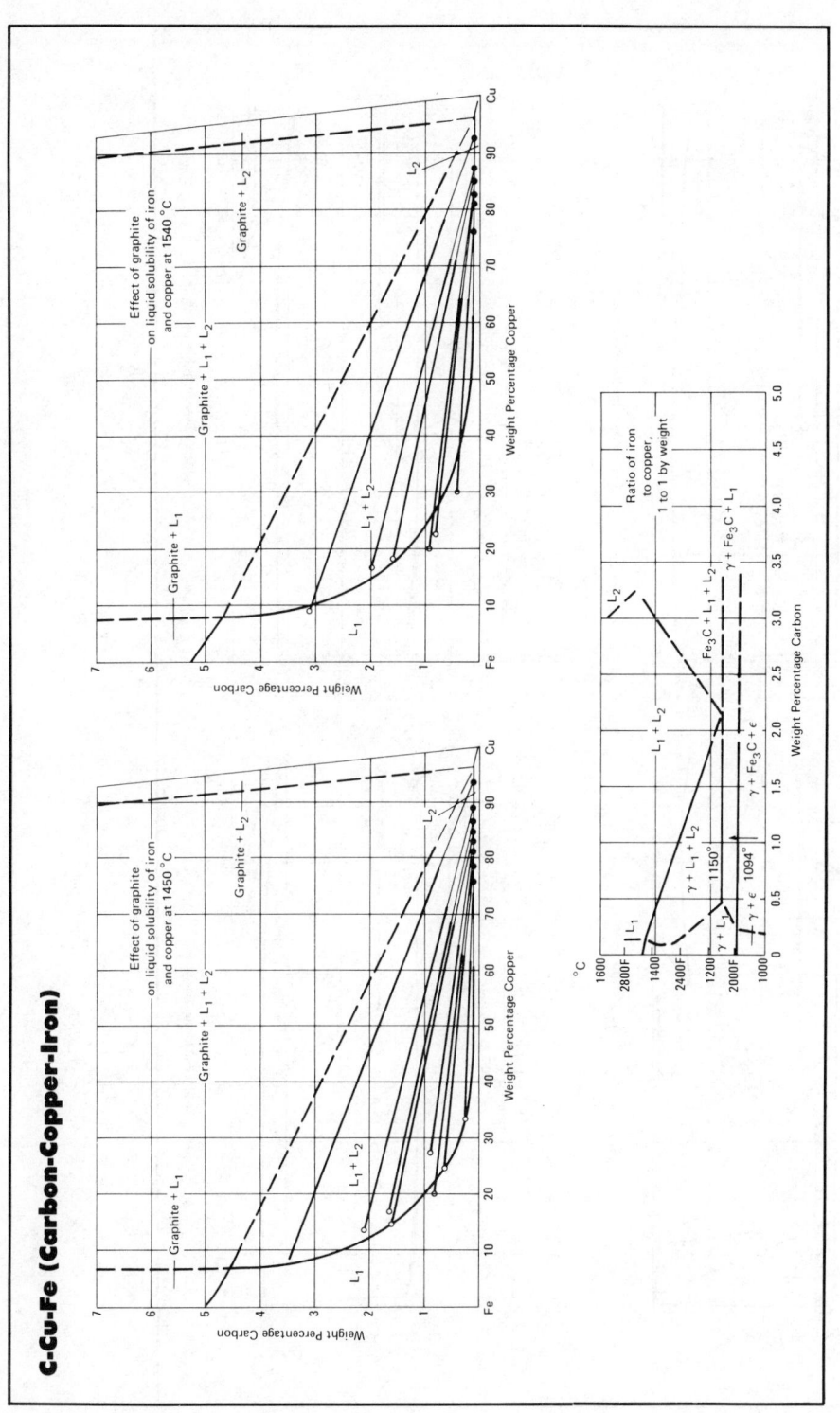

C-Fe-Mn (Carbon-Iron-Manganese)

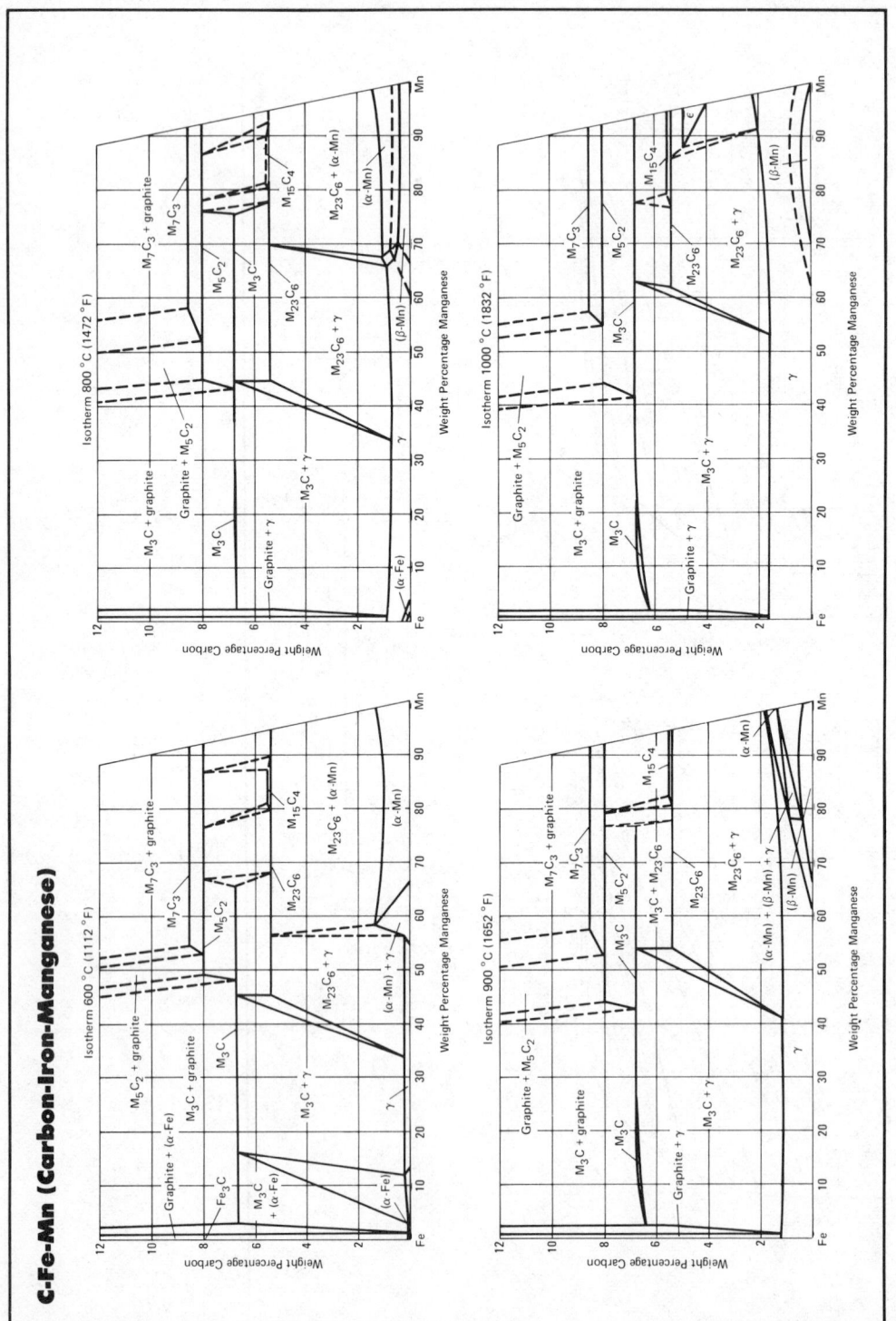

C-Fe-Mn (Carbon-Iron-Manganese) (continued)

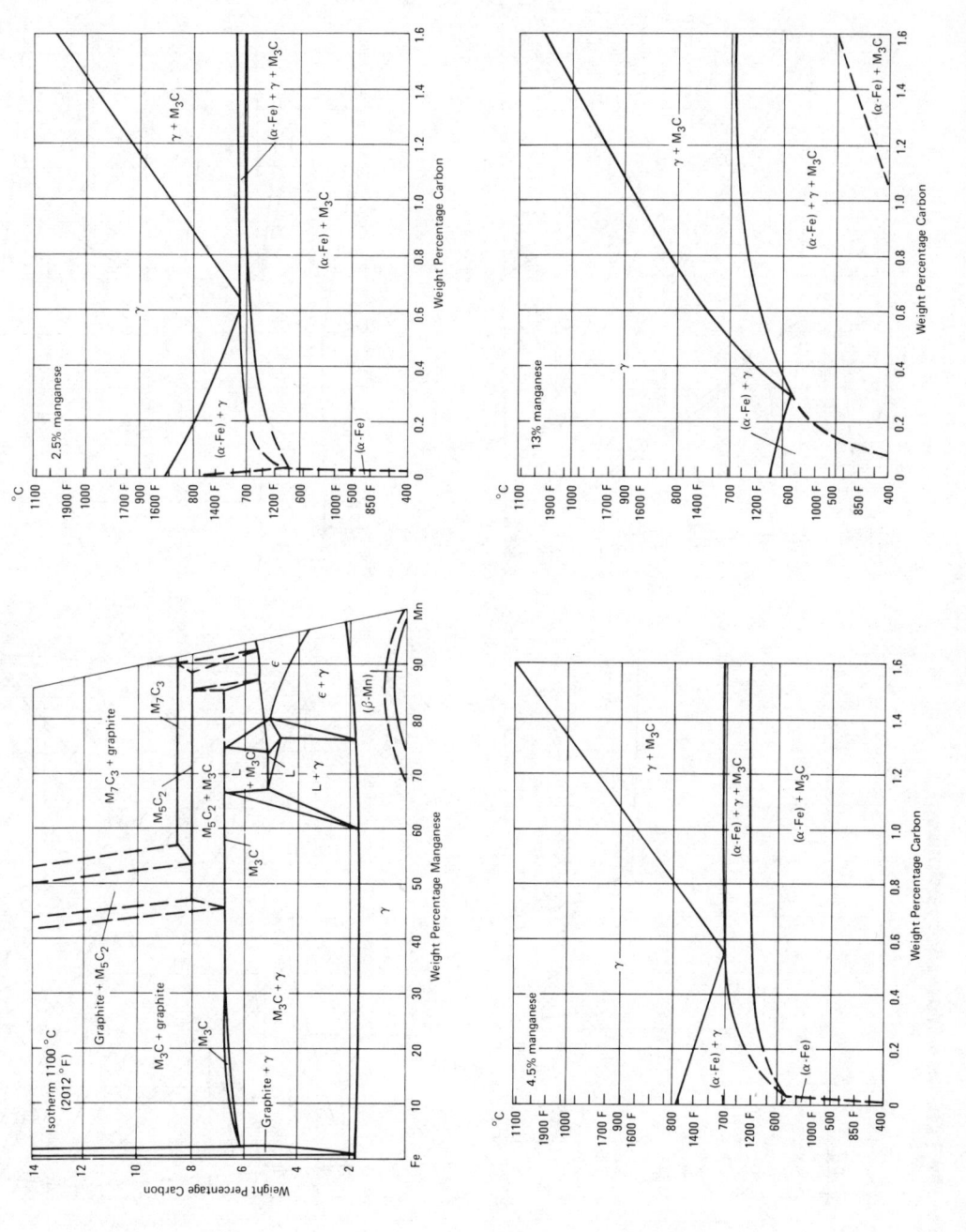

C-Fe-Mo (Carbon-Iron-Molybdenum)

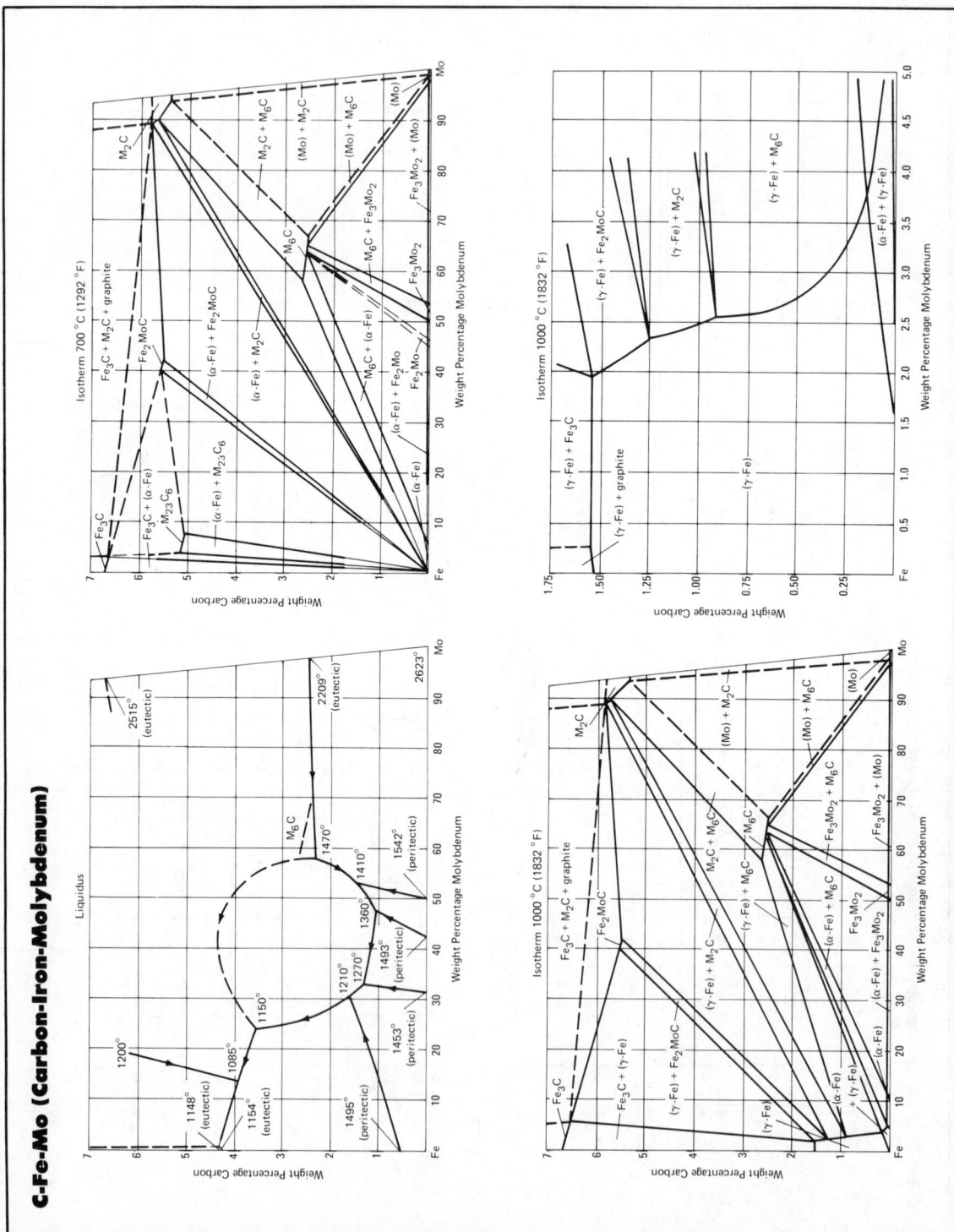

C-Fe-Mo (Carbon-Iron-Molybdenum) (continued)

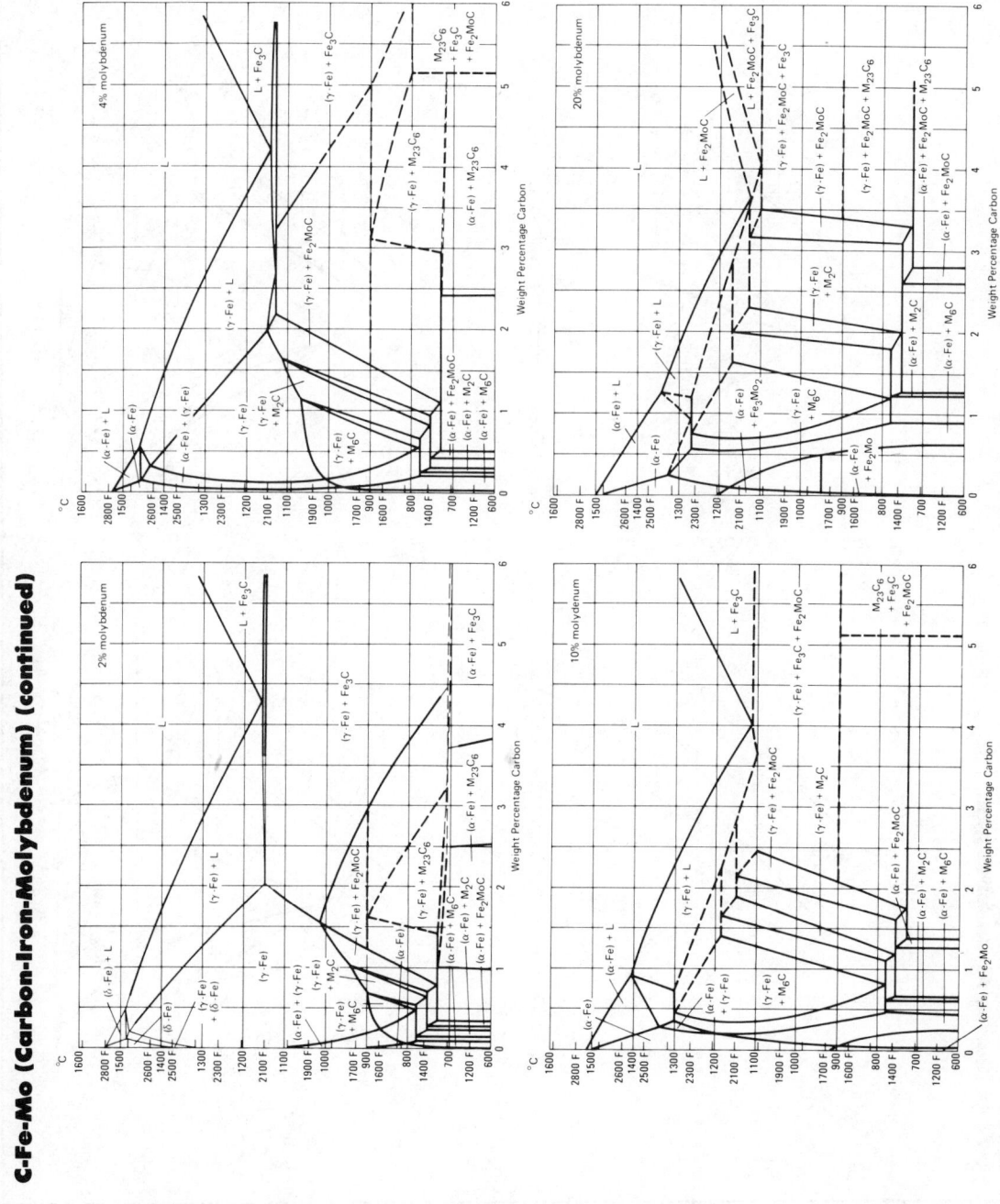

C-Fe-N (Carbon-Iron-Nitrogen)

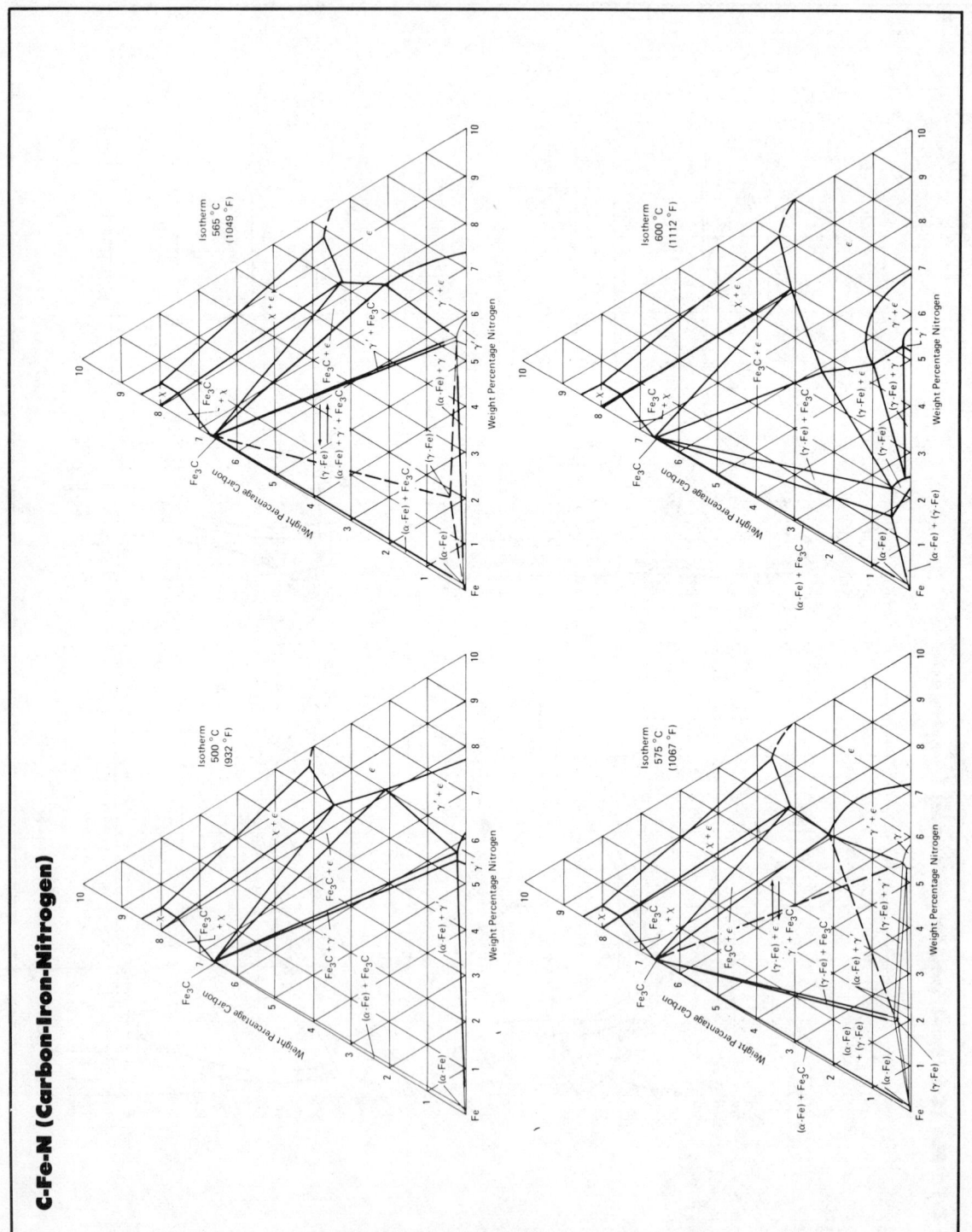

C-Fe-Ni (Carbon-Iron-Nickel)

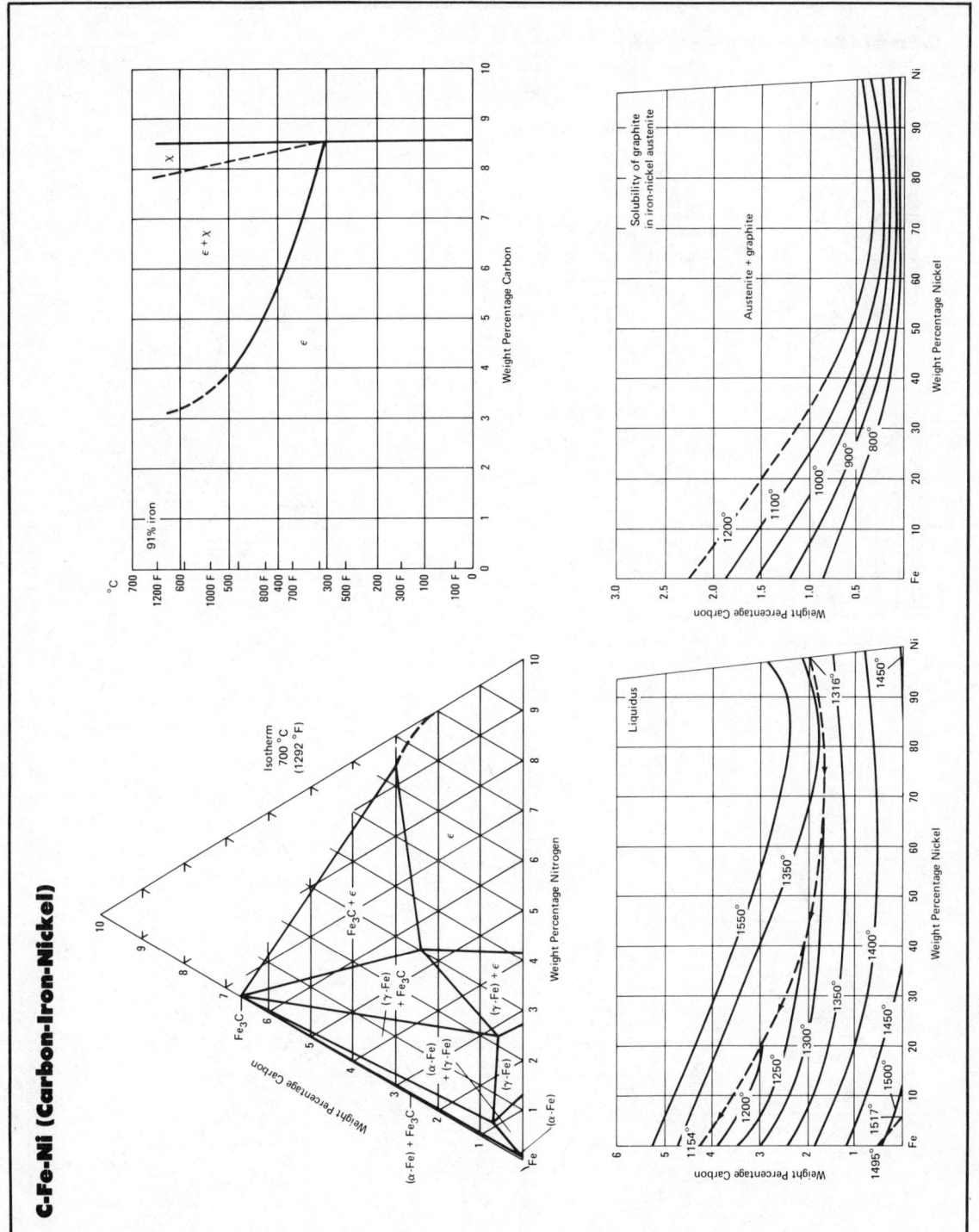

C-Fe-Si (Carbon-Iron-Silicon)

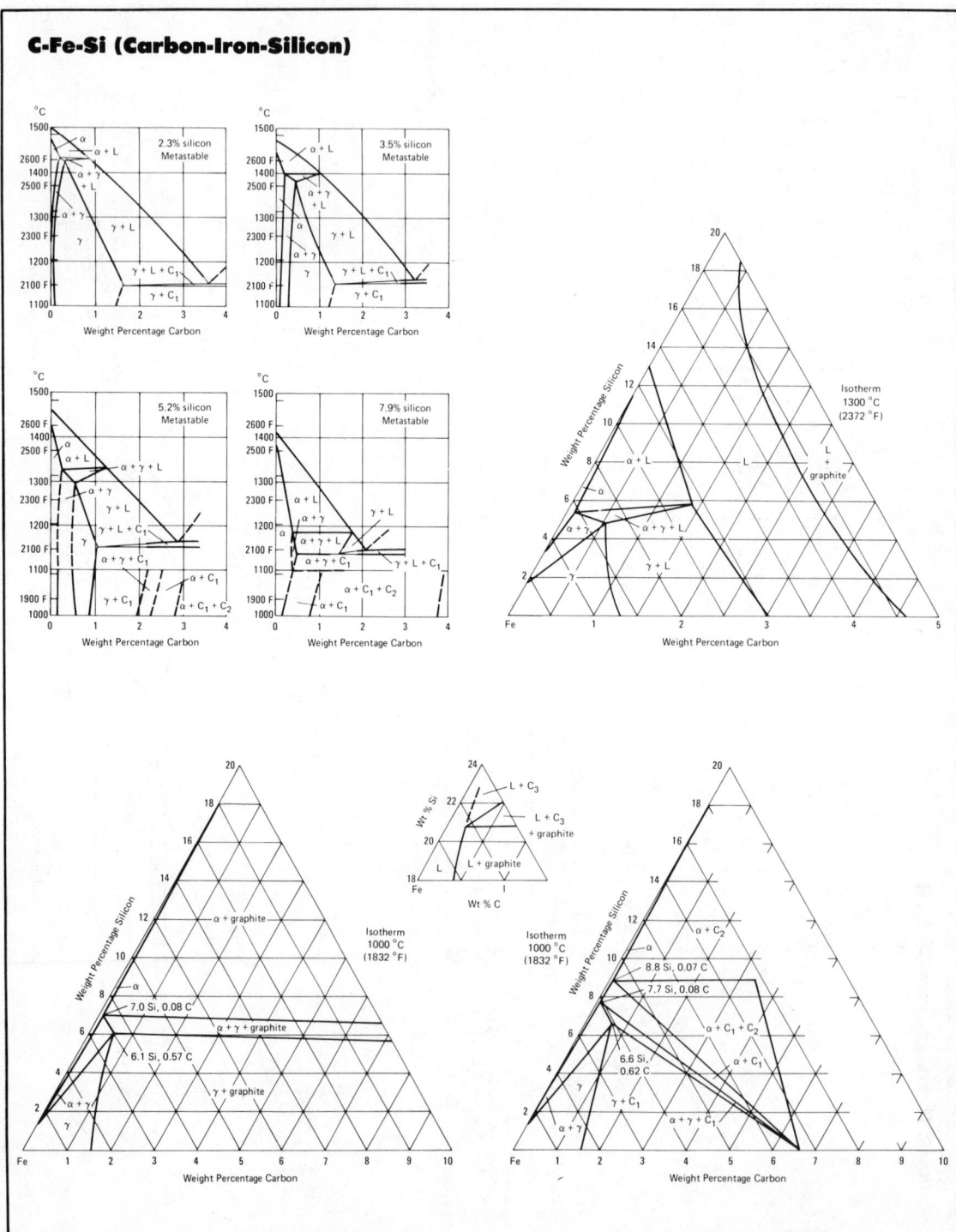

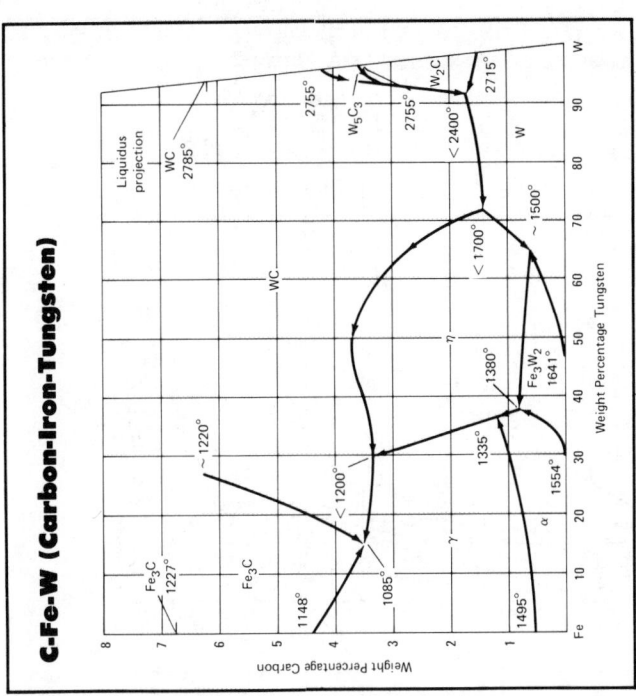

C-Fe-W (Carbon-Iron-Tungsten)

C-Fe-V (Carbon-Iron-Vanadium)

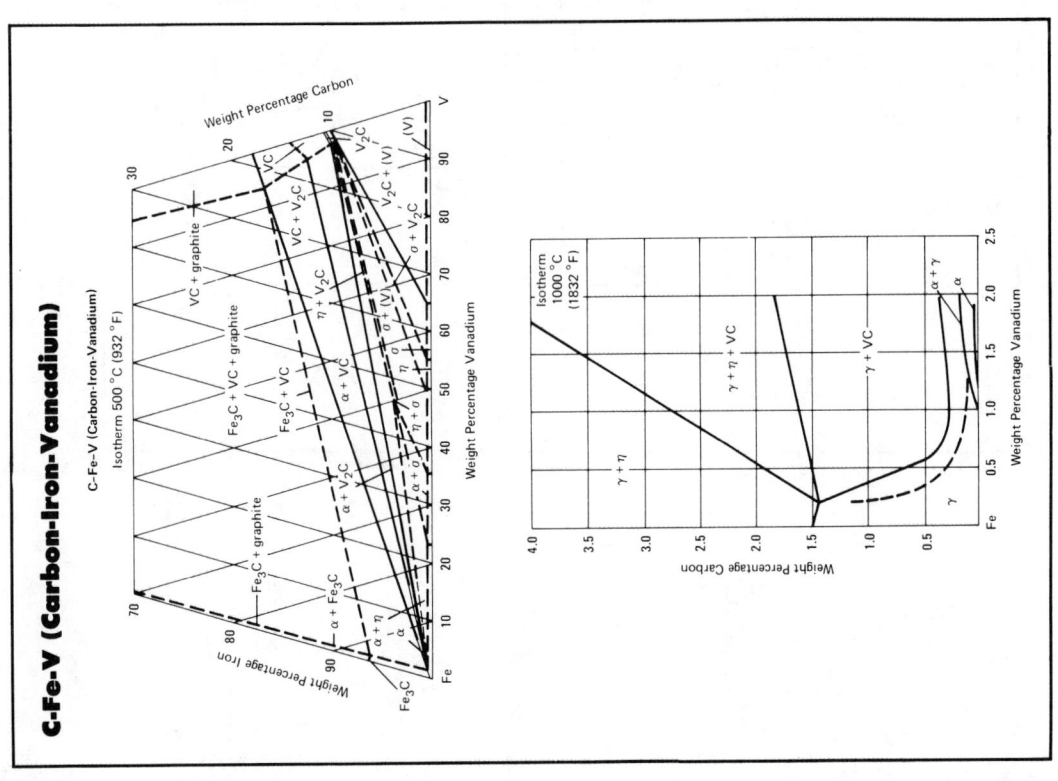

C-Fe-W (Carbon-Iron-Tungsten) (continued)

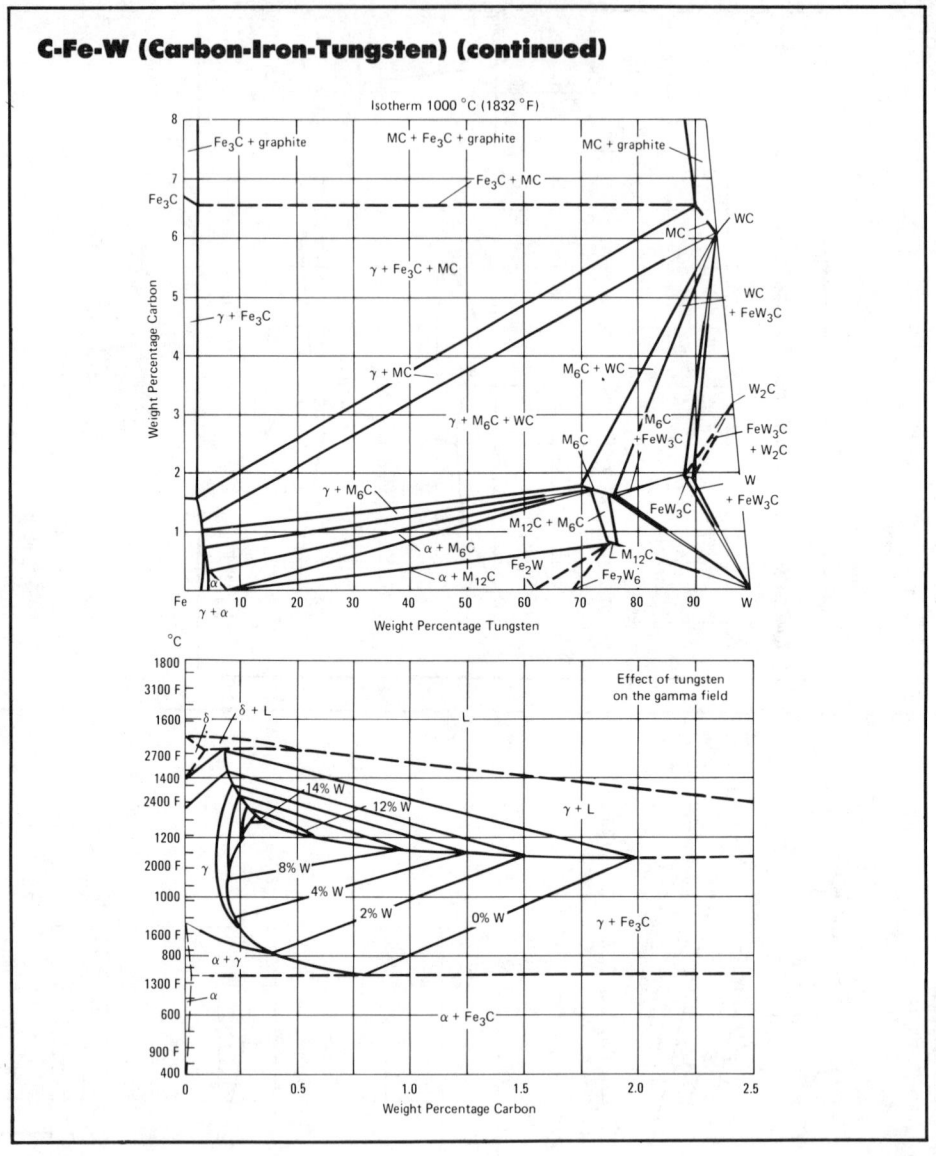

Isotherm 1000 °C (1832 °F)

Effect of tungsten on the gamma field

Co-Cr-Fe (Cobalt-Chromium-Iron)

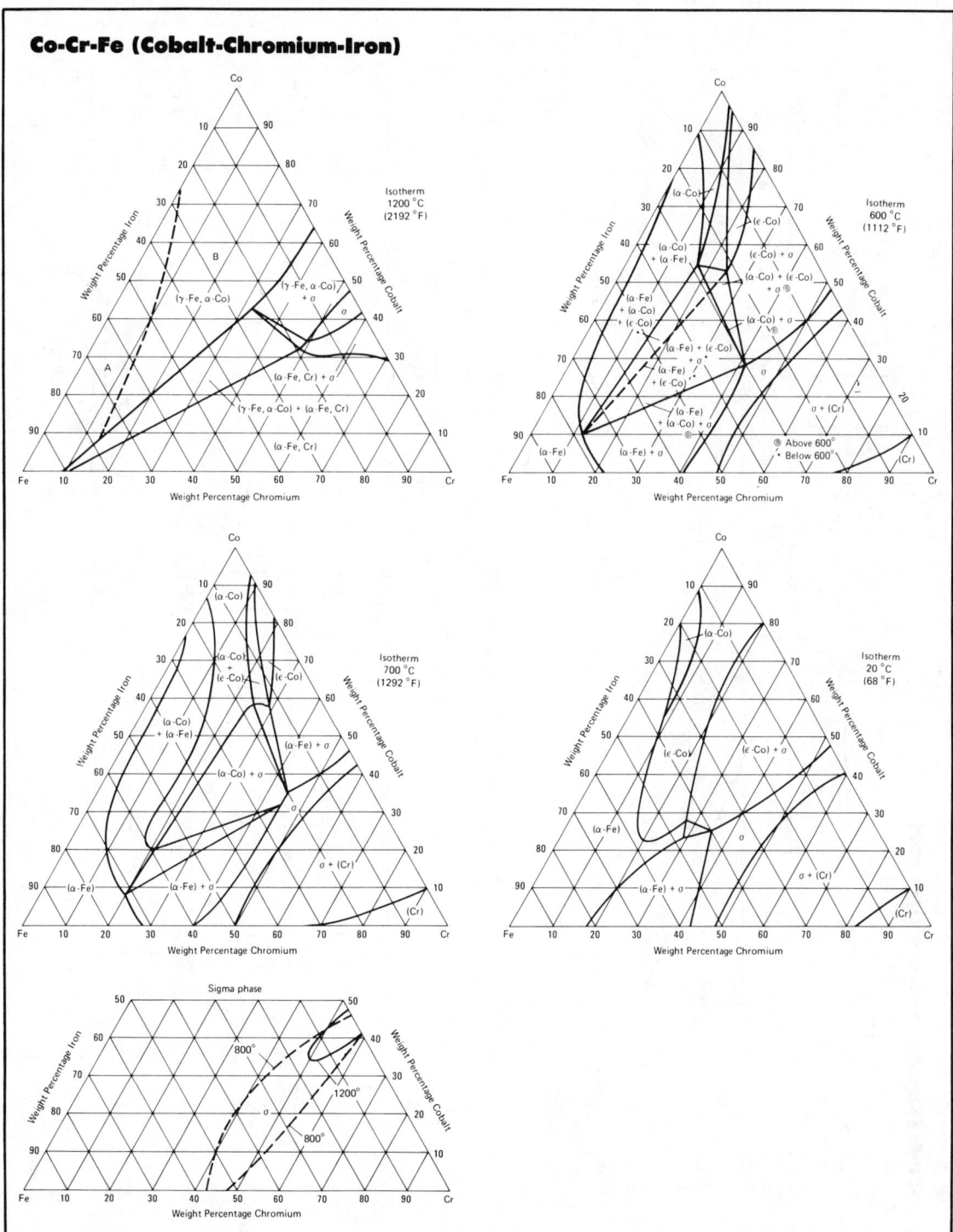

Cr-Fe-N (Chromium-Iron-Nitrogen)

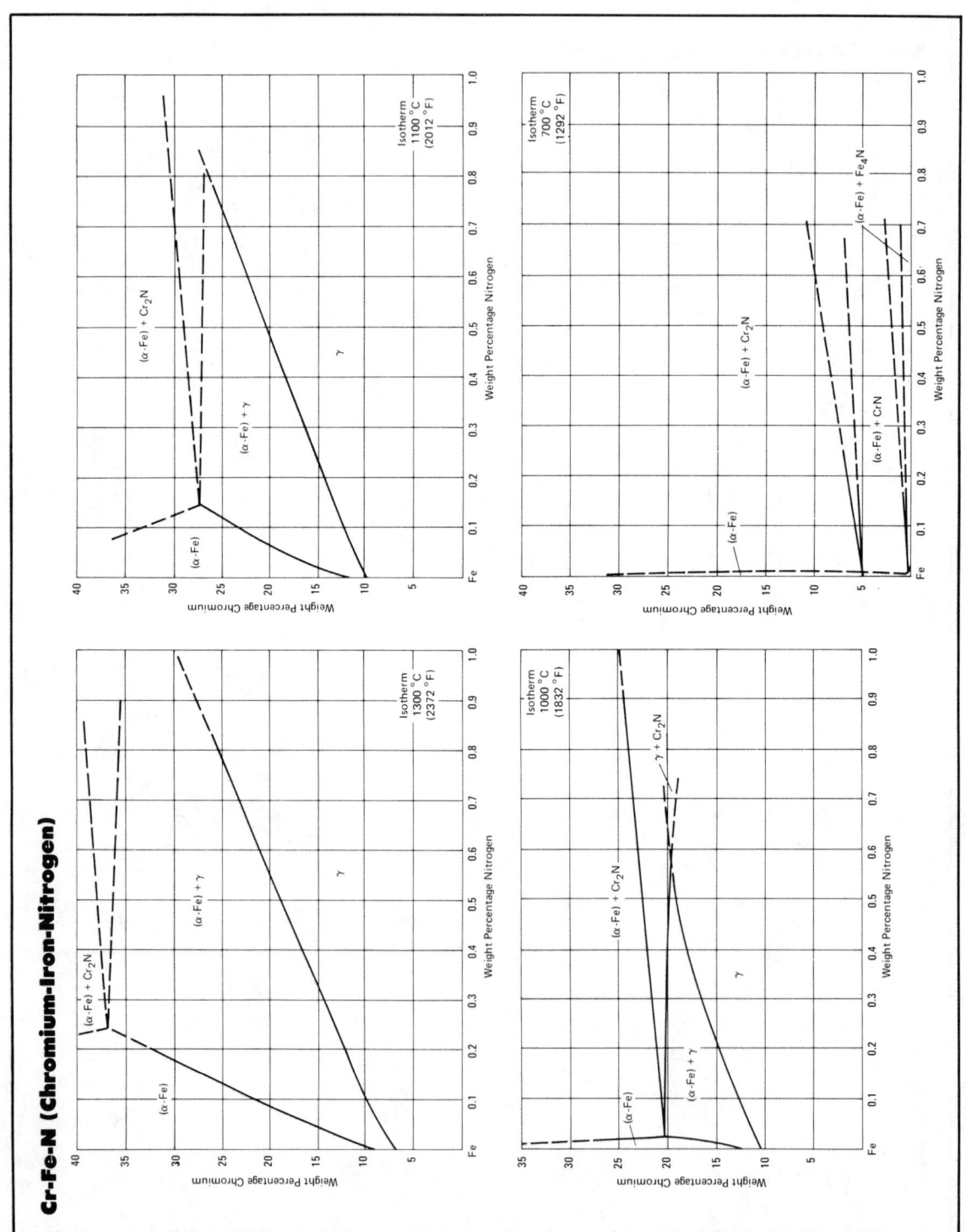

Cr-Fe-N (Chromium-Iron-Nitrogen) (continued)

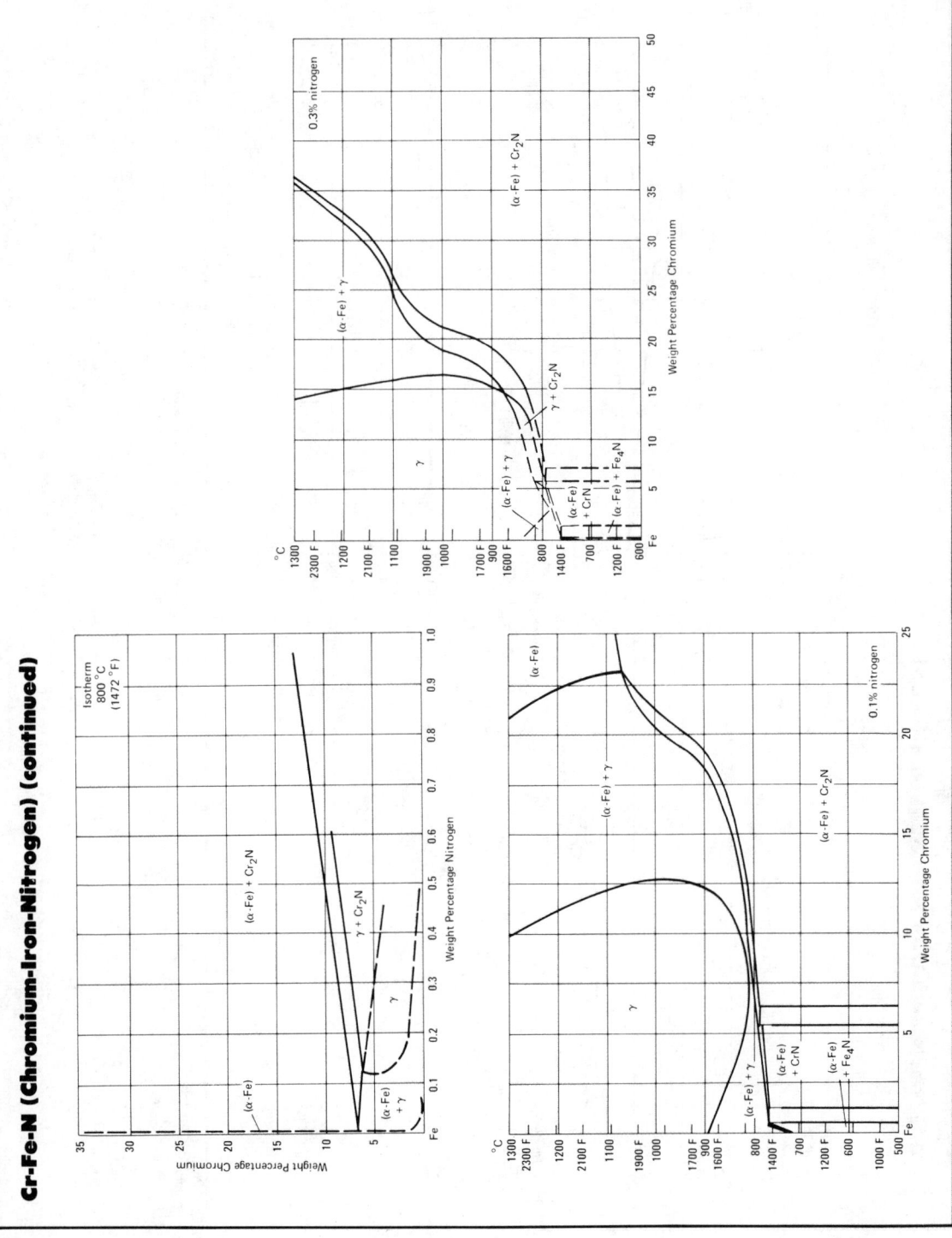

470

Cr-Fe-N (Chromium-Iron-Nitrogen) (continued)

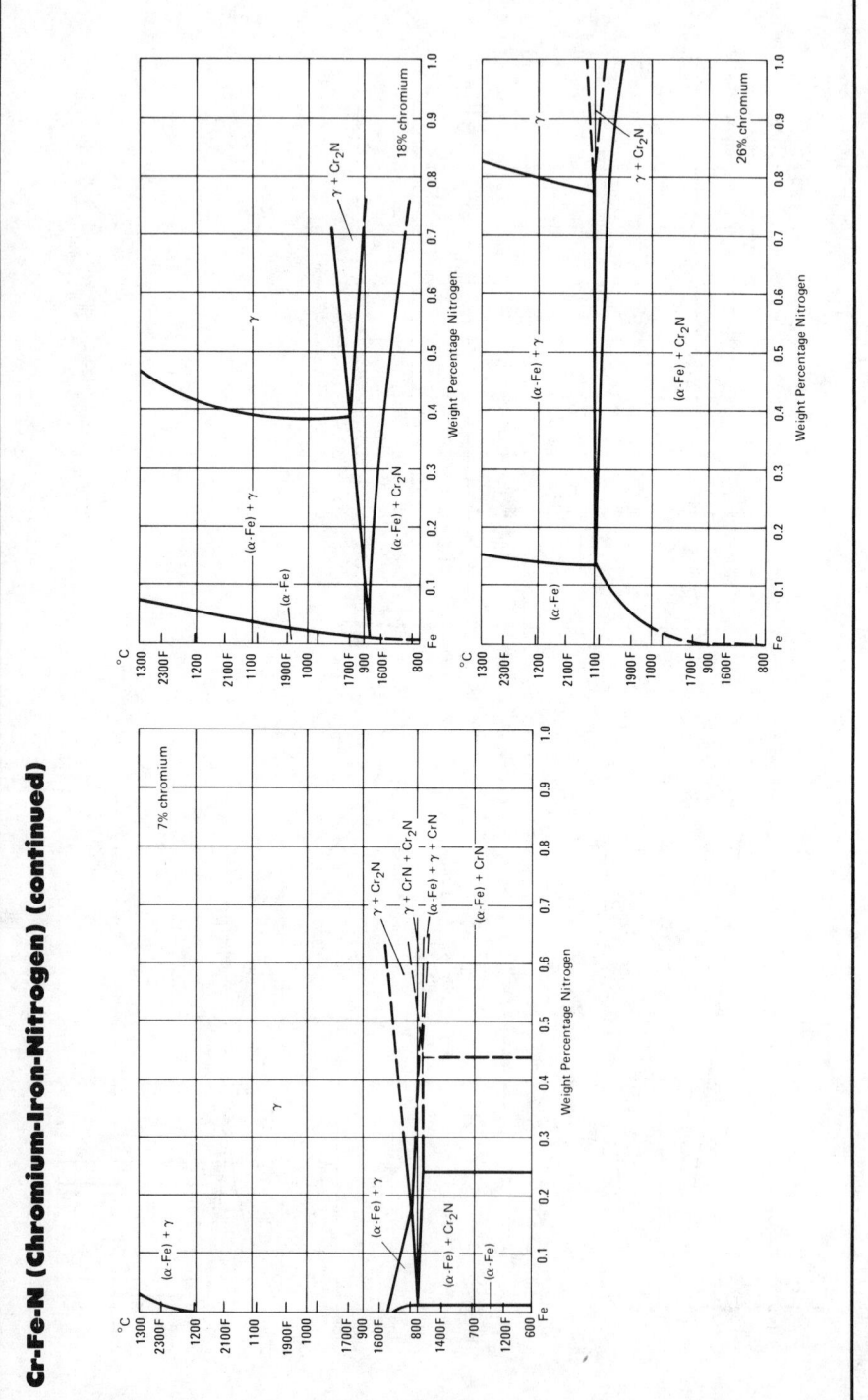

Cr-Fe-Ni (Chromium-Iron-Nickel)

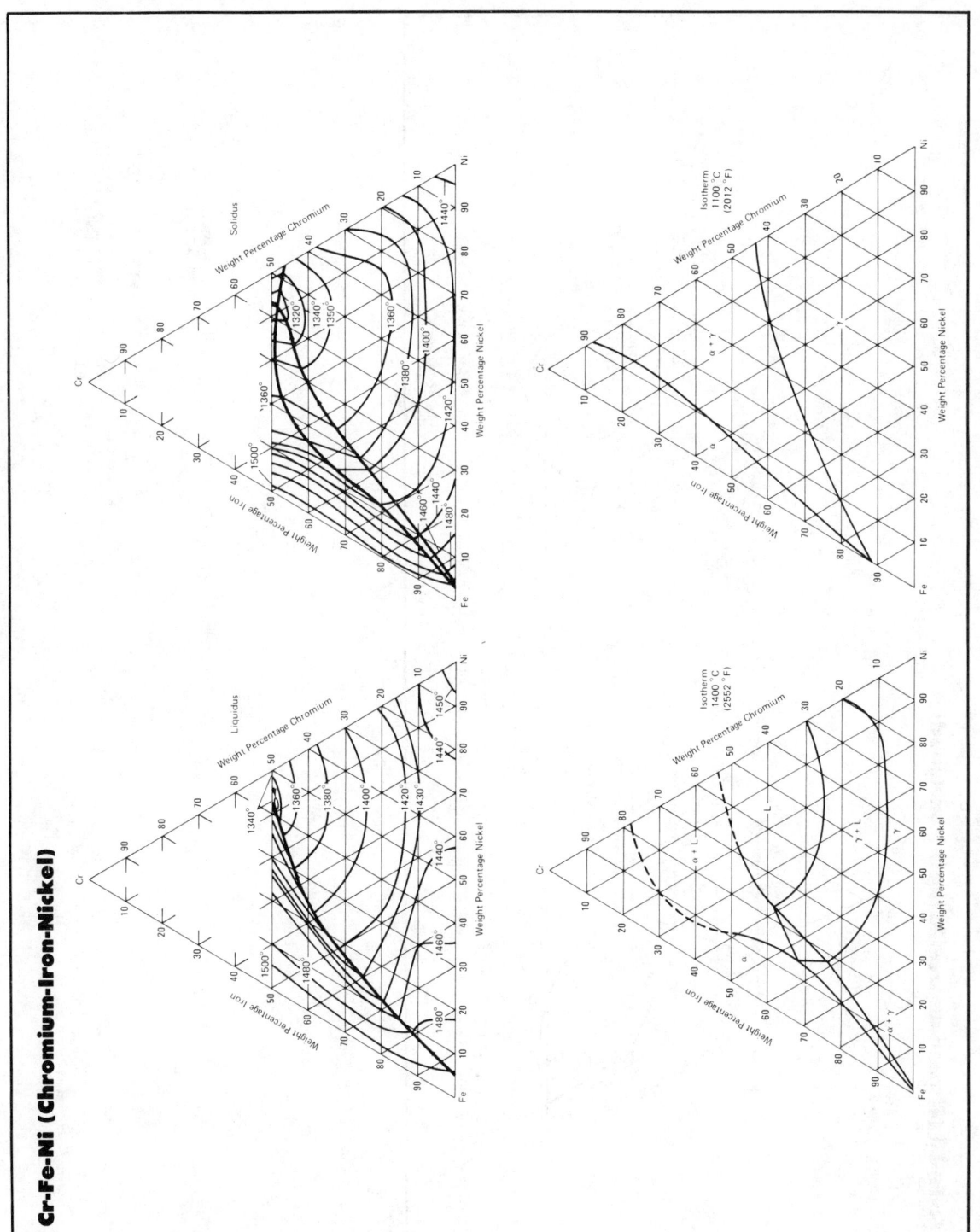

Cr-Fe-Ni (Chromium-Iron-Nickel) (continued)

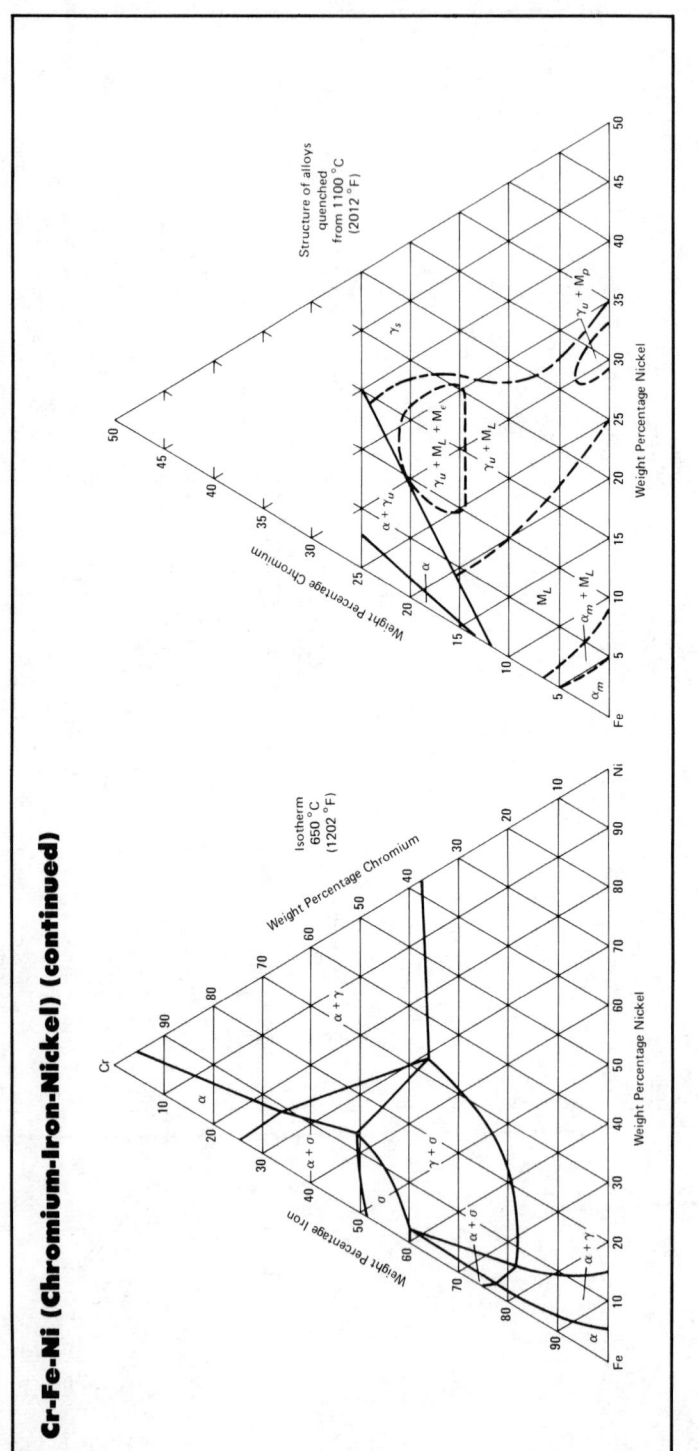

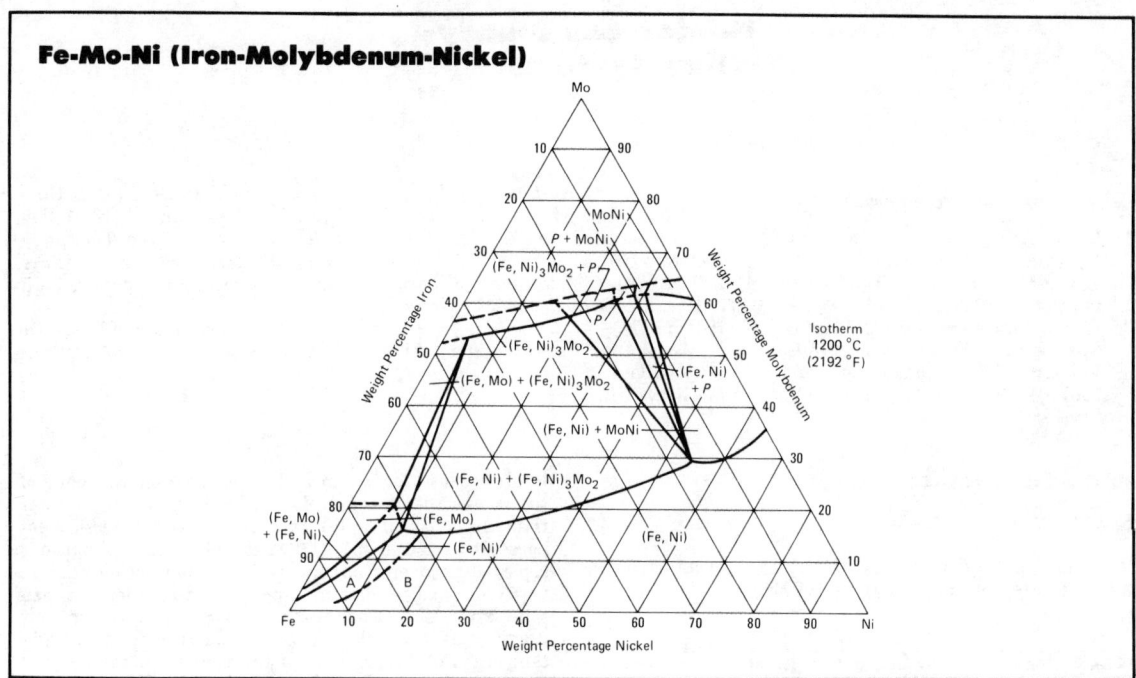

Fe-Mo-Ni (Iron-Molybdenum-Nickel)

Mo

10 — 90

20 — 80

MoNi

P + MoNi

(Fe, Ni)$_3$Mo$_2$ + P

P

(Fe, Ni)$_3$Mo$_2$

(Fe, Mo) + (Fe, Ni)$_3$Mo$_2$

(Fe, Ni) + P

(Fe, Ni) + MoNi

Isotherm
1200 °C
(2192 °F)

(Fe, Ni) + (Fe, Ni)$_3$Mo$_2$

(Fe, Mo)
+ (Fe, Ni)

(Fe, Mo)

(Fe, Ni)

(Fe, Ni)

A   B

Fe   10   20   30   40   50   60   70   80   90   Ni

Weight Percentage Iron

Weight Percentage Molybdenum

Weight Percentage Nickel

Nonferrous Binary Alloy Systems

General References

1 L. A. Willey, Appendix 1, Phase Diagrams, in *Aluminum,* Vol I (K. R. Van Horn, Ed., American Society for Metals, 1967, p 359-381
2 W. B. Pearson, *Handbook of Lattice Spacings and Structures of Metals,* Vol 1 and 2, Pergamon, 1958 and 1967

3 M. Hansen, *Constitution of Binary Alloys,* 2nd ed., prepared with the cooperation of K. Anderko, McGraw-Hill, 1958
4 R. P. Elliott, *Constitution of Binary Alloys, First Supplement,* McGraw-Hill, 1965
5 F. A. Shunk, *Constitution of Binary Alloys, Second Supplement,* McGraw-Hill, 1969

6 R. Hultgren, P. D. Desai, D. T. Hawkins, M. Gleiser and K. K. Kelley, *Selected Values of the Thermodynamic Properties of Binary Alloys,* American Society for Metals, 1973

7 E. Rudy, Compendium of Phase Diagram Data, Part V, Air Force Materials Laboratory Report AFML-TR-65-2, 1969

Al-Cu (Aluminum-Copper)

By L. A. Willey

The diagram is from Willey (Gen Ref 1). Crystal-structure data are from Pearson (Gen Ref 2):

Phase	Formula	Symmetry	Symbol	Prototype
β......	β $AlCu_3$	bcc	$A2$	W
γ.........	...	fcc or cu	$A1$ or $A13$	Cu or β-Mn
γ_1........	...	ord bcc	Related to $D8_{1-3}$	γ brass
γ_2......	Al_4Cu_9	cu	$D8_3$	Al_4Cu_9
χ.........	...	ord bcc	Related to $D8_{1-3}$	γ brass
δ........	...	unknown	...	...
ε_1........	...	unknown	...	...
ε_2......	Al_2Cu_3	pseudocu (?)	...	...
ζ_1........	...	hex	Related to $D8_3$	Al_4Cu_9
ζ_2........	...	mono (?)	...	...
η_1.....	AlCu (HT)	ortho	...	...
η_2.....	AlCu (LT)	eco	...	...
θ.......	Al_2Cu	bct	$C16$	Al_2Cu

Al-Mg (Aluminum-Magnesium)

By L. A. Willey

The diagram is from Willey (Gen Ref 1), with slight modifications in the range 37 to 50% Mg to agree with Bol'shakov et al (1). Bandyopadhyay and Gupta (2) investigated the Al-Mg system between 37 and 54 wt % Mg using thermal, metallographic, dilatometric, and room-temperature and high-temperature x-ray diffraction methods. They report liquidus, solidus, and eutectic temperatures of 6 °C (11 °F) or more lower than those previously observed. These lower temperatures appear un-

likely because, for instance, the congruent melting point of the β phase is placed at 445 °C (833 °F), which is lower than the adjacent well-established Al-rich eutectic temperature at 450 to 451 °C (842 to 844 °F). The β' phase is reported to cover a narrow range of composition near 42 wt % Mg, and to form by a peritectoid reaction between the β and γ phases below 370 °C (698 °F), which confirms previous data of Bol'shakov et al. Based mostly on high-temperature observations of the change in intensity of some x-ray diffraction peaks, and the splitting of some peaks into subpeaks, Bandyopadhyay and Gupta conclude that two transformations occur in the γ phase field in approximately the same area as the single transformation shown on the diagram. The nature of these transformations is not fully resolved, but it appears that they may be more complex than simple lattice distortion with atomic rearrangement. Crystal-structure data are from Hultgren et al (Gen Ref 6):

Phase	Formula	Symmetry	Symbol	Prototype
β.......	β Al_3Mg_2	comp fcc	...	Like Cd_2N
β'	ε $Al_{30}Mg_{23}$	comp rhom	...	Related to R(Mo-Co-Cr)
γ.......	γ $Al_{12}Mg_{17}$	bcc	$A12$	α-Mn
γ'	...	unknown	...	...

REFERENCES

1 K. A. Bol'shakov, P. I. Fedorov and E. I. Smarina, *Russ J Inorg Chem,* Vol 8, 1963, p 734-738
2 J. Bandyopadhyay and K. P. Gupta, *Trans Indian Inst Metals,* Vol 23, No. 4, 1970, p 65-70 (Tech Publ 586)

Al-Mn (Aluminum-Manganese)

By L. A. Willey

The diagram is from Elliott (Gen Ref 4). Crystal-structure data are from Hultgren et al (Gen Ref 6):

Phase	Formula	Symmetry	Symbol	Prototype
β	Al_6Mn	ortho	$D2_h$	Al_6Mn
γ	Al_4Mn	hex	...	...
δ	...	unknown	...	...
ε	φ $Al_{10}Mn_3$	hex	...	Like β (AlMnSi)
$ζ_1$	Al_3Mn	ortho	...	...
$ζ_2$	δ $Al_{11}Mn_4$	tric or cu	...	...
$η_1$	η Al-Mn	hex	...	...
$η_2$	Al_8Mn_5 (?)	bcr	$D8_{10}$	Al_8Cr_5 (LT)
θ	ε Al-Mn	hex	...	...

Al-Si (Aluminum-Silicon)

By L. A. Willey

The diagram is from Willey (Gen Ref 1).

Al-Ti (Aluminum-Titanium)

By L. A. Willey and Harold Margolin

The diagram is from Farrar and Margolin (1), with the Al-rich inset from Willey (Gen Ref 1). Crystal-structure data are from Magneli et al (2):

Phase	Formula	Symmetry	Symbol	Prototype
β	Al_3Ti	tet	DO_{22}	Al_3Ti
γ	AlTi	ord fct	$L1_0$	AuCu I
δ	$AlTi_3$	ord hex	DO_{19}	Ni_3Sn

REFERENCES

1 P. A. Farrar and H. Margolin, Air Force Materials Laboratory Tech Rept AFML-TR-65-69, 1965
2 A. Magneli, L. Edshammar, S. Westman and T. Dagerhamn, University of Stockholm, Inst Inorg Phys Chem Rept 1, 1965 (quoted in Hultgren et al, Gen Ref 6)

Al-Zn (Aluminum-Zinc)

By L. A. Willey

The diagram is from Willey (Gen Ref 1). It generally confirms earlier work, which has been described by Hansen (Gen Ref 3). Recently, however, Goldak and Parr (1), and Presnyakov, Gorban and Chernyakova (2) observed a small discontinuity in an otherwise smooth and continuous lattice parameter curve for alloys near 70 wt % Zn, in the temperature range 340 to 400 °C (644 to 752 °F). This was interpreted as evidence of a narrow two-phase region (of width ~ 1%), and with an intermediate phase at higher Zn contents. Without further supporting data, the intermediate phase was indicated to form by a peritectic reaction at 443 °C (829 °F). A monotectoid reaction at 340 °C (644 °F) at about 69% Zn was also indicated. In the absence of data supporting the x-ray measurements, the earlier form of the diagram has been given.

REFERENCES

1 G. R. Goldak and J. G. Parr, J Inst Metals, Vol 92, 1964, p 230-233
2 A. A. Presnyakov, Yu. A. Gorban and V. V. Chernyakova, Russ J Phys Chem, Vol 35, 1961, p 632-633

Cu-Ni (Copper-Nickel)

By Avinash D. Kulkarni

The solidus and liquidus are from Feest and Doherty (1). Calculations by Elford, Müller and Kubaschewski (2) and Pascoe and Mackowiak (3) are in good agreement. The liquidus of Schürmann and Schulz (4) is up to 7% different from that selected. The miscibility gap was calculated by Elford, Müller and Kubaschewski; it cannot be experimentally verified because of the low temperature. The Curie temperature is from Hansen (Gen Ref 3).

REFERENCES

1 E. A. Feest and R. D. Doherty, J Inst Metals, Vol 99, 1971, p 102-103
2 L. Elford, F. Müller and O. Kubaschewski, Ber Bunsenges Physik Chem, Vol 73, 1969, p 601-605
3 G. Pascoe and J. Mackowiak, J Inst Metals, Vol 98, 1970, p 253-256
4 E. Schürmann and E. Schulz, Z Metallk, Vol 62, 1971, p 758-762

Cu-Pb (Copper-Lead)

By Robert E. Johnson

The diagram is from Hansen (Gen Ref 3).

Cu-Si (Copper-Silicon)

By Donald T. Hawkins

The diagram is from Hansen (Gen Ref 3) and Elliott (Gen Ref 4). Crystal-structure data are from Hultgren et al (Gen Ref 6):

Phase	Formula	Symmetry	Symbol	Prototype
β	ζ Cu-Si	hcp	A3	Mg
γ	β Cu-Si (HT)	bcc (?)	A2	W
δ	γ Cu_5Si	comp cu	A13	β-Mn
ε	δ Cu-Si	comp tet	...	...
ζ	ε $Cu_{15}Si_4$	comp cu	...	...
η, η′, η″	...	comp tet (?)	...	...

Cu-Sn (Copper-Tin)

By Robert E. Johnson

The eutectic composition is from Davey and Happ (1). The remainder of the diagram is from Hansen (Gen Ref 3). Crystal-structure data are from Pearson (Gen Ref 2):

Phase	Formula	Symmetry	Symbol	Prototype
β . . . . · · ·		bcc	$A2$	W
γ	γ Cu_3Sn	fcc	$D0_3$	BiF_3
δ	$Cu_{31}Sn_8$	cu	Like $D8_{1-3}$	γ brass
ε	Cu_3Sn	pseudohex	$A3$	Mg
ζ	$Cu_{20}Sn_6$	trig	· · ·	· · ·
η . . . . · · ·		unknown	· · ·	· · ·
η′	Cu_6Sn_5	hex	$B8_1$	AsNi

REFERENCE

1 T. R. A. Davey and J. V. Happ, *Trans Inst Mining Met,* Ser C, Vol 78, 1969, p 108-110

Cu-Zn (Copper-Zinc)

By Donald T. Hawkins

The diagram is from Hansen (Gen Ref 3), except for the β′ phase boundaries, which were taken from Shinoda and Amano (1). By x-ray studies on alloys annealed for several months, they found some evidence of eutectoid decomposition of part of β′ below 250 °C (482 °F). Massalski and Kittl (2) studied the (Cu)-[(Cu) + β′] phase boundary and found it to lie at slightly lower zinc compositions than shown. They did not find the eutectoid decomposition. From their studies, it appears that a drastic decrease in the solubility limit of the (Cu) phase below 250 °C (482 °F), which was proposed by Shinoda and Amano, is unlikely.

Crystal-structure data are from Pearson (Gen Ref 2):

Phase	Formula	Symmetry	Symbol	Prototype
β	β Cu-Zn	bcc	$A2$	W
β′	β′ Cu-Zn	ord bcc	$B2$	CsCl
γ	γ Cu_5Zn_8	ord bcc	$D8_2$	γ brass
δ	δ $CuZn_3$	ord bcc	$B2$	CsCl
ε	ε Cu-Zn	hcp	$A3$	Mg

REFERENCES

1 G. Shinoda and Y. Amano, *Trans Japan Inst Metals,* Vol 1, 1960, p 54-57

2 T. B. Massalski and J. E. Kittl, *J Australian Inst Metals,* Vol 8, 1963, p 91-97

Mg-Zn (Magnesium-Zinc)

By Donald T. Hawkins

The diagram is from Elliott (Gen Ref 4). Crystal-structure data are from Pearson (Gen Ref 2):

Phase	Formula	Symmetry	Symbol	Prototype
β	Mg_7Zn_3	unknown	· · ·	· · ·
γ	MgZn	hex or ortho	· · ·	· · ·
δ	Mg_2Zn_3	unknown	· · ·	· · ·
ε	$MgZn_2$	hex	$C14$	$MgZn_2$
ζ	Mg_2Zn_{11}	cu	· · ·	· · ·

Pb-Sb (Lead-Antimony)

By Carl DiMartini

The diagram is from Hansen (Gen Ref 3) and Elliott (Gen Ref 4).

Pb-Sn (Lead-Tin)

By Carl DiMartini

The diagram is from Hansen (Gen Ref 3), Elliott (Gen Ref 4), and Shunk (Gen Ref 5).

Ti-V (Titanium-Vanadium)

By Donald T. Hawkins

The diagram is from Rudy (Gen Ref 7).

Nonferrous Binary Alloy
Phase Diagrams

Al-Cu (Aluminum-Copper)

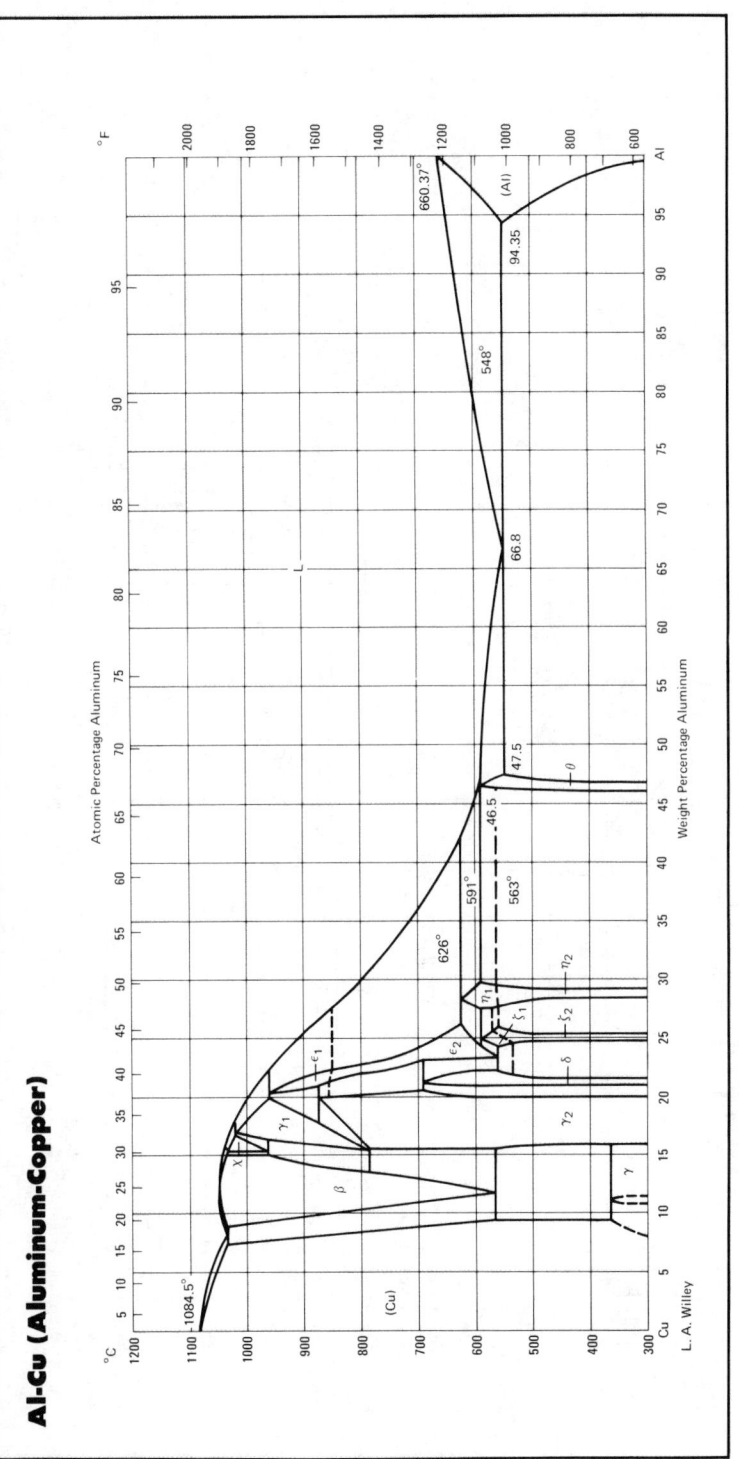

L. A. Willey

Al-Cu (Aluminum-Copper) (continued)

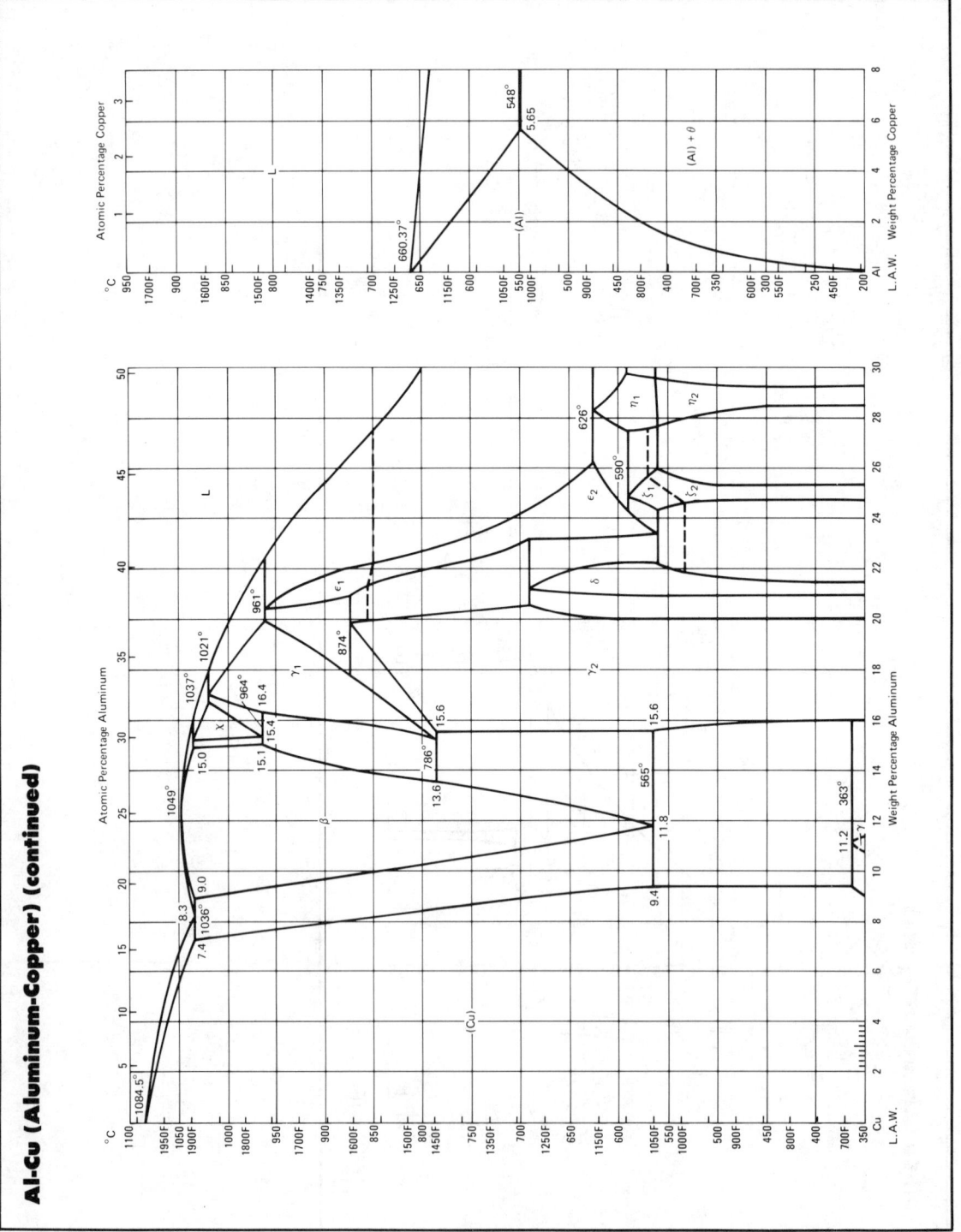

Al-Mg (Aluminum-Magnesium)

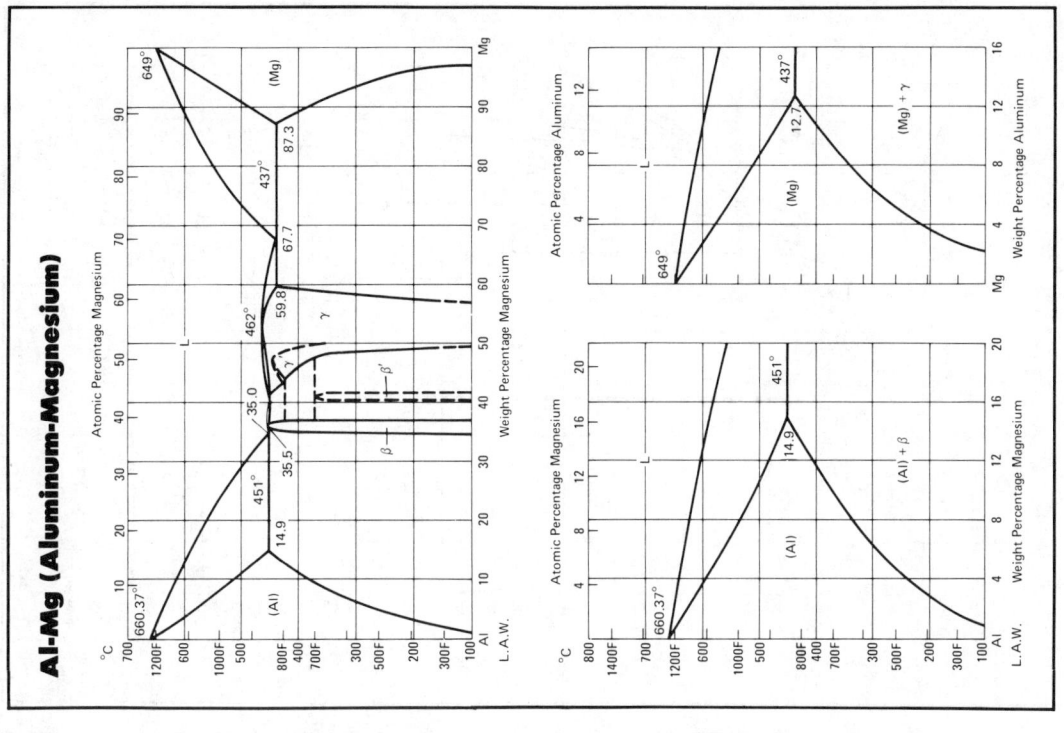

Al-Mn (Aluminum-Manganese)

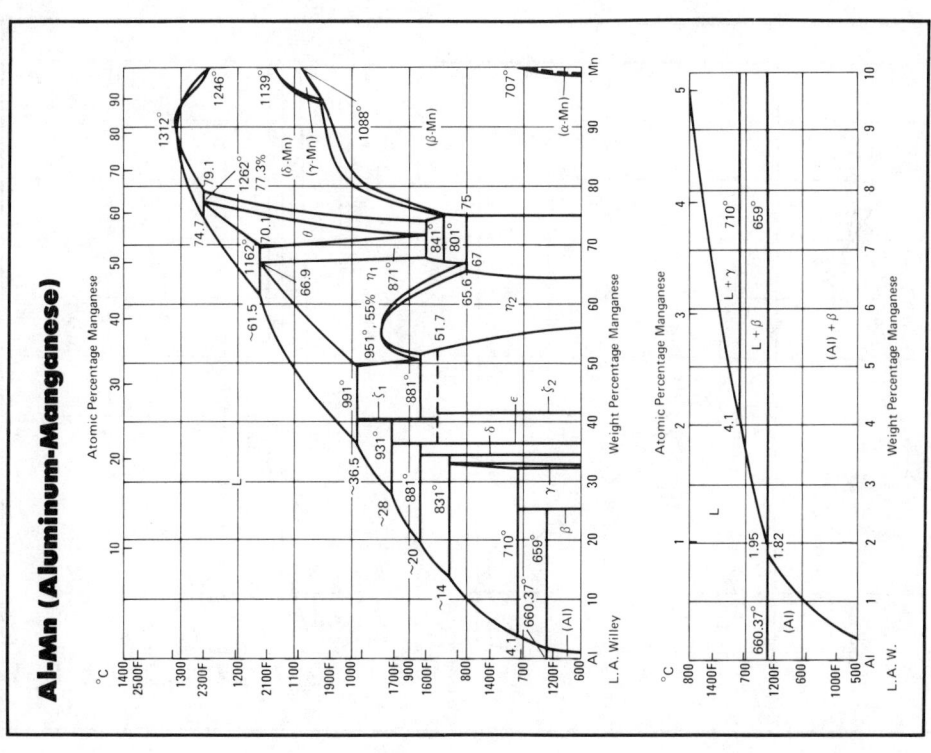

Al-Ti (Aluminum-Titanium)

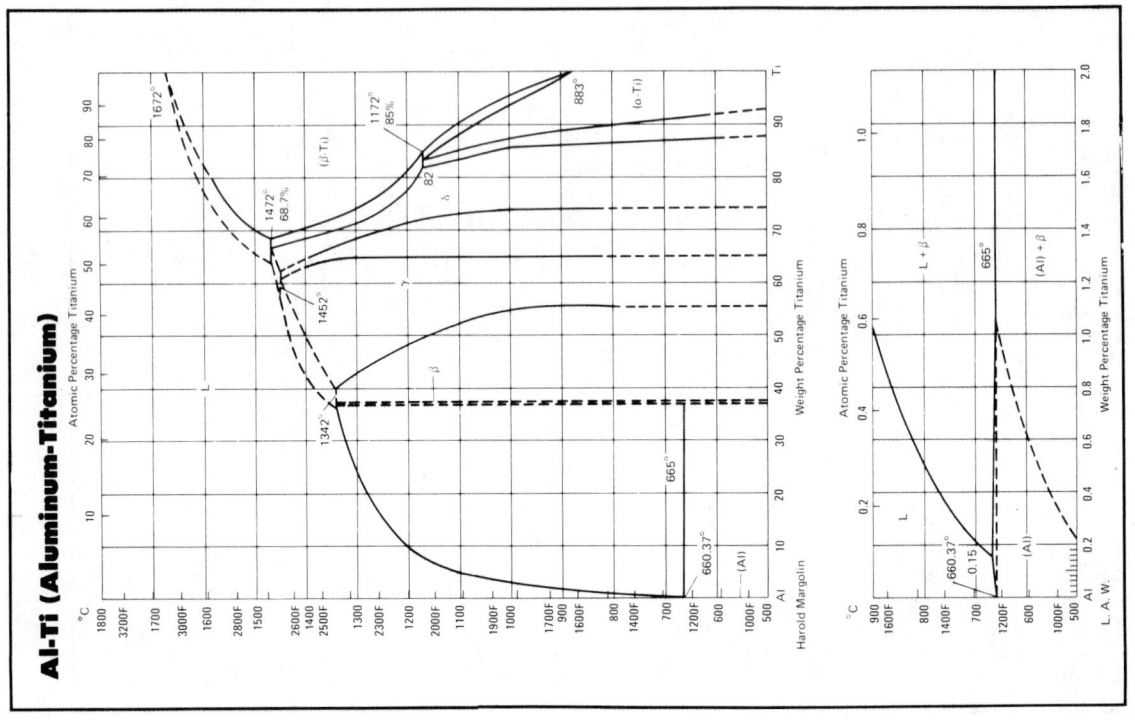

Al-Si (Aluminum-Silicon)

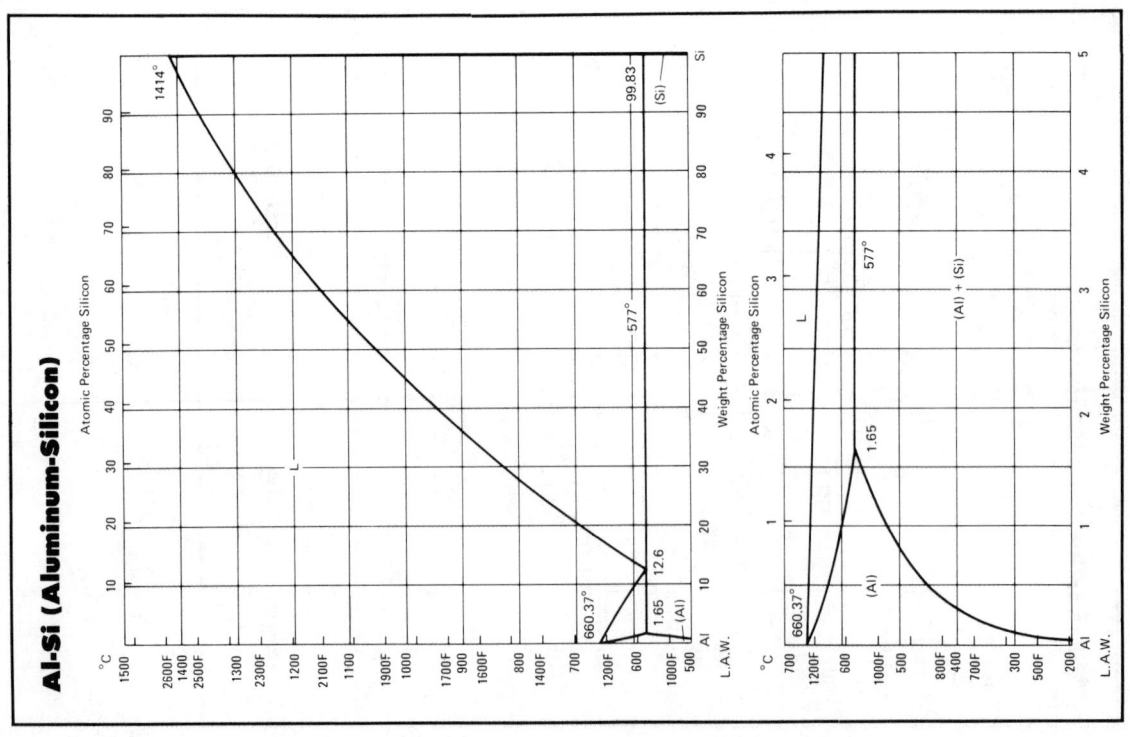

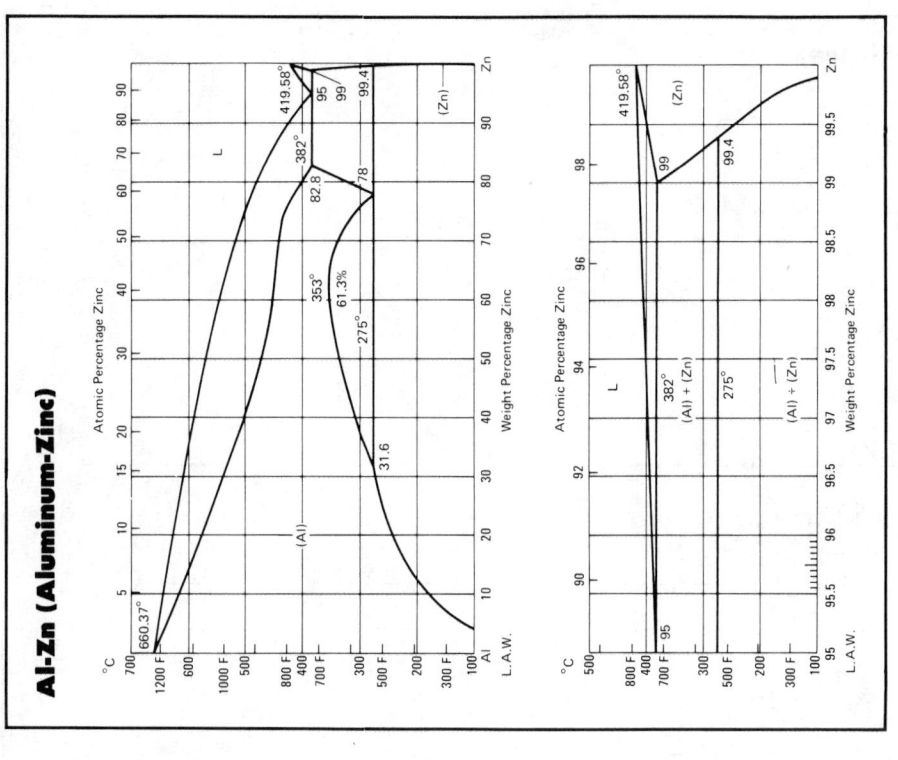

Al-Zn (Aluminum-Zinc)

Cu-Ni (Copper-Nickel)

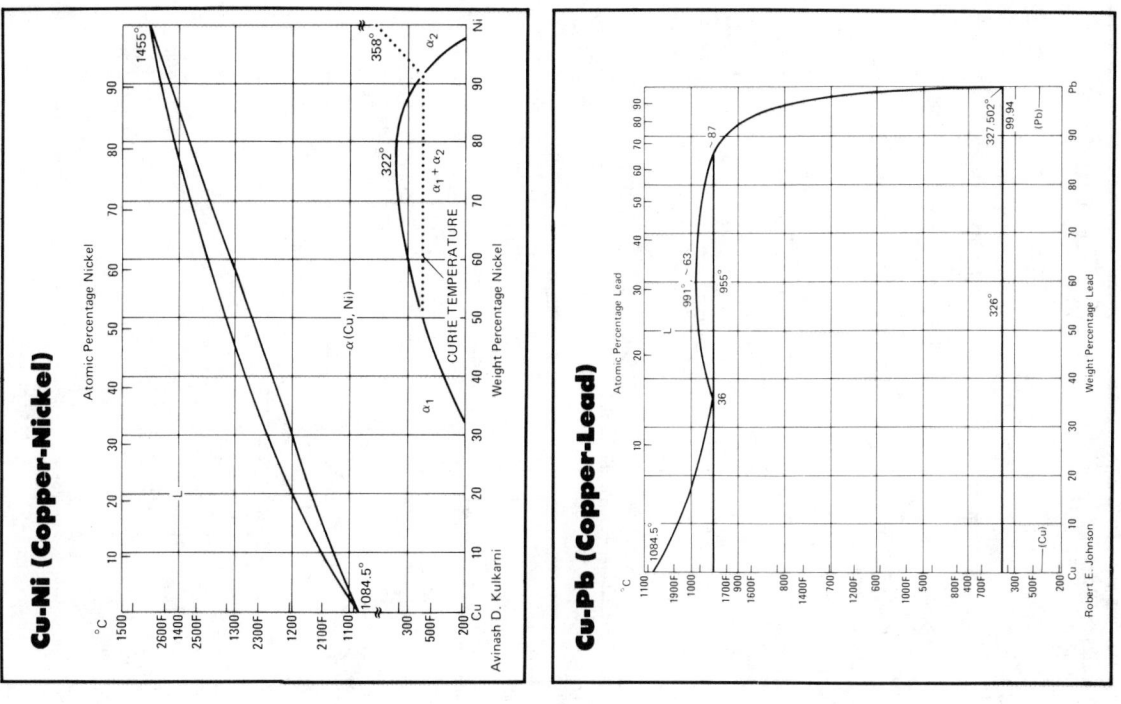

Cu-Pb (Copper-Lead)

Cu-Si (Copper-Silicon)

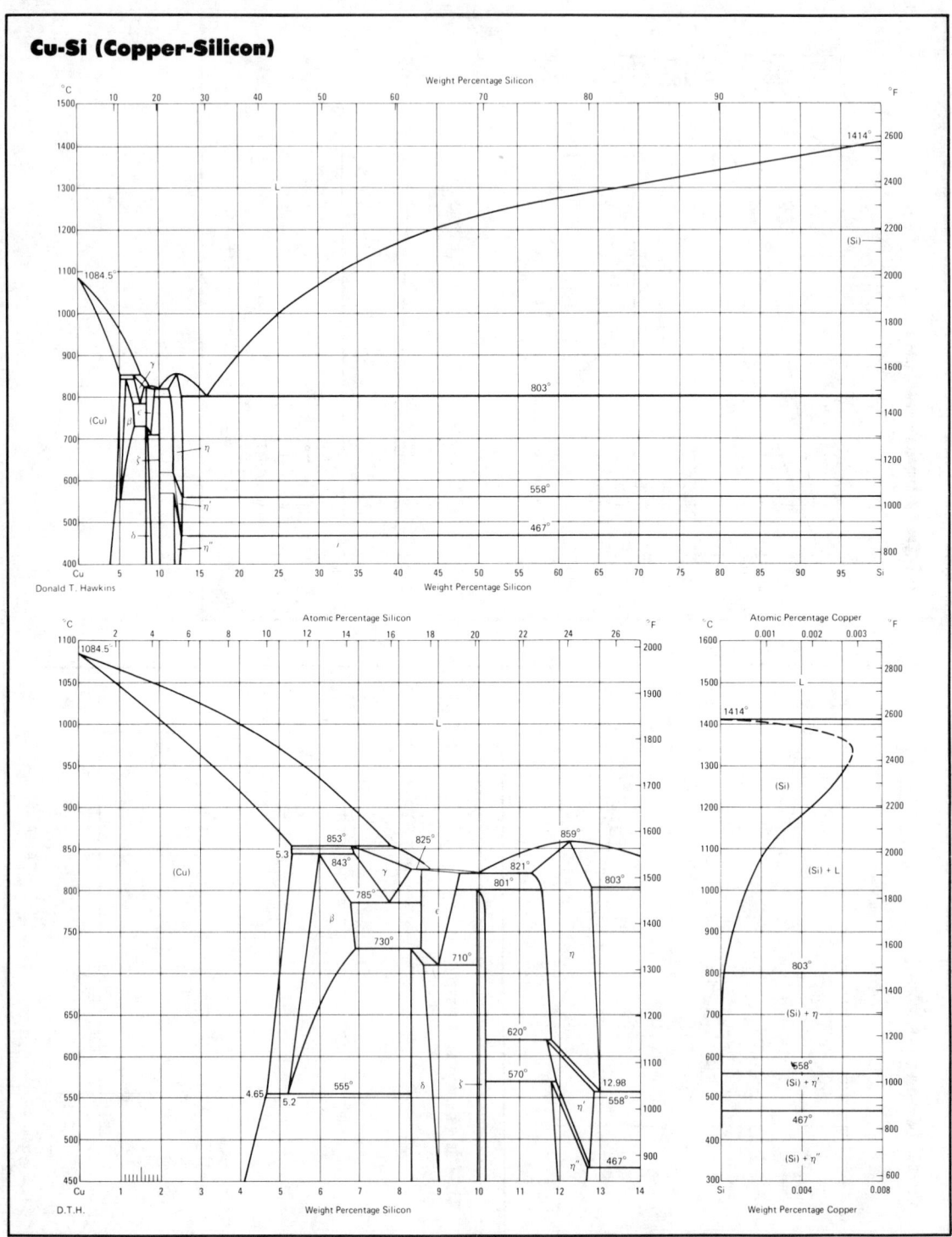

Donald T. Hawkins

D.T.H.

Cu-Sn (Copper-Tin)

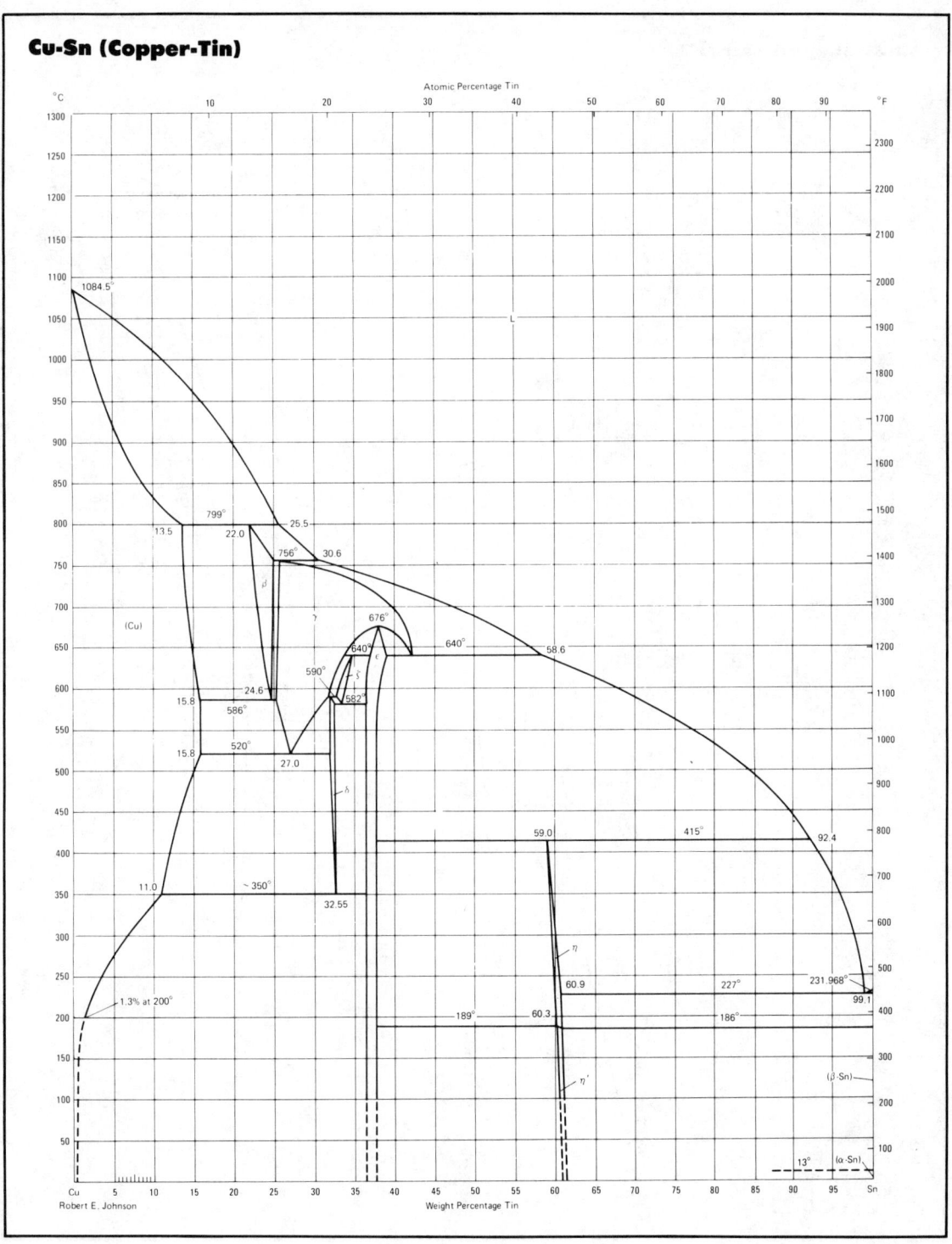

Robert E. Johnson

Cu-Zn (Copper-Zinc)

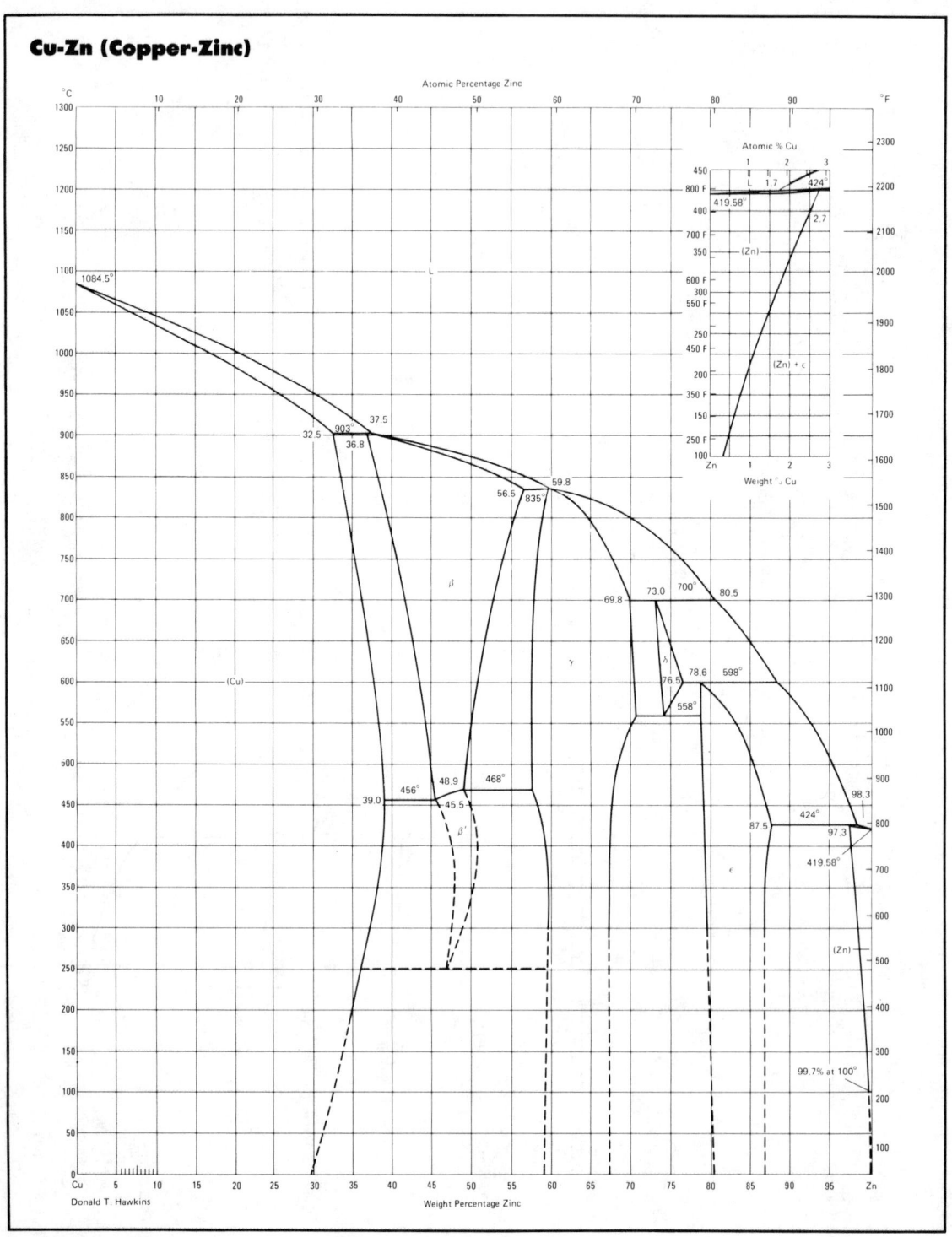

Donald T. Hawkins

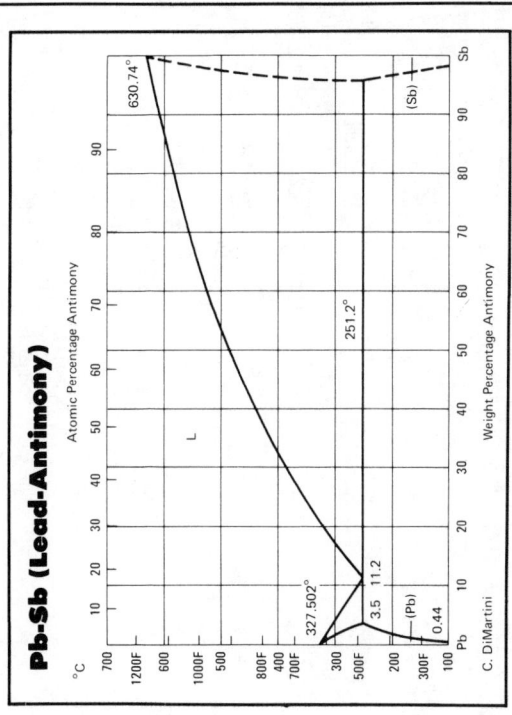

Mg-Zn (Magnesium-Zinc)

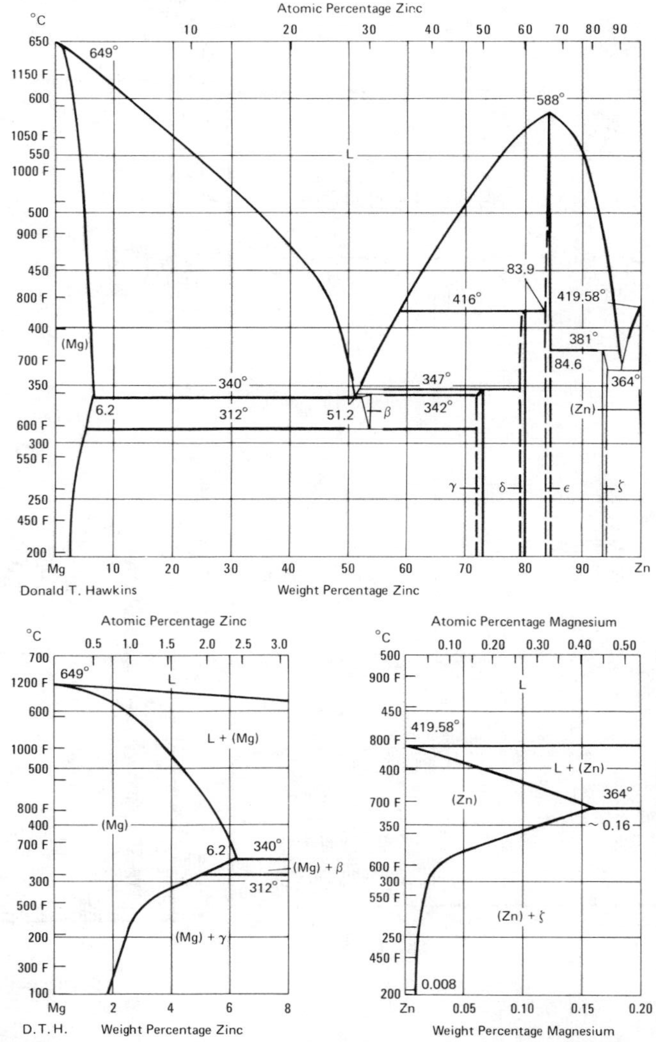

Donald T. Hawkins

D. T. H.

Pb-Sb (Lead-Antimony)

C. DiMartini

Pb-Sn (Lead-Tin)

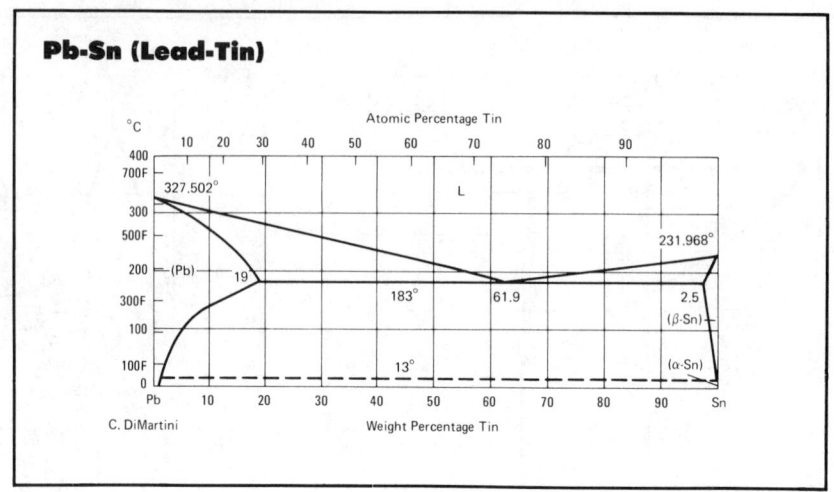

C. DiMartini

Ti-V (Titanium-Vanadium)

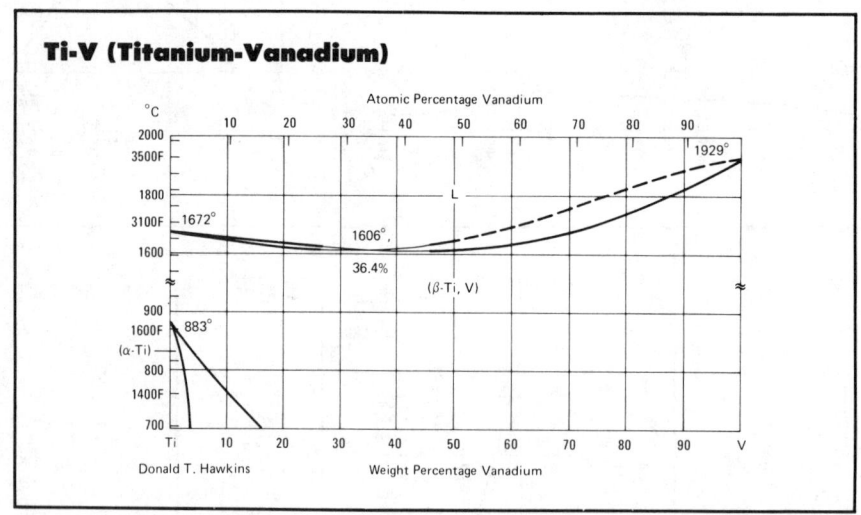

Donald T. Hawkins

Nonferrous Ternary Alloy Systems

General References

1 L. A. Willey, Appendix 1, Phase Diagrams, in *Aluminum,* Vol I, K. R. Van Horn, Ed., American Society for Metals, 1967, p 359-381

2 W. B. Pearson, *Handbook of Lattice Spacings and Structures of Metals,* Vol 1 and 2, Pergamon, 1958 and 1967

3 M. Hansen, *Constitution of Binary Alloys,* 2nd ed., prepared with the cooperation of K. Anderko, McGraw-Hill, 1958

4 R. P. Elliott, *Constitution of Binary Alloys, First Supplement,* McGraw-Hill, 1965

5 F. A. Shunk, *Constitution of Binary Alloys, Second Supplement,* McGraw-Hill, 1969

6 R. Hultgren, P. D. Desai, D. T. Hawkins, M. Gleiser and K. K. Kelley, *Selected Values of the Thermodynamic Properties of Binary Alloys,* American Society for Metals, 1973

7 E. Rudy, Compendium of Phase Diagram Data, Part V, Air Force Materials Laboratory Report AFML-TR-65-2, 1969

Al-Mg-Zn (Aluminum-Magnesium-Zinc)

By L. A. Willey

The Al-Mg-Zn system is a basis for many commercial aluminum-rich and magnesium-rich alloys. Also, zinc-base alloys with aluminum as a major alloying component may contain small percentages of magnesium.

The phases referred to in this compilation (and their formulas) are:

β	Mg_2Al_3	δ	Mg_7Zn_3	θ	Mg_2Zn_{11}
β'	$Mg_3Al_4(Mg_{23}Al_{30})$	ϵ	$MgZn$	τ	$Mg_{32}(Zn,Al)_{49}$
γ	$Mg_{17}Al_{12}$	ζ	Mg_2Zn_3	ϕ	$Mg_5Zn_2Al_2$
		η	$MgZn_2$		

The liquidus diagram shown here is from early investigations by Köster and Wolf (1) and Köster and Dullenkopf (2), modified principally in the region from 0 to 26 wt % Al, 40 to 50 wt % Mg, 28 to 60 wt % Zn, using data from Clark and Rhines (3), Park and Wyman (4), Clark (5), and Yue and Clark (6). Two ternary intermetallic phases, τ and ϕ, crystallize from liquid within this system. In addition, there are three component solid solutions, two binary Al-Mg intermetallic phases, β and γ, and five Mg-Zn intermetallic phases, δ, ϵ, ζ, η and θ, that enter into equilibrium with the ternary liquid. Combinations of these twelve phases take part in four ternary eutectic reactions and eight ternary peritectic reactions. Four binary-type eutectic reactions and one binary-type peritectic reaction are present in ternary quasibinary and pseudobinary systems. The ternary τ phase is first formed by peritectic reaction between liquid and the η phase at 535 °C (995 °F). The τ phase crystallizes primarily within a relatively large range of compositions. The ternary ϕ phase is first formed by peritectic reaction between the τ phase, the γ phase, and liquid at 393 °C (739 °F). The ϕ phase crystallizes directly from liquid within a relatively narrow range of compositions. The η phase forms quasibinary systems with aluminum and with the β phase.

The 335 °C (635 °F) isothermal section is a composite based on investigations by Köster and Wolf (1), Köster and Dullenkopf (2), Köster (7), Fink and Willey (8), Little *et al*

(9), and Clark (5). The ϕ phase, which covers a moderate range of compositions, is shown to be present in equilibrium with (Mg) and the γ and τ phases. The τ phase has a wide range of compositions along a line described by the formula $Mg_{32}(Zn,Al)_{49}$. This phase comes into equilibrium with all phases of the system except the δ, ζ and θ phases of the Mg-Zn binary system, the (Zn) phase, and the (Al') phase. Laves *et al* (10) and Bergman *et al* (11, 12) report the τ phase has a cubic structure with 162 atoms per unit cell; at a composition of 22.1 wt % Mg, 60.9 wt % Zn, 17.0 wt % Al, the cell dimension is reported to be 14.16 A. Raynor and Hume-Rothery (13) report that this dimension increases from a little over 14.00 A at the zinc-rich end to a little less than 14.50 A at the aluminum-rich end of the τ phase, along its boundary with the τ + (Al) phase field. Strawbridge *et al* (14) report that the τ, η and θ phases of the Al-Mg-Zn system are isomorphous with $Mg_{32}(Cu,Al)_{49}$, $CuMgAl$ and $Cu_6Mg_2Al_5$, respectively, of the Al-Cu-Mg system.

An appreciable range of solid solution is present in both the aluminum-rich and magnesium-rich corners of the system. Because maximums of only about 1 wt % Al and less than 0.2 wt % Mg are soluble in zinc, the solid solution range in the zinc corner is extremely small. Details of solid solubility and phase relationships in the aluminum-rich corner of the system are shown in the diagrams for the solidus, from Butchers and Hume-Rothery (15), and for the solvus, from Fink and Willey (8) and Little *et al* (9).

REFERENCES

1 W. Köster and W. Wolf, *Z Metallk,* Vol 28, 1936, p 155-158

2 W. Köster and W. Dullenkopf, *Z Metallk,* Vol 28, 1936, p 309-312, 363-367

3 J. B. Clark and F. N. Rhines, *Trans Met Soc AIME,* Vol 209, 1957, p 425-430

4 J. J. Park and L. L. Wyman, WADC Tech Report No. 57-504, 1957

5 J. B. Clark, *Trans Am Soc Metals,* Vol 53, 1961, p 295-306

6   A. S. Yue and J. B. Clark, *Trans Met Soc AIME,* Vol 221, 1961, p 383-389

7   W. Köster, *Z Metallk,* Vol 39, 1948, p 211-213

8   W. L. Fink and L. A. Willey, *Trans Inst Metals Div AIME,* Vol 124, 1937, p 78-109

9   A. T. Little, G. V. Raynor and W. Hume-Rothery, *J Inst Metals,* Vol 69, 1943, p 423-440, 467-484

10 F. Laves, K. Löberg and H. Witte, *Metallwirtschaft,* Vol 14, 1935, p 793-794

11 G. Bergman, J. L. T. Waugh and L. Pauling, *Nature,* Vol 169, 1952, p 1057-1058

12 G. Bergman, J. L. T. Waugh and L. Pauling, *Acta Cryst,* Vol 10, 1957, p 254

13 G. V. Raynor and W. Hume-Rothery, *Trans Faraday Soc,* Vol 44, 1948, p 29-36

14 D. J. Strawbridge, W. Hume-Rothery and A. T. Little, *J Inst Metals,* Vol 74, 1947-1948, p 191-225

15 E. Butchers and W. Hume-Rothery, *J Inst Metals,* Vol 71, 1945, p 291-311

Pb-Sb-Sn (Lead-Antimony-Tin)

By L. Brewer and S-G. Chang

The diagrams for the Pb-Sb-Sn system are mainly from Iwasé and Aoki (1), corrected to agree with the binary diagrams in this volume. Kogan and Semionov (2) have presented additional data for lead-rich alloys with up to 30 wt % Sb and 20 wt % Sn for temperatures up to 380 °C (716 °F). Their results are shown in vertical sections at 4 and 14 wt % Sn, and by the diagram of the liquidus surface at the lead-rich corner, which is their revision of the corner reported by Weaver (3).

The fields of primary crystallization are: (Sb), the antimony-rich solution; SbSn solution; (β-Sn), the tin-rich solution; and (Pb), the lead-rich solution. There are two invariant points in the system, as shown in the diagram of the liquidus surface: a ternary eutectic at X, where (Sb), SbSn and (Pb) crystallize simultaneously, and a peritectic point at S, where SbSn reacts with liquid to form (β-Sn) + (Pb). Along the line PX, the peritectic reaction of (Sb) with liquid to form SbSn takes place. Another peritectic reaction takes place along the line RS, where SbSn reacts with liquid to form (β-Sn). Eutectic separation of (Pb) and (Sb) takes place along line EX; of (Pb) and SbSn, along line XS; and of (Pb) and (β-Sn), along line SW. A maximum in line XS, the eutectic line of (Pb) and SbSn, at point M is coincident with the intersection of line XS with the pseudobinary line from Pb to SbSn.

Summarizing, the reactions that take place at the invariant points and along the boundary lines are:

X.... $L_x \rightleftarrows$ (Sb) + SbSn + (Pb)	EX... $L \rightleftarrows$ (Pb) + (Sb)
S.... L_s + SbSn $\rightleftarrows$ (Pb) + (β-Sn)	XS .. $L \rightleftarrows$ (Pb) + SbSn
PX .. L + (Sb) $\rightleftarrows$ SbSn	SW .. $L \rightleftarrows$ (Pb) + (β-Sn)
RS .. L + SbSn $\rightleftarrows$ (β-Sn)	

The locations of the invariant points X and S, and point M, as found by the various investigators, are:

Point	Investigator	Composition, wt % Pb	Sb	Sn	Temperature °C	°F
X (eutectic)	Iwasé, Aoki (1)	85.0	11.5	3.5	240.0	464
	Weaver (3)	84.0	12.0	4.0	239.0	462
	Kogan, Semionov (2)	84.3	12.0	3.7	...	...
S (peritectic) ...	Loebe (4)	36.5	3.0	60.5	191.0	376
	Campbell, Elder (5)	40.0	2.5	57.5	189.0	372
	Campbell (6)	40.0	2.5	57.5	189.0	372
	Iwasé, Aoki (1)	40.0	2.5	57.5	189.0	372
	Heyn, Bauer (7)	42.5	4.0	53.5	184.0	363
M (pseudobinary eutectic)	Iwasé, Aoki (1)	80.0	10.0	10.0	245.0	473
	Weaver (3)	80.0	10.0	10.0	246.5	475

Concerning the eutectic in the pseudobinary system, PbSbSn, Loebe (4) and Campbell (6) and earlier investigators took it to be a peritectic point and reported its composition as 80 wt % Pb, 10 wt % Sb, 10 wt % Sn. Iwasé and Aoki (1) first noticed the maximum in line XS at this composition, and Weaver (3), who checked their observation, pointed out the significance of the maximum, and constructed a diagram of the pseudobinary system from thermal data. Earlier investigators may have missed the ternary eutectic because of inadequate etching methods. Weaver used an electrolytic 10% HCl etchant, which clearly distinguished between the (Sb) and SbSn phases.

Solid solubilities of the four phases at the temperatures of the invariant points X and S were determined microscopically by Iwasé and Aoki (1), and they reported the locations of the two-phase and three-phase fields at these temperatures. Their determinations, shown in the isothermal sections at 240 °C (464 °F) and 189 °C (372 °F), were for alloys annealed ten days or more, and would not apply to alloys in the as-cast condition. A small triangular region of liquid extends from the peritectic point S to points on the binary Pb-Sn on either side of the eutectic point W. Horizontally across the diagram, the region is bounded by the two-phase regions (β-Sn) + L and (Pb) + L, which extend on either side of the liquid region to the single-phase regions at the lead-rich and tin-rich corners.

Weaver (3) has given data on the phase fields in the lead corner for as-cast alloys. For as-cast alloys, the boundary between the (Sb) + SbSn + (Pb) field and the SbSn + (Pb) field is closer to the pseudobinary line than it is for annealed alloys.

Additional information on the Pb-Sb-Sn system is given in Ref 8, 9 and 10.

REFERENCES

1   K. Iwasé and N. Aoki, *Kinzoku-no-Kenkyu,* Vol 8, 1931, p 253

2   V. A. Kogan and A. A. Semionov, *Zh Fiz Khim,* Vol 37(4), 1963, p 802

3   F. D. Weaver, *J Inst Metals,* Vol 56, 1935, p 209

4 R. Loebe, *Metallurgie,* Vol 8, 1911, p 7, 33
5 W. Campbell and F. C. Elder, *School Mines Quart,* Columbia Univ., Vol 32, 1911, p 244
6 W. Campbell, *Metallurgie,* Vol 9, 1912, p 422
7 E. Heyn and O. Bauer, *Mitt Materialprüfungsamt Berlin-Dahlem,* Vol 29, 1911, p 29

8 R. A. Morgen, L. G. Swenson, F. C. Nix and E. H. Roberts, *Proc Inst Metals Div AIME, 1928,* p 345
9 O. Bauer and M. Hansen, *Z Metallk,* Vol 26, 1934, p 39
10 E. Hertel and A. Demmer, *Metallwirtschaft,* Vol 10, 1931, p 125

Nonferrous Ternary Alloy Phase Diagrams

Al-Mg-Zn (Aluminum-Magnesium-Zinc)

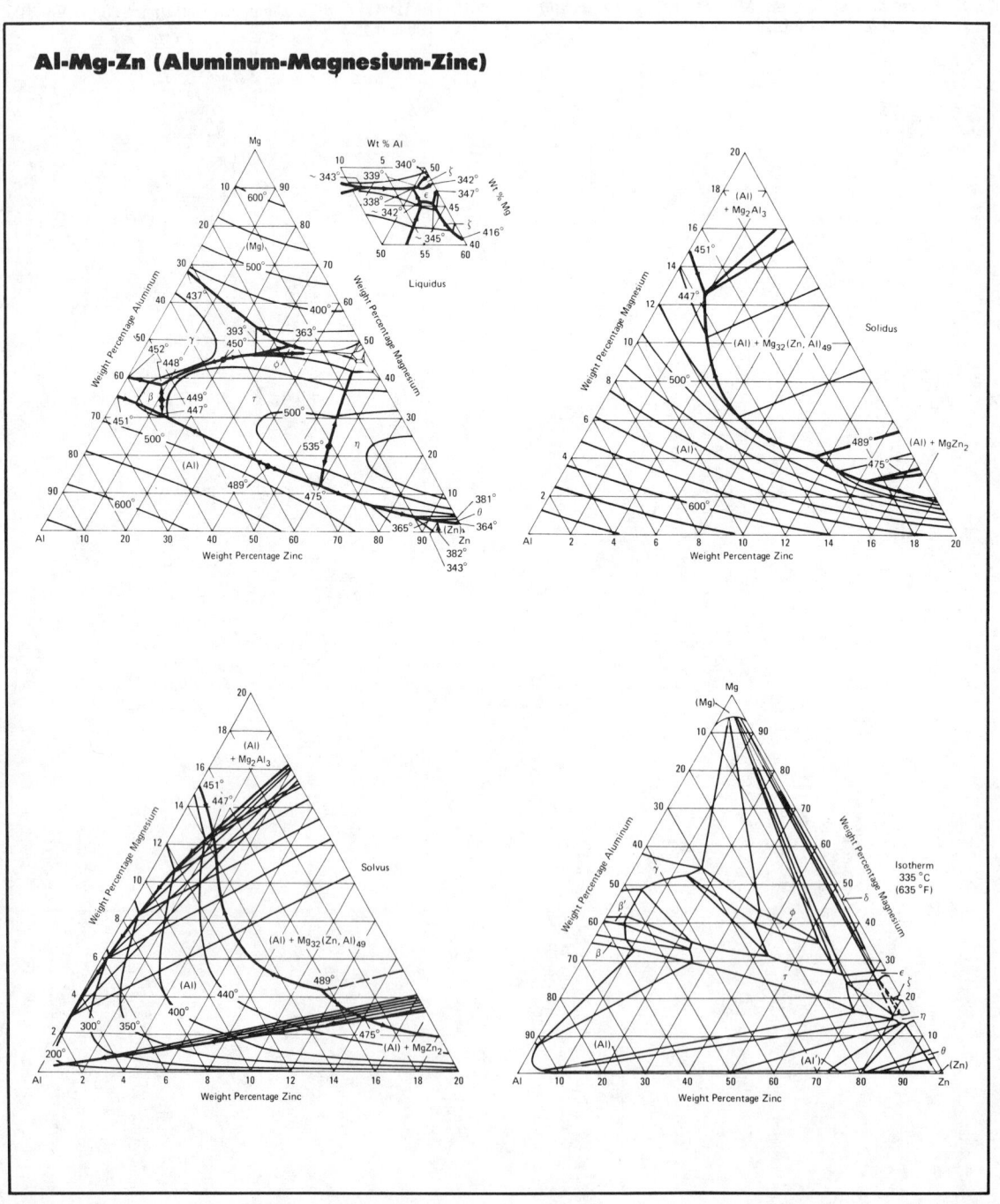

Pb-Sb-Sn (Lead-Antimony-Tin)

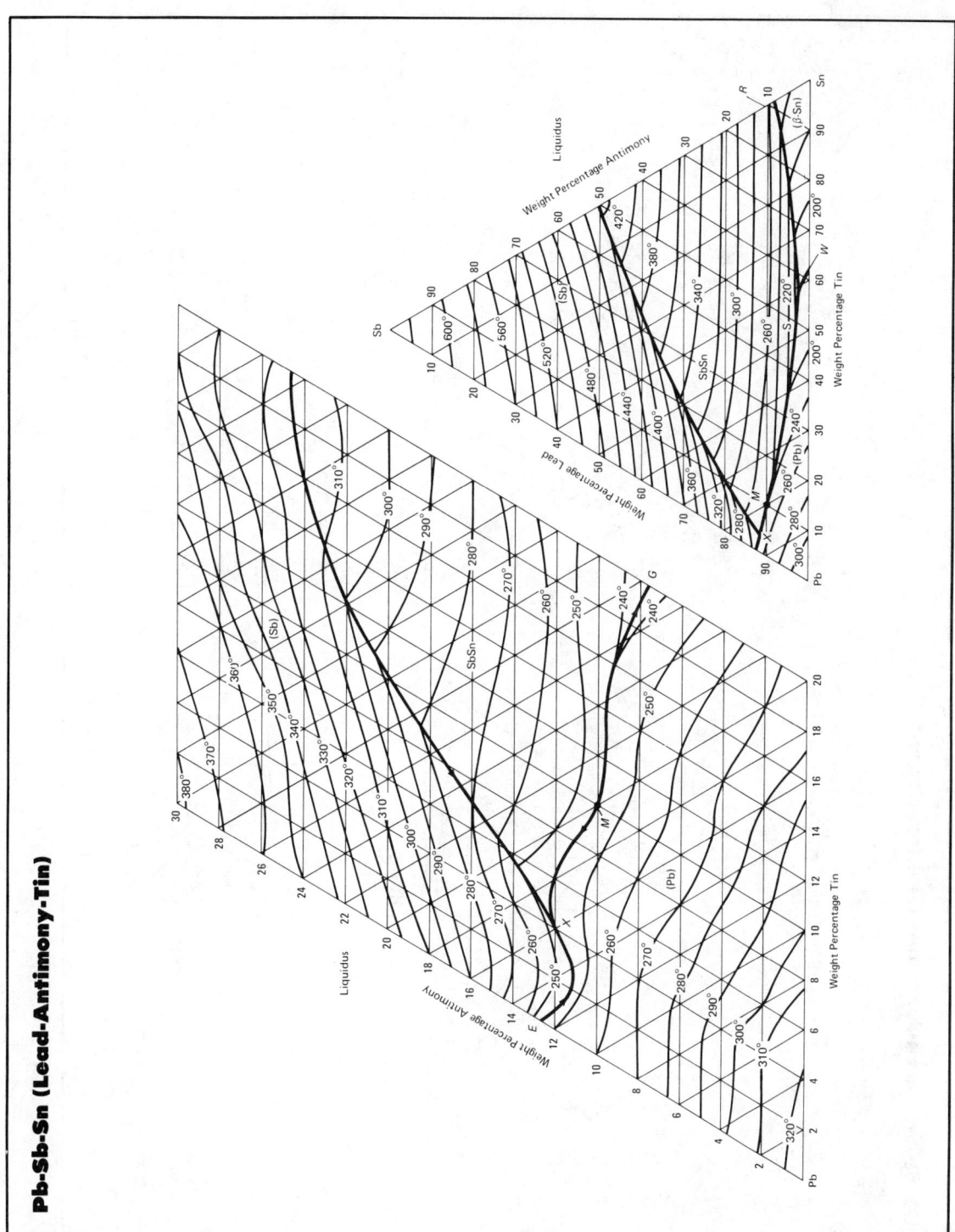

Pb-Sb-Sn (Lead-Antimony-Tin) (continued)

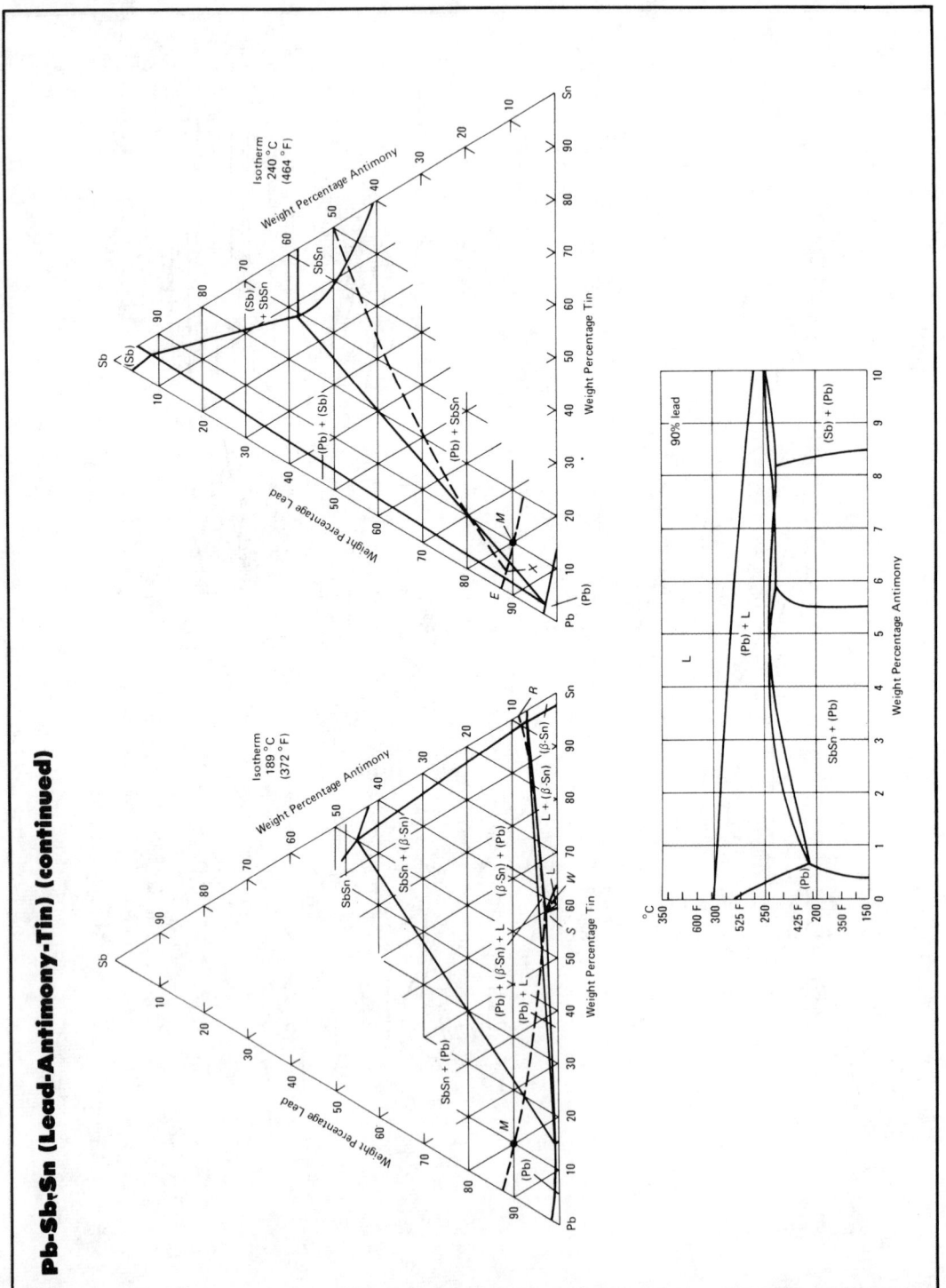

Pb-Sb-Sn (Lead-Antimony-Tin) (continued)

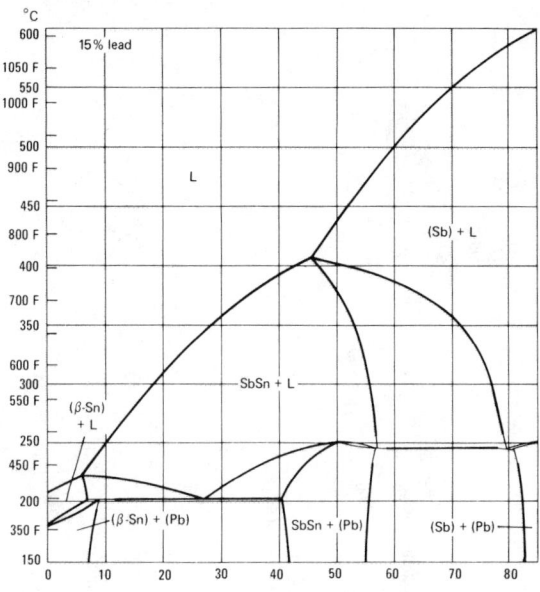

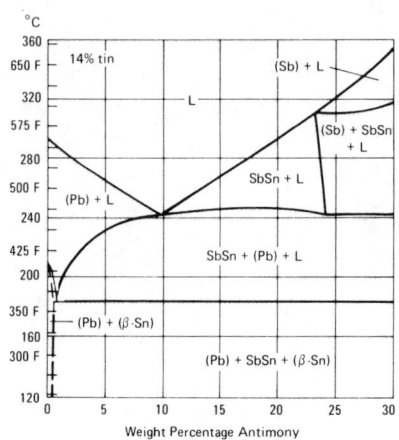

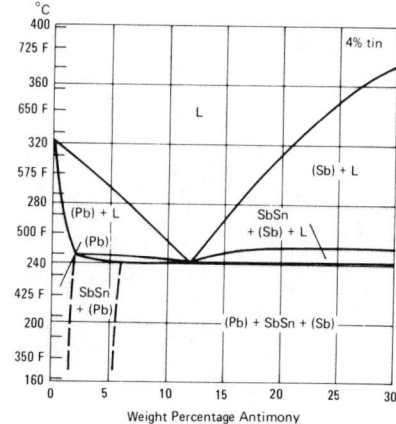

General Engineering Data

Conversion Factors and Tables

Metric Conversion

Units for converting from the English to the metric (SI) system

The Système Internationale d'Unités (SI) is built upon seven base units and two supplementary units. Derived units are related to base and supplementary units by formulas in the right-hand column of the table. Symbols for units with specific names are given in parentheses. The information is adapted from the *Standard for Metric Practice*, ASTM E 380.

Quantity	Unit	Formula
Base SI units		
Length	metre (m)	. . .
Mass	kilogram (kg)	. . .
Time	second (s)	. . .
Electric current	ampere (A)	. . .
Thermodynamic temperature	kelvin (K)	. . .
Amount of substance	mole (mol)	. . .
Luminous intensity	candela (cd)	. . .
Supplementary SI units		
Plane angle	radian (rad)	. . .
Solid angle	steradian (sr)	. . .
Derived SI units with special names		
Frequency (of a periodic phenomenon)	hertz (Hz)	$1/s$
Force	newton (N)	$kg \cdot m/s^2$
Pressure, stress	pascal (Pa)	N/m^2
Energy, work, quantity of heat	joule (J)	$N \cdot m$
Power, radiant flux	watt (W)	J/s
Quantity of electricity, electric charge	coulomb (C)	$A \cdot s$
Electric potential, potential difference, electromotive force	volt (V)	W/A
Electric capacitance	farad (F)	C/V
Electric resistance	ohm (Ω)	V/A
Conductance	siemens (S)	A/V
Magnetic flux	weber (Wb)	$V \cdot s$
Magnetic flux density	tesla (T)	Wb/m^2
Inductance	henry (H)	Wb/A
Celsius temperature	degree Celsius (°C)	K
Luminous flux	lumen (lm)	$cd \cdot sr$
Illuminance	lux (lx)	lm/m^2
Activity (of a radionuclide)	becquerel (Bq)	$1/s$
Absorbed dose	gray (Gy)	J/kg
Dose equivalent	sievert (Sv)	J/kg

Units for converting from the English to the metric (SI) system (continued)

Quantity	Unit
Some common derived units of SI	
Absorbed dose rate	gray per second (Gy/s)
Acceleration	metre per second squared (m/s^2)
Angular acceleration	radian per second squared (rad/s^2)
Angular velocity	radian per second (rad/s)
Area	square metre (m^2)
Concentration (of amount)	mole per cubic metre (mol/m^3)
Current density	ampere per square metre (A/m^2)
Density, mass	kilogram per cubic metre (kg/m^3)
Electric charge density	coulomb per cubic metre (C/m^3)
Electric field strength	volt per metre (V/m)
Electric flux density	coulomb per square metre (C/m^2)
Energy density	joule per cubic metre (J/m^3)
Entropy	joule per kelvin (J/K)
Exposure (X and gamma rays)	coulomb per kilogram (C/kg)
Heat capacity	joule per kelvin (J/K)
Heat flux density, irradiance	watt per square metre (W/m^2)
Luminance	candela per square metre (cd/m^2)
Magnetic field strength	ampere per metre (A/m)
Molar energy	joule per mole (J/mol)
Molar entropy	joule per mole kelvin (J/[mol · K])
Molar heat capacity	joule per mole kelvin (J/[mol · K])
Moment of force[A]	newton metre (N · m)
Permeability (magnetic)	henry per metre (H/m)
Permittivity	farad per metre (F/m)
Power density	watt per square metre (W/m^2)
Radiance	watt per square metre steradian (W/[m^2 · sr])
Radiant intensity	watt per steradian (W/sr)
Specific heat capacity	joule per kilogram kelvin (J/[kg · K])
Specific energy	joule per kilogram (J/kg)
Specific entropy	joule per kilogram kelvin (J/[kg · K])
Specific volume	cubic metre per kilogram (m^3/kg)
Surface tension	newton per metre (N/m)
Thermal conductivity	watt per metre kelvin (W/[m · K])
Velocity	metre per second (m/s)
Viscosity, dynamic	pascal second (Pa · s)
Viscosity, kinematic	square metre per second (m^2/s)
Volume	cubic metre (m^3)
Wave number	1 per metre (1/m)

Multiplication factor	Prefix	Symbol
SI prefixes		
1 000 000 000 000 000 000 = 10^{18}	exa	E
1 000 000 000 000 000 = 10^{15}	peta	P
1 000 000 000 000 = 10^{12}	tera	T
1 000 000 000 = 10^{9}	giga	G
1 000 000 = 10^{6}	mega	M
1 000 = 10^{3}	kilo	k
100 = 10^{2}	hecto(a)	h
10 = 10^{1}	deka(a)	da
0.1 = 10^{-1}	deci(a)	d
0.01 = 10^{-2}	centi(a)	c
0.001 = 10^{-3}	milli	m
0.000 001 = 10^{-6}	micro	μ
0.000 000 001 = 10^{-9}	nano	n
0.000 000 000 001 = 10^{-12}	pico	p
0.000 000 000 000 001 = 10^{-15}	femto	f
0.000 000 000 000 000 001 = 10^{-18}	atto	a

(a) To be avoided where practical

Metric conversion factors

To convert from	To	Multiply by	To convert from	To	Multiply by
angstrom	m	1.0000×10^{-10}(a)	hp(e)	W	7.4570×10^2
atm	Pa	1.0133×10^5	hp(f)	W	7.4600×10^2
Btu(b)	J	1.054×10^3	in.	m	2.5400×10^{-2}
Btu(b)/ft²·h	W/m²	3.1525	in.²	m²	6.4516×10^{-4}
Btu(b)/ft²·h·°F	W/m²·K	5.6745	in.³	m³	1.6387×10^{-5}
Btu(b)·ft/h·ft²·°F	W/m·K	1.7296	in. of Hg(g)	Pa	3.3864×10^3
Btu(b)/ft²·s	W/m²	1.135×10^4	in. of water(c)	Pa	2.4908×10^2
Btu(b)·in./ft²·h·°F	W/m·K	1.4413×10^{-1}	K	°C	$t_{°C} = t_K - 273.15$
Btu(b)·in./s·ft²·°F	W/m·K	5.1887×10^2	kgf	N	9.80665(a)
Btu(b)/lbm·°F	J/kg·K	4.1840×10^3	kgf/mm²	Pa	9.80665×10^6(a)
cal(b)	J	4.1840 (a)	ksi	MPa	6.8948
cal(b)/cm·s·°C	W/m·K	4.1840×10^2(a)	ksi	Pa	6.8948×10^6
cal(b)/g	J/kg	4.1840×10^3(a)	ksi $\sqrt{\text{in.}}$	MPa$\sqrt{m}$	1.089
cal(b)/g·°C	J/kg·K	4.1840×10^3(a)	lb(h)	kg	4.5359×10^{-1}
circ mil	m²	5.0671×10^{-10}	lb/in.³	kg/m³	2.7680×10^4
°C	K	$t_K = t_{°C} + 273.15$	lbf	N	4.4482
degree	rad	1.7453×10^{-2}	lbf·in.	N·m	1.1298×10^{-1}
dyne/cm²	Pa	1.0000×10^{-1}(a)	lbf·ft	N·m	1.3558
°F	°C	$t_{°C} = (t_{°F} - 32)/1.8$	MPa$\sqrt{m}$	MNm$^{-3/2}$	1.0000(a)
°F	K	$t_K = (t_{°F} + 459.67)/1.8$	μin.	m	2.5400×10^8(a)
ft	m	3.0480×10^{-1}	mil	m	2.5400×10^{-5}(a)
ft²	m²	9.2903×10^{-2}	N/m²	Pa	1.0000(a)
ft³	m³	2.8317×10^{-2}	oersted	A/m	79.578
ft of water(c)	Pa	2.9890×10^3	oz/ft²	kg/m²	3.0515×10^{-1}
ft²/h (thermal diffusivity)	m²/s	2.58064×10^{-5}(a)	psi	Pa	6.8948×10^3
ft·lbf	J	1.3558	°R	K	$t_K = t_R/1.8$
ft·lbf/s	W	1.3558	ton(j)	kg	9.0718×10^2
ft/s	m/s	3.0480×10^{-1}	ton(k)	kg	1.0160×10^3
gauss	T	1.0000×10^{-4}(a)	ton/in.²	Pa	1.3786×10^4
gallon(d)	m³	3.7854×10^{-3}	tonne	kg	1.0000×10^3(a)
g/cm³	kg/m³	1.0000×10^3(a)	torr	Pa	1.3332×10^2
g/cm³	Mg/m³	1.0000(a)	Ω/circ mil·ft	Ω·m	1.6624×10^{-9}

(a) Exactly. (b) Thermochemical. (c) At 4 °C (39.2 °F). (d) U.S. liquid. (e) Mechanical (1 hp = 550 ft·lbf/s). (f) Electrical. (g) At 0 °C (32 °F). (g) Avoirdupois. (j) Short; equal to 2000 lbm. (k) Long; 2240 lbm.

Temperature conversions

The general arrangement of this table was devised by Sauveur and Boylston more than 40 years ago. The middle column of figures (in bold-faced type) contains the reading (°F or °C) to be converted. If converting from degrees Fahrenheit to degrees Centigrade, read the Centigrade equivalent in the column headed "°C". If converting from Centigrade to Fahrenheit, read the Fahrenheit equivalent in the column headed "°F". $°C = \frac{5}{9}(°F - 32)$

°F		°C	°F		°C	°F		°C	°F		°C	°F		°C
···	−458	−272.22	···	−358	−216.67	−432.4	−258	−161.11	−252.4	−158	−105.56	−72.4	−58	−50.00
···	−456	−271.11	···	−356	−215.56	−428.8	−256	−160.00	−248.8	−156	−104.44	−68.8	−56	−48.89
···	−454	−270.00	···	−354	−214.44	−425.2	−254	−158.89	−245.2	−154	−103.33	−65.2	−54	−47.78
···	−452	−268.89	···	−352	−213.33	−421.6	−252	−157.78	−241.6	−152	−102.22	−61.6	−52	−46.67
···	−450	−267.78	···	−350	−212.22	−418.0	−250	−156.67	−238.0	−150	−101.11	−58.0	−50	−45.56
···	−448	−266.67	···	−348	−211.11	−414.4	−248	−155.56	−234.4	−148	−100.00	−54.4	−48	−44.44
···	−446	−265.56	···	−346	−210.00	−410.8	−246	−154.44	−230.8	−146	−98.89	−50.8	−46	−43.33
···	−444	−264.44	···	−344	−208.89	−407.2	−244	−153.33	−227.2	−144	−97.78	−47.2	−44	−42.22
···	−442	−263.33	···	−342	−207.78	−403.6	−242	−152.22	−223.6	−142	−96.67	−43.6	−42	−41.11
···	−440	−262.22	···	−340	−206.67	−400.0	−240	−151.11	−220.0	−140	−95.56	−40.0	−40	−40.00
···	−438	−261.11	···	−338	−205.56	−396.4	−238	−150.00	−216.4	−138	−94.44	−36.4	−38	−38.89
···	−436	−260.00	···	−336	−204.44	−392.8	−236	−148.89	−212.8	−136	−93.33	−32.8	−36	−37.78
···	−434	−258.89	···	−334	−203.33	−389.2	−234	−147.78	−209.2	−134	−92.22	−29.2	−34	−36.67
···	−432	−257.78	···	−332	−202.22	−385.6	−232	−146.67	−205.6	−132	−91.11	−25.6	−32	−35.56
···	−430	−256.67	···	−330	−201.11	−382.0	−230	−145.56	−202.0	−130	−90.00	−22.0	−30	−34.44
···	−428	−255.56	···	−328	−200.00	−378.4	−228	−144.44	−198.4	−128	−88.89	−18.4	−28	−33.33
···	−426	−254.44	···	−326	−198.89	−374.8	−226	−143.33	−194.8	−126	−87.78	−14.8	−26	−32.22
···	−424	−253.33	···	−324	−197.78	−371.2	−224	−142.22	−191.2	−124	−86.67	−11.2	−24	−31.11
···	−422	−252.22	···	−322	−196.67	−367.6	−222	−141.11	−187.6	−122	−85.56	−7.6	−22	−30.00
···	−420	−251.11	···	−320	−195.56	−364.0	−220	−140.00	−184.0	−120	−84.44	−4.0	−20	−28.89
···	−418	−250.00	···	−318	−194.44	−360.4	−218	−138.89	−180.4	−118	−83.33	−0.4	−18	−27.78
···	−416	−248.89	···	−316	−193.33	−356.8	−216	−137.78	−176.8	−116	−82.22	+3.2	−16	−26.67
···	−414	−247.78	···	−314	−192.22	−353.2	−214	−136.67	−173.2	−114	−81.11	+6.8	−14	−25.56
···	−412	−246.67	···	−312	−191.11	−349.6	−212	−135.56	−169.6	−112	−80.00	+10.4	−12	−24.44
···	−410	−245.56	···	−310	−190.00	−346.0	−210	−134.44	−166.0	−110	−78.89	+14.0	−10	−23.33
···	−408	−244.44	···	−308	−188.89	−342.4	−208	−133.33	−162.4	−108	−77.78	+17.6	−8	−22.22
···	−406	−243.33	···	−306	−187.78	−338.8	−206	−132.22	−158.8	−106	−76.67	+21.2	−6	−21.11
···	−404	−242.22	···	−304	−186.67	−335.2	−204	−131.11	−155.2	−104	−75.56	+24.8	−4	−20.00
···	−402	−241.11	···	−302	−185.56	−331.6	−202	−130.00	−151.6	−102	−74.44	+28.4	−2	−18.89
···	−400	−240.00	···	−300	−184.44	−328.0	−200	−128.89	−148.0	−100	−73.33	+32.0	±0	−17.78
···	−398	−238.89	···	−298	−183.33	−324.4	−198	−127.78	−144.4	−98	−72.22	+35.6	+2	−16.67
···	−396	−237.78	···	−296	−182.22	−320.8	−196	−126.67	−140.8	−96	−71.11	+39.2	+4	−15.56
···	−394	−236.67	···	−294	−181.11	−317.2	−194	−125.56	−137.2	−94	−70.00	+42.8	+6	−14.44
···	−392	−235.56	···	−292	−180.00	−313.6	−192	−124.44	−133.6	−92	−68.89	+46.4	+8	−13.33
···	−390	−234.44	···	−290	−178.89	−310.0	−190	−123.33	−130.0	−90	−67.78	+50.0	+10	−12.22
···	−388	−233.33	···	−288	−177.78	−306.4	−188	−122.22	−126.4	−88	−66.67	+53.6	+12	−11.11
···	−386	−232.22	···	−286	−176.67	−302.8	−186	−121.11	−122.8	−86	−65.56	+57.2	+14	−10.00
···	−384	−231.11	···	−284	−175.56	−299.2	−184	−120.00	−119.2	−84	−64.44	+60.8	+16	−8.89
···	−382	−230.00	···	−282	−174.44	−295.6	−182	−118.89	−115.6	−82	−63.33	+64.4	+18	−7.78
···	−380	−228.89	···	−280	−173.33	−292.0	−180	−117.78	−112.0	−80	−62.22	+68.0	+20	−6.67
···	−378	−227.78	···	−278	−172.22	−288.4	−178	−116.67	−108.4	−78	−61.11	+71.6	+22	−5.56
···	−376	−226.67	···	−276	−171.11	−284.8	−176	−115.56	−104.8	−76	−60.00	+75.2	+24	−4.44
···	−374	−225.56	···	−274	−170.00	−281.2	−174	−114.44	−101.2	−74	−58.89	+78.8	+26	−3.33
···	−372	−224.44	−457.6	−272	−168.89	−277.6	−172	−113.33	−97.6	−72	−57.78	+82.4	+28	−2.22
···	−370	−223.33	−454.0	−270	−167.78	−274.0	−170	−112.22	−94.0	−70	−56.67	+86.0	+30	−1.11
···	−368	−222.22	−450.4	−268	−166.67	−270.4	−168	−111.11	−90.4	−68	−55.56	+89.6	+32	±0.00
···	−366	−221.11	−446.8	−266	−165.56	−266.8	−166	−110.00	−86.8	−66	−54.44	+93.2	+34	+1.11
···	−364	−220.00	−443.2	−264	−164.44	−263.2	−164	−108.89	−83.2	−64	−53.33	+96.8	+36	+2.22
···	−362	−218.89	−439.6	−262	−163.33	−259.6	−162	−107.78	−79.6	−62	−52.22	+100.4	+38	+3.33
···	−360	−217.78	−436.0	−260	−162.22	−256.0	−160	−106.67	−76.0	−60	−51.11	+104.0	+40	+4.44

Temperature conversions (continued)

°F		°C	°F		°C	°F		°C	°F		°C	°F		°C
107.6	42	5.56	305.6	152	66.67	503.6	262	127.78	701.6	372	188.89	899.6	482	250.00
111.2	44	6.67	309.2	154	67.78	507.2	264	128.89	705.2	374	190.00	903.2	484	251.11
114.8	46	7.78	312.8	156	68.89	510.8	266	130.00	708.8	376	191.11	906.8	486	252.22
118.4	48	8.89	316.4	158	70.00	514.4	268	131.11	712.4	378	192.22	910.4	488	253.33
122.0	50	10.00	320.0	160	71.11	518.0	270	132.22	716.0	380	193.33	914.0	490	254.44
125.6	52	11.11	323.6	162	72.22	521.6	272	133.33	719.6	382	194.44	917.6	492	255.56
129.2	54	12.12	327.2	164	73.33	525.2	274	134.44	723.2	384	195.56	921.2	494	256.67
132.8	56	13.33	330.8	166	74.44	528.8	276	135.56	726.8	386	196.67	924.8	496	257.78
136.4	58	14.44	334.4	168	75.56	532.4	278	136.67	730.4	388	197.78	928.4	498	258.89
140.0	60	15.56	338.0	170	76.67	536.0	280	137.78	734.0	390	198.89	932.0	500	260.00
143.6	62	16.67	341.6	172	77.78	539.6	282	138.89	737.6	392	200.00	935.6	502	261.11
147.2	64	17.78	345.2	174	78.89	543.2	284	140.00	741.2	394	201.11	939.2	504	262.22
150.8	66	18.89	348.8	176	80.00	546.8	286	141.11	744.8	396	202.22	942.8	506	263.33
154.4	68	20.00	352.4	178	81.11	550.4	288	142.22	748.4	398	203.33	946.4	508	264.44
158.0	70	21.11	356.0	180	82.22	554.0	290	143.33	752.0	400	204.44	950.0	510	265.56
161.6	72	22.22	359.6	182	83.33	557.6	292	144.44	755.6	402	205.56	953.6	512	266.67
165.2	74	23.33	363.2	184	84.44	561.2	294	145.56	759.2	404	206.67	957.2	514	267.78
168.8	76	24.44	366.8	186	85.56	564.8	296	146.67	762.8	406	207.78	960.8	516	268.89
172.4	78	25.56	370.4	188	86.67	568.4	298	147.78	766.4	408	208.89	964.4	518	270.00
176.0	80	26.67	374.0	190	87.78	572.0	300	148.89	770.0	410	210.00	968.0	520	271.11
179.6	82	27.78	377.6	192	88.89	575.6	302	150.00	773.6	412	211.11	971.6	522	272.22
183.2	84	28.89	381.2	194	90.00	579.2	304	151.11	777.2	414	212.22	975.2	524	273.33
186.8	86	30.00	384.8	196	91.11	582.8	306	152.22	780.8	416	213.33	978.8	526	274.44
190.4	88	31.11	388.4	198	92.22	586.4	308	153.33	784.4	418	214.44	982.4	528	275.56
194.0	90	32.22	392.0	200	93.33	590.0	310	154.44	788.0	420	215.56	986.0	530	276.67
197.6	92	33.33	395.6	202	94.44	593.6	312	155.56	791.6	422	216.67	989.6	532	277.78
201.2	94	34.44	399.2	204	95.56	597.2	314	156.67	795.2	424	217.78	993.2	534	278.89
204.8	96	35.56	402.8	206	96.67	600.8	316	157.78	798.8	426	218.89	996.8	536	280.00
208.4	98	36.67	406.4	208	97.78	604.4	318	158.89	802.4	428	220.00	1000.4	538	281.11
212.0	100	37.78	410.0	210	98.89	608.0	320	160.00	806.0	430	221.11	1004.0	540	282.22
215.6	102	38.89	413.6	212	100.00	611.6	322	161.11	809.6	432	222.22	1007.6	542	283.33
219.2	104	40.00	417.2	214	101.11	615.2	324	162.22	813.2	434	223.33	1011.2	544	284.44
222.8	106	41.11	420.8	216	102.22	618.8	326	163.33	816.8	436	224.44	1014.8	546	285.56
226.4	108	42.22	424.4	218	103.33	622.4	328	164.44	820.4	438	225.56	1018.4	548	286.67
230.0	110	43.33	428.0	220	104.44	626.0	330	165.56	824.0	440	226.67	1022.0	550	287.78
233.6	112	44.44	431.6	222	105.56	629.6	332	166.67	827.6	442	227.78	1040.0	560	293.33
237.2	114	45.56	435.2	224	106.67	633.2	334	167.78	831.2	444	228.89	1058.0	570	298.89
240.8	116	46.67	438.8	226	107.78	636.8	336	168.89	834.8	446	230.00	1076.0	580	304.44
244.4	118	47.78	442.4	228	108.89	640.4	338	170.00	838.4	448	231.11	1094.0	590	310.00
248.0	120	48.89	446.0	230	110.00	644.0	340	171.11	842.0	450	232.22	1112.0	600	315.56
251.6	122	50.00	449.6	232	111.11	647.6	342	172.22	845.6	452	233.33	1130.0	610	321.11
255.2	124	51.11	453.2	234	112.22	651.2	344	173.33	849.2	454	234.44	1148.0	620	326.67
258.8	126	52.22	456.8	236	113.33	654.8	346	174.44	852.8	456	235.56	1166.0	630	332.22
262.4	128	53.33	460.4	238	114.44	658.4	348	175.56	856.4	458	236.67	1184.0	640	337.78
266.0	130	54.44	464.0	240	115.56	662.0	350	176.67	860.0	460	237.78	1202.0	650	343.33
269.6	132	55.56	467.6	242	116.67	665.6	352	177.78	863.6	462	238.89	1220.0	660	348.89
273.2	134	56.67	471.2	244	117.78	669.2	354	178.89	867.2	464	240.00	1238.0	670	354.44
276.8	136	57.78	474.8	246	118.89	672.8	356	180.00	870.8	466	241.11	1256.0	680	360.00
280.4	138	58.89	478.4	248	120.00	676.4	358	181.11	874.4	468	242.22	1274.0	690	365.56
284.0	140	60.00	482.0	250	121.11	680.0	360	182.22	878.0	470	243.33	1292.0	700	371.11
287.6	142	61.11	485.6	252	122.22	683.6	362	183.33	881.6	472	244.44	1310.0	710	376.67
291.2	144	62.22	489.2	254	123.33	687.2	364	184.44	885.2	474	245.56	1328.0	720	382.22
294.8	146	63.33	492.8	256	124.44	690.8	366	185.56	888.8	476	246.67	1346.0	730	387.78
298.4	148	64.44	496.4	258	125.56	694.4	368	186.67	892.4	478	247.78	1364.0	740	393.33
302.0	150	65.56	500.0	260	126.67	698.0	370	187.78	896.0	480	248.89	1382.0	750	398.89

Temperature conversions (continued)

°F		°C	°F		°C	°F		°C	°F		°C	°F		°C
1400.0	760	404.44	2390.0	1310	710.00	3380.0	1860	1015.6	4370.0	2410	1321.1	5450.0	3010	1654.4
1418.0	770	410.00	2408.0	1320	715.56	3398.0	1870	1021.1	4388.0	2420	1326.7	5468.0	3020	1660.0
1436.0	780	415.56	2426.0	1330	721.11	3416.0	1880	1026.7	4406.0	2430	1332.2	5486.0	3030	1665.6
1454.0	790	421.11	2440.0	1340	726.67	3434.0	1890	1032.2	4424.0	2440	1337.8	5504.0	3040	1671.1
1472.0	800	426.67	2462.0	1350	732.22	3452.0	1900	1037.8	4442.0	2450	1343.3	5522.0	3050	1676.7
									4460.0	2460	1348.9	5540.0	3060	1682.2
1490.0	810	432.22	2480.0	1360	737.78	3470.0	1910	1043.3	4478.0	2470	1354.4	5558.0	3070	1687.8
1508.0	820	437.76	2498.0	1370	743.33	3488.0	1920	1048.9	4496.0	2480	1360.0	5576.0	3080	1693.3
1526.0	830	443.33	2516.0	1380	748.89	3506.0	1930	1054.4	4514.0	2490	1365.6	5594.0	3090	1698.9
1544.0	840	448.89	2534.0	1390	754.44	3524.0	1940	1060.0	4532.0	2500	1371.1	5612.0	3100	1704.4
1562.0	850	454.44	2552.0	1400	760.00	3542.0	1950	1065.6	4550.0	2510	1376.7	5702.0	3150	1732.2
1580.0	860	460.00	2570.0	1410	765.56	3560.0	1960	1071.1	4568.0	2520	1382.2	5792.0	3200	1760.0
1598.0	870	465.56	2588.0	1420	771.11	3578.0	1970	1076.7	4586.0	2530	1387.8	5882.0	3250	1787.7
1616.0	880	471.11	2606.0	1430	776.67	3596.0	1980	1082.2	4604.0	2540	1393.3	5972.0	3300	1815.5
1634.0	890	476.67	2624.0	1440	782.22	3614.0	1990	1087.8	4622.0	2550	1398.9	6062.0	3350	1843.3
1652.0	900	482.22	2642.0	1450	787.78	3632.0	2000	1093.3	4640.0	2560	1404.4	6152.0	3400	1871.1
									4658.0	2570	1410.0	6242.0	3450	1898.8
1670.0	910	487.78	2660.0	1460	793.33	3650.0	2010	1098.9	4676.0	2580	1415.6	6332.0	3500	1926.6
1688.0	920	493.33	2678.0	1470	798.89	3668.0	2020	1104.4	4694.0	2590	1421.1	6422.0	3550	1954.4
1706.0	930	498.89	2696.0	1480	804.44	3686.0	2030	1110.0	4712.0	2600	1426.7	6512.0	3600	1982.2
1724.0	940	504.44	2714.0	1490	810.00	3704.0	2040	1115.6	4730.0	2610	1432.2	6602.0	3650	2010.0
1742.0	950	510.00	2732.0	1500	815.56	3722.0	2050	1121.1	4748.0	2620	1437.8	6692.0	3700	2037.7
1760.0	960	515.56	2750.0	1510	821.11	3740.0	2060	1126.7	4766.0	2630	1443.3	6782.0	3750	2065.5
1778.0	970	521.11	2768.0	1520	826.67	3758.0	2070	1132.2	4784.0	2640	1448.9	6872.0	3800	2093.3
1796.0	980	526.67	2786.0	1530	832.22	3776.0	2080	1137.8	4802.0	2650	1454.4	6962.0	3850	2121.1
1814.0	990	532.22	2804.0	1540	837.78	3794.0	2090	1143.3	4820.0	2660	1460.0	7052.0	3900	2148.8
1832.0	1000	537.78	2822.0	1550	843.33	3812.0	2100	1148.9	4838.0	2670	1465.6	7142.0	3950	2176.6
									4856.0	2680	1471.1	7232.0	4000	2204.4
1850.0	1010	543.33	2840.0	1560	848.89	3830.0	2110	1154.4	4874.0	2690	1476.7	7322.0	4050	2232.2
1868.0	1020	548.89	2858.0	1570	854.44	3848.0	2120	1160.0	4892.0	2700	1482.2	7412.0	4100	2260.0
1886.0	1030	554.44	2876.0	1580	860.00	3866.0	2130	1165.6	4910.0	2710	1487.8	7502.0	4150	2287.7
1904.0	1040	560.00	2894.0	1590	865.56	3884.0	2140	1171.1	4928.0	2720	1493.3	7592.0	4200	2315.5
1922.0	1050	565.56	2912.0	1600	871.11	3902.0	2150	1176.7	4946.0	2730	1498.9	7682.0	4250	2343.3
1940.0	1060	571.11	2930.0	1610	876.67	3920.0	2160	1182.2	4964.0	2740	1504.4	7772.0	4300	2371.1
1958.0	1070	576.67	2948.0	1620	882.22	3938.0	2170	1187.8	4982.0	2750	1510.0	7862.0	4350	2398.8
1976.0	1080	582.22	2966.0	1630	887.78	3956.0	2180	1193.3	5000.0	2760	1515.6	7952.0	4400	2426.6
1994.0	1090	587.78	2984.0	1640	893.33	3974.0	2190	1198.9	5018.0	2770	1521.1	8042.0	4450	2454.4
2012.0	1100	593.33	3002.0	1650	898.89	3992.0	2200	1204.4	5036.0	2780	1526.7	8132.0	4500	2482.2
									5054.0	2790	1532.2	8222.0	4550	2510.0
2030.0	1110	598.89	3020.0	1660	904.44	4010.0	2210	1210.0	5072.0	2800	1537.8	8312.0	4600	2537.7
2048.0	1120	604.44	3038.0	1670	910.00	4028.0	2220	1215.6	5090.0	2810	1543.3	8402.0	4650	2565.5
2066.0	1130	610.00	3056.0	1680	915.56	4046.0	2230	1221.1	5108.0	2820	1548.9	8492.0	4700	2593.3
2084.0	1140	615.56	3074.0	1690	921.11	4064.0	2240	1226.7	5126.0	2830	1554.4	8582.0	4750	2621.1
2102.0	1150	621.11	3092.0	1700	926.67	4082.0	2250	1232.2	5144.0	2840	1560.0	8672.0	4800	2648.8
2120.0	1160	626.67	3110.0	1710	932.22	4100.0	2260	1237.8	5162.0	2850	1565.6	8762.0	4850	2676.6
2138.0	1170	632.22	3128.0	1720	937.78	4118.0	2270	1243.3	5180.0	2860	1571.1	8852.0	4900	2704.4
2156.0	1180	637.78	3146.0	1730	943.33	4136.0	2280	1248.9	5198.0	2870	1576.7	8942.0	4950	2732.2
2174.0	1190	643.33	3164.0	1740	948.89	4154.0	2290	1254.4	5216.0	2880	1582.2	9032.0	5000	2760.0
2192.0	1200	648.89	3182.0	1750	954.44	4172.0	2300	1260.0	5234.0	2890	1587.8	9122.0	5050	2787.7
									5252.0	2900	1593.3	9212.0	5100	2815.5
2210.0	1210	654.44	3200.0	1760	960.00	4190.0	2310	1265.6	5270.0	2910	1598.9	9302.0	5150	2843.3
2228.0	1220	660.00	3218.0	1770	965.56	4208.0	2320	1271.1	5288.0	2920	1604.4	9392.0	5200	2871.1
2246.0	1230	665.56	3236.0	1780	971.11	4226.0	2330	1276.7	5306.0	2930	1610.0	9482.0	5250	2898.8
2264.0	1240	671.11	3254.0	1790	976.67	4244.0	2340	1282.2	5324.0	2940	1615.6	9572.0	5300	2926.6
2282.0	1250	676.67	3272.0	1800	982.22	4262.0	2350	1287.8	5342.0	2950	1621.1	9662.0	5350	2954.4
2300.0	1260	682.22	3290.0	1810	987.78	4280.0	2360	1293.3	5360.0	2960	1626.7	9752.0	5400	2982.2
2318.0	1270	687.78	3308.0	1820	993.33	4298.0	2370	1298.9	5378.0	2970	1632.2	9842.0	5450	3010.0
2336.0	1280	693.33	3326.0	1830	998.89	4316.0	2380	1304.4	5396.0	2980	1637.8	9932.0	5500	3037.7
2354.0	1290	698.89	3344.0	1840	1004.4	4334.0	2390	1310.0	5414.0	2990	1643.3	10 022.0	5550	3065.5
2372.0	1300	704.44	3362.0	1850	1010.0	4352.0	2400	1315.6	5432.0	3000	1648.9	10 112.0	5600	3093.3

Metric stress or pressure conversions

The middle column of figures (in bold-faced type) contains the reading (in MPa or ksi) to be converted. If converting from ksi to MPa, read the MPa equivalent in the column headed "MPa". If converting from MPa to ksi, read the ksi equivalent in the column headed "ksi". 1 ksi = 6.894757 MPa. 1 psi = 6.894757 kPa.

ksi		MPa	ksi		MPa	ksi		MPa	ksi		MPa
0.14504	1	6.895	8.2672	57	393.00	33.359	230	1585.8	114.58	790	...
0.29008	2	13.790	8.4122	58	399.90	34.809	240	1654.7	116.03	800	...
0.43511	3	20.684	8.5572	59	406.79	36.259	250	1723.7	117.48	810	...
0.58015	4	27.579	8.7023	60	413.69	37.710	260	1792.6	118.93	820	...
0.72519	5	34.474	8.8473	61	420.58	39.160	270	1861.6	120.38	830	...
0.87023	6	41.369	8.9923	62	427.47	40.611	280	1930.5	121.83	840	...
1.0153	7	48.263	9.1374	63	434.37	42.061	290	1999.5	123.28	850	...
1.1603	8	55.158	9.2824	64	441.26	43.511	300	2068.4	124.73	860	...
1.3053	9	62.053	9.4275	65	448.16	44.962	310	2137.4	126.18	870	...
1.4504	10	68.948	9.5725	66	455.05	46.412	320	2206.3	127.63	880	...
1.5954	11	75.842	9.7175	67	461.95	47.862	330	2275.3	129.08	890	...
1.7405	12	82.737	9.8626	68	468.84	49.313	340	2344.2	130.53	900	...
1.8855	13	89.632	10.008	69	475.74	50.763	350	2413.2	131.98	910	...
2.0305	14	96.527	10.153	70	482.63	52.214	360	2482.1	133.43	920	...
2.1756	15	103.42	10.298	71	489.53	53.664	370	2551.1	134.89	930	...
2.3206	16	110.32	10.443	72	496.42	55.114	380	2620.0	136.34	940	...
2.4656	17	117.21	10.588	73	503.32	56.565	390	2689.0	137.79	950	...
2.6107	18	124.11	10.733	74	510.21	58.015	400	2757.9	139.24	960	...
2.7557	19	131.00	10.878	75	517.11	59.465	410	2826.9	140.69	970	...
2.9008	20	137.90	11.023	76	524.00	60.916	420	2895.8	142.14	980	...
3.0458	21	144.79	11.168	77	530.90	62.366	430	2964.7	143.59	990	...
3.1908	22	151.68	11.313	78	537.79	63.817	440	3033.7	145.04	1000	...
3.3359	23	158.58	11.458	79	544.69	65.267	450	3102.6	147.94	1020	...
3.4809	24	165.47	11.603	80	551.58	66.717	460	3171.6	150.84	1040	...
3.6259	25	172.37	11.748	81	558.48	68.168	470	3240.5	153.74	1060	...
3.7710	26	179.26	11.893	82	565.37	69.618	480	3309.5	156.64	1080	...
3.9160	27	186.16	12.038	83	572.26	71.068	490	3378.4	159.54	1100	...
4.0611	28	193.05	12.183	84	579.16	72.519	500	3447.4	162.44	1120	...
4.2061	29	199.95	12.328	85	586.05	73.969	510	...	165.34	1140	...
4.3511	30	206.84	12.473	86	592.95	75.420	520	...	168.24	1160	...
4.4962	31	213.74	12.618	87	599.84	76.870	530	...	171.14	1180	...
4.6412	32	220.63	12.763	88	606.74	78.320	540	...	174.05	1200	...
4.7862	33	227.53	12.909	89	613.63	79.771	550	...	176.95	1220	...
4.9313	34	234.42	13.053	90	620.53	81.221	560	...	179.85	1240	...
5.0763	35	241.32	13.198	91	627.42	82.672	570	...	182.75	1260	...
5.2214	36	248.21	13.343	92	634.32	84.122	580	...	185.65	1280	...
5.3664	37	255.11	13.489	93	641.21	85.572	590	...	188.55	1300	...
5.5114	38	262.00	13.634	94	648.11	87.023	600	...	191.45	1320	...
5.6565	39	268.90	13.779	95	655.00	88.473	610	...	194.35	1340	...
5.8015	40	275.79	13.924	96	661.90	89.923	620	...	197.25	1360	...
5.9465	41	282.69	14.069	97	668.79	91.374	630	...	200.15	1380	...
6.0916	42	289.58	14.214	98	675.69	92.824	640	...	203.05	1400	...
6.2366	43	296.47	14.359	99	682.58	94.275	650	...	205.95	1420	...
6.3817	44	303.37	14.504	100	689.48	95.725	660	...	208.85	1440	...
6.5267	45	310.26	15.954	110	758.42	97.175	670	...	211.76	1460	...
6.6717	46	317.16	17.405	120	827.37	98.626	680	...	214.66	1480	...
6.8168	47	324.05	18.855	130	896.32	100.08	690	...	217.56	1500	...
6.9618	48	330.95	20.305	140	965.27	101.53	700	...	220.46	1520	...
7.1068	49	337.84	21.756	150	1034.2	102.98	710	...	223.36	1540	...
7.2519	50	344.74	23.206	160	1103.2	104.43	720	...	226.26	1560	...
7.3969	51	351.63	24.656	170	1172.1	105.88	730	...	229.16	1580	...
7.5420	52	358.53	26.107	180	1241.1	107.33	740	...	232.06	1600	...
7.6870	53	365.42	27.557	190	1310.0	108.78	750	...	234.96	1620	...
7.8320	54	372.32	29.008	200	1379.0	110.23	760	...	237.86	1640	...
7.9771	55	379.21	30.458	210	1447.9	111.68	770	...	240.76	1660	...
8.1221	56	386.11	31.908	220	1516.8	113.13	780	...	243.66	1680	...

Metric stress or pressure conversions (continued)

ksi	MPa	MPa	ksi	MPa	MPa	ksi	MPa	MPa	ksi	MPa	MPa
246.56	1700	...	278.47	1920	...	310.38	2140	...	342.29	2360	...
249.46	1720	...	281.37	1940	...	313.28	2160	...	345.19	2380	...
252.37	1740	...	284.27	1960	...	316.18	2180	...	348.09	2400	...
255.27	1760	...	287.17	1980	...	319.08	2200	...	350.99	2420	...
258.17	1780	...	290.08	2000	...	321.98	2220	...	353.89	2440	...
261.07	1800	...	292.98	2020	...	324.88	2240	...	356.79	2460	...
263.97	1820	...	295.88	2040	...	327.79	2260	...	359.69	2480	...
266.87	1840	...	298.78	2060	...	330.69	2280	...	362.59	2500	...
269.77	1860	...	301.68	2080	...	333.59	2300	...			
272.67	1880	...	304.58	2100	...	336.49	2320	...			
275.57	1900	...	307.48	2120	...	339.39	2340	...			

Metric stress-intensity conversions

The middle column of figures (in bold-faced type) contains the reading (in MPa√m or ksi√in.) to be converted. If converting from ksi√in. to MPa√m, read the MPa√m equivalent in the column headed "MPa√m". If converting from MPa√m to ksi√in., read the ksi√in. equivalent in the column headed "ksi√in.". 1 ksi√in. = 1.098845 MPa√m.

ksi,√in.		MPa,√m	ksi,√in.		MPa,√m	ksi,√in.		MPa,√m	ksi,√in.		MPa,√m	ksi,√in.		MPa,√m
0.91005	1	1.0988	37.312	41	45.051	73.714	81	89.003	110.12	121	132.95	146.52	161	176.91
1.8201	2	2.1976	38.222	42	46.150	74.624	82	90.102	111.03	122	134.05	147.43	162	178.01
2.7301	3	3.2964	39.132	43	47.248	75.534	83	91.200	111.94	123	135.15	148.34	163	179.10
3.6402	4	4.3952	40.042	44	48.347	76.444	84	92.300	112.85	124	136.25	149.25	164	180.20
4.5502	5	5.4940	40.952	45	49.446	77.354	85	93.398	113.76	125	137.35	150.16	165	181.30
5.4603	6	6.5928	41.862	46	50.545	78.264	86	94.497	114.67	126	138.45	151.07	166	182.40
6.3703	7	7.6916	42.772	47	51.644	79.174	87	95.596	115.58	127	139.55	151.98	167	183.50
7.2804	8	8.7904	43.682	48	52.742	80.084	88	96.694	116.49	128	140.65	152.89	168	184.60
8.1904	9	9.8892	44.592	49	53.841	80.994	89	97.793	117.40	129	141.75	153.80	169	185.70
9.1005	10	10.988	45.502	50	54.940	81.904	90	98.892	118.31	130	142.84	154.71	170	186.80
10.011	11	12.087	46.412	51	56.039	82.814	91	99.991	119.22	131	143.94	155.62	171	187.90
10.921	12	13.186	47.322	52	57.138	83.724	92	101.09	120.13	132	145.04	156.53	172	189.00
11.831	13	14.284	48.232	53	58.236	84.634	93	102.19	121.04	133	146.14	157.44	173	190.10
12.741	14	15.383	49.143	54	59.335	85.544	94	103.29	121.95	134	147.24	158.35	174	191.19
13.651	15	16.482	50.053	55	60.434	86.454	95	104.39	122.86	135	148.34	159.26	175	192.29
14.561	16	17.581	50.963	56	61.533	87.364	96	105.48	123.77	136	149.44	160.17	176	193.39
15.471	17	18.680	51.873	57	62.632	88.275	97	106.58	124.68	137	150.54	161.08	177	194.49
16.381	18	19.778	52.783	58	63.730	89.185	98	107.68	125.59	138	151.63	161.99	178	195.59
17.291	19	20.877	53.693	59	64.829	90.095	99	108.78	126.50	139	152.73	162.90	179	196.69
18.201	20	21.976	54.603	60	65.928	91.005	100	109.88	127.41	140	153.83	163.81	180	197.78
19.111	21	23.075	55.513	61	67.027	91.915	101	110.98	128.32	141	154.93	164.72	181	198.88
20.021	22	24.174	56.423	62	68.126	92.825	102	112.08	129.23	142	156.03	165.63	182	199.98
20.931	23	25.272	57.333	63	69.224	93.735	103	113.18	130.14	143	157.13	166.54	183	201.08
21.841	24	26.371	58.243	64	70.323	94.645	104	114.28	131.05	144	158.23	167.45	184	202.18
22.751	25	27.470	59.153	65	71.422	95.555	105	115.37	131.96	145	159.33	168.36	185	203.28
23.661	26	28.569	60.063	66	72.521	96.465	106	116.47	132.87	146	160.42	169.27	186	204.38
24.571	27	29.668	60.973	67	73.620	97.375	107	117.57	133.78	147	161.52	170.18	187	205.48
25.481	28	30.766	61.883	68	74.718	98.285	108	118.67	134.69	148	162.62	171.09	188	206.57
26.391	29	31.865	62.793	69	75.817	99.195	109	119.77	135.60	149	163.72	172.00	189	207.67
27.301	30	32.964	63.703	70	76.916	100.11	110	120.87	136.51	150	164.82	172.91	190	208.77
28.211	31	34.063	64.613	71	78.015	101.02	111	121.97	137.42	151	165.92	173.82	191	209.87
29.121	32	35.162	65.523	72	79.114	101.93	112	123.07	138.33	152	167.02	174.73	192	210.97
30.032	33	36.260	66.433	73	80.212	102.84	113	124.16	139.24	153	168.12	175.64	193	212.07
30.942	34	37.359	67.343	74	81.311	103.75	114	125.26	140.15	154	169.22	176.55	194	213.17
31.852	35	38.458	68.253	75	82.410	104.66	115	126.36	141.06	155	170.31	177.46	195	214.27
32.762	36	39.557	69.164	76	83.509	105.57	116	127.46	141.97	156	171.41	178.37	196	215.36
33.672	37	40.656	70.074	77	84.608	106.48	117	128.56	142.88	157	172.51	179.28	197	216.46
34.582	38	41.754	70.984	78	85.706	107.39	118	129.66	143.79	158	173.61	180.19	198	217.56
35.492	39	42.853	71.893	79	86.805	108.30	119	130.76	144.70	159	174.71	181.10	199	218.66
36.402	40	43.952	72.804	80	87.904	109.21	120	131.86	145.61	160	175.81	182.01	200	219.76

504

Metric energy conversions

The middle column of figures (in bold-faced type) contains the reading (in J or ft·lb) to be converted. If converting from ft·lb to J, read the J equivalent in the column headed "J". If converting from J to ft·lb, read the equivalent in the column headed "ft·lb". 1 ft·lb = 1.355818 J.

ft·lb		J	ft·lb		J	ft·lb		J	ft·lb		J
0.7376	1	1.3558	28.7649	39	52.8769	56.7923	77	104.3980	129.0734	175	237.2681
1.4751	2	2.7116	29.5025	40	54.2327	57.5298	78	105.7538	132.7612	180	244.0472
2.2127	3	4.0675	30.2400	41	55.5885	58.2674	79	107.1096	136.4490	185	250.8263
2.9502	4	5.4233	30.9776	42	56.9444	59.0050	80	108.4654	140.1368	190	257.6054
3.6878	5	6.7791	31.7152	43	58.3002	59.7425	81	109.8212	143.8246	195	264.3845
4.4254	6	8.1349	32.4527	44	59.6560	60.4801	82	111.1771	147.5124	200	271.1636
5.1629	7	9.4907	33.1903	45	61.0118	61.2177	83	112.5329	154.8880	210	284.7218
5.9005	8	10.8465	33.9279	46	62.3676	61.9552	84	113.8887	162.2637	220	298.2799
6.6381	9	12.2024	34.6654	47	63.7234	62.6928	85	115.2445	169.6393	230	311.8381
7.3756	10	13.5582	35.4030	48	65.0793	63.4303	86	116.6003	177.0149	240	325.3963
8.1132	11	14.9140	36.1405	49	66.4351	64.1679	87	117.9562	184.3905	250	338.9545
8.8507	12	16.2698	36.8781	50	67.7909	64.9055	88	119.3120	191.7661	260	352.5126
9.5883	13	17.6256	37.6157	51	69.1467	65.6430	89	120.6678	199.1418	270	366.0708
10.3259	14	18.9815	38.3532	52	70.5025	66.3806	90	122.0236	206.5174	280	379.6290
11.0634	15	20.3373	39.0908	53	71.8583	67.1182	91	123.3794	213.8930	290	393.1872
11.8010	16	21.6931	39.8284	54	73.2142	67.8557	92	124.7452	221.2686	300	406.7454
12.5386	17	23.0489	40.5659	55	74.5700	68.5933	93	126.0911	228.6442	310	420.3036
13.2761	18	24.4047	41.3035	56	75.9258	69.3308	94	127.4469	236.0199	320	433.8617
14.0137	19	25.7605	42.0410	57	77.2816	70.0684	95	128.8027	243.3955	330	447.4199
14.7512	20	27.1164	42.7786	58	78.6374	70.8060	96	130.1585	250.7711	340	460.9781
15.4888	21	28.4722	43.5162	59	79.9933	71.5435	97	131.5143	258.1467	350	474.5363
16.2264	22	29.8280	44.2537	60	81.3491	72.2811	98	132.8702	265.5224	360	488.0944
16.9639	23	31.1838	44.9913	61	82.7049	73.0186	99	134.2260	272.8980	370	501.6526
17.7015	24	32.5396	45.7288	62	84.0607	73.7562	100	135.5818	280.2736	380	515.2108
18.4390	25	33.8954	46.4664	63	85.4165	77.4440	105	142.3609	287.6492	390	528.7690
19.1766	26	35.2513	47.2040	64	86.7723	81.1318	110	149.1400	295.0248	400	542.3272
19.9142	27	36.6071	47.9415	65	88.1282	84.8196	115	155.9191	302.4005	410	555.8854
20.6517	28	37.9629	48.6791	66	89.4840	88.5075	120	162.6982	309.7761	420	569.4435
21.3893	29	39.3187	49.4167	67	90.8398	92.1953	125	169.4772	317.1517	430	583.0017
22.1269	30	40.6745	50.1542	68	92.1956	95.8831	130	176.2563	324.5273	440	596.5599
22.8644	31	42.0304	50.8918	69	93.5514	99.5709	135	183.0354	331.9029	450	610.1181
23.6020	32	43.3862	51.6293	70	94.9073	103.2587	140	189.8145	339.2786	460	623.6762
24.3395	33	44.7420	52.3669	71	96.2631	106.9465	145	196.5936	346.6542	470	637.2344
25.0771	34	46.0978	53.1045	72	97.6189	110.6343	150	203.3727	354.0298	480	650.7926
25.8147	35	47.4536	53.8420	73	98.9747	114.3221	155	210.1518	361.4054	490	664.3508
26.5522	36	48.8094	54.5796	74	100.3305	118.0099	160	216.9308	368.7811	500	677.9090
27.2898	37	50.1653	55.3172	75	101.6863	121.6977	165	223.7099			
28.0274	38	51.5211	56.0547	76	103.0422	125.3856	170	230.4890			

Conversion of inches to millimeters

Inches	Milli-meters	Inches	Milli-meters	Inches	Milli-meters
0.001	0.025	0.290	7.37	0.660	16.76
0.002	0.051	0.300	7.62	0.670	17.02
0.003	0.076	0.310	7.87	0.680	17.17
0.004	0.102	0.320	8.13	0.690	17.53
0.005	0.127	0.330	8.38	0.700	17.78
0.006	0.152	0.340	8.64	0.710	18.03
0.007	0.178	0.350	8.89	0.720	18.29
0.008	0.203	0.360	9.14	0.730	18.54
0.009	0.229	0.370	9.40	0.740	18.80
0.010	0.254	0.380	9.65	0.750	19.05
0.020	0.508	0.390	9.91	0.760	19.30
0.030	0.762	0.400	10.16	0.770	19.56
0.040	1.016	0.410	10.41	0.780	19.81
0.050	1.270	0.420	10.67	0.790	20.07
0.060	1.524	0.430	10.92	0.800	20.32
0.070	1.778	0.440	11.18	0.810	20.57
0.080	2.032	0.450	11.43	0.820	20.83
0.090	2.286	0.460	11.68	0.830	21.08
0.100	2.540	0.470	11.94	0.840	21.34
0.110	2.794	0.480	12.19	0.850	21.59
0.120	3.048	0.490	12.45	0.860	21.84
0.130	3.302	0.500	12.70	0.870	22.10
0.140	3.56	0.510	12.95	0.880	22.35
0.150	3.81	0.520	13.21	0.890	22.61
0.160	4.06	0.530	13.46	0.900	22.86
0.170	4.32	0.540	13.72	0.910	23.11
0.180	4.57	0.550	13.97	0.920	23.37
0.190	4.83	0.560	14.22	0.930	23.62
0.200	5.08	0.570	14.48	0.940	23.88
0.210	5.33	0.580	14.73	0.950	24.13
0.220	5.59	0.590	14.99	0.960	24.38
0.230	5.84	0.600	15.24	0.970	24.64
0.240	6.10	0.610	15.49	0.980	24.89
0.250	6.35	0.620	15.75	0.990	25.15
0.260	6.60	0.630	16.00	1.000	25.40
0.270	6.86	0.640	16.26	...	...
0.280	7.11	0.650	16.51	...	...

Conversion of millimeters to inches

Milli-meters	Inches	Milli-meters	Inches	Milli-meters	Inches
0.01	0.0004	0.35	0.0138	0.68	0.0268
0.02	0.0008	0.36	0.0142	0.69	0.0272
0.03	0.0012	0.37	0.0146	0.70	0.0276
0.04	0.0016	0.38	0.0150	0.71	0.0280
0.05	0.0020	0.39	0.0154	0.72	0.0283
0.06	0.0024	0.40	0.0157	0.73	0.0287
0.07	0.0028	0.41	0.0161	0.74	0.0291
0.08	0.0031	0.42	0.0165	0.75	0.0295
0.09	0.0035	0.43	0.0169	0.76	0.0299
0.10	0.0039	0.44	0.0173	0.77	0.0303
0.11	0.0043	0.45	0.0177	0.78	0.0307
0.12	0.0047	0.46	0.0181	0.79	0.0311
0.13	0.0051	0.47	0.0185	0.80	0.0315
0.14	0.0055	0.48	0.0189	0.81	0.0319
0.15	0.0059	0.49	0.0193	0.82	0.0323
0.16	0.0063	0.50	0.0197	0.83	0.0327
0.17	0.0067	0.51	0.0201	0.84	0.0331
0.18	0.0071	0.52	0.0205	0.85	0.0335
0.19	0.0075	0.53	0.0209	0.86	0.0339
0.20	0.0079	0.54	0.0213	0.87	0.0343
0.21	0.0083	0.55	0.0217	0.88	0.0346
0.22	0.0087	0.56	0.0220	0.89	0.0350
0.23	0.0091	0.57	0.0224	0.90	0.0354
0.24	0.0094	0.58	0.0228	0.91	0.0358
0.25	0.0098	0.59	0.0232	0.92	0.0362
0.26	0.0102	0.60	0.0236	0.93	0.0366
0.27	0.0106	0.61	0.0240	0.94	0.0370
0.28	0.0110	0.62	0.0244	0.95	0.0374
0.29	0.0114	0.63	0.0248	0.96	0.0378
0.30	0.0118	0.64	0.0252	0.97	0.0382
0.31	0.0122	0.65	0.0256	0.98	0.0386
0.32	0.0126	0.66	0.0260	0.99	0.0390
0.33	0.0130	0.67	0.0264	1.00	0.0394
0.34	0.0134	...	...	...	...

Metric length and weight conversion factors

Unit	Inches to millimeters	Millimeters to inches	Pounds to kilograms	Kilograms to pounds
1	25.400 1	0.039 371	0.453 59	2.204 62
2	50.800 1	0.078 742	0.907 19	4.409 24
3	76.200 2	0.118 112	1.360 78	6.613 86
4	101.600 2	0.157 483	1.814 37	8.818 49
5	127.000 3	0.196 854	2.267 96	11.023 11
6	152.400 3	0.236 225	2.721 56	13.227 73
7	177.800 4	0.275 596	3.175 15	15.432 35
8	203.200 4	0.314 966	3.628 74	17.636 97
9	228.600 5	0.354 337	4.082 33	19.841 59
10	254.000 6	0.393 708	4.355 92	22.046 22

Conversion factors and measurements

Equivalents

1 gram = 15 432 grains
1 meter = 39.371 inches or 3.28083 feet
1 millimeter = 0.03937 inch, or 1/25 in. approx
1 metric ton   ⎰ = 2204.6 pounds or
1000 kilograms ⎱ = 0.9842 ton or 2240 pounds
1.016 metric ton ⎰
1016 kilograms ⎱ = 1 ton of 2240 pounds
1 kilogram per sq centimeter = 14.2234 lb per sq in.
1 kilogram per sq millimeter = 1422.32 lb per sq in.
1000 lb per sq in. = ⎰ 0.70308 kilograms per sq mm
                              ⎱ 70.308 kilograms per sq cm

Linear measure

12 inches = 1 foot
3 feet = 1 yard = 36 inches
5½ yards = 1 rod or pole = 16½ feet
40 rods = 1 furlong = 220 yards = 660 feet = ⅛ mile
8 furlongs = 1 statute mile = 1760 yards = 5280 feet
3 miles = 1 league = 5280 yards = 15 840 feet

Square measure

144 square inches = 1 square foot
9 square feet = 1 square yard = 1296 square inches
30¼ square yards = 1 square rod = 272¼ square feet
160 square rods = 1 acre = 4840 square yards
640 acres = 1 square mile = 3 097 600 square yards

Cubic measure

1728 cubic inches = 1 cubic foot
27 cubic feet = 1 cubic yard
144 cubic inches = 1 board foot
128 cubic feet = 1 cord

Liquid measure

4 gills = 1 pint
2 pints = 1 quart = 8 gills
4 quarts = 1 gallon = 8 pints = 32 gills
31½ gallons = 1 barrel = 126 quarts
2 barrels = 1 hogshead = 63 gallons = 252 quarts

Nautical measure

6 feet = 1 fathom
100 fathoms = 1 cable's length (ordinary) = 608 ft (Br.)
   = 607.61 ft (U.S.)
120 fathoms = 1 cable's length (U.S. Navy)
10 cable's lengths = 1 nautical mile = 6080 ft (Br.)
   = 6076.1033 ft (U.S.)
1 nautical mile = 1.1508 statute miles
3 nautical miles = 1 league (marine)
60 nautical miles = 1 degree (of a terrestial great circle)

Advoirdupois weight

27¹¹⁄₃₂ grains = 1 dram
16 drams = 1 ounce = 437½ grains
16 ounces = 1 pound = 256 drams = 7000 grains
100 pounds = 1 hundredweight = 1600 ounces
20 hundredweight = 1 ton = 2000 pounds
112 pounds = 1 long hundredweight
20 long hundredweight = 1 long ton = 2240 pounds

Troy weight

24 grains = 1 pennyweight
20 pennyweights = 1 ounce = 480 grains
12 ounces = 1 pound = 240 pennyweights = 5760 grains

Apothecaries' weight

20 grains = 1 scruple
3 scruples = 1 dram = 60 grains
8 drams = 1 ounce = 24 scruples = 430 grains
12 ounces = 1 pound = 96 drams = 283 scruples
   = 5760 grains

Dry measure

2 pints = 1 quart
8 quarts = 1 peck = 16 pints
4 pecks = 1 bushel = 32 quarts = 64 pints
105 quarts = 1 barrel (for fruits, vegetables, and other dry
   commodities) = 7056 cubic inches

Circular measure

60 seconds (″) = 1 minute (′)
60 minutes = 1 degree (°)
90 degrees = 1 quadrant
4 quadrants = 1 circle of circumference

Roman numerals

1................I		8......................VIII
2................II		9......................IX
3................III		10....................X
4................IV		50....................L
5................V		100..................C
6................VI		500..................D
7................VII		1000.................M

The chief symbols are I = 1; V = 5; X = 10; L = 50; C = 100; D = 500; and M = 1000. Note that IV = 4, means 1 short of 5; IX = 9, means 1 short of 10; XL = 40, means 10 short of 50; and XC = 90, means 10 short of 100. Any symbol following one of equal or greater value adds its value—11 = 2. Any symbol preceding one of greater value subtracts its value—IV = 4. When a symbol stands between two of greater value, its value is subtracted from the second and the remainder is added to the first—XIV = 14; LIX = 59. Of two equivalent ways of representing a number, that in which the symbol of larger denomination preceded is preferred—XIV instead of VIX for 14.

Numerical data

1 cubic foot of water at 4 °C (weight)................62.43 lb	
1 foot of water at 4 °C (pressure)............... 0.4335 lb/in.^{3}	
Velocity of light in vacuum, c................ 186,280 mi/s = 2.998 × 10^{10} cm/s	
Velocity of sound in dry air at 20 °C, 76 cm Hg.......1127 ft/s	
Degree of longitude at equator................... 69.173 miles	
Acceleration due to gravity at sea-level, 40′ Latitude, g ....32.1578 ft/s^2	
$\sqrt{2g}$ 8.020	
Base of natural logs ε 2.718	
1 radian...................................180° ÷ π = 57.3	
360 degrees......................................2π radians	
π 3.1416	
Sine 1′... 0.00029089	
Arc 1°... 0.01745 radian	
Side of square ....0.707 × (diagonal of square)	

Mathematical symbols

× or ·	Multiplied by
÷ or :	Divided by
+	Positive. Plus. Add
−	Negative. Minus. Subtract
±	Positive or negative. Plus or minus
∓	Negative or positive. Minus or plus
= or ∷	Equals
□	Identity
≅	Approximately equal to
≠	Not equal to
>	Greater than
≫	Much greater than
<	Less than
≪	Much less than
≧	Greater than or equal to
≦	Less than or equal to
∴	Therefore
∠	Angle
△	Increment. Decrement
⊥	Perpendicular to
‖	Parallel to
$|n|$	Absolute value of n

a, b, c used for known quantities
x, y, z used for unknown quantities

Mathematical constants

$\pi =$	3.14	$\sqrt{\pi} = 1.77$	
$2\pi =$	6.28	$\sqrt{\dfrac{\pi}{2}} = 1.25$	
$(2\pi)^2 =$	39.5	$\sqrt{2} = 1.41$	
$4\pi =$	12.6	$\sqrt{3} = 1.73$	
$\pi^2 =$	9.87	$\dfrac{1}{\sqrt{2}} = 0.707$	
$\dfrac{\pi}{2} =$	1.57	$\dfrac{1}{\sqrt{3}} = 0.577$	
$\dfrac{1}{\pi} =$	0.318	$\log \pi = 0.497$	
$\dfrac{1}{2\pi} =$	0.159	$\log \dfrac{\pi}{2} = 0.196$	
$\dfrac{1}{\pi^2} =$	0.101	$\log \pi^2 = 0.994$	
$\dfrac{1}{\sqrt{\pi}} =$	0.564	$\log \sqrt{\pi} = 0.248$	

Greek alphabet

Capital	Small	Name	Commonly used to designate
A	α	Alpha	Angles, coefficients, attenuation constant, absorption factor, area
B	β	Beta	Angles, coefficients, phase constant
Γ	γ	Gamma	Complex propagation constant (cap), specific gravity, angles, electrical conductivity, propagation constant
Δ	δ	Delta	Increment or decrement (cap or small), determinant (cap), permittivity (cap), density, angles
E	ε	Epsilon	Dielectric constant, permittivity, base of natural logarithms, electric intensity
Z	ζ	Zeta	Coordinates, coefficients
H	η	Eta	Intrinsic impedance, efficiency, surface charge density, hysteresis, coordinates
Θ	θ	Theta	Angular phase displacement, time constant, reluctance, angles
I	ι	Iota	Unit vector
K	κ	Kappa	Susceptibility, coupling coefficient
Λ	λ	Lambda	Permeance (cap), wavelength, attenuation constant
M	μ	Mu	Permeability, amplification factor, prefix micro
N	ν	Nu	Reluctivity, frequency
Ξ	ξ	Xi	Coordinates
O	o	Omicron	
Π	π	Pi	3.1416
P	ρ	Rho	Resistivity, volume charge density, coordinates
Σ	σ	Sigma	Summation (cap), surface charge density, complex propagation constant, electrical conductivity, leakage coefficient
T	τ	Tau	Time constant, volume resistivity, time-phase displacement, transmission factor, density
Υ	υ	Upsilon	
Φ	φ	Phi	Scalar potential (cap), magnetic flux, angles
X	χ	Chi	Electric susceptibility, angles
Ψ	ψ	Psi	Dielectric flux, phase difference, coordinates, angles
Ω	ω	Omega	Resistance in ohms (cap), solid angle (cap), angular velocity

Note: Use small letter except where capital (cap) is specified.

Miscellaneous Electrical Information

Fusing currents of wires

The current I in amperes at which a wire will melt can be calculated from $I = Kd^2$ where d is the wire diameter in inches and K is a constant that depends on the metal concerned; a wide variety of factors influence the rate of heat loss and these figures must be considered as approximations

AWG B & S gage	Wire diameter (d), in.	Fusing current, A, for wire type:				
		Copper, K = 10 244	Aluminum, K = 7585	German silver, K = 5230	Iron, K = 3148	Tin, K = 1642
40 . . . . .0.0031		1.77	1.31	0.90	0.54	0.28
38 . . . . .0.0039		2.50	1.85	1.27	0.77	0.40
36 . . . . .0.0050		3.62	2.68	1.85	1.11	0.58
34 . . . . .0.0063		5.12	3.79	2.61	1.57	0.82
32 . . . . .0.0079		7.19	5.32	3.67	2.21	1.15
30 . . . . .0.0100		10.2	7.58	5.23	3.15	1.64
28 . . . . .0.0126		14.4	10.7	7.39	4.45	2.32
26 . . . . .0.0159		20.5	15.2	10.5	6.31	3.29
24 . . . . .0.0201		29.2	21.6	14.9	8.97	4.68
22 . . . . .0.0253		41.2	30.5	21.0	12.7	6.61
20 . . . . .0.0319		58.4	43.2	29.8	17.9	9.36
19 . . . . .0.0359		69.7	51.6	35.5	21.4	11.2
18 . . . . .0.0403		82.9	61.4	42.3	25.5	13.3
17 . . . . .0.0452		98.4	72.9	50.2	30.2	15.8
16 . . . . .0.0508		117.0	86.8	59.9	36.0	18.8
15 . . . . .0.0571		140.0	103.0	71.4	43.0	22.4
14 . . . . .0.0641		166.0	123.0	84.9	51.1	26.6
13 . . . . .0.0719		197.0	146.0	101.0	60.7	31.7
12 . . . . .0.0808		235.0	174.0	120.0	72.3	37.7
11 . . . . .0.0907		280.0	207.0	143.0	86.0	44.9
10 . . . . .0.1019		333.0	247.0	170.0	102.0	53.4
9 . . . . .0.1144		396.0	298.0	202.0	122.0	63.5
8 . . . . .0.1285		472.0	349.0	241.0	145.0	75.6
7 . . . . .0.1443		561.0	416.0	287.0	173.0	90.0
6 . . . . .0.1620		668.0	495.0	341.0	205.0	107.0

Electrical formulae

$$\text{Resistance} = \frac{\text{voltage}}{\text{ampere turns}} \times \text{turns} \left(\Omega = \frac{V}{IT} \times T\text{'s} \right)$$

$$\text{Ampere turns} = \frac{\text{voltage}}{\text{resistance}} \times \text{turns} \left(IT = \frac{V}{\Omega} \times T\text{'s} \right)$$

$$\text{Effective turns} = \frac{\text{total resistance}}{\text{resistance of inductive coil}} \times \text{turns of inductive coil}$$

$$\text{Amperes} = \frac{\text{ampere turns}}{\text{turns of inductive coil}} \left(I = \frac{IT}{\Omega} \right)$$

Ohm's law for direct current

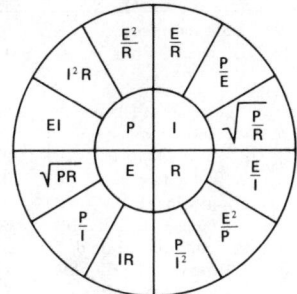

P = power in watts
I = current in amperes
E = electromotive force in volts
R = resistance in ohms

Two resistances in parallel combination:

$$Req = \frac{R_1 \; R_2}{R_1 + R_2}$$

Any number of resistances in parallel combination:

$$\frac{1}{Req} = \frac{1}{r_1} + \frac{1}{r_2} + \cdots \frac{1}{r_n}$$

For calculating capacitances in series combinations, substitute C for R in the above formulas

Ohm's law for alternating current

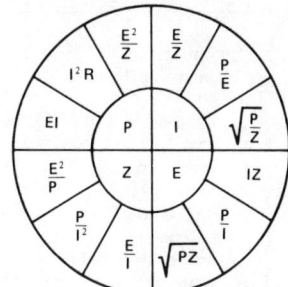

$$f = \frac{1}{2\pi \sqrt{LC}} = \frac{1}{2\pi CXc} = \frac{XL}{2\pi L}$$

$$XL = 2\pi fL$$

$$Xc = \frac{1}{2\pi fC}$$

$$L = \frac{XL}{2\pi f} = \frac{1}{(2\pi f)^2 C}$$

$$C = \frac{1}{2\pi fXc} = \frac{1}{(2\pi f)^2 L}$$

$$Z = \sqrt{R^2 + X^2} = \sqrt{R^2 + (XL - Xc)^2}$$

$Z = R$ when $XL = Xc$

Z = impedance in ohms
XL = inductive reactance in ohms
Xc = capacitive reactance in ohms
L = inductance in henrys
C = capacitance in farads
f = frequency in cycles per second
$2\pi f \approx 377$ for 60 cps

Miscellaneous Factors and Tables

Physical constants

Name and symbol	Value and units
Velocity of light, c	2.997902×10^{10} cm/s
Planck constant, h	6.62377×10^{-27} erg s/molecule
Avogadro constant, N	6.02380×10^{23} molecule mol
Faraday constant, F	96 493.1 C/equivalent
Absolute temperature of ice point, T (0 °C)	273.15 K
Pressure-volume product for 1 mol of gas at 0 °C and zero pressure (PV) $P = 0$ $T = 0$ °C	2271.16 J/mol
Gas constant $R = \dfrac{(PV) \begin{smallmatrix} P = 0 \\ T = 0\ °C \end{smallmatrix}}{T(0\ °C)}$	8.31469 J/mol°
	1.98726 cal/mol°
Boltzmann constant $k = R/N$	1.38031×10^{16} erg/molecule°
Constant relating wave number and energy $Z = Nhc$	11.96171 Jcm/mol
	2.858917 cal cm/mole
Standard atmosphere, atm	1 013 250 dynes/cm^2
Thermochemical calorie	4.1840 J (exact)
	4.18331 J (international)

Miscellaneous conversion factors

To convert from	To	Multiply by
Atmospheres	cm of Hg at 0 °C	76
Atmospheres	gm/cm^2	1033.3
Atmospheres	inches of Hg at 32 °F	29.921
Atmospheres	lb/in^2	14.696
Btu	calories (gram)	252
Calories, gram	Btu	3.968×10^{-3}
Centimeters	angstrom units	1×10^8
Centimeters	feet	0.032808
Cubic cm	cubic inches	0.061023
Cubic cm	gallons (U.S.)	2.6417×10^{-4}
Cubic cm	ounces (U.S. fluid)	0.033814
Cubic cm	pints (U.S. fluid)	0.0021134
Cubic ft	cubic cm	28317
Cubic ft	cubic meters	0.02832
Cubic ft	gallons (U.S.)	7.481
Cubic ft	liters	28.316
Cubic in. (U.S.)	cubic cm	16.3872
Cubic in. (U.S.)	liters	0.016387
Cubic yd (U.S.)	cubic meters	0.7646
Cubic yd of sand	pounds	2700
Feet (U.S.)	centimeters	30.48
Feet (U.S.)	meters	0.3048
Gallons (U.S.)	cubic centimeters	3785.4
Gallons (U.S.)	cubic feet	0.13368
Gallons (U.S.)	cubic inches	231
Gallons (U.S.)	liters	3.7854
Grams	ounces (avoir.)	0.03527
Grams	pounds (avoir.)	0.002205
Horsepower	Btu (mean)/min	42.418
Horsepower	calories, log (mean)/min	10.688
Inches (U.S.)	angstrom units	2.5400×10^8
Inches	mils	1000

To convert from	To	Multiply by
Inches of Hg at 32 °C	atmospheres	0.033421
Inches of Hg at 32 °C	feet of water at 39.1 °F	1.13299
Inches of water at 39.2 °F	inches of Hg	0.073554
Kilograms	ounces (avoir.)	35.274
Kilograms	pounds (avoir.)	2.2046
Liters	cubic feet	0.035316
Liters	gallons (U.S.)	0.2642
Liters	ounces (U.S. fluid)	33.8143
Liters	pints (U.S. liquid)	2.11336
Liters	quarts (U.S. liquid)	1.05668
Meters	angstrom units	1×10^{10}
Meters	feet (U.S.)	3.28083
Meters	inches (U.S.)	39.3700
Microns	angstrom units	1×10^4
Microns	inches	3.937×10^{-5}
Microns	millimeters	0.001
Microns	mils	0.03937
Millimeters	inches (U.S.)	0.03937
Millimeters	microns	1000
Millimeters	mils	39.37
Ounces (avoir.)	grams	28.3495
Ounces (U.S. fluid)	cubic cm	29.5737
Ounces (U.S. fluid)	cubic in.	1.8047
Ounces (U.S. fluid)	liters	0.02957
Pints (U.S. liquid)	cubic cm	473.179
Pints (U.S. liquid)	liters	0.473168
Pounds (avoir.)	grams	453.5924
Square cm	square in.	0.1550
Square in. (U.S.)	square cm	6.5416
Years (leap)	hours	8784

Load conversion table, tsi to psi

tsi	psi	tsi	psi	tsi	psi	tsi	psi	tsi	psi	tsi	psi
10.0	22 400	22.5	50 400	35.0	78 400	47.5	106 400	70	156 800	95	212 800
10.5	23 520	23.0	51 520	35.5	79 520	48.0	107 520	71	159 040	96	215 040
11.0	24 640	23.5	52 640	36.0	80 640	48.5	108 640	72	161 280	97	217 280
11.5	25 760	24.0	53 760	36.5	81 760	49.0	109 760	73	163 520	98	219 520
12.0	26 880	24.5	54 880	37.0	82 880	49.5	110 880	74	165 760	99	221 760
12.5	28 000	25.0	56 000	37.5	84 000	50	112 000	75	168 000	100	224 000
13.0	29 120	25.5	57 120	38.0	85 120	51	114 240	76	170 240	101	226 240
13.5	30 240	26.0	58 240	38.5	86 240	52	116 480	77	172 480	102	228 480
14.0	31 360	26.5	59 360	39.0	87 360	53	118 720	78	174 720	103	230 720
14.5	32 480	27.0	60 480	39.5	88 480	54	120 960	79	176 960	104	232 960
15.0	33 600	27.5	61 600	40.0	89 600	55	123 200	80	179 200	105	235 200
15.5	34 720	28.0	62 720	40.5	90 720	56	125 440	81	181 440	106	237 440
16.0	35 840	28.5	63 840	41.0	91 840	57	127 680	82	183 680	107	239 680
16.5	36 960	29.0	64 960	41.5	92 960	58	129 920	83	185 920	108	241 920
17.0	38 080	29.5	66 080	42.0	94 080	59	132 160	84	188 160	109	244 160
17.5	39 200	30.0	67 200	42.5	95 200	60	134 400	85	190 400	110	246 400
18.0	40 320	30.5	68 320	43.0	96 320	61	136 640	86	192 640	111	248 640
18.5	41 440	31.0	69 440	43.5	97 440	62	138 880	87	194 880	112	250 880
19.0	42 560	31.5	70 560	44.0	98 560	63	141 120	88	197 120	113	253 120
19.5	43 680	32.0	71 680	44.5	99 680	64	143 360	89	199 360	114	255 360
20.0	44 800	32.5	72 800	45.0	100 800	65	145 600	90	201 600	115	257 600
20.5	45 920	33.0	73 920	45.5	101 920	66	147 840	91	203 840	116	259 840
21.0	47 040	33.5	75 040	46.0	103 040	67	150 080	92	206 080	117	262 080
21.5	48 160	34.0	76 160	46.5	104 160	68	152 320	93	208 320	118	264 320
22.0	49 280	34.5	77 280	47.0	105 280	69	154 560	94	210 560	119	266 560

Load conversion table, kg/mm² to psi

kg/mm₂	psi	kg/mm²	psi	kg/mm²	psi	kg/mm²	psi	kg/mm²	psi	kg/mm²	psi
10	14 223	35	49 782	60	85 340	85	120 899	110	156 457	135	192 016
11	15 646	36	51 204	61	86 763	86	122 321	111	157 880	136	193 438
12	17 068	37	52 627	62	88 185	87	123 744	112	159 302	137	194 861
13	18 490	38	54 049	63	89 607	88	125 166	113	160 724	138	196 283
14	19 913	39	55 471	64	91 030	89	126 588	114	162 147	139	197 705
15	21 335	40	56 894	65	92 452	90	128 011	115	163 569	140	199 128
16	22 757	41	58 316	66	93 874	91	129 433	116	164 991	141	200 550
17	24 180	42	59 738	67	95 297	92	130 855	117	166 414	142	201 972
18	25 602	43	61 161	68	96 719	93	132 278	118	167 836	143	203 395
19	27 024	44	62 583	69	98 141	94	133 700	119	169 258	144	204 817
20	28 447	45	64 005	70	99 564	95	135 122	120	170 681	145	206 239
21	29 869	46	65 428	71	100 986	96	136 545	121	172 103	146	207 662
22	31 291	47	66 850	72	102 408	97	137 967	122	173 525	147	209 084
23	32 714	48	68 272	73	103 831	98	139 389	123	174 948	148	210 506
24	34 136	49	69 695	74	105 253	99	140 812	124	176 370	149	211 929
25	35 558	50	71 117	75	106 675	100	142 234	125	177 792	150	213 351
26	36 981	51	72 539	76	108 098	101	143 656	126	179 215	151	214 773
27	38 403	52	73 962	77	109 520	102	145 079	127	180 637	152	216 196
28	39 826	53	75 384	78	110 943	103	146 501	128	182 059	153	217 618
29	41 248	54	76 806	79	112 365	104	147 923	129	183 482	154	219 040
30	42 670	55	78 229	80	113 787	105	149 346	130	184 904	155	220 463
31	44 093	56	79 651	81	115 210	106	150 768	131	186 327	156	221 885
32	45 515	57	81 073	82	116 632	107	152 190	132	187 749	157	223 307
33	46 937	58	82 496	83	118 054	108	153 613	133	189 171	158	224 730
34	48 360	59	83 918	84	119 477	109	155 035	134	190 594	159	226 152

Approximate hourly production

Time to make one piece, s	Gross production per hour, pieces	Gross time per 1000 pieces, h	Time to make one piece, s	Gross production per hour, pieces	Gross time per 1000 pieces, h	Time to make one piece, s	Gross production per hour, pieces	Gross time per 1000 pieces, h
0.5	7200	0.14	26	138	7.22	52	69	14.5
1	3600	0.28	27	133	7.50	54	66	15.0
2	1800	0.55	28	128	7.78	56	64	15.6
3	1200	0.83	29	124	8.06	58	62	16.1
4	900	1.11	30	120	8.33	60	60	16.7
5	720	1.39	31	116	8.62	62	58	17.2
6	600	1.67	32	112	8.90	64	56	17.8
7	514	1.94	33	109	9.17	66	54	18.4
8	450	2.22	34	106	9.45	68	53	18.9
9	400	2.50	35	103	9.73	70	51	19.5
10	360	2.78	36	100	10.00	72	50	20.0
11	327	3.05	37	97	10.30	74	49	20.6
12	300	3.33	38	95	10.56	76	47	21.1
13	276	3.62	39	92	10.83	78	46	21.7
14	257	3.89	40	90	11.11	80	45	22.2
15	240	4.17	41	88	11.39	82	44	22.8
16	225	4.44	42	86	11.67	84	43	23.3
17	212	4.72	43	84	11.94	86	42	23.9
18	200	5.00	44	82	12.22	88	41	24.5
19	189	5.28	45	80	12.50	90	40	25.0
20	180	5.56	46	78	12.78	92	39	25.5
21	171	5.83	47	77	13.05	94	38	26.1
22	164	6.12	48	75	13.34	96	37	26.7
23	156	6.40	49	73	13.61	100	36	27.8
24	150	6.67	50	72	13.9			
25	144	6.95						

Decimal and metric equivalents of fractions of an inch

Fraction of an inch	Equivalents in.	mm	Fraction of an inch	Equivalents in.	mm
1/64	0.015625	0.39687	33/64	0.515625	13.09671
1/32	0.03125	0.79374	17/32	0.53125	13.49362
3/64	0.046875	1.19061	35/64	0.546875	13.89045
1/16	0.0625	1.58748	9/16	0.5625	14.28737
5/64	0.078125	1.98435	37/64	0.578125	14.68419
3/32	0.09375	2.38123	19/32	0.59375	15.08111
7/64	0.109375	2.77809	39/64	0.609375	15.47793
1/8	0.125	3.17497	5/8	0.625	15.87485
9/64	0.140625	3.57183	41/64	0.640625	16.27167
5/32	0.15625	3.96871	21/32	0.65625	16.66859
11/64	0.171875	4.36557	43/64	0.671875	17.06541
3/16	0.1875	4.76245	11/16	0.6875	17.46234
13/64	0.203125	5.15931	45/64	0.703125	17.85915
7/32	0.21875	5.55620	23/32	0.71875	18.25608
15/64	0.234375	5.95305	47/64	0.734375	18.65289
1/4	0.25	6.34994	3/4	0.75	19.04982
17/64	0.265625	6.74679	49/64	0.765625	19.44663
9/32	0.28125	7.14368	25/32	0.78125	19.84356
19/64	0.296875	7.54053	51/64	0.796875	20.24037
5/16	0.3125	7.93743	13/16	0.8125	20.63731
21/64	0.328125	8.33427	53/64	0.828125	21.03411
11/32	0.34375	8.73117	27/32	0.84375	21.43105
23/64	0.359375	9.12801	55/64	0.859375	21.82785
3/8	0.375	9.52491	7/8	0.875	22.22479
25/64	0.390625	9.92175	57/64	0.890625	22.62159
13/32	0.40625	10.31865	29/32	0.90625	23.01853
27/64	0.421875	10.71549	59/64	0.921875	23.41533
7/16	0.4375	11.11240	15/16	0.9375	23.81228
29/64	0.453125	11.50923	61/64	0.953125	24.20907
15/32	0.46875	11.90614	31/32	0.96875	24.60602
31/64	0.484375	12.30297	63/64	0.984375	25.00281
1/2	0.5	12.69988	1	1.0	25.4

Comparison of standard gages(a)

Gage No.	A(a)	B(b)	C(c)	D(d)	E(e)	F(f)
0000000	...	...	0.4900	0.5000	0.500	...
000000	...	0.580000	0.4615	0.4687	0.464	...
00000	...	0.516500	0.4305	0.4375	0.432	...
0000	0.454	0.460000	0.3938	0.4062	0.400	...
000	0.425	0.409642	0.3625	0.3750	0.372	...
00	0.380	0.364796	0.3310	0.3437	0.348	...
0	0.340	0.324861	0.3065	0.3125	0.324	...
1	0.300	0.289297	0.2830	0.2812	0.300	...
2	0.284	0.257627	0.2625	0.2656	0.276	...
3	0.259	0.229423	0.2437	0.2500	0.252	0.2391
4	0.238	0.204307	0.2253	0.2344	0.232	0.2242
5	0.220	0.181940	0.2070	0.2187	0.212	0.2092
6	0.203	0.162023	0.1920	0.2031	0.192	0.1943
7	0.180	0.144285	0.1770	0.1875	0.176	0.1793
8	0.165	0.128490	0.1620	0.1719	0.160	0.1644
9	0.148	0.114423	0.1483	0.1562	0.144	0.1495
10	0.134	0.101897	0.1350	0.1406	0.128	0.1345
11	0.120	0.090742	0.1205	0.1250	0.116	0.1196
12	0.109	0.080808	0.1055	0.1094	0.104	0.1046
13	0.095	0.071962	0.0915	0.0937	0.092	0.0897
14	0.083	0.064084	0.0800	0.0781	0.080	0.0747
15	0.072	0.057068	0.0720	0.0703	0.072	0.0673
16	0.065	0.050821	0.0625	0.0625	0.064	0.0598
17	0.058	0.045257	0.0540	0.0562	0.056	0.0538
18	0.049	0.040303	0.0475	0.0500	0.048	0.0478
19	0.042	0.035890	0.0410	0.0437	0.040	0.0418
20	0.035	0.031961	0.0348	0.0375	0.036	0.0359
21	0.032	0.028462	0.03175	0.0344	0.032	0.0329
22	0.028	0.025346	0.0286	0.0312	0.028	0.0299
23	0.025	0.022572	0.0258	0.0281	0.024	0.0269
24	0.022	0.020101	0.0230	0.0250	0.022	0.0239
25	0.020	0.017900	0.0204	0.0219	0.020	0.0209
26	0.018	0.015941	0.0181	0.0187	0.018	0.0179
27	0.016	0.014195	0.0173	0.0172	0.0164	0.0164
28	0.014	0.012641	0.0162	0.0156	0.0148	0.0149
29	0.013	0.011257	0.0150	0.0141	0.0136	0.0135
30	0.012	0.010025	0.0140	0.0125	0.0124	0.0120
31	0.010	0.008928	0.0132	0.0109	0.0116	0.0105
32	0.009	0.007950	0.0128	0.0102	0.0108	0.0097
33	0.008	0.007080	0.0118	0.0094	0.0100	0.0090
34	0.007	0.006305	0.0104	0.0086	0.0092	0.0082
35	0.005	0.005615	0.0095	0.0078	0.0084	0.0075
36	0.004	0.005000	0.0090	0.0070	0.0076	0.0067
37	...	0.004453	0.0085	0.0066	0.0068	0.0064
38	...	0.003965	0.0080	0.0062	0.0060	0.0060
39	...	0.003531	0.0075	...	0.0052	...
40	...	0.003144	0.0070	...	0.0048	...

(a) Birmingham Wire (BWG) and Stubs' Iron Wire. Birmingham Wire gages used principally for strips, bands, hoops, and wire. (b) American Wire (AWG) and Brown and Sharpe; gages used principally for nonferrous sheets, rod, and wire. (c) U.S. Steel Wire; American Steel and Wire; Washburn and Moen; and Steel Wire. U.S. Steel Wire gages used principally for steel wire, except music wire. (d) U.S. Standard (old); gages used principally for stainless steel sheets. (e) British Imperial Standard Wire (SWG); gages used principally for English legal standard wire gage. (f) Manufacturers' Standard; gages used principally for uncoated steel sheets.

Lengths, Areas, Volumes, Weights

Mensuration: Lengths, Areas, Volumes

In the figures and equations that follow, **a, b, c, d, s** denote lengths, **A** denotes area, **V** denotes volume.

Right triangle

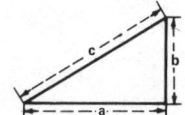

$$A = \tfrac{1}{2}\, ab \qquad a = \sqrt{c^2 - b^2}$$
$$c = \sqrt{a^2 + b^2} \qquad b = \sqrt{c^2 - a^2}$$

Equilateral triangle

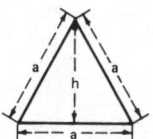

$$A = \tfrac{1}{2}\, ah = \tfrac{1}{4}\, a^2 \sqrt{3}$$
$$h = \tfrac{1}{2}\, a \sqrt{3}$$

Square

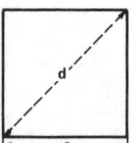

$$A = a^2$$
$$d = a \sqrt{2}$$

Oblique triangle

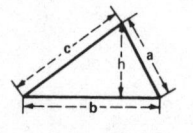

$$A = \tfrac{1}{2}\, bh$$

Rectangle

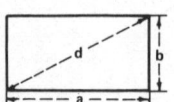

$$A = ab$$
$$d = \sqrt{a^2 + b^2}$$

Trapezoid

One pair of opposite sides parallel

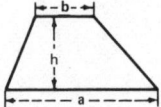

$$A = \tfrac{1}{2}\, h\,(a + b)$$

Parallelogram

Opposite sides parallel

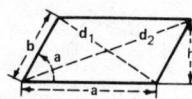

$$A = ah = ab \sin \alpha$$
$$d_1 = \sqrt{a^2 + b^2 - 2\, ab \cos \alpha}$$
$$d_2 = \sqrt{a^2 + b^2 + 2\, ab \cos \alpha}$$

Isosceles Trapezoid

Nonparallel sides equal

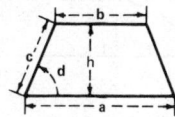

$$A = \tfrac{1}{2}\, h\, (a + b) =$$
$$\tfrac{1}{2}\, c \sin \alpha\, (a + b) =$$
$$c \sin \alpha\, (a - c \cos \alpha) =$$
$$c \sin \alpha\, (b + c \cos \alpha)$$

Cube

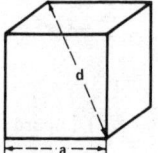

$$V = a^3$$
$$d = a\sqrt{3}$$
Total surface = $6\, a^2$

Ellipsoid

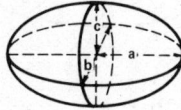

$$V = \tfrac{4}{3}\, \pi abc$$

Circle

C = circumference
α = central angle in radians

$$C = \pi D = 2\, \pi R$$

$$c = R\alpha = \tfrac{1}{2}\, D\alpha = D \cos^{-1} \frac{d}{R} = D \tan^{-1} \frac{l}{2\, d}$$

$$l = 2\sqrt{R^2 - d^2} = 2\, R \sin \frac{\alpha}{2} = 2\, d \tan \frac{\alpha}{2} = 2\, d \tan \frac{c}{D}$$

$$d = \tfrac{1}{2}\sqrt{4\, R^2 - l^2} = \tfrac{1}{2}\sqrt{D^2 - l^2} = R \cos \frac{\alpha}{2} = \tfrac{1}{2}\, l \cot \frac{\alpha}{2} = \tfrac{1}{2}\, l \cot \frac{c}{D}$$

$$h = R - d$$

$$\alpha = \frac{c}{R} = \frac{2\, c}{D} = 2 \cos^{-1} \frac{d}{R} = 2 \tan^{-1} \frac{l}{2\, d} = 2 \sin^{-1} \frac{l}{D}$$

$$A_{(circle)} = \pi R^2 = \tfrac{1}{4}\, \pi D^2 = \tfrac{1}{2}\, RC = \tfrac{1}{4}\, DC$$

$$A_{(sector)} = \tfrac{1}{2}\, Rc = \tfrac{1}{2}\, R^2\alpha = \tfrac{1}{8}\, D^2\alpha$$

$$A_{(segment)} = A_{(sector)} - A_{(triangle)} = \tfrac{1}{2}\, R^2(\alpha - \sin \alpha) = \tfrac{1}{2}\, R\left(c - R \sin \frac{c}{R}\right)$$

$$= R^2 \sin^{-1} \frac{l}{2\, R} - \tfrac{1}{4}\, l\sqrt{4\, R^2 - l^2} = R^2 \cos^{-1} \frac{d}{R} - d\sqrt{R^2 - d^2}$$

$$= R^2 \cos^{-1} \frac{R - h}{R} - (R - h)\sqrt{2\, Rh - h^2}$$

Prism or cylinder

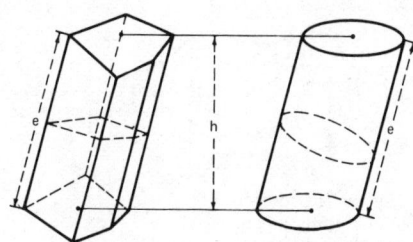

V = area of base × altitude

Lateral area = perimeter of right section × lateral edge

Rectangular parallelopiped

$$V = abc$$
$$d = \sqrt{a^2 + b^2 + c^2}$$
Total surface = $2(ab + bc + ca)$

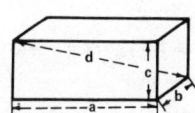

Regular polygon of n sides

All sides equal

All angles equal

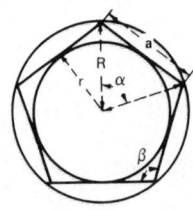

$$\beta = \frac{n-2}{n} 180° = \frac{n-2}{n} \pi \text{ radians}$$

$$\alpha = \frac{360°}{n} = \frac{2\pi}{n} \text{ radians}$$

n	a	r	R	A	
3	$2r\sqrt{3} = R\sqrt{3}$	$\frac{1}{6} a\sqrt{3}$	$\frac{1}{3} a\sqrt{3}$	$\frac{1}{4} a^2\sqrt{3}$	$= 3 r^2\sqrt{3}$
					$= \frac{3}{4} R^2\sqrt{3}$
4	$2r = R\sqrt{2}$	$\frac{1}{2} a$	$\frac{1}{2} a\sqrt{2}$	a^2	$= 4 r^2 = 2 R^2$
6	$\frac{2}{3} r\sqrt{3} = R$	$\frac{1}{2} a\sqrt{3}$	a	$\frac{3}{2} a^2\sqrt{3}$	$= 2 r^2\sqrt{3}$
					$= \frac{3}{2} R^2\sqrt{3}$
8	$2r(\sqrt{2}-1)$	$\frac{1}{2}a(\sqrt{2}+1)$	$\frac{1}{2}a\sqrt{4+2\sqrt{2}}$	$2a^2(\sqrt{2}+1)$	$= 8 r^2(\sqrt{2}-1)$
	$= R\sqrt{2-\sqrt{2}}$				$= 2R^2\sqrt{2}$
n	$2r\tan\frac{\alpha}{2}$	$\frac{a}{2}\cot\frac{\alpha}{2}$	$\frac{a}{2}\csc\frac{\alpha}{2}$	$\frac{na^2}{4}\cot\frac{\alpha}{2}$	$= nr^2\tan\frac{\alpha}{2}$
	$= 2R\sin\frac{\alpha}{2}$				$= \frac{nR^2}{2}\sin\alpha$

Area by approximation

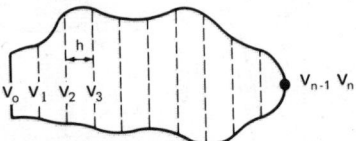

Let $y_0, y_1, y_2, \ldots, y_n$ be the measured lengths of a series of equidistant parallel chords, and let **h** be their distance apart, then the area enclosed by any boundary is given approximately by one of the following rules.

Trapezoidal rule:

$$A_T = h[\frac{1}{2}(y_0 + y_n) + y_1 + y_2 + \cdots + y_{n-1}]$$

Durand's rule:

$$A_D = h[0.4(y_0 + y_n) + 1.1(y_1 + y_{n-1}) + y_2 + y_3 + \cdots + y_{n-2}]$$

Simpson's rule:

$$A_s = \frac{1}{3} h[(y_0 + y_n) + 4(y_1 + y_3 + \cdots + y_{n-1}) + 2(y_2 + y_4 + \cdots + y_{n-2})]$$

where n is even

The larger the value of n, the greater is the accuracy of approximation. In general, for the same number of chords, A_s gives the most accurate, A_T, the least accurate approximation.

Trapezium

No sides parallel

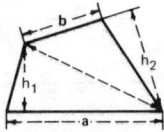

$$A = \frac{1}{2}(ah_1 + bh_2) =$$

sum of areas of two triangles

Cycloid

r = radius of generating circle

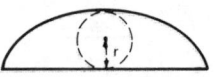

$$A = 3\pi r^2$$

Length of arc (**s**) = 8r

Catenary

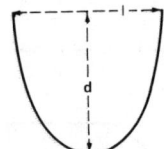

Length of arc (**s**) =

$$l\left[1 + \frac{2}{3}\left(\frac{2d}{l}\right)^2\right] \text{ approx}$$

if **d** is small in comparison with **1**

Torus

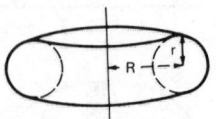

$$V = 2\pi^2 R r^2$$

Surface (**S**) = $4\pi^2 R r$

Ellipse

$$A = \pi ab$$

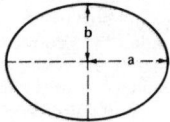

Perimeter (**s**) =

$$\pi(\mathbf{a} + \mathbf{b}) \left[1 + \tfrac{1}{4} \left(\frac{\mathbf{a} - \mathbf{b}}{\mathbf{a} + \mathbf{b}} \right)^2 + \frac{1}{64} \left(\frac{\mathbf{a} - \mathbf{b}}{\mathbf{a} + \mathbf{b}} \right)^4 + \frac{1}{256} \left(\frac{\mathbf{a} - \mathbf{b}}{\mathbf{a} + \mathbf{b}} \right)^6 + \cdots \right].$$

$$\cong \pi \;\; \frac{\mathbf{a} + \mathbf{b}}{4} \left[3(1 + \lambda) + \frac{1}{1 - \lambda} \right] \qquad \lambda = \left[\frac{\mathbf{a} - \mathbf{b}}{2(\mathbf{a} + \mathbf{b})} \right]^2$$

Pyramid or cone

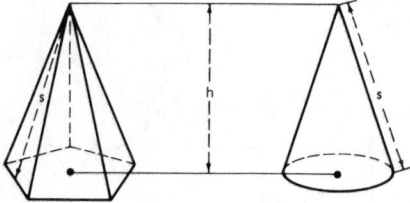

$\mathbf{V} = \tfrac{1}{3}$ (area of base) × (altitude)
Lateral area of regular figure =
½ (perimeter of base) × (slant height)

Frustum of pyramid or cone

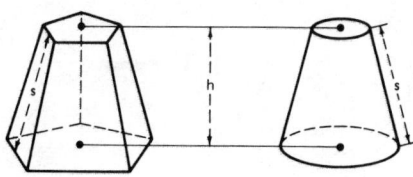

$$\mathbf{V} = \tfrac{1}{3} \left(\mathbf{A}_1 + \mathbf{A}_2 + \sqrt{\mathbf{A}_1 \times \mathbf{A}_2} \right) \mathbf{h}$$

where $\mathbf{A}_1$ and $\mathbf{A}_2$ are areas of bases, and $\mathbf{h}$ is altitude.
    Lateral area of regular figure =
    ½ (sum of perimeters of bases) × (slant height)

Prismatoid

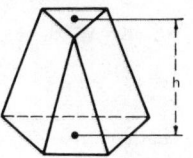

Bases are in parallel planes, lateral faces are triangles or trapezoids

$$\mathbf{V} = \tfrac{1}{6} (\mathbf{A}_1 + \mathbf{A}_2 + 4 \mathbf{A}_m) \, \mathbf{h}$$

where $\mathbf{A}_1$, $\mathbf{A}_2$ are areas of bases, $\mathbf{A}_m$ is area of midsection, and $\mathbf{h}$ is altitude

Sphere

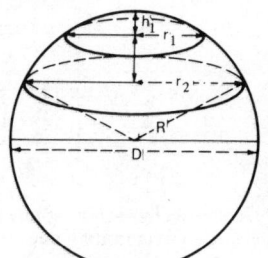

$$\mathbf{A}_{(\text{sphere})} = 4\,\pi\mathbf{R}^2 = \pi\mathbf{D}^2$$

$$\mathbf{A}_{(\text{zone})} = 2\,\pi\mathbf{R}\mathbf{h} = \pi\mathbf{D}\mathbf{h}$$

$$\mathbf{V}_{(\text{sphere})} = \tfrac{4}{3}\,\pi\mathbf{R}^3 = \tfrac{1}{6}\,\pi\mathbf{D}^3$$

$$\mathbf{V}_{(\text{spherical sector})} = \tfrac{2}{3}\,\pi\mathbf{R}^2\mathbf{h} = \tfrac{1}{6}\,\pi\mathbf{D}^2\mathbf{h}$$

$$\mathbf{V}_{(\text{spherical segment of one base})} =$$
$$\tfrac{1}{6}\,\pi\mathbf{h}_1 (3\mathbf{r}_1{}^2 + \mathbf{h}_1{}^2) = \tfrac{1}{3}\,\pi\mathbf{h}_1{}^2(3\,\mathbf{R} - \mathbf{h}_1)$$

$$\mathbf{V}_{(\text{spherical segment of two bases})} =$$
$$\tfrac{1}{6}\,\pi\mathbf{h} (3\,\mathbf{r}_1{}^2 + 3\,\mathbf{r}_2{}^2 + \mathbf{h}^2)$$

Parabola

$$A = \tfrac{2}{3}\, \text{ld}$$

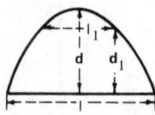

Length of arc (**s**) =

$$\tfrac{1}{2}\sqrt{16\, \mathbf{d}^2} + \mathbf{l}^2 + \frac{\mathbf{l}^2}{8\, \mathbf{d}} \ln \left(\frac{4\, \mathbf{d} + \sqrt{16\, \mathbf{d}^2 + \mathbf{l}^2}}{\mathbf{l}} \right)$$

$$= 1\left[1 + \tfrac{2}{3}\left(\frac{2\, \mathbf{d}}{\mathbf{l}} \right)^2 - \tfrac{2}{5}\left(\frac{2\, \mathbf{d}}{\mathbf{l}} \right)^4 + \cdots \right]$$

Height of segment ($\mathbf{d}_1$) = $\dfrac{\mathbf{d}}{\mathbf{l}^2}(\mathbf{l}^2 - \mathbf{l}_1{}^2)$.

Width of segment ($\mathbf{l}_1$) = $\mathbf{l}\sqrt{\dfrac{\mathbf{d} - \mathbf{d}_1}{\mathbf{d}}}$

Solid (V) or surface (S) of revolution

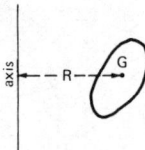

Generated by revolving any plane area (**A**) or arc (**s**) about an axis in its plane, and not crossing the area or arc

$$\mathbf{V} = 2\,\pi\mathbf{RA};\ \mathbf{S} = 2\,\pi\mathbf{Rs}$$

where **R** = distance of center of gravity (**G**) of area or arc from axis

Solid angle

At any point (P) subtended by any surface (S), the solid angle (Ψ) is equal to the portion (A) of the surface of a sphere of unit radius which is cut out by a conical surface with vertex at P and the perimeter of S for base. The unit solid angle (Ψ) is called a steradian. The total solid angle about a point = $4\,\pi$ steradians

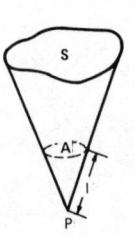

Paraboloidal segment

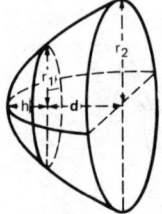

$$\mathbf{V}_{(\text{segment of one base})} = \tfrac{1}{2}\,\pi\mathbf{r}_1{}^2\mathbf{h}$$

$$\mathbf{V}_{(\text{segment of two bases})} = \tfrac{1}{2}\,\pi\mathbf{d}\,(\mathbf{r}_1{}^2 + \mathbf{r}_2{}^2)$$

Weight Formulas and Conversions

Weight formulas

Steel weights are based on 0.2833 lb/in.³. Aluminum weights are based on 0.0979 lb/in.³, which applies to 1100 alloy.

Rounds
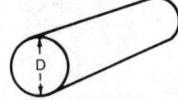

Steel:
Pounds per lineal foot = $2.67036 \times D^2$
Aluminum:
Pounds per lineal foot = $0.9227 \times D^2$
D = size in inches

Squares
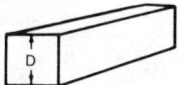

Steel:
Pounds per lineal foot = $3.4 \times D^2$

Aluminum:
Pounds per lineal foot = $1.1748 \times D^2$
D = size in inches

Hexagons
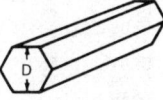

Steel:
Pounds per lineal foot = $2.9446 \times D^2$

Aluminum:
Pounds per lineal foot = $1.0175 \times D^2$
D = size in inches

Tubing
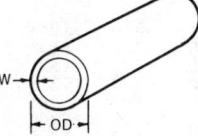

Steel:
Pounds per lineal foot =
$10.68 \times (OD - W) \times W$
Aluminum:
Pounds per lineal foot =
$3.6904 \times (OD - W) \times W$
OD = outside diameter
    to 3 decimal places
W = wall thickness
    to 3 decimal places

Flats
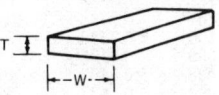

Steel:
Pounds per lineal foot = $3.4 \times T \times W$
Aluminum:
Pounds per lineal foot = $1.1748 \times T \times W$
T = thickness in inches
W = width in inches

Octagons
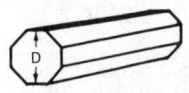

Steel:
Pounds per lineal foot = $2.8166 \times D^2$

Aluminum:
Pounds per lineal foot = $0.9733 \times D^2$
D = size in inches

Circles
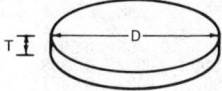

Steel:
Weight of circle in pounds =
$0.22253 \times T \times D^2$
Aluminum:
Weight of circle in pounds = $0.0769 \times T \times D^2$
D = diameter in inches
T = thickness in inches

Rings
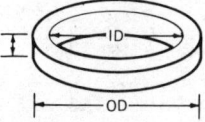

Steel:
Weight of ring in pounds =
$0.22253 \times T \times (OD^2 - ID^2)$
Aluminum:
Weight of ring in pounds =
$0.07690 \times T \times (OD^2 - ID^2)$
OD = outside diameter in inches
ID = inside diameter in inches
T = thickness in inches

Source: Earl M. Jorgensen Co.

Weight conversion factors

To obtain weight of	Density (lb/in.^{3})	Multiply weight of steel by	To obtain weight of	Density (lb/in.^{3})	Multiply weight of steel by	To obtain weight of	Density (lb/in.^{3})	Multiply weight of steel by
Aluminum	0.098	0.3462	6063 Aluminum	0.098	0.3462	Monel	0.307	1.084
1100 Aluminum	0.098	0.3462	7075 Aluminum	0.101	0.3568	**Stainless steels**		
2011 Aluminum	0.102	0.3604	7178 Aluminum	0.102	0.3604	300 series	0.286	1.010
2014 Aluminum	0.101	0.3568	Gold	0.698	2.466	400 series	0.283	1.000
2017 Aluminum	0.101	0.3568	Tungsten	0.697	2.462	Carbon, alloy steels	0.283	1.000
2024 Aluminum	0.100	0.3533	Tantalum	0.600	2.120	Tin	0.264	0.932
3003 Aluminum	0.099	0.3498	Lead	0.410	1.448	Zinc	0.258	0.911
5005 Aluminum	0.098	0.3462	Silver	0.379	1.339	Cast iron	0.258	0.911
5052 Aluminum	0.097	0.3427	Molybdenum	0.369	1.303	Zirconium	0.230	0.812
5056 Aluminum	0.095	0.3356	Copper	0.324	1.144	Titanium	0.163	0.575
5083 Aluminum	0.096	0.3392	Nickel	0.322	1.137	Beryllium	0.067	0.236
5086 Aluminum	0.096	0.3392	Niobium	0.310	1.095	Magnesium	0.065	0.229
6061 Aluminum	0.098	0.3462	Brass	0.307	1.084			

Source: Earl M. Jorgensen Co.

Theoretical weights of carbon-steel bars(a)

Thickness or diameter, in.	Round lb/in.	Round lb/ft	Square lb/in.	Square lb/ft	Hexagon lb/in.	Hexagon lb/ft
1/32	0.0002	0.0026	0.0003	0.0033	0.0002	0.0028
1/16	0.0009	0.0104	0.0011	0.0133	0.0010	0.0115
3/32	0.0020	0.0235	0.0025	0.0299	0.0022	0.0259
1/8	0.0035	0.0417	0.0044	0.0531	0.0038	0.0460
5/32	0.0054	0.0652	0.0069	0.0830	0.0060	0.0719
3/16	0.0078	0.0939	0.0100	0.1195	0.0086	0.1035
7/32	0.0106	0.1278	0.0136	0.1627	0.0117	0.1409
1/4	0.0139	0.1669	0.0177	0.2125	0.0153	0.1840
9/32	0.0176	0.2112	0.0224	0.2689	0.0194	0.2329
5/16	0.0217	0.2608	0.0277	0.3320	0.0240	0.2875
11/32	0.0263	0.3155	0.0335	0.4018	0.0290	0.3479
3/8	0.0313	0.3755	0.0398	0.4781	0.0345	0.4141
13/32	0.0367	0.4407	0.0468	0.5611	0.0405	0.4860
7/16	0.0426	0.5111	0.0542	0.6508	0.0470	0.5636
15/32	0.0489	0.5867	0.0623	0.7471	0.0538	0.6470
1/2	0.0556	0.6676	0.0708	0.8500	0.0613	0.7361
17/32	0.0628	0.7536	0.0800	0.9596	0.0693	0.8310
9/16	0.0704	0.8449	0.0896	1.076	0.0776	0.9317
19/32	0.0785	0.9414	0.0999	1.199	0.0865	1.038
5/8	0.0869	1.043	0.1107	1.328	0.0958	1.150
21/32	0.0958	1.150	0.1220	1.464	0.1057	1.268
11/16	0.1052	1.262	0.1339	1.607	0.1160	1.392
23/32	0.1150	1.380	0.1464	1.756	0.1268	1.521
3/4	0.1252	1.502	0.1594	1.913	0.1380	1.656
25/32	0.1358	1.630	0.1729	2.075	0.1498	1.797
13/16	0.1469	1.763	0.1870	2.245	0.1620	1.944
27/32	0.1584	1.901	0.2017	2.421	0.1747	2.096
7/8	0.1704	2.044	0.2169	2.603	0.1879	2.254
29/32	0.1828	2.193	0.2327	2.792	0.2015	2.418
15/16	0.1956	2.347	0.2490	2.988	0.2157	2.588
31/32	0.2088	2.506	0.2659	3.191	0.2303	2.763
1	0.2225	2.670	0.2833	3.400	0.2454	2.944
1-1/16	0.2512	3.015	0.3199	3.838	0.2770	3.324
1-1/8	0.2816	3.380	0.3586	4.303	0.3106	3.727
1-3/16	0.3138	3.766	0.3995	4.795	0.3460	4.152
1-1/4	0.3477	4.172	0.4427	5.313	0.3834	4.601
1-5/16	0.3833	4.600	0.4881	5.857	0.4227	5.072
1-3/8	0.4207	5.049	0.5357	6.428	0.4639	5.567
1-7/16	0.4598	5.518	0.5855	7.026	0.5070	6.085
1-1/2	0.5007	6.008	0.6375	7.650	0.5521	6.625
1-9/16	0.5433	6.519	0.6917	8.301	0.5991	7.189
1-5/8	0.5876	7.051	0.7482	8.978	0.6479	7.775
1-11/16	0.6337	7.604	0.8068	9.682	0.6988	8.385
1-3/4	0.6815	8.178	0.8677	10.41	0.7515	9.018
1-13/16	0.7310	8.773	0.9308	11.17	0.8060	9.67
1-7/8	0.7823	9.388	0.9961	11.95	0.8626	10.35
1-15/16	0.8354	10.02	1.064	12.76	0.9211	11.05
2	0.8901	10.68	1.133	13.60	0.9815	11.78
2-1/16	0.9466	11.36	1.205	14.46	1.044	12.53
2-1/8	1.005	12.06	1.279	15.35	1.108	13.30
2-3/16	1.065	12.78	1.356	16.27	1.174	14.09
2-1/4	1.127	13.52	1.434	17.21	1.242	14.91
2-5/16	1.190	14.28	1.515	18.18	1.312	15.75
2-3/8	1.255	15.06	1.598	19.18	1.384	16.61
2-7/16	1.322	15.87	1.683	20.20	1.458	17.49
2-1/2	1.391	16.69	1.771	21.25	1.534	18.40
2-5/8	1.533	18.40	1.952	23.43	1.691	20.29
2-3/4	1.683	20.19	2.143	25.71	1.856	22.27
2-7/8	1.839	22.07	2.342	28.10	2.028	24.34
3	2.003	24.03	2.550	30.60	2.208	26.50
3-1/8	2.173	26.08	2.767	33.20	2.396	28.75
3-1/4	2.350	28.21	2.993	35.91	2.592	31.10
3-3/8	2.535	30.42	3.227	38.73	2.795	33.54
3-1/2	2.726	32.71	3.471	41.65	3.006	36.07
3-5/8	2.924	35.09	3.723	44.68	3.224	38.69
3-3/4	3.129	37.55	3.984	47.81	3.451	41.41
3-7/8	3.341	40.10	4.254	51.05	3.684	44.21
4	3.560	42.73	4.533	54.40	3.926	47.11
4-1/8	3.786	45.44	4.821	57.85	4.175	50.10
4-1/4	4.019	48.23	5.118	61.41	4.432	53.18
4-3/8	4.259	51.11	5.423	65.08	4.700	56.36
4-1/2	4.506	54.07	5.738	68.85	4.970	59.63
4-5/8	4.760	57.12	6.061	72.73	5.248	62.98
4-3/4	5.021	60.25	6.393	76.71	5.536	66.44
4-7/8	5.289	63.46	6.734	80.80	5.831	69.98
5	5.563	66.76	7.083	85.00	6.134	73.61
5-1/8	5.845	70.14	7.442	89.30	6.445	77.34
5-1/4	6.133	73.60	7.809	93.71	6.763	81.16
5-3/8	6.429	77.15	8.186	98.23	7.089	85.07
5-1/2	6.732	80.78	8.571	102.85	7.422	89.07
5-5/8	7.041	84.49	8.965	107.58	7.763	93.16
5-3/4	7.357	88.29	9.368	112.41	8.112	97.35
5-7/8	7.681	92.17	9.779	117.35	8.470	101.63
6	8.011	96.13	10.200	122.40	8.833	106.00

(a) Theoretical weight per cubic inch = 0.2833 lb. Theoretical weight per cubic foot = 489.6 lb. For flats, to determine the theoretical weight in pounds per linear foot, multiply the width in inches times the thickness in inches times 3.4.

Source: American Iron and Steel Institute

Diameters, weights and thicknesses of large-diameter steel pipe

Outside diameter	Weight		Thickness	
in.	lb/ft	kg/m	in.	mm
12¾	25.22	37.57	0.188	4.78
	27.20	40.51	0.203	5.16
	28.23	42.05	0.210	5.33
	29.31	43.66	0.219	5.56
	33.38	49.72	0.250	6.35
	37.42	55.74	0.281	7.14
	41.45	61.74	0.312	7.92
	45.58	67.89	0.344	8.74
14	27.73	41.30	0.188	4.78
	29.91	44.55	0.203	5.16
	30.93	46.07	0.210	5.33
	32.23	48.01	0.219	5.56
	36.71	54.68	0.250	6.35
	41.17	61.32	0.281	7.14
	45.61	67.94	0.312	7.92
	50.17	74.73	0.344	8.74
16	31.35	47.29	0.188	4.78
	34.25	51.02	0.203	5.16
	35.38	52.70	0.210	5.33
	36.91	54.98	0.219	5.56
	42.05	62.63	0.250	6.35
	47.17	70.26	0.281	7.14
	52.27	77.86	0.312	7.92
	57.52	85.68	0.344	8.74
	62.58	93.21	0.375	9.52
	67.62	100.72	0.406	10.31
18	35.76	53.26	0.188	4.78
	38.55	57.42	0.203	5.16
	39.86	59.38	0.210	5.33
	41.59	61.95	0.219	5.56
	47.39	70.59	0.250	6.35
	53.18	79.21	0.281	7.14
	58.94	87.79	0.312	7.92
	64.87	96.62	0.344	8.74
	70.59	105.14	0.375	9.52
	76.29	113.63	0.406	10.31
	82.15	122.36	0.438	11.13
	87.81	130.79	0.469	11.91
	93.45	139.19	0.500	12.70
	104.67	155.91	0.562	14.27
	115.98	172.75	0.625	15.88
20	46.27	68.92	0.219	5.56
	52.73	78.54	0.250	6.35
	59.18	88.15	0.281	7.14
	65.60	97.71	0.312	7.92
	72.21	107.56	0.344	8.74
	78.60	117.07	0.375	9.52
	84.96	126.55	0.406	10.31
	91.51	136.30	0.438	11.13
	97.83	145.72	0.469	11.91
	104.13	155.10	0.500	12.70
	116.67	173.78	0.562	14.27
	129.33	192.64	0.625	15.88
22	50.94	75.88	0.219	5.56
	58.07	86.50	0.250	6.35
	65.18	97.09	0.281	7.14
	72.27	107.65	0.312	7.92
	79.56	118.50	0.344	8.74
	86.61	129.01	0.375	9.52
	93.63	139.46	0.406	10.31
	100.86	150.23	0.438	11.13
	107.85	160.64	0.469	11.91
	114.81	171.01	0.500	12.70
	128.67	191.65	0.562	14.27
	142.68	212.52	0.625	15.88
	156.60	233.26	0.688	17.48

Outside diameter	Weight		Thickness	
in.	lb/ft	kg/m	in.	mm
24	63.41	94.45	0.250	6.35
	71.18	106.02	0.281	7.14
	78.93	117.57	0.312	7.92
	86.91	129.45	0.344	8.74
	94.62	140.94	0.375	9.52
	102.31	152.39	0.406	10.31
	110.22	164.17	0.438	11.13
	117.86	175.55	0.469	11.91
	125.49	186.92	0.500	12.70
	140.68	209.54	0.562	14.27
	156.03	232.41	0.625	15.88
	171.29	255.14	0.688	17.48
	186.23	277.39	0.750	19.05
26	68.75	102.40	0.250	6.35
	77.18	114.96	0.281	7.14
	85.60	127.50	0.312	7.92
	94.26	140.40	0.344	8.74
	102.63	152.87	0.375	9.52
	110.98	165.30	0.406	10.31
	119.57	178.10	0.438	11.13
	127.88	190.48	0.469	11.91
	136.17	202.83	0.500	12.70
	152.68	227.42	0.562	14.27
	169.38	252.29	0.625	15.88
	185.99	277.03	0.688	17.48
	202.25	301.25	0.750	19.05
28	74.09	110.36	0.250	6.35
	83.19	123.91	0.281	7.14
	92.26	137.42	0.312	7.92
	101.61	151.35	0.344	8.74
	110.64	164.80	0.375	9.52
	119.65	178.22	0.406	10.31
	128.93	192.04	0.438	11.13
	137.90	205.40	0.469	11.91
	146.85	218.73	0.500	12.70
	164.69	245.31	0.562	14.27
	182.73	272.18	0.625	15.88
	200.68	298.91	0.688	17.48
	218.27	325.11	0.750	19.05
	235.78	351.19	0.812	20.62
30	79.43	118.31	0.250	6.35
	89.19	132.85	0.281	7.14
	98.93	147.36	0.312	7.92
	108.95	162.28	0.344	8.74
	118.65	176.73	0.375	9.52
	128.32	191.13	0.406	10.31
	138.29	205.98	0.438	11.13
	147.92	220.33	0.469	11.91
	157.53	234.64	0.500	12.70
	176.69	263.18	0.562	14.27
	196.08	292.06	0.625	15.88
	215.38	320.81	0.688	17.48
	234.29	348.97	0.750	19.05
	253.12	377.02	0.812	20.62
	272.17	405.40	0.875	22.22
	291.14	433.65	0.938	23.83
	309.72	461.33	1.000	25.40
32	84.77	126.26	0.250	6.35
	95.19	141.79	0.281	7.14
	105.59	157.28	0.312	7.92
	116.30	173.23	0.344	8.74
	126.66	188.66	0.375	9.52
	136.99	204.05	0.406	10.31
	147.64	219.91	0.438	11.13
	157.94	235.25	0.469	11.91
	168.21	250.55	0.500	12.70
	188.70	281.07	0.562	14.27

Outside diameter	Weight		Thickness	
in.	lb/ft	kg/m	in.	mm
32	209.43	311.95	0.625	15.88
	230.08	342.70	0.688	17.48
	250.31	372.84	0.750	19.05
	270.47	402.87	0.812	20.62
	290.86	433.24	0.875	22.22
	311.17	463.49	0.938	23.83
	331.08	493.14	1.000	25.40
34	90.11	134.22	0.250	6.35
	101.19	150.72	0.281	7.14
	112.25	167.20	0.312	7.92
	123.65	184.18	0.344	8.74
	134.67	200.59	0.375	9.52
	145.67	216.98	0.406	10.31
	157.00	233.85	0.438	11.13
	167.95	250.16	0.469	11.91
	178.89	266.46	0.500	12.70
	200.70	298.94	0.562	14.27
	222.78	331.83	0.625	15.88
	244.77	364.58	0.688	17.48
	266.33	396.70	0.750	19.05
	287.81	428.69	0.812	20.62
	309.55	461.07	0.875	22.22
	331.21	493.34	0.938	23.83
	357.44	524.96	1.000	25.40
36	95.45	142.17	0.250	6.35
	107.20	159.67	0.281	7.14
	118.92	177.13	0.312	7.92
	131.00	195.12	0.344	8.74
	142.68	212.52	0.375	9.52
	154.34	229.89	0.406	10.31
	166.35	247.78	0.438	11.13
	177.97	265.09	0.469	11.91
	189.57	282.36	0.500	12.70
	212.70	316.82	0.562	14.27
	236.13	351.72	0.625	15.88
	259.47	396.48	0.688	17.48
	282.35	420.56	0.750	19.05
	305.16	454.54	0.812	20.62
	328.24	488.91	0.875	22.22
	351.25	523.19	0.938	23.83
	373.80	556.78	1.000	25.40
38	125.58	187.05	0.312	7.92
	138.35	206.07	0.344	8.74
	150.69	224.45	0.375	9.52
	163.01	242.80	0.406	10.31
	175.71	261.72	0.438	11.13
	187.99	280.01	0.469	11.91
	200.25	298.27	0.500	12.70
	224.71	324.71	0.562	14.27
	249.48	371.60	0.625	15.88
	274.16	408.36	0.688	17.48
	298.37	444.42	0.750	19.05
	322.50	480.36	0.812	20.62
	346.93	516.75	0.875	22.22
	371.28	553.02	0.938	23.83
	395.16	588.59	1.000	25.40
40	132.25	196.99	0.312	7.92
	145.69	217.01	0.344	8.74
	158.70	236.38	0.375	9.52
	171.68	255.72	0.406	10.31
	185.06	275.65	0.438	11.13
	198.01	294.94	0.469	11.91
	210.93	314.18	0.500	12.70
	236.71	352.58	0.562	14.27
	262.83	391.49	0.625	15.88
	288.86	430.26	0.688	17.48
	314.39	468.28	0.750	19.05

Source: Italsider (IRI-Finsider Group)

Diameters, weights and thicknesses of large-diameter steel pipe (continued)

Outside diameter	Weight		Thickness	
in.	lb/ft	kg/m	in.	mm
40	339.84	506.19	0.812	20.62
	365.62	544.59	0.875	22.22
	391.32	582.87	0.938	23.83
	416.52	620.41	1.000	25.40
42	153.04	227.95	0.344	8.74
	166.71	248.31	0.375	9.52
	180.35	268.63	0.406	10.31
	194.42	289.59	0.438	11.13
	208.03	309.86	0.469	11.91
	221.61	330.09	0.500	12.70
	248.72	370.47	0.562	14.27
	276.18	411.37	0.625	15.88
	303.55	452.14	0.688	17.48
	330.41	492.15	0.750	19.05
	357.19	532.03	0.812	20.62
	384.31	572.43	0.875	22.22
	411.35	612.71	0.938	23.83
	437.88	652.22	1.000	25.40
44	160.39	238.90	0.344	8.74
	174.72	260.25	0.375	9.52
	189.03	281.56	0.406	10.31
	203.78	303.53	0.438	11.13
	218.04	324.77	0.469	11.91
	232.29	346.00	0.500	12.70
	260.72	388.34	0.562	14.27
	289.53	430.87	0.625	15.88
	318.25	474.03	0.688	17.48
	346.43	516.01	0.750	19.05
	374.53	557.86	0.812	20.62
	403.00	600.27	0.875	22.22
	431.39	642.56	0.938	23.83
	459.24	684.04	1.000	25.40
46	167.74	249.85	0.344	8.74
	182.63	272.18	0.375	9.52
	197.70	294.47	0.406	10.31
	213.13	317.46	0.438	11.13
	228.06	339.70	0.469	11.91
	242.97	361.90	0.500	12.70
	272.73	406.23	0.562	14.27
	302.88	451.14	0.625	15.88
	332.95	495.93	0.688	17.48
	362.45	539.87	0.750	19.05
	391.88	583.71	0.812	20.62
	421.69	628.11	0.875	22.22
	451.42	671.79	0.938	23.83
	480.60	715.85	1.000	25.40
48	175.08	260.78	0.344	8.74
	190.74	284.11	0.375	9.52
	206.37	307.39	0.406	10.31
	222.49	331.40	0.438	11.13
	238.08	354.62	0.469	11.91
	253.65	377.81	0.500	12.70
	284.73	424.11	0.562	14.27
	316.23	471.02	0.625	15.88
	347.64	517.81	0.688	17.48
	378.47	563.73	0.750	19.05
	409.22	609.53	0.812	20.62
	440.38	655.95	0.875	22.22
	471.46	702.24	0.938	23.83
	501.96	747.67	1.000	25.40
50	198.56	295.76	0.375	9.52
	214.85	320.02	0.406	10.31
	231.63	345.02	0.438	11.13
	246.53	367.21	0.469	11.91
	264.09	393.37	0.500	12.70
	296.46	441.58	0.562	14.27

Outside diameter	Weight		Thickness	
in.	lb/ft	kg/m	in.	mm
50	329.27	490.46	0.625	15.88
	362.00	539.21	0.688	17.48
	394.14	587.07	0.750	19.05
	426.17	634.79	0.812	20.62
	458.65	683.17	0.875	22.22
	491.05	731.42	0.938	23.83
	522.85	778.78	1.000	25.40
52	206.57	307.69	0.375	9.52
	223.52	332.93	0.406	10.31
	240.98	358.95	0.438	11.13
	257.88	384.11	0.469	11.91
	274.76	409.26	0.500	12.70
	308.46	459.45	0.562	14.27
	342.70	510.44	0.625	15.88
	376.70	561.09	0.688	17.48
	410.14	610.90	0.750	19.05
	443.51	660.61	0.812	20.62
	477.39	711.08	0.875	22.22
	511.06	761.23	0.938	23.83
	544.18	810.56	1.000	25.40
54	232.31	346.03	0.406	10.31
	250.46	373.07	0.438	11.13
	267.88	399.02	0.469	11.91
	285.43	425.15	0.500	12.70
	320.44	477.30	0.562	14.27
	355.95	530.20	0.625	15.88
	391.44	583.05	0.688	17.48
	426.15	634.75	0.750	19.05
	460.82	686.40	0.812	20.62
	496.19	739.07	0.875	22.22
	531.08	791.05	0.938	23.83
	565.52	842.35	1.000	25.40
56	240.84	358.73	0.406	10.31
	259.60	386.80	0.438	11.13
	277.95	414.02	0.469	11.91
	296.09	441.04	0.500	12.70
	333.03	496.06	0.562	14.27
	369.29	550.06	0.625	15.88
	406.18	605.02	0.688	17.48
	442.14	658.58	0.750	19.05
	478.16	712.23	0.812	20.62
	514.70	766.60	0.875	22.22
	551.09	820.86	0.938	23.83
	586.86	874.14	1.000	25.40
58	249.50	371.63	0.406	10.31
	268.59	400.07	0.438	11.13
	287.90	428.83	0.469	11.91
	306.83	457.03	0.500	12.70
	344.43	513.04	0.562	14.27
	382.73	570.08	0.625	15.88
	420.73	626.69	0.688	17.48
	458.16	682.43	0.750	19.05
	495.49	738.04	0.812	20.62
	533.35	794.42	0.875	22.22
	570.07	850.07	0.938	23.83
	608.20	905.92	1.000	25.40
60	258.17	384.54	0.406	10.31
	278.36	414.63	0.438	11.13
	297.91	443.74	0.469	11.91
	317.44	472.83	0.500	12.70
	356.49	531.00	0.562	14.27
	396.10	590.00	0.625	15.88
	435.41	648.55	0.688	17.48
	474.15	706.26	0.750	19.05
	512.95	764.05	0.812	20.62
	552.02	822.24	0.875	22.22

Outside diameter	Weight		Thickness	
in.	lb/ft	kg/m	in.	mm
60	591.14	880.51	0.938	23.83
	629.55	937.72	1.000	25.40
62	266.83	397.45	0.406	10.31
	287.71	428.55	0.438	11.13
	307.92	458.65	0.469	11.91
	328.31	489.02	0.500	12.70
	368.62	549.07	0.562	14.27
	409.30	609.66	0.625	15.88
	450.08	670.42	0.688	17.48
	490.16	730.10	0.750	19.05
	530.15	789.66	0.812	20.62
	570.69	850.05	0.875	22.22
	611.15	910.32	0.938	23.83
	650.88	969.50	1.000	25.40
64	338.77	504.61	0.500	12.70
	380.68	567.03	0.562	14.27
	422.64	629.53	0.625	15.88
	464.78	692.30	0.688	17.48
	506.23	754.04	0.750	19.05
	547.47	815.47	0.812	20.62
	589.49	878.06	0.875	22.22
	631.17	940.13	0.938	23.83
	672.22	1001.28	1.000	25.40
66	349.45	520.51	0.500	12.70
	392.42	584.51	0.562	14.27
	436.41	650.04	0.625	15.88
	479.46	714.16	0.688	17.48
	522.18	777.75	0.750	19.05
	564.80	841.28	0.812	20.62
	608.03	905.67	0.875	22.22
	651.26	970.05	0.938	23.83
	693.56	1033.07	1.000	25.40
68	404.40	602.36	0.562	14.27
	449.85	670.06	0.625	15.88
	494.14	736.03	0.688	17.48
	538.18	801.62	0.750	19.05
	582.14	867.10	0.812	20.62
	626.72	933.50	0.875	22.22
	671.20	999.76	0.938	23.83
	715.04	1065.06	1.000	25.40
70	416.40	620.23	0.562	14.27
	463.26	690.03	0.625	15.88
	508.89	758.00	0.688	17.48
	554.18	825.46	0.750	19.05
	599.53	893.00	0.812	20.62
	645.38	961.30	0.875	22.22
	691.21	1029.57	0.938	23.83
	736.52	1097.05	1.000	25.40
72	428.39	638.09	0.562	14.27
	475.99	709.00	0.625	15.88
	523.66	780.00	0.688	17.48
	570.19	849.30	0.750	19.05
	616.79	918.71	0.812	20.62
	664.06	989.12	0.875	22.22
	711.23	1059.39	0.938	23.83
	757.58	1128.43	1.000	25.40
74	440.41	656.00	0.562	14.27
	489.42	729.00	0.625	15.88
	538.19	801.64	0.688	17.48
	586.19	873.14	0.750	19.05
	634.11	944.52	0.812	20.62
	682.77	1017.00	0.875	22.22
	731.25	1089.21	0.938	23.83
	778.93	1160.22	1.000	25.40

Source: Italsider (IRI-Finsider Group)

Diameters, weights and thicknesses of large-diameter steel pipe (continued)

Outside diameter	Weight		Thickness		Outside diameter	Weight		Thickness		Outside diameter	Weight		Thickness	
in.	lb/ft	kg/m	in.	mm	in.	lb/ft	kg/m	in.	mm	in.	lb/ft	kg/m	in.	mm
76	452.51	674.02	0.562	14.27	84	666.23	992.35	0.750	19.05	92	911.41	1357.55	0.938	23.83
	502.66	748.72	0.625	15.88		720.77	1073.60	0.812	20.62		970.99	1446.30	1.000	25.40
	552.87	823.51	0.688	17.48		776.09	1156.00	0.875	22.22					
	602.21	897.00	0.750	19.05		831.34	1238.29	0.938	23.83	94	560.32	834.60	0.562	14.27
	651.44	970.33	0.812	20.62		885.63	1319.15	1.000	25.40		622.71	927.53	0.625	15.88
	701.60	1045.04	0.875	22.22							685.01	1020.33	0.688	17.48
	751.27	1119.02	0.938	23.83	86	512.35	763.14	0.562	14.27		746.25	1111.54	0.750	19.05
	800.26	1192.00	1.000	25.40		569.35	848.06	0.625	15.88		807.67	1203.03	0.812	20.62
						626.42	933.06	0.688	17.48		869.46	1295.06	0.875	22.22
78	464.63	692.08	0.562	14.27		682.24	1016.20	0.750	19.05		931.43	1387.37	0.938	23.83
	516.01	768.60	0.625	15.88		738.10	1099.40	0.812	20.62		992.34	1478.09	1.000	25.40
	567.55	845.38	0.688	17.48		794.91	1184.02	0.875	22.22					
	618.20	920.82	0.750	19.05		851.36	1268.10	0.938	23.83	96	572.31	852.46	0.562	14.27
	668.78	996.15	0.812	20.62		907.01	1351.00	1.000	25.40		636.05	947.40	0.625	15.88
	720.08	1072.56	0.875	2$.22							699.70	1042.20	0.688	17.48
	771.39	1149.00	0.938	23.83	88	524.30	781.00	0.562	14.27		762.25	1135.38	0.750	19.05
	821.78	1224.05	1.000	25.40		582.76	868.03	0.625	15.88		824.73	1228.44	0.812	20.62
						640.97	954.73	0.688	17.48		888.27	1323.08	0.875	22.22
80	476.36	709.54	0.562	14.27		698.24	1040.03	0.750	19.05		951.44	1417.18	0.938	23.83
	529.34	788.46	0.625	15.88		755.42	1125.20	0.812	20.62		1013.81	1510.07	1.000	25.40
	575.52	857.25	0.688	17.48		813.71	1212.03	0.875	22.22					
	634.48	945.06	0.750	19.05		871.43	1298.00	0.938	23.83	98	584.30	870.32	0.562	14.27
	686.17	1022.06	0.812	20.62		928.31	1382.72	1.000	25.40		649.38	967.26	0.625	15.88
	738.75	1100.37	0.875	22.22							714.38	1064.07	0.688	17.48
	791.54	1179.01	0.938	23.83	90	536.46	799.07	0.562	14.27		778.26	1159.22	0.750	19.05
	843.29	1256.08	1.000	25.40		596.03	887.80	0.625	15.88		842.06	1254.25	0.812	20.62
						655.65	976.60	0.688	17.48		906.81	1350.70	0.875	22.22
82	488.35	727.41	0.562	14.27		714.37	1064.06	0.750	19.05		971.46	1447.00	0.938	23.83
	542.68	808.33	0.625	15.88		772.75	1151.01	0.812	20.62		1035.28	1542.06	1.000	25.40
	596.92	889.12	0.688	17.48		832.12	1239.44	0.875	22.22					
	650.21	968.50	0.750	19.05		891.60	1328.03	0.938	23.83	100	596.30	888.19	0.562	14.27
	703.63	1048.06	0.812	20.62		949.32	1414.02	1.000	25.40		662.72	987.13	0.625	15.88
	757.31	1128.02	0.875	22.22							729.13	1086.04	0.688	17.48
	811.32	1208.47	0.938	23.83	92	548.32	816.73	0.562	14.27		794.26	1183.06	0.750	19.05
	864.29	1287.36	1.000	25.40		609.37	907.66	0.625	15.88		859.39	1280.06	0.812	20.62
						670.33	998.47	0.688	17.48		925.47	1378.50	0.875	22.22
84	500.35	745.27	0.562	14.27		730.24	1087.70	0.750	19.05		991.62	1477.02	0.938	23.83
	556.01	828.19	0.625	15.88		790.22	1177.03	0.812	20.62		1056.36	1573.45	1.000	25.40
	611.61	911.00	0.688	17.48		850.78	1267.25	0.875	22.22					

Source: Italsider (IRI-Finsider Group)

Organizations and Engineering Data Sources

Colleges and Universities with Metallurgy/Materials Curricula

Schools with metallurgy faculties

University of Alabama
University, Alabama 35486
Department of Chemical and Metallurgical Engineering
(205)-348-6450

B.S. Met. Eng.; M.S. Met. Eng.

University of Alberta
Edmonton, Alberta T6G 2G6
Canada
Department of Mineral Engineering
(403)-432-3337

M. Eng.; Met., Met. Eng., B.Sci.; Met. Eng., Mnl. Eng. M.Sci.; Met., Met. Eng., Mnl. Eng. Ph.D.; Met., Met. Eng., Mnl. Eng.

University of Arizona
Tucson, Arizona 85721
Department of Metallurgical Engineering
(602)-626-1361

B.S. Met. Eng.; M.S. Met.; Ph.D. Met.

University of British Columbia
Metallurgical Engineering Department
307-6350 Stores Road
Vancouver, British Columbia,
Canada V6T 1W5
(604)-228-2676

B.A.Sc. Met. Eng.; M.A.Sci. Met. Eng.; M.Sci. Met.; Ph.D. Met.

Brown University
Providence, RI 02912
Division of Engineering
(401)-863-2628

Undergraduate: Matls. Eng. Option; Graduate: Matls. Sci.

University of California
Berkeley, California 94720
Department of Materials
Science and
Mineral Engineering
(415)-642-3801

B.S., M.S., Ph.D., D. Eng.

Abridged from *Metallurgy/Materials Education Yearbook,* American Society for Metals, May, 1981.

Schools with metallurgy faculties (continued)

University of California, Los Angeles
Los Angeles, California 90024
Department of Materials Science and Engineering
(213)-825-5534

M.S., Ph.D. Phys. Met. and/or Cer. and Sci. of Matls.

California Polytechnic State University
San Luis Obispo, California 93407
Metallurgical and Welding Engineering Department
(805)-546-2568

B.S. Met. Eng.

Carnegie-Mellon University
Pittsburgh, Pennsylvania 15213
Department of Metallurgical and Materials Science
(412)-578-2700

B.S. Met. and Mat. Sci.; M.S., M.E.; Ph.D. Met. & Mat. Sci.

Case Western Reserve University
Cleveland, Ohio 44106
Department of Metallurgical and Materials Science
(216)-368-4230

B.S., M.S., Ph.D.

University of Cincinnati
Cincinnati, Ohio 45221
Department of Materials Science and Metallurgical Engineering
(513)-475-3096

B.S. Met. Eng.; M.S. Mat. Sci. and Met. Eng. of M. Eng.; Ph.D. Mat. Sci. and Met. Eng. or D. Sci.

Cleveland State University
Cleveland, Ohio 44115
Department of Chemical Engineering
(216)-687-3500

M.S.; Dr. Eng.

Colorado School of Mines
Golden, Colorado 80401
Department of Metallurgical Engineering
(303)-279-0300 ext. 2770

B. Met. Eng.; M.S.; M.E.; & Ph.D. Met. Eng.

Columbia University
Henry Krumb School of Mines
New York, New York 10027
Division of Metallurgy and Chemical Metallurgy
(212)-280-5049/2905

B.S., M.S., Mnl. Engr., Met. Engr., Ph.D., Eng. Sci. D.

University of Connecticut
Storrs, Connecticut 06268
Department of Metallurgy
(203)-486-4620

Ph.D., Met.; M.S. Met.

Cornell University
Ithaca, New York 14853
Department of Materials Science and Engineering
(607)-256-4135

B.S.; M. Eng. Mat.; M.S.; Ph.D.

University of Dayton
Dayton, Ohio 45469
School of Engineering
Materials Engineering Graduate Program
(513)-229-2241

M.S., D.E., Ph.D., Mat. Eng.

University of Delaware
Newark, Delaware 19711
Materials and Metallurgy Faculty
College of Engineering
(302)-738-2421

B.S., Met. Option; M.S., Ph.D. in Appl. Sci.

Drexel University
Philadelphia, Pennsylvania 19104
Department of Materials Engineering
(215)-895-2322

B.S., M.S., Ph.D. in Mat. Eng.

Evening College
Department of Materials and Metallurgical Engineering
(215)-895-2179/2158

B.S.

Duke University
Durham, North Carolina 27706
Department of Mechanical Engineering and Materials Science
(919)-684-2832

École Polytechnique
C.P. 6079, Succursale A
Montreal (Quebec), Canada H3C 3A7
Department of Metals Engineering
(514)-344-4787

B.A. Sci., M.A. Sci., M. Eng., Ph.D.

University of Florida
Gainesville, Florida 32611
Department of Materials Science and Engineering
(904)-392-1451

B.S., M.S., Ph.D.

Georgia Institute of Technology
School of Chemical Engineering
Metallurgy Program
Atlanta, Georgia 30332
(404)-894-2879

M.S., Ph.D. Met.

Grove City College
Grove City, Pennsylvania 16127
Department of Engineering
(412)-458-6600 ext. 284, 313

B.S.

The Hartford Graduate Center
Hartford, Connecticut 06120
Department of Metallurgy
(203)-549-3600

M.S. Met. Eng., M.S. Eng. Sci.

Harvard University
Cambridge, Massachusetts 02138
Division of Engineering and Applied Sciences
(617)-495-5829

Ph.D.

University of Idaho
Moscow, Idaho 83843
Department of Mineral Engineering
and Metallurgy
(208)-885-6376

B.S., M.S., Ph.D.

**University of Illinois at Chicago
Circle**
Department of Materials Engineering
P.O. Box 4348
Chicago, Illinois 60680
(312)-996-3428

B.S., M.S., Ph.D.

**University of Illinois at Urbana-
Champaign**
Urbana, Illinois 61801
Department of Metallurgy and Mineral
Engineering
(217)-333-1441

B.S. Met. Engr.; M.S. and Ph.D.

Illinois Institute of Technology
Chicago, Illinois 60616
Department of Metallurgy and
Materials Engineering
(312)-567-3050, 3051, 3052

B.S., M.S., Ph.D.

**Iowa State University of Science
and Technology**
Ames, Iowa 50011
Department of Materials Science and
Engineering
(515)-294-1214

B.S., M.S., Ph.D. Cer. E.; B.S. Met. E.,
Met.; M.S., Ph.D. Met.; M.E., MSE.

The Johns Hopkins University
Baltimore, Maryland 21218
Department of Civil Engineering/
Materials Science and Engineering
(301)-338-7126

B.A., B.S., M.A., M.S., Ph.D.

University of Kentucky
Lexington, Kentucky 40506
Department of Metallurgical Engineering and Materials Science
(606)-258-8884

B.S., M.S., Ph.D.

Lafayette College
Easton, Pennsylvania 18042
Department of Metallurgical Engineering
(215)-253-6281

B.S. Met. Eng.

Laval University
Quebec, Canada G1K 7P4
Department of Minerals and Metallurgy
(418)-656-2167

B.Sci., M.S., D.Sci.

Lehigh University
Whitaker Laboratory #5
Bethlehem, Pennsylvania 18015
Department of Metals and Materials
Engineering
(215)-861-4220

B.S., M.S., Ph.D. Met. and Matls.
Eng.

Marquette University
Milwaukee, Wisconsin 53233
Department of Mechanical Engineering
(414)-224-7259

Option in Met. and Matls. Sci., B.S.
M.E., M.S. and Ph.D. Met. and Matls.
Sci.

University of Maryland
College Park, Maryland 20742
Department of Chemical and Mechani-
cal Engineering
Engineering Materials Program
(301)-454-2421

M.S. and Ph.D.

University of Massachusetts
Amherst, Massachusetts 01002
Department of Mechanical Engineering
(413)-545-2505

M.S. and Ph.D. Mech. Eng., Mat. Major; M.S. Manufacturing Eng.

**Massachusetts Institute of
Technology**
Cambridge, Massachusetts 02139
Department of Materials Science and
Engineering
(617)-253-3300

S.B., Mat. Sci. and Eng.; Mat. Eng.,
Met. Eng.; S.M. Sci.D. and Ph.D. in:
Ceramics, Mat. Eng., Mat. Sci., Metal-
lurgy and Polymerics

McGill University
Montreal, Quebec, Canada
Department of Minerals and Metallurgical Engineering
(514)-392-5706

B.E. Met. Eng., M.E. Met. Eng., Ph.D.
Met. and Met. Eng.

McMaster University
Hamilton, Ontario, Canada
Department of Metallurgy and Materials Science
(416)-525-9140 ext. 4293

M.E., M.S., Met.; M.S., Ph.D., Matl.
Sci.

University of Michigan
Ann Arbor, Michigan 48109
Department of Materials and Metallurgical Engineering
(313)-764-7489

B.S., M.S. and Ph.D., Mat. and Met.
Eng.

Michigan State University
East Lansing, Michigan 48824
Department of Metallurgical, Mechanical and Materials Sciences
(517)-335-5141

B.S. Mat. Sc., Mechanics, M.S., Ph.D.
Met. Mechanics and Mat. Sci.

Michigan Technological University
Houghton, Michigan 49931
Department of Metallurgical Engineering
(906)-487-2630

B.S., M.S., Ph.D. Met. Eng.

University of Minnesota
Minneapolis, Minnesota 55455
Department of Chemical Engineering and Materials Science
(612)-373-2300

B.S., M.S., Ph.D. Met. Eng.

University of Missouri-Rolla
Rolla, Missouri 65401
(314)-341-4711

B.S., M.S., Ph.D. Met. Eng.

Montana College of Mineral Science and Technology
Butte, Montana 59701
Department of Metallurgy and Mineral Processing
(406)-496-4208

B.S., M.S. Metallurgical Eng.; M.S. Met.; B.S., M.S. Mineral Processing Eng.

University of Nebraska
Department of Mechanical Engineering
Metallurgy Option 255 NEC
Lincoln, Nebraska 68588
(402)-472-2375

B.S., M.S., Ph.D. M.E., Met. Eng. Option

University of Nevada-Reno
Reno, Nevada 89557
Department of Chemical and Metallurgical Engineering
(702)-784-6961

B.S. Met. Eng., M.S. Met., M.S. Met. Eng., Ph.D. Eng. in cooperation with the College of Engineering

New Mexico Institute of Mining and Technology
Socorro, New Mexico 87801
Department of Metallurgy and Materials Engineering
(505)-835-5229

B.S., M.S., Met. Eng.; B.S. Ph.D. Met.

Polytechnic Institute of New York
Brooklyn, New York 11201
Department of Physical and Metallurgical Engineering
(212)-643-2196

B.S. Met. Eng.; M.S. Met. Eng.; Met. Engr.; Ph.D. Phys. Met., Matls. Sci., Met. Eng.

State University of New York
Stony Brook, L.I., New York 11794
Department of Materials Science and Engineering
(516)-246-6759

B.S. Eng. Sci.; M.S., Ph.D. Matl. Sci.

North Carolina State University
Raleigh, North Carolina 27650
Department of Materials Engineering
(919)-737-2377

B.S., M.S., Ph.D. Matl. Eng.

Northwestern University
Evanston, Illinois 60201
Department of Materials Science and Engineering
(312)-492-3537/3587

B.S., M.S., and Ph.D., Matls. Sci. and Eng.

University of Notre Dame
Notre Dame, Indiana 46556
Department of Metallurgical Engineering and Materials Science
(219)-283-4516

B.S. Met. Eng.; M.S., Ph.D.

Technical University of Nova Scotia
Halifax, Nova Scotia, Canada
Metallurgy Section
Department of Minerals and Metals

Co-op Program—B. Eng./M. Eng. Metallurgy; B. Eng., Ph.D. in either Mining or Met. Eng.

The Ohio State University
Columbus, Ohio 43210
Department of Metallurgical Engineering
(614)-422-2491

B.S. Met. Eng., M.S., Ph.D.

University of Oklahoma
Norman, Oklahoma 73019
School of Chemical Engineering and Materials Science
(405)-325-5811

B.S. Met. Eng., M.S., Ph.D.

The Pennsylvania State University
University Park, Pennsylvania 16802
Department of Materials Science and Engineering
(814)-865-0497

B.S., M.S., Ph.D., Met.; B.S., M.S., Ph.D., Cer. Sci.; B.S., M.S., Ph.D. Plm. Sci.

University of Pennsylvania
Philadelphia, Pennsylvania 19104
Department of Materials Science and Engineering/K1
(215)-243-8337

B.S., M.S., Ph.D. Matl. Sci. and Eng.

University of Pittsburgh
Pittsburgh, Pennsylvania 15261
Department of Metallurgy and Materials Engineering
(412)-624-5300

B.S., M.S., Ph.D. Met. Eng., M.S. and Ph.D. Matl. Eng.

Schools with metallurgy faculties (continued)

Purdue University
School of Materials Engineering
West Lafayette, Indiana 47907
(317)-749-2601

B.S., M.S., Met. Eng.; B.S. Eng.; M.S.
Eng.; M.S.; Ph.D.

Queen's University
Kingston, Ontario, Canada
Department of Metallurgical Engineering
(613)-547-5877

B.S., M.Sc., Ph.D., Met. Eng.; Ph.D.
Met.

Rensselaer Polytechnic Institute
Troy, New York 12181
Materials Engineering Department
(518)-270-6372

B.S. Mat. Eng.; M.E., M.S. Mat.; D.
Eng.; Ph.D.

Rice University
Houston, Texas 77001
P.O. Box 1892
Department of Materials Science
(713)-527-8101, 4993

Ph.D., M.S. Matl. Sci.; M. Matl. Sci.

Rutgers University College of Engineering
Department of Mechanical and Materials Science
P.O. Box 909
Piscataway, New Jersey 08854
(201)-932-2245, 2584

M.S.; Ph.D.

San Jose State University
San Jose, California 95192
Department of Materials Engineering
(408)-277-2446

B.S., M.S. Matl. Engrg.

South Dakota School of Mines and Technology
Rapid City, South Dakota 57701
Department of Metallurgical Engineering
(605)-394-2341

B.S., M.S. Met. Eng.

University of Southern California
University Park
Los Angeles, California 90007
Department of Materials Science
(213)-743-6225

M.S./Ph.D. Matl. Sci.; M.S. Matl. Eng.

Stanford University
Stanford, California 94305
Department of Materials Science and Engineering
(415)-497-2534

M.S., Ph.D. Matl. Sci. and Eng.

Stevens Institute of Technology
Hoboken, New Jersey 07030
Department of Materials and Metallurgical Engineering
(201)-420-5270

B.E.; B.S.; M.S.; Ph.D.

Syracuse University
Syracuse, New York 13210
Department of Chemical Engineering and Materials Science
(315)-423-2831

M.S. Matl. Sci.; M.S. and Ph.D. in Solid State Sci. and Tech.

Texas A & M University
College Station, Texas 77843
Departments of Mechanical and Civil Engineering

B.S. Mech. Eng.—option in Met. Eng.;
M.S., M.E., or Mechanics and
Materials, Ph.D., D. Eng.

The University of Tennessee
Knoxville, Tennessee 37916
Department of Chemical, Metallurgical, and Polymer Engineering
(615)-974-2421

Ph.D. Met. Eng.

University of Toronto
Toronto, Ontario, Canada M5S 1A4
Department of Metallurgy and Materials Science
(416)-978-3012

B. Appl. Sci., M. Appl. Sci., Ph.D. M. Eng.

University of Utah
Salt Lake City, Utah 84112
Department of Metallurgy and Metallurgical Engineering
(801)-581-6386

B.S., M.S., M.E., Ph.D. Metallurgy

College of Engineering
Department of Materials Science and Engineering
(801)-581-6863

B.S., M.S., Ph.D. Matl. Sci. & Eng.

Vanderbilt University
Box 1621, Station B
Nashville, Tennessee 37235
Mechanical Engineering and Materials Science
(615)-322-2413

B.E., M.S., Ph.D.

Schools with metallurgy faculties (continued)

University of Virginia
Charlottesville, Virginia 22901
Department of Materials Science
(804)-924-3264 / 924-3462

M.M.S.; M.S. & Ph.D. Matl. Sci.

Virginia Polytechnic Institute and State University
Blacksburg, Virginia 24061
Department of Materials Engineering
(703)-961-6640

B.S. Matl. Eng.; M.S. Matl. Eng.;
Ph.D. Matl. Eng. Sci.

University of Washington
Seattle, Washington 98195
(206)-543-2600

M.S. Met. Eng.; M.S. Eng.; M.S.,
Ph.D.

Washington State University
Pullman, Washington 99164
Department of Materials Science and
Engineering
(509)-335-8521

B.S. Phys. Met.; M.S. Matl. Sci. and
Eng.; Ph.D. Eng. Sci.

University of Waterloo
Waterloo, Ontario, Canada N2L 3G1
Department of Mechanical
Engineering
(519)-885-1211

Option in Matls. Sci.; M.A.Sc. Mech.
Eng., Ph.D. Mech. Eng.

Wayne State University
Detroit, Michigan 48202
Department of Chemical and
Metallurgical Engineering
(313)-577-3800

B.S., M.S., Ph.D. Met. Eng.

University of Western Ontario
London, Ontario, Canada N6A 5B9
Materials Engineering Group
Faculty of Engineering Science
(519)-679-3304

B.E. Sci., M.E. Sci., Ph.D.

University of Windsor
Windsor, Ontario, Canada N9B 3P4
Department of Engineering Materials
(519)-253-4232

B.A. Sc., M.A. Sc., Ph.D.

University of Wisconsin-Madison
1509 University Avenue
Madison, Wisconsin 53706
Department of Metallurgical
Engineering
(608)-262-1478

B.S., M.S., Ph.D. Met. Eng.; B.S.,
M.S., Ph.D., Min. Eng.

University of Wisconsin-Milwaukee
Milwaukee, Wisconsin 53201
Materials Department
(414)-963-5181

B.S., M.S., M.E., D. Eng.

Wright State University
Dayton, Ohio 45435
Materials Science and Engineering
Program
(513)-873-2403

B.S. Mat. Sci. Eng.

Yale University
New Haven, Connecticut 06520
Department of Engineering and
Applied Science

B.S., M.S., Ph.D.

Youngstown State University
Youngstown, Ohio 44555
Department of Chemical and
Metallurgical Engineering
(216)-742-3020

B.S., M.S. in Met. Eng.

Schools with faculties of ceramics

Alfred University
New York State College of Ceramics
Alfred, New York 14802
Division of Engineering and Science
(607)-871-2448

B.S., M.S., Ceramic Engineering,
Ceramic Science, Glass Science;
Ph.D., Ceramics

University of California at Berkeley
Berkeley, California 94720
Department of Materials Science
and Engineering
(415)-642-3801

University of California at Los Angeles
Los Angeles, California 90024
Materials Department
(213)-825-5473

University of Florida
Gainesville, Florida 32611
Ceramics Division

Georgia Institute of Technology
Atlanta, Georgia 30332
School of Ceramic Engineering
(494)-894-2850

B. Ceramic Eng.; M.S. Ceramic Eng.;
M.S.; Ph.D.; Ph.D. Ceramic Eng.

University of Illinois
Urbana, Illinois 61801
Department of Ceramic Engineering
(217)-333-1770

B.S., M.S., Ph.D. in Ceramic Eng.;
M.S., Ph.D. in Ceramics

Iowa State University of Science and Technology
Ames, Iowa 50010
Department of Materials Science and
Engineering
(515)-294-1214

B.S., M.S., Ph.D. Ceramic Eng.

The University of Missouri at Rolla
Rolla, Missouri 65401
School of Mines and Metallurgy
Department of Ceramic Engineering
(314)-341-4401

B.S., M.S., Ph.D. Cer. Eng.

Massachusetts Institute of Technology
Cambridge, Massachusetts 02139
Department of Materials Science and
Engineering
Ceramics Division

S.M. Cer.; Sci. D. Ph.D. Cer.

New Mexico Institute of Mining and Technology
Socorro, New Mexico 87801
Department of Metallurgy and
Materials Engineering

The Ohio State University
Columbus, Ohio 43210
Department of Ceramic Engineering
(614)-422-2960

B.Sc., Cer. Eng.; M.S., Ph.D.

The Pennsylvania State University
University Park, Pennsylvania 16802
Materials Science and Engineering
Department
Ceramic Science and Engineering
Section
(814)-865-4992

B.S., M.S., Ph.D., Ceramic Sci.

Rutgers, The State University
New Brunswick, New Jersey 08902
Department of Ceramics
(201)-932-2220

B.S., M.S., Ph.D. in Ceramics and
Ceramic Eng.

The University of Utah
Salt Lake City, Utah 84112
College of Engineering
Department of Materials Science and
Engineering
(801)-581-6863

B.S., M.S., Ph.D. Mat. Sci. and Eng.

Virginia Polytechnic Institute and State University
Blacksburg, Virginia 24061

Ceramic Eng.; Option Materials Eng.;
B.S. Cer. Eng.; M.S. Cer. Eng.; Ph. D.
Matl. Eng. Sci.

Schools with faculties of polymer science

University of Akron
Department of Polymer Science
Whitby Hall
Akron, Ohio 44325

(216)-375-7542

Institute of Polymer Science
Auburn Science and Engineering
Center
Akron, Ohio 44325

(216)-375-7500

Case Western Reserve University
Department of Macromolecular
Science
University Circle
Cleveland, Ohio 44106

(216)-368-4184

University of Lowell
Department of Plastics Engineering
Lowell, Massachusetts 01854

(617)-452-5000 ext. 2234

**Massachusetts Institute of
Technology**
Department of Materials Science and
Engineering
Polymer Division
Cambridge, Massachusetts 02139

S.M. Polymers; Sci.D. and Ph.D.
Polymers

University of Southern Mississippi
Department of Polymer Science
Hattiesburg, Mississippi 39401

(601)-266-7186

Pennsylvania State University
Department of Materials Science and
Engineering
University Park, Pennsylvania 16802

B.S., M.S. & Ph.D. in Polymer Sci.

University of Utah
Department of Materials Science and
Engineering
Salt Lake City, Utah 84112

(801)-581-6863

B.S. (Polymer Science option), M.S.,
Ph.D. Mat. Sci. and Eng. with
Specialty in Polymer Sci.

Selected Engineering Data Sources in the United States

Organization	Address and Telephone Number	Area of Interest • Services
Aluminum Association	818 Connecticut Ave. NW Washington, DC 20006 (202)-862-5100	Aluminum alloys and mill products • Publications: abstracts; indexes; shipment statistics; standards; specifications; technical data manuals
American Association of Engineering Societies	345 E. 47th Ave. New York, NY 10017	Technical issues of national and international importance • Resources of the engineering societies and organizations in this national organization
American Ceramic Society	65 Ceramic Drive Columbus, OH 43214 (614)-268-8645	Glass; ceramic-metal systems; electronic and nuclear ceramics; refractories; structural clay products; whitewares; related ceramic materials • Library. Publications: periodicals; books; abstracts
American Chemical Society	1155 16th St. NW Washington, DC 20036 (202)-872-4600	Organic and inorganic chemistry; analytical chemistry; physical chemistry; industrial and engineering chemistry; etc. • Library. Publications: periodicals; books; abstracts. Educational courses
American Council of Independent Laboratories, Inc.	1725 K St. NW Washington, DC 20006 (202)-659-3766	Professional association of independent engineering and scientific laboratories
American Die Casting Institute	2340 Des Plaines Ave. Des Plaines, IL 60018 (312)-298-1220	Zinc, aluminum, magnesium, and brass die casting alloys • Publications: product standards for die castings
American Electroplaters' Society	1201 Louisiana Ave. Winter Park, FL 32789 (305)-647-1197	Electroplating; deposition and finishing of metals • Library. Publications: periodicals; books (symposiums and conference proceedings). Course and educational training aids
American Foundrymen's Society	Golf and Wolf Rd. Des Plaines, IL 60016 (312)-824-0181	Brass; bronze; gray, ductile, and malleable iron; aluminum; magnesium; environmental control; technical and engineering services • Library. Publications: periodicals; annual transactions; handbooks. Current awareness and documentation retrieval
American Institute of Aeronautics & Astronautics	1290 Avenue of the Americas New York, NY 10019 (212)-581-4300	Materials used in aircraft and space applications • Library. Publications: periodicals; abstracts; indexes; reports; meeting papers; proceedings. Information retrieval. Microfiche
American Iron & Steel Institute	1000 16th St. NW Washington, DC 20036 (202)-452-7100	Steels (carbon, alloy, stainless, tool); steel products (bars, rods, wire, tubing, sheets, etc.); heat-resistant alloys • Publications: steel product manuals
American National Standards Institute	1430 Broadway New York, NY 10018 (212)-354-3300	Industrial and commercial standards; development of international standards • Publications: standards
American Powder Metallurgy Institute	105 College Rd. East Princeton, NJ 08540 (609)-452-7700	Powder metallurgy; metal powders • Publications: periodicals; handbook; proceedings; bibliographies; newsletter; quarterly journal

(continued)

American Society for Metals	Metals Park, OH 44073 (216)-338-5151	Metals, alloys and related engineering materials; metalworking processes, fabrication, testing methods ● Library. Publications: periodicals; books (including handbooks, abstracts and conference proceedings). Information retrieval. Literature searches
American Society for Nondestructive Testing	4153 Arlington Plaza Columbus, OH 43228 (614)-274-6003	Methods and equipment for nondestructive testing of materials in raw, semi-finished, and finished forms ● Library. Publications: periodicals; handbook; technical papers; textbooks; conference proceedings
American Society for Quality Control	230 Wells St. Milwaukee, WI 53203 (414)-272-8575	Quality control; reliability; inspection; research and development; statistics; engineering ● Library. Publications: journals; transactions; books; indexes; reports; reprints
American Society for Testing & Materials	1916 Race St. Philadelphia, PA 19103 (215)-299-5400	Standards for materials, products, services and systems ● Publications: standards; indexes; symposiums; proceedings; periodicals
American Society of Mechanical Engineers	345 E. 47th St. New York, NY 10017 (212)-705-7722	Boiler, pressure vessel, and power test codes; equipment safety codes and standards ● Library. Publications: periodicals; books, including abstracts and reviews; standards and specifications
American Welding Society	550 NW LeJeune Rd. Miami, FL 33126 (305)-642-7090	Welding and allied processes; techniques and materials ● Library. Publications: periodical; handbook; books; bibliographies; standards; specifications. Film directory. Educational courses
Assn. of Iron & Steel Engineers	Suite 2350 Three Gateway Center Pittsburgh, Pa 15222 (412)-281-6323	Technology in the iron and steel industry ● Publications: periodical; yearbook; directory of iron and steel plants; standards, specifications, and recommended practices relating to the steel industry
Copper Development Association	405 Lexington Ave. New York, NY 10174 (212)-953-7300	Copper and its alloys ● Publications: standards; production data. Technical data available for specific requests
Defense Technical Information Center	Bldg. 5, Cameron Station Alexandria, VA 22314 (202)-274-6900	Central facility for Department of Defense for scientific and technical reports ● Publications: reports; abstracts; bibliographies. Information retrieval. Microfiche
Forging Industry Association	1121 Illuminating Bldg. Cleveland, OH 44113 (216)-781-6260	Closed-impression die forgings; forging equipment and supplies ● Publication: quarterly journal
Hazardous Materials Control Research Institute	9300 Columbia Blvd. Silver Spring, MD 20910 (302)-587-9390	Control of hazardous materials, substances, and environments ● Conferences, expositions, publications, newsletters, seminars. Non-profit organization promoting establishment and maintenance of a reasonable balance between expanding industrial productivity and an acceptable environment
International Lead Zinc Research Organization	292 Madison Ave. New York, NY 10017 (212)-532-2373	Lead, zinc, and their alloys ● Publications: research digest; books; manuals; abstracts; reports; reprints

(continued)

536

International Magnesium Association	c/o Bell Publicom, 1406 Third National Bldg., Dayton, OH 45402, (513)-223-0419	Magnesium and its alloys • Publication: monthly news letter
Investment Casting Institute	8521 Clover Meadow Dr., Dallas, TX 75243, (214)-341-0488	Precision investment castings • Publication: handbook
Iron Castings Society	20611 Center Ridge Rd., Rocky River, OH 44116, (216)-333-9600	Gray, ductile and malleable cast iron • Publications; newsletter; handbooks; manuals; films; brochures; specification summaries. Advisory service
Iron & Steel Society—AIME	P.O. Box 411, Warrendale, PA 15086, (412)-776-1535	Ferrous metals and alloys • Library. Publications: periodicals; proceedings; symposiums; monographs
Machinability Data Center	Metcut Research Associates, 3980 Rosslyn Dr., Cincinnati, OH 45209, (513)-271-9510	Machining of all types of materials • Library. Publications: handbook; reports. Advisory service. Information retrieval
The Metallurgical Society—AIME	420 Commonwealth Dr., Warrendale, PA 15086, (412)-776-9000	Ferrous and nonferrous metals and alloys • Library. Publications: periodicals; books, primarily proceedings and symposiums; abstracts
Metal Powder Industries Federation	105 College Rd. East, Princeton, NJ 08540, (609)-452-7700	Ferrous and nonferrous metal powders • Library. Publications: periodicals; handbook; proceedings; bibliographies; reviews; standards and specifications
Metal Properties Council, Inc.	345 E. 47th Ave., New York, NY 10017, (212)-705-7693	Engineering properties of metals and alloys • Identifies major unfulfilled needs for reliable data
Metals and Ceramics Information Center	Battelle Columbus Laboratories, 505 King Ave., Columbus, OH 43201, (614)-424-5000	Advanced metals and ceramics • Publications: reviews of recent developments; state-of-the-art reports; technical memoranda; technical notes; handbooks; current awareness bulletins; newsletters. Advisory service. Literature searches. Research investigations
National Association of Corrosion Engineers	P.O. Box 218340, Houston, TX 77218, (713)-492-0535	Materials deterioration and prevention of deterioration • Library. Publications: periodicals; books; proceedings; abstracts; standards and specifications; special reports
National Bureau of Standards	Washington, DC 20234, (301)-921-2318	Standard reference data and calibration; physical, chemical and materials measurement • Publications: reports; standards
National Technical Information Service	U.S. Department of Commerce, Springfield, VA 22161, (202)-487-4600	Government-generated science and technology information (unclassified) • Publications: reports; abstracts; bibliographies. Information retrieval. Microfiche
Non-Ferrous Founders' Society	455 State St., Des Plaines, IL 60016	Nonferrous casting metals and alloys (aluminum, brass, bronze, etc.) • Publications: handbooks; reports; pamphlets
Plastics Technical Evaluation Center (PLASTEC)	U.S. Army Armament R & D Command, Dover, NJ 07801, (201)-328-4222	Plastics, adhesives and composites • Library. Publications: specifications; standards; bibliographies. Information retrieval

(continued)

Rare-Earth Information Center	Institute for Atomic Research, Iowa State University Ames, IA 50011 (515)-294-2272	Physical metallurgy and solid-state physics of rare-earth metals and their alloys ● Library. Publications: newsletter; special reports; bibliographies; reviews; compilations. Information retrieval
Silver Institute	1001 Connecticut Ave. NW Suite 1138 Washington, DC 20036 (202)-331-1227	Silver and silver products ● Library. Publications: newsletters; journal; brochures. Seminars
Society of Automotive Engineers	4000 Commonwealth Dr. Warrendale, PA 15096 (412)-776-4841	Engineering materials for ground, air and space transport, and mobile equipment ● Library. Publications: periodicals; transactions; handbook; abstracts; indexes; technical papers; standards and specifications; unified numbering system
Society of Die Casting Engineers	Triton College Campus 2000 North 5th Ave. P.O. Box 3002 River Grove, IL 60171	Nonferrous die casting alloys; die casting processes; steels and other die materials; design of castings and related equipment ● Publications: periodicals; transactions; textbooks. Educational courses
Society of Manufacturing Engineers	20501 Ford Rd. Dearborn, MI 48128 (313)-271-1500	Processing and fabricating metals, alloys, and related materials ● Library. Publications: periodicals; books (handbooks, manuals, abstracts, etc.); technical papers; proceedings; standards and specifications. Programmed learning courses
Society of Plastics Engineers	14 Fairfield Dr. Brookfield Center, CT 06805 (203)-775-0471	Plastics materials, properties, processing and applications ● Publications: periodicals; books; meeting preprints. Educational courses
Society of the Plastics Industry	355 Lexington Ave. New York, NY 10017 (212)-573-9400	Plastics and their properties ● Technical, marketing, and management programs. Publications: conference proceedings
Steel Founders' Society	20611 Center Ridge Rd. Rocky River, OH 44116 (216)-333-9600	Carbon and alloy steels for castings; casting technology ● Publications: periodicals; handbook
Thermophysical Properties Research Center	Purdue University 2595 Yeager Rd. West Lafayette, IN 47906 (317)-463-1581	Thermophysical properties of materials ● Library. Publications: reports; handbooks. Information retrieval. Advisory service
Tin Research Institute Inc.	1353 Perry St. Columbus, OH 43201 (614)-424-6200	Tin and its alloys; tin coatings and chemicals ● Library. Publications: quarterly and annual report; handbooks; reprints; technical papers; statistics. Films. Advisory service
Naval Publications & Forms Center	5801 Tabor Ave. Philadelphia, PA 19120 (215)-697-3321	Federal and military specifications ● Publications: specifications and related documents
Zinc Institute/Lead Industries Association	292 Madison Ave. New York, NY 10017 (212)-578-4750	Zinc, lead and their alloys; die casting and galvanizing technology; lead in batteries, gasoline, ceramics and paints ● Library. Publications: annual review; technical booklets; reprints. Films. Advisory service

Automated Literature Searching

The storage of technical information in computer data banks is increasing. In this practice, computer terminals, linked to a central database by telephone, are used to search the stored information and retrieve relevant data.

The American Society for Metals maintains Metadex, a database on metals and materials. The contents of this and other databases relevant to metals and materials are described in the following section on DIALOG information services. Because the formats and search techniques for most databases are similar, the Metadex database will be used to illustrate briefly automated literature searching.

Each month ASM editors abstract documents for the Metadex database. These documents include:

• Seminar proceedings
• Journals
• Magazines
• Books
• Technical reports
• Patents
• Dissertations
• Translations

The document, as keyed for input to the system by the editor, consists of this appropriate information:

Distribution tape field No.	Contents
1	Document number
2	Title
3	Abstract, number of references, abstract, author's initials
4	Authors, corporate authors
5a	Conference name
5b	Conference place
5c	Conference date
5d	Publisher name
5e	Publisher address
5f	Publishing date
5g	Journal title
5h	Report number
5i	Date
5j	Number
5k	Volume reference
5l	Pages
5m	Languages
5n	Patent number
5o	Patent country
5p	Patent application date
5q	ISSN or ISBN
6	Indexing terms
7	Alloy indexing terms
8	Cross references, security code

The document information is also compiled into an abstract, similar to the example given here, which is entered into the database file.

Abstracts are arranged in logical subject groupings for publication in the bound volumes of *Metals Abstracts*. The two digit categories are included in the document number and may be used in retrieval. The subject codes and categories for grouping abstracts are:

Subject codes	Categories
11	Constitution
12	Crystal properties
13	Lattice defects
14	Structural hardening
15	Physics of metals
16	Irradiation effects
21	Metallography
22	Testing and control
23	Analysis
31	Mechanical properties
32	Physical properties
33	Electrical and magnetic
34	Chemical and electrochemical properties
35	Corrosion
41	Ores and raw materials
42	Extraction and smelting
43	Refining and purification
44	Physical chemistry of extraction and refining
45	Ferrous alloy production

Subject codes	Categories
46	Nonferrous alloy production
51	Foundry
52	Working
53	Machining
54	Powder technology
55	Joining
56	Thermal treatment
57	Finishing
58	Metallic coating
61	Engineering components and structures
62	Composites
63	Electronic devices
71	General and nonclassified
72	Special publications

Subject codes, standard indexing terms (descriptors), and key words can be used to search for abstracts in the database. For example, the computer can be used to select every abstract which:

• Is categorized in subject code 51 (Foundry)
• Contains the standard indexing term "casting" in the descriptor field
• Contains the phrase "improvements in quality" somewhere in the abstract text

Any of these three commands, which are circled on the sample abstract, would retrieve the abstract and display it on the computer terminal.

Sample abstract

771587 81-510668
Trends in Casting Technology.
Chandler, H E ; Baxter Jr, D F
Met. Prog.   , Jan. 1981,   119, (1),   96-100
Language: ENGLISH
Document Type: ARTICLE

New die-casting applications for Al in the U.S. auto industry will include: engine blocks, cylinder heads, intake manifolds and pistons; the use of cold chamber cast Mg for transmission cases is being considered. Improvements in casting technology include: use of hot isostatic pressing to cast complex Ti components for the aerospace industry and to improve performance of Al castings; investment casting of stainless steels and Ti; spin casting of iron rolls and Al. (improvements in quality) will result from automation, use of the AOD (Ar   O decarburization) process, metal stream inoculation, in-mold measurement of gas pressure and ceramic foam filter for liquid metal filtration. --M.G.S.

Descriptors: Ferrous alloys, (Casting;) Nonferrous alloys, Casting; Automotive components, Casting; Foundry practice
Section Heading: (51.) (FOUNDRY) Journal Announcement: 8107

Selected Databases Available Through DIALOG Information Services, Inc.

The DIALOG Information Retrieval Service, from DIALOG Information Services, Inc., has been operating since 1972. There are 160 databases available on the system.

The databases on the DIALOG system contain in excess of 50 000 000 records. Records, or units of information, can range from a directory-type listing of specific manufacturing plants to a citation with bibliographic information and an abstract referencing a journal, conference paper, or other original source. For more information on DIALOG write:

DIALOG Information Service, Inc.
Marketing Department
3460 Hillview Avenue
Palo Alto, California 94304

The DIALOG service may also be called at these numbers:

- (800) 227-1927, marketing and training
- (800) 227-1960, customer services Toll-free in the continental U.S., except California
- (800) 982-5838, California
- (415) 858-3785, all other locations

The following databases related to metals and the metals industry are available on the DIALOG system:

BHRA Fluid Engineering

British Hydromechanics Research Association
Cranfield, Bedford MK43 OAJ
England

1974 to present, 63 460 records, indexing and abstracting of worldwide information of fluid engineering, including theoretical research and the latest technology and applications on the following subjects: civil engineering hydraulics, industrial aerodynamics, dredging, fluid flow, fluid power, fluid sealing, fluidics feedback, and tribology

Chemical Abstracts

Chemical Abstracts Service
Columbus, Ohio

Maintains five separate databases covering information relating to chemicals and the chemical-processing industry:

- **Chemical Industry Notes (CIN).** 1974 to present, 368 300 records. Extracts 78 worldwide business-oriented periodicals which cover the chemical processing industry
- **Chemical Regulations and Guidelines System (CRGS).** Regulatory materials in effect. Index to U.S. federal regulatory material relating to the control of chemical substances, covering federal statutes, promulgated regulations, and available federal guidelines, standards, and support documents
- **Chemname.** 1 179 796 chemical substances derived from Chemical Abstracts Service (CAS) Registry Nomenclature File, quarterly updates. Contains a listing of chemical substances in a dictionary-type non-bibliographic file
- **Chemsearch.** 183 876 chemical substances, derived from Chemical Abstracts (CA) Search. Companion file to Chemname listing most recently cited substances
- **Chemsis.** 1967 to 1971, 850 000 records, closed file: 1972 to 1976, 1 178 410 records, closed file; 1977 to present, 1 320 880 records, irregular updates. Dictionary, non-bibliographic file containing those chemical substances cited once during a collective index period of chemical abstracts

Claims

Plenum Data Company
Arlington, VA

Seven separate files on patent information:

- **Claims/Chem.** 265 000 U.S. chemical and chemically related patents issued from 1950 to 1962
- **Claims/Citation.** 1 757 000 records. Includes all patent numbers cited in U.S. patents from 1947 to 1981
- **Claims/Class.** 100 743 records. Classification code and title dictionary for all classes and selected subclasses of the U.S. Patent Classification System
- **Claims/U.S. Patents.** 1963 to 1970, 483 000 records. Contains all patents listed in the general, chemical, electrical, and mechanical sections of the "Official Gazette" of the U.S. Patent Office
- **Claims/U.S. Patent Abstracts.** 1971 to present, 745 200 records. Citations and abstracts for all patents classified by the U.S. Patent Office in the areas of aerospace and aeronautical engineering, agricultural engineering, chemical engineering, chemistry, civil engineering, electrical and electronics engineering, electromagnetic technology, mechanical engineering, nuclear science, and general science and technology
- **Claims/U.S. Patent Abstracts Weekly.** Companion to previous file; includes most current weekly update and records from the current month
- **Claims/Uniterm.** 1950 to present, 497 000 records. Access to chemical and chemically related patents. Subject indexing to facilitate retrieval of chemical structures and polymers

Compendex

Engineering Information, Inc.
New York, NY

January 1970 to present, 1 001 000 records. Machine-readable version of the "Engineering Index". Worldwide coverage of approximately 3500 journals, publications of engineering societies and organizations, papers from the proceedings of conferences, and selected government reports and books

DOE Energy

U.S. Department of Energy
Washington, D.C.

1974 to present, 640 000 records. Covers all aspects of energy and related topics from journal articles, report literature, conference papers, books, patents, dissertations, and translations. Topics include: nuclear, wind, fossil, geothermal, tidal, solar, as well as the environment, energy policy, and conservation

Metadex

American Society for Metals
Metals Park, Ohio 44073

Two separate databases on metals and related materials:

- **Metals Abstracts/Alloys Index.** 1966 to present, 300 000 citations produced by the American Society for Metals (ASM) and the Metals Society (London). Provides coverage of international literature on the science and practice of metallurgy. Included are *Review of Metal Literature* 1966 to 1967, *Metals Abstracts* 1968 to present, and since 1974, *Alloys Index*. *Metals Abstracts* include about

40 000 citations each year from about 1200 primary journal sources. *Alloys Index* supplements *Metals Abstracts* by providing access to the citations through commercial, numerical, and compositional alloy designations; specific metallic systems; and intermetallic compounds found within these systems. In addition to specialized topics (including specific alloy designations, intermetallic compounds, and metallurgical systems), six basic categories of metallurgy are covered: materials, processes, properties, products, forms, and influencing factors. Each month about 3500 new documents related to metals technology are scanned and abstracted, with intensive coverage of appropriate conference papers, reviews, technical reports, and books. These sources are international in scope, including the U.S.S.R. and Eastern European nations among the 43 countries covered

MIDAS, Metals Information Datafile on Alloys and Specifications. Available only through System Development Corporation (SDC), Santa Monica, CA (not available through DIALOG). Commencing Fall, 1982. A numerical data bank on the mechanical and physical properties, composition, and specifications of ferrous and nonferrous alloys. Alloys can be located by specification/designation number, trade name, or by composition. When available, data on product form (such as bar or plate) and condition (heat treatment) are also provided. The mechanical and physical properties covered include: tensile, shear, impact, fatigue, crack propagation, hardness, fracture toughness,

specific heat, thermal conductivity, thermal expansion, melting temperature, electrical resistivity, electrical conductivity, and magnetic permeability. The data is obtained from journal articles, conference proceedings, compilations, and technical reports. The source of the data is listed in each record

Non-Ferrous Metals Abstracts
British Non-Ferrous Metals
  Technology Center
Wantage, Oxfordshire, England

1961 to present, 60 000 records. Covers nonferrous metallurgy and technology. Sources include journals, monographs, British patents, reports, standards, and conference papers

NTIS
National Technical Information
  Service
U.S. Department of Commerce
Springfield, VA

1964 to present, 863 500 citations. Consists of government-sponsored research, development, and engineering, plus analyses prepared by federal agencies, their contractors, or grantees

Standards and Specifications
National Standards Association, Inc.
Bethesda, MD

1950 to present, 72 000 records. Access to all government and industry standards, specifications, and related documents which specify terminology, performance testing, safety, materials, products or other requirement, and characteristics of interest to a particular technology or industry (Relevant standards-issuing organizations included in this database are listed in the

following article on selected standards-issuing organizations)

Surface Coatings Abstracts
Paint Research Association of
  Great Britain
Middlesex, England

1976 to present, 50 200 citations. Contains references to research literature on all aspects of paints and surface coatings, including pigments, dyestuffs, resins, solvents, plasticisers, printing inks, testing pollution, and marketing

Weldasearch
The Welding Institute
Cambridge, England

1967 to present, 55 400 records. Primary coverage of the international literature on all aspects of the joining of metals and plastics, and related areas, such as metals spraying and thermal cutting

World Aluminum Abstracts
American Society for Metals
Metals Park, Ohio 44073

1968 to present, 81 600 citations, monthly updates. Provides coverage of the world's technical literature on aluminum, ranging from ore processing (exclusive of mining) through end uses. Includes information abstracted from approximately 1600 scientific and technical patents, government reports, conference proceedings, dissertations, books, and journals. All aspects of the aluminum industry, aside from mining, are covered, including the following major subject areas: aluminum industry in general, and foundry, metalworking, fabrication, finishing, physical and mechanical metallurgy, engineering properties and tests, quality control and tests, and end uses

Selected Standards-Issuing Organizations Related to Metals and Metalworking

Standards issued by these organizations may be searched using the Standards and Specifications database described in the previous article on DIALOG information services.

Aerospace Industries Association
National Aerospace Standards
  Committee
1725 DeSales St., N.W.
Washington, DC 20036

American Bureau of Shipping
65 Broadway
New York, NY 10006

American Chemical Society
1155 16th St. NW
Washington, DC 20036

American Foundrymen's Society
Golf and Wolf Rd.
Des Plaines, IL 60016

American Gear Manufacturers
  Association
1901 N. Fort Myer Dr.
Arlington, VA 22209

American Institute of Steel
  Construction, Inc.
400 N. Michigan Ave., 8th Floor
Chicago, IL 60611

American Iron & Steel Institute
1000 16th St. N.W.
Washington, DC 20036

American National Standards
  Institute
1430 Broadway
New York, NY 10018

American Nuclear Society, Inc.
555 N. Kensington Ave.
LaGrange Park, IL 60525

American Society for Nondestructive
  Testing, Inc.
4153 Arlingate Plaza
Columbus, OH 43228

American Society for Quality
  Control, Inc.
230 W. Wells St.
Milwaukee, WI 53203

American Society for Testing and
  Materials
1916 Race St.
Philadelphia, PA 19103

American Society of Lubrication
  Engineers
838 Busse Hwy.
Park Ridge, IL 60068

American Society of Mechanical
  Engineers, Inc.
345 E. 47th St.
New York, NY 10017

American Vacuum Society
335 E. 45th St.
New York, NY 10017

American Welding Society
550 N.W. LeJeune Road
P.O. Box 351040
Miami, FL 33125

Anti-Friction Bearing Manufacturers
  Association, Inc.
Century Building, Suite 704
1235 Jefferson Davis Hwy.
Arlington, VA 22202

Architectural Aluminum
  Manufacturers Association
35 E. Wacker Dr.
Chicago, IL 60601

Association of Iron and Steel Engineers
Three Gateway Center, Suite 2350
Pittsburgh, PA 15222

Can Manufacturers Institute
1625 Massachusetts Ave., N.W.
Washington, DC 20036

Cast Bronze Bearing Institute, Inc.
221 North LaSalle, Suite 2026
Chicago, IL 60601

Cast Iron Soil Pipe Institute
1499 Chain Bridge Rd., Suite 203
McLean, VA 22101

Composite Can and Tube Institute
1800 M St. N.W.
Washington, DC 20036

Computer Aided Manufacturing
  International, Inc.
611 Ryan Plaza Dr., Suite 1107
Arlington, TX 76011

Concrete Reinforcing Steel Institute
180 N. LaSalle St., Room 2110
Chicago, IL 60601

Copper Development Association, Inc.
405 Lexington Ave., 57th Floor
New York, NY 10017

Diamond Core Drill Manufacturers
  Association
59 E. Main St.
Moorestown, NJ 08057

Diesel Engine Manufacturers
  Association
c/o A.P. Wherry & Associates
712 Lakewood Center North
Cleveland, OH 44107

Equipment and Tool Institute
1545 Waukegan Road
Glenview, IL 60025

Expansion Joint Manufacturers
  Association
25 N. Broadway
Tarrytown, NY 10591

Fluid Controls Institute, Inc.
Plaza 222, U.S. Highway One
P.O. Box 3854
Tequesta, FL 33458

Forging Industry Association
1121 Illuminating Bldg.
55 Public Square
Cleveland, OH 44113

General Aviation Manufacturers
  Association
1025 Connecticut Ave. N.W.
Washington, DC 20036

Heat Exchange Institute
c/o Thomas Associates, Inc.
1230 Keith Building
1621 Euclid Ave.
Cleveland, OH 44115

Hoist Manufacturers Institute
1326 Freeport Rd.
Pittsburgh, PA 15238

Hydraulic Institute
c/o A.P. Wherry and Associates, Inc.
712 Lakewood Center North
Cleveland, OH 44107

Industrial Fasteners Institute
1505 E. Ohio Building
1717 E. 9th St.
Cleveland, OH 44114

Institute of Electrical and
  Electronics Engineers, Inc.
345 E. 47th St.
New York, NY 10017

Institute of Scrap Iron and Steel, Inc.
1627 K St. N.W., Suite 700
Washington, DC 20006

Instruments Society of America
67 Alexander Dr.
P.O. Box 12277
Research Triangle Park, NC 27709

Investment Casting Institute
8521 Clover Meadow Dr.
Dallas, TX 75243

Lead Industries Association, Inc.
292 Madison Ave.
New York, NY 10017

Magnetic Materials Producers
  Association
c/o H.P. Dolan & Associates, Mgrs.
3451 W. Church St.
Evanston, IL 60603

Manufacturers Standardization
  Society of Valve and Fittings
  Industry
5203 Leesburg Pike, Suite 502
Falls Church, VA 22041

Material Handling Institute, Inc.
1326 Freeport Rd.
Pittsburgh, PA 15238

Metal Powder Industries Federation
105 College Rd. E.
Princeton, NJ 08540

Metal Treating Institute, Inc.
1311 Executive Center, Suite 200
Tallahassee, FL 32301

National Association of Architectural
  Metal Manufacturers
221 N. LaSalle, Suite 2026
Chicago, IL 60601

National Association of Corrosion
  Engineers
P.O. Box 218340
Houston, TX 77218

National Association of Pipe Coating
  Applicators
717 Commercial National Bank
  Building
Shreveport, LA 71101

National Association of Recycling
  Industries, Inc.
330 Madison Ave.
New York, NY 10017

National Board of Boiler and
  Pressure Vessel Inspectors
1055 Crupper Ave.
Columbus, OH 43229

National Coil Coaters Association
1900 Arch St.
Philadelphia, PA 19103

National Solid Wastes Management
  Association
1120 Connecticut Ave. N.W., Suite 930
Washington, DC 20036

Pipe Fabrication Institute
1326 Freeport Rd.
Pittsburgh, PA 15238

Plastics Pipe Institute
369 Lexington Avenue, 10th Floor
New York, NY 10017

Refractories Institute
1102 One Oliver Plaza, No. 3760
Pittsburgh, PA 15222

Resistance Welder Manufacturers
  Association
1900 Arch St.
Philadelphia, PA 19103

Semiconductor Equipment &
  Materials Institute
625 Ellis St., Suite 212
Mountain View, CA 94043

Society for Technical Communication
815 15th St. N.W., Suite 506
Washington, DC 20005

Society of Automotive Engineers, Inc.
400 Commonwealth Dr.
Warrendale, PA 15096

Society of Die Casting Engineers, Inc.
Triton College Campus
2000 N. 5th Ave.
P.O. Box 3002
River Grove, IL 60171

Society of Naval Architects and
  Marine Engineers
One World Trade Center, Suite 1369
Washington, DC 20005

Society of the Plastics Industry, Inc.
355 Lexington Ave.
New York, NY 10017

Specialty Wire Association
1101 Connecticut Ave., No. 700
Washington, DC 20036

Spring Manufacturers Institute, Inc.
1211 W. 22nd St.
Oak Brook, IL 60521

Steel Bar Mills Association
1125 W. Lake St.
Oak Park, IL 60301

Steel Founders' Society of America
20611 Center Ridge Rd.
Cast Metals Federation Building
Rocky River, OH 44116

Steel Joist Institute
1703 Parham Rd., Suite 204
Richmond, VA 23229

Steel Structures Painting Council
4400 Fifth Ave.
Pittsburgh, PA 15213

Steel Tank Institute
666 Dundee Rd., Suite 705
Northbrook, IL 60062

Ultrasonic Industry Association, Inc.
c/o Michael Management
P.O. Drawer F
Jamesburg, NJ 08831

Zinc Institute, Inc.
292 Madison Ave.
New York, NY 10017

Industrial Alloy Cross-Referencing

Two problem areas frequently encountered in the purchasing, researching, and selection of industrial alloys are cross-referencing of designations and specifications and matching equivalent alloys from differing countries. Presented below is a list of reference books and other data sources that are useful resources to consult for such information.

Stahlschlussel, Key to Steel, 12th ed., Verlag Stahlschlussel, 1980. Over 40 000 brands and designations of steels. Includes the material numbers of German and other foreign data; the standards of other countries are related and set opposite to the German number. Includes ASTM, Military, Federal QQ, AMS-AISI, and SAE. Tri-lingual: German, French, and English

International Metallic Materials Cross Reference, 1979, General Electric Company. Presents chemical specifications for materials which are roughly equivalent. Includes steel, aluminum, and copper alloys. The alloys are cross-indexed by numerical, alphabetical, national, and International Standards Organization (ISO). Includes Australia, Belgium, Brazil, Canada, France, West Germany, Italy, Japan, Spain, U.K., U.S., U.S.S.R., and seven other European countries

Woldman's Engineering Alloys, 6th ed., 1979, American Society for Metals. 40 000 alloys arranged alpha-numerically by trade name. Provides chemical composition, main applications, mechanical properties, and the producer for each alloy listed

Unified Numbering System for Metals and Alloys, 2nd ed., 1977, Society of Automotive Engineers. Correlates nationally (United States) used numbering systems currently administered by societies, trade associations including (ferrous and nonferrous): AA, ACI, AISI, and SAE, AMS, ANSI, ASME, ASTM, AWS, CDA, QQ and WW, and MIL

Worldwide Guide to Equivalent Irons and Steels, 1979, American Society for Metals. Provides chemical compositions, mechanical properties, available product forms, and designations for over 20 000 irons and steels from 21 countries. Alloys are sorted by descending carbon content, thereby drawing alloys with equivalent compositions close together in the listing

Worldwide Guide to Equivalent Nonferrous Metals and Alloys, 1980, American Society for Metals. Companion to the above volume for nonferrous alloys

Handbook of Comparative World Steel Standards, 1974, International Technical Information Institute, Tokyo. Steel standards and specifications from Japan, U.S., U.K., West Germany, France and the U.S.S.R. tabulated in a convenient, comparative style

Metallic Materials Specification Handbook, 3rd ed., E. & F.N. Spon, 1980. Provides ingredients and properties of a symbol representing a specification or trade name. Includes 50 000 trade names specifications or symbols relating to metals

Soviet Alloy Handbook, Oct 1980, Battelle Metals and Ceramics Information Center, Columbus, Ohio. 358 pages consisting almost entirely of tables of data and cross-indexes

Handbook of International Alloy Compositions and Designations—Volume 1, Titanium, Nov 1976, Battelle Metals and Ceramics Information Center, Columbus, Ohio. 240 pages, 89 references, 33 tables, 13 appendixes

Handbook of International Alloy Compositions and Designations—Volume 2, Superalloys, Dec 1978, Battelle Metals and Ceramics Information Center, Columbus, Ohio. 405 pages, 126 references, 31 tables, 8 appendixes

Handbook of International Alloy Compositions and Designations—Volume 3, Aluminum, Oct 1980, Battelle Metals and Ceramics Information Center, Columbus, Ohio. 863 pages, 4 tables, 7 appendixes

Aluminum Standards and Data, Aluminum Association, Inc., 818 Connecticut Ave. NW, Washington DC 20006

Standards Handbook, Part 7, Data Specifications, Copper Development Association, 405 Lexington Ave., New York, NY 10174

Alloy Digest, Engineering Alloy Digest, Inc., 356 North Mountain Ave., Box 823, Upper Montclair, NJ 07043

Cross Reference to Steels

The following table and its accompanying index cross reference the standard AISI carbon, alloy, stainless, and tool steels to selected chemically similar specifications established by the United States Government and standards organizations from France, Germany, Japan, Sweden and the United Kingdom. It is recommended that this reference be used only as a guide. Any determination of the true equivalency of any two alloys should be made only after carefully comparing their chemical compositions. For further information on the chemical compositions and mechanical properties of the alloys listed in this reference, consult the publications listed in the preceding section, Industrial Alloy Cross-Referencing.

AISI	UNS	FED	MIL-SPEC (MIL)	France (AFNOR)	Germany (DIN)	Japan (JIS)	Sweden (SS$_{14}$)	U.K. (B.S.)
Nonresulfurized Carbon Steels								
1008	G10080	QQ-S-637 (C1008) QQ-S-698 (C1008)	S-11310 (CS1008)		1.0204			
1010	G10100		S-11310 (CS1010)	XC 10	1.1121	S 10 C S 12 C S 9 CK		
1012	G10120		S-11310 (CS1012)					
1015	G10150	QQ-S-698 (C1015)	S-16974	XC 15 XC 18	1.1141	S 15 C S 17 C S 15 CK	1370	
1016	G10160		S-866		1.0419			
1017	G10170		S-11310 (CS1017)	X C 15 X C 18	1.1141	S 15 C S 17 C S 17 CK	1370	
1018	G10180		S-11310 (CS1018)					
1019	G10190							
1020	G10200		S-11310 (CS1020)	CC 20	1.0402		1450	040 A 20 070 M 20
1021	G10210							
1022	G10220		S-11310 (CS1022)		1.1133	SMnC 21		
1023	G10230	QQ-S-700 (C1025)	S-11310 (CS1025)		1.1158	S 25 C S 28 C		
1025	G10250	QQ-S-700 (C1025)	S-11310 (CS1025)		1.1158	S 25 C S 28 C		
1026	G10260							
1029	G10290							
1030	G10300	QQ-S-700 (C1030)	S-11310 (CS1030)		1.1172			
1035	G10350	QQ-S-635 (C1035) QQ-S-700 (C1035)		CC 35	1.0501		1550	060 A 35 080 A 32 080 A 35 080 A 37 080 M 36
1037	G10370							
1038	G10380			X C 38 TS	1.1176			
1038H	H10380			XC 38 TS	1.1176			
1039	G10390			35 M 5	1.1157			120 M 36 150 M 36

AISI	UNS	FED	MIL-SPEC (MIL)	France (AFNOR)	Germany (DIN)	Japan (JIS)	Sweden (SS$_{14}$)	U.K. (B.S.)
1040			S-11310 (CS1040)		1.1186	S 40 C		080 A 40 2 S. 93
1042		QQ-S-635 (C1042)		XC 42 XC 42 TS XC 45 XC 48	1.1191	S 45 C S 48 C	1672	
1043	G10430			CC 45	1.0503		1650	060 A 47 080 H 46 080 M 40 080 M 46
1044	G10440							
1045	G10450	QQ-S-635 (C1045) QQ-S-700 (C1045)		XC 42 XC 42 TS XC 45 XC 48	1.1191	S 45 C S 48 C	1672	
1045H	H10450			XC 42 XC 42 TS XC 45 XC 48	1.1191	S 45 C S 48 C	1672	
1046	G10460							
1049	G10490			XC 48 TS	1.1201	S 50 C	1660	
1050	G10500	QQ-S-635 (C1050) QQ-S-700 (C1050)	S-16974		1.1210	S 53 C S 55 C		
1053	G10530							
1055	G10550	QQ-S-700 (C1055)			1.1209			
1060	G10600		S-16974	CC 55	1.0601			060 A 62
1070	G10700		S-11713 (2)	XC 68	1.1231		1770 1778	
1078	G10780			XC 75	1.1248		1774	
1080	G10800	QQ-S-700 (C1080)	S-16974					
1084	G10840	QQ-S-700 (C1084)			1.0647			
1090	G10900				1.1273			
1095	G10950	QQ-S-700 (C1095)	S-16788 (C1095)		1.1274	SUP 4	1870	060 A 96 En. 44 B

Resulfurized Carbon Steels

AISI	UNS	FED	MIL-SPEC (MIL)	France (AFNOR)	Germany (DIN)	Japan (JIS)	Sweden (SS$_{14}$)	U.K. (B.S.)
1110	G11100	QQ-S-637 (C1110)			1.0702	SUM 11 SUM 12		
1117	G11170	QQ-S-637 (C1117)	S-18411					
1118	G11180	QQ-S-637 (C1118)						
1137	G11370	QQ-S-637 (C1137)						
1139	G11390	QQ-S-637 (C1139)						
1140	G11400	QQ-S-637 (C1140)		35 MF 4	1.0726		1957	
1141	G11410	QQ-S-637						
1144	G11440	QQ-S-637 (C1144)						
1146	G11460	QQ-S-637 (C1146)		45 MF 4	1.0727		1973	
1151	G11510	QQ-S-637 (C1151)	S-20137A					

AISI	UNS	FED	MIL-SPEC (MIL)	France (AFNOR)	Germany (DIN)	Japan (JIS)	Sweden (SS$_{14}$)	U.K. (B.S.)
Rephosphorized and Resulfurized Carbon Steels								
1211	G12110	QQ-S-637 (C1211)						
1212	G12120	QQ-S-637 (C1212)			1.0711	SUM 21		
1213	G12130	QQ-S-637 (C1913)			1.0715	SUM 22		220 M 07
12L14	G12144				1.0718	SUM 22 L SUM 24 L	1914	
1215	G12150	QQ-S-637						
High-Manganese Carbon Steels								
1513	G15130							
15B21H	H15211				1.5523			
1522	G15220							
1522H	H15220				1.1133	SMnC 21		
1524	G15240				1.1160			
1524H	H15240				1.1160			
1526	G15260				1.1161			
1526H	H15260							
1527	G15270				1.1161			
15B35H	H15351							
15B37H	H15371							
1541	G15410			40 M 5	1.1161	SMn 2 H SMn 2 SCMn 3	2120	
1541H	H15410			40 M 5	1.167	SMn 2 H SMn 2 SCMn 3	2120	
15B41H	H15411				1.5527			
1548	G15480				1.1226			
15B48H	H15481							
1551	G15510							
1552	G15520				1.1226			
1561	G15610							
15B62H	H15621							
1566	G15660				1.1260			
Alloy Steels								
1330	G13300		S-16974		1.1165	SMn 1 H SCMn 2		
1330H	H13300				1.1165	SMn 1 H SCMn 2		
1335	G13350		S-16974	40 M 5	1.1167	SMn 2 H SMn 2 SCMn 3	2120	
1335H	H13350			40 M 5	1.1167	SMn 2 H SMn 2 SCMn 3	2120	

AISI	UNS	FED	MIL-SPEC (MIL)	France (AFNOR)	Germany (DIN)	Japan (JIS)	Sweden (SS$_{14}$)	U.K. (B.S.)
1340	G13400		S-16974					
1340H	H13400				1.5069			
1345	G13450				1.0912			2 S 516 2 S 517
1345H	H13450				1.0912			2 S 516 2 S 517
4023	G40230							
4024	G40240							
4027	G40270							
4027H	H40270							
4028	G40280							
4028H	H40280							
4032	G40320	QQ-S-00629 (FS4032)						
4032H	H40320							
4037	G40370							
4037H	H40370							
4042	G40420							
4042H	H40420							
4047	G40470							
4047H	H40470							
4118	G41180							
4118H	H41180							
4130	G41300		S-16974	25 CD 4(S)	1.7218	SCM 2 SCCrM 1	2225	CDS 110
4130H	H41300			25 CD 4(S)	1.7218	SCM 2 SCCrM 1	2225	CDS 110
4135	G41350		S-18733	35 CD 4 35 CD 4 TS	1.7220	SCM 1 SCCrM 3	2234	708 A 37
4135H	H41350			35 CD 4 35 CD 4 TS	1.7220	SCM 1 SCCrM 3	2234	708 A 37
4137	G41370			40 CD 4 42 CD 4	1.7225	SCM 4 H SCM 4	2244	708 A 42 708 M 40 709 M 40
4137H	H41370			40 CD 4 42 CD 4	1.7225	SCM 4 H SCM 4	2244	708 A 42 708 A 40 709 A 40
4140	G41400		S-16974	40 CD 4 42 CD 4	1.7225	SCM 4 H SCM 4	2244	708 A 42 708 M 40 709 M 40
4140H	H41400			40 CD 4 42 CD 4	1.7225	SCM 4 H SCM 4	2244	708 A 42 708 M 40 709 M 40
4142	G41420							
4142H	H41420				1.7223			

AISI	UNS	FED	MIL-SPEC (MIL)	France (AFNOR)	Germany (DIN)	Japan (JIS)	Sweden (SS₁₄)	U.K. (B.S.)
4145	G41450		S-16974					
4145H	H41450							
4147	G41470				1.7228	SCM 5 H SCM 5		
4147H	H41470				1.7228	SCM 5 H SCM 5		
4150	G41500		S-11595 (ORD4150)		1.7228	SCM 5 H SCM 5		
4150H	H41500				1.7228	SCM 5 H SCM 5		
4161	G41610							
4161H	H41610							
4320	G43200							
4320H	H43200							
4340	G43400		S-16974		1.6565	SNCM 8		817 M 40 3111 Type 6 2 S 119 3 S 95
4340H	H43400				1.6565	SNCM 8		817 M 40 3111 Type 6 2 S 119 3 S 95
E4340	G43406		S-5000		1.6562	40 NiCrMo 7 40 NiCrMo 7 KB		Type 8 S 139
E4340H	H43406				1.6562			Type 8 S 139
4615	G46150		S-7493					
4620	G46200		S-7493					
4620H	H46200							
4626	G46260							
4626H	H46260							
4720	G47200							
4720H	H47200							
4815	G48150							
4815H	H48150							
4817	G48170							
4817H	H48170							
4820	G48200							
4820H	H48200							
50B40	G50401				1.7007			
50B40H	H50401				1.7007			
50B44	G50441							
50B44H	H50441							

AISI	UNS	FED	MIL-SPEC (MIL)	France (AFNOR)	Germany (DIN)	Japan (JIS)	Sweden (SS₁₄)	U.K. (B.S.)
5046	G50460							
5046H	H50460							
50B46	G50461							
50B46H	H50461							
50B50	G50501				1.7138	SUP 11		
50B50H	H50501				1.7138	SUP 11		
50B60	G50601							
50B60H	H50601							
5117	G51170							
5120	G51200			20 MC 5	1.7147			
5120H	H51200			20 MC 5	1.7147			
5130	G51300				1.7030			530 A 30 530 H 30
5130H	H51300			32 C 4	1.7033	SCr 2 H SCr 2		530 A 32 530 H 32
5132	G51320			32 C 4	1.7033	SCr 2 H SCr 2		530 A 32 530 H 32
5132H	H51320			38 C 4	1.7034	SCr 3 H		530 A 36 530 H 36 Type 3
5135	G51350			38 C 4	1.7034	SCr 3 H		530 A 36 530 H 36 Type 3
5135H	H51350			42 C 4	1.7035	SCr 4 H		530 A 40 530 H 40 530 M 40 2 S 117
5140	G51400			42 C 4	1.7035	SCr 4 H		530 A 40 530 H 40 530 M 40 2 S 117
5140H	H51400			42 C 2 45 C 2	1.7006			
5150	G51500			42 C 2 45 C 2	1.7006			
5150H	H51500							
5155	G51550			55 C 3	1.7176			
5155H	H51550			55 C 3	1.7176			
5160	G51600							
5160H	H51600							
E51100	G51986				1.3503			
E52100	G52986		S-980 S-7420 S-22141	100 C 6	1.3505			534 A 99 535 A 99
6118	G61180				1.7511			
6118H	H61180				1.7511			

AISI	UNS	FED	MIL-SPEC (MIL)	France (AFNOR)	Germany (DIN)	Japan (JIS)	Sweden (SS₁₄)	U.K. (B.S.)
6150	G61500		S-8503	50 CV 4	1.8159	SUP 10	2230	735 A 50 En. 47
6150H	H61500			50 CV 4	1.8159	SUP 10	2230	En. 47 735 A 50
81B45	G81451							
81B45H	H81451							
8615	G86150		S-866					
8617	G86170			20 NCD 2 22 NCD 2	1.6523	SNCM 21 H SNCM 21		805 H 20 805 M 20
8617H	H86170			20 NCD 2 22 NCD 2	1.6523	SNCM 21 H SNCM 21		805 H 20 805 M 20
8620	G86200		S-16974	20 NCD 2 22 NCD 2	1.6523	SNCM 21 H SNCM 21		805 H 20 805 M 20
8620H	H86200			20 NCD 2 22 NCD 2	1.6523	SNCM 21 H SNCM 21		805 H 20 805 M 20
8622	G86220				1.6543			805 A 20
8622H	H86220				1.6543			805 A 20
8625	G86250		S-16974					
8625H	H86250							
8627	G86270							
8627H	H86270							
8630	G86300		S-16974		1.6545			
8630H	H86300				1.6545			
86B30H	H86301							
8637	G86370							
8637H	H86370							
8640	G86400		S-16974		1.6546			Type 7
8640H	H86400				1.6546			Type 7
8642	G86420							
8642H	H86420							
8645	G86450		S-16974					
8645H	H86450							
86B45	G86451							
86B45H	H86451							
8650	G86500							
8650H	H86500							
8655	G86550							
8655H	H86550							
8660	G86600							
8660H	H86600							
8720	G87200				1.6543			805 A 20

AISI	UNS	FED	MIL-SPEC (MIL)	France (AFNOR)	Germany (DIN)	Japan (JIS)	Sweden (SS₁₄)	U.K. (B.S.)
8720H	H87200				1.6543			805 A 20
8740	G87400	S-6049			1.6546			Type 7
8740H	H87400				1.6546			Type 7
8822	G88220				1.6543			805 A 20
8822H	H88220				1.6543			805 A 20
9260	G92600			60 S 7 61 SC 7	1.0909			250 A 58
9260H	H92600			60 S 7 61 SC 7	1.0909			250 A 58
9310H	H93100							
94B15	G94151							
94B15H	H94151							
94B17	G94171							
94B17H	H94171							
94B30	G94301							
94B30H	H94301							

Water-Hardening Tool Steels (W Series)

AISI	UNS	FED	MIL-SPEC (MIL)	France (AFNOR)	Germany (DIN)	Japan (JIS)	Sweden (SS₁₄)	U.K. (B.S.)
W1	T72301							
W2	T72302	QQ-T-580(W-2)						
W5	T72305							

Shock-Resisting Tool Steels (S Series)

AISI	UNS	FED	MIL-SPEC (MIL)	France (AFNOR)	Germany (DIN)	Japan (JIS)	Sweden (SS₁₄)	U.K. (B.S.)
S1	T41901	QQ-T-570(S-1)		55 WC 20	1.2550			
S2	T41902	QQ-T-570(S-2)						
S5	T41905	QQ-T-570(S-5)						
S6	T41906	QQ-T-570(S-6)						
S7	T41907							

Oil-Hardening Cold Work Tool Steels (O Series)

AISI	UNS	FED	MIL-SPEC (MIL)	France (AFNOR)	Germany (DIN)	Japan (JIS)	Sweden (SS₁₄)	U.K. (B.S.)
O1	T31501	QQ-T-570(O-1)			1.2510			BO1
O2	T31502	QQ-T-570(O-2)		90 MV 8	1.2842			BO2
O6	T31506	QQ-T-570(O-6)						
O7	T31507	QQ-T-570(O-7)						

Medium-Alloy, Air Hardening Cold Work Tool Steels (A Series)

AISI	UNS	FED	MIL-SPEC (MIL)	France (AFNOR)	Germany (DIN)	Japan (JIS)	Sweden (SS₁₄)	U.K. (B.S.)
A2	T30102	QQ-T-570(A-2)		Z 100 CDV 5	1.2363	SKD 12	2260	BA2
A3	T30103	QQ-T-570(A-3)						
A4	T30104	QQ-T-570(A-4)						
A6	T30106	QQ-T-570(A-6)						
A7	T30107	QQ-T-570(A-7)						
A8	T30108	QQ-T-570(A-8)						
A9	T30109	QQ-T-570(A-9)						
A10	T30110	QQ-T-570(A-10)						

AISI	UNS	FED	MIL-SPEC (MIL)	France (AFNOR)	Germany (DIN)	Japan (JIS)	Sweden (SS₁₄)	U.K. (B.S.)

High-Carbon, High Chromium Cold Work Tool Steels (D Series)

AISI	UNS	FED	MIL	France	Germany	Japan	Sweden	U.K.
D2	T30402	QQ-T-570(D-2)			1.2379			BD2
D3	T30403	QQ-T-570(D-3)		Z 200 C 12	1.2080	SKD 1		BD3
D4	T30404	QQ-T-570(D-4)						
D5	T30405	QQ-T-570(D-5)						
D7	T30407	QQ-T-570(D-7)						

Low-Alloy Special-Purpose Tool Steels (L Series)

L2	T61202	QQ-T-570(L-2)			1.2210			
L6	T61206	QQ-T-570(L-6)		55 NCOV 7	1.2713			

Mold Steels (P Series)

P2	T51602							
P3	T51603							
P4	T51604				1.2341			
P5	T51605							
P6	T51606				1.2735			
P20	T51620				1.2330			
P21	T51621							

Hot Work Tool Steels (H Series)

H10	T20810	QQ-T-570(H-10)		32 DCV 28				
H11	T20811	QQ-T-570(H-11)		Z 38 CDV 5	1.2343	SKD 6		BH11
H12	T20812	QQ-T-570(H-12)			1.2606	SKD 62		BH12
H13	T20813	QQ-T-570(H-13)		Z 40 COV 5	1.2344	SKD 61	2242	BH13
H14	T20814	QQ-T-570(H-14)						
H19	T20819	QQ-T-570(H-19)						
H21	T20821	QQ-T-570(H-21)		Z 30 WCV 9	1.2581	SKD 5		BH21
H22	T20822	QQ-T-570(H-22)						
H23	T20823	QQ-T-570(H-23)			1.2625			
H24	T20824	QQ-T-570(H-24)						
H25	T20825	QQ-T-570(H-25)						
H26	T20826	QQ-T-570(H-26)						
H42	T20842	QQ-T-570(H-42)						

Tungsten High-Speed Tool Steels (T Series)

T1	T12001	QQ-T-590		Z 80 WCV 18-04-01	1.3355	SKH 2		BT1
T2	T12002	QQ-T-590						
T4	T12004	QQ-T-590		Z 80 WKCV 18-05-04-01	1.3255	SKH 3		BT4
T5	T12005	QQ-T-590			1.3265	SKH 4A		BT5
T6	T12006	QQ-T-590(T-6)						
T8	T12008	QQ-T-590						
T15	T12015	QQ-T-590			1.3202			BT15

AISI	UNS	FED	MIL-SPEC (MIL)	France (AFNOR)	Germany (DIN)	Japan (JIS)	Sweden (SS$_{14}$)	U.K. (B.S.)
Molybdenum High-Speed Tool Steels (M Series)								
M1	T11301	QQ-T-590(M-1)		Z 85 DCWV 08-04-02-01	1.3346			BM1
M2	T11302	QQ-T-590(M-2)		Z 85 WDCV 06-05-04-02	1.3343	SKH 9	2722	BM2
M3 Class 1	T11313	QQ-T-590(M-3)		Z 90 WDCV 06-05-04-02	1.3342			
M3 Class 2	T11323	QQ-T-590(M-3)		Z 120 WDCV 06-05-04-03	1.3344	SKH 52 SKH 53		
M4	T11304	QQ-T-590(M-4)						
M6	T11306	QQ-T-590(M-6)						
M7	T11307	QQ-T-590(M-7)			1.3348			
M10	T11310	QQ-T-590(M-10)						
M30	T11330	QQ-T-590(M-30)						
M33	T11333	QQ-T-590(M-33)			1.3249			BM 34
M34	T11334	QQ-T-590(M-34)			1.3249			BM 34
M36	T11336	QQ-T-590(M-36)						
M41	T11341	QQ-T-590(M-41)		Z 110 WKCDV 07-05-04-04-02	1.3246			
M42	T11342	QQ-T-590(M-42)		110 DKCWV 09-08-04-02-01	1.3247			BM 42
M43	T11343	QQ-T-590(M-43)						
M44	T11344	QQ-T-590(M-44)						
M47	T11347							
Austenitic Stainless Steels								
201	S20100	QQ-S-766						
202	S20200	QQ-S-763 QQ-S-766 STD-66						
205	S20500							
301	S30100	QQ-S-766	S-5059	Z 12 CN 17.08	1.4310	SUS 301		
302	S30200	QQ-S-763 QQ-S-766 QQ-W-423	S-862					
302B	S30215							
303	S30300		S-862	Z 10 CNF 18.09	1.4305	SUS 303	2346	303 S 21
303Se	S30323		S-862					
304	S30400	QQ-W-423 QQ-S-763 QQ-S-766 STD-66	F-20138 S-862 S-5059 S-23195 S-23196 T-6845 T-8504 T-8506	Z 6 CN 18.09	1.4301	SUS 304	2332 2333	304 S 15 302 S 17 304 S 16 304 S 18 304 S 25 304 S 40 En. 58 E
304L	S30403	QQ-S-763 QQ-S-766	S-862 S-23195 S-23196		1.4306	SUS 304 L SCS 19	2352	304 S 12 304 S 14 304 S 22 S. 536

AISI	UNS	FED	MIL-SPEC (MIL)	France (AFNOR)	Germany (DIN)	Japan (JIS)	Sweden (SS₁₄)	U.K. (B.S.)
S30430	S30430							
304N	30451							
305	S30500	QQ-S-763 QQ-W-423			1.4303	SUS 305 SUS 305 J1		
308	S30800				1.4303	SUS 305 SUS 305 J1		
309	S30900	QQ-S-763 QQ-S-766	S-862		1.4828			
309S	S30908			Z 15 CN 24.13	1.4833			
310	S31000	QQ-S-763 QQ-S-766 QQ-W-423 STD-66	S-862	Z 12 CNS 25.20	1.4841	SUS Y 310		310 S 24
310S	S31008							
314	S31400			Z 12 CNS 25.20	1.4841	SUS Y 310		310 S 24
316	S31600	QQ-S-763 QQ-S-766 QQ-W-423	S-862 S-5059	Z CND 17.11	1.4401	SUS 316 SUH 309 SUS Y 316	2347	316 S 16 316 S 18 316 S 25 316 S 26 316 S 30 316 S 40 316 S 41 En. 58 H
316L	S31603	QQ-S-763 QQ-S-766	S-862	Z 2 CND 17.12	1.4404	SUS 316 L SUH 310	2348	316 S 12 316 S 14 316 S 22 316 S 24 316 S 29 316 S 30 316 S 31 316 S 37 316 S 82 S. 537
316F	S31620							
316N	S31651							
317	S31700	QQ-S-763	S-862		1.4449	SUS 317		
317L	S31703			Z 2 CND 19.15	1.4438		2367	317 S 12
321	S32100	QQ-S-763 QQ-S-766 QQ-W-423	S-862	Z 6 CNT 18.10	1.4541	SUS 321	2337	CDS-20 321 S 12 321 S 18 321 S 22 321 S 27 321 S 40 321 S 49 321 S 50 321 S 59 321 S 87 En. 58 B En. 58 C
329	S32900							
330	N08330							
347	S34700	QQ-S-763 QQ-W-423	S-862 S-23195 S-23196	Z 6 CNND 18.10	1.4550	SUS 347	2338	347 S 17 En. 58 F En. 58 G ANC 3 Grade B

AISI	UNS	FED	MIL-SPEC (MIL)	France (AFNOR)	Germany (DIN)	Japan (JIS)	Sweden (SS$_{14}$)	U.K. (B.S.)
348	S34800	QQ-S-766	S-23195 S-23196		1.4546			347 S 17 347 S 18 347 S 40 S. 130 S. 525 S. 527
384	S38400							

Ferritic Stainless Steels

AISI	UNS	FED	MIL-SPEC (MIL)	France (AFNOR)	Germany (DIN)	Japan (JIS)	Sweden (SS$_{14}$)	U.K. (B.S.)
405	S40500	QQ-S-763	S-862	Z 6 CA 13	1.4002	SUS 405		405 S 17
409	S40900	QQ-S-763 QQ-S-766 QQ-W-423	S-862					
430	S43000	QQ-S-763 QQ-S-766 QQ-W-423	S-862	Z 8 C 17	1.4016	SUS 430	2320	430 S 15
430F	S43020		S-862	Z 10 CF 17	1.4104	SUS 430F	2383	
430FSe	S43023		S-862					
434	S43400			Z 8 CD 17.01	1.4113	SUS 434	2325	434 S 19
436	S43600							
442	S44200							
446	S44600	QQ-S-763 QQ-S-766	S-862					

Martensitic Stainless Steels

AISI	UNS	FED	MIL-SPEC (MIL)	France (AFNOR)	Germany (DIN)	Japan (JIS)	Sweden (SS$_{14}$)	U.K. (B.S.)
403	S40300	QQ-S-763	S-862		1.4024	SUS 43 SUS 403 SUS 416		420 S 29 En. 56 B
410	S41000	QQ-S-763 QQ-W-423	S-862	Z 10 C 13 Z 10 C 14 Z 12 C 13	1.4006	SUS 410	2302	410 S 21 En. 56 A ANC 1 Grade A 3 S. 61 S. 141
414	S41400	QQ-S-763						
416	S41600	QQ-W-423	S-862	Z 12 CF 13	1.4005		2380	416 S 21
416Se	S41623		S-862					
420	S42000	QQ-S-763 QQ-S-766 QQ-W-423	S-862	Z 20 CB	1.4021	SUS 420 JI	2303	420 S 37 CDS-18 EN. 56 C 3 S. 62
420F	S42020							
422	S42200				1.4935	SUH 616		
431	S43100		S-862		1.4057	SUS 431	2321	431 S 29 5 S 80
440A	S44002	QQ-S-763	S-862					
440B	S44003	QQ-S-763	S-862		1.4112	SUS 440 B		
440C	S44004	QQ-S-763	S-862		1.4125	SUS 440 C		

Precipitation-Hardening Stainless Steels

AISI	UNS	FED	MIL-SPEC (MIL)	France (AFNOR)	Germany (DIN)	Japan (JIS)	Sweden (SS$_{14}$)	U.K. (B.S.)
630	S17400		C-24111 S-81506 S-81591					

AISI	UNS	FED	MIL-SPEC (MIL)	France (AFNOR)	Germany (DIN)	Japan (JIS)	Sweden (SS₁₄)	U.K. (B.S.)
631	S17700		S-25043 W-46078		1.4568			
632	S15700				1.4532			
633	S35000							
634	S35500							
660	K66286				1.4980			

Index to the
Cross Reference to Steels

France

Designation — AISI

AFNOR

Designation	AISI
20 MC 5	5120
20 MC 5	5120H
20 NCD 2	8617
20 NCD 2	8617H
20 NCD 2	8620
20 NCD 2	8620H
22 NCD 2	8617
22 NCD 2	8617H
22 NCD 2	8620
22 NCD 2	8620H
25 CD 4(S)	4130
25 CD 4(S)	4130H
32 C 4	5130H
32 C 4	5132
32 DCV 28	H10
35 CD 4	4135
35 CD 4	4135H
35 CD 4 TS	4135
35 CD 4 TS	4135H
35 M 5	1039
35 MF 4	1140
38 C 4	5132H
38 C 4	5135
40 CD 4	4137
40 CD 4	4137H
40 CD 4	4140
40 CD 4	4140H
40 M 5	1335
40 M 5	1335H
40 M 5	1541
40 M 5	1541H
42 C 2	5140H
42 C 2	5150
42 C 4	5135H
42 C 4	5140
42 CD 4	4137
42 CD 4	4137H
42 CD 4	4140
42 CD 4	4140H
45 C 2	5140H
45 C 2	5150
45 MF 4	1146
50 CV 4	6150
50 CV 4	6150H
55 C 3	5155
55 C 3	5155H
55 NCOV 7	L6
55 WC 20	S1
60 S 7	9260
60 S 7	9260H
61 SC 7	9260
61 SC 7	9260H

France (continued)

AFNOR

Designation	AISI
90 MV 8	02
100 C 6	E52100
CC 20	1020
CC 35	1035
CC 45	1043
CC 55	1060
XC 10	1010
XC 15	1015
XC 15	1017
XC 18	1015
XC 18	1017
XC 18 S	1023
XC 25	1023
XC 38 TS	1038
XC 38 TS	1038H
XC 42	1042
XC 42	1045
XC 42	1045H
XC 42 TS	1042
XC 42 TS	1045
XC 42 TS	1045H
XC 45	1042
XC 45	1045
XC 45	1045H
XC 48	1042
XC 48	1045
XC 48	1045H
XC 48 TS	1049
XC 68	1070
XC 75	1078
Z 2 CND 17.12	316L
Z 2 CND 19.15	317L
Z 6 CA 13	405
Z 6 CN 18.09	304
Z 6 CND 17.11	316
Z 6 CNN6 18.10	347
Z 6 CNT 18.10	321
Z 8 C 17	430
Z 8 CD 17.01	434
Z 10 C 13	410
Z 10 C 14	410
Z 10 CF 17	430F
Z 10 CNF 18.09	303
Z 12 C 13	410
Z 12 CF 13	416
Z 12 CN 17.08	301
Z 12 CNS 25.20	310
Z 12 CNS 25.20	314
Z 15 CN 24.13	309S
Z 20 CB	420
Z 30 WCV 9	H21
Z 38 CDV 5	H11

France (continued)

AFNOR

Designation	AISI
Z 40 COV 5	H13
Z 80 WCV 18-04-01	T1
Z 80 WKCV 18-05-04-01	T4
Z 85 DCWV 08-04-02-01	M1
Z 85 WDCV 06-05-04-02	M2
Z 90 WDCV 06-05-04-02	M3 Class 1
Z 100 CDV 5	A2
Z 110 DKCWV 09-08-04-02-01	M42
Z 110 WKCDV 07-05-04-04-02	M41
Z 120 WDCV 06-05-04-03	M3 Class 2
Z 130 WDCV 06-05-04-04	M3 Class 2
Z 200 C 12	D3

Germany (Federal Republic of)

DIN

Designation	AISI
1.0204	1008
1.0402	1020
1.0419	1016
1.0501	1035
1.0503	1043
1.0601	1060
1.0647	1084
1.0702	1110
1.0711	1212
1.0715	1213
1.0718	12L14
1.0726	1140
1.0727	1146
1.0909	9260
1.0909	9260H
1.0912	1345
1.0912	1345H
1.1121	1010
1.1133	1022
1.1133	1522H
1.1141	1015
1.1141	1017
1.1151	1023
1.1157	1039

Germany (continued)

DIN

Designation	AISI
1.1158	1025
1.1160	1524
1.1160	1524H
1.1161	1526
1.1161	1527
1.1165	1330
1.1165	1330H
1.1167	1335
1.1167	1335H
1.1167	1541
1.1167	1541H
1.1172	1030
1.1176	1038
1.1176	1038H
1.1186	1040
1.1191	1042
1.1191	1045
1.1191	1045H
1.1201	1049
1.1209	1055
1.1210	1050
1.1226	1548
1.1226	1552
1.1231	1070
1.1248	1078
1.1260	1566
1.1273	1090
1.1274	1095
1.2080	D3
1.2210	L2
1.2330	P20
1.2341	P4
1.2343	H11
1.2344	H13
1.2363	A2
1.2365	H10
1.2379	D2
1.2510	O1
1.2550	S1
1.2581	H21
1.2606	H12
1.2625	H23
1.2713	L6
1.2735	P6
1.2842	O2
1.3202	T15
1.3246	M41
1.3247	M42
1.3249	M33
1.3249	M34
1.3255	T4
1.3265	T5

Germany (continued)

DIN

Designation	AISI
1.3342	M3 Class 1
1.3343	M2
1.3344	M3 Class 2
1.3346	M1
1.3348	M7
1.3355	T1
1.3503	E51100
1.3505	E52100
1.4002	405
1.4005	416
1.4006	410
1.4016	430
1.4021	420
1.4024	403
1.4057	431
1.4104	430F
1.4112	440B
1.4113	434
1.4125	440C
1.4301	304
1.4303	305
1.4303	308
1.4305	303
1.4306	304L
1.4310	301
1.4401	316
1.4404	316L
1.4438	317L
1.4449	317
1.4512	409
1.4532	632
1.4541	321
1.4546	348
1.4550	347
1.4568	631
1.4828	309
1.4833	309S
1.4841	310
1.4841	314
1.4935	422
1.4980	660
1.5069	1340H
1.5523	15B21H
1.5527	15B41H
1.6523	8617
1.6523	8617H
1.6523	8620
1.6523	8620H
1.6543	8622
1.6543	8622H
1.6543	8720
1.6543	8720H
1.6543	8822
1.6543	8822H
1.6545	8630
1.6545	8630H
1.6546	8640
1.6546	8640H
1.6546	8740
1.6546	8740H
1.6562	E4340
1.6562	E4340H
1.6565	4340
1.6565	4340H
1.7006	5140H
1.7006	5150
1.7007	50B40
1.7007	50B40H
1.7030	5130
1.7033	5130H
1.7033	5132
1.7034	5132H
1.7034	5135
1.7035	5135H

Germany (continued)

DIN

Designation	AISI
1.7035	5140
1.7138	50B50
1.7138	50B50H
1.7147	5120
1.7147	5120H
1.7176	5155
1.7176	5155H
1.7218	4130
1.7218	4130H
1.7220	4135
1.7220	4135H
1.7223	4142H
1.7225	4137
1.7225	4137H
1.7225	4140
1.7225	4140H
1.7228	4147
1.7228	4147H
1.7228	4150
1.7228	4150H
1.7511	6118
1.7511	6118H
1.8159	6150
1.8159	6150H

Japan

JIS

Designation	AISI
40 NiCrMo 7	E4340
40 NiCrMo 7 KB	E4340
S 9 CK	1010
S 10 C	1010
S 12 C	1010
S 15 C	1015
S 15 C	1017
S 15 CK	1015
S 15 CK	1017
S 17 C	1015
S 17 C	1017
S 20 C	1023
S 20 CK	1023
S 22 C	1023
S 25 C	1025
S 28 C	1025
S 40 C	1040
S 45 C	1042
S 45 C	1045
S 45 C	1045H
S 48 C	1042
S 48 C	1045
S 48 C	1045H
S 50 C	1049
S 53 C	1050
S 55 C	1050
SCCrM 1	4130
SCCrM 1	4130H
SCCrM 3	4135
SCCrM 3	4135H
SCM 1	4135
SCM 1	4135H
SCM 2	4130
SCM 2	4130H
SCM 4	4137
SCM 4	4137H
SCM 4	4140
SCM 4	4140H
SCM 4 H	4137
SCM 4 H	4137H
SCM 4 H	4140
SCM 4 H	4140H

Japan (continued)

JIS

Designation	AISI
SCM 5	4147
SCM 5	4147H
SCM 5	4150
SCM 5	4150H
SCM 5 H	4147
SCM 5 H	4147H
SCM 5 H	4150
SCM 5 H	4150H
SCMn 2	1330
SCMn 2	1330H
SCMn 3	1335
SCMn 3	1335H
SCMn 3	1541
SCMn 3	1541H
SCr 2 H	5130H
SCr 2 H	5130H
SCr 2 H	5132
SCr 3 H	5132H
SCr 3 H	5135
SCr 4 H	5135H
SCr 4 H	5140
SCS 19	304L
SKD 1	D3
SKD 5	H21
SKD 6	H11
SKD 12	A2
SKD 61	H13
SKD 62	H12
SKH 2	T1
SKH 3	T4
SKH 4A	T5
SKH 52	M3 Class 2
SKH 53	M3 Class 2
SKN 9	M2
SMn 1 H	1330
SMn 1 H	1330H
SMn 2	1335
SMn 2	1335H
SMn 2	1541
SMn 2	1541H
SMn 2 H	1335
SMn 2 H	1335H
SMn 2 H	1541
SMn 2 H	1541H
SMnC 21	1022
SMnC 21	1522H
SNCM 8	4340
SNCM 8	4340H
SNCM 21	8617
SNCM 21	8617H
SNCM 21	8620
SNCM 21	8620H
SNCM 21 H	8617
SNCM 21 H	8617H
SNCM 21 H	8620
SNCM 21 H	8620H
SUH 309	316
SUH 310	316L
SUH 409	409
SUH 616	422
SUM 11	1110
SUM 12	1110
SUM 21	1212
SUM 22	1213
SUM 22 L	12L14
SUM 24 L	12L14
SUP 4	1095
SUP 10	6150
SUP 10	6150H
SUP 11	50B50
SUP 11	50B50H
SUS 43	403
SUS 301	301
SUS 303	303

Japan (continued)

JIS

Designation	AISI
SUS 304	304
SUS 304 L	304L
SUS 305	305
SUS 305	308
SUS 305 J 1	305
SUS 305 J 1	308
SUS 316	316
SUS 316 L	316L
SUS 317	317
SUS 321	321
SUS 347	347
SUS 403	403
SUS 405	405
SUS 410	410
SUS 416	403
SUS 420 J 1	420
SUS 430	430
SUS 430 F	430F
SUS 431	431
SUS 434	434
SUS 440 B	440B
SUS 440 C	440C
SUS Y 310	310
SUS Y 310	314
SUS Y 316	316

Sweden

SS$_{14}$

Designation	AISI
1370	1015
1370	1017
1450	1020
1550	1035
1650	1043
1660	1049
1672	1042
1672	1045
1672	1045H
1770	1070
1774	1078
1870	1095
1914	12L14
1957	1140
1973	1146
2120	1335
2120	1335H
2120	1541
2120	1541H
2225	4130
2225	4130H
2230	6150
2230	6150H
2234	4135
2234	4135H
2242	H13
2244	4137
2244	4137H
2244	4140
2244	4140H
2260	A2
2302	410
2303	420
2320	430
2321	431
2325	434
2332	304
2337	321
2338	347
2346	303
2347	316

Sweden (continued)

Designation	AISI
SS$_{14}$	
2348	316L
2352	304L
2367	317L
2380	416
2383	430F
2722	M2

United Kingdom

Designation	AISI
B.S.	
040 A 20	1020
060 A 35	1035
060 A 47	1043
060 A 62	1060
060 A 96	1095
070 M 20	1020
080 A 32	1035
080 A 35	1035
080 A 37	1035
080 A 40	1040
080 H 46	1043
080 M 36	1035
080 M 40	1043
080 M 46	1043
2 S. 93	1040
2 S. 117	5135H
2 S. 117	5140
2 S. 119	4340
2 S. 119	4340H
2 S. 130	348
2 S. 516	1345
2 S. 516	1345H
2 S. 517	1345
2 S. 517	1345H
3 S. 61	410
3 S. 62	420
3 S. 95	4340
3 S. 95	4340H
5 S. 80	431
120 M 36	1039
150 M 36	1039
220 M 07	1213
250 A 58	9260
250 A 58	9260H
302 S 17	304
303 S 21	303
304 S 12	304L
304 S 14	304L
304 S 15	304
304 S 16	304
304 S 18	304
304 S 22	304L
304 S 25	304
304 S 40	304
310 S 24	310
310 S 24	314
316 S 12	316L
316 S 14	316L
316 S 16	316
316 S 18	316
316 S 22	316L
316 S 24	316L
316 S 25	316
316 S 26	316
316 S 29	316L
316 S 30	316
316 S 30	316L
316 S 31	316L
316 S 37	316L
316 S 40	316

United Kingdom (continued)

Designation	AISI
B.S.	
316 S 41	316
316 S 82	316L
317 S 12	317L
321 S 12	321
321 S 18	321
321 S 22	321
321 S 27	321
321 S 40	321
321 S 49	321
321 S 50	321
321 S 59	321
321 S 87	321
347 S 17	347
347 S 17	348
347 S 18	348
347 S 40	348
405 S 17	405
409 S 17	409
410 S 21	410
416 S 21	416
420 S 29	403
420 S 37	420
430 S 15	430
431 S 29	431
434 S 19	434
530 A 30	5130
530 A 32	5130H
530 A 32	5132
530 A 36	5132H
530 A 36	5135
530 A 40	5135H
530 A 40	5140
530 H 30	5130
530 H 32	5130H
530 H 32	5132
530 H 36	5132H
530 H 36	5135
530 H 40	5135H
530 M 40	5140
530 M 40	5135H
530 M 40	5140
534 A 99	E52100
535 A 99	E52100
708 A 37	4135
708 A 37	4135H
708 A 40	4137H
708 A 42	4137
708 A 42	4137H
708 A 42	4140
708 A 42	4140H
708 M 40	4137
708 M 40	4140
708 M 40	4140H
709 A 40	4137H
709 M 40	4137
709 M 40	4140
709 M 40	4140H
735 A 50	6150
735 A 50	6150H
805 A 20	8622
805 A 20	8622H
805 A 20	8720
805 A 20	8720H
805 A 20	8822
805 A 20	8822H
805 H 20	8617
805 H 20	8617H
805 H 20	8620
805 H 20	8620H
805 M 20	8617
805 M 20	8617H
805 M 20	8620
805 M 20	8620H
817 M 40	4340

United Kingdom (continued)

Designation	AISI
B.S.	
817 M 40	4340H
3111 Type 6	4340
3111 Type 6	4340H
ANC 1 Grade A	410
ANC 3 Grade B	347
BA 2	A2
BD 2	D2
BD 3	D3
BH 11	H11
BH 12	H12
BH 13	H13
BH 21	H21
BM 1	M1
BM 2	M2
BM 34	M33
BM 34	M34
BM 42	M42
BO 1	O1
BO 2	O2
BT 1	T1
BT 4	T4
BT 5	T5
BT 15	T15
CDS-18	420
CDS-20	321
CDS 105/106	1039
CDS 110	4130
CDS 110	4130H
En. 44 B	1095
En. 47	6150
En. 47	6150H
En. 56 A	410
En. 56 B	403
En. 56 C	420
En. 58 B	321
En. 58 C	321
En. 58 E	304
En. 58 F	347
En. 58 G	347
En. 58 H	316
S. 139	E4340
S. 139	E4340H
S. 141	410
S. 525	348
S. 527	348
S. 536	304L
S. 537	316L
Type 3	5132H
Type 3	5135
Type 7	8640
Type 7	8640H
Type 7	8740
Type 7	8740H
Type 8	E4340
Type 8	E4340H

United States

Designation	AISI
FED	
QQ-S-00629 (FS4032)	4032
QQ-S-635 (C1030)	1030
QQ-S-635 (C1035)	1035
QQ-S-635 (C1042)	1042
QQ-S-635 (C1045)	1045
QQ-S-635 (C1050)	1050
QQ-S-637 (C1008)	1008
QQ-S-637 (C1110)	1110
QQ-S-637 (C1117)	1117
QQ-S-637 (C1118)	1118

United States (continued)

Designation	AISI
FED	
QQ-S-637 (C1137)	1137
QQ-S-637 (C1139)	1139
QQ-S-637 (C1140)	1140
QQ-S-637	1141
QQ-S-637 (C1144)	1144
QQ-S-637 (C1146)	1146
QQ-S-637 (C1151)	1151
QQ-S-637 (C1211)	1211
QQ-S-637 (C1212)	1212
QQ-S-637 (C1213)	1213
QQ-S-637	1215
QQ-S-698 (C1008)	1008
QQ-S-698 (C1015)	1015
QQ-S-700 (C1025)	1025
QQ-S-700 (C1030)	1030
QQ-S-700 (C1035)	1035
QQ-S-700 (C1045)	1045
QQ-S-700 (C1050)	1050
QQ-S-700 (C1055)	1055
QQ-S-700 (C1080)	1080
QQ-S-700 (C1084)	1084
QQ-S-700 (C1095)	1095
QQ-S-763	202
QQ-S-763	302
QQ-S-763	304
QQ-S-763	304L
QQ-S-763	305
QQ-S-763	309
QQ-S-763	310
QQ-S-763	316
QQ-S-763	316L
QQ-S-763	317
QQ-S-763	321
QQ-S-763	347
QQ-S-763	403
QQ-S-763	410
QQ-S-763	405
QQ-S-763	414
QQ-S-763	420
QQ-S-763	429
QQ-S-763	430
QQ-S-763	440A
QQ-S-763	440B
QQ-S-763	440C
QQ-S-763	446
QQ-S-766	201
QQ-S-766	202
QQ-S-766	301
QQ-S-766	302
QQ-S-766	304
QQ-S-766	304L
QQ-S-766	309
QQ-S-766	310
QQ-S-766	316
QQ-S-766	316L
QQ-S-766	321
QQ-S-766	347
QQ-S-766	348
QQ-S-766	420
QQ-S-766	429
QQ-S-766	430
QQ-S-766	446
QQ-T-570	A2
QQ-T-570	A3
QQ-T-570	A4
QQ-T-570	A6
QQ-T-570	A7
QQ-T-570	A8
QQ-T-570	A9
QQ-T-570	A10
QQ-T-570	D2
QQ-T-570	D3
QQ-T-570	D4
QQ-T-570	D5

Designation	AISI
FED	
QQ-T-570	D7
QQ-T-570	H10
QQ-T-570	H11
QQ-T-570	H12
QQ-T-570	H13
QQ-T-570	H14
QQ-T-570	H19
QQ-T-570	H21
QQ-T-570	H22
QQ-T-570	H23
QQ-T-570	H24
QQ-T-570	H25
QQ-T-570	H26
QQ-T-570	H42
QQ-T-570	L2
QQ-T-570	L6
QQ-T-570	O1
QQ-T-570	O2
QQ-T-570	O6
QQ-T-570	O7
QQ-T-570	S1
QQ-T-570	S2
QQ-T-570	S5
QQ-T-570	S6
QQ-T-580	W2
QQ-T-590	M1
QQ-T-590	M2
QQ-T-590	M3 Class 1
QQ-T-590	M3 Class 2
QQ-T-590	M4
QQ-T-590	M6
QQ-T-590	M7
QQ-T-590	M10
QQ-T-590	M30
QQ-T-590	M33
QQ-T-590	M34
QQ-T-590	M36
QQ-T-590	M41
QQ-T-590	M42
QQ-T-590	M43
QQ-T-590	M44
QQ-T-590	M46
QQ-T-590	T1
QQ-T-590	T2
QQ-T-590	T4
QQ-T-590	T5
QQ-T-590	T6
QQ-T-590	T8
QQ-T-590	T15
QQ-W-423	302
QQ-W-423	304
QQ-W-423	305
QQ-W-423	310
QQ-W-423	316
QQ-W-423	321
QQ-W-423	347
QQ-W-423	410
QQ-W-423	416
QQ-W-423	420
QQ-W-423	429
QQ-W-423	430
STD-66	202
STD-66	304
STD-66	310
STD-66	430
MIL SPEC	
MIL-C-24111	630
MIL-F-20138	304
MIL-S-862	302
MIL-S-862	303
MIL-S-862	303Se
MIL-S-862	304

Designation	AISI
MIL SPEC	
MIL-S-862	304L
MIL-S-862	309
MIL-S-862	310
MIL-S-862	316
MIL-S-862	316L
MIL-S-862	317
MIL-S-862	321
MIL-S-862	347
MIL-S-862	403
MIL-S-862	405
MIL-S-862	410
MIL-S-862	416
MIL-S-862	416Se
MIL-S-862	420
MIL-S-862	429
MIL-S-862	430
MIL-S-862	430F
MIL-S-862	430FSe
MIL-S-862	431
MIL-S-862	440A
MIL-S-862	440B
MIL-S-862	440C
MIL-S-862	446
MIL-S-866	1016
MIL-S-866	8615
MIL-S-980	E52100
MIL-S-5000	E4340
MIL-S-5059	301
MIL-S-5059	304
MIL-S-5059	316
MIL-S-6049	8740
MIL-S-7420	E52100
MIL-S-7493 (A4615)	4615
MIL-S-7493 (A4620)	4620
MIL-S-8503	6150
MIL-S-11310 (CS1008)	1008
MIL-S-11310 (CS1010)	1010
MIL-S-11310 (CS1012)	1012
MIL-S-11310 (CS1017)	1017
MIL-S-11310 (CS1018)	1018
MIL-S-11310 (CS1020)	1020
MIL-S-11310 (CS1022)	1022
MIL-S-11310 (CS1025)	1025
MIL-S-11310 (CS1030)	1030
MIL-S-11310 (CS1040)	1040
MIL-S-11595 (ORD4150)	4150
MIL-S-11713 (2)	1070
MIL-S-16788 (C10)	1095
MIL-S-16974	1015
MIL-S-16974	1050
MIL-S-16974	1060
MIL-S-16974	1080
MIL-S-16974	1330
MIL-S-16974	1335
MIL-S-16974	1340
MIL-S-16974	4130
MIL-S-16974	4135
MIL-S-16974	4140
MIL-S-16974	4145
MIL-S-16974	4340

Designation	AISI
MIL SPEC	
MIL-S-16974	8620
MIL-S-16974	8625
MIL-S-16974	8630
MIL-S-16974	8640
MIL-S-16974	8645
MIL-S-18411	1117
MIL-S-18733	4135
MIL-S-20137A	1151
MIL-S-22141	E52100
MIL-S-23195	304
MIL-S-23195	304L
MIL-S-23195	347
MIL-S-23195	348
MIL-S-23196	304
MIL-S-23196	304L
MIL-S-23196	347
MIL-S-23196	348
MIL-S-25043	631
MIL-S-81506	630
MIL-S-81591	630
MIL-T-6845	304
MIL-T-8504	304
MIL-T-8506	304
MIL-W-46078	631
UNS	
G10080	1008
G10100	1010
G10120	1012
G10150	1015
G10160	1016
G10170	1017
G10180	1018
G10190	1019
G10200	1020
G10210	1021
G10220	1022
G10230	1023
G10250	1025
G10260	1026
G10290	1029
G10300	1030
G10350	1035
G10370	1037
G10380	1038
G10390	1039
G10400	1040
G10420	1042
G10430	1043
G10440	1044
G10450	1045
G10460	1046
G10490	1049
G10500	1050
G10530	1053
G10550	1055
G10600	1060
G10700	1070
G10780	1078
G10800	1080
G10840	1084
G10900	1090
G10950	1095
G11100	1110
G11170	1117
G11180	1118
G11370	1137
G11390	1139
G11400	1140
G11410	1141
G11440	1144
G11460	1146
G11510	1151

Designation	AISI
UNS	
G12110	1211
G12120	1212
G12130	1213
G12144	12L14
G12150	1215
G13300	1330
G13350	1335
G13400	1340
G13450	1345
G15130	1513
G15220	1522
G15240	1524
G15260	1526
G15270	1527
G15410	1541
G15480	1548
G15510	1551
G15520	1552
G15610	1561
G15660	1566
G40230	4023
G40240	4024
G40270	4027
G40280	4028
G40320	4032
G40370	4037
G40420	4042
G40470	4047
G41180	4118
G41300	4130
G41350	4135
G41370	4137
G41400	4140
G41420	4142
G41450	4145
G41470	4147
G41500	4150
G41610	4161
G43200	4320
G43400	4340
G43406	E4340
G46150	4615
G46200	4620
G46260	4626
G47200	4720
G48150	4815
G48170	4817
G48200	4820
G50401	50B40
G50441	50B44
G50460	5046
G50461	50B46
G50501	50B50
G50601	50B60
G51170	5117
G51200	5120
G51300	5130
G51320	5132
G51350	5135
G51400	5140
G51500	5150
G51550	5155
G51600	5160
G51601	51B60
G51986	E51100
G52986	E52100
G61180	6118
G61500	6150
G81451	81B45
G86150	8615
G86170	8617
G86200	8620
G86220	8622
G86250	8625

United States (continued)

Designation	AISI

UNS

G86270	8627
G86300	8630
G86370	8637
G86400	8640
G86420	8642
G86450	8645
G86451	86B45
G86500	8650
G86550	8655
G86600	8660
G87200	8720
G87400	8740
G88220	8822
G92600	9260
G94151	94B15
G94171	94B17
G94301	94B30
H10380	1038H
H10450	1045H
H13300	1330H
H13350	1335H
H13400	1340H
H13450	1345H
H15211	15B21H
H15220	1522H
H15240	1524H
H15260	1526H
H15351	15B35H
H15371	15B37H
H15410	1541H
H15411	15B41H
H15481	15B48H
H15621	15B62H
H40270	4027H
H40280	4028H
H40320	4032H
H40370	4037H
H40420	4042H
H40470	4047H
H41180	4118H
H41300	4130H
H41350	4135H
H41370	4137H
H41400	4140H
H41420	4142H
H41450	4145H
H41470	4147H
H41500	4150H
H41610	4161H
H43200	4320H
H43400	4340H
H43406	E4340H
H46200	4620H
H46260	4626H
H47200	4720H
H48150	4815H
H48170	4817H
H48200	4820H

United States (continued)

Designation	AISI

UNS

H50401	50B40H
H50441	50B44H
H50460	5046H
H50461	50B46H
H50501	50B50H
H50601	50B60H
H51200	5120H
H51300	5130H
H51320	5132H
H51350	5135H
H51400	5140H
H51500	5150H
H51550	5155H
H51600	5160H
H51601	51B60H
H61180	6118H
H61500	6150H
H81451	81B45H
H86170	8617H
H86200	8620H
H86220	8622H
H86250	8625H
H86270	8627H
H86300	8630H
H86301	86B30H
H86370	8637H
H86400	8640H
H86420	8642H
H86450	8645H
H86451	86B45H
H86500	8650H
H86550	8655H
H86600	8660H
H87200	8720H
H87400	8740H
H88220	8822H
H92600	9260H
H93100	9310H
H94151	94B15H
H94171	94B17H
H94301	94B30H
K66286	660
N08330	330
S15700	632
S17400	630
S17700	631
S20100	201
S20200	202
S20500	205
S30100	301
S30200	302
S30215	302B
S30300	303
S30323	303Se
S30400	304
S30403	304L
S30430	S30430
S30451	304N

United States (continued)

Designation	AISI

UNS

S30500	305
S30800	308
S30900	309
S30908	309S
S31000	310
S31008	310S
S31400	314
S31600	316
S31603	316L
S31620	316F
S31651	316N
S31700	317
S31703	317L
S32100	321
S32900	329
S34700	347
S34800	348
S35000	633
S35500	634
S38400	384
S40300	403
S40500	405
S40900	409
S41000	410
S41400	414
S41600	416
S41623	416Se
S42000	420
S42020	420F
S42200	422
S42900	429
S43000	430
S43020	430F
S43023	430FSe
S43100	431
S43400	434
S43600	436
S44002	440A
S44003	440B
S44004	440C
S44200	442
S44600	446
T11301	M1
T11302	M2
T11304	M4
T11306	M6
T11307	M7
T11310	M10
T11313	M3 Class 1
T11323	M3 Class 2
T11330	M30
T11333	M33
T11334	M34
T11336	M36
T11341	M41
T11342	M42
T11343	M43
T11344	M44

United States (continued)

Designation	AISI

UNS

T11346	M46
T11347	M47
T12001	T1
T12002	T2
T12004	T4
T12005	T5
T12006	T6
T12008	T8
T12015	T15
T20810	H10
T20811	H11
T20812	H12
T20813	H13
T20814	H14
T20819	H19
T20821	H21
T20822	H22
T20823	H23
T20824	H24
T20825	H25
T20826	H26
T20842	H42
T30102	A2
T30103	A3
T30104	A4
T30106	A6
T30107	A7
T30108	A8
T30109	A9
T30110	A10
T30402	D2
T30403	D3
T30404	D4
T30405	D5
T30407	D7
T31501	O1
T31502	O2
T31506	O6
T31507	O7
T41901	S1
T41902	S2
T41905	S5
T41906	S6
T41907	S7
T51602	P2
T51603	P3
T51604	P4
T51605	P5
T51606	P6
T51620	P20
T51621	P21
T61202	L2
T61206	L6
T72301	W1
T72302	W2
T72305	W5